NATIONAL
GEOGRAPHIC

COMPLETE
BIRDS
OF NORTH AMERICA

SECOND EDITION
FULLY REVISED and UPDATED

COMPLETE
BIRDS
OF NORTH AMERICA

SECOND EDITION
FULLY REVISED and UPDATED

EDITED BY **JONATHAN ALDERFER**
WITH **JON L. DUNN**
MAPS BY **PAUL LEHMAN**

CONTRIBUTING AUTHORS

Jessie H. Barry, Edward S. Brinkley, Steven W. Cardiff, Allen T. Chartier,
Cameron D. Cox, Donna L. Dittmann, Jon L. Dunn, Ted Floyd,
Kimball L. Garrett, Matthew T. Heindel, Paul Hess, Steve N. G. Howell,
Marshall J. Iliff, Alvaro Jaramillo, Ian L. Jones, Tony Leukering,
Mark W. Lockwood, ~~Cardiff~~ E. Quady,
Kurt ~~...~~ ~~...~~van,

NATIONAL GEOGRAPHIC
WASHINGTON, D.C.

CONTENTS

Tennessee Warbler

INTRODUCTION BY JONATHAN ALDERFER

Since the publication of the first edition of this work in 2005 the birding landscape in North America has changed in myriad ways—excellent new books and articles on bird identification have been published, the science of ornithology has amazed us with discoveries of unforeseen evolutionary relationships, new species have been recorded for North America, migration paths have been unraveled, and on and on. The scope of the first edition, ambitious as it was, no longer covered today's birding territory. We hope this new edition of *Complete Birds* will fill that gap with refreshed and reliable information. Herein, you'll find in-depth information on identification, similar species, geographic variation, voice, status and distribution, population, and range, accompanied by an unequaled number of illustrations and maps.

Coverage

This reference describes all species of wild birds reliably recorded in North America—the continent north of Mexico, plus adjacent islands and seas within 200 miles of the coast, excluding Greenland. With a cutoff date of November 2013, the book's last included species was the Common Redstart, a vagrant from the Old World. We also include a variety of exotic species that are established or regularly observed.

Text The text for the first edition was written in 2005 by 24 of the most experienced field ornithologists in North America (see Contributors, page 741), with additional editorial work by Paul Hess and Paul Lehman. The added text for the second edition, including new 12 new family accounts and over 60 new species accounts, was written by Edward S. Brinkley, Paul Hess, and Kimball L. Garrett, with updating and editorial work by Jonathan Alderfer and Jon L. Dunn.

Art and Photographs Most of the artwork in this work comes from the *National Geographic Field Guide to the Birds of North America*, 6th edition (2011). New art of six recently added vagrant species (Providence Petrel, Double-toothed Kite, Rufous-necked Wood-Rail, Chiffchaff, Common Redstart, and Asian Rosy-Finch) was painted expressly for this book. Credits appear on page 726.

Maps All the range maps from the 6th edition field guide have been updated as necessary by Paul Lehman, with over 70 maps adjusted. A map key is located at right. Note that boundaries are drawn where a species ceases to be regularly seen in the proper habitat. Nearly every species will be rare at the edges of its range. The range maps are augmented with a selected group of large-scale specialty maps that show long-range migration routes, vagrancy patterns, or historical data. There are over 800 maps in this edition.

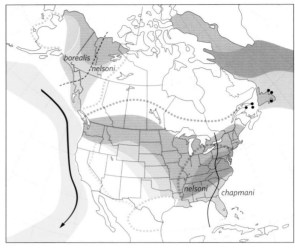

Range Map Symbols

Breeding range, generally in spring and summer	Extent of irregular or irruptive range in some winters
Year-round range	Extent of irregular migration in spring and autumn
Winter range (*if no winter or year-round range is shown, bird winters outside North America*)	Extent of irregular migration primarily in spring
Migration range in spring and autumn	Extent of irregular migration primarily in autumn, or post-breeding dispersal
Migration range primarily in spring	*fulva* Subspecies name and its boundary
Migration range primarily in autumn	Subspecies boundary where only approximate
Principal direction of migration	**Subspecies Maps Only** (*these symbols not shown on sample map above*)
Selected breeding colonies	Subspecies boundary during particular seasons
Extent of irregular breeding range	Subspecies group boundary
Extent of irregular year-round range	Zone of intergradation between two subspecies

Fieldcraft

The art and practice of birding is a skill that develops with time. If you are new to birding, here are a few basic recommendations: First, get out in the field and look at birds with a decent pair of binoculars. Study your field guide before you go, but try leaving it behind. Instead make notes, even if they're just mental ones; soak in what you're seeing, and check the book later. Most expert birders have had a mentor—someone with great field skills who is willing to share knowledge. One of the most important things you can do is learn status and distribution—starting with your home area.

Taxonomic Organization

This book follows the taxonomic sequence and naming conventions adopted by the American Ornithologists' Union (AOU) as of August 2013 and is organized by family, genus, and species. The taxonomy of birds is currently in a period of flux as genetic research uncovers evolutionary relationships and this trend is certain to continue. Getting to know the latest taxonomic sequence is part of a birder's education.

Family A short essay accompanies each of the 89 families, detailing its distinctive characteristics and how those characteristics relate to field identification.

Genus Genera (pl. of genus) are groups of closely related species. The genus name is the first part of a species' scientific name: i.e., *Spizella* in *Spizella passerina,* for Chipping Sparrow. Deciding a bird's genus can be useful in identifying it. A brief text describes many of the genera, but where distinctions are less useful, genera are lumped together or have no text.

Species Taxonomists differ on the definition of "species." With new scientific research, especially in the field of molecular genetics, species limits will continue to be redefined, ultimately rendering a clearer picture of the evolutionary relationships of the birds we watch. In this book, the species' accounts start with a brief overview and follow a standard layout; some accidental species have abbreviated accounts.

Subspecies (Geographic Variation) Some birders find it challenging to observe variation *within* the species. This book attempts to give information on many field-recognizable subspecies, or groups of subspecies. A species that shows no geographic variation is known as **monotypic.** When within-species variation forms two or more distinctive populations, the species is known as **polytypic.** Those populations are given a scientific subspecies name, or **trinomial.** *Zonotrichia leucophrys gambelii* is the trinomial for "Gambel's" White-crowned Sparrow, a distinctive, field-recognizable subspecies of the White-crowned Sparrow. In polytypic species, the first type described is referred to as the **nominate subspecies,** with a name that repeats the second part of its scientific species name. For example, *Zonotrichia leucophrys leucophrys* is the nominate subspecies of

White-crowned Sparrow. Many subspecies have no commonly used English name and are known only by their scientific subspecies epithet. When different subspecies interbreed, the result is known as an **intergrade.** When a species or subspecies varies gradually across a geographical region, it is said to form a **cline,** or to vary clinally. Our authority for subspecies names and ranges is the most recent edition of the *Howard and Moore Complete Checklist of the Birds of the World.*

Plumage Variation

The sex and age of an individual can be determined for many species, most often by observing its plumage. In many species the male and female look quite different. This book illustrates and describes most of those differences. Age determination ranges from simple to complex, depending on the species and how quickly or slowly it reaches adult plumage. Accurate age assessment requires a basic understanding of molt. **Molt** describes the replacement of a bird's feathers: A fresh new feather growing from the same follicle pushes out the old worn feather. All birds molt, and a molt produces a plumage. Birds with old, abraded feathers are referred to as **worn;** birds with new feathers are **fresh.** Most birders have adopted an amalgam of terms for molt and plumage to describe what they are seeing—not a rigorous or scientific use of terminology, but a practical one. Some of the terms used in this book are defined below.

Juvenile Juveniles (birds in **juvenal** plumage) are wearing their first true coat of feathers, the ones in which they usually fledge (leave the nest). Some young birds hold their juvenal plumage only a few weeks, while others such as true hawks, loons, and many other waterbirds hold their juvenal plumage into the winter, or even the following spring.

First-fall or First-winter Most species replace their juvenal plumage in late summer or early fall with a partial molt into this plumage, usually while retaining the major, juvenal wing and tail feathers. All plumages between juvenal and adult are **immature,** an imprecise term. The sequence of immature plumages may continue for some years until adult plumage is attained.

First-summer, Second-winter, etc. These terms describe immature plumages between first-winter and adult. **Life-year** terms may be used, such as first-year, second-year, etc. These are age (not molt or plumage) terms; that is, a first-year bird is in its first 12 months of life. **Calendar-year** refers to the bird's actual calendar-year age; that is, a first-calendar-year bird becomes a second-calendar-year bird after December 31.

Adult Birds are adult when their molts produce a stable cycle of identical (or definitive) plumages that repeats for the rest of their lives. Most adult birds have either one complete molt per year *or* one partial and one complete molt per year. Adult birds with a single

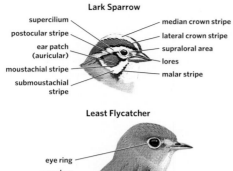

Lark Sparrow

- supercilium
- postocular stripe
- ear patch (auricular)
- moustachial stripe
- submoustachial stripe
- median crown stripe
- lateral crown stripe
- supraloral area
- lores
- malar stripe

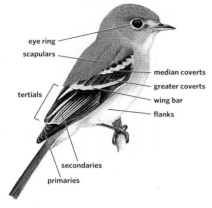

Least Flycatcher

- eye ring
- scapulars
- tertials
- median coverts
- greater coverts
- wing bar
- flanks
- secondaries
- primaries

complete molt per year (usually after breeding) look the same all year, except for the effects of wear. Adults with two annual molts generally undergo a partial (**prealternate**) molt in late winter or early spring—often involving the head, body, and some wing coverts—into breeding (or alternate) plumage. Most of these species later undergo a complete (**prebasic**) molt after breeding into their nonbreeding or winter (or basic) plumage. **High breeding** refers to plumage or bare-parts colors evident only during the brief period of courtship, such as the rich colors of the bill, lores, legs, and feet of many herons. **Eclipse** plumage usually describes the briefly held, female-like plumage of male ducks in summer, during which they molt their flight feathers and become flightless.

Other Variations Two species may interbreed, producing **hybrids** that may look partly like one parent, partly like the other, or may show unexpected characteristics. For a few species this is common; we describe and illustrate these cases. **Morph** describes a regularly occurring variation, usually in plumage coloration, such as the gray-morph Eastern Screech-Owl. Other terms describe abnormal plumages: **Albino** birds are pure white with red eyes; **leucistic** birds have unusually pale plumage or a piebald appearance with scattered white feathers, sometimes in a symmetrical pattern; and **melanistic** birds have excessively dark pigmentation.

Art Labels In most cases the art labels match those in the field guide's 6th edition. Male (♂) and female (♀) symbols are used where the sexes are discernibly different. Many figures are labeled with an age designation.

If a species shows little or no difference between ages or sexes (e.g., storm-petrels), there are no age or sex labels. Some figures are simply labeled (♂) or (♀) when exact age is difficult to assess. Often a subspecies or subspecies group name is included, with geographic designation.

Feather Topography

Knowing the names of the feather groups and how they are arranged is central to the language of birding. The colors and patterns of a bird's feathers are usually its most important field marks: wing bars, scapulars, eye rings, etc. Sometimes the physical proportions are equally important: for example, primary extension past the tertials. Most of the bolded terms below are labeled on the illustrations here.

Head A careful look at a bird's head can often provide a correct identification. At the top of the Lark Sparrow's head (top, left) there is a pale **median crown stripe,** bordered by dark **lateral crown stripes.** The line running from the base of the bill up and over the eye is the **eyebrow** or **supercilium;** in front of the eye is the **supraloral area.** The area between the eye and the bill is the **lores.** Note eye color: The pupil is always black, but the **iris** (pl. irides) can be colored. The **orbital ring** (below, left) is naked flesh around the eye; when contrastingly feathered, the area forms an **eye ring** (above, left), or **eye crescents.** A dark stripe extending back from the eye is a **postocular stripe;** when it extends through the eye it is an **eyeline.** The Lark Sparrow shows a distinctive brown patch behind and below the eye known as the **ear patch,** or **auricular.** Its lower border is the **moustachial stripe;** below that is the **submoustachial stripe;** and below that is the **malar stripe.** Most gulls (below) and some other species have a prominent ridge on their lower mandible, the **gonys,** which forms the **gonydeal angle.** The top of the bill is the **culmen.** Raptors have a bare patch of skin over part of the upper mandible, the **cere** (below). Some species, such as pelicans and cormorants, have a fold of loose skin hanging from the throat, the **gular pouch.**

Wings Folded wings and extended wings look very different, and it's useful to study both aspects. The longest, strongest feathers of the wing, the **flight feathers,** or **remiges,** are covered on both their upper and lower surfaces by shorter, protective feathers called **coverts** (opposite). Each feather's shaft divides its **inner web** from its **outer web.** The outermost 9 or 10 tapering flight feathers comprise the **primaries (P1-P10);** inward from

orbital ring

cere

Common Black-Hawk

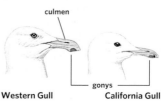

culmen

gonys

Western Gull **California Gull**

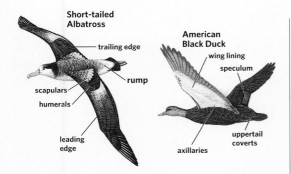

Short-tailed Albatross

trailing edge

rump

scapulars

humerals

leading edge

American Black Duck

wing lining

speculum

uppertail coverts

axillaries

the primaries are a variable number of **secondaries.** The innermost secondaries are a group of feathers known as the **tertials.** On the folded wing, the distance the primaries extend past the longest tertial is the **primary tip projection.** Covering the area where the wing joins the body are the **scapulars.** Rows of wing coverts cover the bases of the flight feathers. From **leading edge** to **trailing edge** they are the **marginal, lesser, median,** and **greater wing coverts.** The greater and median coverts form single rows; the lesser and marginal coverts consist of multiple rows of small feathers. **Wing bars** (opposite) are formed by the pale tips of the greater and median wing coverts. On long-winged species (e.g., albatrosses, above), the **humerals** are well developed. **Mantle** refers to the overall plumage of the back and scapulars. The uniquely colored secondaries of some ducks are referred to as the **speculum** (above). A contrasting bar on the leading edge of the wing is a **carpal bar;** a bar that cuts diagonally across the inner wing is an **ulnar bar.** Dark primaries, ulnar bars, and rump may form an **M-pattern,** prominent on some seabirds. On the underwing, **wing linings** refer to the underwing coverts as a whole; **axillaries** are the feathers of the bird's "armpit."

Tail Most birds have 12 tail feathers, also called **rectrices.** On the folded tail from above, only the two central tail feathers are visible; from below only the two outermost tail feathers are visible. Note the pattern of the undertail and length of the tail projecting past the undertail coverts. Some patterns and the tail's overall shape are visible only in flight.

Size All species accounts include average length, **L,** from tip of bill to tip of tail; some large or soaring species also have wingspan, **WS,** measured from wing tip to wing tip. The perception of size is fraught with variables. Lighting, background coloration, and posture affects size perception, as does the use of birding optics.

Abundance Terms and Codes

Abundance must be considered in relation to habitat. Some species are highly local, found only in specialized habitat. Here they may be common, while nearby they are rare. Ranges and abundance are detailed in the text, as well as extralimital records and population trends. We use the following terms with to describe abundance:

Rare Occurs in low numbers, but annually; e.g., visitors or rare breeding residents.

Casual Not recorded annually, but with six or more total records, *including* three or more in the past 30 years, reflecting a pattern of occurrence.

Accidental Species with five or fewer records *or* species with fewer than three records in the past 30 years.

ABA Abundance Codes Number codes (1–6) developed by the American Birding Association are found after the banding codes on the opening line of each species account. They are general indicators of a species' abundance in the whole of North America, north of Mexico. **Code 1=Common** (usually widespread); **Code 2=Uncommon** (or with a limited range); **Code 3=Rare; Code 4=Casual; Code 5=Accidental;** and **Code 6=Not found** (extirpated, extinct, or exists as a released population that is not yet naturally reestablished). For more information consult the *ABA Checklist: Birds of the Continental United States and Canada* (listing.aba.org/checklist/abachecklist_v7.5.pdf).

Banding Codes These codes, used by bird banders, are four-letter codes usually derived from the first two letters of a bird's two-word name (e.g., TUVU is the code for Turkey Vulture), but this varies with longer and shorter names and to avoid duplication. Banding codes are found after each species' scientific name.

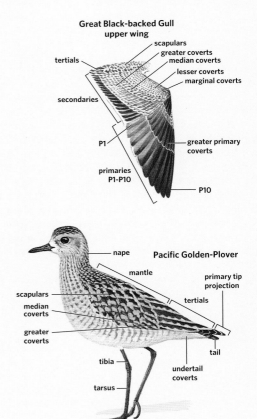

Great Black-backed Gull upper wing

scapulars

greater coverts
median coverts

tertials

lesser coverts
marginal coverts

secondaries

greater primary coverts

P1

primaries P1-P10

P10

Pacific Golden-Plover

nape

mantle

primary tip projection

scapulars

median coverts

tertials

greater coverts

tail

tibia

undertail coverts

tarsus

DUCKS, GEESE, AND SWANS Family Anatidae

Black Scoter, female (second from left) with males (NJ, Mar.)

Anatids are familiar as domesticated species, and all share some basic traits that are distinctive. They have fully webbed feet. Most species have a standard "duck bill" that is flattened and broadened at the end; the major exception is the mergansers, which have long, thin, serrated bills. The bodies of ducks, geese, and swans are rotund, tails are usually short, while the wings are strong and generally pointed. The placement of the legs differs considerably depending on whether the species is a surface swimmer or a diver. The legs of surface species are more centrally placed and allow for a horizontal body stance, while those of divers tend to be set far back on the body, making it difficult for these ducks to walk on land. Whistling-ducks, an exception, are slim and have legs placed far back, allowing them to stand more upright than other ducks.

Plumage The range in variation in plumage is huge; many are very attractive and are kept as ornamental species in captivity. Geese tend to be brownish or subdued colors, but often with attractive patterns, and there is no sexual dimorphism. Swans are often white, although two are black or black necked. Whistling-ducks are attractively patterned, and often show characteristic markings on the flanks; they are sexually monomorphic. Ducks are incredibly variable, but are usually sexually dimorphic, with males showing bright drake plumage with large patches of solid color or pattern; females more cryptic brown plumage. Some species of dabbling ducks, particularly those on islands, have henlike male plumages. Dabbling ducks and some other species show a colorful, and often iridescent, color patch on the secondaries known as the speculum.

Behavior This family is aquatic. Some birds feed from the surface of the water, either filtering food or tipping down to feed on vegetation, while other species are active divers ranging from piscivorous predators to species specializing on shellfish or submerged vegetation. Geese graze on open fields. Courtship behavior is well marked in ducks and is usually easy to observe during the winter and early spring. Geese and swans tend to mate for life, while ducks find new mates each year.

Distribution Anatids occur worldwide, breeding in all continents except Antarctica, where the Black-necked Swan is a vagrant.

Taxonomy This is a large family, with 49 genera and 158 species. It is divided into five subfamilies of which three, Dendrocygninae (whistling-ducks), Anserinae (true geese and swans), and Anatinae (ducks) are found in North America. The family is thought to be most closely related to the screamers (Anhimidae) and the magpie-geese (Anseranatidae); these families and the galliform families form an old lineage (Galloanserae) within modern birds (Neognathae). In fact the Galloanserae is the sister lineage to all other modern birds, the third and most basal lineage being the Palaeognathae (Tinamous and Ratites). Species level taxonomy within N.A. has been quite stable, although currently there is interest in determining adequate species level divisions within the geese. One confounding factor in waterfowl is that they readily hybridize with each other, sometimes quite commonly, and even between genera.

Conservation Six species are extinct, six critically endangered, nine endangered, and twelve vulnerable.

WHISTLING-DUCKS Genus *Dendrocygna*

The eight species of whistling-ducks are found mainly in tropical regions. Slim, long necked, and long billed, they stand rather upright. Their whistling calls inspired their name; some perch on trees, giving them the alternate name "tree-ducks." They fly with shallow wingbeats on rounded wings. The whistling-ducks comprise a subfamily of the Anatidae.

BLACK-BELLIED WHISTLING-DUCK *Dendrocygna autumnalis*   BBWD ■ 1

This gorgeous duck is often found perching on trees or shrubs. Polytypic (2 ssp.; *fulgens* in N.A.). L 21" (53 cm) **IDENTIFICATION** A long-legged, broad-winged, and long-necked duck. **ADULT:** Bright red bill obvious on gray face; the eye is surrounded by a bold white eye ring. Lower neck and breast tawny, with contrasting black belly and flanks. Upperparts warm brown, folded wings show whitish shoulder/

forewings, and dark primaries and tertials. Legs bright pink. **JUVENILE:** Similar to adult but bill grayish and body plumage much duller. Flanks brownish and belly off-white, not black. Wing pattern similar to that of adult, although duller and less extensively white. **FLIGHT:** A warm brown with black hindparts, including rump and lower back, with strikingly large white patch on upper wings from inner greater coverts to outer primary coverts and primary bases. **SIMILAR SPECIES** Likely to be confused only with Fulvous Whistling-Duck, which lacks the red bill and black belly. The large white wing patch of Black-bellied is diagnostic. Juveniles show duller plumage, but note white on the wing and gray face.

VOICE CALL: A multisyllabic, nasal whistle, *pe-che*, or a longer *pe-che-che-ne*. **STATUS & DISTRIBUTION** Fairly common, widespread in neotropics south to central Argentina. **YEAR-ROUND:** Wetlands, those with overhanging woody vegetation preferred. **MIGRATION:** A short-distance migrant, traveling not far into northern Mexico. Arrival on breeding grounds by Apr., southbound movements noted from Aug. to Oct. **VAGRANT:** Casual north of regular range as far north as MN, MI, and ON and to southern CA and NM. Vagrants often appear in small flocks. **POPULATION** Both the population and range have expanded noticeably over the last several decades; now found over much of the Southeast.

FULVOUS WHISTLING-DUCK *Dendrocygna bicolor*   FUWD ■ 1

The Fulvous Whistling-Duck is a long-legged and long-necked largely tawny duck. Monotypic. L 20" (51 cm) **IDENTIFICATION** Bill gray, legs blue-gray. Face, neck sides, breast and belly tawny. Throat and foreneck buffy-white, speckled with black on

lower neck. Top of crown brown, continues as a brown line (broken in male, continuous in female) down the back of the neck to the upper back. Upperparts black with broad cinnamon edges to feathers. Tawny underparts accented by crisp white stripes on flanks, vent white. **FLIGHT:** Appears tawny with blackish wings, including underwings. Lower back and tail black, contrasting with white rump band. **SIMILAR SPECIES** Juvenile Black-bellied has a gray face and large white wing patch. Fulvous Whistling-Duck is always more tawny, with white flank stripes. In flight, Fulvous shows a strong contrast between blackish underwings and tawny body, and a white rump band. In addition Fulvous is narrower

winged than the Black-bellied and has a deeper wing flap. **VOICE CALL:** A two-note, high-pitched, nasal, whistled call, *pt-TZEEW*, accent on second syllable.

STATUS & DISTRIBUTION Fairly common. Widely distributed in New World tropics south to central Argentina, as well as in Africa and the Indian subcontinent. **YEAR-ROUND:** Freshwater grassy wetlands, rice fields commonly used. **MIGRATION:** Some of FL population moves south to Cuba and some TX and LA birds south into Mexico. Arrival on breeding grounds in Feb.–Mar., southbound movements Aug.–late Sept. **DISPERSAL:** Regular northward dispersal after breeding. **VAGRANT:** Casual throughout continent, north to BC, AB, and NS. Vagrant records occur in years of increased postbreeding northward wandering. Records are concentrated in Mississippi Basin and Atlantic Coast. **POPULATION** Numbers have fluctuated noticeably; however, California population and range has seriously decreased and contracted since the mid-1900s, and the species became extirpated as a breeder shortly after 2000.

GRAY GEESE Genus Anser

These are the "standard" or quintessential geese. Our domesticated goose is derived from a member of this genus, the Graylag (*Anser anser*). There are eight species in the genus, found largely in the Palearctic. These geese are brownish with a variable gray wash to the coverts. Bill color and pattern as well as leg color aid in identification.

TAIGA BEAN-GOOSE Anser fabalis   TABG ◼ 3

Taiga Bean-Goose of northern Eurasia has been recorded only a handful of times in North America. It was formerly combined with Tundra Bean-Goose as a single species known as "Bean Goose," *A. fabalis*. The recent (2007) split of Bean Goose has prompted extensive discussions about North American records of bean-geese, most of which have not yet been identified to species despite good photographs or even specimens. Polytypic (3 ssp.). L 30–35" (76–90 cm)

IDENTIFICATION A large brownish goose with long, almost swan-like neck and long, black bill with distinctive pale orange subterminal band. Legs orange. **ADULT:** Brown with head usually noticeably darker and white belly. **FLIGHT:** Coverts show gray cast, contrasting with much darker remiges; dark brown rump separated from dark rectrices by white coverts, recalling uppertail pattern of white-fronted geese.

GEOGRAPHIC VARIATION Subspecies size increases from west to east from smaller *fabalis* west of the Urals to largest *middendorffii* mainly in the Russian Far East. The westernmost subspecies, nominate *fabalis*, breeds from northern Scandinavia east to the Urals. The medium-size *johanseni* (sometimes synonymized with *fabalis*) breeds in western Siberia from the Urals to Lake Baikal, and the large *middendorffii* breeds from Lake Baikal to the Russian Pacific coast and south to northern Mongolia and the Altay. Intergradation between subspecies apparently occurs, but more research is needed. Although some authorities treat Taiga and Tundra Bean-Geese as conspecific, ongoing field and biochemical studies support the distinction of Tundra and Taiga as distinct species.

SIMILAR SPECIES Although measurements overlap between the smallest Taiga Bean-Geese (*fabalis*) and the largest Tundra Bean-Geese (*serrirostris*), most Tundras are considerably smaller than Taigas, with much shorter necks, stubbier bills, and darker heads. Some, however, appear intermediate. Taiga typically shows longer bill with minimal or no "grinning patch," unlike Tundra, and usually has more extensive area of orange in bill, but this feature is highly variable. Bill color and pattern eliminate Graylag and Greater White-fronted Geese; pale belly (lacking speckles or bars) also unlike white-fronted geese.

VOICE CALL: A nasal, sonorous, somewhat grating *gang gang*, usually lower in pitch and more resonant than Tundra.

STATUS & DISTRIBUTION BREEDING: Breeds in taiga and tundra-forest from northern Scandinavia east to western Anadyrland. **WINTER:** From northeastern Europe to eastern China and Japan. **VAGRANT:** AK has certain records from St. Paul I. (Apr. 1946), Shemya I. (3 birds; fall/winter 2007–2008), and Adak I. (May 2009), but other AK records of bean-geese are still being evaluated. Records of single bean-geese from IA/NE (1984-85), QC (1987), NE (1998), and WA (2002) appear to be of Taiga, with some authorities identifying them as the largest subspecies *middendorffii*. A single bean-goose at the Salton Sea, CA, in fall 2010 has not yet been identified conclusively. **POPULATION** Although some populations (especially nominate *fabalis*) are relatively stable or declining slowly, others (especially *middendorffii*) have been declining more rapidly.

long sloped bill with little or no "grinning patch"

adults
middendorffii

overall coloration and marking like Tundra Bean-Goose but much larger and longer necked

TUNDRA BEAN-GOOSE Anser serrirostris   TUBG ◼ 3

Tundra Bean-Goose is a smaller version of the closely related Taiga Bean-Goose and nests farther north, as its English name suggests. Polytypic (2 ssp.). L 28–33" (71–84 cm)

IDENTIFICATION A medium-size to moderately large brownish goose, comparable in size to *gambelli* Greater White-fronted Goose. Relatively compact appearance, a product of the short neck and short, stout bill (especially in smaller western subspecies *rossicus*). Bill blackish with pale orange or yellow subterminal band. Legs orange. **ADULT:** Brown with head usually considerably darker and white belly. **JUVENILE:** As adult, but pattern in flanks and coverts more muted. **FLIGHT:** Impressions of plumage as described for Taiga Bean-Goose, but head often appears contrastingly darker, neck much shorter. **GEOGRAPHIC VARIATION** The easterly nominate subspecies *serrirostris* breeds in the Russian Far East from Khatanga to Anadyrland and Kamchatka; it is relatively larger and longer billed than subspecies *rossicus,* which breeds in areas west of the nominate, in northwestern Siberia from the Taymyr Peninsula west to the Kanin Peninsula. **SIMILAR SPECIES** See Taiga Bean-Goose. Tundra Bean-Goose could be confused with Pink-footed Goose, but the latter has pinkish (not orange) legs and feet and a stubbier, smaller bill with pink patch near the tip. **VOICE** Similar to Taiga Bean-Goose, but some calls higher in pitch and sound less muffled, at least in some

gray on upperwing like Greater White-fronted

adults
serrirostris

darkish head

slightly larger than Greater White-fronted

rather short necked

short black bill with "grinning patch" and yellow-orange subterminal ring

adult

some Alaskan records and one from Salton Sea, CA, have appeared more intermediate between Taiga and Tundra; identification uncertain

Salton Sea record

subspecies. In Tundra Bean-Goose, nominate *serrirostris* is said to have lower-pitched calls than western *rossicus* but similar cadence in call structure. **STATUS & DISTRIBUTION BREEDING:** Tundra north to the Arctic across northern Russia. **WINTER:** From Europe to eastern China and Japan. **VAGRANT:** In AK, this species is recorded mostly as a spring vagrant, with verified specimens and photographs from Bering Sea and Aleutian Is. and probably the Seward Peninsula. Away from AK, there are apparent records from QC (possibly subspecies *rossicus;* 1982) and YT (2000). In autumn 2013, California's first Tundra Bean-Goose turned up at the Salton Sea in Oct. (very near the site of the 2010 bean-goose), and Nova Scotia's first Tundra Bean-Goose was found at Yarmouth in Nov. **POPULATION** Good population trend data are not available for Tundra Bean-Goose, but wintering populations appear to indicate stable or increasing populations of *rossicus* in many areas (also true in recent decades of many other goose species that nest at high latitudes) but declining populations of *serrirostris.*

PINK-FOOTED GOOSE *Anser brachyrhynchus* PFGO ▪ 4

A midsize brownish gray goose similar in structure to the Greater White-fronted, the Pink-footed Goose is a vagrant from Greenland most likely to be found in the Northeast. Monotypic. L 26" (66 cm) **IDENTIFICATION** Legs and feet pink, bill black with extensive pink on terminal half. Juveniles may show dull or even orange tone to legs. **ADULT:** A brownish goose with darker head and white belly. Upperparts tipped white, creating neatly barred appearance. Upperparts show gray wash, contrasting with browner underparts. Tertials crisply fringed white. Flanks edged with narrow white line, which forms a separation with the folded wings. Hind flanks dark barred. **JUVENILE:** As adult, but lacking white edge to flanks and barring on flanks. Legs often orange tinged. **FLIGHT:** A brown goose showing obvious gray wash to the upper wings. The upperparts contrast with the narrow white rump band; the tail is brown with a comparatively wide white border. **SIMILAR SPECIES** Juvenile White-fronted Goose similar, but it has longer pink bill lacking black base. In addition, Pink-footed Geese show obvious white fringes on the tertials. Also very similar to both bean-geese, but the Pink-footed has pink on bill and legs. Bean-geese are browner, lacking grayish wash to upperparts of most Pink-footed Geese. Furthermore, Pink-footed is much smaller than Taiga Bean-Goose with stockier shape; closer in size to Tundra Bean-Goose. In flight, Pink-footed Goose has a more prominent white edge to the tail and more gray on wings. **VOICE CALL:** A high-pitched *ayayak.* **STATUS & DISTRIBUTION** Vagrant, breeds in Greenland, Iceland, and Svalbard, wintering in British Isles and Low Countries of Europe. **VAGRANT:** Casual to Atlantic Canada and northeastern U.S.

more extensive blue-gray on upperwing than Greater White-fronted

broad white tail tip

small, stubby, dark-based bill with pinkish band

comparatively small head, dark neck

some have white rim at bill base

warm cinnamon-buff tinge at base of neck

light silvery or "frosty" gray upperparts on most adults; some are browner

pinkish legs and feet

broad white tip to tail

GREATER WHITE-FRONTED GOOSE *Anser albifrons* GWFG ■ 1

Colloquially known as the "Speckle-belly" due to the variable black barring on the underparts, this is a medium-size grayish brown goose. Polytypic (5 ssp.; 4 in N.A.). L 28" (71 cm)

IDENTIFICATION Legs and feet orange, bill pink, sometimes more orange. **ADULT:** Brownish with a distinctive white band around the bill base, most obvious on the forehead. The underparts are brown with black barring from lower breast to upper belly; belly and vent are white. The flanks are bordered by a white line. Above brown with paler feather tips. **JUVENILE:** Similar to adult, but lacks the white forehead and speckling on underparts; bill has a dark nail. **FLIGHT:** Brown showing a gray wash to the coverts and a narrow white rump band.

GEOGRAPHIC VARIATION Five subspecies recognized, three breeding in N.A. The smaller *sponsa* (breeds in western AK and Russian Far East; winters in Pacific states and western MX) and the larger "Tule Goose," *elgasi* (see sidebar p. 15), breeding between Cook Inlet and Alaska Range, Alaska, and wintering in Sacramento Valley. The large pale *gambelli* from northern Alaska

and northwestern Canada winters in Mexico to Texas. Another subspecies, *flavirostris,* the "Greenland White-fronted Goose," is a vagrant to eastern North America. This form is larger than *sponsa,* darker brown, more heavily barred on underparts, and shows an orange bill. The nominate subspecies breeds in northern Eurasia.

SIMILAR SPECIES White front and barred belly diagnostic in adult, although some adult Pink-footed Geese have limited white feathering at the bill base. Compare immature with other "brown geese" and accidental Lesser White-fronted Goose.

VOICE CALL: A laughing two- to three-syllable *ka-yaluk* or *kaj-lah-aluk,* somewhat grating and high-pitched.

STATUS & DISTRIBUTION Common, also in Eurasia. **BREEDING:** Tundra and to a lesser extent taiga wetlands. **MIGRATION:** Mid-continent populations move south from Aug.–Sept. to stopover areas from eastern AB to western MB; most depart by mid-Oct. for coastal TX, then to sites from Mexico to LA. Northward movements begin in Feb. and track snowmelt, arriving in central AK late Apr.–early

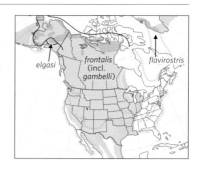

May. Pacific Population of central to southwestern AK, including "Tule Goose," fly south along coast eventually moving inland to stop in Klamath region of southern OR, northern CA in Sept.–Oct. More southern breeders and "Tule Goose" continue to winter in Central Valley of CA, while others continue to central plateau of Mexico. **WINTER:** Wetlands, agricultural areas, short-grass fields. **VAGRANT:** Rare east of range in migration and winter.

POPULATION Generally increasing, Pacific populations declined strongly in the 1970s and '80s, but are regaining in numbers. "Tule Goose" numbers estimated at 7,500.

dark neck

orange bill

Greenland adult
flavirostris

heavier black bars on underparts

orange legs

long, thick-based bill

long, thick dark neck

"Tule Goose" taiga adult
elgasi

gray on upper wing

adult
sponsa

white rump band with dark gray tail base

high-pitched laughing calls heard in flight

immature
sponsa

bill color of immature varies from orangish to pinkish

some white at base of bill by end of first winter

juvenile
sponsa

western AK *sponsa* is small and pale; birds breeding from northern AK (sometimes recognized as *gambelli*) are also pale, but larger

adult
sponsa

fewer black bars (variable)

largely unmarked belly

LESSER WHITE-FRONTED GOOSE *Anser erythropus* LWFG ▪ 5

This medium-size brown goose is accidental from the Palearctic. Monotypic. L 22–26" (55–66 cm)
IDENTIFICATION The Lesser White-fronted is extremely similar to the White-fronted Goose, but stockier in structure, shorter necked, and smaller billed, with longer and narrower wings. The wings extend beyond the tail when folded. The legs and feet are deep orange, bill is pink, orbital ring is yellow. **ADULT:** White "front" extends back onto the crown. Brown below with minimal black barring on the belly. **JUVENILE:** Similar to the adult,

but fresh juveniles lack the white forehead and speckling on the underparts.
SIMILAR SPECIES Smaller, darker, longer winged, shorter necked, and smaller billed than Greater White-fronted. In adults the extensive white on the forehead and the yellow orbital are diagnostic. Juveniles may be separated from the Greater White-fronted by darker body, size and structure, yellow orbital, and bill with pale nail.
VOICE Yelping and high-pitched calls.
STATUS & DISTRIBUTION Accidental, breeds from Scandinavia to Siberia, wintering in a few midlatitude sites.

VAGRANT: One specimen from Attu I., AK, June 5, 1994, and a recent record from St. Paul I., Pribilofs, AK (June 21, 2013).

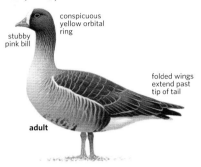

conspicuous yellow orbital ring

stubby pink bill

folded wings extend past tip of tail

adult

GRAYLAG GOOSE *Anser anser* GRGO ▪ 5

This native of Europe and northern Asia is the progenitor of most domestic geese. Often not shy, the domestic forms are larger and heavier bodied, with deeper bellies than their wild ancestors. Many show variable amount of white at base of bill, like the White-fronted. When swimming, the

domestic's rear end floats higher than rest of its body. L 29–32" (74–81 cm)
IDENTIFICATION Wild-type Graylag is large, with thick neck, large head, heavy, pinkish orange to pink bill, and dull pink legs. The dumpy domestic Graylag is a very large, heavy-bodied, thick-necked goose. Domestic forms range from gray-brown to entirely white or multicolored. Hybridization between domestics and sometimes with Canada Goose is frequent, producing a wide away of structures and plumages. Presumed Canada x domestic Graylag often has a large, buffy cheek patch, white eye-ring, brown crown and neck, and some white at base of bill, suggestive of Canada x White-fronted hybrids.
SIMILAR SPECIES Wild form compared to juvenile White-fronted Goose is stockier, thicker necked, and grayer

overall. Compare domestics carefully to White-fronted Goose and white forms to Snow Goose. Graylag is larger, stockier, with deep belly, often practically dragging on the ground. Domestic Graylag bills are usually larger and thicker at base then those of White-fronted Goose. Some white forms, known as "Domestic Goose" can have relatively petite bills, but do not show "grinning patch" of the Snow Goose. Plumage of the Ross's Goose could be confused, but it is significantly smaller and lighter bodied, with a dainty bill. The Graylag is larger and stockier and lacks dark markings on bill of the bean-goose or Pink-footed Goose.
VOICE Loud, raw, deep *honks*. Wide repertoire of calls.
POPULATION The Graylag is increasing in native lands of Europe.

head, neck, and body uniformly gray-brown

large, thick, orangish bill

adult
anser

striking pale gray forewing visible in flight

"Tule Goose" Subspecies of Greater White-fronted Goose

The "Tule White-fronted Goose," subspecies *elgasi* but previously named *gambelli* or *gambeli,* is rare and range-restricted. It has been known largely from the wintering grounds, and only in the 1970s were the breeding grounds determined. More work needs to be done to clarify the status of birds breeding in the Yukon Territory, which may also be *elgasi.* In winter both the "Tule" and the "Pacific" population of Greater White-fronted Geese are found in the Sacramento Valley, sometimes together. It is when they are together that identification is most straightforward. The "Tule"

"Tule Goose" (CA)

is larger, with a longer neck and a noticeably longer and deeper bill. In direct comparison these size and structural features should be apparent. In addition, the "Tule" is darker, particularly on the head and neck; in fact these geese often show a contrasting dark cap that extends as a dark line along the back of the neck. The speckling on the belly is less extensive on the "Tule" than on the typical White-fronted Goose. Finally, there are ecological differences. The "Tule" often feeds in marshes, eating the large tubers of tule *(Scirpus robustus),* while other White-fronted Geese forage in open grassy fields. ▪

SNOW GOOSE AND ALLIES Genus *Chen*

This is a small genus of three species with mainly a North American distribution, sometimes incorporated into *Anser*. Two of the three species are polymorphic, a feature not found in other goose genera. Plumages of all Snow Geese and allies show white tails and largely white heads.

EMPEROR GOOSE *Chen canagica*   EMGO ■ 2

The Emperor Goose is a gorgeous marine goose of the Bering Sea area. Monotypic. L 26" (66 cm)
IDENTIFICATION A small stocky goose with thick and strong legs. Bill pink with black cutting edges and nostril, and legs bright orange. **ADULT:** Distinctive, gray bodied, with each feather neatly tipped dark subterminally and white terminally, giving a characteristic scaly look. Head white, continuing down the back of the neck and contrasting with black throat and foreneck. Sometimes the head may be stained rust with oxides. In flight, the Emperor Goose appears gray, with white tail and white head and hind neck. **IMMATURE:** Browner than adult, but shows scaly appearance and white tail although it has an entirely brown head and neck.
SIMILAR SPECIES Distinctive, although blue morph Snow or Ross's Geese show similar dark body with white face or neck. Emperor Goose has a diagnostic pattern of white head and hind neck contrasting with black throat and foreneck. Barnacle Goose shares grayish scaly appearance, but it has an entirely black breast and neck with only face white and a dark tail.
VOICE CALL: A hoarse and repeated *kla-ha,* although usually remains quiet.

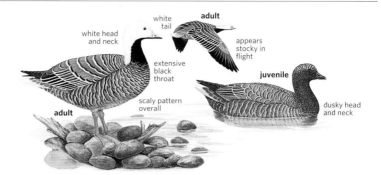

STATUS & DISTRIBUTION Uncommon, also breeds in the Russian Far East. **BREEDING:** In coastal wetlands with tidal influence. **MIGRATION:** In fall, begin staging from mid-Aug.–Oct. on the north shore of the Alaska Peninsula, dispersing to the Aleutian Islands by Nov. Wintering areas largely in Aleutian Islands and south coast of Alaska Peninsula. Birds depart Mar.–early Apr. for staging lagoons on the north shore of the Alaska Peninsula. Birds stage until May, and then fly to nesting areas, largely in the Yukon-Kuskokwim Delta. **WINTER:** Intertidal habitats, from rocky coasts to eelgrass beds. **VAGRANT:** Casual south to CA on the

Pacific coast, with few inland records in the Sacramento Valley.
POPULATION Appears to have decreased between 1960 and the 1980s, but has remained stable since then.

SNOW GOOSE *Chen caerulescens*   SNGO ■ 1

This species is the larger and more common white Arctic goose. There is also a blue morph, which commonly occurs in the mid portion of the continent and is called the "Blue Goose." Polytypic (2 ssp.; both in N.A.). L 26–33" (66–84 cm)
IDENTIFICATION Medium size, with a pink bill and distinctive "grinning patch" along the cutting edge of the bill. Adult's legs pink, immature's legs dusky pink. **ADULT WHITE:** White, with black primaries. Often face is stained rust from oxides in water. **JUVENILE WHITE:** Similar to adult, but gray-brown wash on head, neck, and upperparts; dark centered tertials. In flight shows dusky secondaries. **ADULT BLUE:** Brown body, white head and neck. Wings with gray-blue coverts,

long tertials, inner greater coverts black with bold white fringes. Tail gray with white border. **JUVENILE BLUE:** Dull brown, including head and neck. Pattern of wing and tertials less well developed. **GEOGRAPHIC VARIATION** Interior and western populations known as "Lesser Snow Goose," *caerulescens;* eastern population is the "Greater Snow Goose," *atlantica.* Size overlaps and perhaps better considered monotypic. Blue morph lacking in Pacific and Atlantic wintering populations. **SIMILAR SPECIES** Most similar to Ross's Goose, but adult Snow Geese of both morphs have black "grinning patch" on bill, are larger, longer necked, and longer billed. Blue morph has white

head and neck and blue wing panel; compare to Ross's. Immature white morph Snow Geese are extensively brownish washed above, and in flight shows brownish secondaries. **VOICE CALL:** A barking *whouk, or kow-luk.* **STATUS & DISTRIBUTION** Abundant, casual to Europe, now rare to Japan where formerly numerous. **BREEDING:** Moist tundra. **MIGRATION:** Southbound movements begin late Aug.– early Sept., arriving in wintering areas by early Oct., peaking late Oct. and Nov. Northbound movements begin in Feb., peak early Apr. in southern Canada and reach breeding areas in late May. **WINTER:** Wetlands and agricultural fields.

POPULATION Growing exponentially since 1960s; they are exceeding the carrying capacity at breeding sites, overgrazing sensitive grassy tundra.

ROSS'S GOOSE *Chen rossii* ROGO ▪ 1

The smallest of the white Arctic geese, Ross's Goose has a short, triangular bill. Monotypic but polymorphic. L 23" (58 cm)
IDENTIFICATION Stocky and short necked, with high forehead and short legs. The bill is pink and shows grayish "warts" on the base of the upper mandible. Legs pink. **ADULT:** Entirely white, with black primaries. **JUVENILE:** Similar to adult but variably washed dusky, particularly on the nape and back of neck. The tertials show dark shafts or centers. **HYBRIDS:** With the Snow Goose, intermediate in size and structure and "grinning patch," best identified by comparing with the two parental species. **GEOGRAPHIC VARIATION** Monotypic, but a rare blue morph exists. The origin of this blue morph is controversial and is thought to be due either to introgression with blue-morph Snow Geese or a recurrent mutation of genes controlling feather color. Blue-morph Ross's Geese are

most frequent in wintering flocks in California's Central Valley. They are structurally like typical Ross's Geese. **SIMILAR SPECIES** Like the Snow Goose but smaller, shorter necked, smaller billed, and rounder headed. Ross's lacks "grinning patch" but shows gray warts at bill base. In flight, smaller and shorter necked with more rapid wingbeats than Snow Goose. Blue-morph Ross's Geese differ from blue-morph Snow Geese in showing only a white face, with dark hind crown, nape, and neck; and extensive white on the lower breast and belly. **VOICE CALL:** A high, nasal *hawhh.* **STATUS & DISTRIBUTION** Common. **BREEDING:** Wet tundra. **MIGRATION:** Little is known about boreal staging areas in this species. However, the western population stops over in eastern AB, western SK largely in Sept. From there they move through

western MT, northern ID, eastern OR into the wintering areas in CA by Oct. The eastern population moves through Hudson Bay and stages in eastern SK and MB, before moving south through the Dakotas to wintering areas in central Mexico, eastern TX, and LA. Spring migration begins as early as Feb. and retraces the fall route, following the advance of the snowmelt. Staging in Canadian prairies takes place mid-Apr.–mid-May, and arrival in breeding grounds from end of May to early June Lingers later in wintering grounds than the Snow Goose. **WINTER:** Wetlands and adjacent agricultural areas. **VAGRANT:** Rare but regular east of the Mississippi River; increasing. **POPULATION** Since the mid-1950s Ross's Goose populations have increased dramatically; some estimates give a 10 percent annual increase from the 1950s to the 1990s.

CANADA GOOSE AND ALLIES Genus *Branta*

This is a genus of six species found throughout the Northern Hemisphere, including Hawaii. The birds have distinctive patterns on the face, neck, or breast. Most show a black neck "sock," with contrasting white face or neck patches. All have black bills and legs. Several are high-Arctic breeders; most forage on short grass during the nonbreeding season.

BRANT *Branta bernicla*   BRAN ■ 1

A sea goose that forages largely on marine grasses, the Brant is known as the "Brent Goose" in Great Britain. Polytypic (3 ssp.; 2 in N.A.). L 25" (64 cm)

IDENTIFICATION A smallish, short-billed, short-necked stocky goose. Bill and legs black. **ADULT:** All forms show a black head, neck, and breast; a fishbone pattern white neck collar varies in extent geographically. The upperparts are solid brown, contrasting with a white rump and vent and dark tail. Underparts variable depending on race, varying from pale brown to blackish brown with contrasting white flanks. **IMMATURE:** Similar to adult but lacks neck collar and shows white tips to wing coverts.

GEOGRAPHIC VARIATION There are three named subspecies—nominate *bernicla*, "Dark-bellied Brant," breeding in central and western Russian Arctic and wintering in western Europe; *hrota*, "American Brant," breeding in eastern Arctic of N.A., Greenland, and Svalbard and wintering in Atlantic

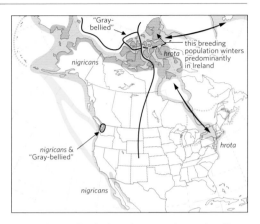

coast of U.S., England, and Ireland; and *nigricans*, "Black Brant," breeding from eastern Russia through western Arctic N.A. including AK to Victoria I., NU, and wintering on Pacific Coast from AK to Mexico. There is a fourth, as yet unnamed, population known as the "Western High Arctic" population or "Gray-bellied Brant." These birds breed mainly on Melville Is. and Prince Patrick Is., NT, and winter in the Puget Sound area and Boundary Bay, BC. The "Pale-bellied Brant" has a small neck collar, divided at front, pale underparts contrasting with black neck "sock," and white belly behind legs. "Black Brant" is darker above and below. It shows little contrast between the dark underparts and the neck sock; pale flanks contrast with

the dark underparts; the neck collar is broad and full and meets in the front of the neck. "Dark-bellied Brant" is a possible East Coast vagrant from Europe. It is dark below, but paler than "Black Brant," lacks the contrasting white flanks, and has a small and broken neck collar. "Gray-bellied Brant" appears to be variable, ranging from pale birds similar to "American

juvenile lacks broken whitish collar

juvenile
hrota

adults

all Brant fly with rapid, shallow wingbeats

hrota

"Gray-bellied Brant" adult

collar usually complete at bottom

underparts variable, usually much paler than *nigricans*; often like *hrota*

white neck collar broken in front

"American Brant"
hrota

pale belly contrasts with black neck

white-fringed coverts on 1st winter and juvenile

1st winter

white vent and black belly

complete white collar

"Black Brant"
nigricans

adult

flanks like adult

blackish breast and belly

grayish brown flanks

indistinct collar
juvenile

dark border to rear flank

frosty white barring on flanks

Brant," to birds with gray underparts and contrasting paler flanks; they have smaller, broken neck collars than the "Black Brant."

SIMILAR SPECIES Entirely black head, neck, and breast; white collar diagnostic. Cackling and Canada Geese have white face patch.

VOICE CALL: A quavering *crrr-oonkkk.*

STATUS & DISTRIBUTION Common, also breeds in Palearctic. **BREEDING:** In salt marshes in low Arctic and moist tundra in high Arctic. **MIGRATION:** Four populations are treated here; two are part of the "Pale-bellied Brant" subspecies, the Atlantic and Eastern High Arctic population. **ATLANTIC BRANT:** Breed largely near the Foxe Basin, NT; stage in James Bay in Sept. and fly nonstop to coastal NY and NJ, arriving in late Oct.–Nov. They move north in Apr. Some use a coastal route to the St. Lawrence River Estuary; others move inland over Lake Ontario and up the Ottawa River Valley (late May). Both routes converge in James

Bay, where they stage before heading to the breeding grounds by early June. Historically more birds moved up the coast, with important staging areas in the Maritimes; these routes have shifted westward. **EASTERN HIGH ARCTIC BRANT:** They leave their breeding grounds in late Aug. and stage in western Greenland in Sept. before continuing to Iceland (mid-Sept.) and wintering sites in Ireland and England by late Sept. or Oct. Spring migration begins in late Apr.; they stage for several weeks in Iceland and fly over the Greenland ice cap directly to breeding grounds by mid-June **BLACK BRANT AND "GRAY-BELLIED BRANT":** They make their way from various breeding areas from the western Canadian Arctic to congregate at Izembek Lagoon staging area on the Alaska Peninsula as early as mid-Sept. Victoria I. breeders fly a coastal route around AK and arrive at the staging grounds by mid-Oct. They then fly directly to the Queen Charlotte Is. and southern BC coast

before continuing south well offshore on the Pacific coast to wintering sites, arriving late Oct.–Nov. A few migrate to Japan for the winter. Northbound migration peaks in Mar. and Apr. on California coast, with staging areas on east coast of Vancouver I. before flight to Alaska Peninsula, peaking there in mid-May, and arriving at breeding grounds mid–late June **WINTER:** Coastal, primarily in large estuaries where beds of eelgrass and other intertidal plants are available for forage. **VAGRANT:** Casual in interior, away from eastern Great Lakes. "Dark-bellied Brant" has been reported but not conclusively documented from the East; it should be looked for in wintering concentrations of "Pale-bellied Brant."

POPULATION "Black Brant" have decreased in last several decades; "Atlantic Brant" have fluctuated markedly and are now stable or increasing. The "Gray-bellied Brant" population is small and of conservation concern.

BARNACLE GOOSE *Branta leucopsis*   BARG ▪ 4

Before bird migration was accepted, legends told that when Barnacle Geese disappeared from wintering sites in western Europe, they turned into barnacles. Monotypic. L 27" (69 cm)

IDENTIFICATION A small, short-necked goose with a stubby bill. Bill and legs black. **ADULT:** White face and black lores contrast with black neck and breast. Underparts white with gray barring on flanks. Upperparts gray, with scaled appearance created by dark subterminal band and white terminal band on each feather. **FLIGHT:** Gray wings and back, with black neck; lower back and tail contrasting with white rump band. **SIMILAR SPECIES** Distinguishable from

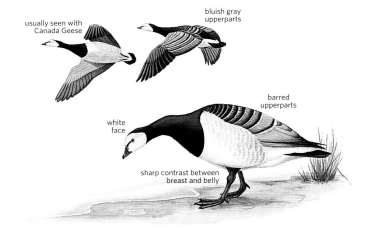

usually seen with Canada Geese

bluish gray upperparts

barred upperparts

white face

sharp contrast between breast and belly

Cackling Goose and Brant by white underparts and grayish wings and back; white face a further distinction from Brant. Hybrids with Cackling or Canada geese are known.

VOICE CALL: A barking *kaw.*

STATUS & DISTRIBUTION Vagrant, status unclear as truly wild birds impossible to differentiate from escaped captives. Breeds on tundra in eastern Greenland, Svalbard, and Novaya Zemlya, winters largely in Ireland and Scotland and the Netherlands. **VAGRANT:** Numerous sightings throughout continent, mostly from

the east; most or many of these pertain to escapes from captivity, particularly well away from East Coast. Records from Maritime Provinces, NF and Labrador, and northeastern U.S. may include wild birds. These sightings include small family groups and singles from October to April, all mixed with other geese, especially Canadas. There is a growing consensus that these likely involved wild birds from Greenland or Svalbard island group.

POPULATION Has increased in last couple of decades.

CACKLING GOOSE *Branta hutchinsii* CACG ■ 1

The Cackling Goose was split from the Canada Goose in 2004. Polytypic (4 ssp.; all in N.A.). L 23–33" (58–84 cm)

IDENTIFICATION A small, stubby-billed goose with short legs, stocky body, and short neck. Comparatively long winged, with primaries extending noticeably past the tail. **ADULT:** Black neck and head with a contrasting white cheek patch that wraps around the throat. Often shows a white neck ring below the black "sock."

GEOGRAPHIC VARIATION Four subspecies recognized. "Richardson's Goose" *(hutchinsii)* breeds from Mackenzie Delta, Northwest Territories, east to western Baffin I. and south to Southhampton I. and McConnell River, Hudson Bay, Nunavut; may breed in western Greenland, winters from eastern NM and northern TX south into highlands of Mexico, also coastal TX to western LA and south to northern Veracruz, Mexico. "Taverner's Goose" *(taverneri)* breeds in coastal wetlands of Seward Peninsula and North Slope, AK, and winters in Columbia River Valley, WA and OR, south to the Central Valley of CA. "Cackling Goose" *(minima)* breeds in the Yukon-Kuskokwim Delta, AK, and winters largely from Willamette Valley, OR, south to Central Valley of CA, previously wintered largely in CA but has shifted north; "Aleutian Goose" *(leucopareia)* breeds on Aleutian and Semidi Islands, AK, and winters in Central Valley, CA, but expanding. "Richardson's" is the palest, with a pale breast form and intermediate in size but with a proportionately long bill. Dark-breasted *minima* is the smallest, with a stubby short bill. "Taverner's" is the largest, with a longer neck and rounder head; intermediate in darkness, but variable. "Aleutian" is larger than *minima,* but smaller than "Taverner's"; a moderately dark form; it typically has a complete white neck collar.

SIMILAR SPECIES Due to the split of Cackling and Canada Geese, we have a vexing problem in field identification, mainly that the largest Cackling Goose (subspecies *taverneri*) is close in size to the smallest Canada Goose (subspecies *parvipes*). In the past, *taverneri* and *parvipes* were grouped together as a single subspecies; however, recent studies clearly show them to be genetically different. On average, *parvipes* Canada is larger than *taverneri* Cackling and has a relatively longer neck and longer

bill. In general, the Cackling Goose has longer and more pointed wings than the Canada Goose, and this difference shows up as a longer primary extension (past tertials) and wing extension (past tail) in the Cackling. The wintering range of *taverneri* is not clearly known, but it is not expected to be regular east of the Cascade Range/Sierra Nevada Mountains. So a troubling intermediate goose in the East is more likely *parvipes,* which migrates and winters largely east of the mountains. The dark-bodied *parvipes* from Anchorage, AK, winters in the Pacific Northwest.

VOICE CALL: Higher pitched and more cackling than Canada Goose.

STATUS & DISTRIBUTION Common. **BREEDING:** Moist coastal tundra. **MIGRATION:** Richardson's flies south from late Aug., staging in central and eastern prairie provinces, peaking in Oct. Northward movements begin in Feb.–Mar. Cackling *(minima)* and Aleutian stage in Alaska Peninsula into Oct., then perform nonstop (2–4 day) flight to OR and CA. They move north by late Apr. **WINTER:** Short-grass fields, including agricultural areas and wetlands. **VAGRANT:** Rare away from regular migratory areas.

POPULATION General increase in various populations during the last decade; the Aleutian subspecies has been removed from the endangered species list.

"Aleutian" *leucopareia*

"Cackling" *minima*

flies with rapid, shallow wingbeats

short neck especially noticeable in flight

rounded head

short neck

stubby bill

calls higher pitched than Canada

"Taverner's"

"Richardson's" *hutchinsii*

overall pale

like *parvipes* Canada but with rounder head and stub-

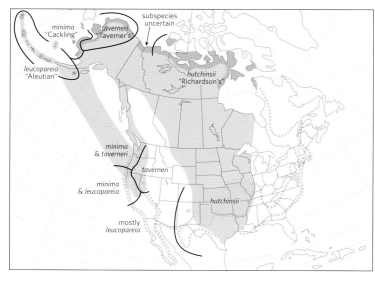

subspecies uncertain

minima "Cackling"

taverneri "Taverner's"

leucopareia "Aleutian"

hutchinsii "Richardson's"

minima & *taverneri*

taverneri

minima & *leucopareia*

hutchinsii

mostly *leucopareia*

CANADA GOOSE *Branta canadensis*   CANG ▪ 1

The "honker" is the common goose in most of North America. Polytypic (7 ssp.; all in N.A.). L 30–43" (76–109 cm)
IDENTIFICATION A large and long-necked goose; legs and bill black. **ADULT:** Black neck and head with a contrasting white cheek patch. Body brown, paler below, often darker on rear flanks. Belly and vent white. **FLIGHT:** Brown above including wings, lower back blackish, as is the tail, contrasting with a white rump band.
GEOGRAPHIC VARIATION Seven subspecies in N.A. Subspecies *canadensis* breeds Ungava Bay east to Newfoundland, wintering on Atlantic seaboard; *interior* breeds west of Ungava Bay through Hudson Bay lowlands to northern MB (Manitoba) and north to southern Baffin Island and southwestern Greenland, wintering throughout the East; *maxima* ("Giant Canada Goose") breeds central MB and MN south to KS and western KY, some resident, others short-distance migrants; *moffitti* breeds south-central BC to western MB south to CO and OK, wintering southern portion of breeding range to southern CA, northern Mexico, and TX; *parvipes* ("Lesser Canada Goose") breeds in boreal forest zone from central AK to northwestern Hudson Bay, wintering eastern WA and eastern OR to northeastern Mexico and eastern TX; *fulva* ("Vancouver Canada Goose") breeds in Copper River Delta and Prince William Sound, AK, wintering in Willamette Valley, OR; *occidentalis* ("Dusky Canada Goose") breeds from Glacier Bay, AK, to northern Vancouver Island, many resident, others winter in Willamette and Columbia Valleys, OR. These subspecies intergrade to various extents and can be thought of as fitting three general groups. The subspecies *canadensis, interior, maxima,* and *moffitti* are very similar. The "Lesser Canada Goose" *(parvipes)* varies in size, but on average is smaller than all other Canada Geese. The "Vancouver" and "Dusky Canada Geese" are very dark and color saturated, the "Dusky" being smaller than the "Vancouver."
SIMILAR SPECIES See Cackling Goose account.
VOICE CALL: Males give a lower pitched *hwonk,* females a higher *hrink.*
STATUS & DISTRIBUTION Abundant,

the birds have been introduced to Western Europe. **BREEDING:** Various freshwater wetlands, golf courses. **MIGRATION:** Complex due to number of different populations and staging areas; however, most wild populations of the Canada Goose are migratory. In modern times "feral" Canada Geese have been introduced in various parts of the continent, many of these being primarily stocks descended from mixes of "Giant Canada Goose" and others. Many of these birds are residents in urban areas, or they perform only minor migrations. Migrant geese tend to leave breeding grounds in Aug.–Sept., peak in Oct., and arrive at wintering areas from mid-Oct.–Nov. Spring movements begin in Feb., peaking in Mar. **WINTER:** various grassy habitats, from urban parks to native wetlands; also agricultural fields.
POPULATION This goose has dramatically increased in number since the

1940s; it is now estimated that nearly five million Canada Geese live in North America, and the birds are considered urban pests in some regions.

smallest Canada; compare to *taverneri* Cackling

"Lesser" *parvipes*

long, black neck

"Atlantic" *canadensis*

honking calls lower pitched than Cackling

long, black neck

flies with slower, deeper wingbeats than Cackling

"Dusky" *occidentalis*

darkest Canada Goose

moffitti

widespread, large western subspecies

SWANS Genus *Cygnus*

Of the seven species of swans, six comprise the genus *Cygnus*. The genus has a worldwide distribution, but only in temperate regions. All *Cygnus* are huge birds, with long necks, long bills, and strong legs. Southern Hemisphere species are black or black necked, but all North American species are entirely white, differing from each other in subtle structural and bill pattern variations.

TRUMPETER SWAN *Cygnus buccinator*   TRUS ■ 1

The long-necked Trumpeter is our native swan of forested habitats. Monotypic. L 60" (152 cm)
IDENTIFICATION Huge swan, with a long, sloping head and bill profile, as well as a relatively thick-based neck. The bare loral skin is as wide as the eye and narrowly encircles the eye. Shape of upper edge of bill, where it meets forehead, comes forward to a create a V-shaped point along the bill's midline. **ADULT:** White, often stained yellowish on head and upper neck. Bill black with orange stripe on lower mandible along cutting edge. Legs black. **JUVENILE:** Appears dirty, pale brownish gray throughout. Bill dull pink with dark base and dark loral skin, also dark on tip and cutting edge.
SIMILAR SPECIES Most likely to be confused with the Tundra Swan. The Trumpeter is larger, has a longer, sloping bill with a V-shaped upper edge. The upper mandible of the Trumpeter is always black, but a few Tundras may lack yellow bill spot. The loral skin of

the Trumpeter is wide, but narrow on the Tundra.
VOICE CALL: A bugling *oh-OH* like an old car horn, second syllable emphasized.
STATUS & DISTRIBUTION Uncommon to rare, currently being stocked in various states and provinces in the East. **BREEDING:** Various freshwater wetlands, at least 300 feet long, needed for takeoff. **MIGRATION:** Migrants leave north by mid-Oct., arriving in south by early Nov. or later; northbound movements begin late Feb., arriving on breeding grounds from Apr. Alaskan population winters mainly in coastal BC and western WA. YT and NT

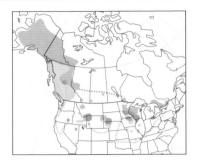

population migrates largely east of Rockies to winter in tristate area of MT, WY, and ID. Southern breeding populations resident or make only local movements. Newly stocked populations in central and eastern N.A. largely resident, but some may move several hundred miles to south. **WINTER:** Freshwater wetlands to grassy and agricultural fields near water bodies. **VAGRANT:** Very rare to CA, casual to NM and TX.
POPULATION Historically suffered a huge population decrease, but conservation efforts have allowed native western populations to increase. A somewhat controversial reintroduction to the central part of the continent has been successful, although former range in eastern North America is very unclear.

white forms
V-shape
on forehead

retains darker
plumage later
in winter than
Tundra

adult

juvenile

TUNDRA SWAN *Cygnus columbianus*   TUSW ■ 1

The Tundra is the widespread and more highly migratory swan in the continent. Polytypic (2 ssp.; both in N.A.). L 52" (132 cm)
IDENTIFICATION Large, shorter-necked swan with rounded forehead, clearly setting off bill profile. The bare skin on the lores narrows to a point before the eye. Upper edge of bill, where it meets forehead is smoothly curved—shallow U-shape when seen from front.
ADULT: White throughout. Bill black

with variable yellow patch at bill base, extending forward from eye; patch rarely absent. Yellowish stripe on lower mandible along cutting edge. Legs black. **JUVENILE:** Dull whitish gray throughout, but some become largely white by midwinter. Bill pink to base, including loral skin; this darkens from base outward as bird ages. Obvious black nostril on pink bill.
GEOGRAPHIC VARIATION Subspecies *columbianus* widespread, *bewickii*

("Bewick's Swan") of Asia and Europe a vagrant to western N.A. It is differentiated from *columbianus* by having much more extensive yellow on bill base, covering more than a third of the bill. Hybrid family groups have been seen in N.A.
SIMILAR SPECIES Most likely to be confused with the larger Trumpeter Swan. Tundra has shorter bill, steeper forehead, and rounder crown. When observed from the front, the upper edge

to the bill, where it meets the forehead, is shaped like a shallow U, not the sharp V of the Trumpeter. Yellow at bill base is characteristic of Tundra, and absent on Trumpeter; the loral skin of Tundra pinches in before the eye. Immatures begin with a pink base to bill and quickly become much whiter-plumaged than most Trumpeters.

VOICE CALL: A loud barking and somewhat gooselike *kwooo.*

STATUS & DISTRIBUTION Common, a subspecies in Palearctic. **BREEDING:** Tundra lakes and ponds. **MIGRATION:** Western population stages on Great Salt Lake, then moves to more coastal wintering sites. Eastern population stages from ND to MN and flies to Atlantic coast wintering areas. Staging areas used in Oct., arrival in wintering sites by late Oct. to mid-Nov. Northbound by late Feb. and early Mar. Eastern population stages around Lake Erie, continuing to MN and Canadian prairies, arrives in Arctic breeding grounds by mid-May. Western population retraces route through Great Salt Lake, or through Klamath region, arrives in breeding grounds of AK in Apr. **WINTER:** Various wetland habitats and agricultural fields.

VAGRANT: Casual to Maritimes and Gulf Coast; rare throughout interior away from regular migration routes.

POPULATION Appears stable now, but is thought to have doubled between the 1960s and '90s.

white on forehead forms a V

white on forehead cuts more straight across

bill has more concave shape than Trumpeter

eye stands out

size of yellow lore spot variable, absent on some

extensive yellow squared off at base of bill

"Bewick's Swan" adult *bewickii*

juvenile *columbianus*

adult *columbianus*

Trumpeter

Tundra

WHOOPER SWAN *Cygnus cygnus*   WHOS   ≡   3

This swan is a vagrant from the Old World. Monotypic. L 60" (152 cm)

IDENTIFICATION Huge, similar structurally to Trumpeter Swan. Bill long; sloping forehead accentuates bill length. Adults have extensively yellow bill base, covering more than half of the bill, at least to the level of the nostril. Much of the underside of the lower mandible is also yellow. The shape of the yellow comes forward to a point. **ADULT:** Body entirely white, sometimes stained yellowish on head and neck. **JUVENILE:** Body grayish brown throughout, bill whitish gray shaped like yellow pattern of adult; nostril, cutting edge, and nail black.

SIMILAR SPECIES In shape and size resembles Trumpeter Swan, although large amount of yellow on bill easily separates the Whooper. Tundra Swan smaller, with shorter bill and steeper

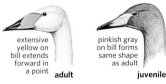

extensive yellow on bill extends forward in a point   **adult**

pinkish gray on bill forms same shape as adult   **juvenile**

forehead as well as small yellow patch on bill base. However, vagrant *bewickii* subspecies, "Bewick's Swan," has more extensive yellow on bill, more like Whooper Swan. Note that yellow on Whooper comes forward to a point and extends past nostril; on "Bewick's" yellow ends more abruptly and does not reach the nostril. Immatures have a bill pattern that suggests that of the adult, helping to identify them.

VOICE CALL: Bugling; often gives three or four calls in a group *kloo-kloo-kloo.*

STATUS & DISTRIBUTION Vagrant, widely distributed from Europe to Asia,

breeding in higher latitudes, wintering at midlatitudes. **BREEDING:** Has bred on Attu I. **WINTER:** Large wetlands and agricultural fields. **VAGRANT:** Rare in winter on central and outer Aleutians and casual elsewhere in western AK and Bering Sea islands. Casual south to OR and northern CA. Many records in eastern N.A. likely pertain to escapes.

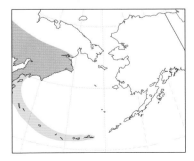

MUTE SWAN *Cygnus olor*   MUSW   ■ 1

This is the quintessential swan, a species introduced from Europe no doubt to give our local parks and ponds a royal touch. However, this species is a great ecological problem here. Monotypic. L 60" (152 cm)

IDENTIFICATION A huge, long-necked swan that holds its neck in a characteristic S-shape with the bill pointing downward. The wings are held partially raised, arched over the back. The bill has a prominent black knob at the base, which is larger in males than females. **ADULT:** Entirely white body, sometimes with yellowish wash to head and neck. Bill bright orange or salmon with black base, cutting edge, and nail, as well as black bulbous knob at base of culmen. Bill color of females duller. Legs black. **JUVENILE:** Entirely brownish gray; bill gray with black base and black around nostril. Older immature becomes progressively whiter, and bill becomes pale pink; black knob slowly enlarges as immature ages.

SIMILAR SPECIES Both Tundra and Trumpeter Swans may be found in established range of the Mute Swan. Aggressive posture with wings raised over body, and black-and-orange bill with black knob diagnostic for Mute Swan. Immatures of all swan species more similar; Mute has black base to bill as in Trumpeter and shows distinctive S-shape to neck; bill begins grayish and turns pinkish orange later on in life, by which time growth of bill knob has started.

Swans in city parks are likely Mute Swan, although reintroduced Trumpeter Swans in the East may also frequent urban parks.

VOICE CALL: As its name suggests, this swan is usually silent. However, it is not mute. Usually calls heard are hisses and other soft alarm calls. Rarely it gives a resonant bugle.

STATUS & DISTRIBUTION Common, introduced from Europe. Native from eastern Europe to western Asia. **YEAR-ROUND:** From small urban ponds and lakes to large wetlands. **MIGRATION:** Mostly resident in North America, although some birds make short-distance movements to larger bodies of water. **WINTER:** May flock together in larger bodies of water, including the Great Lakes. **VAGRANT:** Rare in Mississippi watershed and the southern Atlantic coast; at least some of these birds are dispersing from more northerly established sites. Small populations now locally established in Pacific states. Most birds elsewhere are either feral or local escapes.

POPULATION Booming in northeastern states, with substantial increases in Chesapeake Bay population and continuing expansion of the range particularly to New England. Other populations stable or increasing and of general concern as a management problem. Mute Swans can greatly alter the ecology of wetlands by uprooting emergent vegetation as they feed.

sometimes arches wings

dark knob

dark border at base of bill

adult

juvenile

darker than other juvenile swans

Identification of Young Swans

In their first fall, juvenile swans migrate south in their juvenal plumage, typically accompanying their parents. Adults traveling with their young help in identifying them. During fall and early winter, juvenile swans are largely pale gray to dirty white. On average Tundra juveniles are paler than Trumpeter juveniles, although both Trumpeter and Mute Swans have a rare all-white juvenile plumage. During winter juveniles wear somewhat and also perform a partial body molt, giving them a paler plumage as they age. By late winter Tundras are largely white; other species still grayish. Other than structure, bill patterns are the best distinctions. Juvenile Tundras have a pink bill base; Mutes and Trumpeters have a dark bill base. Mute Swans begin to develop a bill knob early. ■

Trumpeter Swan

Tundra Swan

Mute Swan

PERCHING DUCKS Genera *Cairina* & *Aix*

These genera are represented by two species worldwide. One species from each genus is present in N.A., the Muscovy (*Cairina*) and the Wood (*Aix*). Both are shy species, at home in slow-moving waters surrounded by large trees. They nest in cavities or use nest boxes and have dark glossy plumage. Otherwise, the two species are very different.

MUSCOVY DUCK *Cairina moschata*   MUDU ■ 2

This large, tropical duck is found most often near Falcon Dam in southern Texas. Domestic birds are in parks throughout North America and widely established in Florida. Shy; most active at dusk and dawn. Monotypic. L 26–33" (66–84 cm)
IDENTIFICATION Adults blackish with iridescent greenish and purple highlights. **MALE:** Glossier than females, with

dark reddish knob above bill. **FEMALE:** Duller and smaller; lacks knob on bill. **JUVENILE:** Duller, slowly acquires white wing patch during first winter. **FLIGHT:** Appears massive, with very broad wings with large white patches.
SIMILAR SPECIES Domestic Muscovies usually have white blotches on head and body.
STATUS & DISTRIBUTION Local and

uncommon from Mexico to northern half of S.A. **YEAR-ROUND:** Wild birds present near Falcon Dam; scarce and difficult to find. **BREEDING:** Nests in holes, hollow trees, or earth banks.
VOICE Generally silent.
POPULATION Declining in Mexico due to hunting.

knob

female much smaller than male

slow gooselike flight

highly variable

domestic variety ♂

extensive white under-wing

juvenile ♀

large white wing patch

adult ♂

adult ♂

WOOD DUCK *Aix sponsa*   WODU ■ 1

Wood Ducks are usually found in heavily wooded swamps and are often detected by the loud squeal of the female when taking flight. Monotypic. L 18" (47 cm)
IDENTIFICATION Both sexes have a bushy crest that makes the head appear large and rounded. **MALE:** Stunning, intricate plumage renders it distinctive. **ECLIPSE MALE:** As female, but retains distinctive bill, eye, colors, and throat pattern. **FEMALE:** Distinguished by large white eye-patch, white throat, and head shape. **JUVENILE:** Like female but duller, with spotted belly. **FLIGHT:** Dark wings with white trailing edge; uplifted head,

broad wings, and long rectangular tail create distinctive flight profile.
SIMILAR SPECIES The female is similar to female Mandarin (p. 54), a closely related Asian species common in waterfowl collections and parks, which lacks eye-patch of the Wood Duck.
VOICE CALL: Female flight call a rising squeal, *ooEEK*. Male infrequently gives a high, thin *jeeee*.
STATUS & DISTRIBUTION Widespread and common, generally in low densities. **BREEDING:** Nest in cavities in

hollow trees or nest boxes in wet woods. **MIGRATION:** Spring: Begins moving north in Feb.; peaking in New England in late Mar.; in the Great Lakes in mid-Apr. Fall: Many Wood Ducks disperse widely after breeding and prior to fall migration. Most begin moving south in late Sept.; numbers in southern states increase steadily through Dec. **WINTER:** Found in bottomland forests and swamps. Rather rare over much of Southwest. Rare to casual in southeastern AK.
POPULATION Currently fairly stable, expanding range across northern plains in the last 50 years.

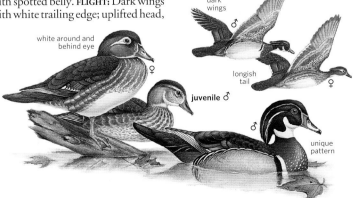

white around and behind eye

♀

dark wings ♂

longish tail ♀

juvenile ♂

♂

unique pattern

DABBLING DUCKS Genus *Anas*

This is a large, diverse group of waterfowl, 16 species of which occur in North America; of these, ten species breed, and six are visitors from the Old World or the Caribbean. Structure and size vary greatly. Males have colorful plumages that are held for most of the year, except in midsummer, when bright plumage is replaced by dull, often called "eclipse plumage." It is usually worn until fall and camouflages males as they replace flight feathers. Females are mottled brown and show little seasonal variation. Most male dabblers and some females show a colorful speculum, an iridescent patch on the secondaries that can be useful for identification.

FALCATED DUCK *Anas falcata*   FADU ■ 4

This stunning East Asian species is closely related to the Gadwall. North American vagrants often mix with flocks of Gadwall or wigeon. Monotypic. L 19" (48 cm)
IDENTIFICATION Structure like that of the Gadwall but with a smaller thin bill and a shaggy head coming to a point on males. **MALE:** Elongated tertials trail in the water. Green-and-white striped throat often hidden; visible when the head is lifted. **FEMALE:** All brown with tan face, pale belly, and dull green speculum with narrow white upper border.
SIMILAR SPECIES Female separated from female Gadwall by longer, thinner, dark bill; crested nape, uniform head color, and subtle dark fringes to the body feathers. Dark thin bill combined with the large head eliminate all other female dabblers.

STATUS & DISTRIBUTION Common in eastern Asia. **VAGRANT:** Rare migrant to western Aleutians; casual on the Pribilofs; accidental along the West Coast south to CA. Evidence suggests that some of the West Coast records represent true vagrants, while some may be escapes from waterfowl collections.
VOICE Generally silent.
POPULATION Stable.

dark speculum with faint pale borders

slight hint of crest

distinctive head shape

white throat with blackish band

all-gray bill

long sickle-shaped tertials

EURASIAN WIGEON *Anas penelope*   EUWI ■ 2

This Old World counterpart of the American Wigeon is regular in small numbers in North America, mainly in the Pacific states and British Columbia. It is typically found in flocks of American Wigeon; individuals often return to the same location for several years. Monotypic. L 20" (51 cm)
IDENTIFICATION Structure is almost identical to that of the American Wigeon. In any plumage the gray axillaries and underwing coverts of the Eurasian Wigeon are diagnostic. The innermost secondary of most Eurasian Wigeons is white, not gray, and is visible as a horizontal white bar separating the tertials from the dark speculum. **MALE:** Distinctive, with bright chestnut head, cream-colored crown, and vermiculated gray back and sides. Some males may show small patches of green around the eye, which falls within the normal variation of the Eurasian Wigeon and does not necessarily indicate hybridization. **FEMALE:** Very similar to female American, but usually has warmer brown heads; some gray morph females have dull gray heads. Females often have pale, unmarked throats and are uniformly dull with no contrast between the head and the breast. Female American often shows contrast between gray throat and warmly colored breast. **IMMATURE MALE:** Adult male plumage acquired slowly during first winter. Most individuals closely resemble adult male by Jan., but retain brown wing coverts. **FLIGHT:** Gray axillaries and underwing coverts are usually visible; white ovals on upper wings of adult males like

rusty cast to head

rufous-morph ♀

head and chest uniform in color

adult ♂
white forewing

dusky underwing and axillaries

cream crown

gray-morph ♀

rufous head

gray-morph ♀

adult ♂

adult ♂

immature ♂

American Wigeon. Females have uniform brown upper wing, lacking white covert bar of the female American. **SIMILAR SPECIES** Compare males to hybrids with the American Wigeon. Most hybrids show obviously intermediate characteristics. Typical hybrids show an American Wigeon head pattern with rusty wash; flanks are a mixture of gray and pink, usually more gray. Head pattern can be variable, so the flanks are the best indication of hybridization. Any wigeon with a mixture of pink and gray flanks is almost certainly a hybrid. **VOICE CALL:** Male whistles a sharp,

high-pitched, single note, wiry *WHEEOOO*, with a distinct similarity to the Gray-cheeked Thrush flight call. The call is more attenuated than that of the American Wigeon and is clearly audible above a noisy group of American Wigeon. Female gives a harsh call like that of the American Wigeon. **STATUS & DISTRIBUTION** Small numbers winter along and near both coasts; more common in the Pacific states and BC. Rare but regular on the Atlantic coast; generally rare to casual elsewhere, with records from almost every state and province. Regular migrant and winter visitor on the western and central

Aleutians. In the lower 48, the earliest arrivals are seen in late Aug. and some individuals linger into May.

AMERICAN WIGEON *Anas americana* AMWI ▭ 1

The American Wigeon is a colorful, noisy dabbler equally comfortable on land as in the water. Monotypic. L 19" (48 cm)
IDENTIFICATION The American Wigeon's large head, chunky body, short legs, and fairly long, pointed tail are characteristic of wigeon. Males have a small blue bill and females are gray. Either sex can have a black ring at the base of the bill that the Eurasian Wigeon lacks, but many female and immature Americans also lack this ring. Axillaries and underwing coverts are white, which, if visible, is a diagnostic difference from the Eurasian. Another characteristic is the color of the innermost secondary; visible just below the tertials, which is gray in the American, usually white in the Eurasian. **MALE:** Sports a white crown, gray face, glossy green eye stripe, pinkish breast and flanks. One common variation has variable amounts of cream on the throat and face. **FEMALE:** Gray-brown overall with orange or pinkish tinged flanks; brown upper wing with a white greater covert bar. **FIRST-WINTER MALE:** Like adult male, but more mottling on the upperwing coverts. **FLIGHT:** White ovals on the upper wings of males are visible from a great distance.
SIMILAR SPECIES Males are distinctive, but see Eurasian Wigeon account for separation of hybrids. The female is separated with caution from the female Eurasian Wigeon by gray head, speckled throat, flanks with orange or pink tones, and colder overall color with clear contrast between head, breast, and flanks. The Eurasian shows uniform coloration on throat and breast, while the American's

white or buffy white crown

adult ♂

white underwing coverts and axillaries

white forewing

adult ♂

eclipse adult ♂

bluish gray bill with dark tip

gray throat contrasts with its warm-colored breast. Note color of innermost secondary.
VOICE CALL: Male gives two- or three-part whistle, *whee-WHOO* or *whee-WHOO-who,* given in flight or on the water; call is heard constantly from any large flock. Call of the female, given infrequently, sounds like a harsh, nasal cough.
STATUS & DISTRIBUTION Abundant and widespread, particularly in the West. **BREEDING:** Farther north than all other dabbling ducks except the Northern Pintail. Nests on tundra pools, river deltas, boreal lakes, and prairie potholes. **MIGRATION:** Winter birds begin leaving in early Feb.; in most states bordering Canada it peaks in the first two weeks of Apr.; birds arriving on the northern breeding grounds in the first two weeks of May. Early fall migrant, beginning in mid-Aug. Great Lakes peak is the end of Oct., and migration is largely over by late Nov. It peaks in Nov. on coast of BC. **WINTER:** Widespread in winter, with scattered areas of high concentrations.

gray face contrasts with cinnamon-buff chest

Found in open marshes, coastal estuaries, wet farm fields, and flocks may even gather at parks and golf courses, where they become fairly tame. Rare but a regular stray to the British Isles and a very rare vagrant to the rest of Europe and the Russian Far East. **POPULATION** Abundant and stable.

GADWALL *Anas strepera* GADW ▪ 1

The Gadwall is generally thought to be nondescript, but the male's intricate plumage is stunning. Monotypic. L 20" (51 cm)

IDENTIFICATION From Mallard, has a steeper forehead and a narrow, shallow-based bill. Both sexes show a mixture of chestnut and black on the upper wing and a white square on the inner secondaries. Males have a more colorful wing pattern, often concealed on swimming birds; in flight, the wing pattern is diagnostic. **MALE:** Tan face, black-and-white scalloped breast that appears gray at a distance, gray flanks, black rump, and long scapulars with faint reddish fringes. **FEMALE:** Dull, mottled brown head and body, with a black bill evenly edged with orange, and a white belly.

SIMILAR SPECIES Female similar to female Mallard in plumage; separated by smaller bill and high, steep forehead; in flight by wing pattern and white belly. Female wigeon has shorter gray bill, rounded head, brighter flanks.

VOICE CALL: Female gives a low, harsh *aack.* Male utters a soft, burping *meep.*

STATUS & DISTRIBUTION Common and widespread across Northern Hemisphere. **BREEDING:** Nests on ponds, lakes, and marshes. **MIGRATION:** Spring: Late migrant; many do not begin to move north until well into Apr. Numbers peak in the Great Lakes and western plains in late Apr. **FALL:** Migration begins in Sept.; Great Lakes peak occurs in early Nov.; OR peak mid- to late Oct. **WINTER:** Marshes of LA are the most important wintering area for Gadwall. Widespread in shallow lakes and marshes.

POPULATION Has expanded rapidly in the East.

small white speculum patch ♂

white belly

chestnut-edged scapulars

mostly gray body ♂

black undertail coverts

steep forehead ♀

even line of orange on sides of bill ♀

AMERICAN BLACK DUCK *Anas rubripes* ABDU ▪ 1

A northeastern species, the American Black is most numerous in coastal salt marshes. Monotypic. L 23" (58 cm)

IDENTIFICATION Sexes similar and structurally identical to the Mallard. The two species are often found in mixed flocks and frequently hybridize. **MALE:** Body blackish with a dark crown, sharply contrasting tan face and yellow-olive bill. **FEMALE:** Like male but slightly paler body with duller olive bill. **FLIGHT:** Speculum is deep purple and can show thin white borders. Dark body contrasts with white underwing.

SIMILAR SPECIES Resembles female Mallard but darker, showing strong contrast between body and face, bill yellow to olive, speculum purple. Hybrids with Mallard may be paler in coloration overall, partially green head, brighter yellow or orange bill, white bordering the speculum, curved uppertail coverts or paler tail feathers. Some hybrids are subtle. Similar to western subspecies of Mottled Duck; see that entry.

VOICE Like Mallard, but slightly lower and harsher.

STATUS & DISTRIBUTION Locally common. **BREEDING:** Nests in a wide variety of wetlands. **MIGRATION:** Spring migration begins in early Feb. and proceeds gradually following the appearance of open water. Arrives in ME in early Apr., northern QC in late May. Fall migrants peak in mid-Atlantic in early Nov., peak in the western Great Lakes in mid-Nov. **WINTER:** Primarily in mid-Atlantic salt marshes. Also use a variety of wetlands throughout the Northeast; some wintering as far north as NF. Casual to western Gulf States and western N.A. Casual to British Isles.

POPULATION Serious declines; replacement by Mallards in deforested portions of its range and overhunting in past decades are the primary causes.

American Black Duck x Mallard hybrid ♂

underwing contrasts strongly with dark body

purplish speculum lacks white forward border

♀

♂

streaked face

MALLARD *Anas platyrhynchos*   MALL ▪ 1

Probably the most widely recognized and widespread duck in North America, the Mallard forms massive, noisy flocks in large marshes and agricultural fields; Mallards also make use of almost any body of water, from a roadside puddle to the Great Lakes. Polytypic (3 ssp.; 2 in N.A.). L 23" (58 cm)

IDENTIFICATION Mallards are large dabbling ducks with heavy bodies and large, rounded heads. Both sexes have bright blue speculums with bold white borders. The central uppertail coverts of the males are curled upwards, a characteristic frequently passed on to hybrid offspring. **MALE:** Distinctive, with bright yellow bill and metallic green head bordered by a white neck ring. Breast is deep chestnut, and rump black with white outer tail feathers. The tertials are very unusually shaped, extremely broad at the base, tapering to a point. **FEMALE:** Nondescript medium brown overall, paler tan face with a streaked, brown crown, indistinct brown eye line, and orange bill with a black saddle. The belly is unmarked and buffy, paler than the body, and the outer tail feathers are whitish. **ECLIPSE MALE:** Like female but the bill remains bright yellow and the crown and breast are darker than female. **FLIGHT:** Wingbeats steady and shallow, slower than other ducks.

GEOGRAPHIC VARIATION Two subspecies in N.A., nominate and the southwestern subspecies *diazi,* "Mexican Duck," which is found in pockets along the Rio Grande River in Texas, New Mexico, and southeastern Arizona. It is 10 percent smaller than *platyrhynchos,* and the sexes are similar, lacking the distinctive plumage of male northern Mallards. Its plumage is darker than female northern Mallards, and it has an all brown tail. The white borders to the speculum are narrower than on *platyrhynchos* but still distinct and obvious. "Mexican Duck" resembles Mottled Duck, but has a grayer more distinctly patterned face, streaked throat, and stronger white borders to the speculum.

SIMILAR SPECIES Female paler than Mottled (especially western subspecies) and American Black Duck, bill brighter orange, tail and rump paler, with distinct white borders to the blue speculum. From the smaller female Gadwall by rounded head shape, lacking a distinct forehead, larger, deeper based bill, iridescent speculum with white borders. Much larger than female teal with a larger, extensively orange bill.

VOICE Female gives series of loud, descending *quacks* for which Mallard is known. Male less vocal; shorter, soft *quack* is often drowned out by the calls of the females.

STATUS & DISTRIBUTION Abundant and widespread across Northern Hemisphere. **BREEDING:** Any place that has both water and cover. Typically nests on the ground in thick grass or shrubs, but also may build nests in hollow trees, on top of duck blinds, or use an old nest of another bird. **MIGRATION:** Driven by the availability of open water. In the spring it is an early migrant, along with Northern Pintail and Cinnamon Teal, appearing north of its winter range as soon as there is open water. Begins in early Feb.; peaks in the Midwest and Great Lakes region in late Mar.; northernmost breeders arrive in AK in early May. In the fall it is among the last to leave the north, retreating only as water freezes. Some begin moving south in Sept.; however, most northern breeders do not move south until Oct.; peak numbers arrive in the Great Lakes and across the northern plains in early Nov. **WINTER:** Spends the winter in a variety of habitats, mostly shallow freshwater. Often very mobile in the winter; moving farther north during warmer periods, then driven back south by cold fronts.

POPULATION The most abundant duck in N.A., population more or less stable.

eclipse ♂

underwing contrasts less with body than Black Duck

blue speculum bordered broadly with white

mostly white tail

dark saddle on orange bill

male with yellow bill

body darker than female Mallard ♂

♀

yellow bill ♂

curled tail feathers

MOTTLED DUCK Anas fulvigula   MODU ▪ 1

This southern look-alike of the American Black Duck is primarily a resident of the coastal marshes of the Gulf of Mexico and Florida. Like American Black Duck, the sexes are similar in appearance. Mottled Ducks retain their pair bond for most of the year, so are most often seen in pairs and do not form large winter flocks. Polytypic (2 ssp.; both in N.A.). L 22" (56 cm)

IDENTIFICATION Structure is very similar to Mallard, but slightly smaller with proportionally shorter wings. Overall color is closer to American Black Duck, especially darker *maculosa*. The blue-green speculum has very narrow white borders. Noticeable black spot at gape. **ADULT MALE:** Dark brown body with golden brown, V-shaped marks on interior of flank feathers. Buffy brown face with an unmarked throat, slightly darker crown and eye line, yellow bill. **ADULT FEMALE:** Like male, with dull orange or olive-yellow bill

with black marking on the culmen. **FLIGHT:** Upper wing appears all dark; underwing coverts flash bright white. Broad winged with slow wingbeats like Mallard.

GEOGRAPHIC VARIATION Two subspecies, nominate in the FL Peninsula and along the coast of GA and SC; *maculosa* found, primarily coastally, from LA and coastal Mississippi to northern Mexico. The two are similar but *maculosa* is darker and more coarsely marked than nominate *fulvigula*. Some authorities consider Mottled Duck to be a subspecies of Mallard.

SIMILAR SPECIES Similar to female Mallard and American Black Duck. Separated from female Mallard by darker overall color; buffy, unstreaked throat; all brown tail; and narrower white border to the speculum. Great care should be taken when separating from American Black Duck. Compared to Black Duck it has a buffier face, no streaking on the throat, paler brown V-shaped markings on the interior of the flank feathers, and a blue-green speculum with narrow white borders. Black Duck has essentially all-dark flanks, a grayer face, streaked throat, and

lacks white borders on the speculum. **VOICE** Females give a loud *quack* like Mallard but slightly softer and not as harsh. Males infrequently give a short, soft *quack* like male Mallards.

STATUS & DISTRIBUTION Closely tied to the Gulf Coast, where it is fairly common but rarely found in large numbers. **YEAR-ROUND:** Essentially nonmigratory, however individuals move in response to food resources and water levels. Large numbers move from coastal marshes to interior rice fields in fall to feed on the ripening rice. Rare away from coastal areas but known to wander and occasionally nest in the interior. Introduced to SC coast. **BREEDING:** Nests in cordgrass meadows, fallow rice fields, and grassy islets in marshes. Rare breeder in northeastern TX, southern AR, and OK. Very rare north to KS. **VAGRANT:** Casual to KY and IL; accidental to southern ON. **POPULATION** Fairly stable.

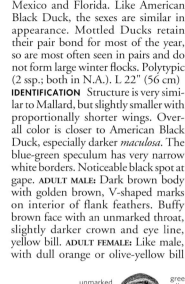

unmarked buffy face and throat
greenish yellow bill
black spot at gape
♂
western Gulf Coast
maculosa

mainly Florida
fulvigula
yellower bill
averages paler overall than *maculosa*
black spot at gape
♂

EASTERN SPOT-BILLED DUCK Anas zonorhyncha   ESPD ▪ 4

This robust, dark brown duck is an east Asian species and a casual vagrant to Alaska. Its structure and habits are identical to the Mallard. It is found on freshwater marshes and lakes, lagoons, and rivers. Despite its name, this recently split species (from *A. poecilorhyncha*) lack the red spots at its bill base that the two related, more westerly taxa exhibit. Monotypic. L 22" (56 cm)

IDENTIFICATION The diagnostic, sharply defined yellow tip on the black bill is visible at great distances. Sexes are similar; female is slightly paler. Head is pale with light gray throat and cheeks, dark crown, dark stripe through the eye, and shorter, dark stripe beginning at the base of the bill. Body is dark brown; speculum is

bluish with white border. Diagnostic white-edged tertials are readily visible when at rest. **FLIGHT:** Large and dark with a heavy body and broad wings, striking white underwing coverts, and dark upper wing with narrow white borders to the blue speculum.

SIMILAR SPECIES Like the American Black Duck and female Mallard; separated

by distinctive bill coloration, paler head, scalloped edges to flank feathers, and pale-fringed tertials.

VOICE CALL: Females give a loud *quack*. Males give raspy *kreep*.

STATUS & DISTRIBUTION Generally widespread and locally common throughout its range in eastern Asia. **VAGRANT:** Casual to western and central Aleutians. Accidental to Kodiak I.

dark facial stripes
distinct yellow tip to black bill
pale neck
mostly dark body
distinct white tertial edges

BLUE-WINGED TEAL *Anas discors*   BWTE   1

This small duck is largely a summer resident only. Monotypic. L 15" (39 cm) **IDENTIFICATION** Both sexes have all-black bills and a pale area at the base of the bill that is crescent shaped in males and oval shaped in females. **ADULT MALE:** Cold blue head with large white crescent on the face; the body is bronze, covered with black spots. **ADULT FEMALE:** Grayish brown with paler face, a distinct dark eye stripe, white eye arcs, and a pale loral spot. **FLIGHT:** Powder blue wing coverts

and green speculums are visible; the wingbeats are rapid. Male has a broad white border to blue forewing. **SIMILAR SPECIES** Larger and colder gray than Green-winged, with a pale loral spot and longer bill. See sidebar below for separation from Cinnamon Teal. **VOICE** Males give a high, whistled *peew;* females give a shrill *quack.* **STATUS & DISTRIBUTION BREEDING:** Nests in grassy clumps near small ponds and small flooded fields. Most dense in the prairie pothole region, where it is the most abundant duck. **MIGRATION:** Relative to other waterfowl, is a late spring and early fall migrant. **SPRING:** Begin arriving along the Gulf Coast in Feb.; peaking in Apr. in the Midwest; migration is largely

over by the end of May. **FALL:** Peaks in the Midwest and Southwest in late Sept.–early Oct.; peaks in the Gulf states in early Oct., but quickly decline later in the month. **WINTER:** Most of the birds winter in S.A. **VAGRANT:** Annual in British Isles; numerous records from western Europe and northwestern Africa. **POPULATION** Stable.

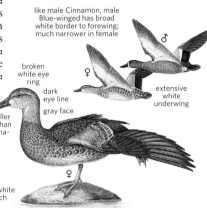

like male Cinnamon, male Blue-winged has broad white border to forewing; much narrower in female

broken white eye ring

extensive white underwing

dark eye line

gray face

smaller bill than Cinna-

white face crescent

vertical white flank patch

Identification of Female Teal: Blue-winged and Cinnamon

These two female teal are very similar, but by paying attention to detail most individuals can be identified with confidence. As with most difficult identifications, it is useful to start with structure. These two species are almost identical in size and proportions. Cinnamon Teal, however, has almost no forehead; instead the head slopes directly into the bill like that of Northern Shoveler. Blue-winged Teal shows a slight rising forehead, so the head appears more square than that of Cinnamon Teal. Bill shape can be a subtle but important difference. Cinnamon Teal has a longer, wider bill that has a spoonlike appearance. The bill of Blue-winged Teal is smaller; it is more similar to the bills of other dabblers, lacking the spatulate look.

When dealing with a difficult teal hard to identify, it is also important to evaluate the overall color of the bird. Blue-winged Teal tend to be pale gray-brown overall and give a cold impression. Cinnamon Teal are slightly darker brown with a strong rusty wash that gives them a much warmer look.

Blue-winged Teal, female (CA, Nov.)

Cinnamon Teal, female (CA, Dec.)

As for plumage, Blue-winged Teal have a sharply defined face pattern, with large white eye arcs, a strong brown eye line, and a clear pale loral spot at the base of the bill. The loral spot is usually connected to the pale male's throat and is reminiscent of the male's white crescent. In contrast, Cinnamon Teal have a diffuse face pattern, usually lacking noticeable eye arcs. The eye line of Cinnamon Teal looks washed out, and the loral spot is buffy. Sometimes the bird completely lacks the loral spot, but when present, it does not have the definition shown by Blue-winged Teal and is usually separated from the throat by some stippling across the malar area. Overall, dark facial markings are more distinct in Blue-winged than in the browner, blended face of the Cinnamon.

By using a combination of these characteristics, most difficult teal will no longer be challenging. However, birds with mixed features are better left unidentified. Male hybrids between these two species occur somewhat regularly, and there are certainly some female hybrids as well. ■

CINNAMON TEAL   *Anas cyanoptera*   CITE ■ 1

This uniquely plumaged teal is exclusively a western species. It is usually found in small, tight groups that move and forage together. Polytypic (5 ssp.; *septentrionalium* in N.A.). L 16" (41 cm)

IDENTIFICATION The head is large and rounded, grading into a long bill reminiscent of Northern Shoveler. Males have intense red eyes in all but first few months of life. **MALE:** Bright cinnamon head and body are distinctive. **FEMALE:** Rusty or golden brown overall with faintly scalloped flanks; variable, but usually small, pale loral spot and faint eye stripe. **FLIGHT:** Powder blue upperwing coverts contrast sharply with bright cinnamon plumage of male. Male has a broad white border to blue coverts. Flight is swift with rapid wingbeats.

SIMILAR SPECIES Female is smaller and darker than female Northern Shoveler

with less distinctly spatulate bill. (See sidebar p. 31 for separation from female Blue-winged Teal.)

VOICE Male makes dry series of *click* notes; female gives shrill *quack.*

STATUS & DISTRIBUTION Locally common in the West. **BREEDING:** Unlike its close relative, the Blue-winged Teal, the Cinnamon largely avoids prairie potholes; most common in the Great Basin and mountain regions of the western U.S. Nests on marshes, ponds, and shallow lakes; often uses highly alkaline water. **MIGRATION:** In spring, begins in mid-Jan. in Southwest; peaks in late Apr. in UT; most breeders are in place by mid-May. In fall, numbers diminish sharply after early Sept. **WINTER:** Uncommon to rare in Southwest. **VAGRANT:** Casual to AK and eastern N.A.; some records may involve escapes.

POPULATION Due to its limited breeding range, it is one of the least numerous species of waterfowl in North America. Little is known about population trends; it appears to be stable, in most of its range, declining slightly in the Pacific Northwest.

buffier face, less evident eye line and eye ring than female Blue-winged

♀

underwing of Cinnamon and Blue-winged Teal more extensively white than Green-winged Teal

♂

larger spatulate bill than Blue-winged

all but youngest males have red-orange eyes

juvenile Cinnamon has slightly more evident eye line than adult Cinnamon

♂

NORTHERN SHOVELER   *Anas clypeata*   NSHO ■ 1

This is an odd-looking, Holarctic species with a distinct spatulate bill. It obtains most of its food from the surface, straining it through its large bill. Often small groups of Northern Shovelers bring food to the surface by swimming rapidly in a circle while swinging their bills side to side. Monotypic. L 19" (48 cm)

IDENTIFICATION On the water, shovelers look front heavy and short necked, with the bill angled downward. Adult males have bright golden eyes, while eye color in young males and females ranges from dull yellow to brown. Upperwing coverts powder blue in both sexes. **MALE:** Distinctive, with bright green head, white breast, reddish brown flanks and

belly. **FEMALE:** Mottled light brown head and body; some have a faint rusty wash. Most have dark bills with orange edges, but a few have all-dark bills. **ECLIPSE MALE:** As female, but has golden eyes and bright orange legs. **FALL MALE:** An intermediate plumage in early fall, with heavily spotted breast and flanks, dull head, and pale facial crescent.

SIMILAR SPECIES The long, spatulate bill is diagnostic.

VOICE Generally silent, courting males give a repeated low, nasal *erp-EERP.*

STATUS & DISTRIBUTION Common. **BREEDING:** Uses a wide variety of shallow wetlands for nesting, such as saline

ponds and sewage-treatment plants. **MIGRATION:** In spring, a late migrant: begins migration in late Mar., peaks in the southern Great Lakes in early Apr., arriving on the prairie breeding grounds mid-Apr.–early May. Fall: Begins in late Aug.; peaks in BC in late Sept. or early Oct.; Great Lakes peaks mid-Oct. **WINTER:** Primary wintering areas CA and the Gulf Coast. Found in marshes, ponds, and bays.

POPULATION Long trend of stability, but population has increased substantially in recent years. Breeding range expanding eastward.

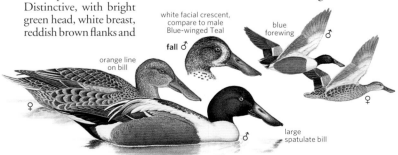

orange line on bill

♀

fall ♂

white facial crescent, compare to male Blue-winged Teal

blue forewing

♂

♀

♂

large spatulate bill

WHITE-CHEEKED PINTAIL *Anas bahamensis*   WCHP ■ 4

This neotropical duck is a vagrant from the West Indies, where it is casual to southern Florida. Sightings even from Florida are often of uncertain origin as the White-cheeked Pintail is popular in captivity; those away from Florida are most likely birds that have escaped from captivity. The White-cheeked Pintail is found on shallow-water habitats in salt water or fresh water, where it feeds by dabbling and tipping up. The large, white patch on the side of the White-cheeked's head and throat, for which it is named, is distinctive.

Polytypic (3 ssp.; *bahamensis* in N.A.). L 17" (43 cm). **IDENTIFICATION** Sexes of this pintail are similar; the female is paler than the male, with a slightly shorter tail. White cheeks and throat contrast with a dark forehead and cap; the blue-gray bill has a red spot near the base. The long, pointed tail is buffy; the underparts are tawny or reddish and are heavily spotted. **FLIGHT:** The bird's slender body and long, pointed tail are evident. Both sexes of White-cheeked Pintail show brown forewing, green speculum bordered on each side with broad buffy edge. The bird's pale, buffy rump and tail contrast with the darker brown body. **GEOGRAPHIC VARIATION** Nominate *bahamensis* from the West Indies and northeastern South America occurs in North America. Subspecies *bahamensis* is brighter and larger than *rubrirostris,* which is from mainland South America, and *galapagensis* of the Galápagos Islands is the palest. **SIMILAR SPECIES** Generally distinctive, not likely to be mistaken. Prominent white cheek patch readily distinguishes the

White-cheeked Pintail from the Northern Pintail. **VOICE** Generally silent. Female gives weak descending *quack* notes. Male produces low whistle. **STATUS & DISTRIBUTION** Widespread and locally common resident in West Indies; small movements between islands occur. **VAGRANT:** Casual to southern FL, majority of records mid-Dec.–late Apr., with bulk of records from Everglades N.P., FL. Accidental in coastal southern TX, perhaps an escape. **POPULATION** Threatened, with moderate declines in the West Indies due to habitat loss, hunting, and introduced predators.

red at bill base / sharply delineated white cheek and throat / buffy pointed tail ♂

NORTHERN PINTAIL *Anas acuta*   NOPI ■ 1

This elegant species is as easily identified by structure as by plumage. The birds frequently gather in huge flocks to feed in grain fields. Monotypic. L 20–26" (51–66 cm). **IDENTIFICATION** The thin grayish bill; long, slender neck; slim body; and long, wedge-shaped tail give pintails a diagnostic shape. **MALE:** Combination of brown head with white neck stripe, white breast, and spikelike tail is distinctive. **FEMALE:** Mottled brown, with a plain, warm buff head. **FLIGHT:** Looks long and lean, with slender neck extended. The wings are thin and pointed, with bold white trailing edges to the secondaries. **SIMILAR SPECIES** Female's combination of structure and unmarked buffy head is unlike any other female dabbler. **STATUS & DISTRIBUTION** Abundant, especially in the West. **BREEDING:** Holarctic species; nests on marshes, lakes, and

tundra pools. **MIGRATION:** In spring, early migrant relative to other waterfowl. Mid-Atlantic peak late Feb.; northern plains peak in early Apr.; Arctic breeders arrive in mid- to late May. In fall, arrive in Southwest mid-Aug. Peak in the mid-Atlantic and Great Lakes late Oct.–early Nov. Pacific Northwest peak mid- to late

Oct. **WINTER:** Prefers open marshes, agriculture fields, and tidal flats. Much of the population winters in CA; also, rarely, in southern AK and the Great Lakes region. **VOICE** Male gives a high, whining *mee-meee* during courtship and short, mellow *proop-proop,* like the Green-winged Teal's but lower pitched and more musical; female gives a hoarse, weak *quack.* **POPULATION** Fluctuates; low during prairie droughts. High in 1950s and 1970s; sharp decline in time of drought around 1990.

♂ / long pointed tail / dark speculum with narrow white trailing edge / ♀ / grayish bill / white stripe extends up neck / long pointed tail ♂

GARGANEY *Anas querquedula*   GARG ■ 4

The Garganey is the highly migratory Eurasian counterpart of the Blue-winged Teal. Monotypic. L 15" (39 cm) **IDENTIFICATION** Structure is very similar to that of Blue-winged. **BREEDING MALE:** Bold white supercilium, brown breast, drooping tertials, and gray flanks. **ADULT FEMALE:** Pale face with strong dark eye stripe and additional diffuse dark stripe across the face starting at the lores. Pale loral spot and white throat. **ECLIPSE MALE:** Like female, but retains the paler gray upper wing of the breeding male. The Garganey holds its eclipse plumage much longer than most ducks, so that from midsummer to late January or February, males look much like females. **FLIGHT:** The wing pattern is distinctive. The male has pale blue forewing; the female's forewing is gray-brown. Silvery gray inner primary webs. The underwing is pale with a strongly contrasting dark leading edge. **SIMILAR SPECIES** Compare female and eclipse male to female Blue-winged Teal. Garganey's wing pattern is diagnostic. On the water look for its double-striped face, pale loral spot separated from the white throat, and all dark scapulars with sharp white fringes. Garganey is larger, with a heavier bill than Green-winged Teal, has a pale throat, and lacks a pale streak on its undertail coverts. **VOICE** Male gives a series of dry clicking noises; female utters a harsh *quack* like Green-winged Teal. **STATUS & DISTRIBUTION** The Garganey is rather common across Europe and Asia. **MIGRATION:** Highly migratory, completely vacating its breeding range. **VAGRANT:** Regular migrant on the western Aleutians; casual on the Pribilofs and in the Pacific states; additional records are widely scattered throughout N.A. Most often found in Apr. and May, when migrating males are most easily recognizable. **WINTER:** Central Africa and southern Asia. **POPULATION** Common but declining in recent decades in eastern Asia, likely resulting in fewer records from western N.A.

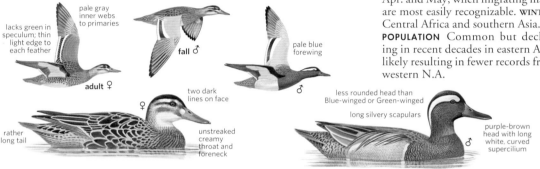

pale gray inner webs to primaries

lacks green in speculum; thin light edge to each feather

fall ♂

pale blue forewing

adult ♀

♂

two dark lines on face

♀

rather long tail

unstreaked creamy throat and foreneck

less rounded head than Blue-winged or Green-winged

long silvery scapulars

purple-brown head with long white, curved supercilium

♂

BAIKAL TEAL *Anas formosa*   BATE ■ 4

The Baikal Teal is an East Asian teal only slightly larger than the Green-winged Teal. Populations had been declining sharply, but the bird appears to have rebounded dramatically in recent years. Most Baikal Teal winter in eastern China and South Korea. The bird favors several lakes, where counts can reach hundreds of thousands. It blends into flocks of Green-winged Teals surprisingly well; however, a glimpse of the elongated, pink-edged scapulars or yellowish cheek patch is sufficient to pick out a male. Although Baikal Teal is rare in captivity, most sightings, especially away from the West, are possibly escapes. Monotypic. L 17" (43 cm) **IDENTIFICATION** Small dabbler with a proportionally large, square head and small bill. **MALE:** Its intricate face pattern and long, ornate scapulars are exquisite. **FEMALE:** Tawny brown with two short stripes on the face; pale loral spot with a brown outline and an extensive pale throat that extends up onto the face. **SIMILAR SPECIES** Female Baikal is very similar to female Green-winged Teal; look for the pale throat, face pattern, and a loral spot stronger than on most Green-winged Teal. However, some female Green-wingeds have strongly patterned faces that appear very similar to that of the Baikal. Even the most strongly patterned Green-wingeds generally do not show the sharply delineated pale loral spot of the Baikal. The upperwing pattern is much like the Green-winged's, but the narrow cinnamon-buff upper border to the green speculum is straight, not wedge shaped; the white trailing edge is broader than on the Green-winged; and the under wing is more extensively gray. **VOICE** Male frequently gives a deep, repeated *wot-wot-wot*. Female gives a soft *quack*. **STATUS & DISTRIBUTION** Locally common in East Asia. **VAGRANT:** Spring overshoot to western and northern AK (May–June) and fall migrant (Sept.–Oct.). Fall migrants recorded on Seward Peninsula, St. Lawrence I., western Aleutians, and Pribilof Is. Western records Dec.–May. **POPULATION** Recent discovery of enormous flocks in South Korea and an increase in Alaska records perhaps indicate a rise in population.

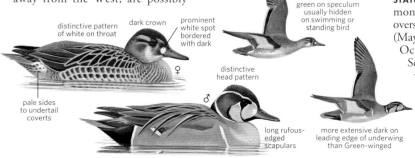

distinctive pattern of white on throat

dark crown

prominent white spot bordered with dark

green on speculum usually hidden on swimming or standing bird

♀

distinctive head pattern

pale sides to undertail coverts

♂

long rufous-edged scapulars

more extensive dark on leading edge of underwing than Green-winged

GREEN-WINGED TEAL *Anas crecca*   GWTE ▪ 1

A tiny, Holarctic dabbler, the Green-winged Teal is frequently seen foraging on mudflats. Polytypic (2 ssp.; both in N.A.). L 14" (37 cm)
IDENTIFICATION Small, with rounded head; short, thin bill. **ADULT MALE:** Chestnut head with green ear patch and a vertical white bar on each side. **ADULT FEMALE:** Brown body and head, black bill with dull orange edges, a pale line on the sides of the undertail coverts, and white belly. **FLIGHT:** Flies with very rapid wingbeats, forming tight bunches that twist and turn erratically. White restricted to central part of underwing, unlike the Blue-winged and Cinnamon. **GEOGRAPHIC VARIATION** The subspecies *carolinensis* is widespread in N.A.; the Eurasian subspecies nominate *crecca* is regular in the Aleutians and the Pribilofs, but a vagrant elsewhere in N.A. It has a white horizontal

bar above the flanks instead of the vertical bar and more white on the face. **SIMILAR SPECIES** The female Green-winged has a smaller bill and white sides to the undertail coverts that separate it from other female teal.
VOICE Males give a sharp, whistled *kreek* (like the Northern Pintail); females give a high, thin *quack*.
STATUS & DISTRIBUTION Common and widespread. **BREEDING:** Wooded ponds, potholes, and tundra pools. **MIGRATION:** In spring, mid-Atlantic peak late Mar. In fall, mid-Atlantic peak mid- to late Oct. Peak late Oct.–early Nov. on BC coast. **WINTER:** Shallow wetlands and marshes, often in large flocks. **VAGRANT:** Eurasian *crecca* is regular to western AK, uncommon to fairly common on

Aleutians and Pribilofs, rare to St. Lawrence I. and mainland western AK, very rare on West Coast, regular to NF in winter; casual elsewhere in N.A.
POPULATION Stable.

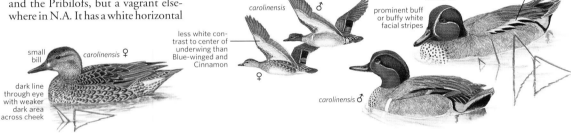

"Eurasian Teal"
crecca ♂

prominent buff or buffy white facial stripes

carolinensis ♂

carolinensis ♂ [flight]

less white contrast to center of underwing than Blue-winged and Cinnamon
♀

small bill

carolinensis ♀

dark line through eye with weaker dark area across cheek

BAY DUCKS Genus *Aythya*

This genus is made up of 12 species worldwide. Five species breed in North America, and two are vagrants. *Aythya* are medium to medium-large diving ducks that often gather together in impressive rafts on coastal bays and large lakes in the winter. *Aythya* ducks regularly hybridize with each other. Hybrids often resemble other members of this genus and can be mistaken for a rare Tufted Duck or Common Pochard.

COMMON POCHARD *Aythya ferina*   CPOC ▪ 3

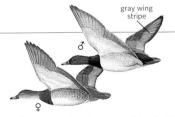

gray wing stripe

♂

♀

This Old World counterpart of the Canvasback, also similar to the Redhead, is a Eurasian species rare in western Alaska. Monotypic. L 18" (46 cm)
IDENTIFICATION Heavily built *Aythya* with a thick neck and a long sloping forehead grading smoothly into the bill. Common Pochard's bill has a dark base and a black tip separated by a thick whitish blue stripe. **MALE:** Patterned as

the Canvasback with chestnut head, black breast, and light gray body. **FEMALE:** Light brown head with pale throat, lores often paler than the face, fairly prominent pale eye ring, brown breast, mottled gray and brown body. **FLIGHT:** Wings uniform gray with no wing stripe. Appears chunkier than the Canvasback with short, rounded wings. **SIMILAR SPECIES** Males separated from the Canvasback by shorter bill with more concave profile, more compact build, shorter neck, slightly grayer back, and lack of black crown. Separated from the Redhead by sloping forehead, bill pattern, darker eye color, and paler back. Females separated from Redheads and Ring-necked Ducks by structure and the presence

of gray on the flanks and back. Differentiated from the Canvasback by structure, bill pattern, pale loral spot, and darker color. Hybrids between Canvasback and Redhead can appear very similar to Common Pochard but lack its unique bill pattern.
VOICE Generally silent.
STATUS & DISTRIBUTION Widespread in Europe and Palearctic Asia. **VAGRANT:** Rare migrant to the western and central Aleutians; casual to the Pribilofs; records from St. Lawrence I., Seward Peninsula, and coastal southern AK. Accidental to southern CA.

head shape and coloration of both sexes resemble Canvasback

♀

pale gray center to bill in both sexes

♂

CANVASBACK *Aythya valisineria* CANV ▪ 1

The Canvasback is perhaps one of the more elegant ducks; its distinctive structure makes identification simple. Monotypic. L 21" (53 cm)
IDENTIFICATION Forehead slopes straight down to a long, dark bill. The largest *Aythya* species, it has a thick neck and long body. **MALE:** Chestnut head with a dark crown and forehead, red iris. The flanks, back, and tertials are very pale, almost white. **FEMALE:** Light brown head and breast with a slightly darker crown; contrasting pale gray flanks and back. **FLIGHT:** Wings are uniform, with only a faint gray wing stripe.
SIMILAR SPECIES Compare to Redhead

and Common Pochard. Canvasback can be distinguished by its sloping forehead and long, uniformly colored bill. The forehead is straight, not concave as in similar species. In all plumages, Canvasback has a dark bill and pale back unlike any other *Aythya* species.
VOICE Generally silent.
STATUS & DISTRIBUTION Locally common south to central Mexico; casual to Honduras. **BREEDING:** Uses a wide variety of freshwater wetlands, from lakes and ponds to large marshes; also found on alkali lakes. **MIGRATION:** Spring migration begins in early Feb.; peaks in the Midwest and Great Plains

in early Apr.; most breeders have arrived by mid- to late May. Fall migration is fairly late. Most begin moving south in Oct.; peaking in the Great Lakes in early Nov.; peaks in early Dec. on the Gulf Coast. **WINTER:** Ranges widely in freshwater lakes and brackish bays and estuaries. **VAGRANT:** Casual on the Aleutian and Pribilof Islands. Accidental to Iceland.
POPULATION Numbers have fluctuated widely due to changing water levels on the breeding grounds and, to some degree, hunting regulations.

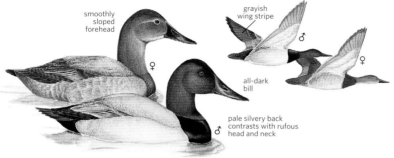

smoothly sloped forehead

grayish wing stripe ♂

all-dark bill

pale silvery back contrasts with rufous head and neck ♂

♀

♀

REDHEAD *Aythya americana* REDH ▪ 1

An attractive duck usually found in small groups except in massive rafts along the barrier islands of southern Texas in winter. Monotypic. L 19" (48 cm)
IDENTIFICATION Large *Aythya*, slightly larger than Greater Scaup, with a short bill, rounded head. **MALE:** Tricolored bill with a black tip, white subterminal ring, and pale blue base. The head is uniformly rufous with a yellow eye. Back and flanks appear smoky gray. **FEMALE:** Brown overall, with paler face and dark crown. Bill is tricolored with a slate gray base. **FLIGHT:** Brown upper wing, with a paler gray stripe on the flight feathers.

SIMILAR SPECIES Male separated from Canvasback by short, tricolored bill and darker back. Female from Canvasback by structure and browner color.
VOICE Generally silent.
STATUS & DISTRIBUTION Fairly common, south to central Mexico: casual to Honduras. **BREEDING:** Highest concentrations in the western prairies; nesting on prairie potholes, seasonal pools, and lakes. **MIGRATION:** In spring, begins leaving the wintering grounds in late Jan., peaking in the southern prairies

in mid-Mar. In fall, migration begins in late Aug., peaking in the southern prairies in mid- to late Oct., arriving on the Gulf Coast in numbers in early Nov. **WINTER:** Widespread but fairly sparse except along the western Gulf Coast, where most of the population winters, especially the Laguna Madre of southern TX and northern Mexico. **VAGRANT:** Accidental in Bermuda, HI, and the U.K.
POPULATION Stable, but eastern breeders are declining.

more rounded and uniform buffy head than female Ring-necked

head and back uniform in color

gray wing stripe

♀

♂

♀

rounded head

pale blended band on bill

darker gray back than Canvasback

♂

RING-NECKED DUCK *Aythya collaris*   RNDU ■ 1

This distinctive duck is usually found in marshy pools and small ponds. Monotypic. L 17" (43 cm)
IDENTIFICATION Angular head, peaking near the rear with the nape slanting forward. Both sexes have a white ring adjacent to a black-tipped bill; male has an additional thin ring at base of bill. **ADULT MALE:** White wedge between the breast and flanks, a unique feature; dark purple head, gray flanks, and black back. **ADULT FEMALE:** Gray face with a white eye ring, darker crown,

and indistinct pale area at the base of the bill; some show a pale thin stripe behind the eye. **FLIGHT:** Rapid, showing a slight gray stripe on the secondaries.
SIMILAR SPECIES Male similar to Tufted Duck, but lacks tuft, has more prominently marked bill, gray flanks separated from breast by a white wedge, and lacks wing stripe. Female similar to female Redhead, but with an angular head, darker back and crown, and more distinctly pale face. Separated from female scaup by gray face contrasting with darker peaked crown and white eye ring, and lack of white wing stripe in flight.
VOICE Generally silent.
STATUS & DISTRIBUTION Widespread and fairly common. **BREEDING:** Nests on freshwater marshes and shallow ponds and lakes. **MIGRATION:** In spring,

begins early Feb., peaking in Great Lakes in early Apr.; most breeders in place by late Apr. or early May. In fall, migration begins in mid-Sept., peaking on Great Lakes in late Oct., Pacific Northwest in late Nov. **WINTER:** Sometimes uses coastal marshes, but usually in shallow freshwater. **VAGRANT:** Annual to western Europe.
POPULATION Stable or increasing.

adult ♂
grayish stripe
♀

peaked head
white eye ring and grayish face ♀
prominent white ring near tip of bill
cinnamon collar is hard to see
black back
vertical white bar on side
adult ♂

TUFTED DUCK *Aythya fuligula*   TUDU ■ 3

An Old World species, regular in Alaska and a vagrant to the rest of N.A., the Tufted Duck is usually found in the company of scaup or sometimes Ring-necked Ducks. Individuals often return to the same locations year after year. Monotypic. L 17" (43 cm)
IDENTIFICATION **MALE:** Its long and shaggy crest is the most striking feature. The bill is similar to that of a scaup but has more black on the tip. From a distance, black back and gleaming white flanks are distinctive. **FEMALE:** Dark brown head, back, and breast, with paler mottled brown flanks. The female shows a variably sized crest and a black tipped bill. **FIRST-WINTER MALE:**

Like the adult male, but has a shorter crest and light gray flanks. **FLIGHT:** Shows an extensive white stripe on the wing, broader than that of the Greater Scaup.
SIMILAR SPECIES First-winter males are similar to hybrids between a scaup and a Tufted Duck or a scaup and a Ring-necked Duck. Scaup x Tufted Duck hybrids are regular, almost as frequent as pure Tufted Ducks; hybrids tend to show a short crest, reduced black on the tip of the bill, a pale gray wash on the flanks, and grayish barring on the

upper back; they lack the faint white spur on the flanks of the Ring-necked x scaup hybrids.
VOICE Generally silent.
STATUS & DISTRIBUTION Abundant and widespread, breeding across northern Palearctic Europe and Asia. **WINTER:** Ponds, bays, and rivers; a number of records have come from ponds or lakes in city parks. In southern Europe and Asia, south to Africa. **VAGRANT:** Regular migrant in AK; rare but regular winter visitor on the West Coast south to southern CA. Casual on the East Coast. Accidental elsewhere.
POPULATION The most numerous of the *Aythya* in the world.

some females whitish around bill
♀
crest
broad black bill tip
dark brown back
black back
1st winter ♂
crest length variable on males and females
extensive and bold white wing stripe
adult ♂
♀
adult ♂
pure white sides

GREATER SCAUP *Aythya marila* GRSC ∎ 1

Very similar to the Lesser Scaup, the Greater Scaup often forms mixed flocks. In coastal bays and inlets or other areas of deep water, the Greater Scaup generally dominates these flocks, while the Lesser Scaup is more numerous in the interior and on shallower lakes and ponds. Polytypic (2 ssp.; *nearctica* in N.A.). L 18" (46 cm)

IDENTIFICATION Structure is the best way to separate and identify scaup. The Greater Scaup has a larger and more rounded head, thicker neck, and larger, thicker bill. The Greater Scaup also has more black on the bill, covering much of the bill tip. **MALE:** Dark head that shows a bright green sheen in bright light. The breast is black; the back is pale gray covered with fine vermiculations; the flanks are white with very light to moderated gray vermiculations.

FEMALE: Dark brown head with a slight reddish tone. There is an extensive white ring encircling the base of the bill. Females in worn plumage show a variably sized pale ear patch. The back of the female is brown, and the flanks are mottled gray-brown. **FIRST-WINTER MALE:** Like females in fall, but with less white surrounding the bill; acquires plumage like adult males by spring, but with more vermiculations on the flanks and heavily worn brown tail feathers. **FLIGHT:** White wing stripe extends almost to the tip of the wing.

SIMILAR SPECIES See sidebar (below). Compare females to other female *Aythya*. Female Ring-necked shows a more angular head, ringed bill, and more defused pale face pattern. Female Redhead is more evenly and paler brown, with more black on the bill, and lacks the white ring surrounding the bill. See Lesser Scaup entry for separation from that species.

GEOGRAPHIC VARIATION North American and east Asian birds are of the subspecies *nearctica*. There is one specimen record of the nominate *marila* from St. Paul I. in AK in late June.

VOICE Generally silent.

STATUS & DISTRIBUTION Widespread and abundant in the Northern Hemisphere. **BREEDING:** Holarctic breeder; nests on tundra pools and lakes. **MIGRATION:** In spring, begins gradually in early Mar. Mid-Atlantic peak mid-Mar to mid-Apr. In the West, most have left the southern part of their winter range by late Mar. Peak in BC the last two weeks of Apr. In fall, departs breeding grounds mid-Sept.– late Oct. The bulk of the population migrates to the Atlantic coast via the Great Lakes. Peak in Great Lakes the second week of Oct.; late Oct.–early Nov. in the mid-Atlantic. Peak in southern BC mid- to late Oct. **WINTER:** Uses bays, estuaries, and deep inland lakes and rivers.

POPULATION Numbers are declining for unknown reasons.

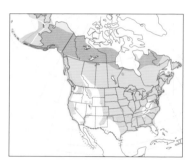

adult *nearctica* ♂

extensive white stripe

nearctica ♀

many females with prominent whitish ear patches

nearctica ♀

larger, rounder head than Lesser

nearctica ♀

1st winter *nearctica* ♂

adult *nearctica* ♂

more black on bill tip than Lesser

Identification of Male Scaup on the Water

Scaup rank highly on the list of most frequently misidentified birds in North America because birders often rush the identification process. While scaup are truly difficult and can cause much frustration, they can be learned, but only with patience and practice. Two keys to learning scaup are studying them carefully when you see the species together and taking time when identifying them.

The most solid field marks for scaup are subtle aspects of structure that can be altered by a bird's behavior and posture; time is often required to correctly assess structure. There are several useful plumage characteristics, but they are variable, and relying on plumage leads to errors.

Greater Scaup, male

Size can be a starting point in identification. In direct comparison, Greater is noticeably larger than Lesser. Head shape is often the key to identifying scaup. Greater has a smoothly rounded head that appears large in comparison with the body; Lesser has a smaller, more angular head shape. Start by looking at the nape, which in Lesser is tilted slightly forward in a straight line until it meets a slight bump on the back of the crown. This bump is caused by several long crown feathers coming to an end; it is frequently referred to as the topknot. At times the topknot forms an obvious bump; in some postures the bump disappears, becoming just a few ruffled feathers on the crown. The head has a strongly peaked

LESSER SCAUP *Aythya affinis*   LESC ■ 1

The most abundant, frequently encountered *Aythya* in North America, Lesser Scaup is smaller than Greater Scaup and usually more common in freshwater. Monotypic. L 16.5" (42 cm)

IDENTIFICATION Lesser Scaup are medium size with small, narrow heads; high, peaked crowns; and thin necks. Their bills are small and thin, with only a small amount of black surrounding the nail. **MALE:** Dark head has a purple gloss in good light, but may also appear green. It has a vermiculated gray back and lightly vermiculated flanks that may appear light gray or white. **FIRST-WINTER MALE:** Like female in fall, but with no white surrounding the bill; acquires plumage like adult male's by spring, but with more vermiculations on the flanks. **FEMALE:** Dark brown head with a moderate white ring enclosing the base of the bill. A few females, in worn plumage, show a small, pale ear patch; back is dark brown, and flanks are a mottled

mix of gray and brown. **FLIGHT:** White wing stripe is strong on the secondaries and very faint on the primaries.
SIMILAR SPECIES See sidebar (below). Compare females to other female *Aythya*. Female Ring-necked has a paler head and a sharp white ring at the base of the bill. Female Lesser Scaup is smaller and darker than female Redhead or Canvasback. Separate with caution from female Greater Scaup by noticeably small size in direct comparison, angular head shape, usually with a slight bump (topknot) at the peak of the nape. Usually shows noticeably less white on the face enclosing the bill; also most have little or no pale ear patch. On average appears slightly more mottled and "messier" than Greater Scaup.
VOICE Generally silent.
STATUS & DISTRIBUTION Abundant and widespread. **BREEDING:** Nest near lakes and pools and large, permanent potholes in the western prairies. **MIGRATION:** In spring, some Lesser Scaup begin to leave their winter range in early Feb., while large numbers remain through

Apr. Mid-Atlantic and Great Lakes peak early Apr., late Mar in the Pacific Northwest. In fall, they remain on the breeding grounds until Sept. South of its breeding range, in the lower 48, the fall peak occurs in the first two weeks of Nov. virtually everywhere. **WINTER:** Found on lakes, reservoirs, and coastal lagoons. The bulk of the population winters near the Gulf of Mexico in FL, LA, and TX. **VAGRANT:** Casual to Bering Sea, north to Greenland, and Europe. Accidental in S.A.
POPULATION No strong population trends. Overall, the Lesser Scaup seems to be declining slightly, but the breeding range has expanded due to the creation of reservoirs.

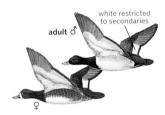

white restricted to secondaries

adult ♂

less black on bill tip than Greater

smaller, more peaked head than Greater

♀

adult ♂

appearance at times; the highest point on the head is near the back of the crown. Viewed directly in front or behind, the head of Lesser is very narrow. On Greater the nape bulges out slightly near the base, then smoothly curves to a slight peak above the eye. Its head does not look as tall as that of Lesser and has a narrow crown that expands below the eye, giving a heavy-jawed appearance. Lesser has a small bill with parallel sides; Greater has a proportionally longer bill that expands near the tip. Greater tends to have more black surrounding the nail, but this is variable enough to be unreliable.

Most plumage characteristics can be used only to support conclusions drawn from structure. However,

Lesser Scaup, male (CA, Jan.)

plumage tends to be more discernable than structure, making it easier to choose which scaup to concentrate on in a large raft.

Greater tends to have gleaming white flanks, while Lesser tends to have slightly gray flanks; a few individuals of both species, however, show the "wrong" flank color. In strong light the color of the head gloss can be obvious. Greater has bright green gloss, but Lesser usually has purple gloss, but it sometimes appears green. Using head gloss can often be misleading.

Even with years of practice and knowledge of all field marks, everyone will misidentify a scaup occasionally, but with care, most scaup can be correctly named. ■

EIDERS Genera *Polysticta* and *Somateria*

Each of the four species worldwide are found in North America. These large, bulky diving sea ducks have dense down feathers that help insulate them from cold northern waters. Females pluck their own down to line nests. Eiderdown is harvested and sold for pillows and blankets. Eiders generally migrate in large flocks. Spectacular movements can be witnessed at points in coastal northwestern Alaska. Most eiders head north to tundra nesting grounds as soon as sea ice breaks up. In years of late breakup in combination with severe storms, many eiders, mainly King, starve to death. In spring 1964, an estimated 100,000 King Eiders, ten percent of the population migrating on the Arctic coast, died.

STELLER'S EIDER *Polysticta stelleri*   STEI ■ 3

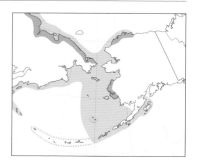

This small, compact sea duck seems inappropriately named as an eider. Structurally and behaviorally it is more reminiscent of a dabbling duck. It lacks the highly developed frontal lobes and bill feathering characteristic of other eiders. Monotypic. L 17" (43 cm)
IDENTIFICATION Head is rectangular, with a flat crown and sharply angled nape. Its long, pointed tail is generally held out of the water. **BREEDING MALE:** Distinctive. Appears mostly white at a distance. **ECLIPSE MALE:** Like adult female, but darker, retaining white in the wing and broad pale tertial tips. **FEMALE:** Easily overlooked in mixed flock. Plain brown face, with pale eye ring, warm cinnamon brown col-

oration overall. Blue speculum and white-tipped tertials diagnostic when present. **JUVENILE:** Gray-brown, plumage wears and fades quickly; thin white speculum borders, tertials dull and short. **FLIGHT:** Flies swiftly in tight flocks, with a long-winged, short-necked, heavy-bellied appearance.
SIMILAR SPECIES Blue speculum with broad white boarders, like a Mallard, in combination with long, curved white-tipped tertials separate Steller's from all other sea ducks. Female distinguished from Harlequin Duck by pale eye ring on warm brown face and lack of white spots on head and shape.
VOICE Noisy in winter flocks; females giving rapid guttural call, inciting females give loud *qua-haa* or *cooay*. Growling and barking noises given by both sexes; does not produce cooing calls like other eiders.
STATUS & DISTRIBUTION Threatened. **BREEDING:** Arrive on AK breeding grounds on Arctic tundra late May–early June. **WINTER:** To southern Bering

Sea, arrive late Oct.–Nov. Steller's Eider favors shallow, sheltered coastal lagoons and bays. **VAGRANT:** Casual outside Alaskan waters to BC, WA, OR, CA. Casual to accidental on East Coast; records from QC, ME, MA, MD, and Baffin I.
POPULATION Marked decline in AK breeding population since 1960s for reasons largely unknown. Population estimates of wintering concentrations along Alaskan Peninsula (includes Russian breeders) are also dropping.

white forewing

adult ♂

blue speculum bordered by white

♀

dark cinnamon-brown color overall

rather small unfeathered, dark bill

♀

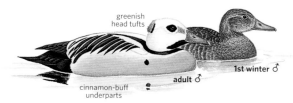

greenish head tufts

cinnamon-buff underparts

adult ♂

1st winter ♂

SPECTACLED EIDER *Somateria fischeri*   SPEI ■ 3

This hardy sea duck is appropriately named for its bold, pale "spectacles," which are apparent in all plumages. The Spectacled Eider has a gradually sloping forehead, with feathers extending down the bill. It lacks the highly developed frontal lobes of Common and King Eiders. Otherwise, Spectacled is structurally similar to Common Eider in flight but significantly smaller, with a bit shorter neck. Monotypic. L 21" (53 cm)
IDENTIFICATION BREEDING MALE: Bold head pattern with big white, black-bordered goggles and orange bill; generally distinctive. **ECLIPSE MALE:**

Resembles breeding plumage, but white and green areas replaced with dusky gray feathers. **FEMALE:** Head appears light overall with dark forehead, due to pale "spectacles" and cheek. Bill is dark, blue-gray. Vertical barring on the flanks, paler body coloration overall than other eiders. **FLIGHT:** Male is only eider with black belly extending to upper breast. Upperwing coverts all white, like on Common Eider. Underwing duskier than on other eiders. Female difficult to identify, but dark forehead contrasts with pale spectacles and cheek

and can be useful in combination with structural characteristics.

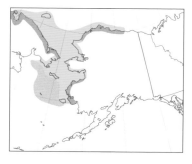

green head with white "goggles"

feathering on all birds extends well out on bill

adult ♂

black extends into breast

female has pale "goggles," but pattern is subdued ♀

pattern of white on upperparts similar to Common Eider

adult ♂

♀

SIMILAR SPECIES Spectacled Eider is smaller than Common and King Eiders. Similar to Common in flight, but more extensive dark breast on males and distinctive head pattern. Female Spectacled Eider is separated from Common and King Eiders by pale cheek and "goggles"; feathering extends down the smaller bill, and lacks the chevron pattern on the flanks of King Eider.

VOICE Typically silent. Male display call is a very soft *hoo-hoo*. Female gives a short, guttural *croak* and a two-syllable clucking call.
STATUS & DISTRIBUTION Threatened. **BREEDING:** Return to AK breeding grounds, coastal tundra near lakes and ponds, first week of May. **MIGRATION:** Depart late June for offshore molting areas. **WINTER:** Bering Sea in large flocks in openings of pack ice, primarily south of St. Lawrence Island. **VAGRANT:** Casual on Aleutian and Pribilof Is. Not prone to wander far from breeding and wintering grounds.
POPULATION Three distinct breeding populations in western and northern AK and in Russia. Russian population is currently much larger than Alaskan populations, which have experienced dramatic declines in the last 50 years.

KING EIDER *Somateria spectabilis*   KIEI ■ 2

The King Eider undergoes spectacular migrations in large flocks to arrive on the breeding grounds with enough time to breed and molt in the short Arctic summer. Monotypic. L 22" (56 cm)
IDENTIFICATION MALE: The King Eider is distinctive in breeding plumage, with bright orange frontal lobes, pinkish red bill, baby blue head, and a green wash on the side of the face. **FIRST-WINTER MALE:** Has a brown head, pinkish or buffy bill, and lacks white wing patches. Full adult plumage is attained by the third winter. **ECLIPSE MALE:** Dark overall, with only a bit of white mottling on mantle and in wing coverts; pinkish bill; and a distinct head shape with thin white postocular stripe. **FEMALE:** Brown overall with dark chevron markings on flanks and scapulars. Bill is dark, appears stubby. **FLIGHT:** Large rectangular head and short neck; stockier than the Common Eider, with faster wingbeats. The Common is larger, with a more horizontal body posture and triangular head. King Eider tends to have the rear dragging below its rectangular shaped head and a shorter neck. Male shows a partly black back and black wings with white patches. Young males have white flank patches. **SIMILAR SPECIES** Males in flight have more black on upperparts than Common Eider. Females can be difficult to separate from Common Eider. Structurally King is smaller, with a flatter crown, rounded nape, and slightly bulging, shorter forehead that does not slope evenly into its bill, which is relatively smaller than Common's.

At rest the bill of King is held more horizontally than that of Common Eider, which is often angled down a bit. Also, note that female Kings have a crescent or V-shaped markings on the flanks and scapulars as opposed to the vertical barring on Common Eider. **VOICE** Males give a low, dovelike *urr urr urr*. Various low croaks given in flight.
STATUS & DISTRIBUTION Fairly common. **BREEDING:** Common on tundra and coastal waters in northern part of range. **MIGRATION:** Move in impressive large flocks at all hours. In spring, begin departing Bering Sea in Apr. and early May. Undergo molt migration 3–4 weeks after end of spring migration. Fall: Rare in MA, generally not seen before late Oct. **WINTER:** Very rare on the Great Lakes, except on Lake Ontario, where it is rare and increasing. **VAGRANT:** Casual to FL and West Coast to southern CA.

Some West Coast individuals linger into summer. Accidental to Gulf Coast.
POPULATION Declining. Food depravation during spring migration may have significant impact on numbers; large percentage of population (e.g., 10 percent) may die of starvation.

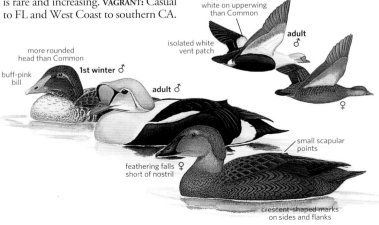

more restricted white on upperwing than Common

adult ♂

isolated white vent patch

1st winter ♂

more rounded head than Common

buff-pink bill

adult ♂

♀

small scapular points

feathering falls ♀ short of nostril

crescent-shaped marks on sides and flanks

COMMON EIDER *Somateria mollissima*   COEI ▪ 1

This most common and widespread eider shows extensive geographic variation. Polytypic (6 ssp.; 4 in N.A.). L 24" (61 cm)

IDENTIFICATION ADULT MALE: Distinctive head pattern; frontal lobes and bill color vary with respect to subspecies. **ECLIPSE MALE:** Dark with white mottling overall. **IMMATURE MALE:** Dark with white breast. Full adult plumage by third year. **FEMALE:** Evenly barred flanks and scapulars. "Pacific Eider" *(v-nigrum)* is colder brown and duller overall compared to eastern subspecies, which is warm reddish brown. In the East, *dresseri* is dark and richly colored; *borealis* is reddish with dark markings; *sedentaria* palest.

GEOGRAPHIC VARIATION There are four subspecies in N.A.: *dresseri, borealis, sedentaria,* and *v-nigrum*. See sidebar below.

SIMILAR SPECIES See King Eider entry.

VOICE Male gives a ghostly *ahoooooo* during courtship. Female gives a grating *krrrr* and other guttural calls.

STATUS & DISTRIBUTION Locally abundant. **BREEDING:** Nests in colonies along marine coasts. **MIGRATION:** Some populations sedentary, others long-distance migrants. In spring, Mar.–mid-June for Arctic breeders. Molt migration in June and July. In fall, peaks in MA Oct.–Nov.; late Nov.–early Dec. in NJ. **WINTER:** "Pacific Eider" winters in Bering Sea to south coastal AK in N.A. **VAGRANT:** In East *dresseri* and *borealis* casual on Great Lakes, rare in winter on coast south to NC, casual south to FL. "Pacific Eider" *(v-nigrum)* rare in BC; accidental to Pacific states south to northwestern CA; rare east to Greenland; *sedentaria* accidental outside Hudson and James Bays.

POPULATION Subspecies *v-nigrum* and *borealis* are declining. On Atlantic, *dresseri* relatively stable.

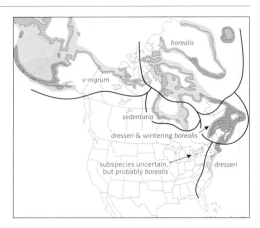

adult ♂
extensive white on back
v-nigrum

♀

feathering extends to nostril
♀ *v-nigrum*
overall color variable
♀ *dresseri*
evenly barred on sides and flanks

white back
flat forehead
1st winter ♂
dresseri
adult ♂

Subspecies Identification of Adult Male Common Eiders

Six Common Eider subspecies are recognized, four of which breed in North America. They fall into two groups, Eastern and Pacific. The Eastern group includes *dresseri* ("Atlantic Eider"), *sedentaria* ("Hudson Bay Eider"), and *borealis* ("Northern Eider"). The Pacific group consists of *v-nigrum* ("Pacific Eider").

Nominate *mollissima,* and *faeroeensis* are found in Europe.

"Pacific Eider," *v-nigrum,* is the largest eider subspecies. Adult males are readily distinguished from the Eastern group by a black V on the chin and a bright orange bill. The frontal lobes are narrow and pointed, and green on the head is extensive, continuing to a narrow line

"Atlantic Eider"
♂ *dresseri*

"Hudson Bay Eider"
♂ *sedentaria*

Genus *Camptorhynchus*

LABRADOR DUCK *Camptorhynchus labradorius*   LABD ▪ 6

This wary sea duck was the first North American endemic to reach extinction. The last reliable report was a specimen collected on Long Island, New York, in 1875. Bill shape suggests the species was a food specialist, but very little is known about its natural history. It is believed to have foraged near sandbars for shellfish. Monotypic. L 22.5" (57 cm)
IDENTIFICATION BREEDING MALE: Head, neck, and chest are white; the crown has a black stripe. FEMALE: Plumage is gray-brown overall, with throat whiter than head, mantle and scapulars slaty blue, and white

white head and breast

dark band across neck

unique bill shape

adult ♂

speculum. **IMMATURE AND ECLIPSE MALE:** Resemble females.
SIMILAR SPECIES Confusion unlikely.
VOICE Unknown.
STATUS & DISTRIBUTION Extinct. BREEDING: Remote areas, probably rocky coasts, islands, or short distance inland, likely Labrador or farther north. WINTER: Atlantic coast from NS south possibly to Chesapeake and Delaware Bays, estuaries, and sandy bays. VAGRANT: In poor weather, reported from rivers as far inland as Philadelphia. Accidental to Montreal, QC, spring 1862.
POPULATION Extinct. Explanations for extinction are largely speculative. The population already was believed to be small in the 1800s. There are no published accounts of causes of mortality aside from shooting and trapping in nonbreeding season for consumption and collection.

Genus *Histrionicus*

HARLEQUIN DUCK *Histrionicus histrionicus*   HADU ▪ 1

This small diving duck is structurally unique, with a steep rounded forehead, stubby bill, and chunky body. Males have a white crescent in front of the eye, a small white circular patch near the ear, and a white vertical stripe on the hind neck; these create the clown-like appearance for which it is named. Monotypic. L 16.5" (42 cm)
IDENTIFICATION BREEDING MALE: Distinctive. Dark blue-gray overall with chestnut flanks; appears dark at a distance. Scapulars and tertials white, white band on breast and neck.

ECLIPSE MALE: Dark sooty brown, but still shows male characteristics. **FIRST-WINTER MALE:** Like adult, but duller overall. Attains adultlike features throughout winter. **FEMALE:** Dark brown overall, pale belly, variably sized white patch in front of the eye, white spot behind eye. **FLIGHT:** Small head, steep forehead, plump body with long tail. Wings are entirely dark, but adult male has small white spots on a few coverts; whitish belly of female is not very noticeable, and it appears dark overall. Flies swiftly, low over water with rapid wingbeats; usually seen in small numbers.
SIMILAR SPECIES Female, compared with female Bufflehead in flight, has darker body and longer, pointed tail. Smaller overall size, smaller bill, steeper forehead than female scoters. Chunky body with proportionally tiny head and bill, round spot on face, and longer tail separate it from female Black Scoters. Darkest of female Long-tailed Ducks are similar, but have different head pattern and more white on body.

---

below the eye. In the Eastern group, *dresseri* is medium size, with an olive-green to greenish yellow bill and broad frontal lobes rounded at the top. It is distinguished from *sedentaria,* which has narrower, shorter frontal lobes. Subspecies *borealis* is found in the northernmost regions of the Atlantic and is the smallest subspecies in N.A. It has

a bright yellow-orange bill; its frontal lobes are narrow, tapered, and rounded; and it has less green on the side of the head than *dresseri.* Subspecies *borealis* resembles *v-nigrum* in bill color, but its overall bill size is smaller, and it lacks the black V on its throat. However, the black V is occasionally seen on *dresseri* and nominate *mollissima.* ▪

"Northern Eider"
♂ *borealis*

"Pacific Eider"
♂ *v-nigrum*

VOICE High-pitched, nasal squeaking; commonly a mouselike squeak, *gia*.
STATUS & DISTRIBUTION Locally fairly common from N.A. to Iceland and Eastern Asia. **BREEDING:** Nests on ground, near fast-flowing rivers, streams, lakes. **MIGRATION:** Not seen moving in large concentrations, short- to intermediate-distance migrant. In spring, moves inland from the coasts, departing East Coast Apr.–mid-May. West Coast departs late Mar. and largely gone by mid-May. Immature and injured birds may stay on winter grounds. In fall, males undergo molt

migration starting in late June. Rare in MA before Nov. **WINTER:** Rocky coastlines. **VAGRANT:** Rare, but increasing on Great Lakes in winter. Casual south to FL; accidental on Gulf Coast to TX. In West occurs in small numbers to northern CA, rare to southern CA, casual to Baja California. Accidental to CO, NV, and AZ.
POPULATION Declining. Pacific population larger than Atlantic. Local extirpations contribute to range reduction. Many studies underway to better understand and conserve Harlequins.

unique face pattern

round head with small bill

adult ♂

♀

white belly

chestnut sides

adult ♂

♀

adult ♂

head spots develop over winter

1st winter ♂

Genus *Clangula*

LONG-TAILED DUCK   *Clangula hyemalis*   LTDU ■ 1

This sea duck, formerly known as the "Oldsquaw," is unique in having three plumages a year. Monotypic. L 6–22" (41–56 cm)
IDENTIFICATION ADULT MALE WINTER: Long tail conspicuous. White head

with gray patch around eye, black band across breast, long gray scapulars, light gray flanks. **EARLY SUMMER MALE:** Head black, still with eye patch, scapulars buffy with black centers. By late summer, head and neck become whiter, flanks darker gray. **FIRST-YEAR MALE:** Gradually attains adultlike features by second fall. **FEMALE:** Crown

and nape blackish brown, breast grayish, flanks whitish, scapulars buff with black centers. By fall whiter head and neck. In summer head and neck dusky overall. **FLIGHT:** Appears very white, with uniformly dark underwings.

winter adult ♂

dark wings

1st fall ♀

long pointed tail

variably white on face

winter ♀

winter ♀

early summer adult ♂

males have pink band on bill

small bill

1st winter ♂

females have grayish bills

winter adult ♂

Identifiable by swift, careening flight. **SIMILAR SPECIES** Female similar to female Harlequin, but Harlequin's dark flanks, steep forehead, rounded head, and facial pattern are distinct. Female Steller's Eider is darker overall with plain face, white-edged speculum, curved tertials. In winter flight, guillemots can cause confusion; they have a bold white wing patch, unlike the black wings of the Long-tailed. **VOICE** Highly vocal. Male utters

nasal, loud, yodeling, three-part, *ahr-ahr-ahroulit.* Both sexes give soft *gut* or *gut-gut* call while feeding. **STATUS & DISTRIBUTION** Common. **BREEDING:** Arctic tundra ponds and marshes. **MIGRATION:** In spring, East coast in late Mar.–early Apr.; West and Great Lakes in late Feb.–May. In fall, U.S. coasts and Great Lakes peak numbers in late Nov. and Dec. **POPULATION** Hard to evaluate; information is lacking.

SCOTERS Genus *Melanitta*

The AOU recognizes four species of scoters; Europeans split White-winged Scoter for a total of five. They are medium to large, with black or blackish plumage and orange or yellow on the bill. They nest in northern Canada and Alaska and winter along both coasts. While scoters migrate nocturnally when crossing land, along the coast they migrate diurnally, often in mixed flocks. During peak migration these flocks may consist of several thousand individuals. Away from the breeding grounds and the Great Lakes, scoters are primarily coastal. Small numbers of all three N.A. breeding species, however, are found throughout the interior every year, most often on lakes and rivers after late fall storms.

SURF SCOTER *Melanitta perspicillata*   SUSC ▪ 1

Surf Scoters are the most common scoter in many areas. Male Surfs have a distinctive clownlike face. Monotypic. L 20" (51 cm)
IDENTIFICATION Forehead and bill are smoothly merged, forming a wedge-shaped head. The bill meets the face in a straight, vertical line; some feathers extend down the culmen toward the nares. Diving birds flick their wings open just as they submerge. **ADULT MALE:** Black with white patches on the nape, forecrown, and base of the bill. Eye pale and bill orange with a yellow

tip. **ADULT FEMALE:** Uniform blackish brown with diffuse pale loral and postocular spots; some also have pale napes. **FIRST-WINTER MALE:** As female in fall, but with a pale belly; acquires much of the black-and-white plumage and orange bill during the first winter. **FIRST-WINTER FEMALE:** As female but with a pale belly. **FLIGHT:** Large, wedge-shaped head, sleek body, and pointed wings give the Surf Scoter a unique silhouette. Forms dense flocks. **SIMILAR SPECIES** First-winter females can have pale cheeks like Black Scoter, but can be separated by head shape. **VOICE** Generally silent. **STATUS & DISTRIBUTION** Common. **BREEDING:** Nesting on shallow lakes, often with rocky shores. **MIGRATION:** In spring, begins in Mar.; peaking in CA and OR in second half of Apr.; mid-Atlantic peak in late Mar. or early

Apr.; movements less noticeable in spring than fall on Atlantic coast. In fall, begins in late Aug., peaking in CA in early Nov.; mid-Atlantic peak mid- to late Oct. **WINTER:** In large flocks along both coasts, usually close to shore. **VAGRANT:** Rare in the interior, where it is primarily a migrant; rare on the Gulf Coast. Casual to U.K. and Greenland.
POPULATION Declining for unknown reasons.

adult ♂

more uniformly dark wings than Black

immature ♀

whitish belly

forward whitish patch vertically shaped

1st winter ♀

feathering out culmen

adult ♂

white head patches

forward whitish patch vertically shaped

white nape on adult female

adult ♀

1st winter ♂

bill shows color by Dec.

immatures of all scoters have pale bellies

WHITE-WINGED SCOTER *Melanitta fusca*   WWSC ■ 1

When migrating, these—the largest and most distinctive of the four scoters—form smaller flocks. Polytypic (3 sp.; 2 in N.A.). L 21" (53 cm)
IDENTIFICATION The white secondaries are obvious on close birds. White secondaries may be visible on swimming birds, but are often hidden. When diving, they flip their wings open even more distinctly than Surf Scoters. **MALE:** Head, breast, and back are black; sides are dark brown. A small white comma mark surrounds the eye; orange tip on bill. **FEMALE:** Dark brown with two faint round spots on the face. **FLIGHT:** Broad wings and heavy flight are reminiscent of eiders; white secondaries diagnostic.
GEOGRAPHIC VARIATION Three subspecies: widespread *deglandi* in N.A.; records of *stejnegeri* from

St. Lawrence I. and Nome. Asian subspecies *stejnegeri* has a more obvious nasal protuberance, different bill color pattern, and black flanks. Western Palearctic adult male *fusca* ("Velvet Scoter," illustrated), casual to Greenland, also has black flanks and largely lacks a bill knob, but note different bill coloration. Females of all subspecies very similar to one another.
SIMILAR SPECIES Compare females to Surf Scoter. Face pattern less distinct, feathers extend farther down bill. White secondaries distinctive if visible.
VOICE Generally silent.
STATUS & DISTRIBUTION Locally common. **BREEDING:** Tundra pools. **MIGRATION:** Early Mar.–May. In fall, peaks in the mid-Atlantic in late Nov. **VAGRANT:**

Rare to casual in the interior and the Gulf Coast. Nominate *fusca* casual to Greenland, but no N.A. records.
POPULATION Has declined for many regions.

juveniles of both sexes show whitish patches; forward patch horizontally shaped

adult ♂ *deglandi*

only scoter with white in wings

1st winter ♂ *deglandi*

1st winter ♀ *deglandi*

feathering

adult ♀ *deglandi*

adult male has white slash under eye

brownish flanks

adult ♂ *deglandi*

adult ♂ *stejnegeri*

"Velvet Scoter" *fusca*

white not always visible on folded wing

adult ♀ *deglandi*

recorded from Greenland

small knob

yellow on sides of bill

adult ♂

black flanks like *stejnegeri*

bill color and white behind eye differs from *deglandi*; note bulbous nostrils

black flanks

COMMON SCOTER *Melanitta nigra*   COSC

Common Scoter, the Eurasian counterpart to Black Scoter, was until recently considered conspecific with that species. Even before the two taxa were elevated to full species in 2010, sea-watching birders in coastal areas from Labrador to Massachusetts had kept a keen eye out for Common Scoter, which breeds as near as Iceland and has been documented as close as Greenland. Monotypic. L 18–19" (45–49 cm)
IDENTIFICATION **ADULT MALE:** Uniformly black in plumage with a distinctive bill. The base of the maxilla appears swollen at the base, then appears oddly indented across the culmen; most of the bill is black, the central portion of the maxilla yellow-orange, including (usually) a variably

narrow stripe in the middle of the swollen portion. **FEMALE:** Closely resembles Black Scoter, being dark brown with contrasting pale cheek and throat. **IMMATURE:** First-year birds resemble females but have paler belly. Young males acquire a bill pattern similar to the adult's during the first winter.
SIMILAR SPECIES Adult male most likely to be confused with Black Scoter, but differences in bill shape and pattern are diagnostic. In addition, there are very some subtle structural differences: The wings of Common are more pointed than those of Black, and Common has a slightly thinner and more rounded head than Black. In these closely related species, females and immatures are not known to be distinguishable in the field; however,

the nostrils are set closer to the base of the bill in Common Scoter.
VOICE Male's call is very similar to Black Scoter's mournful whistle in quality, but male Common Scoter's is lower in pitch and usually (by a factor of four) much shorter than Black's, though Black sometimes gives shorter calls.

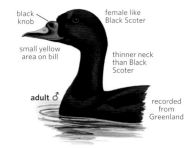

black knob

female like Black Scoter

small yellow area on bill

thinner neck than Black Scoter

adult ♂

recorded from Greenland

STATUS & DISTRIBUTION BREEDING: Adjacent freshwater pools, lochs, lakes, and rivers in tundra and heath habitats from Iceland across much of northern Eurasia, east to the Olenek River in northwestern Siberia. WINTER: Coastal (saltwater) areas from Norway to the British Isles, sparingly south to Portugal and Gibraltar.
POPULATION The species is highly sensitive to disturbances in wintering areas. Oil spills have claimed many Common Scoters, and much of the European wintering population is found near oil extraction operations. Populations in Scotland declined 45 percent between 1995 and 2007, and similar declines are suspected in Iceland for reasons unknown, though a warming climate could be responsible in the southern latitudes of the species' range.

BLACK SCOTER *Melanitta americana*   BLSC   1

The handsome Black Scoter is often located in winter and spring by the male's plaintive, whistling calls, given incessantly, particularly during foggy or calm conditions, as flocks are resting in rafts. Black Scoters migrate in flocks, mixing with other scoters as well as other waterfowl. On late autumn days with heavy migratory passage the spectacle of scoter migration can be spellbinding, with ribbons of birds numbering in the many thousands. Monotypic. L 18–19" (45–49 cm)
IDENTIFICATION Chunky body with small head and bill; at a great distance the bill seems to disappear, leaving the impression of a flat profile. The tail may be held flat but is often held cocked above the water, like a Ruddy Duck's. **ADULT MALE:** Uniformly black with a bright orange knob at the base of the maxilla. **FEMALE:** Dark brown with contrasting pale cheek and throat. First-year birds resemble females but have paler belly. **IMMATURE:** Young males slowly acquire the orange knob during the first winter.
SIMILAR SPECIES Distinctive at close range. Female Ruddy Duck, mostly brownish with a pale cheek, is superficially like female Black Scoter. Distant Black Scoters in flight can be difficult to separate from Surf Scoter, but some are identifiable with careful study of head, wing, and body shape. Black has a pot-bellied appearance compared with Surf, a shorter "hand" with more rounded wingtips, and stronger silver flash on the underside of the primaries (in optimal light). Its rounded head, with steep forehead, differs from the long, sloping profiles of White-winged and Surf. As a Black Scoter's primaries wear, they become strongly translucent, a feature that is most useful in the spring, especially for adult males, and an attribute that also separates Black from the two other American scoters. Unlike White-winged and Surf, Black dives with a small leap, wings held closed. On the water, it often rears out of the water, dips its head down, and performs a rapid series of wingbeats; Surf and White-winged Scoters hold their heads up when stretching their wings. See also Common Scoter.
VOICE Males produce low, mellow whistle and less often a rattling call. Females are said to give squealing and growling calls on the breeding grounds. Adult males also produce wing noise when flying, like other scoters, perhaps not as loud as that of Surf Scoter.
STATUS & DISTRIBUTION BREEDING: On tundra pools, lakes, and rivers in two disjunct parts of N.A.; the western population nests mostly in AK and adjacent YT, while the eastern population is confined mostly to northern QC and Labrador. A small population exists in the Russian Far East. MIGRATION: Usually in large flocks, most visible in fall. In spring, peak in the mid-Atlantic in early Apr., in Canadian Maritime Provinces in early May, and arrival on the breeding grounds in northern QC occurs in the third week of May. In CA, migration begins in early Mar. Southward migration in fall begins in Sept., peaking in the third week of Oct. in the mid-Atlantic, and in early Nov. in BC. Smaller numbers pass through the Great Lakes, and the species can be common on Lake Ontario, where some winter. WINTER: Found on inshore bays and inlets and offshore along both Atlantic and Pacific coasts; very uncommon to very rare along Gulf Coast and in the interior.
POPULATION Based on counts of migrating and wintering birds, Black Scoter appears to be declining sharply in recent decades, but little research has been done in breeding areas to determine what factors might be causing the decline.

yellow-orange knob at base of bill

round head

no feathering out on bill

pale face contrasts sharply with dark cap

dull yellow at bill base

adult ♂

adult ♀

adult ♂

secondaries and primaries paler than rest of wing, especially on adult males

1st winter ♂

adult ♀

BUFFLEHEAD *Bucephala albeola*   BUFF ■ 1

This diving duck has adapted to a wide variety of habitats from sheltered bays, rivers, to flooded fields. It is one of the smallest ducks, with a large puffy head, a steep forehead, and a short stubby bill. Its plumages are suggestive of miniature goldeneyes. It flies fast and with such rapid wingbeats that the wings blur; usually seen only in small flocks. Monotypic. L 13.5" (34 cm) **IDENTIFICATION MALE:** Glossy black back, white below, with large white patch on iridescent purple-green head. **ECLIPSE MALE:** Like female, but darker head with larger white patch; retains white in wing. **FIRST-YEAR MALE:** Like female. **FEMALE:** Dark gray-brown above, dusky white below. Dark head, elongated white patch under eye.

SIMILAR SPECIES Female's very rapid wingbeats and small size can resemble some alcids. Compare female Harlequin; Bufflehead has larger head, shorter tail, and faster wingbeats. Female and juvenile Long-tailed Ducks are larger, with slower wingbeats, longer tail, whiter head. **VOICE** Generally silent, except during courtship displays. **STATUS & DISTRIBUTION** Common and widespread. **BREEDING:** Nests in woodlands near small lakes, ponds. **MIGRATION:** In spring, begins in Feb., arrives on breeding grounds early Apr.–early May. In fall, begins late Oct., with numbers on wintering grounds increasing until early Dec. **WINTER:** Uses a wide range of open-water habitats

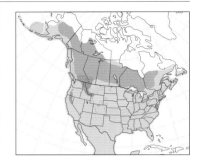

including sheltered bays, rivers, and lakes. **VAGRANT:** Casual to U.K. **POPULATION** One of the few species of ducks whose numbers have increased markedly since the mid-1950s.

extensive white head pattern

adult ♂

small dark bill

1st winter ♂

white cheek spot

adult ♂

♀

COMMON GOLDENEYE *Bucephala clangula*   COGO ■ 1

This common and widespread goldeneye is usually found in or near deep, clear water. Polytypic (2 ssp.; *americana* in N.A.). L 18.5" (47 cm) **IDENTIFICATION** Large triangular head, sloping forehead, and longer bill than the Barrow's, with shallower base. **ADULT MALE:** White circular spot on dark head distinctive. **FIRST-YEAR MALE:** Like female; transition to adult male throughout first winter. White loral spot develops slowly, can appear crescent shaped, causing confusion with the Barrow's. **FEMALE:** Dull brown head, pale gray body, and black bill with yellow tip. Rarely can have mostly yellow bill. **FLIGHT:** Heavy and muscular appearance. Show more white on the upper wing than the Barrow's, but difficult to judge.

GEOGRAPHIC VARIATION Subspecies *americana* is found in N.A.; smaller nominate *clangula* is found across the Old World. **SIMILAR SPECIES** Males have more white on the scapulars than Barrow's and lack the black spur on the breast of Barrow's. Loral spot round and forehead not as steep as Barrow's. Hybrid Common × Barrow's are rare; majority of reports from the East. Male hybrids (that can be identified) typically look almost exactly intermediate between the two species (see illustration). **VOICE** Male gives a soft, short *preent* during display. Female

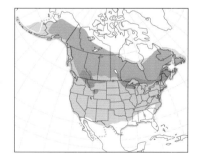

occasionally gives a harsh croak: *gack.* **STATUS & DISTRIBUTION** Common. **BREEDING:** Open lakes with nearby woodlands where nest holes are available. **MIGRATION:** In spring, departs southern part of winter range in late Feb. Peak on Atlantic coast late Mar.–early Apr. In fall, New England and Atlantic coast peaks early Dec. Pacific coast migration late Oct.–early Dec.; CA peak third week of Nov. **WINTER:** Coastal areas, inland lakes, and rivers. **POPULATION** Most data suggests population is relatively stable.

adult ♂

americana

round white spot on face

adult *americana* ♂

mostly white scapulars

♀

1st winter *americana* ♂

americana ♀

pure white sides

BARROW'S GOLDENEYE *Bucephala islandica*   BAGO   ◼   1

North American and Icelandic species. Monotypic. L 18" (46 cm)
IDENTIFICATION Puffy, oval-shaped head, steep forehead, and stubby, triangular bill. **BREEDING MALE:** Bold white crescent; black spur interrupts white breast and flanks. **ADULT FEMALE:** Dark brown head, gray body, yellow-orange bill that becomes black in summer. **FIRST-WINTER MALE:** White facial crescent develops slowly, evident on many by late December. **FIRST-WINTER FEMALE:** Like adult, but with largely black bill.

JUVENILE: Like female; dark eyes and bill, darker and browner overall; difficult to distinguish from Common.
SIMILAR SPECIES Adult female separated from Common Goldeneye by steep forehead and shorter bill. Bill usually all yellow-orange; some female Commons can have a yellow bill, but color is paler, not as orange as Barrow's. Separation of many individuals in fall is very difficult; hybridization complicates.
VOICE Generally silent.
STATUS & DISTRIBUTION Fairly common in West, uncommon to rare in East; northeast to Greenland and Iceland. **BREEDING:** Open lakes and small ponds. **MIGRATION:** In spring, Pacific coast mostly late Mar.–early Apr.; MA by early Apr. In fall, Pacific coast late

Oct.–early Nov.; arrive late Nov–early Dec. in ME and MA. **WINTER:** Sheltered coastal areas, lakes, and rivers. **VAGRANT:** Casual away from Great Lakes and Northeast. Accidental to Southeast.
POPULATION Stable.

adult ♂

slightly less white on wing coverts than Common

many immature males have a shaded crescent on face by late Dec.

1st winter ♂

♀

steeper forehead and stubbier bill than Common

black scapulars with white spots

1st winter ♀

white crescent on face

vertical black spur on side

adult ♂

orange-yellow bill

adult ♀

Barrow's x Common hybrid adult ♂

hybrid adult males have intermediate amount of white on scapulars and white facial mark

---

MERGANSERS   Genera *Lophodytes, Mergus,* and *Mergellus*

The long, thin, serrated bills and slender, long-necked bodies of mergansers aid them in catching fish, crustaceans, and aquatic insects. There are six species worldwide, four in N.A. Fast and direct in flight, mergansers show pointed wings, shallow wingbeats, long neck and bill. They nest in tree cavities, nest boxes, and on ground sheltered by vegetation.

SMEW   *Mergellus albellus*   SMEW   ◼   3

This Eurasian species is a vagrant throughout N.A. Most records are from Alaska and the West Coast. The Smew is kept in captivity in U.S., therefore origin is often questioned for sightings away from Alaska and the West Coast. Monotypic. L 16" (41 cm)

IDENTIFICATION BREEDING MALE: White with black markings, black-and-white wings conspicuous in flight. **FIRST-SPRING MALE:** Dark rufous crown and nape, dark around eye, white cheek, gray body. **FEMALE:** White throat and lower face contrast with reddish head and nape. **FLIGHT:** Male and female have large white ovals on upper wing.
SIMILAR SPECIES White patch on lower face of females and nonbreeding males

separates all other sea ducks. Female-like plumages resemble Common Merganser, but significantly smaller; extensive white lower face and dark smaller, stouter bill are distinctive.
VOICE Generally silent, except during courtship.
STATUS & DISTRIBUTION Rare migrant to west and central Aleutians mid-Mar.–May and primarily Oct. to early winter, where it has been observed in small flocks. **BREEDING:** Northern Eurasia near rivers and small lakes in forested areas. **VAGRANT:** Casual elsewhere in AK and in the Pacific states; accidental in the East, including three records from ON. Most records away from AK Nov.–Mar.
POPULATION In Europe apparently stable after long-term decline.

adult ♂

♀

1st spring ♂

slight crest

adult ♂

short, dark bill

reddish brown crown and nape

white throat

♀

black face

white body with black bars

HOODED MERGANSER *Lophodytes cucullatus* HOME ■ 1

This small, oddly shaped diving duck is frequently found on ponds, wooded sloughs, and streams in small groups or pairs. Monotypic. L 18" (46 cm) **IDENTIFICATION ADULT MALE:** Black-and-white fan-shaped crest, black back, chestnut flanks, white breast with black band, black bill. **FIRST-YEAR MALE:** Like female, but white in crest,

yellow eye, sometimes dark feathers on head, bill blackens. **ECLIPSE MALE:** Dusky-brown crest, dark bill, yellow eye. **FEMALE:** Brownish overall, paler breast, upper mandible dark, lower yellowish. **FLIGHT:** Very quick wing-beats, head usually held low, crest flattened, prominent tail. **SIMILAR SPECIES** Female can be confused with female Red-breasted. Much smaller; thinner dark bill; lower mandible yellow. In flight like Wood Duck, but more slender shape and with faster, shallower wingbeats. **VOICE** Generally silent; but froglike growl often given during courtship displays. **STATUS & DISTRIBUTION** Uncommon in West, common throughout much

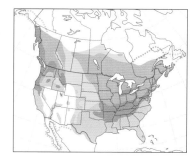

of East. **BREEDING:** Forested areas with nest cavities. **MIGRATION:** In spring, begin departing Southeast early Feb.; remain on coastal areas in Pacific Northwest until mid-Apr. In fall, peak mid-Nov. in New England; numbers increase by late Dec. in FL. Arriving on wintering areas late Oct. in Pacific Northwest. **POPULATION** Stable, possibly increasing in some areas.

adult ♂

very fast wingbeats

♀

crest sometimes raised

very similar to female but pale iris by midwinter

1st spring ♂

crest often flattened

adult ♂

brownish bushy crest

♀

dull yellow bill

chestnut sides

COMMON MERGANSER *Mergus merganser* COME ■ 1

This hardy merganser is regularly seen along raging streams or loafing on chunks of ice in winter lakes. It forms

large single-species flocks in winter, often tightly packed and facing the same direction. Frequently obtains food by diving in deep water; however, also forages in shallow water by swimming with its face submerged looking for prey. Polytypic (3 ssp.; 2 in N.A.). L 25" (64 cm) **IDENTIFICATION** One of the largest ducks, with a heavy body, thick neck,

and long, slender, hooked bill. Sits low in the water, holding its large head erect. Slight crest on nape can be held close to the head so it almost

thick base to bill

1st spring *americanus* ♂

adult *americanus* ♂

americanus

americanus ♀

short crest

adult *americanus* ♂

dark spur on wing

♀

rich rufous head and lower throat contrast sharply with white chin

white underparts often tinged pinkish

thinner bill with more distinct hook than *americanus*

"Goosander" adult ♂
merganser

lacks black spur on wing

disappears. Thicker bill than other mergansers, especially at base. ADULT MALE: Dark head, white breast and flanks, generally distinctive. FEMALE: Bright chestnut head and neck contrast with white chin; the bill is bright red and the body uniformly pale gray. ECLIPSE MALE: Resembles female, but retains adult male's wing pattern. FIRST-WINTER MALE: Resembles adult female, but with paler body and with variable amounts of dark green feathers appearing during the first winter. FLIGHT: Shows large white wing patches with a single black bar across the median coverts, larger on males than females. Has bulkier body and thicker neck than Red-breasted Merganser. Flight is extremely swift, with shallow, rapid wingbeats. **GEOGRAPHIC VARIATION** Three subspecies worldwide; only *americanus* regular in North America. Nominate *merganser* migrant to western Aleutian Islands, found across Eurasia. Males of the nominate subspecies have a

shallower base to the bill, larger hook on bill, and lack the black bar across the white wing patch. Largest subspecies, *comatus,* is found in Central Asia. **SIMILAR SPECIES** Female can be confused with the smaller female Red-breasted Merganser, but has sharply delineated white chin surrounded by chestnut head and throat, unlike diffused pale throat and neck of Red-breasted. Bill has a heavier base than a Red-breasted that meets the head on a straight line. Tends to look paler and cleaner than Red-breasted, which appears more mottled. Robust build with a thick neck and large head, unlike slender Red-breasted. **VOICE** Generally silent, except during courtship or when alarmed. **STATUS & DISTRIBUTION** Common. BREEDING: Nests in woodlands near lakes and rivers. MIGRATION: Short- to intermediate-distance migrant, nocturnal overland movements and diurnal migration along coasts. In spring, leave southern part of wintering range in mid-Feb.;

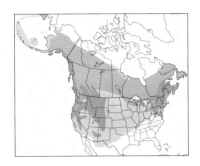

most birds gone by Apr. Great Lakes region, arrive late Mar., peak late Apr. In fall, on Atlantic coast, arrive mid-Nov., peak Dec.–Feb. in VA. On West Coast peak in BC in Nov. WINTER: Found as far north as there is open water; large lakes, reservoirs, rivers, sometimes on coastal bays, estuaries, and harbors. In eastern N.A., regular in winter to Ohio River, casual to Gulf Coast. **POPULATION** Stable or increasing in N.A. Good indicator species to assess contamination levels.

RED-BREASTED MERGANSER *Mergus serrator*   RBME   ▪ 1

This species is most often seen in flocks in coastal wintering areas. Monotypic. L 23" (58 cm) **IDENTIFICATION** Medium size; a slim neck and smaller bill; ragged, two-pronged crest. ADULT MALE: Green head; white collar; brown breast; gray flanks; distinctive. ADULT FEMALE: Gray-brown body with a dull reddish brown head; slender reddish bill; all-white throat and breast. FIRST-YEAR MALE: Similar to female, but paler gray with variable green on head. FLIGHT: White wing patches divided by two black lines; head and neck appear flat. **SIMILAR SPECIES** Female similar to Common, but Red-breasted is slender, with thinner neck, much thinner bill,

full ragged crest, uniformly pale throat. **VOICE** Generally silent, except during courtship. Male gives catlike *yeow-yeow.* Female produces raspy croaking during displays. **STATUS & DISTRIBUTION** Common. BREEDING: Arrive mid- to late May. Nest in woodlands near freshwater or sheltered coastal areas. MIGRATION: Abundant on Great Lakes, where some winter. In fall, pass through Great Lakes first two weeks of Nov., with numbers increasing in mid-Atlantic in early Nov. Peaks mid-Sept.–early Nov. in BC. Uncommon in interior, more common in East as migrant. WINTER: Favors salt water more than other mergansers. Common in coastal regions and Great

Lakes, except Lake Superior. **POPULATION** Declining; reasons unknown.

STIFF-TAILED DUCKS Genera *Nomonyx* and *Oxyura*

Long, narrow, stiff tail feathers serve as a rudder for these diving ducks. Males have a bright blue bill in breeding season. They are found in fresh and brackish water; the Masked prefer ponds with dense emergent vegetation. When actively swimming and diving, the tail is frequently laid on water, not cocked and spread as it is held when resting.

MASKED DUCK *Nomonyx dominicus*   MADU ■ 3

This small tropical stiff-tail most frequently appears in southern TX and north to the upper TX coast. Shy and difficult to see even when present, it spends most of its time hidden in dense aquatic vegetation. Monotypic. L 13.5" (34 cm)

IDENTIFICATION Slightly smaller than Ruddy Duck, with smaller head and shorter bills. Like Ruddy, it has a long tail that can be cocked above the body or trail behind on the water. **BREEDING MALE:** Black face with sky blue bill. Nape and body are dull reddish brown with black-centered flank feathers. **FEMALE:** Warm buffy brown, with a buffy supercilium and two dark stripes on the cheek. **WINTER MALE:** Like female, but with stronger stripes on the face. **FLIGHT:** Large white patches on inner wing. Unlike Ruddy, it takes to flight easily off the water.

SIMILAR SPECIES Duller plumages are similar to female and winter male Ruddy Duck, but overall coloration warmer (buffier) and with

two distinct dark stripes on face. Also shows a pale supercilium, which Ruddy lacks. Ruddy Ducks prefer more open water with less vegetation. Any stiff-tail with white wing patch can instantly be identified as a Masked Duck. Male Ruddy Ducks can show an entirely black face or heavily flecked cheek patch during molt to breeding plumage.

VOICE Generally silent.

STATUS & DISTRIBUTION Rare; sometimes occurs in small numbers after invasions in TX, but then it will go for many years without being recorded at all. **BREEDING:** Nests on lowland ponds, marshes, and rice fields with much floating vegetation. **MIGRATION:** Nonmigratory. **VAGRANT:** Rare in southern and southeastern TX, casual in LA and FL. Accidental in East, records from WI, MA, NC, MD, and VT.

POPULATION Numbers fluctuate in TX, more likely in wet years. Recent invasions in 1930s, late 1960s, early

1970s, and 1990s. Due to the species' secretive nature, preference for highly vegetated ponds, and fact that it is often on private land, it is difficult to assess the population and probably not as rare as is perceived.

breeding ♂

long pointed tail

white wing patches, but species seldom seen in flight

♀

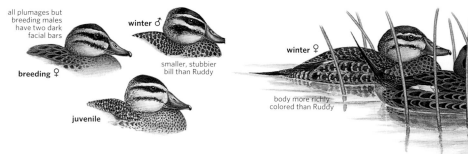

all plumages but breeding males have two dark facial bars

winter ♂

breeding ♀

smaller, stubbier bill than Ruddy

juvenile

winter ♀

blackish head

body more richly colored than Ruddy

breeding ♂

RUDDY DUCK *Oxyura jamaicensis*   RUDU ■ 1

The Ruddy Duck is the only widespread stiff-tail in N.A. Bright breeding plumage is held during summer and dull plumage during winter, unlike other ducks of N.A. Polytypic (2 ssp.; *rubida* in N.A.). L 15" (38 cm)

IDENTIFICATION Small, chunky duck with proportionally large head and bill. Frequently cocks long tail in a fan shape. **BREEDING MALE:** Bright baby blue bill, black crown contrast with white cheeks and chestnut body.

Often with black flecking or almost entirely black cheeks when molting to breeding. **WINTER MALE:** Retains a white cheek and black crown but body becomes dull gray-brown. **FEMALE:** Fairly dull gray year-round. Dark crown; a single dark line crosses buffy cheek. **FLIGHT:** Rarely seen, prefers to dive or run on water to avoid danger. When they do fly, wingbeats are rapid, usually barely clearing the water.

SIMILAR SPECIES Unique structure and plumage, most likely to be confused with Masked Duck. Female Bufflehead is superficially similar, but has a rounded head with a high forehead and smaller bill.
VOICE Generally silent.
STATUS & DISTRIBUTION Common. **BREEDING:** Nests in dense vegetation of freshwater wetlands. **MIGRATION:** Almost exclusively nocturnal. In spring, begins in early Feb. in south;

ends third week of May in northern parts of range. In fall, earlier than other diving ducks, beginning in late Aug., peaks mid-Sept.–late Oct.,

ending in Dec. **WINTER:** Lakes, bays, and salt marshes.
POPULATION Stable or increasing in N.A.

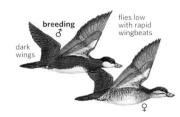

EXOTIC WATERFOWL

Many waterfowl species are brought into North America from other continents for zoos, farms, parks, and private collections. They occasionally escape from captivity. The species shown here are among those now seen all over; some are becoming established in the wild. While some species such as the Tufted Duck, Smew, and Baikal Teal are natural vagrants to North America, they are also kept in captivity. Use caution when assessing the origin of waterfowl and look for marks of formerly captive birds such as leg bands, lack of a hind toe, notches in webbing, and unusual feather wear.

SWAN GOOSE *Anser cygnoides*   CHGO ■ EXOTIC

This native to eastern Asia has a domestic form commonly known as the Chinese Goose, which is the only domestic goose not of Graylag ancestry. There are two types: the slimmer-bodied Chinese Goose (Brown and White) and the larger, heavy-bodied African Goose. The wild form is slim; its long, swanlike bill lacks knob at base. It could potentially reach western AK as a vagrant. Rear end floats higher than rest of body. L 45" (114 cm)

IDENTIFICATION Crown, nape, hind neck dark brown; cheek and sides of neck white. Upperparts dusky brown with buffy edges.
SIMILAR SPECIES Generally distinctive. White form could be confused with other domestic white geese of the Graylag-type, but note prominent orange knob on bill of Graylag-types.
STATUS & DISTRIBUTION Domestic forms found regularly in N.A.
POPULATION Little is known; endangered in its native land of eastern Asia.

"Chinese Goose" domestic type

BAR-HEADED GOOSE *Anser indicus*   BHGO ■ EXOTIC

This native to central Asia is known for its record high-altitude flights over the Himalaya; it was recorded above Mount Everest (29,029 feet). It is particularly tolerant of cold climatic conditions. The Bar-headed Goose is common in captivity in N.A and Europe, where it frequently escapes. L 30" (76 cm)
IDENTIFICATION ADULT: Two conspicuous black bars on rear of white head; white line runs down side of gray neck. Back is pale gray with white-edged

feathers. Bill is yellow with a black nail; orange-yellow legs. **JUVENILE:** Less distinctive than adult, with white face, dark gray crown and hind neck, paler gray body.
SIMILAR SPECIES Generally not confusing.
VOICE Honking flight call.
STATUS & DISTRIBUTION Escapes can be found anywhere in N.A. and western Europe. **BREEDING:** High-mountain lakes of central Asia. **MIGRANT:** Flies over the Himalaya to reach wintering

grounds. **WINTER:** Mountain rivers, lakes, and grassy wetlands primarily in northern half of Indian subcontinent, west to Pakistan and east to northern Burma.

two black bars on head

EGYPTIAN GOOSE *Alopochen aegyptiacus*   EGGO ▪ EXOTIC

adults

This native of tropical Africa is usually found inland on freshwater. It feeds by grazing but also dabbles, swimming with rear end floating higher than rest of the body. Monotypic. L 27" (68 cm) **IDENTIFICATION** **ADULT:** Head and neck buffy with dark brown patch around eye and base of bill. Body buffy to gray-brown, paler on flanks, whitish belly. **JUVENILE:** Duller than adult, with dark crown and hind neck, lacking brown eye patch. **FLIGHT:** Strong and quick flier with slow wingbeats, white wing coverts, green to purple speculum. **SIMILAR SPECIES** Likely to be confused only with Ruddy Shelduck, which is smaller, shorter legged, with orange-chestnut body, and no dark eye patch. **VOICE** Male gives a harsh, breathing sound; female a loud, harsh quacking. **STATUS & DISTRIBUTION** An escape throughout N.A., mostly southern CA and warm climates elsewhere. Feral breeder in Britain; escape in Europe.

RUDDY SHELDUCK *Tadorna ferruginea*   RUSH ▪ EXOTIC

This Afro-Eurasian species (popular in zoos and private collections) is mainly a nomadic species. Monotypic. L 26" (66 cm)

adult ♂

IDENTIFICATION Prominent white wing coverts, green speculum, and ruddy back in flight. **MALE:** Head and neck buffy, with thin black band around neck, buffy orange body. **FEMALE:** Like male, but lacks neck collar and buffier overall. **FLIGHT:** Upperwing coverts white, black primaries, green speculum. Underwing coverts, white, contrasting with dark flight feathers. **SIMILAR SPECIES** Likely to be mistaken only for Egyptian Goose (mainly brown, larger, and longer legged). **VOICE** Frequently gives a loud, nasal trumpeting call in flight. **STATUS & DISTRIBUTION** Common in captivity. A record of a flock of six on July 23, 2000, at Southampton I., NU, may have been wild birds from the Old World; accidental in Greenland. **POPULATION** Long-term decline until recently.

COMMON SHELDUCK *Tadorna tadorna*   COSH ▪ EXOTIC

This native to Europe, Asia, and northern Africa is often known as the "Shelduck." Commonly kept in collections and reported as an escape in N.A., Common Shelduck is as at home on land as it is in water; favors salt or brackish water. Monotypic. L 25" (64 cm) **IDENTIFICATION** Sexes are similar. Bright red bill; green head; broad chestnut band across white breast and belly. **ADULT:** Bold black band around neck. Prominent knob at base of bill, especially on males in breeding season. **JUVENILE:** White face; lacks chestnut band on breast. **FLIGHT:** White wing coverts prominent. **SIMILAR SPECIES** Generally distinctive. **VOICE** Loud calls. Males give musical whistles in flight and perched females give a low and nasal *ak-ak-ak-ak*. **STATUS & DISTRIBUTION** Abundant in western Europe. Recent records from NF and MA have not been accepted as of wild origin.

POPULATION Population increasing in Europe.

adult ♂

MANDARIN DUCK *Aix galericulata*   MADU ▪ EXOTIC

This native of eastern Asia is one of two species in the genus *Aix*. The female is very similar to the female Wood Duck. One of the most popular exotic waterfowl species, it is common in collections. It favors wooded lakes and rivers and is often seen perching in trees. Monotypic. L 16" (41 cm)

compare to female Wood Duck

IDENTIFICATION **BREEDING MALE:** Elaborate and ornate plumage. Pinkish red bill, white band sweeps behind eye, orange "mane," dark green crown, and orange tertial "sails." **ECLIPSE MALE:** Resembles female, but bill is reddish with less distinct eye ring, shaggier crest, and glossy upperparts. **FEMALE:** Resembles female Wood Duck. Pale and grayer overall, with shaggy hind neck. White eye ring with thin white postocular stripe. Bill grayish with pale nail. Wing coverts dull brown (glossy on Wood Duck). **JUVENILE:** Like female, but duller with less distinct facial pattern. Attains adult plumage throughout first winter. **FLIGHT:** Dark with white belly and dark upper wings. **VOICE** Generally silent, except during display. **STATUS & DISTRIBUTION** Eastern Asia. Established in Britain. Feral population in central Sonoma County. **BREEDING:** Russian Far East, China, Japan; nests in tree cavities. **WINTER:** Lowlands of eastern China and southern Japan. **POPULATION** Declining in Asia due to habitat destruction and previous export from China in large numbers.

CURASSOWS AND GUANS Family Cracidae

Plain Chachalaca (TX, Nov.)

chickenlike and their legs are fairly long and strong.
Behavior In general, cracids live in groups; the chacha-lacas are the most sociable. Most species are found moving through the canopy of tropical forests, but chachalacas generally stay closer to the ground.
Plumage Overall plumage varies from olive brown to black. The sexes appear similar in most species with only the curassows showing strong sexual dimorphism.
Distribution A New World family, cracids are found from south Texas and northwestern Mexico through the neotropics to northern Argentina. The family reaches its greatest diversity in northern South America. The chachalacas are very similar; they appear to occupy similar niches to the point that there is virtually no overlap in distribution between the different species.
Taxonomy The family can be informally divided into three major groups: curassows, guans, and chacha-lacas—50 species in 11 genera. The 12 species of chachalaca—characterized by small size (for the family) and plain coloration—are all in the genus *Ortalis*.
Conservation Most chachalaca species are common; none are considered of great conservation concern. Seven guan and six curassow species, however, are considered threatened. Small clutch sizes do not allow their populations to recover from hunting pressures as readily as temperate gallinaceous species. In addition, many of the threatened species depend on primary forest, much of which is seriously threatened by deforestation.

Cracids are a diverse family of primarily arboreal game birds, many of which have elaborate wattles and knobs. They generally feed on leaves and fruits of trees, with some species foraging on the ground for fallen fruits.
Structure They range in size from medium to large and have heavy bodies. Common characteristics include longish necks with small heads, long broad tails, and rounded wings. Their bills are rather

---

CHACHALACAS Genus *Ortalis*

PLAIN CHACHALACA *Ortalis vetula*   PLCH ◼ 2

Although generally secretive, this south Texas bird is very vocal and has become habituated to feeding stations. The Plain Chachalaca is always found in small social groups—numbering 10–15 individuals—as it moves through the understory and on the ground. Well adapted to woodland habitats, it often hops from branch to branch without the use of its wings. The English name of the genus,

chachalaca, is derived from the loud chorus given by groups of these birds. Polytypic (5 ssp. in C.A.; *mccallii* in N.A.). L 22" (56 cm)
IDENTIFICATION Large with a long tail and small head. **ADULT:** Brownish olive above, buffy brown below. Gray head and neck. Long dark tail with a green sheen and broad white tip. Patch of bare skin on throat gray to dull pink, becoming bright pinkish red during breeding season. **JUVENILE:** Similar to adult, but duller and often mottled brown above; tail feathers tipped with pale brown and less rounded.
VOICE Noisy, especially at dawn and dusk; much more vocal during the breeding season. **CALL:** A raucous *cha-cha-lac* given in very loud chorus. Female voice is higher pitched than male's. Various other guttural chatter given, including *krrr* notes.
STATUS & DISTRIBUTION Common in limited N.A. range. **YEAR-ROUND:** Resident in brushy areas and riverine

white tail tips

woodlands of south TX. Successfully introduced to Sapelo Island, GA. World range extends to Costa Rica along the Atlantic coast.
POPULATION The population appears stable throughout its range.

♂

carmine-pink wattle, more pinkish in female, is often hidden

long blackish tail

NEW WORLD QUAIL Family Odontophoridae

Gambel's Quail, male (center) and females (AZ, May)

This family composed primarily of terrestrial game-birds is restricted to the New World; it includes both temperate and tropical members. Many species have crests and head plumes as well as rather ornate body plumage. Omnivorous, they feed on insects as well as mast from a variety of plants.

Structure Of small to medium size, these birds have heavy bodies and rather small heads. All have stout bills with slightly serrated cutting edges, and long toes and short tarsi, which do not develop spurs. A long tail separates the wood-partridges from the rest of the family.

Behavior In general, New World quail are found in pairs during the breeding season and in groups otherwise.

Plumage They generally exhibit sexual dimorphism, although it ranges from strong in some quail to weak in wood-quail to virtually nonexistent in the wood-partridges. Although the overall plumage of most species is dominated by grays and browns, many species are quite ornate. Most species of quail have crests or head plumes or distinctive head plumage.

Distribution Found from southwestern Canada to Paraguay, the family reaches its greatest diversity in western N.A. and Mexico.

Taxonomy This family informally divides into three groups: wood-partridges, quail, and wood-quail. Worldwide, there are 32 species in nine genera recognized.

Conservation Poorly studied, wood-partridges and at least four tropical wood-quail are of concern. BirdLife International lists five species as threatened, with three more as near threatened.

Genera *Callipepla* and *Oreortyx*

SCALED QUAIL *Callipepla squamata*   SCQU ◼ 1

A bird of the desert grasslands, mesquite savanna, and thorn-scrub of the Southwest, the Scaled Quail is generally sedentary, moving short distances to form winter coveys. Fall and winter coveys can have as many as 200 individuals. It prefers to run to escape, but it will fly when startled. Hybrids with Gambel's Quail and Northern Bobwhite have been documented. Polytypic (3 ssp.; 2 in N.A.). L 10" (25 cm)

IDENTIFICATION This quail is plump and short-tailed and has grayish plumage. The bluish gray breast and mantle feathers are edged in black or brown giving it a scaled appearance. **ADULT:** Sexes are similar. The male has a prominent white crest; the female's is smaller and buffier in coloration. **JUVENILE:** Similar to adults but more mottled above and less conspicuous scaling on the underparts.

GEOGRAPHIC VARIATION Subspecies *pallida* inhabits most of Scaled Quail's range in N.A.; *castanogastris* occurs in southern Texas, below the Balcones Escarpment. A chestnut patch on the belly and darker overall plumage,

particularly on the upperparts, separates *castanogastris* from *pallida*. Some authors divide *pallida* into two weakly defined taxa, *pallida* and *hargravi*.

SIMILAR SPECIES Distinctive and unlikely to be confused with any other quail in N.A.

VOICE During the breeding season, both sexes give a location call when separated, a nasal *chip-churr,* accented on the second note. The male's advertising call is a rhythmic *kuck-yur,* often followed by a sharp *ching.*

STATUS & DISTRIBUTION Fairly common year-round resident.

POPULATION Scaled Quail has

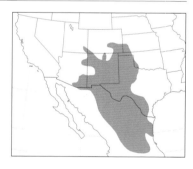

declined dramatically in the United States since the 1940s. In Texas, the species has become increasingly scarce in the Hill Country and Rolling Plains; reasons for the decline are unknown, but they are probably attributable to habitat degradation; still abundant in the Trans-Pecos northward to the Texas Panhandle.

pale buffy top to crest

south Texas ♂
castanogastris

♂ *pallida*

scaly breast

dark chestnut belly patch

juvenile

CALIFORNIA QUAIL *Callipepla californica*   CAQU ▪ 1

Although highly sedentary, the California Quail congregates in large coveys during the fall and winter. It hybridizes with the Gambel's Quail where their ranges overlap. Polytypic (5 ssp.; 4 in N.A.). L 10" (25 cm)

IDENTIFICATION Plump with gray and brown plumage; a prominent teardrop-shaped head plume or double plume in both sexes. **ADULT MALE:** Pale forehead with brown crown and black throat, scaled belly with a chestnut patch, brown upperparts. **ADULT FEMALE:** Similar to male but muted and lacking distinctive facial pattern; head plume smaller. **JUVENILE:** Grayish brown to brown overall and heavily mottled. Belly pale and lacking the scaled appearance of the adults.

GEOGRAPHIC VARIATION Differences are based on coloration and size, which are more pronounced in females. Adult female *canfieldae* (found in east-central CA) and *californica* (the most widespread subspecies) have grayish upperparts; adult female *brunescens* (found in the wetter coastal mountains) have brown upperparts. The *catalinensis* subspecies is endemic to Santa Catalina Island.

SIMILAR SPECIES California

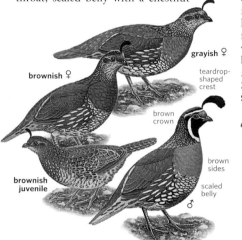

brownish ♀

grayish ♀

teardrop-shaped crest

brown crown

brown sides

scaled belly ♂

brownish juvenile

Scaled x Gambel's hybrid

♂

Quail is similar in structure and size to the Gambel's, but Gambel's lacks the scaled underparts and brown sides and crown.

VOICE CALL: An emphatic *chi-ca-go*; sometimes shortened on only one or two syllables. A variety of grunts and sharp cackles are also made.

STATUS & DISTRIBUTION Common year-round in open woodlands and brushy foothills, usually near permanent water sources. This species has been introduced locally within the general boundaries of the mapped range, including UT.

POPULATION Since 1960, the overall population has declined in the U.S.

GAMBEL'S QUAIL *Callipepla gambelii*   GAQU ▬ 1

Although a common quail of the desert Southwest, Gambel's Quail requires a lot of water. Generally sedentary, it moves short distances in the late summer to form coveys, usually comprising several family groups. It hybridizes with both California and Scaled Quails where their ranges overlap. Polytypic (2 ssp.; nominate in N.A.). L 11" (28 cm)

IDENTIFICATION Plump, short-tailed quail with gray plumage; prominent teardrop-shaped head plume or double plume in both sexes. **ADULT MALE:** Chestnut crown with black forehead and black throat; chestnut sides with whitish underparts with a black belly; gray upperparts. **ADULT FEMALE:** Similar to adult male but muted and lacking distinctive facial pattern. Head plume smaller. **JUVENILE:** Grayish brown overall and heavily mottled. Usually has a short head plume.

SIMILAR SPECIES California Quail is similar in structure and size, but the chestnut crown and sides and lack of scaling on Gambel's easily separate them.

VOICE CALL: A plaintive *qua-el;* and a loud *chi-ca-go-go* similar to California's, but higher pitched and usually four notes, sometimes shortened on only one or two syllables. Also a variety of clucking and chattering calls.

STATUS & DISTRIBUTION Common year-round in desert shrublands and thickets, usually near permanent water sources. Introduced to HI, ID, and San Clemente Island, CA.

POPULATION The numbers appear to be stable over the past 60 years.

chestnut crown

teardrop-shaped crest

black belly surrounded by white ♂

chestnut sides

♀

juvenile

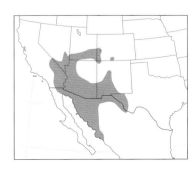

MOUNTAIN QUAIL Oreortyx pictus   MOUQ ▪ 1

Although fairly common in much of its range, the Mountain Quail is very shy and hard to find under most circumstances. This bird is perhaps best seen when males call from rocks or in late summer, and when family groups often forage along roadsides near dense brush. This species seems to occur with greater frequency in mixed evergreen woodlands on mountain slopes than in chaparral or riparian corridors. As its name suggests, it can be found on mountain slopes at elevations as high as 10,000 feet. Polytypic (5 ssp.; 4 in N.A.). L 11" (28 cm)

IDENTIFICATION Plump, short-tailed quail with gray and brown plumage. Two long, thin head plumes present in both sexes. **ADULT MALE:** Brown or grayish brown above; slate blue chest and chestnut belly barred with white on the flanks; slate blue crown with chestnut throat bordered in white. **ADULT FEMALE:** Very similar to adult male, but normally has shorter, brownish plumes and brown mottling on the hind neck in interior populations. **JUVENILE:** Dull version of the adult. Lacks bold flank pattern and the rich coloration of the face and throat pattern. Body more mottled with dark and light edging.

GEOGRAPHIC VARIATION Subspecies differ in general coloration of the back and breast. The interior subspecies—*plumifer, eremophilus,* and *russelli*—are grayer above, generally lacking brown mottling in the nape and breast. The nominate subspecies (now includes *palmeri*) is found in

more mesic habitats and is brown on the nape with some brown mottling on the sides of the breast.

SIMILAR SPECIES Distinctive; no similar species exist within its range. California and Gambel's Quail share some common characteristics, but a reasonable view of the bird should provide a straightforward identification; long head plumes and chestnut throat eliminate both species. The prominent white bars on chestnut flanks are often a key mark in distinguishing this species when seen in heavy cover. A juvenile Mountain Quail is more like a California in plumage, but note its straight head plumes and less intricate belly plumage.

VOICE The male's advertising call is a clear, descending *quee-ark* that can carry up to a mile away. The covey call that is frequently given is an extended series of whistled *kow* or *cle* notes.

STATUS & DISTRIBUTION Locally common in evergreen woodlands, chaparral, brushy ravines, and mountain slopes. Nominate subspecies introduced on Vancouver I. **WINTER:** Some descend to lower elevations, usually on foot.

POPULATION The population in the U.S. appears to be stable in most areas. It has declined significantly in Idaho, where only three small populations persist. Heavy agricultural use of prime habitat in northern Baja California is affecting the only population in Mexico.

chestnut throat bordered by white

long, thin, straight plumes

interior ♀

coastal ♂
pictus

juvenile

white bars on chestnut flanks

Genera *Colinus* and *Cyrtonyx*

NORTHERN BOBWHITE Colinus virginianus   NOBO ▪ 1

slight crest

white supercilium

white throat

buffy supercilium

virginianus ♂

buffy throat

virginianus ♀

juvenile
virginianus

The Northern Bobwhite, found in coveys except during breeding season, is the most widespread and familiar quail in North America. Polytypic (20 ssp.; 4 in N.A.). L 9.7" (25 cm)

IDENTIFICATION Plump, short-tailed quail with reddish brown plumage. **ADULT MALE:** Intricate body plumage of chestnut, brown, and white; blackish plumage on head; white throat and eye line. **ADULT FEMALE:** Similar to adult male but throat and eye line buffy and plumage on head brown. **JUVENILE:** Similar to adult female but body plumage browner, less rufous, and eye line less prominent. **RUFOUS MORPH:** Very rufous body plumage masks markings.

Black face superficially resembles the "Masked Bobwhite." Extremely rare. **"MASKED BOBWHITE":** Similar to eastern birds, but it has a black face with the white eye line greatly reduced and flecked with black. Underparts are entirely rufous with white markings on the flanks; the upperparts have prominent rufous markings, becoming browner on the lower back and rump. **GEOGRAPHIC VARIATION** Complex, with differences sometimes striking in the plumage of the adult male. The species is often broken into four distinct groups, two of which occur in the U.S. The northern group occurs in eastern N.A.; it is made up of four similar subspecies that are differentiated by changes in overall coloration and the width of barring on the underparts. Birds from peninsular FL (Gainesville and south), *floridanus*, are smaller and darker than the widespread nominate subspecies. The subspecies *taylori* (SD to northern TX; introduced locally in WA, OR, and ID) and the more southerly *texanus* (southwestern TX to northern MX) are paler. The second group is the "Masked Bobwhite" *(ridgwayi)* found in southern Arizona

and Sonora; at times it has been considered a separate species.

SIMILAR SPECIES Distinctive; no similar species exist within its range. Montezuma Quail is superficially similar, but its range overlaps with Northern Bobwhite only in south-central Texas.

VOICE Male's advertising call is a whistled *bob-WHITE* or *bob-bob-WHITE*. Other calls include a low whistled *ka-lo-kee* and a variety of clucks.

STATUS & DISTRIBUTION Uncommon to common year-round in a variety of open habitats with sufficient brushy cover. Introduced to northwestern N.A., the Bahamas, several Caribbean islands, and New Zealand. The "Masked Bobwhite" is rare in the Altar Valley of southern AZ as a result of transplanted individuals from remaining populations in Sonora; it is federally listed as endangered.

POPULATION Although still locally common, Northern Bobwhite has declined significantly in the U.S. Manipulation of habitat and the high populations of the introduced fire

ant are believed to be key factors in the decline. Dramatic declines have left the species virtually extirpated in many areas where it was once common. The native population of "Masked Bobwhite" was extirpated from the U.S. by 1900.

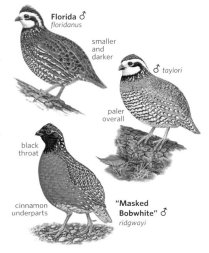

Florida ♂
floridanus

smaller and darker

♂ *taylori*

paler overall

black throat

cinnamon underparts

"Masked Bobwhite" ♂
ridgwayi

MONTEZUMA QUAIL *Cyrtonyx montezumae*   MONQ ■ 2

The Montezuma Quail is perhaps the most attractive quail in North America. The elaborate plumage of the male provides a very cryptic pattern in tall grasses or in dappled sunlight under a shrub. It has highly specialized long claws, used for digging up bulbs and tubers. It remains in pairs all summer, although nesting often does not begin until the monsoon rains of July and August. It is often seen cautiously crossing roads or in grassy areas along roadsides. When chanced upon, it prefers to crouch to remain hidden rather than run or fly away. In winter, it forms into small coveys (rarely more than 15 birds). Populations appear to be as much as 60 percent male, enhancing mate selection in females. Polytypic (5 ssp.; *mearnsi* in N.A.). L 8.7" (22 cm)

IDENTIFICATION Plump, short-tailed, round-winged quail. **ADULT MALE:** Distinctive facial pattern with a rounded brown crest on the back of the head; brown upperparts, but heavily marked with black and tan; dark chestnut breast; black sides and flanks, heavily spotted with white. **ADULT FEMALE:** Brown overall with mostly unmarked pinkish brown underparts; upperparts heavily marked with black and tan; brown rounded crest on the back of the head; face somewhat lighter tan

than remainder of the body. **IMMATURE MALE:** Seen only in late summer. A pale gray face; black sides and flanks, heavily spotted with white. Distinctive facial pattern appears in late fall. **JUVENILE:** Similar to adult female, but lighter in overall color and more heavily mottled with black.

SIMILAR SPECIES Unmistakable; no similar species exist within its range. Northern Bobwhite is superficially similar, but their ranges overlap only in south-central TX.

VOICE Male's advertising call is a descending whistled *vwirrrrr*. Covey call, which is given by both sexes, is a loud descending whistle. Contact calls are a rather quiet *whi-whi*.

STATUS & DISTRIBUTION Uncommon; secretive and local in open juniper-oak or pine-oak woodlands on semiarid slopes.

POPULATION Montezuma Quail was once much more widespread in North America. In TX, it was found throughout all mountain ranges and in the central portion of the state east to San Antonio. In AZ and NM, it was more widespread in the mountain ranges in

the southern half of each state. By 1950, the population had shrunk to its current range. Tall grasses in oak-juniper woodlands are believed to be an important part of habitat quality.

slightly crested look

♀

exotic head pattern

dark buffy crest laid over nape

juvenile

♂

solid blackish brown chest

round white spots on sides and flanks

PARTRIDGES, GROUSE, AND TURKEYS Family Phasianidae

Wild Turkey (ON, Jan.)

Phasianidae is arguably one of the more spectacular families in the world. Many species have elaborate courtship rituals. Most exhibit complex plumage variation that is still poorly understood. And, since many species require large tracts of quality habitat, looking for them takes birders to pristine and magnificent locations.

Structure Well adapted for life on the ground, all family members have strong legs with three long toes and a small hind toe. Adults of many species have one or more sharp spurs on the back of the leg to rake opponents during quarrels. The relatively short wings allow birds to explode off the ground.

Behavior Phasianids are unobtrusive. Most will fly only when danger is imminent; they spend their lives with minimal movement, both in their daily activities and seasonally. During the mating season, the males of most species, particularly those with polygynous systems where males compete for females at leks, engage in a barrage of displays that include drumming, wing whooshing, maniacal cackles, foot stomping, flight displays, and hoots.

Plumage Most species are sexually dimorphic, dramatically so in many pheasants. Males are often larger, and in many species of grouse have special adornments on their head and neck. Otherwise, most species are cryptically colored. Most appear very similar throughout the year, although ptarmigan change dramatically from their cryptic lichen-covered-rock summer plumage to nearly completely white in winter.

Distribution While members of this familiy are distributed across North America, diversity is generally higher away from the Southeast. Most species are sedentary.

Taxonomy Phasianidae includes four distinct subfamilies, which are sometimes designated as full families: pheasants and partridges (Phasianinae), grouse (Tetraoninae), turkey (Meleagridinae) and guineafowl (Numidinae). Of the approximately 180 species worldwide, 11 grouse and the Wild Turkey are native to North America. Attempts have been made to introduce several species of pheasants and partridges (and Helmeted Guineafowl for tick control), but only the Chukar, Gray Partridge, Himalayan Snowcock, and Ring-necked Pheasant are considered established. The sedentary nature of phasianids with little genetic mixing has given rise to extensive regional variation and subspecies. Taxonomy continues to undergo revision: The Gunnison Sage-Grouse was described as a new species in 2000 and the "Blue Grouse" was split into the Dusky Grouse and Sooty Grouse in 2006.

Conservation Habitat loss poses the most serious threat, especially to native grassland grouse. The Wild Turkey's recovery, however, demonstrates that rebounds can occur when conservation is undertaken.

Genus *Alectoris*

CHUKAR   *Alectoris chukar*   CHUK ▪ 2

This introduced Old World species can be challenging to find and is best located by call or by searching near water sources. By late summer it is not unusual to see a covey with one to three adults and 30 to 50 young. Polytypic (14 ssp.). L 14" (36 cm)

IDENTIFICATION Resembles an overgrown short-tailed quail. **ADULT:** Cream-colored face and throat broadly outlined in black. Flanks boldly barred with black. **JUVENILE:**

Similar but smaller, mottled, and no bold black markings; usually with adults. **FLIGHT:** It explodes into the air and then glides away; look for chestnut on spread tail just before landing.

GEOGRAPHIC VARIATION Complex; most birds in N.A. believed to be nominate subspecies, which is the darkest and brownest subspecies.

SIMILAR SPECIES No similar species established in N.A.; however, two released game birds are similar: Redlegged Partridge *(A. rufa)* has a white face and throat; its neck is conspicuously streaked with black, giving the appearance of having a "neck-shawl." Rock Partridge *(A. graeca)* has more black (less white) between the bill and the eye and a whiter throat.

VOICE CALL: A calm *chuck, chuck, chuck* that often becomes progressively louder and drawn out, culminating in a voluminous eruption of *chuckara-chuckara-chuckara.* The alarm call of flushed birds is a loud piercing squeal followed by *whitoo* notes.

STATUS & DISTRIBUTION Uncommon

to fairly common; now established in steep rocky terrain in much of the West. Released birds are possible almost anywhere, even in the East.

POPULATION Widely released in North America starting in the 1930s, the Chukar became established in much of the West by the late 1960s. The largest populations are in WA, OR, NV, ID, UT, and CA.

adults

extensive rufous on outer tail feathers

buffy throat outlined by black

black-and-white barred flanks

juvenile

Genus *Tetraogallus*

HIMALAYAN SNOWCOCK   *Tetraogallus himalayensis*   HISN ▪ 2

The Central Asian Himalayan Snowcock has established a toehold in North America, but seeing this exotic is no small feat. Its numbers are small (only about 1,000 birds) and it lives only in Nevada's high, rugged Ruby Mountains. At dawn, the Himalayan Snowcock calls from its roost site and flies downhill. It spends the day

white areas bordered by chestnut bands

♂

dark belly

walking back up, feeding, resting, and preening. Birds congregate just before dark, then fly or walk to roost. Polytypic (6 ssp.; nominate in N.A.). L 28" (71 cm)

IDENTIFICATION Himalayan Snowcock is large (nearly 2.5 times the size of the Chukar). It is generally grayish overall with rusty-brown streaks on upperparts. Two prominent chestnut stripes on each side of the head outline the whitish face and throat. The predominantly white primaries are conspicuous in flight. **FEMALE:** Smaller and duller than the male with a buff forehead; the area around the eye is grayer; lacks spurs. **JUVENILE:** Similar to the adult female, but smaller and duller and lacks rusty tones to the spotting above. Very young birds lack the chestnut markings on the face.

SIMILAR SPECIES None within restricted range.

VOICE CALL: Generally similar to Chukar; other calls surprisingly reminiscent of a Long-billed Curlew. Clucks and cackles persistently while foraging.

STATUS & DISTRIBUTION Successfully introduced to the high elevations of the Ruby Mountains of northeast NV, where it resides in subalpine and alpine habitats. **MOVEMENTS:** Generally sendentary; may move slightly downslope during severe winters.

POPULATION The Himalayan Snowcock was first released into the Ruby Mountains in 1963; the first brood was seen in 1977. By the 1980s the species was self-sustaining. Limited threats include overgrazing by sheep and pressure from hunters and birders.

Genus *Perdix*

GRAY PARTRIDGE *Perdix perdix*   GRAP ■ 2

Established as a game bird in N.A. since the early 1900s in agricultural areas, this often evasive Old World species is best found at dawn or dusk and when there is snow cover. In fall it forms coveys of 12 to 15 birds. Polytypic (8 ssp.). L 12.5" (32 cm)

IDENTIFICATION Mostly streaked brownish above, grayish below. Broad rufous bars and narrow cream-colored streaks on flanks. Contrasting rufous outer tail feathers often revealed as bird

flicks its tail open. **MALE:** Conspicuous chestnut patch on upper belly and bright rufous-orange head and throat. **FEMALE:** Belly patch absent or reduced; head and throat more buff. **IMMATURE:** Duller in both sexes. **FLIGHT:** Explodes into the air with rapid wingbeats and then alternates glides and rapid wingbeats; flights are typically low to the ground and of short distance.

GEOGRAPHIC VARIATION Clinal. In its Old World range the western subspecies are darker gray and more rufous than subspecies farther east. Little information on introductions, but most believed to be western subspecies (including nominate *perdix*).

SIMILAR SPECIES None. The Gray Partridge's broad rufous bars on the flanks and face pattern are distinctive. Compare female to smaller female California Quail, which overlaps with Gray Partridge in the western part of its range.

VOICE CALL: A harsh *kee-uck,* likened to a rusty gate. Alarm call a rapidly repeated

kuta-kut-kut-kut. Also various clucks.

STATUS & DISTRIBUTION Uncommon in most areas; it inhabits farmland and grassy fields.

POPULATION Gray Partridge is thought to be declining in much of its range in North America, largely due to changes in farming practices.

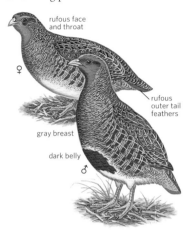

rufous face and throat

♀

rufous outer tail feathers

gray breast

dark belly

♂

PHEASANTS Genus *Phasianus*

RING-NECKED PHEASANT *Phasianus colchicus*   RNEP ■ 1

Native to Asia, the flashy Ring-necked Pheasant has been widely introduced as a game bird in North America. It is frequently seen along roadsides in agricultural areas. Polytypic

(±30 ssp.). L ♂ 33" (84 cm) ♀ 21" (53 cm)

IDENTIFICATION Large. Very long, pointed tail; short, rather rounded wings. **MALE:** Iridescent bronze overall, mottled with brown, black, gray, and rufous. Head a dark glossy green to purplish, with red fleshy eye patches and iridescent ear tufts. Most show a wide white neck ring. **FEMALE:** More buff colored with sparse dark spots and bars on breast and flanks. **FLIGHT:** When flushed, it erupts with powerful

loud whirring wingbeats. The long tail is distinctive even in flight.

GEOGRAPHIC VARIATION Extensive, particularly among males. Many subspecies have been introduced in N.A.; intergrades are numerous. The male "White-winged Pheasant" (not illustrated) with distinctive white upperwing coverts has become established in parts of the West. The Japanese "Green Pheasant," often considered a separate species *(P. versicolor),* is largely extirpated from the areas where it was introduced, mainly the Tidewater region of VA and southern DE.

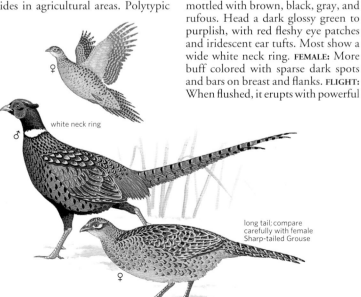

♀

white neck ring

♂

long tail; compare carefully with female Sharp-tailed Grouse

♀

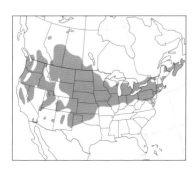

Green Pheasant
versicolor

♂

♀

SIMILAR SPECIES The female resembles a Sharp-tailed Grouse, but the latter is smaller, with heavily scalloped and spotted underparts and has a shorter tail with white outer tail feathers. **VOICE CALL:** Male territorial call is a loud, penetrating *kok-cack*. Both sexes give hoarse, croaking alarm notes. **STATUS & DISTRIBUTION** Locally common to uncommon resident of open country, farmlands, brushy areas, and edges of woodlands and marshes. Local hunting releases account for the presence of some individuals outside their normal range. **POPULATION** While populations fluctuate with weather changes, pheasants are declining in parts of the East. Some populations are preserved through continual introductions.

Genus *Meleagris*

WILD TURKEY *Meleagris gallopavo*   WITU ▪ 1

Writing to his daughter, Benjamin Franklin observed that the Wild Turkey would have made a better national symbol for the United States than the Bald Eagle, proclaiming it "a bird of courage." The Wild Turkey is among our best known birds. After major declines, the Wild Turkey is once again regularly encountered in open woods bordered by clearings, particularly where oaks are prevalent. At night it roosts in trees. There is one other turkey in the world, the Ocellated Turkey *(M. ocellata),* found on the Yucatán Peninsula and in Belize and Guatemala, but the Wild Turkey is the parent stock from which all domesticated turkeys are descended. Polytypic (6 ssp.; 4 in N.A.). L ♂46" (117 cm) ♀37" (94 cm)

IDENTIFICATION The Wild Turkey is the largest game bird in North America, albeit smaller and more slender than its domesticated cousin. **MALE:** Dark, iridescent body, flight feathers barred with white, red wattles, blackish breast tuft, spurred legs; bare-skinned head is blue and pink. **FEMALE AND IMMATURE:** Smaller, duller, and often lack the breast tuft of the male. **GEOGRAPHIC VARIATION** The widespread "Eastern Turkey" *(silvestris)* has tail, uppertail coverts, and lower rump feathers tipped with chestnut. These feathers are tipped a buff white in "Merriam's Turkey" (found in the Great Plains and Rockies). The "Rio Grande Turkey" *(intermedia),* found from Kansas south to Mexico, is intermediate in plumage; in fall and winter it forms huge flocks of up to 500 birds (40–50 typical of other subspecies). The "Peninsular Florida Turkey" *(osceola)* is similar to the "Eastern Turkey," but smaller. Introductions of subspecies outside of native range and presence of escaped and released domestic birds and resultant interbreeding greatly

complicate regional variation. Two other subspecies occur in Mexico. **SIMILAR SPECIES** None. **VOICE CALL:** The distinctive gobble given by males in spring may be heard a mile away. Other calls include various yelps, clucks, and rattles. **STATUS & DISTRIBUTION** Restocked in much of its former range and introduced in other areas; now found in each of the lower 48 states and southernmost Canada. **POPULATION** The Wild Turkey declined throughout its range in the 19th and early 20th century; the loss was generally blamed on overhunting and habitat loss. Since then, the species has rebounded dramatically with an estimated population of nearly four million. Population numbers in Mexico remain extremely low.

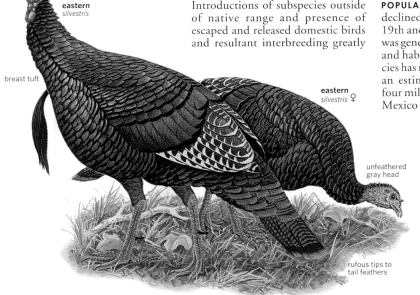

unfeathered reddish head

eastern
silvestris

breast tuft

eastern
silvestris ♀

unfeathered gray head

rufous tips to tail feathers

displaying ♂

western
merriami

PTARMIGAN Genus *Lagopus*

All three ptarmigan species are found in North America: The White-tailed is endemic to the mountains of the West; the Rock and Willow Ptarmigan are circumpolar. They have thickly feathered toes and tarsi, short tails, and cryptic plumages. All but the "Red Grouse" (*scotica* species of Willow Grouse) of the British Isles molt to a near completely white plumage in winter.

WILLOW PTARMIGAN *Lagopus lagopus*   WIPT ■ 1

The largest ptarmigan, the circumpolar Willow is characteristic of willow flats and low dense vegetation near tree line. It has the most widespread distribution of any grouse. Polytypic (19 ssp. worldwide; 7 in N.A.). L 15" (38 cm)

IDENTIFICATION Both sexes have a black tail, predominantly white wings, and white-feathered legs year-round. The red eye combs are more pronounced in males; they may be largely concealed or inflated during courtship or aggression. Both sexes are white in winter, except for black tails. **SUMMER MALE:** Mottled, but predominantly rufous. **BREEDING FEMALE:** Warm brown; very similar to Rock Ptarmigan. **JUVENILE:** Heavily barred

black and buffy yellow. Dark wings for a short period in summer.
GEOGRAPHIC VARIATION Extensive. Differences in both sex and molt complicate subspecies recognition.
SIMILAR SPECIES Willow Ptarmigan is larger than Rock Ptarmigan and has a bigger bill. Vocalizations and habitat are helpful in identification. While Willow will move upslope in mountains, it is restricted to relatively lush patches of vegetation. In winter, male Willow has a relatively plain face and head lacks the distinctive dark eye line of the male Rock.
VOICE In displays, male utters a loud raucous *go-back go-back go-backa go-backa go-backa*. **CALL:** Low growls and croaks; noisy cackles.

STATUS & DISTRIBUTION Widespread sometimes abundant tundra species. Fond of moist locations with dwarf willow and alder thickets. **BREEDING:** Generally prefers wetter, brushier habitat than Rock Ptarmigan. **MOVEMENTS:** Vary between largely resident populations with only short shifts between breeding and wintering ranges, to relatively long-distance migrants. Thousands have been recorded in a day moving through Anaktuvuk Pass, AK. Flocks of up to 2,000 individuals have been recorded in migration. **MIGRATION:** Fall peak at Anaktuvuk in mid-Oct. Spring migration begins as early as mid-Jan., peaking late Apr. **VAGRANT:** Casual in spring and winter to southern ON, southern SK, and northern tier states (e.g., MT, ND, MN, WI, and ME).
POPULATION Substantial cyclic ups and downs in populations over several-year periods. Little data on long-term trends, but apparently stable.

dark rufous head and neck

molting spring ♂

summer ♀

thick bill

winter

summer ♂

summer ♂

black tail

ROCK PTARMIGAN *Lagopus muta*   ROPT ■ 1

The Rock Ptarmigan is the archetypal bird of the arctic. Found in cold and barren windswept tundra, it has been recorded as far north as 75° N in winter, when the region is enshrouded in 24 hours of darkness. Like other ptarmigan it has an unusual sequence of three body molts each year, generally allowing it to remain well camouflaged at all times of the year. Males molt later than females in spring. By early June many are still mostly white and can be spotted from up to a mile away; at the same time the females have completed their

molts and are difficult to spot from a few feet away. Polytypic (26–30 ssp.; 14 in N.A.). L 14" (36 cm)
IDENTIFICATION Like the Willow Ptarmigan, both sexes have a black tail, predominantly white wings, and white-feathered legs year-round. The mottled summer plumage is black, dark brown, or grayish brown; the exact patterning varies with subspecies. **ADULT MALE:** White winter plumage is similar to the Willow Ptarmigan but the black eye line is prominent. **ADULT FEMALE:** Very difficult to distinguish from Willow Ptarmigan, except

by overall size and bill size. **JUVENILE:** Heavily barred black and buffy yellow;

has dark wings for a short period in summer.

GEOGRAPHIC VARIATION Extensive regional variation. Individual variation and complex molt complicates subspecies identification. Subspecies on the Aleutians are larger with heavier bills. Males on westernmost Aleutians are mostly blackish in breeding plumage; central Aleutian males more ochre-toned.

SIMILAR SPECIES Rock Ptarmigan is smaller than Willow Ptarmigan, and with a smaller bill. Use caution when estimating bill size as the dark feathers between the eye and bill on birds in transitional plumages can give the impression that the bill is larger. The presence of dark lores is diagnostic for male Rock Ptarmigan (their absence is not diagnostic).

VOICE In courtship and territorial flight displays, male utters a low rattling *ah-AAH-ah-AAAAH-a-a-a-a.* **CALL:** Low growls and croaks; noisy cackles.

STATUS & DISTRIBUTION Common. **BREEDING:** Generally prefers higher and more barren habitat than the Willow Ptarmigan. **MOVEMENTS:** May withdraw from northernmost summer range, but only irregularly south of the southern boundary of the breeding range. Most make only limited movements, which are largely altitudinal and likely influenced by weather and snow depth. Populations on the Aleutian Islands are resident. **VAGRANT:** Accidental on Queen Charlotte I., BC, and, most remarkably, in northeastern MN.

POPULATION Numbers fluctuate. Some southern populations may have contracted. Most of range is remote, but disturbance from natural resource extraction and climate changes are potential concerns. Populations in the Aleutian Islands declined as a result of non-native arctic fox introductions.

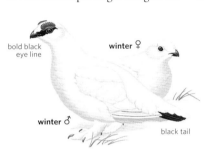

bold black eye line

winter ♀

winter ♂

black tail

slighter bill than Willow

summer ♀

summer ♂

fall ♂

WHITE-TAILED PTARMIGAN *Lagopus leucura*   WTPT ■ 2

The White-tailed Ptarmigan is the smallest grouse in North America, and it is the only ptarmigan regularly found in the lower 48. Its ability to blend in perfectly with its remote alpine tundra environment makes it among the most challenging resident species to find. Territorial during the breeding season, the White-tailed forms flocks of up to 80 birds from late October to late April. During winter, it roosts under the snow. White-tailed Ptarmigan forage largely on buds, stems, and seeds year-round, but they augment this diet with insects in the summer. Polytypic (5 ssp.; all in N.A.). L 12.5" (32 cm)

IDENTIFICATION WINTER: Completely white except for eye combs, dark eye and bill, and fine dark shaft streaks to primaries. **SUMMER MALE:** More white on belly than female; conspicuous coarsely barred brown and black breast feathers form a necklace. **SUMMER FEMALE:** More evenly patterned with barring of buffy yellow and black. **JUVENILE:** Heavily barred black and buffy yellow and have dark wings for a short period in summer.

GEOGRAPHIC VARIATION Clinal among five subspecies; average larger and grayer in the south, smaller and darker in the north.

SIMILAR SPECIES The white tail distinguishes it from both Willow and Rock Ptarmigan in all seasons; however, the tail feathers of all ptarmigan are typically concealed by the uppertail coverts, making it difficult to judge tail color.

VOICE Males give a loud raucous flight call with four distinct syllables, *ku-ku-KII-KIIEUR;* also a variable chattering clucking series of calls given on the ground, *duk-duk-DAAAK-duk-duk; duk-DAK-DADAAK-duk.* **CALL:** Clucks, churrs, and high-pitched chirps.

STATUS & DISTRIBUTION Endemic to mountains of the West where it is found in rocky alpine areas at or above timberline. Uncommon to locally common. Small numbers have been introduced in the Wallowa Mountains, Sierra Nevada, Pike's Peak, and the Uinta Mountains. Re-established into northern NM. **MOVEMENTS:** Depends on severity of winter and snowfall,

slighter bill than Willow

molting fall ♂

summer ♀

summer ♂

but usually moves to lower elevations; during heavy snow years moves to streambeds and avalanche chutes dominated by willows and alders.
POPULATION Numbers fluctuate widely between years, but there is little data on population changes. Local populations may be affected by natural resource extraction, road construction, off-road vehicles, and overgrazing by domestic livestock and elk; these activities have the greatest effect on reducing winter forage, principally willow. Population on Vancouver Island *(saxatilis)* considered vulnerable.

winter

all plumages have all-white tail

FOREST GROUSE Genera *Bonasa, Falcipennis,* and *Dendragapus*

Members of these genera live in relatively forested landscapes. The Spruce and Siberian Grouse of *Falcipennis* are quite similar to the Dusky and Sooty Grouse but lack the unfeathered neck sacs. There is considerable geographic variation among all the genera.

RUFFED GROUSE *Bonasa umbellus*   RUGR   1

This generally solitary species is the most widespread grouse in North America. In spring, the male's deep accelerated drumming often resonates in open woodlands. Displaying males also raise their ruffs and crest. Most active in morning and evening, Ruffed Grouse may be seen along edges of woodland road, or in winter, perched high in deciduous trees feeding on buds, twigs, and catkins. It sometimes visits feeding stations. Polytypic (14 ssp.; all in N.A.). L 17" (43 cm).
IDENTIFICATION Slim, long-tailed species with a small crest (most apparent when bird is alarmed). Black ruffs on side of neck often hard to see. Most populations have two color morphs, red and gray, most apparent in coloration of tail; coloration varies regionally. Barred underparts and tail with many narrow bands and a wide dark subterminal band (center of which is typically broken or blotchy on female). **JUVENILE:** Similar to female, but less well marked and lacks subterminal band; usually seen with female. **FLIGHT:** When alarmed, it explodes into the air with a roar of the wings, but even then tail pattern is usually easily seen.
GEOGRAPHIC VARIATION Complex, with 14 recognized subspecies. Birds in the Pacific Northwest average darker and richer brown; in interior West, paler gray; in the Northeast, generally grayish; in the Southeast, brownish.
SIMILAR SPECIES Spruce, Dusky, and Sooty Grouse are uncrested. Larger Dusky and Sooty Grouse are more uniformly colored, without the barring on the underparts. Spruce Grouse has a shorter tail and neck with more noticeable white spotting below. Note the tail pattern.
VOICE CALL: Both sexes may give nasal squeals and clucks, particularly when alarmed.
STATUS & DISTRIBUTION Fairly common resident of deciduous and mixed woodlands.
POPULATION Generally stable, with populations rising and falling in regular cycles.

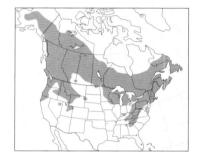

red-morph ♂

male with solid dark tail band

crest

displaying gray-morph ♂

red-morph ♀

female with broken tail band

SPRUCE GROUSE *Falcipennis canadensis*   SPGR   2

While this species may be seen along roadsides, particularly in fall, the Spruce Grouse is often challenging to find. Typically very tame, this species is so well camouflaged that by remaining still it can remain undetected even when observers are only a few feet away. Spruce Grouse are usually solitary, but will gather in small loose flocks in winter. Polytypic (6 ssp.; all in N.A.). L 16" (41 cm)
IDENTIFICATION Short-tailed compact grouse. Over most of range, both sexes have black tail with chestnut tail tip. Male has dark throat and breast, edged with white and red eye combs. Female occurs as red and gray color-morphs; in all females note the heavy dark barring and white spots below. **JUVENILE:** Similar to red-morph female.

GEOGRAPHIC VARIATION Two groups of subspecies: "Taiga Grouse" (*canadensis, atratus, osgoodi, canace,* and *isleibi*) and "Franklin's Grouse" (*franklinii*). Some argue that "Franklin's Grouse" of the northern Rockies and Cascades should be considered a separate species from "Taiga Grouse" found throughout the rest of the range. "Franklin's" has bold white spots on the uppertail coverts, and on males the tail is uniformly dark. Courtship displays also differ: Both give strutting displays where the male spreads his tail, erects the red combs above his eyes, and rapidly beats his wings; some give a series of low hoots. In territorial flight display, "Taiga" male flutters upward on shallow wing strokes; "Franklin's," however, ends the performance by beating his wings together, which produces a loud clapping sound.

SIMILAR SPECIES Compared with Dusky, Sooty, and Ruffed Grouse, Spruce Grouse appears to have a shorter tail. Female Spruce is similar to both Dusky and Sooty Grouse, but it has more heavily marked underparts. Ruffed Grouse has a broad black subterminal tail band.

VOICE Relatively quiet outside the breeding season. **CALL:** Birds in foraging groups give a *sreep.* Female gives a high-pitched call that is thought to be territorial.

STATUS & DISTRIBUTION Rare to fairly common resident of conifer forests, particularly early successional stages with young dense trees. "Franklin's Grouse" is also found in more open subalpine habitats. Usually found near conifers, but when dispersing in fall may be seen in deciduous woods.

POPULATION Population declines in the southern perimeter of the species' range are possibly due to changes in habitat; populations are generally stable in the northern part of the range. Fires are important to renew the species' favored habitat.

red comb

displaying ♂

red-morph ♀

gray-morph ♀

rufous tail band

white tips on uppertail coverts

"Franklin's Grouse" ♂
franklinii

DUSKY GROUSE   *Dendragapus obscurus*   DUGR ▪ 2

Birding visitors to the Rocky Mountains prize the swashbuckling spectacle of a displaying male Dusky Grouse—circling in fluttering flight, strutting with tail fanned, body tipped forward, head drawn in, wings splayed, and neck feathers fanned open, revealing brilliant maroon patches of skin on the sides of the neck. Polytypic. (4 ssp.). L 20" (51 cm)

IDENTIFICATION A relatively large, long-tailed grouse. **MALE:** Medium brownish gray, mottled plumage overall. On each side of the neck, white-based feathers cover patches of reddish bare skin called cervical apteria, exposed during displays. Prominent yellow eye combs, called superciliary apteria, become orange during display. **FEMALE:** Similar to male but upperparts browner, with a more uniformly grayish belly. **JUVENILE:** Similar to female but with fine pale streaks above. Tail is heavily mottled brown.

GEOGRAPHIC VARIATION Four subspecies are named. Nominate *obscurus* ranges from Utah and Wyoming south to Arizona and New Mexico. Subspecies *oreinus,* found from southern Idaho to western Utah, is similar but much paler, and females have more pronounced white edgings in scapulars and wing coverts. The two subpecies found farther north in the Rockies, *richardsonii* and *pallidus,* are sometimes together referred to as "Richardson's Grouse." Both essentially lack the gray terminal tail band found in the other subspecies; *richardsonii* is distinguished from *pallidus* by its darker, blacker white-tipped undertail coverts.

SIMILAR SPECIES See the Identification section of the very similar Sooty Grouse (p. 68) for distinguishing characters of both species. Female Dusky Grouse also resembles female Spruce Grouse, a smaller species, but has less heavily patterned

northern subspecies lack gray tail band

purple air sac

displaying northern Rockies ♂
richardsonii

southern Rockies ♀
obscurus

underparts and a longer tail. Ruffed Grouse has a broad black subterminal tail band and more heavily barred underparts.

VOICE CALL: Males in display give ultra-low-pitched, short hoots, audible only at close range and usually delivered from the ground. Both sexes give various soft clucking sounds; male gives low, nasal, rapid *gr gr gr gr;* female gives odd braying yelp.

STATUS & DISTRIBUTION Fairly common resident of open coniferous and mixed woodlands, brushy lowlands, and mountain slopes. **MOVEMENTS:** Early spring movements are often from lower to higher elevations, into alpine/subalpine habitats, especially for northern populations. Later in the breeding season, some move from relatively open areas to dense coniferous forest. In some

parts of range, these grouse move downslope into sagebrush habitats after breeding.

POPULATION Although historical declines in populations and local extirpations are known, Dusky Grouse numbers currently are not decreasing as rapidly as those of prairie-chickens or sage-grouse, to which they are more closely related than Spruce Grouse or other species.

SOOTY GROUSE *Dendragapus fuliginosus* SOGR ▪ 2

broad gray tail band

displaying coastal ♂ *fuliginosus*

yellow air sac

Sooty Grouse was until recently combined with Dusky Grouse as a single species, "Blue Grouse," but early settlers and naturalists generally distinguished between these birds because of their many distinctions. Sooty inhabits primarily coastal Pacific ranges, especially the Cascades and Sierra Nevada, though some populations are found well into the interior, barely into Nevada. Polytypic (4 ssp.). L 20" (51 cm)

IDENTIFICATION Plumages generally very like Dusky Grouse overall, and in most places, range will determine the identification; hybridization is very limited. Sooty is darker than Dusky, and all subspecies of Sooty have a gray tail band (absent in "Richardson's Dusky"). Compared to Dusky Grouse, Sooty's plumage averages darker. During displays, the male's bare yellow (over most of

range) neck skin is exposed; in southeast AK and north coastal BC, the aptera is red like Dusky. Both sexes of Sooty Grouse have 18 rectrices (versus 20 in Dusky), and in adults, the rectrix tips are rounder (versus more squared in Dusky). In Sooty, the rectrices are also more graduated in length, so that the tail overall looks more rounded than in Dusky. Even the downy chicks are different: grayish in Dusky but reddish above, yellowish below in Sooty.

GEOGRAPHIC VARIATION Four subspecies are named. The northernmost, and very dark, subspecies *sitkensis* is found from southeastern AK to northern BC, including Haida Gwaii (Queen Charlotte I.). Nominate *fuliginosus,* found from BC south to Sonoma and Trinity counties, CA, is rather similar; males of both have fairly narrow gray terminal tail band, but female *sitkensis* are more rufous above than *fuliginosus.* Subspecies *sierrae* is found on the eastern slope of the Cascades in Washington south to Mendocino and Fresno counties, CA, while *howardi* is found south of there, in the southern Sierra Nevada. The latter two subspecies are similar, but *howardi* is paler, with longer, more graduated tail.

SIMILAR SPECIES Very similar to Dusky Grouse; see Identification section for distinguishing characters of both species.

VOICE CALL: Male Sooty in display gives six short hoots, higher in frequency than Dusky's and thus audible about one-half mile away;

Dusky's hooting is hard to hear more than 200 ft. away. Sooty usually displays in tree rather than on ground. Both sexes have repertoire of clucks and wails, similar to Dusky.

STATUS & DISTRIBUTION Uncommon, locally declining resident of coniferous forests, from sea level (in northern portions of range) to high elevations (12,000 ft) near treeline. **MOVEMENTS:** Apparently travel by walking rather than flying between breeding and nonbreeding areas. Males begin trek to breeding grounds in early spring and vacate them beginning in June.

POPULATION Sooty Grouse populations fluctuate, especially on the coast, but long-term declines have clearly occurred in California, where the species is no longer found in the southern portion of its historical range (Mt. Pinos and the Tehachapi Mts.—populations likely exterpated in the 1930s). Some fluctuations in populations appear to be driven in part by cycles of habitat alteration, such as timber operations.

SAGE-GROUSE Genus *Centrocercus*

GREATER SAGE-GROUSE *Centrocercus urophasianus* GRSG ▪ 1

The largest North American grouse, the Greater Sage-Grouse is a highly social denizen of aromatic sagebrush

flats and foothills centered on the Great Basin. Where common, it may form flocks of several hundred

birds. Impressive lekking displays begin in late February or early March and continue into early summer.

Monotypic. L ♂28" (71 cm) ♀22" (56 cm)
IDENTIFICATION Distinctive dark belly and long pointed tail feathers. **MALE:** Yellow eye combs, black throat and bib, large white ruff on the breast. **FEMALE:** Similar to male but smaller. Brown throat and breast, showing only a hint of the male's pattern. **JUVENILE:** Resembles the female but with upperparts finely streaked. **FLIGHT:**

Superficially suggests waterfowl, but note dark belly, long tail, and grouse shape.
GEOGRAPHIC VARIATION Some consider birds in the interior Northwest a separate subspecies (*phaios*), but differences are minor and validity questionable. It hybridizes occasionally with Sharp-tailed and Dusky Grouse.
SIMILAR SPECIES See Gunnison Sage-Grouse. Distinguished from all other

grouse by black belly and larger size.
VOICE CALL: Courting male fans tail and rapidly inflates and deflates air sacs emitting two popping sounds, at times audible up to two miles away; also a variety of wing swishes, coos, whistles, and tail rattles. When flushed, it gives a rapid cackling *kek-kek-kek*.
STATUS & DISTRIBUTION Uncommon to fairly common but local. **MOVEMENTS:** Generally resident, but may engage in local movements, particularly to find sage in times of deep snow. Movements of more than 100 miles have been noted in ID and WY.
POPULATION The species declined during the 20th century; hunting probably reduced numbers early on, but threats to habitat pose the biggest risks today.

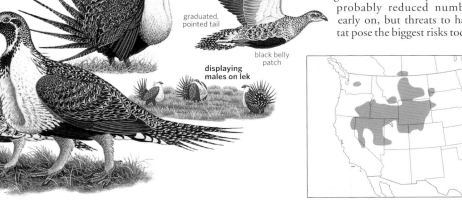

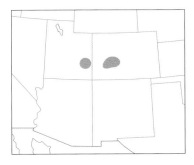

GUNNISON SAGE-GROUSE *Centrocercus minimus*   GUSG ■ 2

This bird was only recognized as a separate species by scientists in the 1990s when differences in size, plumage, and lekking displays were described. Only 2,000 to 4,000 birds remain in less than 193 square miles of land in south-central Colorado and extreme southeastern Utah. Gunnison Sage-Grouse require large expanses of sage with a diversity of forbs and grasses and a healthy riparian habitat, interspersed with grassy open areas, where the males perform

elaborate courtship displays. Monotypic. L ♂22" (56 cm) ♀18" (46 cm)
IDENTIFICATION Roughly two-thirds the size of the Greater Sage-Grouse. Identify by range: Gunnison Sage-Grouse is the only sage-grouse in south-central Colorado and southeastern Utah.
SIMILAR SPECIES Greater Sage-Grouse is very similar. During display, the Gunnison has pronounced, even white bands on the tail, whereas the Greater's tail feathers are more mottled; it also appears noticeably paler at a distance, almost blond in color. Gunnison males display less frequently, and produce different sounds (typically nine air sac plopping sounds during each display, compared with two for Greaters). Gunnison's longer, thick filoplumes are thrown forward over the top of the head, and then flop back down (unlike Greater, where the sparse filoplumes are held erect). Female Gunnison is nearly identical to female

Greater in plumage and behavior.
VOICE CALL: Similar to Greater Sage-Grouse.
STATUS & DISTRIBUTION Endangered. Most in Gunnison Basin of Gunnison and Saguache Counties in CO; six smaller disjunct populations, each with fewer than 300 birds. Fewer than 150 birds in UT.
POPULATION Habitat loss, degradation, and fragmentation have reduced the populations. In recent years this has been exacerbated by drought.

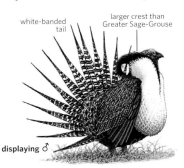

PRAIRIE GROUSE Genus *Tympanuchus*

Endemic to North America, the three species of *Tympanuchus* are found in open grasslands and brushy habitats. They are generally barred or mottled in shades of brown, black, and pale buff. The males have colorful neck sacs and tufts (pinnae) on their head that they inflate and raise during their foot-stomping springtime displays.

SHARP-TAILED GROUSE *Tympanuchus phasianellus*   STGR ■ 2

Early each spring, Sharp-tailed Grouse gather at leks located on rather open elevated sites. The male inflates his purplish neck sacs, rapidly stomps his feet, bows his wings, and shakes his tail rapidly from side to side, producing a distinctive rattling sound. A variety of cackles and low coos accompany the performance. In other seasons, the Sharp-tailed is more difficult to see. During winter, look for it feeding on buds and catkins in deciduous trees early and late in the day. Where ranges overlap, it occasionally hybridizes with Greater Prairie-Chicken and Greater

Sage-Grouse; at least one record of hybridization with Dusky Grouse. Polytypic (6 ssp.; all in N.A.). L 17" (43 cm) **IDENTIFICATION** Similar to prairie-chickens, but underparts scaled and spotted. **ADULT:** Pointed tail, with central rectrices extending far beyond the others. Yellow eye combs (sometimes difficult to see). Small erectile crest most noticeable when bird agitated. Female smaller and with a less contrasting facial pattern. **JUVENILE:** Like female, but shorter central tail feathers and outer tail more brownish (not white). **GEOGRAPHIC VARIATION** Four northern subspecies (*caurus, kennicotti, phasianellus*, and *campestris*) are darker, with heavy markings below and bold pale spotting on upperparts. Two southern subspecies (*columbianus*, and *jamesi*) are paler brown and more uniform. **SIMILAR SPECIES** Greater Prairie-Chicken is always heavily barred below; it also has a rounded tail shape and color to edge. Female Ring-necked Pheasant is superficially similar but is larger with a much longer tail, unfeathered legs, and less heavily marked underparts.

VOICE CALL: Various clucks and peeps; also a three-note *whucker-whucker-whucker* given in flight. **STATUS & DISTRIBUTION** Fairly common in most of range; rare in western portions of range. **BREEDING:** Open and brushy habitats including grasslands, sagebrush, woodland edges, and river canyons. **MOVEMENTS:** Sedentary, but snow cover may induce movements to more wooded areas. **POPULATION** Habitat destruction and fragmentation led to declines throughout much of its southern range; populations generally stable in north. Extirpated from California and Oregon; reintroductions underway in some places.

pointed tail shorter than female Ring-necked Pheasant
displaying ♂
purplish air sacs
grayish plumage overall, heavily mottled and barred with dark
♀

GREATER PRAIRIE-CHICKEN *Tympanuchus cupido*   GRPC ■ 2

The once widespread Greater Prairie-Chicken is now very local in mixed-grass and tallgrass prairie. It is best found and appreciated in early spring, when males gather at hilltop leks; the sounds of their dramatic booming display may carry for nearly a mile. Polytypic (3 ssp.; all in N.A., but 1 is extinct). L 17" (43 cm) **IDENTIFICATION** Heavily barred above and below with dark brown, cinnamon, and pale buff. Barring on underparts broader. **MALE:** Uniformly dark, short, rounded tail; yellow-orange neck sacs (inflated during the breeding season); and long pinnae, erected in displays to form "rabbit ears." **FEMALE:** Subtle version of the male, with smaller pinnae, indistinct eye combs, and barred tail. **JUVENILE:** Resembles female but is smaller with

white shaft streaks on upperparts. **GEOGRAPHIC VARIATION** The two coastal subspecies are darker: "Attwater's Prairie-Chicken" *(attwateri)* of southeastern Texas is nearly extinct, and the "Heath Hen," nominate *cupido,* is extinct but was once found on the coastal plain from Massachusetts to Virginia—the last record was on Martha's Vineyard in 1932. The widespread *pinnatus* is found in the rest of the species' range. **SIMILAR SPECIES** Lesser Prairie-Chicken is best separated by range. Greater Prairie-Chicken is slightly larger, with darker, broader, and more uniform barring on underparts. Sharp-tailed Grouse is similar but is more heavily scalloped and spotted below with a pointed tail. **VOICE** Courting males make a deep

low *ooAH-hooooom* sound known as "booming," often likened to the sound of blowing across a bottle, and also give a variety of wild frenzied cackles and yelps. **CALL:** Both sexes give a variety of clucking notes.

pinnatus

attwateri

short, square black tail

displaying ♂

golden orange air sacs

♀

STATUS & DISTRIBUTION Uncommon and local; "Attwater's Prairie-Chicken" is endangered. **MOVEMENTS:** Poorly understood; show no clear relationship with weather or food availability. Typically less than 25 miles between breeding and wintering areas. **POPULATION** Greater Prairie-Chickens have experienced enormous declines in much of their range. The species was extirpated from Tennessee (1850); Kentucky (1874); Arkansas (1913); Louisiana (*attwateri*, 1919); Texas (*pinnatus*, 1925); Ohio (1934); Indiana (1972); Iowa (1984); Alberta (1965); Manitoba (1970); Ontario (1975); and Saskatchewan (1976). Some of the first conservation

legislation in the United States was passed in 1791 to protect the Heath Hen from market hunting—efforts that ultimately failed. Market hunting was a major component in historic declines. Today, habitat destruction, deterioration, and fragmentation pose threats. Small isolated populations also face genetic, demographic, and environmental uncertainties. Conservation and reintroduction efforts exist in many areas. Some, such as efforts by the Colorado Division of Wildlife and local communities in northeastern Colorado, have been extremely successful in rebuilding Greater Prairie-Chicken numbers.

LESSER PRAIRIE-CHICKEN *Tympanuchus pallidicinctus*   LEPC ■ 2

This shortgrass prairie grouse has a very limited range in the southern Great Plains. It is best looked for during early spring at leks on relatively open sites on exposed hills and ridges. During this time, the differences between Lesser and Greater Prairie-Chicken are best appreciated. Note Lesser Prairie-Chicken's dull orange-red neck sacs, its more maniacal wails and cackles, and slight differences in displays. Monotypic. L 16" (41 cm)
IDENTIFICATION Very similar to Greater Prairie-Chicken; see above for separation from that species. As with the Greater, the Lesser male has a uniformly dark tail while that of the female is barred. Juvenile Lesser is also similar to a Greater but is slightly paler.
SIMILAR SPECIES Best separated from Greater Prairie-Chicken by range; the two species come very close in central Kansas and Oklahoma, but do not overlap. Lesser Prairie-Chicken is slightly smaller than the Greater, with paler barring on the underparts, which is narrower and less uniform in width. The barring typically becomes paler on the center of the belly. On Greater the barring is bold throughout the breast and belly; the upperparts are more finely barred with black edging. There is much individual variation.
VOICE CALL: Similar to Greater Prairie-Chicken; male's courtship calls are higher, more piercing, and even more demonic than the Greater's.
STATUS & DISTRIBUTION Uncommon and local; found in sand sagebrush-bluestem and shinnery oak-bluestem ecosystems. **MOVEMENTS:** Sedentary,

although more likely to be found in agricultural fields during winter.
POPULATION This species has declined by an estimated 97 percent since the 1800s, including a 92 percent reduction in range (78 percent since 1963). Conversion of native habitat to cropland, excessive grazing by livestock, and herbicide control of sand sagebrush and shinnery oak all contributed to habitat loss. Hunting likely played some role in early declines. Droughts have exacerbated all these factors.

displaying ♂

reddish air sacs

♀

weaker barring below than Greater Prairie-Chicken

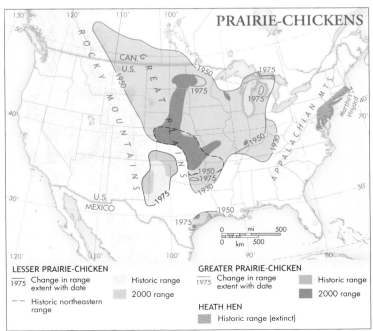

PRAIRIE-CHICKENS

LESSER PRAIRIE-CHICKEN
1975 Change in range extent with date
-- Historic northeastern range
Historic range
2000 range

GREATER PRAIRIE-CHICKEN
1975 Change in range extent with date
Historic range
2000 range

HEATH HEN
Historic range (extinct)

LOONS Family Gaviidae

Pacific Loon, breeding (MB, June)

L oons, usually called "divers" in the Old World, present major identification challenges in both breeding and winter plumages: Swimming birds dive frequently, making them difficult to study, and flying birds rarely allow prolonged views. On swimming birds note the bill shape (stout or slender, straight or with any uptilt), head shape, and overall head and bill proportions and posture. In breeding plumages, note the overall patterns, especially of the head, neck, and back. On winter and juvenal plumages, details of the dark/light patterning on the head and neck are important, in combination with bill shape and color.

Structure Bodies are long and heavy, with large webbed feet set well back; bills are strong and dagger-shaped. Wings are relatively small with ten primaries and 22 to 24 secondaries; tails are small and short, with 16 to 20 rectrices. Males are up to 10 percent larger than females.

Behavior Loons are aquatic birds that swim and dive with proficiency. Birds trying to avoid predators can swim very low in the water, whereas resting birds sometimes ride quite high and buoyantly, and preening birds often roll to the side, exposing their flanks and even belly. Such differences in posture can help determine the extent and pattern of white along a bird's sides. Flight is fairly heavy and direct, with steady, measured wingbeats; the neck held outstretched and often slightly below the body plane; the large feet are visible beyond the short tail.

Plumage Plumages are mostly monochromatic, with breeding adults more boldly patterned than winter adults and juveniles; sexes look similar. Juveniles and first-summer birds resemble winter adults. Second-summer birds resemble breeding adults but with messier, less complete breeding patterns, especially on the head and neck. Adult plumage aspect is attained in the third winter.

Distribution High-latitude breeders throughout the Northern Hemisphere, loons are short- to mid-distance migrants that winter south in the New World to the southern United States and northern Mexico. Non-breeding immatures often remain south in summer.

Taxonomy Five species in one genus, *Gavia,* make up the modern-day loon family; all occur in North America.

Conservation Oil spills present a major threat to non-breeding populations in marine waters. Industrial poisons (such as mercury) may affect both breeding and wintering populations. Human disturbance and acid rain affect nesting Common Loons at more southerly lakes.

Genus *Gavia*

ARCTIC LOON *Gavia arctica*   ARLO ■ 2

This Old World species' breeding range extends into western Alaska, where it is a prized find to be distinguished with care from the much commoner Pacific Loon. Polytypic (2 ssp.; *viridigularis* in N.A.). L 28" (73 cm)

IDENTIFICATION Resembles the Pacific Loon, but averages larger, with a longer and stouter bill, and often a more angular forehead. The Arctic's single best field mark in all plumages is the white flank patch on swimming birds or a white flank bulge on flying birds. The white patch extends up onto the sides of the rump. Otherwise, all age-related variation and molts are like Pacific Loon, except breeding.

BREEDING ADULT: Arctic's hind neck is a slightly duskier, smoky gray, and its throat patch is glossed green (very

hard to see). **WINTER ADULT AND JUVE-NILE:** The dusky chinstrap common on Pacific is usually (perhaps always) lacking on Arctic.

GEOGRAPHIC VARIATION North American records are of the eastern Eurasian subspecies, *viridigularis,* distinguished from nominate *arctica* of western Eurasia by its slightly larger size and a more greenish sheen to the breeding adult's throat patch.

SIMILAR SPECIES Pacific Loon averages smaller, with a rounder head that accentuates the smaller bill, and it lacks well-defined white femoral patches, but a swimming bird can show white along the waterline. Usually this is an untidy patch at about midbody. Arctic Loon has a generally neater patch that flares up the

sides of the rump. Breeding Pacific Loon has a paler crown and nape, and a purple-glossed throat patch bordered by less contrasting white stripes that are separated from the chest striping by a narrow black bar (stripes typically merge on Arctic). A winter/juvenile Pacific often has a distinct dark chinstrap and averages

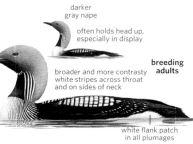

darker gray nape

often holds head up, especially in display

broader and more contrasty white stripes across throat and on sides of neck

breeding adults

white flank patch in all plumages

paler on the hind neck. A winter/juvenile Common Loon is larger and bulkier with a stouter bill, a broken

whitish eye ring, a more irregular dark/white border to the neck sides (suggesting the breeding pattern), and the hind neck is darker than the back, the opposite of Arctic Loon. **VOICE** Similar to Pacific Loon, but lower pitched.

STATUS & DISTRIBUTION Eurasia and western AK. **BREEDING:** Uncommon in N.A., small to large tundra lakes. **MIGRATION:** In Bering Sea, western Aleutians. Small numbers are regularly noted flying northeast past Gambell, St. Lawrence Island in late spring. A few are

seen in the fall. Rare on the Pribilofs. **WINTER:** Casual south along or near Pacific coast to Baja California (also a few summer records of immatures). Accidental in CO.
POPULATION No data.

never has a "chinstrap"

juvenile

winter adult

PACIFIC LOON *Gavia pacifica* PALO ▪ 1

Intermediate in size between the more widespread Red-throated and Common Loons, Pacific Loon is best known as a migrant and winter visitor along the Pacific coast. When flocks migrate north in spring, adults are in their handsome breeding plumage. Monotypic. L 26" (66 cm)
IDENTIFICATION Pacific Loon has a straight, fairly stout bill and usually a fairly smoothly rounded head and hind neck profile. In winter/juvenile plumages, note the well-defined and generally even dark/light border on the head and neck sides, and the contrast between a paler hind neck and darker back. Pacific is told from the slightly larger Arctic Loon in all plumages by its black thigh patch. **BREEDING ADULT:** Head and hind neck are ashy gray with a purple-glossed throat patch bordered by white stripes; upperparts are boldly patterned black-and-white. Bill is black, eyes red. **WINTER ADULT:** Throat and foreneck are white; about 90 percent of birds have a dusky to dark chinstrap, and the blackish upperparts have very muted paler checkering. Bill is pale gray with a dark culmen. **JUVENILE/FIRST-YEAR:**

Juvenile resembles winter adult but paler feather tips to its upperparts create a scaly patterning; its upperwing coverts lack white spots; eyes are duller, brownish red. About half of juveniles lack the dusky chinstrap. Many first-summers look like faded winter adults. **SECOND-WINTER:** Produced by complete second prebasic molt in late summer, resembles winter adult but upperwing coverts lack distinct white spots. Following a molt in spring of the third year, resembles breeding adult but the head and neck can be mottled white, the white back markings average smaller, and the bill may be piebald.
SIMILAR SPECIES Winter adult/juvenile Common Loon is distinctly larger and bulkier, and its stouter bill has a pronounced gonydeal angle. Note Common's broken whitish eye ring, the more irregular dark/white border to its neck sides (suggesting the breeding pattern), and that its hind neck is darker than the back, the opposite of Pacific Loon. Red-throated Loon is lighter in build, often has a paler face, and has an uptilted bill. Winter adults and juveniles are usually paler overall with fine white flecks on the

upperparts and either a more extensive white face or (juveniles) a messier and more extensively dark face and neck than the Pacific. See Arctic Loon.
VOICE Rarely vocal in winter. **CALL:** Includes a low, grunting *awhrr.* **SONG:** Loud, slightly trumpeting, melodic wailing, *hooo-AHHr-uh,* and variations.
STATUS & DISTRIBUTION N.A. **BREEDING:** Fairly common on small to large lakes. **MIGRATION:** Mainly Apr.–early June and Sept.–early Dec. over inshore waters, often in flocks. **WINTER/VAGRANT:** Rare inland throughout West, very rare in Midwest, and casual to East Coast.
POPULATION Stable or increasing.

pale gray nape

faint white vertical streaks on sides of neck

winter adult

some young birds lack "chinstraps"

juveniles

white scaling on scapulars

breeding adult

dark "chinstrap"

sharp and even division between front and rear of neck

winter adult

dark upperparts

COMMON LOON   *Gavia immer*   COLO ▪ 1

This large and widespread species is the most familiar loon throughout N.A. Monotypic. L 32" (81 cm)
IDENTIFICATION Distinctive stout, dark bill with, usually, a distinct gonydeal angle. Head shape fairly angular, often with a distinct forehead bump. In winter/juvenile plumages, white extends over eye and jagged dark/light borders on the neck sides. **BREEDING ADULT:** Head and neck oily green-black with two horizontal white neck bands striped dark; back boldly patterned black-and-white. Bill black, eyes red. **WINTER ADULT:** Throat and foreneck white, upperparts blackish brown with very muted paler checkering. Bill pale gray with dark culmen. **JUVENILE/FIRST-YEAR:** Like winter adult, but paler feather tips on upperparts create scaly effect and eyes duller. Some first-summer birds retain almost all juvenal plumage, which can become faded and pale. Other first-summers attain new head, neck,

and back feathers, so their upperparts have scattered darker feathers and sometimes small white spots. **SECOND-WINTER:** Like winter adult, but upperwing coverts lack distinct white spots.
SIMILAR SPECIES Yellow-billed Loon is slightly larger overall; longer bill has straight culmen, often held slightly raised, accentuating upturned shape; bill pale yellow (breeding) to creamy or pale ivory with a duskier base, but culmen always pale, at least near tip (dark on Common). Winter/juvenile Yellow-billed paler above, especially face, which usually has dark auricular mark; pale upperpart scaling on juvenile Yellow-billed averages more contrasting.
VOICE CALL: Includes a slightly manic, yelping tremolo. **SONG:** Loud, mournful, and eerie wailing cries are far-carrying, *whoo'oo ooh-ooh*.
STATUS & DISTRIBUTION Holarctic

breeder. **BREEDING:** Fairly common, on large lakes. **MIGRATION:** Mar.–May and Sept.–Nov. **WINTER:** Mainly coastal, also on large, inland water bodies.
POPULATION Bulk of world population nests in Canada.

white collar

breeding adult

winter adult

heavy silver-colored bill with dark, slightly curved culmen

uneven line of separation of white and dark on neck

half dark collar

juvenile

Identification of Loons in Flight

Loons are often seen in flight, particularly off coastal headlands during migration. Flight identification of loons takes practice, and even experienced observers have a "distance threshold," beyond which they are not comfortable.

The larger loons, Common and Yellow-billed, usually can be distinguished by their slower, more measured (or gooselike) wingbeats, bigger "front end" (with a thicker neck and larger head and bill), and bigger feet. They tend to fly singly or in small, loose groups, often higher than smaller loons. Specific identification depends mainly on details of bill shape and color.

Red-throated, Pacific, and Arctic Loons are smaller, with quicker wingbeats, have smaller feet, and tend to fly lower over water. Red-throated often looks more gangling, accentuated by a thin neck, often held slightly drooped and then raised briefly, but not all

Red-throateds fly with a drooped neck. Pacific Loon has a steadier flight and looks more compact than the Red-throated, with a thicker neck, often held fairly level. Arctic Loon flies like a heavily built Pacific and has slightly slower wingbeats.

Red-throated Loon usually flies alone or in very small groups; ones and twos often join flocks of Pacifics. Pacific Loons often fly in loose flocks or lines of ten to 50 birds, sometimes hundreds. It's a pretty good bet that if you see a flock of loons, they will likely be Pacifics, although another species could be mixed in. Arctic Loons fly alone or in pairs but also are often seen with Pacific Loons at St. Lawrence Island, AK. Specific identification is always best confirmed by plumage and bill features; beware that Red-throated Loon shows some white coming up the sides of the rump, suggestive of Arctic. ▪

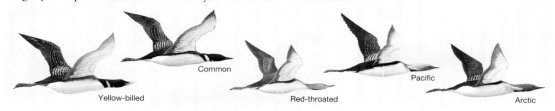

Yellow-billed

Common

Red-throated

Pacific

Arctic

YELLOW-BILLED LOON *Gavia adamsii* YBLO ▪ 2

This spectacular far-northern loon is being found in winter with increasing regularity in the lower 48, but it is still best appreciated when breeding adults are seen on their high-Arctic summer grounds. Monotypic. L 34" (86 cm)
IDENTIFICATION Most Yellow-billed Loons are readily identified, but on problem birds always check the bill shape and color: The bill is long and stout with straight culmen (that is pale, at least distally) and a distinct gonydeal angle; hence the lower man-

dible looks slightly recurved, an effect enhanced by the bill usually being held slightly raised. Winter adults and juveniles typically have a pale-faced aspect with a dark auricular mark. **BREEDING ADULT:** Bill is straw yellow. **WINTER ADULT AND JUVENILE:** Bill is pale creamy yellow to ivory, sometimes with a grayish base.
SIMILAR SPECIES Yellow-billed Loon resembles Common Loon in all plumages, but winter Yellow-billed adults and juveniles are usually paler above. Common Loon is slightly smaller overall with a smaller bill that has a decurved culmen and less exaggerated wedge shape to the lower mandible; Common's bill is black (breeding

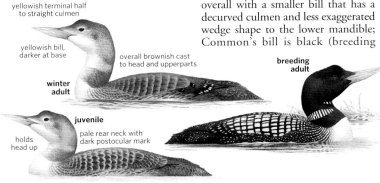

yellowish terminal half to straight culmen

yellowish bill, darker at base

overall brownish cast to head and upperparts

winter adult

breeding adult

juvenile

holds head up

pale rear neck with dark postocular mark

adult) to pale gray with a dark culmen. **VOICE** Similar to that of Common Loon but lower pitched and hollow; tremolo is slower paced.
STATUS & DISTRIBUTION Holarctic breeder. **BREEDING:** Uncommon to fairly common on tundra lakes and rivers. **MIGRATION:** Apr.–early June and late Sept.–Nov. **WINTER:** Rare on Pacific coast south to CA, casual to very rare inland in West, and casual east to Great Lakes region, south to TX and northern GA.
POPULATION No information.

RED-THROATED LOON *Gavia stellata* RTLO ▪ 1

This smallest and most lightly built loon differs from other loons in its relatively subdued breeding plumage and in having a complete molt in late fall. Monotypic. L 25" (64 cm)
IDENTIFICATION Note slender, slightly upturned bill, often held slightly raised. Swimming birds often lack a chest bulge at the waterline, shown by other loons. **BREEDING ADULT:** Head and neck gray with black-and-white hind neck lines; brownish upperparts; dark red throat patch. Bill black; eyes red. **WINTER ADULT:** White face and neck; bill gray. Small, mainly subterminal, white spots and ovals on scapulars. **JUVENILE/FIRST-SPRING:** Resembles winter adult, but face and neck variably smudged dusky (not to be confused with the Pacific Loon's "chinstrap"), scapulars have narrower, chevronlike, subterminal whitish marks, and eyes duller, brownish red. First-summer birds range in appearance from like winter adult to having patches of adultlike gray and dark red on neck. Complete second prebasic molt in late summer produces adultlike winter plumage.

SIMILAR SPECIES Pacific Loon is slightly larger and bulkier with thicker neck, straight bill, white back patches, and has a fuller-chested profile when swimming. Winter/juvenile Pacific has dark cap extending down through eye, and hind neck more extensively dark (beware juvenile Red-throated, but its dark areas are messy, not well defined). **VOICE CALL:** Flight call in summer a low, grunting or barking *ahrr* or *ahrk,* usually in series. **SONG:** A male-female duet of loud, wailing, drawn-out screams, *whEEahhr, heeAHH,* etc.
STATUS & DISTRIBUTION Holarctic breeder. **BREEDING:** Fairly common, mainly on lakes and ponds. **MIGRATION:** Mar.–May and Sept.–Nov., mainly coastal but also overland in East, especially Great Lakes. **WINTER:**

Casual inland and south to Gulf Coast.
POPULATION The Alaska population on western tundra declined about 50 percent between 1977 and 1993.

1st spring

bill slightly upturned and often holds head up

face and neck washed with dusky

juvenile

breeding adult

reddish throat often looks dark

extensive pure white face

winter adult

white speckling on upperparts

GREBES Family Podicipedidae

Least Grebe, adult with chicks (TX, Apr.)

Grebes are fairly small to medium-large diving birds of freshwater and inshore habitats. A tail-less appearance and a fairly long neck distinguish them from most swimming birds; relative to loons, note their fluffy rear ends. Rarely seen in flight, grebes can be identified readily to species, given good views of head shape, face and neck patterns, and bill size, shape, and color.

Structure Grebes are heavy bodied with longish necks and lobed feet set far back on the body. Their beaks vary from somewhat long and pointed to stout and chickenlike. The wings are relatively small and pointed, with 11 primaries and 17 to 22 secondaries; vestigial tail feathers are difficult to distinguish from body feathers.

Behavior Grebes dive for food. Smaller species are mostly found singly or in small groups, but Eared and *Aechmophorus* grebes occur in flocks of hundreds, sometimes thousands. They fly direct with steady, fairly quick wingbeats, the neck held outstretched and feet trailing.

Plumage Seasonally dichromatic plumage, but sexes look similar; juveniles resemble winter adults. Generally, plumages are dark above and light below; two species have golden head plumes in breeding plumage. Adult plumage aspect attained at about one year of age.

Distribution Cosmopolitan, with both resident and migratory species. In North America, grebes occur in all but the far northern latitudes.

Taxonomy About 22 species worldwide in six genera, with seven species in four genera in N.A.

Conservation Pollution, disturbance, and draining of inland lakes affect breeding and wintering populations. Oil spills threaten nonbreeding grebes on marine waters. BirdLife International lists four species as threatened, with one as near threatened.

Genera *Tachybaptus* and *Podilymbus*

LEAST GREBE *Tachybaptus dominicus*   LEGR ■ 2

This well-named, diminutive, and somewhat retiring grebe of southern Texas often holds its neck retracted and its rear end puffed out. Polytypic (4 ssp.; 2 in N.A.). L 8.5–9.5" (21–24 cm)

IDENTIFICATION Bill slender and slightly uptilted. **BREEDING ADULT:** Black face and throat; slate gray sides of neck. Black bill finely tipped white, eyes golden. **WINTER ADULT:** A paler face with a mottled or dusky throat; lower mandible mostly pale grayish to horn. **FIRST-YEAR:** Juvenile head sides striped dark gray and whitish; lower mandible mostly pale horn; eyes paler and duller. Eyes develop golden color over first year. First-summer bird has throat mottled sooty gray, not solidly black as in breeding adult.

GEOGRAPHIC VARIATION Subspecies *brachypterus* of TX averages slightly larger and longer billed than *bangsi*, which is a vagrant to the Southwest from western Mexico.

SIMILAR SPECIES No other species should be confused with Least Grebe in its limited N.A. range; Pied-billed Grebe is nearly twice its size and has a stout, mostly pale bill.

VOICE An overall descending, purring trill, *pc, pc, purrrrrrrrrrrrr;* often starts hesitantly. Also a quacking *kwrek,* and a slightly shrill, emphatic *ehkehr.*

STATUS & DISTRIBUTION Southern TX to S.A. **BREEDING:** Fairly common, but local on vegetated ponds and lakes; nests year-round in southern TX (mainly Apr.–Aug.). **VAGRANT:** Casual visitor to southeastern CA, southern AZ (has bred), southern FL, and the upper TX coast.

POPULATION No data.

juvenile
brachypterus

golden-colored iris

slender bill

winter
brachypterus

purplish gray neck

white throat

breeding adult
brachypterus

PIED-BILLED GREBE *Podilymbus podiceps*   PBGR ▪ 1

This widespread chunky grebe is usually found singly or in small groups on small ponds. It can hide by sinking until only its head is above water. Polytypic (3 ssp.; nominate in N.A.). L 11–13.5" (28–34 cm)

IDENTIFICATION Thick bill diagnostic; fairly large head, and, unlike other North American grebes, lacks distinct white upperwing patches. **BREEDING ADULT:** Face and throat black, neck sides grayer; bill pale blue-gray with sharply defined black medial band. **WINTER ADULT:** Face paler and throat whitish, neck sides warmer and browner; bill pale horn to grayish, without black band. **FIRST-YEAR:** Juveniles have head striped dark gray and whitish, and bill pale horn. Some birds retain vestiges of dark head striping through winter. First-summer plumage resembles breeding adult, but it can have a less solidly black throat and the bill band may be narrower and less complete.

SIMILAR SPECIES Eared and Horned Grebes have slenderer bills, and a more capped appearance in nonbreeding plumages; also see Least Grebe.

VOICE CALL: Includes single clucks and a rapid-paced, slightly nasal, chatter in interactions. **SONG:** Loud series of gulping notes that can fade away or run on into more prolonged variations.

STATUS & DISTRIBUTION N.A. to southern S.A. **BREEDING:** Lakes, small ponds with emergent vegetation. **MIGRATION:** Mainly Mar.–Apr., late Aug.–Oct. in interior regions, less often on saltwater. **VAGRANT:** Casual visitor north to AK.

POPULATION Declining locally, especially in the East.

white rear end · dark eye · ring on thick, whitish bill · striped head · **breeding adult** *podiceps* · black chin · **juvenile** *podiceps* · **downy young** *podiceps* · **winter** *podiceps* · white throat · tawny brown neck

---

Genus *Podiceps*

RED-NECKED GREBE *Podiceps grisegena*   RNGR ▪ 1

This grebe often forms flocks at favored coastal locations in winter. Polytypic (2 ssp.; *holboellii* in N.A.). L 17–20" (43–51 cm)

IDENTIFICATION Distinctive thick neck and chunky head, with overall dusky to dark head and neck. White secondaries and white patagial panel visible in flight. Medium-long, pointed bill with yellow at base. **BREEDING ADULT:** Neck chestnut; throat and auriculars smoky gray, outlined in white. **WINTER ADULT:** Neck dusky; face dusky gray with paler auricular patch. **JUVENILE/FIRST-YEAR:** Head sides striped dark gray and buffy to whitish, neck sides washed rufous. First-summer plumage like breeding adult, but crown browner and foreneck duller.

SIMILAR SPECIES Eared Grebe in winter plumage is somewhat similar, but is distinctly smaller and has a slender dark bill, steep forehead, usually whiter and more contrasting auricular patches, and bright reddish eyes. Eared also often has a puffier rear end and shows more whitish along waterline. A molting Horned Grebe can suggest a breeding Red-necked, but is smaller and has a smaller blackish bill, bright-red eyes, and some indication of golden supercilium. Eared and Horned Grebes also lack a clean-cut white patagial panel on the upper wings.

VOICE Mainly vocal when breeding. Low, slightly gull-like wailing cries often run into slightly metallic, pulsating bickering or chattering series. Also a clipped, harsh note, usually repeated, *kerk! kerk!*

STATUS & DISTRIBUTION Holarctic breeder. **BREEDING:** Favors shallow freshwater lakes. **MIGRATION:** Mainly Mar.–early May, Sept.–Nov. **WINTER:** Rare in interior south of northern tier of states; casual to the Southwest and Gulf Coast.

POPULATION No clear population trend in recent decades, but declines have been noted, especially at edge of range.

holboellii

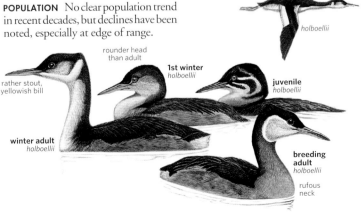

rounder head than adult · **1st winter** *holboellii* · **juvenile** *holboellii* · rather stout, yellowish bill · **winter adult** *holboellii* · **breeding adult** *holboellii* · rufous neck

HORNED GREBE *Podiceps auritus*   HOGR ■ 1

This fairly small but stocky grebe is generally less familiar than the Eared Grebe on the breeding grounds, but it is more widespread in North America in winter, mainly on saltwater, but also on freshwater lakes, where locally numerous. Overall scarcer in the interior West than inland eastern N.A. Polytypic (2 ssp.; *cornutus* in N.A.). L 11.5–13" (29–33 cm)

IDENTIFICATION Medium-length straight bill with fine white tip in all ages, shallow-angled forehead, and flat crown. Undertail coverts rarely fluffed into a high rear end. White secondaries visible in flight. **BREEDING ADULT:** Black head with bushy postocular "horns," black bill, and red eyes; neck and body sides chestnut. **WINTER ADULT:** White head sides and foreneck with blackish cap extending down to eye level; white on face nearly meets on back of head; bill grayish. **FIRST-YEAR:** Juvenile head sides striped

dark gray and whitish, lower mandible mostly pale horn. First-summer plumage averages duller, with paler and less extensive horns.
SIMILAR SPECIES Eared Grebe is more lightly built with a thinner neck and steep forehead that create a pointier, less blocky head profile; its slimmer, upturned bill lacks a white tip. Eared often rides higher on the water, with a puffy rear end. Most winter Eared Grebes are typically much dingier, with the black cap extending down well into the cheeks. Some Eareds, though, have surprisingly bright whitish necks and faces; check bill shape and depth of cap. Breeding Eared has a shaggy fan of golden postocular plumes and black lores. The bold black-and-white pattern of the Clark's and Western Grebes can suggest a winter Horned, but the

former are much larger and longer necked with long and thin yellowish bills. Also see the larger Red-necked Grebe.
VOICE Vocal mainly when breeding. A fairly high-pitched, rapid, pulsating whinny often given in duet, and a slightly hoarse, whistled *híki-sheíhr,* repeated.
STATUS & DISTRIBUTION Holarctic breeder. **BREEDING:** Favors small to medium-size lakes with emergent vegetation. **MIGRATION:** Mainly Mar.–early May, Sept.–Nov. Casual in NF.
POPULATION More than 100,000 Horneds are estimated in N.A., but the numbers are apparently slowly declining with a steady northwestward contraction of breeding range.

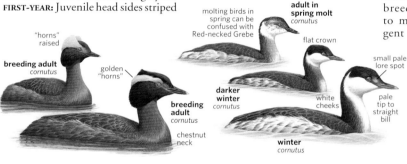

"horns" raised

breeding adult
cornutus

golden "horns"

breeding adult
cornutus

chestnut neck

molting birds in spring can be confused with Red-necked Grebe

adult in spring molt
cornutus

flat crown

darker winter
cornutus

small pale lore spot

white cheeks

pale tip to straight bill

winter
cornutus

EARED GREBE *Podiceps nigricollis*   EAGR ■ 1

This fairly small, mostly western grebe is famed for its fall molt migrations, when tens of thousands of birds swarm at Mono Lake, California, and Great Salt Lake, Utah. It also nests colonially. Known as "Black-necked Grebe" in the U.K. Polytypic (3 ssp.; *californicus* in N.A.). L 11–12.5" (28–31.5 cm)

IDENTIFICATION Distinctive steep forehead and relatively thin neck. Overall

dusky to dark head and neck and slender, slightly upturned bill. White secondaries visible in flight. **BREEDING ADULT:** Black head and neck, expansive postocular fan of golden plumes, black bill, and red eyes; body sides chestnut. **WINTER ADULT:** Dusky head sides and neck with a whitish throat and post-auricular patches; gray bill with dark culmen. **FIRST-YEAR:** Juvenile resembles a winter adult, but its head sides and foreneck are washed buff and the eyes are duller, orangey. In fall, the formative

buffy wash to neck

1st fall
californicus

some quite pale-necked like Horned

often raises and fluffs out rear end

paler winter
californicus

peaked or rounded head

upturned all-dark bill

dark cheek

dusky neck

winter
californicus

downy young
californicus

golden "ears"

breeding adult
californicus

black neck

puffier cheeks, and white almost meets on back of head

winter

Horned   **Eared**

plumage resembles winter adult but the upperparts average browner and the eyes are orange to orange-red. First-summer plumage resembles a breeding adult, but the head and neck average duller, with paler and less extensive head plumes. **SIMILAR SPECIES** Horned Grebe is more thickset and has a shallow-sloping forehead, flatter crown, and a straight, slightly thicker bill tipped white. Horned often rides flatter on the water, without the pronounced puffy rear end of the Eared. Winter

Horned Grebe is typically a much cleaner black-and-white, with a shallow black cap (beware of nonbreeding Eared with bright whitish neck; check bill shape and depth of cap). Molting birds of the two species can look quite similar (mainly Apr. and Sept.): Note bill shape differences and the position of yellow on the head sides. Also see Red-necked Grebe. **VOICE** An upslurred, slightly reedy or plaintive *hoor-EEP* or *eeihr-EK!* Rapid-paced, high piping chittering

in interactions. The two Old World subspecies are similarly plumaged to *californicus* but appear to have very different vocalizations. **STATUS & DISTRIBUTION** Holarctic breeder. **BREEDING:** Common. Favors shallow ponds with high macroinvertebrate productivity; rarely on ponds with fish. **MIGRATION:** Mainly Mar.–May, Aug.–Nov.; rare in East, a few winter in the Southeast. **POPULATION** No demonstrable trend apparent in N.A. over recent decades.

Genus *Aechmophorus*

WESTERN GREBE *Aechmophorus occidentalis*   WEGR ■ 1

Western Grebe looks similar to Clark's Grebe, and the two were once considered conspecific under the name Western Grebe. Flocks in winter can number hundreds, often with Clark's Grebes mixed in. Polytypic (2 ssp.; nominate in N.A.). L 19–25" (48–64 cm) **IDENTIFICATION** Extent of white on face varies with age and season. Intermediate (and unidentifiable) birds are not rare; some may be hybrids. **BREEDING ADULT:** Dusky gray to blackish lores and supraorbital region, yellow bill with dark culmen, and orange-red to red eyes. **WINTER ADULT:** Dusky lores and supraorbital region, often with whitish loral patch. **FIRST-YEAR:** Aging criteria as Clark's. First summer may have pale loral spot. **SIMILAR SPECIES** Bill color best separates Western Grebe (greenish yellow to yellow) from Clark's Grebe (orange-yellow). See Clark's for other differences. In flight, a Red-necked Grebe is more compact with a duskier neck; the upper wing has a more restricted white secondary panel and white patagial patch. Also see the winter Horned Grebe.

VOICE Varied screechy and scratchy calls include a two-note *kreeih chik* or *kree-kreet,* the first part upslurred. Begging calls much like Clark's Grebe. **STATUS & DISTRIBUTION** N.A. to central Mexico. **BREEDING:** Favors freshwater lakes with large areas of open water and bordered by reeds or other emergent vegetation. **MIGRATION:** Mainly Apr.–May and Sept.–Nov. Large numbers of nonbreeding birds oversummer on inshore waters along Pacific coast. **WINTER:** Casual in the East. **POPULATION** See Clark's Grebe.

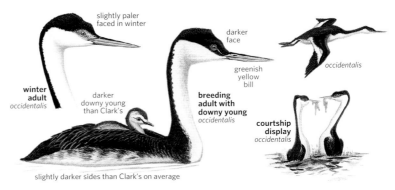

slightly paler faced in winter

darker face

greenish yellow bill

occidentalis

winter adult *occidentalis*

darker downy young than Clark's

breeding adult with downy young *occidentalis*

courtship display *occidentalis*

slightly darker sides than Clark's on average

CLARK'S GREBE *Aechmophorus clarkii*   CLGR ■ 1

A large, striking black-and-white grebe once considered conspecific with the generally commoner Western Grebe. The two species associate readily and hybridize. Polytypic (2 ssp.; *transitionalis* in N.A.). L 18–24" (45–61 cm) **IDENTIFICATION** This grebe has a long, slender neck and a long, slender, pointed bill (noticeably larger on males) with a distinct gonydeal angle. **BREEDING ADULT:** White lores and

supraorbital region, bright orange-yellow bill with dark culmen, orange-red to red eyes. **WINTER ADULT:** White to mostly dusky lores and supraorbital region with whitish loral patch. **FIRST-YEAR:** Juvenile similar to winter adult but crown washed grayish and eyes duller and paler, orangey. After molt, rarely distinguishable from adult, although upperparts can be mixed with worn, brownish feathers. First-summer may have dusky face, lacking

bright white surround to eye, which is paler red through first year. **SIMILAR SPECIES** Western Grebe is best separated by its greenish yellow to yellow bill, but also note its more extensive black crown (extending down to the eyes), broader black hind neck stripe, and overall grayer flanks. Breeding Western Grebe has blackish to dusky-gray lores and surround to eye, but these areas are paler on winter birds, which can have a loral pattern similar to Clark's. Although the upper wing of Clark's averages more extensive white on primaries, it is not diagnostic. The range of variation within each species is not well understood, compounded by hybridization; some birds, mainly in winter, may not be identifiable. In flight, Red-necked Grebe is more compact with a duskier neck and its upper wing has a more restricted white secondary panel and white patagial patch. Also see the winter Horned Grebe.

VOICE CALL: A rather drawn-out, disyllabic *kreeeIH* or *kreeet* among others. Screechy and scratchy. Juveniles solicit adults with high-pitched whistles that can still be given in midwinter.
STATUS & DISTRIBUTION N.A. to central Mexico. **BREEDING:** Fairly common. **MIGRATION:** Mainly Apr.–May and Sept.–Nov. Nonbreeding birds oversummer on inshore waters along Pacific coast. **WINTER:** Fairly common to uncommon on West Coast (only

10–20 percent of "Western Grebe" flocks in CA are Clark's). Accidental in the East.
POPULATION From 1890s to about 1906, tens of thousands of Western/Clark's Grebes were shot for their silky white feathers; today, oil spill mortality, disturbance, habitat degradation, and pesticides continue to keep present populations below historic levels.

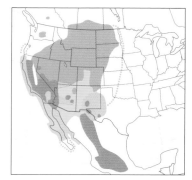

FLAMINGOS Family Phoenicopteridae

American Flamingo (Bahamas, Mar.)

Extremely long necks, a unique bill shape, and pink plumage make flamingos among the most recognizable birds in the world. Nonetheless, flamingos are rare in the United States, occurring regularly at only one site. They are common in captivity, however, which casts doubt on many extralimital reports, and even induces identification issues—not all escapes are species native to North America.
Structure Proportionately, flamingos are the longest necked and longest legged birds in the world. The neck is so long that it must be draped over one side of the body when the bird is at rest. The bill bends down in

the middle, giving the bird a distinctive "broken nose." Webbed feet provide stability on mudflats and allow the flamingos to swim in water too deep for wading.
Plumage All species are unmistakably pink or red with black flight feathers. The color of the feathers is obtained from carotenoid pigments in the food supply; it takes three or more years to reach full adult plumage. Immatures are gray or brown.
Behavior Highly gregarious, flamingos nest, forage, and roost in colonies apart from other birds. Breeding takes place on mud- or marlflats near water; flamingos build unique volcano-shaped nests of mud at the top of which they lay a single egg. They use their curiously shaped bill to strain algae, diatoms, and aquatic invertebrates; when foraging, the head faces directly downward so that the distal half of the bill is parallel with the water. They usually prefer to forage in hypersaline estuaries and lakes; they feed by swinging their bill from side to side as they walk, and may even dabble like ducks in deep water. Their flight is swift and direct, with quick wingbeats and with the neck fully extended. Flocks fly in lines or in V-formation.
Distribution Flamingos are found on all continents except Antarctica and Australia. In North America, only the American Flamingo occurs naturally in extreme southern Florida. Elsewhere along the Gulf and south Atlantic coasts it is a casual visitor; however, many of the records may be of escaped captives. Adding to the confusion, escapes of other flamingo species are regularly observed.
Taxonomy Recent genetic studies have revealed a close relationship to grebes. Flamingos are placed in their

own order, Phoenicopteriformes. There are six species worldwide in three genera. American Flamingo was formerly classified as a subspecies of Greater Flamingo, which included birds from the Old World. The two subspecies were split into two species in 2008: American Flamingo *(Phoenicopterus ruber)* in the New World and Greater Flamingo *(P. roseus)* in the Old World.

Conservation Flamingos have suffered from hunting and disturbance at nesting colonies, but they were largely spared the ravages of the late 18th- to early 19th-century millinery trade because their feathers fade once plucked from their bodies. Populations are recovering now that colonies are protected, but they remain at risk from wetland loss, pollution, disturbance, and other factors.

Genus *Phoenicopterus*

AMERICAN FLAMINGO   *Phoenicopterus ruber*   AMFL ■ 3

This well-known species is common in exhibits, but it has an extremely limited native range in the United States. In Florida Bay—an area within Everglades National Park—one small, mostly wintering flock offers the thrill of seeing flamingos in their natural habitat. At least some of these birds originate from Mexico—one fledgling color-banded in a colony at the Yucatán Peninsula was discovered recently at Florida Bay. Monotypic. L 46" (117 cm) WS 60" (152 cm)
IDENTIFICATION Unmistakable (except for escaped flamingos of other species). **ADULT:** The sexes look similar. Plumage entirely pink (darker below); black flight feathers often not visible at rest. Pink orbital ring and yellow eyes.

Tricolored bill—gray base, pink mid-section, black tip. Pink legs and feet. **JUVENILE:** Grayish white overall with blackish streaking on wing coverts. Eyes dark; mid-section of bill gray; gray or grayish pink legs and feet. Transition to adult plumage occurs gradually over three years; intermediate plumages not well known. **FLIGHT:** All ages show black flight feathers on both upper and lower wings.
SIMILAR SPECIES Roseate Spoonbill shares the same coastal habitats and pink plumage, but its body shape and proportions and foraging behavior differ considerably. Escaped flamingos of other species are rare but present identification problems—be particularly circumspect with extralimital reports. Greater Flamingo *(P. roseus)* from the Old World is mostly whitish with reddish flight feathers, and the base of its bill is pink. Chilean Flamingo *(P. chilensis)* has very pale pink plumage except it is darker on

the neck and has reddish flight feathers; its bill lacks any pink and its legs are gray with pink ankle joints. Lesser Flamingo *(P. minor)* is smaller and has a shorter, thicker neck; the basal half of its bill is maroon, the distal half is red with a black tip.
VOICE Low guttural sounds given when feeding. **FLIGHT CALL:** A goose-like honking.
STATUS & DISTRIBUTION Rare in the U.S.; restricted to southern FL. Locally common in Mexico, Cuba, and the Bahamas. **BREEDING:** Builds a volcano-shaped mound of mud in shallow coastal estuaries. No certain U.S. nesting records. **DISPERSAL:** Prone to postbreeding and nonbreeding dispersal from colonies in Yucatán Peninsula, Cuba, and Great Inagua, Bahamas. **VAGRANT:** Individual flamingos occasionally move north to Gulf or southern Atlantic coastal states, but many extralimital reports may refer to escaped captives. Reported north to KS, MI, ON, QC, NB, and NS. Flamingos in CA, NV, and WA presumably refer to escapes (a variety of species).
POPULATION The species was formerly a locally abundant postbreeding visitor to Florida Bay from Mexican or Greater Antillean breeding grounds; only a small flock remains. Some populations now are increasing with protection. The species numbers some 80,000 in the West Indies; about 40 nonbreeders are semi-resident at Florida Bay.

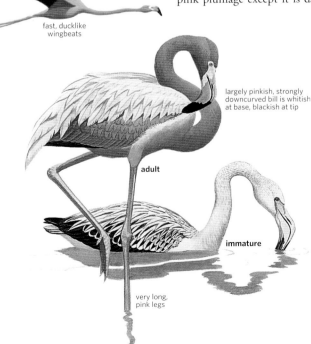

black remiges

fast, ducklike wingbeats

largely pinkish, strongly downcurved bill is whitish at base, blackish at tip

adult

immature

very long, pink legs

ALBATROSSES Family Diomedeidae

Black-footed Albatross (CA, Nov.)

lbatrosses, open ocean birds, spend the majority of their lives at sea, some wandering thousands of miles in search of food. They are characterized by their large size and long, narrow wings; the Wandering and Royal Albatrosses possess the largest wingspans of any bird, measuring just over 11 feet! While the North American species are smaller, all are strikingly large when compared to other seabirds. Although identification can be straightforward, the separation of similar species relies on observation of subtle plumage differences and bill color. Albatrosses live long lives, typically tens of years.

Structure Albatrosses are characterized by their long, narrow wings and thick, heavy bodies. Unlike the procellariids, which have a single tube on the culmen, albatrosses have very large bills with single tubes on either side of the culmen. The bills, made up of individual plates, are sharply hooked. Bill size and coloration can be critical field marks, and bill color often changes with maturation. Albatrosses have large, webbed feet, which sometimes extend past the tail in flight.

Behavior Flight style varies with wind speed and sea condition, but almost always incorporates a large amount of energy-efficient arcing and gliding, especially in stiff oceanic gales. Wingbeats are heavy and labored, and in light winds albatrosses work hard to stay aloft, often sitting on the water in calm conditions. Albatrosses typically feed by surface dipping, alighting on the water and picking prey—including squid, fish, flotsam, offal, and carrion—from the surface. Their acute sense of smell helps in detecting food; many are attracted to fishing boats and chum. Although vocal on breeding grounds, most albatrosses are silent at sea. Most species nest in large colonies on remote oceanic islands; pairs mate for life.

Plumage Albatross plumages are largely variations of black and white, with some having gray or brown tones. The contrasting patterns of black and white are important identification field marks. Some species' plumages change with maturation; others exhibit less variation. Albatrosses show little sexual dimorphism; however, more work is needed. The molt can be complex and protracted, particularly in the larger species; it is not possible to replace all the feathers between breeding cycles, hence the long maturation.

Distribution Albatrosses reach their greatest diversity in the Southern Hemisphere; the doldrums of the equatorial waters presumably represent a barrier to their northward dispersal. In the North Pacific there are three breeding albatrosses: Laysan, Black-footed, and Short-tailed. The Short-tailed Albatross was historically common and is slowly recovering from near extinction. All others occur as vagrants in North America. Other than when they nest, albatrosses are entirely pelagic.

Taxonomy Albatross taxonomy continues to undergo scientific debate. Worldwide 13 to 24 species in two to four genera are recognized. The AOU lists eight species in four genera as occurring in North America: three mollymawks or southern albatrosses, one sooty albatross, one great albatross, and three North Pacific albatrosses.

Conservation BirdLife International lists 16 species as threatened and four as near threatened. Threats include feral predators (esp. rats, pigs, and cats) at nesting sites; long-line fishing at sea; ingestion of floating plastics by adults and the subsequent regurgitation of this material to chicks; and the human harvesting of birds and eggs for food.

MOLLYMAWKS Genus *Thalassarche*

Three of the 11 mollymawk species, medium-size albatrosses of southern oceans, occur as vagrants in North American waters. All are superficially similar in plumage—largely white below with blackish brown backs and upper wings. Important distinctions are the extent of dark margins on the underwings and bill coloration. Superficially all mollymawks resemble an adult Great Black-backed Gull, but are much larger, and have longer wings and larger bills.

SHY ALBATROSS *Thalassarche cauta*   SHAL ▪ 4

This vagrant to the northeastern Pacific resembles an oversize Laysan Albatross. Polytypic (3–4 ssp.; 2–3 in N.A.). L 35–39" (90–99 cm) WS 87–101" (220–256 cm)

IDENTIFICATION A large black-and-white albatross with pale gray mantle. Nominate subspecies *cauta* described. **ADULT:** Head is whitish, with grayer cheeks and dark brow, often appearing white-capped. Its large, yellowish gray bill has a yellow tip. Its wings are dark above and strikingly white below with very narrow dark border and diagnostic "thumbmark"—a small dark spot on the leading edge of the underwing where it meets the body. The mantle is pale gray where it meets the hind neck, turning darker blackish brown distally; also has a white rump and a gray tail. **JUVENILE:** Similar to adult, but its bill is darker grayish blue and has a black tip, and its head is grayer, showing little contrast with mantle. **SUBADULT:** The head becomes whiter with age, typically starting with the nape. The bill often retains a dark tip.

GEOGRAPHIC VARIATION Worldwide three to four subspecies have been described, and they are recognized as full species by some authors. Nominate *cauta* ("White-capped Albatross"), which breeds on islands off Tasmania, is recorded as a specimen from Washington (1951). Sight records from California may pertain to *salvini* ("Salvin's Albatross"), which breeds on the Snares and Bounty Is. off New Zealand and has more extensively dark primaries and a pale gray head, or to *steadi* (not recognized by some authors and nearly identical to nominate *cauta*), which breeds on Auckland, Antipodes, and Chatham Is., New Zealand. Unrecorded *eremita* ("Chatham Island Albatross") is resident around the Chatham Is., New Zealand, and has a dark gray head. Subspecies determination depends on head and bill color and is further complicated by age variation.

SIMILAR SPECIES Compared with Laysan Albatross, Shy Albatross is much larger, with a paler gray mantle, larger gray bill, and diagnostic, dark "thumbmark" on an otherwise much whiter underwing. Longer tail is gray, not blackish.

STATUS & DISTRIBUTION Vagrant to North Pacific. **BREEDING:** Islands off Tasmania and New Zealand. **NONBREEDING:** Disperses east and west from breeding areas across southern oceans, ranging to waters off western S.A. and southern Africa. **VAGRANT:** Casual off Pacific coast from northern CA to AK.

POPULATION Global populations of Shy Albatross were decimated by plume hunters in the 1800s, but they are recovering. Feral pigs threaten some breeding colonies. BirdLife International lists subspecies *salvini* and *eremita* as threatened.

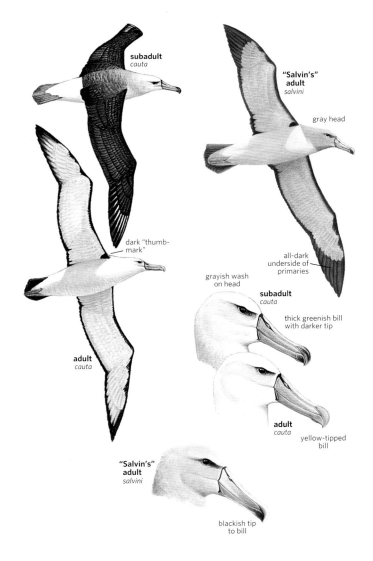

subadult *cauta*

"Salvin's" adult *salvini*

gray head

dark "thumbmark"

all-dark underside of primaries

grayish wash on head

subadult *cauta*

thick greenish bill with darker tip

adult *cauta*

adult *cauta*

yellow-tipped bill

"Salvin's" adult *salvini*

blackish tip to bill

YELLOW-NOSED ALBATROSS *Thalassarche chlororhynchos* YNAL ■ 4

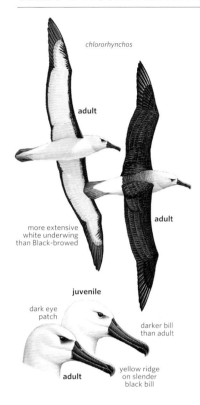

chlororhynchos

adult

adult

more extensive
white underwing
than Black-browed

juvenile

dark eye
patch

darker bill
than adult

adult

yellow ridge
on slender
black bill

The smallest and slimmest mollymawk occurring in North America's Atlantic waters. Polytypic (2 ssp.; nominate in N.A. and *bassi* of the Indian Ocean). L 32" (81 cm) WS 80" (203 cm)

IDENTIFICATION A rather small and slim albatross. ADULT: Gray headed in fresh plumage, otherwise strikingly black-and-white. Upper wings blackish with white primary shafts; back grayer; underwings white, with thin black margins, widest at the leading edge. Gray tail. At a distance, bill appears black; at close range, yellow ridge along the length of the culmen and orange-reddish tip visible. JUVENILE: All dark bill. White head, lacking the adult's grayish wash. May show wider dark margins on the leading edge of the underwing, causing confusion with adult Black-browed Albatross.

SIMILAR SPECIES Black-browed Albatross will cause the most confusion. Adults differ in structure, underwing pattern, and most notably bill coloration. The more heavily built Black-browed has a broader dark leading edge on the underwings and a yellowish bill with an orange tip. Immature Black-browed's dark bill is actually grayish

with a notably dark tip. Juvenile and immature Black-broweds appear gray-headed, often with a grayish collar, turning whiter with age. Juvenile and immature Yellow-noseds appear white-headed, becoming grayer with age. Head patterns can overlap; caution is needed. An immature Northern Gannet has a longer pointed head and tail, and a pointed, not blunt-tipped, bill. Adult Great Black-backed Gulls have also been confused with this species and Black-browed Albatross.

STATUS & DISTRIBUTION BREEDING: Islands in southern oceans; nominate *chlororhynchos* breeds on Tristan da Cunha and Gough Islands. NONBREEDING: Wanders north to subtropical and subantarctic waters of the Atlantic and Indian Oceans. VAGRANT: Casual offshore from the Gulf of Mexico to the Canadian Maritimes; several records from shore or inland. Remarkably, there are a handful of records from onshore or inland where recorded in the interior as far as ON.

POPULATION The global population is apparently stable, but BirdLife International lists *chlororhynchos* as near threatened.

BLACK-BROWED ALBATROSS *Thalassarche melanophris* BBAL ■ 5

The large and stocky Black-browed is a striking bird. The adult's bright yellow-orange bill and dark brow mark give it a fierce countenance. Polytypic (2 ssp.; nominate in N.A.) L 35" (89 cm) WS 88" (224 cm)

IDENTIFICATION Heavily built with thick neck and large-bodied jizz. ADULT: Unique combination of yellow-orange bill with bright orange tip and whitish head with dark brow. Underwings darkest on leading edge, generally characterized by broad dark margins. JUVENILE: Grayish to dark horn-colored bill, typically with dark blackish tip. Grayish head, turning whiter with age, with darkest gray on the hind neck often connecting across the breast to form a partial collar. Underwing often darker than adult; can appear all-dark at a distance. SUBADULT: Bill changing from dark to mostly pale with dark tip. Underwing becoming mostly adultlike and head mostly white with dark brow.

SIMILAR SPECIES See Yellow-nosed Albatross—the major identification problem in the western North

Atlantic. Black-browed is larger and bulkier than Yellow-nosed, with differing underwing pattern and head and bill coloration.

STATUS & DISTRIBUTION Circumpolar in southern oceans. BREEDING: Nominate *melanophris* nests around tip of S.A. on subantarctic oceanic islands; *impavida* off New Zealand. NONBREEDING: North to waters off Peru and off Brazil. VAGRANT: Casual in N. Atlantic Ocean; most records are from European waters. Several

records off the East Coast; one very well documented immature off VA (Feb. 6, 1999).

POPULATION Black-browed is the most abundant albatross worldwide, but BirdLife International lists it as near threatened. Incidental drowning from long-line fishing is the likely cause of local declines.

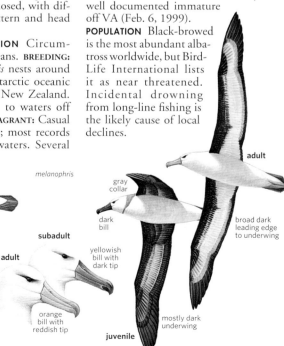

adult

rather thick
orange bill

melanophris

subadult

adult

orange
bill with
reddish tip

gray
collar

dark
bill

yellowish
bill with
dark tip

adult

broad dark
leading edge
to underwing

mostly dark
underwing

juvenile

SOOTY ALBATROSSES Genus *Phoebetria*

LIGHT-MANTLED ALBATROSS Phoebetria palpebrata LMAL ■ 5

This elegant, rakish albatross with long, narrow wings flies with effortless grace. Monotypic. L 31–35" (79–89 cm) WS 72–86" (183–218 cm) **IDENTIFICATION** Very distinctive. **ADULT:** Large. Dark head with prominent white eye crescents. Strikingly pale mantle and body. Long, dark, and wedge-shaped tail. Bluish purple cutting edge to the upper mandible (sulcus). **JUVENILE:** Similar to adult, but with browner plumage, less

prominent eye crescents, and gray or brownish sulcus. **SIMILAR SPECIES** Confusion with Black-footed Albatross is possible, but Light-mantled has a longer tail, prominent white eye crescents, and very different jizz. Related Sooty Albatross (*P. fusca,* unrecorded in N.A.) has an all-dark body and a yellow sulcus. **STATUS & DISTRIBUTION** Circumpolar in southern oceans. **BREEDING:** Subantarctic islands. **VAGRANT:**

Accidental, one record off northern CA (Cordell Bank, July 17, 1994) was well documented with photos. **POPULATION** Global populations appear stable, but losses occur due to drowning from long-line fishing.

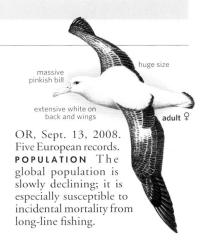

dark head contrasts with pale gray mantle

white eye crescents

long, dark, wedge-shaped tail

adult

GREAT ALBATROSSES Genus *Diomedea*

WANDERING ALBATROSS Diomedea exulans WAAL ■ 5

This huge seabird's wingspan reaches over 11 feet. Polytypic (5–7 ssp.). L 42–53" (107–135 cm) WS 100–138" (254–351 cm) **IDENTIFICATION** Massive pinkish bill and white underwings with narrow dark trailing edge and primary tips. **ADULT:** Extensively white back and wings. **JUVENILE:** Dark chocolate brown with conspicuous white face. **SUBADULT:** Complex plumage maturation takes up to 15 years. Becomes white first on the mantle, body, and head, eventually spreading across the upperwing coverts.

SIMILAR SPECIES Superficially similar to Short-tailed Albatross (see species for details). Very similar to Royal Albatross (*D. epomophora,* unrecorded in N.A.) of Southern Hemisphere. **STATUS & DISTRIBUTION** Circumpolar in southern oceans. **BREEDING:** Subantarctic islands. **NONBREEDING:** Disperses north after breeding, but typically stays south of Tropic of Capricorn. **VAGRANT:** Accidental. Two records in N.A.: one perched on a CA sea cliff, Sea Ranch, Sonoma Co., July 11–12, 1967; one at Perpetua Bank off central

OR, Sept. 13, 2008. Five European records. **POPULATION** The global population is slowly declining; it is especially susceptible to incidental mortality from long-line fishing.

massive pinkish bill

huge size

extensive white on back and wings

adult ♀

NORTH PACIFIC ALBATROSSES Genus *Phoebastria*

Of the four species in this genus, three occur in the North Pacific. Laysan and Black-footed are small for albatrosses, while Short-tailed is much larger. Waved Albatross breeds on the Galápagos Islands and has not been recorded in North America. All take advantage of cold, nutrient-rich waters for foraging.

LAYSAN ALBATROSS Phoebastria immutabilis LAAL ■ 2

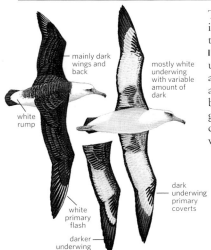

mainly dark wings and back

white rump

mostly white underwing with variable amount of dark

dark underwing primary coverts

white primary flash

darker underwing

The only regular white-bodied albatross in the eastern North Pacific. Monotypic. L 32" (81 cm) WS 79" (200 cm) **IDENTIFICATION** No age-related or sexual differences in plumage, but black areas turn paler with wear. Blackish above; white below. Distinctive dark brow above the eye; auriculars washed gray. Yellowish to pinkish bill with a dark tip. Variably dark underwings with central white area (mostly white

underwings exceptional). Black tail. **HYBRID:** Rare hybrids with a Black-footed Albatross usually appear as darker versions of the Laysan's plumage pattern; often washed gray on the body and head. On darker hybrids the white facial markings of a typical Black-footed are exaggerated, giving the face a prominent white appearance. **SIMILAR SPECIES** Easily confused with the larger and rarer Shy Albatross, which

rather thin pinkish bill with dark tip

dusky face

has a whiter underwing with a dark "thumbmark," a grayish mantle, and a longer, gray tail. Laysan lacks the yellowish wash on the head of an adult or older subadult Short-tailed Albatross. **STATUS & DISTRIBUTION BREEDING:** Atolls in the northwestern HI islands and on Kauai; recently established in small numbers in the Revillagigedo Archipelago and Isla de Guadalupe, Mexico. Regular to cold AK waters in summer. **NONBREEDING:** Moves mostly north and east into the Northern Pacific Gyre; found widely scattered

in the eastern Pacific. Most commonly observed from CA to WA in deep water Sept.–Feb. **VAGRANT:** Casual in spring in desert Southwest (southeastern CA and southwestern AZ), where (presumably) individuals traveling north up the Gulf of California continued overland. **POPULATION** Stable and likely increasing, recovering from plume hunting and other human exploitation at the end of the 19th century. Threats include introduced predators at breeding colonies and the ingestion of plastic that is regurgitated to nestlings.

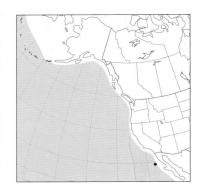

BLACK-FOOTED ALBATROSS *Phoebastria nigripes*   BFAL ◾ 1

The Black-footed is the only regularly occurring dark albatross in the eastern North Pacific. Monotypic. L 32" (81 cm) WS 80" (203 cm)
IDENTIFICATION Typically appears overall dark brown at all ages and seasons; worn individuals can have paler brown wing coverts. Variable amounts of white at bill base and on uppertail and undertail coverts, occasionally onto the lower belly. Black tail. Grayish black bill, sometimes paler with yellow or pink tones. **ADULT:** Older bird has a whiter face, sometimes even appearing white headed due to wear and sun bleaching, and has more white on the undertail coverts. **JUVENILE:** Like adult, but generally darker in fresh plumage with less white at bill base. White typically restricted to uppertail coverts. **HYBRID:**

variable white in uppertail coverts

darkish bill

white primary flash

dark underwing

some with pale head and under tail

older adult

Rarely with Laysan Albatross (see that species).
SIMILAR SPECIES A mostly dark-plumaged juvenile or young subadult Short-tailed Albatross differs from a Black-footed by its much larger size and large pink bill. An exceptionally pale, adult Black-footed has been confused with a subadult Short-tailed, but note the different distribution of white plumage areas and the Short-tailed's pink bill.
STATUS & DISTRIBUTION Uncommon to locally common year-round in northeast Pacific waters. **BREEDING:** Atolls in the northwestern HI islands and on islands off Japan. **NONBREEDING:** Disperses widely during summer. Moves largely north and east in the

Northern Pacific Gyre, reaching the Aleutians. Rare in Bering Sea. Most numerous in deep water off CA, OR, and WA from June–Aug.; some occur year-round.
POPULATION The global population is declining, threatened by human disturbance, habitat loss, introduced predators, long-line fishing, and oil pollution.

SHORT-TAILED ALBATROSS *Phoebastria albatrus*   STAL ◾ 3

A prize among pelagic birds of the West, this large albatross with its striking "bubblegum pink" bill is becoming more regular in North American waters. Once nearly extinct, the species has partly recovered in recent years

and has been found with increasing regularity in North America, particularly off the coast of Alaska. Our most variably plumaged species, the Short-tailed Albatross—sometimes known as "Steller's Albatross"—has

a succession of age-related plumages similar to the great albatrosses found in the southern oceans. It is obviously larger and more robust than Black-footed Albatross, with which it often appears in association off the West

Coast. Most recent records off the West Coast are of chocolate brown juveniles, especially away from the Aleutians, AK. Monotypic. L 36" (91 cm) WS 87" (215–230 cm)

IDENTIFICATION Where it occurs with other North Pacific species, the Short-tailed's large size, massive pink bill with pale bluish tip, and dark humerals are distinctive in all plumages. **ADULT:** Plumage pattern similar to that of great albatrosses of Southern Hemisphere. Largely white both ventrally and dorsally with golden wash on head. White scapulars and upperwing coverts contrast with black humerals, primary coverts, and outer greater upperwing coverts creating a somewhat pied pattern on the upper wing. Underwings white with narrow dark margin. **JUVENILE:** Recently fledged juvenile has dark bill with traces of pink. Timing of color change uncertain (possibly within first two months at sea), but bill soon pink. Fresh plumage wholly dark brown with small variable white around eye and below bill, but often appearing cleanly dark-headed. Worn, older juveniles more white-headed overall. Underwings dark. **SUBADULT:** Slow change (up to ten years) from the mostly brown juvenile-like plumage of younger subadult, to the whiter adultlike plumage of older subadult. Scapulars, upperwing coverts, and back change from brown to white as bird ages. Underwing changes from all dark to white. Older subadult can appear nearly adultlike but often retains dark nape patch on the otherwise unmarked yellowish head.

SIMILAR SPECIES Adult Short-tailed is not likely to be confused with any other North Pacific albatross. The combination of dark underwings in juvenile and early subadult plumages and smaller size separate it from a Wandering Albatross, though it's unlikely the species occur in the same waters. Recently fledged juveniles with dark bills are very similar to Black-footed Albatross, and older juveniles and younger subadults are superficially similar. Note the bright pink bill on the older birds, visible at long range. On all ages, note much larger size. Black-footed Albatross always lacks white on the upper wing unlike the pied upper wing of a subadult Short-tailed. Hybrids between Black-footed and Laysan Albatrosses can resemble a Short-tailed; however, the hybrids are smaller and lack white on the upper wing.

STATUS & DISTRIBUTION Rare but increasing. **BREEDING:** Historically multiple islands off Japan; currently restricted to Torishima and Minami-kojima. Single pairs have occurred in albatross colonies on Midway and have bred there. **NONBREEDING:** Disperses to surrounding waters and follows the trade winds to the northeast Pacific and southern Bering Sea and rarely farther south off the west coast of N.A.

POPULATION Hunting by Japanese plume hunters and several volcanic eruptions severely reduced this species by 1900; additional exploitation pushed it to the brink of extinction by the early 1930s. Now fully protected on their breeding islands—although eggs and chicks still suffer predation by introduced rats, and future volcanic eruptions threaten nesting habitat—the Short-tailed population is slowly recovering; thus the higher percentage of juveniles and subadults than in a stable population. The population has rebounded to around 4,000 individuals in 2013. Records off the west coast of North America are becoming more regular; historically, large numbers of Short-taileds occurred during the nonbreeding season from the Aleutians to California.

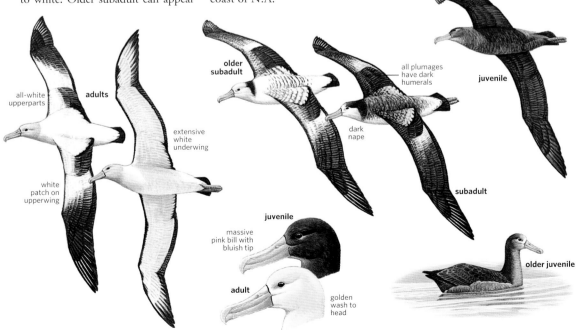

SHEARWATERS AND PETRELS Family Procellariidae

Pink-footed Shearwater (CA, Oct.)

The identification of shearwaters and petrels—the procellariids—is challenging. Almost all North American observations are made at sea, where rocking boats, great distances, and generally drab plumages can exasperate even an experienced seabirder. Birders should focus on the light-dark plumage patterns and the body proportions. Distant identifications of shearwaters and petrels must often be based on a bird's unique flight style and shape, and these identification skills are gained only through experience.

Structure Heavy bodies, short tails, and long, narrow wings are the standard for procellariidae. Their bills are made up of individual plates and are sharply hooked. The nostrils are situated in tubes on the upper surface (culmen) of the bill—hence the term *tubenoses.* Both bill size and color can be useful field marks. Shearwaters and petrels have webbed feet.

Behavior Shearwaters and petrels can be found singly or in large groups, with the shearwaters being more prone to flocking behavior than the others. Their flight is usually a series of wingbeats interspersed with stiff-winged glides. The speed, duration, and depth of the wingbeats and the frequency and duration of the glides define a species' flight style. Body and wing proportions are also major variables. Wind speed and a bird's activity will alter its flight style. In traveling flight, the bird's feet are often hidden, tucked into the belly feathers. Feeding styles vary from surface dipping to deep diving. Prey items range from small copepods to squid and fish, and many species are attracted to fishing boats and chum. Their acute sense of smell helps them to detect food. Procellariids are generally silent at sea.

Plumage Procellariids appear monochromatic. Their plumages are generally variations on black, gray, brown, or white; some are tinged with blue. Male, female, and immature plumages look very similar. Many species have dark upperparts and light underparts. Some species have distinct color morphs (e.g., light or dark), and several species have individuals that show intermediate characteristics between light and dark, further complicating identification. All undergo a single annual molt after breeding. In our spring, most austral breeders will be noticeably in molt; fresh-plumaged birds of the same species are likely to be juveniles. Subadults and non-breeders may molt earlier than breeders, complicating the issue. When worn and bleached, dark colors become paler and browner. Light conditions also affect color perception.

Distribution Shearwaters and petrels reach their greatest diversity in the Southern Hemisphere, but they inhabit all oceans. With few exceptions they come ashore only to breed; most nest in colonies on remote oceanic islands. Many nest in burrows underground. Only two procellariids—the Northern Fulmar and the Manx Shearwater—breed in North America; the others only migrate through or spend their nonbreeding season in North American waters. A number of species are casual or accidental. Species in this family are known to wander, and many records of wayward individuals come from both the Pacific and the Atlantic Oceans.

Taxonomy Taxonomy continues to undergo revision, and ornithologists often differ on species limits, especially in the genera *Pterodroma* and *Puffinus.* Worldwide between 70 and 80 species in 12 to 14 genera are recognized. The American Birding Association lists 30 species in six genera as occurring in North America. Informally, they are often subdivided into three groups: fulmars (1 sp.), gadfly petrels (14 sp.), and shearwaters (15 sp.).

Conservation Many nesting sites are threatened by feral predators, especially rats, pigs, and cats. Long-line fishing drowns birds at sea. The human harvesting of birds and eggs for food is a threat to some species.

FULMARS Genus *Fulmarus*

NORTHERN FULMAR *Fulmarus glacialis*   NOFU ▪ 1

The Northern Fulmar is the stocky, familiar seabird of winter pelagic trips off both coasts. In the Atlantic and in the high Arctic, light-morph birds predominate and appear gull-like in plumage. Along the West Coast, dark-morph birds are more common and are often confused with dark shearwaters and petrels. Polytypic (3 ssp.; all in N.A.). L 19" (48 cm) WS 42" (107 cm)
IDENTIFICATION Polymorphic, appearing in light, dark, and intermediate plumages. Key features are its stocky build, very narrow wings, and acrobatic flight style in strong winds. Ages and sexes similar. **LIGHT MORPH:** White below and gray dorsally; white head with dark eye smudge. Pale primary bases appear as wing patch on upper wing. Rump white or gray. Tail gray. Underwing mostly white. **DARK MORPH:** Plumage wholly chocolate brown to brownish gray, often with darker smudge around eye. Underwing coverts concolorous with body, but primaries appear paler. Upperwing may or may not have pale primary patch. **INTERMEDIATE:** Any shade of color between dark brown and white. Most often appears light, pearly gray with darker upperwing. Blotchy birds often seen on West Coast. **FLIGHT:** Flies on stiff, outstretched wings with little bend at the wrist, arcing high off the water in strong winds. In light winds, it flies direct and low to water; its stiff, short wingbeats alternate with accomplished glides.
GEOGRAPHIC VARIATION The polymorphic *rodgersii* of the Pacific has a darker tail that contrasts with the rump and a more slender bill (on average). The nominate *glacialis* describes Atlantic high-arctic breeders, predominately the light morph; the polymorphic *auduboni* describes

intermediate morph
rodgersii

light morph
rodgersii

white flash on inner primaries

Pacific birds have dark tails that contrast with paler rump

variable, some individuals even paler

light morph
rodgersii

Pacific birds' plumage more variable than Atlantic birds

Atlantic dark morph
glacialis

heavier bill than Pacific birds

dark morph
rodgersii

uniformly dark

Atlantic light morph
glacialis

tail blends with rump in both Atlantic morphs

heavy yellowish bill

light morph
rodgersii

Atlantic populations breeding in low Arctic and boreal areas. Both Atlantic subspecies have tails similar in color to the rump and the uppertail coverts.
SIMILAR SPECIES Light morphs are similar to large gulls. Note the fulmar's stocky overall shape; relatively short, narrow wings; and complicated bill structure with multiple plates and tube on the culmen. Dark morphs can be confused with dark shearwaters, which have thinner, darker bills and longer, broader wings. Confusion with dark *Pterodroma* petrels is a problem, but note the small, stocky, dark bills of those species and their more dynamic flight on longer, crooked wings.
VOICE Generally silent at sea when alone; feeding flock can be noisy. At breeding colonies, it makes a variety of cackling and croaking sounds.
STATUS & DISTRIBUTION Common.

BREEDING: Breeds on sea cliffs along the coasts of the north Pacific and north Atlantic as well as on high-arctic islands. **NONBREEDING:** Moves south in most years to winter off the coast of New England and off the West Coast south to Mexico. Numbers fluctuate annually, and in some invasion years hundreds are visible from shore along the West Coast. Found year-round off the West Coast, but typically few are present during the summer months.
POPULATION Increasing in the Atlantic and stable in the Pacific. Breeding colonies are threatened by introduced predators, pesticides, and oil pollution.

rodgersii   *glacialis*

PETRELS Genus *Pterodroma*

This wide-ranging genus includes 31 species worldwide, of which ten occur in North America. In many species, a notable M-pattern appears across the upperwing, formed by dark primaries, primary coverts, greater upperwing coverts, and in some cases rump and uppertail coverts. Their unique flight style is striking at sea, typically when they cover large tracts of ocean on dynamic, high arcs over the sea. Most nest in burrows underground on oceanic islands, spending the majority of their lives at sea.

GREAT-WINGED PETREL *Pterodroma macroptera*   GWPE = 5

This vagrant gadfly petrel from the southern oceans has been recorded casually off California. Polytypic (2 ssp.). L 16" (41 cm) WS 38" (97 cm)
IDENTIFICATION Subspecies *gouldi*, which the California birds were thought to be, is described and illustrated. Large for a petrel; wholly dark except pale gray around bill and on chin and throat. Overall uniform blackish brown, but can appear darker on head and breast; thick, stout, black bill. Ages and sexes alike,

primary flash fainter than Murphy's

uniform dark underwing

white evenly distributed around bill

gouldi

overall color browner, less gray, than Murphy's

but dark brown fades paler with wear. **FLIGHT:** More languid wingbeats than other Pterodromas, but still performs dynamic arcs in strong winds bounding high above the ocean's surface.
SIMILAR SPECIES Most similar to Murphy's Petrel, which is smaller overall with quicker wingbeats. Murphy's has prominent pale-based flight feathers and grayer plumage. Pale feathering around the bill base is more restricted to area below bill on Murphy's. Murphy's legs and feet are pink, not black. Providence

Petrel is very similar to Great-winged but has more prominent whitish primary bases, dark-tipped greater primary coverts, and a subtle M-pattern across the upperwing.
STATUS & DISTRIBUTION BREEDING: Oceanic islands from Australasia to S. Atlantic; *gouldi* off New Zealand. **NONBREEDING:** Movements poorly understood; in general disperses to surrounding waters. **VAGRANT:** Casual off central and southern CA in late summer and fall (about 5 records).
POPULATION Populations largely stable or increasing. Taken for food by native Maoris; commercial harvesting has been banned. Threatened at breeding colonies by introduced cats and rats.

PROVIDENCE PETREL *Pterodroma solandri*   PRPE = 4

This gadfly petrel performs an annual transequatorial migration from breeding areas in Australia to the North Pacific during the boreal summer. Monotypic. L 17.5" (44 cm) WS 41" (104 cm)
IDENTIFICATION A medium-size, mostly brownish, stocky gadfly petrel. Relatively heavy bill and broad wings. Upperparts dark gray-brown overall, with mantle, greater coverts, and rump more grayish, contrasting with remainder of upperparts to create dark carpal-ulnar M-pattern; whitish plumage encircling bill. Underwing dark brown except for silvery white bases of primaries with distinct dark tips to the pale-based primary coverts (the best field mark).

SIMILAR SPECIES Murphy's Petrel is smaller, with shorter, narrower wings, smaller bill, and different underwing pattern; in Murphy's, the white in the face tends to be most extensive in the throat, less so above the bill, the reverse of Providence.
STATUS & DISTRIBUTION BREEDING: Lord Howe I. and Phillip I., off Norfolk I. Adults begin to arrive in late February; last fledglings depart in late November. **NONBREEDING:** Dispersal northward into the central North Pacific, where mostly recorded Sept.–Apr. In N.A. waters, ranges north into

dark crescent formed by tips of greater primary coverts

white around bill most prominent on lores

heavy bill

western Bering Sea: 11 were studied during research cruises in the Aleutians Aug. 4–25, 1994, and 25 were documented Sept. 15, 2011, close to Russian waters, where previous records (e.g., 18 on Sept. 23, 2006) indicate that the species is likely regular. A probable Providence Petrel was photographed 32 miles west of Tofino, BC (Oct. 6, 2009).
POPULATION About 32,000 pairs remain; listed as vulnerable. Formerly nested on Norfolk I.; extirpated by 1800.

HERALD PETREL *Pterodroma arminjoniana*   HEPE = 3

This gadfly petrel is rare but regular in warm Gulf Stream waters off the East Coast. A medium-size and polymorphic petrel, the Herald lacks the strong M-pattern on the upperwing shown by many congeners: Its upperparts are generally a smooth velvety brown. The Herald's dark morph is more commonly observed in North America. Polytypic (2 ssp.; nominate

in N.A.). L 15.5" (39 cm) WS 37.5" (95 cm)
IDENTIFICATION Polymorphic; the three morphs vary mainly in body color; all lack bold contrasts across the upperwing. Underwing coloration is variable, but generally matches morph type (e.g., dark morph has darker underwings overall). Flight feathers whitish with dark trailing edge; underwing

coverts largely blackish often with paler lesser coverts (usually lacking on dark morphs). Upperwing brown, often with subtle, darker brown M-pattern (usually absent on dark morph). **LIGHT MORPH:** White below and brownish gray above; brownish head and neck sides often give dark-hooded look. Flanks dusky brownish; belly and undertail coverts bright white. **DARK MORPH:** Plumage

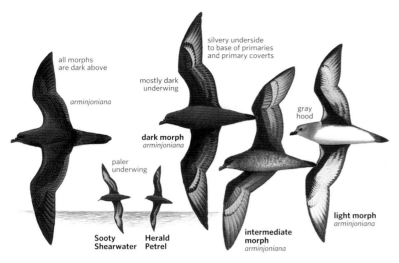

all morphs are dark above

arminjoniana

silvery underside to base of primaries and primary coverts

mostly dark underwing

dark morph
arminjoniana

gray hood

paler underwing

Sooty
Shearwater

Herald
Petrel

**intermediate
morph**
arminjoniana

light morph
arminjoniana

wholly chocolate brown to brownish gray, often with darker head and upper breast. Underwing coverts concolorous with body, but primary bases flash white below due to pale bases of primaries and greater primary coverts. **INTERMEDIATE:** Ventrally the body can be a mix of dark brown and white, generally palest on belly. **FLIGHT:** Acrobatic and agile, arcing high above the water on stiff set wings, angled back at the wrist. Direct flight low with stiff, measured flapping. **GEOGRAPHIC VARIATION** Pacific *heraldica* given full-species status by many authors; also known as the "Trinidad Petrel." **SIMILAR SPECIES** Light morph similar to Fea's Petrel, but it lacks the bold upperwing pattern and pale tail of Fea's. The dark morph can be confused with Sooty Shearwater, but it has white restricted to primary bases on the underwing and lacks the pale underwing coverts of the Sooty. **STATUS & DISTRIBUTION** Rare visitor to Gulf Stream. **BREEDING:** Nests on islands off Brazil: Trinidad and Martin Vaz Is. **NONBREEDING:** Disperses to surrounding waters. Rare but regular as far north as VA in deep offshore waters from May–Sept. **VAGRANT:** Accidental inland after hurricanes as far north as PA and NY. **POPULATION** Small population size and limited breeding range make nominate *arminjoniana* especially susceptible to drastic declines; BirdLife International considers it vulnerable.

MURPHY'S PETREL *Pterodroma ultima*   MUPE ▪ 3

Murphy's Petrel is the dark *Pterodroma* most likely to be observed off the West Coast. Although it superficially resembles dark shearwaters and dark-morph Northern Fulmars, Murphy's Petrel flies on stiff, angled wings with dynamic agility. Monotypic. L 16" (41 cm) WS 38" (97 cm)

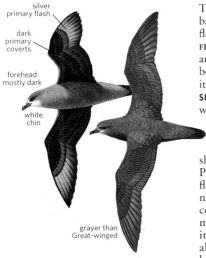

silver primary flash

dark primary coverts

forehead mostly dark

white chin

grayer than Great-winged

IDENTIFICATION Small, dark, thick-necked, and large-headed. Bill is all dark, slender for a *Pterodroma,* and short. Plumage is overall gray-brown, with a subtle dark M-pattern across upperwing. It appears white-throated, with a few white feathers on the forecrown. The upperwing is grayish brown, concolorous with upperparts. The underwing is dark, with white bases to primaries showing a white flash below. Ages and sexes are alike. **FLIGHT:** Murphy's flies with dynamic arcs above the horizon with stiff wings bent back at the wrist. In light winds, its flight is low with stiff wingbeats. **SIMILAR SPECIES** Similar to Great-winged Petrel; see that species account for details. Providence Petrel, breeding in the South Pacific, is also very similar and should be considered. The larger Providence shows a prominent white flash in the underwings and prominent dark tips to the greater primary coverts. Providence has a bigger, more bulbous bill, and the white on its face is more evenly distributed above and below the bill (mostly below in Murphy's).

STATUS & DISTRIBUTION Rare visitor to West Coast. **BREEDING:** Breeds on islands in the central south Pacific. **NONBREEDING:** Poorly known; thought to disperse to tropical Pacific waters, but reaches waters off CA; mainly seen over deep water mid-Apr.–May; a few later into summer. **VAGRANT:** Offshore north to OR. **POPULATION** Little is known about breeding populations and nesting habits. BirdLife International considers the Murphy's Petrel as near threatened.

MOTTLED PETREL *Pterodroma inexpectata* MOPE ▪ 2

An elusive gadfly petrel of the north Pacific, this species is most easily identified by its striking wing pattern and its diagnostic gray belly. Given any reasonable view, this species is one of few *Pterodroma* petrels that can be considered easy to identify. Monotypic. L 14" (36 cm) WS 32" (81 cm)

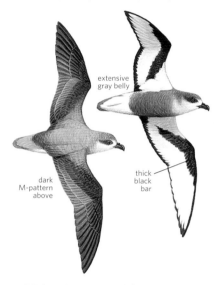

extensive gray belly

dark M-pattern above

thick black bar

IDENTIFICATION Medium-size and agile; largely white below; bold black ulnar bar across the underwing; diagnostic grayish black belly patch contrasts with the white chest and paler undertail coverts. Can appear largely gray below; white throat and restricted white upperbreast; dark neck sides connect with gray belly and flanks. Wear and molt can make dark belly somewhat paler. Underwing white; dark outer leading edge and black ulnar bar across the underwing coverts extend to the axillaries; dark trailing edge. Upperwing gray, with black M-pattern. Tail gray with white outer tail feathers. Gray head; dark eye smudge; stout black bill. Juvenile has pale-tipped upperparts, imparting a more scaled appearance. **FLIGHT:** Arcs high over the water on stiff, crooked wings bent at the wrist. **SIMILAR SPECIES** Dark belly patch contrasting with white chest and undertail is unique among *Pterodroma* species. Smaller Cook's Petrel lacks the contrasting pattern below, appears largely white on the underwings, and has a bolder dark M-pattern across the upperwing.

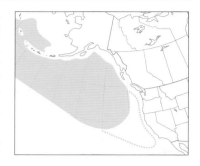

VOICE Generally silent at sea.
STATUS & DISTRIBUTION BREEDING: Nests on islands off New Zealand. NONBREEDING: Transequatorial migrant. Spends nonbreeding season in the north Pacific; many off the Aleutians and in the Gulf of Alaska (mostly May–Oct.). Rare and irregular farther south far off the West Coast (mostly mid-Nov.–mid-Apr.). Many have been found onshore, dead or exhausted. VAGRANT: Remarkable specimen rec. in upstate NY (Apr. 1880).
POPULATION Considered near threatened by BirdLife International.

HAWAIIAN PETREL *Pterodroma sandwichensis* HAPE ▪ 4

This rarely encountered gadfly petrel was recently split from the very similar Galápagos Petrel *(P. phaeopygia)*. Both taxa were previously classified as a single species, the Dark-rumped Petrel. Monotypic. L 17" (43 cm) WS 39" (99 cm)
IDENTIFICATION Medium-sized; sleek; elegant pattern of clean white below, black above. Blackish gray, uniform upperparts. Dark tail and rump; occasionally flecked white on uppertail coverts. Head largely dark; white forehead; dark neck sides form partial collar. Underwing largely white; darker

flight feathers; restricted black ulnar bar. Axillaries white; dark patch visible when arcing high into the wind. Ages and sexes alike. **FLIGHT:** Strong flight on long wings bent back at wrist; quick deep flaps followed by high arcs above the sea.
SIMILAR SPECIES Very similar in the field to Galápagos Petrel. Note Hawaiian Petrel's lighter build and cleaner white underparts; white on the face extends up in a point behind the ear coverts.
VOICE Generally silent at sea. At breeding colonies, gives several call types including a resonant *A'uuuuu'A'uu'A'uu'A'* presumably the origin of their Hawaiian name, the "'Ua'u."
STATUS & DISTRIBUTION BREEDING: Once nested on all the Hawaiian Is. Now primarily restricted to small colonies on Maui and Hawaii; small numbers possibly remain on three other islands. NONBREEDING: Some disperse north and east, where small numbers are seen annually off CA; has occurred as far north as BC.
POPULATION Small global population estimated at 5,000 pairs. Decimated

by habitat loss and predation by introduced predators (inc. humans) and feared extinct until rediscovery in 1948. Designated as vulnerable by BirdLife International and as endangered by U.S. government.

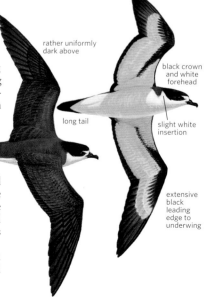

rather uniformly dark above

black crown and white forehead

long tail

slight white insertion

extensive black leading edge to underwing

COOK'S PETREL *Pterodroma cookii* COOP ◼ 3

Cook's Petrel is a small, sharply patterned gadfly petrel of the Pacific. It is often grouped with the Stejneger's Petrel and several other very similar extralimital species that are referred to collectively as Cookilaria petrels. Monotypic. L 10.8" (27 cm) WS 26" (66 cm)

IDENTIFICATION Combination of small size, pale gray upperparts with striking dorsal M-pattern, and clean white underparts is unlike most other *Pterodromas*. Tail gray above with dark, central tip; white outer tail feathers difficult to see under

field conditions. Head largely grayish; dark eye patch; white throat. Clean white underparts, including underwing; thin, dark mark at wrist. Ages and sexes alike. **FLIGHT:** Dynamic but erratic compared to other *Pterodromas,* with quick direction changes.

SIMILAR SPECIES In N.A., Stejneger's Petrel is the most similar, but it is darker above with a less distinct M-pattern and a darker head, imparting a capped or hooded appearance. **VOICE** Generally silent at sea. **STATUS & DISTRIBUTION** BREEDING: Nests on three islands off New Zealand. NONBREEDING: Moves northeast to eastern Pacific. It is rare in deep water well off the West Coast from spring through fall (fairly common some summers). VAGRANT: Accidental in Alaskan waters. Has been found several times at the Salton Sea in mid-summer (presumably birds traveling north across the desert from north end of the Gulf of California).

POPULATION Global population threatened from predation of

introduced mammals and from a native rail, the Weka. Removal of cats and Wekas from some breeding islands has produced promising results. BirdLife International designates it as endangered.

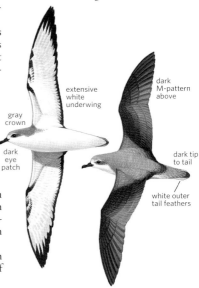

extensive white underwing

dark M-pattern above

gray crown

dark eye patch

dark tip to tail

white outer tail feathers

STEJNEGER'S PETREL *Pterodroma longirostris* STPE ◼ 4

Similar to Cook's Petrel in its overall appearance, the dark-capped Stejneger's Petrel has been recorded just a few times off the West Coast. The first North American record was in 1979, off California. Monotypic. L 11" (28 cm) WS 23" (58 cm)

IDENTIFICATION A small, slim *Pterodroma* showing a dark M-pattern on its gray wings and upperparts, with clean white underparts. It has a notably black-capped (or half-hooded) head; the black half-hood encompasses the eye and contrasts strongly with the gray mantle and pale forehead. Normally the dark half-hood is very noticeable, but it can be difficult to see on birds flying away. The white of the throat hooks up slightly behind the eye, accentuating an intrusion of dark feathers just behind it. The tail is long and appears uniformly dark; the white in the outer tail feathers is restricted and difficult to see under field conditions. Underwings are clean white, with a thin, dark mark at wrist (similar to the Cook's Petrel). Ages and sexes are alike. **FLIGHT:** Dramatic and bounding, Stejneger's Petrel arcs high off

the water in strong winds. It is prone to erratic maneuvers, though somewhat less so than Cook's Petrel. **SIMILAR SPECIES** Similar to Cook's Petrel and to other unrecorded Cookilaria petrels. When comparing it to Cook's Petrel, note Stejneger's black cap, its darker gray upperparts, its blacker outer wing, and its darker tail. Structurally, Stejneger's Petrel is slightly smaller and shorter-winged. **VOICE** Generally silent at sea. **STATUS & DISTRIBUTION** Vagrant to North America. BREEDING: Nests on the Juan Fernandez Is. off Chile. NONBREEDING: Disperses to surrounding waters with movements north of the Equator. It has been suggested that Stejneger's Petrel makes a loop migration through the northern Pacific, passing through waters east of Japan as it moves north, and through the eastern Pacific (closer to our West Coast) as it moves south in Oct. and Nov. VAGRANT: Casual off CA; fewer than ten records from midsummer through fall (mostly Nov.) in deep offshore waters. More regular much farther offshore, beyond the 200-mile limit, but confusion with

similar species confounds status. Also a remarkable specimen record from coastal TX: one salvaged at Port Aransas Sept. 15, 1995.

POPULATION Population likely stable. Threatened at breeding colonies, mostly by feral cats. BirdLife International has designated Stejneger's Petrel as vulnerable.

extensive white underwing

dark cap and nape

more uniform tail than Cook's

BERMUDA PETREL *Pterodroma cahow* BEPE ■ 3

Once thought to be extinct, this endangered species (also known as the "Cahow") was rediscovered breeding on islets off Bermuda. In N.A. it has been recorded only from Gulf Stream waters off North Carolina. In recent years, records (first recorded in N.A. in 1996) have increased. Its rebounding population may account for this, but increased observer awareness is also a factor. Monotypic.

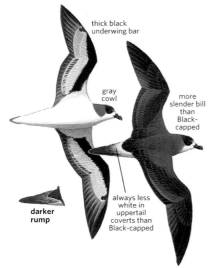

thick black underwing bar

gray cowl

more slender bill than Black-capped

darker rump

always less white in uppertail coverts than Black-capped

L 15" (38 cm) WS 35" (89 cm)

IDENTIFICATION Distinctive small, slim structure and dark-hooded appearance (typically lacking pale collar). Generally dark across the back, sometimes blackish with faint M-pattern when fresh. Pale uppertail coverts usually restricted. Dark blackish gray head, typically lacking white collar; pale throat and dark neck sides. Underparts white. Underwing white, with dark flight feathers and bold dark ulnar bar diagonally across the underwing coverts. Ages and sexes similar. **FLIGHT:** Dynamic high arcing flight typical of gadfly petrels.

SIMILAR SPECIES Most similar to Black-capped Petrel in plumage, but lacks that species' distinct white collar and bright white uppertail coverts; generally lacks the white-tailed appearance of that species when viewed dorsally. Head darker than Black-capped's, with a more cowled appearance. In flight more like the smaller western Atlantic gadflies; overall smaller and slimmer than Black-capped.

VOICE Generally silent at sea; gives eerie nocturnal flight calls at breeding colonies.

STATUS & DISTRIBUTION Casual in N.A. **BREEDING:** One of the rarest seabirds in the world, breeds only on islets off

Bermuda. **NONBREEDING:** Disperses to surrounding waters. **VAGRANT:** Casual in warm Gulf Stream waters off NC mainly May–Aug.; accidental MA (June 28, 2009), but recent satellite tagging indicates that Bermuda Petrels may regularly range as far north as waters off Atlantic Canada and even Ireland.

POPULATION Slaughtered for food during the 17th century, leading to its near extinction. Rediscovered—after a 300-year absence—breeding on islets off Bermuda. World population now estimated to be about 500 birds (2012). Protection of colony on Nonsuch I., Bermuda, largely responsible for this species' recovery. Considered endangered by BirdLife International.

BLACK-CAPPED PETREL *Pterodroma hasitata* BCPE ■ 2

Black-capped Petrel is the most commonly occurring gadfly petrel in North America (found regularly in warm Gulf Stream waters off the East Coast north to VA). One can expect to encounter numbers of these petrels on a typical pelagic trip off Cape Hatteras in May to October. Its highly acrobatic flight style is distinctive, and its bright, white plumage details make it identifiable at

long range. Polytypic (2 ssp.; nominate in N.A.). L 16" (41 cm) WS 37" (94 cm)

IDENTIFICATION Large and stocky; with dark blackish upperwings, a bold white collar (sometimes subdued or brownish), and a very prominent and broad white rump band. Upperwings dark blackish with indistinct, dark M-pattern. "White-faced" birds have a mostly white head with a smaller black cap, broad white collar, and a thin, white supercilium that makes the dark eye stand out on the face; "dark-faced" birds have a more extensive cap that surrounds the eye and usually a brownish collar; intermediate types occur. The two types may represent different breeding populations—more study is needed. Large, stocky, dark bill. White below, with white underwings and variable dark ulnar bar across the underwing coverts. Ages and sexes similar; molt can cause unusual splotching on dark upperwing. **FLIGHT:** Dynamic, arcing high above the water; rarely flapping

on stiff wings angled back at wrist. Arcs higher above the water than shearwaters.

GEOGRAPHIC VARIATION Two subspecies generally recognized, but some authorities consider them separate species: nominate restricted to breeding on Hispaniola, with the largest historic colonies in Haiti; recent information

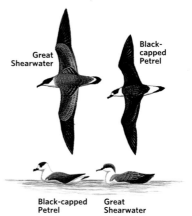

Great Shearwater

Black-capped Petrel

Black-capped Petrel

Great Shearwater

suggests possible breeding in Cuba and other small Caribbean islands. The all-dark *caribbaea* ("Jamaican" Petrel), bred only on Jamaica; it was last seen in the 1800s and is now feared extinct. **SIMILAR SPECIES** Most similar to Bermuda Petrel, but larger and stockier with bold white collar and broad white band across rump and uppertail coverts. **VOICE** Generally silent at sea. **STATUS & DISTRIBUTION** Common in restricted range. **BREEDING:** Mainly on Hispaniola. **NONBREEDING:** Common over deep Gulf Stream waters off NC, May–Oct.; rare in winter. Perhaps regularly ranges to VA and FL. Disperses to surrounding waters north of the Gulf Stream, rarely to the Canadian Maritimes, and south to Brazil. **VAGRANT:** Casual inland to the Great Lakes and other large bodies of water after hurricanes. **POPULATION** Global population dwindling; likely just a few thousand breeding pairs persist. Breeding areas in Haiti decimated by habitat destruction, with

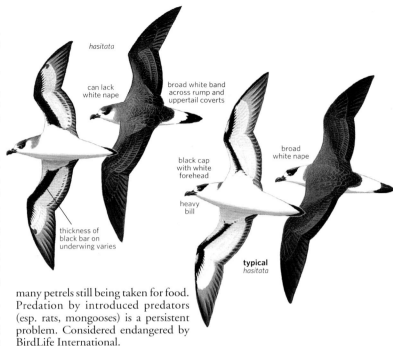

many petrels still being taken for food. Predation by introduced predators (esp. rats, mongooses) is a persistent problem. Considered endangered by BirdLife International.

FEA'S PETREL   *Pterodroma feae*   FEPE ▪ 3

Part of a confusing species complex (including Zino's Petrel), this infrequent visitor to Gulf Stream waters is one of the more strikingly plumaged tubenoses in the region. Polytypic (2 ssp.; both in N.A.). L 14" (36 cm) WS 37" (94 cm)
IDENTIFICATION Small gadfly with bold dorsal M-pattern; blackish underwings; pale uppertail and rump. Upperwing grayish, often with distinct darker gray M-pattern; some

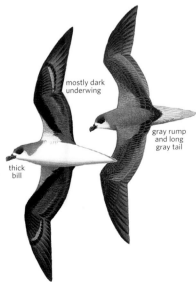

have brownish tinge above. Head generally dark, imparting a grayish hooded appearance; throat pale. Tail distinctive: grayish to occasionally whitish above; pale rump typically paler than mantle. White below; dark gray sides of neck typically do not connect in breast band. Ages, sexes similar. **FLIGHT:** Typical gadfly style, arcing high over the water in strong winds; direct flight on stiff, clipping wingbeats.
GEOGRAPHIC VARIATION Subspecies taxonomy unclear; recent information suggests that the two currently recognized taxa might be separate species. Nominate *feae* nests on the Cape Verde Is.; the slightly larger-billed *deserta* nests on Bugio I. in the Desertas Is. (part of the Madeira Archipelago). Fea's Petrel may also nest in the Azores.
SIMILAR SPECIES Unlike any other species in the western Atlantic, but similar to the accidental Zino's Petrel (*P. madeira*). Zino's Petrel (one of the world's rarest seabirds, endemic to the island of Madeira) is smaller overall, notably so in the field; but perhaps most importantly, it is smaller billed. Zino's may also show extensive white on the underwing, unlike Fea's. A Zino's Petrel was well photographed off Hatteras, NC (Sept. 16, 1995) and was recently (Dec. 2013) accepted by

the ABA Checklist Committee, too late for inclusion in this volume.
VOICE Generally silent at sea.
STATUS & DISTRIBUTION Rare. **BREEDING:** Colonies on islands in the eastern Atlantic. **NONBREEDING:** Disperses to surrounding waters. Rare from May–Sept. in warm Gulf Stream waters off Hatteras, NC (100+ recs.). **VAGRANT:** Accidental inland after hurricanes and north in the Gulf Stream to the Canadian Maritimes (1 rec.).
POPULATION Global population fewer than 1,000 pairs. Suffers predation from introduced predators, direct persecution from humans, and indirect habitat loss due to feral grazers. BirdLife International considers it near threatened (and Zino's endangered).

BULWER'S PETREL *Bulweria bulwerii*   BUPE ▪ 5

This unusual petrel—which recalls an oversize, dark storm-petrel with very long wings and an exceptionally long, pointed tail—is accidental in North American waters. The genus *Bulweria* contains only one other species, the larger Jouanin's Petrel (*B. fallax*) of the Indian Ocean. Monotypic. L 10" (26 cm) WS 26" (66 cm)

IDENTIFICATION A medium-size and all-dark petrel with a long, pointed

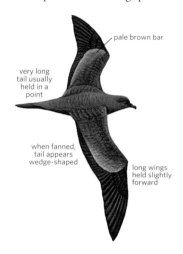

pale brown bar

very long tail usually held in a point

when fanned, tail appears wedge-shaped

long wings held slightly forward

tail, recalling a large storm-petrel. The long, pointed wings are dark brown above and below, with a bold tawny carpal bar extending forward to the leading edge of the wing. The tail is long, dark, and pointed; its wedge shape is concealed except during twisting aerial maneuvers. The head and the underparts are wholly dark brown; the bill is heavy and black. Bulwer's Petrel appears dove-headed, lacking the thick neck and heavy build of most *Pterodroma* petrels. Ages and sexes are alike. **FLIGHT:** Buoyant, erratic, and zigzagging with long wings slightly bowed and held well forward; usually close to the water's surface; in calm conditions, sometimes a few wingbeats then a short glide. Bulwer's has the lowest wing-loading of any tubenose. Unlike *Pterodroma* petrels, it does not normally arc high over the water, nor does it flutter like storm-petrels. **SIMILAR SPECIES** Bulwer's Petrel is most similar to Black Storm-Petrel, but note its larger size and medium brown coloration as well as its long, wedge-shaped tail; Black Storm-Petrel has a deeply forked tail. Dark-morph Wedge-tailed Shearwater has a similar tail, but it is much larger

and has a different flight style: slow, languid wingbeats, as well as prolonged soaring with bowed wings angled forward to the wrist, then swept back; usually stays close to the water.

VOICE Generally silent at sea.

STATUS & DISTRIBUTION Vagrant to N.A. **BREEDING:** Breeds on oceanic islands in the three major oceans, including the islands of Macaronesia in the eastern Atlantic and the Hawaiian Is. (The majority of the world's population nests on Nihoa, in the northwest chain of HI.) These are the two likely sources for North American vagrants. **NONBREEDING:** At-sea distribution is poorly understood, but it disperses offshore to surrounding tropical waters. **VAGRANT:** Two photographed records: off Monterey Bay, CA (July 26, 1998) and off Oregon Inlet, NC (Aug. 8, 1998); also an accepted sight record from off Cape Hatteras, NC, July 1, 1992. There are other sight records from VA to FL, and there are a least a dozen sight reports from the West Indies.

POPULATION Stable, but populations are experiencing predation at breeding colonies by introduced mammalian predators.

The five species of *Procellaria* petrels inhabit southern oceans, mostly at higher latitudes—two species are vagrants to N.A. waters. They are robust tubenoses with rather broad wings and short, thick bills. *Procellaria* petrels are larger than shearwaters but smaller than giant-petrels or albatrosses and occupy an intermediate ecological niche between them.

WHITE-CHINNED PETREL *Procellaria aequinoctialis*   WCPE ▪ 5

Among petrels, White-chinned is a bruiser, built for foraging in some of the planet's roughest seas. Heavy-billed and amply proportioned, this species is not likely to be overlooked among the more delicate shearwaters and gadfly petrels observed on North American pelagic trips. Monotypic. L 21.5" (55 cm) WS 55.5" (140 cm)

IDENTIFICATION A very stout-bodied, wide-winged *Procellaria* petrel, entirely chocolate brown in plumage except for a small amount of white in the "chin" (interramal space at the base of the mandible) that gives the species its English name. Heavy bill is yellowish, with each section (or plate) of the bill bordered narrowly in black.

SIMILAR SPECIES Both the comparably built Westland Petrel (*P. westlandica;* unrecorded in the northern hemisphere) and the more delicately built Parkinson's Petrel lack the white chin and have dusky-tipped bills, but close studies are required to discern these details.

VOICE At sea, occasionally gives grating squeal or squawk when competing for food.

STATUS & DISTRIBUTION BREEDING: On South Georgia, Prince Edward Is., Crozet Is., Kerguelen Is., Auckland I., Campbell I., Antipodes Is., and Falklands; when breeding and afterward, ranges widely at sea, from Antarctic pack ice to subtropical latitudes.

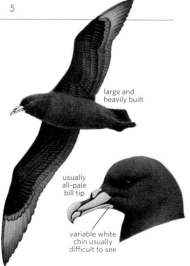

large and heavily built

usually all-pale bill tip

variable white chin usually difficult to see

VAGRANT: In N.A. waters, five records of single birds: near Rollover Pass, TX (Apr. 27, 1986); off Bar Harbor, ME (Aug. 24, 2010); Half Moon Bay, CA (Oct. 18, 2009); near San Miguel I., CA (Sept. 6, 2011); and Cordell Bank, off Marin Co., CA (Oct. 16, 2011). **POPULATION** Incidental mortality in long-line fisheries is exceptionally high in this species, which will likely hasten the population declines, as seen in albatrosses and other large tubenoses worldwide. It is classified currently as vulnerable.

PARKINSON'S PETREL *Procellaria parkinsoni*   PAPE ▪ 5

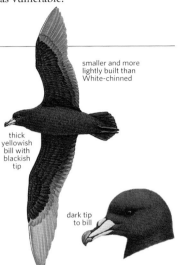

smaller and more lightly built than White-chinned

thick yellowish bill with blackish tip

dark tip to bill

Simply called "Black Petrel" in its native New Zealand, Parkinson's Petrel is named for botanical illustrator Sydney Parkinson, who traveled on Captain James Cook's first voyage. These petrels are frequent patrons at fishing vessels, where they feed on by-catch and offal. Monotypic. L 18" (46 cm) WS 45" (115 cm) **IDENTIFICATION** A sturdy, blackish brown petrel, with black legs and feet. Comparatively elegant for a *Procellaria* petrel, in flight recalling a large shearwater. Stout bill is greenish yellow, with each part of the bill bordered narrowly in black, which has a dusky tip. **SIMILAR SPECIES** Compared to its beefier, more bull-headed congeners, White-chinned and Westland Petrels (the latter not recorded from N.A.), Parkinson's has disproportionately longer, lankier wings, noticeable in flying birds; on the water, the wings project well past the tail, giving a different profile than White-chinned or Westland. Parkinson's might be passed over for a smaller Flesh-footed Shearwater, which has pink legs and feet and a pink bill that lacks the neat black plate outlines of *Procellaria* bills. **VOICE** Generally silent at sea. **STATUS & DISTRIBUTION** BREEDING: Only on Great Barrier I. and Little Barrier I., New Zealand. Adults arrive in October; fledglings depart by June. WINTER: Ranges widely over Pacific waters from southern Mexico to northern Peru and west to Australia. VAGRANT: Two records of single birds in N.A. waters: 20 miles northwest of Point Reyes, CA (Oct. 1, 2005) and near Heceta Bank, OR (Oct. 22, 2005).

POPULATION Listed as vulnerable. The small population of 5,000 appears stable.

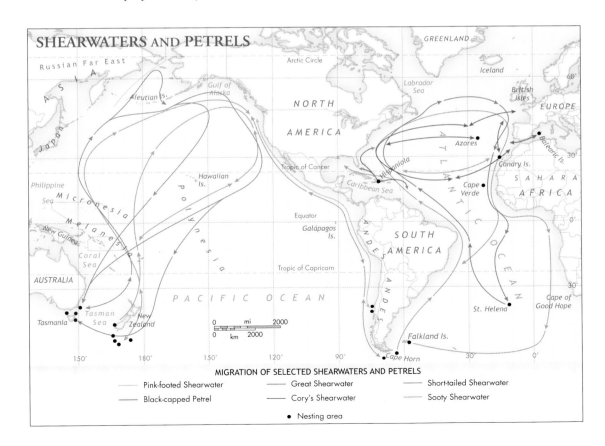

MIGRATION OF SELECTED SHEARWATERS AND PETRELS

— Pink-footed Shearwater — Great Shearwater — Short-tailed Shearwater
— Black-capped Petrel — Cory's Shearwater — Sooty Shearwater
● Nesting area

LARGE SHEARWATERS Genus *Calonectris*

These shearwaters are typically large and long-winged; their slow, languid wingbeats are quite unlike the shorter, shallow wingbeats of the *Puffinus* shearwaters. They are able to arc high off the water, when they appear uncharacteristically buoyant for their size. Three species are now recognized: Cory's and Cape Verde (Atlantic) and Streaked (Pacific).

STREAKED SHEARWATER *Calonectris leucomelas* STRS 4

This rare wanderer from the western Pacific is a much sought after specialty of West Coast fall pelagic trips. Monotypic. L 19" (48 cm) WS 48" (122 cm)

IDENTIFICATION A large pale shearwater giving a dirty white-headed appearance; overall dark above and white below. Variable head pattern, but almost always appears white at a distance; at close range streaked with brown, heaviest on nape and palest around bill and eye. The bill is bluish or yellowish gray (rarely pinkish) with a darker tip. Dark brownish upperparts, often appearing scaly due to pale fringes on the upperwing coverts and mantle feathers; the upperparts become paler and browner when worn and bleached. Long, dark tail. The uppertail coverts are variable: they can be all-dark or white, forming a pale U-shape on the rump. The underwings are largely white with a broad rear border and wing tip formed by the dark flight feathers. Just out from the wrist, the mostly dark, median primary coverts form a conspicuous dark patch on the underwing. The legs and feet are fleshy pink. Ages and sexes alike. **FLIGHT:** Agile, with long, broad-based wings. Flies with slow, languid wingbeats unlike most shearwaters; arcs high above the sea surface when traveling.

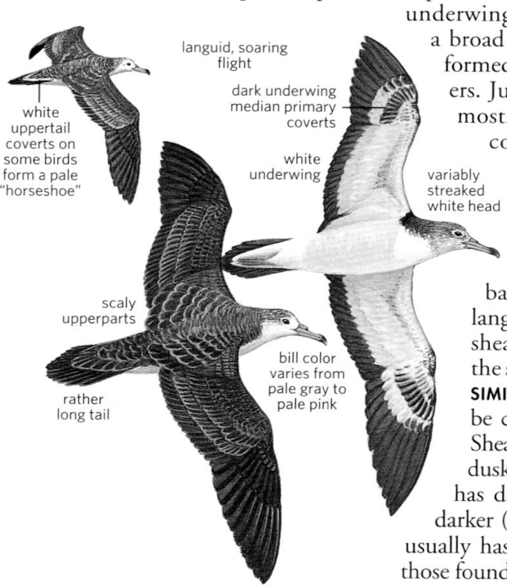

languid, soaring flight
white uppertail coverts on some birds form a pale "horseshoe"
dark underwing median primary coverts
white underwing
variably streaked white head
scaly upperparts
bill color varies from pale gray to pale pink
rather long tail

SIMILAR SPECIES Most likely to be confused with Pink-footed Shearwater. Pink-footed appears dusky headed with a pink bill; has dark undertail coverts and darker (not scaly) upperparts; and usually has darker underwings than those found on Streaked.

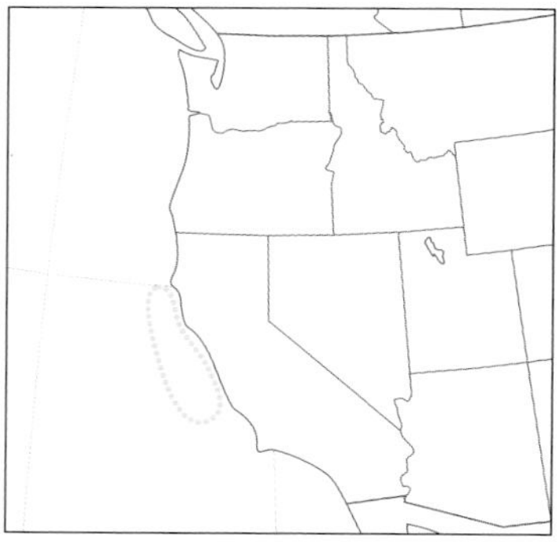

VOICE Generally silent at sea, but gives excited cackling in feeding groups.

STATUS & DISTRIBUTION Casual off the West Coast. **BREEDING:** Nests abundantly on islands off Japan and Asia. **NONBREEDING:** Migrates south to waters off the Philippines, Indonesia, and eastern Australia. **VAGRANT:** Casual in the eastern Pacific. Most records from off central CA (Sept.–Oct.); one Sept. record from off OR. Two inland records: one captured at Red Bluff, CA (Aug. 1993) and one near Medicine Bow, WY (June 2006).

POPULATION Large global population stable. Threatened at breeding colonies by introduced mammalian predators.

CORY'S SHEARWATER *Calonectris diomedea* COSH 1

Cory's Shearwaters, known among fishermen as "tuna ducks" for their habit of following and foraging above schools of tuna, are the largest of the world's shearwater species. Often found in mixed flocks with Great Shearwater in summer and fall off the Atlantic coast, Cory's dive deeply in pursuit of baitfish and squid, using wings and tail to propel and steer them with remarkable grace underwater. Polytypic (2 ssp.; both in N.A.). L 18" (46 cm) WS 46" (117 cm)

IDENTIFICATION A large shearwater, sometimes appearing gull-like in shape, with long, lanky wings and rather heavy yellow bill with dusky mark near the tip. Upperparts medium to pale brownish, often palest on the mantle and slightly grayish in the head. Upperwing sometimes shows carpal-ulnar M-pattern. Underparts and underwing coverts bright white. Pinkish legs and feet. Molting birds, frequently seen in summer months, show bluish gray, fresh flight feathers contrasting with worn brownish ones; may show whitish "wing bars" due to dropped coverts. **FLIGHT:** In low winds, can appear somewhat ungainly in flight, with deep, laborious flaps and rather loping flight style. In higher winds, far more agile, sweeping up and down in sinusoidal arcs, sometimes steeply.

GEOGRAPHIC VARIATION With a few exceptions, colonies of the smaller nominate subspecies *diomedea* (called "Scopoli's Shearwater") are located on islands within the Mediterranean Sea, whereas colonies of the larger subspecies *borealis* (called Cory's Shearwater) are

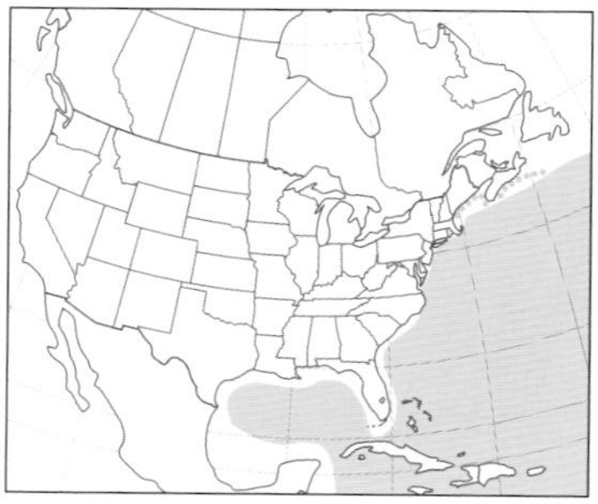

on oceanic islands from the Azores south to the Canary Is. Distinguishing these subspecies, both of which occur commonly off the East Coast, is possible for many individuals, though it is helpful to take photographs to confirm identification: In the underwing, *borealis* shows a clean break between the white of the primary coverts and the dark primaries, whereas in nominate *diomedea,* the bases of the primaries show some white, especially the outer primaries, producing a different look in the "hand" from below. "Scopoli's" also averages smaller, slimmer of bill, and narrower of wing.

SIMILAR SPECIES Distant Cory's might be confused with Great Shearwater, which usually appears darker above and cleanly dark-capped; Great flies more boldly and snappily, on outstretched wings that look more

pointed than those of Cory's. See also Cape Verde Shearwater.

VOICE Bleating calls frequently heard from actively foraging and squabbling birds in flocks.

STATUS & DISTRIBUTION The birds observed in N.A. waters are thought, based on their presence during the nesting season in the eastern Atlantic and on their state of molt, to be nonbreeding birds, largely immatures. **BREEDING:** Nests on islands in the eastern Atlantic and Mediterranean. Adults return to nesting areas in April; the last fledglings depart

by October. **WINTER:** Largely in the South Atlantic off Africa and South America. **VAGRANT:** Casual inland after hurricanes. Accidental off central and southern CA (two records; one bird spent two summers on the Coronodos Is., off Baja California Norte).

POPULATION The global population stands at about 500,000 birds. The species appears to be declining, probably from a combination of factors that threaten many pelagic seabirds: introduced predators, fishing operations, overexploitation of marine resources.

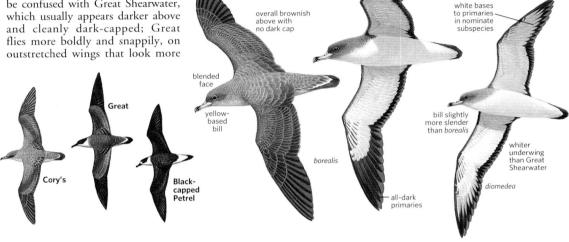

overall brownish above with no dark cap

white bases to primaries in nominate subspecies

blended face

yellow-based bill

bill slightly more slender than *borealis*

borealis

whiter underwing than Great Shearwater

Great

Cory's

Black-capped Petrel

all-dark primaries

diomedea

CAPE VERDE SHEARWATER   *Calonectris edwardsii*   CVSH   5

This large shearwater resembles a smaller version of Cory's Shearwater, with which it was considered conspecific for most of the 20th century. Endemic to the Cape Verde Is. off Africa, it is still a little-known species, particularly in the nonbreeding season. Monotypic. L 15.5" (39 cm) WS 39.5" (101 cm)

IDENTIFICATION A large shearwater, nearly as large as Cory's, brown above and white below; longest uppertail coverts usually show some white. Bill slim, olive-gray with dusky tip, some showing dull pinkish or olive-yellow tones at close range.

SIMILAR SPECIES Cory's Shearwater is very similar in plumage but averages bulkier and larger in all measurements; the slightly smaller "Scopoli's Shearwater" (currently classified as the nominate subspecies of Cory's) is closer to Cape Verde in proportions but still has broader wings and greater bulk. In direct comparison to Cory's and "Scopoli's," Cape Verde has a

notably smaller head and slimmer bill showing little or no yellow. The head often looks darker than in Cory's, with the dark of the crown contrasting more with the white underparts, unlike in Cory's, in which these colors are less contrasting, more diffuse where they meet. Capped appearance is never so strong as seen in smaller Great Shearwater, which is also darker above and has dark markings in underwing coverts and a dark patch on the belly.

VOICE Nasal, braying calls heard from birds at sea, especially when foraging in flocks.

STATUS & DISTRIBUTION BREEDING: Nests in summer only in the Cape Verde Is. First adults arrive in late February; last fledglings depart in November. **WINTER:** Ranges at sea off western Africa near breeding grounds to waters off Brazil and possibly Argentina, but very little known about range in nonbreeding season. **VAGRANT:** The first N.A. record was of

one found 30 miles southeast of Hatteras Inlet, NC (Aug. 15, 2004); the only subsequent record is off Worcester, Co., MD (Oct. 21, 2006).

POPULATION About 10,000 pairs persist on five or six islands; listed as Near Threatened.

smaller than Cory's and has a slightly longer tail

slim olive-gray bill

more contrast on face than Cory's

SHEARWATERS Genus *Puffinus*

This large, complex genus (15–20 sp.) comprises all the remaining shearwaters in the world. Ongoing debate has resulted in the recognition of 40–49 taxa. In N.A., the Manx is the only breeding representative; the remaining species visit offshore waters during migration or dispersal. In the eastern Pacific, six species occur regularly; another four have been recorded as vagrants. In the western Atlantic, four species occur regularly, and two have been recorded as vagrants. These species are known to wander, and individuals of several species have turned up multiple times in the wrong ocean altogether! All nest in burrows and are pelagic when not breeding.

PINK-FOOTED SHEARWATER *Puffinus creatopus*   PFSH ■ 1

Pink-footed Shearwater is a large, pale-bellied species of offshore Pacific waters and is easily seen on pelagic trips from late spring through fall. Monotypic. L 19" (48 cm) WS 43" (109 cm)
IDENTIFICATION Large; lacks strong plumage contrasts; appears mostly grayish above and pale below. Heavy pink bill with dusky tip; legs and feet also pink, often visible against the dark undertail coverts in flight. Upper-wing coloration ranges from gray to brownish, with odd white patches visible when molting. Underparts and underwings are quite variable: some authors consider it polymorphic, with light, dark, and intermediate morphs. Head generally dusky gray-brown, occasionally wholly dark on the darkest individuals; typically has a whitish throat. Whitish underparts, often dusky tinged, typically not appearing bright white below (beware strong sunlight); palest on belly. Underwing variable, but consistently shows dark flight feathers and pale underwing coverts, with variable dark markings on the axillaries. Dark individuals can have extensive grayish brown flanks and undertail coverts. **FLIGHT:** Generally flaps with slower, more labored wingbeats

than other shearwaters, arcing and gliding high off the water in strong winds. Wings are usually angled back slightly at the wrist, showing a long-handed appearance unlike most other *Puffinus* shearwaters.
SIMILAR SPECIES Darker individuals can be confused with Flesh-footed Shearwater (esp. distant, swimming birds). Flesh-footed is similar in shape, size, and flight style, but it never shows a pale belly or throat.
VOICE Generally silent at sea.
STATUS & DISTRIBUTION Common in offshore Pacific waters from spring through fall (a few north to off south-eastern AK), rare in winter. **BREEDING:** Nest colonially on oceanic islands off Chile. **NONBREEDING:** Generally moves north from breeding areas across the Equator to spend the austral winter in N.A. Pacific waters.
POPULATION Threatened at breeding sites by introduced mammalian predators, including coatis. Considered vulnerable by BirdLife International.

FLESH-FOOTED SHEARWATER *Puffinus carneipes*   FFSH ■ 3

This large, all-dark shearwater is generally rare during fall in the throngs of shearwaters gathering to feed in offshore Pacific waters. Monotypic. L 17" (43 cm) WS 41" (104 cm)
IDENTIFICATION All-dark plumage, slow-flapping, languid wingbeats, and pale bill are good clues to its identity. Upperparts wholly chocolate brown, lacking contrasting patterns. Underparts and underwing coverts dark brown; primaries flash silvery in strong sunlight (compare to the Sooty Shearwater). Dark head, with a stout, pink to pinkish horn bill with dark tip. Pink legs and feet, visible against the dark undertail coverts in flight. **FLIGHT:** Flaps slowly and deliberately, lacking the hurried wingbeats of most other shearwaters. Arcs and glides high over sea surface in strong winds, with wings bent back slightly at the wrist.
SIMILAR SPECIES Most similar in size, shape, and flight style to the Pink-footed Shearwater, but differs in being wholly dark below. Larger than Sooty and Short-tailed Shearwaters,

lacking silvery underwing coverts, and always showing a thick, pinkish bill. Dark-morph Wedge-tailed Shearwater is more delicately built and longer tailed, with a thinner dark bill, a wedge-shaped tail, and dark flight feathers. Also compare carefully to Parkinson's and White-chinned Petrels and to first-winter Heermann's Gulls.

VOICE Generally silent at sea.

STATUS & DISTRIBUTION Rare (sometimes uncommon) off the West Coast. **BREEDING:** Nests on oceanic islands off Australia and New Zealand and in the Indian Ocean. **NONBREEDING:** In the Pacific, migrates north across the Equator. Small numbers move east to the eastern Pacific, occurring from BC to Baja California (Aug.–Dec., rarely through spring); rare south of Point Conception, CA.

POPULATION Threatened largely by habitat loss and introduced mammalian predators.

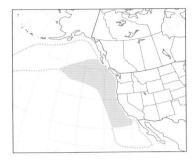

silvery flight feathers

blackish wing linings

pink bill with dark tip

pink feet often hard to see

GREAT SHEARWATER *Puffinus gravis*   GRSH ▪ 1

This strikingly plumaged large shearwater of the Atlantic often attends boats, where it forages for refuse as well as for handouts. Monotypic. L 18" (46 cm) WS 44" (112 cm)

IDENTIFICATION Large and dark above; more neatly patterned than other Atlantic shearwaters. Brownish black upperwings; pale tips on the upperwing coverts and mantle create a scaled appearance. Secondaries and greater upperwing coverts washed pale gray when fresh. Dark tail, with white uppertail coverts often forming a bold, U-shaped rump patch. Unique head pattern: The dark cap extends just below the eyes and is set off by white collar. The rear edge of the dark cap and a brownish half-collar near the wings frame a prominent white triangular shape on the side of the head; often prominent on distant birds, this is completely lacking on the Cory's Shearwater. Dark bill is long and thin. Clean white underparts, with gray belly patch (often hidden, lost in shadow, or occasionally absent; but diagnostic if

seen). The pale-centered underwings are broadly framed in dark with variable dark markings on the axillaries and coverts. Does not molt in North American waters. **FLIGHT:** Although a large shearwater, the Great Shearwater flies with stiff wingbeats like the smaller *Puffinus* shearwaters and unlike the slow, languid wingbeats of Cory's Shearwater. Great Shearwater arcs and glides on stiff wings that are not crooked back at the wrist, but held straight out, generally staying close to the water.

SIMILAR SPECIES Most similar to Black-capped Petrel, but it lacks the clean white underparts and dark ulnar bar as well as the white-tailed appearance (when viewed dorsally) of that species. The black cap of Black-capped Petrel is variable, but usually more restricted than on Great Shearwater; and most Black-cappeds show a prominent white forehead and a broader white collar. At a distance, Great Shearwater often stays lower to the water than Black-capped Petrel and, noticeably, holds its wings more straight out. Great Shearwater is smaller and more boldly patterned than Cory's Shearwater, with a dark cap and pale collar.

VOICE Generally silent at sea, but gives a whining drawn-out call when excited during foraging. Often vocalizes upon landing among other feeding seabirds.

STATUS & DISTRIBUTION Common off the Atlantic coast from May through late fall. **BREEDING:** Nests on oceanic islands in the S. Atlantic Ocean. **NONBREEDING:** Moves north to spend the austral winter in the North Atlantic, where large numbers congregate in late summer. **VAGRANT:** Casual inland after hurricanes. Accidental off central CA.

POPULATION Large global population.

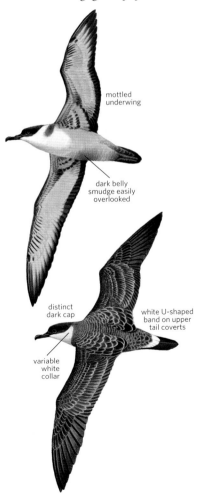

mottled underwing

dark belly smudge easily overlooked

distinct dark cap

white U-shaped band on upper tail coverts

variable white collar

BULLER'S SHEARWATER *Puffinus bulleri*   BULS ▪ 2

This strikingly plumaged Pacific species can be mistaken for a rare *Pterodroma* petrel due to the bold M-pattern across the surface of its wings. Widely considered the most elegant shearwater in North America, its plumage, structure, and flight style are unusually graceful. Monotypic. L 16" (41 cm) WS 40" (102 cm)

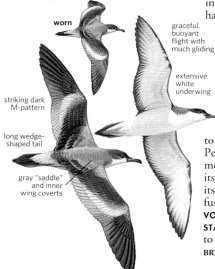

worn

graceful, buoyant flight with much gliding

striking dark M-pattern

extensive white underwing

long wedge-shaped tail

gray "saddle" and inner wing coverts

IDENTIFICATION Distinctly black-capped head, with a clean demarcation between the dark cap and light sides of face and throat. Dove gray upperparts set off by a bold, blackish M-pattern, creating a black-on-gray pattern unique among shearwaters. Gleaming white underwings and ventral body; the underwings are neatly outlined in black. Some fresh-plumaged birds have a very subdued M-pattern; worn birds are browner above. Ages and sexes alike. **FLIGHT:** Elegant, with smooth, unhurried wing-beats and elongated appearance. In direct flight uses slow, measured wingbeats; in strong winds arcs high like a *Pterodroma*, flapping little, if at all.
SIMILAR SPECIES Most similar to *Pterodromas*, especially the Cook's Petrel, but note the larger size and more rangy appearance of the Buller's, its slower wingbeats (if flapping), and its long, thin bill. Unlikely to be confused with other shearwaters.
VOICE Generally silent at sea.
STATUS & DISTRIBUTION Uncommon to fairly common off the West Coast. **BREEDING:** Nests colonially on a few

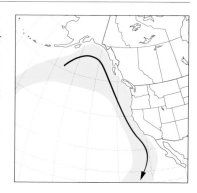

islands off New Zealand. **NONBREEDING:** Moves north and east of breeding islands to feeding areas in the North Pacific. Uncommon; sometimes numerous or in large flocks, in late summer and fall off central CA northward to the Gulf of Alaska; rare south of Point Conception. **VAGRANT:** Accidental inland at the Salton Sea (Aug. 6, 1966) and off NJ (Oct. 28, 1984).
POPULATION Historic numbers impacted by human exploitation and predation by introduced predators. BirdLife International considers it vulnerable.

SOOTY SHEARWATER *Puffinus griseus*   SOSH ▪ 1

The commonest dark shearwater in both oceans, but much more abundant in the Pacific, this transequatorial migrant reaches our waters during the austral winter. Spectacular concentrations occur during mid-summer off

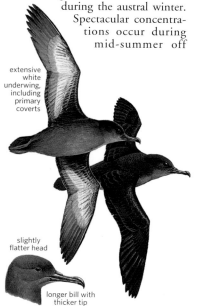

extensive white underwing, including primary coverts

slightly flatter head

longer bill with thicker tip

central California, where hundreds of thousands can often be seen from shore. Monotypic. L 18" (46 cm) WS 40" (102 cm)
IDENTIFICATION Medium-sized, dark shearwater with silvery underwings. Bill long and black. Underwing pattern distinctive, with pale silvery coverts and dark flight feathers. Lighting can affect the visibility of this character: In strong morning light, underwings can appear very white, but variable, depending on light. **FLIGHT:** Varies with wind speed like all seabirds, but generally intersperses glides with bursts of clipped, stiff-winged flaps. Flapping often appears hurried. Arcs high over the water in strong winds when it can be confused with dark *Pterodroma* petrels.
SIMILAR SPECIES Most similar to the Short-tailed Shearwater (see sidebar opposite). Told from other dark shearwaters by silvery underwing coverts, dark bill, and heavy structure.
VOICE Generally silent at sea.
STATUS & DISTRIBUTION Abundant in nearshore waters off West Coast;

fairly common off East Coast. **BREEDING:** Nests colonially on oceanic islands off southern South America, Australia, and New Zealand. **NONBREEDING:** Undertakes long, transequatorial migration in both oceans moving north to spend the austral winter (May–Aug.) in North Pacific and North Atlantic waters; rare at other times. **VAGRANT:** Recorded eight times at the Salton Sea during summer; accidental southwestern AZ.
POPULATION Large global population is still exploited as a human food resource in New Zealand.

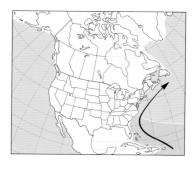

SHORT-TAILED SHEARWATER *Puffinus tenuirostris*   SRTS ▪ 2

This dark shearwater visits the West Coast primarily during late fall and winter, where separation from the more common Sooty Shearwater is problematic. Spectacular concentrations of up to a million birds have been seen off parts of Alaska, where this species congregates in late summer to feed in the nutrient-rich, shallow waters of the Bering Sea. Monotypic. L 17" (43 cm) WS 39" (99 cm)

IDENTIFICATION Dark and somewhat delicately built. Largely dark underwings and usually a pale chin. Bill structure and head shape are important identification features. Upperparts wholly chocolate brown,

lacking strong contrasts. Underwings variable, typically dark brownish, but many have pale silvery underwing coverts (like Sooty Shearwater). Typically round headed, imparting a dove-like appearance. Bill short, thin, and all-dark. Head often appears dark capped due to the pale feathers around the bill, on the throat, and often on the lower face. **FLIGHT:** Flaps in short bursts, stiffly from the shoulder. Does not appear long winged; rather its wings often look short and narrow (stick-like) for its relatively bulky body. Rarely shows much angle at the wrist, usually holding wings completely outstretched. Arcs high above water in strong winds.
SIMILAR SPECIES Notoriously similar to Sooty Shearwater (see sidebar below). Told from other dark shearwaters by overall shape, size, and plumage pattern.
VOICE Generally silent at sea.
STATUS & DISTRIBUTION Common off AK in summer, less common farther south in winter. **BREEDING:** Nests colonially on islands of Australia. **NONBREEDING:** Moves north during the austral winter (our

summer) to waters off AK. Most abundant off the Alaska Peninsula, the Aleutians, and in the Bering Sea. Some move farther south along the West Coast in fall and winter, where it can be found regularly off WA, OR, and central CA. Scarcer in southern CA and Baja California waters.
POPULATION Large global population; classified as least concern by BirdLife International.

some can show pale on inner wing but note darker primary coverts

worn Sooty and Short-tailed Shearwaters have a brownish cast

darkish underwing but variable

more rounded head

short, thin bill

can have paler chin

Identification of Sooty and Short-tailed Shearwaters

Separating Short-tailed and Sooty Shearwaters is one of the most difficult field problems off the West Coast. Key field marks for separating the two are the shape and coloration of the head and bill, and the underwing pattern; however, these are not diagnostic. A combination of features is necessary to confirm identification, and typically any candidate is considered a Sooty Shearwater until proven otherwise. Probability should weigh into any identification: Sooty Shearwaters greatly outnumber Short-tailed Shearwaters during spring, summer, and early fall; but during late fall and winter, they can occur in relatively even numbers.

Note the dove-like head shape of Short-tailed; it is more delicate, with a rounded crown and steeper forehead than Sooty's flatter forehead. The bill of Short-tailed is finer, shorter, and slimmer overall. The bill of Sooty

Sooty Shearwater

Short-tailed Shearwater (CA, Nov.)

is thicker at the base and tip (with a "pinched-in" center) and longer, with a more prominent nail.

Generally, the underwing of Short-tailed Shearwater is darker than the Sooty, and any individual with dark underwing coverts is automatically a strong candidate for the Short-tailed. But variation exists in this feature. On Short-taileds with paler underwings, there exists a subtle difference in the distribution of light and dark: Typically Short-tailed's underwing is palest in the center (on the median secondary coverts), whereas Sooty is palest farther out on the underwing (on the median primary coverts), creating a strong contrast between these pale feathers and the dark primaries. Additionally, on Sooty, the pale feathers of the secondary coverts have dark tips that often form thin, dark rows on the inner wing. ∎

WEDGE-TAILED SHEARWATER *Puffinus pacificus* WTSH ▪ 4

This rare visitor is always a surprise when encountered on West Coast pelagic trips. The provenance of individuals recorded in North America is a matter of some debate, as breeding populations exist on islands off Mexico and in Hawaii. Its graceful shape and flight style immediately separate it as something unique among Pacific shearwaters. Monotypic. L 18" (46 cm) WS 40" (102 cm)

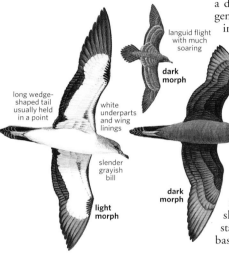

long wedge-shaped tail usually held in a point

white underparts and wing linings

slender grayish bill

light morph

languid flight with much soaring

dark morph

dark morph

IDENTIFICATION Polymorphic, occurring in both light and dark morphs. A fairly large, lanky shearwater characterized by a slim build and a long, wedge-shaped tail. Head appears small; the long, thin, blue-gray bill has a dark tip. Pinkish legs and feet. LIGHT MORPH: White below and brownish above with pale tips on the upperwing coverts and scapulars giving a scaly appearance. White throat, chest, and belly, but head can appear largely dark at a distance. Underwing variable, but generally light; some with darker leading edge and dark line across the underwing coverts connecting with the axillaries. Dark flight feathers. DARK MORPH: Upperparts as in light morph, but often slightly darker. Underwing and ventral body wholly dark, with dark flight feathers. Some intermediate birds occur: dark morphs with grayer underparts or paler underwings, and light morphs with a grayish breast band or gray below with a white throat. FLIGHT: Extremely graceful, with slow, languid wingbeats; usually stays low to the water. Holds broad-based wings slightly forward, bent

back at the wrist and bowed down.
SIMILAR SPECIES Light morph is most similar to Buller's Shearwater in size and shape, but note Wedge-tailed's longer tail, bulging secondaries, and plainer, gray-brown back. The dark morph is similar to other dark shearwaters in plumage; note unique shape and flight style; most similar to Flesh-footed Shearwater, but note Wedge-tailed's longer tail, lighter build, thinner black bill, and darker flight feathers.
VOICE Generally silent at sea.
STATUS & DISTRIBUTION Vagrant to N.A. waters. BREEDING: Nests colonially on oceanic islands in the tropical Pacific (including HI and Isla San Benedicto, Revillagigedo Is., off western Mexico) and Indian Oceans. NONBREEDING: Disperses to surrounding tropical waters; migration poorly known. VAGRANT: Casual; fewer than ten records for the Pacific coast, most from CA; one inland record at the Salton Sea (July 31, 1988). Both color morphs have been recorded.
POPULATION Global populations severely impacted historically by human exploitation, predation pressure from introduced mammalian predators, and habitat loss.

TOWNSEND'S SHEARWATER *Puffinus auricularis* TOSH ▪ 5

The endangered Townsend's Shearwater is a small tropical species with two distinctive, subspecies that nest in Mexico and Hawaii, where the latter is known as "Newell's Shearwater." Polytypic (2 ssp., *newelli* in N.A.). L 13" (33 cm) WS 33" (83 cm)
IDENTIFICATION A small shearwater, blackish above, white below, with white partly wrapping up toward the rump at the flank, just past the trailing edge of the wing. A small amount of white also extends from the throat to the rear of the auriculars.
GEOGRAPHIC VARIATION Nominate *auricularis* of Mexico and *newelli* ("Newell's Shearwater") of Hawaii are treated by most authorities as separate species. "Newell's" averages larger, 13.8–15" versus 12.6–13.8". In nominate Townsend's, the undertail coverts are blackish, whereas "Newell's" has white undertail coverts with blackish tips. "Newell's" also shows a cleaner, more distinct division of black from white plumage at the neck and breast.

SIMILAR SPECIES Manx Shearwater is larger and more heavyset, with short tail and long white undertail coverts that nearly extend to the tail tip. The longer-billed Black-vented shows diffuse border of dark/light plumage in the head and neck.
VOICE Generally silent at sea.
STATUS & DISTRIBUTION BREEDING: Nominate Townsend's nests only on Socorro I. in the Revillagigedo Is. off Mexico and disperses to adjacent waters after breeding. "Newell's" breeds only in the main Hawaiian Is., with about 90 percent nesting in the mountains of Kauai. Adults arrive to court in May; the last fledglings depart in October. Foraging is primarily around seawater fronts in the Equatorial Counter Current. Post-breeding dispersal at sea mostly south and east of the nesting areas, possibly as far as Clipperton I. VAGRANT: One "Newell's" wrecked on land at Del Mar, CA (Aug. 1, 2007).
POPULATION In Mexico, nominate Townsend's Shearwater has declined

"Newell's Shearwater"
newelli

plumage sharply black and white

overall similar to Manx Shearwater, but more dark on undertail coverts and longer tail

from about 500 pairs in the 1990s to fewer than 100 pairs in 2008; it is critically endangered. Population estimates for "Newell's" range from about 16,000 to 19,000 total pairs, but these numbers, and other sources of data, suggest a 60 to 70 percent decline in the population since the 1970s. The species is classified as endangered. Part of the rapid decline in "Newell's" is associated with the impact of Hurricane Iniki (1992). Each summer, an average of 350 "Newell's Shearwaters" die in collisions with manmade structures.

BLACK-VENTED SHEARWATER *Puffinus opisthomelas*   BVSH ▪ 2

This small shearwater occurs during the fall and winter months in the nearshore waters off southern California. Monotypic. L 14" (36 cm) WS 34" (86 cm) **IDENTIFICATION** Dingy; stocky build. Dark undertail coverts; individuals with white undertail coverts have been recorded. Upperwings largely brownish; can appear blackish when fresh; fade substantially with wear. Head generally dusky grayish brown, lacks strong contrast between dark cap and paler throat. Underparts variable: typically whitish, often washed on the flanks with dusky gray-brown. Ages and sexes alike.

FLIGHT: In direct flight, flaps with stiff hurried wingbeats, wings held straight out with little bend at the wrist. In strong winds, arcs high over the water. **SIMILAR SPECIES** Distant birds resemble Pink-footed Shearwater, but fly with faster wingbeats and are generally seen in flocks. Cautiously separated from much rarer Manx: Note Manx's cleaner white appearance below, more contrasting black-and-white plumage, white undertail coverts, pale "ear-surround." Harsh lighting can make Black-vented appear more black-and-white. The accidental Townsend's Shearwater (subspecies *newelli*) is very similar, but has much more contrasting, black-and-white plumage. **VOICE** Generally silent at sea. **STATUS & DISTRIBUTION** Common within limited West Coast range. **BREEDING:** Oceanic islands off Baja California. **NONBREEDING:** Disperses primarily north after breeding; large

numbers spend the winter (Oct.–Mar.) off southern CA; rare north to BC. **POPULATION** Small global population and restricted breeding range; considered vulnerable by BirdLife International.

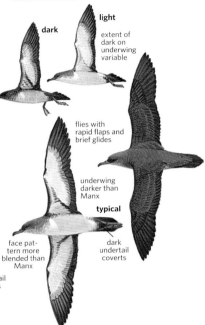

dark

light

extent of dark on underwing variable

flies with rapid flaps and brief glides

underwing darker than Manx

typical

face pattern more blended than Manx

dark undertail coverts

dark

dark bill

dark undertail coverts

MANX SHEARWATER *Puffinus puffinus*   MASH ▪ 2

This is one of a group of small black-and-white shearwaters that are confusing to identify and were once considered subspecies. Monotypic. L 13.5" (34 cm) WS 33" (84 cm) **IDENTIFICATION** Small, stocky build; sharply demarcated black-and-white plumage. Upperwings typically blackish, occasionally fade browner. Distinctive head pattern: eye encompassed by dark cap, contrasting sharply with white throat and a pale area that wraps up behind the ear coverts (the pale "ear surround"). Clean white underparts. Underwing variable, often pure white coverts contrast with blackish flight feathers; occasionally a dark intrusion across coverts connects with axillaries. **FLIGHT:** Wingbeats stiff, hurried; in strong winds arcs high above water. Wings held straight out with little or no bend at the wrist. **SIMILAR SPECIES** Most similar to Audubon's Shearwater in the western Atlantic. Note Audubon's lighter build, longer tail, whiter face, and dark undertail coverts. Manx in the eastern Pacific (where rare) most closely

resembles Black-vented Shearwater. In comparison, Manx flight style slightly more languid with more soaring. **VOICE** Generally silent at sea. **STATUS & DISTRIBUTION** Regular off the Atlantic coast; rare off the West Coast. **BREEDING:** Oceanic islands in the North Atlantic; in N.A. only on islands off NF and MA. **NONBREEDING:** In winter (Sept.–May) moves south to waters off the east coast of S.A.; some stay as far north as NC. Some birds (presumed nonbreeding) summer as far south as NC, where rare. Rare off the West Coast (100+ records for CA); breeding not yet confirmed in Pacific. **VAGRANT:** Casual inland to the Great Lakes; accidental western MT (June 2004). **POPULATION** Large global population; considered least concern by BirdLife International.

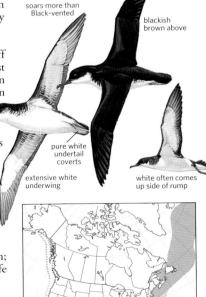

soars more than Black-vented

blackish brown above

pure white undertail coverts

extensive white underwing

white often comes up side of rump

white wraps around ear coverts

white undertail coverts

AUDUBON'S SHEARWATER   *Puffinus lherminieri*   AUSH ■ 1

This small black-and-white shearwater resides in the tropical oceans and is closely related to Manx and Barolo Shearwaters. Polytypic (at least 2 ssp.; both in N.A.). L 12" (30 cm) WS 27" (69 cm)

IDENTIFICATION Small, fast flying; solidly blackish above, white below. Blackish brown upperparts; thin pale tips to greater upperwing coverts when fresh. Head variable, but typically dark capped; sharply demarcated from the white cheek and throat, bordered behind with a dark half-collar; some show white

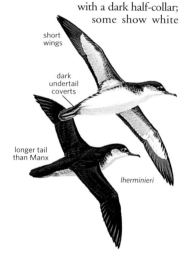

short wings

dark undertail coverts

longer tail than Manx

lherminieri

completely encircling the eye. Medium-length bill is thin and dark. Underparts bright white, including underwings; occasionally with variable dark bar across underwing coverts to axillaries. Flight feathers grayish below, black above. Tail long and dark; undertail coverts typically blackish brown, giving a "dipped in ink" appearance; occasionally shows mixed white feathers in the undertail coverts. Legs and feet typically pink, but can be blue or a mix of both. Ages and sexes alike. **FLIGHT:** Flight is low and fluttering with buoyant glides on slightly arched wings

GEOGRAPHIC VARIATION Taxonomy is in a state of flux. The nominate subspecies from the Caribbean is the most likely to be encountered based on proximity. The status of the smaller *loyemilleri*, which nests in the Caribbean off Panama, is uncertain. Subspecific identification at sea is impossible due to lack of identification criteria.

SIMILAR SPECIES Most similar to Manx. Note Audubon's brownish cast to upperparts, longer tail, and dark undertail coverts. Also similar to Barolo Shearwater, but larger, with darker face, larger bill, longer tail, and dark undertail coverts. In the Pacific, Galápagos

Shearwater *(P. subalaris)*—a potential vagrant—is smaller and darker, and lacks the half-collar (recorded north to off southwestern Mexico).

VOICE Generally silent at sea.

STATUS & DISTRIBUTION Common in warm Gulf Stream waters typically along temperature breaks. **BREEDING:** Nearest populations nest on islands in the Caribbean. **NONBREEDING:** Caribbean breeders disperse north into the Gulf Stream (May–Sept.). Uncommon in Gulf of Mexico and rare off New England. **VAGRANT:** Casual inland after hurricanes (exceptionally to western KY).

POPULATION Considered least concern by BirdLife International, but numbers off the East Coast have declined.

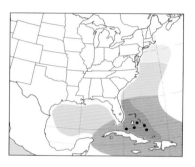

BAROLO SHEARWATER   *Puffinus assimilis*   BASH ■ 5

Until recently the Barolo Shearwater was considered a subspecies of Little Shearwater *(P. assimilis),* and the current taxonomy may undergo further revision. Barolo Shearwater is known in North America from a few records off the East Coast. Monotypic. L 11" (28 cm) WS 25" (64 cm)

IDENTIFICATION Very small shearwater with relatively short and slightly rounded wings. Upperparts largely blackish, but a pale gray panel on the upper secondaries is usually visible. Dark-capped head; bold white face, eye completely encircled with white. Tiny, all-dark bill. White underparts, including underwing coverts and, most importantly, the undertail coverts. Grayish blue legs and feet. **FLIGHT:** Flies low to the water on hurried wingbeats and short glides; sometimes raises head upward at the end of glides.

SIMILAR SPECIES Most similar to Audubon's Shearwater but structurally smaller, more compact, and stockier, with shorter, more rounded wings.

Barolo Shearwater further differs from Audubon's in having a white face and white undertail coverts. The treatment of the taxon *boydi* (breeds and possibly resident on Cape Verde Is.; unrecorded N.A.) is unresolved. Some authorities treat it as a subspecies of Audubon's; others treat it as a subspecies of Barolo Shearwater; and some treat it as a full species. Birds of the taxon *boydi* have a dark face and lack the white eye-surround of Barolo Shearwater and are unlikely to be separable from Audubon's Shearwater in the field based on known plumage characters.

VOICE Generally silent at sea.

STATUS & DISTRIBUTION Vagrant to N.A. **BREEDING:** Breeds on islands in the eastern North Atlantic, from the Azores east to the Canary Is. **NONBREEDING:** Poorly known; some likely sedentary, others dispersing to surrounding waters. **VAGRANT:** Casual. Recorded about six times in N.A. (several other unsubstantiated reports); most recently from off Sable I., NS (Sept. 23, 2003, and two birds on

Sept. 24, 2003) and MA (Aug. 25, 2007, and July 29, 2011). May prove to be annual in late summer and fall in the Gulf Stream, MA, north to off the Maritimes.

POPULATION Several small populations declining due to human exploitation as a food resource; also impacted by habitat loss and introduced predators.

flies with stiff, shallow wingbeats

white extends prominently over eye

white undertail coverts

extensive white underwing

pale gray creates two-tone upperwing

STORM-PETRELS Family Hydrobatidae

Wilson's Storm-Petrel (NC, Aug.)

The small size, quick flight low over the water, and monochromatic, generally dark plumages of storm-petrels challenge the abilities of many birders. For identification purposes, first know that in general Northern Hemisphere storm-petrels have longer wings and shorter legs, whereas those from the Southern Hemisphere have shorter, rounded wings and long legs; there is some overlap in the tropics. Secondly, focus on the exact distribution of white on the rump and, perhaps more important, on flight style. Not all birds are identifiable, and sufficient views are a must. Many species will closely approach boats when attracted by chum and fishing refuse; they have well developed olfactory systems and use their sense of smell to locate food sources—and even individual nest burrows and crevices. At these times, close scrutiny is possible.

Structure Most storm-petrels are small birds, lightly built with long narrow wings. They have large heads, with slim bodies and tails that range from short to quite long. Some species have long legs that dangle below the body and long feet, which they patter on the surface of the water.

Behavior Flight pattern varies with each species, but most stay low to the water due to their small size and foraging techniques. They often appear swallowlike when searching for small prey items on the surface, but some of the larger species act more like small shearwaters, gliding and arcing on stiff wings. Several species have exaggerated, languid wingbeats like those of a nighthawk. Flight style changes significantly with activity and wind conditions. Many species are more dynamic in direct flight, incorporating more arcing and gliding when covering large tracts of open ocean. Mixed-species flocks of up to 10,000 birds occur off the California coast during early fall. Prey items are generally phytoplankton or other small invertebrates, but many species will occasionally take small fish if available. Most species only make nocturnal visits to their breeding colonies and are rarely observed in the immediate vicinity.

Plumage For the most part, storm-petrels are dark brown to blackish, but several species are gray above and pale below. Many have extensively pale rumps, other species are all-dark. Most dark storm-petrels show a pale diagonal bar across the upper wing, formed by the buffy and often faded greater upperwing coverts; its extent and prominence can aid field identification. Ages and sexes generally look alike, but plumages fade considerably with wear. Most birds complete a single annual molt on the nonbreeding grounds at sea. One can see a wide variety of molt states within a single species in a day, presumably due to age and fitness.

Distribution Storm-petrels occur in all oceans. In North America, at least nine species occur in the eastern Pacific; in the western Atlantic three white-rumped species commonly occur (Wilson's, Leach's, and Band-rumped), and at least four other species have been recorded as vagrants. Only Wilson's regularly undertakes a transequatorial migration from subantarctic breeding areas into North American waters. Black and Least Storm-Petrels regularly migrate into the Southern Hemisphere during their nonbreeding season. Migratory routes are poorly understood. It is believed that many species disperse into surrounding waters during the nonbreeding season. Some undertake northward dispersal events in the eastern Pacific; others disperse southward from breeding colonies.

Taxonomy Worldwide there are 23 species in seven genera, including the newly discovered Pincoya Storm-Petrel *(Oceanites pincoyae)* from Chile and the rediscovered New Zealand Storm-Petrel *(Oceanites maorianus)*. The taxonomy of some species is still a matter of debate. The AOU lists 14 species in five genera as occurring in North America.

Conservation Storm-petrels are subject to avian (especially skuas) and mammalian predation. The introduction of non-native predators to breeding islands has resulted in the decimation of some breeding colonies; the Guadalupe Storm-Petrel *(Oceanodroma macrodactyla)* from Guadalupe I. off Baja California has presumably gone extinct. BirdLife International lists one species as critical, two as endangered, and two as near threatened.

Genus *Oceanites*

WILSON'S STORM-PETREL *Oceanites oceanicus* WISP ▪ 1

The small, dark, white-rumped Wilson's Storm-Petrel is abundantly observed in the Atlantic. Its distinctive feeding style includes pattering on the surface with long, dangling legs and feet. Polytypic (3 ssp.; nominate in N.A.). L 7.3" (18 cm) WS 16" (41 cm)

IDENTIFICATION Overall dark plumage, with bold white rump extending well down onto the undertail coverts. Well-defined pale carpal bar, frosty gray when fresh, fading to pale tan when worn, on the upper wing. Tail shape variable, often appearing squared in direct flight, but can appear slightly notched or even rounded during feeding. Long dark legs with yellow-webbed feet distinctive, often visible during feeding behavior. Ages and sexes look similar. Adults undergo molt during spring and summer in North American waters. **FLIGHT:** Flies low to the sea surface, occasionally arcing up into the air over wave tops. Direct flight with fluttery wingbeats interspersed with short glides. Feet extend past the tip of the tail. When feeding, holds its wings above the horizontal, fluttering while pattering the surface with its feet; often appears to be "walking on water."

GEOGRAPHIC VARIATION Nominate *oceanicus* is proportionately smaller than the extralimital *exasperatus* and *chilensis*, but the three are inseparable under field conditions.

SIMILAR SPECIES Wilson's has similar plumage to other North Atlantic storm-petrels, but Leach's and Band-rumped Storm-Petrels have distinctly longer wings and different flight styles. Leach's flight is nighthawk-like; Band-rumped's is shearwater-like. European Storm-Petrel is very similar, but is smaller and lacks both the foot projection and prominent pale carpal bar of the Wilson's. See the European Storm-Petrel account for more.

VOICE Generally silent at sea, but gives a faint peeping call when excitedly feeding in groups.

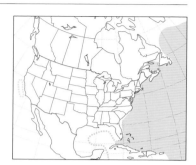

STATUS & DISTRIBUTION Common. **BREEDING:** Nests colonially on oceanic islands in the southern oceans. **NON-BREEDING:** Common offshore in the Atlantic May–Sept.; regularly seen from shore at favored locations. Rare in the Gulf of Mexico and rare off the CA coast late summer and fall. **VAGRANT:** Casual elsewhere on West Coast and inland after hurricanes.

POPULATION The global population of this species, considered one of the most abundant seabirds, is likely stable.

European

molt gap

dropped greater coverts

old (faded) outer primaries

new inner primaries

Wilson's different stages of wing molt from spring to summer

feet project beyond tail

short "arm"

white extends well below tail

long legs

feet with yellow webs

typical foot-pattering behavior

extensive white on sides of undertail

Genus *Pelagodroma*

WHITE-FACED STORM-PETREL *Pelagodroma marina* WFSP ▪ 3

This much sought-after species occurs rarely in the western Atlantic, where it is now seen almost annually in late summer. Polytypic. (6 ssp.; 2 in N.A.) L 7.5" (19 cm) WS 17" (43 cm)

IDENTIFICATION Where it occurs in the western Atlantic with other dark storm-petrels, this small, dynamic bird is distinctive in having primarily whitish underparts and a bold white facial pattern with dark auriculars and dark half-collar. The clean white underwing coverts contrast with the dark flight feathers. The brownish mantle and darker upperwing coverts and remiges create a pale saddled look; contrasting pale rump. Long tarsi dangle below the body during feeding and extend past the tail in direct flight. Ages and sexes look alike. **FLIGHT:** Its unique foraging technique, where it appears to hop across the water in pogo-stick fashion, its long legs dipping into the water as it moves across the sea surface, is unlike any other storm-petrel. It appears to spring off the surface of the water while rapidly changing direction with its wings almost always held fully extended and often seeming slightly bowed down.

It flaps with stiff, short wingbeats, and skips across the water with short glides. It flies in direct flight with similar stiff, short wingbeats followed by long glides on outstretched wings. **GEOGRAPHIC VARIATION** The majority of North American records are of the Cape Verde Islands *eadesi*. It has a whiter forehead and pale collar compared with the darker faced *hypoleuca* of the Salvages (Ilhas Salvagens). **SIMILAR SPECIES** White-faced is unlikely to be confused with any other storm-petrel in the western Atlantic due to combination of distinctive plumage features and feeding behavior. Potentially it could be confused with a winter-plumaged phalarope while sitting on the ocean, but note the White-faced's different posture, wing shape, and flight style when flushed. **VOICE** Generally silent at sea. **STATUS & DISTRIBUTION** Very rare in N.A. **BREEDING:** Widely distributed on oceanic islands across the southern oceans and, north of the Equator, in the eastern Atlantic on the Cape Verde Islands and Salvages. **NONBREEDING:** Disperses to waters north and west of breeding grounds, reaching the continental shelf of N.A. in late summer and fall. **POPULATION** Least concern, but suffers predation by native and introduced predators, as well as habitat loss on breeding islands.

pushes off water and changes direction

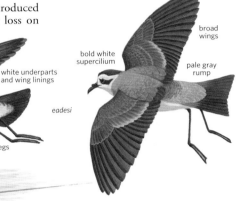

white underparts and wing linings

long bill

very long legs

bold white supercilium

broad wings

pale gray rump

eadesi

Genus *Hydrobates*

EUROPEAN STORM-PETREL *Hydrobates pelagicus*   EUSP   4

This very rare nonbreeding visitor from the eastern Atlantic usually associates with Wilson's Storm-Petrels and is easy to overlook. Polytypic (2 ssp.; N.A. records presumably of nominate). L 5.5–6.5" (14–17 cm) WS 14–15.5" (36–39 cm) **IDENTIFICATION** Tiny size and blackish overall. Bold, clean, white rump patch extends marginally onto rump sides. The white greater underwing coverts are best looked for when it feeds with wings held high, otherwise they are difficult to see in flight. Upper wing typically dark, lacking the pale carpal bar of other Atlantic storm-petrels; juveniles can have pale-tipped greater upperwing coverts. Head and bill are small and delicate. Feet and legs are dark. Ages and sexes look alike. **FLIGHT:** Flies with flexing bursts recalling a large swift or Spotted Sandpiper; changes direction frequently. Wings rarely appear to break the horizontal in direct flight. **SIMILAR SPECIES** Most likely to be confused with the Wilson's Storm-Petrel.

In a flock of Wilson's, it can be picked out by its smaller size and darker upper wings lacking a pale carpal bar, lack of foot projection, longer narrower wings with rounded tip, different flight style, and white greater underwing coverts. **VOICE** Generally silent at sea. **STATUS & DISTRIBUTION BREEDING:** Nests on oceanic islands in the northeastern Atlantic and Mediterranean.

NONBREEDING: Transequatorial migrant; wintering largely in southwest African waters. **VAGRANT:** More than ten North American records. Single 1970 specimen record from NS; all recent records (mid-May to early June) from off Cape Hatteras, NC. **POPULATION** Stable.

smaller and darker overall than Wilson's

white underwing bar

short legs

lacks obvious carpal bar

long "arm"

BLACK-BELLIED STORM-PETREL *Fregetta tropica* BBSP ▪ 5

This widespread southern ocean species has a distinctive flight style and plumage, but despite its name, the black belly can be relatively hard to see. Polytypic (2 ssp.; N.A. records presumably of nominate). L 8" (20 cm) WS 18" (46 cm)

IDENTIFICATION Black-and-white pattern; distinctive among other western Atlantic storm-petrels in being mostly white below. The upperparts are dark, with pale-tipped greater upperwing coverts; the rump is white. It has a variable dark line that connects the dark chest with dark undertail coverts; the dark line is absent on some birds. The white underwing coverts contrast with dark flight feathers. Long legs and feet project past the tail. Ages and sexes look alike. **FLIGHT:** The bird's distinctive foraging technique of splashing its breast into the water and then

springing off forward with its long legs is unique. In direct flight it flies low across the water, rarely flapping. **SIMILAR SPECIES** When seen well, it is unlikely to be confused with other storm-petrels in the western Atlantic. The combination of white underparts, dark stripe down belly, and distinctive foraging action are unique. It is separated from the similar-looking White-bellied Storm-Petrel (*F. grallaria*) by its darker upperparts, typically black belly stripe, and long feet that project past the tail. Its flight style is similar to White-faced Storm-Petrel; however, it splashes into the water with its breast and does not pogo-hop with its feet. **VOICE** Generally silent. **STATUS & DISTRIBUTION BREEDING:** Nests on subantarctic islands. **NONBREEDING:** Disperses to tropical seas north of breeding areas, reaching the

Equator in the Atlantic. **VAGRANT:** Four N.A. records (late May–mid-Aug.) from Gulf Stream waters off NC, all since 2004. **POPULATION** It suffers predation by native and introduced predators.

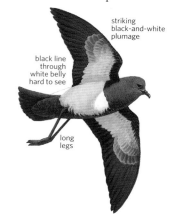

striking black-and-white plumage

black line through white belly hard to see

long legs

FORK-TAILED STORM-PETREL *Oceanodroma furcata* FTSP ▪ 2

The Fork-tailed Storm-Petrel is found year-round in the northern Pacific, but only rarely south of central California. Polytypic (2 ssp.; both in N.A.). L 8.5" (22 cm) WS 18" (46 cm)

IDENTIFICATION Only grayish storm-petrel in the northeastern Pacific. Underparts pale gray with darker grayish black underwing coverts. Head grayish, with dark blackish mark through eye. Upperparts gray, with

pronounced M-pattern across wings. **FLIGHT:** Wingbeats shallow and measured; flight relatively direct. **GEOGRAPHIC VARIATION** West Coast *plumbea* is smaller and darker than nominate *furcata* of Asia and the Aleutians. **SIMILAR SPECIES** In the northeastern Pacific, confusion is unlikely with other storm-petrels, which are largely blackish brown. It could be confused with winter-plumaged phalaropes while sitting on the water, but note differences in posture, structure, and flight style. **VOICE** Generally silent at sea. A raspy

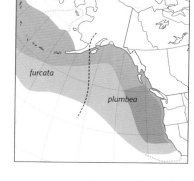

furcata

plumbea

ana, ana, ana given at the nesting colonies. **STATUS & DISTRIBUTION** Uncommon to fairly common. **BREEDING:** Nests on islets from AK south to northern CA. **NONBREEDING:** Disperses south and west after breeding. Rare off southern CA; accidental inland in CA (Yolo Co.). **POPULATION** Least concern; losses from native and introduced predators on some breeding islands.

prominent pale carpal bar whitens near body

distinctive fork to tail

rounded wing tip

pearly gray with blackish wing lining

dark mask can suggest a phalarope

RINGED STORM-PETREL *Oceanodroma hornbyi*   RISP ■ 5

This most distinctive of the world's storm-petrels is also the most mysterious: its breeding grounds are still unknown. In most of the world, it is still known as "Hornby's Storm-Petrel," in honor of Adm. Sir Phipps Hornby of the British Royal Navy. Monotypic. L 7.8–8.7" (20–22 cm) **IDENTIFICATION** Very strikingly patterned medium-size storm-petrel with rather long bill; flight swooping and erratic, with rapid wingbeats. Underparts white except for the broad band of dark gray across the breast. Upperparts grayish, palest on mantle and uppertail coverts, browner in marginal, lesser, and humeral coverts and scapulars, with a starkly contrasting cream-colored carpal bar (median and greater coverts). A thin band of white across the nape, extending to the throat, separates the gray back from the blackish cap. Remiges and rectrices very dark gray; tail strongly forked. **SIMILAR SPECIES** No storm-petrel or other seabird known from the New World shows Ringed's distinctive

pattern of dark gray, brown, cream, and white, but a poorly seen or ill-lit Fork-tailed Storm-Petrel could cause momentary confusion. Polynesian Storm-Petrel *(Nesofregetta fuliginosa)* is similar in pattern but has a completely dark face, brown rather than gray upperparts, white underwing, dark undertail coverts, and less fork in the tail. **VOICE** Generally silent at sea. **STATUS & DISTRIBUTION** In the eastern Pacific Ocean, large numbers have been observed in the cold Peru Current off northern Chile and Peru, with a few observed north to Ecuadorean waters. The breeding grounds of this species have never been discovered, and so nothing is known of its nesting ecology and populations, though at-sea records suggest that it probably nests in Chile somewhere between 20° and 25° S. Fresh-plumaged juveniles are recorded mostly in June and July, indicating that nesting likely occurs between Dec. and June. Mummified adults and fledglings have been discovered

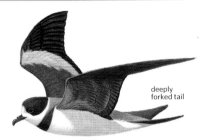

deeply forked tail

striking pied plumage with black cap and breast band

in the extremely arid Atacama Desert, up to 30 miles from the ocean and at elevations up to 5,000 feet. **VAGRANT:** In N.A. waters one certain record: one well-photographed bird about 25 miles south-southwest of Santa Rosa I., CA (Aug. 2, 2005); another about 50 miles off Coos Co., OR (May 3, 2007), was seen only very briefly, but accepted by the state records committee, a decision questioned by many. One specimen record at Isla Gorgona, Colombia (July 1979), is the only other north Pacific record.

SWINHOE'S STORM-PETREL *Oceanodroma monorhis*   SSTP ■ 5

only all-dark Atlantic storm-petrel

pale carpal bar

white base to primary shafts visible at close range

A rather mysterious species in the Atlantic Ocean, single Swinhoe's Storm-Petrels were discovered on islands around Madeira and England in the late 20th century and have since been observed on other islands and at sea elsewhere in the North Atlantic. As yet, no breeding pairs have been discovered, so these could represent prospectors from Indian Ocean populations. Monotypic. L 8" (20 cm) WS 18" (46 cm)

IDENTIFICATION A dark-rumped, medium-size storm-petrel, comparable in size to Leach's and similarly proportioned, with forked tail and head and bill larger than Wilson's. Long, angular wings recalling Leach's or Band-rumped. Plumage entirely deep brown, darkest in flight feathers, except for broad, buffy carpal bar in upperwing. In the upperwing, the outer 4–5 primaries show white shafts at the base, visible at close range. Flight remarkably different from Leach's, typically loping and unhurried, with relatively shallow wingbeats. **SIMILAR SPECIES** In the Atlantic, no similar species, but beware Leach's Storm-Petrels with extensive dark brown plumage in central uppertail coverts, showing whitish uppertail coverts only on the sides; in all plumages, Leach's lacks white primary shafts. Bulwer's Petrel larger, with larger bill, longer wings, and very long tail, also lacks white primary shafts. In the eastern North Pacific, where Swinhoe's is a plausible vagrant (reported once off AK), dark-rumped Leach's are common, and many other dark-rumped species must also be

considered. All of these differ in structure, flight style, and plumage from Swinhoe's, but extensive study would be required to confirm identification. Because viewing conditions at sea can be challenging, many "dark-rumped storm-petrels" observed in the Atlantic and elsewhere are commendably left unidentified. **VOICE** Generally silent at sea. **STATUS & DISTRIBUTION** BREEDING: Islands in the western North Pacific and northern Indian Oceans, mostly off Russia, Japan, Korea, and Taiwan; since the 1990s, small numbers have been detected on islands off Europe and Africa, from Scandinavia to the British Isles and south to Great Salvage I. (near Madeira) and the Azores, as well as in the Mediterranean (Italy, Israel). Post-breeding dispersal still little known. **VAGRANT:** Off Hatteras, NC, single birds have been documented four times (Aug. 20, 1993; Aug. 8, 1998; June 2, 2008; and June 6, 2009). One was documented 192 miles southeast of Cape Sable I., NS (Aug. 9, 2010), just inside N.A. waters. **POPULATION** The Asian population is believed to be in steep decline.

LEACH'S STORM-PETREL *Oceanodroma leucorhoa* LESP ▪ 1

The widespread and variable Leach's is easily confused with several congeners. Essentially a Northern Hemisphere species, it occurs in both oceans from the Equator to the Arctic. Polytypic. (4 ssp.; 3, possibly 4, in N.A.) L 8" (20 cm) WS 18" (46 cm)

IDENTIFICATION Dark brownish black below. Underwing coverts and flight feathers dark brown. Upper wing dark brown with prominent tawny carpal bar extending to the leading edge of the wing. Rump pattern variable, typically lacks a clean white appearance, often smudged gray through the middle, resulting in a "split-rumped" appearance in all Atlantic and most North Pacific populations; it can range from all-white to all-dark in more southerly Pacific breeders. Entirely dark-rumped birds regularly occur in southern California waters. Ages and sexes similar, but molt and wear can result in plumage fading. **FLIGHT:** Very active. Flies with deep, rowing wingbeats incorporating frequent direction changes, bounding from side to side and up and down. Arcs and glides over wave tops on bent wings. In strong winds it may fly more shearwater-like. When flushed, can give short, hurried wingbeats but quickly returns to languid deep wingbeats.

GEOGRAPHIC VARIATION Four subspecies. Nominate *leucorhoa* breeds in the North Atlantic and Pacific Oceans. It is the largest and the least variable in terms of plumage. Individuals in the southern part of its range begin to show intermediate rump characteristics, yet, totally dark-rumped individuals are rare. The subspecies *chapmani* ("Chapman's") breeds off Baja California. It has a high proportion of dark-rumped individuals, slightly rounder wing tips, and a longer and more deeply notched tail. The subspecies *socorroensis* ("Townsend's") breeds in summer on Isla de Guadalupe, off Mexico. It is typically smaller and darker than northern populations, with a shorter tail and a shallow tail fork. Most *socorroensis* show solid white to partly white rumps, but a few are dark-rumped. Some apparent white-rumped *socorroensis* have been photographed well off southern CA in summer. The subspecies *cheimomnestes* ("Ainley's") is not well known. It breeds in winter on Isla de Guadalupe and may range north into southern CA waters. It is mid-size and variably white-rumped. Field identification of subspecies in the Pacific is problematic, but individuals can be characterized as white-rumped, intermediate, or dark-rumped.

SIMILAR SPECIES Separated from other similarly plumaged storm-petrels by nighthawk-like flight style, typically split-rumped appearance, and pale carpal bars reaching the leading edge of the wing. In the Pacific, a dark-rumped individual is easily confused with Ashy and Black Storm-Petrels. Differs from the Ashy in having darker underwings, darker overall plumage, and deeper flight style; differs from the Black by smaller size, less notably forked tail that often appears shorter, and less shearwater-like flight behavior.

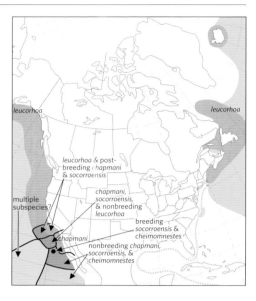

In the Atlantic and Gulf, see the Band-rumped Storm-Petrel. In the Pacific some dark-rumped individuals may be unidentifiable under field conditions.

VOICE Generally silent at sea. Gives a series of generally eerie sounding nocturnal vocalizations including growls and grunts, as well as a low owl-like purring, at breeding colonies.

STATUS & DISTRIBUTION Common at sea around breeding colonies; usually scarce elsewhere. Prefers deep water; rarely seen from shore. **BREEDING:** In burrows on offshore islets. **NONBREEDING:** Generally disperses to surrounding waters, but some populations undertake long migratory movements across the Equator to tropical oceans. **VAGRANT:** Casual inland during hurricanes.

POPULATION Widely distributed and numerous.

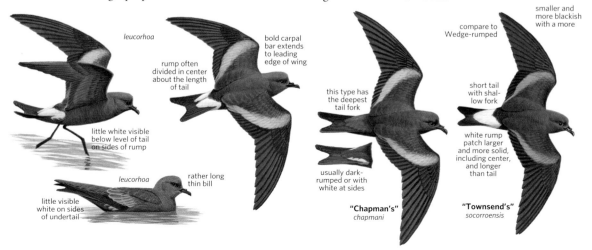

leucorhoa

rump often divided in center about the length of tail

little white visible below level of tail on sides of rump

leucorhoa

rather long thin bill

little visible white on sides of undertail

bold carpal bar extends to leading edge of wing

this type has the deepest tail fork

usually dark-rumped or with white at sides

"Chapman's"
chapmani

compare to Wedge-rumped

smaller and more blackish with a more

short tail with shallow fork

white rump patch larger and more solid, including center, and longer than tail

"Townsend's"
socorroensis

BAND-RUMPED STORM-PETREL *Oceanodroma castro* BSTP ■ 2

This blackish, long-winged storm-petrel flies with a clean-cut white rump patch on stiff bowed wings and often glides like a small shearwater. Monotypic (but see Geographic Variation). L 9" (23 cm) WS 17" (43 cm)
IDENTIFICATION Upper wings blackish brown. Indistinct carpal bar typically does not reach leading edge of wing; worn individuals in late summer/early fall can show a more prominent bar. Shallow-forked tail, often looks square in direct flight. U-shaped rump clean white, extending partially onto rump sides. **FLIGHT:** Flies on shallow, steady, measured wingbeats. Wings have a slightly bowed down appearance.
GEOGRAPHIC VARIATION Often treated as monotypic, but different populations sometimes treated as subspecies, even full species. Up to four cryptic species have been proposed just for the North Atlantic breeders based on vocalizations and timing of breeding (up to two populations nest on the same island, but at different seasons). North American birds are believed to mainly involve widespread "Grant's" from eastern North Atlantic islands. Birds with slightly broader rump bands may be "Cape Verde" that nest on the Cape Verde Is.
SIMILAR SPECIES Darker overall than Leach's, with a less forked tail and a pure white rump pattern (rare in Atlantic Leach's). At least some adult "Grant's" are undergoing primary molt in spring, unlike Leach's in the Atlantic. From Wilson's Storm-Petrel, it lacks the foot projection of that species and its wings are longer and thinner. Wilson's has a fluttery flight style.
STATUS & DISTRIBUTION BREEDING: Nests on the Azores, Berlengas, Canary Is., Madeira Is., Salvages, Cape Verde Is., and farther south in the Atlantic; on HI and other islands in the Pacific. **NONBREEDING:** Disperses west from eastern Atlantic breeding colonies to the Gulf Stream from FL to VA late May–early Sept., rarely as far north as MA. Also regular over deep water in Gulf of Mexico. **VAGRANT:** Casual inland after hurricanes in East.
POPULATION Introduced mammals have reduced Atlantic island numbers.

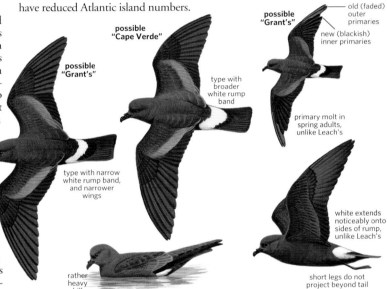

possible "Grant's"
possible "Cape Verde"
possible "Grant's"
old (faded) outer primaries
new (blackish) inner primaries
type with broader white rump band
primary molt in spring adults, unlike Leach's
type with narrow white rump band, and narrower wings
white extends noticeably onto sides of rump, unlike Leach's
rather heavy bill
short legs do not project beyond tail

WEDGE-RUMPED STORM-PETREL *Oceanodroma tethys* WRSP ■ 4

This diminutive vagrant appears white-tailed in flight. Polytypic (2 ssp.). L 6.5" (17 cm) WS 13.3" (34 cm)
IDENTIFICATION Small and dark. White rump patch extends well down uppertail coverts and even the undertail coverts, giving bird the appearance of being white-tailed; the dark corners are visible at close range. On the water, folded wings mostly hide the white rump. Underparts blackish brown; pale-centered dark underwings distinctive. Upperparts dark with pale carpal bar restricted primarily to the wing's inner half. **FLIGHT:** Deep, strong wingbeats; direct flight higher above water than most storm-petrels.
GEOGRAPHIC VARIATION Nominate *tethys* breeds on the Galápagos Is.; smaller *kelsalli* on islets off Peru.
SIMILAR SPECIES Separation from Wilson's Storm-Petrel is problematic, but note Wedge-rumped's extensive triangular white rump patch, longer, pointed wings, and lack of foot projection.
VOICE Generally silent at sea.
STATUS & DISTRIBUTION Vagrant to waters off CA. **BREEDING:** Visits colonies during the day, unlike other storm-petrels. **NONBREEDING:** Disperses to tropical eastern Pacific waters off Colombia and Ecuador and north to Panama. **VAGRANT:** Seven records from central and southern CA (July–Oct.); one *kelsalli* specimen (Jan.).
POPULATION Least concern.

triangular white rump almost reaches end of tail
kelsalli
small size and relatively dark plumage
kelsalli
white on sides of undertail often hidden by wings

ASHY STORM-PETREL *Oceanodroma homochroa* ASSP ▪ 2

Endemic to California and waters off northwestern Baja California, the Ashy Storm-Petrel is best identified by its direct, but fluttery flight style, gray-brown plumage, and pale silvery

molting tail looks more like Least

molting fall adult

distinctly long tailed with deep tail fork

slightly paler gray on uppertail coverts

some pale visible in center of underwing, but can be hard to see

paler than Black Storm-Petrel with ashy-gray plumage

underwing coverts. It can be very similar to a dark-rumped Leach's Storm-Petrel. Monotypic. L 8" (20 cm) WS 17" (43 cm)

IDENTIFICATION Midsize, all-dark storm-petrel with medium brown plumage washed gray when fresh. The unique pale underwing coverts can appear silvery at sea, but they are often difficult to see under field conditions. Overall plumage is paler than other dark storm-petrels, but difference is subtle and subject to wear and fading. Ages and sexes alike. **FLIGHT:** Steady, measured wingbeats, rarely raising wings above the horizontal, and rarely gliding. **SIMILAR SPECIES** Separate Ashy from Leach's by its silvery wing linings and stiff, steady wingbeats and direct flight style. Compared with Black and Least Storm-Petrels, Ashy has shallower wingbeats and paler plumage; also note its proportionally longer tail length, especially compared with Least. **VOICE** Generally silent at sea. Around breeding colonies it makes eerie barking and yelping noises and purring calls from inside burrows.

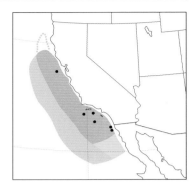

STATUS & DISTRIBUTION Rare to locally common. **BREEDING:** Nests colonially on islets off the coast of CA north to the Farallon Is. and south to the Coronados Is. off Tijuana, Mexico. **NONBREEDING:** In early fall, it congregates in large numbers in Monterey Bay and Cordell Bank, where much of the world population stages. Main wintering area(s) unknown; rarely recorded in winter off CA. **POPULATION** This species' low numbers (approx. 5,000 pairs) and restricted range make it especially susceptible to accidental oil spills and predation from introduced predators.

BLACK STORM-PETREL *Oceanodroma melania* BLSP ▪ 2

The Black Storm-Petrel is the largest and darkest regularly occurring dark-rumped storm-petrel in the eastern Pacific. Monotypic. L 9" (23 cm) WS 19" (48 cm) **IDENTIFICATION** Appears more substantial at sea than other dark, eastern Pacific storm-petrels. Entirely blackish brown, with extensive pale carpal bar formed by tawny greater upperwing coverts. Tail deeply forked. **FLIGHT:** Wingbeats deep and deliberate, Black

Tern-like, rising high above the horizontal before each downstroke. Glides frequently and arcs over wave tops like a small shearwater. **SIMILAR SPECIES** Separation from other dark-rumped Pacific storm-petrels is difficult, especially from a dark-rumped Leach's. Note the distinctive flight style of Black Storm-Petrel, which lacks the nighthawk-like, bounding quality of Leach's. Black Storm-Petrel is noticeably larger than Least Storm-Petrel and appears to have a longer tail and wings. It is darker overall than Ashy Storm-Petrel, with longer wings and more deliberate flight style. It is similar to the paler Tristram's (one N.A. rec.), which has a more deeply forked tail. Also similar to Markham's and Matsudaira's Storm-Petrels (unrecorded in N.A.). **VOICE** Generally silent at sea. A shrieking repeated call given in flight over colonies at night. **STATUS & DISTRIBUTION** Uncommon to common during late summer and early fall off CA coast. **BREEDING:**

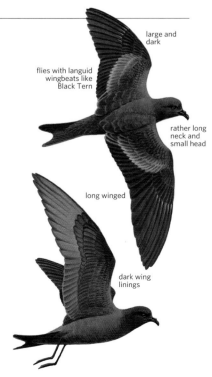

large and dark

flies with languid wingbeats like Black Tern

rather long neck and small head

long winged

dark wing linings

Colonially on islands off western Mexico and in the Gulf of California. A very few nest on rocks off Santa Barbara I., CA. **NONBREEDING:** Regular postbreeding visitor along the CA coast in late summer, more

during warm-water years, but generally not occurring farther north than central CA. Most birds winter off C.A. and south to Peru. **VAGRANT:** Casual to northern OR; also inland at the Salton Sea and along the lower

Colorado River, particularly after strong cyclonic storms.
POPULATION Least concern, but predation by cats and rats at breeding colonies and the limited availability of breeding habitat adversely affect its population.

LEAST STORM-PETREL *Oceanodroma microsoma* LSTP ■ 3

The diminutive, all-dark Least Storm-Petrel is endemic to Mexican waters and disperses northward to near-shore California waters during the late summer and early fall. Its overall appearance has been described as batlike due to its unusual shape, and its flight is quick and erratic on deep wingbeats. Monotypic. L 5.3" (13 cm) WS 15" (38 cm)
IDENTIFICATION Completely dark brownish black and relatively featureless in terms of plumage. Wings relatively long with slightly rounded tip. Often appears almost tail-less in flight due to short wedge-shaped (tapered) tail. Dark

upper wing, with a tawny carpal bar primarily on the inner wing. **FLIGHT:** Wingbeats unusually deep, strong, and direct for a bird of its small size; little or no gliding. Wings held high above body while feeding on the surface.
SIMILAR SPECIES Least differs from other all-dark storm-petrels in its small size, flight style, and shorter wings and tail. It superficially resembles Black Storm-Petrel, but it is much smaller with much shorter wings and tail. Except for Least, other all-dark storm-petrels have notched tails.
VOICE Generally silent at sea.
STATUS & DISTRIBUTION Uncommon in N.A. most years; common during warm-water events. **BREEDING:** Colonially on islands off west coast of Baja California and on islands in the northern Gulf of California. **NONBREEDING:** Disperses north to southern CA waters during warm-water years (sometimes a few to central CA); in other years majority stay south of North American waters. **VAGRANT:** Casual inland at the Salton Sea (hundreds in 1976 after tropical storm Kathleen) and elsewhere in the desert Southwest, particularly

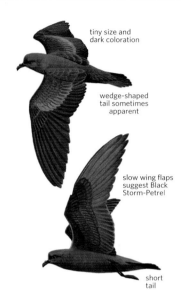
tiny size and dark coloration / wedge-shaped tail sometimes apparent / slow wing flaps suggest Black Storm-Petrel / short tail

after strong cyclonic storms.
POPULATION Least concern, but restricted by limited breeding habitat available on offshore islands and losses from introduced mammalian predators at breeding colonies.

TRISTRAM'S STORM-PETREL *Oceanodroma tristrami* TRSP ■ 5

A species of the subtropical western Pacific Ocean, Tristram's Storm-Petrel breeds as close to North America as the northwestern Hawaiian Is. The species' name commemorates 19th-century English naturalist Henry Tristram. Monotypic. L 10" (25 cm) WS 22" (56 cm)
IDENTIFICATION A very large, bulky, long-winged storm-petrel with deeply forked tail. Plumage rich brown overall, with prominent buffy carpal bars in upperwing and subtle gray sheen to head and rump, often visible in favorable light at close range. In moderate winds, flight tends to be direct rather than erratic, with shallow flaps.
SIMILAR SPECIES Tristram's is larger and paler (especially head and rump) than more delicately proportioned Black Storm-Petrel, which also has a more shallowly forked tail and deeper,

more decisive wingbeats. Markham's and Matsudaira's storm-petrels, unrecorded in N.A., should also be considered. Markham's is also smaller and more delicately proportioned and lacks gray in the rump; Matsudaira's is larger than Tristam's but rather lightly built, with shallower tail fork, generally more buoyant flight style. Both typically show white shafts in outer primaries dorsally, not present in Tristram's.
VOICE Generally silent at sea.
STATUS & DISTRIBUTION BREEDING: Breeds in the Hawaiian archipelago on Nihoa, Necker, French Frigate Shoals, Laysan, and Pearl and Hermes Reef, as well as on islets of the Bonin

and Izu Is. off Japan. The first adults return to court in late Oct., and the last fledglings depart in June. Post-breeding dispersal is little known, but recorded east of Japan in summer. **VAGRANT:** One captured in a mist net at Southeast Farallon I., CA (Apr. 22, 2006) represents the only firm N.A. record.
POPULATION The world population stands at around 10,000 pairs; the species is listed as near threatened because of its small population. The species formerly nested on Volcano Is. (Japan).

larger and grayer than similar Black Storm-Petrel with paler carpal bar and more deeply forked tail

TROPICBIRDS Family Phaethontidae

Red-billed Tropicbird (Galápagos Is., Jan.)

Tropicbirds are pelagic seabirds with long central tail streamers and direct flight styles. For field identification, focus on the distribution of black on the wings (especially on the primaries and primary coverts, which are visible only from above) and flight style; the bird's size, bill color, and tail streamers can be difficult to assess from a distance.

Structure Tropicbirds have heavy, pointed red to yellowish bills and long central rectrices forming spectacular streamers in adults. The pointed wings angle back and the body appears front-heavy in flight. Small, black fully

webbed feet, set well back on the body, make tropicbirds walk with clumsy lunges.

Behavior Flight is direct with species-unique wingbeats; glides are usually brief. On oceangoing trips, you may see them making a brief pass over the boat, or less frequently resting on the water in the distance, with tail streamers held in a high arc. They are found singly or in small, loose groups. They plunge-dive for prey, preferring flying fish and squid.

Plumage Tropicbirds are mainly bright, satiny white with some black markings; the sexes look similar but juveniles are more heavily barred with black than adults. Molts are prolonged and poorly studied; adult tail streamers are often broken.

Distribution Found over most tropical and subtropical oceans, tropicbirds nest on oceanic islands or offshore islets or stacks. None nest in our area. They are uncommon to casual visitors to our southern waters, and vagrants (often storm-related) occur well north of their regular range, onshore and inland.

Taxonomy The world's three tropicbird species form a uniform group without close relatives and were recently classified as comprising a separate order, Phaethontiformes. Geographic variation is minor.

Conservation Tropicbirds face threats from introduced predators on nesting islands; numbers are likely about the same from North American waters, and sightings have tended to increase with better offshore coverage.

Genus *Phaethon*

WHITE-TAILED TROPICBIRD *Phaethon lepturus*   WTTR ▪ 3

The White-tailed, the smallest and most graceful tropicbird, has the most buoyant flight of the tropicbirds. Its long, white tail streamers are especially prominent in flight, undulating and ribbonlike. The distinctive upperwing pattern is only hinted at from underneath, and can be difficult to judge from above in harsh light. Polytypic. (5 ssp.; 2 in N.A.) L 15" (38 cm) Tail streamers 12–25" (30–64 cm) WS 37" (94 cm)
IDENTIFICATION ADULT: The large, black diagonal upperwing bar of adults is unique; it extends from the longest scapulars and tertial centers out the wing to near the wrist joint. The outer

wing is white, with black on outer four or five visible primaries; the black appears cut off by the white primary coverts halfway along the outer wing. The remainder of the upperparts and underparts are white with no barring; the few black flank markings are not generally visible in the field. There is

a small black patch through the eye. The long to very long white tail streamers are ribbonlike (they may have a pink or golden tinge) with black shafts. **JUVENILE:** Lacks tail streamers and is heavily barred with black on upperparts and upperwing coverts, but still shows sharp contrast between black outer primaries and white primary coverts. Black around eye does not extend back around nape. The bill is a dull yellow. Older immatures have lost most dorsal barring and have brighter yellow bills. **FLIGHT:** It is more graceful and buoyant than other tropicbirds; it appears

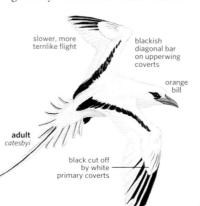

slower, more ternlike flight

blackish diagonal bar on upperwing coverts

orange bill

adult
catesbyi

black cut off by white primary coverts

greenish yellow bill

Pacific adult
dorotheae

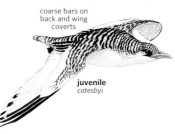

coarse bars on back and wing coverts

juvenile
catesbyi

slim-bodied and flies with rapid tern-like wingbeats. The suggestion of a black W-pattern above is distinctive.

GEOGRAPHIC VARIATION There is slight geographic variation in size, bill color, and extent of black in the primaries. Caribbean and Atlantic records refer to *catesbyi,* with a deep orange bill (can approach red in color) and slightly more black on wing tips (showing black on five visible primaries). On Pacific *dorotheae* note greenish yellow bill and black on only four primaries.

SIMILAR SPECIES Separating White-tailed from Red-billed Tropicbird in the Atlantic is the main problem. Differences in flight styles can be very helpful (see above). Bill color is unreliable, and the diagnostic upperwing pattern may be hard to see. Adult Red-billeds have barred (not white) upperparts and more black in the outer wing, extending through the primary coverts closer to the bend of

the wing. Juvenile Red-billeds show an upperwing pattern like adults, but species differences are subdued by extensive dorsal barring in both species. Juvenile Red-billed also has a longer black eye patch that extends to or across the nape. Red-billed and Red-tailed Tropicbirds are much larger, with broader wings and larger bills than White-tailed.

VOICE Generally silent at sea. Grating *keek keek* series and shrill whistled notes given mainly on breeding grounds.

STATUS & DISTRIBUTION Rare in our area. **BREEDING:** Pantropical; nearest colonies on Bermuda, Bahamas, and Greater Antilles. **NONBREEDING:** Regular visitor (mostly adults) mainly May–Sept., to Gulf Stream off NC; also formerly around Dry Tortugas, FL; casual elsewhere off Atlantic coast north to NS and in the Gulf of Mexico. Very rarely onshore after hurricanes. **VAGRANT:** One onshore record

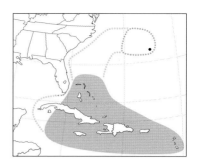

from southern CA (only certain record of *dorotheae,* which breeds as near as HI), and inland records for AZ (likely of Caribbean origin), PA, and NY.

POPULATION Stable. The declines on Bermuda and in the West Indies have been partly offset by conservation measures (predator control, artificial nest sites); there are about 2,500 pairs on Bermuda and 3,500 pairs in the West Indies.

RED-BILLED TROPICBIRD *Phaethon aethereus* RBTR ■ 3

Red-billed Tropicbirds range off both of North America's southern coasts, but are most numerous off CA, where they are most often seen on long-range day trips beyond the Channel Islands (Aug.–Sept.). The combination of barred upperparts, red bill, and long white tail streamers is diagnostic for adults. Polytypic (3 ssp.; *mesonauta* in N.A.). L 18" (45 cm) Tail streamers 12–20" (30–50 cm) WS 44" (112 cm)

IDENTIFICATION Large tropicbird with obvious black in primaries and barred upperparts in all plumages. **ADULT:** Mainly white with fine black barring above from nape to uppertail coverts; extensive black in the primaries and black in the longer primary coverts of upper wing. A long black mark extends from in front of the eye to the nape. Very long, thin white tail streamers lack a black

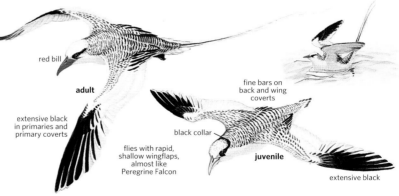

red bill

adult

extensive black in primaries and primary coverts

flies with rapid, shallow wingflaps, almost like Peregrine Falcon

black collar

fine bars on back and wing coverts

juvenile

extensive black

shaft. Bill bright red. **JUVENILE:** Finely barred with black above. Black mark through eye wraps around nape, unlike juveniles of other tropicbirds. Tail shows black tip; no tail streamers. Bill dull yellowish to orange. Older immatures have only sparse spots on crown and short white tail streamers. **FLIGHT:** Flies with rapid, stiff wingbeats, different from the more buoyant flight of White-tailed and the slower more powerful strokes of Red-tailed.

SIMILAR SPECIES White-tailed Tropicbird poses an identification problem in the Atlantic and Gulf of Mexico; see that species for separation. In the Pacific, a casual Red-tailed

Tropicbird also has a deep red bill, but whiter plumage; all ages of Red-tailed look completely white-winged at any distance. Red-billed superficially resembles large Royal (often seen well at sea) and Caspian Terns; but note the tropicbird's more rapid wingbeats, elongated central tail feathers (adults), black barring on the upperparts, dark tertials, and contrasting white secondaries and inner primaries.

VOICE Generally silent at sea. Grating ternlike notes are given mainly around the nesting colonies.

STATUS & DISTRIBUTION Rare in our area, but double figures have been seen on long day trips off southern

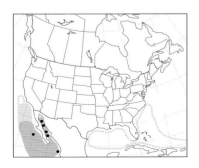

CA. **BREEDING:** Tropical and subtropical eastern Pacific, southern Atlantic, and northwestern Indian Oceans. Nearest colonies in northern Gulf of California and Puerto Rico/Virgin Is. **NONBREEDING:** Regular at sea off southern (casually northern) CA, mainly May–Sept.; esp. frequent in waters off Channel Islands, CA. Rare off Southeast and casual in Gulf of Mexico (May–Sept.). **VAGRANT:** North to WA (one record in 1945), New England, along Gulf Coast (including several storm-wrecked birds), and inland in southern CA and AZ. **POPULATION** Worldwide, the Red-billed is the least numerous tropicbird. Nearby, some 1,800–2,500 pairs breed in the West Indies (mainly Lesser Antilles) and 500–1,000 pairs breed in the Gulf of California.

RED-TAILED TROPICBIRD *Phaethon rubricauda*   RTTR ■ 4

This large, broad-winged, robust tropicbird flies with slower and more powerful wingbeats than the other tropicbirds. Casual in North America, most records are from far off the California coast. Polytypic (4 ssp.; likely *melanorhynchos in N.A.*). L 18" (45 cm) Tail streamers 12–15" (30–38 cm) WS 44" (112 cm)

IDENTIFICATION ADULT: Entirely white (adults may show a faint rosy tinge below) except for a black mark through the eye, black centers to the tertials (a hint of which may show from below), and black shafts to the primaries (not visible from below); the black centers to some flank feathers are not usually visible in the field. The central tail feathers are very thin, stiff, and red (with black shafts); though diagnostic, they tend to disappear against a blue sky or water background. The tail streamers frequently break or are lost. Bill is bright red. **JUVENILE:** It is heavily barred with black above, spotted black on the crown and nape; black in the primaries extends to vanes adjacent to shafts, thus slightly more extensive than in adults. The tail appears mainly white, wedge-shaped; lacks streamers. The bill is black, changing after fledging to yellow. Older immatures retain limited spotting on head and some black in the primary vanes adjacent to shaft, and show short, whitish or pinkish tail streamers. The bill becomes orange to orange-red in the second year. **FLIGHT:** Strikingly white in flight, it appears large, heavy, and broad winged.

GEOGRAPHIC VARIATION There is minor variation in size between subspecies; *melanorhynchos* breeds in the north-central Pacific Ocean, including in the Hawaiian Is.

SIMILAR SPECIES At-sea distribution in eastern Pacific overlaps with Red-billed Tropicbird; note Red-billed's extensive black in the primaries, long white tail streamers, and more rapid wingbeats.

VOICE Generally silent at sea. Grating *ack,* often repeated, around nesting islands.

STATUS & DISTRIBUTION Very rare (but perhaps annual) in our area. **BREEDING:** Tropical Indian and Pacific Oceans; nearest colonies are on HI, but a few may breed on islands far off southwestern Mexico; absent from Atlantic Ocean. **NONBREEDING:**

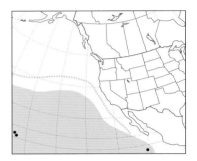

Ranges widely over tropical and subtropical Pacific and Indian Oceans. Probably regular more than 100 miles off CA coast, most records Aug.–Jan. Recorded on the Farallones. **VAGRANT:** One onshore record at a large tern colony at Bolsa Chica Ecological Reserve, Orange County, CA (July 10, 1999); remains of one found at Buttle Lake Park on central Vancouver I., BC (June 1992).

POPULATION The Pacific Ocean breeding population consists of fewer than 15,000 pairs. Exploitation for human food threatens some important central Pacific colonies. Introduced mammalian predators also impact many breeding populations.

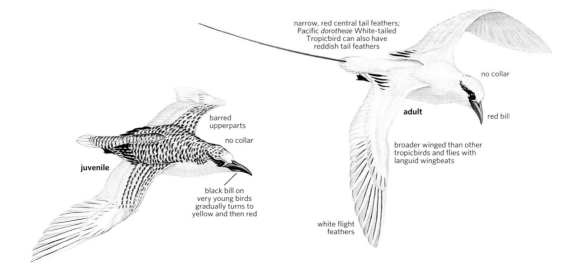

narrow, red central tail feathers; Pacific *dorotheae* White-tailed Tropicbird can also have reddish tail feathers

no collar

adult

red bill

barred upperparts

no collar

juvenile

black bill on very young birds gradually turns to yellow and then red

broader winged than other tropicbirds and flies with languid wingbeats

white flight feathers

STORKS Family Ciconiidae

Wood Stork (FL, May)

Storks superficially resemble other wading birds (especially ibises), but are larger and bulkier. They lack the breeding plumes characteristic of herons and egrets. The only species seen regularly in the United States is the Wood Stork, which forages in any freshwater habitat; saline habitats are used more frequently in the northern parts of their U.S. range.

Structure Storks are large to huge birds, with long necks and legs and heavy bills that may be straight or slightly upturned or downcurved. The tails are short and the wings long and rounded.

Plumage Nearly all species have a combination of white and black feathering. Many have featherless faces or heads to aid in foraging in deeper water or scavenging at carcasses. Soft parts are reddish or orange in most species.

Behavior Most species are colonial; the Jabiru and a few other species nest solitarily or in small groups. Tactile foragers, storks stand or walk slowly in marshes or lakes with their bills open; the mandibles close instantly upon contact with prey. They feed on a wide variety of aquatic animals, primarily fish; some Old World species also scavenge carrion. Unlike wading birds, but similar to cranes, storks fly with the head and neck fully extended. They soar to great heights on thermals, allowing them to efficiently travel great distances from nesting colonies.

Distribution Found on all continents except Antarctica, storks number 19 species in six genera. They are most widespread in the Old World, especially Africa and Asia. Only the Wood Stork and Jabiru occur in North America.

Taxonomy Although storks were once thought to be closely related to New World vultures (Cathartidae) they are now placed in the revamped order Ciconiiformes with a new position following the tropicbirds (Phaethontiformes). Storks are separated into three subfamilies: the Mycteriini (Wood Storks and openbills; 6 sp.); Ciconini ("typical" storks; 7 sp.); and Leptoptilini ("giant" storks; 6 sp.).

Conservation BirdLife International lists three species as endangered and two as vulnerable, with two others as near threatened.

Genus *Jabiru*

JABIRU *Jabiru mycteria*   JABI ■ 4

This huge stork—the largest flying bird in the Americas—is an uncommon resident of tropical freshwater wetlands, where it feeds on fish, mud eels, and other aquatic vertebrates. Monotypic. L 52" (132 cm) WS 90" (229 cm)

IDENTIFICATION Unmistakable, with a massive, upturned bill. **ADULT:** Sexes similar. Plumage entirely white, including wings and tail. Unfeathered head and neck black, with conspicuous red band at base of neck. Black eyes, bill, legs, and feet. **JUVENILE:** Downy feathers on head and neck shed shortly after fledging. Upperparts pale gray, edged grayish brown. Inner webs of primaries pale brown. **IMMATURE:** Scattered brownish feathers on upperparts; red patch on lower neck paler.

SIMILAR SPECIES Wood Stork is smaller, with a slightly downcurved bill, entirely black neck, and black flight feathers on the upper and underwings.

VOICE Generally silent; rattles bill when disturbed, primarily at nest.

STATUS & DISTRIBUTION Uncommon. Native from southeastern Mexico to Bolivia, Argentina, and Uruguay. **BREEDING:** Usually singly but sometimes in loose colonies; nest usually in crown of tall palm or in mangroves. **DISPERSAL:** Nonmigratory, but prone to summer-fall postbreeding dispersal. **VAGRANT:** Casual stray to northern Mexico (Veracruz) and TX (±8 reports); accidental to OK, LA, and MS.

POPULATION Near threatened in C.A. but more numerous in S.A.

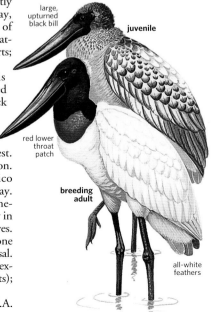

large, upturned black bill

juvenile

red lower throat patch

breeding adult

all-white feathers

Genus *Mycteria*

WOOD STORK *Mycteria americana*   WOST ■ 1

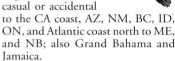

extensive black on flight feathers

The Wood Stork is the largest wading bird regularly encountered in N.A. Flocks forage in fresh water with the bill held open, ready to close on contact with prey, primarily fish. Monotypic. L 40" (102 cm) WS 61" (155 cm)

IDENTIFICATION **ADULT:** Distinctive, large white wading bird with black flight feathers and a tail with green or purple iridescence. The head and upper neck are featherless, "scaly," and dark gray. Dark bill is wide at the base and slightly decurved. Legs

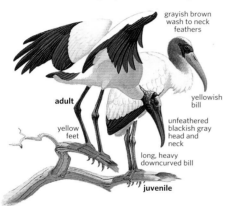

grayish brown wash to neck feathers

adult

yellowish bill

yellow feet

unfeathered blackish gray head and neck

long, heavy downcurved bill

juvenile

are dark; feet pale pinkish. **BREEDING ADULT:** The feet turn pink. Develops plumelike undertail coverts. **IMMATURE:** Adult plumage attained in fourth year. Neck and most of head covered with grayish feathering, lost by the second year. Head and upper neck become "scaly." Straw-colored bill becomes blackish. **FLIGHT:** Black flight feathers contrast strikingly with white underparts and wing linings; black tail partially obscured by legs.

SIMILAR SPECIES In flight, Wood Stork is distinguished from American White Pelican by legs protruding beyond tail, color and shape of head and neck, and irregular flock formation.

VOICE Silent except at nests.

STATUS & DISTRIBUTION Locally common. **BREEDING:** Colonially, typically in tall trees. Locally from coastal SC and GA throughout the FL Peninsula. Also in the American tropics and West Indies. **DISPERSAL:** Nonmigratory, but prone to postbreeding dispersal north and west. **VAGRANT:** Still regular to the south end

of the Salton Sea, where now there are few (formerly large numbers); casual or accidental to the CA coast, AZ, NM, BC, ID, ON, and Atlantic coast north to ME, and NB; also Grand Bahama and Jamaica.

unfeathered head and neck

adult

POPULATION Although listed as endangered because of a sensitivity to local water levels when nesting, the species' population in the U.S. is stable at about 5,000 pairs. Colonies extirpated from TX, LA, and AL.

FRIGATEBIRDS Family Fregatidae

Magnificent Frigatebird, male (FL, Apr.)

Consummate aerialists, frigatebirds are lightly built tropical seabirds with long, angular wings and distinctive, long, deeply forked tails. Only the Magnificent Frigatebird is expected in North America; two other species occur accidentally. Identification is fraught with difficulties.

Structure Light wing loading allows for long periods of effortless flight. The tail is deeply forked, and the small legs and feet are useful only for perching. The long bill is deeply hooked at the tip; the male's gular pouch is bare and red, and can be greatly inflated during courtship.

Behavior Soars great distances over warm waters, occasionally flapping with deep, slow wingbeats; does not alight on the water, but selects a perch such as rocks, shrubs, rigging, and guy wires. Frigatebirds pluck fish, squid, or offal from near the water's surface, often attending seabird feeding-flocks or concentrating around fish cleaning operations. They frequently pirate prey from pelicans, boobies, gulls, terns, and other seabirds through deft aerial chases. Breeding colonies are on shrubs, mangroves on oceanic islands, or coastal bays.

Plumage Adult males are mostly black; females show more white on the underparts and juveniles are even more extensively white below and on the head. Plumage transitions to adult, taking three or more years, are complex and variable.

Distribution Frigatebirds are found throughout tropical waters, but show great potential for vagrancy well beyond their normal ranges (accidental north to NF and AK).

Taxonomy Five species in one genus. The family is now placed in the recently created order Suliformes, which also includes the boobies and gannets (Sulidae), cormorants (Phalacrocoracidae), and darters (Anhingadae).

Conservation Introduced exotic species and human disturbance to nesting colonies have caused strong declines in many populations.

Genus *Fregata*

MAGNIFICENT FRIGATEBIRD *Fregata magnificens* MAFR ■ 1

Our only "expected" frigatebird, the Magnificent is numerous in southern Florida, but occurs widely along the Gulf Coast and more rarely on the Atlantic and Pacific coasts and in the interior. Monotypic. L 40" (102 cm) WS 88" (224 cm)

IDENTIFICATION Frigatebirds are unmistakable. **ADULT MALE:** He is entirely blackish overall, with a slight purple gloss; the gular pouch (inflated in display) is orange to red. The bill and feet are gray. **ADULT FEMALE:** The black on the head comes to a point on the chest; the lower breast is white and the upper belly is constricted in the center. The bill is gray, the feet are pink. **JUVENILE:** Blackish, with a white head, chest, and belly patch, and brown wing coverts. Complex plumage maturation takes four to six years.

SIMILAR SPECIES See Great and Lesser Frigatebirds.

STATUS & DISTRIBUTION Locally common. **BREEDING:** Colony on Dry Tortugas Is., FL.; nearest Mexican colonies in Baja California Sur. **DISPERSAL:** Along Gulf of Mexico coast and throughout Gulf of California. Rare but regular, mainly June–Sept., on CA coast and inland at Salton Sea; fewer in recent years. **VAGRANT:** Casual (mainly summer and fall) north on coasts to southern AK, NF, and through interior of continent.

POPULATION There are about 200 pairs on the Dry Tortugas; large colonies exist in western Mexico.

white head

juvenile

displaying adult ♂

adult ♂

long, thin, angled wings

all-dark head

white chest

long forked tail, usually closed

adult ♀

GREAT FRIGATEBIRD *Fregata minor* GREF ■ 5

Closely similar to the Magnificent Frigatebird, Great Frigatebird vagrants have been recorded three times in North America. Polytypic (5 ssp.; N.A. ssp. unknown). L 37" (95 cm) WS 85" (216 cm)

IDENTIFICATION ADULT MALE: Differs from Magnificent in the brown bar on the upperwing coverts, pinkish feet, and (often) whitish scallops on axillars. **ADULT FEMALE:** Pale gray throat (black in the Magnificent), red orbital ring, and a rounder (less tapered) black belly patch in front. **JUVENILE:** Rust wash on head and chest; pink feet

STATUS & DISTRIBUTION Accidental. **BREEDING:** Extensive range in Pacific and Indian Oceans; large colonies in northwestern Hawaiian Is. and in east Pacific from off western Mexico to Ecuador. West Atlantic subspecies *(nicolli)* on Trinidade and Martín Vaz Is. is nearly extinct. **VAGRANT:** One specimen from OK (Nov. 3, 1975); adults have been photographed in CA in Mar. (Farallon Is.) and Oct. (Monterey Bay).

POPULATION Nearest source colonies include up to 10,000 pairs in Hawaiian Archipelago and about 165 pairs on Revillagigedo Is. off western Mexico.

adult female most distinctive with dark head, white throat, and red orbital ring

all plumages similar to Magnificent Frigatebird

adult ♀

LESSER FRIGATEBIRD *Fregata ariel* LEFR ■ 5

The smallest frigatebird, only a bit larger than the largest gulls, Lesser Frigatebird has been recorded four times in North America. Polytypic (3 ssp.; N.A. ssp. unknown). L 30" (76 cm) WS 73" (185 cm)

IDENTIFICATION In all plumages, white extends from the sides into the axillars. **ADULT MALE:** Differs from the Magnificent Frigatebird by its much smaller size and the white flank patch that extends into the axillars. **ADULT FEMALE:** It has a black head and chest (like Magnificent), but the white breast sides extend into the axillars. **JUVENILE:** Pale rusty head; a white axillar spur is present.

STATUS & DISTRIBUTION Accidental: four widely scattered records, three in July (ME, WY, CA) and one in Sept. (MI). Nearest breeding colonies are in S. Atlantic at Trinidade I. and Martín Vaz I., east of Brazil. Widespread in southwest and central Pacific and Indian Oceans. Regular north to western Pacific. Casually north to Ussuriland and Amurland, Russian Far East.

white spur extends from flanks into axillaries

adult ♂

smallest frigatebird

BOOBIES AND GANNETS Family Sulidae

Northern Gannet (NF, July)

Boobies and gannets are supreme plunge-divers, entering the water from great heights to catch prey up to 30 feet underwater. Flight silhouette is distinctive, with long pointed wings and a body pointed both fore and aft. All are generally pelagic in our area, but gannets are readily seen from shore, as is the occasional booby. For field identification note especially the distribution of dark and light in the plumage, foot and bill color, and overall size. At least some plumages of each species can resemble one or more other species.

Structure Sulids are very large (gannets) to medium-size seabirds with long, pointed wings set well back on the body, wedge-shaped tails, and moderately long, thick necks; the pointed bills lack a hook on the tip and have serrated cutting edges. All sulids have thick, short legs and fully webbed feet. The bare gular pouch, orbital ring and loral area, and the forward-directed eyes give the birds a distinctive countenance.

Behavior The steady flap-and-glide flight may be low over water or (esp. gannets) well above the water surface. The plunge-diving foraging Behavior is directed at fish and squid. Sulids ride buoyantly on the water and swim well. Masked and Blue-footed Boobies generally perch on beaches, rocks, and gravel bars; Brown and especially Red-footed Boobies perch on bare tree limbs, buoys, and ship's rigging (often riding long distances). Gannets usually roost at sea. In most boobies sexes differ in voice, with males (except Red-footed) giving wheezy whistles, and females giving quacking calls; usually silent at sea.

Plumage Most species as adults are largely white below and variably all-dark to white above. The primaries are dark; other flight feathers are variably dark to white. Juveniles are generally darker than adults; plumage maturation takes three years (up to six for gannets). Red-footed Booby has distinct color morphs. Bare parts are often brightly colored, especially in breeding birds, and may differ between sexes and ages. Gannets show more extensive feathering on face.

Distribution Boobies are found in tropical and subtropical waters worldwide; gannets occupy colder waters in the North Atlantic, southern Africa, and Australasia. Nonbreeders disperse to coastal or pelagic waters. Gannets are migratory.

Taxonomy The Sulidae is now placed in the recently created order Suliformes, which also includes frigatebirds (Fregatidae), cormorants (Phalacrocoracidae), and darters (Anhingadae). The family comprises two closely related genera, *Sula* (seven species of booby, five occurring in or near N.A.) and *Morus* (three gannets, one in N.A.). Boobies show weak to moderate geographic variation; a subspecies of Masked Booby—the Nazca Booby—was accorded full-species status in 2000.

Conservation Nesting sulids have long been exploited for human food, and introduced predators may also decimate colonies. Gannets are largely protected and well recovered from past declines, but few booby colonies enjoy effective protection. BirdLife International classifies the Abbott's Booby *(S. abbotti),* nesting only on Christmas Island (in the Indian Ocean), as endangered.

BOOBIES Genus *Sula*

MASKED BOOBY *Sula dactylatra*   MABO ▪ 3

The Masked Booby is the largest of the boobies. It is also the most highly pelagic, staying well at sea except when breeding. Polytypic (4–7 ssp.; 2 in N.A.). L 32" (81 cm) WS 62" (158 cm)

IDENTIFICATION Large with a heavy bill and relatively short tail. Facial skin is bluish black; legs olive-yellow to blackish. **ADULT:** White with black trailing edge to wing, mostly black tail, black tips to scapulars; yellow bill. **JUVENILE:** Dark brown hood with white collar, back and upperwing coverts brown with pale edges, and underparts and most of wing linings white; bill gray to dull yellowish. **SUB-ADULT:** Replaces brown body feathering with white during its second and third years; bill shows yellow by second year. **GEOGRAPHIC VARIATION** Subspecies differ slightly in size and bare part colors. Nominate *dactylatra,* occurring in the Caribbean, is smaller than the Pacific subspecies. Birds recorded in California are likely *californica* (although central Pacific *personata* is very similar).

SIMILAR SPECIES Adults can suggest the much smaller, white-morph Red-footed Booby (see species) and larger Northern Gannet, which lacks black on its secondaries and tail, has a blue-gray bill, and shows a yellow wash on its head and neck. Near-adult gannets with all-black secondaries have been reported, resembling the pattern of an adult Masked, but

they have had all-white scapulars and white in the tail. Juvenile Masked can be confused with juvenile or first-winter Northern Gannet; however, note the gannet's darker underwing (only the axillars are white), lack of white collar, and feathered gular area. See Nazca Booby.

STATUS & DISTRIBUTION Very rare breeder and uncommon to rare non-breeding visitor in N.A. **BREEDING:** Small numbers have nested in Dry Tortugas Islands, FL, since the mid-1980s; closest large colonies in the Caribbean are off the Yucatán Peninsula, with small colonies in the northern Caribbean. A small Pacific colony is located at Alijos Rocks west of Baja California; larger colonies are on the Revillagigedo archipelago and especially Clipperton Island off southwestern Mexico. **NONBREEDING:** Regular at Dry Tortugas Islands, with up to double figures in a day; rare visitor elsewhere off the

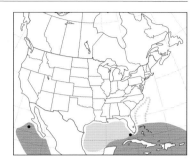

Atlantic coast north to Outer Banks, NC, and in the Gulf of Mexico west to TX. **VAGRANT:** Eighteen records for CA, plus 14 juvenile Masked/Nazca; recorded throughout year, but mostly in summer.

POPULATION Least concern.

all-white wing linings

adult
dactylatra

mostly white underwing

black mask

subadult
dactylatra

adult
dactylatra

juvenile
dactylatra

dark extends to body

dull yellow-green bill

juvenile
dactylatra

white collar on most young birds

subadult
dactylatra

adult
dactylatra

NAZCA BOOBY *Sula granti*   NABO

Recently split from Masked Booby, this eastern Pacific endemic ranges north to Mexico; all Masked-type boobies along the West coast should be scrutinized for this species. Monotypic. L 32" (81 cm) WS 62" (158 cm)

adult with orange-pink (not yellow) bill and more orange eye

all plumages very similar to Masked Booby

adult

IDENTIFICATION Resembles Masked Booby in all plumages. **ADULT:** Orange-pink bill and more orange (not yellow) iris. **JUVENILE:** Extent of plumage variation unknown. Juvenile Nazca averages paler dorsally than Masked (gray-brown, not dark chocolate brown) and lacks or has an inconspicuous white collar (most Masked juveniles show a broad white neck collar).

SIMILAR SPECIES Nazca Booby averages a shorter and thinner bill, shorter legs, and longer wings and tail; it often shows more extensive white at base of tail feathers. Adult Masked shows a yellow to yellow-green bill (not orange-pink) and a yellow (not orangish) iris.

STATUS & DISTRIBUTION Common in its limited range. Potential free-flying vagrant to N.A. **BREEDING:** Main colonies are on the Galápagos and off western Colombia; about 200 breed on Clipperton Island far off southwestern Mexico, and a few breed on the Revillagigedo archipelago and possibly Alijos Rock southwest of Baja. **DISPERSAL:** Generally inshore, north to west coast of Mexico (a few into the Gulf of California). **VAGRANT:** One May record of a juvenile landing on a ship off northwest Baja California and riding into San Diego Bay, CA, has not been accepted by the state records committee (origin questionable).

BLUE-FOOTED BOOBY *Sula nebouxii* BFBO ■ 4

This aptly named booby with sky blue feet is an inshore species. Polytypic (2 ssp.; nominate in N.A.). L 32" (81 cm) WS 62" (158 cm)

IDENTIFICATION Large, very long tailed, long-billed booby with dirty brown and white patterning and distinctive blue (to blue-gray) legs and feet. Central tail feathers mostly whitish. **ADULT:** Streaked head and neck diagnostic. Brownish upperparts with white barring; white patches on upper back and rump. Underwing with white linings divided by dark median covert bar. Pale yellow eyes

streaked head
white on upper back
white scalloping on lower back
long attenuated bill thins near tip
adult *nebouxii*
white uppertail coverts
bright blue feet
white central tail feathers

(dark ring around pupil in female); dark gray bill; blackish facial skin. **JUVENILE:** Head, neck more solidly and extensively brown than in adult, whitish dorsal patches more limited. Grayish blue feet; dull grayish bill; gray-brown eyes.

SIMILAR SPECIES Compare juveniles: Juvenile Brown Booby is smaller and solidly dark above; dark brown breast contrasts sharply with slightly paler underparts; underwings lack dark median covert bar, tail all dark. Juvenile Masked Booby usually shows a complete white collar, shorter all-dark tail, and more extensively white underwing.

STATUS & DISTRIBUTION Irregular vagrant in N.A. **BREEDING:** Common on islets in Gulf of California south to the Galápagos Islands and Peru. **NONBREEDING:** Rare and irregular (mainly mid-July–mid-Oct.) to Salton Sea, with occasional larger irruptions. Casual to coastal and central CA, southern NV, and western AZ usually during irruption years. Two major "flight years" since 1977: a small irruption in 1990 and the largest ever recorded in 2013. **VAGRANT:** Records north to OR, WA, and BC; and single records from NM and a long-staying bird in central TX.

POPULATION Least concern.

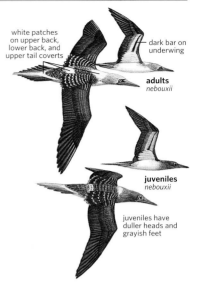

white patches on upper back, lower back, and upper tail coverts
dark bar on underwing
adults *nebouxii*
juveniles *nebouxii*
juveniles have duller heads and grayish feet

BROWN BOOBY *Sula leucogaster* BRBO ■ 3

A dark brown and white bird of tropical and subtropical waters, the Brown Booby occurs rarely but regularly in southern North American waters, mainly in southern Florida and California. Polytypic (4 ssp.; 2 in N.A.). L 30" (76 cm) WS 57" (145 cm)

IDENTIFICATION Moderately small. Sharp contrast between dark chest and paler belly and underwing coverts; contrast subtle in the darkest juveniles. Bill and foot colors vary with population, age, and season (brighter at onset of breeding). **ADULT:** Upperparts, tail, head, and breast dark brown, meeting bright white underparts in a sharp line across the lower breast. Axillars and underwing secondary coverts white.

Legs and feet pale green to yellowish. On female, yellow to pale flesh-colored bill, yellow facial skin, and a dark spot in front of the eye. On male, dusky yellowish bill, blue facial skin. **JUVENILE:** Dusky brown head, breast, upperparts; underparts dull whitish with dusky scalloping. Axillars and underwing secondary coverts dull whitish to mottled pale gray, but showing some contrast with remaining underwing. Bill bluish gray. **SUBADULT:** Increasingly white on belly, wing linings.

GEOGRAPHIC VARIATION Nominate *leucogaster* (Caribbean and tropical Atlantic) occurs in our southeastern waters. In *brewsteri* (Gulf of California to islands well off southwestern Mexico), males are distinctive with extensive whitish frosting on the head and upper neck; juveniles are often very dark on the underparts.

SIMILAR SPECIES Compare juveniles: Juvenile Brown, even the darkest

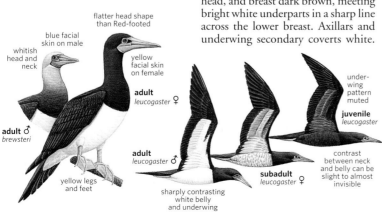

flatter head shape than Red-footed
blue facial skin on male
whitish head and neck
yellow facial skin on female
adult *leucogaster* ♀
adult *brewsteri* ♂
adult *leucogaster* ♂
yellow legs and feet
sharply contrasting white belly and underwing
subadult *leucogaster* ♀
contrast between neck and belly can be slight to almost invisible
underwing pattern muted
juvenile *leucogaster*

brewsteri, shows a sharp line of contrast between the breast and slightly paler lower breast and belly, but it can be very subtle, even invisible on distant birds. A paler Red-footed Booby juvenile is mostly tan below

with a dark band across the chest; a darker juvenile is more uniform, but it still shows a chest band. Red-footed's underwings are darker overall and more uniformly dark. Legs and feet can appear pinkish in both species, but head shape differs subtly; Brown has a more continuous contour from the culmen to the forehead, Red-footed has a more concave culmen and more rounded forehead. Juvenile Masked and Nazca Boobies are larger than a Brown, have a dark hood limited to the head and neck, and show white on underwing primary coverts. Juvenile Northern Gannet is much larger, spotted with white, and has a dark underwing with only the axillars white.

STATUS & DISTRIBUTION Rare in N.A. **BREEDING:** Common in most tropical and subtropical seas including Gulf of California and parts of Caribbean. Recently established as a breeder on the Los Coronados Is. off northern Baja California. **NONBREEDING:** Rare to our southern coasts; regular to Dry Tortugas, FL; very rare on TX coast. Now annual along and off the CA coast (100+ recs. for CA, casual inland from Salton Sea area and the lower Colorado River). **VAGRANT:** Casual on Atlantic coast (mostly July–Sept.), north to MA (twice) and NS, usually after southerly storms. Accidental to OR and Niagara River, NY/ON (Oct. 7–Nov. 2013).
POPULATION Least concern.

RED-FOOTED BOOBY *Sula sula*   RFBO ▪ 4

This smallest and relatively graceful booby is widespread in tropical oceans, but occurs only marginally in North America. Many of the records, especially from Pacific waters, have been of birds observed from boats. Adults show an array of color morphs, but they always have striking red feet. Polytypic (3 ssp.; 2 in N.A.). L 28" (71 cm) WS 60" (152 cm)
IDENTIFICATION Red-footed Booby is relatively slender with long, narrow wings. The forehead is more rounded than in other boobies, and the small bill is slightly concave along the culmen. **ADULT:** Plumage variable. In all morphs, the bill is pale blue, the facial skin pink, the orbital ring pale blue, and the feet brilliant red. **WHITE MORPH:** White to creamy white throughout with black primaries and secondaries (innermost secondaries white), black median

primary coverts on underwing, pure white scapulars; tail white or black (tipped white). **BROWN MORPH:** Brown throughout, but lighter tan on head and underparts and darker dusky on flight feathers; underwings entirely dark. **WHITE-TAILED BROWN MORPH:** Similar, but rump, vent, tail coverts, and tail white. **JUVENILE:** Brownish throughout, but usually paler (even creamy tan) on head, neck, and underparts, with dark chest band. Bill blackish, feet dull pinkish. **SUBADULT:** Approaches respective adult color morph; white morph subadult often shows brown mantle (retained in some adults). Bill pink with dark tip, gradually changing to blue-gray; feet gradually brighten; definitive adult plumage attained in third year.
GEOGRAPHIC VARIATION Caribbean birds (nominate *sula*) and eastern Pacific birds (*websteri*) differ little; all-

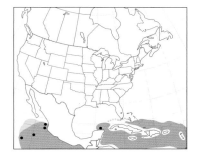

brown morph is rare in the Caribbean.
SIMILAR SPECIES White morph adult with black tail resembles the much larger Masked and Nazca adults, but note Red-footed's blue-gray (not yellow or orange) bill, pink and pale blue facial skin, and bright red feet. In flight, the black on Red-footed's wings does not reach the body (scapulars and

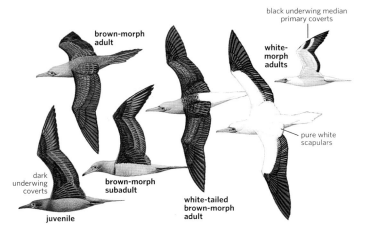

innermost secondaries are white) and there is a black "comma" mark on the primary coverts. See Brown Booby for juvenile separations.
VOICE Generally silent in our area.
STATUS & DISTRIBUTION Pantropical distribution. Casual in our area.

BREEDING: Well-vegetated, tropical islets; nearest colonies off western Mexico and in the eastern Caribbean (off Yucatán), and rarely the Bahamas. **DISPERSAL:** Nonbreeders generally well out to sea. **VAGRANT:** Casual but most regular on Dry Tortugas Is. and Gulf

Stream waters off FL; has occurred north to SC on Atlantic coast; one record off upper TX coast. Casual in CA, along the coast and well offshore (19 recs., mainly Aug.–Nov., north to San Francisco area).
POPULATION Least concern.

GANNETS Genus Morus

NORTHERN GANNET   Morus bassanus   NOGA ■ 1

This species is the only common sulid along the Atlantic coast. Readily seen from shore in winter and on migration, it often flies in low lines or may soar high above the ocean. It executes spectacular aerial feeding dives, wings folded back just before hitting the surface. Monotypic. L 37" (94 cm) WS 72" (180 cm)
IDENTIFICATION Larger and with longer wings than a booby. Adult plumage attained in about four years. **ADULT:** Mainly white with black primaries; golden buff wash on head. Pale gray bill, black facial skin, mostly black feet. **JUVENILE:** Slate brown, spotted with white, with white uppertail coverts; underwings dark with white axillars. **SUBADULT:** Dark body, wing coverts, secondaries, and rectrices gradually replaced by white, with much variation within age classes. Third-year birds often

have a checkered or "piano key" look to the wings and a yellowish head. Central tail feathers and secondaries are usually the last dark feathers replaced.
SIMILAR SPECIES A distant immature flying low over the water can suggest a large shearwater or even an albatross. See Masked Booby for differences between immatures.
STATUS & DISTRIBUTION Common to abundant. **BREEDING:** Colonially on uninhabited islets and rocky cliffs, mainly Gulf of St. Lawrence and off NF. **MIGRATION:** Atlantic coast, mainly Mar.–May, Oct.–Dec. **WINTER:** Atlantic coast from MA to FL, and Gulf of Mexico, mainly Nov.–Apr. **VAGRANT:** Casual on Great Lakes; accidental elsewhere in Northeast, Midwest, once (Sept.) on Victoria I.,

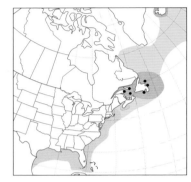

NT. Recent sight records off northern AK and one long-staying adult at Farallones Is., CA.
POPULATION Least concern.

CORMORANTS Family Phalacrocoracidae

Double-crested Cormorant, immatures (NJ, Jan.)

ormorants (including "shags") are long-necked diving birds represented in North America by the widespread Double-crested Cormorant and five more localized species. For identification, concentrate on shape and color of bill and gular region, overall size, tail length, neck thickness, flight silhouette, and distinctive markings of breeding adults.

Structure Cormorants have long necks, long, heavy bodies, stiff tails, and long bills strongly hooked at the tip. Their feet are fully webbed for underwater propulsion but serve also for perching on branches, cables, and cliff faces.

Behavior Cormorants pursue fish in shallows or exploit schooling fish in deeper water. The dense, wettable body plumage reduces buoyancy; after a diving bout cormorants perch with spread wings to dry. Flight involves steady, shallow wingbeats; cormorants briefly glide but rarely soar. Cormorants nest in small to very large colonies on sandy or rocky islands, cliff faces, or mangroves and other shrubs or trees. Their grunting calls are usually heard only around the colonies.

Plumage North American species are mainly black as adults, often with a green or purple gloss and, in some species, scaly patterning on the upperparts; many extralimital species are white below. Cormorants variously show wispy plumes, crests, bright gular pouches, and white flank patches early in the breeding season. The sexes look similar. Juveniles and immatures are duller and browner than adults and lack ornaments. Adult plumage is attained in two to four years.

Distribution Cormorants frequent coastlines and interior lakes and wetlands worldwide. Some species are quite sedentary, others are somewhat migratory. Four North American species are strictly marine; two others also occur inland.

Taxonomy The cormorant family is now placed in the recently created order Suliformes, which also includes frigatebirds (Fregatidae), boobies and gannets (Sulidae), and darters (Anhingidae). All the ±39 species of cormorants worldwide are in the genus *Phalacrocorax*.

Conservation Many cormorant species have declined (one is extinct; at least ten are threatened), but North American populations are generally stable or expanding.

Genus *Phalacrocorax*

BRANDT'S CORMORANT *Phalacrocorax penicillatus* BRAC ■ 1

Common along the Pacific coast, this species is also the most numerous cormorant well offshore. Monotypic. L 35" (89 cm) WS 48" (122 cm)

IDENTIFICATION Fairly large with a short tail. Always shows pale buff on throat. **BREEDING ADULT:** Black throughout (with slight gloss) except for buff throat feathering; gular pouch bright blue; thin white plumes on neck, back. **NONBREEDING ADULT:** Similar to breeding, but no gloss or plumes and pouch dull gray. **JUVENILE:** Dark brown above; lighter brown below with band of pale buff on throat. Adult plumage attained third year. **FLIGHT:** Straight neck, head only slightly elevated.

SIMILAR SPECIES Double-crested and Pelagic are found within range of Brandt's. Double-crested has a yellow to orange gular pouch and facial skin, deeper bill, and scaly upperparts (adults); in flight it shows a thick neck with elevated head and longer, more pointed wings. Much slimmer Pelagic has a longer tail, smaller head, and thinner bill; juveniles are darker and all plumages lack buff throat patch.

STATUS & DISTRIBUTION Common.

BREEDING: Colonially on offshore islands. **NONBREEDING:** General northward postbreeding move July–Oct., and southward move in fall and winter; during these seasons found north to southeastern AK. **VAGRANT:** Accidental in interior CA.

POPULATION Populations fluctuate greatly, but overall numbers are stable.

buffy band under bill in all plumages

juvenile

blue skin

winter adult

breeding adult

appears shorter tailed than Pelagic

NEOTROPIC CORMORANT *Phalacrocorax brasilianus* NECO ▪ 1

Resembling a small, long-tailed version of the Double-crested Cormorant, this aptly named, primarily Neotropical species is fairly common in North America along the Texas and western Louisiana coasts and locally inland in Texas and New Mexico. Polytypic (2 ssp.; *mexicanus* in N.A.). L 26" (66 cm) WS 40" (102 cm)

IDENTIFICATION The combination of the bird's characteristic field marks— small size, relatively long tail, short bill, black feathered lores, and dull yellow gular pouch bordered by white feathering and distinctly pointed at the rear—identify most Neotropic from Double-crested Cormorants, though the two species are most easily told when together. Easily separated in flight by proportional differences

described above. **BREEDING ADULT:** Blackish throughout with a slight bluish gloss; the back feathers and wing coverts are olive-gray with black borders. The gular pouch varies from brownish to yellow, but it is always bordered at the rear by a distinct band of white feathering coming to a point behind the gape. Short, thin white plumes are concentrated on the sides of the head and neck. **NONBREEDING ADULT:** Resembles a breeding bird, but lacks the white plumes, body plumage is less glossy, and gular pouch is a paler yellow with a slightly less distinct white border. **JUVENILE:** Brown throughout, often considerably paler on the head, neck, and breast. Gular pouch is a pale yellow; the white border is indistinct or absent. Older immatures gradually attain the dorsal patterning, blacker body plumage, and white throat border of adults.

GEOGRAPHIC VARIATION North American birds *(mexicanus)* are considerably smaller than nominate birds found ranging from Panama through South America.

SIMILAR SPECIES To distinguish immature Neotropic from the closely similar Double-crested Cormorant, see the sidebar below. It is unlikely that

other small, long-tailed cormorants (e.g., the Pelagic) will be confused with Neotropic.

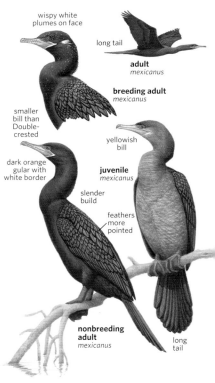

wispy white plumes on face

long tail

adult
mexicanus

breeding adult
mexicanus

smaller bill than Double-crested

yellowish bill

dark orange gular with white border

juvenile
mexicanus

slender build

feathers more pointed

nonbreeding adult
mexicanus

long tail

Identification of Immature Double-crested and Neotropic Cormorants

The Double-crested Cormorant occurs widely, especially in winter, within the more limited U.S. range of the Neotropic, and vagrant Neotropics are often seen with Double-cresteds; therefore opportunities for close comparison of the two species are often available.

Some characters that readily distinguish adults, such as the shape of the gular pouch, the distinct white throat border of Neotropic, the absence of bare yellow or orange supraloral skin in Neotropic, and the twin curled crests of Double-crested can be absent, obscure, or difficult to assess in juveniles and younger immatures. Thus close scrutiny of facial features in combination with careful attention to (and ideally comparison of) size and shape characters are important.

The yellowish gular pouch of Neotropic comes to a point behind the gape of the bill, with the throat feathering coming forward to a point under the pouch. In Double-crested the gular pouch is squared or rounded at the rear, with less feathering protruding forward underneath. This shape difference holds for all postnatal plumages. In all Double-cresteds there is a patch of yellow-orange facial skin in front of the eye and above the dark lores; this area is dark and feathered in

the Neotropic. (But beware: Some individuals show a thin yellowish streak above and in front of the eye.)

Double-crested Cormorants average larger than our *mexicanus* Neotropics in all standard measurements except tail length, but the smallest Double-crested can overlap Neotropic in bill length. Neotropics have slightly slimmer bodies and smaller heads. The most distinctive shape difference is the relatively much longer tail of Neotropic (the tail is 60 percent of the wing length in Neotropic, but less than 50 percent in Double-crested); in flight the tail length of Neotropic roughly equals the length of the neck (with head), whereas the tail is shorter than the neck in the Double-crested. ∎

yellow-orange above dark lores

juvenile Neotropic

more acute angle at gape

more rounded shape at gape

juvenile Double-crested

STATUS & DISTRIBUTION Common and widespread from Mexico through South America. **BREEDING:** Fairly common in main U.S. range (coastal TX and southwestern LA); found in coastal bays, inlets, and freshwater lakes and ponds. More local inland on reservoirs, lakes, and rivers; breeds in southern NM and southern OK. **NONBREEDING:** Northeast and north-central TX birds withdraw south in winter. Regular and increasing visitor to southern AZ and southeastern CA. **VAGRANT:** Rare to casual in southern NV, CO, central Great Plains, and the Midwest; accidental to southern ON and MD.

POPULATION After suffering severe declines in the 1960s, the populations in the United States have largely rebounded, but there remains strong yearly variation in colony site use and population size.

DOUBLE-CRESTED CORMORANT *Phalacrocorax auritus* DCCO ▪ 1

The Double-crested is North America's most widespread and familiar cormorant, found in both marine and freshwater habitats. It often flies higher than other cormorants, often in goose-like V-formations. Polytypic (5 ssp.; 4 in N.A.). L 32" (81 cm) WS 52" (132 cm)

IDENTIFICATION Stocky, with medium-length tail, thick neck, and moderately thick bill; yellow to orange gular pouch, rounded or square at rear; bare supraloral skin yellow to orange; lores dark. **BREEDING ADULT:** Black; coverts and back feathers with grayish centers, black fringes. Facial skin orange; mouth lining blue. Nuptial crests black, white, or peppered. Bill blackish above, spotted with pale below. **NONBREEDING ADULT:** Similar to breeding adult, but crests lacking and facial skin often more yellow. **JUVENILE:** Dark gray-brown above with paler centers to back feathers and wing coverts; underparts color varies with individual and feather wear, but breast always paler than belly. Throat and breast dark brownish gray, becoming paler, even nearly white, with wear and fading (flanks and lower belly darker). Bill extensively yellow to pale orange. Older immatures resemble nonbreeding adults but plumage duller and browner.

GEOGRAPHIC VARIATION Subspecies differ in overall size and in color and shape of crest in breeding adults. Nominate *auritus* (widespread from the Great Basin and Rocky Mountain region east through central and eastern N.A.) is moderately large with black crests; *floridanus* (FL) is our smallest subspecies, with dark crests; *cincinnatus* (coastal AK to northern BC) is largest, with straight white crests; and *albociliatus* (Pacific coast from southern BC to western Mexico, and presumably inland to the Great Basin) is large and usually shows white to partially white crests. Subspecies range limits in interior western North America are not clear.

SIMILAR SPECIES Neotropic Cormorant is similar (see sidebar opposite) as are Great and Brandt's Cormorants (see those species). Very pale-breasted immatures might be mistaken at a distance for loons or, when perched, boobies.

STATUS & DISTRIBUTION Common to abundant. **BREEDING:** A wide range of

lakes, marshes, coastal estuaries, and offshore islands; colonially on rocky or sandy islands, marsh vegetation, waterside trees, or artificial structures. **MIGRATION:** Northern breeders move north mainly Mar.–Apr., return south mainly Aug.–Nov. **WINTER:** Mainly coastal habitats, but also on inland lakes, reservoirs, and fish farms. **VAGRANT:** Some summer wandering north of regular range (e.g., Great Slave Lake). Casual to Bering Sea region.

POPULATION Currently there are about 350,000 breeding pairs in North America. Most populations declined greatly from the 1800s through the early 1900s because of persecution and introduced predators. Major declines occurred again from the 1950s to about 1970 because of pesticide impacts. Numbers have strongly rebounded since the 1970s in the interior and Atlantic coast regions. Some populations are now controlled through management (especially in the Great Lakes region).

short tail

adult

wispy black crest

breeding adult

juvenile

dark belly

pale breast

yellow-orange stripe above dark loral stripe

orangish gular pouch

winter adult

wispy white crest

western breeding adult

mostly dark below

some 1st years have white bordering gular pouch

some 1st years have paler bellies

2nd year

1st year

1st year

GREAT CORMORANT *Phalacrocorax carbo*   GRCO   1

This species, North America's largest cormorant, has a huge worldwide range, but in North America it breeds only in easternmost Canada and Maine; nonbreeders occur southward along the Atlantic coast in winter. Polytypic (7 ssp.; *carbo* in N.A.). L 36" (94 cm) WS 63" (160 cm)

IDENTIFICATION Large size, large blocky head, short tail, and limited yellow gular pouch bordered by white feathering. **BREEDING ADULT:** Black with scaly upperparts; broad white throat patch bordering small lemon yellow gular pouch and facial skin; thin white plumes on sides of head and bold white patch on flanks. **NONBREEDING ADULT:** Similar to breeding, but white throat patch more limited and white head plumes and flank patches absent. **JUVENILE:** Gray-brown head, neck and breast contrast with white belly; flanks, wing linings, and upperparts dusky; pale feathered border to small yellow facial skin area. Older immatures progressively darker on body; definitive plumage attained in third year.

SIMILAR SPECIES Double-crested Cormorant, the only other cormorant in Great Cormorant's North American range, is smaller with a slightly slimmer neck and bill. A breeding adult Double-crested lacks Great Cormorant's white feathering on the face and flanks and has more extensive and deeper orange-yellow facial skin. A nonbreeding adult is more similar, but note that the Great Cormorant's limited pale yellow facial skin is pointed, not rounded at rear, and its broad pale feathering bordering the facial skin points forward on throat toward bill. Most immature Great Cormorants are readily told from immature Double-crested Cormorants by their two-toned underparts: dark on the breast and flanks, whitish on the belly; a few Greats are more uniformly dark below and best told by facial characters (above) and grayish (not extensively orangish) bill. Also compare with much smaller and longer-tailed Neotropic Cormorant.

STATUS & DISTRIBUTION Fairly common. **BREEDING:** Maritime habitats; nests colonially on ledges, plateaus on rocky islands, coastal points. **NONBREEDING:** A few immatures summer to NJ, rarely farther south. **WINTER:** Seacoasts, bays, and lower portions of major rivers; regular south to NC and SC, very rarely to FL. **VAGRANT:** Rare and irregular inland in Northeast as far as Lake Ontario. Casual in Gulf of Mexico (western FL, MS); one record for WV. **POPULATION** Only about 5,500 pairs breed in North America, 80 percent of these in Quebec and Nova Scotia. Largely recovered from 19th-century declines, populations are stable.

large blocky head

yellowish throat pouch

white

olive-bronze sheen on back

nonbreeding adult
carbo

white belly

juvenile
carbo

long, heavy bill

white on throat more diffuse

white neck feathers soon lost

dark neck

dark throat pouch

breeding
carbo

white flank patch

breeding adult

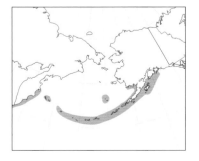

RED-FACED CORMORANT *Phalacrocorax urile*   RFCO   2

A close relative of the smaller Pelagic Cormorant, this species is found only from southwest Alaska west through the Pribilofs and Aleutians to the northern Sea of Japan. It is not especially gregarious and nests in small colonies on cliff faces and steep rocky islands. Monotypic. L 31" (79 cm) WS 46" (117 cm)

IDENTIFICATION Extensive yellow to pale yellowish gray on bill in all plumages. A bit stockier in build than Pelagic, with a slightly thicker bill. **BREEDING ADULT:** Glossy black with browner wings, large white flank patch; bill yellowish with blue base; bright red pouch and facial skin surrounding eyes and extending across forehead; thin white neck plumes. **NONBREEDING ADULT:** Similar to breeding, but body less glossy, red face and blue bill base duller, and white neck plumes and flank patch lacking. **JUVENILE:** Dark brown, slightly paler

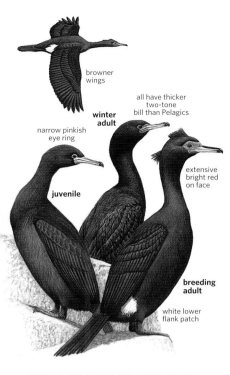

browner wings

all have thicker two-tone bill than Pelagics

winter adult

narrow pinkish eye ring

juvenile

extensive bright red on face

breeding adult

white lower flank patch

on breast; facial skin dull yellow to pinkish (but forehead feathered); some yellow on bill.

SIMILAR SPECIES Closely resembles Pelagic Cormorant, but Red-faced is about 20 percent heavier with a stocky build, including a thicker neck and blockier head, apparent at rest and in flight; also slightly thicker bill. The Red-faced Cormorant's bill is extensively pale (usually yellow), whereas the bill of Pelagic Cormorant is slate gray or blackish. In adult Red-faced, the dark brown wings contrast with black body (Pelagic is more uniformly black). The facial skin of adult Red-faced Cormorant is a bright red and extends across forehead and broadly surrounds the eyes; in Pelagic, the red is deeper in color and limited to the lower face (forehead is feathered black). Juveniles are best told apart by the pale yellowish bill and face

of the Red-faced (Pelagic is dark in these areas).

STATUS & DISTRIBUTION Common year-round. **YEAR-ROUND:** Nests on sea cliffs with kittiwakes and alcids, commonly on the Pribilofs and Aleutians; some fall and winter dispersal within the general breeding range. It is less common and more local in the remainder of the AK range. **VAGRANT:** Casual in AK north to Norton Sound and southeast to Sitka. Recorded twice (Apr., June) on Queen Charlotte Is., BC.; one May sighting in northwestern WA—though accepted by the state rarities committee, it should perhaps be considered equivocal. Similarly, sightings from St. Lawrence Is. are not substantiated; some were clearly of Pelagic Cormorants. Well out-of-range claims of this species should be substantiated with photos. **POPULATION** About 50,000 birds breed in Alaska; populations are apparently stable, but little trend data exists.

PELAGIC CORMORANT *Phalacrocorax pelagicus*   PECO   1

Despite its name, this small, slender cormorant frequents rocky coastlines and is rarely seen far at sea. Polytypic (2 ssp., both in N.A.). L 26" (66 cm) WS 39" (99 cm)

IDENTIFICATION Slender neck (held straight in flight), small head, very thin blackish bill, and relatively long tail. Plumage entirely dark, except for white flank patches of adults early in the breeding season. **BREEDING ADULT:** Shiny black, glossed purple and green, except for large white patch on flank and scattered thin white plumes on the neck. Short crests on crown, nape. Deep red facial skin limited to area around chin and eyes. **NONBREEDING ADULT:** Similar to breeding, but no flank patch, neck plumes, or crests. Body plumage slightly less glossy;

red facial skin less evident. **JUVENILE:** Blackish brown above, only slightly paler on underparts; our darkest juvenile cormorant. Adult plumage attained second year.

GEOGRAPHIC VARIATION Nominate *pelagicus* breeds from southern British Columbia through Alaskan and northeast Asian range; *resplendens,* from southern British Columbia to extreme northern Baja California, Mexico, is smaller, with a more slender bill.

SIMILAR SPECIES See Red-faced Cormorant. Along the Pacific coast, Pelagic Cormorant can be confused with Brandt's Cormorant; the latter is larger, with a heavier body and neck, larger head, thicker and longer bill, and shorter tail. Brandt's also shows a buff throat (dark in all Pelagic Cormorant plumages).

STATUS & DISTRIBUTION Common, but rarely in large groups. Regularly forages up rivers and in bays well away from the immediate coast. **BREEDING:** Colonially on steep rocky slopes or cliff faces on coastlines and islands. **NONBREEDING:** Generally resident, but northernmost populations withdraw south for winter. Winters south to Pacific coast of Baja California. **VAGRANT:** Casual on northern coast of

thin neck

all have slender, dark bill

winter adult

limited dark red facial skin

juvenile

uniformly dark brown, including chin

breeding adult

white lower flank patch

AK and on leeward Hawaiian Is. Accidental east of Sierra Nevada at Silver Lake, Mono Co., CA (Dec. 8, 1976). **POPULATION** The North American population of about 130,000 birds (mostly in AK) is generally stable.

pelagicus

resplendens

DARTERS Family Anhingidae

Anhinga, breeding male (FL, Apr.)

Darters resemble slender, long-tailed cormorants. They are often seen loafing or sunning on waterside perches.

Structure Darters have a thin neck and small head with a pointed bill and small gular pouch; the heronlike neck is kinked (serving as a hinge for quick strikes at prey). The long broad wings, thin-based tail, and straight neck give flying darters a crosslike appearance. The short legs have large, fully webbed feet.

Behavior Highly aquatic, they swim with the body submerged; the wettable plumage reduces buoyancy, but it requires long periods of drying and sunning with wings and tail spread. Flight consists of several flaps and short glides; darters often soar on thermals. They use their sharply pointed bills to spear fish and other prey. They are seen singly or in small, loose groups and nest in waterside trees and mangroves, often alongside herons, ibis, cormorants, and storks.

Plumage Adult males are largely black with silver-white markings on the upperwing coverts, back, and scapulars; in Old World taxa the head and neck have brown and white markings. Females and immature males are extensively gray-brown to buff on the head, neck, and breast.

Distribution Darters occur in warmer regions of the New World from the southeastern U.S. through tropical South America, sub-Saharan Africa, South Asia, and Australia/New Guinea. Some populations are migratory or move in response to changing water conditions.

Taxonomy The darter family is now placed in the recently created order Suliformes, which also includes frigatebirds (Fregatidae), boobies and gannets (Sulidae), and cormorants (Phalacrocoracidae). The entire darter family consists of the Anhinga in the Americas and a widespread Old World group variously considered either one species or three.

Conservation Populations are generally stable, although organochlorine residues and the draining of wetlands have caused some local declines.

Genus *Anhinga*

ANHINGA *Anhinga anhinga*   ANHI ■ 1

Anhingas inhabit warm southern wetlands. Polytypic (2 ssp.; *leucogaster* in N.A.). L 35" (89 cm) WS 45" (114 cm)
IDENTIFICATION Unique. Long, snakelike neck, small head, sharply pointed yellow bill, and long fan-shaped tail with pale tip. **ADULT MALE:** Black throughout with extensive silver-white panel on upperwing coverts and streaks on lance-like back and scapular feathers. Breeding birds show wispy white plumes on sides of head and neck, green lores, and orange gular pouch. **ADULT FEMALE:** Like male, but with grayish tan head, neck, and breast (richer cinnamon on lower breast). **JUVENILE:** Similar to adult female but duller, with reduced white on scapulars, coverts.

SIMILAR SPECIES A soaring cormorant can suggest an Anhinga, but note Anhinga's longer tail and different bill shape. Escaped African Darters have been seen in southern CA; they lack a pale tail tip and have brown greater secondary coverts; males have rufous throat and neck and white stripe below cheek (females hint at this pattern).

VOICE Generally silent away from nesting colony. Clicking and rattling calls sometimes heard from perched birds.

STATUS & DISTRIBUTION Fairly common. **BREEDING:** Wooded swamps, lakes, slow-moving rivers, and mangrove estuaries, with highest densities in FL, LA, southeast GA; formerly bred to OK, IL, MO, OK, and KY. **WINTER:** Northernmost populations withdraw south in Oct; return in Mar. **VAGRANT:** Casual to Great Plains, Great Lakes states, southern ON, NJ, MD, CA, AZ, CO, and NM.

POPULATION Least concern.

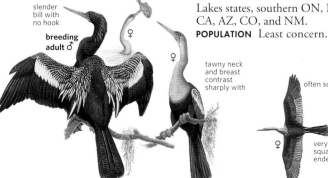

slender bill with no hook

breeding adult ♂

♀

♀

tawny neck and breast contrast sharply with

often soars

♀

very long, square-ended tail

PELICANS Family Pelecanidae

Brown Pelican (CA, Feb.)

Pelicans are instantly recognizable very large fish-eating birds with long, hooked bills and comically extensive gular pouches. In North America, the Brown Pelican inhabits subtropical and temperate coastlines and the American White Pelican breeds in the interior western and central parts of the continent and winters on the southern coastlines. Identification is straightforward, so concentrate on age determination, breeding condition, and scrutiny of flocks for individually marked birds.

Structure Pelicans are huge waterbirds with long, broad wings, short tails, long necks, and long, hook-tipped bills. The gular pouch can greatly expand as the mandibles bow outward to form a large fishing "net" or scoop. The long bill rests on the foreneck in flight and often when at rest. The legs are short and stout; the feet are webbed. The skeleton is lightweight, and pelicans ride buoyantly on the water. Despite such adaptations for weight reduction, American White Pelicans may weigh up to 20 pounds; the smallest Brown Pelicans (West Indian populations) only weigh about eight pounds.

Behavior Pelicans spend a lot of time loafing on beaches, lakeshores, low islets or (Brown Pelican) piers, barges, rocky islets, and mangroves. They are highly gregarious, often feeding or loafing in groups of hundreds or even thousands; American White Pelicans breed in colonies on low islets on large lakes, while Brown Pelicans, a marine species, breed on rocky coastal islands, vegetated barrier islands, or mangroves. American White Pelicans feed by swimming, often in coordinated groups, and dipping for fish with the pouched bill. Brown Pelicans rarely feed this way; instead they plunge-dive from as high as 50 feet, hitting the water with a slight leftward rotation, and scoop up as much as 2–2.5 gallons of water, filling the pouch underwater. Mergansers, cormorants, and other fish-eating birds often accompany feeding flocks of White Pelicans, and Heermann's and Laughing Gulls frequently attend diving Brown Pelicans to steal captured fish. Pelicans have a distinctive flap-and-glide flight, often in lines or V's; they can soar for long periods, and migrating flocks of American White Pelicans are often seen high within thermals. Pelicans are generally silent, but they do hiss and grunt in breeding colonies.

Plumage Worldwide, six pelican species are predominantly white, with variably dark flight feathers; Brown and Peruvian Pelicans are darker bodied. Molts are prolonged and complex; definitive adult plumage is reached in about three years. Bare parts can be brightly colored, especially during courtship season.

Distribution Pelicans are found on inland lakes and coastlines over much of North America, southern Eurasia, Africa, Australasia, and the northern and western coasts of South America. The North American species differ in habitat; Brown Pelican is almost exclusively marine (it is the only pelican species normally seen at sea, though generally absent from waters more than about 100 miles from land). All species undergo postbreeding dispersal, and some are truly migratory.

Taxonomy The Pelecanidae family is placed in the recently reconfigured order Pelecaniformes (herons and ibises and spoonbills are now also included in this order). Worldwide the Pelecanidae family has only one genus with eight species. The Peruvian Pelican *(P. thagus)* has been recently split from the widespread smaller Brown Pelican.

Conservation The fish diet magnifies problems of organochlorine contamination of aquatic environments. Evidence of Brown Pelican reproductive failures owing to eggshell thinning mediated by the DDT metabolite DDE, providing a clear cause and effect, was a key motivation in the banning of DDT and related pesticides. Populations that crashed between the 1950s and early 1970s have generally recovered, and the species was removed from the U.S. Endangered Species List in 2009. Human disturbance of nesting colonies is also an important and persistent threat.

AMERICAN WHITE PELICAN   *Pelecanus erythrorhynchos*   AWPE ■ 1

The huge American White Pelican has black primaries and outer secondaries and a yellow to orange bill. Breeding on inland lakes, this pelican is the only one likely to be seen away from seacoasts in most of North America. Monotypic. L 62" (158 cm) WS 108" (274 cm)

IDENTIFICATION Unmistakable in all plumages. **ADULT:** Entirely white except for black primaries and outer secondaries. Bill, gular pouch, and bare facial skin orange-yellow, brightening to deep orange-pink in breeding birds; legs and feet yellowish, becoming orange in breeding birds. Early breeding season adults have pale yellowish plumes on crest and center of breast, and a fibrous knob up to two inches high two-thirds of the way out the culmen; after egg-laying a supplemental molt results in dusky gray feathering on crown and nape. Sexes similar. **JUVENILE:** Similar to adult but head, neck, and especially upperwing coverts marked with dusky gray; bill and feet duller yellow. Older immatures resemble adults, but lack yellowish plumes and supplemental crown markings.

SIMILAR SPECIES A distant flying bird could be confused with the smaller Wood Stork, also largely white with black on the flight feathers. Other white-plumaged pelicans could occur as escapes; of these only the Great

White Pelican (*P. onocrotalus*) shares the all-white body plumage, sharp black and white wing pattern, and yellow pouch. Great White Pelican is even larger, has black bill sides, and always lacks the bill knob.

STATUS & DISTRIBUTION Common. **BREEDING:** Lakes in inland west and prairie regions; largest colonies in western and southern Canada. Small resident population in coastal TX. **MIGRATION:** Arrive breeding colonies Mar.–May; fall movements protracted. **WINTER:** Mainly from central CA and Gulf states to southern Mexico on coastal bays and estuaries, but also on large inland lakes; nonbreeders may summer well outside breeding range. **DISPERSAL:** Irregular postbreeding wandering brings small numbers to much of the Northeast (to Canadian Maritimes) and Atlantic coast. **VAGRANT:** Accidental in southeastern AK and far northern Canada.

POPULATION The population is stable or increasing since the 1960s. More than 20,000 pairs nest in the U.S., more than 50,000 in Canada.

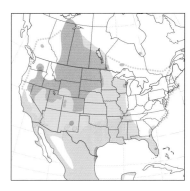

extensive black on wing

nonbreeding adult

immature

mottled coverts

grayish on crown and nape

chick-feeding adult

yellowish tuft

knob

breeding adult

nonbreeding adult

BROWN PELICAN   *Pelecanus occidentalis*   BRPE   ■   1

The small, dark Brown Pelican of marine coasts feeds mainly by plunge-diving. Although still considered endangered in parts of its range, it has generally rebounded well from severe declines caused by pesticide residues. Polytypic (5 ssp.; 3 in N.A.). L 48" (122 cm) WS 84" (213 cm)

IDENTIFICATION Diagnostic typical pelican form and extensively dark plumage. Sexes look similar, but age classes differ in plumage and body feather shape (narrower and more pointed in adults). Adults have white heads and dark bellies; juveniles the opposite. **BREEDING ADULT:** Upperparts appear gray (feathers slaty with silvery center streaks); underparts dark brown, streaked with silver on sides, flanks. Head and neck mainly white, with dark brown to chestnut hindneck stripe and foreneck (encircling yellow or whitish chest patch) and yellowish crown. Eyes white. Gular pouch dark greenish gray (bright red at base in courting *occidentalis*); bill with extensive pink or orange toward tip, and whitish area near base during chick-feeding stage. **NONBREEDING ADULT:** Similar, but entire neck whitish, eyes dark by late summer, bright colors on pouch bill less evident. **JUVENILE:** Mainly brown on head, chest, and upperparts, but white from breast through remaining underparts; greater underwing coverts pale grayish white, forming a long stripe below. Eyes dark. **IMMATURE:** Year-old birds have some pale feathering on head and neck and some dark flecking on underparts; third-year birds resemble adults but show some whitish mottling on belly and hint of pale underwing stripe.

GEOGRAPHIC VARIATION The Pacific coast subspecies *californicus* is distinguished from the Atlantic and Gulf coast *carolinensis* by its larger size and (in adults early in breeding season) the bright red rather than blackish green gular pouch. The small, darker-bellied nominate *occidentalis* (from the West Indies) has been collected once in coastal northwest FL.

SIMILAR SPECIES Unmistakable. Adults can look pale above when seen in strong lighting and momentarily may be mistaken for American White Pelicans. Even a heavily marked juvenile White Pelican is still mainly white and thus not likely to be confused.

STATUS & DISTRIBUTION Common. **BREEDING:** Atlantic and Gulf coast colonies nest mainly on low vegetated islands (mangroves in FL). Northernmost colonies are in MD and on

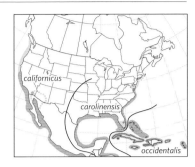

northern Channel Is., CA. **DISPERSAL:** Postbreeding dispersal May–Oct. north to DE, NJ, WA, then south to breeding regions by late fall. Regular postbreeding visitor to Salton Sea, where common; in 1990s, some breeding attempts occurred. Elsewhere, irregular in Southwest, but in some years widespread. **VAGRANT:** Rare (mainly late summer, fall) to NS, BC; recorded rarely but widely in the interior to ID, ND, and Great Lakes region.

POPULATION The species has recovered since the 1970s and is no longer listed as federally endangered. The West Coast population was the last to be removed from the U.S. Endangered Species List—this occurred in 2009 when its estimated population had grown to about 11,000 breeding pairs.

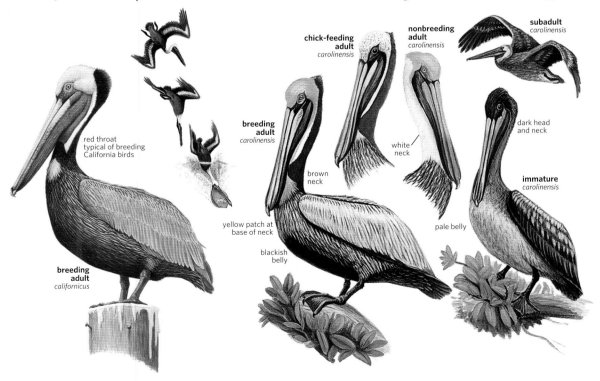

red throat typical of breeding California birds

breeding adult
californicus

chick-feeding adult
carolinensis

breeding adult
carolinensis

yellow patch at base of neck

blackish belly

brown neck

nonbreeding adult
carolinensis

white neck

subadult
carolinensis

dark head and neck

pale belly

immature
carolinensis

HERONS, BITTERNS, AND ALLIES Family Ardeidae

Yellow-crowned Night-Heron (NJ, Apr.)

One of the most distinctive and graceful bird families, the Ardeidae played a major role in the formation of the American conservation movement. Wading bird populations were decimated in the late 19th and early 20th centuries, when it was fashionable for women to wear hats adorned with feathers, wings, or even entire stuffed birds. Hundreds of thousands of wading birds were slaughtered for the fine plumes—aigrettes—grown by some species for courtship displays. The loss of the adults during the breeding season meant that their nestlings starved, causing the populations of most species to plummet within a few decades. Ardeids are called wading birds in North America, a term that should not be confused with the British term "waders," which refers to shorebirds.

Structure Some ardeid species are small and chunky, but most species have moderate to large, slender bodies with long necks and legs—the legs of most species extend beyond the tail when in flight. The wings are generally long and rounded, the tail short. The long, dagger-shaped bill aids in capturing aquatic prey. The neck, which is held in an S-curve while birds are resting, may be extended fully when foraging, and is drawn toward the body during flight.

Plumage Herons generally have a dull blue or gray plumage, while egrets typically show a plumage that is entirely white. Bitterns are plumaged cryptically, with brown or buffy streaking. Many species grow elongated plumes from the head, breast, or back that are used in elaborate courtship displays. Also associated with courtship ("high breeding") is the intensifying of soft-part colors of the lores, bill, and legs of many species.

Behavior Most species nest colonially; nightly roosts or breeding colonies may contain thousands of individuals of several species. Wading birds use several foraging strategies that vary greatly by species. Many species stand in or near water and wait for prey to pass by—usually fish and small crustaceans, but snakes and other vertebrates are taken by larger species. Others walk slowly, perhaps also stirring up the water with their feet to flush prey. The Reddish Egret has a particularly active foraging behavior.

Distribution Cosmopolitan. The largest numbers of wading birds occur in areas with abundant wetlands and mild climate. The Everglades is perhaps the area most associated with wading birds in North America, but large populations occur along the entire Gulf Coast and in parts of California.

Taxonomy Recent higher-level taxonomic revisions now place the family Ardeidae in the order Pelecaniformes with pelicans (Pelecanidae) and ibises and spoonbills (Threskiornithidae). Species limits remain uncertain for some taxa. Worldwide, 60 to 65 species in 16 to 21 genera are recognized. According to the AOU, 20 species in 10 genera occur in North America. They are often divided into three groups: bitterns (3 sp.), herons and egrets (15 sp.), and night-herons (2 sp.).

Conservation BirdLife International lists three species as endangered and five as vulnerable—all from the Old World. In North America, the end of the millinery trade in the early 1900s allowed many wading bird populations to recover—a recovery that continues for especially sensitive species such as the Reddish Egret. Wetland loss or degradation, pollution, colony disturbance, and overharvesting of prey, however, now threaten some populations.

BITTERNS Genera *Botaurus* and *Ixobrychus*

In contrast to other wading birds, which are showy and feed in the open, bitterns are secretive marsh dwellers with cryptic, streaked plumage that allow them to blend in with their surroundings. Bitterns are solitary species, but they can be common in an ideal habitat. Three species occur in North America, one as an accidental.

AMERICAN BITTERN *Botaurus lentiginosus*   AMBI   ■   1

Cryptic in plumage and mostly crepuscular in habits, the American Bittern is a denizen of marshes with thick vegetation. It attempts concealment by pointing its bill upward; it may sway its head and neck back and forth to mimic wind-rustled vegetation. It is detected primarily by its eerie "pumping" vocalizations, given primarily on the breeding grounds. Monotypic. L 28" (71 cm) WS 42" (107 cm)
IDENTIFICATION A large brown stocky heron with a proportionately thick neck and short legs. **ADULT:** Sexes similar. Brown upperparts finely flecked with black. Brown head with darker cap, white supercilium, chin, and throat, and wide blackish malar streak. Yellow eyes and mostly yellow bill; legs and feet pale greenish or yellowish. White or pale underparts with bold rufous streaking. **JUVENILE:** Blackish malar stripe absent. **FLIGHT:** Strong and direct; may appear hunchbacked. Wings somewhat pointed. On upperwings, blackish flight feathers contrast with brown coverts; underwings sooty.
SIMILAR SPECIES Immature night-herons are smaller, lack the dark malar stripe, have rounded wings and reddish eyes, and are often found in open habitats. The contrasting upperwing surface of an immature Yellow-crowned Night-Heron is similar to American Bittern, but its bill is short and black.
VOICE SONG: Distinctive, resonant "pumping" *oonk-a-lunk* (rarely heard

on wintering grounds). **CALL:** A hoarse *wok* or *wok-wok-wok* when flushed.
STATUS & DISTRIBUTION Uncommon. **BREEDING:** Non-colonially in freshwater wetlands with tall emergent vegetation; also found in saline and brackish marshes in migration and winter. Throughout southern Canada and northern U.S.; rarely in southern U.S. **MIGRATION:** Partially migratory; some birds probably resident in Pacific region and in the southern portion of breeding range. **WINTER:** Pacific region as far north as Puget Sound, southern U.S., Mexico, Cuba, and possibly C.A. **VAGRANT:** Prone to extralimital wandering; reported from Bermuda, Canary Is., Great Britain, Greenland, Iceland, Lesser Antilles, Norway, and Spain.
POPULATION American Bittern numbers may be declining severely due to wetland loss and degradation.

often seen motionless with head up
juvenile
adult
blackish malar stripe
dark brown streaks
adult
wings slimmer and more pointed than night-herons
dark primary coverts, primaries, and secondaries contrast with paler upperwing coverts

YELLOW BITTERN *Ixobrychus sinensis*   YEBI   ■   5

This Old World relative of the Least Bittern has occurred once in N.A. Monotypic. L 15" (38 cm) WS 21" (53 cm)
IDENTIFICATION ADULT: Sexes similar. Head and neck buffy with black cap; brown back with black tail. Underparts white with buffy streaking on neck and breast, extending to flanks. **JUVENILE:** Plumage browner, boldly streaked with black on upperparts (except rump); neck also streaked. **FLIGHT:** Black primary coverts and flight feathers. Adults with buffy wing patch on upper wing. Juveniles with brown secondary coverts heavily streaked with black.
SIMILAR SPECIES The highly migratory Schrenk's Bittern (*I. eurhythumus*), breeding north to the Russian Far East, is distinguished by its larger size, dark rufous upperparts (solid in male, light-spotted in female, and similar in juvenile), and less contrasting upperwing pattern.
VOICE Guttural grunt.
STATUS & DISTRIBUTION Locally common. **BREEDING:** Wetlands, even tall trees. From India to Japan through Malay Peninsula. **MIGRATION:** Partially migratory. **WINTER:** Much of breeding range (withdraws from Japan and China) to New Guinea and Guam. **VAGRANT:** One specimen record in N.A. (Attu I., AK, May 17–22, 1989). Also accidental to Australia and Christmas I.
POPULATION Not threatened.

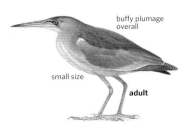

buffy plumage overall
small size
adult

LEAST BITTERN *Ixobrychus exilis*   LEBI ■ 1

The Least Bittern, the world's smallest heron, is easily overlooked. It is usually associated with cattails, bulrush, or sawgrass in freshwater marshes, but it also occurs in mangroves. It is very adept at clambering up reed stems, aided by its small size, long toes, and extremely narrow body. Like its larger cousin, the American Bittern, the Least Bittern attempts to avoid detection by "freezing" in place, with its head and bill pointed skyward. Knowledge of its vocalizations will aid in finding this secretive, solitary species. Polytypic (5 ssp.; nominate in N.A.). L 13" (33 cm) WS 17" (43 cm)

IDENTIFICATION Tiny with bright buff wing patches, a dark crown, and two white or buff scapular stripes that contrast with a dark mantle. Whitish underparts are variably streaked on the foreneck with buff. Eyes, loral skin, and bill are mostly yellow; legs are greenish in front and yellow in back. **ADULT MALE:** Crown, back, rump, and tail are a glossy black. The loral skin turns reddish pink during courtship. **ADULT FEMALE:** Similar, but upperparts are dark brown rather

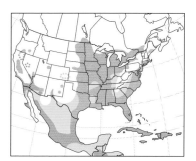

than black; foreneck has fine, dark streaking. **JUVENILE:** Similar to female, but colors are more muted. Streaking on foreneck is browner and bolder, extending onto sides of neck; buffy wing coverts have dark centers. **DARK MORPH:** Very rare; known as "Cory's Least Bittern." It was best known from Ashbridge Bay, Toronto, ON, in the early 20th century, but these marshes were filled decades ago. It also bred at Lake Okeechobee, FL. This distinct color morph may not have been recorded since the mid-20th century. In all ages, chestnut replaces pale areas on upperparts and wing patches, as well as on upperwing coverts. In adults, the white scapular lines are lacking. **FLIGHT:** Flies with quick wingbeats low over reed beds, dropping abruptly into them. All plumages (except "Cory's") show a large buffy patch on the upper wing that contrasts strongly with the dark flight feathers and deep rufous greater coverts.

SIMILAR SPECIES A juvenile Green Heron is also small with streaked underparts, but its wings are always entirely dark.

VOICE Quite vocal with a varied repertoire. **ALARM:** Several calls, including *quoh* and *hah*. **CALL:** A rail-like *tut-tut-tut* or *kek-kek-kek-kek* heard year-round (often mistaken for a King or Clapper Rail). **SONG:** Male only gives three or four short, low *coo* notes in rapid succession.

STATUS & DISTRIBUTION Uncommon to locally common. **BREEDING:** Dense

marshland vegetation, especially cattails; rare in mangrove swamps. Very local in interior West. **MIGRATION:** Partially migratory. Arrives in breeding areas Apr.–May; departs Aug.–Sept. **WINTER:** Resident in southernmost N.A. breeding areas. Majority of N.A. birds winter in C.A. and the Caribbean. **VAGRANT:** Casual north to southwest Canada and NF. Accidental to the Azores, Bermuda, Clipperton I., and Iceland.
POPULATION Not threatened.

contrasting buffy wing coverts

exilis

adult ♂

juvenile

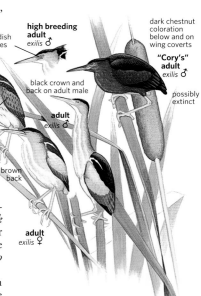

high breeding adult *exilis* ♂

reddish lores

dark chestnut coloration below and on wing coverts

"Cory's" adult *exilis* ♂

possibly extinct

juvenile *exilis*

black crown and back on adult male

adult *exilis* ♂

dark-centered coverts

brown back

adult *exilis* ♀

Genus *Tigrisoma*

BARE-THROATED TIGER-HERON *Tigrisoma mexicanum*   BTTH ■ 5

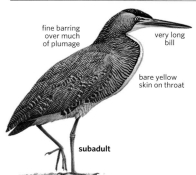

fine barring over much of plumage

very long bill

bare yellow skin on throat

subadult

Named for their extensively barred, cryptic plumages, tiger-herons are solitary nesters and foragers. Bare-throated is unique in the group in having an unfeathered throat, a useful field mark at close range. Monotypic. L 28–33" (70–80 cm)

IDENTIFICATION A very heavy-set, short-legged heron with heavy bill. **ADULT:** Blackish cap, gray face, bare yellow throat, and finely barred neck. Flight feathers blackish. Legs slaty olive. **JUVENILE:** Boldly barred

and spotted, including remiges, with cinnamon-buff and brown; neck often with trace of adult pattern.

SIMILAR SPECIES Superficially similar to American Bittern, juvenile Black-crowned and Yellow-crowned Night-Herons, but these species have smaller bills, yellowish legs, and lack gray neck barring.

VOICE CALL: When flushed, birds give a heronlike *wah* or *hauk*, often in rapid succession.

STATUS & DISTRIBUTION Resident of

coastal habitats from Mexico (found along coastal slopes north to southern Tamaulipas and southern Sonora) to northwestern Colombia. No migration is known in the species, but local movements are likely, especially during drought cycles. The only record north of Mexico is of a single subadult at Bentsen-Rio Grande Valley State Park, Hidalgo Co., TX (Dec. 21, 2009– Jan. 20, 2010). **POPULATION** Trends unknown.

Genus *Ardea*

GREAT BLUE HERON *Ardea herodias* GBHE ◼ 1

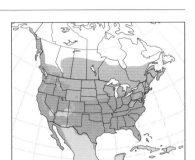

The largest and most widely distributed wading bird in N.A., the Great Blue Heron is found in a range of wetland and upland habitats. It feeds on crustaceans, vertebrates, and small mammals. The white morph, "Great White Heron," is sometimes considered a distinct species. Polytypic (4 ssp.; 3 in N.A.). L 46" (117 cm) WS 72" (183 cm)
IDENTIFICATION BLUE MORPH: Blue-gray upperparts; grayish neck; large, daggerlike bill. **NONBREEDING ADULT:** Sexes similar. Bluish gray back and wings with scapular plumes and a black shoulder patch; blackish tail. Neck gray or rusty gray with fine black streaking on the underside; short, gray plumes on lower neck and breast. Head and crown white, divided by a wide black stripe that extends to black occipital plumes. Yellow eyes; bluish gray lores; mostly yellow bill. Legs and feet brown, brownish green, or greenish black. Underparts blackish. **BREEDING ADULT:** Soft-part coloration intensifies but varies regionally. Lime green or blue lores; yellow, orange, or reddish bill; reddish or greenish yellow legs. **JUVENILE:** Darker overall without plumes; black crown; blackish lores and upper mandible. **FLIGHT:** Strong, slow, deliberate wingbeats. Gray-ish coverts contrast with dark flight feathers. **"GREAT WHITE HERON":** Size, shape, and behavior like Great Blue, but white in all plumages; legs and feet dull yellow. **INTERMEDIATE MORPH ("WURDEMANN'S HERON"):** Intergrade between blue and white morphs; similar to blue morph but head white and neck pale with inconspicuous streaking on underside.
GEOGRAPHIC VARIATION Nominate subspecies is found over most of North America. Pacific Northwest *fannini* has darker plumage and a shorter bill; white *"occidentalis"* and intermediate morphs are restricted to southern Florida.
SIMILAR SPECIES Closely resembles Gray Heron from the Old World. The "Great White Heron" is distinguished from Great Egret by its larger size, larger, thicker bill, and entirely dull yellow legs and feet.
VOICE Relatively silent away from nest. **CALL:** A load, hoarse *kraaank*.
STATUS & DISTRIBUTION BLUE MORPH: Common. **BREEDING:** Colonially (rarely solitary) in all lower 48 states, AK, all but northern Canadian provinces, West Indies, and American tropics. **MIGRATION:** Partially migratory; withdraws from northern portion of breeding range, except along coasts. **WINTER:** Along coasts and in southern portion of breeding range. **VAGRANT:** Wanders widely; casual to southwestern AK and the Arctic coast, Azores, Clipperton I., Greenland, HI, and Spain. **WHITE MORPH:** Uncommon. Restricted in U.S. to southern FL (±850 pairs resident) and also resident on Cuban cays. Strays along the coasts north to TX and NJ/southern NY, and inland to PA and KY.
POPULATION Numbers are stable.

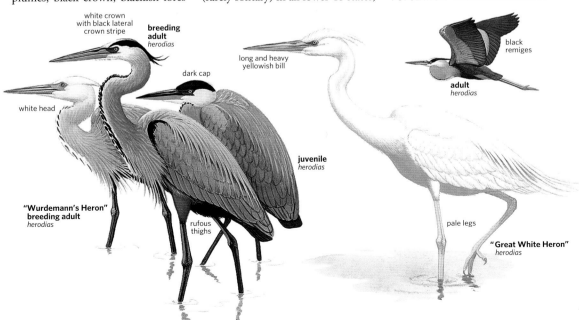

white crown with black lateral crown stripe

breeding adult *herodias*

dark cap

long and heavy yellowish bill

black remiges

adult *herodias*

white head

juvenile *herodias*

"Wurdemann's Heron" breeding adult *herodias*

rufous thighs

pale legs

"Great White Heron" *herodias*

GRAY HERON *Ardea cinerea*   GRAH ◼ 5

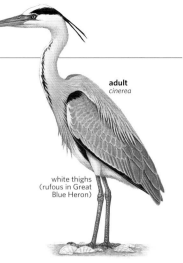

adult
cinerea

white thighs
(rufous in Great
Blue Heron)

A familiar species of town and countryside across Eurasia and much of Africa, Gray Heron is the Old World counterpart to Great Blue Heron of N.A., which it closely resembles. Polytypic (4 ssp.). L 33–40" (84–102 cm) **IDENTIFICATION** A large heron, medium slaty gray above, white crown, whitish below, with two rows of neat, dark stippling down most of neck to breast; mostly grayish bill and legs. **ADULT:** Breeding birds have bright yellow bill, with single black plumes extending behind each eye; legs also partly yellow. **JUVENILE:** Sides of neck and head grayish brown; dark crown. **GEOGRAPHIC VARIATION** Nominate *cinerea* is found through most of the Palearctic, Africa, India, and Sri Lanka. Other subspecies include East Asian *jouyi,* increasingly threatened *firasa* on Madagascar, and *monicae,*

restricted to islands off Mauritania, which some authorities consider a full species. **SIMILAR SPECIES** Only likely to be confused with larger, longer-necked, and longer-legged Great Blue Heron. In all plumages Gray Heron lacks rufous thighs of Great Blue; in flight, leading edge of wing shows prominent white area, rather than rufous. Gray Heron is paler above and on the neck than Great Blue. **VOICE** CALL: An abrupt, resonant *ka-ark!* and raspy *yeehr!* often given in flight. Also a screeching *rraank!* similar to Great Blue Heron but higher in pitch. **STATUS & DISTRIBUTION** BREEDING: Vast Old World range. **WINTER:** Mostly in breeding range; northern breeders fully migratory, moving southward in autumn. **VAGRANT:** Casual to western

AK (Aleutians and Pribilofs) and Newfoundland. Also over 20 records from Greenland and rather widely recorded from West Indies. **POPULATION** In most areas, populations appear to be stable.

GREAT EGRET *Ardea alba*   GREG ◼ 1

This tall, stately egret is found in a variety of freshwater, brackish, or saline habitats over much of the United States. It feeds primarily on fish and aquatic invertebrates, but is also found locally in fields and pastures, stalking small mammals. The plumes of Great Egrets were among the most sought after for the millinery trade, resulting in severe population declines of this once again common species. Polytypic (4 ssp.; *egretta* in N.A.; Asian *modesta* and possibly Eurasian *alba* are vagrants). L 39" (99 cm) WS 51" (130 cm) **IDENTIFICATION** Large, white egret with a long, slender neck. **NONBREEDING ADULT:** Plumage is entirely white. Eyes, lores, and bill are yellow; legs and feet are black. **BREEDING ADULT:** Develops long, graceful plumes (aigrettes) on the back that extend beyond the tail. Unlike many other herons

and egrets, there are no plumes on the head or neck. The lores flush lime green and the bill turns orangish. **IMMATURE:** Similar to a nonbreeding adult. **GEOGRAPHIC VARIATION** In breeding plumage, Eurasian *alba* and slightly smaller Asian *modesta* have black bills. **SIMILAR SPECIES** Most similar to the white morph of Great Blue Heron ("Great White Heron") but somewhat smaller and less bulky, with thinner bill, and black legs and feet; lacks the

short head plumes of adult "Great White Heron." The bill of the white morph Reddish Egret is either all dark (immature) or bicolored with a pink base (adults). Other egrets are much smaller and lack the combination of yellow bill and black legs and feet. **VOICE** Generally silent except when nesting or disturbed, when it may utter *kraak* or *cuk-cuk-cuk* notes. **STATUS & DISTRIBUTION** Found on all continents except Antarctica. **BREEDING:** In shrubs or trees in colonies with other wading birds, along the Atlantic, Gulf, and Pacific coasts, the Mississippi River floodplain, and locally elsewhere in the interior. Also in extreme southern Canada, the West Indies, C.A., and S.A. **MIGRATION:** Partially migratory; birds mostly resident along coasts and in southern U.S. Postbreeding dispersal may carry birds north of regular summer range. **WINTER:**

lime green
facial skin

high breeding adult
egretta

very long
neck

yellow facial
skin

long
yellow bill

breeding
modesta

Asian *modesta* is slightly
smaller than Eurasian
alba (not shown);
both have black bills in
breeding plumage

nonbreeding
egretta

dark legs
and feet

long
plumes

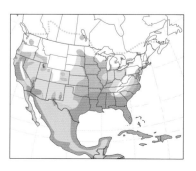

Locally in the West and throughout the Southeast. VAGRANT: Casual to southeastern AK and Atlantic Canada. Asian *modesta* casual to western and central Aleutians, AK. A black-billed bird from VA may have been *alba,* on distribution probability.
POPULATION Although decimated by the plume trade in the late 1800s and early 1900s, the species has largely recovered. Some populations are still increasing.

Genus *Mesophoyx*

INTERMEDIATE EGRET *Mesophoyx intermedia* INEG ■ 5

The serviceably, if not poetically, named Intermediate Egret is indeed intermediate in size between the smaller and larger white egrets, for which it is easily mistaken. It is accidental to North America, with just two records from the Aleutian Is., Alaska. Polytypic (3 ssp.). L 22–28" (56–72 cm)
IDENTIFICATION An all-white egret with dark legs and feet and a yellowish bill with a dark tip. In N.A. context, Intermediate is a bit larger than a Snowy Egret and well smaller than a Great Egret. ADULT: In high breeding condition soft parts change color variably among the different subspecies: The bill becomes reddish, red and yellow, or black; loral skin becomes green or yellow; iris color changes in some subspecies to red; and legs in some subspecies take on yellow or pinkish tones.
GEOGRAPHIC VARIATION Medium-size, nominate *intermedia* is found from southeastern Asia and western Indonesia to Japan, northern populations are migratory; largest, *brachyrhyncha,* inhabits Africa south of the Sahara and is not migratory; and smallest, *plumifera,* is found from east Indonesia to New Guinea and Australia and is also not migratory.
SIMILAR SPECIES Smaller Snowy and Little Egrets have yellow-soled feet and always have longer, slenderer black bills. Great Egret has a much longer neck, usually with a kink in the middle, and a longer bill in which the culmen aligns with the flat top of the head (Intermediate shows a "bump" at the forehead). When foraging, Snowy Egret tends to chase prey actively, while Intermediate Egret forages more like Little Blue Heron, with neck extended forward, watching downward, moving slowly or scarcely at all before seizing prey.
VOICE A soft buzzing call is said to be unique among herons, but more typical egret-like calls (*grak, glok, kroo, kraa*) are heard from flying and nesting birds.

STATUS & DISTRIBUTION MIGRATION: Mostly resident in its fragmented Old World range. Japanese breeders are migratory and depart in mid-autumn, wintering in Borneo and the Philippines. VAGRANT: Two records, certainly of the nominate subspecies, from AK: Buldir I. (May 30, 2006) and Shemya I. (Sept. 28, 2010).
POPULATION Significant declines have been documented in urbanized Japan.

dark-tipped yellow bill

breeding adult
intermedia

Genus *Egretta*

This genus contains 13 to 14 species. Four species (all called egrets) are white, six (mostly called herons) are blue or blackish, and three or four others contain both white and dark morphs. The species inhabit a variety of freshwater and saline environments.

CHINESE EGRET *Egretta eulophotes* CHEG ■ 5

This elegantly plumed egret, a rare Asian species, has strayed to North America just once. It is related to Snowy and Little Egrets. Monotypic. L 27" (65 cm) WS 41.5" (105 cm)
IDENTIFICATION Similar to the Snowy Egret in size and plumage. Sexes and ages similar. Plumage entirely white. Yellow eyes; yellow-green lores; dusky bill with pale base to the lower mandible. Yellow-green legs and feet. BREEDING ADULT: Develops shaggy plumes from the nape, breast, and back. Soft-part colors intensify: yellow bill, turquoise lores, black legs, and yellow feet.
SIMILAR SPECIES Snowy and Little Egrets are very similar, but Chinese Egret has shorter legs. In nonbreeding plumage, identify Chinese Egret by its yellow-green legs and feet. In breeding plumage, the combination of its turquoise lores, entirely yellow bill, and black legs with yellow feet is distinctive.
VOICE Undescribed.
STATUS & DISTRIBUTION Considered endangered by BirdLife International, with ±1,800–2,500 birds remaining. BREEDING: Colonially on islands off Korea, China, and possibly the Russian Far East. DISPERSAL: Wanders widely from northern Japan south. WINTER: From the Malay Peninsula to the Philippines and northern Borneo. VAGRANT: One accidental record in N.A., on Agattu I., AK (June 16, 1974); also accidental to Myanmar and Sulawesi.
POPULATION Nearly exterminated by plume hunters, populations continue to decline, primarily from destruction of wetlands.

shaggy crest in breeding plumage

orange-yellow bill with bluish lores

breeding adult

black legs and yellow feet

LITTLE EGRET *Egretta garzetta* LIEG ■ 4

This Old World counterpart to the Snowy Egret inhabits a variety of freshwater and saline environments. It feeds primarily on small fish but takes other prey. It has two color morphs, a typical white morph (all N.A. records) and a rare dark morph.

lores often greenish gray, unlike Snowy

nonbreeding adult
garzetta

bill appears longer than Snowy

two long plumes

breeding adult
garzetta

always all-black legs

yellow feet

A few ornithologists consider the Little Egret conspecific with the Western Reef-Heron *(E. gularis)*. Polytypic (3 ssp.; nominate in N.A.). L 24" (60 cm) WS 36" (91 cm)
IDENTIFICATION Medium-size egret, very similar to Snowy Egret. Sexes similar. White morph described. **NONBREEDING ADULT:** Plumage entirely white. Yellow eyes, black legs, and yellow feet. The lores are usually bluish gray, but can be greenish, pale yellow, or whitish. **BREEDING ADULT:** Two white occipital plumes, as well as numerous plumes on lower neck and back. Lores and feet flush reddish, but can be yellow. **IMMATURE:** Pale base to lower mandible, gray lores, brownish green legs.
SIMILAR SPECIES Little Egret is most readily told from the Snowy Egret in nonbreeding plumage by its blue-gray rather than yellow lores; in breeding plumage, it has only two or three occipital plumes. Little Egret also has a flatter crown, a somewhat thicker bill, duller yellow feet, and more upright, longer necked posture (often appearing larger than Snowy Egret). Snowy Egret

habitually forages in a hunched manner and tends to move around more frenetically. Mostly white, immature Little Blue Heron often has dark tips to the primaries, and greenish yellow legs and feet. A nonbreeding Cattle Egret differs by its yellow bill and dark legs and feet.
VOICE Usually silent. **CALL:** A low guttural *kraak*.
STATUS & DISTRIBUTION Common. Widespread in Eurasia and Australasia. Rare but increasing in the West Indies (±20 pairs breed at Barbados). **VAGRANT:** Casual along the Atlantic coast north to Maritimes (most frequently recorded in spring and summer) and also to Puerto Rico and Lesser Antilles. Accidental to QC and the western Aleutians, AK (male in breeding plumage found dead on Buldir I., May 27, 2000).
POPULATION Little Egret is widespread in the Old World. New World populations have been increasing since the first report in the West Indies in 1954. Colonization of the New World probably originated from birds in West Africa flying across the Atlantic Ocean. Little Egrets will presumably be found with greater frequency in the U.S. as the West Indies population increases.

WESTERN REEF-HERON *Egretta gularis* WERH ■ 5

This Old World species has recently colonized the West Indies and has strayed several times to the United States and Atlantic Canada. It has two color morphs, white and dark, with rare intermediates; only the dark morph has been recorded from the New World. The Western Reef-Heron is sometimes considered conspecific with the Little Egret. Polytypic (2 ssp.; nominate in N.A.). L 23.5" (60 cm) WS 37.5" (95 cm)
IDENTIFICATION Dark morph described. **NONBREEDING ADULT:** The sexes are similar. Entirely slate gray overall, with a white chin and throat. Eyes yellow; lores and bill dusky yellow. Legs black and feet yellow. **BREEDING ADULT:** Develops two long, wispy, slate occipital plumes. Soft-part colors intensify: Lores flush red and feet turn black. **IMMATURE:** Overall brown or dark gray-brown plumage with a white chin and throat.
SIMILAR SPECIES Adult Little Blue

Heron is entirely dark; in breeding plumage it has a shaggy head and neck plumes and entirely dark legs and feet. Tricolored Heron has an entirely white belly. Rare dark morph of Little Egret (no N.A. records) is very similar, but it often has a dark throat. White morph Little Egret is best distinguished by bill structure. The reef-heron's bill has a subtle decurvature toward the tip.
VOICE Usually silent. **CALL:** A throaty squawk.
STATUS & DISTRIBUTION Occurs coastally from West Africa east to India; vagrant to southern Europe, possibly from deliberate releases. Recent arrival in the New World, beginning in mid-1980s. About ten records from the West Indies, primarily Barbados. Six records in N.A. (1983 and 2005–2007) from Atlantic Canada and the Northeast, perhaps the more recent records could

have involved only one individual visiting a number of locations.
POPULATION Widespread in the Old World, the Western Reef-Heron is increasing in the West Indies.

overall slaty-gray with white chin and throat

dark-morph adult
gularis

yellow feet

SNOWY EGRET   *Egretta thula*   SNEG ▉ 1

The delicate and graceful Snowy Egret occurs throughout the New World and is found in a wide range of freshwater and saline habitats. It feeds on a variety of prey such as fish, aquatic invertebrates, and even snakes and lizards. Monotypic. L 24" (61 cm) WS 41" (104 cm)

IDENTIFICATION Medium-size white heron with a slender build and a thin black bill. **NONBREEDING ADULT:** The sexes look similar. The plumage is entirely white. Eyes and lores yellow; legs black and feet distinctly yellow ("golden slippers"). **BREEDING ADULT:** Develops elegant plumes on its crown, nape, foreneck, and back. Some soft-part colors intensify: Lores flush red, and feet flush orange or reddish. **JUVENILE:** Bill has a pale gray base; lores are grayish green.

SIMILAR SPECIES Little Egret closely resembles Snowy Egret. The Snowy is distinguished by its yellow lores, more rounded crown, and shorter bill, narrower at the base. It also has numerous breeding plumes, compared to only the two or three occipital plumes found on Little Egret. When foraging, the Little's posture is more upright, in a manner similar to a Great Egret. Mostly white, immature Little Blue Heron has a thicker, more bicolored bill, greenish yellow legs and feet, often has dark gray tips to the primaries, and feeds more deliberately and slowly.

VOICE Generally silent except away

from breeding colony. **CALL:** A harsh, raspy *aah-raarrh*.

STATUS & DISTRIBUTION Common. **BREEDING:** Colonially with other wading birds. Locally from eastern OR and CA across the interior West; primarily coastally along the Atlantic and Gulf coasts from ME south to S.A.; also in the West Indies. **MIGRATION:** Partially migratory; most inland breeders in northern part of range withdraw. **WINTER:** Coastally from OR and NJ south to S.A.; also all of FL, along the entire Gulf Coast, and the West Indies. **DISPERSAL:** Wanders irregularly north to south Canada from BC to NF; also to HI, Bahamas, Lesser Antilles, Bermuda, and the Galápagos. Accidental to southeastern AK, Clipperton I., Iceland, and South Georgia I.

POPULATION Its numerous breeding plumes made it among the most sought after species for the millinery trade in the late 1800s and early 1900s, resulting in the near extinction of the species, but numbers have since recovered. The species' adaptability to a range of environmental conditions has allowed it to expand its range beyond its original distribution in the past several decades.

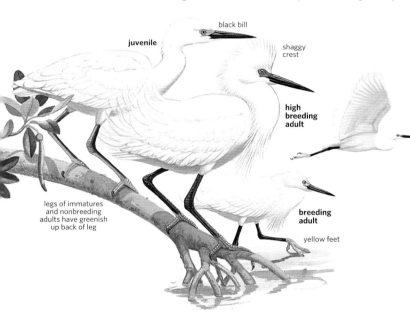

juvenile

black bill

shaggy crest

high breeding adult

breeding adult

yellow feet

legs of immatures and nonbreeding adults have greenish up back of leg

---

Identification of White Egrets

Separating various white egret species is fairly straightforward as long as a few features are kept in mind. Great Egret is the largest species; Cattle Egret, the smallest, has proportionately short legs. Young juvenile Cattle Egrets can be tricky to identify when their bills are dark. With their sharply separated pink-based bill and shaggy head plumes during the breeding season, white-morph Reddish Egrets are easily identified. In contrast, immatures and winter adults have entirely dark bills. The "dancing" feeding behavior is an excellent differentiating characteristic for the species at all ages and times of year. Immature Little Blue Herons during their first fall and winter closely resemble Snowy Egrets, but they have duller lores and a somewhat thicker and more obviously bicolored bill. Most, but not all, have some dusky in the wing tips, often only visible in flight. They also have more uniformly greenish yellow legs and feed more deliberately than Snowy Egrets.

In flight, Great Egret is very long-necked and flies with slow wingbeats. Much smaller Snowy Egret has much more rapid wingbeats. The small, short-legged Cattle Egret appears short and round-winged and flies with rapid wingbeats. ∎

LITTLE BLUE HERON   *Egretta caerulea*   LBHE ■ 1

Unique among herons, the Little Blue has a nearly all-white juvenal plumage and an all-dark adult plumage. Found in freshwater and saline habitats, it forages for aquatic invertebrates, fish, and amphibians by walking slowly or by standing still, with neck bent forward at a 45-degree angle. Monotypic. L 24" (61 cm) WS 40" (102 cm)
IDENTIFICATION Medium-size heron. **ADULT:** Sexes similar. Dark slate blue body with purple head and neck. Yellow eyes, dull greenish lores, blue-gray bill with black tip; gray to greenish gray legs and feet. **BREEDING ADULT:** Bright reddish purple head and neck with long lanceolate plumes on the crest and back. Lores and base of bill turquoise; legs and feet black. **JUVENILE:** Entirely white, but often with dark gray tips to most of the primaries. **FIRST-SPRING:** Adult plumage acquired gradually by second summer or second fall, resulting in splotchy white and blue ("calico") plumage.
SIMILAR SPECIES Dark-morph Reddish Egret is larger and paler overall, with shaggy head and neck plumes, and a pink-based, black-tipped bill in the breeding adult. White-morph Reddish Egret is larger with pale lores and dark legs and feet; it also has a different foraging behavior. Tricolored Heron has white underparts in all plumages. Immature Little Blue is distinguished from Snowy Egret by its stouter, more bicolored bill, gray-green lores, greenish yellow legs and feet, and usually dark tips to the primaries. Cattle Egret has a stout yellow bill.

VOICE Generally silent away from breeding colony. **CALL:** A harsh, croaking *aarr-aarrh*.
STATUS & DISTRIBUTION Common. **BREEDING:** Colonial. Along Atlantic coast from NY (local north to southern ME) through FL, along entire Gulf coast to northern S.A., up the Mississippi River drainage to southern IL and IN, and in southern Great Plains. Along Pacific coast from Sonora and Baja California south, and locally at San Diego, CA; rare and sporadic breeder to ND and SD. **MIGRATION:** Partially migratory; withdraws from northern portion of breeding range. **WINTER:** Along entire Gulf Coast, along Atlantic coast to VA, and from extreme southwestern CA south along Pacific coast. **DISPERSAL:** Widely after breeding; casual to BC,

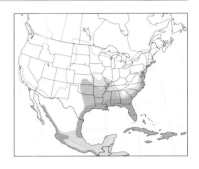

HI, MT, NF, and WA. **VAGRANT:** Accidental to the Azores, Chile, and western Greenland.
POPULATION Threatened by wetland degradation or loss and colony disturbance, but some range expansion has occurred since the 1970s.

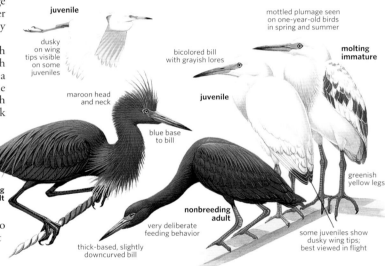

juvenile

dusky on wing tips visible on some juveniles

maroon head and neck

breeding adult

blue base to bill

mottled plumage seen on one-year-old birds in spring and summer

bicolored bill with grayish lores

molting immature

juvenile

greenish yellow legs

nonbreeding adult

very deliberate feeding behavior

thick-based, slightly downcurved bill

some juveniles show dusky wing tips; best viewed in flight

TRICOLORED HERON   *Egretta tricolor*   TRHE ■ 1

This colorful, slender-necked heron of southern affinity is a graceful and active hunter. When feeding, it often dashes about in the shallows pursuing prey with wings slightly raised, and may change direction swiftly and with precision; it often wades in deeper water than most other herons. Polytypic (2–3 ssp.; *ruficollis* in N.A.). L 26" (66 cm) WS 36" (91 cm)
IDENTIFICATION Medium-size heron with a long, slender neck and long, thin bill. **NONBREEDING ADULT:** Slate-gray head, neck, back, wings, and tail; back covered with elongated purplish maroon scapular feathers. White underparts, throat, and neck stripe.

Reddish brown eyes, yellow lores, mostly yellow bill, and grayish yellow legs and feet. **BREEDING ADULT:** A few long, white occipital plumes and elongated purplish plumes on the lower neck. Soft-part coloration intensifies: red eyes; turquoise lores and base of bill; and maroon, orange, or pink legs and feet. **IMMATURE:** Rich chestnut head and neck; scapulars and wing coverts edged with chestnut. **FLIGHT:** Swift and direct. White underparts and wing linings contrast with dark flight feathers.
SIMILAR SPECIES Distinctive.
VOICE Relatively silent away from breeding colony. **CALL:** A raspy *aaah*.
STATUS & DISTRIBUTION Uncommon to

common. **BREEDING:** Colonial, usually on islands. Along the entire Atlantic and Gulf Coasts from NY (locally to southern ME) to northern Mexico,

rufous on neck and wing coverts

juvenile
ruficollis

long bill

long, slim neck

white belly

breeding adult
ruficollis

and throughout FL and most of LA, and occasionally inland to KS and AR; along the Pacific coast from Baja California south to northern S.A.. Irregular breeder inland in SC, ND, and SD. **MIGRATION:** Withdraws from

northern portions of breeding range; some U.S. breeders winter in the West Indies. **WINTER:** Along the Atlantic coast from VA south, peninsular FL, and the Gulf Coast; now casual to southern CA (formerly more regular). **VAGRANT:** Casual to northern CA, OR, the Southwest, and southern Canada from MB to NF. Accidental in the Azores.

POPULATION Notable increases in breeding and wintering populations have been recorded along the Atlantic coast from North Carolina since the 1950s. However, Florida populations are showing a recent rapid decline, perhaps due to wetland loss or degradation.

REDDISH EGRET   *Egretta rufescens*   REEG   1

North America's rarest and most restricted heron, the Reddish Egret is confined as a breeder to the Gulf coast. It has two distinct color morphs; the dark morph predominates in North America. It is also notable for its curious, active foraging behavior, quickly pursuing fish in shallows—"dancing" around with wings flapping. It feeds on fish and aquatic invertebrates. Polytypic (2 ssp.; both in N.A.). L 30" (76 cm) WS 46" (117 cm)
IDENTIFICATION Large wading bird restricted to shallow estuaries. **NON-BREEDING DARK MORPH:** Entirely rufous head and neck with short, "shaggy" plumes. Remainder of body dark gray; some dark morphs have scattered white feathers on wings or body. Pale eyes, bill entirely dark, blackish legs and feet. **BREEDING ADULT DARK MORPH:** Elongated plumes on head, neck, and back. Basal portion of bill pink, lores violet-blue, legs

and feet bluish gray. **IMMATURE DARK MORPH:** Plumage brownish gray; wing coverts edged with rufous. Pale eyes; black bill. **WHITE MORPH:** Plumage entirely white. Soft part colors as in dark morph.
GEOGRAPHIC VARIATION Two weakly differentiated subspecies. Nominate *rufescens* is found in the U.S. Southeast and the West Indies; *dickeyi*, with browner head and neck and darker plumage, is found from Baja California to Central America (it wanders north to CA and the Southwest).
SIMILAR SPECIES The dark morph is distinctly larger than Little Blue Heron, and has black legs and feet; note differences in feeding behavior. Compared to a nonbreeding white-morph Reddish Egret, an immature Little Blue has greenish legs and feet and many have dusky wingtips. Nonbreeding white-morph Reddish Egret is distinguished from Snowy Egret by

its larger size, all-black legs and feet, and lack of yellow lores.
VOICE Generally silent. Low *raaaah* and other notes when foraging or disturbed.
STATUS & DISTRIBUTION Uncommon to locally common. **BREEDING:** Colonially with other wading birds; typically islands in coastal estuaries. Currently ±2,000 pairs in the U.S., most in TX. **DISPERSAL:** Nonmigratory, but disperses along Gulf Coast and southern Atlantic coast to NC during spring and summer; some move inland. **WINTER:** Perhaps some withdrawal from northern parts of Gulf Coast; a few regularly move north from Baja California to southern CA (most are found along the coast or at the Salton Sea). **VAGRANT:** Casual north to central CA, NV, WY, CO, MI, IL, NY, and MA. All western records are of dark morphs.
POPULATION Continues to reclaim historic breeding range in Florida, although some colonies are threatened from disturbance by boaters; coastal development has greatly reduced foraging areas in Florida.

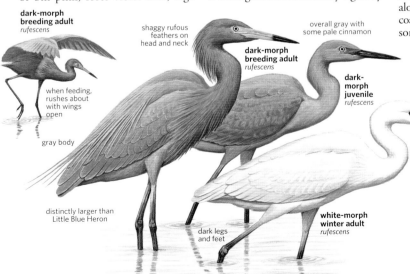

dark-morph breeding adult *rufescens*

when feeding, rushes about with wings open

gray body

distinctly larger than Little Blue Heron

shaggy rufous feathers on head and neck

dark-morph breeding adult *rufescens*

overall gray with some pale cinnamon

dark-morph juvenile *rufescens*

dark legs and feet

only birds in breeding plumage have pink-based bills; bill otherwise blackish in all birds in fall and winter

white-morph winter adult *rufescens*

CATTLE EGRET *Bubulcus ibis*   CAEG ■ 1

This species often feeds in fields, well away from water. There, it habitually follows cattle and other large grazing animals to forage on insects and other invertebrates that the animals flush up, and has learned to follow tractors and other machinery in the same way. Polytypic (2 ssp.; both in N.A.). L 20" (51 cm) WS 36" (91 cm)

IDENTIFICATION Nominate subspecies described. Small, stocky, white egret, with a short, thick neck and relatively short and stocky bill. **ADULT:** Plumage entirely white. Yellow bill; black legs and feet. **BREEDING ADULT:** Bright orange-buff plumes develop on the crown, nape, lower back, and foreneck. (Aberrant individuals may be extensively peach-colored.) Bill becomes two-toned with a reddish base and yellow tip; lores flush purplish pink; and legs and feet become dark red. **JUVENILE:** Like nonbreeding adult, but with black bill.

GEOGRAPHIC VARIATION Accidental (AK) *coromandus* from Asia is larger and longer-necked and has a rich cinnamon wash on head and neck in breeding plumage.

SIMILAR SPECIES Snowy Egret is slimmer with an entirely black bill and black legs with yellow feet. Great Egret is much larger with a longer neck and bill and much longer black legs. Immature Little Blue Heron has a gray bill with a black tip, dark-tipped primaries, and greenish gray legs and feet.

VOICE Emits a coarse nasal *rick-rack* around the nest or roost.

STATUS & DISTRIBUTION Common to abundant. **BREEDING:** Colonially with other wading birds in shrubs or small trees, often on islands. **MIGRATION:** Partially migratory, withdrawing from the northern part of breeding range. **WINTER:** In the southern U.S., especially in southeastern CA, along the entire Gulf Coast, and throughout the FL Peninsula; also in the West Indies and

American tropics. **DISPERSAL:** Wanders well north. **VAGRANT:** To AK and the northern Canadian Provinces. One record to AK of the Asian subspecies *coromandus* at Agattu, Aleutians (June 19, 1988).

POPULATION First documented in the New World in the West Indies in the 1930s, the species colonized Florida by 1941. Its arrival in the New World probably originated with birds from Africa crossing the Atlantic during favorable westerly winds. By the early 1970s, Cattle Egrets had colonized most of the continental United States. Most expansion has slowed or stopped, and major population declines have been seen in the Northeast (and possibly elsewhere) since the peak in the 1970s and early '80s.

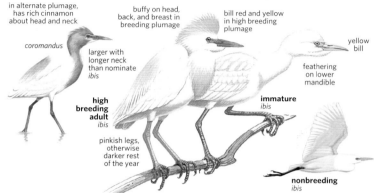

in alternate plumage, has rich cinnamon about head and neck

coromandus

larger with longer neck than nominate *ibis*

buffy on head, back, and breast in breeding plumage

bill red and yellow in high breeding plumage

yellow bill

feathering on lower mandible

immature *ibis*

high breeding adult *ibis*

pinkish legs, otherwise darker rest of the year

nonbreeding *ibis*

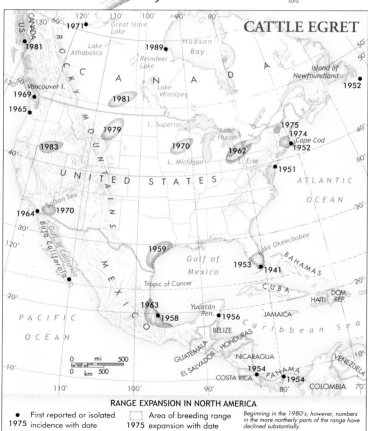

CATTLE EGRET

RANGE EXPANSION IN NORTH AMERICA

● First reported or isolated
1975 incidence with date

▢ Area of breeding range
1975 expansion with date

Beginning in the 1980's; however, numbers in the more northerly parts of the range have declined substantially.

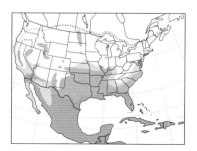

Genus *Ardeola*

CHINESE POND-HERON   *Ardeola bacchus*   CHPH ▪ 5

The Chinese Pond-Heron is the most northeasterly occurring of the three pond-herons. Monotypic. L 18" (46 cm) WS 34" (86 cm) **IDENTIFICATION NONBREEDING ADULT:** Short and stocky with entirely white wings, rump, and tail. Head, neck, breast, and back dark. Yellow bill with black tip; yellow eyes, legs, and feet. **BREEDING ADULT MALE:** Head, neck, and upper breast bright chestnut. Base of bill bluish. Lower breast slaty colored. Long plumes. **BREEDING ADULT FEMALE:** Plumes less showy, foreneck pale; no slaty patch on breast. **IMMATURE:** Head and breast buffy, streaked with dark

brown; mantle brown. **FLIGHT:** White wings, rump, and tail contrast vividly with dark body. **SIMILAR SPECIES** None; dark body and white wings and tail are distinctive. **VOICE** Undescribed; other pond-herons squawk. **STATUS & DISTRIBUTION BREEDING:** Common in China and Indochina. **WINTER:** Throughout much of S.E. Asia. Has strayed to Japan (where increasing) and Korea. **VAGRANT:** Casual. Three summer records of breeding-plumaged birds from western AK islands: Attu I., St. Paul I., and St. Lawrence I.

POPULATION Declining in some breeding areas.

chestnut head and neck in breeding plumage

breeding adult ♂

white wings in all plumages;

Genus *Butorides*

GREEN HERON   *Butorides virescens*   GRHE ▪ 1

The Green Heron (found throughout N.A. and C.A.) was once taxonomically combined with the Striated Heron (*B. striata,* found throughout S.A. and much of the Old World, and recorded once on Bermuda) as the "Green-backed Heron," based on reported widespread hybridization where their ranges overlap in the southern West Indies, southern Central America, and northern South America. Their status as separate species was restored after it was learned that hybridization was not as widespread as was previously thought. The Green Heron is generally found alone or in family groups, occurring in many wetland habitats ranging from freshwater to saline, and open to wooded. It feeds primarily on small fish, but also takes a variety of aquatic invertebrates. The Green Heron is one of the few North American birds to use tools to forage: It places a leaf, feather, piece of bread, or other object on the surface of the water. When a fish swims in to investigate, the heron grabs it. Polytypic. (4 ssp.; 2 in N.A.) L 18" (46 cm) WS 26" (66 cm) **IDENTIFICATION** Small and stocky, with short legs, a thick neck, and a small crest. **NON-BREEDING ADULT:** Sexes similar. Crown and short crest blackish, sides of head and neck rufous. White chin, throat, underside of neck, and breast.

Yellow-orange eyes; dark lores. Blackish yellow bill with greenish yellow base. Yellow legs and feet. Greenish gray back, wings, rump, and tail; upperwing feathers edged with buff. Gray belly and undertail coverts. **BREEDING ADULT:** Soft-part colors intensify: The lores flush bluish black; the bill turns largely black; the legs and feet flush orange. **JUVENILE:** Sides of face, neck, and breast heavily streaked with brown. Upperparts browner; upperwing feathers marked with buff edges and whitish spots at the tips. **FLIGHT:** Underwings of adult are uniformly dark gray; those of immature are paler with contrasting pale feather edges. **GEOGRAPHIC VARIATION** The larger, paler subspecies *anthonyi* is found along the Pacific coast of Mexico and in the U.S. Southwest; subspecies *virescens* occurs in east and central North America, Mexico, the West Indies, and all of Central America.

SIMILAR SPECIES The only other small wading bird regularly encountered in North America is the smaller Least Bittern, which shows white scapular lines and bright buffy wing coverts in all typical plumages. **VOICE** Sharp *skeow* given when migrating at night or when flushed (often accompanied in flight by defecating). **ALARM CALL:** A series of *kuk* notes; raises crest and flicks tail. **STATUS & DISTRIBUTION** Common. **BREEDING:** From ND east to southern NS south to TX through FL. Also, mostly coastally from Puget Sound south through Baja California east to western AZ. Rare and sporadic through much of the interior. **MIGRATION:** May be seen casually through much of U.S. **WINTER:** Many birds resident along coasts and in the Deep South; distribution in C.A. more widespread. **VAGRANT:** Has strayed to the Azores, U.K., Greenland, and HI. **POPULATION** Numbers are stable.

short yellowish legs

adult *virescens*

brownish above

adult *virescens*

streaked neck

juvenile *virescens*

dark cap

BLACK-CROWNED NIGHT-HERON Nycticorax nycticorax   BCNH   1

This stocky heron feeds day and night on fish and aquatic invertebrates and even upland prey. Polytypic (4 ssp.: 2 in N.A.). L 25" (64 cm) WS 44" (112 cm) **IDENTIFICATION NONBREEDING ADULT:** Black back; pale gray upperwings, rump, and tail. Black crown and nape; white forehead, face, neck. Two or three white occipital plumes. Red eyes; black bill; yellow legs and feet. **BREEDING ADULT:** Legs and feet bright pink

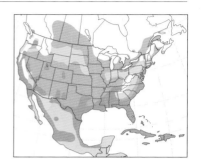

or red. Occipital plumes lengthen. **JUVENILE:** Dark brown above and paler below, with upperpart feathers tipped with large white spots, and underparts heavily streaked. Orange eyes; yellow bill with a dark culmen and tip. **FIRST-SUMMER:** Dark brown crown, back, and upperwings. Face and neck pale with diffuse light brown streaking; belly and undertail coverts white. **SECOND-SUMMER:** Like adult, but white forehead, hind neck, and underparts washed with brown or gray. **FLIGHT:** Fast and direct. Wings short and rounded; only the foot projects beyond tail. **GEOGRAPHIC VARIATION** Subspecies *hoactli* is widespread in N.A.; nominate *nycticorax* from Eurasia is very similar. **SIMILAR SPECIES** Larger, solitary American Bittern (rarely seen in open) has a conspicuous blackish malar stripe; in flight, it shows rather

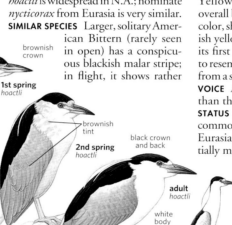

pointed wings, and dark flight feathers contrast with brown coverts. Juvenile Black-crowned is told from juvenile Yellow-crowned Night-Heron by overall browner plumage, paler crown color, shorter legs, and extensive greenish yellow on slightly slimmer bill. By its first summer, an immature begins to resemble an adult, and is readily told from a similarly aged Yellow-crowned. **VOICE** A harsh *wock*, lower pitched than the Yellow-crowned. **STATUS & DISTRIBUTION** Locally fairly common. **BREEDING:** Colonial. Also in Eurasia and Africa. **MIGRATION:** Partially migratory; withdraws from most of breeding range. **WINTER:** Along Pacific coast, entire Gulf Coast, along Atlantic coast from MA south. **VAGRANT:** Casual to AK, Clipperton I., Greenland, HI. Eurasian subspecies recorded from western Aleutians and Pribilofs, AK. **POPULATION** Thought to be stable.

YELLOW-CROWNED NIGHT-HERON Nyctanassa violacea   YCNH   1

Ranging from coastal beaches to inland cypress swamps, this species is more likely to be found nesting alone or in scattered pairs compared to other colonial heron species. It forages for aquatic invertebrates. Polytypic. (5 ssp.; 2 in N.A.) L 24" (61 cm) WS 42" (107 cm) **IDENTIFICATION** Longer neck and more upright posture than Black-crowned Night-Heron. **NONBREEDING ADULT:** Head black with bold buffy white cheek patch, pale yellow crown, and a few wispy occipital plumes. Largely blue-gray; feathers on back and upperwings black with wide blue-gray borders. Orange eyes; short, thick, black

bill; yellowish green legs and feet. **BREEDING ADULT:** Eyes scarlet; lores dark green; legs scarlet or bright orange. **JUVENILE:** Mostly grayish brown above with small white tips to feathers on upperparts; head and neck heavily streaked. **FIRST-SUMMER:** Like juvenile, but darker crown and mostly brown upperparts. **FLIGHT:** Slower wingbeats than Black-crowned; feet and a portion of the legs extend beyond tail. **GEOGRAPHIC VARIATION** Nominate subspecies occurs in central and eastern U.S. and Atlantic coast of C.A. Subspecies *bancrofti* has paler plumage, narrower dorsal streaking, and a

larger bill, and is found from northwestern Mexico to the Pacific coast of C.A. and the West Indies; recently

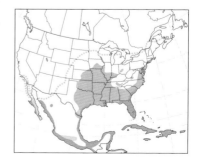

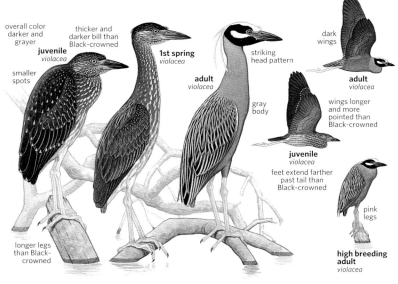

established in the San Diego region, CA.
SIMILAR SPECIES Juvenile is told from juvenile Black-crowned by somewhat darker and grayer plumage with less conspicuous white markings on upperparts; shorter, slightly thicker, entirely dark bill; and longer legs.
VOICE Harsh *wock,* higher pitched than Black-crowned Night-Heron.
STATUS & DISTRIBUTION Uncommon to fairly common, and local. **BREEDING:** Primarily coastal, from MA through FL to TX; a few inland north to MN, OH, and WI. **MIGRATION:** Most interior populations are migratory. **WINTER:** In the U.S., primarily central and south FL, but north to NC and west to TX. **VAGRANT:** To central CA, AZ, NM, ND, and southern Canada from SK to NS; also Clipperton I.
POPULATION Thought to be stable.

IBISES AND SPOONBILLS Family Threskiornithidae

Roseate Spoonbill (TX, Mar.)

A lthough easily distinguished from each other by bill shape, plumage coloration, and foraging behavior, ibises and spoonbills are closely related; hybridization between the two groups has been documented in the Old World.
Structure Threskiornithids are medium to large birds with relatively long necks and legs. Bill shape defines this family: long, downcurved, and slender in ibises, and long, straight, and flattened, with a broad, spoon-shaped distal end in spoonbills. Flight is strong and direct on moderately long, rounded wings; flocks of ibises often soar.
Plumage Species are mostly unicolored, typically white in the spoonbills (but pink in Roseate Spoonbill) and white or blackish in the ibises. Immature plumage is dusky or mottled with brown in ibises, and whitish in spoonbills.
Behavior Colonial breeders, ibises and spoonbills nest

and roost with other wading birds. They tend to forage in flocks away from other species. Tactile feeders, they feed on fish and aquatic invertebrates (especially crabs) in shallow fresh or saltwater. Spoonbills feed most often by sweeping their bills back and forth, while ibises tend to probe to locate prey. Ibises also forage visually in uplands, feeding on terrestrial invertebrates such as grasshoppers.
Distribution The family is represented on all continents except Antarctica, reaching their greatest diversity in the tropics. Worldwide, there are about 32 species in 13 genera (26 ibises in 12 genera; six spoonbills in one genus).
Taxonomy Recent higher-level taxonomic revisions now place the family Threskiornithidae in the order Pelecaniformes with the pelicans (Pelecanidae) and herons (Ardeidae). The species-level taxonomic status of ibises in North America has been debated for centuries. Glossy and White-faced Ibises have often been considered conspecific, but hybridization has only recently been documented. Also often considered conspecific are White and Scarlet Ibises, based on hybridization in South America, where the two species occur sympatrically, as well as in Florida, where Scarlet Ibises were unsuccessfully introduced in the 1960s.
Conservation BirdLife International lists four species as critical, three as endangered, and one as vulnerable, with two others as near threatened. The *Plegadis* ibises are expanding beyond previous range limits, but all threskiornithids remain at risk from colony disturbance and wetland loss or degradation.

SPOONBILLS Genus *Platalea*

ROSEATE SPOONBILL *Platalea ajaja* ROSP ▪ 1

The Roseate Spoonbill's pink and red plumage makes it one of the most flamboyant wading birds. Monotypic. L 32" (71 cm) WS 50" (127 cm)

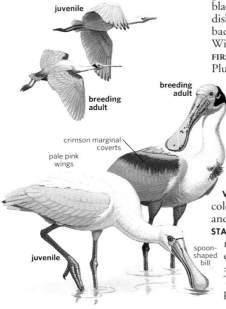

juvenile

breeding adult

breeding adult

crimson marginal coverts

pale pink wings

juvenile

spoon-shaped bill

IDENTIFICATION Unmistakable. All ages have a large, flat, spoon-shaped bill and pink wings and back. **ADULT:** Head featherless, yellowish or greenish gray with black collar. Bill grayish with black markings; eyes, legs, and feet reddish. Red patch on white breast; lower back pink; rump red. Tail orange. Wings pink with wide red carpal bars. **FIRST-WINTER:** Head fully feathered. Plumage white except for wings and tail, which are entirely pale pink. Yellowish bill. **SECOND-WINTER:** Like adult, but plumage paler; wings lack red carpal bars. **SIMILAR SPECIES** Flamingos share the spoonbill's pink plumage, but have much longer and thinner necks and longer legs, as well as very different bill shape and foraging posture. **VOICE** Silent except at breeding colonies; various low, guttural grunts and clucking. **STATUS & DISTRIBUTION** Locally common. **BREEDING:** Colonially on estuarine islands, rarely inland. ±5,000 pairs breed in FL, LA, and TX. **DISPERSAL:** Nonmigratory, but prone to summer-fall postbreeding

dispersal along Gulf and S. Atlantic coasts; many disperse inland. Movements farther north usually made by juveniles. **WINTER:** Resident in coastal southern FL, LA, and TX. **VAGRANT:** Wanders regularly to northeast TX, southeast OK, AR, and casually north to OH, PA, and NY and west to CO and UT. Casual to southern CA (once to central CA) and the Southwest from colonies in northwestern Mexico, but numbers reduced in recent decades. **POPULATION** Recovering in range and numbers from declines during the millinery trade and associated disturbance at colonies.

Genus *Eudocimus*

SCARLET IBIS *Eudocimus ruber* SCIB ▪ 5

This wading bird graces wetlands in the New World tropics. Monotypic. L 23" (58.5 cm) WS 36" (91 cm) **IDENTIFICATION** Unmistakable. **BREEDING ADULT:** Entirely brilliant red with black tips to outermost four primaries. Red legs and bill with blackish distal half. **NONBREEDING ADULT:** Pinkish bill. **IMMATURE:** Head and neck margined brown and gray. Brown upper back and wings; white lower back and rump tinged with pink. Underparts white tinged with

pinkish buff; dark gray legs and feet. **SIMILAR SPECIES** Roseate Spoonbill is larger with a spoon-shaped bill and pink and white plumage, rather than red. Juveniles indistinguishable from juvenile White Ibis until pink feathering develops. **VOICE** Usually silent. **FLIGHT CALL:** Low, grunting *hunk-hunk-hunk*. **STATUS & DISTRIBUTION** Common. Resident in northern S.A. and Trinidad. **BREEDING:** Colonial. **DISPERSAL:** Nonmigratory but prone to dispersal.

VAGRANT: Perhaps to FL but valid occurrence clouded by issue of a few (mostly lost) 19th-century specimens and, more recently, by possibility of escapes. Observations elsewhere in U.S. presumed to be escapes. **POPULATION** Declining, but still common.

adult

WHITE IBIS *Eudocimus albus* WHIB ▪ 1

Historically one of the most abundant wading birds in N.A., the White Ibis remains numerous in parts of the Southeast despite recent substantial declines. It is found in virtually all wetland types and is often seen foraging in grassy fields or along roadways. Despite often breeding in coastal habitats, it

feeds primarily on aquatic crustaceans (especially crabs) in freshwater marshes, or on insects in fields. Monotypic. L 25" (64 cm) WS 38" (97 cm) **IDENTIFICATION ADULT:** Sexes similar. Plumage entirely white except for the black tips of the four outermost primaries. Soft parts orange

or orange-red; decurved bill with a blackish tip. Eyes pale blue. **BREEDING ADULT:** Soft part coloration intensifies: face, bill, legs, and feet bright red or red-orange. **JUVENILE:** Head and neck streaked with dark brown. Dark brown mantle, upperwings, and tail; white rump. Underwings white with blackish band on trailing half of flight feathers. **FIRST-SUMMER:** Juvenile-like head and neck pattern with increasing white

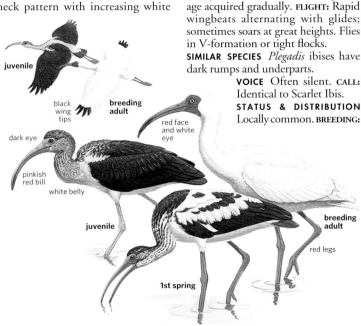

feathering on the back, wings, and finally head and neck as adult plumage acquired gradually. **FLIGHT:** Rapid wingbeats alternating with glides; sometimes soars at great heights. Flies in V-formation or tight flocks. **SIMILAR SPECIES** *Plegadis* ibises have dark rumps and underparts. **VOICE** Often silent. **CALL:** Identical to Scarlet Ibis. **STATUS & DISTRIBUTION** Locally common. **BREEDING:**

Colonially, with other wading birds or in large single-species colonies. Along Atlantic coast from VA south; along the entire Gulf Coast. **MIGRATION:** Partially migratory in northern parts of breeding range. Elsewhere highly nomadic and prone to long-distance vagrancy. Regular visitor to northern TX, OK, AR. **VAGRANT:** To southern CA and the Southwest, north to SD, MI, and southern Canada; also to the Bahamas, Puerto Rico, Bermuda, and Clipperton I. **POPULATION** Declines from habitat loss.

DARK IBISES Genus *Plegadis*

GLOSSY IBIS   *Plegadis falcinellus*   GLIB ▪ 1

The predominantly eastern Glossy Ibis inhabits a variety of wetlands, from freshwater to saline. Its range now overlaps with its western counterpart, the White-faced Ibis. Monotypic. L 23" (58 cm) WS 36" (91 cm) **IDENTIFICATION BREEDING ADULT:** Rich chestnut-red on head, neck, upper back, and underparts. Lower back and tertials, glossed mostly with purple, wing coverts mostly with green. Uniformly grayish bill tinged horn-colored or dull red, thin and downcurved. Dark gray facial skin bordered by pale bluish skin above and below the eye; brown eyes; dark gray legs and feet with dark maroon ankle joints. **NONBREEDING ADULT:** Dark brown head and neck

feathers edged with white, appearing finely streaked. **JUVENILE:** Sooty brown on the head, neck, and underparts with variable amounts of white splotching on the throat and chin. By late fall, many juveniles have acquired the pale blue stripes bordering the facial skin. **FLIGHT:** Flocks fly in a weak V-formation; birds appear all-dark at a distance. **SIMILAR SPECIES** White-faced Ibis is very similar; see sidebar page 152. **VOICE CALL:** A sheeplike *huu-huu-huu-huu.* Flocks sometimes emit a subdued chattering or grunting, particularly on taking flight. **STATUS & DISTRIBUTION** Locally common. **BREEDING:** Coastally from southern ME south to VA and locally south along the Atlantic coast to coastal and inland locations in southern FL and locally along the Gulf Coast west to LA. Widespread in the Old World. **WINTER:** Mainly along the Gulf Coast, throughout FL, and along Atlantic coast north to SC. **VAGRANT:** A few birds now found annually far

north and west to southern Canada, WY, MT, NM, AZ, CA, and WA. **POPULATION** Rare in North America through the 1930s, but its numbers have increased since that time.

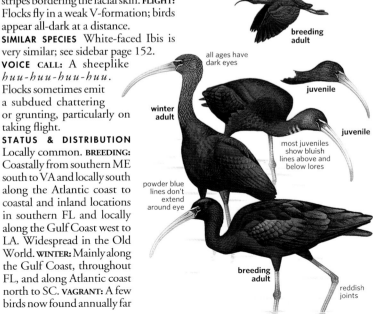

Identification of *Plegadis* Ibises

Separating the Glossy Ibis and White-faced Ibis has become even more important because both species have undergone substantial range expansions in recent years. White-faced Ibises are now annual along the Atlantic and eastern Gulf coasts, and in the Midwest, while Glossy Ibises now occur regularly to the Great Plains, and some have reached the Pacific coast. The breeding ranges of both species overlap in Louisiana and perhaps elsewhere. Hybridization in the wild has been documented; thus, the identification of *Plegadis* ibises can no longer be taken for granted. Scrutinize vagrants carefully; some cannot be identified to species with certainty.

Distinguishing *Plegadis* ibises in breeding plumage is easier than in other plumages. Breeding adult Glossy Ibises have brown eyes, dark lores with narrow pale blue borders that do not extend behind the eyes or under the chin, and gray legs and feet with contrasting reddish ankle joints. In contrast, breeding adult White-faced Ibises have red eyes, pink or red loral skin with wide white borders that extend behind the eyes and under the chin, and more extensive pinkish legs and feet.

The identification of *Plegadis* ibises in nonbreeding and immature plumages requires close views in good light, and even then, some immatures cannot be identified with certainty. Nonbreeding adult Glossy Ibises retain from breeding plumage the dark eyes and dark gray facial skin with a narrow but well-defined pale blue border above and below the eye. The legs are entirely grayish. Adult White-faced Ibises retain their red eyes and pale pinkish facial skin, but lose the white outline around it. Juvenile *Plegadis* ibises cannot always be identified conclusively in the field. However, by their first fall, many juvenile Glossy Ibises have acquired the pale blue facial stripes, and many juvenile White-faced Ibises have acquired a reddish cast to their eyes and a touch of pink facial skin by their first fall. Leg color and bill color between the two species are too variable to be used as useful field marks. ■

WHITE-FACED IBIS *Plegadis chihi*   WFIB ■ 1

The White-faced Ibis is the quite similar, western counterpart to the eastern Glossy Ibis, mostly replacing it west of the Mississippi River. Monotypic. L 23" (58 cm) WS 36" (91 cm)
IDENTIFICATION Tall bird with long legs and a long decurved bill. **BREEDING ADULT:** Rich chestnut maroon plumage on the head, neck, upper back, and underparts. The lower back, tertials, and median wing coverts are a glossy iridescent purple; the lesser coverts and remainder of the wing are mostly metallic bronzy green. The eyes are red; a narrow but prominent white band of feathering extends around the eyes and under the chin, bordering reddish or pinkish facial skin. The grayish bill has a reddish suffusion, particularly near the tip. The legs and feet are largely reddish brown. **NONBREEDING ADULT:** Less glossy than breeding adult; chestnut overall. The dark brown head and neck feathers are edged with white, appearing finely streaked; the throat and foreneck are a paler brown with less streaking. The legs are dull brownish or brownish gray and, the bill is gray; both of these features lack the reddish tones of breeding season. **JUVENILE:** It is duller overall with less sheen than the adult and sooty brown on the head, neck, and underparts, with variable amounts of white splotching on the throat and chin. Legs and bill are dull brownish or brownish gray. **SIMILAR SPECIES** Glossy Ibis is very similar; see sidebar above.
VOICE Sound identical to Glossy Ibis. **CALL:** A sheeplike *huu-huu-huu-huu*. Flocks sometimes emit a subdued chattering or grunting.
STATUS & DISTRIBUTION Locally common. **BREEDING:** Colonially, usually in dense marsh vegetation in low shrubs or trees, often on islands. Locally

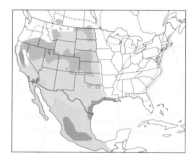

from OR, ID, AB, MT, ND, and SD south to CA, TX, and through Mexico; casual breeder in FL. The largest concentrations are found year-round near the Salton Sea, with a peak population estimated in the tens of thousands. **MIGRATION:** Withdraws from northern part of breeding range; migrates primarily south through the Great Basin and the Colorado River Valley; migrants in the desert southwest usually seen singly or in small groups. TX and LA populations mainly resident. **WINTER:** Southern CA, southwestern AZ, and coastal TX and LA. **VAGRANT:** Disperses widely, but casually to the Midwest and the East Coast from NY and MA to FL. Accidental to Clipperton I. and HI.
POPULATION White-faced Ibis populations declined during the 1970s, but, in recent years, the breeding range and populations have expanded in the West. Populations in Texas and Louisiana have shown declines.

Labels on illustration:
red iris
pinkish skin on lores
often with faint whitish supraloral line; iris red by late winter
winter adult
facial skin grayish, sometimes with some pinkish by fall
juvenile
breeding adult
red facial skin bordered by white feathered margin
extensively reddish legs

NEW WORLD VULTURES Family Cathartidae

Turkey Vulture (FL, Nov.)

N ew World vultures share their naked heads and the practice of feeding on carrion with Old World vultures. Usually found in warm climates, they have a unique habit of urinating down their legs, letting the evaporative process cool their bodies in the process. Banding and tagging studies must therefore use wing tags instead of the standard metal or plastic leg band, which would interfere with that cooling process, and in some cases harm the bird.

Behavior Usually solitary nesters, New World vultures are otherwise gregarious in roosting, searching for food, and gang-feeding at carcasses. Bills are hooked for tearing into flesh, but their feet are relatively weak, not adapted for killing or grasping prey as in hawks and eagles. A single bird spiraling down out of the sky toward food will attract the attention of other birds from a great distance. At the roost site in the morning, they often will perch spread-winged in the sun before leaving to forage. Eggs are laid on bare surfaces, such as cliff ledges, caves, hollow logs, abandoned buildings, and the ground.

Plumage All species are primarily black on the wings, body, and tail, with some pattern of white or silver on the underwings. All share the naked head, in order to feed on carcasses without fouling head feathers. Immatures usually have dark skin on the head, with the adult's skin lightening or becoming more colorful. The King Vulture *(Sarcoramphus papa),* distributed from central Mexico to northern Argentina, is unique in having extensively white plumage as an adult.

Distribution Primarily warm-weather species, New World vultures are colonizing more northerly areas, perhaps as a consequence of global warming, or perhaps due to increased availability of winter roadkills.

Taxonomy Recent genetic data strongly refute this family's close relationship to the storks (Ciconiidae). Some authorities place the New World vultures in their own order, although the AOU retains them in Acciptriformes (Hawks, Kites, Eagles, and Allies). Worldwide there are seven species in the New World vultures family, three of which occur in the continental United States.

Conservation Farmers occasionally persecute the Black Vulture for aggressive behavior toward newborn farm animals. The California Condor is on the U.S. Endangered Species List, and is the focal point of federal and state programs where captive-bred individuals are released into prospective habitats to re-establish it as a viable wild species.

Genus *Coragyps*

BLACK VULTURE   *Coragyps atratus*   BLVU ■ 1

The gregarious Black Vulture roosts, feeds, and soars in groups, often mixed with Turkey Vultures. A carrion feeder that will bully a Turkey Vulture away from a carcass, it occasionally kills smaller live prey. Polytypic (3 ssp.; nominate in N.A.). L 25" (64 cm) WS 57" (145cm) **IDENTIFICATION ADULT:** Glossy black feathers can show iridescence in the right light. Whitish inner primaries

soars on flat wings; wingflaps are rapid and shallow

adult
atratus

extensive white base to outer primaries

pale grayish legs

short tail

atratus

often hard to see on the folded wing. Whitish legs contrast with dark gray head color. Skin of head wrinkled; bill dark at base and tipped ivory or yellowish. **JUVENILE:** Black body and wing feathers usually duller, less iridescent. Skin of head smooth, darker black than an adult. **FLIGHT:** Conspicuous white or silvery patches at base of primaries that contrast with black wings, body, and tail. Whitish legs extend almost to tip of relatively short tail. Soars and glides with wings held in a slight dihedral. If seen at a distance, the quick, shallow, choppy wingbeats interspersed with glides are usually enough to make an identification. **SIMILAR SPECIES** The Turkey Vulture shows silvery inner secondaries and a pronounced dihedral while in flight, along with a deeper, more fluid wingbeat. **VOICE** Hisses when threatened. **STATUS & DISTRIBUTION** Abundant in the Southeast, expanding up the East Coast

into southern New England. Less common in southern Great Plains, local in southern AZ. **BREEDING:** Nests in a sheltered area on the ground, including abandoned buildings. **MIGRATION:** Sedentary, northern breeders may migrate with Turkey Vultures to warmer winter territory. **VAGRANT:** Casual to CA, northern New England, and southern Canada; accidental to southwestern NM, BC, and southern YT. **POPULATION** The species adapts well to human presence, feeding on roadkills and at garbage dumps.

Genus *Cathartes*

TURKEY VULTURE   *Cathartes aura*   TUVU ■ 1

The most widespread vulture in North America, the Turkey Vulture is locally called "buzzard" in many areas. A Turkey Vulture standing on the ground can, at a distance, resemble a Wild Turkey. It is unique among our vultures in that it finds carrion by smell as well as by sight. When threatened, it defends itself by vomiting powerful stomach acids. Polytypic (4 ssp.; 3 in N.A.). L 27" (69 cm) WS 69" (175 cm) **IDENTIFICATION** Overall, it is black with brownish tones, especially on the

feather edges. Legs are dark to pinkish in color, the head unfeathered. **ADULT:** The red skin color of the head contrasts with the ivory bill and dark feather ruff on the neck. **JUVENILE:** The skin of the head is dark, the bill dark with a pale base. **FLIGHT:** A large dark bird that flies with its wings held in a noticeable dihedral. Usually rocks side-to-side, especially in strong winds. Underneath, the silvery secondaries and primaries contrast with the black wing coverts, giving a two-toned look to the wing. The tail is relatively long. **GEOGRAPHIC VARIATION** Three subspecies are found in North America *(aura, meridionalis, septentrionalis).* Only a few minor differences in size and overall tone separate them. **SIMILAR SPECIES** The Black Vulture has a quicker, shallow flap, shorter tail, and obvious white patches at the base of the primaries on its shorter, broader wings. The Zone-tailed Hawk and Golden Eagle will mimic the wing dihedral when hunting, and

adults

long tail

two-toned underwing

soars on slight dihedral; wingflaps are slow and deep

reddish head

grayish head

adult

juvenile

dark-morph Swainson's Hawks and Rough-legged Hawks also can fly in a dihedral, but at closer range all have feathered heads and different wing shapes. Zone-tailed also shows banded tail and yellow cere and legs.

VOICE Hisses when threatened. **STATUS & DISTRIBUTION** Year-round in southern U.S., migrates into northern U.S. and Canada. **BREEDING:** Nests on the ground, using a shallow cave, hollow log, or thick vegetation.

MIGRATION: Northern populations migratory, some heading to C.A. **WINTER:** Increasing numbers in snowbound states. **VAGRANT:** Casual to AK, YT, NT, and NF. **POPULATION** Stable.

Genus Gymnogyps

CALIFORNIA CONDOR *Gymnogyps californianus* CACO ■ 6

The largest raptor in North America, the California Condor is unmistakable when seen flying with any other bird. It will soar for hours on thermals in search of carrion, rarely flapping, and can cover vast amounts of territory. The wingbeat is slow and deep, the wingtips appearing almost to touch at the bottom of the stroke. Monotypic. L 47" (119 cm) WS 108" (274 cm) **IDENTIFICATION** **ADULT:** The body is black; the head shows bare red-orange to yellow skin. Large areas of white are visible on the underwings. Legs are whitish. **JUVENILE:** Body and legs are similar to adult, but the head is dark. White on the underwings is mottled, less extensive. **FLIGHT:** Wings held flat or with the tips slightly higher, with large splayed primaries ("fingers") evident. The dark tail is relatively short. Seen from above, the wings have a thin white diagonal line at the greater coverts. **SIMILAR SPECIES** The Turkey Vulture is much smaller, with no white on underwings, and flies with a more pronounced dihedral. The smaller Golden Eagle has a feathered head and longer tail; the juvenile has restricted white patches at base of primaries. A juvenile Bald Eagle always flies with flat wings and shows irregular white on underwings, but is smaller and has a feathered head.

VOICE Hisses and croaks. **STATUS & DISTRIBUTION** California Condor fossil evidence has been found through much of western North America and even southern Canada. By the early 1900s the species was restricted to a relatively small area in southern California. Many condors died by feeding on poisoned carcasses set out by farmers to control coyotes. The U.S. listed the species as endangered in 1967. As the population dwindled, the remaining 22 wild birds were live-trapped (the last on April 19, 1987) and placed into a captive breeding program. After successfully building up the captive breeding population, two captive-bred condors were released into the wild in 1992. Today, free-flying birds (over 230 in 2013) are gaining a toehold in ranges in CA, the Grand Canyon area of AZ and UT, and northern Baja California. Some released condors have produced young in the wild (starting in 2002), but about 200 birds remain in the captive breeding program. **YEAR-ROUND:** Nonmigratory. **BREEDING:** Solitary nesters, usually on a cliff face in a shallow cave; easily disturbed by man. Fledged chicks

are looked after for up to eight months. **POPULATION** Even though the Condor Recovery Plan has achieved a large measure of success, the California Condor is still considered critically endangered, with a total population (captive and free flying) in May 2013 of 435 birds.

adult

huge size

orange head

adult

juvenile

whitish wing linings

wings are long, broad, and of rather uniform width

unlike Turkey Vulture soars with no distinct dihedral

white wing linings contrast with black flight feathers

short tail

adult

white tips to greater coverts and white on edges of secondaries visible from above

OSPREY Family Pandionidae

Osprey (Scotland, July)

Having been shuffled back and forth between its own monotypic family and a subfamily within the more diverse Accipitridae, this widespread fish-catching specialist is now generally afforded family rank. **Structure** A large, long-winged, hawklike bird, it has a narrow, rounded head that lacks the bony "brows" of most other large raptors. The long legs have strong talons and remarkable spikelike pads on the toes that help Ospreys grip slippery fish prey; the outer toe can be reversed to better hold prey.

Behavior Consummate fish-catchers. With a strong, gull-like flapping flight, Ospreys are less reliant on rising air currents than most other large raptors and more readily cross large water gaps.

Plumage See species account.

Distribution Ospreys breed widely through North America, locally in the Bahamas and Cuba, and in Eurasia and northern Africa, as well as the Australasian region. They are nearly or entirely absent from most oceanic island groups, New Zealand, and Madagascar.

Taxonomy The family includes only one genus and species with four moderately divergent subspecies; some authorities have considered the subspecies *cristatus* of the Australasian region and the pale-headed *ridgwayi* of the Caribbean to be a distinct species. Recognition of the family Pandionidae as distinct from the other diurnal birds of prey is based on studies of DNA, chromosomes, and morphology.

Genus *Pandion*

OSPREY *Pandion haliaetus*   OSPR ■ 1

Known to many as the "Fish Hawk," the Osprey is almost exclusively a fish-eater, capturing its prey by plunging into the water feet-first. Polytypic (4 ssp.; *carolinensis* in N.A.). L 22–25" (56–64 cm) WS 58–72" (147–183 cm)

IDENTIFICATION Large raptor; dark above and white below with a white head and a prominent dark eye stripe. **ADULT:** Upperparts are uniformly dark, eyes yellow. Most have some streaking on the upper breast, with larger females usually showing heavier markings, forming a "necklace." **JUVENILE:** Upperparts fringed with pale buff, giving it a "scaly" appearance. **FLIGHT:** Long, narrow wings held back at the "wrist," dark carpal patches conspicuous; wings slightly arched when soaring. **SIMILAR SPECIES** Immature Bald Eagles show a dark-light pattern on the head, underwings, and underparts. Their larger size and straight-wing silhouette set the eagles apart.

VOICE Series of loud, whistled *kyew*'s.

STATUS & DISTRIBUTION YEAR-ROUND: Southern U.S. birds tend to be non-migratory. BREEDING: A large stick nest is built, usually in a dead treetop, but increasingly on man-made structures. MIGRATION: Postbreeding dispersal occurs in late summer, and migrants are common through Sept. at most hawk-watch sites.

POPULATION Controls on pesticides (chiefly DDE) have resulted in rebounding populations over much of range.

dark "wrist"

pale wing linings

barred flight feathers

gull-like flight

carolinensis

long angled wings

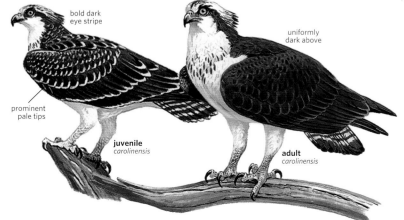

bold dark eye stripe

prominent pale tips

uniformly dark above

juvenile *carolinensis*

adult *carolinensis*

HAWKS, KITES, EAGLES, AND ALLIES Family Accipitridae

Bald Eagle, adult (left) and subadult (AK, Mar.)

Most non-birdwatchers can identify a large, hook-billed, fierce-looking bird as a "hawk" or an "eagle," even if they do not know much more than that. The North American members of family Accipitridae include a wide range of body sizes, structures, flight styles, prey selection, and habitat preference.

Structure Common features of this group are a hooked beak, strong feet with curved talons for grasping and in some cases killing, and sharp eyesight for locating prey at a distance. The buteos have long, wide, rounded wings for soaring, as do the eagles. The kites and harrier have slimmer wings with pointed or rounded wing tips and long tails to help with hovering or maneuvering. Short, wide wings and long tails give the accipiters the ability to make quick turns in close quarters. Most juveniles have shorter wing feathers and longer tails than adults. Females are usually larger than males, and among species that inhabit a wide range of latitudes, northern breeders are usually larger than southern birds.

Behavior Not surprisingly, birds with different body and wing types employ different hunting strategies and prey selection. The soaring buteos and eagles spy prey from afar and make long glides or dives from above on small mammals, reptiles, and ground birds. The three pointed-wing kites actively flap and all hover—except Mississippi Kites—in search of small

prey on the ground, mostly mammals, reptiles, and insects; the two other kites are snail-eating specialists that course low over their habitats in search of their slow-moving meals. The accipiters are bird-catchers, speeding through the trees and brush to capture their prey on the wing. All species except the harrier build a stick nest, usually in a tree, but western species may use a rock ledge. Harriers use grasses and sedges in their ground nest, which is usually concealed in longer grasses or reeds. Most breeders are site-faithful, and the larger birds will reuse a nest site for many years, adding to the nest structure. Most species are migratory, with the northern nesters following well-known migratory paths, passing regular hawk-watch sites to the delight of birders. The southwestern species are more sedentary, but may wander outside of the breeding season. Mississippi and Swallow-tailed Kites are becoming regular wanderers in late spring up to the northeastern states, upper Midwest, and southern Ontario, and it is theorized that global warming may be contributing to possible range expansions.

Plumage Immatures are distinct from adults, with most species acquiring adult plumage in one year; the eagles take four years. Body feathers are replaced annually, as are flight feathers on the smaller hawks. The eagles and many buteos typically do not replace all their flight feathers in one year, creating visible patterns of old and new feathers. All show significant feather wear by spring, bleaching out colors on the upperparts.

Distribution The species in this group inhabit virtually every habitat in North America, from tundra to desert, mountain forests to coastal marshes. A species' relative abundance is usually tied to its habitat and prey preference. All are regular breeders in North America (though not necessarily common) except two of the sea-eagles, Double-toothed Kite, Crane Hawk, and Roadside Hawk, which are vagrants.

Taxonomy The birds in this group comprise 14 genera in N.A. The genus *Buteo* is best represented, with ten species. The Osprey was recently placed in its own family, Pandionidae (previous page).

Conservation Worldwide, 34 species are listed as threatened, including Old World vultures, 25 as near threatened. The fish-eating birds have suffered population declines in the past due to pesticide-laden food sources. After banning most of the offending substances for use in North America, increases in breeding populations have been encouraging. Species that winter in Central America and South America, where some of these compounds are still in use, are still considered at risk, but to a lesser degree than previously. Traditionally, raptors have suffered most from indiscriminate shooting, although in recent years, statistics on banded raptors found dead have seen a drop in the percentage killed by gunshot, especially in the eastern states.

KITES Genera *Chondrohierax, Elanoides, Elanus, Rostrhamus, Harpagus,* and *Ictinia*

Kites are a loosely-related group of raptors, six of which occur in North America. They are all of different genera, and do not share many common physical characteristics. Our kites are found in the warmer regions of North America, and two are long-distance migrants, with Swallow-tailed and Mississippi kites wintering as far south as South America.

HOOK-BILLED KITE *Chondrohierax uncinatus* HBKI ■ 3

A tropical hawk that is a decidedly uncommon resident in the Rio Grande Valley area in Texas. Most sightings are of birds taking flight in the morning; they rarely venture up in the air later in the day. Polytypic (4 ssp.; *aquilonis* in N.A.). L 18" (46 cm) WS 36" (91 cm)
IDENTIFICATION Featuring a long, slim body with long, square tail, the wing silhouette is unique: Wide, rounded primaries and bulging secondaries taper down to a narrow wing base, giving a diagnostic paddle-shaped wing. Dark morph is rare in the U.S. **ADULT MALE:** Dark gray with light barring across the belly, tail with two wide

bands, white eye color. **ADULT FEMALE:** Brown with heavy rufous barring on underside, rufous on underwings, two tail bands, white eye. **BLACK MORPH:** Rarely seen in U.S.; adult is all black except for a single white tail band and tail tip. **JUVENILE:** Like adult female, but with thinner barring underneath, no rufous in wings, three narrow bands on tail, brownish eyes. **FLIGHT:** Slow, deep wingbeats for a bird of its size; will occasionally soar on thermals.
SIMILAR SPECIES Gray Hawk is not as strongly barred underneath, has dark eyes. Red-shouldered Hawk shows crescents at the base of the primaries, narrow white tail-bands.

VOICE Loud rattling notes given near its nest when disturbed or during courtship.
POPULATION The U.S. population is small but seemingly stable, as long as snails are available.

black-morph adult

black-morph juvenile

aquilonis

juvenile *aquilonis*

paddle-shaped wings

hooked bill

rufous wing linings

adult ♀ *aquilonis*

barred grayish underparts

white collar

variable dark barring below

whitish underparts

adult ♀ *aquilonis*

juvenile *aquilonis*

rufous collar

adult ♂ *aquilonis*

barred rufous underparts

SWALLOW-TAILED KITE *Elanoides forficatus* STKI ■ 1

usually seen in flight; white body and wing linings contrast with black flight feathers

long forked black tail

adults *forficatus*

Elegant and graceful in flight, this species readily soars but does not hover. It feeds mostly on insects, lizards, snakes, and frogs, usually taken on the wing. Polytypic (2 ssp.; *forficatus* in N.A.). L 23" (58 cm) WS 48" (122 cm)
IDENTIFICATION Head, underparts, and wing linings pure white; otherwise blackish. **ADULT:** Has red eyes. **JUVENILE:** Brownish eyes, the shorter tail is less deeply forked, flight feathers have lighter tips. **FLIGHT:** Unmistakable.
VOICE Mostly silent.
STATUS & DISTRIBUTION Very uncommon. **BREEDING:** Breeds in FL, along the Gulf Coast into eastern TX, up the Atlantic coast to SC. Builds a flimsy stick nest in a treetop, often exposed to view or on a lone tree. **MIGRATION:** Most pass through the FL Keys, increasing numbers through southern TX hawkwatch sites. **WINTER:** Radio-tagged

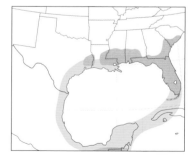

individuals from the Orlando, FL, area have been satellite-tracked to S.A. and back. **VAGRANT:** Increasing frequency of records into the upper Midwest and northeastern U.S.
POPULATION Very small (approx. 1,000 pairs in U.S.), but appears stable. Historical range much larger (north to MN, IL, OH).

WHITE-TAILED KITE *Elanus leucurus* WTKI ■ 1

A nearly all-white bird often seen hovering above a field or marsh in the Pacific states, Texas, and Florida. Often seen perched along roadsides on telephone wires or dead snags, when hunting they slowly descend (it can hardly be considered a dive) on small rodents with their wings held up in a deep V. Polytypic (2 ssp.; *majusculus* in N.A.). L 16" (41 cm) WS 42" (107 cm) **IDENTIFICATION ADULT:** The white head, tail, and undersides contrast with black upperwing coverts (the "shoulders," formerly called "Black-

shouldered Kite"), and gray crown, back, and upper flight feathers. Below, the black carpal patch contrasts with gray primaries and white secondaries and coverts. Deep red eyes are framed with dark feathers. **JUVENILE:** Best told by brownish feathers on the back and crown, rufous wash across the breast, brown to orange eye. The tail has faint gray subterminal band, flight feathers with white tips. The darker body feathers are molted within a few months, giving the bird a more adult appearance, but still with the retained flight feathers. **FLIGHT:** Fast, shallow wingbeats interspersed with glides, pausing to hover tail-down, wings held high when hunting. Glides with a slight dihedral. **SIMILAR SPECIES** Mississippi Kites are darker on the tail and body, immatures are heavily streaked and do not hover. An adult male Northern Harrier has a white rump above a darker tail, no carpal patches on underwing, and a low, swooping flight. **VOICE** Calls include various whistled notes. **STATUS & DISTRIBUTION YEAR-ROUND:** Resident in coastal CA to the Sierra

Nevada, grasslands of southeastern AZ and into TX, uncommon in FL, rarely in southern NM. **BREEDING:** Builds a small nest in a solitary tree or brushline. **VAGRANT:** Casual north to BC and MN, east to the East Coast.

adults *majusculus* — habitually hovers when foraging — dark spot near "wrist" — long whitish tail — juvenile *majusculus* — buffy on chest

black shoulders — adult *majusculus*

SNAIL KITE *Rostrhamus sociabilis* SNKI ■ 2

Formerly known as "Everglades Kite," this bird is a specialist that feeds on apple snails, using its hooked beak to extract its prey. With paddle-shaped wings, it flies low over marshes in search of its quarry. Often see perched on bushes in marshes. Polytypic (4 ssp.; *plumbeous* in N.A.). L 17" (43 cm) WS 46" (117 cm) **IDENTIFICATION** Long wings and white-

tipped dark tail with large white patch at base. **ADULT MALE:** Dark gray overall, appearing black in poor light; flight feathers darker than back and wing coverts. **ADULT FEMALE:** Dark brown above, thin white streaking on underparts, white markings around red eye. **JUVENILE:** Brown eye, tawny markings on head and upperparts, undersides tawny with dark streaks. **FLIGHT:** Slow, graceful wingbeats, wings held with tips cupped. **SIMILAR SPECIES** Female and juvenile Northern Harriers have white rump, no white in the tail, and fly with wings in a dihedral.

VOICE Mostly silent; occasionally a nasal grating sound. **STATUS & DISTRIBUTION YEAR-ROUND:** Resident in southern FL up to the Orlando area. Also resident on Cuba. **BREEDING:** Nests in low trees or bushes, often surrounded by water. Will nest colonially. **VAGRANT:** Two records from TX likely are of birds from Mexico. **POPULATION** Endangered and intensely managed (approx. 500 pairs in FL). Stable, but dependent on freshwater marshes for snails.

adult *plumbeous* ♂ — whitish supercilium and chin — long, thin curved bill — strong eye line — white upper-tail coverts — heavy streaks on underparts — juvenile *plumbeous* — adult *plumbeous* ♀

white undertail — gray tail tip — adult *plumbeous* ♂

DOUBLE-TOOTHED KITE *Harpagus bidentatus* DTKI ■ 5

adult *fasciatus*
checkered primaries
two notches on bill
rufous barring on underparts
puffy white undertail coverts
shape and structure recall an *Accipiter*
underparts with bars and streaks on subadult
subadult *fasciatus*
banded tail

The Double-toothed Kite's scientific name, *bidentatus* (two teeth), derives from the notches on the cutting edge of the maxilla. This feature, as in shrikes and falcons, is used to dispatch prey quickly with a bite to the neck. Polytypic (2 ssp.; likely northern *fasciatus* in N.A.). L 13.5" (34 cm) WS 27" (69 cm) **IDENTIFICATION** A small compact raptor recalling an *Accipiter* in shape, with relatively long tail and rounded wings. **ADULT:** Underparts barred vivid rufous on breast, grading to gray and white barring on belly, contrasting with fluffy white feathering in undertail coverts. Underwing coverts white; remiges barred. Upperparts a soft gray. **IMMATURE:** Upperparts brownish with pale-spotted scapulars. Under-

parts buffy white with brown streaks. **GEOGRAPHIC VARIATION** Nominate ssp. (most of northern South America) has more intensely rufous and more densely barred underparts than *fasciatus* (central Mexico through Colombia). **SIMILAR SPECIES** A soaring Double-toothed Kite may bear an uncanny resemblance to an *Accipiter*. At close range, the dark line down the middle of the throat and fluffy white leg tufts and tail coverts distinguish Double-toothed from all other New World raptors. **VOICE** Sharp, high-pitched whistled *tswee!* in series or singly, sometimes ending with a short, rising flourish, *yip!* Some calls sound uncannily like those of pewees (*Contopus*). **STATUS & DISTRIBUTION** BREEDING:

Lowland and mid-elevation forests **MIGRANT:** Not known as a long-distance regular migrant. **VAGRANT:** Subadult at High I., TX (May 4, 2011), the only record for the U.S.

MISSISSIPPI KITE *Ictinia mississippiensis* MIKI ■ 1

This small, pointed-winged kite looks more like a falcon than any other of our kites. A buoyant flier, it soars on flat wings, often high up in the air on thermals, catching and eating insects on the wing. Monotypic. L 14.5" (37 cm) WS 35" (89 cm) **IDENTIFICATION ADULT MALE:** Dark gray overall, lighter head with red eyes, dark primaries and tail. Seen from above, light secondaries form a bar across the wings. **ADULT FEMALE:** Like male, but darker head. **JUVENILE:** Dark brown eyes in a gray-brown head, with wide, creamy superciliary line and gray

cheeks. Back and wings are dark brown with buffy edges; scapulars have white spots. Underparts are heavily streaked, and the dark tail has multiple thin white bands. **SUBADULT:** Body plumage similar to adult's, but with a blend of juvenile and adult feathers, especially on tail and flight feathers in late summer/first spring. **FLIGHT:** It does not hover. The pointed wings are notable in that the outer primary is much shorter than

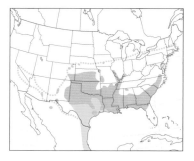

the next one. Tail is square-tipped, usually flared in flight. Underwing coverts are gray in adults, mottled in juvenile. **SIMILAR SPECIES** Most other kites are whiter. Adult Peregrine Falcon is larger, but with similar flight silhouette; its moustache mark and more powerful flight are diagnostic. **VOICE** Downward whistle, given mainly on breeding grounds. **STATUS & DISTRIBUTION** BREEDING: Central and southern Great Plains states, the South, and up the Atlantic coast into the Carolinas. Isolated colonies in NM and AZ. A stick nest built in a tall tree may be part of a loose colony of up to 20 pairs. **MIGRATION:** Most migrate Aug.–early Sept., stragglers into Oct., often in large groups through coastal TX. Return late Mar.–late Apr. **WINTER:** Well down in S.A. **VAGRANT:** Casual, with spring and summer (has nested) records from southern New England, the mid-Atlantic, and the upper Midwest. Accidental to NF. **POPULATION** Slowly spreading north.

never hovers
adult ♂
whitish secondary patch
dark wings
adult ♀
black tail
pale gray head
juvenile
banded tail held through 1st summer
adult ♂

SEA-EAGLES Genus *Haliaeetus*

These birds are very large raptors with a preference for fish; there are eight species worldwide, one breeding in North America, and two casual visitors. Large, wide wings allow them to soar effortlessly as they search for prey that can also include waterfowl and carrion. They are skilled at kleptoparasitism, stealing prey from Ospreys, hawks, and falcons.

BALD EAGLE *Haliaeetus leucocephalus*   BAEA   1

The official symbol of the United States, the Bald Eagle, while primarily a fish-eater, will also take waterfowl and occasionally mammals, but most often prefer to steal kills from other raptors, as well as feed on carrion. Monotypic. L 31–37" (79–94 cm) WS 70–90" (178–229 cm)

IDENTIFICATION A large, dark raptor with bulky body, large head and neck. Females larger than males. Adults have longer wings than immatures. Flight feathers take two to three years to replace. Head and neck project about one-half the length of the tail. **ADULT:** The pure white head and tail contrast with the dark brown body, wings, and legs. The bill and eyes are yellow. **IMMATURE:** There are four distinct stages, corresponding to each year of life. **JUVENILE:** Uniformly dark brown, with white on underwing coverts and axillaries; trailing edge of the secondaries is even. Beak and eye are dark. **SECOND-YEAR:** The eye turns gray-brown or whitish; the head gets a whitish superciliary line. The back and belly become speckled with varying amounts of white. Trailing edge of wings is irregular due to new (shorter) and retained older (longer) secondaries. **THIRD-YEAR:** The cheeks and head become whiter, with a contrasting dark eye stripe; bill becomes yellow; eye color turns yellow. Back and belly still with white mottling, underwing coverts not as white as before. Secondary feathers are even in length. **FOURTH-YEAR:** Body and wing coverts mostly dark, but with white spots throughout. Head white with dark eye line, bill yellow. Tail is usually white, but often has a black terminal band. **FLIGHT:** Soars on flat wings; wingbeats are heavy, slow and powerful.

SIMILAR SPECIES Turkey Vulture glides in a dihedral and rocks side-to-side when turning. Its two-tone underwings are visible in flight. Golden Eagle at all ages has dark axillaries, dark wing linings; head and neck project less than half the tail length. Immature Golden has white at the base of the primaries, white base of tail.

VOICE It has a remarkably weak twittering or whistled call, often delivered in a staggered rhythm.

STATUS & DISTRIBUTION Throughout most of N.A., always near water. Southern birds are nonmigratory. **BREEDING:** Long-lived, they build impressive stick nests in the crotch of a tall tree, adding to the structure every year. **MIGRATION:** They follow coastlines, rivers, and mountain ridges in fall. **DISPERSAL:** Southern juveniles wander to the Great Lakes in early summer. **WINTER:** Northern birds keep to the coastlines or south of the freeze line to find open water. Hydroelectric dams are a favorite location. **POPULATION** The species formerly was common throughout the continent; pesticides in the food chain dramatically reduced numbers in the 1950s and '60s. Conservation efforts highly successful, with nesting reported in most states. In 2007, the Bald Eagle was removed from the U.S. endangered species list.

2nd year

juvenile

whitish underwing coverts and axillaries

note Osprey-like face pattern

3rd year

longer head projection than Golden

larger bill than Golden

juvenile

white head

adults

white tail

tail shorter than Golden

WHITE-TAILED EAGLE *Haliaeetus albicilla* WTEA ■ 4

This large fish-eating eagle from Eurasia most closely resembles the Bald Eagle. Females are larger than males; immatures take four to five years to attain adult plumage. Monotypic. L 26–35" (66–89 cm) WS 72–94" (183–239 cm) **IDENTIFICATION** Perched bird has wing tips that reach the tip of the tail. **ADULT:** Dark brown body and wings, white tail slightly wedge-shaped, head color creamy brown, blending into darker brown upper breast. Bill yellow. Undertail coverts are dark brown. **IMMATURE:** Generally follows the age patterns of Bald Eagle, but with less white on underwings and axillaries; tail longer and less wedge-shaped with variably dark feathers with white centers or white mottling.

Wings are shorter and wider than those of adults. **FLIGHT:** Shows seven emarginated (notched) primaries or "fingers" in flight. Caution—raptors can occasionally have "extra" primaries, so this character should be used in conjunction with other field marks. **SIMILAR SPECIES** Bald Eagle in flight has six emarginated primaries. Adult has white head, tail, and undertail coverts, dark chest. Immatures have white axillaries (wingpit), more white on underwing coverts, dark band on tip of tail. **STATUS & DISTRIBUTION** Widespread across Eurasia **VAGRANT:** Most N.A. records are from western AK, predominantly from the western Aleutians,

where it nested on Attu I. One record from central Aleutians and another on St. Lawrence I., AK. Accidental to MA.

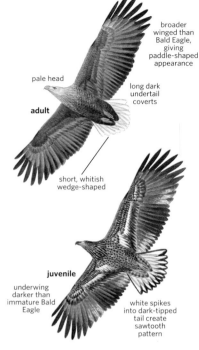

broader winged than Bald Eagle, giving paddle-shaped appearance

pale head

long dark undertail coverts

adult

short, whitish wedge-shaped

palish head

overall scaly appearance

short tail

adult

juvenile

underwing darker than immature Bald Eagle

white spikes into dark-tipped tail create sawtooth pattern

STELLER'S SEA-EAGLE *Haliaeetus pelagicus* STSE ■ 4

A sea-eagle of eastern Asia, Steller's is a vagrant to Alaska. Sexes are similar; females are larger than males. Monotypic. L 33–41" (84–104 cm) WS 87–96" (221–244 cm) **IDENTIFICATION** Huge eagle with a very long, wedge-shaped tail, wide wings, and a massive bill. **ADULT:** Striking white upper and lower wing coverts, tail coverts, tail, and leg feathering. Head is dark with grayish streaks; bill is bright yellow. Neck, body, and flight feathers are dark brown. **IMMATURE:** Bill yellow with dusky tinges, head brown with pale streaking on neck and breast. Underwing coverts brown streaked with white, white axillaries, flight feathers dark with light bases to primaries. Leg feathers and undertail dark, wedge-shaped white tail flecked with dark. **FLIGHT:** Soars with wings in a strong dihedral, secondaries often wider than those of other eagles. **SIMILAR SPECIES** Adults are unique, immatures are like Bald Eagle but

with longer, wedge-shaped tails and huge yellow bill. **STATUS & DISTRIBUTION VAGRANT:** Fewer than ten widely distributed records from AK, as far southeast as Juneau. A probable Bald Eagle x Steller's Sea-Eagle hybrid was photographed in BC during the winter of 2004–2005.

paddle-shaped appearance

white leading edge to wing

massive size

huge orange bill

long white acutely wedge-shaped tail

adult

juvenile with mostly orange bill

white axillaries

huge bright orange bill

white shoulders

juvenile

long wedge-shaped tail is white in adult, dark tipped in juvenile

white primary patch

adult

white leggings

EAGLES Genus *Aquila*

Worldwide, there are ten species in this genus, only one of which occurs in North America: the Golden Eagle. Large raptors with wide wings, all are fierce hunters of mammals and birds, and many are prized for falconry. The human descriptive term "aquiline nose" refers to the hooked beak of *Aquila* eagles.

GOLDEN EAGLE *Aquila chrysaetos* GOEA ■ 1

This large, dark eagle of open areas in the U.S. West and Canada is a fierce hunter, preying on small to medium-size mammals but also taking ground birds, reptiles, insects, and even carrion in winter. Females are larger than males. Polytypic (5 ssp., *canadensis* in N.A.). L 30–40" (76–102 cm) WS 80–88" (203–224 cm)

IDENTIFICATION Overall coloration of all ages is a uniform dark brown, with varying amounts of golden feathers on nape and crown. Legs are feathered down the tarsis to the toes. Bill appears tricolored: dark tip, lighter base, and yellow cere. **JUVENILE:** Has noticeable white base to tail, cleanly separated from dark terminal band, inner primaries with white patch visible in flight, sometimes even from above. Amount of white in the wing varies by individual, not simply age. Wings are wider and shorter than those of older birds. **SUBADULT:** Second-, third-, and fourth-year plumages are distinctive. Second-year birds will have an irregular trailing edge to the wing as new,

shorter feathers mix with retained juvenal feathers. New tail feathers have a less clear-cut boundary between white tip and dark base. Upperwing coverts become tawny, giving a diagonal bar across the wing. Third- and fourth-year birds lose the white base of tail and white on primaries also disappears. **ADULT:** All-dark plumage with no white in wings or tail. Adult male has two or three dark gray bands on brown tail, while adult females have a single gray region in center of tail. **FLIGHT:** Powerful flaps can accelerate it surprisingly fast; it often power-dives after prey. Soars with wings flat or in a dihedral, and sometimes flies with and mimics Turkey Vultures.

SIMILAR SPECIES Immature Bald Eagle always has white by the body at the base of the wing, a longer head and shorter tail projection in flight, and a larger, bicolored bill. Turkey Vulture has silvery flight feathers underneath and a naked head.

VOICE A rather faint and thin *kee-yep* or *yep,* sometimes in a series;

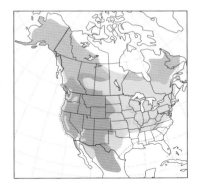

usually silent away from nesting area.
STATUS & DISTRIBUTION Found throughout northern Europe, Asia, and N.A. **BREEDING:** Builds large stick nest on a high ledge or cliff, occasionally in a large tree. **MIGRATION:** Northern nesters move south late (late Oct.–Dec.). Southern birds (mostly adults) are sedentary. **WINTER:** Casual to the Gulf Region.
POPULATION Formerly persecuted by farmers and ranchers. Breeding numbers are thought to be slowly declining in the western U.S. but stable in Canada and AK.

golden nape

adult
canadensis

juvenile
canadensis

short head projection

whitish wing patch

adult
canadensis

whitish tail base

dark or faintly barred tail

HARRIERS Genus *Circus*

There are 13 species in this genus; one or more species occurs on every continent except Antarctica. Only the long-winged, slim-bodied Northern Harrier is found in North America (known as the "Hen Harrier" in the Old World). Harriers have facial discs that help focus the sounds made by hidden rodents.

NORTHERN HARRIER *Circus cyaneus* NOHA ▪ 1

A long-winged, long-tailed hawk coursing low over a grassy field or marsh is likely to be this species—formerly known as "Marsh Hawk." Females are larger than males. Adult males and females are dissimilar in coloration; juveniles are similar to females. Polytypic (2 ssp.; *hudsonius* in N.A., nominate accidental in AK). L 17–23" (43–58 cm) WS 38–48" (97–122 cm)
IDENTIFICATION Long, slim, rounded wings, and a long tail with white on the rump. The head is small, with well-defined facial discs like an owl's. **ADULT MALE:** Often gives the appearance of a gull, with pale gray mantle above, white wings and body below, and black wing tips. Eyes yellow. **ADULT FEMALE:** Brown back and wings above, head with yellow eyes, brown on neck. Underparts are lighter brown with distinct streaking on breast and belly. **JUVENILE:** Similar to adult female, with darker brown head, undersides washed with rufous, streaking less heavy and less extensive. Undersides of secondaries usually darker than primaries. Juvenile male has yellowish to gray eyes; female's eyes dark brown.

FLIGHT: Wings held in a dihedral, rocks side-to-side in quartering flight over marshes. Occasionally soars high on set wings, at which time the long tail and slim wings are unique.
SIMILAR SPECIES Rough-legged Hawk has dark carpal patches on underwing, white at base of tail. Swainson's Hawk has pointed wings, shorter tail. Turkey Vulture is larger and darker.
VOICE Usually silent, except around nest or during courtship when both sexes give a fast series of chattering *kek* notes.
STATUS & DISTRIBUTION Fairly common. **BREEDING:** Nests on ground in tall reeds or grasses. Nesting location and success are dependent on local rodent populations. **MIGRATION:** Prefers to follow coastlines, but will ride thermals along ridges. East Coast migration is composed mostly of juveniles. Unafraid

to cross open water. **WINTER:** Ranges from coastal and southern U.S. into Mexico. Often associates with Short-eared Owls. **VAGRANT:** A wing (salvaged and retained) tentatively identified as of nominate Old World subspecies; found on Attu I., AK (June 1990).
POPULATION Widely dispersed. Breeding has declined in many regions as agricultural areas have disappeared, especially in the eastern states.

white underwing with dark wing tip and trailing edge

adult *hudsonius* ♂

roundish owl-like head

gray above

juvenile *hudsonius*

darker secondaries

barred remiges

adult *hudsonius* ♀

unstreaked belly

adult *hudsonius* ♂

streaked belly

long tail

juvenile *hudsonius*

whitish rump

long, thin wings with rounded tips

adult *hudsonius* ♂

juvenile *hudsonius*

ACCIPITERS Genus *Accipiter*

Three species of these short-winged, long-tailed raptors occur in North America, 46 worldwide. Primarily bird-catchers, they hunt by ambush, flying rapidly around and even through brush to grab their prey with long legs and slender toes. Females are often visibly larger than males. All start out with yellow eyes as immatures, changing to orange, then red.

NORTHERN GOSHAWK *Accipiter gentilis*   NOGO ■ 1

Our largest accipiter, the Northern Goshawk is found in northern forests and in the montane West. Aggressive around the nest, it noisily attacks humans that approach too closely. It preys on birds and small mammals. Sexes are alike, females larger; adults distinct from juveniles. Polytypic (10 ssp.; 3 in N.A.). L 21–26" (53–66 cm) WS 40–46" (102–117 cm)

IDENTIFICATION A large, thick-bodied hawk with wide wings. The long tail with graduated-length feathers is wedge-shaped when folded. **ADULT:** Dark gray above, light gray or white superciliary line, lighter below with fine barring, fluffy white crissum. Red eye. **JUVENILE:** Brown above with heavy mottling, often with a checkerboard effect. Undersides light, with thick, dark streaks, including crissum. Dark tail bands are wavy, with white borders between bands. **FLIGHT:** From above, adult has darker flight feathers than coverts, from below looks more pointy-winged than other accipiters. White-appearing superciliary easily visible, giving head a striped appearance. Juvenile from above has a pale diagonal band across wing coverts, irregular tail bands. From below, heavy streaking on breast and belly. When soaring, the wings taper out to the tip.

GEOGRAPHIC VARIATION Subspecies *langi,* found from coastal southeastern AK to Vancouver I., BC, is heavily barred and darker than widespread *atricapillus.* Subspecies *apache,* from Southwest mountains, is larger and darker than *atricapillus.*
SIMILAR SPECIES Adult Cooper's Hawk has no eyebrow, rufous barring on chest. Juvenile Cooper's has larger-appearing head, is more finely streaked underneath with an all-white undertail, and has longer, rounded tail with white tip. In flight, immature Red-shouldered Hawk has buffy crescents at base of primaries.
VOICE Around the nest, a loud, accelerating *kek-kek-kek-kek* and a wailing *kee-ah.* Young birds remain in the nesting area for weeks after fledging and are very vocal—an excellent way to locate this secretive species in summer.
STATUS & DISTRIBUTION Widespread, but never common, in mature northern forests and mountains. **BREEDING:** Stick nest is placed high, usually in main fork of a deciduous tree. **MIGRATION:** Late-season migrant; numbers

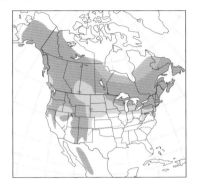

cyclical. **WINTER:** Usually only juveniles winter in the Northeast. Fall passage adults usually work their way back north before winter's end. Mountain residents sometimes head to lower altitudes. **VAGRANT:** During invasion years, a few may be found as far south as Gulf Coast.
POPULATION In eastern U.S., trend toward reforestation has created more habitat, with migration counts indicating a population increase.

prominent pale supercilium

long wings for an accipiter

juvenile ♀

juvenile *atricapillus*

long, pale gray supercilium contrasts sharply with dark auriculars

streaked undertail coverts

adult *atricapillus* ♂

gray barred underparts

thin, pale, wavy bands border dark bands

SHARP-SHINNED HAWK *Accipiter striatus*   SSHA ■ 1

Our smallest accipiter, the "Sharpie" is a jay-size hawk that frequents backyard bird feeders in winter, bursting from nearby bushes to snatch a small bird. Sharpies serve a needed function of keeping wild bird populations healthy and wary. Sexes look alike, females are larger than males, adults differ from imma-

juvenile
velox

shorter square tail

curved leading edge

small head

juvenile
velox ♀

adult
velox ♂

thin legs

tures. Polytypic (10 ssp.; 3 in N.A.). L 10–14"(25–36 cm) WS 20–28" (51–71 cm)
IDENTIFICATION A small, round-winged, long-tailed hawk of woods, edges, and mixed habitat. Tail tip appears squared-off; tail bands are wide and straight. Head is rounded, with a distinct "notch" in the profile from crown to beak. Eye appears more centered in head. **ADULT:** Blue-gray crown and nape are same color as back, creating a "hooded" effect against buffy cheeks. Eyes are red-orange (females) to deep red (males). Underparts are barred rufous; undertail is white. **JUVENILE:** Brown back feathers have rufous tips, white spots on wing coverts. Eye is pale yellow. White underparts are marked with blurry, reddish brown lines and spots down through the belly. **FLIGHT:** Quick, choppy wingbeats interspersed with short glides. Almost no "flex" to the wing. When gliding or soaring, wings are held forward with wrists bent, head barely projecting in front of wings.
GEOGRAPHIC VARIATION Subspecies *velox* is common throughout most of N.A.; *perobscurus,* found on islands of British Columbia, is darker and more heavily barred; *suttoni* of Mexico, into Arizona and New Mexico, is lighter, with fainter barring.

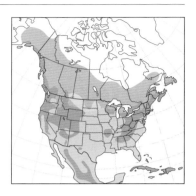

SIMILAR SPECIES Cooper's Hawk is noticeably larger, although small males in western populations approach large female Sharpies in length, but have a longer, rounded tail and flatter head. Adult Cooper's has a dark cap, not hooded as in Sharpie.
VOICE A high, chattering *kew-kew-kew* is heard around the nest; otherwise it is mostly silent.
STATUS & DISTRIBUTION Widespread in northern and western forests. **BREEDING:** Prefers coniferous and mixed forests; makes a small stick nest usually high and close to the trunk. **MIGRATION:** Commonest accipiter at hawk-watches; adult birds prefer to follow mountain ridges, while many juveniles end up following the coasts. Juveniles migrate first, followed by adults. **WINTER:** Throughout much of N.A. and into Mexico.
POPULATION Never abundant, but population appears to be stable.

Juvenile Sharp-shinned versus Juvenile Cooper's

Hawk-watchers in the 1960s and 1970s would get into heated arguments over whether a passing juvenile accipiter was a Sharp-shinned or Cooper's Hawk (henceforth, Sharpie and Cooper's). Since then, watchers and hawk banders have gradually worked out many of these field-identification problems.

Physically, Sharpies are grackle-size, noticeably smaller than Cooper's, which are about the length of a crow. This is handy when a bird is seen in a backyard or perched on an object of known size. Sharpies have a rounded head, with a noticeable "notch" in the profile where the forehead meets the bill. Cooper's has a flatter crown, especially if the hackles are raised,

Sharp-shinned Hawk, juvenile (NJ, Oct.)

and the profile from crown to bill is smooth and continuous. The eye of a Cooper's appears larger in the head and nearer to the bill. Both can show a pale superciliary line, both have pale yellowish eyes. Back and wing coloration are similar—brown with rufous feather tips that slowly wear down through the winter and spring. The scapulars often show large white spots. The tail is brown with dark, evenly spaced bands. The undersides are white, with Sharpie showing blurry, reddish brown streaking and spotting down through the belly, while Cooper's has thinner, darker breast streaks that thin out or stop at the belly. Both have a white undertail, although some western

COOPER'S HAWK *Accipiter cooperii*   COHA   1

The "chicken hawk" of colonial America, this medium-size accipiter is a common sight at home bird feeders across the country, swooping in to nab an unwary dove or jay. Females are larger and bulkier than males, juveniles differ from adults. Monotypic. L 14–20" (36–51 cm) WS 29–37" (74–94cm)

IDENTIFICATION The long tail is rounded at the tip, also the relatively short wings and flat-topped head are good field marks. Eye is close to the beak. Crown merges with forehead and bill in a smooth line. **ADULT:** Blue-gray upperparts, crown is darker and contrasts with lighter nape and buffy cheeks, giving the look of wearing a "beret." Eye color is orange to red. Undersides with rufous barring, undertail is white. **JUVENILE:** Brown above, with rufous edges and white spots on upperwing coverts. Tail long, with straight bands and wide, white tip that wears down by spring. Head usually buffy, eyes pale yellow. Undersides are white with thin brown streaks, white undertail. **FLIGHT:** Wings typically held straight out from body, head and neck projecting forward. This, along with tail length, make a "flying cross" appearance. Shallow, quick wingbeats alternate with short glides.

GEOGRAPHIC VARIATION Hunting in more open country, western populations are smaller, with longer wings and shorter legs than eastern birds.

SIMILAR SPECIES Northern Goshawk is usually larger, heavier appearing, and has relatively shorter tail and longer wings. Sharp-shinned Hawk is smaller and has a square tail.
VOICE A low *keh-keh-keh* uttered around nest, occasionally mimicked by jays. Juvenile gives a squeaky whistle.
STATUS & DISTRIBUTION Widespread through U.S. and southern Canada, more commonly seen in suburbs, probably due to reforestation in the East. **BREEDING:** Nests in a variety of forest types, preying on small- to medium-size birds and small mammals, hunting from perches under the canopy. **MIGRATION:** Increasing numbers seen at eastern hawk-watches probably due to better identification skills. **WINTER:** Juveniles winter farther north than adults; eastern birds move to the southern states, western birds as far south as Mexico.
POPULATION Common and increasing throughout the continent.

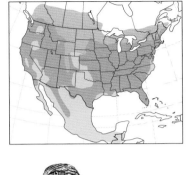

larger head, longer neck, and tawny nape

juvenile ♀

adult ♂

head projects

long rounded tail

straight leading edge

juvenile

Cooper's show thin streaks there.

When viewing a perched bird from the front, the tail can be diagnostic—Cooper's has a proportionally longer, rounded tail, created by the feathers decreasing in length from inner to outer. Sharpie's tail feathers are almost the same length, creating a square tail tip. Seen from behind, Cooper's tail has a broad white terminal band, while Sharpie's tail only has a thin white edge. By spring, much of the white has worn off a Cooper's tail tips, especially the longest (central) feathers. In flight, they have different characteristics, or "jizz." The shorter wings of a Sharpie allow it to flap more quickly in a rapid, choppy motion with almost

Cooper's Hawk, juvenile (NJ, Sept.)

no flexing. Cooper's wings are much longer, and thus its wingbeats have a more fluid motion, almost like a wave traveling out the wing. Both fly in bursts of flapping and gliding. When gliding, the Sharpie pushes its wings forward, cocking the wrists. As a result, the head barely protrudes in front of the leading edge of the wing. Cooper's holds its wings almost straight out from its body, giving it a noticeable head and neck projection. This, along with the longer, rounded tail, gives a Cooper's the appearance of a "flying cross." As in all aspects of bird identification, the more birds you see in the field, the easier it is to name them. ■

NEAR-BUTEOS Genera *Geranospiza, Buteogallus,* and *Parabuteo*

These raptors are somewhat similar in appearance to those in genus *Buteo,* with long, wide, rounded wings. All are neotropical species that are found in the arid southwestern United States.

CRANE HAWK *Geranospiza caerulescens*   CRHA ■ 5

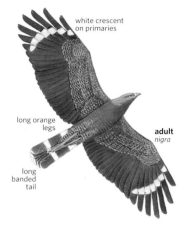

white crescent on primaries

long orange legs

long banded tail

adult
nigra

This mid-size tropical hawk has a lanky body, small head, and long tail. The Crane Hawk is unusual for having very long, double-jointed legs that allow it to reach into tree crevices and nest holes in search of prey. Polytypic (6 ssp.; *nigra* in N.A.). L 18–21" (46–53 cm) WS 36–41" (91–104 cm) **IDENTIFICATION** Dark gray body; broad and rounded wings; long, slightly rounded tail with two bright-white tail bands; the legs are orange. When the bird is perched, the wing tips barely reach the base of the tail. **ADULT:** Slate gray, often with fine barring on leg feathers and belly. **JUVENILE:** Overall browner and slightly paler than adult, white on forehead and cheeks. **FLIGHT:** Note the white crescents on the primaries of the long wings; flight feathers are darker than underwing coverts. Wingbeats are loose and floppy; usually soars for only for brief periods, often with tail spread (normally closed in steady flight).

GEOGRAPHIC VARIATION Subspecies *nigra* occurs up the eastern edge of Mexico, and is also resident on Mexico's west coast from Sinaloa south; *livens* is found in northern Sinaloa and southern Sonora. **SIMILAR SPECIES** Common Black-Hawk is stockier, has broader wings, shorter tail, and yellow legs. Also compare to rare black-morph Hook-billed Kite. **VOICE** A loud, plaintive whistle *wheeeooo.* **STATUS & DISTRIBUTION** YEAR-ROUND: Found from Mexico to Peru and southern Argentina. Prefers tropical lowlands, usually near water. VAGRANT: Accidental; a Crane Hawk was present from December 20, 1987, to March 17, 1988, at Santa Ana National Wildlife Refuge in the Rio Grande Valley, TX.

COMMON BLACK-HAWK *Buteogallus anthracinus*   COBH ■ 2

This dark raptor is from the Southwest. Juveniles are different from adults; females are slightly larger than males. Polytypic (5 ssp.; nominate in N.A.). L 21" (53 cm) WS 50" (127 cm) **IDENTIFICATION** A large, all-dark raptor with long, wide wings and a short tail. **ADULT:** Body and wings black, cere and legs are orange-yellow. Adult females can show some white below the eye. Tail has a broad white band and narrow white tip. **JUVENILE:** Dark brown overall, with rufous markings above and heavy dark streaks on buffy breast and underparts. The head is patterned with dark crown, eye line, and malar mark contrasting with buffy supercilium and throat. The tail is white with numerous dark, wavy bands, and is proportionally longer than in an adult. **FLIGHT:** Wide wings make the tail appear very short in comparison to other dark buteos. Adult has small, whitish crescent at base of primaries; tail band is very evident. Immature has large buffy patch at base of primaries; tail appears lighter

juvenile
anthracinus

finely banded remiges

banded tail

short broad wings

whitish patch

adult
anthracinus

short tail with single, broad, white band

extensive orange-yellow in lores

adult
anthracinus

juvenile
anthracinus

stocky body shape with short tail

longer tail than adult

with dark terminal band. Soars on flat wings.

SIMILAR SPECIES Zone-tailed Hawk soars with a dihedral and has a slimmer wing shape. With close views, Zone-tailed has a white forehead, gray lores. Dark-morph buteos will show an underwing contrast between flight feathers and coverts, and have longer-appearing tails.

VOICE Call is a series of loud whistles, falling in intensity near the end.

STATUS & DISTRIBUTION A tropical hawk of C.A. and S.A., an uncommon to rare breeder in western TX, NM, AZ, and southwestern UT, very susceptible to human disturbance around the nest. Always nests near water, usually a year-round stream or river. Usually hunts from a perch. **VAGRANT:** Casual north to CA, southern UT, southern NV, CO, and southern TX.

POPULATION Probably stable. About 250 pairs breed in the United States, mostly in Arizona. The species is threatened by habitat destruction and human disturbance.

HARRIS'S HAWK *Parabuteo unicinctus*   HASH   1

Also known as "Bay-winged Hawk" in South America, this large, dark raptor of the Southwest is noted for its habit of cooperative hunting in the deserts and scrub areas it inhabits. Working in groups of up to a dozen birds, individuals will take turns flushing prey out of heavy brush into the waiting talons of the others. A call from a perched "lookout" bird will summon others from their concealed locations. Females are larger than males; adults differ from juveniles. Polytypic (2 ssp.; *harrisi* in N.A.). L 21" (53 cm) WS 46" (117 cm)

IDENTIFICATION A long-tailed, long-legged, dark raptor with long, wide wings. It is often seen on a dead snag, telephone pole, or similar open perch. The dark tail has a broad white terminal band and a white base with white undertail coverts. The wing tips reach halfway down the tail. **ADULT:** Uniformly dark head, body, and wings, with chestnut wing coverts and leg feathers. The facial skin, cere, and legs are yellow. **JUVENILE:** Dark brown body with lighter streaks on head and neck, whiter streaking on belly than on breast. The tail has fine dark barring, as do the flight feathers. The white at the base of the tail is not as extensive as in an adult. **FLIGHT:** Very active hunter, flying with quick, shallow wingbeats, wings held slightly downward, often cupped. Its short, broad wings and long tail allow it to be very maneuverable; it can hover for short periods. From below, adult has chestnut coverts and dark flight feathers; immature shows barred gray secondaries, barred primaries with dark tips and lighter bases. Soars with flat wings. Wing silhouette is broad, often called "paddle shaped."

GEOGRAPHIC VARIATION Some authorities recognize two N.A. subspecies—*harrisi* (TX) and *superior* (AZ)—but they are virtually identical; the nominate subspecies occurs in S.A.

SIMILAR SPECIES Dark-morph buteos have two-tone underwings with silvery-tone flight feathers. Zone-tailed Hawk soars with wings in a dihedral, has dark undertail coverts, dark tail tip. Juvenile Northern Harrier at a distance, when compared to a juvenile Harris' Hawk, has slimmer, longer wings, a white rump (not the base of the tail), and glides with a dihedral.

VOICE Call is a long, harsh, grating *eeaarr*.

STATUS & DISTRIBUTION Common, permanent resident in southern TX, into southern NM and southern AZ; recent records (inc. nesting) in southern CA are exceptional. Occasionally expands breeding range, probably in response to prey availability. **BREEDING:** A complex social structure including multiple adults and pairings is still being investigated by researchers. Breeding groups consist of a dominant pair, related helpers that are offspring of that pair, unrelated helpers, and sometimes additional females with an intermediate position of dominance. **MIGRATION:** Mostly sedentary. **WINTER:** Often gathers in large groups to roost and hunt. **VAGRANT:** A popular falconer's bird; any records outside of their normal range should be treated with caution.

POPULATION Apparently stable, but under pressure from habitat (esp. mesquite) destruction, falconry, and electrocution from power lines.

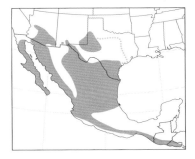

short rounded wings

juvenile *harrisi*

adult *harrisi*

heavily streaked below with blackish brown

white undertail base and tip

chestnut wing linings

adult *harrisi*

chestnut wing coverts

chestnut thighs

white undertail coverts

long tail with white tip

BUTEOS Genus *Buteo*

Called "buzzards" in many parts of the world except North America, there are 29 species in this genus—11 in North America; all are regular breeders except Roadside Hawk, which is a casual visitor into south Texas. They soar on wide, rounded wings, hunting small mammals, lizards, snakes, and the occasional unwary bird. They also perch-hunt along treelines or under the canopy. Sexes are similar in appearance, females are slightly larger than males, and juveniles have different plumage than adults. During fall migration, buteos travel in groups, riding thermal updrafts along mountain ridges or coastlines. In late September, nearly one million buteos pass over Texas on their way south.

ROADSIDE HAWK *Buteo magnirostris*   ROHA   ▪   4

A small tropical hawk, the Roadside Hawk is common in much of Mexico, but only a casual vagrant to Texas. Females are slightly larger than males; juveniles are slightly different from adults. Polytypic (at least 14 ssp.; *griseocauda* in N.A.). L 14" (36 cm) WS 30" (75 cm)

IDENTIFICATION A small, long-tailed, long-legged raptor, it frequently perches and hunts from fence posts, telephone poles, or trees alongside fields and roads. Wing tips reach halfway down the long tail, which has a buffy or whitish U along the uppertail coverts. Legs are long and slim. **ADULT:** Gray to brownish on head, back, and wings; large, pale yellow eyes. The long tail has 3–4 dark bands. Undersides have a brown bib and barred belly. **JUVENILE:** Browner than adults, with a wide superciliary line

and darker eyes. The chest has vertical streaks, the belly horizontal barring. The tail has 4–6 dark bands. **FLIGHT:** The wings are wide and rounded, with barred flight feathers, rufous on inner primaries. Stiff, rapid wingbeats and the long tail make it look very accipiter-like; glides with wings bowed.

GEOGRAPHIC VARIATION Populations farther south show more rufous in the wings and grayer plumage.

SIMILAR SPECIES See Broad-winged Hawk and Gray Hawk. Barred belly and dark bib on adult, vertical streaks on breast of juvenile, plus long, evenly banded tail are distinctive.

VOICE In U.S., birds have been silent, but in Mexico one of the Roadside's characteristic sounds is a drawn-out complaining scream, delivered from a perch.

STATUS & DISTRIBUTION Common resident of open and semi-open country. One of the most widespread raptors of the Neotropics, it world range stretches from Mexico to Ecuador and northern Argentina. **VAGRANT:** Casual to the Rio Grande Valley, TX, with seven accepted records and multiple sightings during the winter of 2004–2005.

POPULATION Apparently stable.

RED-SHOULDERED HAWK *Buteo lineatus*   RSHA   ▪   1

A common hawk of wet deciduous woodlands, it is the noisiest of the buteos, especially during spring courtship. A perch-hunter of the forest understory, it feeds on frogs, snakes, lizards, and small mammals. Females are larger than males, sexes are similar in appearance, and juveniles differ from adults. Polytypic (5 ssp., all in N.A.) L 15–19" (38–48 cm) WS 37–42" (94–107 cm)

IDENTIFICATION Medium-size buteo with rounded wing tips that do not

reach the tip of the tail. **ADULT:** Brown above with lighter feather edges and some streaking on head. Rufous on the upperwing coverts gives the "red shoulders." The primaries are barred or checkered black and white, the dark tail has three white bands. Underparts are rufous with white barring. **JUVENILE:** Mostly brown above, with less rufous on the shoulders than an adult. Undersides are buffy with variable dark streaks, the brown tail has multiple narrow bands. **FLIGHT:** All

ages show a distinct light crescent at the base of the primaries. Soars on flat wings held forward, glides with wings cupped, giving a "hunched" appearance. Wingbeats are quick and shallow, interspersed with quick glides.

GEOGRAPHIC VARIATION Eastern subspecies (*lineatus*) descibed above, with adult showing dark streaks on breast. Juvenile tail from below is light with dark bands. Southeastern subspecies (*alleni*) and Texas subspecies (*texanus*) are very similar to each other, with adults showing no dark streaks on breast; *texanus* is slightly brighter rufous. Juveniles have darker underside of tail, heavier markings on underparts. South Florida subspecies *(extimus)* is the palest subspecies. Adults are pale gray above, pale rufous underneath, with the head appearing light at a distance. Juveniles are less rufous, have thinner streaking underneath. California subspecies *(elegans)* is the brightest subspecies, with juveniles appearing more like adults. Adults have unbarred rufous on the breast, wider tail bands. Juveniles have similar upperwings to the adult, with the most rufous on the shoulders of any subspecies, and a white crescent on the primaries. Underparts are barred rufous and buffy. The dark tail shows whitish bands.

SIMILAR SPECIES Broad-winged Hawk adult has rufous on breast, but no barring on primaries, only one or two tail bands; in flight shows more pointed wings. Juveniles are similar, but Red-shouldered shows three bands on the folded secondaries; Broad-winged lacks the pale crescent at the base of the primaries. Juvenile Northern Harrier is rufous underneath, but long tail, thin wings, and flight style are very different. A perched juvenile Northern Goshawk shows heavy streaking underneath and a long tail, but dark-banded tail pattern and shorter wings will help the identification process.

VOICE A loud, repeated *KEE-ahh,* often in groups of 8–10 repetitions.

STATUS & DISTRIBUTION A widespread breeder throughout East and into southern Canada. Found throughout the South, into eastern and southern TX. CA subspecies is found into OR and southern WA. **BREEDING:** Nests in deciduous woodlands, usually river bottoms, near lakes or swamps. Coexists in similar habitat with the Barred Owl. **MIGRATION:** Northern birds migrate to southern states and into Mexico. Southern subspecies are non-migratory. **VAGRANT:** There are records of eastern birds west to CO; *elegans* wanders casually to AZ, ID, and NM. **POPULATION** Stable, as far as is known.

BROAD-WINGED HAWK *Buteo platypterus* BWHA ▪ 1

juvenile
platypterus

black
moustachial
streak

rufous-
brown
barred
underparts

adult
platypterus

This bird made hawk-watching famous—thousands of birders gather to watch the annual fall migration of Broad-winged Hawks. The hawks start migrating in September in New England, traveling down Appalachian ridges on their way to wintering grounds in South America. They occur in light and dark (rare) morphs; juveniles different from adults. Polytypic (6 ssp.; nominate in N.A., others resident in West Indies). L 16" (41 cm) WS 34" (86 cm)

IDENTIFICATION One of the smallest North American buteos. **LIGHT-MORPH ADULT:** Head, back, and wings are brown, throat is white, wing tips dark, dark tail with one wide white band. A second, thinner band may be visible on the fanned tail. Undersides are white with brown or rufous barring across breast, less on the belly. Some individuals may have a solid-colored dark breast, giving the bird a dark bib. **JUVENILE:** Brown above like adult, but with pale superciliary line on head, dark malar stripe; brown tail has multiple darker bands, widest band at tip. Underparts are white with dark streaking on breast and belly, but amount of streaking is highly variable, sometimes almost absent. **DARK MORPH:** All-dark body with dark wing coverts and silvery flight feathers, the dark tail has a wide white band. The juvenile is dark with variable light streaking on body and wing coverts. **FLIGHT:** Wing tips more pointed than those of the other common buteos, and trailing edge is almost straight. Adult has pale wing linings and flight feathers contrasting with dark primary tips, a wide dark band along

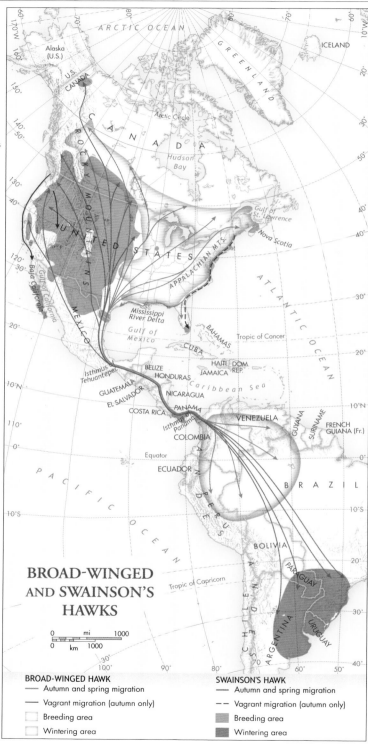

**BROAD-WINGED
AND SWAINSON'S
HAWKS**

	mi	
0		1000
0	km	1000

BROAD-WINGED HAWK
—— Autumn and spring migration
—— Vagrant migration (autumn only)
☐ Breeding area
☐ Wintering area

SWAINSON'S HAWK
—— Autumn and spring migration
- - - Vagrant migration (autumn only)
▨ Breeding area
■ Wintering area

the trailing edge of the wing. Juveniles have slightly longer tails, but the same wing silhouette with the trailing edges

not as dark. Backlit wings show a light rectangle at the base of the primaries. Underwing coverts are variably

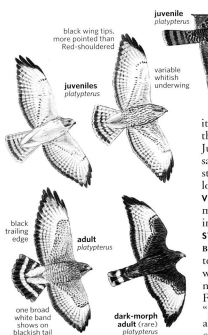

black wing tips, more pointed than Red-shouldered

juveniles
platypterus

juvenile
platypterus

variable whitish underwing

black trailing edge

adult
platypterus

one broad white band shows on blackish tail

dark-morph adult (rare)
platypterus

streaked, as is the belly. Wingbeats are stiff; soars on flat wings.

SIMILAR SPECIES In flight, juvenile Red-shouldered Hawk shows light crescents at base of primaries and longer tail. Perched, it has a brown tail with dark bands; three bands on folded secondaries. Juvenile Cooper's Hawk can show the same overall markings, but no malar stripe; shorter, barred wings and much longer tail give a different shape.

VOICE Thin, whistled *kee-eee*, often mimicked by Blue Jays, rarely given in migration.

STATUS & DISTRIBUTION Common. **BREEDING:** Nests in eastern woodlands to eastern TX and MN, in Canada west to AB and BC. Rare dark morph nests in western Canada. **MIGRATION:** Famed for migrating in groups (called "kettles"), generally utilizing updrafts along mountain ridges. Reluctant to cross open water. During the last six

days in Sept., typically more than 700,000 Broad-wings pass over Corpus Christi, TX. Very small numbers are regularly seen in fall (fewer in spring) at favored hawk-watching sites in the West. **WINTER:** Small numbers, usually juveniles, in southern FL and southern TX; rare in CA. **VAGRANT:** Dark morph birds casually seen in the East and West, in migration.

POPULATION Stable, as far as is known.

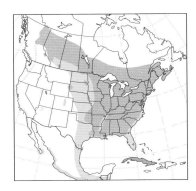

GRAY HAWK *Buteo plagiatus*   GRHA   2

This bird was called the "Mexican Goshawk" in older literature due to its overall gray coloration, longish tail, and barred undersides. Sexes look alike; females are larger and juveniles are different from adults. Formerly placed in the genus *Asturina* and recently split from the polytypic Gray-lined Hawk *(B. nitidus)* of Central and South America. Monotypic. L 17" (43 cm) WS 35" (89 cm)

IDENTIFICATION A small raptor with wide, rounded wings and long tail; all ages have uppertail coverts that form a

distinct white U. **ADULT:** Overall gray, with fine white barring on breast and belly. White wing linings have fine gray barring, appearing whitish at a distance. Undertail coverts white; tail has two distinct white bands. **JUVENILE:** Dark brown body; distinct head pattern with light supercilium and cheek contrasting with dark eye line and malar stripes. Undersides heavily streaked; tail has multiple stripes that grow progressively wider toward the tip. Tail is longer than adult's. **FLIGHT:** Accipiter-like; short, choppy wingbeats are interspersed with flat glides. From underneath, adult shows black tips to primaries.

SIMILAR SPECIES Juvenile Broad-winged Hawk lacks the face pattern and upper-tail U-mark and has more pointed wings, a shorter tail, and flies with slower wing beats. Juvenile Red-shouldered Hawk has buffy crescent at base of primaries.

VOICE Calls include a loud, descending whistle.

STATUS & DISTRIBUTION Resident in the Lower Rio Grande Valley; breeds in Big Bend N.P. and into southeastern AZ, rarely to southwestern NM. Nests in large trees, usually near water. **VAGRANT:** Accidental in winter to southwestern CA.

POPULATION Population stable, possibly expanding. About 80 pairs breed in the U.S.

distinct head pattern with dark eye line, dark malar, and pale cheek

juvenile

fine gray bars on underparts

adult

longer tail than Broad-winged

juveniles

whitish uppertail coverts

adult

banded black-and-white tail

SHORT-TAILED HAWK   *Buteo brachyurus*   STHA ■ 2

This small buteo of the tropics occurs in two color morphs—light and dark. Usually catches its prey below the forest canopy, but most often seen soaring high in the sky. Polytypic (2 ssp.; *fuliginosus* in N.A.). L 15" (39 cm) WS 35" (89 cm)

IDENTIFICATION Perched, the wing tips reach the tail tip. LIGHT-MORPH: Adult dark above, including head and cheeks, giving a "helmeted" appearance. Light below, usually white or creamy-white on throat, body, and undertail coverts. Tail from below has wide, dark terminal band and lighter, thin bands. Juvenile has more patterning on face, brownish secondary coverts, variable streaking on sides of breast; tail from below is light with many thin bands and narrow terminal band. Both show white underwing coverts with darker, barred flight feathers, black tips, and paler base to primaries. DARK-MORPH: Adult uniformly dark brown, flight feathers lighter than coverts, but with dark trailing edge, lighter bases to primaries. Tail is lighter than body with dark terminal band. Juveniles have heavily marked bodies and underwing coverts; tail has fine bands and dark tip. FLIGHT: A highly aerial species, Short-tailed Hawk shows a pointed-wing silhouette, especially when gliding. It typically soars at high altitudes when hunting, and dives or glides down when prey is sighted. When wind conditions allow, it often kites for considerable periods of time. White body and wing coverts stand out from darker flight feathers. White spot on forehead and lores is often visible.

SIMILAR SPECIES Broad-winged Hawk is similar size, but has lighter flight feathers and shorter wings when perched. Juveniles are spotted or streaked underneath. Light-morph Swainson's Hawk is larger, has two-tone underwings, but flight feathers are more uniform in color.

VOICE Usually silent, especially in nonbreeding season. Has a high-pitched, drawn-out, slightly descending, two-syllable call.

STATUS & DISTRIBUTION Widespread throughout C.A. and S.A., uncommon resident in southern FL. Formerly rare, more recently uncommon in southeastern AZ, southwestern NM, and southern TX. BREEDING: Only recorded in FL and recently in the Chiricahua Mts., AZ; a failed nesting attempt with a Swainson's Hawk in southern TX is interesting.

POPULATION Stable in FL.

dark helmet

rarely seen perched

light-morph adult *fuliginosus*

uniform white underparts

whitish patch at base of outer primaries

light-morph adult *fuliginosus*

white wing linings

dark-morph adult *fuliginosus*

black border to trailing edge

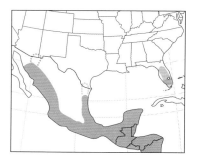

Head Details of Ferruginous, Red-tailed, Swainson's, and Rough-legged Hawks

When viewing a perched raptor closely through a spotting scope, here are a few ways to identify Red-tailed, Ferruginous, Swainson's, and Rough-legged Hawks.

The buteos display a bewildering amount of plumage variation, so on perched birds it often helps to look carefully at structural differences rather than at feathers. Concentrate on the bird's head shape, the length and visibility of the gape (the fleshy edges of the mouth visible on the side of the head), and the bill size. Additionally, most juveniles have light-colored eyes, while adults have dark or darker eyes.

Among the large buteos, Ferruginous Hawk has the largest head and bill, and the head usually appears

Ferruginous Hawk

Red-tailed Hawk

flat-topped. This hawk's yellow gape is very visible and long, extending past the centerline of the eye. The large bill appears taller (deeper) than its length, and the cere is yellow.

Red-tailed Hawk has a rounder forehead, but the

ZONE-TAILED HAWK *Buteo albonotatus*   ZTHA ▪ 2

This dark raptor of the Southwest closely mimics a Turkey Vulture both in appearance and flight, and as a result is easily overlooked. Monotypic. L 20" (51 cm) WS 51" (130 cm) **IDENTIFICATION** Appears blackish at a distance, has barred flight feathers, and the tail shows a distinct pattern. When perched, wing tips equal or exceed tail length. **ADULT:** Uniformly black with a grayish cast (the blackish Turkey Vulture has a brownish cast); yellow cere and legs. The tail of the adult male has one broad, white, mid-tail band and narrower, white inner band; the adult female's tail also has a mid-tail band, but with three or four inner bands. **JUVENILE:** Black overall, dull yellow legs and cere (brighter by fall); tail is dark from above, light below with many thin dark bands and a dark tip. **FLIGHT:** Long, dark wings are two-tone, with heavily barred flight feathers. Adults show a dark trailing edge to the wing.

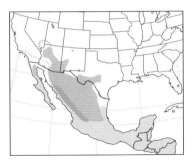

Wingbeats are deep, but not as floppy as a vulture's. Head and banded tail project farther than a vulture's. Usually glides and tips side-to-side in a dihedral, just like a Turkey Vulture, but can also glide on flat wings; drops from low glide onto small birds, rodents, lizards, and fish. **SIMILAR SPECIES** Common Black Hawk in flight has shorter, broader wings and a shorter tail; when perched it has longer legs, a bigger bill, and wing tips do not reach tip of tail. Juvenile dark-morph Red-tailed Hawk has light panels at base of primaries; juvenile White-tailed Hawk usually has a white chest patch and a pale crescent on the upper tail coverts. **VOICE** Gives a loud screaming call, not too different from Red-tailed, but more of a whistle and less harsh, especially at the end. **STATUS & DISTRIBUTION** Occurs throughout C.A., into S.A. Northern birds are migratory. **BREEDING:** Uncommon and local in wooded hills, mountains, and mesas often near watercourses from central and western TX through AZ, now also in southern CA and has bred in southern TX. **MIGRATION:** South into Mexico and beyond. **WINTER:** Rare in the Rio Grande Valley, TX, southern CA, and southern AZ. **VAGRANT:** Records for NV and UT, east to LA, and to NS. **POPULATION** Slowly expanding in CA, northern AZ, and NM.

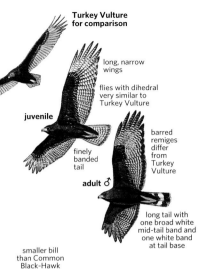

Turkey Vulture for comparison

long, narrow wings

flies with dihedral very similar to Turkey Vulture

juvenile

finely banded tail

barred remiges differ from Turkey Vulture

adult ♂

long tail with one broad white mid-tail band and one white band at tail base

smaller bill than Common Black-Hawk

adult

Swainson's Hawk

Rough-legged Hawk

Swainson's Hawk has a rounder and smaller head than either Ferruginous or Red-tailed Hawks. This hawk's visible gape does not reach the leading edge of the eye. The gape and cere are yellow and the lores are light. Most individuals, except dark-morph adults, have a pale throat framed in dark and a whitish forehead.

Rough-legged Hawk has the smallest bill and the roundest appearing head of all four large buteo species. The gape extends to between the leading edge and the centerline of the eye. Both gape and cere are yellow in adults, grayish in juveniles. Many Rough-legs have a dark spot within a pale area located on the nape. ▪

head still has a flat crown. Its bill is slightly smaller and longer than that of the Ferruginous Hawk. The gape reaches past the leading edge of the eye. Both gape and cere range in color from greenish yellow in juveniles to bright yellow in adults.

SWAINSON'S HAWK   *Buteo swainsoni*   SWHA ▪ 1

smaller bill than Red-tailed

light-morph juvenile

often with darker lateral breast patches

light-morph juvenile

light-morph juvenile

brown breast band

light-morph adult

light-morph adult

whitish underwing coverts contrast with dark flight feathers

broad dark subterminal band

A large hawk with long, narrow pointed wings, it is found in open grasslands and agricultural areas. Widely distributed across the West and highly variable in plumage. Females are larger, and juveniles and subadults differ from adults. Feeds mostly on small mammals and insects, primarily grasshoppers and caterpillars. Monotypic. L 21" (53 cm) WS 52" (132 cm)

IDENTIFICATION Individuals grade evenly from light to rufous to dark morphs. When perched, adults' long wings extend past the wing tips; a juvenile's wings almost reach the tail tip. Year-old birds are similar to juveniles, but have wide subterminal bands on tail, flight feathers. **LIGHT-MORPH ADULT:** Dark above and dark-headed with white on the forehead,

lores, and throat. Dark breast contrasts with white or lightly barred underparts, forming a bib. Tail is gray with dark subterminal band and many thinner dark bands, and a light U on the uppertail coverts. Underwings are distinctly two-tone with white coverts and darker gray flight feathers and dark gray trailing edges. **LIGHT-MORPH JUVENILE:** Dark back and wings with buffy feather edges, buffy head with a pale superciliary line, buffy cheek, and dark malar line. Underparts are light-colored with variable streaking, usually with dark patches on the sides of the breast. Tail is brown with narrow bands, darker tip, and white undertail coverts. **INTERMEDIATE (RUFOUS) MORPH:** Above, similar to light-morph adult, but underparts are rufous, either evenly colored or with a darker chest than the rufous underparts. Throat is white, as are the undertail coverts. Juveniles are similar to light-morph except for buffy underparts with heavier streaks. **DARK MORPH:** Adults are overall dark above and below, varying from black to dark brown. Undertail coverts always lighter, and underwing coverts can feature whitish or rufous mottling, paler flight feathers with dark trailing edge. Juveniles from above are similar to other morphs, but have darker wing coverts and heavy, dark mottling on body. **FLIGHT:** Best told by the combination of long, slim, pointed wings with the two-tone underwing coloration, relatively long tail, and dark bib on light-morph adults. Often soars with wings in a dihedral; wingbeats

relatively quick and light for a large buteo; often hovers when hunting. **SIMILAR SPECIES LIGHT MORPH:** Short-tailed Hawk is smaller, has secondaries darker than primaries. Immature has darker head, shorter wings. Adult White-tailed Hawk has no bib, wider wings with darker primaries than secondaries, white tail with black subterminal band. Juvenile has dark body and underwing coverts contrasting with a white chest patch. Red-tailed Hawk has shorter wings, dark patagial mark on the leading edge of the evenly colored underwing; most have a dark belly band. Perched, its wing tips do not reach the end of the tail. **RUFOUS AND DARK MORPHS:** All other dark adult buteos have dark undertail coverts and silvery, not gray, flight feathers. **VOICE** A drawn-out scream, usually heard near nest site. **STATUS & DISTRIBUTION BREEDING:** From the Great Plains westward to central CA, north to Canada and locally into AK and YT, and south into Mexico. **MIGRATION:** In fall, migrates in huge flocks through the western states, often descending into fields to feed on insect swarms. They pass through southern TX in late Sept. and early to mid-Oct. Rare fall migrant in the East, almost always juveniles. **WINTER:** Primarily in Argentina (see map p. 172). A small but regular population winters in southern FL, in the Rio Grande Valley, TX—especially immatures around the sugarcane fields—and a few in the Sacramento Valley, CA. **POPULATION** Overall stable, but with declines in CA.

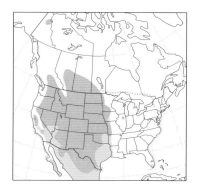

dark-morph adult

intermediate-morph adult

WHITE-TAILED HAWK   *Buteo albicaudatus*   WTHA ■ 2

This large, striking raptor found in open areas and scrubby habitat in southern Texas has an adult plumage and three immature plumages: juvenile, second-year, and third-year. A dark-morph exists in southern populations, but not in the United States. When the White-tailed Hawk is perched, the wing tips reach past the tip of the tail, even in juveniles. Females are larger than males. It feeds on small mammals, insects, reptiles, and even birds. It will gather at a prairie fire as well as burning sugarcane fields to pursue displaced prey. Polytypic (3 ssp.; *hypsopodius* in N.A.). L 20" (51 cm) WS 51" (130 cm)

IDENTIFICATION ADULT: Uniform gray on back, head, and wings, with a rufous patch on the lesser coverts forming an easily visible shoulder patch. Throat is white, as are the underparts, though some individuals will show fine barring underneath. The white lower back and uppertail coverts lead to the distinctive white tail with a black subterminal band and multiple faint bands. **JUVENILE:** Above, dark brown to blackish on head, back, and wings, with rufous feather edges on upper coverts. Head is dark with whitish areas on cheek, and dark throat; most show a white patch on the breast. Amount of dark on belly and white on chest varies greatly among individuals, with some heavily mottled to completely dark, but with lighter undertail coverts. Tail is light gray with many fine dark bands, contrasting with a white U on the uppertail coverts. Wings are narrower, and tail is longer, than adult's. **SECOND-YEAR:** A blend of juvenile and

adult features, including wing and tail proportions of an adult and coloration of a juvenile. Head is more uniformly dark, the back and wings are dark with a rufous shoulder patch. Chest is whitish with variable streaking, undersides barred with black or rufous. Uppertail coverts are white, rump is black, and tail is gray with fine bands and a dark subterminal band. Below, undertail coverts are white with variable mottling. **THIRD-YEAR:** Similar to adult plumage overall, but with blackish head, throat, neck, and back; tail is a mix of gray and white feathers, rump is a mix of black and white. Underneath, whitish barred with rufous. **FLIGHT:** Long, pointed wings pinch in at the body. In flight, often soars in a dihedral; will hover while hunting. The wingbeats of White-tailed are heavier than those of Swainson's Hawk. On adults, the white underwing coverts contrast with darker flight feathers, secondaries being lighter than primaries. Its namesake tail with wide black subterminal band is unique among buteos. All immature plumages show variable dark streaking on the underwing coverts, lighter at bases of flight feathers than at tips. A white chest spot is usually visible.

SIMILAR SPECIES In flight, adult light-morph Swainson's Hawk has two-tone underwings, but primaries and secondaries are the same color; uniform dark trailing edges to wing and dark bib on chest. Wing does not pinch in at the body like White-tailed. Dark-morph Swainson's Hawk is similar to juvenile White-tailed, but has a more noticeably barred tail with a wide,

dark terminal band, and does not show the white chest patch or white uppertail coverts. Ferruginous Hawk has all-white undersides, but legs are dark, underwings are uniformly white, tail lacks dark subterminal band. All other dark buteos are distinguished from juvenile White-tailed by the silvery flight feathers and lack of white U on the uppertail coverts.

VOICE Rarely heard except when disturbed at nest site.

STATUS & DISTRIBUTION Widespread in similar habitat throughout S.A., into C.A, and locally into coastal southern TX. Birds in TX are resident from just west of Galveston and Houston to Brownsville on coastal prairie and agricultural land, especially large private ranches. **VAGRANT:** Casual along the Gulf Coast as far as LA and well inland in TX.

POPULATION The species is listed as threatened in Texas, mostly due to its limited range and lack of knowledge of its breeding dynamics. The small Texas population of 200 to 400 pairs appears to be stable.

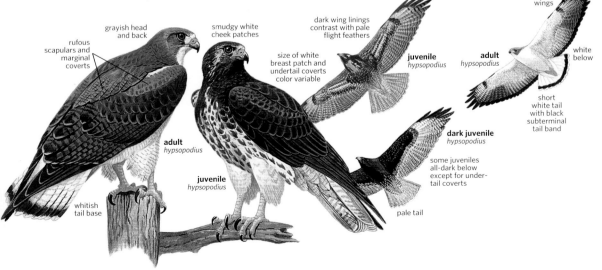

grayish head and back

rufous scapulars and marginal coverts

smudgy white cheek patches

dark wing linings contrast with pale flight feathers

size of white breast patch and undertail coverts color variable

long pointed wings

juvenile *hypsopodius*

adult *hypsopodius*

white below

adult *hypsopodius*

juvenile *hypsopodius*

short white tail with black subterminal tail band

dark juvenile *hypsopodius*

some juveniles all-dark below except for undertail coverts

whitish tail base

pale tail

RED-TAILED HAWK *Buteo jamaicensis* RTHA ▪ 1

The Red-tailed Hawk's widespread breeding range makes it the "default raptor" in most of the U.S. and Canada. It utilizes a wide range of habitats, from wooded to open areas, farmland to urban settings. Red-tails come in a variety of color morphs, from pale to rufous to dark; average size varies from north (largest) to south (smallest). Prey species include rodents and small mammals, snakes, occasionally birds, even carrion. They are prone to albinism, occasionally appearing totally white. Polytypic (14 ssp.; 6 in N.A.). L 22" (56 cm) WS 50" (127 cm)

IDENTIFICATION A large, chunky, short-tailed raptor, with a large head and bill. All show a diagnostic dark patagial mark on the leading edge of the underwing, more easily seen in light-morph birds. The bulging secondaries give the wing a sinuous trailing edge. Juveniles have more slender wings and longer tails than adults, but when seen perched the wing tips usually fall short of the tail. Most adults have a reddish tail, varying in intensity by color morph and subspecies. **LIGHT-MORPH ADULT:** Occurs in all subspecies. Brown head is offset by darker malar stripe, white throat (except in some western populations and Florida). Back and wings are brown, with white spots on scapulars, giving the appearance of a whitish V on the back. Chest and underparts white, crossed by a belly band of spots or streaks (darkest in Western and Florida subspecies, missing in *fuertesi*). Tail is orange to brick-red with a dark terminal band and often with multiple thinner bands (Western). **LIGHT-MORPH JUVENILE:** Brown head with a dark malar stripe, often with a lighter superciliary line. Throat is white (Eastern) to streaked darker (Western). Dark back and wings are mottled with white on the scapulars, forming a light V. Primaries are lighter than the secondaries, giving a two-tone look to the wing. Below, white undersides are separated by a belly band of heavy dark streaks. Tail is brownish with multiple thin, dark bands, often with a slightly wider terminal band. Underwings are light, coverts occasionally washed with rufous (Western), flight feathers tipped dark and lightly barred, and a pale rectangle at the base of the primaries. **DARK ADULTS (WESTERN BIRDS):** Darker brown above, usually without white spots on wing coverts. Dark belly is

offset by rufous to black chest. Tail is rufous, with thin dark bands and wider subterminal band. On rufous morph, undertail coverts are unbanded rufous, uppertail coverts barred brown.

Underwing coverts are rufous with variable dark barring, patagial mark is visible. Darkest birds are uniformly dark; the dark undertail coverts are barred rufous. Tail is dark rufous with

white wing linings with sparse, black markings

adult light-morph "Harlan's"

mostly white below

white head with variable gray

tail variable with smudgy dark terminal band

adult dark-morph "Harlan's"

black wing linings with variable spotting

adult "Krider's"

light rufous tail often with white base

pale underparts and wing linings

moderate to faint patagial bar

pale head

variable white on forehead and around eye

slightly paler scapulars

adult dark-morph "Harlan's"

white marks on black underparts

extensive pale above

pale head

adult "Krider's"

pale underparts

pale underparts

juvenile *calurus*

heavy bill

white chest with heavily mottled belly

adult *calurus*

belly band

warm buffy wash to underparts

rufous tail

dark-morph
adult
calurus

pale breast
contrasts with
darkish head and
variably marked
belly band

adult
fuertesi

dark
patagial bar

uniform pale
underparts

adult
borealis

pale rufous tail with
narrow or no dark
subterminal band

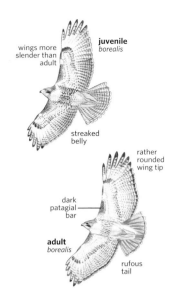

wings more
slender than
adult

juvenile
borealis

streaked
belly

rather
rounded
wing tip

dark
patagial
bar

adult
borealis

rufous
tail

thin barring and a wider dark subterminal band. **DARK JUVENILES:** Above, similar to light-morph with white speckles on secondaries, often darker on the head and throat. Variable below, they are heavily streaked with rufous across chest; dark belly band with white or rufous streaks. Underwing coverts are mottled rufous or dark, with patagial mark often hard to discern. Tail is brown with many darker bands, like light-morph Western. **"KRIDER'S RED-TAIL":** A pale, Great Plains color morph of *borealis*, treated by some as a separate subspeces, *kriderii*. Head whitish with little or no malar stripe. Back and upperwings heavily mottled with white, underparts almost pure white with reduced patagial mark on underwing. Adult has orangish tail

fading to white at the base; immature has light tail banded with dark bars. It is considered a morph rather than a subspecies as the majority of individuals within the described breeding range do not resemble "Krider's." Definitions of what constitute a valid subspecies vary, but one widely accepted rule is that 75 percent of one population of subspecies must differ from 100 percent of an adjacent subspecies population. **"HARLAN'S RED-TAIL":** In *harlani,* the dark-morph is by far the most numerous. The adult is blackish, similar to other dark-morph birds, but with variable white mottling or streaking on chest and belly. Tail is gray with wide, dark subterminal band. Rare light-morph adult "Harlan's" similar to "Krider's," with light head but dark malar stripe, darker wings and flight feathers. Tail has a gray subterminal band and gray mottling fading to white at tail base. **FLIGHT:** Wingbeats are heavy, usually slow. Red-tails glide with wings level, occasionally soar in a slight dihedral. Immatures show a light rectangle at the base of the primaries. **GEOGRAPHIC VARIATION** Eastern *borealis* is found west through the Great Plains, Western *calurus* west from the Great Plains, north into Canada, and south

to the range of "Fuertes's" *fuertesi* in Arizona, New Mexico, southern Texas. Alaskan *alacensis* along the Alaskan coast, to the Pacific Northwest, "Harlan's" *harlani* from northwestern Canada into southern Alaska, and *umbrinus* in peninsular Florida. **SIMILAR SPECIES** Red-shouldered Hawk has light crescents at base of primaries in all plumages. Adult Ferruginous Hawk has mostly white underparts with dark legs, slimmer wing silhouette. Juveniles lack patagial mark, have dark crescent at the end of underwing coverts and pale primaries. Dark Ferruginous has white comma at end of underwing coverts, unbanded tail. Unlike Ferruginous and especially Rough-legged Hawk, which habitually hover while hunting in flight, Red-tailed seldom hovers unless there is a stiff breeze. **VOICE** A husky scream, rising then dropping in pitch *shee-eeee-arrr.* **STATUS & DISTRIBUTION** Found across N. A., into Mexico, and across the Caribbean. **MIGRATION:** Late fall along mountain ridges, the Great Lakes, fewer along the East Coast. Reluctant to cross open water; late July–Aug. movement along the south shore of Lake Ontario is primarily juveniles in postbreeding dispersal. Spring peak is Mar.–early Apr. along the south shores of the Great Lakes. **WINTER:** Many northern breeders winter in the southern U.S., subspecies mixing together. **VAGRANT:** Casual to Bermuda and NF. **POPULATION** Stable.

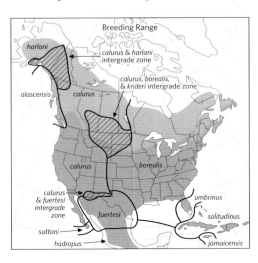

Breeding Range

harlani

calurus & harlani
intergrade zone

calurus, borealis,
& krideri intergrade zone

alascensis

calurus

calurus

borealis

calurus
& fuertesi
intergrade
zone

fuertesi

umbrinus

solitudinus

suttoni

hadropus

jamaicensis

FERRUGINOUS HAWK   *Buteo regalis*   FEHA   ▪ 1

The largest of our buteos, the Ferruginous Hawk is sometimes called the "Ferruginous Rough-leg" for its feathered tarsi. Sexes are similar, although the females are larger; immatures are different from adults. Their variations include light, rufous, and dark morphs, light birds being far more commonly found. Ferruginous Hawks often perch on the ground. They generally feed on small mammals (prairie dogs are among their favorite prey), snakes, insect swarms, and occasionally birds taken on the ground. The species is notable for gathering in winter roosts. Monotypic. L 23" (58 cm) WS 56" (142 cm)

IDENTIFICATION Large-headed, big-chested raptor with leg feathering that reaches to the toes. Wings are long and tapered with pale flight feathers. The light tail is of moderate length. Perched birds show the wing tips almost reaching the tip of the tail, shorter on juveniles. Head features a large bill and an enormous yellow gape reaching past the centerline of the eye (see sidebar pp. 174–175).

LIGHT-MORPH ADULT: Head is variably colored whitish or gray with rufous streaks and dark eye line. Back and upperwings are chestnut with dark markings and primary coverts, and

long gape extends back under rear of eye

palish head with whitish supercilium and dark postocular

juvenile

yellow cere and gape

juvenile

pale head

extended gape

juvenile

juveniles whiter below than adults

mostly white below

feathered legs

adult

rufous underparts

dark-morph adult

whitish tail lacks bands

rufous leggings

basal third of tail is white

variable rufous feathering in underwing coverts

adults

all show broad whitish crescents

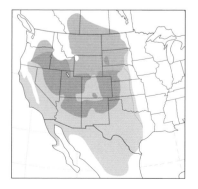

darker wing tips. Primaries are white at the base. Underparts are white with variable rufous barring on belly and flanks; legs barred rufous. Underwing coverts are variable, white to heavily barred, with light flight feathers almost unbarred, and dark crescent ("comma") at wrist. Tail is plain, varying from white to gray, often with a rufous wash. **LIGHT-MORPH JUVENILE:** Dark brown back and upperwings are similar to many Red-tailed juveniles, but the head shows a large gape, dark eye line, and lighter cheeks. Primaries have light bases that make a noticeable white spot on the upper wings in flight. White undersides are usually sparsely marked, and leg feathers are white. Underwings are mostly white with scattered dark markings, and with a dark comma at wrist. Light tail has light bands on outer half and whitish base. **RUFOUS AND DARK-MORPH ADULTS:** Very uncommon with variations from rufous to very dark. Dark brown or gray head, back, and wings, outer primaries show gray. Unbanded tail is gray above, silvery on underside. Undersides are dark, with variable amounts of white or rufous streaking. Underwings have silvery flight feathers and dark brown or rufous coverts, and white comma at wrist. Legs are dark, barred brown, or rufous. **DARK JUVENILES:** Dark brown above, head and breast slightly lighter and more rufous than belly; underwing coverts dark with white comma at wrist, dark primary tips; silvery flight feathers

with light barring. Note dark forehead. Tail is gray with darker bands.
FLIGHT: From above, white at base of primaries is very visible, as is the light tail. Long, pointed wings are often held in a dihedral. Wingbeats are slow and powerful, interspersed with gliding. Hunts by soaring at high altitudes followed by long dives or low-level pursuit-flights near the ground to flush and ambush prey.
SIMILAR SPECIES Light-morph Red-tailed Hawks have a dark patagial mark on the underwing that Ferruginous Hawks lack, and they are less rufous or chestnut above. Other dark-morph buteos have different wing shapes, lack the pale area on the upper wing at the base of the primaries, and have more patterned tails. Dark-morph Rough-legged Hawk has a light forehead.
VOICE Quite vocal during the breeding season, especially when near its nesting site or during confrontations with other raptors. Gives harsh alarm calls, *kree-a* or *kaah*. May call to signal alarm or location, to beg for food, or while engaged in territorial defense.
STATUS & DISTRIBUTION Uncommon. **BREEDING:** Breeds in the western U.S. into Canada and south to northwestern TX, NM, and AZ in open, dry country, often hilly. The large stick nest is built on a lone tree or on a rocky outcrop, and usually reused over many years. **MIGRATION:** Northern birds migrate into southwestern U.S., TX, and northern Mexico, and

west to CA. Fall movements are at their peak in Oct.; in spring, adults move north in Feb.–Mar., juveniles tend to migrate later. Southern breeders are more or less sedentary.

VAGRANT: Casual east to KY, IN, OH, VA, NJ, and FL, most regularly in MN and WI.
POPULATION Never abundant, the Ferruginous Hawk is susceptible to

habitat loss and to human disturbance when nesting; otherwise, it seems stable. In parts of its range it can suffer from programs to poison ground squirrels or prairie dogs.

ROUGH-LEGGED HAWK *Buteo lagopus* RLHA ■ 1

The most northerly of our buteos, the Rough-legged Hawk is a winter visitor to the lower 48 states from its Arctic breeding grounds. The numbers of visiting birds vary geographically and from year to year, most likely in response to the abundance or lack of prey within the birds' range. Rough-leggeds have small bills and have the smallest feet and toes of the large buteos. They often perch on thin and delicate branches, which would not support Red-tailed Hawks and other large species of buteo. Rough-legs feed heavily on voles, lemmings, and other small rodents and occasionally on small birds captured on the ground, including Snow Buntings, shorebirds, and others. Females are larger than males and have different plumage characteristics; juveniles are different from adults; all occur in light and dark morphs. Polytypic (4 ssp.; *sanctijohannis* breeds in N.A., Asian *kamtschatkensis* was recently collected on Shemya I., western Aleutians). L 22" (56 cm) WS 56" (142 cm)
IDENTIFICATION All have their legs feathered down to the toes. When perched, the wing tips reach past the tip of the tail in adults, but just reach it in juveniles. Most show a white or light forehead. **ADULT:** All morphs have a dark trailing edge to the underwings and a subterminal band on the tail. Light-morph shows dark carpal patches

on underwing and variably marked underwing coverts. Light-morph male has breast more heavily marked than its belly (which can be almost white), dark gray back and wings, light tail with multiple dark bands and a wide, dark subterminal band. Light-morph female has a browner back and upper wings and a more heavily marked belly than the male, usually with a lighter band between breast and belly, and the undertail has a large, dark subterminal band. Dark-morph adults have a black carpal patch against dark brown coverts and silvery flight feathers with a dark trailing edge. Males are uniformly dark, with three or four light bands on a dark tail. Females show a single dark band on the light undertail. **JUVENILE:** Light-morph has markings similar to light-morph adult female, but with less dark markings on the underwing coverts, breast, and legs, including a narrower dark band on the trailing edge of the wing. The tail has a single, diffuse dark terminal band and a light base. Head often appears whitish. Seen from above, primaries have a light patch at the base; coverts are darker. Uppertail coverts are light. Dark-morphs are similar to dark adult females, with narrower dark trailing edges to the underwings and tail, sometimes with lighter heads. **FLIGHT:** Long wings flap slowly and frequently are

held in a dihedral. When gliding, wings can push forward into a loose W-shape, like an Osprey. Juveniles show a large white patch at the base of the primaries, especially from above. Wing tip is rather blunt and shows five "fingers."
SIMILAR SPECIES Ferruginous Hawk light-morph lacks the dark carpal patch on the underwing and breast streaking. Dark-morph juveniles resemble Rough-legs but have a white comma at the wrist and dark foreheads. Dark "Harlan's Red-tail" usually has white markings on the breast or head, a blurry terminal band on the undertail, and gray on the uppertail.
VOICE During breeding season gives a soft, plaintive courting whistle. Alarm call is a loud screech or squeal.
STATUS & DISTRIBUTION Fairly common. **BREEDING:** Throughout the Holarctic region, nesting on tundra and less commonly in subalpine forests. Nest is built on a rocky outcropping, on the ground, or in a lone tree, if present. **MIGRATION:** The entire breeding population leaves the tundra to winter in lower Canada and the U.S., but the timing and abundance of the migration is determined by the abundance of their food source. Few arrive in northern states as early as late Sept.; most arrive in late Oct. and Nov. Spring migration sees adults moving north in Mar., followed by juveniles in Apr. and into May. **WINTER:** Immatures often winter farther south than adults; adult males winter farther south than females. A few winter as far south as north-central Mexico, and casually to southern TX, the Gulf Coast, and northern FL.
POPULATION Appears to be stable.

adult
sanctijohannis ♀

adult
sanctijohannis ♂

— squarish carpal patch

small bill

dark-morph adult
sanctijohannis ♂

juvenile with pale head and blackish belly

white tail base with broad, black subterminal band

adult males with multiple blackish bands at tail base

feathered tarsus

juvenile
sanctijohannis

RAILS, GALLINULES, AND COOTS Family Rallidae

Purple Gallinule (FL, Mar.)

One of the most widely distributed families of terrestrial vertebrates, this group includes some of our most familiar birds as well as some of our most mysterious and difficult to observe species. Most are small- to medium-size ground dwellers and are easy to identify if well seen or heard. However, because rails are secretive and often inhabit dense, relatively inaccessible, wetland vegetation, observation can be difficult and requires patience. Typically, an observation involves a brief glimpse as a bird runs, jumps, or swims or is accidentally flushed. Gallinules and particularly coots tend to be less secretive, more ducklike, and often frequent open water or marsh edge.

Structure All species have proportionately short and rounded wings, short tails, and strong legs and feet. Laterally compressed bodies allow rails and gallinules to move effortlessly, almost rodentlike, through dense vegetation. Long slender toes help distribute weight and allow some species to walk on floating vegetation. The more aquatic coots are ducklike, heavier, wider bodied, with lobed toes for swimming. Bill shapes vary from short, stubby, and chickenlike to long and slender, mainly reflecting diet: more herbivorous/omnivorous (taking vegetation, seeds, or snails) versus taking a higher proportion of invertebrates such as crabs, respectively.

Behavior Most rails are adapted to living in low, dense vegetation on moist substrates. Thus, most occur in association with water: fresh water or brackish, salt marshes, mangrove swamps, bogs, wet meadows, or flooded fields. All species can swim; some can sink beneath the water's surface. Most can dive and use their wings underwater to escape predators. Coots and gallinules are the most aquatic, forage while swimming, and must patter across the water to get airborne. Rails rarely forage in the open, bur they do they can be surprisingly tame and oblivious. Most rails and gallinules prefer to run to avoid danger, but if startled will flush, fly a short distance with legs dangling, then drop back into cover. Once airborne, most species are remarkably strong fliers, and many are capable of sustained long-distance migration or vagrancy. As a group rails are very vocal, especially during the breeding season and at night. Vocalizations are usually loud and diagnostic. Calls vary from soft cooing to harsh and monotonous series of mechanical sounds; some species engage in duets.

Plumage Somber, cryptic colors provide camouflage; a few species are brightly colored. Sexes are usually similar, and plumage generally does not vary seasonally, although some species have a slightly brighter breeding plumage. Chicks are covered in black down, some adorned with colorful plumes or naked skin on the head. Down is replaced by juvenal plumage while birds are still substantially smaller than adults, and juvenal plumage is fairly quickly replaced by first basic plumage. These immatures resemble adults, but typically have duller plumage and soft parts. The postbreeding molt is complete; the rapid loss of flight feathers results in temporary flightlessness in some species.

Distribution Worldwide except polar regions and waterless deserts. Many oceanic islands have been colonized by rails, many of which have evolved into new species, and some of which have become flightless.

Taxonomy About 133 extant species. Family relationships have undergone a number of reclassifications; genetic studies are ongoing. Our Common Gallinule was recently split from the Old Word Eurasian Moorhen.

Conservation About 25 percent of extant species is globally threatened; 17 species are almost certainly extinct since about 1600. Many oceanic species and populations, especially flightless species, are extinct or seriously threatened by introduced predators. Habitat loss has caused declines in many continental species.

YELLOW RAIL *Coturnicops noveboracensis* YERA ■ 2

This secretive and mysterious species is rarely found away from dense cover and is not easily flushed, preferring to run and hide. It relies on its cryptic plumage for camouflage, but when flushed the wings show a conspicuous white secondary patch. Polytypic (2 ssp.; nominate in N.A.). L 7.3" (18 cm)
IDENTIFICATION ADULT: Small. Sexes similar. Back feathers black broadly edged yellow-buff with narrow white terminal crossbars. Crown and nape blackish, dark stripe below eye. Broad supercilium; rest of face, with darker brown feather tips. Throat and belly white, sides and flanks black with narrow white bars. Tips of greater coverts and trailing edge of inner wing extensively white. Nonbreeding males and most females have blackish olive bill; breeding males, some breeding females have paler yellow bills; legs brownish to greenish. **IMMATURE:** Darker; face, neck, breast, and sides less buffy, with heavier blackish brown and white barring, white spots on crown.
SIMILAR SPECIES Sora similar in shape (juvenile Sora also similar in coloration) but larger and underwings barred with gray; no white in secondaries. Corn Crake much larger, no white in secondaries. Black Rail smaller, more uniformly dark, grayer, no white in wing.
VOICE Sounds like two stones tapped together, *tic-tic tic-tic-tic,* four or five notes repeated. Usually vocalizes only on breeding grounds, mainly at night.
STATUS & DISTRIBUTION Uncommon, local. **BREEDING:** Sedge- or grass-dominated fresh water or brackish marsh. **MIGRATION:** Nocturnal. Spring: Apr.–May. Fall: late Sept.–Nov. **WINTER:** Marsh, grassy fields, and rice fields along S. Atlantic and Gulf coastal plains, NC to TX; also a few in coastal Bay Area marshes on central CA. **VAGRANT:** Casual or accidental from southeastern AK (heard only), BC, AZ, NM; also Labrador, NF, Bermuda, and Bahamas. **POPULATION** Probably declining; secretive behavior complicates estimates.

upperparts blackish and buff with thin, horizontal whitish bars | juvenile *noveboracensis* | *noveboracensis*

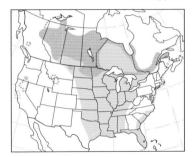

noveboracensis — extensive white secondary patch

BLACK RAIL *Laterallus jamaicensis* BLRA ■ 2

Our tiniest rail is rare, local, and usually only heard. If flushed, it flies a short distance then drops out of sight. Best chance of viewing North America's most secretive species is during unusually high tides in some California salt marshes, when flooding forces rails to marsh edge. Polytypic (5 ssp.; 2 in N.A.). L 6" (15 cm)
IDENTIFICATION ADULT: Generally dark gray, with short black bill, red eyes, dark brown hind crown, dark chestnut nape, white speckling on upperparts, white barring on flanks, lower belly, undertail coverts, and underwing coverts. Female paler gray throat and breast. **IMMATURE:** Similar to female but more brownish, less dorsal spotting.
GEOGRAPHIC VARIATION Two subspecies breed in N.A., three others resident in S.A. Smaller, more brightly colored *coturniculus* occurs year-round in CA, southwestern AZ, and northwestern Baja California. Larger, duller nominate occurs in eastern U.S., eastern Mexico, and Caribbean slope of C.A.
SIMILAR SPECIES Downy chicks of other rail species are small and uniformly black, but have disproportionately large legs and feet. Yellow Rail, almost as small, is dark above but more patterned and paler below, with extensive white on wing. Sora larger, paler, browner, with stouter, paler bill, and extensive white on leading edge of wing and under wing.
VOICE *Kik-kee-do, kik-kee-derr,* or *kik-kik-kee-do,* mainly at night during breeding season; also gives a series of *grr* notes, heard year-round.
STATUS & DISTRIBUTION Rare to locally uncommon. **BREEDING:** Shallow freshwater or salt marshes, wet meadows. **MIGRATION:** In spring, arrives mid-Mar.–mid-May. Departs early Sept.–early Nov., peak mid-Sept.–mid Oct. **WINTER:** Local along southern Atlantic and Gulf coasts, NC to TX, casual north to NJ and south to Greater Antilles (may breed Cuba), C.A., S.A. **VAGRANT:** Rare to casual inland in eastern N.A. north to CO, MN, Great Lakes region, ON, QC, ME, CT, RI; accidental to Bermuda.
POPULATION Declining. Generally considered threatened; endangered in Arizona.

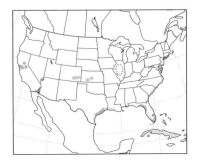

tiny size with white speckling | dark red iris | maroon nape | extremely secretive | dark bill | *jamaicensis* | slaty gray face and underparts

CORN CRAKE *Crex crex*   CORC ▪ 4

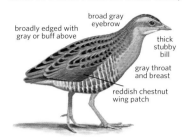

broadly edged with
gray or buff above

broad gray
eyebrow

thick
stubby
bill

gray throat
and breast

reddish chestnut
wing patch

Once a regular vagrant, the Corn Crake has been recorded only five times in North America since 1928, most recently in 2002. Secretive, it prefers dense grass, runs to avoid danger, and is not easily flushed. Monotypic. L 10.8–12" (27–30 cm)

IDENTIFICATION Relatively large. Bill stubby, pale. Shoulders chestnut, back feathers blackish brown broadly edged buff, flanks banded reddish brown and buff. **MALE:** Gray breast and superciliary stripe. **FEMALE AND IMMATURE:** Similar, browner, less gray on face, breast.

SIMILAR SPECIES Yellow Rail smaller with white-tipped secondaries, lacks chestnut shoulders.

VOICE Loud, rasping *crex crex*.

STATUS & DISTRIBUTION Common but increasingly localized in Old World. **BREEDING:** Meadows; northern Europe to Siberia. **MIGRATION:** In fall, Aug.–Nov. In spring, late Mar.–May. **WINTER:** Eastern Africa. **VAGRANT:** Casual/accidental (fall), northeastern N.A. coast (Baffin I. to MD); recent records from Bermuda and Guadeloupe.

POPULATION Vulnerable.

TYPICAL RAILS Genus *Rallus*

Considerable size variation exists among the six New World species, but all have relatively long, slender, slightly decurved bills and rusty breasts; most have barred flanks. All have adapted to a semiaquatic existence and are often seen foraging along habitat edge.

CLAPPER RAIL *Rallus longirostris*   CLRA ▪ 1

The "clapper call" is a familiar sound of our coastal salt marshes, and this large rail can occasionally be observed foraging for crabs, other invertebrates, and seeds along the edges of tidal channels, mudflats, or ditches. It prefers to run into vegetation rather than fly when startled, but will readily flush, fly a short distance with legs dangling, then drop back into cover. It swims across deeper tidal channels. Polytypic (±20 ssp.; 8 in N.A.). L 12.8–16.4" (32–41 cm)

IDENTIFICATION ADULT: Sexes similar in plumage, but males about 20–25 percent larger. Bill long, dark, with paler reddish to brown base of lower mandible (brighter on male), eye reddish brown, legs long and brownish red to gray. Dark brown crown and nape, brown to blackish back feathers edged with gray, olive, or brown. Cheek gray, sometimes bordered brownish gray. Throat white. Color of breast varies geographically from buff to deep rufous to olive-gray. Flanks and undertail coverts gray to black, heavily barred with white. Plumage is darker and brighter in fall when fresh, becomes progressively paler and duller into spring and summer. Some subspecies exhibit considerable individual color variation, referred to as olive and brown morphs to describe color of feather edges on back. **JUVENILE:** Similar to adult but center of breast more extensively white contrasting with darker gray sides, white barring on flanks less conspicuous, and underparts varyingly mottled with blackish gray feather tips. Fully feathered juveniles can be considerably smaller than adults.

GEOGRAPHIC VARIATION About 20 subspecies can be broken into three groups based on size, coloration, and genetics. Recently it has been proposed that these three groups (two occur in the U.S.) be considered separate species. 1.) The *obsoletus* group consists of four western

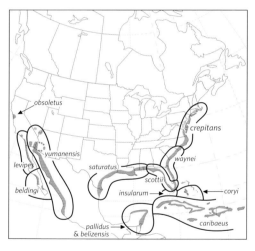

subspecies; all have richly colored plumage, bright rufescent brown underparts, and a brownish gray cheek. The three subspecies found in U.S. include: "California Clapper Rail"—*obsoletus*—of San Francisco Bay (largest); slightly smaller "Light-footed Clapper Rail"—*levipes*—of coastal southern CA south into Baja California; and "Yuma Clapper Rail"—*yumanensis*—of the lower Colorado River drainage and Imperial Valley (dullest of western group). 2.) The *crepitans* group includes 11 subspecies. The five species in the U.S. are variably colored saltwater birds that intergrade with each other along the Atlantic and Gulf coasts: "Northern Clapper Rail"—*crepitans*—of the northern Atlantic Coast from New England

coastal southern
California
levipes

dark and
rich rusty
neck

obsoletus from San
Francisco Bay Area is
similar but is slightly
less richly colored below

south Atlantic
coast
waynei

all subspecies
have brown
centered feathers
on upperparts
with grayish edges

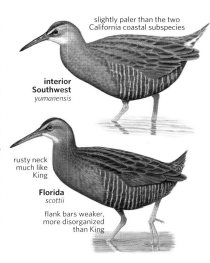

slightly paler than the two
California coastal subspecies

**interior
Southwest**
yumanensis

rusty neck
much like
King

Florida
scottii

flank bars weaker,
more disorganized
than King

to North Carolina (grayest on chest and back); "Wayne's Clapper Rail"—*waynei*—of the southern Atlantic Coast from North Carolina to northeastern Florida (slightly smaller and darker); "Florida Clapper Rail"—*scottii*—along most of the mainland Florida coast (darkest upperparts); "Mangrove Clapper Rail"—*insularum*—of the Florida Keys (smaller, paler); and "Louisiana Clapper Rail"—*saturatus*—of the Gulf Coast from western Florida to Mexico (more rufescent below and most similar to King Rail). Remaining subspecies of the *crepitans* group occur in eastern Mexico, northern Central America, and the West Indies. 3.) The *longirostris* group consists of six subspecies from northern South America; all are small,

pale, thick-billed birds of mangrove habitats.

SIMILAR SPECIES Virginia Rail smaller and more delicate even compared to half-size juvenile Clapper. Some Clapper subspecies very similar to King Rail, but King Rail is larger, overall browner, has browner cheeks, back edgings, and brighter chestnut wing coverts. (See sidebar below).

VOICE CALL: Most typically a series of ten or more loud *kek* notes, accelerating then slowing (King Rail similar but series usually deeper, shorter, slower, more evenly spaced); also shorter *kek-kek-kek,* other squawks, grunts. Calls day or night, but most vocal at dawn and dusk during breeding season.

STATUS & DISTRIBUTION Uncommon

Atlantic and Gulf Coast Clapper Rail Versus King Rail

Along the Atlantic and Gulf coasts, Clapper and King Rails breed in close proximity. King Rails prefer freshwater habitats; Clappers prefer salt or brackish marshes. Vegetation type offers a clue to salinity and, thus, which rail is likely to occur at a site. Grasses, sedges, and especially cattails are indicators of the freshwater haunts of King Rails. Mangroves, cordgrass, needle rush, and pickerelweed are typical of the preferred saline habitats of Clappers. In some areas King Rails inhabits one side of a road and Clappers the other. In addition, in some areas King Rails breed in brackish marsh, and the occasional Clapper has been found in freshwater marsh. Wanderers/migrants of both species can be found away from typical habitat, and hybrids have been reported from DE to LA. Atlantic coast *crepitans* (New England to northeastern NC) and *waynei* (southeastern NC to northeastern FL) are largely gray-breasted, and more easily distinguished from a King Rail. However, Gulf Coast subspecies *insularum* (FL Keys), *scottii* (most of FL coast), and *saturatus* (central and western Gulf Coast) have cinnamon brown chests and are more similar to King Rail. Some individuals are grayer chested like Atlantic Clappers; whether grayness is age-related or a plumage polymorphism is not known. Considering the potential for individual or regional variation, identifications should be based on a combination of characters. King Rail averages larger, more brightly colored, and overall browner. Clapper averages smaller, grayer, and shows a greater degree of individual, seasonal, and regional variation. King Rail's bright chestnut upper-

"Gulf Coast" Clapper Rail (TX, Apr.)

King Rail (TX)

wing coverts stand out compared to the browner coverts of the Clapper. The pale edges of back feathers are more consistently brown in King Rail and gray to olive in Clapper Rail. The Clapper has extensively gray cheeks; gray is more limited in King Rail. Kings tends to have cleaner white throats, sharply contrasting with rusty breast; Clappers average less contrast. Adult Kings never have grayish wash on sides and across the breast that is typical of Clappers. Juveniles of both species are grayer below and almost identical on the underparts, but dorsal coloration is similar to adult. The barring pattern on the flanks is variable between species, individuals, and ages, and although the King tends to be more conspicuously barred black and white, some individuals are duller. Dark and pale morphs have been described for the King, darker birds with more olive-edged back feathers, which could further complicate identification. Consider lighting, distance, and seasonal changes in plumage that occur due to wear and fading. Rails occupying "wrong" habitat for their appearance or lacking definitive identification characters are sometimes referred to as "Cling Rails." First generation hybrids intermediate in size and coloration would be difficult to distinguish with certainty from a pure Clapper Rail. Variation shown by Clapper Rails may be the result of introgression with King Rails, and some authorities consider them conspecific. However, recent work suggests splitting the Clapper Rail into three species (two in U.S.) and the King Rail into two species (one in the U.S.). ∎

to common. **YEAR-ROUND:** Coastal salt marsh, mangroves; freshwater marsh in interior Southwest. Also patchily distributed West Indies, Mexico, C.A., northern S.A. **MIGRATION:** Northern Atlantic Coast populations at least partially migratory, generally scarce in winter north of NC. Arrives north

Atlantic coast mid-Mar.–Apr. Departs late Aug.–Nov. Other populations considered nonmigratory, but individuals will undertake erratic post-breeding movements. **VAGRANT:** Western subspecies casual to south-central AZ, Farallon Is., extreme northwestern coastal CA, and inland away from breeding areas.

In East casual to NE, TN, WV, central VA, western PA, central NY, NH, VT, ME, NB, PE, NS, NF, and Bermuda. **POPULATION** Western subspecies are endangered. Eastern populations are stable but have experienced declines, vulnerable to wetlands loss. It is a game bird in many states.

KING RAIL *Rallus elegans*   KIRA ■ 1

King Rails occasionally forage in the open along marsh edges or in roadside ditches; in spring and summer parents venture into the open with their black downy chicks in tow. If startled, individuals run with head lowered in line with the body and disappear into vegetation. Polytypic (3 ssp.; nominate in N.A.). L 15.2–19.2" (38–48 cm)
IDENTIFICATION ADULT: Sexes similar in plumage; males average larger. Very large, heavy rail, but considerable individual size variation. Bill long with yellowish orange base, eye orange-brown, legs and feet grayish brown to grayish olive. Plumage overall rusty-brown in coloration, mantle heavily streaked brownish black and edged with brown to olive. Upperwing coverts are conspicuously chestnut on folded wing. Crown is dark brown, cheek mostly brown or with a small patch of gray. Throat white contrasting with bright rusty breast, belly buffy white, flanks blackish to dark brown barred with white, undertail coverts mottled dark brown and white. Plumage is brightest in fall, and becomes much duller, paler, and abraded by spring and summer. **JUVENILE:** Heavily mottled with gray below and is darker and less conspicuously striped above. Downy chick like the Clapper Rail, black with a bicolored bill, pale at tip and around nostril.
GEOGRAPHIC VARIATION Nominate *elegans* is closely related to the subspecies *ramsdeni* from Cuba. A recent genetic paper finds that the King Rail subspecies from the interior of central Mexico *(tenuirostris)* is more closely

related to the western subspecies of Clapper Rail than it is to the two other subspecies of King Rail. In which case, it might soon be elevated to full species rank.
SIMILAR SPECIES Virginia Rail superficially similar but much smaller and has pure gray face. Briefly glimpsed chicks might be confused with the Black Rail, but have large legs and feet and downy plumage. (For separation from Clapper Rail, see sidebar previous page.)
VOICE CALL: Most typically a series of ten or fewer loud *kek* notes, fairly evenly spaced; also shorter *kek-kek-kek*, other squawks, grunts. Calls day or night, but most vocal at dawn and dusk during breeding season.
STATUS & DISTRIBUTION Common. **BREEDING:** Freshwater and brackish marshes, rice fields. **MIGRATION:** Nocturnal, partial migrant. Spring: Apr.–May. Fall: Sept.–Oct. **WINTER:** Mainly near Gulf Coast, coast of eastern Mexico south to Veracruz, and Cuba, occasionally farther north

within the breeding range; typically absent from areas with regular freezes. **VAGRANT:** Casual/accidental west and north to western TX, southeastern NM, CO, ND, MN, ME, MB, ON, QC, NB, NS, NF, and PE.
POPULATION Serious declines in northern breeding range since mid-1900s. Vulnerable in Canada; considered endangered, threatened, or of concern in 12 states. Game bird in many eastern states.

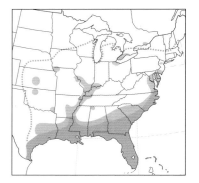

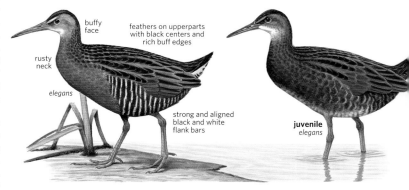

buffy face

feathers on upperparts with black centers and rich buff edges

rusty neck

elegans

strong and aligned black and white flank bars

juvenile
elegans

VIRGINIA RAIL *Rallus limicola*   VIRA ■ 1

Almost as widespread as Sora, Virginia Rails are seen less frequently as they forage in dense vegetation. Polytypic (4 ssp.; nominate in N.A.). L 8.8–10.8" (22–27 cm)

IDENTIFICATION ADULT: Sexes similar. Relatively small with reddish-brown eye, long reddish bill, reddish legs, and rusty chest and belly. Crown and nape brown, contrasting with

gray cheek. Upperparts brown edged with rufous. Throat white, flanks black barred white, undertail coverts mottled black, rusty, and white. **JUVENILE:** Similar but duller above, underparts extensively mottled black-ish or brownish, legs and iris brown. **IMMATURE:** Like adult but buffy-white on lower breast and belly. **SIMILAR SPECIES** King and Clapper Rails generally similar but much larger. Blackish chicks could be confused with Black Rails. **VOICE** Year-round, descending series of *oink* notes; breeding calls include *kid-kid-kiDik-kiDik-kiDik-kiDik.* **STATUS & DISTRIBUTION** Common. **BREEDING:** Primarily freshwater marsh, less frequently brackish or salt marsh. Other populations resident in south-ern Mexico, C.A., northwestern S.A. **MIGRATION:** Mostly migratory. Arrives in spring Mar.–May, peak early Apr.– early May. Departs mid-Aug.–Nov., peak mid-Sept.–mid Oct. **WINTER:** Western and southern U.S. south to central Mexico; regular in salt marshes. **VAGRANT:** Rare to Bermuda, casual or accidental to AK, Cuba, Greenland. **POPULATION** Possibly declining. Game bird in many states.

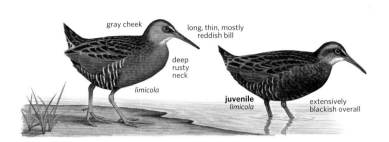

gray cheek

long, thin, mostly reddish bill

deep rusty neck

limicola

juvenile *limicola*

extensively blackish overall

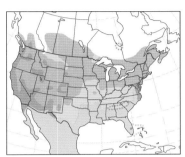

Genus *Porzana*

SORA *Porzana carolina*   SORA ▪ 1

The calls of the Sora are some of the most familiar sounds of the fresh-water marsh, and with patience or luck one can frequently be observed as it ventures into the open to forage along marsh edges or the shoreline of a ditch, walking deliberately with bobbing head and cocked tail. Sora runs to avoid danger, but will also flush when nearly underfoot, fly a short distance, and then drop from sight into dense vegetation. Soras commonly swim across deeper chan-nels and will occasionally even forage coot-style in the water. Monotypic. L 7.6–10" (19–25 cm) **IDENTIFICATION ADULT:** Stubby yel-low bill with black around base and on throat to center of upper breast. Breast gray barred white, belly white, sides barred white, brown, and black, undertail coverts buffy-white. Crown

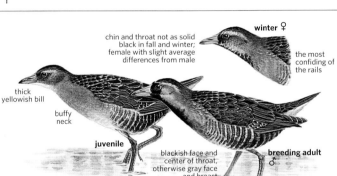

winter ♀

chin and throat not as solid black in fall and winter; female with slight average differences from male

the most confiding of the rails

thick yellowish bill

buffy neck

juvenile

blackish face and center of throat, otherwise gray face and breast

breeding adult ♂

and nape rich brown with black cen-ter, contrasting with gray face and neck, and white spot behind eye. Rest of upperparts brown, back and tail with black feather centers and white fringes. In flight, leading edge of inner wing conspicuously white, underwing coverts barred gray and white. Legs and feet are yellowish-green, toes long. Female similar but bill duller, black on face and throat less extensive, slightly smaller. **JUVENILE:** Paler with white throat, buffier face and breast, lacks black on face, bill darker; immature gradually acquires black on face and throat, and gray on neck and breast. **SIMILAR SPECIES** Duller, buffier juve-nile superficially similar to Yellow Rail. Sora is larger, upperparts have white streaks (not buffy stripes and white bars) and in flight, lacks white on trailing edge of wing. Black Rail smaller, uniformly darker, with all-dark bill. **VOICE** Year-round gives a descend-ing whinny, *Whee-hee-hee-hee-hee-hee,* and a sharp *keek*; when breeding, an upslurred *soo-rah.* **STATUS & DISTRIBUTION** Common. **BREEDING:** Shallow freshwater marsh; occasionally brackish or salt marsh. **MIGRATION:** In spring, late Mar.–mid-May, peak mid-Apr.–early May. In fall, late July–early Nov., peak Sept.–Oct. **WINTER:** Vegetated shallow wet-lands in CA and southern Atlantic and Gulf Coast regions to northern S.A. **VAGRANT:** Casual/accidental to east-central AK, Queen Charlotte Is., southern Labrador, Bermuda, Green-land, western Europe, and Morocco. **POPULATION** Generally stable, but declining in central U.S. from habitat loss. Game bird over much of the East.

Genera *Aramides, Neocrex,* and *Pardirallus*

RUFOUS-NECKED WOOD-RAIL   Aramides axillaris   RUWR ■ 5

Though gaudy in color, this species blends remarkably well into its shadowy environment and is most easily detected by its incisive calls. Monotypic. L 11–12.5" (28–32 cm) **IDENTIFICATION** A brightly colored, chicken-size rail unlikely to be mistaken for any other species. **ADULT:** Underparts a velveteen chestnut except for whitish throat and black or slaty gray vent. Mantle a grayish blue contrasting with olive coverts and tertials and blackish rump. Legs and iris coral red. Heavy, rather long bill green-yellow, becoming orange-yellow at base. **FLIGHT:** Seldom seen in flight. Brilliant rufous remiges

contrast with black-and-white barred underwing coverts. **JUVENILE:** Underparts mostly grayish, upperparts olive, with buffy gray head, darker gray crown; bill mostly dusky. **SIMILAR SPECIES** None. **VOICE** A short series of high-pitched, sharp *pik!* calls, delivered mostly at night or very early or late in the day. **STATUS & DISTRIBUTION** Uncommon permanent resident of mangrove swamps, marshes, lagoons, but also in forest undergrowth to 6,000 feet elevation, from western Mexico and the Yucatán Peninsula south to Panama, along the Caribbean coast of S.A. to Suriname, and south along

the Pacific coast of S.A. to northern Peru. Also on Trinidad and larger islands off Honduras. **VAGRANT:** One at Bosque del Apache N.W.R., NM (found July 7, 2013, and remained for more than a week).

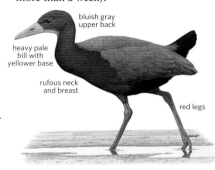
bluish gray upper back / heavy pale bill with yellower base / rufous neck and breast / red legs

PAINT-BILLED CRAKE   Neocrex erythrops   PBCR ■ 5

This secretive and poorly known vagrant from South America has been encountered only twice in North America, both times in the 1970s during winter. Polytypic (2 ssp.; possibly both in N.A.). L 7.2–8" (18–20 cm) **IDENTIFICATION ADULT:** Superficially like a small, uniformly colored Sora. Most of upperparts are dark olive-brown. Forecrown, supercilium, and face are gray. Throat whitish, rest of underparts dark gray; sides, flanks, undertail coverts, and underwing coverts barred white. Bill yellow-green with bright orange base, legs red. **JUVENILE:** Duller. **GEOGRAPHIC VARIATION** Two sub-

species, nominate *erythrops* of coastal Peru and the Galápagos Is. is a vagrant to TX. A Virginia specimen is tentatively identified as *olivascens,* patchily distributed in S.A. east of the Andes; darker, with less white on throat. **SIMILAR SPECIES** Sora is more patterned above and below, lacks orange base to bill, has yellow-green legs. Black Rail is smaller, has dark bill and legs, chestnut nape, white spots on upperparts. **VOICE** Guttural, froglike *qur'r'r'r'rk,* or *pip* notes. **STATUS & DISTRIBUTION** Poorly understood, secretive; abundance varies locally from rare to common. Also

recorded from Costa Rica and Panama, but status unclear. **YEAR-ROUND:** Marsh, pasture, rice fields, overgrown drainage ditches, damp thickets in scrub or woodland. **MIGRATION:** No clear migratory pattern. **VAGRANT:** Accidental in TX (south of College Station; Feb. 17, 1972) and VA (near Richmond; Dec 15, 1978).

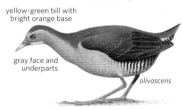
yellow-green bill with bright orange base / gray face and underparts / *olivascens*

SPOTTED RAIL   Pardirallus maculatus   SPCR ■ 5

This amazing tropical rail vagrant was detected in the U.S. twice in the 1970s. Polytypic (2 ssp.; *insolitus* in N.A.). L 10–11.2" (25–28 cm) **IDENTIFICATION ADULT:** Somewhat larger and heavier than Virginia Rail. Upperparts are blackish brown with

black head and body with white speckling / long, slender greenish-yellow bill with red spot at base / adult *insolitus*

white spots, head and breast blacker with white spots, remainder of underparts black banded with white. Bill is long, slender, greenish yellow with red spot at base of lower mandible; eyes, legs, and feet red. **JUVENILE:** Polymorphic, with dark, pale, and barred morphs. All have duller legs and bill, brown eyes. **GEOGRAPHIC VARIATION** Middle American *insolitus* is darker above than West Indian and South American *maculatus* and is spotted rather than streaked with white. **SIMILAR SPECIES** Virtually unmistakable based on size, plumage pattern, soft parts colors. Beware juvenile Virginia, King, and Clapper Rails, which

have varying amounts of blackish or gray plumage. **VOICE** Sharp *geek,* a screech preceded by a grunt, and a series of pumping *wumph* sounds, like a starting motor. **STATUS & DISTRIBUTION** Common. **YEAR-ROUND:** Vegetated wetlands; patchily distributed southern Mexico to Costa Rica *(insolitus);* and Greater Antilles (except Puerto Rico), Trinidad, eastern Panama, and south to northern Argentina *(maculatus).* **VAGRANT:** Accidental in Beaver Valley, PA (Nov. 12, 1976), Brownwood, TX (Aug. 9, 1977), Juan Fernández Is. off Chile, and at sea in S. Atlantic off Brazil. **POPULATION** Unknown.

PURPLE SWAMPHEN *Porphyrio porphyrio*   PUSW ■ EXOTIC

First detected in 1996, this nonnative species quickly colonized southern Florida after escaping from captivity. Despite eradication attempts that resulted in the shooting of more than 3,100 birds (Oct. 2006–Dec. 2008), the effort showed few signs of reducing the population and was discontinued.

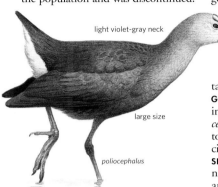

light violet-gray neck

thick red bill

large size

poliocephalus

Polytypic (13 ssp.). L 18–20" (45–50 cm) **IDENTIFICATION ADULT:** Sexes similar, females somewhat smaller. Unmistakable. Suggestive of a huge Purple Gallinule, with thick, reddish bill, broad reddish frontal shield, red eye, massive reddish legs. Body plumage generally purplish blue; wings, lower throat, upper breast paler greenish blue; sides of head, neck, upper throat gray, tinged light blue. Undertail coverts white. **JUVENILE:** Duller, grayer plumage; bill, frontal shield, and legs are darker grayish. **GEOGRAPHIC VARIATION** Most Florida individuals resemble gray-headed *poliocephalus* group, occurring from Turkey to southern Asia, which includes subspecies *poliocephalus, caspius,* and *seistanicus.* **SIMILAR SPECIES** Adult Purple Gallinule smaller with azure frontal shield and yellow legs.

VOICE Dull moans or mooing.
STATUS & DISTRIBUTION Common in native range. Locally common in southern FL. A possible disperser from FL was photographed in GA (Nov. 2009). **YEAR-ROUND:** Shorelines of artificial lakes and natural marshlands.
POPULATION Continues to increase in FL.

PURPLE GALLINULE *Porphyrio martinicus*   PUGA ■ 1

This colorful rail prefers floating or emergent vegetation, using long toes for balance or climbing. More secretive than the Common Moorhen, it spends less time in the water. Monotypic. L 13" (33 cm)
IDENTIFICATION ADULT: Purplish blue head and underparts. Upperparts green, wing coverts turquoise. Undertail coverts bright white. Bill red with yellow tip; frontal shield pale blue; legs, feet yellow. **JUVENILE:** Upperparts brownish to greenish, underparts buffy, whiter on throat, belly, undertail. Bill is greenish brown; legs, feet dull greenish. Downy chick black, legs and feet dark, bill red, tip black, black and pink subterminal bands. Full adult plumage is aquired during the second winter.

SIMILAR SPECIES Coots and Common Moorhen blackish gray, larger, spend more time on water. Purple Swamphen is much larger with red shield and legs.
VOICE Sharp *puckk;* cackling, grunting.
STATUS & DISTRIBUTION Common. **BREEDING:** Freshwater marsh with dense floating or emergent vegetation. Breeds casually north to IL, OH, MD, DE. **MIGRATION:** Primarily a trans-Gulf spring migrant, late

Mar.–May, peak mid-Apr.–early May; occurs rarely north of breeding range. Migrates in fall late Aug.–Oct., peak Sept., stragglers into Nov. **WINTER:** Primarily areas of year-round occurrence, peninsular FL, eastern and central-western Mexico to S.A. Rare Gulf Coast. **VAGRANT:** Many records well north of eastern breeding range to southern Canada; casual to Southwest and CA; accidental to U.K.
POPULATION Stable.

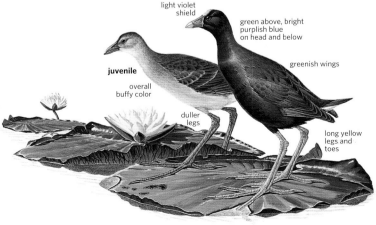

light violet shield

green above, bright purplish blue on head and below

greenish wings

juvenile

overall buffy color

duller legs

long yellow legs and toes

COMMON GALLINULE *Gallinula galeata* COGA ▪ 1

This species forages while swimming, picking at floating vegetation. In protected areas where not hunted, gallinules are often tame and confiding, sometimes entering backyards to feed and rest. Polytypic (7 ssp.; *cachinnans* in N.A.). L 12.8–14" (32–35 cm)

IDENTIFICATION ADULT: Head, neck, and breast are blackish or gray, underparts charcoal gray with white stippling down side to flanks. Upperparts olive brown. Central undertail coverts black but otherwise white. Bill lipstick red, neatly tipped with yellow; frontal shield and eyes red. Legs bright yellow, cherry red at base, with very long toes. **JUVENILE:** Paler gray below, throat and belly white, bill dusky, and legs greenish. Downy chick is black, with yellow and orange head plumes; frontal shield is red, orange-yellow bill is red tipped. **GEOGRAPHIC VARIATION** Seven subspecies named, but only *cachinnans*

recorded in N.A., where it ranges south to western Panama. Subspecies *pauxilla* nests from eastern Panama to northern S.A., and farther south, subspecies *galatea* and *garmani* have been recognized. The Caribbean has two endemic subspecies, *barbadensis* on Barbados and *cerceris* elsewhere. In Hawaii, subspecies *sandvicensis* is endemic and was formerly treated as a full species, "Hawaiian Gallinule." **SIMILAR SPECIES** See American Coot, Purple Gallinule, Eurasian Moorhen. **VOICE** Most commonly heard is a loud, trumpeting "cackle" call, a series of harsh, staccato *ka* notes, rising in pitch, followed by falling, slower series of *kreh* notes (sometimes likened to call of Laughing Gull). This laughing vocalization is unknown in any subspecies of Eurasian Moorhen; also gives clucking calls and screeches. **STATUS & DISTRIBUTION BREEDING:** Fresh to slightly brackish marshes with emergent vegetation; range highly fragmented. **MIGRATION:** Partial migrant, mainly in East. In spring, mid-Mar.–mid-May. In fall, mid-Aug.–early Nov. **WINTER:** Generally vacates

eastern N.A. to southeastern coast and south. **VAGRANT:** Casual in southern Canada beyond breeding range, and to Greenland (6 records!). **POPULATION** Generally stable, but considered threatened or a species of concern in some northern states due to local declines and extirpations. This species and Purple Gallinule are hunted for food in many southern states.

darker bill

winter *cachinnans*

red bill with yellow tip

breeding *cachinnans*

bronze-brown back

juvenile *cachinnans*

thin whitish stripe

EURASIAN MOORHEN *Gallinula chloropus* EUMO ▪ 5

Eurasian Moorhen, a familiar bird in castle moats, city parks, and fish farms across its enormous Old World range, has only recently been recognized as a species distinct from Common Gallinule of the New World. Polytypic (5 ssp.). L 10.5–12" (27–31 cm)

IDENTIFICATION Plumages very similar to Common Gallinule. Adult gallinules show neat yellow tips to the red bill. Gallinules have a reddish brown tone to upper parts, especially distinct in the mantle; moorhens show more olive tones in the mantle. The two have differently shaped frontal shields: Moorhen's is rounded at top and widest in middle, whereas gallinules' are widest at the top and often look squared off or even notched (with 2 "lobes") at the top. The condition of the frontal shield,

however, relates to breeding condition, so a narrow shield rounded at the top is not diagnostic of a moorhen, especially in autumn, but a notched shield probably rules out a moorhen. In adult moorhens, the iris tends to be deep ruby red, while gallinules usually have maroon irides, but this feature is variable. Immature Eurasian Moorhen is almost identical to Common Gallinule. **GEOGRAPHIC VARIATION** Nominate *chloropus,* likely involved in the AK record, is found throughout Eurasia and northern Africa. **VOICE CALL:** Lacks the trumpeting cackle of Common Gallinule. Most commonly heard calls are a rapid staccato *krrrr!* and a rough, sharp *krek,* frequently repeated in series. Many other abrupt clucking sounds, shrieks, and grating

sounds, some similar to Common. **STATUS & DISTRIBUTION** Widespread and quite common across much of Europe, tropical Africa, and much of Asia, including oceanic islands. **VAGRANT:** Has reached Iceland, the Faeroe Is., Svalbard, and Tristan da Cunha. An immature male at Shemya I., AK (Oct. 12–14, 2010) was collected and determined by DNA analysis to represent a Eurasian Moorhen.

shorter bill and more rounded top to frontal shield

adult breeding *chloropus*

record from western Aleutians, a juvenile,

COOTS Genus *Fulica*

EURASIAN COOT *Fulica atra*   EUCO ■ 5

This vagrant to the far northern corners of North America looks and acts very much like our American Coot. Polytypic (4 ssp.; nominate in N.A.). L 14.4–15.2" (36–38 cm)

IDENTIFICATION ADULT: Larger than American Coot with all-white bill and frontal shield. The all-black—rather than white-sided—undertail coverts are diagnostic. Legs and feet greenish. **JUVENILE:** Similar, but much paler below.

SIMILAR SPECIES The American Coot averages smaller, is less uniformly blackish with more head-body contrast, has white on undertail coverts, partial dark ring on bill, and usually has a dark spot on its frontal shield. **VOICE** *Kowk* and double *kek-wock.*
STATUS & DISTRIBUTION Common and widespread in Old World. **BREEDING:** Freshwater marshes. **MIGRATION:** Northern Eurasian populations migratory. In fall, migrates mid-Aug.–Nov. In spring, migrates late Feb.–May. **VAGRANT:** Casual to Greenland, accidental in fall and winter in AK, QC, Labrador, and NF.
POPULATION Stable.

all-white frontal shield and bill

atra

black undertail

AMERICAN COOT *Fulica americana*   AMCO ■ 1

Whether swimming on a lake, grazing on a golf course, diving for submerged vegetation, or wading along a pond, this comical, chickenlike rail is one of our most familiar waterbirds. Coots have to run along the surface to take flight, but once in the air flight is strong and fast. Can occur in large flocks. Polytypic (2 ssp.; nominate in N.A.). L 12.8–17.2" (32–43 cm)

IDENTIFICATION ADULT: Sexes similar, males slightly larger. No other species has a short, whitish, chickenlike bill, overall gray coloration, and, in flight, white trailing edge to secondaries. Head and neck black, body somewhat paler blackish gray, center of belly usually paler, undertail coverts black with white outer patches. Bill and frontal shield mostly white with brownish-red partial ring on bill and spot in center of frontal shield. Size and coloration of frontal shield variable: Some individuals have more yellowish white shield without darker center. Eyes are red, legs and lobed toes greenish yellow. **JUVENILE:** Duller and paler, with more whitish underparts. Bill dusky, legs more olive. Downy chick is black above, dark gray below, with stiff curly orange and yellow fluff on forehead, chin, and lores; fleshy red skin on crown, blue skin above eye, and bright red bill with black tip.

SIMILAR SPECIES Immature Common Gallinule similar but smaller, more slender, has white stripe along sides, browner back, and more slender bill. Caribbean Coot *(F. caribaea)* of West Indies more uniformly blackish below, has larger, broader, yellowish white frontal shield. Some American Coots lack brownish red center in frontal shield and may suggest Caribbean Coot based on coloration of frontal shield alone. Caribbean Coot, sometimes regarded as conspecific, not satisfactorily documented in our area. Eurasian Coot similar but averages larger, overall blacker, with an all-white frontal shield and bill, and has entirely black undertail coverts.

VOICE Assortment of grunting and cackling notes, including an emphatic *puck,* crowing *croooah,* and *punk-unk-punk-uh-punk-unk-uh.* Vocal day or night.

STATUS & DISTRIBUTION Abundant. **BREEDING:** Freshwater wetlands with emergent vegetation, occasionally slightly brackish marshes. Breeds south through most of Mexico and locally in C.A., Colombia, and West Indies. **MIGRATION:** Northern populations migratory; others resident, may migrate during severe winters. Spring: Feb.–May. Fall: Late Aug.–Nov. **WINTER:** Northern interior populations move to coasts and southern half of continent. Occurs on brackish estuaries in winter. **VAGRANT:** Rare in HI; casual to western AK, northern and eastern Canada, Greenland, Iceland, and Ireland.

POPULATION Currently stable, but has declined historically. Considered a game bird in many states.

whitish bill with subterminal band

adult *americana*

some lack dark top to shield

juvenile *americana*

americana

lobed toes

swims like a duck most of the time

SUNGREBES Family Heliornithidae

Sungrebe, female (Costa Rica, Apr.)

These aquatic and generally unfamiliar birds are found in the New World tropics (the Sungrebe) and in tropical Africa and southeastern Asia (the two finfoots).

Structure These are superficially grebelike swimming birds, with lobed feet; slender bodies; long necks; and sharp, pointed bills. In size, they are roughly like gallinules or coots, but more slender. Unlike grebes or coots, they have a long, broad tail (which is stiffened in the finfoots but not the Sungrebe).

Behavior These shy swimming birds inhabit vegetated swamps, ponds, and rivers, where they are found in low densities. They are more prone to flee by moving into vegetation rather than flying. Though principally aquatic, they can run on land and climb into trees. All three species are generally solitary or found in pairs during the breeding season. The nest of twigs and reeds is built within streamside vegetation. Sungrebes have a very short incubation period (about 11 days), and the altricial young can be transported by the adult male in flaps of skin under his wing—even in flight. Young finfoots are more precocial, capable of swimming when very young. Sungrebes and finfoots feed mainly on aquatic invertebrates and to some extent on small amphibians and fish.

Plumage All are generally gray-brown to brown above and pale below (barred in the African Finfoot); the outstanding visual characters include longitudinal neck stripes in all species, colorful beaks, and colorful legs and feet (orange or green in finfoots, banded yellow and black in the Sungrebe). Males and females differ slightly in plumage.

Distribution Sungrebes occur from northeastern Mexico to northern Argentina. The African Finfoot is found widely in wetter areas of sub-Saharan Africa, while the Masked Finfoot is restricted to Southeast Asia.

Taxonomy Worldwide there are three species, each in a monotypic genus. The family belongs within the "core" Gruiformes (along with cranes, rails, limpkins, and trumpeters), but its exact placement is still debated; the most recent molecular work suggests a sister relationship to the rails, and particularly to the rails known as flufftails (which, if this sister-group relationship should be verified, would be given family rank, Sarothruridae).

Conservation All species are probably declining through loss of forested wetlands.

Genus *Heliornis*

SUNGREBE *Heliornis fulica*   SUNG ▪ 5

A small and somewhat secretive aquatic species of tropical America has occurred just once in North America. Monotypic. L 11–12" (28–31 cm)

IDENTIFICATION Unlikely to be confused with any other species. Resembles some combination of large rail and duck, but with long tail and rather heavy, moderately long bill. Toes lobed, banded black and yellow. Usually observed on the water, swimming with slow, jerking movements of head and neck. Body plumage a rich auburn brown; tail blackish. The head and neck have a black-and-white pattern of stripes that recall a grebe

chick. Female has rufous cheek, white in male. **FLIGHT:** Rapid and direct, with the long tail extending far past the trailing edge of the wing.

VOICE A deeply resonant, nasal, monkey-like *eeyaaa*, repeated in short series; both softer and harsher *coo* and *cooo-ah* calls; and clucking sounds.

STATUS & DISTRIBUTION Permanent resident in its range in coastal lagoons, slow-moving rivers, lakes and ponds, usually with forested or vegetated banks as far north as central Veracruz (a few to southern Tamaulipas) and through lowland South America as far south as northern Argentina.

Migration is unknown in the species, but Sungrebes do make movements in search of appropriate habitat and are strong fliers. **VAGRANT:** One female at Bosque del Apache National Wildlife Refuge, NM (Nov. 13–18, 2008) was most unexpected. The species has been found as a vagrant to Trinidad, 12 miles off the coast of South America. **POPULATION** No reliable data.

ear coverts tawny-buff (white on male)

distinctive head pattern

♀

LIMPKIN Family Aramidae

Limpkin (FL, Mar.)

This family consists of a single large, wading-type species, limited in North America to Florida. The Limpkin superficially resembles herons and ibises, but it is only distantly related to them—its closest relatives are the cranes.

Structure Long neck, bill, and legs, a short tail, and long, rounded wings.

Behavior Loosely colonial, perhaps in part due to clumped distribution of prey abundance or availability. Active throughout the day and night, it forages by slowly walking in water; prey is detected either visually or by contact with bill. It nests in a variety of freshwater sites, ranging from mounds of vegetation built in marshes to stick nests built high in trees. Despite its name, the Limpkin does not limp; rather, it walks slowly while flicking its tail. It flies with its neck fully extended and held below the body plane, creating a distinctive hunchbacked profile. Its flight style—quick upstrokes followed by slower downstrokes—is similar to cranes. When flying short distances, the Limpkin dangles its legs below the body, but in sustained flight it draws the legs up.

Distribution Restricted to the New World, primarily the tropics.

Taxonomy Limpkin taxonomy is much debated. It is currently placed within the order Gruiformes, which comprises ten families, including rails, cranes, and sungrebes. The Limpkin formerly was considered two species: one found in North America, Central America, and the Caribbean and the other in South America.

Conservation Overall stable in its neotropical range, estimated at more than one million birds.

Genus *Aramus*

LIMPKIN *Aramus guarauna*   LIMP ≡ 2

This Florida species feeds nearly exclusively on freshwater mollusks, primarily apple snails and clams. The tip of the Limpkin's lower mandible is curved to the right to facilitate extracting snails from their shells. Polytypic (4 ssp.; *pictus* in N.A.). L 26" (66 cm)

IDENTIFICATION Large, brown, heron-like bird with dense streaking over much of its body. Long, slender, slightly downcurved bill is yellow with a blackish culmen and tip. **ADULT:** Wholly dark brown, paler on face, chin, and throat. Most feathers, especially those on the neck, have white markings; those on the upper back and upperwing coverts show large white triangles. Dark eyes; blackish legs and feet. **JUVENILE:** White markings are narrower, appearing more streaked than spotted.

SIMILAR SPECIES American Bittern adults have a conspicuous dark malar streak and more pointed wings in flight. Immature night-herons have orange or red eyes, shorter, straight bills, shorter necks, and yellow legs. *Plegadis* ibises have wholly darker plumage.

VOICE Loud, varied, and distinctive; mostly by male except when breeding pairs duet. Most common calls are a drawn-out *kreow* and a *kow*.

STATUS & DISTRIBUTION Uncommon and local; in the U.S. now restricted to peninsular FL, where generally sedentary. The N.A. subspecies (*pictus*) is also found in the Bahamas, Cuba, and Jamaica. **VAGRANT:** Casual throughout Southeast and accidental along the Atlantic and Gulf states from TX to NJ and NS.

POPULATION Hunting nearly extirpated the Florida population by the early 1900s. The population has since recovered to about 5,000 pairs and is apparently stable, despite continuing wetland loss. Recent extirpation from the state's eastern panhandle is unexplained and alarming.

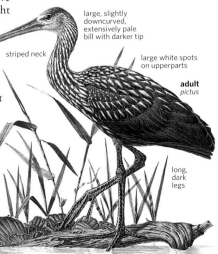

large, slightly downcurved, extensively pale bill with darker tip

striped neck

large white spots on upperparts

adult
pictus

long, dark legs

CRANES Family Gruidae

Sandhill Cranes (NM, Oct.)

Cranes are evocative of wild and open habitats in temperate regions around the world. In North America they present limited identification challenges, although distance and lighting can play into determining features accurately. Note overall plumage coloration and contrasts, details of head pattern, and the extent and contrast of any black on the remiges (best seen in flight); with experience, voice can be useful, but be aware that vocal variation can also be related to age and, in Sandhill Cranes, to subspecies.

Structure Cranes are large, heavy-bodied, and long-necked. They have a distinct tertial bulge, or "bustle," which is conspicuous when they stand. Their wings are long and broad, with ten primaries and 18 to 21 secondaries (including 4–5 elongated tertials); their short, square tails have 12 rectrices. Their bills are moderately long, straight, and pointed, and their legs long and sturdy, with unwebbed feet. Males average 5 to 10 percent larger than females. Juveniles do not attain full growth until nearly a year old; they often look noticeably smaller than adults.

Behavior Cranes are terrestrial and social birds of open habitats, where they pick and probe for food in soil and marshes; they are not known to perch in trees. Their flight is graceful. With their necks and legs outstretched and holding their wings mostly above the body plane, they fly with stiff wingbeats with a relatively quick and slightly jerked upstroke. Migrating cranes are diurnal and often fly high, typically in V-formations or well-spaced lines, and they frequently glide and sail on fairly flat wings, without flapping. Cranes are perennially monogamous; courtship involves spectacular leaping and "dancing" by pairs and groups, accompanied by much calling. Extended parental care means that families remain together for nine to ten months, through migration and winter. Although cranes nest in territorial pairs, they form large flocks in migration and winter, when they commute from safe roost sites to feeding areas. They often feed in agricultural land, and roost on islands or gravel bars in lakes and rivers. Their far-carrying calls are loud cries with a distinctive rolled or throaty quality, sometimes likened to bugling. Cranes

show strong site fidelity to their migration staging grounds and to their wintering grounds.

Plumage Their plumage is gray to white overall, usually with darker to blackish primaries, and often with variable rusty blotching on the neck and upperparts. Adults have bare forecrowns patterned red and black; juveniles have mostly cinnamon-brown feathered heads and necks. Adult prebasic molts occur mostly on or near the breeding grounds, whereas first-cycle preformative molts take place on nonbreeding grounds. Cranes have high wing loading and shed their primaries synchronously to become flightless for a short period—adults do this while nesting; prebreeding immatures also molt their remiges during the summer months, in remote areas with adequate food. Adult plumage aspect is attained by the second prebasic molt at about one year of age, although second-year birds are often distinguishable by their less developed and duller bare crowns and the presence of a few retained and worn juvenal feathers, such as secondaries.

Distribution Cosmopolitan, with both resident and migratory species.

Taxonomy Worldwide, there are 15 species in four genera; in North America, three species (two breeding, one vagrant from Eurasia) in one genus. Two other species, the Demoiselle Crane (*Anthropoides virgo*) and the Hooded Crane (*Grus monacha*) are not (yet) accepted on the official list.

Conservation Several crane species rank among the world's most threatened birds due to habitat modification, pollution, and hunting. Their faithfulness to traditional sites in migration and winter makes them particularly vulnerable to localized habitat loss (e.g., 80 percent of the mid-continent Sandhill Cranes stage in spring along Nebraska's Platte River). Most populations of Sandhill Crane are stable or increasing—some states even permit hunting. The Whooping Crane, however, is one of the rarest birds in North America; conservation measures have brought it back from only 15 to 16 birds in the 1940s to more than 200 individuals today.

SANDHILL CRANE *Grus canadensis*   SACR ▣ 1

This crane is the only one seen in most of N.A. The sight of many thousands staging in spring along the Platte River in NE is one of the great wildlife spectacles in N.A. Polytypic (6 ssp.; 5 in N.A.). L 34–48" (86–122 cm) WS 73–90" (185–229 cm)

IDENTIFICATION ADULT: Gray overall with variable rusty blotching on neck, body, and upperwing coverts. Naked forecrown dark red. **JUVENILE/FIRST-WINTER:** Feathered head, neck, and upperparts cinnamon to rusty. Preformative molt changes over first winter, with forecrown becoming unfeathered midwinter through summer; plumage becomes variegated with newer gray and rusty feathers. After second prebasic molt, resembles an adult. **GEOGRAPHIC VARIATION** See sidebar below. **SIMILAR SPECIES** Whooping Crane is larger and mostly white with contrasting black wing tips. See Common Crane. Great Blue Heron, often referred to by non-birders as a "crane," is rare in groups of more than 10 to 20 birds and usually flies with its neck pulled in with steady downbeats of arched wings. A distant resting heron has a longer neck, often kinked; longer bill; and lacks a bustle. **VOICE** Deep, far-carrying, rolled or rattled honking cries, *k'worrrh*, *grr-rowh* or *grrrah-uu*, and *ah grruu*. In duets of mated pairs, female calls higher and shriller. "Lessers" average higher-pitched calls. Immatures have very different, high, slightly reedy trills through at least first year. **STATUS & DISTRIBUTION BREEDING:**

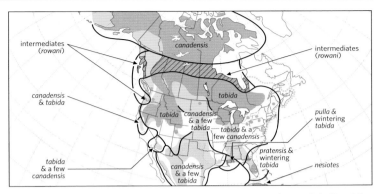

Common to locally rare (Apr./May–Sept.) in a variety of open, undisturbed habitats in N.A., Russian Far East, and Cuba (where resident). **MIGRATION:** Mainly Feb.–Apr./May and Sept.–Nov. **WINTER:** Farmland and open country,

usually near marshes and shallow lakes or playas in N.A. and south to northern Mexico. Very rare on East Coast. **VAGRANT:** Accidental in Europe and HI; very rare Japan. **POPULATION** Mississippi subspecies *pulla* is endangered; its population is 75 to 80 percent captive-produced.

red crown

intermediate-size *rowani* breeds from northern BC to James Bay, ON

adult *rowani*

stained adult *rowani*

juvenile *rowani*

all cranes have tertial "bustles"

neck extended in flight

dark dusky remiges

adult *rowani*

Subspecies Variation in Sandhill Cranes

Six subspecies of Sandhill Crane are recognized, the northern populations being migratory and the southern ones resident. The smallest subspecies is nominate *canadensis*—the "Lesser Sandhill" or "Little Brown Crane"—which breeds from Alaska (and eastern Siberia) across Arctic Canada to Baffin Island, and winters mainly from California to Texas. All other populations (collectively termed "Greater Sandhill Crane") are larger, but *rowani* of Canada bridges the size difference between *canadensis* and the largest subspecies, *tabida,* which breeds from the prairie provinces through the West and the northern Midwest. Two large subspecies are resident locally in the Southeast: paler *pratensis* in southeast

Georgia and Florida, and darker *pulla* in southern Mississippi. Subspecies *nesiotus* is resident on Cuba and Isla de la Juventud.

The most obvious field distinction of the subspecies is their size, however, males of a given population average 5 to 10 percent larger than females. Furthermore, a large-scale study of mid-continent populations found traditional criteria for distinguishing *canadensis, rowani,* and *tabida* to be unsatisfactory. Therefore, although size differences among Sandhill subspecies can be quite striking, as within wintering flocks in California, only extremes may be safely identifiable. ∎

COMMON CRANE Grus grus   COMC ▪ 4

juvenile
lilfordi

yellowish
bill

black neck
with white
stripe

adult
lilfordi

This Old World species, a vagrant to North America, is usually found with migrant Sandhill Cranes. Polytypic. (2 ssp.; ssp. in N.A. unknown)

L 44–51" (112–130 cm)
WS 79–91" (201–231 cm)

adult
lilfordi

blackish
remiges

IDENTIFICATION Averages larger and slightly heavier in build than Sandhill Crane. Distinctive black-and-white head-and-neck pattern on adult; contrasting black remiges on all ages. **ADULT:** Gray overall, sometimes with pale cinnamon wash to chest and back. Naked crown black with red band behind eye; black foreneck, which contrasts with white auriculars and hind neck. **JUVENILE/FIRST-WINTER:** Feathered head, neck, and upperparts extensively cinnamon; forecrown becomes unfeathered midwinter through summer, head and neck develop muted adult pattern.

SECOND-YEAR: Resembles adult by first summer, but some may have a partially feathered forecrown and retain some abraded juvenal wing feathers.
SIMILAR SPECIES See Sandhill Crane.
VOICE Far-carrying, trumpeting *krooh* and *krooah,* and a harsher *kraah.* Immature gives high reedy *peep* or *cheerp* calls through its first year.
STATUS & DISTRIBUTION Eurasia. **VAGRANT:** Casual in migration and winter from AK to Great Plains and to IN, QC, CA, and NV; of about 20 records, some may pertain to escapes. Mixed Common and Sandhill Crane pairs with hybrid offspring have been seen in eastern N.A. Also has been recorded from Greenland.

WHOOPING CRANE Grus americana   WHCR ▪ 2

This large, tall-standing, white icon of endangered species management is best known on its winter grounds in coastal Texas. It breeds in northwestern Canada. Monotypic. L 50–55" (127–140 cm) WS 87" (221 cm)
IDENTIFICATION Occasional cinnamon-tipped upperwing coverts may be shown by birds of any age. **ADULT:** Large, white overall with naked red crown, black face, dark red malar. Has dark greenish bill with orange-yellow base; pale yellow eyes. Black wing tips, usually concealed at rest, contrast strikingly in flight. **JUVENILE/FIRST-WINTER:** Feathered head, neck, and upperparts are extensively cinnamon; eyes are olive. Over first winter, forecrown becomes unfeathered from midwinter through summer, and the neck and upperparts variegated with growth of new white feathers. **SECOND-YEAR:** After second prebasic molt in first summer,

the Whooping resembles an adult, but some second-winter birds may have a duller and partially feathered forecrown and retain some abraded juvenal wing coverts or secondaries. The eye averages duller, and the bill base is a duller yellow. Occasional cinnamon-tipped underwing coverts may be shown by birds of any age.
SIMILAR SPECIES Unlikely to be confused. Wood Stork, rare on the Whooping's winter grounds along the Texas coast, is smaller with a longer, decurved bill and a blackish head and neck; in flight it also has black secondaries. Be aware that in some light Sandhill Cranes can look silvery, but not truly white.
VOICE Loud, trumpeting or bugling cries that carry more than a mile: *k'raah-hu k'raah* and *k'raah kah, kah;* higher pitched and less throaty than Sandhill Crane's. In duets of mated pairs, female calls are higher and shriller. Immatures have strikingly different, reedy whistles through at least their first winter.
STATUS & DISTRIBUTION North America, formerly wintered south to central Mexico. **BREEDING:** Rare (May–Sept.), in freshwater marshes found in boreal forests. **MIGRATION:** In spring, mainly late Mar–early May; in fall, mid-Sept.–mid-Nov. Flight path is along a 50- to 100-mile-wide corridor in a fairly direct route across the Great

Plains between breeding and wintering grounds. Includes brief fall staging in southern SK, and irregular spring and fall stopovers in central NE, KS, and OK. Casual recently in migration west to southeast BC and CO, and east to IL and AR. **WINTER:** Restricted to estuarine marshes and shallow coastal bays in southeast TX, arriving late Oct.–mid-Nov., departing late Mar.–Apr.
POPULATION Endangered. Only self-sustaining population breeds in and around Wood Buffalo N.P., AB, and winters in coastal TX in and around Aransas N.W.R.

red crown

adult

adult

black wing tips

juvenile

THICK-KNEES Family Burhinidae

Double-striped Thick-knee (Mexico, Mar.)

Thick-knees are large shorebirds with a unique cryptic plumage. Their crepuscular and nocturnal habits are also different from other shorebirds. They earned their name for rather thick intertarsal joints (knees).

Structure Thick-knees are large and heavily built like plovers. The head and eyes are large and the bill is stout and straight. Wings and tail are fairly long and the legs are exceptionally long.

Behavior Thick-knees are terrestrial and can run quite fast. Their typical gait is ploverlike with runs punctuated by abrupt stops. Flight is typically low over the ground with long legs projecting well beyond the tail. Wingbeats are silent and rapid. During the day they often sit or stand in the shade of a bush or tree, often with other thick-knees. Most of their activity is between dusk and dawn. They feed on insects, reptiles, and amphibians. Their nests are simple scrapes on the ground.

Plumage The sexes look alike, although the females average slightly smaller than the males. Their highly cryptic plumage camouflages them while they rest during the day. Juveniles are duller and can be separated from adults; adult plumage is achieved in approximately a year.

Distribution The Burhinidae family is composed of nine species that are found primarily in the tropics and the Old World. Two species live along rivers or coastlines, while the other species typically nest in savanna or other grassland habitats.

Taxonomy There are seven species in the genus *Burhinus;* two species are from the New World, but only one is found in the Northern Hemisphere. Thick-knees were once thought to be more closely related to the bustards (Otididae) of the Old World, and indeed they superficially resemble bustards, but skeletal, biochemical, and other studies have proved thick-knees to be shorebirds.

Conservation More information is needed, although some species might actually benefit from forest destruction as it opens habitats they could not otherwise use. BirdLife International lists one species as near threatened.

Genus *Burhinus*

DOUBLE-STRIPED THICK-KNEE *Burhinus bistriatus*   DSTK   5

The large head and eye, cryptic plumage, and unique behavior separate this accidental species from all other shorebirds. The terrestrial Double-striped Thick-knee can run quite fast; indeed, it seems to prefer running over taking flight. When it does fly, it is fast and low to the ground, with legs stretched behind the tail, showing white in the wings. The Double-striped Thick-knee becomes most active after sundown; its large eyes enable it to hunt in the dark for insects and other small animals. Polytypic (4 ssp.; presumably nominate from Mexico in U.S.). L 16.5" (42 cm)

IDENTIFICATION Brown upperparts; a gray-brown face, neck, and chest; a white belly; and a brown tail. A dark lateral crown stripe borders the top edge of its bold white supercilium. **ADULT:** The brown upperparts are edged cinnamon or buff in fresh plumage. Its face, neck, and chest are gray-buff and streaked darker. Its yellow bill has a dark tip. The legs and eyes are bright yellow. **JUVENILE:** It is duller and grayer overall than an adult with greenish yellow legs. The streaking along the neck is narrower and darker. In addition to the white supercilium and dark lateral crown stripe, a dark stripe below the supercilium extends to the nape. **FLIGHT:** Striking. The upper wing is dark with the outer primaries marked with a short white bar. A second white bar covers the base of the inner primaries. The underwings are white.

SIMILAR SPECIES Unique and offers no identification challenge. Other congeners are extremely unlikely to occur here naturally.

VOICE CALL: Loud and far-carrying barking or cackling *kah-kah-kah.*

STATUS & DISTRIBUTION Fairly common to common from southern Mexico to northern S.A.; also on Hispaniola.

BREEDING: Arid and semiarid savannas and grasslands from Mexico to Brazil. **VAGRANT:** One record for U.S., King Ranch, TX (Dec. 5, 1961). A tame individual in AZ was determined to have been transplanted from Guatemala.

dark lateral crown stripe borders broad white supercilium

large yellow eye

long yellowish legs

bistriatus

STILTS AND AVOCETS Family Recurvirostridae

Black-necked Stilt, female (CA, Apr.)

Stilts and avocets are found around the world, primarily in warmer climates. Typically pied black-and-white, they walk gracefully on long legs, particularly the stilts.

Structure These large shorebirds are slender. The bill is long and thin, and typically black; the bill of the avocet is thicker at the base and upturned. The neck is often held in a gentle curve or pulled into the shoulders. Wings are long and pointed; in flight, the neck is outstretched and the legs trail behind the tail. Males are larger than females in stilts and equal in size in avocets.

Behavior Typically feed along the water's edge or in deeper water beyond the reach of other shorebirds. They pick at the surface, but they can also grab invertebrates well below the surface. On land, their gait is typically graceful when walking (it can be brisk with long strides),

but they do appear gangly when they run. Their flight is direct, and avocets appear to be stronger fliers than stilts. Both genera swim; avocets do so well and more frequently than stilts. These are highly social birds, particularly so in the nonbreeding season, when they might form large flocks. They are usually colonial nesters and mob predators with a zeal beyond most shorebirds. Unlike most shorebirds, they are quite conspicuous when breeding, out in the open and very noisy instead of the more common cryptic strategy.

Plumage Both stilts and avocets have a striking black-and-white pattern. Differences between the sexes are slight, but noticeable. Seasonal differences are marked in avocets, but slight in stilts. The wings are dark in stilts and mixed black-and-white in avocets.

Distribution Species are scattered around the world, primarily in warmer climates. Most breeding populations withdraw to (or close to) the coasts in winter.

Taxonomy The taxonomy of this family continues to be debated. The stilts (*Himantopus*) have been placed into as few as one, and as many as eight, species. Avocets (*Recurvirostra*) have also been subject to taxonomic revision, but the recognition of four species spread across the world is generally accepted. The stilts and avocets share behavioral traits, and the third genus in the family (*Cladorhynchos*), with one representative (Banded Stilt) from Australia, is intermediate between the two genera. Stilts and avocets have hybridized in captivity and in the wild in California.

Conservation Hunting and trapping sharply reduced populations in the 19th century, but populations are in the process of reclaiming some of the former breeding areas along the mid-Atlantic coast. Pollutants and habitat loss remain threats. BirdLife International lists the Black Stilt as critically endangered.

STILTS Genus *Himantopus*

BLACK-WINGED STILT *Himantopus himantopus* BWST ■ 5

A vagrant from the Old World, the Black-winged Stilt picks at the water with its needlelike bill for prey. Polytypic (2 ssp.; nominate in N.A.). L 13" (33 cm)

IDENTIFICATION Black back and wings; white underparts; long, bright pink legs. **ADULT:** Head and neck vary from entirely white to extensively black, although a portion of the hind neck will remain whitish or pale gray-brown. Extent of black on head and neck not diagnostic of sex, although males tend to be more white headed. **BREEDING MALE:** Back glossy, pink flush to underparts at onset of breeding season. **FEMALE:**

Duller than male, lacking gloss and pink flush. **JUVENILE:** Gray-brown back; thin, white trailing edge to wing (visible in flight); legs duller and paler.

SIMILAR SPECIES Black-necked Stilt always has entire hind neck black and has white spot above and behind eye. **VOICE CALL:** A repeated *kik, kik, kik* like Black-necked Stilt.

STATUS & DISTRIBUTION Accidental. **BREEDING:** Nominate subspecies breeds in Europe and locally in Asia and the Middle East. **VAGRANT:** Three spring records for western Aleutians and Pribilof Is. This species increasingly occurs in Japan and Korea, with

rising breeding records, and there have been a few extralimital records in the Russian Far East.

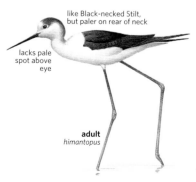

like Black-necked Stilt, but paler on rear of neck

lacks pale spot above eye

adult
himantopus

BLACK-NECKED STILT *Himantopus mexicanus* BNST 1

This graceful wader with an elegant posture is highly social. Polytypic (2 ssp.; nominate in N.A.). L 14" (36 cm)

IDENTIFICATION Boldly pied, black above and white below; long pink or red legs. The dark head has a white spot above and behind the eye. **BREEDING MALE:** Glossy black back, red legs, and occasional pink flush at beginning of breeding season. **NONBREEDING MALE:** Duller legs; gloss and pink flush absent. **FEMALE:** Back has brown tones. **JUVENILE:** Paler than adult. Buffy edges to upperparts; grayish pink legs. Inner primaries and secondaries tipped white, visible in flight.

SIMILAR SPECIES See accidental Black-winged Stilt.

VOICE CALL: A loud, piercing *kek kek kek,* particularly when disturbed while breeding. A loud *keek,* recalling a Long-billed Dowitcher, is a common contact note.

STATUS & DISTRIBUTION Locally common. **BREEDING:** Marshy areas and shallow ponds with emergent vegetation. Range is spreading north in interior states and along Gulf and Atlantic coasts.

MIGRATION: Short to medium-distance migrant, usually away from inland areas during winter. Spring migrants primarily Mar.–Apr., but locally as early as late Feb., and into May. Fall migration protracted with many areas seeing movement in July, but most areas peak Aug.–Sept., trickling into early Nov. **WINTER:** Central and southern CA, Gulf and Atlantic coasts, and southern FL. **VAGRANT:** Rare or casual across southern Canada and various places in the Northeast and Great Lakes.

POPULATION Despite recent expansions, the species has not recovered from 19th-century losses.

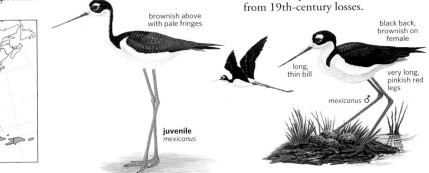

brownish above with pale fringes

black back, brownish on female

long, thin bill

very long, pinkish red legs

mexicanus ♂

juvenile
mexicanus

AVOCETS Genus *Recurvirostra*

AMERICAN AVOCET *Recurvirostra americana* AMAV 1

Avocets feed in water by sweeping their bills back and forth, and on mudflats, where they peck at the surface for prey. Monotypic. L 18" (46 cm)

IDENTIFICATION Black scapular stripes on upperparts; white underparts; black-and-white wings. Long, thin bill is recurved; male's bill is longer and straighter, female's has more of an upward tilt. Bluish gray legs. **BREEDING:** Rusty colored head and neck, with white eye ring and white at base of bill. **NONBREEDING:** Pale gray head and neck. **JUVENILE:** Pale cinnamon wash on head and neck; pale fringes on tertials and coverts.

VOICE CALL: Loud, piercing *wheet* or *kleep,* usually repeated when agitated, higher pitched than the call of a stilt.

STATUS & DISTRIBUTION Fairly common. **BREEDING:** Shallow ponds, marshes, and lakeshores. **MIGRATION:** Anywhere in the West and along the Atlantic (mostly DE and south) and Gulf coasts. Spring migration primarily mid-Mar.–May. A few found in the Great Lakes region with arrivals mid-Apr.–mid-May, dozens in exceptional years. Fall migration July–Oct. with adults through early Aug., followed by juveniles mid-Aug.–mid-Oct. Lingerers into Nov. **WINTER:** Coastal and locally, inland CA, locally along Gulf Coast, and southern FL. Rarely inland near more northerly breeding sites (e.g., UT, NV, OR). **VAGRANT:** North to southern AK and southern Canada.

POPULATION Largely stable.

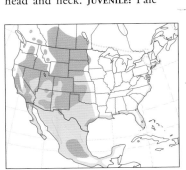

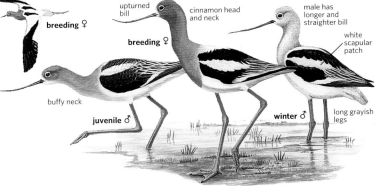

breeding ♀

upturned bill

cinnamon head and neck

male has longer and straighter bill

breeding ♀

white scapular patch

buffy neck

juvenile ♂

winter ♂

long grayish legs

OYSTERCATCHERS Family Haematopodidae

American Oystercatcher, adult (left) with juvenile (NY, July)

Oystercatchers are large, chunky birds usually tied to the coast. Their bills, which they use like a chisel during feeding, are long and bright orange or red.

Structure These are bulky shorebirds, larger than most species. The stout and straight bill flattens laterally toward the tip. Oystercatcher legs are short and thick, and typically pink or red.

Behavior Oystercatchers frequent sandy beaches or rocky shores. They tend to run rather than take flight to escape danger. When they stand upright, they often tuck the neck in and they hold the bill below horizontal. Flight, usually low to the surface, is powerful and direct on deep, rapid wingbeats with rounded wings.

To feed, the oystercatcher plunges its bill into the sand for prey or uses its bill to chisel mollusks off surfaces and pry them open. The birds feed singly, or in small, noisy flocks on coastal beaches and mudflats. Unlike most shorebirds, they feed their young, and the young often stay with the parents through the first winter.

Plumage Oystercatchers are black or pied (black and white). Closer views show substantial brown in the upperparts plumage, with only the Eurasian Oystercatcher adult being truly black. Juveniles are paler, with duller soft part coloration.

Distribution Oystercatchers live around the world's coasts. Some species are less tied to the immediate shore; most are not expected away from the ocean.

Taxonomy The family is composed of 12 species all in the genus *Haematopus*. However there is considerable taxonomic debate about the worldwide number of oystercatcher species—as few as four according to some authorities. The pied species and the black species look very different, but hybridization occurs, resulting in intermediate plumages. In the U.S., two species are mostly resident and one is accidental.

Conservation Beach development impacts oystercatchers, as do spills of harmful pollutants (e.g., the *Exxon Valdez* spill in AK); yet, they quickly colonize dredged areas. BirdLife International lists one species as endangered and another as near threatened.

Genus Haematopus

EURASIAN OYSTERCATCHER *Haematopus ostralegus*   EUOY ■ 5

This unexpected vagrant occurs in the Old World. Polytypic (3 ssp.; probably 2 in N.A.). L 16.5" (42 cm)

IDENTIFICATION Black head, neck, and upperparts contrast with the white rump, wing stripe, and remainder of the underparts. Eyes, eye rings, and

white extends up back

bold white wing stripe

breeding adult *ostralegus*

bill are bright red. **ADULT BREEDING:** Legs pinkish red. **ADULT NONBREEDING:** A white bar extends across the lower throat. **JUVENILE:** Browner above, with a narrower tail band, than the adult. Dark-tipped bill, brown eyes, and grayish legs; bare parts get brighter over first year. **FLIGHT:** White wing stripe extends almost to tip, with pale primary shafts on outer primaries. Base of tail is white.

GEOGRAPHIC VARIATION Subspecies include nominate *ostralegus* of W. Europe (likely the subspecies of birds seen in NF) and *osculans* of the Russian Far East (likely the subspecies of the bird seen in AK). The subspecies *osculans* has a slightly shorter wing

stripe and dark outer primary shafts.

SIMILAR SPECIES Similar to American Oystercatcher but adult Eurasian is black above, lacking brown, has red eyes, and its legs are brighter.

VOICE CALL: Series of loud, piercing whistles, similar to congeners.

STATUS & DISTRIBUTION BREEDING: Eurasia, from Iceland to Kamchatka, more local in Asia. **MIGRATION:** Southern populations largely resident, but northern populations move south. **WINTER:** Southern Europe, N. Africa, Arabian Peninsula, and coastal Asia. **VAGRANT:** Accidental. Three records from NF (2 in spring, 1 fall) and one in spring from Buldir I., Aleutians, AK.

AMERICAN OYSTERCATCHER *Haematopus palliatus*   AMOY   ◼ 1

Found along sandy beaches, oyster bars, bays, and mudflats. Polytypic (5 ssp.; 2 in N.A.). L 18.5" (47 cm)

IDENTIFICATION Striking. Black head with contrasting brown back, white underparts, large red-orange bill, flesh-colored legs, and yellow eyes. **JUVENILE:** Whitish fringes on upperparts give a scaly appearance. Duller bill with a dusky tip, dusky eyes, and grayish flesh-colored legs. **FLIGHT:** Bold white wing stripe and white base of tail.

GEOGRAPHICAL VARIATION Five subspecies, including nominate *palliatus* (along the Atlantic coast) and *frazari* (casual in southern CA from nearby breeding grounds in Baja California). Some *frazari* appear identical to *palliatus,* but most show some dark mottling on lower breast and uppertail

coverts (white on *palliatus*); white wing stripe shorter. This subspecies is perhaps best considered as an intermediate population.

SIMILAR SPECIES See Eurasian Oystercatcher. Hybridization between *frazari* and Black Oystercatcher creates a variety of plumages.

VOICE CALL: Loud *wheep* or *whee-ah,* often rising into an excited chatter.

STATUS & DISTRIBUTION Fairly common. **BREEDING:** Populations extend south through both coasts of Mexico, C.A., and S.A. Expanding northward in the East. **MIGRATION:** Poorly understood but some northern Atlantic pop-

ulations withdraw in Sept., returning Mar.–Apr. Populations south of VA are resident and augmented in winter by northern birds. Very rare in southern CA. **VAGRANT:** Flock of three immatures at Salton Sea, CA; sight record from ID; and a record from ON.

POPULATION Recent expansion in the Northeast.

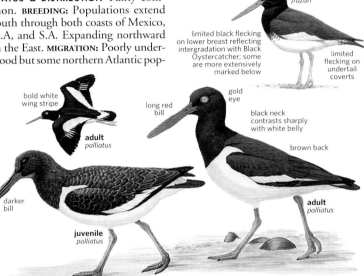

adult
frazari

limited black flecking on lower breast reflecting intergradation with Black Oystercatcher; some are more extensively marked below

limited flecking on undertail coverts

bold white wing stripe

long red bill

gold eye

black neck contrasts sharply with white belly

brown back

adult
palliatus

adult
palliatus

darker bill

juvenile
palliatus

BLACK OYSTERCATCHER *Haematopus bachmani*   BLOY   ◼ 1

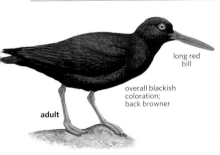

long red bill

overall blackish coloration; back browner

adult

The Black Oystercatcher is a dark, large, chunky shorebird likely to be seen only along the rocky coastline of the Pacific. It is typically found hunting for food on tide-exposed rocks, where it blends into the dark rocks, only to be revealed by its large, red-orange bill. It demonstrates tremendous site fidelity, with pairs typically returning to the same territory for many years. Monotypic. L 17.5" (45 cm)

IDENTIFICATION All-dark body washed with brown on the upperparts and

belly. Yellow eyes with a red eye ring; legs pinkish. Sexes similar, except females are slightly larger and bill is slightly more orange than male's. While subtle, these differences might be evident in pairs. **JUVENILE:** Outer half of bill is dusky, legs are grayish pink, and eyes are dusky. Fine, pale edges to upperparts in fresh plumage.

SIMILAR SPECIES Unmistakeable in N.A., but beware of hybrids between Black and American Oystercatchers in northern Baja California and southern CA: The resulting offspring can look similar to either parent. See the American Oystercatcher account for more detail.

VOICE CALL: Similar to American Oystercatcher. Loud, piercing *wheep* or *whee-ah* notes can accelerate into an excited chatter.

STATUS & DISTRIBUTION Common. **BREEDING:** Rocky shores and islands along the Pacific coast from the Aleutians to Baja California. **MIGRATION:** Largely

resident, but some local movements, not all of which are understood. Many northern breeding areas are abandoned as some birds move south. Autumn numbers in BC peak in Nov., with vacated territories reoccupied in Mar.–Apr. **WINTER:** Territories are abandoned and flocks are formed during winter, but range largely within breeding range. **VAGRANT:** A Jan. record of a "distressed" bird found at 3,400 feet in the Washington Cascades is hard to explain.

POPULATION Human disturbances have eliminated local populations in the Pacific Northwest.

LAPWINGS AND PLOVERS Family Charadriidae

Black-bellied Plover (NY, Nov.)

irds of wetlands, grasslands, and shorelines, plovers are gregarious. They are known for undertaking long migrations, but some species are sedentary.
Structure Plovers have medium to long legs, a rather sturdy frame, a short neck, and a round head. The bill is short; thick at the base, it usually narrows and then expands at the tip. Most plovers have long, pointed wings, excellent for rapid flight and energy-efficient for long migrations. The less migratory lapwings have more rounded wings. The pectoral muscles of plovers are relatively large; this fact and their ability to store large amounts of fat allow them to migrate long distances over the ocean.

Behavior Plovers hunt by running and abruptly stopping. After a few seconds, if no prey is taken, they run again. They pick small invertebrates from either wet or dry surfaces. They rarely perch; most plovers lack the hind toe that would help them do so. Flight is generally rapid for "true" plovers, but floppy in the broad-winged lapwings. The nest is typically a scrape, and the precocial young feed themselves. Some species have exaggerated "broken-wing" displays to distract potential predators. Quite gregarious, they form flocks for migration and roosting; however, they are very territorial during breeding season. Some species will defend feeding territories in winter and during migration.

Plumage Sexual differences are slight, with males usually brighter than females. Plovers typically molt their contour/body feathers in spring. Most species are in breeding plumage when they migrate, a few start in winter plumage and molt on the way north. Many species have black facial markings, which are boldest during breeding season and may play a role in territorial displays. Most Arctic breeders start molting head feathers when incubation begins. Fall migrants usually show some signs of molt, but some species start migrating before much molt is evident. Juveniles do not distinctly differ from winter-plumaged adults, other than look far more crisp in early fall. Some first-summer birds look similar to breeding adults, whereas others look like winter-plumaged birds. First-years have varying migration strategies, sometimes remaining on the wintering grounds, moving north to an intermediate point, or moving all the way to the breeding grounds.

Distribution Worldwide the family numbers more than 65 species—17 have been recorded in North America—inhabiting the tropics, mountains, deserts, and a variety of wetlands. Plovers are both resident and some of the longest-distance migrants in the avian world.

Taxonomy Generally the family is divided into lapwings and "true" plovers, but some work suggests that the four Arctic-breeding *Pluvialis* species belong in their own subfamily. The ten genera include some rather unique plovers; however, North American birds are quite similar. There are few controversial placements at the species level.

Conservation Historically, plovers were hunted in large numbers (with tens of thousands killed on a spring migration day); they are now protected. Today, loss or degradation of habitat and human disturbance are the biggest threats.

LAPWINGS Genus *Vanellus*

Lapwings are generally more strikingly colored than other plovers. Their large, boldly marked, broad wings often have spurs at the carpal joint. While most lapwings are sedentary and found in the tropics, some species live at higher latitudes and are migratory. They are quite noisy when approached, particularly during breeding.

NORTHERN LAPWING *Vanellus vanellus* NOLA ■ 4

A vagrant from Eurasia, the Northern Lapwing is closely tied to wet or dry fields. Monotypic. L 12.5" (32 cm) **IDENTIFICATION** Black and white from a distance, it is actually quite colorful. The face, throat, and upper breast are black; the back has a glossy green sheen. Most of the auriculars and adjacent areas are gray; remaining underparts are white, with pale cinnamon undertail coverts. Wings are broad with white tips, obviously rounded; wing linings are white. Flight is slow and floppy. White rump and base of tail contrast with the tail's black tip. Wispy but prominent crest is unique. **NONBREEDING ADULT:** Most

of the gray near the face of a breeding adult is replaced with buffy tones; black is reduced and throat is white. Upperparts have pale buff edges to the feathers. **JUVENILE:** Duller than adult, with a shorter crest, and bolder edges to upperparts. Legs are gray. **SIMILAR SPECIES** None. Unique. **VOICE CALL:** Whistled *pee-wit*, with the second note higher. **STATUS & DISTRIBUTION** Eurasian species. Casual in the Northeast and Atlantic provinces in late fall and winter. Accidental elsewhere in the East and at other seasons; recorded south to FL, west to OH; one record from Shemya I., western Aleutians, AK (Oct. 12, 2006).

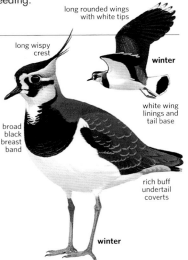

Genus *Pluvialis*

The largest North American plovers, members of this tundra-breeding group are plump and tall. Their breeding plumage is impressive—black bellies (less black in females) with white- or gold-spangled upperparts—whereas their basic plumage is dull (usually grayish or dingy brown). They feed in the typical plover fashion and usually fly and roost in flocks.

EUROPEAN GOLDEN-PLOVER *Pluvialis apricaria* EUGP ■ 4

The European Golden-Plover, as its name suggests, is a vagrant from the Old World. Monotypic. L 11" (28 cm) **IDENTIFICATION** Short tertial extension emphasizes long primaries. **BREEDING MALE:** Pure white sides of breast quite broad; sides, flanks, and undertail coverts more purely white, and gold spotting denser than on Pacific Golden-Plover. **BREEDING FEMALE:** Duller than male; less black below. **NONBREEDING ADULT:** Brownish, with yellow-gold spangles above; dingy whitish below. **JUVENILE:** Supercilium

and auriculars rather indistinct; heavily barred below. **FLIGHT:** Uniform brown upperparts, including rump and tail; bright white underwings distinctive. **SIMILAR SPECIES** See other golden-plovers. **VOICE CALL:** Mournful, evenly pitched two-note whistle. **STATUS & DISTRIBUTION** Eurasian species. **BREEDING:** Greenland to western Siberia. **VAGRANT:** Irregular spring migrant to NF; casual elsewhere in Atlantic Canada and along the East Coast of U.S., south to DE. **WINTER:** Europe to N. Africa; one exceptional specimen record for southeastern AK.

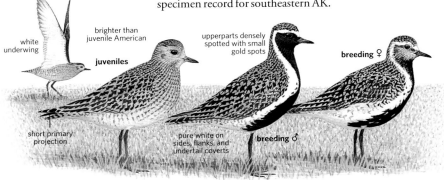

AMERICAN GOLDEN-PLOVER *Pluvialis dominica*   AMGP ▪ 1

Golden-plovers are most frequently seen as migrants, often in flocks. Monotypic. L 10.3" (26 cm) **IDENTIFICATION** American Golden-Plover is plump, but not as large as the Black-bellied Plover. Early spring arrivals are in nonbreeding plumage; they molt over the course of spring, sometimes not finishing until they reach the breeding grounds. Primaries on the long wings extend noticeably beyond the tail and tertials, the latter usually end closer to the base of the tail. **BREEDING MALE:** Gold spotting on most upperpart feathers. White forecrown extends into the supercilium and then broadens into a neck stripe, which wraps to the sides of the breast, extending slightly toward the breast's center, but goes no farther down the lower sides and flanks. Underparts are usually all black. **BREEDING FEMALE:** Like male, but duller; white flecks may be present throughout the underparts. **NONBREEDING ADULT:** Molts on its way south, but early arrivals in southern Canada and the lower 48 are likely to be in partial breeding plumage; later in fall, it usually retains some black flecks below. Full nonbreeding plumage is dull brown above and dingy below,

with minimal marking, except for the supercilium. **JUVENILE:** Grayish brown overall, with a bold whitish supercilium. Upperparts are gray with gold-notched feathers. Underparts dingy, with pale brown barring or mottling on the sides of breasts and flanks. Postocular spot is dark, but diffuse at its edges. **FLIGHT:** Rather uniform brown above, including the rump; indistinct white wing stripe. From below, underwings pale gray with no black in axillaries. **SIMILAR SPECIES** American Golden-Plover is similar in all plumages to the Pacific Golden-Plover (see that species). In nonbreeding and juvenal plumage, American Golden-Plover is also similar to Black-bellied Plover in those plumages. Compared to Black-bellied, the smaller American has a more slender bill, bolder supercilium, and darker crown; in flight, it appears uniform brown above, and the pale gray underwings lack black axillaries. **VOICE CALL:** Shrill two-note *queedle* or *klee-u,* with the second note shorter and lower pitched. **STATUS & DISTRIBUTION** Common. **BREEDING:** Dry tundra. **MIGRATION:** In spring, primarily through interior U.S. Early arrivals late Feb., but more

typically Mar. in southern states, late Mar. in southern Great Lakes, with peak late Apr.–mid-May. Fall migrants typically off the Atlantic coast. Adults first detected early July, but more typically mid-Aug, peak into Sept. Juveniles peak mid-Sept.–mid-Oct., smaller numbers into Nov., and exceptionally Dec. **WINTER:** Primarily in S.A., winter status in U.S. uncertain; some records likely pertain to very late migrants or misidentifications. **VAGRANT:** Casual spring migrant and rare fall migrant in the West; fall records are nearly all juveniles. **POPULATION** Hunting in the 19th century took a heavy toll on this species; the numbers have rebounded but current studies are needed.

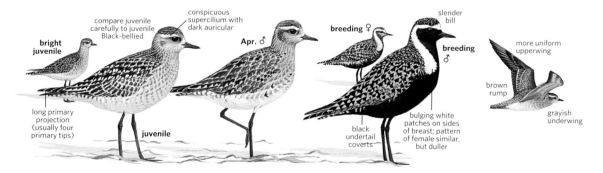

bright juvenile

compare juvenile carefully to juvenile Black-bellied

conspicuous supercilium with dark auricular

Apr. ♂

breeding ♀

slender bill

breeding ♂

more uniform upperwing

long primary projection (usually four primary tips)

juvenile

black undertail coverts

bulging white patches on sides of breast; pattern of female similar, but duller

brown rump

grayish underwing

PACIFIC GOLDEN-PLOVER *Pluvialis fulva*   PAGP ▪ 2

Pacific Golden-Plover, a powerful flier, is known for long, transoceanic migrations. It prefers sandy beaches more than the American Golden-Plover does; on migration, an interior bird is most likely encountered in fields or around ponds. Monotypic. L 9.8" (25 cm) **IDENTIFICATION** Primaries do not extend much beyond the tertials; the latter are long and often extend near the tip of the tail. Adults in spring are usually in breeding plumage;

only second-year birds are expected in nonbreeding plumage in early to mid-spring. Migrant fall adults in July already show signs of molt. **BREEDING MALE:** White forecrown extends into the supercilium and then becomes a stripe that extends to the neck and along the flanks, where black spots or stripes mix with the white. Underparts are black from face and throat through to belly, but not the undertail coverts. **BREEDING FEMALE:** Similar to male, but duller, with black more limited

and mottled. **NONBREEDING ADULT:** Grayish brown above with yellowish edges to feathers, and pale yellow supercilium; dingy whitish or with buff tones below. **JUVENILE:** Rather bright, with distinctly buffy-yellow tones. Spots on upperparts; brown streaks on breast and neck. Bold, dark postocular spot. **FLIGHT:** From above, uniform brown, including rump and tail, and indistinct wing stripe; pale gray underwings from below. Feet usually project noticeably beyond tail.

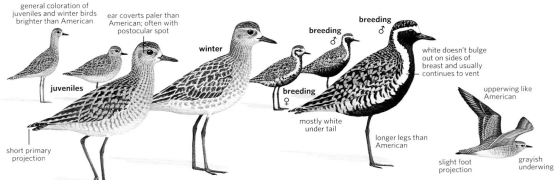

general coloration of juveniles and winter birds brighter than American

ear coverts paler than American; often with postocular spot

winter

breeding

breeding ♂

breeding ♂

white doesn't bulge out on sides of breast and usually continues to vent

juveniles

breeding ♀

upperwing like American

short primary projection

mostly white under tail

longer legs than American

slight foot projection

grayish underwing

SIMILAR SPECIES American Golden-Plover is similar in all plumages. In breeding plumage, male American has a broad white stripe that stops at the sides of the breast, and wholly black underparts—a pattern unmatched by Pacific. However, a breeding-plumaged, female American has white flecks along the flanks, similar to male Pacific. Juvenile and winter Pacifics typically appear brighter than juvenile and winter Americans, which are grayish with a bold white supercilium and a dusky, rather poorly defined postocular spot. American has dusky barring or mottling on dingy white underparts; this barring can extend well past the legs. Pacific is usually buffier, almost yellow in some cases, with a bold buffy supercilium and a dark, but small postocular spot. Proportions in golden-plovers are subtly different and can be useful field marks. The tertials are long in Pacific and the primary extension is shorter, whereas the tertials are shorter in American, with a longer primary projection. Pacific shows one to three primaries extending beyond the tertials, whereas American shows four or five. Fresh-plumaged juveniles are easiest to judge. Pacific Golden-Plover also looks to have longer legs (they project noticeably behind the tail in flight) and a longer and heavier bill than American.

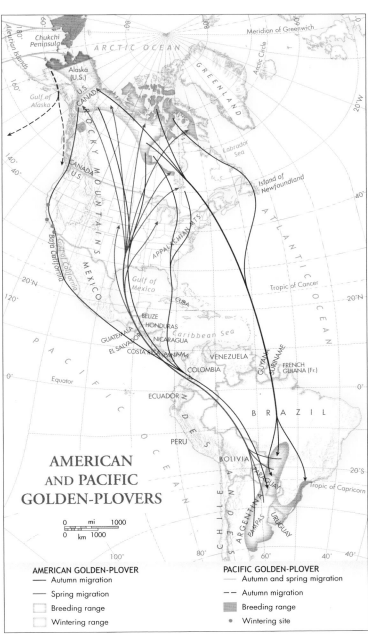

AMERICAN AND PACIFIC GOLDEN-PLOVERS

0 mi 1000
0 km 1000

AMERICAN GOLDEN-PLOVER
— Autumn migration
— Spring migration
☐ Breeding range
☐ Wintering range

PACIFIC GOLDEN-PLOVER
— Autumn and spring migration
-- Autumn migration
☐ Breeding range
• Wintering site

VOICE CALL: A loud, rich *chu-weet* with the last note higher pitched, like a Semipalmated Plover. STATUS & DISTRIBUTION Common breeder, rare transient. BREEDING: Prefers moister habitats than the American Golden-Plover, including the coast and river valleys. Russian Far East to western

AK. MIGRATION: Movement from wintering grounds typically mid-Apr.–mid-May, with arrivals on breeding grounds late May, or June where snowmelt is later; arrivals into southwestern AK as early as late Apr. In fall, adults migrate primarily July–Aug., with lingerers into Sept. Juveniles are most numerous

Sept.–Oct., lingerers into Nov. WINTER: Primarily southern Asia to Pacific islands; small numbers on West Coast and in central CA. VAGRANT: Casual in late summer on East Coast. POPULATION Some southern wintering populations (e.g., in Australia) are thought to have declined.

BLACK-BELLIED PLOVER *Pluvialis squatarola* BBPL ■ 1

While lacking the more colorful tones of the related golden-plovers, the Black-bellied Plover is a handsome bird in breeding plumage. It feeds singly or somewhat spread out, but collects into flocks for roosting. Polytypic (3 ssp.; 2 ssp. in N.A., not field-identifiable). L 11.5" (29 cm) IDENTIFICATION Larger plover than its congeners. Bill bulky. BREEDING MALE: Frosty crown and nape, black-and-white barred upperparts and tail. Black face, throat, and breast to belly; remainder of underparts white. BREEDING FEMALE: Less black; varies from rather dull to almost as bright as males. NONBREEDING ADULT: Drab gray-brown above, with only a hint of a supercilium; dingy below, with grayish brown streaks on neck and streaks or mottling on sides of breast and flanks. JUVENILE: Like adult, but upperparts spotted whitish, with buff tone; fresh juvenile can have pale buff spots. Breast more heavily streaked.

FLIGHT: White uppertail coverts, barred white tail, and bold white wing stripe from above; black axillaries from below. SIMILAR SPECIES Winter and juvenile American Golden-Plover and, to a lesser extent, Pacific Golden-Plover are similar. See those species. Structurally, Black-bellied's bill is larger, with a bulbous tip, and its overall size is larger, more robust. VOICE CALL: A drawn out, three-note whistle, *wee-er-ee;* the second note is lower pitched. STATUS & DISTRIBUTION Common. BREEDING: Dry Arctic tundra. MIGRATION: In spring, southern states (where they also winter) peak mid-Apr.–early May; late Apr. peak in NW; Great Lakes peak mid to late May; lingerers into June, nonbreeders linger all summer. In fall, adults early July, typically late July–Aug., some linger to late Sept.; juveniles first arrive late Aug. (exceptionally earlier), and peak mid-

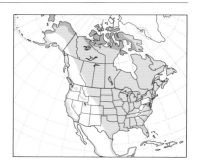

to late Sept., lingerers to mid-Nov., exceptionally later. WINTER: Common on coasts to S.A., Central and Willamette Valleys in CA and OR, and Salton Sea. Otherwise, rare to accidental elsewhere in the interior. POPULATION Numbers are generally stable, with little fluctuation. Nineteenth-century hunting did not victimize the Black-bellied Plovers as it did the American Golden-Plovers.

bright juvenile

short primary projection

winter

large bill

breeding ♀

silver-gray on crown and upperparts

white vent and undertail coverts

breeding ♂

juvenile

RINGED PLOVERS Genus *Charadrius*

Charadrius contains the most species of plovers. Ringed plovers are brown or sandy-colored above and white below, and often have one or two breast bands. Many species have a white collar and/or black face markings during the breeding season. Many species reside in the tropics, while others breed at high latitudes and migrate long distances.

LESSER SAND-PLOVER *Charadrius mongolus* LSAP ■ 3

The Lesser Sand-Plover's breeding plumage differs substantially from its nonbreeding plumage—quite

unlike most of the smaller plovers. This vagrant from Asia was formerly called "Mongolian Plover." Always a

welcome sight, it is likely found on a beach or mudflat, as opposed to an irrigated field, like larger plovers.

Polytypic (5 ssp.; 1 N.A.). L 7.5" (19 cm)

IDENTIFICATION Mid-size plover, but larger than the plovers with which it would most likely associate in the United States. Lesser lacks a pale collar, and has a dark bill and dark legs. **BREEDING MALE:** Unmistakable. Bright rufous breast extends more broadly at the sides, usually bordered above by a narrow black line. Facial pattern is bold—black mid-crown, cheek, and lores contrast with a white forecrown and throat. **BREEDING FEMALE:** Duller than male, but still colorful. The facial markings are brown instead of black. **NONBREEDING ADULT:** Lacks the rufous

breast band and black facial markings. Most birds show a white supercilium and have broad grayish patches on the sides of the breast. **JUVENILE:** Broad, buffy wash across the breast, sharply contrasting with the white throat. Upperparts are edged with buff. Legs might be paler, greenish gray.

GEOGRAPHIC VARIATION N.A. records believed to be *stegmanni,* which has a white forehead and breeds farther northeast than others.

SIMILAR SPECIES Lesser Sand-Plover presents no identification challenge when it is in breeding plumage. In other plumages it might recall a Wilson's Plover, but note Wilson's pale

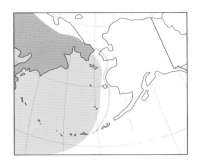

legs, pale collar, and longer, heavier bill. Semipalmated Plover can appear to lack a collar when it is hunched over; its soft parts often retain some orange color, it is smaller, and it lacks the broad patches at the sides of the breast. Greater Sand-Plover, an extremely unlikely vagrant, can be very similar to Lesser Sand-Plover in nonbreeding plumages. Most Lessers have dark legs, but juveniles can be quite pale-legged; the larger bill of the Greater is always the best distinguishing feature.

VOICE CALL: A short, rapid trill.

STATUS & DISTRIBUTION Asian visitor. Rare but regular in migration in western and northwestern AK, casual in summer and it has bred. Casual along West Coast in fall. Accidental elsewhere, primarily eastern North America (recorded in IN, ON, RI, NJ, LA).

POPULATION Stable.

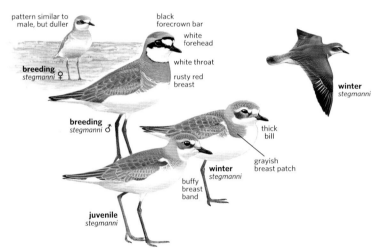

pattern similar to male, but duller

breeding *stegmanni* ♀

black forecrown bar

white forehead

white throat

rusty red breast

breeding *stegmanni* ♂

winter *stegmanni*

thick bill

winter *stegmanni*

grayish breast patch

buffy breast band

juvenile *stegmanni*

GREATER SAND-PLOVER *Charadrius leschenaultii* GSAP ▪ 5

Closely related to Lesser Sand-Plover, the Greater Sand-Plover is a most unexpected vagrant from the Old World. Polytypic (3 ssp.). L 8.5" (22 cm)

IDENTIFICATION BREEDING MALE: Similar to Lesser Sand-Plover, but its rufous breast band is not as broad or intense. It also has black facial markings, but it lacks the dark border to the upper breast, and there's not as much contrast with the crown and throat. **BREEDING FEMALE:** Similar to male but duller and lacks black. **NONBREEDING ADULT:** Rufous and black colorations are absent, like a dull breeding female. Brown breast band can be complete or incomplete. **JUVENILE:** Like a nonbreeding adult, but it has pale fringes on the upperparts. **FLIGHT:** Broad wing stripe is visible, and the toes project beyond the tail tip.

SIMILAR SPECIES Can be difficult to

separate Greater and Lesser Sand-Plovers, particularly from juvenile Lessers that have pale legs. Greater Sand-Plover is a larger version of the Lesser: Its bill is up to 50 percent larger; its body is 10 percent larger than the average Lesser, with legs 20 percent longer. In combination, it looks more elongate, with a long bill. Greater's legs are pale greenish gray, sometimes yellowish green; only a juvenile Lesser will show pale legs. Greater's larger bill, broader white wing stripe, and toes projecting beyond the tail in flight are the best features. And lastly, Greater's underwings are a cleaner white; Lesser Sand-Plover has a dark mark in the under primary coverts.

STATUS & DISTRIBUTION Accidental. Two records: Bolinas Lagoon, Marin Co., CA (Jan. 29–Apr. 8, 2001) and Duval Co., FL (May 14–26, 2009).

Given the breeding range for this species (Middle East to Central Asia) and that it winters in coastal areas of eastern Africa and Southeast Asia to Australia, two records in N.A. are remarkable.

POPULATION The nominate subspecies appears stable, while other subspecies have shown local declines due to wetlands destruction and runoff from irrigation impacting the breeding grounds.

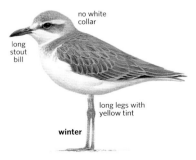

no white collar

long stout bill

long legs with yellow tint

winter

COLLARED PLOVER *Charadrius collaris*   COPL ■ 5

The Collared Plover is common from Mexico to South America; the presence of one in inland Texas was unexpected. Monotypic. L 5.5" (14 cm)
IDENTIFICATION This small plover is pale brown above and white below.

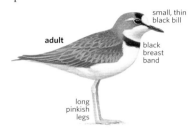

small, thin
black bill

adult

black
breast
band

long
pinkish
legs

It lacks the white collar that most small plovers have. Black breast band is complete and narrow. Black bill is small and notably thin; legs are pinkish, or dull flesh, and proportionately long. **ADULT:** Forecrown is dark, often with a rusty border; auriculars, nape, and sides of breast often have distinct rusty tinges, especially in males. **JUVENILE:** Breast band is incomplete. Upperparts have pale rusty edges. **FLIGHT:** White wing stripe and white edge to outer tail are visible.
SIMILAR SPECIES Collared Plover is most similar to Snowy Plover, but it lacks the white collar. An adult Collared

has a complete breast band, is slightly darker above, and has proportionately longer, pinkish legs, whereas a Snowy has a partial breast band and dark legs. **VOICE CALL:** A sharp *pit,* sometimes repeated.
STATUS & DISTRIBUTION Generally resident from S.A. to Mexico, as far north as southern Tamaulipas and Sinaloa, where it inhabits coastal and inland regions. **VAGRANT:** Accidental, with one record from Uvalde, TX (May 9–11, 1992).
POPULATION Stable, although the population appears to be expanding in South America.

SNOWY PLOVER *Charadrius nivosus*   SNPL ■ 1

The Snowy Plover might be the most overlooked shorebird. In addition to being rather pale, it blends in with its surroundings, and on beaches, it often rests in furrows, cryptically hidden in the sand. A fast runner, it will run to escape disturbances before it will take flight. It generally eats invertebrates that it picks off the sand. Polytypic. (2 ssp.; nominate in N.A.) L 6.3" (16 cm)
IDENTIFICATION Pale sandy brown above, with dark ear patch, and partial breast band that extends from the sides of the breast. White collar on nape is complete. Bill is thin and dark; legs are dark or grayish. **BREEDING MALE:** Ear patch, partial breast band, and forehead are black. Crown and nape may have a buffy orange tint. **BREEDING FEMALE:** Where the breeding male is black, the breeding female is often brown and lacks the warm buff tones to the crown and nape. **NONBREEDING ADULT:** Like a dull breeding female, with no black in the plumage. **JUVENILE:** Like a nonbreeding adult, with pale edges to the upperparts; it quickly fades and looks like an adult. Legs are pale and have a greenish hue. **FLIGHT:** White wing stripe is often

conspicuous. Outer tail has bold white edges, and a dark subterminal bar on the tail contrasts with a paler base and rump.
GEOGRAPHIC VARIATION Gulf Coast birds from Florida, recognized as *tenuirostris* by some, are paler dorsally (also show genetic differences) compared with *nivosus,* the darker nominate subspecies from the interior West to the Pacific. Those from farther west on the Gulf Coast, from Texas for example, while classified as *tenuirostris,* look more like *nivosus.* Some authors consider these the same subspecies, but differences are usually apparent. Snowy Plover was recently split from the Kentish Plover *(C. alexandrinus)* of the Old World.
SIMILAR SPECIES Females and juveniles resemble Piping Plover, the only plover that shares Snowy's pale coloration, although Piping Plover is paler than western Snowies. Note Snowy's thinner, black bill and darker legs.
VOICE CALL: Low *krut* or *prit,* which can be rolled into a trill. Also a soft, whistled *ku-wheet,* second note higher.
STATUS & DISTRIBUTION Uncommon and declining on Gulf Coast. **BREEDING:** Barren, sandy

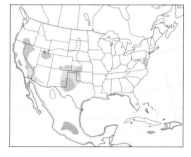

beaches and, inland, alkali playas and marsh edges. **MIGRATION:** Some populations resident, others migrate short to medium distances. In spring, arrive early Mar.–Apr. in CA, slightly later in TX. In fall, adults move from July, more frequently Aug.; migration completed by early Sept.; with stragglers into Oct. **WINTER:** Primarily coastal, but irregularly in winter in southeastern CA, southern AZ, NM, western TX. Recently found in the Rio Grande Valley of south TX in relatively large numbers. **VAGRANT:** Accidental to western AK (Kentish Plover?). Casual across much of N.A., including the Great Lakes and along the southern Atlantic coast.
POPULATION U.S. federal list of endangered species lists the western *nivosus* as threatened. Snowy populations are also under threat in the Southeast. Human disturbance, as well as habitat loss and degradation, play a role in the diminished numbers of this species.

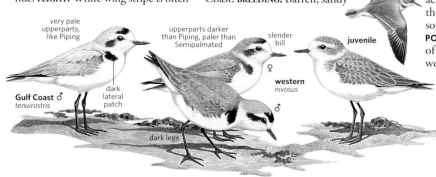

very pale
upperparts,
like Piping

Gulf Coast ♂
tenuirostris

dark
lateral
patch

dark legs

upperparts darker
than Piping, paler than
Semipalmated

slender
bill

♀

western
nivosus

♂

juvenile

WILSON'S PLOVER *Charadrius wilsonia* WIPL ■ 1

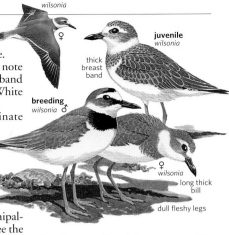

wilsonia

juvenile
wilsonia

thick
breast
band

breeding ♂
wilsonia

♀
wilsonia

long thick
bill

dull fleshy legs

This species is found most frequently on beaches. Polytypic (4 ssp.; 2 in N.A.). L 7.8" (20 cm)
IDENTIFICATION Forecrown is white, with a narrow supercilium. Upperparts are brown, with a broad, white collar; underparts are white, except for a broad breast band. Bill is long, very heavy, and black; legs are pale, but occasionally have yellow or orange tones. **BREEDING MALE:** Breast band is black, and there is some black in the lores and on the crown; ear patch is cinnamon-buff. **BREEDING FEMALE:** Black parts of

male are usually replaced by brown. **NONBREEDING ADULT:** Similar to a dull breeding female. **JUVENILE:** Resembles female, but note scaly-looking upperparts; breast band is usually incomplete. **FLIGHT:** White wing stripe is indistinct.
GEOGRAPHIC VARIATION Nominate *wilsonia,* occurring on the Atlantic and Gulf coasts, has a broader white forehead and narrower breast band compared with the *beldingi* of Mexico's Pacific coast (vagrant to the Pacific coast states).
SIMILAR SPECIES The smaller Semipalmated has a much smaller bill. See the sand-plover species.
VOICE CALL: Sharp, hard *whit.*
STATUS & DISTRIBUTION Fairly common, but declining on barrier islands. **BREEDING:** Sandy beaches and mudflats, although has occurred inland in south TX and once at Salton Sea. **MIGRATION:** Where a summer resident, arrives generally in Mar., but as early as Feb. Fall departures usually in Sept.,

but lingerers into Nov. **WINTER:** FL, otherwise rare along southern Atlantic coast and Gulf Coast; most likely in southernmost TX. **VAGRANT:** Casual to CA, OR, WA, on Atlantic coast north to the Maritimes, and inland to the Great Lakes and Great Plains.
POPULATION Development, disturbance, and habitat destruction are known to impact this species.

wilsonia

beldingi

PIPING PLOVER *Charadrius melodus* PIPL ■ 2

The beautiful and endangered Piping Plover is primarily found on beaches. Pale above, it blends well with its sandy home. Polytypic (2 ssp., both in N.A.). L 7.3" (18 cm)
IDENTIFICATION Pale sandy brown above, with white collar and forecrown. Orange legs are often the field mark most likely to bring attention. **BREEDING MALE:** Black, narrow breast band wraps around the back, below the white collar. Black bar bordering the forecrown nearly connects between the eyes. Obvious orange base to short black bill. **BREEDING FEMALE:** Like male, but collar and bar on crown are paler,

usually not black. **NONBREEDING ADULT:** Bill is black, and black from breeding plumage is lost. **JUVENILE:** Like nonbreeding adult, but with pale edges to the upperparts, difficult to see. **FLIGHT:** Distinct white wing stripe; conspicuous white rump.
GEOGRAPHIC VARIATION Breast band is variable, can be incomplete, especially in eastern populations. Inland subspecies *circumcinctus* has darker facial features than the nominate.
SIMILAR SPECIES In nonbreeding plumage, distinguish it from Snowy Plover by its thicker bill and orange legs.
VOICE CALL: A clear *peep-lo;* also

piping notes during flight display.
STATUS & DISTRIBUTION Endangered, generally uncommon. **BREEDING:** Sandy beaches, lakeshores, dunes. Rare and declining breeder around the Great Lakes. **MIGRATION:** Rare between breeding and wintering ranges. Departures from winter range peak early Apr.–early May. Arrivals in north Atlantic and Great Plains mid-Apr.–mid-May. Departures from breeding ground as early as July, but mostly Aug.–early Sept., lingerers to Nov. **WINTER:** Atlantic and Gulf coasts. **VAGRANT:** Accidental in fall to OR and south end of Salton Sea, CA, and in winter to coastal southern CA.
POPULATION Only a few thousand Pipings are thought to remain. Development, habitat degradation, and water management practices contribute to their decline.

circumcinctus

melodus

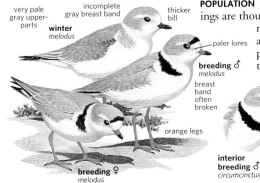

very pale
gray upperparts

incomplete
gray breast band

thicker
bill

winter
melodus

paler lores

breeding ♂
melodus

breast
band
often
broken

orange legs

breeding ♀
melodus

darker lores

complete
breast
band

whitish
uppertail
coverts

interior
breeding ♂
circumcinctus

COMMON RINGED PLOVER *Charadrius hiaticula*   CRPL ▪ 2

Common Ringed Plover is an Old World species, but its breeding range extends into North America. Its behavior and preferred habitat are like those of the Semipalmated Plover, to which it is closely related. It frequents beaches and mudflats, and runs in typical plover fashion in search of food. Polytypic (3 ssp.; 2 in N.A.). L 7.5" (19 cm)

IDENTIFICATION Brown upperparts from crown to back, except for white collar. Dark lores and auriculars offset by white forecrown and throat, the latter continuous with the white collar. Rather distinct white supercilium. Pale yellow eye ring incomplete, or missing entirely. White underparts, except for a broad breast band. Webbing, while difficult to assess in the field, is important: Common Ringed has no webbing between the outer two toes, and only slight webbing between the inner toes. **BREEDING MALE:** Mid-crown bar, lores, auriculars, and breast band

black. Black of lores joins bill at or just below gape. Bill has extensive orange to the base. **BREEDING FEMALE:** Generally paler than male with black parts often replaced partly with brown, but can be difficult to sex. **NONBREEDING ADULT:** Like a dull adult female, with no black on the head. Breast band possibly incomplete. Supercilium broader and more likely to extend in front of the eye than in breeding plumage. Duller legs; bill mostly black. **JUVENILE:** Like nonbreeding adult, except duller. Obvious pale edges and darker submarginal lines create a scaly appearance above through, at least, early fall. **FLIGHT:** White wing stripe quite obvious (much more so than in Semipalmated).

GEOGRAPHIC VARIATION Subspecies *psammodroma* breeds in northeastern Canada; *tundrae* in western AK.

SIMILAR SPECIES Almost identical to Semipalmated Plover (see sidebar below).

VOICE CALL: A soft, fluted *pooee,* lower pitched and less strenuous than Semipalmated's; repeated in a series of notes during display flight.

STATUS & DISTRIBUTION BREEDING: An almost-annual breeder on St. Lawrence Island, AK; regular on Baffin Island, as well as throughout Eurasia. **MIGRATION:** Long-distance migrant, wintering to Africa. Very rare migrant on western AK islands and also recorded on Seward Peninsula. Casual to eastern Canada and New England; accidental CA (Yolo Co.). There is little doubt this species is more numerous than this pattern suggests. While conceivable that most of the breeding birds that creep into North America retrace their steps and winter in the "correct" hemisphere, it is also conceivable that this species passes through either coast annually, only to be overlooked. **POPULATION** Populations of both subspecies are stable.

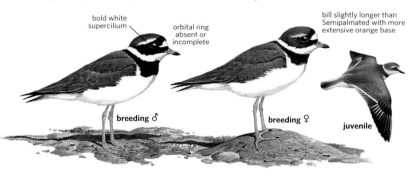

bold white supercilium

orbital ring absent or incomplete

bill slightly longer than Semipalmated with more extensive orange base

breeding ♂

breeding ♀

juvenile

Common Ringed Versus Semipalmated Plover

These two plover species are very similar. Undoubtedly Common Ringed Plover appears in North America more frequently than it has been detected; however, the subtle nature of this identification requires caution. Only consider a vagrant plover if the diagnostic calls are noted. Ideally, use a combination of other characters to confirm the identification.

The calls are straightforward. Semipalmated gives an emphatic, upslurred *chu-weet,* with the emphasis on the second syllable. Common Ringed gives a more plaintive, flute-like *poo-ee.* Both species have minimal webbing between the inner toes; Semi has fairly obvious webbing between the outer toes; Common Ringed lacks

Semipalmated Plover, breeding adult

webbing between the outer toes. If you see webbing in the outer toes, identification is straightforward, but if you do not see webbing, is it because there was no webbing, or was the view insufficient? Verifying this mark is difficult.

Other than calls and webbing, identification is much more complex. The key characters to look at are the supercilium, orbital ring, lore-gape junction, white forecrown, bill, and breast band.

1. Supercilium. Sexing the bird in question is important. Adult male Common Ringed has a distinct white supercilium, fainter in breeding female. In adult male Semipalmated the supercilium is faint or absent, but the breeding female has a bolder supercilium than the male

SEMIPALMATED PLOVER *Charadrius semipalmatus* SEPL ▪ 1

One of our most common plovers, it prefers wet fields and is often found in freshly irrigated fields that attract large numbers of shorebirds. Often flocks when roosting, but the flock spreads out when the birds feed. Monotypic. L 7.3" (18 cm)

IDENTIFICATION Very similar to Common Ringed Plover. Brown upperparts with white collar, supercilium, fore-crown, and throat. Although difficult to assess under most field conditions, this species shows obvious webbing between its two outer toes (and a little webbing between the other toes). Bill is short and has an orange base; legs are orange. **BREEDING MALE:** Lores, auricu-lars, and mid-crown bar are all black, as is the breast band; soft part colors are bright orange. The dark of the lores joins above the gape. An eye ring is usu-ally obvious, orange or yellow in color. Supercilium usually not present in this season. **BREEDING FEMALE:** Similar to male, but facial markings and breast band might be brown; more likely than male to have a partial supercilium. **NONBREEDING ADULT:** Like a dull female with no black on facial markings; breast band brown and sometimes incom-plete; dark bill, duller legs. **JUVENILE:** Pale fringes and dark subterminal marks give bird a scaly appearance; darker legs than in adults. **SIMILAR SPECIES** Semipalmated's dark back separates it from Piping and Snowy Plovers; its much smaller bill separates it from Wilson's Plover.

Almost identical to Common Ringed Plover (see sidebar below). **VOICE CALL:** Distinctive. Whistled, upslurred *chu-weet*, the second syllable higher pitched and emphatic; repeated in a series during breeding. **STATUS & DISTRIBUTION** Common. **BREEDING:** Beaches, lakeshores, riv-ers, tidal flats. **MIGRATION:** A spring migrant on both coasts and through the center of the continent. Migration generally starts mid-Mar., but usually early Apr. before southern states see a lot of migration. Most migration is mid-Apr.–mid-May in southern and middle latitudes; peaks at Great Lakes mid- to late May. Lingers to June and some nonbreeders stay south of the breeding range, complicating depar-ture dates. In fall, much of the eastern movement takes place to the northeast Atlantic and then south. First arrivals usually detected in mid-July, but can be earlier. Most adults move through

late July–mid-Aug., when the juveniles start migrating in numbers. Most states have passed the peak by early Sept., with mostly juveniles through that month. Lingerers occur into Oct., and in some cases Nov. **WINTER:** A coastal bird, with only a few inland locales where it is rare; accidental elsewhere. Winters to S.A. **POPULATION** Numbers are stable.

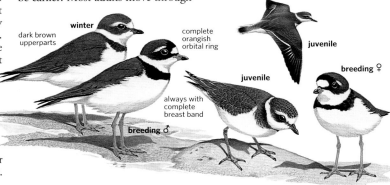

dark brown upperparts
winter
complete orangish orbital ring
juvenile
juvenile
breeding ♀
always with complete breast band
breeding ♂

---

and can closely resemble Common Ringed in this feature.

2. Orbital ring. Semi shows an obvious yellow, or almost orange, orbital ring. Common Ringed shows either a faint yellow orbital ring or none at all.

3. Lore-gape junction. A very difficult mark to accurately assess in the field. In Common Ringed, the white from the throat meets the black of the lores at the gape, whereas in Semi, this junction is above the gape.

4. White forecrown. The white forecrown shape is subtly different on the two species. The rear border is rather straight in Semi and pointed toward the nape on Common Ringed. Furthermore, the white usually runs

Common Ringed Plover, breeding adult

to the eye on Common Ringed; the eye sits well within the dark area on Semi.

5. Bill. Common Ringed has a bill that is a little longer and of more even thickness throughout its length (thinner at the base) than Semi. Furthermore, adult Com-mon Ringed has a more extensive orange base.

6. Breast band. Thicker on Com-mon Ringed and, while variable, normally thinner on Semi. Posture affects this character, so its use should be as supportive evidence, not definitive.

Unless the diagnostic calls are heard, a claim of a vagrant will likely not pass muster without using multiple characters for confirmation. ∎

LITTLE RINGED PLOVER   *Charadrius dubius*   LRPL ■ 5

This rather solitary Old World plover has been recorded as a vagrant in North America. Polytypic (3 ssp.; northernmost breeder *curonicus* in N.A.). L 6" (15 cm)

IDENTIFICATION A small plover, notably smaller than Semipalmated Plover. Brown upperparts; white collar. Lores, forehead, auriculars, and breast band dark, contrasting with white forecrown and white bar behind the dark crown. Conspicuous yellow orbital ring in all seasons. Rather dull yellow legs. On a standing bird, note the long tertials, which almost reach the tail tip. **BREEDING MALE:** Facial markings black; pale base to the lower mandible. **BREEDING FEMALE:** Like male, but auriculars brown. **NONBREEDING ADULT:** Dull brown above; no black facial markings. **JUVENILE:** Often a yellow-buff tint to paler areas on head and throat. Scaly appearance due to pale edges and dark subterminal marks. Dull yellow orbital ring. **FLIGHT:** Uniform brown above; no wing stripe.

SIMILAR SPECIES Little Ringed Plover most closely resembles Semipalmated Plover, but it is smaller, with a yellow orbital ring, and it lacks a wing stripe in flight. **VOICE CALL:** A descending *pee-oo* that carries a long way. **STATUS & DISTRIBUTION** Old World plover, breeds from U.K. to Russian Far East and south to southern Asia and Australasia. Winters from Africa to southern Asia. **VAGRANT:** Casual in spring to western Aleutians.

POPULATION Population is generally expanding as Little Ringeds are opportunistically able to take advantage of man-made developments—more than offsetting declines resulting from habitat loss or agricultural runoff.

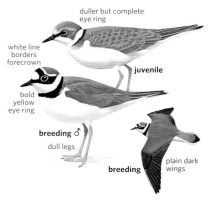

duller but complete eye ring

white line borders forecrown

juvenile

bold yellow eye ring

breeding ♂

dull legs

breeding

plain dark wings

KILLDEER   *Charadrius vociferus*   KILL ■ 1

The Killdeer is North America's most well known plover, although many people know the Killdeer without understanding its family affiliation. It can be common around human developments, frequently seen on playing fields, parking lots, and other unnatural habitats. The Killdeer's "broken-wing" display is famous and known by many non-birders. It feeds in fields and in a variety of wet areas, but rarely along the ocean shore, and in general, it is not too prevalent on mudflats. The Killdeer often forms flocks after breeding in late summer. It is noisy and reacts quickly to any perceived disturbance. Polytypic (3 ssp.; nominate in U.S.). L 10.5" (27 cm)

IDENTIFICATION Brown upperparts turn orange at the rump and uppertail coverts, but this only shows in flight or during its "broken-wing" display. Forecrown is white, as is a short but distinct supercilium, and there is a white collar. Lores are dark; this coloration continues and broadens at the auriculars. Underparts are white, except for two bold black breast bands. Legs are pale, usually dull pinkish, but sometimes with a yellow-green cast. Orbital ring is narrow, but a bright orange-red. The Killdeer is the largest of the "ringed" plovers and looks long-tailed. Sexes generally look similar in adult plumage, and there is little seasonal change. Males average more black on the face than females, but this is variable. Fresh feathers, typically in late summer, have rusty edges to them. **JUVENILE:** Plumage is paler than adult's, with pale edges to the upperparts. Downy young have only one breast band, but they quickly grow out of this plumage. **FLIGHT:** A particularly bold white wing stripe marks the inner wing. In addition, the bright reddish orange rump is visible and the tail has obvious white corners, with dark central rectrices.

SIMILAR SPECIES Double breast band is distinctive; as is its loud, piercing call. A downy young Killdeer has one breast band and might be identified as Wilson's Plover by overeager birders. **VOICE CALL:** Loud, piercing *kill-dee* or *dee-dee-dee*.

STATUS & DISTRIBUTION Common. **BREEDING:** Open ground, usually on gravel, including in cities. **MIGRATION:** Early spring migrants show up in the middle latitudes with the first bit of warmth after mid-Feb. Peak in Great Lakes mid- to late Mar., with most migrants having passed through by mid-Apr. In fall, numbers build July–Aug., sometimes as early as late June. Migration peaks Aug., with numbers decreasing during Sept. Many birds linger until Nov., or later if warm weather persists. **WINTER:** While most winter populations are well established, some vary according to the extent of snow cover. **VAGRANT:** Can occur north of breeding range. **POPULATION** Stable. Has adapted well to human encroachment.

long tail and reddish orange rump

vociferus

vociferus

two breast bands

vociferus

MOUNTAIN PLOVER *Charadrius montanus*   MOPL ■ 2

The Mountain Plover's tan coloration blends in with its barren environment. If not for the white underparts, usually noted only when it moves, sightings would be few indeed. Gregarious in winter, and is usually found in short grass or bare dirt fields. Monotypic. L 9" (23 cm)

IDENTIFICATION Uniform brown upperparts from collar through rump. White forecrown and narrow, somewhat indistinct whitish supercilium. Underparts generally whitish, brightest near belly; breast has a dingy brown wash. Black bill; pale pinkish to gray legs. **BREEDING ADULT:** Black lores and forecrown. Underparts mostly white. **NONBREEDING ADULT:** In fresh plumage (seen on early fall migrants), rufous edges to

scapulars and coverts, fading by early winter in most birds. Buffy tinge on breast more extensive at this season, as is the supercilium. **JUVENILE:** Paler than adult with buff edges to the scapulars and coverts. **FLIGHT:** Brown above, with a hint of a white wing stripe, particularly toward the middle of the wing. Underwing white, visible as bird banks low above the ground. Tail has a pale tip and a dark subterminal bar.

SIMILAR SPECIES Except for the closely related Caspian and Oriental Plovers from Asia, no other species quite looks like a Mountain Plover, but inexperienced birders might mistake it for one of the more numerous American Golden-Plovers when that species arrives in early spring. In nonbreeding plumage in early spring, American Golden-Plovers share a rather brown-and-white plumage, but they are darker above, with pale spots and a bolder supercilium, and have gray underwings and dark legs.
VOICE CALL: Harsh *krrr* note.
STATUS & DISTRIBUTION Local and declining. **BREEDING:** Plains and short-grass prairies. **MIGRATION:** Wintering

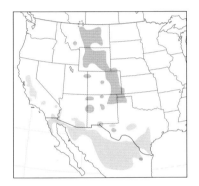

populations thin out late Feb., and withdrawal usually complete by late Mar., early Apr. in TX and CA. First CO arrivals early Mar., those in MT to mid-Apr. Fall dispersal starts July and ends in late Nov. (peak mid-Oct.–early Nov.). Winter populations generally do not arrive until late Oct. **WINTER:** Primarily southern CA, southern AZ, southern and central TX, and Mexico. **VAGRANT:** Rare to casual to Pacific coast; casual in Pacific Northwest; accidental in eastern N.A.
POPULATION Numbers have declined severely with the conversion of grasslands to farmland.

EURASIAN DOTTEREL *Charadrius morinellus*   EUDO ■ 4

A unique vagrant from Eurasia, the Eurasian Dotterel is interesting for its beauty and biology. Rather tame, it allows bird-watchers to approach quite closely. Unlike other plovers, female Eurasian Dotterels have bolder plumage and males tend the nest. Typically, the female only helps rear the young when gender ratios are out

of balance. Monotypic L 9" (23 cm)
IDENTIFICATION Mid-size plover with a small bill; bold white supercilium in all plumages. **BREEDING ADULT:** Gray upperparts have buff edges. Dark crown sets off a bold white line that starts above and extends well behind the eye, wrapping around to the back of the head. White bar separates neck from reddish brown to dark underparts. **NONBREEDING**

ADULT: Similar to breeding bird, but underparts are paler, grayish throughout; white line at the breast is less distinct. **JUVENILE:** Like nonbreeding adult, but has bold, colorful edgings to many feathers above and is buffy below; the breast band is somewhat obscured. The bold white supercilium extends around the entire head.
SIMILAR SPECIES Eurasian Dotterel recalls the golden-plovers, but the supercilium that wraps around the head and the white line on the breast are diagnostic, and Eurasian Dotterel also has pale legs.
VOICE CALL: Soft *put, put,* repeated.
STATUS & DISTRIBUTION Eurasian species; very rare, sporadic breeder in northwestern AK on open, rocky tundra; rare late spring migrant to St. Lawrence I., AK (fewer in recent years). **VAGRANT:** Casual to Aleutian Islands and along West Coast in the fall. Accidental in winter in CA and one record in adjacent Baja California.

JACANAS Family Jacanidae

Northern Jacana (Costa Rica, Aug.)

Jacanas are atypical shorebirds; they look more like rails. Their bold plumage, unique behavior, and rarity in the United States make them a sought-after species. They are easy to identify; the only difficulty posed is pronouncing the name. Most people prefer ZHA-sah-na, from a local tribe in Brazil.

Structure Their long legs and very long toes and toenails are unique. The wings are rounded; some species have a pointed spur at the carpal joint, used for displays and fighting. The bill has a frontal shield.

Behavior Jacanas walk with a high-stepping gait, and frequently do so across floating vegetation. Their long toes and toenails distribute their weight over the vegetation so that they do not sink. They can also swim. Jacanas are quite conspicuous and do not hide in vegetation. Weak fliers, jacanas usually fly only for short distances; in flight, their necks are outstretched and their legs trail behind their tails. They often raise their wings upon landing or in various displays. Jacanas are fiercely territorial. In most species, the female is polyandrous, taking several males in her territory. While the female may help with nest building (on floating vegetation), the males sit on the eggs and care for the young. Jacanas usually feed on aquatic insects, but they also eat small fish and plants.

Plumage The sexes look alike, although the females are larger than the males.

Distribution Eight species of jacanas are found worldwide, primarily in tropical regions.

Taxonomy New World jacanas are in the genus *Jacana*. Some authors consider the Northern Jacana to be conspecific with the Wattled Jacana (*Jacana jacana*) of southern Central and South America; they overlap in Panama.

Conservation The destruction of wetlands through drainage or overgrazing has eliminated populations. One species from Madagascar is near threatened.

Genus *Jacana*

NORTHERN JACANA *Jacana spinosa*   NOJA   ▬   4

This conspicuous bird walks on long legs and is as adept at walking along ditches and grassy lakeshores as across lily pads. It often raises its wings, revealing yellow flight feathers; a spur sticks out from the middle of the wing and

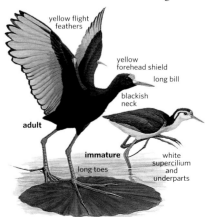

yellow flight feathers

yellow forehead shield

long bill

blackish neck

adult

immature

long toes

white supercilium and underparts

is occasionally visible during these displays. Monotypic. L 9.5" (24 cm)

IDENTIFICATION ADULT: Chestnut with a glossy black head, neck, breast, and upper back. A pale blue cere separates the yellow frontal shield from the yellow bill. **JUVENILE:** White below and brown or olive-brown above with cinnamon edges in fresh plumage. Dark hind neck and crown and dark postocular stripe contrast with whitish buff supercilium. Frontal shield and spur are tiny. One-year-olds are mottled with chestnut, brown, and black; dark bill has a yellow base.

SIMILAR SPECIES Unique. An immature gallinule, perhaps the closest confusion species, has a different shape and coloration.

VOICE CALL: A cackling, harsh *ka-ka-ka-ka* or *jik-jik-jik-jik;* similar to a large rail.

STATUS & DISTRIBUTION Rare and irregular visitor. **BREEDING:** Common nesting species in Mexico and C.A.; has

bred in TX, but not recently. **MIGRATION & WINTER:** Over 30 records, most from freshwater marshes and ponds near the coast of southern TX, primarily Nov.–Apr. A few records are well inland. Casual in southern AZ. Accidental near Marathon, TX.

POPULATION There was a small resident population in Brazoria Co., TX, from 1967 to 1978.

SANDPIPERS, PHALAROPES, AND ALLIES Family Scolopacidae

Long-billed Dowitchers (far right and lower center) and Short-billed Dowitchers (NJ, May)

The large, diverse Scolopacidae family represents the heart of the shorebirds. The general similarity of shorebirds and the inherent identification difficulties keeps many birders from enjoying their distinctive traits. Most sandpiper identifications can be simplified once their basic structure and plumage are understood.

Structure This group has a great variety of shapes, from the large, long legs and long bill of a curlew to the short bills and short legs of sandpipers. Scolopacids are distinct from plovers in having longer bills. Most sandpipers forage primarily by feel; their bills have many receptors to aid in the location of prey. Species that are transequatorial migrants have long wings and primary projection.

Behavior Most sandpipers feed and roost in or near water. They often migrate and winter in mixed-species flocks. During breeding season, however, they are territorial with little tolerance for perceived intruders; many sandpipers are territorial in migration also. Many species perform breeding displays that seem unusual for shorebirds. These displays might include slow, floppy flights with rather well-developed songs and perching in tops of trees. Feeding behavior varies, depending on shape and length of bills and legs. Most scolopacids pick at or near the water's edge, some feed in the water, and others forage in forest litter. Most sandpipers eat invertebrates, such as worms or small bugs, and some swallow mollusks whole—their strong gizzards break up the shell—and yet others eat small fish. The family as a whole is highly migratory; some species migrate long distances over the ocean, but most follow coasts, major waterways, or simply move over land. When sleeping, shorebirds typically stand on one leg and tuck the other into their feathers.

Plumage Understanding molt and plumages is the key to identifying difficult shorebirds. Shorebirds have at least three distinct plumages: breeding, nonbreeding, and juvenile. Transitional or second-year plumages are more complex, but these three plumages provide a good baseline. Some species show sexual dimorphism, but many show little or none. The species have different molt strategies. (The timing of molt might serve as an identification clue.) After breeding, adults generally molt into their winter plumage; most species will be pale and more unmarked below and plainer above. As spring approaches, many species begin to molt into a more colorful plumage, often with red and orange tones and streaks, spots, or splotches on the underparts. Birds less than a year old molt into their first breeding plumage. It is quite variable; sometimes they look like alternate adults, but more frequently they have a dull, incomplete plumage, with a mix of new and old feathers, closer to basic plumage. In most species juveniles leave their natal grounds in a fresh plumage that is quite distinct from either adult plumage, but a few species molt out of many juvenal feathers before moving south. Most sandpiper species retain a few juvenal feathers for months, allowing birds to be aged well into late fall or winter; seeing these feathers can be accomplished under good viewing conditions.

Distribution Of the almost 90 species worldwide, most breed in the Northern Hemisphere and migrate to temperate or tropical areas or, often, well into the Southern Hemisphere. Species nest in a variety of habitats, such as marsh, prairies, tundra, and boreal forests; in winter, most species take advantage of exposed mudflats at intertidal wetlands, although a few use bogs and marshes.

Taxonomy The genera *Catoptrophorus* (Willet) and *Heteroscelus* (tattlers) have been merged into genus *Tringa*. More recently, genetic studies have resulted in the merging of several monotypic genera into genus *Calidris*.

Conservation All shorebirds face the continuing loss of wetlands habitats. Many species have narrow breeding ranges, and many use key staging areas as feeding stops on migration. In the event of some catastrophe, such as an oil spill, or the degradation of a habitat that impacts food sources at staging areas, populations could be rapidly put at risk.

TRINGINE SANDPIPERS Genera *Xenus, Actitis,* and *Tringa*

This tribe of the sandpiper family includes the genera *Xenus, Actitis,* and *Tringa.* Most of these birds are of medium size, although they range from small to rather large. Most of them feed by picking, but those with longer legs will wade in water to probe or take small fish. Many of these sandpipers have gray or gray-brown as their primary color, but breeding and juvenile plumages can be attractive. A majority of these birds breed in the northern latitudes, typically in marshy openings within boreal forests; only the Willet is a common breeder in the lower 48. Most *Tringa* sandpipers migrate long distances, wintering as far south as South America and Australasia.

TEREK SANDPIPER *Xenus cinereus* TESA ■ 3

The Terek Sandpiper, a vagrant from the Old World, has a unique, frenetic personality. When hunting, it uses a jerky, start-and-stop run to cross mudflats. It teeters like a Spotted Sandpiper. Monotypic. L 9" (23 cm) **IDENTIFICATION** Gray or brownish upperparts; white underparts. Long upturned bill. Short orange-yellow legs. **BREEDING ADULT:** Dark-centered scapulars form two dark lines on the back; faint streaking on the breast. **NONBREEDING ADULT:** Reduced black lines on back and breast streaking. Dark lores and faint supercilium both bolder. **JUVENILE:** Like nonbreeding, but darker brown above, with buffy edges to most of the upperparts. **FLIGHT:** Distinctive wing pattern with dark leading edge, grayer median coverts, dark greater coverts, and broad white tips to the secondaries. **SIMILAR SPECIES** Unmistakable. On a roosting bird note Terek's short legs and thick-based upturned bill, unlike any other species. **VOICE CALL:** A series of shrill whistled notes on one pitch, usually in threes. **STATUS & DISTRIBUTION** Rare migrant on outer Aleutians. Casual on Pribilofs, St. Lawrence I., and in Anchorage area; accidental to coastal BC, CA, and MA (June rec.).

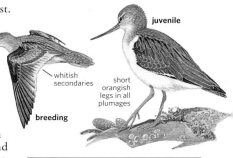

juvenile

whitish secondaries

short orangish legs in all plumages

breeding

thin dark scapular stripe

long upturned bill

breeding

COMMON SANDPIPER *Actitis hypoleucos* COSA ■ 3

Old World equivalent to Spotted Sandpiper, Common Sandpiper has not been recorded away from western Alaska. Monotypic. L 8" (20 cm) **IDENTIFICATION** This species is a small, short-legged sandpiper with a long tail, which it bobs. Typically seen near freshwater, its flight is low over the water, with shallow, flicking wingbeats. The bill is gray and the legs are greenish. **BREEDING ADULT:** Brown above, with dark streaks on the upperparts and dark spots along the edge of the tertials; it is white below, with brown on the upper breast, which is finely streaked. **NONBREEDING ADULT:** Similar to breeding birds, but it lacks extensive streaking. **JUVENILE:** Similar to nonbreeding birds, but upperparts are well marked; pale fringes and dark subterminal lines give a scaled effect; and tertials have barring around the entire feather. **FLIGHT:** Broad white wing stripe extends to the base of the wing and is adjacent to a bold white trailing edge to the inner secondaries. **SIMILAR SPECIES** Juvenile and nonbreeding plumages resemble the nonbreeding and juvenile Spotted Sandpiper. Note Common's longer tail; on juvenile, the barring on the edge of the tertials extends along the entire feather. In flight, Common shows a longer white wing stripe and longer white trailing edge. **VOICE CALL:** A clear *twee-wee-wee,* higher than the Spotted Sandpiper. **STATUS & DISTRIBUTION** Breeds in Eurasia. Rare but regular migrant, usually in spring, on the outer Aleutians, Pribilofs, and St. Lawrence I. Accidental to Seward Peninsula, AK.

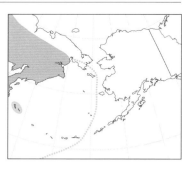

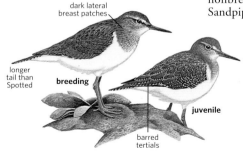

dark lateral breast patches

longer tail than Spotted

breeding

juvenile

barred tertials

juvenile

longer and bolder white wing stripe than Spotted

SPOTTED SANDPIPER *Actitis macularius*   SPSA ■ 1

North America's most widespread breeding sandpiper, the Spotted Sandpiper is found in freshwater habitats. It has a unique and bold breeding plumage and an interesting biology. The polyandrous females arrive early, find and defend territories, and attract males, who take the lead parental role. Its flight is low to the water, on shallow, flicking wingbeats that seem jerky. It bobs and teeters constantly, whether walking along the shore or just standing. Monotypic. L 7.5" (19 cm)
IDENTIFICATION Short-legged, horizontal-looking sandpiper; brown above and white below. Bill is pale at base, varying from flesh to yellow. Legs vary from greenish to yellow. **BREEDING ADULT:** Unmistakable; bold spotting throughout underparts, and dark bars above. Bill is nearly entirely pale; legs bright yellow. **NONBREEDING ADULT:** Dark patches on sides of breast;

white of underparts forms a wedge above the dark wing. Spots are absent or reduced; barred wing coverts contrast with the rather uniform brown back. Bill and legs duller. **JUVENILE:** Similar to nonbreeding, with faint edges to upperparts, but barred wing coverts still contrast with back. Barring on the edge of the tertials, absent on some juveniles, extends no farther than halfway along each feather. **FLIGHT:** White wing stripe is evident, but it does not extend to body; narrow white trailing edge to wing.
SIMILAR SPECIES Compared to Common, Spotted has a shorter tail, and in flight shows a shorter white wing stripe and a shorter white trailing edge. Also Common's shallow wingbeats are not as exaggerated as Spotted's. On juveniles, Spotted shows a greater contrast between wing coverts and scapulars, which are more boldly marked on Common.

Common's tertials are barred; if bars are present on Spotted, they do not extend around the entire feather.
VOICE CALL: Shrill *peet-weet;* a series of *weet* notes in flight. Lower pitched than calls of the Solitary and Common Sandpipers.
STATUS & DISTRIBUTION Common and widespread. **BREEDING:** Sheltered streams, ponds, lakes, or marshes. **MIGRATION:** In spring, peak late Apr. in the southern states (e.g., FL, TX), first three weeks of May in Great Lakes, and after mid-May in the West. Females arrive mid-May in MN, a week before males. In fall, some adults leave late June, peak mid- to late July through Aug. Juveniles predominate in Sept., with stragglers Oct.–mid-Nov. **WINTER:** Most winter in C.A. and S.A. Uncommon to fairly common in coastal southern states; rare to southern edge of breeding range. **VAGRANT:** To Russian Far East and Europe.
POPULATION Stable, but sensitive to the health of rivers and lakes.

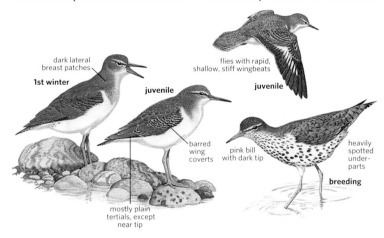

dark lateral breast patches
1st winter
juvenile
flies with rapid, shallow, stiff wingbeats
juvenile
barred wing coverts
pink bill with dark tip
heavily spotted underparts
breeding
mostly plain tertials, except near tip

SOLITARY SANDPIPER *Tringa solitaria*   SOSA ■ 1

This mid-size sandpiper is usually seen singly or in small groups. Typically encountered around small pools, wet grassy areas, or creeks, it bobs its entire body or, more rarely, just the tail. It calls frequently as it flies overhead and is reliably vocal when flushed. Upon landing it often holds its wings up, as if getting its balance, allowing looks at the underwings. Polytypic (2 ssp.; both in N.A.). L 8.5" (22 cm)
IDENTIFICATION Dark brown above, with spots, and white below; lower throat, breast, and sides streaked with blackish brown; undertail coverts might be streaked. Dark bill, often with a pale base. Yellowish to greenish legs.

White eye ring in all plumages; bold white supraloral line does not extend beyond the eye. **BREEDING ADULT:** Dark brown, with extensive spotting on upperparts and streaks on the neck and breast. Pale base to bill can make the bill look two-toned. Legs yellowish. **NONBREEDING ADULT:** Duller, with streaking more obscure, and fewer pale spots above. Bill usually dark; greenish legs. **JUVENILE:** Dark upperparts, with bold white spots, more marked than nonbreeding adult. Plain brown head and breast, with minimal or no streaking. **FLIGHT:** Dark central tail feathers, white outer feathers barred with black, darkish underwing.

GEOGRAPHIC VARIATION Nominate *solitarius* breeds from interior British

Columbia to Labrador; more common in the East as a migrant. Adult has dark lores and juvenile's upperparts are spotted white. Subspecies *cinnamomea* breeds from Alaska to Hudson Bay; more common in the West. It is larger and paler; adult has streaked lores and juvenile's upperparts are spotted buffy. **SIMILAR SPECIES** Most similar to Green and Wood Sandpipers. Bolder white eye ring, shorter olive legs, tail pattern, and call distinguish it from Lesser Yellowlegs. **VOICE CALL:** Shrill *peet-weet*, sometimes three notes, higher pitched and more emphatic than calls of Spotted Sandpiper. **STATUS & DISTRIBUTION** Fairly common. **BREEDING:** Shallow backwaters, taiga pools, and bogs. Uses abandoned passerine nests in small trees, unique for North American shorebirds.

MIGRATION: Journey in spring starts late Mar., but migrants have been noted late Feb. Peak mid-Apr. in TX, late Apr.–mid-May in Great Lakes, mid-Atlantic, etc. Stragglers seen late May south of Canada. In fall, adults arrive late June–early July in Great Lakes and mid-Atlantic, mid-July in TX, etc. Juveniles predominate in West during fall, peak in Aug.

Numbers drop in Sept., with most gone by early Oct., stragglers to Nov. **WINTER:** Primarily the tropics but rarely in the U.S., primarily southern TX. **VAGRANT:** W. Europe, Bering Sea, the Azores, and S. Africa. **POPULATION** Remarkably little data, perhaps owing to the largely remote breeding grounds and the bird's solitary nature.

breeding *solitaria*

legs often yellowish in breeding plumage

juvenile *solitaria*

white eye ring

dark grayish brown breast

short greenish legs

GREEN SANDPIPER *Tringa ochropus* GRSA ■ 4

The Green Sandpiper is a vagrant sandpiper from the Old World. Monotypic. L 8.8" (22 cm) **IDENTIFICATION** Dark above with variable spotting; white below, with breast streaking. Straight, dark bill of medium length; greenish legs. **BREEDING ADULT:** Dark olive-gray tone to upperparts, with white spots on many scapulars and coverts. **NONBREEDING ADULT:** Brown tone to upperparts, less spotting above; neck and breast brown with indistinct streaks. **JUVENILE:** Like nonbreeding adult, but more heavily spotted above, streaks on breast on white background. **FLIGHT:** Blackish

wing linings, white rump, with barring primarily near tips of tail; outer rectrices with little or no barring. **SIMILAR SPECIES** Green Sandpiper resembles a Solitary Sandpiper in plumage, behavior, and calls. Note Green Sandpiper's white rump and uppertail coverts, with less extensively barred tail; darker wing linings; and lack of solidly dark central tail feathers. The similar Wood Sandpiper has more spotting above, more barring on tail, and paler wing linings. **VOICE CALL:** *Peet-o-weet*. Like Solitary

Sandpiper's call, but mixes in notes with different pitch. **STATUS & DISTRIBUTION** Eurasian breeder; winters to central Africa, Indian subcontinent, and Southeast Asia. **VAGRANT:** Casual in spring on outer Aleutian Islands; accidental to Pribilof Islands and St. Lawrence Island.

Solitary Sandpiper breeding

dark center to uppertail coverts and tail, barred outer tail feathers

Green Sandpiper breeding

pure white uppertail coverts and mostly white outer tail feathers

both Solitary and Green have dark wing linings, but Green's are slightly darker

breeding

plumage very similar to Solitary

short olive legs

juvenile

GRAY-TAILED TATTLER *Tringa brevipes* GTTA ■ 3

This vagrant sandpiper from Asia is very similar to the Wandering Tattler in behavior and preferred habitat, but it is more likely to be found on sandy beaches, ponds, or mudflats. Monotypic. L 10" (25 cm)

IDENTIFICATION Gray-tailed Tattler is generally similar to Wandering Tattler in plumage and structure, although it has slightly shorter wings, with the primaries barely extending beyond the tail. Whitish superciliums are more distinct

and meet on forehead. Dark bill might have a pale base; the nasal grooves usually do not extend more than halfway to the tip. **BREEDING ADULT:** Upperparts are pale gray and unmarked. Underparts have fine barring that does not extend beyond the flanks; often there are a few dark marks on the undertail coverts. **NONBREEDING ADULT:** Pale gray above and unmarked white below, except for grayish on the sides and breast. **JUVENILE:** Similar to nonbreeding plumage, but with a white belly and sides, and pale spots on the upperparts.

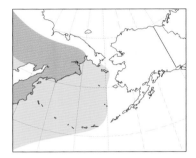

SIMILAR SPECIES Gray-tailed Tattler closely resembles Wandering Tattler; however, its upperparts are slightly paler gray and the barring on the underparts is finer and less extensive in a breeding adult. Juvenile Gray-tailed has less extensive gray on the sides and belly and is more heavily marked above; some are tinged brownish above. The shorter nasal groove of the Gray-tailed can be difficult to accurately assess in the field. Best distinction is voice. **VOICE CALL:** Most frequent vocalization an ascending *too-weet,* similar to Common Ringed Plover. **STATUS & DISTRIBUTION** Breeds in Asia. Regular spring and fall migrant on outer Aleutians, Pribilofs, and St. Lawrence I. **VAGRANT:** Casual visitor to northern AK; accidental in fall to WA and CA.

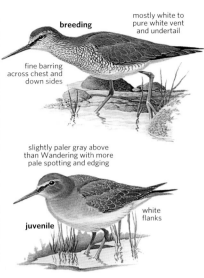

breeding

mostly white to pure white vent and undertail

fine barring across chest and down sides

slightly paler gray above than Wandering with more pale spotting and edging

juvenile

white flanks

WANDERING TATTLER *Tringa incana* WATA 1

One of the shorebirds of the rocky Pacific coast, the Wandering Tattler is admired more for its habits than for its coloration. A chunky gray bird, it can easily go unnoticed as it feeds on rocks exposed by the tide; bobbing as it moves. Its call can be heard over crashing waves, and it is often a clue in detecting its presence. It is generally seen singly or in small groups. Monotypic. L 11" (28 cm)

IDENTIFICATION Long wings and short legs contribute to a horizontal look. Wings extend noticeably beyond the tail tip. Bill is rather long and black, at times showing a pale base; the nasal groove is often visible, extending more than halfway to the tip. Legs are dull yellow, sometimes with a greenish tone. Overall, uniformly dark gray above and white below. **BREEDING ADULT:** Underparts heavily barred, all the way through the sides to the undertail coverts. White supercilium is flecked with gray and

is most prominent in front of the eye. Legs are at their brightest, usually yellow. Some first-year birds, with an incomplete plumage, can be difficult to age. **NONBREEDING ADULT:** Unbarred below with a gray wash on breast, sides, and flanks; uniform gray above. Legs duller, more greenish yellow. **JUVENILE:** Like nonbreeding bird, but with pale fringes on upperparts feathers or alternating light and dark spots along the edges of those feathers. **SIMILAR SPECIES** Wandering Tattler closely resembles Gray-tailed Tattler and is best distinguished by voice. In breeding plumage, Gray-tailed's finer barring does not extend beyond the flanks, with fewer bars on undertail coverts and flanks. Juvenile Gray-tailed is more heavily marked above, with more extensive spotting and notches. In all birds there is a

difference in the length of the nasal groove; Wandering's extends more than halfway to the bill tip. Use caution in assessing this feature—it is extremely hard to see. **VOICE CALL:** Rapid series of clear, hollow whistles, *twee, twee, twee;* all on one pitch. **STATUS & DISTRIBUTION** Fairly common. **BREEDING:** On or near gravelly stream banks. **MIGRATION:** Mostly along the coast or over the ocean. In spring, peaking early Apr.–mid-May in southern CA, mid-Apr.–early May on Pacific islands. Common by early May in southern AK, on breeding grounds by late May. In fall, adults depart July, peak late July–mid-Aug. in OR. Juveniles predominate late Aug.–Sept. Smaller numbers move through Oct. **WINTER:** CA to Ecuador, Pacific islands, and in Australasia. **VAGRANT:** Casual inland in the West during migration; accidental farther east (TX, IL, ON, MN). **POPULATION** No recent data.

breeding

extensively barred below

juvenile

gray flanks

breeding

heavily barred undertail

winter

dark gray flanks

rather short, dull yellow legs

SPOTTED REDSHANK   *Tringa erythropus*   SPRE ▪ 4

The Spotted Redshank, an Old World vagrant to North America, is unique among the *Tringa* for a molt of its underparts to blackish feathers and for its bill that droops toward the tip. Its behavior involves the expected picking and probing near the water's edge, although this species is an impressive wader and seems to take to swimming more readily than its congeners. Monotypic. L 12.5" (32 cm)
IDENTIFICATION Long, mostly dark bill droops at the tip. **BREEDING ADULT:** Dark plumage overall with white

spots on dark brown upperparts; black below, with females more likely to have a black breast or belly interrupted with white spots. Legs are long and dark red, although they can be almost black. Bill is long and black, with red at the base. **NONBREEDING ADULT:** Upperparts a rather uniform gray with a dingy whitish breast; the remaining underparts are white. Red-orange to red legs are much brighter than in a breeding bird. **JUVENILE:** Darker upperparts than nonbreeding bird and somewhat variable. Most juveniles are dark brown with white spots on the upperparts and underparts that are brownish gray; some juveniles can be quite dark, almost approaching breeding females in coloration. Supercilium

is bold and the legs are paler than on a nonbreeding bird. Base of the mandible is orange-red, not as bright as in a nonbreeding bird. **FLIGHT:** Dark above, with a white wedge extending up the back, and a barred rump and tail; light or dark below, depending on the plumage; and white wing linings. Legs and feet extend well beyond the tail.
SIMILAR SPECIES Unlikely to be confused with other sandpipers. Drooping bill, red legs (orangish in some yellowlegs), and white wedge up the back are unlike any other species. Common Redshank, other than having red legs, is not very similar: It is shorter, smaller billed, and browner above, and has a bold white secondary patch in flight.
VOICE CALL: Loud rising *chu-weet*. Similar to Pacific Golden-Plover's call; more strident than Semipalmated Plover's call.
STATUS & DISTRIBUTION Breeds on tundra and northern boggy or marshy openings within open forests in Eurasia. A very rare spring and fall visitor to Aleutian and Pribilof Islands. **VAGRANT:** Casual on both coasts in migration and winter; accidental elsewhere with scattered records across N.A.

juvenile

extensively barred below

extensive pale spotting above

long red-based bill in all plumages with slight droop near tip

breeding
black overall

very dark red legs

white wedge up back

winter

winter

orange legs

GREATER YELLOWLEGS   *Tringa melanoleuca*   GRYE ▪ 1

The tall and somewhat robust Greater Yellowlegs is a common bird and serves as an excellent example to learn about shorebird plumages. It feeds along the water's edge, although it will partially submerge itself and feed in belly-deep water. On shore, it may walk and pick at the surface, or run short distances when in an active feeding mode. It often bobs the front part of its body. Monotypic. L 14" (36 cm)
IDENTIFICATION Long yellow legs, a long neck, and a rather thick and often upturned dark bill. **BREEDING ADULT:** Throat and breast with heavy, blackish streaks; sides and belly with blackish spots and bars. On the upperparts, many feathers are black centered, and the tertials have dark barring. Legs tend to bright yellow with orange tones. Bill is usually dark, but it can have a slightly paler base to the lower mandible. **NONBREEDING ADULT:** Rather pale and

uniform, being gray above and white below. Upperparts have dark notches on the outer edges of feathers and pale fringes; neither are seen easily from a distance or if worn. Legs are paler and the bill usually has an obvious pale base. **JUVENILE:** Like nonbreeding, but upperparts have bold white spots on scapulars and coverts and lack the dark marks of the adult; neck has dark streaking. Bill base averages more pale gray than in adults. **FLIGHT:** Rather plain upperparts with gray wings and white rump and whitish gray tail; legs and feet extend beyond the tail.
SIMILAR SPECIES Similar to Lesser Yellowlegs, but they are usually easy to distinguish when together. Greater Yellowlegs has 50 percent more bulk, and is taller and more robust. Its bill is also thicker and longer (longer than the head) and often upturned, compared to a narrower, shorter, and straighter bill of Lesser Yellowlegs.

Adult Greater molts earlier in fall, so early migrants might have dropped flight feathers; Lesser molts after migration. Greater is more heavily and extensively marked below in breeding plumage and is streaked on neck (not washed) in juvenal plumage. Grayish-based bill differs from Lesser's all-dark bill, except in breeding plumage. Structure and

differing calls are the best methods of identifying these species. Greater Yellowlegs usually gives three or more *tew* notes, Lesser more commonly gives two; the tone is more important than the number of notes. Common Greenshank, a vagrant from the Old World, is similar in size and shape.
VOICE CALL: Loud, slightly descending series of three or more *tew* notes. This ringing call can be heard for long distances. **SONG:** Rolling *whee-oodle.*
STATUS & DISTRIBUTION Common. **BREEDING:** Muskeg and open or sparsely wooded areas. **MIGRATION:**

Migrates on a broad front, but in most areas, in smaller flocks than Lesser Yellowlegs. In spring, usually a week or so earlier than Lesser Yellowlegs. More southerly states see peak passage late Mar.–mid-Apr., but numbers seen into early May. Great Lakes and mid-Atlantic peak mid-Apr.–mid-May, might be seen as early as early Mar. First breeding arrivals late Apr. or early May, depending on latitude. In fall, the migration is protracted late June–late Nov. First adults arrive in late June,

but most adults arrive late July–late Aug., although southern states (e.g., TX) are a little later. Juveniles account for the majority of migrants by late Aug., peaking most areas late Aug. into early Oct. Some remain into winter if conditions are mild enough; this is one of our hardiest shorebirds. **WINTER:** Coastal and, unlike Lesser, some inland areas in the U.S. south to S.A. **VAGRANT:** North of breeding grounds. Casual in Europe, primarily in fall, Japan, and S. Africa.
POPULATION Numbers appear stable.

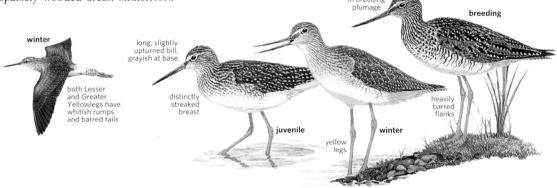

COMMON GREENSHANK *Tringa nebularia*   COMG   3

This species is a vagrant from the Old World. It might be overlooked since it is similar to the Greater Yellowlegs, but it is readily identifiable. Monotypic. L 13.5" (34 cm)
IDENTIFICATION Rather large and robust *Tringa* with a long neck. Black tip of the relatively long, slightly upturned bill contrasts with a gray base. Greenish legs. **BREEDING ADULT:** Brownish gray above, with black centers to many scapulars. White under-

parts, with light to moderate streaking on the breast and neck. **NONBREEDING ADULT:** Streaking absent or restricted, with a rather pale head and foreneck. Upperparts lack black; feathers have dark and light notches on outer feather edges. **JUVENILE:** Like nonbreeding, but boldly marked above; white edges create a distinctly striped appearance. **FLIGHT:** A white rump and a wedge continuing up into the lower back is distinctive.

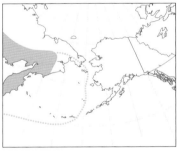

SIMILAR SPECIES Looks similar to Greater Yellowlegs, but the adult is less heavily streaked and the juvenile has striped, not spotted, upperparts. In all plumages, the legs are greenish, and in flight, a white wedge extends up to middle of back.
VOICE CALL: A loud *tew-tew-tew* on same pitch is very similar to Great Yellowlegs, but does not descnd in pitch.
STATUS & DISTRIBUTION Eurasia. Rare migrant to Aleutian and Pribilof Is., very rare to St. Lawrence I., casual to eastern Canada, and two CA records. Most records from spring, but some in fall and winter.

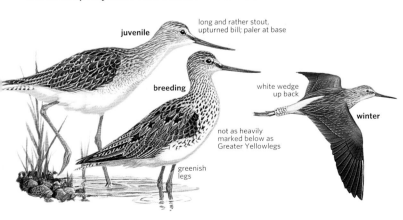

WILLET *Tringa semipalmata*   WILL ▪ 1

The Willet's shape—large, plump, rather long legs, a thick bill—is unique for a shorebird. It picks and probes for prey, feeding along the shore, sometimes in the water, and often bobs its body like a yellowlegs. It often occurs in small flocks during migration or while feeding; it roosts in large groups with other large shorebirds. Polytypic (2 ssp; both in N.A.—see sidebar below). L 15" (38 cm)

IDENTIFICATION Generally gray or brownish gray above and whitish below. Stout, straight, dark bill, gray or brown at base. Gray legs, with possible dull olive or blue tones. **BREEDING ADULT:** Heavily streaked on neck, barred on breast and sides; white belly. Upperparts might have extensive black feathers. **NONBREEDING ADULT:** In general, unmarked gray above and white below. **JUVENILE:** Brownish upperparts, with obvious white or buff spots and fringes. **FLIGHT:** Striking black-and-white wing pattern diagnostic in all plumages.

SIMILAR SPECIES Unmistakable in breeding plumage and in flight. A roosting nonbreeding bird might recall a yellowlegs, but the Willet's legs are always dull, the bill larger, and the shape plumper.

VOICE CALL: In nesting areas, *pill-will-willet,* higher and faster in eastern birds. Also *kip* or *yip,* sometimes repeated.

short pinkish-based bill
early Mar. in molt
western
eastern
heavily barred below
brownish cast to upperparts
breeding
eastern *semipalmata*
contrasting scapulars
juvenile

striking wing pattern in flight
winter

western *inornata*
paler above
western Willets are larger and have a longer, darker bill
less heavily marked below
breeding
worn breeding
plainer and paler upperparts
juvenile

winter

Separation of Eastern and Western Willets

It is possible that the two Willet subspecies—the western *inornata* and the eastern *semipalmata*—may be different species.

The "Eastern Willet" is smaller by approximately 10 percent with a shorter bill and shorter legs, by up to 15 percent. It is more heavily barred below in breeding plumage with more pinkish at the base of the bill; *inornata* has more gray on the breast in nonbreeding plumage. A juvenile "Eastern's" wing coverts are less noticeably marked, compared to a "Western."

The "Western" is paler in all plumages. With its longer legs, the "Western" is more apt to be seen wading in belly-deep water. Its bill is longer, slightly thinner at the base, and more likely upturned; the combination of longer legs, wading in water, and an upturned bill might recall a godwit more than eastern birds.

"Eastern" Willet, breeding (NY)

"Western" Willet, breeding (TX, Apr.)

The white wing stripe is broader in the "Western."

Subspecies can be differentiated by their songs; all calls and songs are lower pitched in the "Western Willet."

And finally, distribution is important: "Eastern Willets" are not recorded inland, nor are they present in winter; their winter range is poorly known but presumably is mainly in S.A. ▪

STATUS & DISTRIBUTION Fairly common. **BREEDING:** Freshwater and alkaline marshes in the West; salt marshes in the East. **MIGRATION:** In spring, "Western Willet" (*inornata*) occurs by late Mar. in inland parts of southern states. Peak late Apr.–early May at Great Lakes. "Eastern" (*semipalmata*) generally returns to breeding areas along Gulf Coast and FL in Mar.; in mid-Atlantic mid- to late Apr. In fall, *inornata* starts moving in late June; adults peak mid-July–mid-Aug., juveniles numerous by late July; most migrants through by early Sept., with stragglers into Oct., rarely later. Most

semipalmata depart early, and arriving *inornata* outnumber them by Aug. **WINTER:** Subspecies *inornata* winters on coastal beaches and at the Salton Sea and Owens L., CA. **VAGRANT:** Casual to YK, northern BC, Hudson Bay, and southern QC. Accidental to Europe. **POPULATION** Hunting in the 1900s devastated Atlantic populations, but they seem to have rebounded.

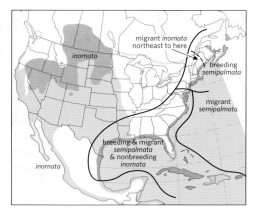

LESSER YELLOWLEGS *Tringa flavipes*   LEYE   1

This dainty version of the Greater Yellowlegs looks almost delicate in comparison. It picks its way along the edge of the water in a deliberate fashion, taking prey with its thin, straight bill, and sometimes bobs the front part of its body in a jerky fashion. Monotypic. L 10.5" (27 cm)
IDENTIFICATION Long yellow legs and a short, thin, dark bill. Generally gray above, white below. **BREEDING ADULT:** Breast is finely streaked, and sides and flanks show fine, short bars. Legs can have an orange tone in high breeding plumage. **NONBREEDING ADULT:** Upperparts are gray or gray-brown, with alternating dark and light notches on the feather edges; these are difficult to see from a long distance or when worn. Underparts are white, except for the dingy breast, which has obscure streaks. **JUVENILE:** Similar to nonbreeding, except upperparts have bold white spots on coverts and scapulars. **FLIGHT:** Gray wings are rather uniform; white rump and whitish tail. Legs and feet extend well beyond the tail.
SIMILAR SPECIES Similar to Greater Yellowlegs but smaller—a comparison

easily made when the two are together. Lesser's all-dark bill is shorter, thinner, and straighter. The calls are different: Lesser utters fewer notes, which are, more importantly, not as long or as piercing as Greater's and do not descend in pitch. Stilt Sandpiper can look similar in flight, but its bill is longer and curved. They usually feed differently, especially when with their frequent companions—dowitchers. Wilson's Phalarope can be superficially similar outside of breeding plumage, but feeds differently; it has smaller, shorter legs, is paler above, and has a white breast.
VOICE CALL: *Tew* notes, higher and shorter than Greater's; notes usually given singly or double, but can be given in threes. **SONG:** *Wheedle-ree,* given with an undulating flight display.
STATUS & DISTRIBUTION Common. **BREEDING:** Tundra or taiga woodland openings, generally farther north than the Greater Yellowlegs. **MIGRATION:** Common in the Midwest and fairly common in the East;

uncommon in far West. More likely to be seen in large numbers (e.g., hundreds) than Greater Yellowlegs, especially during spring. Spring migration is usually a week later than for Greater Yellowlegs. Southern states peak in Apr.; Great Lakes and mid-Atlantic peak late Apr.–mid-May. In fall, the first adults move south in late June and peak late July–mid-Aug. Juveniles generally first arrive in early Aug. and peak late Aug.–Sept. Much more numerous in fall than in spring along the Pacific coast. Lessers have a more defined peak than Greaters and have generally left by mid-Oct., with lingering birds into Nov., exceptionally later. **WINTER:** Most winter in S.A. or C.A. Local along the coasts, especially in TX, but always outnumbered by Greaters. Away from the coast, use caution when identifying winter Lessers. **VAGRANT:** North and west of breeding range, primarily May–June; annual to Europe, primarily Aug.–Oct.; fall records from Japan and China, winter records from New Zealand and Australia.
POPULATION Stable.

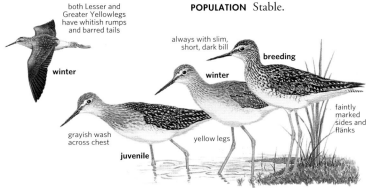

both Lesser and Greater Yellowlegs have whitish rumps and barred tails

always with slim, short, dark bill

breeding

winter

winter

faintly marked sides and flanks

grayish wash across chest

yellow legs

juvenile

MARSH SANDPIPER Tringa stagnatilis   MASA ■ 5

The Marsh Sandpiper is a vagrant from the Old World. It is a small and rather delicate *Tringa*, with very long legs and a long, thin bill. Its gait is rather deliberate as it picks its way around watery locales, and it is more stiltlike in many ways. Monotypic. L 8.5" (21 cm)

IDENTIFICATION Gray-brown above and white below; greenish legs; long, thin, dark bill, almost needlelike. Face pattern different from most *Tringa*, lacking a distinct loral stripe and having a supercilium that extends beyond the eye. **BREEDING ADULT:** Irregular dark markings on plain brown scapulars create somewhat mottled upperparts, alternating black and brown. Tertials and many wing coverts are brown, with black barring. Head, neck, and breast are liberally streaked or spotted with brown, largely eliminating the supercilium. Remaining underparts are white, although some light spotting or barring might be present on the flanks and undertail coverts. **NONBREEDING ADULT:** Plain gray-brown on the upperparts, with the head and back slightly paler than the wings. Scapulars and wing coverts have thin, whitish borders when fresh. Underparts are white below, generally lacking streaks. **JUVENILE:** Similar to nonbreeding, except for darker upperparts, with bold white or buff spots on scapulars and coverts. **FLIGHT:**

Gray upperparts, with slightly darker wings, contrast with the white tail, rump, and lower back. Legs extend well beyond the tail in flight.

SIMILAR SPECIES Only vaguely similar to Lesser Yellowlegs, but the two species share long legs and straight, thin bills. Marsh Sandpiper has a thinner, longer bill, giving it a more dainty look. In flight, the white on the lower back is more like that of Common Greenshank, whereas Lesser Yellowlegs has white restricted to the rump and tail. Somewhat similar to both Stilt Sandpiper and Wilson's Phalarope in nonbreeding plumage. Stilt Sandpiper has a slightly downcurved bill and, in flight, a white wing stripe

and no white on the back. Wilson's Phalarope has a rather needlelike bill and no wing stripe, but it has a dark back. It also looks more compact as its legs are much shorter than those of Marsh Sandpiper; as a result, its legs do not extend as far beyond the tail in flight.

VOICE CALL: *Tew,* repeated like a Lesser Yellowlegs.

STATUS & DISTRIBUTION VAGRANT: Casual from western and central Aleutians, and the Pribilofs, AK; accidental to southern CA (near Mecca, Oct. 26, 2013); accidental to Ensenada, Baja California, Mexico (Oct. 2011).

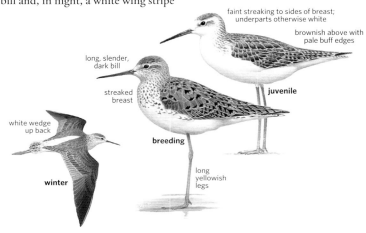

faint streaking to sides of breast; underparts otherwise white

brownish above with pale buff edges

long, slender, dark bill

streaked breast

juvenile

white wedge up back

breeding

long yellowish legs

winter

WOOD SANDPIPER Tringa glareola   WOSA ■ 2

The Wood Sandpiper is a vagrant from the Old World, somewhat intermediate between Lesser Yellowlegs and Solitary Sandpiper. Along with Green and Solitary Sandpipers, Wood Sandpiper is a more compact *Tringa*, with obviously shorter legs. It frequents edges of wetlands, often where

there is grass. It methodically picks food from the surface as it ambles along. It is often first detected when flushed from cover: It rises up steeply while giving its shrill calls; in spring, it frequently breaks into the full display vocalizations uttered in a hovering display flight. Monotypic. L 8" (20 cm)

IDENTIFICATION Dark upperparts are heavily spotted, and underparts are white, with brownish streaks on breast and neck. Has prominent whitish supercilium, particularly bold in front of the eye, which contrasts with a dark lore. An eye ring is usually obvious. Bill is rather short and dark;

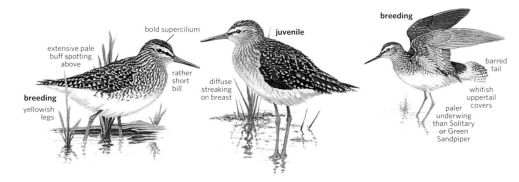

bold supercilium

juvenile

breeding

extensive pale buff spotting above

rather short bill

diffuse streaking on breast

barred tail

whitish uppertail covers

breeding

yellowish legs

paler underwing than Solitary or Green Sandpiper

legs vary from yellow to greenish yellow. **BREEDING ADULT:** Upperparts are boldly checked with white; the breast and neck are heavily streaked. Bill often shows a pale base; legs more often with yellow tone. **NONBREEDING ADULT:** Duller than breeding bird, with upperparts lacking the black-centered feathers. In fresh plumage,

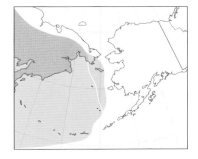

there are pale fringes on most scapulars and white spots on the coverts. **JUVENILE:** Like nonbreeding adult, but upperparts are darker, with more extensive spotting and more streaking on the breast. Crown is distinctly dark, set off from the supercilium. Legs are paler, more greenish. **FLIGHT:** Brown above and whitish below, with plain wings, a white rump, and a whitish tail, moderately barred, including the outer tail feathers. Wing linings are pale gray.
SIMILAR SPECIES Wood Sandpiper resembles a Lesser Yellowlegs in plumage coloration, but its shape is similar to Green and Solitary Sandpipers. Solitary Sandpiper has dark central rectrices and is also darker, more gray, or grayish olive, as opposed to brown. Wood Sandpiper is distinguished from a Green Sandpiper by its

paler wing linings, smaller white rump patch, and more densely barred tail. Compared to Lesser Yellowlegs, note Wood Sandpiper's eye ring, shorter bill, and the presence of a supercilium. In addition, Lesser Yellowlegs is taller, with more leg extending beyond the tail in flight.
VOICE CALL: Loud, sharp whistling of three or more notes. Similar to the call of Long-billed Dowitcher.
STATUS & DISTRIBUTION Eurasia. Casual breeder on the outer Aleutian Islands. **MIGRATION:** Fairly common spring migrant to the western and central Aleutians; uncommon on the Pribilof Is.; rare on St. Lawrence I. Rare in fall. **VAGRANT:** Casual to BC, OR, CA, and northeastern N.A. Nearly all records are of juveniles in fall, except for one at Ridgecrest, Kern County, CA (May 23–24, 2007).

COMMON REDSHANK *Tringa totanus*   COMR ▪ 5

This species is a vagrant from the Old World, where it is one of the most numerous and familiar shorebirds of northwestern Europe. Multiple birds have occurred at the same location and time, suggesting that the right weather pattern (a nor'easter) is likely to deposit more Common Redshanks in the future. Polytypic (6 ssp.; probably *robusta* in N.A.). L 11" (28 cm)
IDENTIFICATION Slightly smaller than its relative, the Spotted Redshank. Bill is short and straight, almost stout, and red at the base. Legs are of medium length and vary from orange to red. Common Redshank is generally brownish above and white with streaking below. **BREEDING ADULT:** Brown above with heavy streaking; whitish background below, usually with streaks on the breast that turn into barring on the flanks. White eye ring is usually evident, but dark lores are obscured.

NONBREEDING ADULT: Rather plain brown above, with some feathers having dark notches on the feather edges; these notches are less evident than on other *Tringa* and might not be visible without good views. Breast and sides are brownish, with some spots or streaks; belly and undertail coverts white, with some barring. **JUVENILE:** Like nonbreeding, but more heavily marked above, with scapulars and coverts having bold white spots. The breast is whiter and streaked; remainder of underparts paler. Legs paler, with a yellow tone to the orange; bill has minimal color to the base of the mandible. **FLIGHT:** A white dorsal wedge is visible; particularly striking is a distinctive broad white trailing edge to the secondaries and inner primaries that looks like a white patch.

SIMILAR SPECIES Given a good view, no other species looks like a Common Redshank. The larger, longer-legged, and longer- and thinner-billed Spotted Redshank is grayer in all plumages (less brownish) and is much paler in winter plumage. Lesser and Greater Yellowlegs, with reddish-orange legs, are not infrequently confused with Redshanks.
VOICE CALL: Loud *twek-twek* and a mournful, liquid whistle, *teu,* with a distinctly lower ending.
STATUS & DISTRIBUTION Breeds in Eurasia, as close as Iceland. Breeding range is more southerly than its congeners, and wintering range is more northerly; the combination makes the Common a less likely candidate for a wide pattern of vagrancy; however, it is casual to NF (five recs.; four in spring 1995). Numerous records for Greenland. Even though it breeds from portions of the Russian Far East, unrecorded for AK.

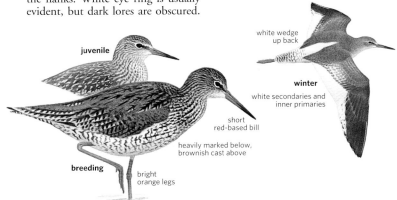

juvenile

white wedge up back

winter

white secondaries and inner primaries

short red-based bill

heavily marked below, brownish cast above

breeding

bright orange legs

These waders have long legs and, usually, long bills. Most plumages are brown with little change between seasons, ages, or sexes. Females tend to be larger than males in most species, with larger bills. The bills grow over the first year. Most species stay near water (less so during breeding season), feeding on crabs, worms, and more, but insects are also a key part of the diet for some species.

UPLAND SANDPIPER *Bartramia longicauda* UPSA ■ 1

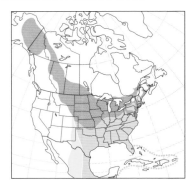

The Upland Sandpiper is typically found in fields, where its head and neck are visible above the grass; on the breeding grounds, it perches on posts. It often bobs the rear portion of its body, and, upon landing, will typically hold its wings up momentarily. It is one of few shorebirds that seems disinterested in water. Monotypic. L 12" (30 cm)

IDENTIFICATION Small, aberrant curlew is an elongate species with a long, thin neck; long legs; long tail; and

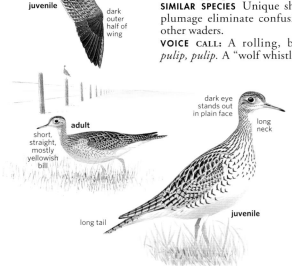

long wings. Bill is short and straight; legs and bill are both yellow; the bill is dark at the tip and atop the ridge. The head is small and dove-like, with large, dark eyes. Generally brown overall; dark upperparts with buff edges; brown streaks on the foreneck that turn into chevrons on the breast and flanks. Belly and undertail coverts white. **ADULT:** Upperparts look barred, whereas the tertials are entirely barred. **JUVENILE:** Can look very similar to adult. Upperparts look less barred due to pale fringes on dark feathers; tertials are edged in buff. Some juveniles have a rich tawny color to the sides of the neck in fresh plumage. **FLIGHT:** Blackish primaries contrast with mottled brown upperparts.

SIMILAR SPECIES Unique shape and plumage eliminate confusion with other waders.

VOICE CALL: A rolling, bubbling *pulip, pulip.* A "wolf whistle," given

on breeding grounds, sounds distinctly human.

STATUS & DISTRIBUTION Rare to fairly common. **BREEDING:** Tallgrass prairies (less common in short-grass prairies); increasingly restricted to airports in the East. **MIGRATION:** More likely found in dirt or short grassy fields, including sod farms. In spring, the first arrivals appear mid-Mar., with peak late Mar.–mid-Apr. in south, mid-Apr.–early May farther north. Fall movement begins in July, with peak late July–mid-Aug. in north, Aug.–mid-Sept. farther south, stragglers to Oct., rarely to Nov. **WINTER:** Primarily southern S.A. **VAGRANT:** Casual on West Coast and in the Southwest in migration; casual in Europe in fall. Accidental in Australia and Guam.

POPULATION Seriously declining in the eastern parts of its range. The loss of habitat due to grassland conversion to agricultural fields has had a greater impact on the population than the hunting pressures of the 19th century.

LITTLE CURLEW *Numenius minutus* LICU ■ 5

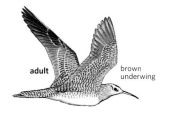

This aptly named, diminutive curlew is a vagrant from Asia. Monotypic. L 12" (30 cm)

IDENTIFICATION Small curlew with a short, slightly curved, dark bill with a pale base. Legs are grayish but variable; they can be flesh-colored or greenish. The dark crown (occasionally showing pale mid-crown) is offset by a pale buff supercilium. A dark line or patch behind the eye is interrupted in front of the eye, leaving the lores at least partially pale. Upperparts are composed of dark scapulars and coverts that are broadly edged buff. Below, there are faint streaks on a buff breast, and slight barring on the sides and flanks.

It is very difficult to age this species other than in fall, when assessing the degree of feather wear will reveal juveniles, who look fresh in early fall. On adults, the tertials are usually distinct, showing brown with narrow, dark bars, whereas juveniles show dark feathers, almost black,

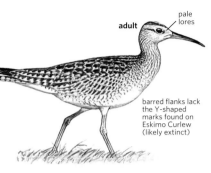

adult
pale lores

barred flanks lack the Y-shaped marks found on Eskimo Curlew (likely extinct)

with buff notches along the edge of the feather. However, this feature appears sufficiently variable to limit its use as a diagnostic tool. **FLIGHT:** It appears uniform brown above, with only slightly darker primaries. The underwings are brown. The toes barely extend beyond the tip of the tail.

SIMILAR SPECIES Similar to the larger Whimbrel, but with a shorter and only slightly curved bill, mostly pale lores, and buffy tones. Size and short bill of Little Curlew are like Upland Sandpiper, but note the curve to Little Curlew's bill and its stronger head pattern. If you are lucky enough to encounter a Little Curlew, you must consider the likely extinct Eskimo Curlew. Eskimo Curlew has a bolder head pattern, including dark lores, more

heavily barred breast and flanks, with bold Y-shaped marks on the flanks (just slight brown barring on Little Curlew); also has overall more cinnamon coloration, including the wing linings.

VOICE Vagrants here are mostly silent. **CALL:** Musical *quee-dlee* and a loud *tchew-tchew-tchew*.

STATUS & DISTRIBUTION BREEDING: Russian Far East **WINTER:** Mainly in northern Australia. **VAGRANT:** Casual in fall to coastal CA (four accepted records from early Aug. to mid-Oct. that involve both juveniles and adults, the last of which was Sept. 1994); two spring records at Gambell, St. Lawrence I., and one spring sighting (flight only) for WA. **POPULATION** Small and disjunct; threatened due to habitat loss in migration and winter range.

ESKIMO CURLEW *Numenius borealis*   ESCU ▪ 6

The Eskimo Curlew is probably extinct; the last confirmed record was of an adult female shot by a hunter in Barbados on Sept. 4, 1963 (specimen at the Academy of Natural Sciences of Philadelphia). There have been several sightings in the intervening years, but none with definitive documentation. Given the species' remote summer and winter habitats, it is conceivable, but not likely, that this species still exists. Monotypic. L 14" (36 cm)

IDENTIFICATION Like Little Curlew or a small Whimbrel, but its head pattern is more muted than the Whimbrel and the central crown stripe is indistinct or lacking entirely. The lores are entirely dark, like Whimbrel, but unlike Little Curlew. Upperparts are dark with buff edges; underparts are buff, with streaks on the neck and dark chevrons down the sides. The wings are long and project past the tail. **FLIGHT:** Entirely brown above, with slightly darker primaries. The underwing is cinnamon but looks darker in flight.

SIMILAR SPECIES Differs from Whimbrel and a buffy Bristle-thighed Curlew in its smaller size, smaller, less curved bill, more muted head pattern, and cinnamon underwings. (Bristle-thighed Curlew has a cinnamon-rust rump and tail.) This identification would be more problematic with a fresh juvenile Whimbrel, as its bill can be substantially shorter than an

adult's. In comparison with the Little Curlew, Eskimo is slightly larger, with a more curved bill, completely dark lores, more heavily marked flanks, and darker underwings.

VOICE CALL: Poorly known, but includes clear whistles.

STATUS & DISTRIBUTION This species is probably extinct, although it is federally listed as endangered. **BREEDING:** Nested on Arctic tundra and wintered in S.A., but its full, former breeding range is not known. The only nests were found in the vicinity of Fort Anderson, NT, during the summers from 1862–1866. **MIGRATION:** This species used to be found in large flocks, with northbound migration being primarily mid-continent, and southbound birds being largely transoceanic, taking the Atlantic Ocean to S.A. Their stopover points hosted large numbers. During spring movements, migrants were found Mar.–Apr. In fall, Eskimo Curlews would stage along Canada's Atlantic coast (esp. Labrador), fattening up on berries for their transoceanic flights, typically Aug.–Sept. Records on the U.S. Atlantic coast were rather sparse, particularly south of New England. **VAGRANT:** Records from AK and Russian Far East; the British Isles in the 19th century.

POPULATION This species, once so numerous, declined precipitously during the two decades following the Civil War. By 1940, it was thought possibly extinct, but then

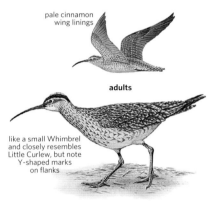

pale cinnamon wing linings

adults

like a small Whimbrel and closely resembles Little Curlew, but note Y-shaped marks on flanks

up to two were photographed on Galveston I., TX, in Mar. and Apr. between 1959 and 1962. The two primary reasons for its decline are 19th-century market hunting pressures, which impacted almost all shorebirds, and, uniquely for the Eskimo Curlew, the extinction of a key prey item on the Great Plains (a grasshopper). The former breeding grounds are hard to access. The two most likely places to search for this species during migration are along the upper Texas coast from late March through mid-April and along the Labrador coast August through September. However, given the highly social nature of this species during migration and winter, it is unlikely to have survived in such small numbers.

WHIMBREL *Numenius phaeopus* WHIM ◼ 1

The Whimbrel is most frequently encountered along the coasts during migration. It forms large flocks away from the breeding grounds. Whimbrels pick at the ground for food while walking and often form large flocks with other large shorebirds, such as godwits and curlews. Polytypic (3 ssp.; all in N.A.). L 17.5" (45 cm)

IDENTIFICATION Medium-size curlew, rather cool brown in color, with bold head stripes. Black bill, with a little pale at the base, is moderate in size and curvature; legs are blue-gray. Crown has a broad dark lateral stripe, with a pale mid-crown stripe. Dark eye line, including the lores, sets off the paler supercilium. Upperparts are dark brown with pale notches along many of the scapulars and coverts. Neck and breast are lightly streaked, with barring on sides; remainder of underparts are dingy whitish buff. **JUVENILE:** Can be difficult to age this species, but juveniles typically have more extensive pale edging and notches on the upperparts and a shorter bill. There is often a greater contrast between the scapulars and coverts on young birds compared to adults. **FLIGHT:** Uniform cool brown above, including dark rump; dark underwings. Feet do not project beyond the tail.

GEOGRAPHIC VARIATION North American *hudsonicus* is described above. European nominate *phaeopus* (casual to East Coast) has a white rump and underwings and coarser dark markings on the breast; Asian *variegatus* (rare to casual in Pacific region) has a whitish, variably streaked rump and underwings. Both *phaeopus* and *variegatus* share a whiter ground color to the underparts, compared to the browner ground color of *hudsonicus*. Many characters of *phaeopus* and *variegatus* overlap; identification should be made carefully. Presume that a vagrant on the East Coast is *phaeopus* and one on the West Coast is *variegatus*.

SIMILAR SPECIES Compared to Long-billed Curlew, Whimbrel is smaller, with a smaller bill and bold head stripes, and its plumage never matches the warmth of the Long-billed. In Alaska, Bristle-thighed Curlew is similar. It has a bold head pattern and, when worn, might lose some of its buff coloration on the body as well as the buff notches to the upperparts, increasing the potential for confusion with the Whimbrel. But even a worn Bristle-thighed has a distinctively buffier tail and rump than the uniform brown of a *hudsonicus* Whimbrel, or compared to the dull gray-and-white barring of a *variegatus* Whimbrel. See also Little Curlew.

VOICE CALL: Series of hollow whistles on one pitch, *pi-pi-pi-pi.*

STATUS & DISTRIBUTION Fairly common. **BREEDING:** Open tundra. **MIGRATION:** Starts late Mar., exceptionally in early Mar. Peak in most southern states from mid- to late Apr.; peak in mid-Atlantic and Great Lakes states mid-May; western states slightly earlier. In fall, adults move first, typically early to mid-July. Juveniles peak mid-Aug.–mid-Sept., stragglers in Great Lakes and mid-Atlantic to mid- or late Oct.

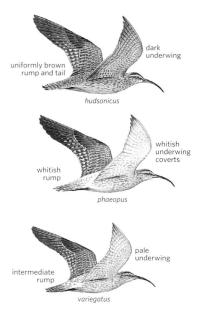

uniformly brown rump and tail

dark underwing

hudsonicus

whitish rump

whitish underwing coverts

phaeopus

intermediate rump

pale underwing

variegatus

In West, much scarcer in the interior in fall. **WINTER:** Beaches and coastal wetlands from CA and VA south to S.A. **VAGRANT:** European *phaeopus* is casual to East, mainly Atlantic coast. Asian *variegatus* is a rare migrant on islands in western AK, casual elsewhere in Pacific region. N.A. *hudsonicus* is casual in Europe, Africa, and Australasia.

POPULATION Stable.

juvenile *hudsonicus*

strong dark lateral crown and eye stripes

decurved bill

adult *hudsonicus*

BRISTLE-THIGHED CURLEW *Numenius tahitiensis* BTCU ◼ 2

Lucky is the birder who finds the Bristle-thighed Curlew on migration, as it is rarely found away from its remote Alaskan breeding grounds or its winter home on ocean islands. The species' common name derives

from the bare shafts on the thigh feathers; reasons for this characteristic are not clear. Monotypic, although there are mensural differences between the two breeding populations. L 18" (46 cm)

IDENTIFICATION Medium-size curlew, like a bright Whimbrel. Bold head stripes include a dark eye line that connects to the bill and a pale supercilium, bordered above by a broad, dark lateral crown stripe. Upperparts

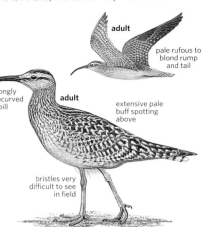

adult

pale rufous to blond rump and tail

strongly downcurved bill

adult

extensive pale buff spotting above

bristles very difficult to see in field

are dark brown with bold, buff edges and notches on the scapulars and coverts. Underparts are buff, rather bright, with streaking on the neck and breast, and bars on the sides. Tail and rump are strong buff, with dark bars on the tail. Worn adults can lose much of the buff edges to the upperparts, and the underparts can become faded; however, rump and tail remain distinctly buffy. Bill is dark, but much paler on the lower mandible than on most curlews, extending almost to the tip; legs are blue-gray. Stiff feathers on the thighs and flanks are hard to see except at close range and in good light. **JUVENILE:** Like adult except in early fall, when plumage is uniform

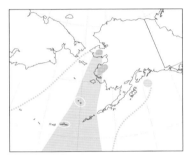

and fresh, compared to more worn adult. As winter progresses, juveniles look darker than adults, retaining buff coloration on only rump and tail. They generally do not return to the breeding grounds for three years. **FLIGHT:** Brown above, with noticeable buff edges. More important, the bright buff rump and tail are usually evident. **SIMILAR SPECIES** Possible confusion with Whimbrel, but Bristle-thighed is brighter buff above and below, and has a bright buff rump and tail. Calls different. **VOICE CALL:** Loud whistled *chu-a-whit,* somewhat recalling a Black-bellied Plover, but louder, with more-distinct syllables; somewhat humanlike. On the breeding grounds, loud, sharp whistles accompany impressive display flights. **STATUS & DISTRIBUTION** Local and uncommon. **BREEDING:** Endemic AK breeder; nests on hills with uneven, hummock tundra. **MIGRATION:** Spectacular flights covering 2,000–4,000 miles with minimal or no stops. In spring, through the Pacific islands in Apr., clearing out of HI early to mid-May on average. Arrive in AK early to mid-May, but as early as late

Apr. In fall, adults stage July–Aug. in AK, where they fatten up on berries prior to moving south. Juveniles move south mid-Aug.–early Sept. **WINTER:** Pacific islands. **VAGRANT:** Casual to the Pacific coast. A weather-related occurrence accounted for the 13 birds found May 6–25, 1998, in northern CA, OR, and WA. Prior to this event, there were eight records (6 spring, 2 fall) for OR, WA, and BC; the two fall records (OR) are sight records. **POPULATION** The population is vulnerable because of its relatively small size and restricted range.

FAR EASTERN CURLEW *Numenius madagascariensis* FECU ▪ 4

The Far Eastern Curlew, a vagrant from Asia, is the largest curlew. Monotypic. L 25" (64 cm) **IDENTIFICATION** Dull coloration, generally brown with streaks, but its size, and extremely long, curved bill are noteworthy. Face pattern is indistinct, with slightly darker lores being the most noticeable feature. Upperparts are grayish brown with warmer tints on occasion; the neck and crown are streaked. Underparts are pale brown with dark streaks on the neck and breast, extending

down the sides. **JUVENILE:** Fresh birds differ from adults by their buffier breast and belly, with less streaking, and they are also more boldly marked above, with buffy fringes to all feathers. As feathers wear, juveniles can be difficult to separate from adults. **FLIGHT:** Wing linings are dingy with dark barring, but in many lights look gray or gray-brown. Upperparts are uniform brown with the rump the same color as the back.

SIMILAR SPECIES Large size and lack of head pattern leave the Long-billed and Eurasian Curlews as the only identification challenges. The more likely problem is Long-billed along the Pacific coast, where a vagrant Far Eastern might occur. Long-billed is slightly smaller, with warm cinnamon tones on the body and, in flight, on the underwings. Far Eastern has heavier streaking than Long-billed. Eurasian and Far Eastern Curlews look very similar at rest (Eurasian slightly paler), but in flight note Eurasian's white rump, lower back, and underwings. **VOICE CALL:** A *curlee,* similar to other curlews. **STATUS & DISTRIBUTION BREEDING:** Marshes and wet meadows of Russian Far East. **WINTER:** Southeast Asia and Australasia. **VAGRANT:** Casual in spring and early summer on the Aleutian and Pribilof Islands. Accidental coastal BC (Sept. 1984). **POPULATION** Vulnerable; numbers are declining due to loss of habitat during migration and on winter grounds.

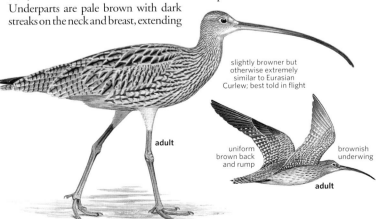

slightly browner but otherwise extremely similar to Eurasian Curlew; best told in flight

adult

uniform brown back and rump

brownish underwing

adult

SLENDER-BILLED CURLEW *Numenius tenuirostris*   SBCU ▦ 6

This Old World species is likely extinct. Monotypic. L 15" (39 cm)
IDENTIFICATION The size of a Whimbrel, but its plumage is more like a Eurasian Curlew. Rump is white; tail is white with dark bars. White underwings are visible in flight. Bill is slender, even at the base.
SIMILAR SPECIES Not particularly similar to Whimbrel or Long-billed Curlew—the expected North American curlews. Slender-billed has a short bill and bright white underparts, similar to Eurasian Curlew,

but Eurasian has a longer and thicker bill, and has chevron-shaped marks rather than round black spots on its flanks.
STATUS & DISTRIBUTION BREEDING: The only nests were found in southwestern Siberia near Tara, north of Omsk, between 1914 and 1924. WINTER: Thought to winter in parts of western and northwestern Africa VAGRANT: A specimen record from Crescent Beach, Lake Erie, ON, "about 1925."
POPULATION Critically endangered, if not already extinct, due to habitat

loss and hunting. The last fully credible record was from 1995 in Morocco.

white rump and underwings are visible in flight

like a small Eurasian Curlew, but note heart-shaped spots on flanks

adult

EURASIAN CURLEW *Numenius arquata*   EUCU ▦ 4

This species is a vagrant from Eurasia. Polytypic (2 ssp.; nominate in N.A.). L 22" (56 cm)

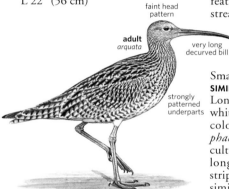

faint head pattern

adult
arquata

very long decurved bill

strongly patterned underparts

IDENTIFICATION BREEDING ADULT: Dark upperparts with black-centered feathers, edged with buff. Heavily streaked below on a whitish background; streaks extend to flanks. Long, strongly downcurved bill. NONBREEDING ADULT: Grayer than breeding. JUVENILE: Smaller bill, buffier above.
SIMILAR SPECIES Distinguished from Long-billed Curlew by white rump, white wing linings, and cooler brown color. Separation from a vagrant *phaeopus* Whimbrel is more difficult; Eurasian Curlew is larger, has a longer bill, and lacks dark head stripes. Far Eastern Curlew is very similar, but Eurasian is slightly paler

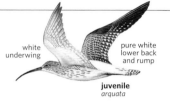

white underwing

pure white lower back and rump

juvenile
arquata

above and obviously different in flight.
VOICE CALL: A sharp *cur-lee*.
STATUS & DISTRIBUTION BREEDING: Eurasia. WINTER: Southern Europe, Africa, and Asia. VAGRANT: Casual on East Coast (about 7 records, none recent), primarily fall and winter, from NF to NY. One summer record from NU.
POPULATION Near threatened; local declines a result of habitat loss.

LONG-BILLED CURLEW *Numenius americanus*   LBCU ▦ 1

barred remiges

cinnamon underwing coverts and flight feathers

adult

Long-billed Curlews are typically seen in large roosting flocks along the coast or feeding in fields. They often roost with other large shorebirds, such as godwits and Willets. In the breeding season they inhabit grasslands and shrub-steppe and eat insects and birds; however, during the nonbreeding season most move to water areas and eat crabs and worms. Polytypic (2 ssp.; both in N.A.). L 23" (58 cm)
IDENTIFICATION Large, warm brown shorebird—not well marked, but distinctive. Upperparts are cinnamon-

brown, with a slightly darker crown due to faint streaking. Upperpart feathers have dark stripes or bars. The face is rather uniform with only a hint of a buff supercilium and slightly darker lores. Underparts are buffy, with streaking on neck and breast. Wing coverts are warm cinnamon with fine bars. Very long bill curves strongly down and the basal half is pale; long legs are gray. JUVENILE: Shorter, less curved bill. Dark wing coverts with broad buff edges result in a striped look.
FLIGHT: Cinnamon-buff wing linings are distinctive in all plumages. The toes extend slightly beyond the tail.
GEOGRAPHIC VARIATION Nominate *americanus* breeds in the southern part of the species' range, north to Nevada, Idaho, Wyoming, and South Dakota. The slightly smaller *parvus* breeds north of *americanus;* these size

differences are difficult to appreciate in the field.
SIMILAR SPECIES Most likely to be confused with a Whimbrel, but it is larger, warmer brown, with buff and cinnamon tones, and lacks the dark head stripes of a Whimbrel. The bill of Long-billed Curlew is longer and more curved, but be aware that juveniles have shorter bills. Eurasian Curlew has white on the rump, tail, and lower back, in addition to white underwings. Far Eastern Curlew presents more of a problem, but it is even larger and plain brown above, lacking any cinnamon tones. The plain brown underwings of Far Eastern are duller than Long-billed, and its bill is long, even in comparison to Long-billed Curlew. A Marbled Godwit, roosting with its bill tucked, will look smaller with unmarked or barred underparts, and has dark legs,

common. The southernmost breeding curlew, with the northernmost winter range. **BREEDING:** Nests in wet and dry uplands. **MIGRATION:** Found on wetlands, grainfields, and coasts. In spring, migrants as early as mid-Feb., but more typically Mar. Peak mid- to late Apr. for UT and the Northwest. Most migrants leave wintering grounds by early May. In fall, first arrivals occur in July, occasionally late June. In most areas they peak in Aug., and almost all birds are gone from their breeding grounds by this time, peaking in southern states in Sept. **WINTER:** Pacific, Gulf, and Southeast coasts, Central and Imperial Valleys of CA, southern AZ, central TX, and Mexico. **VAGRANT:** Casual to mid-Atlantic and Great Lakes in late spring (late May–mid-June) and fall (mid-July–Oct.).

POPULATION Local declines, particularly in eastern parts of its range, due to habitat loss.

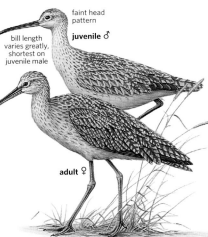

faint head pattern

juvenile ♂

bill length varies greatly, shortest on juvenile male

adult ♀

unlike the paler, gray legs of Long-billed.
VOICE CALL: Loud, musical, ascending *cur-lee.* On breeding grounds it repeats *cur-lee,* followed by sharp, descending whistles.
STATUS & DISTRIBUTION Fairly

GODWITS Genus *Limosa*

Godwits are large with long legs and long, usually upturned bills. Most godwits undergo a substantial change from breeding to nonbreeding plumage, with some sexual dimorphism. They feed by probing and stitching while standing in water or along its edge; they roost with other large shorebirds. All four godwits occur in North America.

MARBLED GODWIT *Limosa fedoa*   MAGO ■ 1

The Marbled Godwit, our largest and most widespread godwit, is numerous along winter beaches, where it feeds by probing into sand and mud. At this time of year it roosts in coastal wetlands with other large shorebirds, typically Willets and curlews. Polytypic (2 ssp.; both in N.A.). L 18" (46 cm)
IDENTIFICATION Large, tawny-brown shorebird with a long, slightly upcurved, bicolored bill. Upperparts have dark feather centers, with buff bars and spots; breast and sides variably barred. **NONBREEDING ADULT:** Bill with brighter, more extensive pink base. Much less barring on underparts. **JUVENILE:** Like nonbreeding, but wing coverts less heavily

marked, contrasting with upperparts. **FLIGHT:** Distinctive cinnamon inner primaries and wing linings.
GEOGRAPHIC VARIATION Nominate *fedoa* is the interior breeder; the slightly smaller *beringiae* population breeds in AK and migrates south as far as southwest AK and northwest CA.
SIMILAR SPECIES A worn Marbled that has lost some of its color could be confused with the smaller and shorter-legged Bar-tailed Godwit. It is easily separated in flight by the lack of cinnamon in the wing. A roosting Long-billed Curlew with its bill tucked has a similar coloration, but Marbled is smaller, has darker legs, and is either unmarked or has some barring on

the sides (versus the breast streaking of a curlew).
VOICE CALL: Loud *ker-ret;* also *widica widica widica.*
STATUS & DISTRIBUTION Common. **BREEDING:** Grassy meadows, near lakes and ponds. **MIGRATION:** In spring, peak mid-Apr.–mid-May in most areas, a little earlier farther south. In fall, adults first appear mid-June, but more typically during July. Most areas peak Aug.–mid-Sept., lingerers annually to Oct., exceptionally to Nov. or even Dec. Rare but regular in the East, primarily late July–late Sept. **WINTER:** South to S.A. Casual to southwest BC and Central Valley of CA. **VAGRANT:** Rare, but annual in Great Lakes; casual in mid-Atlantic in spring, late Apr.–late May, more numerous in fall; also to HI.
POPULATION Stable.

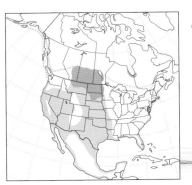

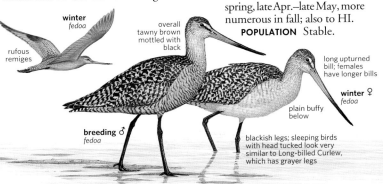

winter *fedoa*

rufous remiges

overall tawny brown mottled with black

long upturned bill; females have longer bills

winter ♀ *fedoa*

plain buffy below

breeding ♂ *fedoa*

blackish legs; sleeping birds with head tucked look very similar to Long-billed Curlew, which has grayer legs

BLACK-TAILED GODWIT *Limosa limosa* BTGD ▪ 3

Black-tailed Godwit is a vagrant from the Old World. Polytypic (3 ssp.; 2 in N.A.). L 16.5" (42 cm)

IDENTIFICATION Straight, or only slightly upcurved, long, bicolored bill. Tail mostly black to tip, white uppertail coverts. Buff or whitish supercilium and dark loral line. Diagnostic flight pattern. **BREEDING MALE:** Chestnut head, neck, and breast; black upperparts with warm rufous edges to scapulars and coverts. White belly and undertail coverts; heavily barred sides and flanks. **BREEDING FEMALE:** Similar to male but paler, sprinkled with white. **NONBREEDING ADULT:** Gray or grayish brown above; grayish breast but otherwise whitish below, lacking barring. **JUVENILE:** Brown upperparts, with warm cinnamon edges to

feathers; pale orangish hindneck and streaked crown. The dark tertials are notched with buff. Pale cinnamon neck and breast; remainder of underparts white. **FLIGHT:** Broad wing stripe across all secondaries and most primaries; conspicuous white wing linings. **GEOGRAPHIC VARIATION** Three subspecies. Subspecies *melanuroides* (from Asia) and *islandica* (breeds in Iceland) are similar in breeding plumage, but an *islandica* breeding male has deeper and more extensive reddish color below; *melanuroides* is darker above in nonbreeding plumage, with a shorter bill and tarsus. Alaska and West Coast birds are likely *melanuroides;* Atlantic birds are likely *islandica*.

SIMILAR SPECIES Black-tailed shares a bold white wing stripe and black-and-

white tail with Hudsonian Godwit; however, note Black-tailed's more extensive wing stripe on the inner portion of the wing and white underwings (black in Hudsonian). In breeding plumage, Hudsonian has dark streaks and lacks reddish color on the hindneck. On juveniles, Black-tailed is warmer colored, with an orange-reddish tint to the hindneck.

VOICE CALL: Generally silent.

STATUS & DISTRIBUTION BREEDING: Wet meadows from Iceland to Russian Far East. **MIGRATION:** Rare, but fairly regular spring migrant on western and central Aleutians; casual to Pribilofs and along Atlantic coast. **WINTER:** Southern Europe to Australia. **VAGRANT:** Casual on Atlantic coast in migration and winter. Accidental inland to ON and south to LA and TX.

POPULATION Stable.

juvenile *melanuroides* — bold and extensive white wing stripe — white wing lining — white base to black tail — juvenile *melanuroides* — winter *melanuroides* — long straight bill; all godwits have extensively pink-based bills, but only Black-tailed has a straight bill — pale chestnut neck — barred sides and flanks — breeding *melanuroides* ♂ — breeding *melanuroides* ♀ — breeding *islandica* ♂ — extensively reddish below

HUDSONIAN GODWIT *Limosa haemastica* HUGO ▪ 1

The striking Hudsonian Godwit is particularly sought after. Its breeding grounds are remote, and it has few reliable migration stopovers. Monotypic. L 15.5" (39 cm)

IDENTIFICATION Long, slightly upcurved, bicolored bill. Black tail,

tipped white, with white uppertail coverts. Pale supercilium and dark loral line. **BREEDING MALE:** Dark upperparts, with buff edges; dark chestnut underparts, finely barred. **BREEDING FEMALE:** Larger and grayer, with reduced chestnut below. Tertials less patterned than on male, often with small notches. **NONBREEDING ADULT:** Rather plain, with gray-brown upperparts, lighter on the breast. Some adults seen in fall migration are in a transitional plumage. **JUVENILE:** Like nonbreeding, but darker above, with buff feather edges to mantle and upper scapulars; lower scapulars and tertials with buff notches. **FLIGHT:** Bold white wing stripe, thin on the secondaries and broader on inner primaries; black underwing coverts and black tail with white base.

SIMILAR SPECIES Nonbreeding adult

Hudsonian resembles Black-tailed Godwit but has dark wing linings and a narrower white wing stripe. Breeding Hudsonian is chestnut below with barring, whereas Black-tailed has chestnut on the hindneck, with minimal barring below. Barred underparts on breeding Marbled Godwit might recall Hudsonian. In flight, Hudsonian's striking wing pattern might recall a Willet, but the Willet has a shorter bill and a paler tail.

VOICE CALL: Generally silent.

STATUS & DISTRIBUTION Uncommon and local. **BREEDING:** From AK to the Hudson Bay, but in very disjunct pockets. **MIGRATION:** Spring route primarily through central and eastern Great Plains. Migrate generally Apr.–May. Peak migration in TX and KS second half of Apr.–early May.

Arrives on AK breeding grounds by late May; in Churchill, ON, by early to mid-June. In fall, migrates much farther east; large staging areas from SK to James and Hudson Bays, with most birds flying southeast to the Atlantic and then to S.A. Great Lakes and mid-Atlantic coast to New England; arrives by early July, peak late Aug.–early Oct., lingers to Nov., exceptionally to Dec. **WINTER:** Southern S.A., unexpected

in N.A.; one sight record for GA. Any winter records should eliminate Black-tailed Godwit, which is more likely in midwinter. **VAGRANT:** Casual to very rare Great Lakes and mid-Atlantic in spring (mid-May–early June); casual in western states in

spring (mid-May–mid-June) and fall (nearly all juveniles, mid-Aug.–mid-Oct.); annual in New Zealand; casual HI, Australia. Accidental in Europe and S. Africa.
POPULATION Stable, but knowledge of breeding range is incomplete.

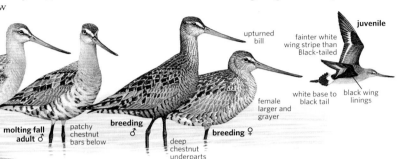

BAR-TAILED GODWIT *Limosa lapponica* BTGO ■ 2

The Bar-tailed is primarily an Old World godwit. It has an impressive migration: The Alaska-breeding subspecies *baueri* stages near breeding grounds and flies nonstop, thousands of miles, to its wintering grounds in New Zealand. Polytypic (5 ssp.; 2 in N.A.). L 16" (41 cm)
IDENTIFICATION Slightly upcurved, long, bicolored bill; brown crown and hindneck, with dark streaks; pale supercilium and dark eye line. Distinctive black-and-white barred tail. **BREEDING MALE:** Reddish brown underparts with some white near the belly, no barring; streaks limited to sides of breast. Dark scapulars with rufous edges. **BREEDING FEMALE:** Larger and paler than male; some females mostly whitish below. **NONBREEDING ADULT:** Gray-brown upperparts with paler buff edges, giving a striped appearance. Grayish brown on foreneck and breast. **JUVENILE:** Like nonbreeding, but more obviously marked: buffy spots and notches on upperparts; wing coverts with dark streaks. Neck and breast with buff tones and light streaks; barring might be present on sides. **FLIGHT:** Weak wing stripe; tail barring visible. White or barred underwings.

GEOGRAPHIC VARIATION Nominate *lapponica,* a vagrant on the Atlantic coast, has a white rump, white wing linings, and brown-barred axillaries; Alaska-breeding *baueri* has brown-and-white barring on rump and underwing. Also, *baueri* is more boldly marked, and the female is more likely to be richly colored in breeding plumage.
SIMILAR SPECIES Resembles Marbled Godwit but lacks cinnamon tones. (Some worn Marbleds can be more problematic, but they still retain cinnamon flight feathers.) Note Bar-tailed's shorter bill, shorter legs, bolder face pattern, white or barred underwing coverts, and barred tail. Bar-tailed's rufous breeding plumage might recall Hudsonian Godwit, but Hudsonian's underparts are more heavily barred and males are deeper chestnut.
VOICE CALL: Generally silent.
STATUS & DISTRIBUTION BREEDING: Tundra from Scandinavia to AK. **MIGRATION:** In spring, AK arrivals mid- to late May, with earliest arrivals late

Apr., and arrivals to north AK into early June. In fall, adults typically depart late July–mid-Aug.; juveniles by Sept. **WINTER:** Europe, Africa, Asia, and Australasia; a few records for the Pacific coast. **VAGRANT:** Rare to casual in spring and, mostly, fall on Pacific coast, particularly in Pacific Northwest. Casual along Atlantic coast in spring, summer, and fall.
POPULATION Stable, but vulnerable due to its limited staging and wintering sites.

TURNSTONES Genus *Arenaria*

Turnstones are named for their habit of flipping over stones and other material with their short, pointed bills in search of food. These short-legged, almost chunky birds are quite social, chattering as they feed and flock along the coasts. In flight, the black-and-white pattern on the back, wings, and tail is impressive.

RUDDY TURNSTONE *Arenaria interpres*   RUTU ■ 1

Ruddy Turnstones frequent shorelines, picking through rocks, shells, seaweed, or other flotsam in search of food. Polytypic (2 ssp.; both in N.A.). L 9.5" (24 cm)
IDENTIFICATION Dark breast and neck on otherwise white underparts; demarcation from black to white on bib uneven, more extensive toward the sides; white throat. Median coverts quite long, hanging down over the wing, like long scapulars. **BREEDING MALE:** Striking, bold black-and-chestnut upperparts; mostly white head, with dark streaks on crown. Bright orange legs. **BREEDING FEMALE:** Duller and browner than male; less rufous in the upperparts. **NONBREEDING ADULT:** Dull brown above and relatively unmarked. **JUVENILE:** Like nonbreeding, but with pale edges on upperparts that give the back a scaly appearance. Once edges wear, remaining upperparts look darker than nonbreeding.

Legs paler orange, some near pinkish. **FLIGHT:** Complex pattern on back and wings: white wing stripe, white back, and white humeral bar separated by rufous or brown patches. Dark tail has white base.
GEOGRAPHIC VARIATION Nominate *interpres* breeds from Greenland eastward to northwest Alaska; it averages more black and less rufous than *morinella,* which breeds from northeast Alaska through Canada.
SIMILAR SPECIES Black Turnstone is similar, particularly in flight, but it has more contrast, lacking brown or rufous patches. Bright orange legs of breeding Ruddy are diagnostic, but this mark is less reliable in winter. Use instead the white in the throat, brown in the plumage, or an uneven dark breast bib to identify a Ruddy.

Some juvenile Ruddies look fairly dark overall.
VOICE CALL: A low-pitched, guttural rattle.
STATUS & DISTRIBUTION Common. **BREEDING:** Coastal tundra. **MIGRATION:** Generally rare inland, except Great Lakes region and Salton Sea. In spring, western populations (primarily *interpres*) peak mid-Apr.–mid-May, with some lingering to June. Eastern populations *(morinella)* move primarily from late Apr., with peak mid-May–early June along mid-Atlantic and Great Lakes. In fall, adults seen in late July, primarily Aug. Juveniles arrive late Aug., peak in Sept., with most gone by early or mid-Oct, depending on region, lingerers to Nov., exceptionally later. **WINTER:** Along the coasts, south to S.A. Accidental to Great Lakes. **POPULATION** Stable. Old World populations show slight increases.

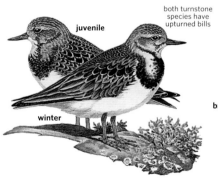

juvenile

both turnstone species have upturned bills

winter

harlequin head and breast pattern

striking wing pattern

breeding ♂

extensive rufous above; female duller

breeding ♂

orange-red legs

BLACK TURNSTONE *Arenaria melanocephala*   BLTU ■ 1

nearly solid slate black head and chest

winter

dark red legs, much darker than Ruddy

A dichromatic bird from the rocky shores of the Pacific coast, the Black Turnstone is social outside of the breeding season, wintering in small flocks. It has a restricted range, in contrast to the circumpolar Ruddy Turnstone. Monotypic. L 9.3" (24 cm)
IDENTIFICATION Dark upperparts, dark throat and upper breast, and white underparts found in all plumages. Legs are reddish brown; some birds show pinkish brown. **BREEDING ADULT:** Head and entire upperparts are black; eyebrow and lore spot are

white. White spotting is visible on sides of neck and breast; dark breast is separated from white underparts in a rather even line. Black median wing coverts are long, hanging over the edge of folded wing, edged with white. **NONBREEDING ADULT:** Browner, but generally looks black. Some scapulars and coverts are edged white; the white on the breast, neck, and head is absent. **JUVENILE:** Like nonbreeding; median coverts are not as large and are edged with white. Legs are often paler, with more of a flesh or

orange tone. **FLIGHT:** Striking black-and-white pattern: white wing stripe, white back, and white scapular bar separated by black. Tail is black, with white base and tips.
SIMILAR SPECIES Wing pattern and habits recall Ruddy Turnstone, but Ruddy has brown or rufous upperparts, an uneven black bib, a white throat, and, usually, bright orange

legs. Surfbird, with which Black Turnstone is often seen, shares the Black's general dark coloration above, white below, with black-and-white tail; however, it is paler above, lacks white patches on the back and scapulars, and has a thicker bill and yellow legs.
VOICE CALL: Includes a guttural rattle, higher than the call of the Ruddy.
STATUS & DISTRIBUTION Locally common. **BREEDING:** Coastal tundra. **MIGRATION:** In spring, movement on the Pacific coast Mar.–early May; peak in southern AK mid-May, arriving on breeding grounds mid-May–early June. First fall migrants appear early July in BC, generally mid-July in OR and WA; adults move earlier, with juveniles arriving in early Aug., peaking late Aug.–Sept. **WINTER:** On rocky coasts from southern AK south. **VAGRANT:** Casual in eastern WA, eastern OR, and interior CA, primarily May and late Aug.–early Sept.; more regular at Salton Sea. Accidental

records east to AZ, NM, AB, MT, and WI. Extralimital records from western Mexico and Russian Far East.
POPULATION Stable, but potentially vulnerable due to its restricted range.

striking wing pattern

winter

white lore spot

white spots on sides breast

breeding

CALIDRINE SANDPIPERS Genus *Calidris*

Calidris sandpipers—the smallest are commonly referred to as peeps or stints—make up this varied group. Plumages tend to be bright in breeding and juvenal birds, rather colorless and unmarked in nonbreeding birds. Most species are rather small with little difference in size or plumage between the sexes. Identifications within this group can be difficult. Most species pick along the shore and mudflats or in shallow water; some species feed in slightly deeper water. Although many species migrate and roost in large flocks, they can be very aggressive and territorial. Most species undergo long migrations, breeding in the Arctic and wintering to South America.

GREAT KNOT   *Calidris tenuirostris*   GRKN ■ 4

The Great Knot is a vagrant from Asia. Monotypic. L 11" (28 cm)
IDENTIFICATION Large and chunky. Generally gray above; white and spots below. **BREEDING ADULT:** Black breast with large spots extending to flanks. Extensive rufous on scapulars. **NONBREEDING ADULT:** Gray-brown above; white with limited streaks or spots below. **JUVENILE:** Dark back feathers

edged with brown and white; buffy wash and distinct spotting below. **FLIGHT:** Faint wing stripe; dark primary coverts and white uppertail coverts.
SIMILAR SPECIES Larger than Red Knot, with longer bill; less rufous on back, none on head and breast on breeding birds. In flight, wing stripe is fainter, and uppertail coverts are whiter than in Red Knot. Remarkably similar in breeding plumage to Surfbird (with which it has hybridized), but has a longer, thinner black bill, as well as dark gray or green legs, not yellowish as in Surfbird.
VOICE CALL: Generally silent.

STATUS & DISTRIBUTION Breeds in Russian Far East. **WINTER:** Primarily Australia through Indonesia; smaller numbers in southeast and south-central Asia. **VAGRANT:** Casual spring migrant in western AK; accidental in fall. Accidental in OR (fall); also to western Palearctic.
POPULATION Stable, but there is major concern over hunting and habitat loss in China and Korea, which the Great Knot uses on migration.

juvenile

upperparts much more patterned than Red Knot

extensive spotting below

bold arrow-shaped spots below

extensive rufous on scapulars

larger and longer-billed than Red Knot

breeding plumage suggests breeding plumage of Surfbird

breeding

RED KNOT *Calidris canutus* REKN ■ 1

Thousands of Red Knots collect at staging areas during migration. The Red Knot feeds along sandy beaches and mudflats, often with dowitchers. Polytypic (6 ssp.; 3, possibly 4, in N.A.). L 10.5" (27 cm)

IDENTIFICATION Chunky and short-legged for a *Calidris*. Medium-length black bill. Legs vary from black to olive or gray. **BREEDING ADULT:** Dappled brown, black, and chestnut above, with buffy chestnut face and breast; entirely or mostly chestnut below. Males average brighter and females average more white on the belly. Worn adult looks darker above, once paler chestnut edges wear. **NONBREEDING ADULT:** Pale gray upperparts and head; white underparts, except breast, which has grayish wash or spots, and fine markings along sides. **JUVENILE:** Like nonbreeding, but with more distinct spotting below and scaly-looking upperparts. Pale fringes and dark subterminal lines on scapulars and coverts. Fresh juveniles often have buff wash on breast. **FLIGHT:** White wing stripe and paler rump with gray barring.

GEOGRAPHIC VARIATION There is uncertainty over the winter distribution of some subspecies and the subspecific limits in the Alaska populations. Of six subspecies, *islandica,* which breeds in Greenland and northern Canadian islands and winters in Europe, is more chestnut above and below.

Breeding south of *islandica* in Canada, *rufa* has less chestnut above and is paler below. It is believed to winter in South America and possibly the southeast United States. Alaska *roselaari* is similar to *rufa,* only darker. It winters from the southern United States to South America. Subspecies *rogersi* breeds on Chukchi Peninsula and may occur in nearby western AK.

SIMILAR SPECIES In breeding plumage, the Red Knot's uniform color is similar to a dowitcher, but the knot has a shorter bill, paler crown, and (in flight) a whitish rump, finely barred with gray. Nonbreeding Red Knot also differs by structure, bill, and tail pattern.

VOICE CALL: Generally silent, but *cur-wit* sometimes given when taking flight.

STATUS & DISTRIBUTION Uncommon to fairly common. **BREEDING:** Dry tundra. **MIGRATION:** Rare to casual migrant in the interior. Often missed in coastal areas not used as staging grounds; may be common in one place, but only in small numbers between the staging

area and the next stop. In spring, earlier in the West than in the East, peaking on Pacific coast late Apr.–mid-May. Mid-Atlantic peak mid-May–early June. In fall, adults seen mid-July, peak late July–mid-Aug. Juveniles peak late Aug.–Sept. Smaller numbers to mid-Oct.; lingerers into Nov. **WINTER:** Coasts, south to Argentina. **VAGRANT:** HI.

POPULATION Severe declines in *rufa* from mid-Atlantic in migration. Species is threatened at staging areas where it relies on a single food source (e.g., horseshoe crab eggs at Delaware Bay).

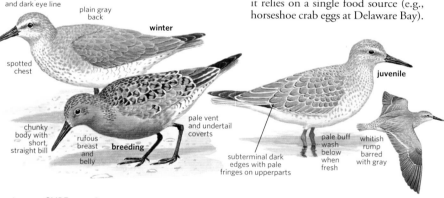

whitish supercilium and dark eye line

plain gray back

winter

spotted chest

chunky body with short, straight bill

rufous breast and belly

breeding

pale vent and undertail coverts

juvenile

pale buff wash below when fresh

whitish rump barred with gray

subterminal dark edges with pale fringes on upperparts

SURFBIRD *Calidris virgata* SURF ■ 1

A Pacific coast rocky shorebird, the rather chunky Surfbird is often seen in the company of Black Turnstones. It picks invertebrates from rocks, although rarely it will feed on mud-

flats and sandy beaches. Monotypic. L 10" (25 cm)

IDENTIFICATION Generally gray above and on the throat, white below. Stout dark bill with a yellow base.

Short yellowish green legs. **BREEDING ADULT:** Head and underparts heavily streaked and spotted with dusky black. Upperparts edged with white and chestnut; scapulars rufous and black. **NONBREEDING ADULT:** Gray spots along sides. White edges to wing coverts. **JUVENILE:** Like nonbreeding, but upperparts have pale edges and darker subterminal markings, creating scaled appearance; chest paler and mottled. **FLIGHT:** White wing stripe and conspicuous black band at end of white tail and rump.

SIMILAR SPECIES Breeding Surfbird could be taken for a Great Knot; however, Great Knot has a longer,

winter

short, sturdy bill, always with greenish yellow at base

arrow-shaped streaks on lower sides and flanks

juvenile

more typical sandpiper bill and dark legs. Surfbird's short, thick bill and bold flight pattern are unique.

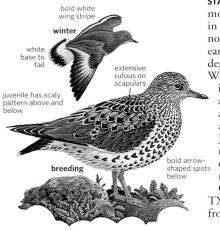

bold white wing stripe

winter

white base to tail

extensive rufous on scapulars

juvenile has scaly pattern above and below

breeding

bold arrow-shaped spots below

VOICE CALL: Generally silent, but chatter notes, like those of a turnstone, often given in flight. **STATUS & DISTRIBUTION** Fairly common. **BREEDING:** Mountain tundra in AK and YT. **MIGRATION:** Spring northward movement noted Mar. to early May along coast, with most birds departed by the end of Apr. Prince William Sound, AK, is a major staging area early to mid-May. In fall, adults first noted south of breeding areas in late June or early July; most adult movement July–early Aug. Juveniles start moving early Aug.; most movement from BC south is over by late Sept. **WINTER:** Reefs and rocky beaches from southern AK to Chile. **VAGRANT:** Casual in spring on TX coast, Salton Sea, and points east from FL to AB; accidental inland in

fall (e.g., sight record Aug. 18, 1979, at Presque Isle, PA). **POPULATION** Stable, but vulnerable since a high percentage of the population uses Prince William Sound, AK, as a staging area.

RUFF *Calidris pugnax* RUFF ▪ 3

The Ruff male has a spectacular breeding plumage. The males gather in leks and display for females. Monotypic. L 10–12" (25–31 cm)
IDENTIFICATION Tall with long legs, plump body, small head, and small bill. Back feathers are often raised when feeding. **BREEDING MALE:** Neck ruff colors black, rufous, or white. Legs may be yellow, orange, or red; bill has pale base, black tip. Some males are dull and small, resembling females. **BREEDING FEMALE ("REEVE"):** Lacks the ruff, is smaller, and has a variable amount of black below. **NONBREEDING ADULT:** Legs and bill usually duller than in breeding, but some birds still with bright tones. Scapulars and coverts

dark, with thin, pale edges; tertials often with bars or internal markings. Underparts dingy whitish with mottling sometimes present on the neck and sides of the breast. Typically with white feathering at base of bill. **JUVENILE:** Like nonbreeding, but bolder edges to the upperparts, buffy breast, and tertials lack internal markings. **FLIGHT:** Distinctive U-shaped white rump band in all plumages; white underwings.
SIMILAR SPECIES Combination of small bill, long legs, and plump body is distinct. Upland Sandpiper is superficially similar, but more marked on

upperparts, has streaks on neck and straight bill, and lacks white ovals on sides of rump. Compare juvenile Ruff with Sharp-tailed Sandpiper, which has a different face pattern. Buff-breasted Sandpiper is smaller and less erect and lacks any facial markings (juvenile Ruffs usually have a postocular line) and white rump ovals.
VOICE CALL: Silent away from breeding grounds.
STATUS & DISTRIBUTION Old World; has bred in AK. **MIGRATION:** In spring, rare in western AK and on Atlantic coast, casual elsewhere, primarily late Mar.–late May, into June in AK. In fall, as above, but more regular in West, primarily late July–Aug. (adults), and Sept.–early Nov. (juveniles); juveniles more numerous than adults in West, opposite in East. **WINTER:** Rare CA, accidental OR. **POPULATION** Stable.

white U-shaped band

juvenile ♀

scaly upperparts

buffy neck and breast

juvenile ♀

displaying breeding males

♀

summer molting ♂

back feathers often raised

summer molting ♀

leg color variable: greenish to bright orange

small head and plump body

whitish lores and forehead

winter ♂

often with colored bill base

Calidris Plumage and Structural Details

At first glance, *Calidris* sandpipers seem incredibly difficult to distinguish. Indeed, a sure sign that birders have truly become enthralled with identification is their love for this particular group of shorebirds. Part of the learning process for correctly identifying *Calidris* sandpipers is refining the identification keys. Instead of looking at the bird as a whole, it is helpful to focus instead on specific characteristics to guide the identification. Three critical aspects in identifying *Calidris* are the bird's structure, the bird's feather topography, and the proper aging of sandpipers.

Begin by considering the bird's structure. Most of the *Calidris* species have a subtly unique structure. The key characters to study to assess structure are bill length and shape, leg length, and primary length. Most sandpipers share a medium-length bill that is relatively straight and which thins toward the end. However, the bill's length relative to the head is variable, as is its thickness and the amount of droop toward the tip. Leg length on its own would not be a good identification mark since it only varies a little; however, used in combination with other features, one can assess whether a sandpiper is shorter, chunkier, or more elongate. Primary projection is also only subtly different, but it helps in both the objective assessment of whether the primaries project past the tail, as well as the subjective determination of the shape of the bird.

The second characteristic to take careful note of when attempting to identify *Calidris* sandpipers is the bird's feather topography. Identifications within this genus will greatly accelerate once there is a working knowledge of the various feather groups. While patterns on the crown, breast, and elsewhere are important, the key patterns to look for and understand are those on the mantle, the scapulars (both upper and lower), the upperwing coverts, and the tertials. You also need to understand what happens to these feathers over the course of the year. The mantle is the upper back; the small feathers here often have bright colors or edges. Near the edge of the mantle, many *Calidris* species have a row of feathers

that are boldly edged white, creating a mantle line. The adjacent feather group is the upper scapulars. Starting at the top, many species have the first row of upper scapulars edged white, creating a scapular line. Some species show both of these lines boldly, creating a V along the sides of the upperparts. The lower scapulars are usually marked slightly differently than the upper scapulars, especially in breeding adults. These scapulars often hang down and cover the wing coverts, a critical feather group when trying to determine the identification of our smallest sandpipers, the peeps or stints.

Finally, be careful to determine the proper aging of the sandpipers present. Aging a *Calidris* sandpiper is typically easy and necessary before one can hope to identify it. There are straightforward, basic age guidelines. Most identification problems occur in fall. In most species, we see juveniles from the end of July through most of October in crisp, often bright plumage. Their feathers are fresh and uniform, often with bright rusty or buff edges and often appearing scaly. In contrast, worn adults in fall have passed their peak appearance, although those seen in July and early August might still be mostly in breeding plumage. By late summer, most adults are molting into nonbreeding plumage, and many of their more colorful breeding feathers on the upperparts are wearing off the bright edges. This wearing results in a darker, more blotchy appearance, in contrast to the uniform appearance of juveniles. When adults complete their molt, most species go into a drab, rather unmarked nonbreeding plumage. Juveniles also molt later in fall (a few species molt earlier) and then look very similar to adults; however, close views of the coverts or tertials usually betray their youth. Sandpipers molt into breeding plumage in spring. Some do so before migrating; others do so during the passage. Many birds that are less than a year old do not migrate all the way north or molt into pristine breeding plumage. Instead, these birds molt into first alternate (breeding) plumage. ∎

Least Sandpiper, juvenile (NJ, Aug.)

Baird's Sandpiper, juvenile (CA, Sept.)

BROAD-BILLED SANDPIPER *Calidris falcinellus* BBIS ▪ 4

Broad-billed Sandpiper is an Old World vagrant. Polytypic (2 ssp.; probably both in N.A.). L 7" (18 cm)

IDENTIFICATION Plump body, short legs, and long bill form a distinctive profile. Thick bill droops noticeably at the tip, at times looking like it has an unnatural kink. The Broad-billed is more likely to probe for food than peck, but it does both. Split supercilium is present in all plumages. Legs vary from olive to almost black. **BREEDING ADULT:** Black scapulars edged buff or rufous; coverts typically edged buff. Underparts white except for the brownish and heavily streaked breast; streaks extend down the sides. **NONBREEDING ADULT:** Gray-brown above with some scapulars having dark centers, giving a somewhat mottled look. White below with variable breast streaking. **JUVENILE:** Scapulars and coverts edged buff or rufous, with bold mantle and scapular lines. Breast lighter, more lightly streaked than breeding plumage.

GEOGRAPHIC VARIATION Nominate *falcinellus* from Europe and west Russia perhaps pertains to the Atlantic record. Asian *sibirica*, which averages more rufous above, is thought to pertain to AK records.

SIMILAR SPECIES While Dunlin is superficially similar—it shares a rather plump appearance, with a downcurved bill—no species matches Broad-billed's combination of plumage and structure.

VOICE CALL: Ascending trill, *brreeet.*

STATUS & DISTRIBUTION Breeds in Eurasia. **WINTER:** Arabian Peninsula to Australia. **VAGRANT:** Casual fall migrant on western and central Aleutians and Pribilofs, AK; accidental in fall in coastal NY. All N.A. sightings have been of juveniles.

POPULATION Stable.

breeding

short legs

split supercilium in all plumages

bright and strongly patterned upperparts on juveniles

broad-based bill with drooped tip

juvenile

TEMMINCK'S STINT *Calidris temminckii* TEST ▪ 3

Temminck's Stint, an Old World vagrant, has a preference for freshwater habitats. It is usually found singly or in small numbers, even where it is common. Monotypic. L 6.3" (16 cm)

IDENTIFICATION Rather brown, horizontal-looking peep. White outer tail feathers are distinctive in all plumages, but hard to see. Bill is dark and short, with a slight droop; legs are yellowish, sometimes with a green tint. Tail is quite long, extending past the wings. **BREEDING ADULT:** Generally brownish upperparts. Scapulars and coverts are dark with rufous or buff edges; although appearing uniform from a distance, they can be colorful. There is little of a face pattern, with no supercilium and only slightly darker lores; white eye ring is usually rather distinct. Breast is brown with dark streaks. Worn adults are brown with dark splotches above. **NONBREEDING ADULT:** Upperparts and breast are plain gray-brown; the remainder of underparts are white. Some shaft streaks are visible on the scapulars. **JUVENILE:** Like nonbreeding, but feathers of upperparts have diagnostic dark, subterminal edges and buffy fringes.

SIMILAR SPECIES Breeding Temminck's resembles in plumage and shape the larger Baird's Sandpiper; however, note Temminck's yellow or greenish yellow legs and tail that extends beyond the primaries, whereas Baird's has dark legs and primaries that project past the tail and tertials. Eager birders might focus on the white outer tail feathers, but this mark, while diagnostic, must be used with caution. Outer tail feathers of other *Calidris* can appear pale.

VOICE CALL: Repeated, fast, dry rattle, almost recalling a longspur; routinely calls when flushed. Call might be the first clue to its identity.

STATUS & DISTRIBUTION BREEDING: Eurasian species in wet tundra from Scandinavia to Siberia. Rare spring and fall migrant on the Aleutians and Pribilofs; very rare on St. Lawrence I. One Sept. record of a juvenile from BC and a late fall record from WA. **WINTER:** Mediterranean, Africa, and southern Asia.

POPULATION Stable; large winter populations in south-central Asia.

juvenile

white outer tail feathers

breeding

general coloration suggests Baird's

greenish yellow legs

dark breast

juvenile

juvenile

dark subterminal fringes

short primary projection

white outer

LONG-TOED STINT   *Calidris subminuta*   LTST ■ 3

This Old World vagrant prefers fresh-water with vegetation and is usually seen singly or in small numbers, even where common. It often appears erect, high-stepping on mudflats. Its long legs and toes, however, are of marginal use for identification. Monotypic. L 6" (15 cm)

IDENTIFICATION Brown peep with a reduced wing stripe. Short and slightly drooped black bill, usually with a pale

greenish base. Yellow legs, with either brown or green tones at times; rela-tively long, so toes extend beyond tail in flight. Dark underwings, unique among stints. **BREEDING ADULT:** Bright above, recalling a miniature breed-ing Sharp-tailed Sandpiper. Crown, auriculars, mantle, scapulars, and ter-tials with rufous edges; lower scapulars often edged white. Greater wing coverts edged buffy. Breast usually buff with fine dark streaks. Rather bold face pat-tern, with dark lores and a split super-cilium. In early fall, when feather edges are worn, bird looks dark brown above. **NONBREEDING ADULT:** Gray-brown above, looking mottled with dark centers to some scapulars and streaks on the crown and neck. Gray-brown breast with faint streaking. Face pattern stays bold. **JUVENILE:** Upperparts and scapulars edged broadly with rufous and white. Bold white mantle line; less distinct scapular line. Median wing coverts edged white. Breast is streaked, but usu-ally on a grayish, less frequently buffy, back-ground. Supercilium stops short of the bill, result-ing in a noticeably dark forehead. **SIMILAR SPECIES** Most likely to be confused with the much more expected Least Sandpiper. Dis-tinguish it in all plumages by

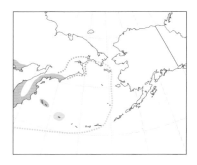

its dark forehead and region above lores; bolder, split eyebrow, broaden-ing behind eye; white-edged median coverts; pale base to the mandible; and darker underwing coverts. Also, Long-toed's breast is often grayish or whitish with fine streaks, whereas Least's is brown or buff with streaks. In breed-ing and juvenal plumage, Long-toed's median upperwing coverts are edged whitish and contrast with the rest of the bright upperparts.

VOICE CALL: *Purp,* lower pitched than Least Sandpiper.

STATUS & DISTRIBUTION Asian breeder. Casual in spring to Pribilofs and St. Lawrence I.; can be fairly common on western Aleutians, but casual in fall. **WINTER:** Australasia and southeast Asia. **VAGRANT:** Accidental elsewhere on West Coast in fall (single records of juveniles from coastal OR and CA).

POPULATION Believed to be stable, but poorly studied.

Illustration labels:
greenish lower mandible base on straight bill
bold white mantle V
juvenile
yellow legs
distinct supercilium broadens behind eye
split supercilium
buffy wash with spots on sides of neck and breast
breeding

LEAST SANDPIPER   *Calidris minutilla*   LESA ■ 1

A common peep throughout the con-tinent, the Least Sandpiper prefers freshwater habitats with vegetation. While common in winter, including in inland habitats, this species does not amass in the large flocks of West-ern and Semipalmated Sandpipers. Monotypic. L 6" (15 cm)

IDENTIFICATION This brown peep is smaller and darker than a Western or Semipalmated, the other two common peeps in N.A. Bill is dark, short, and obviously drooped. Yellow legs can be dull or obscured by mud. **BREEDING ADULT:** Black and brown above, with rufous and buffy edges to the scapulars and all coverts; some feathers are tipped white. Breast is brown with conspicu-ous streaks; remaining underparts are white. Supercilium is bold and meets the base of the bill, so that its forehead

is white. **NONBREEDING ADULT:** Upper-parts and breast are brown, with some dark centers to the scapulars. Super-cilium, while present, is less distinct than in breeding plumage. **JUVENILE:** Like breeding plumage, but very crisp; upperparts are edged rufous and buff, some feathers are tipped white. White mantle line is bold; scapular line less distinct. Usually has a strong buffy wash across the finely streaked breast.

SIMILAR SPECIES Least Sandpiper is not too similar to Western and Semipalmated Sandpipers; however, these three species are the common peeps and the first step in learning sandpiper identification. Beginning birders often emphasize the yellow legs of the Least, compared to the darker legs of the other two species, but the identifications are sufficiently

straightforward to eliminate the need to base an identification on a mark that can be altered by mud or lighting. Least Sandpiper is darker and browner overall than Western and Semipalmated and noticeably smaller when seen with either. In nonbreeding plumage, Least has

a prominent brown breast band. Long-toed Stint, a potential vagrant from Asia, is similar; see that species account. In fall, the first arrival of juveniles and their crisp, bright plumage might suggest vagrant stints to the eager birder, but their upperparts pattern, drooped bill, yellow legs, and buffy, streaked breast should cinch the identification. **VOICE CALL:** A high *kree* or *jeet*. **STATUS & DISTRIBUTION** Common. **BREEDING:** Moist tundra and wet

habitats in boreal forest; breeds farther south than other peeps. **MIGRATION:** Spring movement in the West starts in late Mar., peaks in Apr. (CA) to as late as mid-May (WA). Migration in the East averages 1–2 weeks later. In fall, adults move as early as late June, and are common on both coasts and the Great Lakes by mid-July. Most have migrated by the end of Aug. Juveniles first appear in late July, but more typically there is a push in mid-Aug.; they can remain

common through Sept. Where they do not winter, stragglers seen into Nov., rarely later. **WINTER:** Across southern U.S., as far south as Chile. This is the hardiest and most numerous and widespread winter peep in the interior of N.A. **VAGRANT:** Europe, Russian Far East, Azores, and HI. **POPULATION** Stable, but the numbers are hard to assess given the Least Sandpiper's remote breeding areas and the lack of large migrating or winter flocks.

winter

brownish breast band

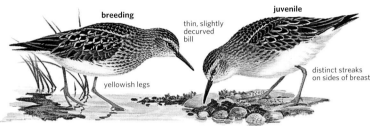

breeding

thin, slightly decurved bill

yellowish legs

juvenile

distinct streaks on sides of breast

LITTLE STINT *Calidris minuta* LIST ■ 4

This Old World vagrant peep is likely found with other small sandpipers during fall migration. Monotypic. L 6" (15 cm) **IDENTIFICATION** Short, black bill has a fine tip. Legs are black; feet do not have webbing. Primaries clearly extend beyond the tail tip. **BREEDING ADULT:** Bright, with dark upperparts brightly fringed with rufous. Yellowish mantle lines appear whitish when worn. Throat and underparts are white; a warm brownish orange wash on the breast contains bold spotting, particularly on the sides of the breast. **NONBREEDING ADULT:** Gray-brown above, with some scapulars having dark feather centers. White below, with some light streaks on the sides of the breast. **JUVENILE:** Most juveniles average brighter than other peeps. Upper scapulars are edged rufous, lower scapulars are dark, usually black, with paler edges. Tertials are dark, typically with bright rufous edges. Both mantle and scapular lines are bold. Crown is dark with a distinctly forked supercilium. Streaking and spotting on the sides of the breast is distinct. **SIMILAR SPECIES** See Red-necked Stint for separation of breeding birds. Breeding Sanderling is superficially similar and the only North American peep that would have such a colorful throat and breast combination;

since this plumage is seen only briefly away from the breeding grounds, unsuspecting birders might jump to an identification as Red-necked or Little Stint, but a Sanderling is bigger, bulkier, with a longer bill and dark, boldly marked tertials. It also lacks a hind toe. Juveniles must be distinguished from Red-necked Stint and Semipalmated Sandpiper. Best distinguished from juvenile Red-necked by its extensively black-centered wing coverts and tertials usually edged with rufous, but it is also usually brighter with bolder mantle lines, and the dark markings on sides of the breast tend to be sharper and more spotted than streaked. Semipalmated Sandpiper has a darker crown, unforked supercilium, and is usually paler than a Little Stint. On a bright Semipalmated, also note its more tubular

bill; in comparison, Little Stint has sharper breast markings, and bolder mantle and scapular lines. Although not as similar, a first fall juvenile Least Sandpiper can often confuse birders due to its bright plumage (see Least Sandpiper). **VOICE CALL:** Single *kip,* often given when flushed. **STATUS & DISTRIBUTION BREEDING:** Eurasia. **MIGRATION:** Very rare on islands in western AK in spring and fall; casual to AK mainland. **WINTER:** Africa and Asia. **VAGRANT:** Casual, primarily in Pacific states (annual in recent years) and on East Coast; breeding adults and juveniles recorded. Accidental in spring and early winter (in winter plumage). **POPULATION** Stable.

coverts and tertials in juvenal and breeding plumage with darker centers than Red-necked and with colored edges

breeding

white mantle V bolder than Red-necked

split supercilium

dark spots on sides of breast with underlying buffy wash

juvenile

RED-NECKED STINT *Calidris ruficollis* RNST ▪ 3

Red-necked Stint is an Old World vagrant. Monotypic. L 6.3" (16 cm) **IDENTIFICATION** Small peep. Black legs, feet unwebbed. Rather short black bill. Primaries extend slightly beyond tail tip. **BREEDING ADULT:** Rufous throat, upper breast, head, and mantle can be bright; paler in some birds, to pinkish orange, with throat entirely white. A necklace of dark streaks extends below the throat onto the white lower breast, with a few spots or streaks on the sides. **NONBREEDING ADULT:** Gray-brown above with dark shaft streaks on scapulars. White below, with some streaks at sides of breast. **JUVENILE:**

Scapulars vary from bright to pale rufous, with white tips. White scapular and mantle lines. Drab tertials, edged in pale buff, some with a faint rust tinge. Plain brown wing coverts with dark anchor-shaped shaft streaks. Crown seems darkest in the center; partial, indistinct forked supercilium. **SIMILAR SPECIES** Breeding Little Stint has breast spotting contained within the color of the breast and always has a white lower breast, usually with extensive rufous-edged tertials and upperwing coverts. Juvenile Red-necked differs from juvenile Little by its plainer gray wing coverts and tertials, dusky streaks rather than darker spots at breast sides, and less distinct mantle and scapular lines; and from juvenile

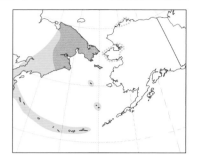

Semipalmated and shorter billed Western by its plainer coverts and tertials; Semipalmated also has a darker crown and less contrast between the scapulars and wing coverts. Away from AK, only one certain record of a juvenile (CA). See Sanderling. **VOICE CALL:** A squeak somewhat like that of Western Sandpiper. **STATUS & DISTRIBUTION BREEDING:** Russian Far East to (rarely) west AK. **MIGRATION:** Regular migrant on western AK islands. **VAGRANT:** Breeding adults casual on both coasts and interior late June–late Aug.; lack of juvenile records in this area, although some probably pass through undetected. Accidental in CA and DE in spring. **WINTER:** Australasia and S.E. Asia.

tertials and coverts rather dull in juvenal and breeding plumage

juvenile

breeding

breeding

rich brick red throat

dark streaks on breast

some have duller rufous that is confined to lower throat

SANDERLING *Calidris alba* SAND ▪ 1

The Sanderling is the familiar sandpiper watched by beachcombers. It seemingly chases waves out to sea, only to be chased back by the next wave, its legs blurring in speedy retreat. Monotypic. L 8" (20 cm) **IDENTIFICATION** Larger than other peeps. Black bill and legs. Lacks hind toe. Prominent white wing stripe in flight; black primary coverts contrast with remainder of wing. **BREEDING ADULT:** Not acquired until late Apr. or later;

head, mantle, and breast are rusty. Tertials often with markings inside the edge, unlike any other peep. **NONBREEDING ADULT:** Palest sandpiper in winter. **JUVENILE:** Blackish above, with pale edges near tip; looks checkered. Breast washed buff when fresh. **SIMILAR SPECIES** Bright breeding-plumaged birds have been most frequently mistaken for Red-necked Stint, but Sanderling is larger, with bolder wing stripe, and lacks dark necklace of streaks below rusty breast. Nonbreeding birds are paler than any congener and larger.

VOICE CALL: A sharp *kip*. **STATUS & DISTRIBUTION** Common. **BREEDING:** Dry tundra of Canadian islands, Eurasia, and, rarely, AK. **MIGRATION:** Spring peak on Pacific coast late Apr.–late May; Great Lakes and mid-Atlantic two weeks later. In fall, adults arrive south mid-July, peaking late July–Aug.; juveniles late Aug.–Sept.; stragglers into Nov. **WINTER:** Coastlines to southern S.A. **POPULATION** Possibly declining.

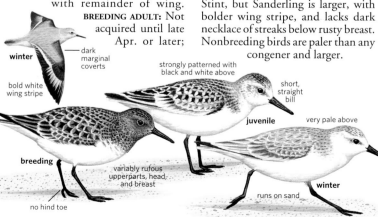

winter

dark marginal coverts

bold white wing stripe

strongly patterned with black and white above

short, straight bill

juvenile

very pale above

breeding

variably rufous upperparts, head, and breast

winter

runs on sand

no hind toe

WESTERN SANDPIPER *Calidris mauri* WESA ■ 1

Western Sandpiper is the abundant western counterpart of the Semipalmated Sandpiper. It is found in large flocks during migration. Monotypic. L 6.5" (17 cm)

IDENTIFICATION Black legs; tapered bill, of variable length, with a slight droop at the tip. **BREEDING ADULT:** Bright rufous wash on crown and auriculars; rufous base to scapulars. White underparts, with streaks and arrow-shaped spots across the breast and along sides and flanks. **NONBREEDING ADULT:** Gray-brown above and white below, with faint streaks on breast. **JUVENILE:** Extensive rufous edges on upper scapulars. Gray-brown

wing coverts with pale edges. **SIMILAR SPECIES** Easily confused with Semipalmated in basic plumage, but Semipalmated (unexpected in winter) has a shorter, more tubular bill, and slightly more contrasty face. See sidebar below for separation of juveniles. In winter, Western might recall Dunlin as both share a bill that droops toward the tip; however, larger Dunlin has a longer bill with a more pronounced droop, is browner, and is more heavily marked on the throat and breast. A winter Least Sandpiper is browner and smaller, with darker throat and upper breast and yellowish or greenish legs. **VOICE CALL:** A high, raspy *jeet*. **STATUS & DISTRIBUTION** Abundant. **BREEDING:** Wet Arctic tundra from

Russian Far East to northwest AK. **MIGRATION:** Throughout N.A., except very rare to casual in ND, MN, and eastern Canada. Movement in spring by late Mar., peaking in WA early Apr.–early May, with lingerers to late May. Casual on East Coast in spring. In fall, first adults late June, most July. First juveniles arrive late July (later in the East), peak Aug.–early Sept., lingerers into Nov. **WINTER:** Primarily coastal to S.A. **VAGRANT:** Europe, Japan, and Australasia. **POPULATION** Stable.

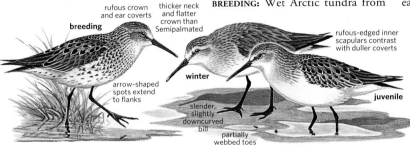

rufous crown and ear coverts

breeding

thicker neck and flatter crown than Semipalmated

rufous-edged inner scapulars contrast with duller coverts

winter

juvenile

arrow-shaped spots extend to flanks

slender, slightly downcurved bill

partially webbed toes

Separation of Juvenile Semipalmated and Western Sandpipers

Many birders forsake identifying juvenile Semipalmated and Western Sandpipers since separating them can be complex. Nevertheless, given good views and a reasonable understanding of their character differences, almost all of them can be identified.

A feature useful in separating them is the size and shape of the bill. Typically, Semipalmateds have an obviously short, tubular bill, whereas Westerns have a long bill, noticeably drooped, with a narrow tip. While there is overlap—and eastern Semipalmateds have longer bills than western Semipalmateds— eastern female Semipalmateds that have long bills still have a stubbier tip. Another good character is the subtle difference in head pattern. The Semipalmated's bold supercilium contrasts with darker auriculars and a darker crown, imparting a capped appearance. The Western has dark lores, but the crown and

Semipalmated Sandpiper, juvenile (Sept.)

Western Sandpiper, juvenile (Sept.)

auriculars are not as dark, nor the supercilium typically as pale.

The scapulars and coverts also subtly differ. Most Semipalmateds look brown above, with pale edges giving them a buffier, more uniform look. Westerns look lighter above and have bold rufous edges to the upper scapulars. The Western's lower scapulars are gray with dark "anchors" near the tip, whereas the Semipalmated's are more uniform. Bright Semipalmateds can cause a problem, as some of them have brighter rust edges to the upper scapulars; worn or duller Westerns duplicate the problem, but, in both cases, the anchors on the wing coverts, head pattern, and bill shape are consistent characters.

Finally, keep in mind that Western Sandpipers molt early in fall, while migrant Semipalmated Sandpipers maintain their juvenal plumage, and that the species have different calls. ■

SEMIPALMATED SANDPIPER *Calidris pusilla* SESA ▪ 1

Semipalmated Sandpiper is abundant. During migration it congregates in large flocks at wetland habitats in the eastern half of the continent. Monotypic. L 6.3" (16 cm)

IDENTIFICATION Short to medium length, tubular-looking bill with rather blunt tip; female's bill larger than male's. Black legs; webbing on toes visible under ideal conditions. **BREEDING ADULT:** Dark upperparts, a mixture of black and brown, with only a tinge of rust on crown, auriculars, and scapular edges; usually lacking spotting on flanks. **NONBREEDING ADULT:** Gray-brown above; largely white below. **JUVENILE:** Strong supercilium; dark crown and ear coverts; uniform brown upperparts, typically edged buff, but variably brighter, edged with rufous.

SIMILAR SPECIES Separating juvenile Western from juvenile Semipalmated can be difficult (see sidebar p. 243). The process of identifying juvenile

Little and Red-necked Stints must first eliminate the more likely Semipalmated; see those species accounts. **VOICE CALL:** A short *churk*, very different from Western Sandpiper.

STATUS & DISTRIBUTION Abundant. **BREEDING:** Tundra from west AK to Atlantic. **MIGRATION:** In spring, first arrivals late Mar., but most in southern states Apr.–mid-May. Peak in mid-Atlantic and Great Lakes mid-May–early June. Rare in West late Apr.–late May. In fall, adults as early as late June, peaking mid-July–mid-Aug.; very rare in West south of WA. Fall juveniles peak in Aug., with numbers declining by early Sept., and largely departed by early Oct. In West, juveniles are much more numerous than adults and mainly seen in Aug. Migration a little later in eastern N.A. A few linger into early Nov. **WINTER:** Mexico (a few) to S.A. Casual in southern FL.

VAGRANT: Casual in Europe.
POPULATION Local declines noted.

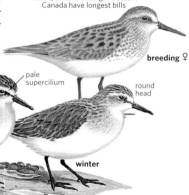

female's bill averages larger than male's; birds breeding in eastern Canada have longest bills

breeding ♀

dark ear coverts

pale supercilium

round head

more uniform upperparts than juvenile Western

breeding

short, straight, tubular bill

partially webbed toes

juvenile

winter

WHITE-RUMPED SANDPIPER *Calidris fuscicollis* WRSA ▪ 1

The White-rumped Sandpiper's larger size and longer wings separate it from the smaller peeps. It is often seen along shore edges or on mudflats. Monotypic. L 7.5" (19 cm)

IDENTIFICATION Long wings, with primaries extending noticeably beyond

the tertials and tail. Black bill, noticeably pale reddish at the base. White, unmarked rump typically only visible during flight; some birds hold their wings lower, allowing a view on a perched bird. **BREEDING ADULT:** Gray-brown upperparts with pale edges on the scapulars and coverts. Distinct whitish supercilium. Crown, auriculars, mantle, and upper scapulars might have some rufous tones. Mantle and scapular lines, if present, are indistinct. Black streaks on breast extend to the flanks. Worn fall birds are remarkably plain; they usually have only remnant streaking along the flanks. **NONBREEDING ADULT:** Gray-brown above and on neck, giving it a hooded look; upperparts have shaft streaks or dark inner markings. Whitish below, with indistinct breast streaks that extend along

flanks. **JUVENILE:** Upperparts brighter than breeding adult, with rather broad rufous edges to crown, mantle, scapulars, and tertials. White mantle and scapular lines rather distinct. Rather bold white supercilium, contrasting with dark lores.

SIMILAR SPECIES Similar to Baird's Sandpiper in structure, but is grayer overall, has an entirely white rump, a bolder, whiter supercilium, and a pale base to the lower mandible. In flight, Curlew and Stilt Sandpipers both show a white rump; however, they show longer legs sticking out past the tail and larger, more curved bills. **VOICE CALL:** A high-pitched *jeet*.

STATUS & DISTRIBUTION Fairly common. **BREEDING:** Moist tundra from north AK to Baffin Island. **MIGRATION:** Late spring migrant, with majority of the population moving through the middle of the U.S. and southern Canada. Although first arrivals reach southern states after mid-Apr., they peak late May–early June; small numbers are still moving through southern states in early June, the Great Lakes area in mid-June, with a few to late June. In fall, most birds take an Atlantic route to their wintering grounds, moving east to Atlantic Canada or the northeast U.S. Adults are found south of the breeding grounds in late July, but most adults migrate Aug.–early Sept. Juveniles are very late migrants, rarely recorded south of Canada before late Sept.; they peak in Oct. and are seen regularly into Nov., strays into Dec. Juveniles yet to be documented in the western states. **WINTER:** Southern S.A. **VAGRANT:** Casual in the West, late May–mid-June; a few fall adults late July–Oct. Rare on the Azores; casual in Europe; regular in Britain and Ireland (with many more records than Pacific states). **POPULATION** Stable.

juveniles not yet recorded in West, not seen in Northeast until late Sept.

juvenile

breeding

dull rufous edges to scapulars

fall migrant adults often in transitional plumage; adults appear in Northeast by early Aug.

fall-molting adult

streaking extends to flanks

long wings and primary projection

juvenile

white rump normally only visible in flight

BAIRD'S SANDPIPER *Calidris bairdii*   BASA ▪ 1

This long-winged sandpiper often feeds up from the shoreline, on drier surfaces, where it pecks for food. Monotypic. L 7.5" (19 cm)
IDENTIFICATION Long primary tips, projecting beyond the tail and tertials when standing. Brown upperparts; breast a lighter brown, with light streaks. Thin dark bill; dark legs. **BREEDING ADULT:** Mixture of brown and buff on dark scapulars and coverts; plain tertials. Rather indistinct pale supercilium; dark lores. In worn plumage, adults look blotchy black and brown above. **NONBREEDING ADULT:** Paler than breeding plumage; rather nondescript light brown on upperparts, slightly paler brown (or gray-brown) on breast. **JUVENILE:** Pale edgings on back give a distinctly scaled appearance. No mantle or scapular lines. Light brown or buffy breast contrasts with white belly.
SIMILAR SPECIES Distinguished from Least Sandpiper, our other brown sandpiper, by its larger size, longer wings, longer and straighter bill, lack of rufous in the plumage, and black legs. Some juvenile Semipalmated Sandpipers can recall a Baird's with their brown plumage and pale feather edges, but Semipalmateds are less elongated, lacking the obviously long wings of Baird's, and breast is not as notably brown. In flight, Pectoral Sandpiper can look and sound somewhat like a Baird's. Pectoral has bolder white sides to the rump and a clearer demarcation at the breast; also, the brighter edges to its upperparts are often visible.
VOICE CALL: A low, raspy *kreeep*.
STATUS & DISTRIBUTION Fairly common. **BREEDING:** Dry tundra from northeast Russian Far East and west AK to Baffin Island and Greenland. **MIGRATION:** In spring, through mid-continent, from TX north through the Great Plains; rare in spring west of the Rockies, east of western Great

Lakes. Early birds documented in early Mar. (exceptionally Feb.), but most pass late Mar.–mid-May. Lingerers found to early June. In fall, most migration is mid-continent, but uncommon at this season to both coasts, usually juveniles. Adults first noted early to mid-July, a few in late June. Many adult Baird's fly thousands of miles in just 5–6 weeks, so they are less likely encountered than juveniles, although adults do pass through the Great Plains. In fall, most observers see juveniles, which peak mid-Aug.–mid-Sept. Numbers drop until rare after early Oct. on West Coast, after late Oct. elsewhere, lingerers to Nov. or even later on occasion. **WINTER:** S.A. **VAGRANT:** HI, Europe, Australasia, and South Africa. **POPULATION** Stable.

like White-rumped, long wings extend beyond tail

shaped like White-rumped, but browner

whitish fringe on upperparts

juvenile

juvenile

breeding

STILT SANDPIPER *Calidris himantopus* STSA ■ 1

whitish rump
and grayish tail

feet project
well beyond tail

distinct pale
supercilium

juvenile   juvenile

winter

chestnut
cheek and
crown

barred
underparts

long,
slightly
down-
curved bill

breeding

molting
juvenile

paler, smaller, and
slimmer than dowitchers;
faintly streaked below

long,
greenish
legs

The Stilt Sandpiper frequently feeds with dowitchers in belly-deep water, its head down and rear end up, as it probes into the mud for food. Monotypic. L 8.5" (22 cm) **IDENTIFICATION** Long-legged with a long, slightly downcurved bill; legs are yellow-green. **BREEDING ADULT:** Chestnut patches on sides of head and rear of crown contrast with dark gray upperparts and pale eyebrow. Heavy barring on the breast thins out somewhat on the belly and lower flanks. Males average brighter than females. **NONBREEDING ADULT:** Gray-brown above including crown, auriculars, and lores, contrasting with white supercilium. Underparts are mostly whitish, with some gray on the breast. **JUVENILE:** Plumage similar to nonbreeding plumage, but scapulars and coverts have broad pale edges.

When plumage is fresh, a buff wash might be present on the breast; dusky breast streaking, and some spotting on the flanks is possible, too. Migrants are often seen with molted scapulars. **FLIGHT:** White uppertail coverts (except in breeding plumage), pale tail, and faint white wing stripe; legs extend noticeably beyond tail. **SIMILAR SPECIES** Curlew Sandpiper is similar, but Stilt Sandpipers has a somewhat straighter bill and yellow-green legs that are noticeably longer. Also, the underparts are always more marked, and in flight, Stilt lacks a prominent wing stripe. See dowitchers. **VOICE CALL:** A low, hoarse *querp,* but often silent. **STATUS & DISTRIBUTION** Common. **BREEDING:** Tundra from AK to

Hudson Bay. **MIGRATION** Primarily migrates mid-continent in spring; first arrivals late Mar., but typically peak in southern states mid-Apr.–early May; closer to mid-May in KS, and the latter half of May in the Prairie Provinces. Very rare in the West, except at south end of Salton Sea where it can be fairly common. Common in fall on East Coast; rare but regular in West, primarily juveniles. Adults first seen early or mid-July, peak in KS late July, generally completing migration in early Aug. Juveniles first arrive in mid-Aug., peaking in Sept. Small numbers into Oct., rarely into Nov. **WINTER:** Scattered areas in southern U.S. (inc. Salton Sea, southern TX, southwestern LA, and southern FL); Mexico to S.A. **VAGRANT:** Europe, Asia, and Australia. **POPULATION** Stable. Possible recent western shift of breeding range.

CURLEW SANDPIPER *Calidris ferruginea* CUSA ■ 3

Curlew Sandpiper is a vagrant from the Old World. Monotypic. L 8.5" (22 cm) **IDENTIFICATION** More elegant than somewhat similar Dunlin. Long, black legs. Medium-length, black, rather evenly downcurved bill.

BREEDING MALE: Distinctive rich chestnut underparts and mottled chestnut back; grayish wing coverts. Scattered bars on sides and across uppertail coverts; white where bill joins face. **BREEDING FEMALE:** Slightly paler. Many

birds show patchy spring plumage or molt to winter plumage, showing grayer upperparts and partly white underparts. **NONBREEDING ADULT:** Gray-brown above, with white edges when fresh. White uppertail coverts unbarred. **JUVENILE:** Appears scaly above; shows rich buff across breast.

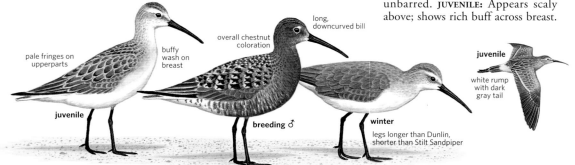

pale fringes on
upperparts

buffy
wash on
breast

overall chestnut
coloration

long,
downcurved bill

juvenile

white rump
with dark
gray tail

juvenile

breeding ♂

winter

legs longer than Dunlin,
shorter than Stilt Sandpiper

FLIGHT: White rump conspicuous; white wing stripe rather distinct. **SIMILAR SPECIES** In fall, young birds are in full juvenal plumage; compare with young Dunlins, which are mostly in winter plumage. Dunlin has a dark rump, shorter legs, and its bill is unevenly curved at the tip. Stilt Sandpiper has longer, greenish legs that in flight extend farther beyond its paler tail. Also note Stilt's fainter wing stripe and faint streaking on underparts.

VOICE CALL: A soft, rippling *chirrup*. **STATUS & DISTRIBUTION BREEDING:** Siberia and Russian Far East; has bred in northern AK. **WINTER:** Africa, Asia, Australasia. **VAGRANT:** Rare on East Coast (primarily mid-Atlantic), casual elsewhere. Spring records primarily late Apr.–late May. In fall, adults most likely encountered mid-July–mid-Aug., juveniles mid-Aug.–mid-Oct, a few records into Nov. **POPULATION** Threats due to loss of habitat in migration and winter.

DUNLIN *Calidris alpina* DUNL ■ 1

Dunlins frequently flock in large groups on the coast during winter and migration. They often feed in water, although they will also feed along the shore, probing into the mud for prey. Polytypic (10 ssp.; 6 in N.A.). L 8.5" (22 cm) **IDENTIFICATION** Mid-size sandpiper with a short neck; it appears hunchbacked. It has a sturdy, black bill, curved at the tip, and black legs. **BREEDING ADULT:** Reddish back is distinctive. Whitish, finely streaked underparts with a conspicuous black belly patch. **NONBREEDING ADULT:** Gray-brown above and on the breast; remainder of underparts are whitish. Molt takes place prior to arrival on winter grounds. **JUVENILE:** Like nonbreeding plumage, but with rufous on crown and edges of upperparts; a bold white mantle line is usually evident. Neck and breast are buffy with dark streaks that extend to the belly, where there is often a partial belly patch. This plumage is not expected south of the breeding range, as molting in N.A. subspecies occurs near the breeding grounds prior to migration. **FLIGHT:** Note dark center to rump and white wing stripe. **GEOGRAPHIC VARIATION** Subspecies *hudsonia* breeds in the Canadian Arctic and winters in eastern North America; it is marked by streaks on its flanks. Subspecies *pacifica* breeds in western Alaska and winters primarily on the West Coast. Asian *articola* breeds in northwestern AK, and Asian *sakhalina* migrates regularly through the central and western Aleutians. Greenland *arctica* and west Palearctic *alpina* have been recorded casually or accidentally on the East Coast **SIMILAR SPECIES** Rock Sandpiper in breeding plumage has a similar black patch, but it is on the chest, not the belly. In nonbreeding plumage, see Curlew and Western Sandpipers.

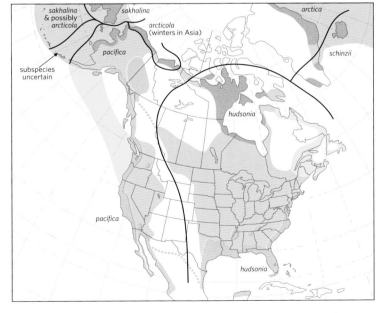

VOICE CALL: A harsh, reedy *kree.*
STATUS & DISTRIBUTION BREEDING: Abundant. **MIGRATION:** Peak in spring early to mid-May in mid-Atlantic and Great Lakes, 1–2 weeks earlier in the West; lingering birds into early June. Generally a late fall migrant, perhaps in part due to its molt prior to migration. Early arrivals usually in early Sept., only rarely earlier, with peak numbers mid-Oct.–mid-Nov. Rare on Great Plains. **WINTER:** On both coasts, across the Southeast, and south to Mexico. **VAGRANT:** To S.A.
POPULATION Local declines are suggested.

SPOON-BILLED SANDPIPER   *Calidris pygmeus*   SBSA ■ 4

This Asian vagrant is perhaps the most sought-after sandpiper due to its rarity and distinctive bill structure. It molts into breeding plumage late, so spring migrants are typically pale. The Spoon-billed is stintlike in its behavior and often forages in shallow water. Monotypic. L 6" (15 cm)
IDENTIFICATION The spoon-shaped bill is diagnostic, but the shape is hard to see at some angles; bill is longer than that of most peeps. **BREEDING ADULT:** Rufous head, throat, and upper breast; reddish edges to upperpart feathers. Dark spotting along lower edge of orange breast color, spreading onto the white lower breast and sides. **NONBREEDING ADULT:** Pale grayish above, paler than most congeners. White supercilium forked and quite broad in front of eye; darker auriculars. **JUVENILE:** Supercilium forked, as in nonbreeding, leaving white forehead that contrasts boldly with dark cheek patch and lores. Upperparts edged white or buff, sometimes rufous. Mantle and scapular lines, if present, usually indistinct. White underparts,

distinct supercilium
dark cheek
juvenile
breeding plumage similar to breeding plumage of Red-necked Stint
breeding
very distinctive spoon-shaped bill harder to discern at a distance

although a buff wash across the breast expected in fresh plumage.
SIMILAR SPECIES Similar in breeding plumage to Red-necked Stint, which usually has a hint of a white supercilium, although Spoon-billed averages more spotting on the underparts. Juvenile Spoon-billed has a darker cheek patch than juvenile Red-necked, emphasized by its forked supercilium. While no other sandpiper has a similar-looking bill, beware of mud creating a misleading blob.
VOICE CALL: A quiet *wheet.*

STATUS & DISTRIBUTION Endangered. **BREEDING:** Tundra in Russian Far East. **WINTER:** Southeast Asia. **VAGRANT:** Casual migrant in AK (no recent records); one fall record of a breeding adult from Vancouver region, BC (July 30–Aug. 3, 1978)
POPULATION Declining and endangered; only a couple hundred birds are thought to remain. Low population, disturbance on breeding grounds, hunting on winter grounds, and habitat loss at migration sites create a perilous existence for this species.

ROCK SANDPIPER   *Calidris ptilocnemis*   ROSA ■ 2

This western counterpart of the Purple Sandpiper is usually seen on rocky shores, often with Black Turnstones and Surfbirds. Polytypic (4 ssp.; 3 in N.A.). L 9" (23 cm)
IDENTIFICATION Shape like Purple Sandpiper; dark bill with pale yellow-green at base; yellow-olive legs. **BREEDING ADULT:** Rufous crown and edges to mantle and scapulars. Supercilium and dark on head variable. Black patch and gray streaks on lower breast; spots on flanks and belly. **NONBREEDING ADULT:** Dark gray upperparts and breast, with distinct spotting on flanks. **JUVENILE:** Like breeding, but paler, with buffy edges to scapulars and coverts; buff color to breast. Molts prior to reaching wintering grounds. **FLIGHT:** White wing stripe and all-dark tail.
GEOGRAPHIC VARIATION Nominate *ptilocnemis* (Pribilofs) is larger and has paler chestnut above, less

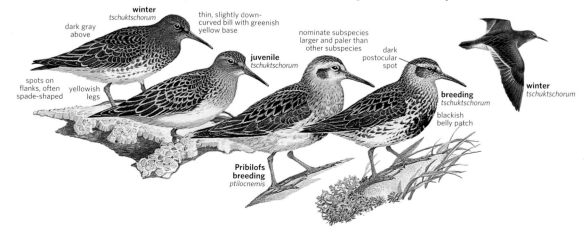

winter
tschuktschorum
dark gray above
thin, slightly down-curved bill with greenish yellow base
juvenile
tschuktschorum
nominate subspecies larger and paler than other subspecies
dark postocular spot
spots on flanks, often spade-shaped
yellowish legs
breeding
tschuktschorum
blackish belly patch
winter
tschuktschorum
Pribilofs breeding
ptilocnemis

black below, and a bolder white wing stripe. Aleutian *couesi* has a darker head and breast and more obscure auricular patch. The brightest breeding bird is *tschuktschorum,* from St. Lawrence I., the Seward Peninsula, and elsewhere.

SIMILAR SPECIES Separate a winter Rock from a Purple Sandpiper by its range and duller bill base and legs. Separate from a Surfbird by its longer bill, smaller size, and different tail pattern.

VOICE CALL: Rough *kreet,* like a Purple's.

STATUS & DISTRIBUTION Fairly common. **BREEDING:** Tundra from west

AK to Russian Far East. **WINTER:** Most birds arrive WA, OR, and northern CA late Oct.–late Nov. Generally leave vicinity of breeding grounds Sept.–Oct. Wintering birds depart by late Apr.–mid-May. Nominate subspecies winters Cook Inlet, AK. **VAGRANT:** Casual coastal southern CA. Accidental interior BC (Atlin, Oct. 29, 1932; specimen) **POPULATION** Stable.

PURPLE SANDPIPER *Calidris maritima*   PUSA   1

The stocky Purple Sandpiper is primarily noted in the Northeast, along rocky coastlines and jetties in winter. Monotypic. L 9" (23 cm)

IDENTIFICATION Long, slightly curved bill, with an orange-yellow base. Yellow legs. **BREEDING ADULT:** Seen only occasionally. Tawny crown streaked with black. Mantle and scapulars edged from white to tawny-buff; rufous usually restricted to scapulars. Breast and flanks spotted with blackish brown. **NONBREEDING ADULT:** Dark gray or slate upperparts, throat, and breast; blurred spots along sides and flanks; white belly. Purple tinge to upperparts difficult to see and also shared by Rock Sandpiper. **JUVENILE:** Like breeding, but less streaked on head and neck, more gray. Upperparts edged buffy or white. Completes molt prior to arriving on wintering grounds.

SIMILAR SPECIES In flight and in winter, Purple resembles a Rock Sandpiper. Separate by range; Rock has more distinct spotting on the breast and sides, less colorful base to the mandible.

VOICE CALL: A low, rough *kweet* or similar note, given in flight.

STATUS & DISTRIBUTION Fairly common. **BREEDING:** Tundra, often with gravel. **MIGRATION:** Typically reach wintering grounds in the Northeast in late Oct. (exceptionally mid-Sept.). Regular in fall on Great Lakes, esp. eastern lakes in late fall. Can linger in spring, well into May. Return to breeding grounds late May–early June. The few Great Lakes spring records range late Mar.–mid-May. **WINTER:** Rocky shores and jetties, often with Ruddy Turnstones and Sanderlings. **VAGRANT:** South of normal winter range along the coast, and

annual to scattered inland areas of the eastern U.S. Casual in winter to FL and Gulf Coast, some have lingered through Apr. Accidental in fall at Pt. Barrow, AK, and southwest UT. **POPULATION** Declines have been noted in Quebec.

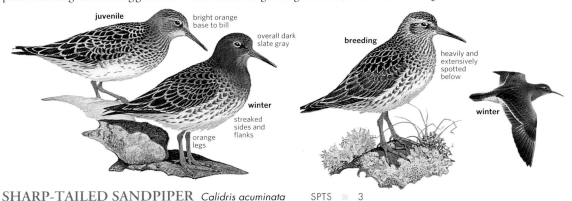

juvenile

bright orange base to bill

overall dark slate gray

breeding

heavily and extensively spotted below

winter

streaked sides and flanks

orange legs

winter

SHARP-TAILED SANDPIPER *Calidris acuminata*   SPTS   3

The Sharp-tailed Sandpiper is an Old World vagrant. Monotypic. L 8.5" (22 cm)

IDENTIFICATION Similar to Pectoral Sandpiper, and often seen with that

species during migration. Dark bill might have pale base; legs greenish. **BREEDING ADULT:** Rufous crown and edges to scapulars and tertials. Buffy breast, spotted below with dark

chevrons on flanks, usually extending to undertail coverts. Distinct white eye ring, rather indistinct supercilium, and dark lores. **NONBREEDING ADULT:** Paler than breeding adult, lacking most rufous and buff tones. Some spots and chevrons on underparts. JUVENILE: Like breeding adult on upperparts, with rufous cap and edges to scapulars and tertials;

breeding adult

ruddy crown and bold white eye ring

extensive dark chevrons on sides and flanks

streaks on undertail coverts

juvenile

broader rufous tertial edges than Pectoral

juvenile Sharp-tailed (center) with juvenile Pectorals

juvenile

bold supercilium

extensive buff on breast below streaking

streaks on sides continue faintly across upper breast

bold supercilium. Bold white mantle and scapular lines. Underparts more lightly marked than on a breeding bird; rich buff breast lightly streaked on upper breast and sides; streaked undertail coverts.

SIMILAR SPECIES Most sightings in N.A. are of juveniles, distinguished from juvenile Pectoral by a bolder white supercilium that broadens behind the eye; bright buffy breast lightly streaked on upper breast and sides only; streaked undertail coverts; brighter rufous cap and edging on upperparts; and call.

VOICE CALL: Mellow two-note whistle, to-wheet.

STATUS & DISTRIBUTION Breeds in Russian Far East, where adults migrate inland and juveniles move to the coast. Casual spring, and irregularly fairly common fall migrant in western AK. Rare in fall and casual in spring along entire Pacific coast, with a few inland records from YK south to CA. Accidental in spring and casual in fall across rest of continent. WINTER: Australia.

POPULATION Stable.

PECTORAL SANDPIPER *Calidris melanotos*   PESA ∎ 1

A favorite among most birders, the Pectoral Sandpiper is found in wet fields, marshy ponds, and similar wetlands. It is often seen in small groups, and its migration is quite long. Monotypic. L 8.8" (22 cm)

IDENTIFICATION Very unusual within the genus *Calidris,* the male Pectoral (like male Ruff) is notably larger than the female. This upright sandpiper has a dark bill with a pale base and yellow legs that might have greenish tones. BREEDING ADULT: Upperparts are brown, with rufous or buff edges. Below, prominent streaking on breast, darker in male, contrasts with clear white belly. NONBREEDING ADULT: Plumage paler and more nondescript than breeding plumage. JUVENILE: Brighter above than the breeding plumage, with bold white mantle and scapular lines; scapulars and coverts are broadly edged in rufous and white. Buffy breast is lightly streaked. FLIGHT: Weak

wing stripe but bold white uppertail coverts, separated by a black bar.

SIMILAR SPECIES Especially compare with juvenile Sharp-tailed Sandpiper (see that account). See also Baird's Sandpiper. In flight, Pectoral's bold white sides of the rump recall a Ruff, but they are not as extensive and not oval-shaped (see Ruff for more detail).

VOICE CALL: A rich, low *churk.* SONG: A deep hoot, with chest distended.

STATUS & DISTRIBUTION Fairly common to common (Midwest). BREEDING: Moist tundra from central Siberia to Hudson Bay. MIGRATION: Common in Midwest; fairly common on East Coast; scarcer from western Great Plains to West Coast. In spring, although the majority of migrants do not reach us until Apr., they are numerous in Mar. in several states, and as early as late Feb. Peak migration late Apr.–early May in Great Lakes and mid-Atlantic coast

(where less numerous), with lingering birds to late May. Very rare west of Rocky Mountains. In fall, they are more numerous along both coasts than in spring, with juveniles accounting for most West Coast records. Adults first arrive early to mid-July but have been seen in late June. Adults decline by late Aug., when the first juveniles arrive, peaking mid-Sept.–mid-Oct. This species often lingers, with a few noted into Nov. and later, even to Dec., in Great Lakes area. WINTER: Casual in the southern states; most winter in southern S.A.; annual in Australasia. VAGRANT: Annual in northwestern Europe; casual elsewhere in Europe, south to Africa.

POPULATION Stable.

breeding ♂

darker breast

breeding ♀

has pale Vs on back and scapulars, is less scaly appearing above than juvenile Baird's

long wings and primary projection

yellowish legs

juvenile

pectoral band of streaks

juvenile

BUFF-BREASTED SANDPIPER *Calidris subruficollis*   BBSA ■ 1

This elegant species is often encountered in plowed or grassy fields. Males join leks in search of mates, using bold wing displays. Monotypic. L 8.3" (21 cm)
IDENTIFICATION Dark eye prominent on buffy face; underparts paler buff.

juvenile

dark crescent

displaying adult

spots on sides of breast

whitish under-wing

juvenile

feathering extends out lower mandible

dark eye stands out in blank face

buffy underparts

breeding adult

yellow legs

Pale orange-yellow legs. Dark bill. **ADULT:** Dark scapulars and coverts, with rather broad buff edges when fresh. **JUVENILE:** Like adult, but scapular edges duller and more narrow, look distinctly scaled. Wing coverts with dark subterminal marks, in addition to the pale fringe. Paler below than adult. Legs not as bright. **FLIGHT:** Bright white wing linings; darker underwing primary coverts. **SIMILAR SPECIES** Unique, but see Ruff. **VOICE CALL:** Generally silent, but utters a low *tu*. **STATUS & DISTRIBUTION** Locally fairly common. **BREEDING:** Tundra. **MIGRATION:** Via interior of N.A. in spring. Arrivals typically early Apr., with peak late Apr.–mid-May, peak in northern Great Plains mid-May; few records to Great Lakes and Atlantic, or in the West. In fall, adults reverse migration through mid-continent; juveniles more likely to wander, very rare on the West Coast, very uncommon in East, at both drying wetland edges and sod farms. Migrate from late July, but peak late Aug.–late Sept., lingerers through Oct. **WINTER:** Southern S.A. **VAGRANT:** Europe, Africa, Asia, and Australia. **POPULATION** Stable.

BUFF-BREASTED AND WHITE-RUMPED SANDPIPERS

mi 1000
km 1000

BUFF-BREASTED SANDPIPER
— Autumn migration
— Spring migration
☐ Breeding range
☐ Wintering range

WHITE-RUMPED SANDPIPER
-- Autumn migration
-- Spring migration
☐ Breeding range
☐ Wintering range

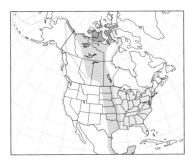

DOWITCHERS Genus *Limnodromus*

The dowitchers are medium-size, chunky, dark shorebirds, with long, straight bills and distinct pale eyebrows. They feed in mud or shallow water, probing with a rapid jabbing motion. In flight they show a white wedge from barred tail to middle of back. Separating the species is easiest with juveniles, difficult with breeding adults, and very difficult in winter.

SHORT-BILLED DOWITCHER *Limnodromus griseus* SBDO ▪ 1

The Short-billed Dowitcher is often encountered in large flocks along the coasts during winter or migration. The Short-billed and Long-billed Dowitcher present very difficult identification challenges. Polytypic (3 ssp.; all in N.A.). L 11" (28 cm)

IDENTIFICATION Rather long, gray bill (shorter in eastern birds); females have longer bills. Alternating light and dark bars, usually more light than dark, on tail. Primaries usually extend a bit past the tail. **BREEDING ADULT:** Dark upperparts with orange feather edges. Underparts a mixture of orange and white, with spots and bars (vary by subspecies). **NONBREEDING ADULT:** Uniform brownish gray above, feathers without darker center. White below with barring on the sides that contrasts with upperparts; gray breast with fine speckling visible at close range. **JUVENILE:** Dark scapulars and coverts, boldly edged orange; tertials, and some coverts, with intricate internal markings and loops. Dark crown contrasts boldly with white supercilium. Orange-buff neck and upper breast; color sometimes extends farther through the underparts. **FLIGHT:** White trailing edge to the secondaries, white wedge from tail through lower back. Tail barred dark gray and white.

GEOGRAPHIC VARIATION Three subspecies: nominate *griseus* (breeds northeast Canada), *hendersoni* (central and western Canada), and *caurinus* (AK). Most Short-billeds show some white on the belly, especially *griseus,* which also has a heavily spotted breast and may have densely barred flanks. The Pacific *caurinus* is intermediate, but more similar to *griseus*. In *hendersoni,* which is mostly cinnamon buff below, the foreneck is much less heavily spotted than in the other subspecies. In juveniles, *caurinus* averages duller, with more narrow markings, although *griseus* is quite similar. Both are duller than *hendersoni,* which can be quite rufous, with broad edges and more buffy below.

SIMILAR SPECIES See Long-billed Dowitcher and Red Knot.

VOICE CALL: A mellow *tu tu tu,* repeated in a rapid series as an alarm call. More likely to remain silent than Long-billed Dowitcher. **SONG:** A rapid *di-di-da-doo* year-round.

STATUS & DISTRIBUTION Common. **BREEDING:** Muskegs in boreal forest. **MIGRATION:** Common along the Atlantic coast *(griseus)*; from the eastern Plains to Atlantic coast from NJ south *(hendersoni)*; and along the Pacific coast *(caurinus)*; a few *griseus* are seen on eastern Great Lakes

pattern of breeding *caurinus* intermediate between *griseus* and *hendersoni*

breeding *caurinus*

worn breeding *griseus*

some white on lower belly

numerous black spots on sides of breast

extensive buffy markings above

spots on sides of breast

white belly

breeding *griseus*

breeding *hendersoni*

more extensively colored underparts than *griseus* with fewer dark markings

faint spots on breast

molting juvenile

bold internal marks on tertials and greater coverts

juvenile

winter

both dowitcher species have white stripe up back

internal bars and stripes

juvenile tertials

griseus

winter *hendersoni*

in late spring, and a few *hendersoni* move through some interior western states. Spring movement for *griseus* begins mid-Mar., peaks late Apr.– early June, particularly mid-May. In the interior, *hendersoni* peaks on the Great Lakes and Great Plains mid-May; *caurinus* a bit earlier, with CA peak mid-Apr., WA late Apr., and

southern AK ±10 May. Fall migration begins earlier than for Long-billed; in three distinct waves, led by adult females, as early as late June; then adult males (adults complete most of migration by Aug.), followed by juveniles, typically arriving in early Aug., peaking through early Sept., with small numbers into early Oct.;

lingerers to early Nov. They molt when they reach wintering grounds; late juveniles still in their juvenal plumage. **WINTER:** Coastal habitats to S.A., almost unknown away from the coasts. **VAGRANT:** A few records for Europe.
POPULATION There are indications of a decline in *griseus*.

LONG-BILLED DOWITCHER *Limnodromus scolopaceus* LBDO ▪ 1

The Long-billed Dowitcher, like the Short-billed Dowitcher, can be seen in large flocks in winter and on migration. It has a preference for fresh-water, although it does frequent coastal estuaries. Monotypic. L 11.5" (29 cm)
IDENTIFICATION The size of the bill is not a particularly helpful feature for identification purposes since all dowitchers have rather long bills. In addition, the female Long-billed's bill is notably longer than the Short-billed's, but the male's bill is not. The tail is barred dark and white, with the dark bars wider than the white bars; the primaries do not extend beyond the tail. **BREEDING ADULT:** The underparts are entirely reddish below, except for whitish edges to fresh feathers that wear off in early spring. The foreneck is heavily spotted and the sides are usually barred. The black scapulars have rufous markings and show white tips in spring. As this plumage wears, usually in July and

August, the bird looks uniform rufous below and a mixture of rufous and black above, making identification difficult. **NONBREEDING ADULT:** The scapulars have dark centers, giving a more mottled appearance; the upperparts are gray-brown. The breast is rather dark gray and is unspotted. **JUVENILE:** It is dark above; the dark crown moderately contrasts with the supercilium. The tertials and greater wing coverts are plain, with thin gray edges and rufous tips; some birds show two pale spots near the tips. The underparts are gray, with only hints of warmth on the breast.
SIMILAR SPECIES Separating the Long-billed from the Short-billed Dowitcher ranges from the straight-forward to the very difficult. A familiarity with variation, molt, and vocalizations is needed and will result in a high percentage of identifications; many dowitchers, however, will likely be left unidentified. Start with the age, and consider which

Short-billed subspecies need to be eliminated. In full breeding plumage, the rufous underparts of a Long-billed are easy to distinguish from those of a breeding *griseus* and many *caurinus,* but they are similar to those of *hendersoni*. Look for white tips to the scapulars of the Long-billed and small dots restricted to the sides of breast of *hendersoni,* whereas Long-billed has spots and bars on the underparts. Note the lighter tail in Short-billed, owing to the wider white bars. This tail feature is hard to assess, further compounded by *griseus* having wider dark bars than the other Short-billed subspecies; *griseus,* however, also shows wavy internal bars on rectrices that are lacking on Long-billed. Breeding Long-billeds have rufous on their central tail feathers, unlike any Short-billed Dowitcher. Adult Long-billeds go to favored locations (often gathering in large numbers) in late summer to molt; Short-billeds molt when they reach winter grounds. In nonbreeding plumage, Long-billeds are darker, with a darker and browner breast that lacks the fine spotting shown by Short-billeds. Juvenile

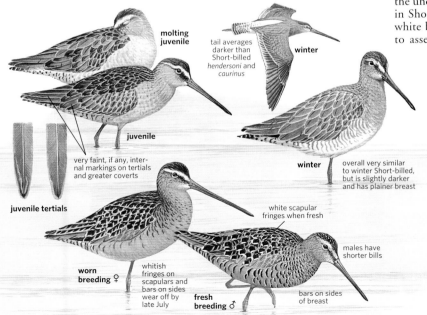

molting juvenile

tail averages darker than Short-billed *hendersoni* and *caurinus*

winter

very faint, if any, internal markings on tertials and greater coverts

juvenile

juvenile tertials

winter

overall very similar to winter Short-billed, but is slightly darker and has plainer breast

white scapular fringes when fresh

worn breeding ♀

whitish fringes on scapulars and bars on sides wear off by late July

fresh breeding ♂

males have shorter bills

bars on sides of breast

Long-billeds are darker above and grayer below than Short-billeds; they also lack the internal feather markings found on the tertials and greater coverts of Short-billeds. One other species could cause some confusion: the Red Knot in breeding plumage. It has a shorter bill and shorter, darker legs, however, and lacks the white wedge up the back.

VOICE CALL: A sharp, high *keek*, given singly or in a rapid series. **SONG:** Like the Short-billed Dowitcher.

STATUS & DISTRIBUTION Common. **BREEDING:** Tundra from northeast Russian Far East to northwest Canada; a restricted and more northerly range than Short-billed. **MIGRATION:** Common in western half of continent; generally uncommon in the east in fall, rare in spring. Spring migration is earlier than the Short-billed's, arriving in nonwintering areas late Feb.–mid-Mar., peaking Great Plains and Great Lakes late Mar.–early May, before most Short-billeds. Arrive breeding grounds mid- to late May. Fall migration begins later than Short-billed's, in mid-July (West) or late July (East). Juveniles migrate later than adults; casual in lower 48 before Sept. Dowitchers seen inland after mid-Oct. are almost certainly Long-billed. **WINTER:** South to C.A., casual to northern S.A. **VAGRANT:** Europe (where annual), Asia, and HI.

POPULATION Population is stable; appears to be expanding west in Siberia.

SNIPES Genera *Lymnocryptes* and *Gallinago*

Snipes are stout, long-billed shorebirds that frequent bogs, marshes, and other similar wetlands. Some of the 19 species are poorly known, isolated on islands around the world. Snipes are generally solitary, but they can collect into flocks during migrations. Their heavily striped, cryptic patterns allow them to blend into their surroundings. They probe the mud for food, and the males' display flights during breeding season are spectacular.

JACK SNIPE *Lymnocryptes minimus*   JASN ■ 4

The Jack Snipe, a Eurasian vagrant, is most frequently detected by hunters. Monotypic. L 7" (18 cm)

some dark oily green above
stripes create V's on mantle
split pale eyebrow
streaked flanks
short bill with pale base
bobs while feeding

IDENTIFICATION The Jack Snipe is small and chunky. Secretive, it is reluctant to flush. Its flight is low, short, fluttery, on rounded wings; it is somewhat rail-like. It bobs while feeding. The short bill has a pale base. The tail is dark and wedge-shaped. It has a bold head pattern with a split supercilium; no median crown stripe. The broad, buffy back stripes are striking, and there is some dark, oily green iridescence on the scapulars. The underparts are pale, with streaking on both the breast and flanks; there is no barring.

SIMILAR SPECIES The Jack Snipe has the general coloration and shape of a snipe, but it is much smaller.

VOICE CALL: Generally silent.

STATUS & DISTRIBUTION **BREEDING:** Forest bogs of northern Eurasia. **WINTER:** Europe, Africa, and Asia. **MIGRATION:** A vagrant migrant. Hunters have shot a few Jack Snipes, which suggests that they are more frequent than the few records would suggest. But their secretive behavior, resistance to flushing, and immediate return to cover once they are flushed make detection difficult. Furthermore, most birders likely do not consider this species when they are flushing snipes and might overlook the occasional vagrant. **VAGRANT:** Three late-fall records for CA, two for OR, and singles for Kenai Peninsula, AK, and Labrador. A sight record for WA in early Sept. (extraordinary if correct). Single spring records from St. Paul I. and St. Lawrence I.

POPULATION Declines in the 20th century that resulted from habitat degradation appear to have stabilized.

WILSON'S SNIPE *Gallinago delicata*   WISN ■ 1

The Wilson's is generally seen singly, standing in wet, grassy habitats, although flocks of dozens can be encountered during winter. It probes the mud with its long bill, much like a dowitcher. This species was called the Common Snipe until the recent decision to re-split Wilson's from that species, restoring both to full-species rank. Monotypic. L 10.3" (26 cm)

IDENTIFICATION The Wilson's is a stocky shorebird with a very long dark-tipped bill, pale at the base; the legs are greenish gray. The head is boldly striped with pale lines above and below the auriculars and in the middle of the crown, all separated by dark brown. The upperparts are dark brown with pale buff or white lines. Below, the breast is brown with dark streaks, and the flanks are heavily barred. The outer rectrices are barred black-and-white, with no trace of orange. **FLIGHT:** It often explodes from the ground straight up into the air. The general impression is a brown breast and white underparts, along with dark gray underwings. The toes do not extend beyond the tail in flight.

SIMILAR SPECIES It is very difficult to separate the Wilson's from a Common Snipe; see that species account.

More likely to be confused with dowitchers in most of North America, as they share a chunky, long-billed

profile. Dowitchers, however, do not share the extensive head stripes or light buffy lines on the upperparts, and snipes lack the dowitcher's white wedge up the back. See also American Woodcock.

VOICE CALL: A raspy, two-note *ski-ape,* given in rapid, zigzagging flight when flushed. On nesting grounds, the male delivers loud *wheet* notes from perches. In swooping display flight, vibrating outer tail feathers make quavering hoots, commonly referred to as "winnowing," similar to the song of the Boreal Owl.

STATUS & DISTRIBUTION Common, but overlooked. **BREEDING:** Marshes. **MIGRATION:** In spring, southern states see movement in late Feb.–early Mar.

Peak Pacific Northwest to Great Lakes late Mar.–late Apr., lingerers recorded to June. Fall migration begins as early as mid-July, but peak mid-Sept.–late Oct. **WINTER:** To northern S.A. **VAGRANT:** Casual in U.K. and the Azores.

POPULATION There have been local declines, but the population appears stable overall.

faint trailing edge

displaying

strongly striped head

dark under-wing

bold pale stripes on upperparts

heavily barred sides and flanks

underwing

COMMON SNIPE *Gallinago gallinago* COSN ▪ 3

The Common Snipe, a visitor from the Old World, is so similar to the Wilson's Snipe that it could easily be overlooked. Polytypic (2 ssp.; nominate in N.A.). L 10.5" (27 cm)

IDENTIFICATION The long bill has a black tip and a pale base; the legs are greenish gray. The bold head pattern has alternating light and dark lines from the light midcrown stripe through the auriculars. The upperparts are dark, with buffy lines running along the back from the head toward the tail. The tertials are heavily barred, all the way to the base. The tail is composed of 14 feathers, the outer feather on each side is broad, and the inner web of these feathers has an orange coloration. Below, the breast is brown with dark streaks; the flanks are barred. **FLIGHT:** Like Wilson's, the Common is likely to explode up from the ground and fly straight up in the air. As it flies overhead, it gives the impression of a dark-breasted bird with a white belly. The underwing has substantial white, although juveniles have more gray than adults. There is a rather broad white trailing edge to the secondaries, easily visible in flight. **SIMILAR SPECIES** The Common Snipe is very similar to the Wilson's; close scrutiny is needed. There are features that average different between the two species. The Common Snipe has a paler, buffier color overall than the Wilson's and slightly fainter flank markings; however, sufficient variation renders these marks suggestive, not diagnostic. The broader white trailing edge to the secondaries and

paler white-striped underwings are consistent features separating the two species. The outer tail feathers are also diagnostic, but you'll need great views of a preening bird or a well-timed photograph; the Common Snipe has one broad feather, with orange on the inner web, versus the Wilson's two narrower feathers, the outer of which lacks orange. **VOICE CALL:** A two-note *ski-ape,* similar to the Wilson's. **SONG:** Male flight display notes distinctly lower pitched than the Wilson's. **STATUS & DISTRIBUTION BREEDING:** Throughout northern Eurasia; it has bred in western Aleutian Is. **MIGRATION:** Regular migrant to the western Aleutian Is. (where it has also been found casually in winter); rare to the central Aleutian and Pribilof Is.; casual to St. Lawrence I. **WINTER:** Southern Europe, Africa, and S.E. Asia. **VAGRANT:** Accidental winter records for: Newfoundland (photograph), Labrador

(specimen), and Riverside, CA (specimen taken by a hunter in Dec. 2011). Given the difficulty in identifying this species, a review of hunters' snipe collections might prove valuable to assess whether this species is more numerous in N.A. **POPULATION** Stable.

bolder white trailing edge

underwing

whitish panel on underwing coverts

gallinago

warmer background color to head, breast, and sides

flank barring fainter than Wilson's

bold pale stripes on upperparts

PIN-TAILED SNIPE *Gallinago stenura*   PTSN ▪ 5

The Pin-tailed Snipe is a vagrant from Asia. Monotypic. L 10" (26 cm)
IDENTIFICATION The Pin-tailed's bill is relatively short and thick for a snipe. It has a bold head pattern with alternating dark and light bars from the pale mid-crown stripe through the auriculars. The expansion of the buff supercilium in front of the eye is broad, leaving a small dark lore stripe and a small dark stripe at the top of the bill. The upperparts are dark with rufous internal markings and rufous-and-white edges. There is a bold white mantle line; it does not have a series of parallel lines. Many of the scapulars and coverts have white around the tip of the feathers, extending to both sides, leaving a slightly more scaled look. In hand, the razor-thin outer tail feathers are diagnostic; this feature is very difficult to see in the field, except perhaps on a preening bird. **FLIGHT:** There is a fairly obvious buffy secondary covert patch; while not the intensity of the patch of a Least Bittern, for example, it is unlike that of the other two snipe species. The secondaries lack pale edges. Below, the underwings are uniformly dark, and the feet distinctly project past the tail.
SIMILAR SPECIES The Pin-tailed most likely needs to be separated from the Common Snipe in western Alaska. Pin-tailed is chunkier, shorter billed, and shorter tailed. On a perched bird, note the barred secondary coverts and even-width pale edges on the inner and outer webs of the scapulars that give a scalloped look. Also, the Pin-tailed's face pattern subtly differs with a particularly broad supercilium in front of the eye. The underwing is the same between the Pin-tailed and Wilson's Snipe, but Wilson's lacks toe projections in flight; the noticeable buffy patch on the upper wing of the Pin-tailed is unmatched by any snipe, except the larger Swinhoe's Snipe *(G. megala)*, which is unrecorded in N.A. but breeds in the nearby Russian Far East. No reliable field marks are known that separate these two species.
VOICE CALL: A high *squak,* sounding more like a duck or a pig.
STATUS & DISTRIBUTION The Pin-tailed breeds across much of Russia. **WINTER:** Primarily Southeast Asia. **VAGRANT:** Accidental, two certain records from western Aleutian Is.
POPULATION Stable.

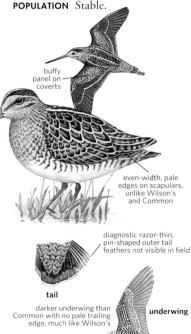

buffy panel on coverts

rather short bill

even-width, pale edges on scapulars, unlike Wilson's and Common

diagnostic razor-thin, pin-shaped outer tail feathers not visible in field

tail

darker underwing than Common with no pale trailing edge, much like Wilson's

underwing

SOLITARY SNIPE *Gallinago solitaria*   SOSN ▪ 5

Like many of the world's large snipe and woodcock, Solitary Snipe nests in remote, inhospitable habitats and is very little known. Nesting above tree line in bogs, swamps, and river valleys, the species moves into foothills and lowlands in winter. Polytypic (2 ssp.; presumably 1 in N.A.). 12" (29–31 cm)
IDENTIFICATION In a genus of cryptic birds, Solitary Snipe is distinctive. Its large size, potbellied shape, and long bill recall a large woodcock as much as a snipe, but the plumage pattern is clearly that of a snipe. **ADULT:** Upperparts overall quite dark, with dark brown and vivid red tones in mantle, scapulars, tertials, and upperwing coverts; mantle bordered by thin whitish "back braces." Underparts paler, with much barring in the flanks and a distinct brownish ginger tone in breast. Single black eye line; brown line lower on face. **FLIGHT:** Dark upperwings lack white trailing edge or other pattern.
GEOGRAPHIC VARIATION Two similar subspecies: *japonica* and nominate *solitaria.* Subspecies *japonica,* presumed to be responsible for N.A. records, is more rufous above, with narrower pale back braces (and generally more uniformly toned upperparts) and a longer bill.
SIMILAR SPECIES Compare to other medium-size snipe: Wilson's, Common, and Pin-tailed. Larger Asian snipe, even though unrecorded in N.A., should also be considered carefully, including Latham's *(G. hardwickii)* and Swinhoe's *(G. megala).* Solitary's single dark eye line extending from the base of the bill to the nape appears to be distinctive; other snipe show a variably thin dark line through the lores that either expands into two parallel lines behind the eye or broadens into a dark streak.
VOICE SONG: Displaying birds give a long series of *choka* calls in flight over territory. **CALL:** An annoyed, low, nasal *kentsh.*
STATUS & DISTRIBUTION BREEDING: High-elevation wetlands of Kazakhstan, Kyrgyzstan, and Mongolia east through northern India and China and parts of Russia east to Kamchatka. The breeding range of *japonica* is not well known but apparently includes part of the Kamchatka Peninsula; the breeding range of nominate *solitaria* covers much of Central Asia. **WINTER:** Mostly in wetland areas near breeding range, at lower elevations; regularly found in winter in Japan, the Koreas, northern India, Pakistan, Iran, and eastern China; *japonica* appears to winter largely in Japan. **VAGRANT:** Accidental; two (1 certain) N.A. records, both from AK: St. Paul I. (Sept. 10, 2008; supported only by marginal photographs) and specimen, Attu I. (May 24, 2010); also accidental to the Commander Is. (Russia).

overall dark coloration, including near solid area on breast sides

very long bill

WOODCOCKS Genus *Scolopax*

Six species. Two are widespread: the American in eastern North America and the Eurasian in the Old World (casual to N.A.). Four others are range-restricted and are found on islands from southern Japan to Indonesia. Chunky and long-billed forest dwellers, all species probably have elaborate flight displays at dawn and dusk, though the island-restricted Asian species are poorly known.

EURASIAN WOODCOCK *Scolopax rusticola*   EUWO ▪ 5

A vagrant from Eurasia. Monotypic. L 13" (33 cm)

IDENTIFICATION The Eurasian Woodcock is a chunky, large-headed bird, with a long, dull pinkish bill that has a darker tip. Like all woodcocks, it has short round wings and a short tail. It is cryptically colored: earthy brown, combined with bars and stripes that allow it to blend into the forest floor. The hindcrown is barred with black; the lores are dark. The upperparts are dark brown, with paler brown markings and tips; there are pale lines on the mantle. The underparts are heavily barred, particularly the breast and flanks. The wing coverts and neck are also barred.

SIMILAR SPECIES The Eurasian Woodcock is similar to the American Woodcock, which is the expected species. The Eurasian Woodcock is considerably larger, browner, and darker with heavily barred underparts with less contrasty pale scapular lines.

VOICE CALL: North American birds have been silent.

STATUS & DISTRIBUTION Eurasian breeder. WINTER: South to N. Africa and Southeast Asia. VAGRANT: A few records, nearly all old—six specimens taken 1859–1890—primarily from the Northeast, inland to QC, PA, and, remarkably, AL. The only N.A. record since 1890 is one photographed at Goshen, NJ (Jan.

2–9, 1956). The species is casual to Greenland, including a specimen as recent as 1974.

POPULATION Millions of Eurasian Woodcocks are hunted annually in Eurasia, yet there is no clear trend in the population.

larger and darker than American Woodcock

heavily barred underparts

AMERICAN WOODCOCK *Scolopax minor*   AMWO ▪ 1

A bird of moist woodland floors and brushy marsh and field edges, this species is usually found singly as it probes and picks through soil and leaves for worms or other prey. It is secretive and seldom seen until flushed; it flies up abruptly and the wings make a twittering sound. Males have a spectacular display flight, typically performed at dawn and dusk. Monotypic. L 11" (28 cm)

IDENTIFICATION The American Woodcock is a chunky, atypical shorebird with rounded wings and a short tail. The long bill is dull pink with a darker tip. The large eyes stand out on a rather buffy face. Bold crossbars mark the hindcrown. Upperparts are dark with obvious pale mantle lines and upper scapulars with pale pinkish or buff edges. Underparts are buffy and unmarked. JUVENILE: It is very much like an adult; it is only separable early in its plumage, when the juvenile has a gray-brown throat and neck that contrasts with the white chin. FLIGHT: A stubby bird in flight with a long bill held downward, round wings, and a short tail. It looks dark brown above and warm buffy-orange below.

SIMILAR SPECIES Most North American birders will likely need to separate the American Woodcock from the Wilson's Snipe. Both species are chunky, with long bills, and share a generally brown coloration. Both also feed by probing in wet, vegetated areas, but woodcocks favor wet woods while snipe are found in marshes or wet fields. The snipe has stripes along its head as opposed to the bars that cross the head of a woodcock, and the woodcock's eyes are isolated on its pale face. The woodcock is also more colorful with rich buffy, unbarred underparts. In flight, the woodcock's rounder wings are noticeable.

VOICE CALL: Males vocalize during spring courtship, mostly at dawn and dusk. A loud buzzy *peent* is repeatedly given from the ground between display flights. When airborne, a long series of twittering notes is given in zigzagging flight, louder just before descent.

STATUS & DISTRIBUTION Common but local. BREEDING: Moist woodlands. MIGRATION: In spring, woodcocks move early; they are the earliest arriving and breeding northern shorebirds, with northern arrivals by late Feb. to early Mar. The first fall migrants are detected late Aug.–early Sept., but most migration is noted in Oct.–Nov. VAGRANT: Casual in West to MT, eastern CO, eastern NM. Accidental in southeastern CA, two fall records.

POPULATION Recent declines might be a result of loss of second-growth forests.

chunky, with short, rounded wings

dark bands on crown and nape

large eyes

rich buffy underparts

PHALAROPES Genus *Phalaropus*

These elegant shorebirds have partially lobed feet and a dense, soft plumage. Feeding on the water, phalaropes often spin like tops, stirring up larvae, crustaceans, and insects. Females are larger and more brightly colored than males, and the sexes reverse roles: females do the courting and males incubate the eggs and care for the chicks. In fall, they rapidly molt to winter plumage (esp. Wilson's and Red); many are seen in transitional plumage farther south. Red-necked and particularly Red Phalaropes spend the winter months on the open ocean (pelagic). Although some phalaropes stay farther north, most migrate to the tropics, and many winter well south of the Equator.

WILSON'S PHALAROPE *Phalaropus tricolor*   WIPH ■ 1

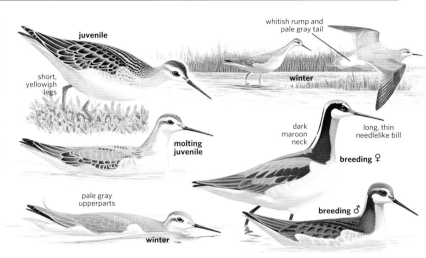

Wilson's Phalarope is a splendid species with beautiful breeding plumage and an impressive biology. Large flocks are encountered during migration. It is the only phalarope that breeds south of Canada. This species is the largest and most terrestrial of the phalaropes. It will swim in circles on shallow ponds, evoking laughter or surprise from those who have not seen it before; this spinning creates a vortex that delivers food to the surface. It also feeds on land, especially juveniles, where it is even more humorous when it walks in circles in a hunched-forward, almost awkward gait. Monotypic. L 9.3" (24 cm)

IDENTIFICATION The Wilson's long, thin bill is black and almost needlelike. **BREEDING FEMALE:** Has a bold black and rufous stripe on the face and neck. The warm orange breast contrasts with the white throat. The mantle and lower scapulars are rufous, separated by gray upper scapulars. The crown is whitish or pale gray; the whitish supercilium usually does not extend beyond the eye. **BREEDING MALE:** Like the female, only usually duller; the crown is dark gray. The upperparts have less contrast and are more mottled, having often dark-centered feathers with pale or rufous edges. Some males are quite dull, almost like nonbreeding plumage, except for the mottled upperparts and the dark, almost black legs. **NONBREEDING ADULT:** The legs are paler, olive to yellow. It is gray above, including gray crown and gray postocular stripe. The upperparts are edged white in fresh plumage. **JUVENILE:** The back is brown with buffy feather edges and the breast is buffy. It quickly molts out of this plumage; most juveniles seen south of breeding grounds are gray backed and identifiable as juveniles by the broad buffy edges to the dark wing coverts and tertials. **FLIGHT:** White uppertail coverts, whitish tail, and the absence of a white wing stripe are important characters. **SIMILAR SPECIES** The lack of the

"phalarope mark" through the eye; the long, thin bill, and the white uppertail coverts, whitish tail, and absence of white wing stripe distinguish juvenile and winter Wilson's from other phalaropes. They are more easily confused with a nonbreeding Stilt Sandpiper, but phalaropes have a straight (not curved) bill and less of a wing stripe, and their short legs do not extend well beyond the tail, as they do on the Stilt. The Lesser Yellowlegs, occasionally noted swimming, is larger and its bill is thicker, and Wilson's has a notably whiter breast. The Marsh Sandpiper, a vagrant from Eurasia, has a longer bill and longer legs; in flight it shows a white wedge up the back.

VOICE CALL: A hoarse *wurk* and other low, croaking notes.

STATUS & DISTRIBUTION Common. **BREEDING:** Grassy borders of shallow lakes, marshes, and reservoirs. **MIGRATION:** Common to abundant in western N.A.; uncommon to rare in the east. Late Mar.–mid-May, spring migrants pass through CA, peak late Apr.–early May. Farther north and East, extending to the Great Lakes, peak is mid-May, although they are rare, and uncommon at best. Adult females are the earliest fall migrants,

with the first arrivals in early June. Large flocks stage at key areas (e.g., Mono Lake, CA, and Great Salt Lake, UT), where hundreds of thousands molt into nonbreeding plumage. Juveniles, seen as early as early July, peak mid-July–Aug., lingerers into Oct., exceptionally later. **WINTER:** Casual in southern CA; recently found in southern TX, sometimes in small flocks. Most winter in S.A., on alkaline lakes in the Andes. **VAGRANT:** Annual in western Europe; casual to S. Africa, Australasia, the Galápagos, and the Falklands.

POPULATION Declines due to habitat loss and drought.

RED-NECKED PHALAROPE *Phalaropus lobatus*   RNPH   ▪   1

This species winters at sea, but migrating flocks occur at inland locales. Monotypic. L 7.8" (20 cm) **IDENTIFICATION** Small, relatively thin, dark bill. **BREEDING ADULT:** Dark above with buff stripes. Chestnut on front and sides of neck, less prominent in male. More prominent supercilium on male. **NONBREEDING ADULT:** Gray above with whitish stripes on the upperparts. Dark patch. **JUVENILE:** Like winter adult but darker above, with bright buff stripes. **FLIGHT:** White wing stripe, whitish

stripes on back, dark central tail coverts. **SIMILAR SPECIES** The Red Phalarope is similar in winter plumage. The smaller Red-necked has pale stripes on the slightly darker upperparts and a thinner, all-black bill, compared to the thicker bill of a Red. In fall, young Red-neckeds retain their dark juvenal plumage much longer than young Reds. **VOICE CALL:** A high, sharp *kit,* often given in a series. **STATUS & DISTRIBUTION** Common. **BREEDING:** Tundra. **MIGRATION:**

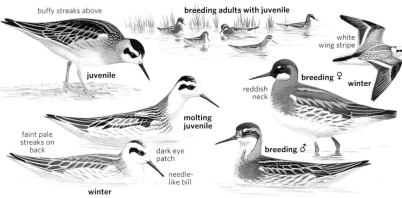

buffy streaks above

breeding adults with juvenile

juvenile

white wing stripe

breeding ♀

winter

reddish neck

molting juvenile

faint pale streaks on back

dark eye patch

breeding ♂

needle-like bill

winter

Common inland in West and off West Coast; rare in Midwest and East; uncommon off East Coast. In spring, migration peaks first half of May, although might appear by early Apr. and linger to June. In fall, in the lower 48, adults are seen early Jul.–mid-Aug., juveniles from early Aug.–late Oct. (mostly in full juvenal plumage). **WINTER:** Chiefly at sea in Southern Hemisphere, but records in southern CA mainly and southern TX. **POPULATION** Recent declines from the Bay of Fundy.

RED PHALAROPE *Phalaropus fulicarius*   REPH   ▪   1

Usually highly pelagic. Monotypic. L 8.5" (22 cm) **IDENTIFICATION** Short bill, thickest of three phalarope species. **BREEDING FEMALE:** Black crown; white cheek patch. Underparts rich chestnut. Upperparts black with buff or whitish lines. Bill mostly yellow, with a dark tip. **BREEDING MALE:** Duller; not as crisp. **NONBREEDING ADULT:** Pale, uniformly

smooth gray above; white below. Bold dark eye patch. Dark bill sometimes with small pale base. **JUVENILE:** Dark upperparts with buffy edges and streaks, peachy buff wash on neck. Most juveniles seen in southern Canada and U.S. have largely molted into winter plumage. **FLIGHT:** White wing stripe, bolder than Red-necked. **SIMILAR SPECIES** Similar to the Red-

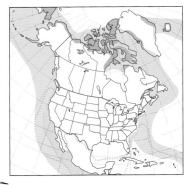

juvenile

molting fall adults

bolder wing stripe than Red-necked

winter

white cheek

molting juvenile

breeding ♀

plain gray upperparts

dark eye patch

red underparts

thicker bill than Red-necked

breeding ♂

winter

necked Phalarope; see that species. **VOICE CALL:** A sharp *keip.* **STATUS & DISTRIBUTION** Common. **BREEDING:** Tundra ponds. **MIGRATION:** In spring, Pacific migration from mid-Apr., but peaks late May. Generally uncommon off East Coast, peaking late May. In fall, adult females as early as late July, stages Aug.–Sept. (e.g., Bay of Fundy); juveniles migrate into Nov. Movements in late fall and early winter off both coasts irregular. **WINTER:** At sea to S.A. **VAGRANT:** Very rare to casual inland, mostly in late fall.

COURSERS AND PRATINCOLES Family Glareolidae

Oriental Pratincole, nonbreeding (Japan, Sept.)

Although ternlike in many aspects, pratincoles are indeed shorebirds. This family, which includes the terrestrial coursers, is unlike any North American family.

Structure Generally mid-size shorebirds, pratincoles have long, pointed wings; long, forked tails; and short legs. The combination of the three gives them a very horizontal, elongated look when perched. The bill is short. Coursers are upright birds with long legs and short toes adapted for running in deserts and brushlands.

Behavior Pratincoles are highly aerial, much more so than other shorebirds; unusual for waders, they rarely enter the water. Their flight is ternlike and can be high and fast. Quite gregarious, they often migrate in large flocks and nest colonially. They nest on bare ground, frequently in plains or desert-like habitats. They feed on insects (their main source of food) in the air, although they will chase insects on the ground. Their gait is ploverlike, alternating running and abruptly stopping.

Plumage In both subfamilies the sexes are alike. Pratincoles molt prior to fall migration, resulting in a more muted plumage, blurring the crisper pattern present in spring. Plumages are mostly cryptic combinations of browns, gray, chestnut, black, and white, but flight patterns can be striking.

Distribution This Old World family is found primarily in southern Europe, Africa, and Asia, with one aberrant pratincole in Australia. There are only three records of pratincoles (of two different species) from the Americas and no records of any coursers.

Taxonomy The 18 species in this family are placed in two subfamilies, the coursers (Cursoriinae, with ten species) and the pratincoles (Glareolinae, with eight species).

Conservation Land reclamation, pesticides, and development have had an impact on some members of this family. BirdLife International lists Jerdon's Courser from India as critically endangered.

Genus *Glareola*

ORIENTAL PRATINCOLE *Glareola maldivarum*   ORPR ■ 5

breeding adult

no white trailing edge to secondaries

chestnut underwings

short tail

The Oriental Pratincole is an unexpected vagrant from the Old World. Its graceful and unique flight would be a welcome sight to any North American birder, particularly given the few records to date. Where common, it is often seen in flight, as it does most of its feeding on the wing; its flight is powerful, very much like a tern, or even a large swallow or martin. On the ground, it might chase insects in ploverlike fashion, but more likely will sit or stand rather motionless until taking flight. Monotypic. L 9" (23 cm)

IDENTIFICATION Adults are brownish above and on breast, but their white rump and deeply forked black tail make for an impressive sight. When there is adequate light, chestnut underwing coverts may be viewed, but when shaded they look dark. **BREEDING:** Both males and females usually have a warmer orange tint to the brown. A black line extends below eye and frames buffy throat with a semicircle. Bill is black, with red restricted to base. **NONBREEDING:** Colors are all muted. Breast is a duller brown. Bill base is a duller red. Black frame to throat is more diffuse. **JUVENILE:** It looks similar to winter adult, but it is duller still, with a pale pink base to mandible, a whitish throat, and pale edges to most of upperparts.

SIMILAR SPECIES Oriental Pratincole is unlike any bird expected in this hemisphere, but there is one record of the similar Collared Pratincole *(G. pratincola),* a bird that wintered in Barbados. Adult Oriental lacks white trailing edge on its secondaries, has darker upperparts (so outer primaries do not boldly contrast), less red on bill, and a shorter tail with a shallower fork.

VOICE CALL: Gives a harsh *kik-kik-kik* or *chik* notes that recall a *Sterna* tern.

STATUS & DISTRIBUTION Fairly common, but some local populations declining. **BREEDING:** Much of Asia to northern and northeastern China, and Russian Far East. **WINTER:** Southern part of breeding range south to northern Australia. **VAGRANT:** Two spring records from islands in western AK: Attu I. (May 19–20, 1985) and St. Lawrence I. (June 5, 1986). Also recorded from northwestern Europe.

SKUAS AND JAEGERS Family Stercorariidae

Long-tailed Jaeger, adult (AK, June)

The skuas and jaegers are superficially gull-like birds and were, until recently, united with them (along with the terns and skimmers) in the family Laridae.
Structure All are long-winged birds capable of rapid and powerful flight. The four skuas are heavily built with relatively broad wings and a short, squared tail with very slightly projecting central rectrices; the three jaegers are more slender in build and as adults have long, projecting central tail feathers. The bills are gull-like, with a strongly hooked tip and a slight to pronounced gonydeal angle. The legs are short and stout compared with those of gulls, with relatively strong claws. Females are slightly larger than males.
Behavior Generally solitary, but small groups may congregate in prime foraging areas in the nonbreeding season. When nesting in tundra regions jaegers take small rodents, birds, nestlings, and even berries. Skuas are predatory, feeding on chicks, eggs, and adults of a variety of seabirds. Nonbreeding jaegers scavenge prey, "pirate" prey from terns or other foraging seabirds, and catch their own fish or other prey. All species are monogamous, with long-term pair bonds in some species. The nest is a simple scrape; from the usual clutch of two eggs, often only the first-hatched chick survives.
Plumage Generally clad in variations of browns and grays. Skuas are entirely dark on the head and body, some species showing reddish or golden tones or pale shaft streaks. Jaegers are mostly pale below, although all three species are polymorphic and may show entirely dark body plumages. All flash white at the bases of the primaries, conspicuous in flight. Age variation is pronounced in jaegers, subtle in skuas.

Distribution The three jaegers breed in high-latitude northern regions, on tundra or coastlines; they migrate southward, mainly over the open ocean or coastlines, but with lesser movements over continental regions, and spend the winter in marine habitats from north temperate regions (a minority of birds) to subtropical, south temperate, or southern high-latitude regions. Three of the four skuas breed in islands and coastlines in the cold southern oceans south to Antarctica; the fourth (Great Skua) breeds in the North Atlantic. Many Great Skuas winter south into the subtropical Atlantic, but Brown and Chilean Skuas appear to remain within the southern oceans for the winter. The spectacular migration of South Polar Skua brings it from Antarctic nesting areas northward into the North Pacific and North Atlantic. All three jaegers and two skuas occur in North America.
Taxonomy It has long been known that jaegers and skuas are closely related to the superficially somewhat similar gulls and terns, on the one hand, and the diving auks, murres, and puffins (alcids) on the other hand. Recent molecular work confirms that the skuas are the sister group to the alcids, necessitating their removal from the family Laridae. The largest jaeger, Pomarine, shares some structural, plumage, and behavioral traits with the larger skuas, and in fact it is more similar in its mitochondrial DNA to Great Skua than to other jaegers; it appears to be the sister to the clade of larger skuas.
Conservation Populations of most species appear stable, although populations of some skua species on certain islands have been eliminated or reduced.

GREAT SKUA   *Stercorarius skua*   GRSK ▪ 3

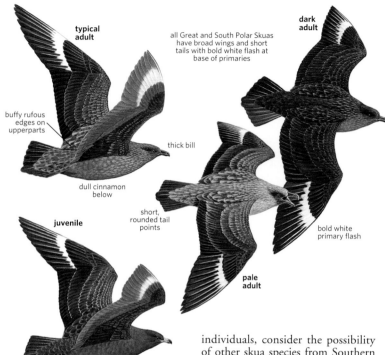

This sought-after visitor to the North Atlantic is often associated with winter mid-Atlantic pelagic trips (where it associates with feeding assemblages around fishing boats), but it can also be found in the western North Atlantic in summer, especially off Canada. Monotypic. L 21–24" (53–61 cm) WS 51–55" (130–140 cm)

IDENTIFICATION Large, bulky; stout bill, broad wings, barely (or not) projecting central rectrices. **ADULT:** Ginger-brown overall; buff streaking on face and neck; rufous to buff mottling and streaking on back, and, to a lesser extent, upperwing coverts; often slightly darker cap or face. Underwing coverts mottled dark brown; white flash across primary bases usually bold and striking in flight. **JUVENILE & FIRST-YEAR:** More uniform overall and often darker. Head and body vary from dark rufous-brown (as shown) to fairly cold brown (suggesting juvenile South Polar, but fresh in fall), without pale neck streaking. Note pale tips or U-shaped subterminal markings on scapulars and upperwing coverts. Grayish bill base. First-year birds in May–June can be largely in bleached juvenile plumage and look paler and more uniform than winter birds, inviting confusion with South Polar.

SIMILAR SPECIES South Polar averages smaller and more lightly built, with more slender bill and narrower wings. Differences difficult to judge at sea, as all skuas are "big" and are not always

seen well. Adult and older immature South Polars differ from adult Greats in uniformly dark upper wings and cold-brown, usually fairly dark head and body, with paler hind neck. Main problem is separating bleached first-summer Great from South Polar: Note Great's evenly worn (and more tapered) juvenal outer primaries, which can be very bleached at tips; dark brown (vs. blackish) underwing coverts, and any remaining juvenal upperwing coverts, with bleached U-shaped markings. Adult South Polars generally are colder brown on head and body; contrasting pale hackles on neck (less uniformly bleached on head and neck than on Great). Many birds not seen well should be recorded as "unidentified skua species." With aseasonal, atypical-looking, or vagrant

individuals, consider the possibility of other skua species from Southern Hemisphere, especially Brown Skua (a few recs. from Maritimes and mid-Atlantic coast probably are this species). Also see smaller Pomarine Jaeger.
VOICE Likely to be silent in N.A.
STATUS & DISTRIBUTION Breeds in northwestern Europe; ranges in north and tropical Atlantic; winters south to Brazil. **MIGRATION & WINTER:** Uncommon; offshore, rarely in sight of land. Some Great Skuas (from Iceland) occur off Atlantic Canada mainly Sept.–Mar.; smaller numbers June–Aug. Farther south, off East Coast of U.S., recorded mainly Dec.–Feb./Mar., south regularly to NC, possibly to FL. Northbound first-summers (from tropical winter grounds) probably also occur May–June, during same time as migrant South Polars.
POPULATION Increased in Britain through 1900s; probably stable (around 6,000 pairs) in Iceland.

SOUTH POLAR SKUA   *Stercorarius maccormicki*   SPSK ▪ 2

This polymorphic bird breeds closer to the South Pole than any other bird species. It is highly migratory, however, and visits both of our coasts.

Monotypic. L 21" (53 cm) WS 52" (132 cm)
IDENTIFICATION The smallest skua, its small size can allow for confusion with

larger female Pomarine Jaegers, particularly dark juveniles. Stocky, thick necked, potbellied. Thick, strong, hook-tipped bill. In many respects, intermediate between Great Skua and Pomarine Jaeger, although clearly a skua, with characteristic short-tailed appearance, broad wings, and large white primary flash visible from above and below. Typically, older ages show a contrasting pale nape; paler birds have a small pale blaze at base of upper mandible. **PALE ADULT:** Distinctive, pale golden brown ("blonde") on head and body, contrasting with dark brown wings and mantle. Narrow golden streaks on back. Face darker, contrasting with pale golden nape. Often strongly developed pale blaze at bill base. Black bill and legs. In flight, pale body contrasts strongly with dark underwings. From above, pale nape and upperback contrast with darker rear quarters and wings. This look (pale in front and dark at back) is characteristic of this morph. **DARK ADULT:** Dark brown throughout; uniform, lacking much streaking on upperparts. Cold brown, lacks rufous or warm tones. Dark, unicolored face contrasts with paler nape, which is narrowly streaked golden. Body dark, so there is no contrast with dark underwing; paler nape most obvious area of any contrast, other than the wing flash. **INTERMEDIATE ADULT:** Intermediate between the two extremes. Body somewhat paler than dark underwings. **JUVENILE:** Similar to dark adult, but gray bill base contrasts with blackish tip; tarsus also gray. Plumage crisp and evenly worn;

lacks streaking or other markings. Nape paler but lacks golden streaking. Unicolored, as in dark adult; lacks warm tones; evenly colored. Body slightly paler than dark underwings. **HYBRID:** Frequently hybridizes with Antarctic (Brown) Skua in Antarctic Peninsula. Distribution of these hybrids during nonbreeding season is not known.

SIMILAR SPECIES Great Skua is extremely similar to juvenile and dark-morph adult South Polar Skuas. South Polars lack rufous or cinnamon tones on plumage and look more uniform in general pattern, but they show a pale nape. Paler South Polars show contrast between pale body and darker underwings. While in our waters, South Polars tend to be in obvious molt, whereas adult Great Skuas molt later (starting Sept.). However, young Great Skuas molt in spring to summer, so lack of midsummer molt is useful only for identification of adult Great.

VOICE Silent while in our region.

STATUS & DISTRIBUTION Uncommon on Atlantic and Pacific Oceans; more frequent on Pacific. Strongly pelagic, extremely unlikely to be seen from shore. **BREEDING:** Antarctica. **MIGRATION:** Hypothesized clockwise migration: arriving earliest on west side of northern oceans, moving later to the east side, before moving south. In Pacific Ocean, rare in spring; numbers increasing through summer,

peak Aug.–Oct. In Atlantic, most common May–July, numbers dropping later on in summer and autumn. Unclear whether all age groups move into Northern Hemisphere; evidence suggests that only younger birds arrive here. Pale-morph adults, or similarly plumaged birds, are extremely rare in N.A., most being darker individuals. **WINTER:** Breed during our winter (southern summer), so winter here during our summer months. They seem to always be on the move during their nonbreeding season, thus they really migrate through our area. **VAGRANT:** Casual in AK. Accidental in interior: ND and TN (after Hurricane Katrina).

POMARINE JAEGER *Stercorarius pomarinus* POJA ▬ 1

This bulky brute with a commanding presence is the largest jaeger. Large individuals recall skuas, while smaller ones are deceptively similar to Parasitic Jaeger. Monotypic. L 21" (53 cm) WS 48" (122 cm)

IDENTIFICATION A large, broad-winged, potbellied, polymorphic jaeger with a thick bull neck—a Rottweiler among jaegers. Adult central tail extensions are broad and twisted, with distinctive spoon-shaped tips. Bill is thick and strong. Outer 4–6 primaries have white shafts. Upper wings unicolored; no contrast between wing coverts and secondaries. As in Parasitic Jaeger, underwings show a white flash at base of the primaries, but a second flash on underside of primary coverts ("double wing flash") is unique to Pomarine Jaeger. This "double flash" is present in juveniles and most adults. **SUMMER PALE-MORPH ADULT:** Dark brown above; wings with variable pale flash at base of primaries. Black cap contrasts with yellow neck; breast white with broad, dark mottled breast band and dark mottled flanks. Males show reduced breast band, rarely lacking it altogether. White underparts and dark vent. Bicolored bill with orange-pink base and dark tip. **SUMMER DARK-MORPH ADULT:** Chocolate

throughout; slightly blacker cap, pale wing flash. **WINTER ADULT:** Pale birds show dark cap; variably barred underparts; coarsely black-and-white barred rump. Tail streamers lacking or less well developed than in summer. **JUVENILE & FIRST-YEAR:** Variable. Brownish, barred below and on underwings. Lacks capped effect. Juveniles are brown to chocolate, lacking warm tones. Head unicolored, not streaked; nape unstreaked. Primaries dark to tip. Tail shows short, broad, rounded extensions on central rectrices. All juveniles (even dark birds) show pale uppertail coverts. **IMMATURE:** Similar to winter adult, but underwings with variable amount of white barring.

SIMILAR SPECIES Most similar in size to Parasitic Jaeger, but shapes differ: Pomarine appears broader-winged, more potbellied, thicker-necked, and shorter-tailed (not including streamers). Breeding pale-morph adult Pomarines easily identified by spoon-tipped, twisted central tail feathers; generally more mottled and broader breast band; bicolored bill; extension of dark cap below gape; brighter yellow face coloration. Dark morph adults separated by structure and tail spoons. In fall, Pomarine is the only jaeger expected to show wing molt while in our waters; the two

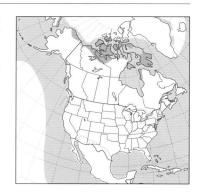

other jaegers tend to molt south of N.A. and later in the season. In winter, adult Pomarine is separable from Parasitic by shape, more heavily barred body plumage, and thicker barring on rump. Growing broad-tipped tail streamers are not pointed as on Parasitic. Overall, Pomarine is less warmly colored than Parasitic, lacks head streaking, and consistently shows coarse black-and-white barring on rump; this is less obvious on Parasitics and is not present in darker morphs. Pomarines have dark primary tips, while Parasitics' primary tips are fringed with pale. Long-tailed Jaegers are much smaller and slimmer, with narrower and longer wings as well as a longer tail. Adult Long-taileds (no

How to Look at Jaegers

Jaegers breed in the Arctic or alpine tundra, and they are pelagic during the nonbreeding season. They are strong and powerful on the wing—falconlike—flying lazily along and suddenly bursting quickly to high speed for a chase. In aerial pursuit they are relentless and maneuverable. Given their aerobatics and highly aerial nature, shape and size are important in separating the three Jaeger species. But this is confounded by the fact that females are substantially bigger and bulkier than males in all species, causing a certain amount of overlap in size. Even so, in shape the three differ on average.

Pomarine Jaeger is bulkiest and appears potbellied and very deep at chest. The wing is broad, tail is short, and head is big. Often it appears as if there is more body before wing than behind wing. Parasitic Jaeger is slimmer and longer, and its wing base is nearly equal to the tail length, with roughly equal amounts of bulk before and after wing. Long-tailed Jaeger is petite, with a long and narrow wing. Its wing base is less than the tail length, and there is less in front of wing than behind wing. A Long-tailed in flight may remind an observer of a tern, which is graceful in nature. The bill is long and thin in Parasitic; Long-tailed has a short

Long-tailed Jaeger, juvenile (Europe, Aug.)

bill, which makes it appear thick, while Pomarine has a thick, stout bill that is pale at the base.

Identifying jaegers is tricky—sometimes impossible. Age adds to the variability, and thus far we

dark morph) show long and pointed tail streamers as well as a small, crisply demarcated black cap. Pomarines at all ages show at least four white shafts on outer primaries; only two on Long-tailed Jaegers. Juvenile Long-taileds are paler and grayer than Pomarines: The palest extremes are contrastingly white-headed, whereas Pomarines are solidly dark-headed. Other than the dark extremes, juvenile Long-taileds show an unbarred white belly, but Pomarines are consistently barred on the belly.

VOICE LONG CALL: A series of *yowk* notes, roughly two per second, given for several seconds. Not heard south of breeding areas.

STATUS & DISTRIBUTION Common in our area. Numerically it is more common than Parasitic Jaeger, but it is less likely to be seen from land and is therefore encountered less often. **BREEDING:** Nests in wet tundra, often near coast. **MIGRATION:** Pelagic on both coasts. Adults move south before juveniles. First arrivals

dark-morph breeding adult

double white wing flash

light-morph breeding adult

long twisted tail feathers, often longer than shown, with thick tip

usually with breast band, which some adult males lack

light-morph breeding adult

thick two-toned bill

dark below gape

thick two-tone bill

light-morph juvenile

dark primaries

strong bars on vent and under-tail coverts

light-morph 1st summer

very short and rounded central tail points

light-morph juvenile

double white wing flash

all nonadult jaegers have checkered underwing coverts

Long-tailed Jaeger, juvenile (CA, Sept.)

know little about aging jaegers, as they are difficult to study when at sea. There appear to be four general age types: juvenile/first-year, second-year, third-year, and adult. Adults show entirely dark underwings.

Juveniles/first-years have variable tones of brown, show strong barring on underwings, and lack dark caps. Second-year immatures show some barring on underwing, particularly on axillaries, and a dark cap; their body plumage is browner than adults'. Third-year immatures are often indistinguishable from adults, but they show a few barred feathers on the wing linings.

Polymorphism as well as the complexity of maturation creates a bewildering array of plumages, often shared by the different species. Note the following highly useful or diagnostic features. When encountering an adult jaeger, look for tail streamers, cap shape, number of white primary shafts, presence and strength of breast band, color of bill and legs, contrast between coverts and secondaries, and strength of yellow on face. On first-year birds, look for overall warmth of plumage color, pale-headed look, presence of white tips on primaries, shape of central tail feather tips, contrast of pale nape, streaked or unstreaked nape, number of white primary shafts, and presence of pale bases to under primary coverts. ■

in July; peak movements Sept.–early Oct., with some wintering in North American waters. Juveniles arrive after Oct. Northbound birds peak late Apr.–late May, arriving in Arctic early to mid-June. Rare in autumn migration on Great Lakes, where a few linger into Dec. WINTER: Most common wintering jaeger, found as far north as central CA, central FL, and Gulf of Mexico. Main wintering areas farther north than either of the other two jaegers, with concentrations in Caribbean and northern S.A. VAGRANT: Casual throughout interior of continent, mainly juveniles in fall. POPULATION Stable; no data on trends.

PARASITIC JAEGER *Stercorarius parasiticus* PAJA ■ 1

This is the standard jaeger, the one to learn as a basis for comparison. It is also the one most likely to be seen from land during migration, often harassing terns. Monotypic. L 19" (48 cm) WS 42" (107 cm)
IDENTIFICATION Polymorphic; medium size; intermediate between other species. Slim yet powerful. Adult central tail extensions are narrow and pointed, not long ribbon-like streamers. Narrow, long bill. Outer 3–5 primaries have white shafts. Upper wings unicolored; no contrast between wing coverts and secondaries. SUMMER PALE-MORPH ADULT: Dark brown above, pale below with dark vent. Black cap; paler forehead; contrasting pale yellow neck. Most adults have a breast band. DARK MORPH: Chocolate overall with a blacker cap. WINTER ADULT: Shows dark cap; variably barred breast and flanks; complete breast band. Tail streamers less well developed than in summer. JUVENILE & FIRST-YEAR: Variable.

Brownish; barred below and on underwings. Lacks dark cap. Juveniles light brown to blackish with cinnamon fringes, but usually cinnamon brown. Streaked pale nape patch. Densely barred throughout; pale primary tips. Paler juveniles show pale uppertail coverts. IMMATURE: Similar to winter adult, but underwings barred white.
SIMILAR SPECIES Compared to Long-tailed, not as slim and ternlike; has shorter tail, longer bill. Compared to Pomarine, slimmer, longer tailed, narrower winged. Breeding pale adults identified by short, pointed tail extensions; narrow, even (unmottled) breast band; black bill. In all ages upper wings are evenly dark above; coverts not paler than secondaries. Parasitics show three or more white primary shafts on outer wing. Juvenile Parasitics are warm colored, showing rusty or rufous tones, unlike more chocolate Pomarine or the colder gray-brown

Long-tailed. Nape paler than head and shows streaking, unlike Pomarine. Above, paler individuals showing pale barred rump have narrower, wavier bars than Pomarine. Primaries pale fringed, unlike other species.
VOICE LONG CALL: A series of 3–4 bisyllabic notes, roughly one per second, rising in pitch. Only in breeding areas.
STATUS & DISTRIBUTION Common in

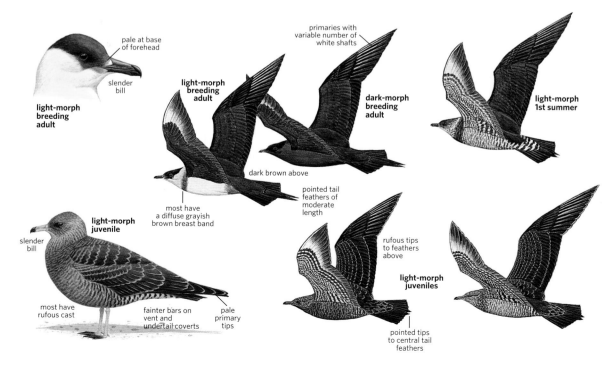

pale at base of forehead

slender bill

light-morph breeding adult

light-morph breeding adult

primaries with variable number of white shafts

dark-morph breeding adult

light-morph 1st summer

dark brown above

most have a diffuse grayish brown breast band

pointed tail feathers of moderate length

slender bill

light-morph juvenile

rufous tips to feathers above

light-morph juveniles

most have rufous cast

fainter bars on vent and undertail coverts

pale primary tips

pointed tips to central tail feathers

our area. **BREEDING:** In Arctic, wet and moist tundra, and coastal wetlands. **MIGRATION:** Pelagic. Small numbers pass through Great Lakes region in autumn. Adults move south before juveniles. First arrivals July; peak late Aug.–late Sept. Juveniles arrive late Aug., at least six weeks before the first juvenile Pomarines appear well south. Northbound birds peak late Apr.–late May; arrive in Arctic late May to mid-June. **WINTER:** Rare in southern CA, FL, and Gulf of Mexico; most winter in S.A. Found closer to shore than other jaegers, where they habitually harass tern flocks. **VAGRANT:** Casual throughout interior of continent.

LONG-TAILED JAEGER *Stercorarius longicaudus*   LTJA ■ 1

The smallest and most elegant jaeger, Long-tailed is locally common on its northern breeding grounds (where it has been nicknamed "tundra kestrel" for its hovering) and is sought after on pelagic trips in the lower 48. Its flight is relatively buoyant and graceful. Unbroken tail streamers add 6–8" (15–20 cm) to length of breeding adult. Monotypic. L 14.5–16" (37–41 cm) WS 37–41" (94–104 cm)

IDENTIFICATION Wings relatively narrow based; tail projection behind wings usually longer than width of wing base, even without streamers. Bill relatively small but appears thick; legs have pale markings at all ages (dark on adults of other species). Wing molt occurs south of N.A. On upper wing of all ages, outer 2–3 white primary shafts most prominent (typically only outer two shafts). Adult lacks dark (or intermediate) morph. **BREEDING ADULT:** Neat black cap; no breast band; belly to vent dusky. On upper wing, blackish remiges (primaries and secondaries) contrast with medium brown-gray coverts; no white flash on underwing. Tail streamers long and finely pointed (often shed in fall). **WINTER ADULT:** Unlikely in N.A. Dusky chest band (partially shown by some fall migrants); barred tail coverts; short tail streamers. **JUVENILE & FIRST-SUMMER:** Polymorphic. Underwing has bold white flash lacking in older ages. Most show bold pale barring on underwings, bold whitish barring on

two to three white primary shafts in nearly all plumages

breeding adult

no white wing flash

gray upperparts contrast sharply with blackish flight feathers

very long, pointed central tail feathers

1st summer

light-morph juvenile

long slender wings

pale tips to feathers above

long tail has rounded central tail feathers with pale tips

dark-morph juvenile

stubby bill

some with pale belly and dark breast

clean blackish cap

stubby bill

dark primaries

breeding adult

strong, straight bars on vent and undertail coverts

typical juvenile

uppertail coverts. Juvenile has bluntly pointed tail streamers; primary tips (at rest) with little or no whitish edging; pale tips to upperwing coverts. First-summer is more capped and lacks pale tips to upperwing coverts; tail streamers have needle-like tips. Dark morph blackish brown overall, usually with bold whitish bars on tail coverts. **SECOND-SUMMER:** Resembles adult but tail streamers average shorter, underwing with variable pale barring; a few have dark body and messy whitish belly.

GEOGRAPHIC VARIATION Breeding adults from N.A. average less extensive dusky on belly than those in northern Europe. Previously considered polytypic.

SIMILAR SPECIES See other jaeger accounts. As a rule, Long-tailed may be likened to Mew Gull, Parasitic likened to Ring-billed Gull, and Pomarine likened to California Gull. Juvenile Long-taileds lack warm reddish tones often shown by Parasitic.

VOICE Shrill, mewing chippers *kyi-kyi-kyik*, etc., and high, clipped yelps. Mostly silent away from breeding grounds.

STATUS & DISTRIBUTION Holarctic breeder, winters mainly off S.A. and S. Africa. **BREEDING:** Common on dry tundra, late May–Aug. **MIGRATION:** Mostly well offshore and not likely to be seen from land. Mainly May and late July to early Oct., stragglers into Nov.; immatures off coasts June–July. Casual to rare inland (mainly fall) and off the Gulf Coast.

POPULATION Most abundant and widespread jaeger in Arctic.

AUKS, MURRES, AND PUFFINS Family Alcidae

Atlantic Puffin, breeding (ME, July)

Spectacular and with many species easily observed at their remote breeding colonies, alcids pose a challenge to scrutinize and identify at sea, where they spend most of their time—much of it submerged. A typical birder's view is of distant birds in flight, often in rough sea condition. Identifications are best based on pattern of dark and light colored plumage, bill shape (if visible), and subtle differences in body shape and flight style.

Structure Heavy compact bodies, dense waterproof plumage, very short tails, and small, short wings are standard. Bills are highly variable in shape and coloration, from dagger-shaped in murres, guillemots, and some murrelets, to hatchet-shaped (puffins), to tiny (auklets, Dovekie, and some murrelets). The legs are short and the feet, which lack a hind toe, are fully webbed; they are brightly colored in some species.

Behavior Alcids normally fly close to the sea surface in a very rapid and direct manner with continuous rapid wingbeats; a few smaller species zigzag among wave tops. Most species (except *Cepphus* guillemots and *Brachyramphus* murrelets) usually remain well offshore except when attending breeding colonies. Solitary sick, weak, or oiled birds sometimes enter bays and harbors. Large numbers occasionally fly by headlands, especially just after dawn, driven close to land during onshore winds. "Wrecks," in which hundreds of birds occur far inland, sometimes take place after late fall and early winter storms. Alcids forage by

wing-propelled pursuit diving with three feeding preferences: schooling fish (murres, puffins), bottom fish (*Cepphus* guillemots), and zooplankton (murrelets, auklets, puffins). The clutch consists of one or two eggs. Alcids are extraordinarily variable in breeding habits: Chicks may depart the colony two days after hatching (*Synthliboramphus* murrelets), when half grown (murres), or remain in their burrows until full-size (puffins, auklets). Adults of some species may be seen at sea with small chicks that resemble miniature adults. Most species breed colonially (except *Brachyramphus* murrelets) on cliff ledges, rock crevices, earth burrows, surface scrapes (Kittlitz's Murrelet), and even mossy tree limbs (Marbled Murrelets, in parts of their range)—invariably in areas where terrestrial predators are absent or scarce. A few species have entirely nocturnal colony attendance. Dexterity on land varies, with some species clumsy and unable to stand upright (murrelets), and others agile (auklets and puffins).

Plumage Alcids have black-and-white or dull gray-and-brownish plumage. A few display spectacular nuptial plumes during the breeding season. Sexes look alike. Most undergo minor seasonal changes in coloration; some change drastically between winter and summer (*Cepphus* guillemots and *Brachyramphus* murrelets), and a few look the same year-round (auklets). Least Auklet is strikingly polymorphic.

Distribution Alcids are restricted to cold seas of the Northern Hemisphere, with the notable exception of three *Synthliboramphus* murrelet species that inhabit warm subtropical waters off southern California and Mexico. Most alcids come ashore only to breed on remote islands and exposed headlands. Long-billed Murrelet is a casual vagrant to North America from Asia.

Taxonomy Worldwide there are 24 living species. Twenty species in ten genera breed in North America, with diversity highest in the North Pacific. Two murrelet species have been split in recent years: Marbled Murrelet into Marbled and Long-billed Murrelets (1997) and Xantus's Murrelet into Scripps's and Guadalupe Murrelets (2012).

Conservation Many species' breeding areas are protected in parks and nature reserves. The greatest threat has been the introduction of predators (rats and foxes) onto breeding islands (now diminishing due to management). Increasing frequencies of both chronic and catastrophic oil pollution events affect all species. Healthy alcid populations are not compatible with intensive gill net fisheries. Four species have experienced severe declines and have threatened or endangered status: Marbled Murrelet (loss of old-growth nesting habitat, gill netting), Scripps's and Guadalupe Murrelets (introduced predators on breeding islands), and Kittlitz's Murrelet (oil spills, gill netting, and unknown factors). Most Atlantic species have been depleted by hunting, drowning in gill nets, and oil pollution at sea. Great Auks were hunted to extinction by the mid-19th century.

DOVEKIES, MURRES, AND AUKS Genera *Alle, Uria, Alca,* and *Pinguinus*

Crisp black-and-white plumages characterize this group, which includes the largest alcids and a small planktivorous species. There were five species; one was extinct by the mid-19th century. All species in these genera occur in the Atlantic (murres also in the Pacific). Males leave the breeding colonies with their single, partly grown chick and provision it at sea.

DOVEKIE *Alle alle*   DOVE ■ 2

Dovekie is by far the smallest alcid in the North Atlantic—nearly half the size of Atlantic Puffin. A swimming bird typically adopts a distinctive neckless posture with its head held low to water. An active bird often "drags" its wings on the surface between dives. Monotypic. L 8.3" (21 cm)
IDENTIFICATION Entirely black and white. Upperparts are black, with distinctive scapular stripes, formed by narrow white margins on scapulars;

underparts are white. Bill is short, blending evenly with forehead to give a bull-headed look; at close range, bill shape is noticeably short, deep, and broad with a strongly curved culmen. Feet are blackish. **BREEDING ADULT:** Face, throat, neck, nape, and upper breast (bib) are entirely black, except for a small highly contrasting white spot "headlight" above eye. **WINTER ADULT:** White extends to throat, sides of nape, and neck, forming an incomplete white collar. **FLIGHT:** Football-shaped body, small size, and rapid "buzzing" wingbeats are distinctive. Secondaries are tipped with white; underwings are blackish. Flies low over sea, zigzagging among waves.
SIMILAR SPECIES Given a decent view, it is unlikely to be confused with any

Atlantic alcid. Starlinglike in size, it is closest in appearance to a juvenile Atlantic Puffin, which is 50 percent larger and has a different bill and head shape, and a dark face, throat, and neck. In the Bering Sea, where it occurs in summer with Least, Parakeet, and Crested Auklets, Dovekie is closest in size to latter two species, but differs in body shape and crisp black-and-white plumage. Parakeet Auklet has broad wings, a potbellied body shape, a large red bill, and a shallow fluttering wingbeat. Dovekie is more than twice the size of a Least Auklet.
VOICE Highly vocal. A high-pitched *ha-keek,* frequently given at sea when flocks are present. At breeding colonies, a variety of maniacal laughlike chattering and screeching, given when perched and in flight.
STATUS & DISTRIBUTION Millions breed in high arctic Greenland, Norway, and Russia. In N.A. a few hundred breed along the Canadian side of the Davis Strait and presumably in AK near the Bering Strait (Little Diomede I. and St. Lawrence I.). Casual summer visitor to other Bering Sea islands. **BREEDING:** Colonially in crevices on scree slopes on high Arctic islands. **MIGRATION:** Moves away from breeding areas in Sept.; returns by late May–June. **WINTER:** Abundant off Atlantic Canada, especially on Grand Banks, arriving Nov. Occasionally large "wrecks" of hundreds of birds are blown inland (usually late fall). Uncommon and irregular off New England (occasionally in large numbers) to NC, Dec.–Mar. **VAGRANT:** Casual south to FL, Great Lakes, and bodies of water near coast.
POPULATION No trends are apparent.

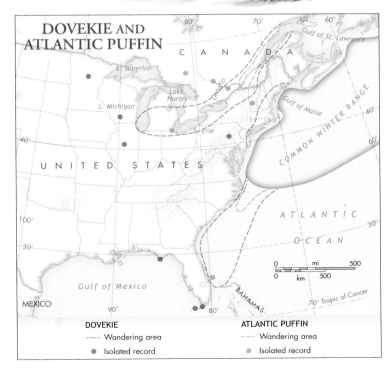

COMMON MURRE *Uria aalge*   COMU ■ 1

Common Murre is the largest living alcid. Its flight is rapid and direct, usually close to the sea surface. Small flocks of Common Murres moving to and from feeding locations often fly in lines or "trains." On land, Common Murre stands nearly vertically upright, but it rests its weight on the full length of the tarsi. It swims buoyantly, when not foraging, with head erect and tail cocked. Polytypic (5 ssp.; 3 in N.A.). L 17.5" (45 cm)

IDENTIFICATION Long dagger-shaped bill, slender neck (contracted in flight), and distinctive head profile of nearly a straight line from crown through culmen (the angle of the gonys is not prominent). Upperparts are nearly uniform dark brown; underparts are white except for sparse dark streaking on flanks. Trailing edge of inner wing is white. Underwing is white, variably mottled with brown on tips of greater underwing coverts; axillaries are usually heavily marked with brown. Bill, legs, and feet are blackish. Eyes are dark brown. **BREEDING ADULT:** Head and neck are blackish brown. Shape of black-and-white border on neck forms a smooth, U-shaped curve. Bridled morph birds, which vary in numbers at different Atlantic colonies, have a prominent white spectacle-like facial mark. **WINTER ADULT:** Similar to a breeding bird, except that its throat, sides to neck, sides of nape, and face are white, save for a dark streak extending from behind eye

toward nape. First-winter bird has a shorter bill and mottled, less contrasting, facial plumage.

GEOGRAPHIC VARIATION Nominate *aalge* is widespread in Atlantic. The two Pacific subspecies (*inornata* and *californica*) are larger and lack the bridled morph. Two additional subspecies occur in the western Atlantic.

SIMILAR SPECIES Common Murre differs strongly from sea ducks, loons, and grebes in its distinctive symmetrical artillery shell-like flight shape, with trailing feet. All plumages of Common Murre appear brownish dorsally (especially in strong light), whereas Thick-billed Murre and Razorbill have blacker backs. Bill shape (long and pointed, lacking a prominent gonydeal angle in Common Murre) is useful in close views, but first-winter Thick-billeds and Commons have more similar bills. The presence of obvious flank streaking— visible on swimming birds—is a good indicator of Common Murre. In winter, distinguish Common Murre from a Thick-billed by its pale face with white extending onto sides of nape, crossed by a prominent dark line extending from behind eye. Molting and subadult Common Murres may have freckled throats and intermediate head patterns; the best way to separate them from Thick-billed Murres is by assessing structural differences. Most populations of Common Murre attain breeding plumage

earlier (often by late winter) than Thick-billed, and this difference can be helpful in picking out the unusual bird. In breeding plumage, Thick-billed has a prominent white gape stripe, which Common Murre lacks; Thick-billed's neck border of black-and-white is V-shaped, whereas Common's neck border is U-shaped. And lastly, Common Murre has a more upright stance when standing and a thinner head and neck than Thick-billed.

VOICE Highly vocal near breeding colonies, giving a variety of harsh grating *arggggh* calls. Not usually vocal at sea and in winter.

STATUS & DISTRIBUTION Common on both coasts, never seen inland. **BREEDING:** Nests in dense colonies on bare cliff ledges and rocky shelves on islands. **MIGRATION:** Moves offshore after departing from breeding colonies Aug.–Sept. **WINTER:** Occurs farther offshore than Thick-billed Murre. **VAGRANT:** Casual on Atlantic coast south of Cape Cod, MA, and much rarer than Thick-billed Murre in those areas; on Pacific coast recorded rarely as far south as Baja California.

POPULATION Various pressures—gill net fishing, oil spills and chronic petroleum pollution, toxic chemicals, and introduced predators (e.g., foxes) on breeding islands—continue to severely affect some Common Murre populations on Pacific coast.

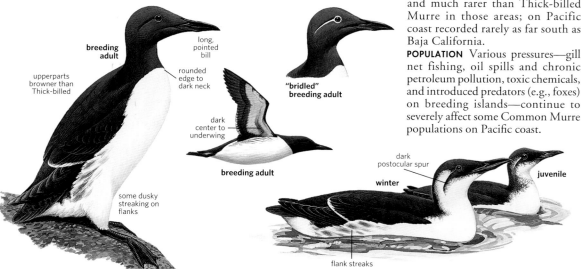

breeding adult

long, pointed bill

rounded edge to dark neck

upperparts browner than Thick-billed

dark center to underwing

"bridled" breeding adult

some dusky streaking on flanks

breeding adult

dark postocular spur

winter

juvenile

flank streaks

THICK-BILLED MURRE *Uria lomvia*   TBMU   1

breeding adult
arra

distinct white line on cutting edge of upper mandible

white forms inverted V into blackish neck

breeding adult
arra

bill slightly less heavy than adult

1st winter
arra

black back

dark face

short, thick bill with curved culmen

faint gape stripe

winter adult
arra

white flanks

Thick-billed Murre is a large black-and-white alcid restricted to cold seas. Its flight is rapid and direct. On land, it stands nearly upright but leans forward slightly, usually resting its weight against a cliff face (it breeds on narrow cliff ledges). Polytypic (4 ssp.; 2 in N.A.). L 18" (46 cm)

IDENTIFICATION Bill relatively thick with curved culmen and notice-able angle of gonys, thick neck, and head shape with relatively promi-nent forehead. Upperparts nearly blackish, head and neck very dark brown, underparts immaculate white.

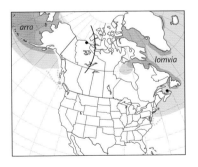

Secondaries white tipped. **BREEDING ADULT:** Prominent white stripe along gape. Upward, V-shaped point at cen-ter of neck formed by black-and-white border. Underwing white, variably mottled with dark gray on underwing coverts. Bill and feet blackish. Irises dark brown. **WINTER ADULT:** Similar to breeding, except that throat and front of neck white with mottled border of entirely dark face. Gape stripe less prominent. First-winter birds have a shorter bill and mottled, less contrast-ing facial plumage.

GEOGRAPHIC VARIATION Pacific sub-species *arra* has a proportionately longer bill with a less strongly curved culmen than nominate *lomvia* of the north Atlantic.

SIMILAR SPECIES Thick-billed's dis-tinctive, bulky, thick-necked artillery shell–like flight shape, with trailing feet, differs strongly from sea ducks, loons, and grebes. It is similar to Com-mon Murre, but in winter it is distin-guished by its darker face, with dark mottling extending well below eye and onto sides of neck, and blacker back. In summer, Thick-billed has a promi-nent white gape stripe and a V-shaped white point in black center of neck, both lacking in Common Murre. In flight, Thick-billed is noticeably blacker above than Common Murre.

A first-winter bird resembles a juvenile Razorbill, but it lacks that species' pointed tail and more extensive white on face. In flight at a distance, Thick-billed Murre is confusable at all ages and seasons with a Razorbill, which has a deeper bill (usually visible) and much longer, graduated tail (surpris-ingly hard to see).

VOICE Highly vocal near breeding colonies, giving a variety of harsh grat-ing *arggggh* calls. Not usually vocal at sea or in winter.

STATUS & DISTRIBUTION BREEDING: Arctic colonies (numbering millions of individuals) south to NF (uncom-mon) and BC (rare). Nests colonially on narrow cliff ledges. MIGRATION: Moves mainly south after departing colonies Aug.–Sept., returning May–June. WINTER: Atlantic birds winter farther south than Pacific birds, occur-ring regularly to mid-Atlantic states (where they are much more likely than Common Murres). VAGRANT: In Atlantic, casual south as far as FL, and inland on Great Lakes. On Pacific coast, casual south of Canada to cen-tral CA (many records from Monterey Bay); once to southern CA.

POPULATION Although Atlantic popu-lations are large (numbering in the millions), hunting and oil pollution do affect them.

RAZORBILL *Alca torda*   RAZO   1

Razorbill is a large Atlantic alcid with a massive head accentuated by a thick neck and an extraordinary, deep, laterally compressed bill with curved culmen. Its flight is rapid and direct, usually close to sea surface. On land, it stands nearly upright and

walks like a penguin. Compared to a murre, it breeds in a wider variety of habitats and regularly forages closer to shore in bays and estuaries. It swims with its head and long tail angled up. Polytypic (2 ssp.; nominate in N.A.). L 17" (43 cm)

IDENTIFICATION Long, graduated tail, extending beyond wing tips on swimming bird, unique among alcids. Upperparts jet black; under-parts white. Trailing edge of second-aries white. White flanks extend onto sides of rump. Underwing mostly white. Mouth interior yellow; bill and feet blackish; eyes dark brown. **BREEDING ADULT:** Face and throat black. Narrow white stripe from top of bill base to eye. Prominent vertical white stripe on grooved bill. **WINTER ADULT:** Similar to breeding, except that throat, front of neck, and ear coverts white, bill stripe less prominent, and face darker. First-winter with similar plumage, but all-blackish bill is shorter and more slender. **SIMILAR SPECIES** Bulky flight shape distinguishes it strongly from sea ducks (beware Long-tailed Duck, which shows extensive white on flanks in flight like Razorbill). Similar to Thick-billed Murre, but in winter it can be told by its large bill and paler face, with white extending well above and behind eye onto sides of nape. Razorbill's facial plumage resembles a

slightly smaller bill

whitish cheek with dusky postocular line

whitish cheek

black hood and upperparts

long pointed tail

thick bill

unique thick bill with white band

immature
torda

breeding adult
torda

winter adult
torda

breeding adult
torda

Common Murre in winter, but lacks the sharply defined dark postocular spur, and Razorbill always has a much deeper bill and is blacker above. In summer, Thick-billed has a prominent white gape stripe and Common Murre entirely lacks white marks on its bill. First-winter Razorbill has a small bill and resembles juvenile Thick-billed, but it has a pointed tail and more extensive white on its face. In flight at a distance, a Razorbill can be confused with a murre, but its bill is deeper (usually visible), white extends onto sides of rump, and its graduated tail is longer, with feet tucked underneath.
VOICE Highly vocal near breeding colonies, giving a variety of harsh growling calls.

STATUS & DISTRIBUTION Locally common at colonies in Gulf of St. Lawrence, NF, and Labrador. **BREEDING:** Variety of habitats, including rock crevices and cliff ledges on coastal islands. **WINTER:** Most of Atlantic population winters on Grand Banks and in Gulf of Maine. Occurs regularly to Long Island; rare off mid-Atlantic states. Winter movements unpredictable, and periodic irruptions send birds much farther south than normal (e.g., flock to off Miami Beach in the winter of 2012–2013). **VAGRANT:** Casual south as far as southern FL and inland on Great Lakes; accidental LA.
POPULATION Razorbill is the least common Atlantic alcid; populations are recovering.

GREAT AUK *Pinguinus impennis*   GRAU ■ 6

Extinct. Great Auk was the original "penguin"—a massive, flightless, black-and-white Atlantic alcid. It was hunted to extinction by the middle of the 19th century. It had a rapid, wing-propelled underwater flight. On land it was fearless, stood upright, and bred in large colonies on a few low islands. Its weight is estimated at about ten pounds. Monotypic. L 31" (80 cm)
IDENTIFICATION Black and white. Resembled a giant Razorbill, with similar deep, laterally compressed bill with numerous concentric grooves. Wings tiny. **BREEDING ADULT:** Face and throat black, oval white face patch between eye and bill. Faint white stripes on grooved bill. **WINTER ADULT:** Similar to summer, but white face patch absent.
STATUS & DISTRIBUTION Extinct; about 80 extant specimens of birds and eggs.

Only three known breeding sites in N.A.: Funk I. (largest colony) and Penguin Is., NF, and Bird Rocks (Rochers aux Oiseaux), Gulf of St. Lawrence, QC. Five colonies in eastern Atlantic (2 in Iceland, 1 in the Faeroes, 1 on St. Kilda, and 1 on Orkney). Highly colonial, forming tightly packed colonies; a single egg laid on bare rock. **WINTER:** Atlantic Canada south to New England, mostly offshore, casual to SC. Remains have been found in prehistoric middens as far south as FL.
POPULATION Great Auk was slaughtered for food, oil, bait, and feathers. The last birds at Funk I. (where an estimated 100,000 pairs bred) were killed in about 1800. The last pair was collected at Eldey Stack, Iceland, on June 3, 1844.

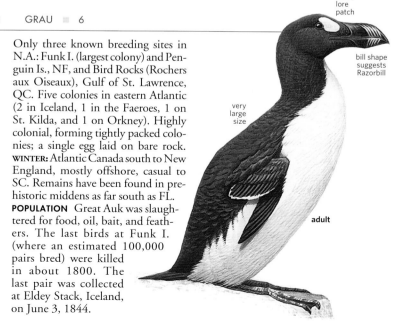

white lore patch

bill shape suggests Razorbill

very large size

adult

GUILLEMOTS Genus *Cepphus*

Medium-size alcids, guillemots show a conspicuous white wing patch year-round. Thin necks and relatively small bills and heads make them look delicate. They fly low with shallow fluttering wingbeats, rather rounded wings, and a heavy-sterned appearance. They usually forage in shallow waters year-round, but some species move offshore in winter.

BLACK GUILLEMOT *Cepphus grylle*   BLGU ▪ 1

Black Guillemot's distinctive features include white, oval wing patches with a smooth outline and gleaming white underwings. Brilliant red feet set off its stunning black-and-white summer plumage. Agile on land, it stands upright and walks on tarsi and toes or on toes alone. Polytypic (5 ssp.; 2 in N.A.). L 13" (33 cm)
IDENTIFICATION Black and white; underwings white. Legs and feet bright red (pink in winter). Eyes dark brown; bill black with mouth interior red. **BREEDING ADULT:** Unmistakable in Atlantic. Completely sooty black except for a broad, crisply defined white patch on greater and lesser coverts and white underwings, excepting tips of flight feathers. **WINTER ADULT:** Very different. Underparts, uppertail coverts, rump, neck, and head white, with sparse dusky streaking and mottling around eye, on crown, and on back of neck. Mantle, scapulars, and uppertail coverts variably mottled. Individuals with patchy, mixed breeding and winter plumage occur late fall and early spring. **JUVENILE:** Similar to winter adult, but more extensively mottled with gray-brown overall; barred wing

coverts (retained through first summer); dull-colored feet.
GEOGRAPHIC VARIATION Subspecies differences are most notable in winter plumage. In N.A., Arctic subspecies *mandtii* is much whiter overall than East Coast breeding *arcticus*. There are three additional subspecies from Iceland, the Faroe Islands, and the Baltic Sea.
SIMILAR SPECIES In all plumages, Black Guillemot's distinctive upperwing patch separates it from all other alcids except Pigeon Guillemot (in northern Bering Sea, the small area of overlap); however, Black Guillemot lacks the dark bar seen in Pigeon Guillemot's wing patch. Black Guillemot's smaller size and faster wingbeats are useful identification cues to separate it from Pigeon Guillemot, especially from locations (e.g., St. Lawrence I.) where both species occur together. Pigeon Guillemots also appears plump. Black's white underwings and axillaries remain the best field mark. A distant bird might be confused with a White-winged Scoter, but a Black Guillemot has faster wingbeats, a pear-shaped silhouette, and is smaller.

VOICE Variety of peeping and thin high-pitched screams and whistles, given near breeding colonies.
STATUS & DISTRIBUTION Common and widespread. **BREEDING:** Small colonies in rock crevices on low rocky islands and in coastal cliffs. **MIGRATION:** Most birds move only short distances from breeding areas. **WINTER:** Wherever ice-free water, even in high Arctic. Occurs regularly south to Long Island. **VAGRANT:** Casual on Atlantic coast south of New England and in AK south of Bering Strait. Also casual inland (Great Lakes); accidental to interior and southeast AK. Interior records all appear to be *mandtii*.
POPULATION No trends are apparent.

winter plumages of *mandtii* paler than Pigeon Guillemot

immatures have barred coverts

1st winter

mandtii

winter adult

winter adult and juvenile *arcticus* average darker than comparable plumages of *mandtii*

winter adult

arcticus

juvenile

smaller and sleeker, with quicker wingbeats than Pigeon Guillemot

mandtii

winter adult

breeding adult

solidly white, oval wing patch

white underwing in all plumages

both guillemots have red legs

white wing patch

breeding adult *arcticus*

PIGEON GUILLEMOT *Cepphus columba* PIGU ■ 1

This Pacific species is similar to Black Guillemot, but it has a dark bar in its white wing patch. Like Black Guillemot, it is often seen from shore. It breeds in small colonies in crevices on low rocky islands and in coastal cliffs, sometimes under wharves, and forages in shallow waters year-round. Polytypic (3–5 ssp.; 2–4 in N.A.). L 13.5" (34 cm)

IDENTIFICATION Black and white. Its conspicuous white wing patch has a black, wedge-shaped intrusion on its lower (or outer) edge. On a flying bird, the patches appear almost divided into two white crescents. Some birds, particularly in western Alaska, have an additional dark wing bar. Underwings uniformly dark. Eyes dark brown; bill is

black; mouth interior bright red. Legs and feet bright red (pink in winter). **BREEDING ADULT:** Completely sooty black except for the wing patch. **WINTER ADULT:** Whitish head and underparts, with dusky markings on crown and back of neck. **JUVENILE:** Similar to winter adult, but more extensively mottled with gray-brown overall, including a dark cap; dull grayish pink feet.
GEOGRAPHIC VARIATION Nominate *columba* (Bering Sea, coast and islands) and *kaiurka* (central and outer Aleutians) have more extensive dark feathering on their wing patches. Some authorities merge *eureka* (CA to OR) and *adiantus* (WA to central Aleutians) with nominate *columba*. Extralimital *snowi* (Kuril Islands) is darkest, sometimes lacking white wing patches.
SIMILAR SPECIES In all plumages distinguished from Black Guillemot by dark underwings and axillaries, and its oval wing patch always has a dark bar,

which Black Guillemot lacks. In the northern Bering Sea, where it overlaps in distribution with Black Guillemot, its larger size, plumper body, and slower wingbeats might also be apparent. Juvenile can be confused with the smaller Marbled Murrelet.
VOICE Variety of thin high-pitched screams and whistles, normally given near breeding colonies.
STATUS & DISTRIBUTION Common and widespread near rocky Pacific coastlines. **BREEDING:** Nests in cavities under rocks and driftwood, and in cliff crevices. **MIGRATION:** Mostly resident, but retreats from areas of heavy ice in Bering Sea. **WINTER:** CA and OR populations apparently move north. **VAGRANT:** Casual to southernmost CA and northwestern Baja California (Islas Coronados).
POPULATION No trends are apparent.

breeding adult

dark underwing
compare to
Marbled Murrelet

dark cap

juvenile

mottled coverts

winter adult

breeding adult

juvenile can have pale center to underwing, which causes confusion with Black Guillemot

dark covert bar often obscured when swimming

MURRELETS Genus *Brachyramphus*

These murrelets are unique among alcids in having a cryptic summer plumage, which camouflages them at their open solitary nests on tree limbs and mountaintops. In fall they molt into black-and-white plumage. A streamlined body form and long, slender wings give them a very fast, direct flight style. They lack agility on land and cannot stand upright.

LONG-BILLED MURRELET *Brachyramphus perdix* LBMU ■ 3

This enigmatic Asian species has occurred as a vagrant in widespread, inland locations across North America and along Pacific coast. In 1997, it was split from the very similar Marbled Murrelet. Monotypic. L 11.5" (29 cm)
IDENTIFICATION In all plumages, underwing coverts whitish. Eyes dark brown; bill, legs, and feet black. **BREEDING ADULT:** Extensively mottled dark grayish brown overall except for whitish throat, with mantle feathers and scapulars thinly edged with buff. **WINTER ADULT:** Black-and-white plumage similar to that of murres;

however, conspicuous white scapular patches contrast strongly with otherwise dark upperparts. Blackish

cap over most of face to well below eye. Nape, sides, and back of neck, mantle, rump, and upperwing coverts

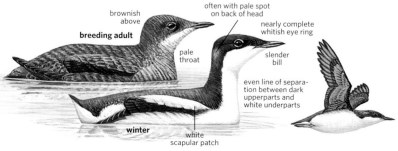

brownish above

breeding adult

often with pale spot on back of head

nearly complete whitish eye ring

pale throat

slender bill

even line of separation between dark upperparts and white underparts

winter

white scapular patch

blackish, except for faint pale oval patch on side of nape. **JUVENILE:** Similar to winter adult, but dusky brownish barring on breast and flanks.

SIMILAR SPECIES Along the Pacific coast, must be carefully separated from Marbled Murrelet. Long-billed is 20 percent larger and has a longer and thinner bill. In winter, it has more extensive dark plumage on lores, nape, and back of neck—entirely lacking pale collar of a winter Marbled. It also lacks dark barring on sides of breast that forms a projecting bar in Marbled and has whitish underwings (dark in Marbled). Long-billed's summer plumage is generally less rufous, more grayish brown overall than Marbled's, with a whiter throat. Beware of confusing Long-billed Murrelet with guillemots (especially juveniles), which are larger and have longer bills and white wing patches.

VOICE Needs study.

STATUS & DISTRIBUTION Asian species. **BREEDING:** Northern Japan, Sea of Okhotsk, and Kamchatka Peninsula; nests in trees. **MIGRATION:** Highly migratory. **WINTER:** Normally near Japan. **VAGRANT:** Casual in N.A. (late summer–early winter), with more than 50 records. Has strayed as far as NF, MA, NC, and FL. Accidental in Europe (Switzerland). With increased observer awareness, a growing proportion of sightings are from the Pacific coast.

POPULATION Trends are unknown.

MARBLED MURRELET *Brachyramphus marmoratus* MAMU ◼ 1

The small, slender-bodied, fast-flying Marbled has a cryptic, speckled breeding plumage adapted to its atypical solitary nesting habits. At sea, it keeps its head tilted up and its tail cocked nearly vertical; its long-necked profile is accentuated by a shallow sloping forehead. It is fequently seen in "pairs" at sea. Monotypic. L 10" (25 cm)

IDENTIFICATION Bill longish, slender, and pointed. Wings narrow and pointed, seeming to blur with high wingbeat frequency; underwings blackish; tail entirely blackish year-round. Eyes dark brown; bill, legs, and feet black. **BREEDING ADULT:** Extensively mottled overall with dark chocolate brown. Mantle feathers thinly edged with rufous; scapulars fringed with white and rufous. Some birds paler with extensive white spotting—pairs often differing in plumage. **WINTER ADULT:** Black and white, recalling a miniature murre. Blackish cap over most of face to well below eye, except for a white loral spot. Neck and lower sides of nape white, giving a white-collared appearance, accentuated by dark barring on side of breast. Whitish scapulars contrast with dark upperparts. Some birds (presumably immatures) in winterlike plumage during summer. **JUVENILE:** Similar to winter adult, but dusky brownish barring on breast and flanks.

SIMILAR SPECIES Kittlitz's Murrelet in breeding plumage has paler, graybrown mottling that does not extend onto lower belly, vent, or undertail coverts and in all seasons shows a pale face with a contrasting dark eye, tiny bill, and hard-to-see white outer tail feathers. Kittlitz's also has a similar black-and-white winter plumage, but its black cap is more restricted, giving the impression of a much whiter face from a distance. Ancient Murrelet lacks white scapulars, shows more extensive black on head and neck, and has bright white underwing coverts.

VOICE Easily identifiable, loud penetrating *keer* calls. Noisy at sea yearround, with a complex vocal repertoire.

STATUS & DISTRIBUTION Locally common in AK and BC, less numerous south of Canada. **BREEDING:** South of AK in large old-growth stands, sometimes far inland. In coastal southern and southwestern AK on ground on steep mountainsides. Often seen close to land, frequenting fjords, deep bays, saltwater lagoons, and even coastal lakes. **MIGRATION:** Some movement away from breeding areas; no evidence of long-distance migration. **WINTER:** Most birds remain near breeding areas; some movement offshore in winter. **VAGRANT:** Very rare to casual south of breeding range to northern Baja California. Apart from regular use of inland lakes near breeding

areas, no confirmed records far inland.

POPULATION Numbers are rapidly declining in southern part of breeding range due to loss of old-growth nesting habitat. The U.S. and Canada list the species as threatened. BirdLife International lists it as endangered.

winter

breeding adult

juvenile

cap dips below eye

white collar

white scapular patch

white lores

winter

mottled brown chin and throat

breeding adult

overall chocolate brown

KITTLITZ'S MURRELET *Brachyramphus brevirostris* KIMU ■ 2

Kittlitz's appears similar to a short-billed Marbled Murrelet—especially in wing and body shape and behavior—but it has a grayish breeding plumage and is more extensively white in winter. It is frequently seen in "pairs" at sea. Monotypic. L 9.5" (24 cm)

IDENTIFICATION Outer tail feathers white year-round. Underwing coverts dark brownish gray. Bill very short, almost invisible; eyes dark brown; bill, legs, and feet black. **BREEDING ADULT:** Buffy upperparts and breast extensively mottled with grayish brown; mantle feathers and scapulars thinly edged buff. Face pale, with distinct dark eye. Belly, vent, and undertail coverts white. **WINTER ADULT:** Black and white. Blackish cap restricted to crown and forehead only. Face, neck, and nape white, giving a white-headed appearance. Blackish barring across breast, forming a nearly complete band. Back blackish, slaty gray appearance in good light, with contrasting whitish scapulars. Some birds (presumably immatures) in winter-like plumage during summer, others intermediate between summer and winter

plumage. **JUVENILE:** Similar to winter adult, but upperparts grayer, with dusky barring on breast and flanks for months after fledging.

SIMILAR SPECIES In breeding plumage, Kittlitz's is similar to Marbled Murrelet, but it has paler gray-brown mottling that does not extend onto lower belly, vent, or undertail coverts, and in all seasons it shows a short bill and a paler face with a contrasting dark eye. White outer tail feathers are hard to see—beware of white overlapping upper and undertail coverts on Marbled. Dark underwings contrast more with pale underparts on Kittlitz's than on Marbled. Ancient Murrelet lacks white scapulars, shows more extensive black on head and neck, and has white underwing coverts.

VOICE Typical call is a nasal grunt. Apparently less vocal than Marbled Murrelet.

STATUS & DISTRIBUTION Uncommon to fairly common, but sharply declining in N.A. **BREEDING:** Solitary ground-nester on high mountainsides among glaciers and snowbeds; often seen close to land, frequenting fjords, deep bays, and saltwater

lagoons. **MIGRATION:** Some movement away from breeding areas; no evidence of long-distance migration. **WINTER:** Poorly known; many birds apparently winter offshore. **VAGRANT:** Accidental to southwestern BC and southern CA.

POPULATION Kittlitz's numbers are rapidly declining in southeast and south-central AK, possibly due to climate change and competition with hatchery-raised salmonid populations. BirdLife International lists it as critically endangered. It is likely the bird species most seriously affected by the massive *Exxon Valdez* oil spill in 1989; many birds also drown in salmon gill nets.

winter

breeding adult

white outer tail feathers

white belly

underwing contrasts darker to body, more so than breeding adult Marbled

breeding adult

short, stubby bill

juvenile

extensive white face

winter

color overall paler than Marbled

MURRELETS Genus *Synthliboramphus*

Five species in the genus: three warm-water species off California and Mexico, one widespread in the North Pacific, and one restricted to waters around central Japan. They're slender-bodied and pointed-winged. They lay a clutch of two eggs and take their precocial chicks to sea two days after hatching. "Xantus's Murrelet" was split into two species in 2012: Scripps's and Guadalupe.

SCRIPPS'S MURRELET *Synthliboramphus scrippsi* SCMU ■ 2

Scripps's Murrelet spends most of its life at sea, where it is wary and difficult to approach in larger boats but is often approachable in very small craft such as kayaks. When flushed, Scripps's, like its relatives Craveri's and Guadalupe Murrelets, lifts directly off water, usually resettling a short distance away. Also like

other *Synthliboramphus* murrelets, Scripps's lays just two eggs and takes its young to sea when they are two days old. The young are fed and reared far from land. At sea, all of these species are encountered singly or in small numbers, often with flightless young chicks that dive to avoid boats. Monotypic. L 9.8" (25 cm)

IDENTIFICATION A remarkably small seabird, resembling a tiny murre in plumage. Underparts bright white, including sides of breast. Upperparts slaty black, with very thin white crescents above and below eye. Bill slim, short, and black. Legs and feet gray, with black webs. Iris dark brown. **JUVENILE:** Plumage like adult's;

chicks at sea have adults in attendance. **FLIGHT:** Wings narrow and pointed; flies with rapid wingbeats. Underwing coverts brilliant white; bases of remiges very pale, creating impression of a mostly white underwing.

SIMILAR SPECIES Craveri's Murrelet is very similar, but underwing appears mostly dark: Its underwing coverts are sooty, remiges blackish. Sometimes, when resting on water, *Synthliboramphus* murrelets raise their body vertically and flap their wings, revealing the underwings. Also note that black of face extends to lower mandible in Craveri's, whereas in Scripps's black extends to a point above the

gape. Craveri's also has a spur of dark plumage on side of breast and a longer bill than Scripps's. In basic plumage, Marbled and Long-billed Murrelets are also black and white, but both show conspicuous white patches on scapulars. See Guadalupe Murrelet.

VOICE At breeding colonies, many birds vocalize simultaneously, giving high-pitched, rolling twitters. Very vocal at sea, unlike most alcids. **CALL:** A repeated piping whistle; also high-pitched chips.

STATUS & DISTRIBUTION Uncommon. Never seen in large numbers except when gathering near colonies during breeding season; far more likely to be seen in nearshore waters than Guadalupe Murrelet. **BREEDING:** Colonially on arid islands in rock crevices and under dense shrubs; breeding begins in Feb. and concludes by Aug. **DISPERSAL:** Postbreeding dispersal apparently mostly to the north in summer and fall, with largest numbers recorded over outer continental shelf waters west of central CA, smaller numbers from northern CA to WA. Northernmost records are from Queen Charlotte Sound, north of Vancouver I., BC.

POPULATION Populations in California have declined historically, mainly due to introduced predators (cats, rats, mice) on breeding islands; the species is listed as threatened there and is a candidate for the federal list as well. The Channel Is. are the species' stronghold in the U.S., with perhaps 2,000–4,000 breeding individuals, and most of the rest of the population is found off Baja California, in Islas Los Coronados and Islas San Benito.

white underwing coverts

slight grayish cast above

white chin

GUADALUPE MURRELET *Synthliboramphus hypoleucus* GUMU ■ 3

Guadalupe Murrelet is an endemic Mexican species, nesting almost entirely on a few predator-free islets off Isla Guadalupe, about 150 miles from the Baja California coast. Guadalupe Murrelet was until recently combined with Scripps's Murrelet as a single species, "Xantus's Murrelet." Recent studies of genetics and vocalizations indicate that Guadalupe and Scripps's merit status as separate species, and they both occur on Islas San Benito. Monotypic. L 9.8" (25 cm)

IDENTIFICATION Plumage almost identical to Scripps's Murrelet, its closest relative, but facial pattern quite

winter range unknown

different. Rather than having an even interface of black and white plumage from the bill to nape, as in Scripps's, the white in face of Guadalupe arcs both above the eye and well into auriculars. Below the eye, Guadalupe has a large amount of white that is very narrowly bordered in black, visible at close range. The white-cheeked impression of Guadalupe is discernible at several hundred yards—remarkable, given the small size of the bird.

GEOGRAPHIC VARIATION None. A few individuals that appear intermediate in facial pattern between Guadalupe and Scripps's Murrelets occur, but their significance is unclear.

VOICE Very unlike Scripps's; a cricket-like rattle, more like Craveri's, and also frequently heard at sea.

STATUS & DISTRIBUTION Uncommon to rare away from colonies. **BREEDING:** Colonially on arid islands in rock crevices; breeding begins in Feb. and concludes by Aug. **DISPERSAL:** Postbreeding dispersal both south and north, and usually found

well offshore. **VAGRANT:** Accidental to WA, two birds about 50 miles offshore (July 19, 2003). Two seen 17 miles west of the tip of Moresby I., Haida Gwaii, BC (Aug. 2, 1994), represent the only record for Canada. A single bird found nesting at Santa Barbara I., CA, is the only U.S. breeding record.

POPULATION Currently listed as endangered in Mexico. Removal of cats and rodents from Isla Guadalupe would benefit species immensely. As with Scripps's Murrelets, oil spills in the vicinity of nesting colonies pose a constant danger, but both are also vulnerable around fishing operations, both because of disorientation by lighting (and thus collision with vessels) and entanglement in nets.

white crescent around eye

slight grayish cast above

white chin

CRAVERI'S MURRELET *Synthliboramphus craveri* CRMU ■ 3

Craveri's Murrelet occurs as a rare post-breeding visitor to California, where it is more regularly observed in warm-water years. Good looks are necessary to separate Craveri's from the very similar Scripps's Murrelet and somewhat similar Guadalupe Murrelet. In California it is normally seen singly or in twos at sea. Monotypic. L 8.5" (22 cm)

IDENTIFICATION Black, with slight brownish cast above and white below. Black cap extends on face below gape onto throat immediately underneath bill. Auricular area, nape, and back and sides of neck are black, extending onto sides of lower breast and giving a partial collared appearance. There are faint white crescents above and below eyes. Underparts, except sides of breast, are immaculate white. Underwing coverts

and underside of flight feathers are dark brownish gray. Eyes are dark brown. Slender, pointed bill is black; legs and feet gray (with black webs). Wings are narrow and pointed; it flies with fast wingbeats. No seasonal variation. **JUVENILE:** Tiny, downy black-and-white chicks taken to sea by adults at two days of age. Fully grown juveniles resemble adults.

SIMILAR SPECIES Very similar to Scripps's Murrelet. Craveri's has a more extensive black cap, black on sides of breast forming a partial collar, blackish to grayish underwings (sometimes with a whitish center, but never as white as the underwings of Scripps's or Guadalupe Murrelets), and a relatively longer and thinner bill—all difficult to see. It is best identified at sea if it raises its wings. Potentially helpful is Craveri's rather long tail that is often cocked up to a point, unlike Scripps's and Guadalupe Murrelets. Guadalupe Murrelet has white extending up over

eye, lacking in both Craveri's and Scripps's. All black-and-white murrelets off the West Coast require close scrutiny to confirm identification: Marbled Murrelet is differently patterned black-and-white in winter, with conspicuous white scapular patches; Ancient Murrelet has a short pale bill, lacks white scapulars, and shows more extensive blue-gray upperparts.

VOICE Vocal at sea. Typical call is a cicada-like rattle, rising to reedy trilling when agitated; also high-pitched chips.

STATUS & DISTRIBUTION Generally rare. **BREEDING:** Nests colonially on arid islands off both sides of Baja California. **DISPERSAL:** Some movement north after breeding season, thus rare in late summer and fall off southern and central CA, especially in El Niño years. **POPULATION** Vulnerable; introduced predators (cats, rats) threaten breeding colonies.

bill longer and thinner than Scripps's

upperparts appear darker than Scripps's

dark under bill

variable dusky underwing coverts

dark breast spur

ANCIENT MURRELET *Synthliboramphus antiquus* ANMU ■ 2

Ancient Murrelet is a strikingly marked, active, social murrelet with strictly nocturnal activity ashore at breeding colonies. Like other *Synthliboramphus* murrelets, it has an elongate cylindrical body and narrow, pointed wings. In flight, it holds its head above the plane of its body during taking off and maneuvering; it tumbles abruptly into sea when alighting. It flies with fast wingbeats. It normally stays in small groups offshore, sometimes coming close to land in bays and harbors. On water, it often appears neckless and flat-crowned. It lacks agility on land and cannot stand upright. Monotypic. L 10" (25 cm)

IDENTIFICATION Ornate plumage. Entire back, rump, uppertail coverts, and upperwing coverts are blue-gray, contrasting with jet-black

flanks and hood and immaculate white underparts. Distinctive short, laterally compressed bill has a black base and yellowish pink tip. Eyes are dark brown; legs and feet gray (with black webs). **BREEDING ADULT:** Black hood extends across upper breast, forming a distinctive bib that contrasts with white sides of neck. Variable, silvery white plumes edge crown and nape (hence the name Ancient), with a second narrow band of similar plumes crossing upper back from sides of breast. Underwing coverts are white. **WINTER ADULT:** White plumes on crown and upper back are reduced, and black bib is smaller, less distinct with light flecking or barring. **JUVENILE:** Tiny, gray-and-white downy chicks are taken to sea by adults at two days of age. Fully grown juveniles resemble

winter adults, but bib is less distinct. **SIMILAR SPECIES** Ancient Murrelet is easily identified by its size, shape, and coloration. Marbled and Kittlitz's Murrelets have different bill shapes; show distinct black-and-white appearance in winter; have dark underwings

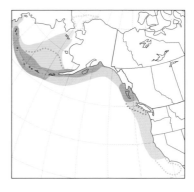

and white scapular patches; and lack the black bib.

VOICE Vocal at sea, typically a loud *chirrup*. Varied repertoire of chattering, chips, and harsh calls given nocturnally at breeding colony.

STATUS & DISTRIBUTION Common, especially when gathering near breeding colonies and at favored wintering areas. **BREEDING:** Colonially on predator-free islands in earth burrows (rarely rock crevices) on forested (BC and southeast AK) and grassy (Aleutian Islands) slopes. **MIGRATION:** Some

movement south after breeding season. **WINTER:** Generally rare but regular off WA, OR, and northern CA; very rare and irregular to southern CA. **VAGRANT:** Casual inland on large lakes in late fall and winter, east to MA and as far south as LA, but most records are from the northern tier of states, many centered around the Great Lakes. Accidental off northwestern

Baja California. One record from U.K.

POPULATION Numbers are declining drastically due to introduced predators (rats, raccoons, foxes) on breeding islands. The population is recovering rapidly in the Aleutian Islands owing to the removal of foxes. Raccoons and rats seriously threaten the population in the Queen Charlotte Islands, BC. Canada lists it as a species of special concern.

AUKLETS Genera *Ptychoramphus* and *Aethia*

The five small alcids in these genera have chunky bodies and short wings; they are not agile on land. Most species are restricted to the Bering Sea and attend colonies (some with millions of birds) in daylight; Cassin's Auklet is more widespread and nocturnal. All species are planktivorous and raise a single chick to full-size in a crevice or burrow.

CASSIN'S AUKLET *Ptychoramphus aleuticus*   CAAU   1

Cassin's Auklet is a small, short-necked auklet with a dull plumage that remains fairly uniform at all ages and seasons. Its flight appears weak; it usually takes off from sea with difficulty at the approach of a vessel. At sea it is often seen in small to moderate aggregations year-round. Strictly nocturnal at breeding colonies. Monotypic. L 9" (23 cm)

IDENTIFICATION Dark brownish gray, except for white or grayish white belly and undertail coverts. Face and crown slightly darker than rest of upperparts. Small white crescents above (larger) and below (smaller) each eye. Black bill larger relative to other auklets' bills;

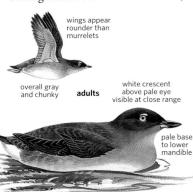

triangular in profile with a straight or even slightly concave culmen, a sharp-pointed tip, and a broad base, with a pale base to lower mandible. White eyes; blue-gray feet with blackish webs. **JUVENILE:** Like adult, but brown or brownish gray eyes.

SIMILAR SPECIES Cassin's pale lower mandible, white iris, and white, eye crescents are visible only at close range. Ancient Murrelet differs greatly in shape (relatively long pointed wings and slender body) and coloration; Rhinoceros Auklet is much larger and has a massive, usually light-colored bill. Near the Aleutians, Cassin's Auklets may be seen with other similar auklets. Whiskered Auklet is smaller and darker (nearly black), and its red bill is much shorter. Crested Auklet is larger and evenly dark colored, and has longer, more pointed wings and a blunt, orange bill. Parakeet Auklet is also larger, shows a bright red bill, and has a more contrasting dark-gray and white body coloration.

VOICE Not vocal at sea, but sometimes makes a grating *krrrk* when alarmed.

STATUS & DISTRIBUTION Common, the most widespread auklet. **BREEDING:**

From the Aleutian Is., AK, to central Baja California. Nests colonially on predator-free islands and isolated coastal cliffs in earth burrows (also rock crevices). **DISPERSAL:** Moves offshore during nonbreeding season. **WINTER:** Mostly offshore, near breeding areas. Some southward movement along West coast. **VAGRANT:** No inland records.

POPULATION Once decimated by introduced predators (rats, foxes), Cassin's Auklet is now recovering rapidly in the Aleutians since the removal of foxes. Rats and raccoons still threaten the species in BC.

PARAKEET AUKLET *Aethia psittacula*   PAAU ■ 2

One of the larger auklets, Parakeet Auk-let is vaguely cootlike, with a chunky body shape, long neck, rounded wings, and relatively large feet. It breeds in small colonies. At sea it is normally seen singly or in small groups. Its wings appear broad and rounded compared to other auklets; its flight is strong and direct. Monotypic. L 10" (25 cm)

IDENTIFICATION Blackish above and white below; minimal seasonal varia-tion. Bill bright red, with oval upper mandible and upcurved lower man-dible. White eyes, giving face a blank expression. Underwing dark gray. Large bluish gray feet with a greenish tinge and black webbing. **BREEDING ADULT:** Upperparts blackish gray, with blackish barring variably extending onto throat, neck, and upper breast, so when resting on sea can appear dark overall. Slender white plumes extend from behind eye. **WINTER ADULT:** Underparts uniformly

white, extending onto center of breast and throat. Bill dull-colored, white facial plumes reduced. **JUVENILE:** Similar to winter adult, but blackish bill and grayish blue eyes.

SIMILAR SPECIES Unmistakable when observed ashore, where facial orna-ments and bill shape are visible. In Gulf of Alaska (off southeastern AK and coastal BC), Parakeet Auklet could be mistaken for Rhinoceros Auklet, which has a longer, pointed bill, a less chunky body shape, and more blended (not black-and-white) plumage. Parakeet Auklet somewhat resembles Least Auklet, but it is twice as large. It is closest in size to a Crested Auklet, but easily distinguished by its white underparts and rounded wings. In flight, it resembles Least and Cassin's Auklets, but size, bill shape, slower wingbeats, lack of white scapulars, and broader wings are use-ful characters. Compared to Dovekie, it has broad wings, slow, fluttery wing-beats, and prominent red bill.

VOICE Highly vocal. At sea, birds taking flight give short high-pitched

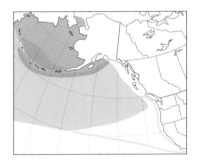

squeals. At colonies, birds give repeti-tive high-pitched whinnying calls.

STATUS & DISTRIBUTION Locally com-mon on Bering Sea and south-central AK islands. **BREEDING:** Colonially in rock crevices, often along cliff-tops. **MIGRATION:** Moves south to avoid winter ice in Bering Sea, dispersing farther offshore than other auklets. **WINTER:** Most birds well offshore in North Pacific, somewhat regularly to latitude of central CA. **VAGRANT:** Casual in winter near coast from BC to CA. Accidental in HI and Europe (Sweden).

POPULATION Least concern.

breeding adults

unique bill shape

dark

orange-red bill

juvenile

winter adult

breeding adult

dark throat and breast

white belly

white underparts

winter adult

LEAST AUKLET *Aethia pusilla*   LEAU ■ 2

Least Auklet is a tiny, chubby alcid, barely larger than a sparrow. Gregari-ous, active, and noisy, it breeds in a few large colonies. Although capable of rapid takeoff from sea, it flies in a characteristic side-to-side weaving style. On land, it stands erect, usu-ally on its toes; it is very agile on its rocky breeding habitat. Monotypic. L 6.3" (16 cm)

IDENTIFICATION Polymorphic. Vari-able seasonally. Short, rounded wings.

Upperparts blackish with white patches on scapulars. Underwing coverts pale. Bill small, nearly black to bright red with a black base, usually with a straw-colored tip; small knob-shaped bill ornament on adult. White eyes. Bluish gray feet with black web-bing. **BREEDING ADULT:** Underparts variably marked with irregular dark spotting, from almost black to nearly unmarked white; throat white with small black chin; scapulars variably

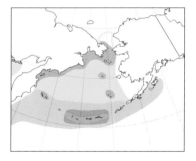

stubby dark red bill

breeding adults

light

white throat

dark

breeding adults

extensive whitish on underwing

winter adults

white underparts

juvenile

faint whitish scapular line

white. Short, evenly scattered, white plumes on forehead, extending onto crown and lores and from behind eye. **WINTER ADULT:** Underparts uniformly white; no ornaments (bill black, no bill knob, only a trace of white facial plumes); and white scapular patches more prominent. **JUVENILE:** Similar to winter adult, but eyes grayish and scapulars less white. **IMMATURE:** Brown forehead with sparse plumes. Dull-colored bill and smaller bill knob than breeding adult, bluish gray eyes.

SIMILAR SPECIES Tiny size and dumpy body shape easily eliminates all other alcids, except other auklets. Least Auklet usually shows white scapulars, which are lacking in all other auklets. It is closest in size to Whiskered Auklet, which is blacker, slightly larger, has somewhat longer wings, and shows pale feathering only on vent (only the darkest Leasts approach Whiskered's dark coloration). Superficially, Least Auklet is most similar to Parakeet Auklet, which is much larger and has slower wingbeats. **VOICE** Highly vocal. At sea, birds taking flight give short high-pitched

squeaks. At colonies, birds give a variety of high-pitched, high-frequency chattering. **STATUS & DISTRIBUTION** Locally abundant in Bering Sea. **BREEDING:** Nests colonially in rock crevices on boulder beaches, talus slopes, boulder fields, lava flows, and cliffs. **MIGRATION:** Moves south to avoid winter ice in Bering Sea. **WINTER:** Poorly known, probably well offshore. **VAGRANT:** Rare inland in western AK in late fall and winter after storms. Accidental in arctic Canada and on West Coast south of AK (one record from central CA). **POPULATION** Stable, but introduced rats at Kiska I. threaten its largest colony.

WHISKERED AUKLET *Aethia pygmaea*   WHAU ■ 2

Although relatively small—only Least Auklet is smaller—Whiskered Auklet is perhaps the most ornamented, living seabird while in breeding plumage. This sought-after species is most often seen at sea, where it flocks together in tidal rips near colonies. Its wings are relatively pointed, allowing for a strong and direct flight style. It breeds at many small colonies throughout the Aleutian Islands, where activity

is mainly nocturnal. On land, it has a forward-tilting posture. Monotypic (Asian ssp. *camtschatica* is sometimes recognized). L 7.8" (20 cm) **IDENTIFICATION** Uniformly dark appearance on water. Upperparts blue-black; underparts blackish gray, except for light gray or whitish lower belly and vent. Flight feathers blackish; underwing coverts dark gray. White eyes; bluish gray legs and toes, with black webbing. **BREEDING ADULT:** Unmistakable at colonies. A slender,

highly variable, forward-curving black forehead crest, and white facial plumes including a showy V-shaped, antenna-like ornament originating on lores, and long, white auricular plumes—all difficult to observe at sea. Bright red bill with a straw-colored tip. Although bird

breeding adult

winter adult

whitish belly

juvenile

winter adult

subdued head pattern in winter adult and especially juvenile

breeding adults

small dark red bill

thin, curled crest

three white head plumes

appears small and black at sea, white vent is visible when bird takes flight. **WINTER ADULT:** Identical to summer (breeding) plumage, but ornaments much reduced and bill is dull red. **JUVENILE:** Similar to winter adult, but

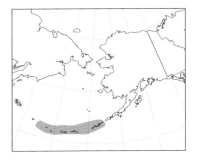

feather ornaments absent, bluish gray eyes, and blackish bill. **SIMILAR SPECIES** Most resembles Crested Auklet, which is larger and grayer above with a more prominent bill. Crested is the only other small alcid that is similarly dark in overall coloration, but it lacks whitish vent region. Whiskered Auklet is closest in size to Least and Cassin's Auklets, which have whiter underparts and dumpier body shapes. Compared to Least Auklet, Whiskered lacks white on scapulars and underparts. **VOICE** Highly vocal. At sea and at breeding colonies, gives a thin kitten-like *mew*. At breeding colony, gives a variety of high-pitched mewing and trilling calls.

STATUS & DISTRIBUTION Locally common on Aleutian Is. only. **BREEDING:** Colonially in rock crevices on coastal cliffs, beach boulders, and talus slopes on predator-free islands. **DISPERSAL:** Resident near breeding islands. **WINTER:** Mostly in large flocks in Aleutian inter-island passes. **VAGRANT:** In AK, casual outside of breeding range, including a specimen record from Gambell, St. Lawrence I. (July 9, 1931). Also recorded from Tatik Point and Bristol Bay. **POPULATION** Once decimated by introductions of exotic mammalian predators onto Aleutian Islands; now recovering since the removal of foxes from many islands.

CRESTED AUKLET *Aethia cristatella*   CRAU ▪ 2

The extremely gregarious Crested Auklet occurs in large flocks year-round; it is rarely seen alone. It breeds in a few large colonies on Bering Sea islands. Its wings are relatively long and pointed compared to other auklets; its flight is strong and direct. On land, it stands erect and is very agile. It has a distinctive citrus-like plumage odor, which is sometimes noticeable at sea. Monotypic. L 9" (23 cm) **IDENTIFICATION** Uniformly dark. White eyes; legs and toes bluish gray with black webbing. **BREEDING ADULT:** Spectacularly ornamented. Bright orange bill with peculiar, curved gape plates, white auricular plumes, and forward-curving, black forehead crest. Larger, more strongly hooked bills on males, thus sexes are distinguishable in the field, unlike other alcids.

WINTER ADULT: Bill plates shed during breeding season, leaving a small dull orange bill by late summer. Black crest and white plumes reduced. **JUVENILE:** Similar to winter adult, but blackish bill, tiny crest, and bluish gray eyes. **SIMILAR SPECIES** Crested Auklet most resembles Whiskered Auklet, which is smaller, blacker above, and shows pale feathering on belly and vent, as does Cassin's Auklet. Crested is closest in size to Parakeet Auklet, which has white underparts, a chunkier body shape, and less pointed wings. Often seen with Least Auklets, which are much smaller and usually show white on scapulars and underparts. Compare first-summer birds, with reduced bill and feather ornaments, to Whiskered Auklet. **VOICE** Highly vocal. At sea, gives a sharp bark like the yap of a small dog. At breeding colony, gives a variety

of trumpeting, cackling, and hooting calls. **STATUS & DISTRIBUTION** Locally abundant in Bering Sea. **BREEDING:** Colonially in rock crevices and caves on talus slopes, lava flows, and cliffs on volcanic islands. **MIGRATION:** Moves south to avoid heavy ice in Bering Sea. **WINTER:** Mostly in a few, food-rich, Aleutian inter-island passes. **VAGRANT:** Rare in Gulf of Alaska east of Kodiak I.; accidental inland in western AK and along West Coast to northwestern Baja California. Accidental to Iceland. **POPULATION** Least concern.

thick, curled crest

breeding adult

red-orange bill plates

breeding adult

uniform dark gray underparts

winter adult

short crest

1st summer

juvenile

white postocular line

winter adult

much smaller bill than breeding adult; loses bill plates

RHINOCEROS AUKLET AND PUFFINS Genera *Cerorhinca* and *Fratercula*

Puffins (3 sp.) are medium-size alcids with big heads, elaborate facial ornamentation, and short, rounded wings. Rhinoceros Auklet has attributes intermediate between puffins and auklets. All are colonial and carry multiple fish externally in their bill to a single chick, which is raised to full-size in a burrow or rock crevice.

RHINOCEROS AUKLET *Cerorhinca monocerata* RHAU ■ 1

Although related to puffins, this alcid's body shape, different facial ornaments, and posture give it an auklet-like appearance. Often seen in large numbers close to shore, it is nocturnal or crepuscular at colonies. Monotypic. L 15" (38 cm)

IDENTIFICATION Upperparts grayish brown. Underparts similar, except for whitish belly and undertail coverts. Yellowish eyes; dull yellowish legs and feet. **BREEDING ADULT:** Two sets of prominent white facial plumes, one extending from gape, other from above eye. Thick, dagger-shaped, orange bill, with decurved culmen, and single, whitish "rhino" horn at base. **WINTER ADULT:** Very similar, but facial plumes much reduced. Bill horn shed in fall, leaving birds with smaller, dull orange bill with blackish base. **JUVENILE:** Similar to winter adult, but no head plumes, darker eyes, and smaller, shallower bill, blackish initially, gradually changing to dull yellowish orange.

SIMILAR SPECIES Recalls a murre as much as a puffin, having a smaller head and more symmetrical body than a puffin. Larger Tufted Puffin lacks white underparts in all seasons, but compare with paler form of a juvenile Tufted, which can show a pale belly. Horned Puffin always has pure white underparts. In flight, Rhinoceros Auklet resembles a grayish and white murre, but its head and bill are disproportionately larger. Cassin's Auklet is smaller and paler with a dumpier body shape, has a more wobbly flight style, and is less approachable. Parakeet Auklet is smaller and more pure white below, especially in winter. Also note markedly different bill shape.

VOICE At and near breeding colony at night, gives sonorous *arr-aarrrgh* calls.

STATUS & DISTRIBUTION Common and widespread along West Coast. **BREEDING:** Colonially in earth burrows on grassy and forested slopes of predator-free offshore islands. Center of breed-

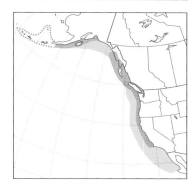

ing abundance is BC; also breeds in Japan. **MIGRATION:** Moves south along West Coast Sept.–Oct.; returns to breeding colonies Mar.–Apr. **WINTER:** Along entire West Coast, regularly south to central Baja California. A few oversummer south of breeding range. **VAGRANT:** Casual near central Bering Sea islands in summer.

POPULATION Least concern.

white plumes on face | "horn"
bright yellow-orange bill
immature
smaller yellowish bill
breeding adult
plumes reduced and no horn
winter adult
breeding adult

ATLANTIC PUFFIN *Fratercula arctica* ATPU ■ 1

This medium-size, large-headed Atlantic alcid has a remarkably ornamented face and bill. It flies direct, and it easily lands on and takes off from flat ground. Very agile on land, it stands erect and on its toes, rather than on tarsi like most large auks; it walks with a hunched posture. Its short neck and heavy, rounded body give it an overall chubby appearance. Polytypic (3 ssp.; nominate in N.A.) L 12.5" (32 cm)

IDENTIFICATION Upperparts blackish with lighter-colored facial disk, black neck collar; underparts immaculate white. Deep, laterally compressed bill with multiple concentric grooves (1–5,

increasing with age). **BREEDING ADULT:** Colorful bill with dark bluish gray crescent-shaped base outlined with pale yellow stripes, remainder bright reddish orange. Pale orange rictal rosettes at gape. Dark eye surrounded by thin red orbital ring, within a clownlike triangular dark gray wattle. Bright reddish orange legs and feet. White facial disks with silvery gray clouding; upperparts silky black fading to brownish black during breeding season. Dark face and winter-type bill on some (first-summer?) birds. Second- and third-summer birds have less brightly colored, more triangle-shaped bills with fewer grooves, and

uniquely shaped and colored bill
pale gray face
dark collar
breeding adult *arctica*
red-orange legs

duller orange feet, compared to adults. Leucistic birds occur. **WINTER ADULT:** Very similar to breeding plumage, but bill plates, rosette, and eye ornaments shed in fall, leaving birds with smaller, dull reddish bill with constricted base and blackish face. Feet yellowish or orange-brown. **JUVENILE:** Similar to winter adult, but bill black, gradually changing to dull reddish, smaller, much shallower and daggerlike.

GEOGRAPHIC VARIATION Northward cline of increasing body size.

SIMILAR SPECIES Unmistakable in summer. In winter, its dirty blackish face, blackish underwings, short, triangular bill, and front-heavy flight shape, resulting from its large head and short neck, distinguish it from murres and guillemots. Superficially, it is similar to much smaller Dovekie, which has a weaving flight style and a very short, stubby bill.

VOICE At breeding colony, gives a variety of low sonorous growling calls, mostly from within burrows.

STATUS & DISTRIBUTION Locally abundant in Atlantic Canada. **BREEDING:** Colonially in burrow networks on grassy slopes of offshore islands. **MIGRATION:** Moves south out of Arctic Sept.–Oct.; returns to breeding areas in late Apr. **WINTER:** Offshore to Grand Banks, far off Canadian Maritime provinces. **VAGRANT:** Rare in coastal waters south to New England, casual south of VA and inland on Great Lakes. Accidental to Talan I., Russian Far East.

POPULATION Least concern.

breeding adult

dark underwing

winter adult

darker face

smaller bill

juvenile

HORNED PUFFIN *Fratercula corniculata* HOPU ■ 1

Horned Puffin is a sister species to Atlantic Puffin, only it is larger, with an even more exaggerated bill and facial ornaments. Its flight is direct, typically high (50–150 ft.) above sea surface, and its feet are prominent. Agile on land, it stands erect and on its toes, walking with a forward-hunched posture. Its short neck and heavy, rounded body give it an overall stocky appearance. Monotypic. L 15" (38 cm)

IDENTIFICATION Upperparts blackish; underparts immaculate white. Bill very large and deep, laterally compressed. **BREEDING ADULT:** Bill mostly bright lemon yellow with bright reddish orange tip. Pale orange rosettes at gape. Dark eye surrounded by thin red eye ring, within clownlike black wattle that includes a "horn" extending vertically from top of eye. Gleaming white facial disks; upperparts satiny black fading to brownish black during breeding season. Dark faces and winter-type bills on a few birds at colonies. Bright reddish orange legs and feet. Second- and third-summer birds have shallower, triangular bills, and duller orange feet, compared to older birds. **WINTER ADULT:** Bill plates, rosette, and eye ornaments shed in fall, leaving bill with smaller dull, greenish brown base, reddish tip, and grotesquely constricted base. Plumage relatively unchanged, but face becomes grayer, sooty black on lores. Yellowish or orange-brown legs and feet. **JUVENILE:** Similar to winter adult, but bill black, gradually changing to dull reddish, smaller, much shallower, and more daggerlike.

SIMILAR SPECIES Horned Puffin is unmistakable in summer. In winter, its large head and short neck produce a front-heavy flight shape, which together with its dirty blackish face, blackish underwings, and triangular bill, identify it as a puffin. Larger Tufted Puffin lacks pure white underparts in all seasons.

"horn"

white face

breeding adult

white below

much smaller and darker bill

immature

darker bill than breeding

breeding adult

juveniles and winter adults darker faced

winter adult

Rhinoceros Auklet also lacks pure white underparts, being more indistinctly mottled grayish brown except for whitish belly; it also has a smaller head and more symmetrical body shape in flight.

VOICE At breeding colony, gives low sonorous growling calls, mostly from within breeding crevices.

STATUS & DISTRIBUTION Abundant and widespread in western AK; locally common in southeastern and south-central AK; rare breeder in BC. **BREEDING:** Colonially in rock crevices (less commonly in earth burrows) on cliffs and rocky slopes of remote islands. **MIGRATION:** Moves out of northern Bering Sea Sept.–Oct., disperses widely into N. Pacific; returns May–June. **WINTER:** Almost entirely far offshore in N. Pacific, south to latitude of CA. **VAGRANT:** Casual, mainly in late spring, along west coast to southern CA. Casual in summer along Arctic coast east of Barrow, AK. Casual in HI. Accidental inland.

POPULATION Least concern.

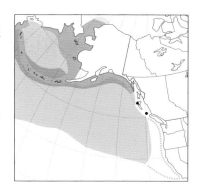

TUFTED PUFFIN *Fratercula cirrhata*   TUPU ■ 1

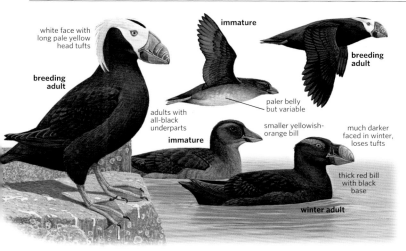

white face with long pale yellow head tufts

breeding adult

adults with all-black underparts

immature

immature

paler belly but variable

smaller yellowish-orange bill

breeding adult

much darker faced in winter, loses tufts

thick red bill with black base

winter adult

The large, dark Tufted Puffin has a huge bill and spectacular golden-blond head tufts in summer. It takes flight from sea with difficulty, sometimes paddling across surface with wings, feet trailing, to escape approaching vessels; once airborne, though, its flight is direct, typically high (50–150 ft) above sea surface, and its feet are very prominent. Its huge head, short neck, and heavy body give it a burly appearance. Social on land, it breeds in large colonies. It stands erect on its toes and is quite agile on land, walking with a forward-hunched posture. Monotypic. L 16" (40 cm)

IDENTIFICATION Entirely sooty brownish black except for pale face. Deep, laterally compressed bill, with deep grooves (2–4, increasing with age). Yellowish white eyes; red eye ring. **BREEDING ADULT:** Colorful bill, basal half olive green, remainder bright reddish orange, and orange gape rosettes. Face gleaming white, with long, blond,

tuftlike plumes that curve backward. Plumage fades to brownish black during season. Second- and third-summer birds have shallower bills with fewer grooves, and duller orange feet. **WINTER ADULT:** Bill plates, rosette, and eye ornaments shed in fall, leaving bill with smaller blackish base, reddish tip, and grotesquely constricted base. Plumage similar to breeding, but head completely blackish, except for hint of pale buff behind eyes and along rear edges of crown. **JUVENILE:** Similar to winter adult, but underparts not as dark, with indistinct light brown areas on belly and throat (variable). Bill blackish, changing to dull orange, smaller than winter adult's.

SIMILAR SPECIES Tufted Puffin is unmistakable in summer. In winter, its huge head and heavy body create a ponderous, front-heavy flight shape, which together with its dirty blackish face, blackish underwings, and triangular bill, identify it as a puffin. Smaller Horned Puffin has

a crisp black-and-white body in all seasons. Rhinoceros Auklet also has whitish underparts, but it is paler and more grayish brown overall than juvenile Tufted; it also has a smaller head and a more symmetrical body shape in flight.

VOICE At colony, gives low sonorous growling calls, most from burrows.

STATUS & DISTRIBUTION Abundant and widespread breeder in western AK; locally common in south-central and southeast AK and BC; uncommon to rare and local south to northern CA; casual to southern CA. **BREEDING:** Colonially in earth burrows (less commonly rock crevices) on grassy slopes and vegetated cliffs of remote islands. **MIGRATION:** Moves out of north Bering Sea in Oct., disperses offshore mainly into north Pacific; returns May–June. **WINTER:** Seldom seen; mostly offshore, but less pelagic than Horned Puffin. **VAGRANT:** Accidental in HI. Single specimen claimed from ME, said to have been shot during winter of 1831–32 and used by J. J. Audubon to paint illustration in his *Birds of North America*. Accidental to the U.K. (Norfolk). **POPULATION** Decreasing in CA, otherwise no trends are apparent.

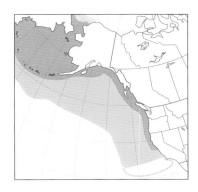

GULLS, TERNS, AND SKIMMERS Family Laridae

Black-headed Gull (center) with Bonaparte's Gulls, nonbreeding adults (NJ, Feb.)

Highly successful and cosmopolitan, gulls are very well known as a group. Even people with scant interest in birds can identify a "seagull." The other subfamilies (terns and skimmers) are less well known, but their relationship to the gulls is not difficult to ascertain. Easily identified to family, they are not so easily identified to species. Discussions of the identity of a gull or tern are classically lengthy. The difficulty in species-level identification is due in part to similarity among closely related species, but also to complex age-related changes and to hybridization in gulls. Species identification is achieved by looking at structure. Soft-part colors are also important. Immatures are usually more difficult to identify than adults, especially when worn, faded, or bleached. These indivduals often provoke the most long-winded debates!

Structure All show webbed feet, but they vary in wing structure and bill shape. Gulls are medium to large in size, with longish blunt-tipped bills. They have long pointed wings; short, square tails; and strong, medium-length legs. The narrow-winged terns are smaller, usually with notched or strongly forked tails, sharply pointed bills, and short legs. Skimmers are the most specialized of the group. They are very long-winged and short-legged, but they have a laterally compressed bill, in which the lower mandible is much longer than the upper.

Behavior Flight is direct and strong, with much gliding in some species. Oceangoing gulls and some terns may arc up over the waves, especially under windy conditions, as tubenoses do. Gulls, terns, and skimmers roost in flocks, often mixed-species flocks. Gulls swim well, as do some species of tern, although terns prefer to perch on floating material. Skimmers can't take off from a swimming position, so they roost only on land. All members tend to breed in colonies, although some are solitary. Foraging varies, with the skimmer having the most impressive strategy: It flies over calm water, trailing its lower mandible in the water. When it senses a fish, it slams its bill closed, capturing the prey. Large gulls tend to be generalists, and a scavenging nature has allowed them to do well in urban areas. Smaller gulls specialize in seizing small prey on the surface. Terns fish, usually by plunge diving and spearing fish with their pointed bills.

Plumage Gulls and terns are usually gray above and white below, with notable exceptions. In summer, terns have black caps; large gulls have white heads, and small gulls, dark heads. Soft-part colors vary.

Distribution This family is as cosmopolitan as they come. There are Arctic and Antarctic species, tropical and temperate species, and pelagic to montane species. Gulls are more diverse in temperate zones, while terns vary more in tropical and subtropical areas. Arctic Terns have record-long migrations.

Taxonomy Although jaegers and skuas were previously placed in this family, recent genetic work has found them to be more closely related to alcids than to gulls—they are now placed in their own family. Other research has resulted in an increase in genera for both gulls and terns. Gull genera in N.A. increased from six to nine; changes include: placing Little Gull in the genus *Hydrocoloeus;* Bonaparte's, Black-headed, and Gray-hooded Gulls in *Chroicocephalus;* and Laughing and Franklin's Gulls in *Leucophaeus.* Tern genera in N.A. increased from three to eight; changes include: placing Sooty, Bridled, and Aleutian Terns in the genus *Onychoprion;* Least Tern in *Sternula;* Gull-billed Tern in *Gelochelidon;* Caspian Tern in *Hydroprogne;* and Royal, Sandwich, and Elegant Terns in *Thalasseus.* The linear sequence of the genera has also been adjusted in both gulls and terns. Species-level taxonomy is equally complicated, and several gulls may be split in the future, for example, Herring Gull and perhaps Mew Gull.

Conservation BirdLife International classifies six species as vulnerable, four as endangered (Black-billed Gull, Peruvian Tern, Black-bellied Tern, and Black-fronted Tern), and Chinese Crested Tern as critically endangered. In N.A., Red-legged Kittiwake is listed as vulnerable; the other species listed above do not occur in the U.S. and Canada. In California the Least Tern (subspecies *browni*) is given legal protection.

SMALLER GULLS Genera *Creagrus, Rissa, Pagophila, Xema, Hydrocoloeus, Rhodostethia,* and *Chroicocephalus*

The gulls in these genera are generally small and ternlike, except for the vagrant Swallow-tailed Gull, an outlier species that is large and feeds nocturnally. The presence of a dark hood is not a taxonomically informative character. Males are larger overall and bigger-billed than females. An understanding of molt is helpful for identification. Most of the species in these genera are two-cycle (or two-year) species that attain adult plumage after their second prebasic molt.

SWALLOW-TAILED GULL *Creagrus furcatus*   STGU ■ 5

in flight, wing pattern similar to Sabine's Gull

very long drooped bill with pale tip

breeding adult

Among the most beautiful of the world's gull species, Swallow-tailed Gull is equally remarkable for its nocturnal habits: It feeds only at night, on vertically migrant squid and fish. No other gull, or species of seabird, is fully nocturnal. Monotypic. L 23" (58 cm)

IDENTIFICATION A very distinctive, medium-size hooded gull. **ADULT:** Upperparts medium gray, underparts white shading to pale gray on breast in some. Long, forked tail is white. Long black, slightly decurved bill has whitish tip. **BREEDING:** Sooty black head with large white spot at base of maxilla (smaller white spot at base of

mandible); orbital ring vivid scarlet. **NONBREEDING:** Dingy white head, dusky orbital ring. **FLIGHT:** Striking dorsal pattern: Blackish outer primaries contrast starkly with white outerwing coverts and secondaries, which contrast with gray mantle and innerwing coverts. Scapulars show neat, thin white edges. **IMMATURE:** Pattern similar to adults, but mantle and innerwing coverts mottled, rectrices black-tipped, head whitish, with blackish around eye and grayish brown ear spot.

SIMILAR SPECIES The striking dorsal pattern is superficially similar to those of Sabine's Gull and Large-billed Tern: The tern has a very large yellow bill and entirely black primaries; the gull, also found in pelagic habitats, has black, not white, outer primaries.

VOICE Calls kittwake-like, including a sharp, grating, rising *kweek* and longer, rattling gurgling sound.

STATUS & DISTRIBUTION BREEDING: Asynchronous breeding, throughout the year, on rocky shores of

Galápagos Is., mostly on Hood, Tower, and Wolf Is., with fewer on the other islands; small numbers nest on Malpelo I. off Colombia in small numbers (50 pairs). **NONBREEDING:** Highly pelagic, found in deepwater areas off the coasts of Ecuador, Peru, and northern Chile often far from shore—most numerous from 3° north to 12° south, with smaller numbers south to central Chile, about 33° south. **VAGRANT:** Accidental, two N.A. records: one photographed at Pacific Grove, CA (June 6–7, 1985), also seen the next day at Moss Landing, CA; and one sight record 15 miles west of Southeast Farallon I., CA (Mar. 3, 1996). A few have been observed off the Pacific coasts of Panama, Costa Rica, and Nicaragua.

POPULATION Global population estimated at 35,000 individuals in 2004; numbers at colonies fluctuate, probably because of prey availability, which is profoundly affected by changes in ocean temperatures, as during El Niño/Southern Oscillation cycles.

Parts of a Gull

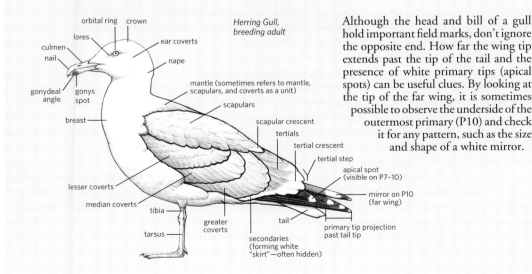

Herring Gull, breeding adult

orbital ring   crown
lores
culmen
nail
gonydeal   gonys
angle   spot
breast
ear coverts
nape
mantle (sometimes refers to mantle, scapulars, and coverts as a unit)
scapulars
scapular crescent
tertials
tertial crescent
tertial step
apical spot (visible on P7–10)
mirror on P10 (far wing)
lesser coverts
median coverts
tibia
tarsus
greater coverts
tail
secondaries (forming white "skirt"—often hidden)
primary tip projection past tail tip

Although the head and bill of a gull hold important field marks, don't ignore the opposite end. How far the wing tip extends past the tip of the tail and the presence of white primary tips (apical spots) can be useful clues. By looking at the tip of the far wing, it is sometimes possible to observe the underside of the outermost primary (P10) and check it for any pattern, such as the size and shape of a white mirror.

BLACK-LEGGED KITTIWAKE *Rissa tridactyla*   BLKI ▪ 1

This gull is buoyant and agile. It feeds mainly in flight by shallow plunge dives or by picking from surface. Polytypic (2 ssp.; both in N.A.). L 17–18" (43–46 cm) WS 37–41" (94–104 cm) **IDENTIFICATION** A three-cycle gull. Medium-length bill; slightly cleft tail; short legs. **BREEDING ADULT:** Medium gray upperparts; upper wing's paler primaries contrast sharply with black wing tip (lack white mirrors); whitish underwing, contrasting black wing tip. Plain yellow bill; black legs. **WINTER ADULT:** Attains dark ear spot, dusky washed hind neck (sometimes on nonbreeding summer adults); primaries molted June–Jan. **JUVENILE & FIRST-SUMMER:** Juvenile has smoky-gray washed nape, blackish ear spot, blackish hind collar (often lost by midwinter on Atlantic birds). Note black M-pattern on upper wings, whitish trailing edge; black tail band. By spring, dark bill mostly yellow; dusky flesh-hued legs blackish. Unlike most juvenile gulls, upperparts gray. Primaries molted late May–Oct. **SECOND-YEAR:** Resembles adult, but more extensive black on outer primaries' outer webs; usually some black on alula; bill sometimes dark tipped. **GEOGRAPHIC VARIATION** Nominate subspecies breeds in North Atlantic; *pollicaris* in North Pacific. The latter averages larger, longer-billed, darker above, more extensive black wing tip, and later molt. **SIMILAR SPECIES** Adult distinctive. Note medium gray upperparts, paler primary bases accentuating "ink-dipped" black wing tips, short black legs, plain yellow bill. In Bering Sea, see Red-legged. On a distant first-year,

upperwing pattern could suggest a Sabine's Gull, but note kittiwake's black ulnar bar, white underwings with small black wing tip. Juvenile Sabine's Gull has brownish hind neck and back; older plumages lack black tail band. Similar upperwing pattern on first-year Little Gull, but it is much smaller with shorter wings; slender bill; relatively longer, pinkish legs. Note Little Gull's dark secondary bar and cap, and lack of black hind collar. **VOICE** Named for rhythmic *ketewehk ketewehk* call; flocks' laughing or honking calls can suggest small geese. **STATUS & DISTRIBUTION** Holarctic breeder, winters in northern seas. **BREEDING:** On sea cliffs. Returns Apr.–May, departs Aug.–Sep. **MIGRATION & WINTER:** Rarely seen from shore. Marked year-to-year variation in southward movements; winter wrecks periodically, with birds blown onshore and inland. Peak fall movements off BC in Sept.; off CA in mid-Nov. First arrivals to New England usually Oct., most Nov.–Dec. Northbound Pacific movement mainly mid-Mar.–Apr. In Atlantic, move north from southern

areas Feb.–Mar.; most depart New England mid-Apr. Very rare on Gulf Coast in winter. Nonbreeders oversummer irregularly south to CA and New England. **VAGRANT:** Rare to casual (mainly late Oct.–Jan.) in interior. **POPULATION** Atlantic populations increased and range expanded since 1960s. Populations in N.A. Arctic show no trends.

RED-LEGGED KITTIWAKE *Rissa brevirostris*   RLKI ▪ 2

This attractive local gull of the Bering Sea region is best known (and readily seen) as a summer resident on the Pribilof Islands. Monotypic. L 15.5–17" (39–43 cm) WS 33–35" (84–89 cm) **IDENTIFICATION** A three-cycle gull with a relatively short, stubby bill; short legs; big dark eyes. Tail slightly cleft (most marked on juvenile). **BREEDING ADULT:** Pale, slaty-gray upperparts; upper wing with narrow white trailing edge, black wing tip (lacking white mirrors); underwings with dusky-gray primaries, black wing tip. Plain yellow bill, bright red legs. **WINTER ADULT:** Attains dusky ear

spot, hind-neck collar (sometimes on summer adults); primaries molted July–Feb. **JUVENILE & FIRST-SUMMER:** Juvenile has smoky-gray wash to nape, blackish ear spot, mottled slaty hind collar. Note upper wings' broad white trailing edge. Dark bill mostly yellow by spring; dusky flesh legs, orange-red by spring. Unlike most juvenile gulls, upperparts gray; lacks black on tail; primaries molted late May–Oct. **SECOND-YEAR:** Resembles adult, but black on outer webs of outer primaries more extensive; bill sometimes tipped dark. **SIMILAR SPECIES** Black-legged

Kittiwake is slightly larger with more sloping forehead, longer bill, paler upperparts. Adult has black legs, whitish underwings; hence, black wing tips contrast more above and below. Head-on in flight, Black-legged's white marginal coverts show up, unlike Red-legged's gray. First-year Black-legged has a black ulnar bar and tail band. Upperwing of first-year Red-legged suggests a Sabine's Gull, but white trailing triangle less extensive, bases of outer primaries gray. Sabine's Gull is smaller and ternlike, with slender bill. **VOICE** Squeakier than Black-legged.

STATUS & DISTRIBUTION BREEDING: On sea cliffs in Bering Sea region; only four known breeding locations (Pribilof Is., Buldir I., Bogoslof Is., and Commander Is.) Returns to colonies Apr.–May; departs Sept.–Oct. **WINTER:** Mainly found near pack ice in Bering Sea; a few probably to North Pacific. **VAGRANT:** Casual (late Nov.–Mar., late June–mid-Aug.) south to WA and OR, also Japan; exceptionally to YT (Oct.), CA (3 Feb. recs.), and NV (July). **POPULATION** In 1970s, estimated 232,000 birds worldwide. Since the 1970s, populations of Red-leggeds have declined on Pribilof Islands and increased on Aleutian Islands.

base of primaries uniform with mantle, which is darker gray than Black-legged

underside of primaries medium gray, except for black tips

breeding adult

no dark M-pattern

broad white trailing edge

no tail band

1st summer

breeding adult

shorter bill than Black-legged

very short coral red legs

juvenile

pinkish legs

IVORY GULL *Pagophila eburnea*   IVGU ▸ 3

This sought-after icon of the high Arctic (*Pagophila* means "ice-loving") strays very rarely into the lower 48. It can be confiding and usually associates only loosely with other gulls except when scavenging. Rarely, and only very briefly, it alights on water. Monotypic. L 17–19" (43–48 cm) WS 36–38" (91–97 cm)
IDENTIFICATION A two-cycle gull with a medium-size, slightly tapered bill; well-developed claws; fairly short legs. **ADULT:** All-white; no seasonal variation. Primaries molted Mar.–Sept., mostly before breeding (when molt suspended). Dark eyes; narrow orbital ring black to red in winter (rarely noticeable), red on breeding birds. Gray-green bill, orange to yellow-orange tip; black legs, feet. **JUVENILE & FIRST-WINTER:** Variable sooty blotching on face and throat; scattered dusky spots on rest of head, back, underparts. Highly variable blackish distal spots on upperwing coverts, tertials, flight feathers.

Some mostly white overall with dark face, black spots largely restricted to flight feathers; others have extensive dark spotting on upperparts. Grayish orbital ring; bill darker gray-green basally than adult's, with yellow-orange tip and often dusky distal marks from late winter to spring. Attains all-white plumage by complete second prebasic molt (mainly Apr.–Aug.).
SIMILAR SPECIES Unmistakable. Beware leucistic or albino individuals of any gull species. Some first-summer Glaucous and Iceland Gulls can bleach to almost all white but are larger with longer pink legs, larger bills (pink-and-black on Glaucous, dark with a flesh base on Iceland), and typical large-gull structure.
VOICE Rarely heard away from breeding grounds. Calls include high, shrill to slightly grating whistles, unlike other gulls.
STATUS & DISTRIBUTION Holarctic breeder on bare ground in high Arctic; mainly around pack ice in winter. **BREEDING:** Uncommon and local (June–Sept.); in N.A. (Canada) breeds

on Ellesmere, Seymour, Devon, Baffin, and Perley Is. **MIGRATION & WINTER:** Tied to pack ice. Main passage late Oct.–Dec. south into Bering Sea and Davis Strait, where most winter (Dec.–Mar./Apr.); irregularly NF. Moves north from Bering Sea late Mar.–Apr.; immatures linger through May in most years (when seen at St. Lawrence I.). **VAGRANT:** Southern occurrences mainly Dec.–Feb. In West, casual BC, exceptionally southern CA and CO. Better known in East, where casual through Great Lakes region, exceptionally TN. Casual from Maritimes south to NJ.
POPULATION Late-winter counts of over 35,000 birds in Davis Strait suggest that the estimate of 2,400 breeding birds in Canadian Arctic is low, although the origin of former numbers is not known (may include Greenland and Old World birds). Recent data suggest major declines in Canadian populations, perhaps linked to shrinking extent of pack ice.

adult

greenish bill with yellow-orange tip

pure white

short black legs

black markings on primary tips and near tail tip

dusky face

1st winter

long, narrow, rather pointed wings

adult

SABINE'S GULL *Xema sabini*   SAGU ■ 1

This striking gull often occurs with Long-tailed Jaeger and Arctic Tern. The Sabine's tern-like flight is buoyant and graceful. Polytypic (2–4 ssp.; 2 in N.A.). L 12.5–14" (32–36 cm) WS 33.5–35.5" (85–90 cm)

IDENTIFICATION Small, two-cycle gull; slender bill, forked tail. Striking upperwing pattern at all ages. **BREEDING ADULT:** Dark slaty hood; black neck ring; lacks white eye crescents; hood often spotted white in fall migrants. Bright white triangle on trailing edge of upper wings; whitish underwing; contrasting black outer primaries; medium-gray inner coverts; dusky secondary bar. Underparts often flushed pinkish. Yellow-tipped black bill; blackish to dusky flesh-hued legs. **WINTER ADULT:** Unlikely in N.A. White head; slaty hind collar; ear spot; dark mottling on nape; neck sides washed smoky. Dusky flesh legs. **JUVENILE & FIRST-SUMMER:** Juvenile (plumage kept through fall migration) has dark gray-brown crown, hind neck, upperparts; scaly pale-edged back feathers with dark subterminal crescents. Wing pattern like adult's, but with dark brown coverts; tail has black distal band. Black bill, flesh legs. Complete preformative molt in winter produces plumage like winter adult's; initially gray back contrasts with brown upperwing coverts. After partial first prealternate molt, first-summer notably variable: Some have slaty-gray hind neck, mostly white head with dark ear spot; others have dark-slaty hood with whitish spotting. Black bill usually has variable yellow tip.

GEOGRAPHIC VARIATION Nominate (breeds northern AK, Arctic Canada, to Greenland; migrant off both coasts) is smallest, palest subspecies; *woznesenskii* (breeds western AK; migrates off Pacific coast) larger, upperparts average darkest. Two other described subspecies (breeding northern Russia; possibly migrating off Pacific coast) are intermediate. Fall migrants (off CA) include smaller, paler adults and larger, darker adults; appreciably different when seen together. Sometimes considered monotypic.

SIMILAR SPECIES Distinctive, but see first-year kittiwakes. Rarely seen juvenile Ross's have dark-brown upperparts and bold white trailing edge to upper wing similar to Sabine's, but tail strongly graduated with black-tipped central rectrices, bill smaller, white postocular area more extensive. Some heavily pigmented juvenile Franklin's Gulls (mainly July–Aug.) have dark brownish head and hind neck, but larger; stouter bill; blackish legs; bolder white tips to primaries; different wing pattern in flight.

VOICE Grating ternlike *kyeerr,* mainly given on summer grounds.

STATUS & DISTRIBUTION Holarctic breeder, winters mainly off S. Africa and western S.A. **BREEDING:** Fairly common (late May–Aug.) in N.A., on low-lying tundra and marshy areas, often near coast. **MIGRATION:** In fall, mainly late July–Oct.; stragglers into Nov., exceptionally Dec. Main passage off West Coast mid-Aug.–late Sept.; rare in interior (mainly Sept.); rare to casual off Atlantic coast. In spring, mainly late Apr.–May off West Coast; casual at best in interior; casual off Atlantic coast (late Apr.–mid-June). Nonbreeders oversummer locally off west coast, rarely in northwest Atlantic. **WINTER:** Accidental Lake Erie and FL.

POPULATION Little data available; difficult to survey remote nesting populations.

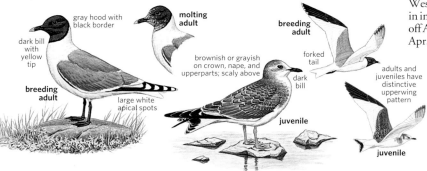

gray hood with black border

molting adult

breeding adult

dark bill with yellow tip

breeding adult

large white apical spots

brownish or grayish on crown, nape, and upperparts; scaly above

dark bill

forked tail

adults and juveniles have distinctive upperwing pattern

juvenile

juvenile

LITTLE GULL *Hydrocoloeus minutus*   LIGU ■ 3

This is the smallest gull, most often seen with flocks of Bonaparte's Gulls. In its fluttery and ternlike flight, it dips down to pick from the water's surface. Monotypic. L 11–11.5" (28–29 cm) WS 27–29" (69–74 cm)

IDENTIFICATION Small, three-cycle gull; slender bill, fairly short legs. **BREEDING ADULT:** Black hood lacks white eye crescent. Pale gray upper wings, smoky gray underwings; note white trailing edge out to tips. Red legs. **WINTER ADULT:** White head with dark cap and ear spot. **JUVENILE &** **FIRST-WINTER:** Juvenile has blackish brown hind neck collar and back; pale-edged scapulars. Soon molts pale gray back of first-winter. Upper wings have bold blackish M-pattern, dark secondary bar. Tail white with black distal band. First-summer can attain partial dark hood and white tail. **SECOND-WINTER:** Resembles winter adult, but some have black wing-tip markings.

SIMILAR SPECIES Bonaparte's is slightly larger and longer-billed, with longer and more pointed wings; first-winter

lacks dark cap, has upper wings with white leading wedge and bolder black trailing edge. Ross's is similar to Little in most plumages, but has shorter bill, long and graduated tail; winter adult lacks dark cap; first-winter has bright white trailing edge on upper wing, black tail tip prominent on long central rectrices. First-winter Black-legged Kittiwake shares bold, black M-pattern

on upper wings, but is markedly larger, with longer, more pointed wings, bigger bill (pale based by winter), dark legs. Note kittiwake's whitish secondaries and lack of dark cap, often with blackish collar through winter.

VOICE Nasal *kek* occasionally given in migration and winter.

STATUS & DISTRIBUTION Eurasia and eastern N.A. **BREEDING:** Rare and local (May–Aug.), solitary pairs or small groups in marshes from Great Lakes region to Hudson Bay. **MIGRATION:**

Mainly Aug.–Nov. and Mar.–May, primarily Great Lakes region, rare New England. **WINTER:** Nov.–Mar., mostly mid-Atlantic coast; movements closely tied to Bonaparte's Gull. Casual north to Maritimes and in southeast. **VAGRANT:** Casual west to Pacific coast, annual to CA.

POPULATION Colonization of N.A. in the 1900s (first bred 1962) mirrored by expansion in western Europe. In N.A., population is in low hundreds, stable or increasing.

ROSS'S GULL *Rhodostethia rosea*   ROGU ■ 3

This sought-after "pink gull" of the high Arctic rarely graces the lower 48. Vagrants are usually found with flocks of Bonaparte's Gulls. Its flight is buoyant and strong, with deep wing beats; it dips down to pick food from water. Monotypic. L 12.5–14" (32–36 cm) WS 32–34" (81–86 cm)

IDENTIFICATION Small, two-cycle gull; long, graduated tail; small bill. All post-juvenal plumages can be strongly flushed pink on head and underparts. **BREEDING ADULT:** Narrow black neck ring. Pale-gray upper wings; smoky-gray underwings, broad white trailing edge to secondaries and inner primaries. Orange-red legs. **WINTER ADULT:** White head with dark ear spot; sometimes broken neck ring; duller legs. **JUVENILE & FIRST-WINTER:** Juvenile has sooty-brown crown, hind neck, and back; pale-edged scapulars. Soon molts into adultlike head and pale-gray back of first-winter. Upper wings have bold blackish M-pattern, broad white trailing edge. White tail, black tips to elongated central rectrices. Flesh legs. First-summer can attain black neck ring, all-white tail; orange-red legs. **SIMILAR SPECIES** Distinctive, but at a distance might be confused with Little Gull, which is similar in most plumages. Little Gull has slightly longer bill; shorter, blunter wings; and "normal" squared tail. Adult Little has narrower white trailing edge to wings extending to wing tip; winter adult has dark cap. First-winter has duller and

less extensive whitish trailing triangle on upper wing (with dark secondary bar); black tail tip forms an even band. See juvenile Sabine's Gull. Most small gulls can attain a pink flush on their head and underparts, but rarely as intense as most Ross's.

VOICE Usually silent. In summer, mellow yapping *p-dew* and ternlike notes.

STATUS & DISTRIBUTION Holarctic breeder, wintering near pack ice. **BREEDING:** Rare and local (June–Aug.); solitary pairs and small colonies at marshes and small tundra lakes. First confirmed N.A. breeding in late 1970s. **MIGRATION & DISPERSAL:** Post-breeding dispersal starts July–Aug. from breeding grounds in Siberia; move east to western Beaufort Sea by late Sept.–early Oct.; followed by late Oct.–Dec. passage south into Bering Sea. Best known in fall from Point Barrow, AK (peak numbers late Sept.–mid-Oct.). In spring, small numbers range east irregularly (late May–June) to northeast Bering Sea. **WINTER:** Distribution probably linked to pack ice; main winter grounds (Dec.–Apr.) apparently Bering Sea south to waters north of

Japan. **VAGRANT:** Casual central Canada (mainly spring) from southern MB east to ON and in lower 48 (mainly winter), mostly in Midwest and Northeast. Accidental in migration and winter south to CO, MO, DE, in Northwest (south to OR and eastern ID), and southern CA (Salton Sea).

POPULATION Estimated 20,000–40,000 birds occur in northern Alaska waters in fall. No data on trends.

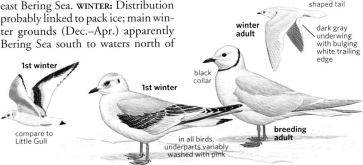

BONAPARTE'S GULL *Chroicocephalus philadelphia* BOGU ▪ 1

This attractive gull is the only small gull that is common and widespread in North America. In migration and winter, it occurs in flocks of up to a few thousand birds, feeding over tidal rips, power station outflows, sewage ponds, and so on. Bonaparte's has a graceful, buoyant flight, which is rather ternlike. Monotypic. L 13–13.5" (33–34 cm) WS 31.5–34" (80–86 cm)

IDENTIFICATION Small, three-cycle gull with slender black bill, white leading wedge on upper wing, and translucent underside to primaries. **BREEDING ADULT:** Slaty black hood. Bold white leading wedge on pale gray upper wing; underwing has translucent primaries with narrow black tips. Black bill, orange-red legs. **WINTER ADULT:** White head with dark ear spot; legs paler, flesh pink to reddish pink. **JUVENILE & FIRST-WINTER:** Juvenile's hind neck and back are mottled cinnamon-brown, soon molting to pale gray in fall. Upper wing has dark brown ulnar bar, black trailing edge, and less prominent white leading wedge than adult's, with mostly dark outer primary coverts. Pale flesh legs; bill black or with dull flesh base (not striking); legs pink to orange. First-summer has variable head plumage:

Some are like winter adult, others attain a partial blackish hood. **SECOND-WINTER:** Resembles winter adult, but with black marks on primary coverts, alula, and sometimes on tertials and tail; legs often paler in early winter.

SIMILAR SPECIES Distinctive and generally common; note wing pattern, fine black bill. See rare Black-headed Gull and Little Gull, often found among flocks of Bonaparte's Gulls.

VOICE Chatters and single *mew* calls all have relatively low, rasping, or buzzy tone; distinct from other gulls' calls.

STATUS & DISTRIBUTION N.A.; winters to northern Mexico. **BREEDING:** Common (May–Aug.), in solitary pairs and loose colonies, around lakes and marshes in boreal forest zone; nests usually placed in conifers. Has nested locally in eastern QC, probably ME. **MIGRATION:** In fall, first birds reach Great Lakes region and Bay of Fundy in July–Aug., but main passage in East is late Oct.–Nov., when rare north to NF. On Pacific coast, a few early migrants occur late July–Aug.; main migration late Oct.–Nov. In spring, mainly Mar.–May. Northbound influxes typically start late Mar. on Great Lakes, mid-May in Prairie Provinces. On Pacific coast, northbound influxes start late Mar.; peak movements Apr.–mid-May.

Nonbreeders oversummer irregularly in Great Lakes region, locally in New England and Pacific states. **WINTER:** Bulk of population spends winter period (Oct.–Mar.) on Lake Erie and Lake Ontario and along Atlantic coast from MA south through mid-Atlantic states. During early winter most birds are on Great Lakes, but once lakes start to freeze, birds move toward Atlantic coast. Casual north to NF. In west, fairly common but local (Nov.–Mar.) in Pacific states, casual to southeastern AK. **VAGRANT:** Casual (late May–June, Aug.) to Bering Sea islands. Also Europe, Africa, HI, Japan.

POPULATION No data on trends. This species' largely inaccessible breeding range makes it difficult to survey.

blackish hood with gray cast

breeding adult

postocular spot

winter adult

1st winter

largely pale inner primaries

dark carpal bar

1st winter

winter adult

white primaries above and below with thin black tips

GRAY-HOODED GULL *Chroicocephalus cirrocephalus* GHGU ▪ 5

This gull has occurred twice in North America. The adult has a gray hood. Also known as "Gray-headed Gull." Polytypic (2 ssp.). L 16" (41 cm) WS 43" (109 cm)

IDENTIFICATION A slim, three-cycle gull. Large for a hooded gull, but much smaller than, say, Ring-billed. Longish bill, long legs. **SUMMER ADULT:** Hood distinctive: pale gray with darker border. Medium gray

mantle; otherwise body white, tail white. Whitish eye, red orbital ring. Dark red bill, orange-red legs. Wing pattern distinctive: white wedge on outer primaries interrupted by large black wing tip; large mirrors on P9 and P10. Dark gray underwing; blacker on primaries with obvious mirrors showing. **WINTER ADULT:** Hood replaced by dark ear spot and smudgy area around eye, both connected by stripes on

crown. Bicolored bill: red base, black tip. **FIRST-YEAR:** Dark ear spot, smudgy area around eye as in winter adult. Medium gray mantle; gray wings, dark tertials; brown centers to lesser and median coverts. Pinkish yellow bill with dark tip; brownish orange legs. In flight, outer wing as in adult: white outer wedge with large black wing tip, but no mirrors. Inner wing gray, brown bar diagonally across inner

wing, contrasting darker secondary bar. Tail white with black tail band. **SECOND-YEAR:** As adult, but smaller primary mirrors.
GEOGRAPHIC VARIATION Subspecies *cirrocephalus* in South America is slightly larger and paler than *poiocephalus* in Africa.
SIMILAR SPECIES Summer adult is easily identified by gray hood and wing pattern. First-year easily confused with Black-headed Gull of similar age. Gray-hooded is larger, with a longer bill and longer legs; shows slightly darker mantle. In flight,

Gray-hooded has a solid black wing tip. Black-headed shows white on primaries, reaching slot-like toward tips. Gray-hooded's primaries are entirely dark on underside. Outer two primaries are white on Black-headed, producing long white underwing stripes.
VOICE Similar to Black-headed Gull, but deeper and rougher.
STATUS & DISTRIBUTION Accidental to N.A. Native to Africa and S.A. **YEAR-ROUND:** Estuaries and coastal concentrations of small gulls. **VAGRANT:** Two records: Apalachicola, FL (Dec.

26, 1998) and Coney Island, NY (July 24–early Aug., 2011).
POPULATION Combined African and South American population estimated to be 50,000 pairs.

breeding adult
cirrocephalus

distinctive black in wing-tip pattern (all plumages)

grayish hood with dark border

red legs and bill

BLACK-HEADED GULL *Chroicocephalus ridibundus*   BHGU   ▪   3

This Old World counterpart of Bonaparte's Gull (with which it is often found) is generally rare in N.A. Polytypic (2 ssp.; both in N.A.). L 15–16" (38–41 cm) WS 35.5–39" (90–99 cm)
IDENTIFICATION Medium-small, two-cycle gull. Medium-long red to pink bill; dark underside to middle primaries. **BREEDING ADULT:** Chocolate brown hood. Bold white leading wedge on pale gray upper wing; dark middle primaries on underwing, narrow white wedge on outer primaries. Deep red bill and legs. **WINTER ADULT:** White head, dark ear spot; paler red bill and legs. **JUVENILE & FIRST-WINTER:** Juvenile has mottled cinnamon-brown hind neck and back, soon molting to pale gray in fall. Upper wing has cinnamon-brown ulnar bar, dark trailing edge,

less prominent white leading wedge than on adult. Pale peach to pinkish-orange bill and legs; black-tipped bill. First-summer head plumage variable: some like winter adult's; others with a solid dark-brown hood, like breeding adult's; legs and bill pink to reddish.
GEOGRAPHIC VARIATION Often considered monotypic. Breeding populations in N.A. presumably nominate (Europe). The subspecies *sibiricus* (eastern Asia) ranges to western N.A., averages larger (esp. longer bill) and molts later.
SIMILAR SPECIES A reasonably distinctive species. Only likely to be confused with smaller and much commoner Bonaparte's Gull. Attention to a few key points (e.g., bill size, bill color, pattern on underside of primaries) should separate these species. Bonaparte's is smaller and daintier with relatively

narrower wings; a more slender black bill; appreciably shorter legs; white underwings with translucent primaries; and slightly darker upperparts—thus white leading wedge on adult upper wing contrasts more. Breeding adult Bonaparte's has a blackish hood (attained a month or so later than Black-headed Gulls attain hoods in the East), orange-red legs; winter adult has a darker smoky-gray hind-neck wash, peach-pink legs. First-winter Bonaparte's averages a darker ulnar bar and has dark streaks on outer primary coverts (mostly white on Black-headed).
VOICE Varied screams and a grating *meeahr*, higher and less buzzy than those of Bonaparte's Gull.
STATUS & DISTRIBUTION Eurasia and eastern N.A.; winters to Africa. **BREEDING:** Rare (May–Aug.) in N.A., where first found breeding in 1977. Colonial or in solitary pairs; on open or partly vegetated ground near water, usually in association with other waterbirds. Mainly NF; smaller numbers in Gulf of St. Lawrence region, QC; has nested, or attempted to, in NS, ME, MA; also possibly northwest IA or

dark inner primaries

dark brown hood

breeding adult

1st winter

dark slate underside to primaries, paler in immature

winter adult

long reddish bill

winter adult

1st winter

winter adult

orangish pink with dark tip

1st summer

slightly paler above than Bonaparte's

adjacent MN. **MIGRATION & DISPERSAL:** In western AK, rare in spring (mainly May–June, a few into summer) and casual in fall (late Aug.–Oct.), mainly to western Aleutians and Pribilofs, a few to St. Lawrence I. Casual in Great Lakes region and west to Great Plains. **WINTER:** Mainly Oct.–Apr. Locally fairly common in NF; uncommon to rare along St. Lawrence Seaway and Atlantic coast south to NJ; very rare in NC; casual farther south along Atlantic coast to FL and (mainly Nov.–Mar.) from TX east through Gulf Coast states. **VAGRANT:** Casual (mainly Sept.–Apr.) along Pacific coast from south-coastal AK to CA. Accidental to HI.

POPULATION Marked increase in Europe from 1800s, especially in 1900s, when spread across north Atlantic. First bred in Iceland in 1911; has bred in western Greenland since the 1960s.

LARGER GULLS Genera *Leucophaeus* and *Larus*

The gulls in these two genera are the larger, "typical" gulls. Laughing and Franklin's Gulls are in the genus *Leucophaeus* and have dark hoods in breeding plumage. Most species are in the genus *Larus,* which includes the widespread Herring Gull and other large species that typically have white heads as breeding adults (the so-called "large white-headed gulls"). Adult plumages are patterned in gray, black, and white, often boldly; juveniles and younger immatures have much brownish in their plumage and are less distinctly marked. Males are larger overall and bigger billed than females; this can be striking in "large white-headed gulls" and is an important consideration in identification. In addition, hybrids are not uncommon, another factor to consider when identifying an atypical individual.

All of these species are either three- or four-cycle gulls with a variety of immature plumages that complicate identification. An understanding of molt is helpful for identification, as gulls can have up to four years of immature plumages. A four-cycle (or four-year) gull is one that attains adult plumage by its fourth prebasic molt. Adults typically have a complete molt in fall and a partial molt in spring; this molt pattern starts with the second prebasic molt at about one year of age and is repeated for life. However, molts in the first year are more variable: Most first-cycle large gulls have a single, protracted molt from fall through spring.

LAUGHING GULL *Leucophaeus atricilla*   LAGU ■ 1

This dark-hooded species is the common and familiar "parking lot gull" of Atlantic and Gulf coast beaches. Polytypic (2 ssp.; 1 in N.A.). L 15–17" (38–43 cm) WS 38–42" (97–107 cm). **IDENTIFICATION** Medium-size, three-cycle gull with long, pointed wings. Relatively long bill often looks slightly droop tipped. **BREEDING ADULT:** Blackish hood, white eye crescents. Upperparts slaty gray with black wing tips; small white outer primary tips worn off by summer. **WINTER ADULT:** Whitish head, dusky auricular smudge. **JUVENILE & FIRST-WINTER:** Juvenile head, neck, chest, and back brown; upperparts with scaly buff edgings. Broad black distal band on tail; underwings dusky brownish. Back, neck, and chest molt to gray in fall; head to whitish with a dark mask. First-summer can attain partial hood. **SECOND-WINTER:**

has dark "earmuffs," but no half hood like Franklin's

winter adult

long bill with slightly thicker tip

breeding adult

dark underside to primaries

breeding adult

long wings

2nd winter

2nd winter

1st winter

tail band includes outer tail feathers

dark "ear muffs," but no hood

gray across breast and down sides and flanks

overall brownish color

juvenile

1st winter

broad, complete tail band

Resembles winter adult, but more black in wing tips, often some black on tail, neck sides grayish.
GEOGRAPHIC VARIATION North American populations are subspecies *megalopterus,* larger than *atricilla* of Caribbean.
SIMILAR SPECIES Franklin's Gull is slightly smaller, with a shorter, undrooping bill; all plumages have bold white eye-crescents. Adult Franklin's has a white medial band inside black wing tip. In spring, adult Franklin's bright pink flush is often striking among white-breasted Laughing Gulls. Winter Franklin's has a blackish half-hood. First-winter

has whitish underwings; its narrower black tail band does not reach tail sides. First-summer Franklin's Gull can suggest adult Laughing, but Franklin's smaller black underwing tip contrasts with pale gray primary bases.
VOICE Laughing and crowing calls.
STATUS & DISTRIBUTION North America to Caribbean; winters to northern South America **BREEDING:** Common (Apr.–Aug.), colonial on sandy and rocky islands, salt marshes. Breeding attempted in southern Great Lakes region, where it has hybridized with Ring-billed Gull. **MIGRATION & DISPERSAL:** In Northeast, peak

movements late Aug.–Sept. (lingerers Dec.), and Apr.– early May (arrivals Mar.); rare north to NF. Casual in Pacific coast states, except at south end of the Salton Sea, CA, where fairly common (mainly June–Oct.; has bred). Casual inland elsewhere in West. **WINTER:** Aug.–Apr. (some oversummer in much of winter range). Casual inland in Midwest. **VAGRANT:** To Europe, Africa, Japan, and Australia.
POPULATION East Coast colonies rebounded after near extirpation in late 1800s by eggers and plume hunters. Current U.S. population is fairly stable.

FRANKLIN'S GULL   *Leucophaeus pipixcan*   FRGU   ■   1

Sometimes known as the "prairie dove," this dark-hooded gull is mainly a bird of the mid-continent. It is unusual in having two complete molts a year. Monotypic. L 14–15" (36–38 cm) WS 35–38" (89–97 cm)
IDENTIFICATION Medium-size, two-cycle gull. **BREEDING ADULT:** Blackish hood with thick white eye crescents; slaty gray upperparts. Black subterminal band on outer primaries is framed by large white primary tips and a white medial band. Red bill; dark reddish legs. **WINTER ADULT:** Head is whitish with a blackish half-hood and thick white eye crescents. Black bill has an orange tip. **JUVENILE & FIRST-WINTER:** Note its hood, like that found on winter adult. Juvenile has brown hind neck and back; back has pale edgings. Tail has black distal band not extending to sides of tail; underwings whitish overall with blackish wing

tips. Back molts to gray in fall, hind neck becomes whitish. After complete molt in winter, first-summer resembles winter adult, but its wing tip lacks white medial band; usually looks like winter adult after complete second prebasic molt in fall.
SIMILAR SPECIES See Laughing Gull.
VOICE Pleasant yelping and laughing chatters often heard from migrant flocks.
STATUS & DISTRIBUTION Breeds interior N.A., winters Pacific coast of S.A. **BREEDING:** Common (May–Aug.) in N.A., colonial in freshwater marshes. **MIGRATION:** In fall mainly Aug.–Nov., south through Great Plains to TX, lingerers into Dec. First spring arrivals Mar., peak movements late Apr.–early May. Uncommon to very rare to Pacific coast; casual north to Bering Sea. Uncommon to rare in Midwest and ON; casual to very rare

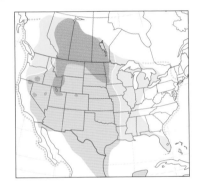

along Atlantic coast from NF to FL. **WINTER:** Rare to casual north to TX. Occasional winter records elsewhere in U.S. **VAGRANT:** To Europe, Africa, Japan, and Australia.
POPULATION Breeding range shifted west and south through 1900s, which—in combination with interannual colony fluctuations—makes population trends difficult to ascertain.

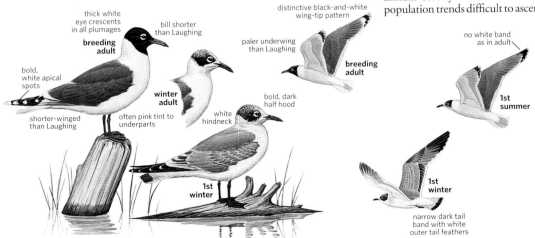

thick white eye crescents in all plumages

breeding adult

bold, white apical spots

shorter-winged than Laughing

bill shorter than Laughing

winter adult

often pink tint to underparts

white hindneck

1st winter

distinctive black-and-white wing-tip pattern

paler underwing than Laughing

breeding adult

bold, dark half hood

no white band as in adult

1st summer

1st winter

narrow dark tail band with white outer tail feathers

BELCHER'S GULL *Larus belcheri* BEGU ▪ 5

This attractive gull was previously known as the Band-tailed Gull. It was lumped with the threatened Olrog's Gull of coastal southeastern S.A. Monotypic. L 20" (51 cm) WS 49" (124 cm)

IDENTIFICATION A medium-size, stocky, four-cycle gull with long thick legs. Long, thick, parallel-sided bill lacks an expansion at the gonydeal angle. **SUMMER ADULT:** Black mantle; white head, underparts; pale gray neck; white tail with wide black tail band, narrow white tip. Tertial, scapular crescents narrow but obvious. Bright yellow bill; black subterminal bar on both mandibles; red tip, rarely missing or restricted to upper mandible. Dark eye, red orbital ring; bright yellow legs. In flight shows extensive black underwing tip. **WINTER**

ADULT: Similar to summer adult, but with dark hood and white eye crescents. **JUVENILE & FIRST-YEAR:** Brown with contrasting dark head, neck, breast; white eye crescents; white belly and undertail coverts. Juveniles with pale-fringed mantle feathers; older birds' mantle gray-brown. Uniform, dark brown wings. Blackish tail contrasts with whitish rump and uppertail coverts. Dull pink legs. Distinctive bill: pinkish or yellowish white with crisp black tip (outer third of bill); red nail. **SECOND-WINTER:** Blackish mantle, like adult. Browner wings, often with blackish inner median covert bar. Extensive brownish hood extends to lower breast. Tail band (like adult's) developed by this age. Bill characteristically bi-colored; red tip now obvious. Bright yellow legs. **SECOND-SUMMER:** Similar, but white head and neck. **THIRD-YEAR:** Very like adult, but brownish wash to wings.

SIMILAR SPECIES Adults identified by crisp black tail band. Black-tailed shows pale eyes, dark gray mantle, thinner bill, slimmer body, longer wings. In comparison to Lesser Black-backed or Yellow-footed, Belcher's has a red-tipped bill, darker mantle, dark eye, no white primary tips. Strongly hooded look of first-year plumages distinctive. Second-year told from same-aged Kelp by brighter yellow legs, red bill tip, and structural differences.
VOICE Undescribed.
STATUS & DISTRIBUTION Casual in N.A., native to Peru Current. **YEAR-ROUND:** Strictly a coastal species, often on rocky coastlines. **VAGRANT:** One record from CA and four from FL. The FL records are unusual as this is a Pacific Gull, suggesting that vagrants cross over C.A.
POPULATION Estimated to be fewer than 100,000 pairs.

thickish bill with black subterminal band and red tip

dark eye

breeding adult

distinct brown hood in winter plumage

darker mantle than Black-tailed

smoky brown head and breast

breeding adult

long yellow legs

winter adult

2nd winter

pinkish legs

1st winter

broad black tail band

BLACK-TAILED GULL *Larus crassirostris* BTGU ▪ 4

This Asian vagrant has the potential to occur nearly anywhere. Adults are slim, with a black tail band and a characteristic red-tipped bill. Monotypic. L 18" (46 cm) WS 47" (119 cm)
IDENTIFICATION A medium-size, long-billed, four-cycle gull with long, narrow wings. Long bill is characterized by being nicely parallel-sided, lacking an expansion at the gonydeal angle. **SUMMER ADULT:** Dark gray mantle. White head, neck, and underparts; white tail has a wide black tail band and narrow white tip. Tertial and scapular crescents narrow but obvious. Bright yellow bill, black subterminal bar on both mandibles, red tip. Yellow eye, orange-red orbital ring. Bright yellow legs. In flight shows extensive black wing tip, to P5 or P4; there are no mirrors on outer primaries.

WINTER ADULT: Similar to summer, but head and neck crisply streaked on face and nape. **JUVENILE & FIRST-YEAR:** Chocolate brown with paler facial area at bill base; contrasting white

breeding adult

pale eye

long bill with black subterminal band and red tip

breeding adult

winter adult

overall smooth brown color with whitish in face

broad, bold black tail band

long wings

rather dark slaty gray upperparts

1st winter

yellowish legs

pink-based bill with dark tip

2nd winter

eye crescents. Juvenile scaly-looking mantle due to crisp buff edges; first-winter birds more gray-brown above with darker feather centers. Unicolored dark brown wings. Blackish tail contrasts with whitish rump and uppertail coverts. Bill distinctive: pale pink with crisp black tip (outer third). Pinkish legs. **SECOND-YEAR:** Dark gray mantle; browner wings. Brown head with white eye crescents and face. Black tail with crisp white tip. Bill greenish at base. **THIRD-YEAR:** As adult, but black on tail more extensive. **SIMILAR SPECIES** Adults are identified by their crisp black tail band and their

pale eyes. The vagrant Belcher's Gull also has a tail band; however, it shows dark eyes, blacker mantle, thicker bill, and bulkier body. Can be told from adult California Gull by a red-tipped bill, darker mantle, pale eye, and very small white primary tips. First-year Black-tailed resembles both California and Laughing Gulls. California shares bi-colored bill, but Black-tailed is more uniformly colored and has a contrasting and nearly unmarked white rump, well-defined white eye crescents, and a longer and slimmer shape. First-year Laughing Gull shares long and slim shape, eye crescents, and

uniform-looking plumage, but it has a dark bill and dark legs. **VOICE LONG CALL:** Mewing and deep. **STATUS & DISTRIBUTION** Casual in N.A. Native to temperate Asian coast. **YEAR-ROUND:** Prefers rocky coastal habitats, but should be looked for any place where medium-size gulls congregate. Most recs. in N.A. Mar.–Aug. **VAGRANT:** Over 20 recs., most in AK. Other records in BC, WA, CA, MB, TX; several sightings around Lake Michigan, MD, VA, NJ, NY, RI, VT, NS, and NF. Other records include Sonora, Mexico, and Belize. This gull can truly show up anywhere!

HEERMANN'S GULL *Larus heermanni* HEEG ■ 1

A striking Pacific gull, entirely dark-bodied and white-headed when breeding—like a photo negative of our smaller dark-hooded and pale-bodied gulls. Monotypic. L 19" (48 cm) WS 51" (130 cm) **IDENTIFICATION** A medium-size, four-cycle gull, neither stocky nor slim, neither large-billed nor small-billed. Bill slightly drooped. **SUMMER ADULT:** Dark body, paler on underparts. Snow-white head contrasts strongly. White head obtained in late fall to early winter and lost by midsummer. Tertial and scapular crescents narrow, but obvious. Deep red bill with black tip; eyes dark with red orbital ring; black legs. In flight, all dark gray, but slightly paler on secondary coverts than flight feathers; shows a narrow white trailing edge extending to inner primaries. No mirrors or white primary tips. Black tail has white tip and contrasts with pale gray rump and uppertail coverts. **WINTER ADULT:** Crisply streaked head; otherwise similar to summer adult. **JUVENILE & FIRST-YEAR:** Dark brown; scaly above due to pale feather edges; older birds gray-brown on mantle. Dull pink bill with dark tip. In flight, wings are brown with darker secondaries and primaries. **SECOND-YEAR:** Body dark gray; wings may be washed brown. Dark brown hood shows white eye crescent. Tertial crescents white; bill reddish at base. **THIRD-YEAR:** Much like adult, but head with more extensive brown wash. **SIMILAR SPECIES** First-year birds may be confused with young California Gulls, but Heermann's is much more uniform above and has blackish legs.

Young often mistaken for juvenile jaegers. Ironically, some Heermann's Gulls show a white patch on greater primary coverts, not primaries as on jaegers. Heermann's Gulls separated by slower, less powerful flight; unbarred body and underwings; and lack of primary flash. **VOICE CALL:** A nasal *ahhh.* **LONG CALL:** Short, deep, and nasal. **STATUS & DISTRIBUTION** Common. Over 90 percent breed on Isla Raza in Gulf of California, so essentially it is a Mexican breeding species that migrates north to our region. **YEAR-ROUND:** Coastal, preferring sandy beaches. **BREEDING:** Very rare breeder in CA. **MIGRATION:** In central CA appears as early as late May in some years, late June in others; reaching Pacific Northwest in July, peaking late July–Sept. During warm-water years arrival is earlier and birds travel farther north. Moves south Sept.–Nov.; largely absent north of Monterey, CA,

Jan.–May. **VAGRANT:** Casual to southeastern AK. Very rare in lower Colorado River Valley, into AZ. Casual farther east; records from NV, NM, UT, WY, OK, TX, MI, OH, FL, VA, and ON. **POPULATION** Estimated at over 300,000 breeding individuals; populations increased steadily from the 1970s to the 1990s.

red bill with small black tip

white head

breeding adult

gray underparts

2nd winter

streaked head

winter adult

dark wings

pink-based bill with dark tip

1st winter

breeding adult

broad black tail band

overall sooty brown color

MEW GULL *Larus canus*   MEGU ■ 1

The Mew Gull is the smallest white-headed gull in N.A. Polytypic (4 ssp.; 3 in N.A.). L 16–17" (41–43 cm) WS 41–44" (104–112 cm)
IDENTIFICATION Medium-small, three-cycle gull with rounded head, relatively small and slender bill, dark eyes in all ages. Adult has large white mirrors on P9–P10. North American Mew Gull described first; other subspecies ("Common" and "Kamchatka Gulls") follow.
BREEDING ADULT: Upperparts medium gray; yellow legs and unmarked bill. **WINTER ADULT:** Head and neck with smudgy brown washing. **JUVENILE & FIRST-WINTER:** Overall brownish; back becomes gray in first winter. Tail mostly dark brown (variable); uppertail coverts barred brown; underwing coverts brownish. **SECOND-WINTER:** Resembles winter adult, but more black on wing tips; tail usually with black markings. **"COMMON GULL":** Average larger, longer billed. **ADULT:** Slightly paler gray upperparts, more black on wing tip than on Mew (but much variation); winter head markings more spotted; bill usually has dusky ring. **JUVENILE & FIRST-WINTER:** Whiter on head, body, and underwing coverts than Mew; uppertail coverts and tail base mostly white with clean-cut black tail band. **"KAMCHATKA GULL":** Largest and bulkiest subspecies, with longer and stouter bill than Mew's.

Adult upperparts average darker, wing tip with more black (but much variation), and eyes often pale (can be pale on Mew); coarser winter head markings. **JUVENILE & FIRST-WINTER:** Mostly white uppertail coverts and tail base; clean-cut black tail band narrowest at sides. Tail typically paler at base; broad, cleaner-cut blackish distal band (pattern matched by some Mews).
GEOGRAPHIC VARIATION Population in N.A. is *brachyrhynchus* (Mew), which is smaller than nominate *canus* group ("Common" includes slightly larger *heinei*) of Eurasia, and *kamtschatschensis* ("Kamchatka") of eastern Asia.
SIMILAR SPECIES Ring-billed is larger with flatter head and deeper, blunter bill; adult pale gray above with pale eyes, neat black bill ring. Juvenile and first-winter paler and more contrasty overall, with whitish underparts (variably spotted dusky) and underwings, and whiter tail base. "Common" and "Kamchatka" more similar to Ring-billed: on first-year Ring-billeds, upperwing coverts have broader, pale-notched tips (often creating a more checkered, rather than scaly,

pattern); tertials tend to be darker with wider and more notched pale edgings; note tail patterns.
VOICE Slightly shrill mewing calls in series; also clipped *kek* in winter.
STATUS & DISTRIBUTION Following refers to Mew Gull unless noted. **BREEDING:** Common (May–Aug.), from tundra and inland marshes to sea cliffs, usually colonial. **WINTER:** Mainly coastal, also inland, commonly (Oct./Nov.–

breeding adult
brachyrhynchus

1st winter
brachyrhynchus

2nd winter
brachyrhynchus

extensive gray on P8

black tail

brownish underwing

overall brownish compared to juvenile Ring-billed

adult
kamtschatschensis

adult
canus

winter adult
brachyrhynchus

brownish wash

2nd winter
brachyrhynchus

most with dark eye

short, slender, yellow bill

breeding adult
brachyrhynchus

darker above than Ring-billed

filled-in brown coverts with pale fringes

juvenile
brachyrhynchus

more black on P8 than *brachyrhynchus*, like *kamtschatschensis*

more black on P8 than *brachyrhynchus*

long wings

kamtschatschensis

head and breast pattern on *canus* more spotted, less washed with brown as in *brachyrhynchus*

often with darker ring on bill

winter adult

1st winter

larger size

pale base to tail

extensively dark tail

1st winter
brachyrhynchus

winter adult
"Common Gull"
canus

black tail band contrasts sharply with white tail base

covert pattern like *brachyrhynch*

1st winter
canus

Mar./Apr.). Locally inland in BC and Pacific states. Casual (mainly Nov.–Mar.) in interior west, Great Plains, TX, and Great Lakes region. **VAGRANT:** Mew casual on Bering Sea islands, accidental on Atlantic coast. "Common" rare in Atlantic Canada (mainly late Oct.–Apr.), most frequent to NF; casual south to mid-Atlantic coast. "Kamchatka" rare spring and casual fall migrant in western Aleutians; casual on Bering Sea islands. **POPULATION** Mew: No data. "Common": European range increasing in past 50 years (Iceland colonized in 1955).

RING-BILLED GULL *Larus delawarensis*   RBGU ■ 1

This is the common "seagull" across much of N.A., yet it is rare offshore. Adaptable and bold feeder, soars to catch insects, scavenges at dumps and parks, and even plucks berries from trees! Monotypic. L 17–20" (43–51 cm) WS 44.5–49" (113–124 cm)
IDENTIFICATION Medium-size, three-cycle gull. On adult, note yellow legs, pale eyes, and "ring bill." **BREEDING ADULT:** Pale gray upperparts; black wing tip with white mirrors on outer 1–2 primaries. Staring pale-yellow eyes set off by red orbital ring; yellow legs. **WINTER ADULT:** Head and neck with fine dusky streaking and spotting. **JUVENILE & FIRST-WINTER:** White head and underparts with dark spots and chevrons (gradually lost through winter). Fresh juvenile has neat, scaly upperparts; back becomes pale gray by late fall. Underwings mostly white. Tail variable: base whitish; broad blackish distal band; tail band sometimes broken. Dark brown eyes, flesh pink bill with dark tip, flesh pink legs. **SECOND-WINTER:** Like winter adult, but more black on wing tips (wing tip all black at rest); tail may have black markings; eyes pale.
SIMILAR SPECIES Larger California Gull (commonly occurs alongside Ring-billed in the West) is a four-cycle gull with a stouter bill and dark eyes in all ages. Adult California has slightly darker medium gray upperparts (white scapular and tertial crescents contrast, unlike Ring-billed), a red gonys spot as well as a black bill band. Legs often duller and more greenish in winter. California has a larger and blunter-based black wing-tip area and larger white mirrors; in flight from below, a dusky-gray subterminal secondary band that offsets white trailing edge (underwings more evenly white on Ring-billed). First-year California is brownish overall, unlike Ring-billed; but second-winter California is similar to first-winter Ring-billed: In addition to noting size and structure, note California's darker gray back; browner greater coverts; legs often have a greenish hue. Second-winter Herring can also suggest a first-winter Ring-billed, but is much larger with stouter bill, dark brownish greater coverts, and usually messier and browner appearance overall. Also see smaller Mew Gull.
VOICE Mewing calls and laughing series, higher pitched and less crowing than California Gull's.
STATUS & DISTRIBUTION Midlatitude N.A., winters to Middle America. **BREEDING:** Common and usually colonial, on low, sparsely vegetated islands in lakes. Arrives late Mar.–May, departs July–Aug. **DISPERSAL:** Post-breeding dispersal begins by late June, with juveniles recorded by mid-July as far south as Salton Sea, CA. Casual north to central AK, accidental Arctic coast of AK. **MIGRATION:** Nearly throughout N.A. where common to abundant in many regions. Mainly Aug.–Oct. and Mar.–mid-May. Oversummering nonbreeders regular along Pacific, Atlantic, and Gulf coasts, local elsewhere. **WINTER:** Main arrival in most of U.S. Sept.–Oct., with most departing Feb.–Apr. Rarely north to NF, casually to south-coastal AK. **VAGRANT:** Casual Europe, W. Africa, HI; accidental to Amazonian Brazil.
POPULATION Largely disappeared from Great Lakes region and other areas during late 1800s due to human persecution. Recolonization

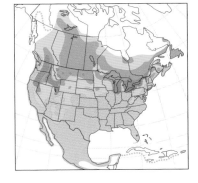

occurred by 1920s, and populations in Great Lakes/St. Lawrence River region exploded during 1960s and 1970s. Today Ring-billed may be the most populous gull in N.A., with an estimated three to four million individuals (most nest in Canada).

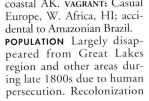

2nd winter

paler underwing than Mew

1st winter

spots and streaks on head

breeding adult

winter adult

pale underside to secondaries

outer wing pattern differs from Mew

small mirror

may have tail markings

black subterminal band

pale eye

2nd winter

breeding adult

pale gray above

yellow legs

dark eye

1st winter

tail pattern variable

pinkish legs

juvenile

small dark centers to coverts

WESTERN GULL *Larus occidentalis*   WEGU   ▪ 1

A bulky, dark-backed Pacific coast gull with a bright banana yellow bill. Polytypic (2 ssp.; both in N.A.). L 25" (64 cm) W 58" (147 cm)

IDENTIFICATION A stocky, four-cycle gull with moderately long legs; relatively short, broad wings; and a bill that expands noticeably at the gonydeal angle to give a "blob-ended" appearance. **ADULT:** White head, neck, and body contrast with a dark gray mantle. Consistently darker-mantled than commonly occurring large gulls on West Coast. Bright yellow bill with a reddish spot on the gonydeal angle. Eye varies from yellow to heavily speckled with dark; yellow orbital ring; and pink legs. In flight shows a broad white trailing edge to wing and one mirror on P10. **JUVENILE & FIRST-WINTER:** Dark and uniform-looking, with dark tail and underparts as well as a solid, unstreaked dark mask. Typically shows a paler area behind dark ear coverts, which extends forward to neck sides. Pale rump contrasts with darker tail and mantle. Dusky pink legs, with dark on front of tarsi. Dark inner primaries; no pale inner wing panel; secondaries neatly tipped white, creating a noticeable but narrow whitish trailing edge to wings. **FIRST-SUMMER:** Similar to first-winter, but more uniform and gray-brown on upperparts; slightly paler on head, neck, and underparts. **SECOND-YEAR:** Whitish head and body with darker streaking on nape and breast sides and a dark wash on belly. Dark gray mantle, contrasting with browner wings, although median coverts usually gray at this age. Tertials have wide white tertial crescent. Black tail contrasts with white rump and uppertail coverts. Usually pink-based bill color, with darker subterminal

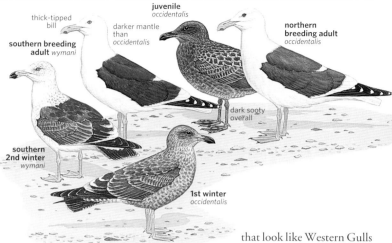

southern breeding adult *wymani*

thick-tipped bill

juvenile *occidentalis*

darker mantle than *occidentalis*

northern breeding adult *occidentalis*

dark sooty overall

southern 2nd winter *wymani*

1st winter *occidentalis*

band and white tip. Dull pink legs. **THIRD-YEAR:** Much as adult, but retains some dark on tertials, greater primary coverts, and tail. Primaries tend to have small or no obvious white tips; head streaking is noticeable. Dark subterminal band on bill. **GEOGRAPHIC VARIATION** The subspecies *occidentalis* breeds north from Monterey County, California, and *wymani* breeds south of Monterey County. The more southern *wymani* differs in having darker upperparts and a tendency to show pale eyes; other ages are not separable.

SIMILAR SPECIES Adult Yellow-footed Gull has yellow legs; a thicker, even more blob-ended bill; and darker upperparts than *occidentalis*. First-year Yellow-footed resembles second-year Western, but its head and underparts are whiter, its bill shows a yellow base to lower mandible, and its legs begin showing yellow tones. Adult Slaty-backed is darker above and more slate-gray than *occidentalis*, lacking the blue-gray of Western Gull. Winter head streaking on Slaty-backed Gull is extensive, particularly around its bright yellow eye, and Slaty-backed shows a complex wing pattern with a "string of pearls" on primaries to P8, dividing gray primary bases from black wing tip. Western Gull also has a darker underwing than Slaty-backed. First-year Slaty-backed has pale inner primaries and pale underwings with narrow dark trailing edge to wing tips, whereas Western Gull is an even brown. Hybrids with Glaucous-winged Gull are common and intermediate in appearance. Individuals

that look like Western Gulls with odd features may be hybrids. Hybrid features include pinkish orbital rings, a paler mantle, and markings on head in winter.

VOICE LONG CALL: Typical of a large *Larus,* but faster paced and higher pitched than Glaucous-winged's.

STATUS & DISTRIBUTION Abundant. **YEAR-ROUND:** Strongly coastal, venturing inland in San Francisco Bay Area, but otherwise usually on or near the coast or offshore. In recent decades, has spread well inland in the Los Angeles metropolitan area. **BREEDING:** Colonial, usually on rocky islands. **DISPERSAL:** Western Gulls of all ages disperse north after breeding season, as far as BC. Movements begin in July, peak Aug.–Sept., with some remaining into winter. Southbound movements peak in Mar. **VAGRANT:** Very small numbers are now found at the Salton Sea. Very rare in AK; accidental to IL and TX. **POPULATION** No data available.

occidentalis

wymani

1st winter

no pale window unlike 1st-winter Herring

southern winter adult *wymani*

southern 3rd winter *wymani*

Large White-headed Gull Basics

The best way to learn these gulls is to spend time looking at the common species and study how they change throughout the year. Learning the variations caused by age, molt, wear, and plain old individual variation of the common species is key in gaining the confidence to tackle finding the less common species.

Consider these three general topics: structure, adult features, and age-related changes. Structure is shorthand for saying shape and size. Relative size can be assessed by comparing the gull to well-known and wide-ranging standard species such as Ring-billed Gull and Herring Gull, the first being a medium-size gull and the latter a large gull. Shape is also best learned by comparison. For example, a Lesser Black-backed Gull is a long-winged species. You could quantify this by looking at comparative lengths on the bird, such as how far primaries extend past the tail, and compare this to a fixed length, like tarsus or bill length. In general, however, the best way to assess shape differences is to compare the gull in question to the common species around it. Gulls are seldom by themselves, so it is not difficult to make comparisons. When a gull is called "bulky" or "big billed" in the accounts, these qualitative descriptions are in reference to other gull species of similar size.

Herring Gull, juvenile (NJ, Jan.)

Learning the adult features of the gulls in your area is important for two reasons. First, adults can often be identified by simple features such as leg color, mantle color, and so forth. In general, adults are not as difficult to separate, and they are less variable than other ages. Learning the structure by looking at the easier-to-identify adults simplifies identification of the more variable immatures. Second, as immatures age, they become more and more

Herring Gull, winter adult (NJ, Feb.)

like adults, so knowing the terminal features allows you to interpret the field marks on an immature more clearly. Apart from noting structure when looking at adult gulls, one should concentrate on mantle shade, eye color, orbital ring color, leg color, streaking pattern (on winter birds), and wing-tip pattern (with particular reference to the extent of black and the number and size of mirrors). On a perched bird, the mirror on the outermost primary (P10) is often visible on the wing tip on the opposite side of the bird, where the underside of the feather is visible.

The third topic to try to master—and it is the most difficult—is how gulls change due to age. To understand these changes, study the beginning stage (the juvenile) and the end point (the adult, as noted above). All other plumages are intermediate between the end points. Juveniles are generally brown, with paler markings and edgings on the mantle and wings. On juveniles and first-year birds, concentrate on bill pattern, tail pattern, presence of secondary bar and paler inner primaries, underwing pattern, extent of banding on greater coverts, and extent of pale markings on tertials. Note that while leg colors vary from pinks to yellow in adults, juveniles of even the yellow-legged species show pink legs. The juvenal is the first plumage obtained after the down, and it all grows in at the same time. This gives the juvenile an even appearance, for all feathers are the same age. In older age classes, newer and fresher feathers always contrast with older and more worn feathers.

Plumage maturation is caused by molts, which can be complete (all feathers replaced) or partial (largely body feathers with no wing and tail feathers). Adult gulls have a complete molt after breeding and a partial molt before breeding. During the complete molt, the first feathers to change are the median coverts, inner greater coverts, and inner primaries. The tail and secondaries change when primaries are half done. Molt ends with outer primaries, inner secondaries, and subscapulars. In larger gulls, the molt out of the juvenile plumage lasts a long time—often into the first spring—and it is a partial molt that involves the head, underparts, and mantle. Most noticeable is the change of the upperparts, with new scapulars replacing the juvenal feathers. Note that early molted feathers are more juvenal-like (brown) and late-molted feathers are more adultlike (gray-brown). By spring, the second molt starts with new median coverts, inner greater coverts, and inner primaries. This is a complete molt. After the first year, the immature bird's molts match up to the adult's. There is a complete molt in summer/fall and a partial molt in late winter.

Here are some key points to remember: First-year gulls are generally brownish and have the juvenal wings and tail. Second-year birds have gone through a complete molt, so they have a new wing and tail, patterned closer to an adult's. In the three-year gulls (which obtain adult plumage in the third year), this second-year plumage is adultlike with some immature features. In four-year gulls, second-year immatures have an adultlike mantle contrasting with a browner wing. Third-year plumage is obtained through a complete molt, and it looks often very much like an adult except for minor features, such as showing some black on the tail or on the greater primary coverts, smaller primary mirrors, more extensive black on primaries, and smaller white primary tips. Plumage changes owing to the partial molts of immatures are minor. Mainly, the head whitens out in spring. Finally, hybridization, feather wear, and bleaching can confound these issues, and not all immature gulls will be identifiable. ■

YELLOW-FOOTED GULL   *Larus livens*   YFGU ◼ 2

A marine gull endemic to the Gulf of California, Yellow-footed Gull has a very thick bill. Monotypic. L 27" (69 cm) WS 60" (152 cm)

IDENTIFICATION A large, stocky, dark-backed gull with a thick and extremely blob-ended bill shape. A three-cycle gull, unlike other gulls of its size. **ADULT:** White head, neck, and body contrasting with a dark gray mantle. Bright yellow bill with a reddish gonys spot. Yellow eye with yellow to orange-yellow orbital ring; yellow legs. In flight it shows a broad white trailing edge to wing and one mirror on P10. **JUVENILE:** A gray-brown gull, brownish above with contrasting white belly and vent. White head and neck with fine but blurry streaking throughout. Dark bill with yellow patch at base of lower mandible; dull pink legs. Dark tail contrasting with whitish rump and uppertail coverts. Wings brownish and uniform, lacking pale inner primary panel; secondaries are neatly tipped white, creating a noticeable but narrow whitish trailing edge. **FIRST-WINTER & FIRST-SUMMER:** Wings and tail as in juvenile, but head and

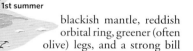

body showing more white, less streaking. Obtains a largely solid, dark-gray mantle starting in first winter, unlike other large gulls; mantle is well developed by summer. First-summer birds may begin showing yellow tones to legs. **SECOND-WINTER:** Similar to adult, with solid slate gray mantle and wings; dark on tail; often a darker secondary bar. Head is variably streaked; bill is pale pinkish yellow with a dark terminal third. Legs are yellowish by this stage. **SECOND-SUMMER:** Similar, but body, head, and neck average whiter. **SIMILAR SPECIES** Most likely to be confused with Western Gull, particularly the darker subspecies *wymani* (found south of Monterey, CA). Adult Yellow-footed Gull differs from Western Gull in its generally thicker bill and yellow legs. But beware, some adults have yellowish legs and have been mistakenly called Yellow-footed Gulls on the southern CA coast. Once first-year Yellow-footeds obtain a gray back, they closely resemble second-year Westerns. These are perhaps best separated by looking for any yellow tones that may be present on legs and for Yellow-footed's generally thicker bill. In addition, Yellow-footed's dark bill tends to show a distinctive yellow patch at base of lower mandible. Second-year Western shows a strongly bicolored bill, with an extensive pinkish yellow base and a dark terminal (half to third). The vagrant Kelp Gull is similar, but records of its appearance are farther east than the range of Yellow-footed Gull. Adult Kelp Gull can be differentiated by its

blackish mantle, reddish orbital ring, greener (often olive) legs, and a strong bill not so obviously blob-ended as on Yellow-footed Gull. First-year Kelp Gull is not so contrastingly white below and has a thinner bill, and any adult-type mantle feathers are blackish, not slate gray.

VOICE LONG CALL: A series of *keow* notes, starting with a slower and longer note and speeding up somewhat toward the end of the series, typical of a large *Larus*. Slower tempo and lower frequency than Western Gull's corresponding call.

STATUS & DISTRIBUTION Uncommon. Breeds in Mexico, restricted to Gulf of California. In the U.S. only as a nonbreeding visitor. **BREEDING:** Breeds on various rocks and small islands in Gulf of California (from 31° N to 24°30′ N). **DISPERSAL:** Largely resident, but disperses north to Salton Sea, CA, mainly June–Aug., fewer in Sept. and only individuals by mid-Oct. Rare to very rare in winter and spring. Disperses south as far as Oaxaca during this period as well. Flocks in Salton Sea may number in the hundreds. **VAGRANT:** Casual to CA coast in Orange, Los Angeles, and San Diego (several recs.) Counties. Otherwise, there are 2–3 records from NV and Mono and San Bernardino Counties, CA.

POPULATION Estimated to be 20,000 pairs, and no population trends exist from Mexico. Yellow-footed Gull was first recorded from Salton Sea in the mid-1960s. The numbers built up to about 50 by 1970, and now the nonbreeding visiting population is in the hundreds, confirming a recent expansion to this nonbreeding site.

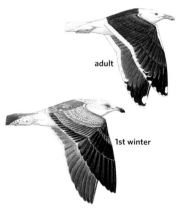

CALIFORNIA GULL *Larus californicus* CAGU ■ 1

A midsize gull of the Western interior, the California Gull is closely tied to saline lakes when breeding. Polytypic (2 ssp.; both in N.A.). L 21" (53 cm) WS 54" (137 cm)

IDENTIFICATION A medium- to large-size, long-legged, long-billed, and long, narrow-winged four-cycle gull. Bill characterized by being nicely parallel sided, lacking an expansion at the gonydeal angle. **SUMMER ADULT:** A dark gray-mantled gull, darker than Ring-billed Gull. White head, neck, underparts, and tail. Bright yellow bill with a red gonys spot and a black subterminal band. During breeding period black bill band is reduced in size; on some birds it may be nearly absent. Eye dark, with bright red orbital ring, carmine gape. Greenish yellow legs. Wing pattern is distinctive: This gull shows extensive black on primaries, particularly so on P8 and P7, giving black wing tip a nearly square-cut shape. Mirrors on P9 and P10 are large. **WINTER ADULT:** Similar to breeding adult, but head and neck streaked brown, concentrated on back of neck, nape, and lower neck sides; often with a dark postocular streak. Throat and front of neck are unstreaked. **JUVENILE:** Usually pale, grayish brown, although some cinnamon brown, and often whitish on center of breast and belly. Bill all dark. Wings lack paler inner primaries, and they show dark-based coverts. Tail is largely dark, and legs are pink. **FIRST-YEAR:** Like juvenile, but bill bicolored with pink base and black tip. Juvenal scapulars replaced by variable patterned feathers, but tend to show a solid gray-brown center and shaft streak and a large buffy gray tip with a narrow blackish terminal fringe. Summer bird has whiter head and worn and faded wings,

contrasting with newer mantle. **SECOND-YEAR:** Dark gray mantle; browner wings with marbled pattern on coverts and tertials, often some gray inner median coverts present. White head, neck, and underparts with streaks concentrated on ear coverts, nape, and particularly breast sides. Bicolored bill, with grayish to gray-green base; legs similarly greenish gray. Wing pattern at this age shows blackish outer primaries clearly contrasting with paler gray inner primaries; dark secondary bar. Tail remains blackish, but now contrasts with white rump and uppertail coverts. In second-summer, head and body much more whitish; may obtain more adultlike soft-part colors. **THIRD-YEAR:** Like adult, but retains dark on greater primary coverts and tail. White mirrors on primaries not as well developed.

GEOGRAPHIC VARIATION Nominate subspecies breeds east to CO, UT, and ID. Subspecies *albertaensis* farther east and north in NT, AB, SK, NB, ND, and SD; intermediates in MT. It is larger, larger billed, and paler on mantle than *californicus,* and it has a tendency to show less black on primaries and larger mirrors, with the tip of P10 often with entirely white.

SIMILAR SPECIES Distinguished from adult Herring Gull by darker mantle, dark eyes, greenish legs, and black and red on bill tip. Second-year birds similar, but note California Gull's darker mantle, dark eyes, grayish bill base and legs, as well as structural differences. (See sidebar p. 305.)

VOICE LONG CALL: A series of *kyow* notes; first two are longer and more drawn out. Call is higher pitched than

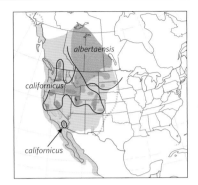

corresponding call of Herring Gull.

STATUS & DISTRIBUTION Abundant. **BREEDING:** Colonies on flat islands, some on saline lakes. **MIGRATION:** Moves to coast after breeding. In summer and early fall, shows a generally northward movement. Southward movements begin in fall and winter, reaching southernmost winter range in midwinter, before moving north again. Interior birds, *albertaensis,* appear to move farther south than *californicus* and return slightly later in spring. **WINTER:** Shifts to Pacific coast with large numbers also offshore; also very common in interior near coast during winter. **VAGRANT:** Casual throughout interior to East Coast, appears to be increasingly regular as a vagrant to the East.

POPULATION Estimated between 500,000 and one million individuals. Population in U.S. estimated to have doubled since 1930. Increases ongoing: For example, in San Francisco Bay, breeders increased from 400 to over 21,000 in last 20 years.

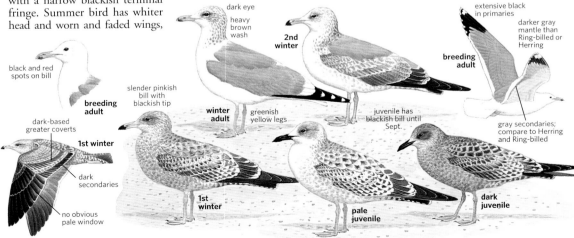

black and red spots on bill

breeding adult

dark-based greater coverts

1st winter

dark secondaries

no obvious pale window

dark eye

heavy brown wash

2nd winter

slender pinkish bill with blackish tip

breeding adult

winter adult

greenish yellow legs

1st winter

pale juvenile

juvenile has blackish bill until Sept.

extensive black in primaries

darker gray mantle than Ring-billed or Herring

breeding adult

gray secondaries; compare to Herring and Ring-billed

dark juvenile

HERRING GULL *Larus argentatus*   HEGU ▪ 1

The most widespread pink-legged gull in N.A, the Herring is common in the East and mainly a winter visitor in the West. Hybrids can be locally common (mainly in the West). Polytypic (4 ssp.; 3, possibly 4, in N.A.). L 22–27" (56–69 cm) WS 54–60" (137–152 cm)

IDENTIFICATION Large gull with four plumage cycles; sloping head and fairly stout bill with distinct (but not bulbous) gonydeal expansion. Pink legs at all ages. Subspecies *smithsonianus* described unless otherwise noted. **BREEDING ADULT:** Pale gray upperparts; black wing tip with white mirrors on outer one or two primaries. (In West, most have mirror only on outermost primary.) Pale yellow eyes (can be dusky, esp. in West); yellow-orange orbital ring; yellow bill, reddish gonys spot. **WINTER ADULT:** Dusky streaking and smudging on head and neck. Duller bill and legs; bill often develops black subterminal mark; orbital ring can be dark to pinkish. **JUVENILE & FIRST-WINTER:** Fresh juvenile sooty brown overall; neat, scaly upperparts; strong dark barring on tail coverts. Inner primaries form pale panel on upper wing. Tail is mostly blackish (but some with extensive whitish hue at base). **VARIABLE FIRST-WINTER MOLT:** Some (esp. East Coast breeders) soon attain new barred and mottled back feathers; others (esp. West Coast winter populations) retain most or all juvenal plumage through winter. Head often bleaches to whitish. Blackish bill soon develops dull flesh color on base; rarely flesh pink with clean-cut black tip by midwinter. **SECOND-WINTER:** Resembles first-winter, but back usually with pale gray feathers (from none to a solid, pale gray saddle); tail mostly black. Pink bill with black distal third (more adultlike in second summer); eyes sometimes become pale. In second-summer, head and body whiter; bill rarely like adult, but usually with black distal mark. **THIRD-WINTER:** Highly variable. Some resemble second-winter; others resemble winter adult, but have more black on wing tips, some black on tail.

Best aged by adultlike pattern of inner primaries. (Second-winter's inner primaries resemble first-winter's.) Bill usually pink with black distal band; eyes pale. In third-summer, head and body mostly white like breeding adult; bill like adult's or with some black marks. **ADULT *VEGAE*:** Upperparts medium gray; eyes dark (mostly) to pale (rarely); reddish orbital ring; rich pink legs. **JUVENILE & FIRST-WINTER *VEGAE*:** Head and body paler overall than *smithsonianus,* with sparser dark barring on tail coverts; tail whitish based with broad blackish distal band. Older ages best distinguished by medium gray tone of upperparts. **ADULT *ARGENTATUS* & *ARGENTEUS*:** Not safely told from *smithsonianus.* **FIRST-WINTER *ARGENTATUS* & *ARGENTEUS*:** Paler

striking pale eye

extensive streaking

pale gray upperparts

winter adult
smithsonianus   pink legs

often pale headed

1st winter
smithsonianus

juvenile
smithsonianus

often more adultlike

pale eye by late 2nd winter

2nd winter
smithsonianus

1st winter
smithsonianus

3rd winter
smithsonianus

apical spots small or absent

some dark in tail

1st winter
smithsonianus

obvious pale window; compare to California Gull

dark tail

pale underside to secondaries

breeding adult
smithsonianus

white tail base

1st winter
argenteus

1st winter
vegae

more checkered upperparts

paler than *smithsonianus*

darker mantle color than *smithsonianus*

winter adult
vegae

breeding adult
vegae

white tail base

1st winter
vegae

and more checkered above and whiter on rump and tail base than first-winter *smithsonianus*. Resembles *vegae*, from which perhaps not safely told except by (presumed) distribution.

GEOGRAPHIC VARIATION Recent work indicates that full-species status may be warranted for several taxa in the traditional Herring Gull complex, including "American" Herring Gull (*smithsonianus*), "European" Herring Gull (*argentatus* of Scandinavia and *argenteus* of Iceland, U.K., and northwest Europe), and "Vega Gull" (*vegae* of east Asia).

SIMILAR SPECIES Adult fairly distinctive, but see Thayer's Gull, which is smaller with more slender bill, shorter legs; eyes often dark; wing tips slaty blackish at rest and mostly pale from below, showing much more white on outer primaries than do Herrings in the West. Beware "diluted" adult Herrings, which may be hybrids with Glaucouswinged or Glaucous Gulls. Juvenile and first-winter Herrings are extremely variable, but not like any other regular large gull in the East. Main problem in the West is separation of small female Herring from dark Thayer's. Besides different structure, Thayer's outer primaries are more extensively pale on inner webs, creating venetian-blind pattern on spread outer primaries; unlike Herring's more solidly dark wing tip. (Some Herring x Glaucouswinged hybrids are very like Thayer's in plumage, but bigger billed.) See sidebar for separation from California Gull. On older immatures, note overall shape, bill structure, and upperwing pattern. Some Herring x Glaucous-winged hybrids look like adult "Vega Gull" but are often bulkier, with slightly paler upperparts and wing tips. First-year "Vega Gull" is told from bulkier Slatybacked by more solidly blackish outer primaries; narrower, blacker tail band. Bleached first-summer birds are not always identifiable.

VOICE Varied. Long call has slightly honking or laughing quality. Call of "Vega Gull" is lower pitched, harsher.

STATUS & DISTRIBUTION The following refers to *smithsonianus* unless otherwise noted. **BREEDING:** Breeds N.A. Common (May–Aug.); colonial or in scattered pairs on coastal islands, islands in lakes, on buildings. Since late 1980s has bred in LA, TX. Subspecies *vegae* is fairly common on St. Lawrence I. (May–Sept.). **MIGRATION:** Nearly throughout N.A., where common to abundant in many regions. Mainly Oct.–Nov. and Feb.–Apr. Subspecies *vegae* rare to casual in Aleutians. **DISPERSAL:** Subspecies *vegae* is rare to uncommon to Arctic coast of AK and Bering Sea islands (mainly June–Sept.) and uncommon (Aug.–Sept.) to Seward Peninsula. **WINTER:** Coastal, inland, and offshore. In West, uncommon along much of immediate coast in range of more numerous Western and Glaucous-winged Gulls. Wintering birds arrive continent-wide by Oct., with marked increase during mid- to late Oct. Departs most wintering areas by May, but some nonbreeders oversummer along Pacific, Atlantic, and Gulf coasts; locally elsewhere. **VAGRANT:** Casual to Europe and HI. Subspecies *vegae* possibly accidental in TX and CA. Subspecies *argentatus* and/or *argenteus* is casual (mainly Nov.–Apr.) in the East, most records from NF. **POPULATION** In N.A., recovered after egging and feather hunting in late 1800s. (U.S. pop. only 8,000 pairs in 1900.) East Coast population greater than 100,000 pairs in mid-1980s and spreading south. Numbers in Northeast and Atlantic Canada now declining; linked to egging and competition with expanding Great Black-backed Gull populations.

Separating First-winter Californias From First-winter Herrings

First-winter California and Herring Gulls look quite similar—specifically Herrings that show a crisply bicolored bill. They share a brown body plumage, dark tail, and pink legs. Though they overlap in size, these gulls do consistently differ in structure. The long-winged California is slimmer, with a long, even-sided tubular bill. The Herring is bulkier, showing a deeper belly, thicker neck, and proportionately shorter, broader wings; the bill tends to expand a bit at the gonydeal angle. Most first-winter Herrings have a dark bill or a variable, but ill-defined pinkish base to the bill; very few show a crisply set-off black tip on a pink bill, like California's. Any black "bleeding" back toward the gape along the cutting edge suggests a Herring. In flight, wing patterns differ. Californias do not show strikingly paler inner primaries, thus they look more evenly patterned. They also show dark greater coverts with

California Gull, juvenile (CA, Sept.)

Herring Gull, 1st winter (CA, Nov.)

paler tips, creating a second dark bar on the wing—the other dark bar being the secondaries. Herrings have strikingly paler inner primaries, and due to more banding on the greater coverts, they lack the second dark bar. Many first-winter Herrings show a whitish head that contrasts with the dark body. Californias that have a whitish head also show a similarly pale neck and breast, with contrasting dark streaks on the breast sides and nape. At rest, California's dark greater coverts, with their whitish markings concentrated at the tip, create a whitish "skirt" pattern. The white markings also concentrate on the inner greater coverts, forming a "covert crescent" inside of the tertial crescent. These patterns are not shown by most Herrings. Finally, mantle molt begins early on California Gulls, and by the first winter they show a complex, messy mosaic of feather patterns. Herrings often look less messy above. ∎

YELLOW-LEGGED GULL *Larus michahellis* YLGU ■ 4

This vagrant from Europe has recently been recognized as a distinct species within the Herring Gull complex, and its status in N.A. is masked by identification difficulties. Polytypic (2 ssp; both may have occurred in N.A.). L 21–26" (53–66 cm) WS 52–58" (132–147 cm)

IDENTIFICATION Large gull with four plumage cycles; intermediate between Herring and Lesser Black-backed. Fairly stout bill, slight to distinct (but not bulbous) gonydeal expansion. **BREEDING ADULT:** Medium gray upperparts; black wing tip with white mirrors on outer 1–2 primaries. Yellow legs; pale lemon eyes, reddish orbital ring. Orange-red gonydeal spot often extends to upper mandible. **WINTER ADULT:** Dusky head and neck streaking concentrated in half-hood and on lower hind neck. Streaking most distinct in late fall; by midwinter most birds are white headed. Bill quite bright yellow, rarely with black subterminal marks. **FIRST-WINTER & SUMMER:** Brownish overall; whitish ground color to head and underparts. Head, neck, and chest often bleach to mostly whitish with variable dusky brown streaking and spotting by spring. White tail coverts sparsely barred dark brown. Bright white base to tail, with sparse blackish markings and broad blackish distal band—striking in flight. Slightly paler inner webs to inner primaries form an indistinct paler panel on spread upper wing. Blackish bill can show dull flesh base by spring; legs flesh pink. **SECOND-YEAR:** Second-winter resembles first-winter, but back usually has some medium gray feathers; tail ranges from extensively black to variably mixed with white; inner primaries average slightly paler, but do not form an obvious pale panel. Some second-winters *(atlantis?)* have a distinctly dark-hooded appearance. In second-summer, head and body whiter; usually a solid medium gray saddle. Brownish to pale lemon eyes; reddish orbital ring in summer.

Through winter, bill typically pinkish basally, blackish distally, with a pale tip and sometimes a blush of red at gonys. In summer, bill typically yellow with reddish gonydeal smudge and black distal band. Legs flesh in winter, yellowish by second summer. **THIRD-YEAR:** Resembles winter adult, but more black (less white) on wing tips; some have black on tail.

GEOGRAPHIC VARIATION Taxonomy complex. Two widely recognized taxa in western Europe: smaller and shorter-legged *atlantis* of Azores and larger, longer-legged *michahellis* of Mediterranean region; both may reach N.A., but most, if not all, appear to be *atlantis*. Other populations in western Europe may deserve recognition as subspecies, but more study is needed. Adult *atlantis* is darker backed, and immatures are more brownish overall, suggesting Lesser Black-backed. Adult *michahellis* is paler backed, and immatures are more whitish on head and body, suggesting Great Black-backed.

SIMILAR SPECIES Note adult's medium gray upperparts (intermediate between Herring and Lesser Black-backed) and yellowish legs. Be aware that Herrings can have yellow legs (mainly in spring) and that Herring x Lesser Black-backed hybrids may closely resemble Yellow-legged. (Hybrid adults typically have extensive dusky head streaking in winter, unlike Yellow-legged.) Lesser Black-backed averages a more slender bill, has narrower and relatively longer wings, and is slightly to distinctly darker above; it has less contrasting and slightly less extensive black wing tips (longer gray basal tongues on P8 and P9) and heavier dusky head and neck streaking in winter. Herring has less extensive black wing tips that often have more white (at least in Northeast, where Yellow-leggeds are most frequent), flesh pink legs, and yellow-orange orbital ring; in winter, Herring has heavier

dusky head and neck streaking and often a duller and more pinkish bill, with more distinct dark distal marks, but no red on upper mandible or gape (shown by many Yellow-leggeds). California is on average smaller and lighter in build with more slender bill, dark eyes; winter adult typically with heavier dark hind neck markings, often more greenish legs. First-winter Yellow-legged is most similar to Lesser Black-backed and perhaps not always separable. Subspecies *michahellis* is larger and bulkier than Lesser Black-backed, with blockier head, stouter bill, and longer legs; tail base averages more extensively white (and less barred) at sides; and head and underparts often whiter overall. Also see Great Black-backed Gull. Subspecies *atlantis* is structurally more similar to Lesser Black-backed, but is on average bulkier and broader winged; outer rectrices usually unbarred at base (usually barred on Lesser Black-backed). Herring Gull is generally darker and browner overall with heavy dark barring on tail coverts, mostly black tail, pale inner primary panel, and often paler-based bill. On older immatures, note overall shape, bill structure, medium gray tone of upperparts, and upperwing pattern.

VOICE Harsher and more grating than Herring Gull; more similar to voice of Lesser Black-backed.

STATUS & DISTRIBUTION Western Europe to N. Africa. **VAGRANT:** Casual (mainly Oct.–Apr.; also June and Aug. records from QC) to eastern N.A., from Atlantic Canada (mainly NF) south to mid-Atlantic coast. Accidental to TX. Specimen from QC referred to *atlantis;* provenance of other birds uncertain.

POPULATION In western Europe, *michahellis* has increased since 1970s; no trends reported for *atlantis* (more than 8,000 pairs in 1990s).

1st winter *michahellis*

stouter bill than Herring

yellow legs

winter adult *michahellis*

darker gray above than Herring

no pale window

1st winter *michahellis*

white tail base

most if not all North American records appear to be darker-mantled *atlantis*

breeding adult *michahellis*

large red spot on lower mandible

winter adult *atlantis*

LESSER BLACK-BACKED GULL *Larus fuscus*   LBBG ▪ 2

This European species has "colonized" N.A. in the last 60 years, but breeding birds have yet to be found here. It can occur anywhere large gulls congregate and is often seen with Herrings. Polytypic (3 ssp.; 2 in N.A.). L 21–25" (54–64 cm) WS 52–58" (132–147 cm) **IDENTIFICATION** Subspecies *graellsii* described and illustrated. Large four-cycle gull slimmer in build than Herring, with relatively longer and narrower wings and more slender bill. **BREEDING ADULT:** Slaty gray upperparts; black wing tip with white mirrors on outer 1–2 primaries. Yellow to orange-yellow legs; pale lemon eyes; reddish orbital ring. Bright yellow bill with orange-red gonydeal spot. **WINTER ADULT:** Dusky head and neck, streaking often concentrated around eyes. **FIRST-WINTER & SUMMER:** Brownish overall; whitish ground color to head and underparts. Many with whiter head and underparts that contrast with dark upperparts. White tail coverts sparsely barred dark brown. Tail has bright white base with blackish barring and variable, clean-cut blackish distal band. Inner primaries on upper wing not appreciably paler than outers; dark-based greater coverts often form a dark band. Dark eyes; blackish bill can show dull flesh-hued base by spring; flesh pink legs. **SECOND-WINTER & SUMMER:** Second-winter resembles first-winter, but some feathers on back are medium gray; tail ranges from extensively black to variably mixed with white. In second-summer, head and body whiter; usually a solid slaty gray saddle. Brownish to pale lemon eyes. Through winter, bill typically blackish with pale creamy tip; in summer, often brightens to yellow with a reddish gonydeal smudge and black distal band. Legs flesh hued in winter, yellowish

by second summer. **THIRD-YEAR:** Resembles winter adult, but more black (and less white) on wing tips; some have black on tail; bill blacker in winter.
GEOGRAPHIC VARIATION Records in N.A. refer mainly to paler-backed *graellsii* of western Europe, with fewer records of darker-backed *intermedius,* which breeds in northwest Europe. Many birds intermediate in appearance, however, and not safely assigned to subspecies.
SIMILAR SPECIES Adult and older immature Lesser Black-backeds in the East are distinctive among Herring and larger Great Black-backed Gulls. First-winters, however, can be overlooked easily among Herrings. Note Lesser Black-backed's smaller size, relatively longer and narrower wings, more slender black bill, and whiter ground color to head and underparts (esp. tail coverts, which have sparser dark barring); in flight, upperwing pattern and tail/uppertail-covert pattern distinctive. Primary molt in first summer is later than Herring's. (Some Lesser Black-backeds in June are just starting primary molt, when Herrings are well advanced.) Main problems arise with separation from rarer species and hybrids. Kelp is larger and bulkier with broader wings and a stouter bill (deeper at gonys than base, the reverse of a typical Lesser Black-backed). Adult Kelps are blackish above; legs often more greenish yellow. First-winter Kelp perhaps not distinguishable from Lesser Black-backed by plumage, but tail often has broader blackish distal band; note also bill shape, broader wings. Kelp × Herring Gull hybrids can resemble Lesser

Black-backed closely, but are bulkier in build with a stouter bill (deepest at gonys); black underwing tip of hybrid adult contrasts more. Yellow-footeds larger and bulkier with much deeper, bulbous-tipped bill and broad wings; adults lack distinct dusky head markings in winter. Adult California is paler above, but structure similar to a small Lesser Black-backed; note California's dark eyes, often more greenish legs, wing-tip pattern.
VOICE Deeper, hoarser than Herring's.
STATUS & DISTRIBUTION Northwestern Europe, wintering to Africa. Has increased dramatically in N.A., especially since 1980 and continues to do so; first N.A. record in 1934 in NJ. **WINTER:** Rare to locally fairly common in east (mainly Sept.–Apr.); max. counts are from mid-Atlantic region south to FL. **VAGRANT:** Casual to rare (mainly Sept.–Apr.) in interior and to West Coast, north to AK.
POPULATION Large increase in *graellsii* and *intermedius* populations since the mid-1900s.

darker than *graellsii*

even longer winged

winter adult *intermedius*

long winged

dark double bar

1st winter *graellsii*

no pale window

extensive dark underwing

heavily spotted head and neck, dark around striking pale eye

dark slaty gray above

slender bill with large reddish spot on gonys

breeding adult *graellsii*

winter adult *graellsii*

yellow legs

2nd winter *graellsii*

winter adult *graellsii*

rather small white mirror on P10

darker greater coverts

1st winter *graellsii*

THAYER'S GULL *Larus thayeri* THGU ■ 2

The identification and taxonomy of Thayer's—a high Arctic gull that winters mainly on the Pacific coast—has been, and remains, controversial. Sometimes Thayer's is lumped with Iceland Gull. Monotypic. L 23" (58 cm) WS 55" (140 cm)

IDENTIFICATION A four-cycle, medium-size, short-legged, small-billed, potbellied, and somewhat long-winged gull. Relatively thin bill, even in width; not expanding at the gonydeal angle. In all ages shows a largely pale underwing, with dark trailing edge of outer primaries. **SUMMER ADULT:** Medium gray mantle; white head, neck, underparts, and tail. Dull yellow bill, sometimes with greener base; red gonys spot. Eye color is variable: most are dark eyed, yet a few appear pale eyed, particularly in sunny conditions; dark red to purplish orbital ring. Bright pink legs. Distinctive variable wing pattern: Outer 5–6 primaries are black (or blackish) with extensive white tongues on inner and sometimes outer vanes, which breaks up black into thin strips—a venetian-blind effect. The mirror on P9 is confluent with white tongue. **WINTER ADULT:** Like summer adult, but head and neck streaked. **JUVENILE:** Retains full juvenal plumage through most of winter. Variable, ranging from dark, Herring-like birds to lighter extremes; exacerbated by bleaching and wear by late winter and early spring. Classic juvenile evenly warm brown, with mantle and wings checkered with pale buff or whitish markings. Primaries dark brown but not blackish; show crisp pale fringes. Dark-centered tertials with variable pale markings on tips. Typically, body is palest, tertials darker, and wings and tail darkest; this stepped gradation is distinctive. Dark bill. In flight, there is a noticeable dark secondary bar as well as paler inner primaries. Bright pink legs. **FIRST-SUMMER:** Similar to juvenal, but has a tendency to bleach out and become very pale, contrasting with newer mantle. **SECOND-YEAR:** Medium gray mantle; browner wings; greater coverts and tertials marbled or vermiculated. Extent of blackish primaries maximal at this age, contrasting with

paler inner primaries. Tail mostly dark but contrasts with whitish rump and uppertail coverts. Bill usually bicolored with pinkish base and dark tip; some retain darker bill. **THIRD-YEAR:** Very much as adult, but some dark on greater primary coverts and tail. **HYBRIDS:** Thayer's and Iceland Gulls are in a complex relationship and hybridize to an unknown degree; "Kumlien's" Iceland Gull could be thought of as intermediate population between Thayer's and nominate Iceland Gull.

SIMILAR SPECIES Thayer's Gull is most often confused with Herring Gulls of corresponding ages. Structural differences are useful: Thayer's shows a smaller bill, steeper forehead, shorter legs, and more potbellied appearance. Adult Thayer's identified by reduced black on upper wings, extensive white tongues, extensively pale underwings, darker eye (when present), reddish orbital ring, greener bill, pinker legs, and often slightly darker mantle. First-year Thayer's are paler than Herrings, and they have paler underwings with dark trailing edge, more uniform upperwing pattern, and noticeably paler tertials than primaries while perched. First-year Thayer's intergrade in these characters with "Kumlien's," and some individuals can't be identified. However, typical adult Thayer's shows black, not gray, on outer primaries; black reaches to P5. First-year Thayer's show dark-centered tertials and solidly dark tail and secondary bar. **VOICE** Long call higher pitched and quicker than that of Herring Gull. **STATUS & DISTRIBUTION** Common

Pacific coast, some adjacent valleys in winter; rare in continental interior. **BREEDING:** Colonial, on steep coastal cliffs in western Arctic. **MIGRATION:** Fall migrants reach central CA by late Oct., larger numbers present Dec. and remain until early Apr. **WINTER:** Variety of habitats, from coastal beach and estuaries, to inland garbage dumps and lakes. **VAGRANT:** Casual to easternmost N.A.

POPULATION Estimated at 6,300 pairs; one of North America's least common gulls.

1st winter

dark secondary bar

pale underwing

brownish on outer webs from above

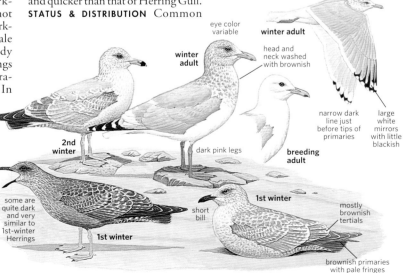

eye color variable

winter adult

winter adult

head and neck washed with brownish

narrow dark line just before tips of primaries

large white mirrors with little blackish

2nd winter

dark pink legs

breeding adult

some are quite dark and very similar to 1st-winter Herrings

1st winter

short bill

1st winter

mostly brownish tertials

brownish primaries with pale fringes

ICELAND GULL *Larus glaucoides*   ICGU ▪ 2

Best known as a bird of Atlantic Canada and the Northeast in winter, the Iceland Gull is an extremely variable-looking gull whose taxonomy remains unclear: Most North American records are of the subspecies *kumlieni,* commonly known as the "Kumlien's Gull." Its behavior is much like other large gulls', but it also forages around sea ice in the northern parts of its winter range. Polytypic (2 ssp.; both in N.A.). L 20–24" (51–61 cm) WS 51–56" (130–142 cm)

IDENTIFICATION A medium-large four-cycle gull with a relatively small and slender bill; all ages have flesh pink legs and lack black in wing tips. **BREEDING ADULT:** Clean white head and neck. "Kumlien's" wing-tip pattern highly variable: Most birds have dark gray markings on outer four or five primaries and a large white tip to P10. Birds with reduced gray markings are not rare, though, and a few birds even appear to have all-white wing tips. (Darker gray restricted to outer webs of P9–P10 often not discernable in the field.) Birds with darkest and most extensive wing-tip markings approach Thayer's in pattern, and some may be hybrids with that species. The nominate has white wing tips that lack gray markings, and its upperparts are slightly paler than "Kumlien's" (more like Glaucous). Eyes usually pale but can be dark; orbital ring reddish pink to purplish. Yellow bill with orange-red gonys spot. **WINTER ADULT:** Head and neck variably mottled and streaked dusky; duller bill often greenish based, rarely with dark subterminal marks. **FIRST-WINTER:** Pale overall; wing tips vary from medium brown to whitish; upper wing lacks contrasting dark secondary

bar. Bill black or with variable dull flesh hue basally. Upperparts, including tertials, finely patterned. (Much juvenal plumage retained into winter.) **SECOND-WINTER:** Resembles first-winter, but back often has some pale gray; bill usually flesh with broad black distal band; eyes can be pale. **THIRD-WINTER:** Resembles winter adult, but upper wings and tail usually washed brownish; bill duller, flesh to yellow with a black subterminal band.

GEOGRAPHIC VARIATION Breeding population in N.A. is *kumlieni* ("Kumlien's Gull"). Adult *kumlieni* typically have variable gray wing-tip markings, unlike pure white wing tips of nominate *glaucoides* (Iceland Gull, which, ironically, breeds in Greenland). Thayer's and Iceland Gulls hybridize to an unknown degree. The remote Arctic breeding grounds of both taxa are difficult to study. Moreover, there have been substantial allegations of fraud directed at the previous study that the current taxonomy is based on.

SIMILAR SPECIES Separating dark-winged "Kumlien's" from Thayer's is compounded by presumed hybridization between them (and some birds are best left unidentified). Thayer's is on average slightly larger and longer billed, and most are readily distinguished from "Kumlien's." Adult Thayer's have slightly darker upperparts, blackish gray (not slaty gray) wing-tip markings; first-winters have dark brown (not medium brown) wing tips and secondary bar, coarse-patterned tertials. Some bleached first-years in spring are mostly white, but note dark secondary bar (usually protected from bleaching). Glaucous is usually much larger, with bigger and deeper bill and relatively short wing projection

beyond tail; wing tips always white or with faint dusky subterminal marks in first winter. First-winter Glaucous has brightly bicolored pink-and-black bill. Adult Glaucous has yellow to orange orbital ring (which can be flesh pink in winter).

VOICE Shriller than Herring Gull's; rarely vocal in winter.

STATUS & DISTRIBUTION Subspecies *kumlieni* breeds in eastern Arctic Canada, winters northeast N.A. Nominate breeds Greenland, winters to northwest Europe. Following applies to *kumlieni* unless stated. **BREEDING:** Fairly common, but local (June–Aug.) on sea cliffs. Presumed to hybridize locally with Thayer's. **WINTER:** Fairly common Atlantic Canada (arriving late Oct.–Nov., departing Apr.–May; very rare in summer), smaller numbers south to NJ, inland to eastern Great Lakes. Rare (mainly Dec.–Mar.) south to NC; casual to FL, Gulf Coast, Great Plains. **VAGRANT:** Casual (Sept.–Oct.) along Arctic coast of AK and south (mainly Dec.–Mar.) into the Northwest and south to CA (annual in recent years). Nominate probably rare but regular (Nov.–Mar.) in Atlantic Canada, accidental (Oct.–Jan.) in the West (AK, YT, CA).

POPULATION Estimates of 5,000 pairs of *kumlieni* and 40,000 pairs of nominate *glaucoides,* but no data on trends.

1st winter
glaucoides

winter adult
kumlieni

breeding adult
glaucoides

slightly paler above than *kumlieni*

a more heavily marked bird

1st winter
kumlieni

dark pink legs

white wing tips

kumlieni has variable gray in primary tips; some all-whitish like *glaucoides*

averages paler than Thayer's, with paler primaries and mottled, less solid tertials; variable tail pattern is sometimes mottled, sometimes with a band, like Thayer's

1st winter
glaucoides

a paler bird

2nd winter
kumlieni

winter adult
kumlieni

SLATY-BACKED GULL *Larus schistisagus* SBGU ▪ 3

This marine gull of the Asian North Pacific regularly visits Alaska. Monotypic. L 25" (64 cm) WS 58" (147 cm) **IDENTIFICATION** A large, long-necked, short-legged, potbellied, four-cycle gull. Strong bill is even in thickness, not showing a bulge toward the tip. Paradoxically, immature plumages show some resemblance to pale-winged gulls such as Thayer's and Glaucous-winged Gulls, although adults have dark backs and wings. **SUMMER ADULT:** Slate gray, appearing blackish in some lights. Dark mantle is a colder gray than Western's blue-gray mantle. White head, neck, underparts, and tail. Yellow bill, red gonys spot. Gleaming yellow eye with a red orbital ring. Legs are pink, often bright bubble-gum pink. Wing pattern distinctive: Slate gray of primary bases separated from black primary tips by a series of white tongues or spots, which begin at P4 or P5 and extend to P8; this line of white spots is referred to as a "string of pearls." Outer primaries are black to the base; mirrors on both or missing on P9. Underwing is similarly distinctive, appearing tricolored. White wing linings contrast with dark gray (or in some lights pale gray) underside of secondaries and primaries and a restricted black area at tip of primaries separated from gray underwing by the "string of pearls." White trailing edge on secondaries is broad, with inner secondaries often looking entirely white. **WINTER ADULT:** Head and neck are densely but crisply streaked, concentrating around eye and also lower nape. **JUVENILE & FIRST-WINTER:** Variable and difficult to characterize at this age. Tend to

look uniform, lacking contrast; in this way they resemble young Glaucous-winged Gulls, although primaries, tail, and secondaries are darker than rest of plumage. Pale underwings show a dark trailing edge to primaries. Upper wings show a large pale inner primary patch, which extends out at least to P8, unlike on other similar large gulls. Other features include uniform or poorly marked greater coverts, dark centered tertials without much pale patterning, a dark tail, black bill, and bright pink legs. **FIRST-SUMMER:** Similar to first-winter, but has a tendency to bleach out and become very pale, with darker tertials, tail, and primaries. **SECOND-WINTER:** At this age, dark slate-gray of mantle is obvious, contrasting with pale brownish wings. Coverts are uniform and pale brown; folded wing shows very little contrast or pattern whatsoever. Tertials are dark centered, with a well-developed white tertial crescent; folded primaries are blackish, contrasting with paler coverts. In flight, underwings are still pale, with a contrasting dark trailing edge to primaries. Dark tail contrasts with white uppertail coverts and rump. Head streaking is as on adults, concentrated into a dark patch around eye and highlighting the now obviously pale eye. Bill shows a pale base, but it is still largely dark. **SECOND-SUMMER:** Similar, but body, head, and neck average whiter; wing coverts are often incredibly bleached, whitish, and contrast strongly with dark mantle. **THIRD-YEAR:** Looks much as adult, but retains some dark on tertials, greater primary coverts, and tail. **SIMILAR SPECIES** Adults likely to be confused with Western Gull, but Slaty-backed has a darker, cold gray mantle and a strongly streaked head,

especially around gleaming yellow eye. Wing pattern of Slaty-backed—particularly the tricolored nature of the underwings—is not found on Western Gull, which is more evenly dark below. Slaty-backed is also shorter legged, potbellied, and longer necked and lacks a blob-ended bill. Identification of first-year Slaty-backed is still uncertain, but useful features are structure, uniform greater coverts, and extensive pale on primaries and underwings. Immature Herring x Glaucous-winged hybrids may show some or all of these features. After the second year, dark mantle, pale eye, and streaking pattern make identification more straightforward. **VOICE LONG CALL:** Slower and deeper than Western Gull, resembling Glaucous-winged. **STATUS & DISTRIBUTION** Breeds in northeast Asia; rare in western AK; casual south of AK, although records throughout continent. **BREEDING:** One record from Cape Romanzof, AK. **VAGRANT:** Annual to Pacific Northwest and south to central CA; other records from YT, AB, southern CA, NV, CO, ID, WI, MO, ON, NY, TX, and FL.

heavily spotted head and neck, dark around striking pale eye

bill of even thickness

blackish slate above

winter adult

broad white tertial crescent

broad white edge to secondaries

breeding adult

2nd summer

deep pink legs

white spots form "string of pearls"

dark mantle contrasts with white coverts

1st summer

1st summer

rather plain greater coverts

primaries like Thayer's Gull

GLAUCOUS-WINGED GULL *Larus glaucescens* GWGU ▦ 1

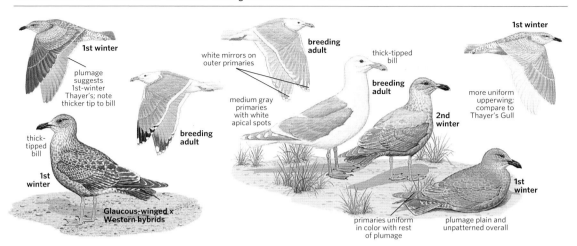

1st winter

plumage suggests 1st-winter Thayer's; note thicker tip to bill

white mirrors on outer primaries

breeding adult

thick-tipped bill

breeding adult

medium gray primaries with white apical spots

1st winter

more uniform upperwing; compare to Thayer's Gull

thick-tipped bill

breeding adult

2nd winter

1st winter

Glaucous-winged x Western hybrids

primaries uniform in color with rest of plumage

plumage plain and unpatterned overall

1st winter

This thick-billed Pacific gull is the only large gull that shows primaries similar in darkness to the body. Monotypic. L 26" (66 cm) WS 58" (147 cm)

IDENTIFICATION A large, stocky, pale-winged four-cycle gull with broad wings, longish legs, and a thick, blob-ended bill. **SUMMER ADULT:** Gray primaries nearly unicolored with gray upperparts. White head, neck, body, and tail. Dark eye, pinkish orbital ring, pink legs, and a yellow bill with red spot at the gonys. In flight, wings look gray above; primaries slightly darker, showing white tongues, a "string of pearls," and a mirror on P10. **WINTER ADULT:** Heavily marked head, most typically finely barred or vermiculated with dark. Often bill becomes a duller yellow. **JUVENILE & FIRST-WINTER:** Pale and uniform, with grayish brown primaries similar in darkness to upperparts. Markings on coverts and tertials reduced, often looking vermiculated. Black bill, dusky pink legs with dark anterior tarsus. In flight looks uniform, lacks darker secondary bar or contrasting pale rump; dark tail. Underwings pale, primaries translucent. **FIRST-SUMMER:** Much paler, bleached, worn than first-winter. Primaries may appear whitish at this age. New scapular and upperparts feathers gray. **SECOND-WINTER:** Gray mantle contrasts with browner wings. Coverts and tertials often very uniform dull brownish gray, lacking obvious pale patterning. Gray-brown tail now contrasts with whiter rump and uppertail coverts. Bill begins to show a pink base at this time, but many retain largely dark

bill into older age classes. **SECOND-SUMMER:** Similar, but body, head, and neck average whiter. **THIRD-WINTER & THIRD-SUMMER:** Much as adult, but retains some dark on tertials, wing coverts, greater primary coverts, and tail. Primaries tend to have small or no white tips. Bill has at least a dark subterminal band; sometimes terminal half is dark. **HYBRIDS:** Hybridizes commonly with Western (in WA) and also with Herring (in southwest and south-central AK).

SIMILAR SPECIES No other large gray-mantled gull shows primaries that are uniform in darkness with upperparts. Some "Kumlien's" Iceland Gulls may show similar grayish primaries, but petite-billed Iceland is much smaller and longer-winged than Glaucous-winged. In addition, older Iceland Gulls tend to show pale eyes. Hybrids with Western Gull, informally known as "Olympic Gulls," range from nearly identical to parental types to intermediate in appearance. A Glaucous-winged that shows a yellow orbital ring, noticeably darker primaries than upperparts, a darker than average mantle color or, in younger ages, a dark secondary bar, and contrasting paler rump or well-marked coverts is likely a hybrid. Hybrids with Herring Gull may resemble Thayer's in plumage features, but note that they are larger billed, bulkier, and longer legged, and often show barred or vermiculated head markings in winter. Juvenile hybrids begin molting their upperparts early in winter; Thayer's retains full juvenal plumage into late winter or early spring.

VOICE LONG CALL: A typical loud

series of evenly spaced *haaaw* notes. Slower pace and lower pitch than Western's.

STATUS & DISTRIBUTION Abundant. Also breeds in Asian Pacific, south to Japan. **YEAR-ROUND:** Rocky coasts preferred. **BREEDING:** Colonial, often uses small rocky islands, or rooftops in Seattle and Vancouver, BC. **MIGRATION:** Migrates south after breeding; first adults arriving in central CA as early as late Aug. Numbers peak in midwinter; most are gone from central CA by early Apr. First-year birds migrate longer distances than adults. **WINTER:** Some winter near breeding areas; others move south as far as Baja California Sur, Mexico. Some winter inland within 50–80 miles or so of coast. It is noted annualy at Salton Sea. **VAGRANT:** Casual to interior with recs. east to IL; accidental in NF, Canary Islands, and Morocco, so its occurrence is possible anywhere.

POPULATION Estimated 200,000 breeders in N.A.; increase in last 50 years likely due to garbage dumps and waste from industrial fishing.

GLAUCOUS GULL *Larus hyperboreus*   GLGU ■ 1

This northern gull basically looks like a larger white-winged Herring Gull. Even more aggressive than that species, it scavenges at dumps and harbors, often with gatherings of other large gulls. Polytypic (4 ssp.; 3 in N.A.). L 22–29" (56–74 cm) WS 56–63" (142–160 cm)

IDENTIFICATION A large four-cycle gull. Stout but not bulbous-tipped bill often expands slightly at culmen base; can be relatively parallel-edged and "slender" on immature females. All ages have flesh pink legs and lack black in wing tips. **BREEDING ADULT:** Clean white head and neck. Pale gray upperparts; pure white wing tips. Pale yellow eyes (dusky flecking on some may indicate hybridization); orange-yellow orbital ring can be orange-red on *pallidissimus* subspecies. Yellow bill; orange-red gonys spot. **WINTER ADULT:** Light to moderate dusky mottling and streaking on head and neck. Duller bill often has pinkish base and dark subterminal marks; orbital ring often fades to flesh pink. **FIRST-WINTER:** Whitish overall; variable pale brownish patterning. Wing tips white to creamy, often with neat dusky subterminal spots or chevrons on outer primaries. Flesh pink bill with clean-cut black distal third; dark eyes. Often retains most or all of juvenal plumage through winter; it can be bleached white overall by spring, when newly molted body feathers look contrastingly dark. **SECOND-WINTER:** Resembles first-winter, but back usually with some pale gray by spring; bill usually has distinct pale tip; eyes can be pale. **THIRD-WINTER:** Resembles winter adult, but upper wings and tail usually washed brownish. Bill duller, flesh to yellowish with a black subterminal band; often some reddish on gonys.

GEOGRAPHIC VARIATION Smallest and darkest-backed *barrovianus* (AK); larger and paler-backed *leuceretes* (Canada, Greenland, and probably Iceland); nominate *hyperboreus* (northwestern Eurasia) is smaller and darker than *leuceretes;* largest and palest-backed *pallidissimus* (northeast Asia). The subspecies *pallidissimus* occurs in northwestern Alaska and breeds on St. Lawrence and St. Matthew Islands; at the latter location it can be seen with *barrovianus*. Immatures not identifiable to subspecies in the field.

SIMILAR SPECIES Glaucous Gull is mostly distinctive, but smaller females may be confused with Iceland Gull. Also beware leucistic or albino individuals of other species (check structure, bill pattern). Iceland's bill is smaller, shorter, and more slender (often greenish based on winter adult); relatively longer wing projection beyond tail: At rest, Glaucous's tail tip usually falls between tips of P7 and P8, Iceland's between tips of P6 and P7. Adult Iceland has a reddish to purplish pink orbital ring. First-winter Iceland lacks Glaucous's clean-cut, pink-and-black bill, but second-winter Iceland's bill can be similar. Glaucous-winged Gull has a more bulbous-tipped bill and darker wing tips. Bleached first-year Glaucous-winged can be whitish overall, including primaries. Note blackish bill on Glaucous-winged (second-winter Glaucous-winged can have a Glaucous-like bill, but its wing tips are grayish). Adult Glaucous-winged is slightly darker above, with gray wing-tip markings. Southern sightings of Glaucous Gull should be double-checked for hybrids with Herring or Glaucous-winged; good views often needed for this. (Flight views of a "classic" Glaucous at even medium range are

insufficient to establish purity.) Herring Gull hybrids can look very like a Glaucous, but their outer primaries have dusky markings (a ghosting of Herring's pattern). Glaucous-winged hybrids can look even more like a Glaucous; some not safely identifiable in field. On first-winters, a bulbous-tipped bill and messy bill pattern are clues; on adults, check eye color and orbital ring color. (Faint gray wing-tip markings may be visible only in the hand.)

VOICE Similar to Herring's, but slightly hoarser and lower pitched.

STATUS & DISTRIBUTION Holarctic breeder, winters to midlatitudes. **BREEDING:** Common on coastal islands, sea cliffs, tundra lakes. Arriving late Apr.–June; mostly departing through Sept. Hybridizes with Herring Gull in AK and Canada and with Glaucous-winged in western AK. **MIGRATION:** Mainly late Oct.–mid-Dec., and in spring Feb.–Apr.; some linger through summer in Canada. **WINTER:** Mainly Nov.–Apr. Rare (mainly late Dec.–Mar.) south of northern tier of states; very rare to southern CA and Gulf states.

POPULATION Poorly known (AK population perhaps more than 100,000 individuals); may be fairly stable in N.A.

2nd winter

winter adult

larger and chunkier than Iceland

winter adult

very pale gray above

shorter wing tip projection past tail than Iceland

pale eye and bill tip in 2nd winter

pinkish legs

white wing tips

breeding adult

1st winter

1st winter

long pink bill with blackish tip

1st winter

color of 1st-winter birds varies; some by mid- to late winter are chalky white

GREAT BLACK-BACKED GULL *Larus marinus*   GBBG ■ 1

The largest gull in the world, the Great Black-backed is an efficient scavenger and predator. It commonly preys on ducklings and can snatch adult puffins in flight! Monotypic. L 25–31" (64–79 cm) WS 60–65" (152–165 cm)

IDENTIFICATION A huge four-cycle gull with very long, broad wings and a low-sloping forehead. Its very stout bill (which is notably smaller on females) has a swollen gonys. **BREEDING ADULT:** Slaty blackish upperparts (browner when worn, in summer) blend into black wing tips; large white tip to P10 merges with a large white mirror on P9. Pale flesh to flesh pink legs; olive to dull, pale yellow eyes; a reddish orbital ring. Yellow bill with an orange-red gonydeal spot. **WINTER ADULT:** Inconspicuous, fine dusky head and neck streaking concentrated mainly on hind neck; birds look white headed at any distance. Bill is duller than in summer, often pinkish at base and with blackish subterminal marks; orbital ring can be pinkish. **FIRST-WINTER & SUMMER:** Head and underparts whitish overall (often bright white by spring) with relatively sparse brownish mottling and streaking. Upperparts contrasty and checkered. Often retains most juvenile plumage through winter. White uppertail coverts and tail; broad black distal tail band typically broken up by internal, narrow white barring; tail base and uppertail coverts with fairly sparse dark bars. Poorly to moderately contrasting pale panel on inner primaries on upper wing. Dark eyes; dull, flesh to flesh pink legs. By spring, black bill usually develops a dull, flesh base. **SECOND-WINTER & SUMMER:** Second-winter often looks very similar to first-winter, but primary tips more rounded; greater coverts plainer and browner, typically with fairly fine markings (unlike boldly barred first-cycle coverts); and bill has large pale tip. A few slaty back feathers appear on some birds. By spring, back has some to many slaty feathers; eyes paler brown. In summer, orbital ring can brighten to orange; some birds develop yellow bill with reddish gonydeal smudge and black subterminal band. **THIRD-WINTER & SUMMER:** Some third-winters resemble second-winters (but note bill pattern and adultlike inner primaries); others much more adultlike, but with black on tail, less white in

wing tips, some brownish on wing coverts. In winter, bill typically flesh pink with broad black subterminal band and creamy tip. In summer, bill usually brightens and may be indistinguishable from adult's, but typically has dark distal marks.

SIMILAR SPECIES Note very large size and stout bill with swollen gonydeal expansion. First-winter and second-winter separated from smaller first-winter Herring by much whiter head and underparts and more boldly checkered upperparts; in flight, note white rump and tail base with broken black tail band. On older immatures and adults, note slaty blackish on upperparts and pale, flesh legs. Adult Kelp has blacker upperparts, limited white in wing tip, and yellowish legs. Adult Great Black-backed x Herring hybrid has been mistaken for Western, but that species is relatively bulkier and broader winged, with a more rounded head and a more bulbous-tipped bill.

VOICE Very deep calls, much lower pitched than Herring Gull.

STATUS & DISTRIBUTION Northeast N.A., northwestern Europe. **BREEDING:** Common (Apr.–Aug.); in small colonies or scattered pairs on rocky islands, beach barriers, locally on

rooftops; fairly common on eastern Great Lakes, rare on western. **MIGRATION:** Mainly late Aug.–Nov. and Mar.–Apr.; small numbers of nonbreeders oversummer in winter range. **WINTER:** Mainly Oct.–Mar., at beaches, fishing harbors, dumps, etc. **VAGRANT:** Casual to very rare (mainly Nov.–Mar.) on Great Plains, and north (May–July) to Hudson Bay and NV. Accidental in AK, the Northwest, and southern CA.

POPULATION Dramatic increase in North America since early 1900s, with range expansion both south along Atlantic coast and inland through Great Lakes. Increased in U.S. by about 17 percent per year from 1926 to1965.

checkered tail with subterminal dark band

1st winter

winter adult

long stout bill with thick tip

pale eye

dark on bill

P10 all-white at tip

blackish above

bold white apical spots

breeding adult

pink legs

blotchy black above

3rd winter

paler below than Herring with strongly checkered upperparts

1st winter

KELP GULL *Larus dominicanus* KEGU ■ 4

one white
mirror

breeding
adult

This stocky, longish-legged, black-backed gull is widespread in the Southern Hemisphere, including western and southeastern S.A. It has bred in Louisiana and has hybridized with Herring Gull, complicating identification. Polytypic (2 ssp.; nominate in N.A.). L 23" (58 cm) WS 53" (135 cm) **IDENTIFICATION** A four-cycle gull that shows an advanced maturation schedule: Blackish upperparts come in as early as first summer; becomes entirely blackish above by second summer. Bulky, large bill, and long greenish legs. Kelps have thick bills that lack an obvious expansion at the gonydeal angle. Adults have a black mantle, showing little difference in darkness of primaries and mantle; only Great Black-backed approaches this darkness. **ADULT:** Adults show a mix of blackish upperparts; bright yellow bill with red gonys spot; reddish orbital ring; variable but usually pale eye; greenish to yellow-green legs; and one mirror on P10. White head, neck, underparts, and tail. Legs vary in color,

being yellow-green during breeding season but becoming olive to olive-gray when not breeding. **FIRST-YEAR:** Brownish, with well-streaked whiter head and underparts. Juvenile mantle brown with neat pale fringes, many of which are replaced in first-winter by dark gray-brown feathers with darker shaft streaks; extent of gray-brown back and lightening of head and underparts continues into first-summer. Tertials are dark with narrow paler fringes; greater coverts dark at bases. Tail is entirely dark. In flight, wings lack paler inner primaries and have dark secondaries and greater coverts. **SECOND-YEAR:** Upperparts blackish by this age, contrast with browner wings. White head and body; moderate streaking on face, breast sides, flanks. Dark post-ocular line retained in even the whitest individuals. Pale pink-based bill by this age; dark eyes; grayish pink to grayish green legs. Tail is largely blackish or white with a broad black terminal band, contrasting with white rump. **THIRD-YEAR:** Much like adult, but it

retains some dark markings on greater primary coverts and sometimes on tail. **SIMILAR SPECIES** The adult Great Black-backed is larger and has dull pink legs, large white tip to P10, and a second mirror on P9. First-year Kelp Gulls differ from Great Black-backeds of similar age by their largely dark tail, more uniformly patterned upperparts, and dark-based greater coverts. Lesser Black-backed Gulls (*graellsi* and *intermedius*) are smaller, slimmer, thinner billed, and much longer winged; adults are paler above than Kelp, which retains its unstreaked head all year long. First-year birds separated on structure and Kelp's darker tail. Kelp x Herring hybrids are similar in mantle color to Lesser Black-backed Gull, but they differ in stockier structure and larger bill. **VOICE LONG CALL:** Similar to Herring Gull's, although faster in tempo. **STATUS & DISTRIBUTION** Casual in N.A., widespread in Southern Hemisphere. **BREEDING:** Colonial; has bred and hybridized with Herrings on the Chandeleur Islands of LA. **VAGRANT:** Casual in eastern N.A. (including breeding records off LA coast). Otherwise, records have been accepted from TX, FL, IN, and a famous long-staying individual in MD. **POPULATION** No estimates exist from S.A., but recent range expansion northward suggests an increasing population.

plumage suggests Lesser Black-backed but chunkier with bigger bill

heavy bill

Kelp x Herring hybrid winter adult

slaty gray

molting juvenile

winter adult

2nd summer

greenish legs

blackish upperparts as dark or darker than Great Black-backed

NODDIES Genus *Anous*

Noddies are tropical terns—three species worldwide—with dark plumage, white caps, and long, wedge-shaped tails. Two species occur in N.A., but only the Brown Noddy breeds here (Dry Tortugas, FL).

BROWN NODDY *Anous stolidus* BRNO ■ 2

In North America, this distinctive all-dark tern nests only on the Dry Tortugas, Florida, where it builds a nest in low bushes. Its flight is steady and graceful, typically low over the water. Here flocks often feed with other tropical seabirds, milling and swooping to pick food from the surface. Brown Noddy does not usually dive for food and rarely alights on the sea. Polytypic (5 ssp.; nominate in N.A.). L 14–16" (36–41 cm) WS 31–35" (79–89 cm)

IDENTIFICATION Medium-size with medium-long bill and long, graduated tail. No seasonal variation in appearance (unlike most terns). **ADULT:** Plumage dark brown overall (inc. underwings) with contrasting ashy-white forecrown offset by black lores and narrow white subocular crescent; ashy-gray hind crown and nape. On resting birds, primaries contrastingly blacker than upperparts, which, in fresh plumage, have gray cast often lost by summer. Black bill; dark gray legs.

Prebasic molt begins April to June, then suspends during breeding, and is completed at sea by March, in time for return to colonies. **IMMATURE:** Juveniles dark brown overall with a neat white supraloral line that continues over bill base; forecrown is dull ashy gray, and brownish hind crown does not contrast with hind neck. Fresh fall birds have subtly paler brown tips to upperwing coverts and scapulars. Protracted complete post-juvenal molt starts in fall and continues into

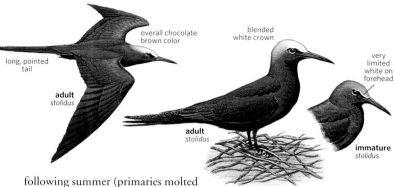

overall chocolate brown color

long, pointed tail

adult
stolidus

blended white crown

very limited white on forehead

adult
stolidus

immature
stolidus

following summer (primaries molted mainly Feb.–Aug.). **FIRST-SUMMER:** Not illustrated. Forecrown mostly ashy white, often with some brownish smudging. Some birds' crown mostly ashy white; others still show a contrasting white supraloral line and have little ashy white on crown. Brownish to gray-brown hind crown merges with nape. Many spring birds still have some contrastingly pale, bleached (juvenal) upperwing coverts. **GEOGRAPHIC VARIATION** Birds in N.A. are of nominate subspecies, *stolidus*. Subspecies *ridgwayi* of west Mexico (not separable in field from *stolidus*) is a potential hurricane-assisted vagrant to the West Coast or the Southwest. **SIMILAR SPECIES** Unlikely to be confused when in the Tortugas, but vagrants elsewhere may be puzzling. Juvenile and first-year Sooty Terns are slightly bulkier overall with a forked

tail, fine pale spotting on upperparts in fresh plumage, contrasting white underwing coverts, and a white central belly and undertail coverts. Black Noddy is smaller and blacker overall (primaries not contrastingly blacker at rest) with a more extensive and contrasting white cap on adults and immatures; its bill is distinctly thinner (obvious in direct comparison), and its white subocular crescent is shorter (lying mostly behind eye's midpoint). At sea, Bulwer's Petrel (accidental in N.A.) can suggest a small noddy, with its dark brown plumage, long wings and tail, and low flight over the water; it lacks any pale head markings, but has a pale-brown ulnar bar on upper wing. Its wingbeats are stiffer, and its flight is more wheeling, with frequent weaving glides. Also compare dark

juvenile jaegers, which have white primary flashes.
VOICE Mostly silent except around colonies, where it gives varied guttural barks and a braying *keh-eh-eh-ehr*. First-year birds give higher, whistled calls.
STATUS & DISTRIBUTION Pantropical. **BREEDING:** Fairly common but very local, arriving (at first nocturnally) at colony mid-Jan.–Feb. and departing through Oct., with breeding season mainly May–July. **DISPERSAL & VAGRANT:** Casual in summer along Gulf Coast and north to Outer Banks, NC; accidental north to New England. **WINTER:** FL population presumably winters at sea in tropical Atlantic.
POPULATION Florida population is believed to be stable, fluctuating between 1,000 and 2,000 pairs.

BLACK NODDY *Anous minutus*   BLNO ■ 3

In North America, this species is most frequently recorded on the Dry Tortugas, where it associates with Brown Noddies. Its flight is typically low over water, but is often quicker than Brown Noddy's, with more fluttery wingbeats. Polytypic (7 ssp.; 1 in N.A.). L 12–13.5" (30–34 cm) WS 26–29" (66–74 cm)
IDENTIFICATION Medium-small bird with a proportionately long and distinctly slender bill and a long, graduated tail. No seasonal variation

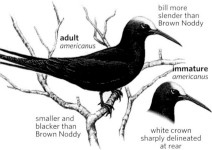

bill more slender than Brown Noddy

adult
americanus

immature
americanus

smaller and blacker than Brown Noddy

white crown sharply delineated at rear

in appearance (like Brown Noddy). **ADULT:** Plumage blackish overall (inc. underwings) with a contrasting white crown offset by black lores, and a narrow white subocular crescent; hind neck is ashy gray. On resting birds, primaries not contrastingly blacker than upperparts. Molts poorly known, but likely similar to Brown Noddy's. **IMMATURE:** Resembles adult, but worn first-summer birds often slightly browner overall, sometimes with bleached (juvenal) upperwing coverts. Rear border to white cap contrasts more sharply with blackish hind neck.
GEOGRAPHIC VARIATION North American birds are presumably of the subspecies *americanus,* which breeds on islands off the north coast of Venezuela.
SIMILAR SPECIES Brown Noddy is larger and bulkier with a thicker (and thus often slightly shorter-looking) bill and dark brown plumage overall. Brown Noddy's primaries are

also contrastingly blacker at rest. However, plumage tones can be difficult to evaluate in bright sunlight: On adults, look for Black Noddy's smaller size and, especially, its long, slender bill; also note more extensive white cap (beware that bright light can cause Brown Noddy to show a large white cap), shorter white subocular crescent, and grayish tail. Immature Black Noddy is more distinctive, with a large and contrasting white cap unlike any plumage of Brown Noddy's.
VOICE Rarely vocal away from colonies; likely to be mostly silent in N.A. Calls include guttural growls, *ahrrr* and *garrr;* first-year birds have a high, piping *swee*.
STATUS & DISTRIBUTION Tropical Atlantic and Pacific Oceans. **VAGRANT:** Rare or casual spring-summer visitor to Dry Tortugas, FL. Accidental TX coast (mid-Apr.–July). Only a few N.A. records over the last two decades.

Genus *Onychoprion*

SOOTY TERN   *Onychoprion fuscatus*   SOTE ■ 2

This handsome, colonial-nesting tropical tern is largely pelagic and comes ashore only to breed. The Florida population spends most of its nonbreeding time off West Africa, where young birds remain for their first one to two years of life. Sooty Terns have a graceful, buoyant flight and swoop down to pick food from the surface; they do not plunge dive. They will rest on the sea, often in flocks. Polytypic (8+ ssp.; 2 in N.A.). L 14–15.5" (36–39 cm) WS 34–36.5" (86–93 cm)

IDENTIFICATION Medium size with fairly broad wings and a deeply forked tail. **BREEDING ADULT:** Above black, with a broad, triangular white forehead patch; white below. Tail black with a white outer edge. In flight, note dark underside to remiges contrasting with white underwing coverts. Black bill and legs. **WINTER ADULT:** Not illustrated, but may be encountered in fall (esp. storm-blown birds). Resembles breeding adult, but hind neck mottled pale gray, back feathers often have broad whitish tips, and forehead patch less neatly defined. **JUVENILE:** Sooty brownish black overall with white-spotted back and upperwing coverts; white vent to undertail coverts. In flight, note contrast between white underwing coverts and dark body. **FIRST-SUMMER:** Lacks juvenile's whitish spotting, but back feathers often tipped whitish, and underparts have variable dusky mottling.
GEOGRAPHIC VARIATION Atlantic breeding populations are the nominate subspecies *fuscatus,* the breeding adults of which have white underparts with little or no smoky clouding on belly and vent. East Pacific populations (casual in southern CA) are the subspecies *crissalis;* adults average smaller and have pale gray clouding on their underparts.
SIMILAR SPECIES Bridled Tern is slightly smaller and lighter in build, with narrower wings, a longer tail, and a more buoyant and lighter flight; it often stands on driftwood, weed patches, and other floating objects but rarely alights on the sea (which Sooty does). Only adultlike plumages of Bridled Tern are likely to be confused with Sooty's. Breeding adult Bridled Tern is dark brownish gray above, with remiges often appearing darkest part of upperparts (opposite of Sooty Tern) and with long tail looking mostly white (striking even at long range). Also note that underside of primaries are white based on Bridled (all dark on Sooty); Bridled's paler hind neck often appears as a hind collar, and its white forehead patch is narrower and more chevron shaped, projecting past eye. Nonbreeding plumages differ in much the same ways, but Bridled has a shorter tail.
VOICE Adult has a nasal barking or laughing *ka-waké* or *ke wéh-de-wek* given year-round. Juveniles give high reedy whistles into their first winter: e.g., *wheeir.*
STATUS & DISTRIBUTION Pantropical.

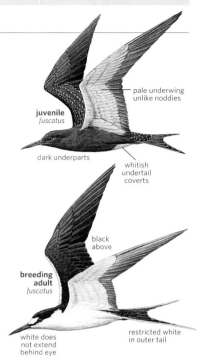

juvenile *fuscatus*
pale underwing unlike noddies
dark underparts
whitish undertail coverts
black above
breeding adult *fuscatus*
white does not extend behind eye
restricted white in outer tail

BREEDING: Large colony on Dry Tortugas, FL, where birds arrive by Mar., depart late July–Sept.; also nests on islets off LA and TX. **DISPERSAL:** Regular in summer (esp. July–Sept.) north to NC. **VAGRANT:** Casual to southern CA. Storm-blown birds (mainly late summer and fall) to Great Lakes and north to Maritimes; accidental to Attu I., AK.

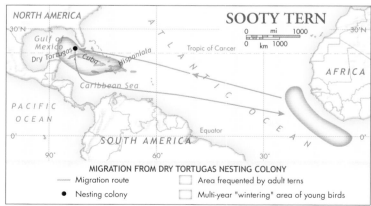

SOOTY TERN

MIGRATION FROM DRY TORTUGAS NESTING COLONY
— Migration route   Area frequented by adult terns
● Nesting colony   Multi-year "wintering" area of young birds

BRIDLED TERN   *Onychoprion anaethetus*   BRTE ■ 2

This handsome tropical tern occurs in North America mainly as a nonbreeding pelagic summer visitor to warm offshore waters. Bridled Terns have a graceful, buoyant flight and swoop down to pick food from the surface; they do not plunge dive. Often found resting on driftwood, sargassum weed mats, and even the backs of turtles, they rarely if ever alight on the sea. Unbroken tail streamers add 2 to 2.5 inches (5–6 cm) to length of a breeding adult. Poly-

typic (6+ ssp.; 2 in N.A.). L 12.5–14"
(32–36 cm) WS 31–33.5" (79–85 cm)
IDENTIFICATION Medium size; long,
deeply forked tail. **BREEDING ADULT:**
Black crown; chevron-shaped white
forehead patch; gray-brown upperparts;
whitish hind collar; outer rectrices
mostly white. **WINTER ADULT:** May be
encountered in fall (esp. storm-blown
birds). Resembles breeding adult, but
back feathers tipped pale gray; forehead patch less neatly defined; whitish streaking in forecrown. **JUVENILE:**
Pattern overall resembles adult's, but
much less well defined. White forehead
and short supercilium are framed by
black auricular mask and dark-streaked

crown; upperparts boldly barred pale
buff to whitish. **FIRST-SUMMER:** Commonly seen in waters of N.A. Variable
appearance intermediate between juvenile and winter adult. Upperparts have
little or no whitish barring; head whitish overall, with an indistinct blackish
auricular mask.
GEOGRAPHIC VARIATION East Coast
birds are *melanopterus,* breeding adults
of which have white underparts. East
Pacific breeding populations (accidental
in southern CA) are of subspecies *nelsoni,* breeding adults of which average
larger than *melanopterus* and have pale
gray clouding on their underparts.
SIMILAR SPECIES See Sooty Tern.
VOICE Not especially vocal except
around the nest, where adults give a
mellow *kowk-kowk* or *kwawk kwawk,*
and a harder *kahrrr.*
STATUS & DISTRIBUTION Pantropical. **BREEDING:** Very local in FL Keys
(Apr.–Aug.). **DISPERSAL:** Regular offshore visitor (mainly May–Oct.) in
Gulf of Mexico and shoreward edge of
Gulf Stream north to NC, rarely to NJ.
VAGRANT: Storm-blown birds (mainly
late summer) north to Atlantic Canada.
Accidental to southern CA and AR.

grayish brown
cast above

extensive whitish
underside to
primaries

**breeding
adult**
melanopterus

extensive white
in outer tail
when spread

white
extends
behind eye

pale collar

pale
head

grayish brown
above

1st summer
melanopterus

head pattern
diluted

pale-edged
upper parts

juvenile
melanopterus

ALEUTIAN TERN *Onychoprion aleuticus*   ALTE ■ 2

This handsome tern breeds on western Alaskan coasts. Its calls are unique.
Monotypic. L 11–12" (28–30 cm)
WS 29.5–32" (75–81 cm)
IDENTIFICATION Deeply forked tail.
On all plumages, note dark secondary
bar on underwing. **BREEDING ADULT:**
Black cap with white forehead, medium
gray above and below. Black legs and
bill. **JUVENILE & FIRST-SUMMER:** Fresh
juveniles are relatively dark above, with
cinnamon to rufous edging and barring on upperparts; cinnamon wash
bleeds onto chest and sides (soon fades
to whitish). Bill pinkish with dark culmen and tip; legs pinkish. Presumably

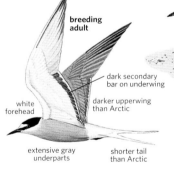

**breeding
adult**

dark secondary
bar on underwing

darker upperwing
than Arctic

white
forehead

extensive gray
underparts

shorter tail
than Arctic

brownish,
with scaly
pattern above

undergoes complete molt in first winter
(similar to Common Tern), and first-
summers may reach Alaskan waters but
go unreported. First-summers probably
resemble winter adult in having white
lores and white forecrown; note dark
underwing secondary bar and medium
gray upperparts.
SIMILAR SPECIES Arctic Tern is smaller
and paler overall. Broader-winged Aleutian has white forehead and shorter tail
streamers; note voice. Breeding adult
Common of East Asian *longipennis* are
slightly slimmer in build with a longer

bill; they lack a black underwing secondary bar and have a full black cap.
VOICE Adult's piping whistled calls have
a mellow to slightly rolled quality: *piiu*
and *piirr-i-u* or *piiu-pi-pip,* and so on.
STATUS & DISTRIBUTION BREED-
ING: North Pacific. Fairly common,
but local; arriving southern AK (inc.
Kodiak I.) and in Aleutians and Bering
Sea mid-May–early June. Departs colonies mid-Aug.–late Aug. **MIGRATION:**
Casual off Queen Charlotte Is., BC
(mid-May–early June). **WINTER:** Winter
grounds not well known, likely in the
South Pacific near Australia and the
Phillipines; regular migrant off Hong
Kong. **VAGRANT:** Accidental to U.K.
POPULATION Ironically, not known to
breed in the Aleutians before 1962, with
subsequent increases perhaps due to
removal of non-native foxes.

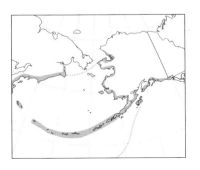

These terns are compact with dark underparts in breeding plumage. They usually feed by dipping down to the water's surface for insects or chasing insects in the air (hawking). Although all three species in the genus have occurred in N.A., only the Black Tern is a common breeder; the other two are vagrants.

BLACK TERN *Chlidonias niger*   BLTE   1

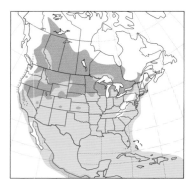

This handsome small tern occurs widely as a migrant and summer visitor at lakes and marshes across North America, where it breeds in freshwater habitats. Its winter range is pelagic. Black Terns have an easy, slightly floppy flight as they patrol back and forth and swoop down to pick food from the surface; they also soar like giant swallows and catch insects in flight, but they do not plunge dive like typical terns. Migrants occur singly or, locally, in flocks of hundreds. Black Terns perch readily on posts and wires, and their nests are mats of floating vegetation. Polytypic (2 ssp.; 1 in N.A.). L 9–9.7" (23–25 cm) WS 23.5–25.5" (60–65 cm)

IDENTIFICATION Small with a fairly short, slender black bill and medium-length legs; tail with shallow cleft. **BREEDING ADULT:** Distinctive, with a black head and body, slaty-gray upperparts, and white undertail coverts. Underwing coverts are smoky gray. Bill is blackish, legs dark reddish. Prebasic molt starts in mid-summer, when white spots appear on head and body; underparts of fall adults are mostly white or blotched black and white. **WINTER ADULT:** Head and underparts are white with a dark-streaked cap connected to a blackish ear spot and forming "headphones"; note dark bar on chest sides and slaty-gray mottling on flanks. **JUVENILE:** Resembles winter adult, but wings are uniformly fresh in fall; back and tertials have variable dark-brown distal markings that create a subtly mottled aspect. **FIRST-SUMMER:** Resembles winter adult, but underparts of some birds have scattered black spots.

GEOGRAPHIC VARIATION North American birds comprise New World subspecies *surinamensis*, which differs from nominate *niger* of the Old World (possible vagrant to the East) in deeper black head and body of breeding adults and gray flanks of winter adults and juveniles.

SIMILAR SPECIES Distinctive; often strikingly small when seen perched among other terns, being barely larger than a Least Tern. See vagrant White-winged and Whiskered Terns.

VOICE A piping, slightly reedy *peep* or *pseeh* and a quiet *kriih* given by birds in feeding flocks; a quacking *kek* in alarm; and a shrill, slightly grating *kehk* given by birds scolding near the nest.

STATUS & DISTRIBUTION North America and western Asia. **BREEDING:** Arrives on nesting grounds late Apr.–May, departing late July–Aug. **MIGRATION:** Mainly mid-Apr.–May and late July–Sept. Uncommon to rare along and off Pacific coast. Casual visitor north to AK and Atlantic Canada. **WINTER:** Over Pacific Ocean waters from southern Mexico to Peru. Casual to CA. **VAGRANT:** In Europe, *surinamensis* is casual.

POPULATION Declining in most areas, mainly because of habitat degradation and loss. North American population in early 1990s was a third of what it was in late 1960s.

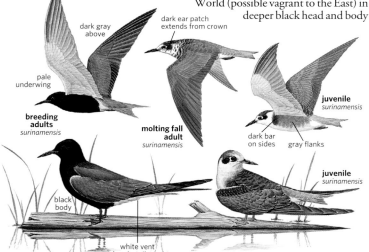

WHITE-WINGED TERN *Chlidonias leucopterus*   WWTE   4

This small tern of Eurasia (where it is more evocatively called "White-winged Black Tern") is a casual visitor to North America, and breeding-plumaged adults are among the most handsome of birds. White-winged Terns might be found anywhere Black Terns occur and readily associate with them. (The two species have hybridized in QC.) Whited-winged's flight is similar to Black Tern's, swooping down to pick food from the water as well as soaring to catch insects in flight. They perch readily on posts and wires. Monotypic. L 9–9.5" (23–24 cm) WS 22–23" (56–59 cm)

IDENTIFICATION Small with a fairly short, slender bill and medium-length legs. Wings are slightly more rounded than Black Tern's, and tail has only a shallow cleft. **BREEDING ADULT:** Unmistakable: Black head, body, and underwing coverts contrast with white "wings," tail coverts, and tail. Outer two to three primaries are dark (basic) feathers that contrast with newer middle primaries attained by the prealternate molt. Bill is blackish to dark red; legs and feet are orange-red. Prebasic molt starts in midsummer, when white spots appear on head and body. Molting adults in fall often lack most or all of the black, except for some underwing coverts. WINTER ADULT: Head, underparts, and underwings are white with a dark-streaked cap and a blackish ear spot often separated from cap by a white supraorbital area; back and tail are pale gray. Note dusky secondaries contrasting with whitish upperwing coverts. JUVENILE: Not recorded in North America. Resembles winter adult, but wings are uniformly fresh in fall; its dark brown saddle contrasts with pale upperwings and uppertail coverts, and it has duller, pinkish legs. A protracted complete molt in first winter (primaries molted Jan.–Aug.; sometimes later on vagrants) produces first-summer plumage. FIRST-SUMMER: Not illustrated, but resembles winter adult. SECOND-SUMMER: Not illustrated, but resembles breeding adult. Some have white spots on underwing coverts, and others have strongly piebald head and body in spring. SIMILAR SPECIES Black Tern is slightly larger with more pointed wings, a slightly more cleft tail, a slightly longer bill, and slightly shorter legs.

All plumages of Black Tern have smoky-gray underwing coverts; its upper wings, back, and tail are essentially concolorous dusky gray, but marginal coverts of breeding adults can be paler and look whitish in some lights. Confusion is most likely in winter plumages, but note White-winged's paler upperwings with contrasting dark secondaries, its pale gray rump and tail, its whitish underwings, its lack of a dark mark on sides of neck, and its lack of dark mottling on flanks. On perched birds, note Black Tern's dark mark on sides of neck and its smokier gray upperparts; in direct comparison, Black Tern's slightly longer (and thus finer-looking) bill and shorter legs may be appreciated. White-winged's overall pale gray

winter plumage recalls many typical terns, but note its small size, short bill, marsh-tern flight, and shallowly cleft tail. If a suspected White-winged does not show the classic suite of features, the possibility of a hybrid with Black Tern should be considered. Also see Whiskered Tern. **VOICE** Mostly silent away from the breeding grounds, White-winged Tern may give a hoarse *kesch,* slightly deeper and harsher than Black Tern's call. **STATUS & DISTRIBUTION VAGRANT:** Casual visitor (mainly late May–Aug.) to widely scattered locales in the eastern half of North America, especially along the Northeast and mid-Atlantic coast, but only a few N.A. records over the last two decades. Accidental in the West (AK and CA).

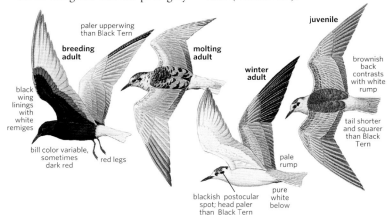

breeding adult — paler upperwing than Black Tern

molting adult

juvenile — brownish back contrasts with white rump

winter adult

tail shorter and squarer than Black Tern

black wing linings with white remiges

bill color variable, sometimes dark red — red legs

pale rump

blackish postocular spot; head paler than Black Tern — pure white below

WHISKERED TERN *Chlidonias hybrida*   WHST ■ 5

breeding adult *hybrida*

stout, dark red bill

dark gray underparts contrast with white cheeks

This Old World marsh tern is an accidental visitor to eastern North America. Whiskered Tern is somewhat intermediate in appearance between typical terns (such as Common) and marsh terns. Its flight is similar to Black Tern's but somewhat heavier, with less floppy wingbeats. Like Black Tern, Whiskered swoops down to pick food from the water, rather than diving. Polytypic (6 ssp.; presumed nominate in N.A.). L 9.5–10" (24–25 cm) WS 26.5–28.5" (67–72 cm)

IDENTIFICATION Medium-size tern with a medium-short and relatively stout bill and medium-length legs. Wings are fairly broad and bluntly pointed; tail is shallowly cleft. **BREEDING ADULT:** Black cap and smoky-gray underparts strongly set off white cheeks (or "whiskers"). Underparts are similar in tone to upperparts, except undertail coverts are whitish and rump and tail are smoky gray. In flight, smoky-gray underbody contrasts with whitish underwings and undertail coverts. Bill and legs are deep red. Prebasic molt usually starts in fall, later than most other marsh terns. **WINTER ADULT:** Head and underparts are white with a black postocular patch that merges into blackish streaking on hind crown; note pale gray rump and tail. Bill and legs are blackish. **JUVENILE:** Unlikely in North America. Resembles winter adult, but wings are uniformly fresh in fall and it has a dark brown saddle

with broad cinnamon bars and pinkish legs. A protracted complete molt in first winter produces first-summer plumage. **FIRST-SUMMER:** Resembles winter adult, but primaries in molt mainly January through August. **SIMILAR SPECIES** Separated from Common and Arctic Terns by typical marsh-tern flight; smaller size; and more compact shape, with a pale gray rump and shallowly cleft tail. Also, breeding adult Whiskered Tern has darker body plumage and more contrasting white cheeks. Winter adult White-winged Tern is smaller and more lightly built, with narrower wings and a shorter, finer bill; its black ear spot is often more distinct and separate from dark cap. **VOICE** Vagrants are usually silent; flight call a rasping *krehk*. **STATUS & DISTRIBUTION VAGRANT:** Accidental (July–Aug.), with two records from 1990s from mid-Atlantic coast (NJ and DE).

TYPICAL TERNS Genera *Sternula, Gelochelidon, Sterna, Hydroprogne, Thalasseus,* and *Phaetusa*

Terns resemble small, angular gulls with pointed bills, shorter legs, and forked tails. In breeding plumage, some have outermost tail feathers elongated into streamers. Typical terns are pale gray to black above and white below, often with a black cap. First-summer birds resemble winter adults. Second-summers are variable—some look like first-summers, others like breeding adults. Many species hover and dive after fish.

LEAST TERN *Sternula antillarum*   LETE ▪ 1

Colonies of this tiny tern "compete" with humans for beach space. Consequently, the species is declining in much of its range. Its very small size is diagnostic among North American terns. Its flight is fast and direct, with fairly deep, hurried wingbeats and frequent hovering before plunge diving for fish. Polytypic (4–5 ssp.; 3 in N.A.). L 8–9" (20–23 cm) WS 19–21" (48–53 cm)

IDENTIFICATION The smallest tern, with a relatively long bill and forked tail. All plumages have a pale smoky-gray rump and tail. **BREEDING ADULT:** Black cap has clean-cut white forehead chevron; outer 1–3 primaries form a contrastingly blackish upperwing wedge against fresher, pale gray middle primaries. Yellow legs; yellow bill tipped black. **WINTER ADULT:** Seen (Aug.–Sept.) before birds leave U.S. Lores and crown become white, with variable dark streaking on crown; bill becomes black. Dark-mottled lesser coverts form patagial bar on upper wing. Best told from first-summer (and juvenile) by strong contrast of old outer 1–3 primaries on upper wing. (Adult prebasic primary molt mainly June–Dec., followed by prealternate primary molt including 7–9 inner primaries). **JUVENILE & FIRST-SUMMER:** Resembles winter adult, but fresh-plumaged juvenile has variable brownish barring and tipping on back and upperwing coverts; bill base often flesh to yellowish. They undergo variably extensive primary molt in first winter and spring, followed by prealternate molt in spring that includes inner primaries. Most frequent first-summer

plumage resembles winter adult's, but with darker, blackish patagial bar and blackish outer 4–5 primaries, forming a contrasting upperwing wedge suggesting Sabine's Gull; bill usually black. Some birds resemble breeding adults, but less clean-cut chevron on forehead, forecrown speckled white, and yellow bill more extensively black distally.
GEOGRAPHIC VARIATION Nominate *antillarum* (breeds on East Coast and Gulf Coast) averages larger and paler than both *browni* (of CA) and, especially, *anthalasso* (of the interior).
SIMILAR SPECIES Unmistakable by its small size. Little Tern *(Sternula albifrons),* its Old World counterpart, is very similar to Least Tern and should be sought on East Coast; all Little Tern plumages have contrasting white rump and tail, unlike pale smoky gray of Least Tern.
VOICE High reedy chippering; chatters with frequent disyllabic calls: *chi-rit* and *k-rrik;* a reedier *kree-it* or *kreet;* a clipped *k'rit;* and longer series, *kik kirvee,* etc.
STATUS & DISTRIBUTION U.S. south to Mexico and Caribbean. **BREEDING:** Fairly common but local, arriving at southern breeding areas late Mar.–

Apr., northern sites through May; fall movements start June–Aug. **MIGRATION & DISPERSAL:** Most birds depart U.S. during Aug., a few linger into early Sept. Casual north to southern Canada. **WINTER:** Poorly known, presumably off coasts from Mexico and Caribbean to northern S.A. **VAGRANT:** Accidental HI. Either a Least Tern, or more likely a Little Tern (on distribution), was photographed on Buldir I. (July 4–6, 2005), the western Aleutians.
POPULATION Pacific coast *(browni)* and interior *(anthalasso)* populations are endangered in U.S.; main threats being habitat modification and human disturbance.

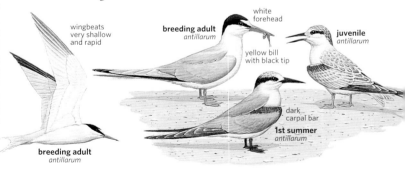

breeding adult
antillarum

white forehead

yellow bill with black tip

juvenile
antillarum

wingbeats very shallow and rapid

dark carpal bar

1st summer
antillarum

breeding adult
antillarum

GULL-BILLED TERN *Gelochelidon nilotica*   GBTE ▪ 1

This distinctive, medium-size tern of salt marshes and beaches has a smooth flight as it sweeps over open areas such as fields or salt flats, picking its prey (insects, small crabs, etc.) from the ground or the water's surface. It does not hover and dive into the water. Polytypic (6 ssp.; 2 in N.A., weakly defined). L 13–14" (33–36 cm) WS 35–38" (89–97 cm)

IDENTIFICATION All plumages have ghostly pale-gray upperparts, lacking a contrasting white rump and tail. Stout bill and relatively long legs are black. **BREEDING ADULT:** Black cap becomes spotted white in late summer. **WINTER ADULT:** Note distinctive white-headed appearance with dark postocular mask or smudge; prima-

ries in molt July–Feb. **JUVENILE & FIRST-YEAR:** Resembles winter adult, but juvenile in fall is fresh plumaged, with variable buff wash and brownish subterminal marks on back; dark mask is often less distinct; and bill can have pinkish base into fall. Protracted complete molt (with primaries molted Jan.–Aug.) produces first-summer plumage, much like winter adult's. **SECOND-YEAR:** Second prebasic molt averages later than in adults, so outer primaries fresher and paler in second summers than in adults. **SIMILAR SPECIES** Unlikely to be confused; note Gull-billed's behavior. Winter and juvenile Forster's Terns have a bolder black mask, are smaller and slimmer in build, and dive in the water for food; on perched birds, note Forster's short, orange-red legs. **VOICE** Nasal but mellow laughing and barking calls, usually 2–3 syllables: *keh-wek* or *ku-wek,* and *keh-w-wek* or *kit-u-wek;* sharper than calls of Black Skimmer. Juvenile has high piping whistles given into first winter. **STATUS & DISTRIBUTION** Warmer climates worldwide. **BREEDING:** Fairly common but local in N.A., present Mar./Apr.– Aug. at most colonies. **MIGRATION:** Mainly Mar.–Apr.

and Aug.–Sept. **WINTER:** Casual north to NC. **VAGRANT:** Casual in interior N.A. except Salton Sea, CA, where it breeds. **POPULATION** Locally erratic, but overall population in N.A. probably stable, although western population perhaps fewer than 600 pairs (inc. Mexico). In San Diego, CA, preys on chicks of the endangered Snowy Plover and the Least Tern, hence Gull-billed Terns have been shot in attempts to adjust the "balance" of nature.

SANDWICH TERN *Thalasseus sandvicensis*   SATE ▦ 1

This medium-size tern nests in dense colonies. Its flight is strong and graceful, with wings held slightly crooked. Polytypic (3 ssp.; 2 in N.A.). L 13.5–14.5" (34–37 cm) WS 34–36" (86–91 cm) **IDENTIFICATION** Slender, yellow-tipped black bill, black legs, shaggy crest. **BREEDING ADULT:** Black cap solid mainly Mar.–June, spotted white late summer. **WINTER ADULT:** White forecrown, primaries molted mainly Aug.–Mar. **JUVENILE & FIRST-YEAR:** Like winter adult, but in fall juvenile has variable blackish subterminal marks on back, dark secondary bar. Black bill may have yellowish side patches (soon lost); develops yellow tip by spring. Complete molt (primaries Dec.–Aug.) produces first-summer plumage like

winter adult's. **SECOND-YEAR:** Second prebasic molt averages later than adult's, so second-summer's outer primaries fresher and paler. White-spotted forecrown of breeding spring birds may occur in all ages. **GEOGRAPHIC VARIATION** "Cayenne Tern" *(eurygnatha)* of Caribbean and S.A. has a yellow to yellow-orange bill; birds with partially yellow bills are common in Caribbean. The subspecies *acuflavida* breeds in N.A. and similar nominate *sandvicensis* in the Old World. **SIMILAR SPECIES** In West and on Gulf Coast, beware hybrids with Elegant Tern. **VOICE** Scratchy, penetrating *kree-ik* or *krrík;* slightly sharp, yelping *kehk;* and slightly reedy *ki-i wii-wii.* Begging young (into first-winter) has high, whistled *sree* or *sri-sree.* **STATUS & DISTRIBUTION** Europe, eastern N.A., eastern S.A. **BREEDING:** Common

in N.A., arriving colonies Apr.–early May, departing Aug.–Sept. **MIGRATION:** Spring mainly Apr.–May. Fall and postbreeding movement north on East Coast mainly Aug.–Oct. **VAGRANT:** Casual southern CA (where hybridized with Elegant), north to Atlantic Canada, inland to Great Lakes. "Cayenne Tern" accidental in NC. **POPULATION** Stable.

COMMON TERN *Sterna hirundo*   COTE ▪ 1

Fairly common along coasts, offshore, and inland, Common Tern is steady and graceful in flight, with smooth wingbeats, hovering briefly before plunge diving for fish. Unbroken tail streamers add 1.5 inches (4 cm) to breeding adult's length. Polytypic (3 ssp.; 2 in N.A.). L 11.5–12.5" (29–32 cm) WS 29.5–32.5" (75–83 cm) **IDENTIFICATION** Medium size; medium-long bill; forked tail. Breeding tail streamers fall about even with wing tips at rest. **BREEDING ADULT:** Solid black cap; black-tipped red bill (mostly dark on early spring migrants, rarely all-red in late summer); red legs. Pale smoky gray underparts; white rump and tail with dark outer web to outer rectrices. Pale gray upper wings usually have distinct dark wedge on trailing edge of primaries. **JUVENILE & FIRST-FALL:** Fresh juveniles have variable gingery wash and barring on back and upper-wing coverts (often fading by fall); pinkish red legs, bill base. Upper wing has contrasting blackish patagial bar, dusky gray secondary bar. Protracted complete molt in first winter (primaries Jan.–July) produces first-summer.

FIRST-SUMMER: Rare in N.A. Forecrown, underparts white; black bill. Note dark patagial bar, dark secondaries. **SECOND-SUMMER:** Some resemble first-summer, but lack worn juvenal outer primaries; bill usually reddish basally. Others resemble breeding adult, but forecrown white or spotted white, underparts paler, tail streamers shorter, bill duller. **GEOGRAPHIC VARIATION** Nominate in N.A., with red bill and legs. Breeding *longipennis* adults (East Asia) have darker plumage overall; bill and legs blackish.

SIMILAR SPECIES Arctic Tern is slightly smaller with shorter neck, slightly more rounded crown, narrower wings, shorter bill, much shorter legs (striking on birds at rest). Primaries are evenly translucent when backlit (only inner primaries look translucent on Common Tern); outer primaries have narrower blackish tips, never show a dark wedge like Common. In flight, Arctic Tern has narrower, proportionately longer wings; shorter head-and-neck projection forward of the wings; generally snappier, less floppy

wingbeats. Breeding adult Arctic has all-red bill, duskier gray underparts, tail streamers that project beyond wings at rest. Juvenile Arctic Tern has a less contrasting dark patagial bar, which on upper wing, blends into gray coverts and whitish secondaries. First-summer Arctic best told by structure. Also see Forster's and Roseate Terns; compare *longipennis* subspecies with Aleutian Tern (see p. 317).

VOICE Calls include a sharp *kik* and *kik-kik;* a grating *krrrih;* slightly drawn-out, disyllabic *eeeahrr;* also a rapidly repeated, grating *kehrr kehrr kehrr* in breeding-season chases.

STATUS & DISTRIBUTION Holarctic breeder. **BREEDING:** Common in N.A., arrives at colonies Apr.–May, departs Aug.–Sept. **MIGRATION:** Mainly Apr.–May, July–Oct. Subspecies *longipennis* rare (mainly spring) on islands of western AK. **WINTER:** Off coasts from Middle America to S.A. Very rare Gulf Coast, southern CA. The subspecies *longipennis* winters in western Pacific south to Australia. **POPULATION** Historical fluctuating trends difficult to assess.

red bill with blackish tip

breeding adult
hirundo

dark carpal bar

1st summer
hirundo

grayish below

red legs

no tail projection past wing tips

blackish bill

2nd summer
hirundo

extensive dark gray below

blackish legs

breeding adult
longipennis

juvenile
hirundo

dark gray secondaries

1st fall
hirundo

dark in outer primaries, often present as a wedge, particularly evident in late summer and early fall

breeding adult
hirundo

ARCTIC TERN *Sterna paradisaea*   ARTE ▪ 1

This medium-small, northern-breeding tern winters in Antarctic waters (see map) and migrates mainly well offshore, with birds rarely seen on coasts and inland. The Arctic's flight is quick and graceful, with snappy wingbeats. It hovers briefly before plunge diving for fish. Unbroken tail

streamers add 1.5 to 2 inches (4–5 cm) to the length of breeding adults. Monotypic. L 11.8–13" (30–33 cm) WS 30–33" (76–84 cm) **IDENTIFICATION** Medium-small tern with a fairly short and slender bill, deeply forked tail, and notably short legs. Breeding adult's tail streamers

project well beyond wing tips at rest. **BREEDING ADULT:** Solid black cap; red bill and legs. Smoky gray underparts may offset white cheeks; has white rump and tail with dark outer webs to outer rectrices. Pale gray upper wings with translucent primaries, narrow dark trailing edge to outer primaries. **JUVENILE:** Fresh juveniles have variable brown wash and barring on back and upperwing coverts (often fading by fall). Upper wing has contrasting dark gray patagial bar and whitish secondaries and inner primaries. Protracted complete molt in first winter (primaries Jan.–May) produces first-summer plumage. **FIRST-SUMMER:** Forecrown and underparts white, or the underparts have gray smudging;

upper wing has dark gray patagial bar; bill black to dull reddish.

SIMILAR SPECIES See Common Tern; see also Aleutian, Roseate, and Forster's Terns. Best distinguishing features for a perched Arctic are its relatively small bill and very short legs. On flying birds, note Arctic's short neck, its long and narrow wings, its evenly translucent primaries with narrow blackish tips, and its generally snappy flight.

VOICE Calls include a high, clipped *kiip;* a shrill grating *keeahr,* higher and drier than Common's; and a shrill, rapid-paced *ki-ki-kehrr* by scolding birds at colony.

STATUS & DISTRIBUTION Holarctic breeder. **BREEDING:** Common in N.A.,

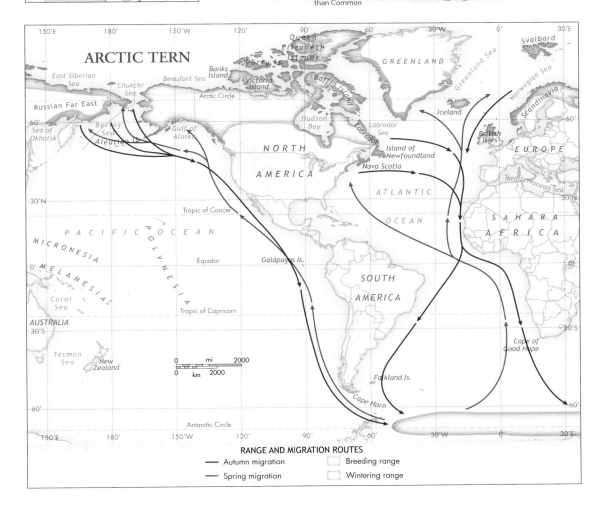

adults in flight have less head projection and longer tail projection than Common

razor-thin black border

juvenile

white secondaries

breeding adult

breeding adult slightly darker and more extensively gray below than Common

1st summer

short bloodred bill

juvenile

breeding adult

shorter legs than Common

ARCTIC TERN

RANGE AND MIGRATION ROUTES
— Autumn migration
— Spring migration
☐ Breeding range
☐ Wintering range

arriving at colonies from mid-May (New England, southern AK) to late June (high Arctic), departing late July–Aug. **MIGRATION:** Mainly late Apr.–early June, late July–mid-Oct.

Mainly casual inland. **WINTER:** Antarctic pack ice and adjacent waters, where rich food enables rapid complete molt. **POPULATION** Recent declines in many areas, from Alaska to New England,

perhaps linked to food shortages in northern oceans and reflecting the combination of changing ocean conditions and human over-exploitation of fisheries.

FORSTER'S TERN *Sterna forsteri*   FOTE   ▪   1

This medium-size tern is rare offshore, but a familiar sight in interior and coastal habitats. Its flight is steady and graceful with strong, smooth wingbeats, hovering briefly before plunge diving for fish. Unbroken tail streamers add 2.5 inches (6 cm) to breeding adult's length. Monotypic. L 12.5–14" (32–36 cm) WS 30–33" (76–84 cm)
IDENTIFICATION Slightly bulkier, heavier billed, and longer legged than Common. It is the only medium-size "white tern" likely seen molting its outer primaries in N.A. Breeding tail streamers project well beyond wing tips at rest (but often broken in mid-summer). **BREEDING ADULT:** Solid black cap (lost by late Aug.), orange-red bill with large black tip, orange-red legs. White underparts, contrasting pale-gray upperparts; pale-gray tail has white outer edges. **WINTER ADULT:** Note diagnostic bold black auricular mask with limited dusky wash on nape; blackish bill. Primaries molted July–Nov. **JUVENILE & FIRST-WINTER:** Resembles winter adult, but fresh juvenile has gingery wash and barring on head, back; first-winter has dark tertials, duskier upsides to primaries, duskier legs than adult. In first-summer, legs brighter orange-red; bill can be orange-red with blackish culmen and tip. First-summer's protracted complete molt (primaries Apr.–Sept.) produces second-winter plumage. **SECOND-YEAR:** Resembles adult, but some second-summers have white-spotted forecrown; outer primaries darker and more worn than on adults, often forming an upper-wing wedge recalling Common. **SIMILAR SPECIES** Common Tern is slightly smaller and slimmer with more slender bill and appreciably shorter legs; all Common Tern

plumages have white tail with contrasting dark outer webs to outermost rectrices. Breeding adult Common is pale gray below; tail streamer tips fall about equal with, or shorter than, wing tips at rest; deeper red bill with smaller black tip; upper wing often shows dark wedge or contrast on trailing edge of duller primaries (but beware second-summer Forster's). Juvenile and first-summer Commons have dark patagial bar on upper wing; black partial cap extends solidly around nape. Arctic Tern is smaller with shorter neck, smaller bill, short legs, relatively longer and narrower wings, and quicker, more clipped wingbeats; all Arctic plumages have white tail, contrasting dark outer webs to outermost rectrices. Breeding adult Arctic is smoky gray below, all-red bill, primaries translucent but not silvery above. Juvenile Arctic Tern has dark patagial bar on upper wing; black partial cap extends solidly around nape. First-summer Arctic Tern has fresh primaries (replaced in winter); cap extends solidly around nape. Also see Roseate Tern.
VOICE Hard, clipped dry *kik* or *krik;* slightly grating, shrill *krrih* or *kyiih*

and *kyerr kyerr;* in breeding-season chases, gruffer, slightly shrill rasping series of *zzhi-zzhi-zzhi*.
STATUS & DISTRIBUTION N.A. south to northern Mexico. **BREEDING:** Common, arrives at southern breeding sites Apr., northern sites (Canada, Great Lakes) mainly late Apr.–May. Departs northern breeding areas through Sept. **MIGRATION & DISPERSAL:** Fall migration mainly Aug.–Oct., ranging north to Maritimes. **WINTER:** Southern U.S. south to Middle America and Caribbean. **VAGRANT:** Increasingly detected in Europe since 1980s.
POPULATION Apparently fairly stable.

ROSEATE TERN *Sterna dougallii*   ROST   ▪   2

This tern feeds offshore; flies with fairly stiff, almost hurried wingbeats; is distinct from other like-size terns. Unbroken tail streamers add 2 inches

(5 cm) to breeding adult's length. Polytypic (5 ssp.; nominate in N.A.). L 12–13" (30–33 cm) WS 26–28" (66–71 cm)

IDENTIFICATION Relatively short wings; long tail. Breeding tail streamers project well beyond wing tips at rest. Bill black most of year. **BREEDING ADULT:** Solid

black cap; bill develops red basally in summer, can be red with black tip by August; rosy underparts flush usually subdued. Dark primary wedge visible at rest. **JUVENILE & FIRST-SUMMER:** Juvenile has variable blackish subterminal marks on back and tail (recalls juvenile Sandwich); forehead often dusky. Protracted complete molt produces first-summer plumage with white forehead, pale gray upperparts, dark patagial bar.
SIMILAR SPECIES Breeding adult in flight quite distinct from Arctic, Common, and Forster's Terns. Note Roseate's dark upperwing wedge on outer few primaries, lack of dark trailing edge to primaries, whitish underparts. Juvenile

patterned like Sandwich. First-summer Roseate told from first-summer Common by lack of dark secondary bar, finer bill.
VOICE A scratchy *krrízzik* or *kír-rik*, often doubled, suggesting Sandwich; a rasping *rrahk* or *ahrrr;* a mellow *ch-dik* or *ch-weet;* and a chippering *cheut cheut.*
STATUS & DISTRIBUTION Atlantic, Indian, western Pacific Oceans. **BREEDING:** Uncommon and local in N.A., arriving colonies Apr.–May. Departs breeding areas through Sept. **MIGRATION:** Rare records Atlantic coast

south of NJ, mostly off NC (late May, late Aug.–Sept.). **WINTER:** Mainly northeast coast of S.A.
POPULATION Atlantic subspecies declining. Northeast breeding population listed as endangered in U.S. (about 3,500 breeding pairs) and threatened in Canada (about 125 breeding pairs).

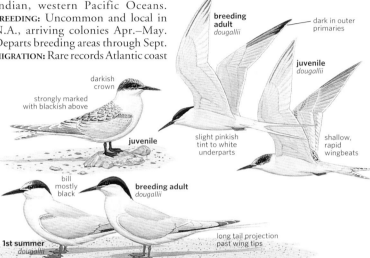

breeding adult
dougallii

dark in outer primaries

juvenile
dougallii

darkish crown

strongly marked with blackish above

juvenile

slight pinkish tint to white underparts

shallow, rapid wingbeats

bill mostly black

breeding adult
dougallii

1st summer
dougallii

long tail projection past wing tips

CASPIAN TERN *Hydroprogne caspia* CATE ■ 1

This, the largest tern in the world, is widespread in interior and coastal habitats but is rarely seen offshore. Its graceful flight is steady and powerful, with fairly shallow wingbeats and wings not held strongly crooked. Monotypic. L 20–22" (51–56 cm) WS 46–51" (117–130 cm)
IDENTIFICATION The heavy-bodied Caspian is notably thickset and lacks tail streamers in its breeding plumage. Very stout red bill, with a black distal mark and pale tip. Extensive blackish panel on underside of outer primaries is diagnostic of all plumages. **BREEDING ADULT:** Black cap becomes streaked white in fall. **WINTER ADULT:** Note densely flecked black-and-white crown, which blends into a broad black auricular mask; primaries molted Aug.–Feb. **JUVENILE & FIRST-YEAR:** Resembles winter adult, but juvenile fresh-plumaged in fall, with variable blackish subterminal marks on back and tail. Paler bill, orange to orange-red, with a dark subterminal mark; legs often yellowish, becoming

dark by winter. A protracted complete molt (primaries molted Jan.–Aug.) produces first-summer plumage much like winter adult's. **SECOND-YEAR:** Second prebasic molt averages later than in adults, so outer primaries fresher and paler in second-summers than in adults; forecrown often streaked whitish.
SIMILAR SPECIES Distinctive. The smaller and more lightly built Royal Tern is a coastal species; its more slender (but still stout) bill is orange, without a dark tip (rarely orange-red in spring). In flight, note Royal's narrower wings, often held more crooked; its deeper wingbeats; a mostly whitish underside to primaries; and a more deeply forked tail (with streamers

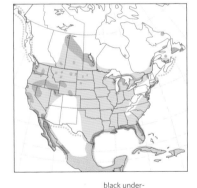

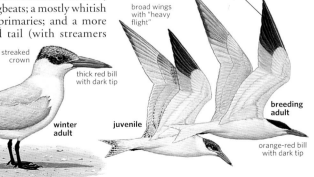

streaked crown

thick red bill with dark tip

winter adult

juvenile

black under-surface to primaries

broad wings with "heavy flight"

breeding adult

orange-red bill with dark tip

in breeding plumage). Winter and immature Royal Terns have an extensively white ("bald") forecrown never shown by Caspian.

VOICE Adult has a distinctive, loud rasping *rraah* or *ahhrr* and a drawn-out, upslurred *rrah-ah-ahr* in chases and when diving at colony intruders.

Begging young (into first-winter) give a high whistled *ssíuuh*.

STATUS & DISTRIBUTION Worldwide except S.A. **BREEDING:** Common in N.A.; arrives at southern nesting sites Mar.–Apr. and northern sites (Canada, Great Lakes) mainly mid-Apr.–May, departs mainly Aug.–

Sept. **MIGRATION:** Mainly Mar.–May and late July–Oct.; stragglers occur through Nov. **WINTER:** Uncommon southern CA.

POPULATION Stable or increasing. Major increases on Pacific coast, in interior West, and in Great Lakes region have taken place in last 50 years.

ELEGANT TERN *Thalasseus elegans*   ELTE ▪ 1

This medium-size Pacific-coast tern nests in dense colonies and disperses north in late summer after breeding, with largest movements in warm-water years. Its flight is strong and graceful, with the wings held somewhat crooked. Unbroken tail streamers add 1.5 to 2 inches (4–5 cm) to the length of breeding adults. Monotypic.

L 14.5–16" (37–41 cm) WS 37–39.5" (94–100 cm)
IDENTIFICATION Note slender bill, which sometimes looks almost droop-tipped, and long, shaggy crest. All plumages can be flushed strongly pink on underparts. **BREEDING ADULT:**

Black cap solid Jan.–June; spotted white in summer. Orange bill varies from bright orange-red with paler tip to uniformly mustard-yellow. Black legs rarely mottled or solidly orange. **WINTER ADULT:** Extensive white forecrown rarely has thin white postocular crescent separating eye from black crest; primaries molted July–January. **JUVENILE & FIRST-YEAR:** Resembles winter adult, but juvenile fresh plumaged in early fall, with

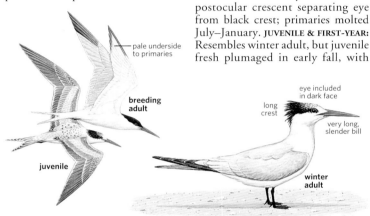

Identifying Royal versus Elegant Terns

Although similar in plumage, these orange-billed terns are not difficult to identify when other terns are present for comparison: Royal and Caspian Terns standing together don't look greatly different in size, and both have stout bills and black caps that do not droop down the nape. But when an Elegant is with a Caspian, it is strikingly smaller, standing little more than half the Caspian's height. With Royal and Elegant Terns together, it is clear the Royal is larger and stands taller.

Bill shape is probably the most important character. Although the Royal's and Elegant's bills are the same length, the Royal's is much thicker, and its gonydeal angle is farther out. The Royal's large, stout bill recalls the Caspian Tern's. The Elegant, however, has a slender bill that varies from being longer, with a

Royal Tern, winter adult (FL, Feb.)

Elegant Tern, winter adult (CA, July)

slightly drooped tip, to being notably shorter and straighter, without any droop. This variation is due mainly to sex: Males average longer bills. The Royal's bill is uniform orange—paler orange on first-cycle birds and orange-red on some spring adults. The Elegant's bill varies in color from reddish orange with a yellowish tip, to uniform orange and bright mustard yellow.

Other pointers include the bright-pink flush that Elegants of any age can have on their underparts (which are lacking or faint on the Royal); Elegant's longer crest (beware that apparent crest length can vary with posture); and its winter head pattern. The Royal Tern typically has a distinct white break separating the eye from the black postocular, whereas any break on the Elegant Tern is usually small and inconspicuous. ∎

dark gray centers to greater coverts and tertials, dark subterminal tail marks, and dark secondary bar. Legs are often yellowish, usually becoming dark by winter. Protracted complete molt (primaries Oct.–July) produces first-summer plumage much like winter adult's. **SECOND-YEAR:** Second prebasic molt averages later than adult's, so outer primaries fresher and paler in second-summer than in adult. White spotting on forecrown of breeding-plumaged spring birds may occur in all ages. **SIMILAR SPECIES** See sidebar opposite for separation from Royals. Some

Elegants' bills have dark smudging; this pattern rarely suggests Sandwich. These may be hybrids with Sandwich or normal variants of Elegant. **VOICE** A slightly screechy *rreeah* or *rreahk,* and *krreéh* or *krreíhr;* and a rough, grating *ehrrk.* Calls scratchy and grating; given incessantly by flocks of hundreds. Similar to Sandwich's, but higher and screechier than Royal's. Begging young (into first-winter) give high, piping *sii, siip-siip.* **STATUS & DISTRIBUTION** Breeds northwestern Mexico and southern CA. Winters Pacific coast of S.A. **BREEDING:** Local southern CA, Mar.–July.

MIGRATION & DISPERSAL: Common postbreeding visitor (mainly July–Oct.) to central CA, irregularly to southernmost BC. **WINTER:** Casual southern CA. **VAGRANT:** Casual inland in Southwest. Accidental on Atlantic and Gulf Coasts (where it has hybridized with Sandwich in FL), and Niagara River, NY/ON (found Nov. 20, 2013). **POPULATION** Increase on U.S. Pacific coast since 1970s may reflect changing ocean conditions rather than population change. Colonized in southern CA since 1950s. Considered near threatened by BirdLife International.

ROYAL TERN *Thalasseus maximus*   ROYT ▪ 1

A large tern of southern coasts, the Royal is rarely seen inland. It nests in dense colonies, often with other species. Its flight is strong and graceful, the wings held somewhat crooked. Unbroken tail streamers add 2 inches (5 cm) to the length of breeding adults. Polytypic (2 ssp.; nominate in N.A.). L 17–19" (43–48 cm) WS 41–45" (104–114 cm) **IDENTIFICATION** Note stout orange bill of all ages. **BREEDING ADULT:** Black cap solid in Jan.–May, becoming spotted white in early summer. Bill deep orange, rarely orange-red; legs black. **WINTER ADULT:** Extensive white forecrown often has a white postocular crescent separating eye from black crest; primaries molted July–Feb. **JUVENILE & FIRST-YEAR:** Resembles winter adult, but juvenile is fresh-plumaged in early fall, with dark gray centers to greater coverts and tertials, dark subterminal tail marks, dark secondary bar. Bill paler orange; legs often yellowish, becoming dark by winter. Protracted complete molt (primaries molted Nov.–Aug.) produces first summer plumage much like

winter adult's. **SECOND-YEAR:** Second prebasic molt averages later than in adults, so in second-summer plumage, outer primaries are fresher and paler than adults'. White spotting on forecrown of breeding-plumaged spring birds may occur in all ages. **GEOGRAPHIC VARIATION** California populations average larger; they breed and molt earlier, and are mostly resident or short-distance migrants. Atlantic coast populations average smaller, breed and molt later, and are medium- to long-distance migrants. **SIMILAR SPECIES** See sidebar opposite for separation from Elegant Tern. Caspian Tern is larger, more heavily built, and broader winged, with a stouter red, dark-tipped bill; in flight, note Caspian's diagnostic dark underside to outer primaries. Winter and immature Caspians have dark-streaked crowns and never show extensive white forecrown typical of Royal Terns. **VOICE** A fairly deep grating *ehrreh* and *rreh'k* or *rreh-eh,* lower and less shrieky than calls of Elegant and Sandwich Terns; a higher and shriekier *krriéh* or *rriehk;* a yelping or clucking *krehk* and *kehk;* and a more laughing *kweh-eh-eh.*

Begging young (into first-winter) give a high whistled *see-ip.* **STATUS & DISTRIBUTION** Mainly New World, also W. Africa. **BREEDING:** Common in N.A. **MIGRATION & DISPERSAL:** Spring migration mainly late Feb.–Apr.; postbreeding northward movement and fall migration on East Coast mainly late June–Nov., regularly to New England, casually Atlantic Canada. **VAGRANT:** Casual in interior and to northern CA. **POPULATION** Generally holding constant. Formerly common into early 1900s as nonbreeding visitor (mainly Sept.–Mar.) north to San Francisco Bay (from Mexican colonies). Conversely, has colonized southern CA as a breeding bird since 1959.

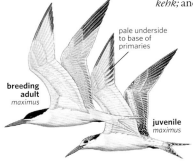

pale underside to base of primaries

breeding adult *maximus*

juvenile *maximus*

bill entirely orange

dark eye stands out in whitish face

short crest

winter adult *maximus*

juvenile *maximus*

LARGE-BILLED TERN *Phaetusa simplex*   LBTE ■ 5

This striking, aptly named species nests in freshwater habitats of South America and moves to coastal regions there in the nonbreeding season.

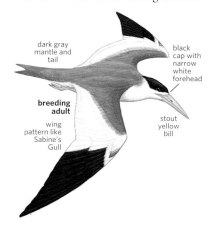

dark gray mantle and tail

black cap with narrow white forehead

breeding adult

wing pattern like Sabine's Gull

stout yellow bill

Its occurrence in North America is accidental, although vagrants have also been found in Bermuda and the Caribbean. They have a graceful flight and feed by plunge diving and hawking for insects in flight. Polytypic (2 ssp.). L 14–15" (36–38 cm) WS 36–38" (91–97 cm)

IDENTIFICATION Proportionately long bill is stout and yellow; legs are medium length and olive to greenish yellow; relatively short tail is shallowly forked. **BREEDING ADULT:** Black cap separated from bill base by a white forehead band. At rest, upperparts are dusky gray, with black primaries and a broad white band along bottom edge of wing. In flight, upperparts display a striking pattern recalling Sabine's Gull: bold white triangular panels on wings

set off by dusky gray lesser upperwing coverts and black primaries and greater primary coverts (inner primaries are whitish). Underwings are mostly white, with dark wing tips. **NONBREEDING ADULT:** Resembles breeding adult, but forecrown mottled white. **JUVENILE & FIRST-YEAR:** Resembles adult, but back and upperwing coverts mottled brownish, cap mottled whitish, and bill is duller yellowish.

SIMILAR SPECIES Should be unmistakable. More likely to be confused (at a distance) with Sabine's Gull than with any North American tern.

VOICE Vagrants are likely to be silent.

STATUS & DISTRIBUTION South American freshwater species. **VAGRANT:** Accidental, three records in late spring and summer from IL, OH, and NJ.

This pantropical genus (3 sp.) is sometimes treated as a separate family, Rynchopidae. Resembling large, angular terns, they are distinguished by long, laterally compressed bills with projecting lower mandibles. They feed in flight, mostly at night, by skimming with an open bill, then snapping it shut on contact with fish and crustaceans.

BLACK SKIMMER *Rynchops niger*   BLSK ■ 1

This species favors coasts with sandy beaches and lagoons, but it has also colonized the Salton Sea. Its flight is strong and buoyant, with smooth wingbeats mainly above the body plane. Flocks often fly in fairly tight formation, at times wheeling like shorebirds. Males are larger than females and often noticeable in the

field (male bills average 10–15 percent longer). Polytypic (3 ssp.; nominate in N.A.). L 17–18" (43–46 cm) WS 45–49" (114–124 cm)

IDENTIFICATION Striking. Long, laterally compressed bill; fairly short legs; very long wings; fairly short, forked tail. **BREEDING ADULT:** Crown, hind neck, and upperparts solidly black, contrasting sharply with white forehead, lores, foreneck, and underparts. Black upper wings have a white trailing edge to secondaries and inner primaries; tail is mostly white with a black central stripe. Whitish underwings grade to dusky on remiges, with blacker wing tips. Black-tipped, bright red bill; red legs. Prebasic molt starts in late summer on the breeding grounds

(when a few inner primaries may be replaced before suspending molt for migration) or on the winter grounds. Primary molt completes Mar.–May, overlapping with prealternate molt of hind neck feathers. **WINTER ADULT:** Resembles breeding adult, but white hind neck, sometimes with a little dusky mottling, and bleached upperparts can look browner in early winter. **JUVENILE & FIRST-YEAR:** Juvenile crown and hind neck heavily streaked buffy and whitish; back and upperwing coverts blackish brown; broad buff to whitish edgings create a bold, scaly

juvenile
niger

white fringes above

very long wings

winter adults
niger

pale collar in winter

breeding adult
niger

"skimming"

striking black-and-white plumage

pattern. Bill duller, less extensive red basally; legs paler, flesh to pale orange-red. Protracted, complete molt starts Sept.–Dec. (primaries molted mainly Jan.–Aug.); produces plumage resembling winter adult's by end of the first summer. Crown is dark sooty-brown to blackish with paler feather edgings; hind neck is white or mottled blackish; bill and legs average paler than on adults.
GEOGRAPHIC VARIATION Birds in N.A. are of the nominate subspecies, with white underwings and an extensively white tail. The nomadic Amazonian-

breeding *cinerascens,* a potential vagrant to N.A., has dusky-gray underwings, reduced white on trailing edge of wings, and a dark tail.
SIMILAR SPECIES Unmistakable.
VOICE A nasal, slightly hollow laughing *kyuh* or *kwuh* and a disyllabic *k'nuk* or *k'wuk,* with calling flocks producing chuckling choruses at times; alarm call a more drawn-out *aaawh.* Juvenile call higher pitched and squawkier than adult's.
STATUS & DISTRIBUTION Warmer regions of the Americas. **BREEDING:** Fairly common to common but local,

arriving at northern colonies late Apr., departing Aug.–Sept. **MIGRATION:** Mainly Mar.–Apr. and Sept.–Nov. **WINTER:** Southern U.S. to S.A. **DISPERSAL & VAGRANT:** Casual inland in coastal states, also (mainly summer) casual in the Southwest, Great Plains, Great Lakes region, and (mainly after storms) Atlantic Canada.
POPULATION Atlantic, Gulf Coast colonies fluctuate greatly: was considered declining in 1970s; some evidence of recent stabilization. West Coast population has increased markedly in last 30 years.

PIGEONS AND DOVES Family Columbidae

Inca Dove (TX, Mar.)

Species in this family are familiar to birders and non-birders alike. The larger species are usually called pigeons, the smaller ones doves; there is no scientific or taxonomic distinction between the two groups. Pigeons and doves feed chiefly on grains, seeds, and fruits. Nests are usually flimsy structures, and most clutches are limited to two white eggs.
Structure This group varies in size from the huge crowned-pigeons of New Guinea to the towhee-size ground-doves of the Americas. Most are plump with short, round-tipped wings and have tails varying from short and square-tipped to long and pointed. All have small rounded heads, and virtually all have short legs and similarly shaped short bills.

Behavior Pigeons and doves are strong, fast fliers (although a few species fill a terrestrial role in dense jungles and fly less frequently); only a few species are long-distance migrants, but many disperse randomly in search of food. While most eat various forms of vegetable matter—primarily seeds and fruit, but also leaves and flowers—some species have been known to eat invertebrates, such as worms or insect larvae. The species covered herein feed primarily on the ground; many species in other parts of the world are arboreal. Their nest-building skills seem poor, but such untidy nests, usually appearing to be little more than several sticks connecting tree branches, might allow them to rebuild quickly in the event of a nesting failure.

Plumage Pigeons and doves vary in color from the near patternless grays of some Old World pigeons to the highly patterned oranges and greens of fruit-doves. Many species show iridescence on the hind neck; many also show a pinkish suffusion to the underparts. Most juveniles have pale-fringed feathers and lack the neck markings of adults. Some species have spectacular head plumage, such as elaborate crests and long, pointed feathers.

Distribution Except for Antarctica, pigeons and doves are present on all continents, with the greatest variety in the tropics. Most are found in woodlands varying from open to dense, but members of this family have evolved to exploit habitats from rather barren deserts to sea cliffs and areas near the tree line in high mountain ranges.

Taxonomy As with all large families, the taxonomy is in a constant state of change, evident from the fact that the AOU as recently as 2003 recognized the "splitting" of the American pigeons, *Patagioenas,* from the Old World pigeons, *Columba.* About 310 species in 40 genera are recognized worldwide, with 21 species in eight genera recorded in the United States and Canada.

Conservation Almost a third of the world's pigeons and doves are threatened to some degree, with a few already extinct. Predators, particularly rats, threaten island species, such as some of the fruit-doves or large terrestrial pigeons, which evolved with little competition and few natural enemies. Wholesale clearing and fragmentation of forests eliminates cover and food for many species, though a few may benefit. Overhunting, for food and for sport, has played a major role in some extinctions, such as the extermination of the Passenger Pigeon. Surprisingly, trapping for the pet trade appears to have little or no impact.

OLD WORLD PIGEONS Genus *Columba*

ROCK PIGEON *Columba livia*   ROPI ■ 1

This highly variable city pigeon is familiar to all urban dwellers. Multicolored birds were developed over centuries of near domestication. Polytypic (12 ssp.; nominate in N.A.). L 12.5" (32 cm)

IDENTIFICATION A medium-size compact pigeon with long wings and a short tail. Birds most closely resembling their wild ancestors are gray with head and neck darker than back, and a prominent white rump. Black tips on greater coverts and secondaries form bold black bars on inner wing, and there is a broad black terminal band on tail. **ADULT MALE:** Metallic green and purple iridescence on neck and breast; iris orange to red; orbital skin blue-gray; bill grayish black; and feet dark red. **ADULT FEMALE:** Like male, but iridescence on neck and breast more restricted and subdued. **JUVENILE:** Generally browner, lacks iridescence; orbital skin and feet gray. **VOICE CALL:** Gives a soft *coo-cuk-cuk-cuk-cooo.*

STATUS & DISTRIBUTION The Rock Pigeon was introduced from Europe by early settlers; it is now widespread and common throughout the U.S. and southern Canada, particularly in urban settings. Gregarious and forming large flocks, it feeds on handouts and grains during the day in city parks and open fields and roosts on buildings at night. Flocks or otherwise displaced pigeons can be found far from civilization. **BREEDING:** Nest is loosely constructed of twigs and leaves, primarily on structures such as window ledges, bridges, and in barns; lays two white eggs. **POPULATION** Primarily associated with human development and dependent on people for food and shelter.

in flight, often gives a loud wing flapping noise

ancestral natural coloration

white rump

color variations
many other variations are seen

AMERICAN PIGEONS Genus *Patagioenas*

SCALY-NAPED PIGEON *Patagioenas squamosa* SNPI ▪ 5

orange orbital ring

dark maroon head and neck, but in most lights appears all dark

adult ♂

A large, dark pigeon of the West Indies, the Scaly-Naped Pigeon has been recorded only twice (old historical records in North America); larger than a Rock Pigeon. Monotypic. L 13.8" (35 cm)

IDENTIFICATION This species appears entirely dark slate-gray, or even blackish, when seen perched or in flight from a distance; there are no obvious markings on wings or tail. **ADULT MALE:** Head and upper breast dark maroon; feathers on sides of neck are more reddish, tipped black, forming diagonal lines and creating the scaled appearance that gives this species its name. Iris orange-red; orbital skin orange; bill dark red with pale yellowish tip; and feet dark red. **ADULT FEMALE:** Same as male, but slightly subdued. **JUVENILE:** Duller with rusty-brown fringes on scapulars.

SIMILAR SPECIES White-crowned Pigeon is slightly smaller, a little darker, and normally shows white on head; however, white on young birds is reduced and washed with grayish brown, so could be overlooked. Red-billed Pigeon is similar, also with a red-based, two-toned bill. But Red-billed lacks any trace of dark-edged nape feathers, and ranges of the two species are widely separated.

VOICE CALL: An emphatic *cruu, cruu-cru-cruuu,* the first syllable soft with a pause before the last three syllables, which sound like "who are you?"

STATUS & DISTRIBUTION Resident, primarily arboreal, where found most commonly in primary forest. Found throughout most of the West Indies; notably absent from the Bahamas and Jamaica; now uncommon to rare on Cuba. Occurs individually or in small flocks. **BREEDING:** Nest is loosely built of twigs lined with grass and found in trees, including palms, but occasionally on the ground. Usually lays one white egg. **VAGRANT:** Single birds collected at Key West, FL, Oct. 24, 1896, and May 6, 1929, are the only two recorded in N.A.

POPULATION While some have adapted to more urban areas, the overall population is declining, with threats of local extirpations, due to deforestation and intensive hunting.

WHITE-CROWNED PIGEON *Patagioenas leucocephala* WCPI ▪ 2

White-crowned Pigeon is the large, dark pigeon seen in the Florida Everglades and Florida Keys. Flocks of varying sizes commute daily from nesting colonies in coastal mangroves to feed inland, at times flying many miles. Monotypic. L 13.5" (34 cm)

IDENTIFICATION A large, square-tailed slate black pigeon with a conspicuous white crown. **ADULT MALE:** Crown is pure white; sides of neck are iridescent green merging into an iridescent purple on hind neck. Iris is whitish, with pale blue-gray orbital skin. Bill is dark reddish with a white tip; feet are dull red. **ADULT FEMALE:** Like male, but slightly duller. **JUVENILE:** Browner; rusty brown fringes on coverts and scapulars. White on crown is reduced and washed with grayish brown; it can be difficult to see at times. Iris is brownish, with brownish gray orbital skin.

SIMILAR SPECIES Some Rock Pigeons can appear black, but will normally show white on rump and lack white on crown. Check all birds carefully for the slim possibility of a vagrant Scaly-naped Pigeon from the West Indies.

VOICE CALL: A loud, deep *coo-cura-coo,* or *whoo-ca-cooo* repeated several times; a soft *coo-crooo* is believed to be given from the nest.

STATUS & DISTRIBUTION Fairly common, but declining throughout the West Indies and the east coast of the Yucatán Peninsula. In Florida, a fairly common resident in the Keys and Everglades. Casual north to St. Lucie and Lee Counties. **BREEDING:** A colonial nester. Nest is a loose platform of twigs, normally in mangroves, but sometimes in dry scrub and trees, and occasionally on the ground; bears one or two white eggs. **VAGRANT:** A sight record for TX.

POPULATION This species is declining dramatically throughout much of the West Indies due to clearing of hardwood forests, severe overhunting, harvesting of nestlings for food, and introduced predators.

white crown

pale eye

red bill with pale tip

dark slaty gray body

♀

RED-BILLED PIGEON *Patagioenas flavirostris*   RBPI ■ 2

The Red-billed Pigeon is a large, all-dark pigeon seen flying in pairs or in small groups along the Rio Grande in southern Texas, but in recent decades, this is a diminishing sight. Polytypic (4 ssp.; nominate in N.A.). L 14.5" (37 cm)

IDENTIFICATION The species generally appears entirely dark when seen from any distance. **ADULT MALE:** Head, neck, most of underparts, and lesser wing coverts are dark maroon; rest of plumage is dark blue-gray, with a blackish tail. Paler gray area on wing coverts forms a bar that might be visible in flight in good light. Iris is reddish orange, with bordering red orbital ring. Bill is pale yellow,

or whitish with dark red at base. Feet are dark red. **ADULT FEMALE:** Female's coloring is similar to male's but generally duller. **JUVENILE:** Like female but its maroon feathering is rustier and paler.

SIMILAR SPECIES Within its expected area of occurrence, there are no likely identification problems. Other columbids do not have rather uniform blue-gray plumage, but have bars on secondaries or tail. In addition, they lack distinctly two-toned bill of Red-billed Pigeon.

VOICE CALL: A distinctive long, high-pitched *cooooo* followed by two to five loud *up-cup-a-coo*'s given in the early spring and summer; also a

single, swelling *whoo* often repeated several times.

STATUS & DISTRIBUTION Widespread throughout the lowlands of Mexico and C.A. Locally uncommon to rare, and declining along the lower Rio Grande (mostly in western Hildago and Starr Counties, fewer upriver to northern Webb and southern Maverick Counties) in TX; rare and very local in winter. Casual north to Nueces and Kerr Cos. Arboreal; perches in tall trees above brushy understory; seldom comes to the ground except to drink. **BREEDING:** Nest is a loose platform of twigs built well above the ground; normally lays one white egg.

POPULATION Deforestation along the Rio Grande in the 1920s greatly reduced the Red-Billed Pigeon's numbers there, and the species is now protected in TX.

pale eye with red orbital ring

red bill with pale yellow tip

maroon head, neck, and wing coverts; otherwise slate gray

flavirostris

flavirostris

BAND-TAILED PIGEON *Patagioenas fasciata*   BTPI ■ 1

The Band-tailed is a large, heavily built gray pigeon that frequents the forests and woodlands of the West. It is most often seen in flocks of varying sizes in rapid flight. Its

white band at top of nape

yellow-based bill

wings make a loud clapping sound when flushed. Polytypic (8 ssp.; 2 in N.A.). L 14.5" (37 cm)

IDENTIFICATION Larger, with a longer tail than a Rock Pigeon; blue-gray on upper parts with contrasting blackish gray flight feathers; a blackish gray tail with a broad, pale gray terminal band. Paler gray greater coverts show as a broad wing stripe when in flight. **ADULT MALE:** Gray on head, and breast tinged pinkish; narrow white half-collar across upper hind neck, with iridescent greenish below. Narrow orbital skin purplish; bill yellow with a black tip; and feet yellow. **ADULT FEMALE:** Like male, but pink color somewhat subdued, and with less iridescent green. **JUVENILE:** Paler than adults, with narrow whitish fringes on breast and coverts; half-collar reduced or obscured.

GEOGRAPHIC VARIATION Nominate

subspecies *fasciata* breeds in the Southwest from UT and CO south into Mexico; *monilis* (often not recognized) breeds in the Pacific states from BC (uncommonly in southeast Alaska) to northern Baja California, Mexico. They are not separable in the field. Six other subspecies are found in Middle and South America, the southern group from Costa Rica south being distinctive with

rather long tail with broad pale tail band

a richer coloration overall and an all-yellow bill. **SIMILAR SPECIES** Rock Pigeons have blackish tails; most have black markings on wings and conspicuous white rumps; at close quarters, bill lacks yellow; and feet are reddish rather than yellow. **VOICE CALL:** A low-pitched *whoo-whoo* delivered several times. **STATUS & DISTRIBUTION** Locally common in low-altitude coniferous forests in the Pacific Northwest, and in oak or oak-conifer woodlands in the Southwest; presence dependent on availability of food; increasingly common in suburban gardens and

parks. **BREEDING:** Nest is a platform of twigs lined with grasses placed in a tree well above the ground; bears one white egg. **MIGRATION & WINTER:** Most birds breeding in the Southwest winter in Mexico, and most breeding in the Pacific Northwest move south into CA in winter. **VAGRANT:** Casual across southern Canada east to NS and New England; also along the Gulf Coast from TX to western FL. **POPULATION** The Pacific population was formerly threatened by over-hunting, but with the introduction of controls, the population is recovering.

OLD WORLD TURTLE-DOVES Genus *Streptopelia*

ORIENTAL TURTLE-DOVE *Streptopelia orientalis*   ORTD ■ 4

This large, heavy dove of Asia, appearing almost pigeonlike when in flight, has strayed to North America. Polytypic (6 ssp.; nominate in N.A.). L 13.5" (34 cm) **IDENTIFICATION** Generally ashy gray, but with warm brown scaly pattern above, formed by broad chestnut fringes on blackish tertials and wing coverts. Flight feathers blackish; underside of wings dark gray; and blunt-tipped tail blackish with broad gray terminal band. **ADULT MALE:** Crown pale gray; underparts grayish brown merging into gray on undertail coverts; conspicuous black-and-white stripes on sides of neck; iris orangish; orbital skin purplish; bill blackish with a trace of dark purple at base; feet dark reddish. **ADULT FEMALE:** As male, but slightly duller and browner on underparts. **JUVENILE:** Paler; fringes on tertials and coverts narrower; neck markings obscured. **GEOGRAPHIC VARIATION** The large and highly migratory nominate

subspecies *orientalis* (breeds central Siberia to Russian Far East, throughout Japan, the Himalaya, and southern India) has reached North America. **SIMILAR SPECIES** European Turtle-Dove is smaller, overall paler, and averages a paler terminal band on tail. **VOICE CALL:** Nominate subspecies gives a four-phase *deh-deh co-co*,

with the last two notes lower-pitched. **STATUS & DISTRIBUTION** Present throughout much of Asia; northern birds highly migratory, withdrawing to S.E. Asia and India in winter. **VAGRANT:** Casual to Aleutians and Bering Sea region in spring, summer, and fall (recorded north to St. Lawrence I.). Accidental YT, Vancouver I., BC, and CA (twice).

tail tips vary from whitish to pale gray

orientalis

rufous fringes give scaly pattern above

orientalis

orientalis

fringes more grayish below

black-and-white streaked neck patch

orientalis

gray rump

EUROPEAN TURTLE-DOVE *Streptopelia turtur*   EUTD ■ 5

orange eye

black-and-white stripes on neck

turtur

scapulars and wing coverts with black centers and orange-brown edges

From biblical verses in the Song of Songs to Shakespeare's poems, folk music, Christmas carols, and spirituals, turtle-doves have been celebrated as symbols of devotion. Through much of its range, it is also a heavily hunted species. Polytypic (4 ssp.). L 10" (26 cm) **IDENTIFICATION** In N.A. context, this slim dove is very distinctive.

ADULT: Vivid orange-brown fringes to scapulars and coverts; black-and-white-striped patch on side of neck. Pale orange eye, surrounded by red orbital skin, stands out starkly in pale gray head. In flight, wedge-shaped tail is strikingly patterned, with grayish brown coverts over black rectrices with broad white tips. **JUVENILE:** Colors far more muted than in adult; lacks neck patch of adult.

GEOGRAPHIC VARIATION Four subspecies recognized: The range of nominate *turtur* extends from Europe to western Siberia; three others occur from northern Africa to western China.
SIMILAR SPECIES In N.A., beware escaped doves and pigeons from aviary collections. Most similar to vagrant Oriental Turtle-Dove, recorded only in the West, but that species is larger overall, with a longer tail; adults have far less rufous in upperwing coverts than European.
VOICE SONG: A rather unmusical, insistent purring *trrrrrrrr,* repeated in series, with some shorter than others and some rising slightly in tone.
STATUS & DISTRIBUTION BREEDING: From the British Isles and Europe to northern Africa and southwestern Asia; winters in sub-Saharan Africa.
VAGRANT: Records of single birds from Monroe Co., FL (Apr. 9–11, 1990)—this record origin questioned by some; St. Pierre I., France (May 15–20, 2001); and Tuckernuck I., MA (July 19, 2001). Iceland has more than 200 records, most from autumn, and so Newfoundland would be a likely place for an autumn vagrant.
POPULATION Has declined in parts of Europe, especially the U.K.

AFRICAN COLLARED-DOVE *Streptopelia roseogrisea* AFCD ▓ EXOTIC

A domesticated form of the African Collared-Dove, also known as Ringed Turtle-Dove, can be encountered as an escaped cage bird almost anywhere. Monotypic. L 10.5" (27 cm)
IDENTIFICATION A very pale Mourning Dove-size bird with a blunt-ended tail. It has a whitish head and neck, and is pale buff above with an obvious black collar on hind neck. Whitish below, including undertail coverts. There are many color variations in adults; a pale buff variant is commonly encountered, but others, including peach-colored and rather uniform whitish birds, are also seen. Primaries are only slightly darker than rest of wing, providing a more uniform look. Tail is black at base, visible from below; black typically falls short of longest undertail coverts. Tail has an entirely white outer web on outermost rectrices, and a broad whitish terminal band.
SIMILAR SPECIES Eurasian Collared-Dove is superficially similar, but most birds can be identified with reasonable views. Given the huge range expansion of Eurasian Collared-Dove, combined with the generally poor ability of African Collared-Dove to survive in the wild, most encounters away from known African Collared-Dove populations will likely pertain to collared-doves. Beware of hybrids between the two species, which may show a mixture of plumage characters. Otherwise, African Collared-Doves are smaller and paler. Their primaries are paler, contrasting less with rest of wing and adjacent coverts. In addition, inner wings, visible in flight, are rather uniform in African Collared-Dove, versus contrasting gray secondaries of Eurasian Collared-Dove. Undertail coverts of African Collared-Dove are white, versus gray, and outer webs of outer rectrices are white in African Collared-Dove, versus black in Eurasian Collared-Dove. Black at base of tail, visible from below, falls short of the longest undertail coverts and is generally hard to see in African Collared-Dove, whereas it extends beyond the longest undertail coverts and is quite evident.
VOICE CALL: A rolling bisyllabic *kooeek-krrroooo,* noticeably softer than that of Eurasian Collared-Dove.

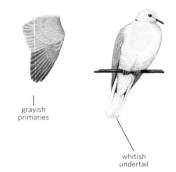
grayish primaries

whitish undertail

STATUS & DISTRIBUTION An escape that can be locally common. Small populations have persisted where fed, but they do not do well in the wild; no known viable wild populations are known in North America. They can be encountered well away from cities, but these are presumably birds that have recently escaped or been released. In its native range, a common species in the Sahel, from Senegal to the Red Sea, including southwestern Saudi Arabia and Yemen.

SPOTTED DOVE *Streptopelia chinensis* SPDO ▓ 2

This Asian dove was introduced into urban areas of southern California about 100 years ago. Polytypic (7 ssp.; nominate in N.A.). L 12" (31 cm)
IDENTIFICATION This is a large and dark dove with broad, rather round-tipped wings. Tail is long, dark, and broad at the end with white tips on outer three to four rectrices. **ADULT MALE:** Gray on top of head bleeds into face. Dark brown upperparts have feathers finely fringed with pale brown, and flight feathers are black. Underparts are a warm dark pinkish brown, merging into gray on undertail coverts. Upper wing in flight has gray secondaries; below, underside of wings are dark. Broad black collar has prominent white spots. Iris is reddish

chinensis
black-and-white spotted collar
chinensis
pinkish underparts
adult
chinensis
juvenile
chinensis
rectangularly shaped tail shows extensive white from below

orange, bordered by narrow, dark red orbital skin. Bill is blackish, and feet are reddish. **ADULT FEMALE:** Same as male. **JUVENILE:** Browner with broader, buffer fringes on feathers of upperparts; collar obscured or missing.

SIMILAR SPECIES Spotted Dove is unique, at least compared to the regular doves of its area. In most cities, it will most likely be compared with Mourning Dove, and it differs in the following ways: Spotted is a large, chunky dove, with broad, round wings and a broad, squared-off tail. In contrast, Mourning Dove has a more slender look, with a more pointed wing and a thinner tail, with an obviously pointed central tail. Also, Spotted is a dark brown bird, lacking paler buff and sandy tones of

Mourning Dove. If Spotted Dove's distinctive black nape, pitted with white spots is visible, identification is simple.

VOICE CALL: A rather harsh *coo-coo-croooo* and *coo-crrooo-coo,* with the emphasis respectively on the middle and last notes of the calls.

STATUS & DISTRIBUTION Spotted Dove occurs naturally throughout Southeast Asia and India; it was introduced into Los Angeles in the early 1900s, then spread throughout much of urban southern CA from Santa Barbara and Bakersfield south to Baja California. Now limited to a small section of south Los Angeles and Avalon (Santa Catalina I.); a few still found in Bakersfield. Reasons for Spotted Dove's decline and extirpation from most areas are not

known, but predation from Cooper's Hawks (now nesting widely in urban areas) is a likely factor. **BREEDING:** The nest is a flimsy platform of twigs, usually built in a tree or shrub, but occasionally found on buildings or even on telephone poles; lays two white eggs.

EURASIAN COLLARED-DOVE *Streptopelia decaocto*   EUCD ◼ 1

A fairly recent arrival to North America, this large pale dove can now be found across the U.S. It flaps on broad wings, and often soars briefly, with wings extended slightly above horizontal as it seemingly floats down to a landing. Polytypic (2 ssp.; nominate in N.A.). L 12.5" (32 cm)

IDENTIFICATION A large, pale gray-buff dove with a black collar, noticeably larger than Mourning Dove. There is also a naturally occurring cream-colored variant, and this species is known to hybridize with African Collared-Dove, so plumage variation will occur. Tail is fairly long and blunt-ended. **ADULT MALE:** Head is an unmarked, pale buff-gray, while upperparts are a darker buff-brown, tinged gray; a conspicuous black collar can be seen on hind neck. The primaries are noticeably darker than rest of wing, appearing blackish; secondaries are gray and contrast with blackish primaries and brown wing coverts in flight. Undersides of wings are pale. Underparts are a paler buff-gray merging into gray on undertail coverts. A dark gray tail has obvious black at base when seen from below; black extends beyond undertail coverts. This black includes outer webs of outer rectrices, and tail has a broad, pale buff-gray terminal band. A reddish brown iris borders narrow grayish-white orbital skin. Blackish bill has gray at base, and feet are dull reddish. **ADULT FEMALE:** Similar. **JUVENILE:** Paler; buff fringes on feathers of upperparts; black collar obscured or missing.

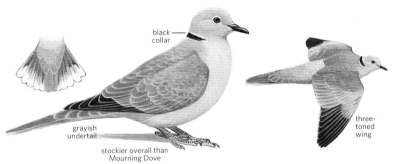

black collar

grayish undertail

stockier overall than Mourning Dove

three-toned wing

SIMILAR SPECIES Domesticated (in U.S.) African-Collared Dove—feral birds were formerly known as "Ringed Turtle-Dove"—is smaller, shorter-tailed, and noticeably paler; it has far less contrast between flight feathers and rest of wing; undertail coverts are white with black at base of tail more restricted, and outer webs of outer rectrices white. In addition, call is distinctly different.

VOICE CALL: A monotonous repeated, trisyllabic *kuk-koooo-kook,* slightly nasal, with emphasis on middle note is most habitually given during the long breeding season; also a harsher *kwurrr* is given frequently in flight throughout the year.

STATUS & DISTRIBUTION A Eurasian species introduced to the Bahamas, which spread to FL in the late 1970s. Explosive expansion across the continent in the late 20th century. The species is now resident in most of

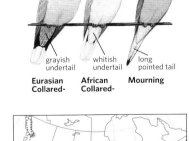

grayish undertail

whitish undertail

long pointed tail

Eurasian Collared-

African Collared-

Mourning

the U.S. (has reached southeast AK) and western Canada. Its westward expansion follows a similar expansion from its original range in Asia all the way to the Atlantic coast of Europe. **BREEDING:** Nest is a flimsy construction of twigs placed in trees, particularly palm trees, but occasionally on man-made structures; normally two white eggs, occasionally more; three to six broods a year from the same nest. **MIGRATION:** Not a migrant in the true sense. Basically year-round resident. But individuals move great distances, thus enabling the species to quickly expand its range across North America.

POPULATION Continues to expand in both numbers and range. Small numbers present in some areas may have escaped or been released from captivity by dove breeders, but most birds are thought to represent genuinely wild colonizers.

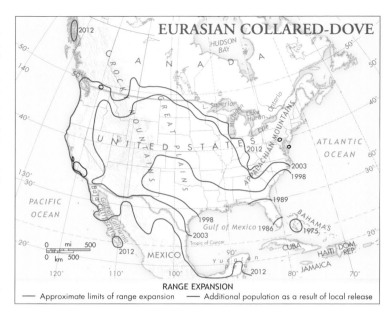

EURASIAN COLLARED-DOVE

RANGE EXPANSION
— Approximate limits of range expansion — Additional population as a result of local release

---

AMERICAN DOVES Genera *Zenaida* and *Ectopistes*

WHITE-WINGED DOVE *Zenaida asiatica*   WWDO ▪ 1

Flocks of these doves, with large white wing patches, and short square-ended tails, are a common sight in summer in many areas near the Mexican border. Polytypic (3 ssp.; 2 in N.A.). L 11.5" (29 cm)

IDENTIFICATION A little larger than Mourning Dove; generally brownish gray above and a paler gray below; white ends on greater coverts form prominent patches on wings, contrasting sharply with blackish flight feathers; blackish square-ended tail with prominent white terminal band. Prominent black crescent framing lower edge of auricular. On perched bird, wing patch shows only as a thin white line along leading edge of folded wing. **ADULT MALE:** Grayish or grayish brown head and neck, with neck and breast washed lightly with pink; iris reddish brown; orbital skin bright blue; bill black; and feet bright red. **ADULT FEMALE:** Similar to male. **JUVENILE:** Paler on head; narrow pale gray fringes on scapulars and wing coverts.

GEOGRAPHIC VARIATION Most recent treatment recognizes three subspecies, but there have been varied interpretations. Until recently, it was considered conspecific with West Peruvian Dove (*Z. meloda*) of South America. The two subspecies in N.A. are nominate *asiatica*, from western Texas eastward along the

Gulf Coast to Florida, and *mearnsi* in the Southwest, east to New Mexico. Differences are weak and clinal, with *asiatica* on average showing a less grayish tone to the brown plumage; but the two subspecies are likely not separable in the field. The third subspecies (*australis*) is found from Costa Rica to Panama.

VOICE CALL: A drawn-out *who-cooks-for-you* cooing; has many variations.

STATUS & DISTRIBUTION Breeds in southern tier of the U.S., from southeastern CA to the Gulf Coast of TX, and from southern FL south through C.A. to Panama. **BREEDING:** Nest is a fragile platform of twigs in medium-height brush; nominate birds typically nest in colonies, while western birds are more solitary; lays two white eggs. **MIGRATION:** Primarily a summer resident in the U.S., with most migrating into Mexico in winter. Increasing numbers are remaining through

the year, establishing isolated resident populations scattered across the U.S. from southeastern CA to FL; regular visitor to the Gulf Coast from LA to FL. **VAGRANT:** Casual on the East Coast north to the Maritime Provinces, in the interior to the Canadian border, and along the West Coast north to extreme southeastern AK.

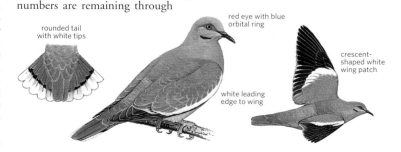

rounded tail with white tips

red eye with blue orbital ring

crescent-shaped white wing patch

white leading edge to wing

ZENAIDA DOVE *Zenaida aurita*   ZEND ■ 5

This shorter-tailed and somewhat darker version of the Mourning Dove is found throughout the West Indies. Polytypic (3 ssp.; *zenaida* in N.A.). L 10.5" (27 cm)
IDENTIFICATION About the same size as a Mourning Dove; warm brown above (averages a richer brown than Mourning Dove), with black spots on scapulars and tertials, and slightly paler cinnamon-brown below. A black crescent frames lower edge of auricular. Flight feathers are blackish; secondaries are tipped white, creating a white trailing edge on inner wing in flight, showing as a white patch on inner secondaries on folded wing, the single best field mark; short tail. **ADULT MALE:** Bronze iridescence on hind neck; iris dark brown; orbital skin pale blue; bill black; and feet bright red. **ADULT FEMALE:** Slightly duller than male and with much less iridescence on hind neck. Most females are a warm shade of brown, but some are grayer (similar to Mourning Dove). **JUVENILE:** Duller with buff fringes on both scapulars and coverts.

SIMILAR SPECIES Differs from Mourning Dove in having white on trailing edge of secondaries, and in having a shorter, rounded, gray-tipped tail.
VOICE CALL: A gentle cooing, very similar to that of Mourning Dove: *coo-oo, coo, coo, coo,* with second syllable rising sharply.
STATUS & DISTRIBUTION A common resident of open woodlands throughout the West Indies and along the east coast of the Yucatán Peninsula. **BREEDING:** Nest is usually found in a bush or tree; may have up to six

broods per year. **VAGRANT:** Said to be resident on small islands off the Florida Keys during the time of Audubon, but only two extant 19th-century specimens. Strictly accidental in the U.S. since 1900, with up to two on Plantation Key (Dec. 18, 1962–Mar. 1963) and singles on Key Largo (June 19–22, 1988, and May 3–6, 2002) being the only unequivocal records.

MOURNING DOVE *Zenaida macroura*   MODO ■ 1

This familiar dove, with its slim body and tapered tail, is the most common and widespread dove in most of North America. Wings make a fluttering whistle when the bird takes flight. Polytypic (5 ssp.; 3 in N.A.). L 12" (31 cm)
IDENTIFICATION Head and underparts unmarked pale pinkish brown, but with black crescent framing lower edge of auricular; upper parts darker and grayer brown; prominent black spots on coverts and tertials, and flight feathers contrasting darker; long pointed tail dark, with black subterminal spots and bold white tips on all but the central rectrices. **ADULT MALE:** Iridescent blue and pink on hind neck, with pinkish bloom extending onto breast; iris blackish; orbital skin pale blue; bill dark; and

feet red. **ADULT FEMALE:** Similar to adult male, but with reduced iridescence and pinkish bloom. **JUVENILE:** Generally darker and browner; pale buff-gray fringes on most of the feathers give the bird a "scaly" appearance; dark crescent below auricular extends forward toward base of bill; cheek area pale.
GEOGRAPHIC VARIATION Subspecies *carolinensis* breeds in the East, *marginella* breeds in the West, and nominate *macroura* from the West Indies recently invaded the Florida Keys; not separable in the field.
VOICE CALL: A mournful *oowoo-woo-woo-woo.*
STATUS & DISTRIBUTION Common throughout the U.S. and southern Canada south through C.A. Prefers

open areas, including rural and residential areas, avoiding thick forests; normally feeds on the ground. **BREEDING:** Nest is a loose platform of twigs, flimsy enough that the eggs are frequently visible from below. **MIGRATION:** Highly migratory, with birds breeding at the northern limit of the range believed to winter in Mexico, but those breeding farther south moving less, with birds present all year in the southern half of the U.S. **VAGRANT:** Casual to AK, northern Canada, the Azores, and Great Britain.
POPULATION The species is a well-managed game bird, with about 45 million killed by hunters in North America each year.

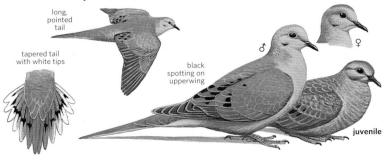

PASSENGER PIGEON *Ectopistes migratorius* PAPI ▪ 6

Extinct. Formerly the most abundant bird in North America, it occurred in huge flocks said to "blacken the sky" in the early 1800s. By the 1870s breeding numbers had been reduced to small scattered colonies, and the last wild bird was recorded in Ohio in 1900. The species became extinct when the last remaining bird in captivity at the Cincinnati Zoo died on Sept. 1, 1914. Monotypic. L 15.8" (40 cm) **IDENTIFICATION** Like Mourning Dove, this species had long, broad wings and a very long graduated tail, but it was substantially larger. **ADULT MALE:** Generally blue-gray above and pinkish below, brightest on breast, with black markings on coverts, and contrasting darker primaries and secondaries; white in tail restricted to outer pair of remiges; bill black; and tarsi and feet coral-red. **ADULT FEMALE:** Browner above and paler below than adult male, lacking pinkish coloration on breast; bare-part colors as in male, but duller. **JUVENILE:** Similar to female, but black markings on wings are obscured or lacking, with feathers of head, breast, and mantle fringed with buff-gray. **VOICE CALL:** Variously reported to give a series of harsh *keek* notes, often ending with a *keooo*. **STATUS & DISTRIBUTION** Extinct. **BREEDING:** Formerly in vast colonies in the deciduous woodlands stretching across the northern U.S. and southern Canada from the Great Plains east to the Atlantic. **WINTER:** The woodlands of the southeastern U.S., south along the Gulf coast of Mexico. **VAGRANT:** Casual in Cuba during the middle of the 19th century; also north to northern Canada, and west to BC and NV; records from Great Britain and France possibly involved genuine vagrants. **POPULATION** Destruction of old-growth deciduous forest and overhunting led to rapid declines and extinction of this once-abundant dove.

male bluish above, pinkish below

female duller

adult ♂

long tail

GROUND-DOVES Genus *Columbina*

INCA DOVE *Columbina inca* INDO ▪ 1

The conspicuously long-tailed, small dove is encountered in urban areas of the Southwest near the Mexican border. Monotypic. L 8.3" (21 cm) **IDENTIFICATION** A small gray dove with black fringes on feathers of both upper and underparts forming an obvious "scaled" appearance. In flight, shows chestnut on upper and underside of wing like Common Ground-Dove; however, also shows prominent white edges on long tail. **ADULT MALE:** Crown and face pale blue-gray and breast tinged lightly with pink; iris reddish; narrow eye ring blue-gray; bill blackish; and feet bright red. **ADULT FEMALE:** Similar, but duller on head and breast. **JUVENILE:** Duller with a slight brownish tinge overall, and "scaling" less noticeable.
SIMILAR SPECIES Its long tail provides a very different shape from that of the ground-dove, but beware of the possibility of an Inca Dove regrowing a lost or molted tail; it can be mistaken for a ground-dove. Scaling of upperparts is unique.
VOICE CALL: A long series of disyllabic *kooo-poo* that can be interpreted as "no hope."
STATUS & DISTRIBUTION Fairly common resident across the southern U.S. from southern NV to western LA; primarily around human habitation and in city parks. Some urban

populations have declined sharply (e.g., Tucson, AZ). Terrestrial, feeding on the ground; fairly tame and easily observed. **BREEDING:** Nest is a small fragile floor of twigs placed in a low bush or shrub; lays two white eggs; two or three broods each year. **MIGRATION:** Resident, but species expanding range northward, especially in the far west. **VAGRANT:** Casual to southern UT, NE, and AR; also recorded in ND and ON.

long tail with white edges

rufous primary patches

scaly pattern on head and body

COMMON GROUND-DOVE *Columbina passerina* COGD ▪ 1

This very small, short-tailed dove lives in open areas throughout most of the southern United States. Polytypic (18 ssp.; 2 in N.A.). L 6.5" (17 cm) **IDENTIFICATION** The smallest of the doves, with short rounded wings and a short, square-ended tail; generally grayish brown above, with underparts paler and washed with pink; prominent blackish spots on coverts and tertials; dark centers to feathers on head and breast create a "scaled" effect; short, square-ended tail blackish; fine white tips on outer two or three rectrices; bright chestnut panel in primaries and entirely rufous underwing all visible in flight. **ADULT MALE:** Crown paler and grayer than rest of upper parts, with iridescent blue-gray on nape and hind neck; face decidedly pinkish, with pinkish tone extending down onto under-parts; blackish spots on wings show iridescent maroon sheen; iris reddish brown; orbital ring blue-gray; bill blackish with bright pink on basal third; and feet bright red. **ADULT FEMALE:** Similar to adult male, but lacking iridescence on hind neck, and pale gray-brown below instead of pinkish; colors on soft parts

less intense. **JUVENILE:** Noticeably browner; scapulars and wing coverts fringed buff-gray; black spots on wings obscured or missing. **GEOGRAPHIC VARIATION** Nominate *passerina* is found in the Southeast, and *pallescens* in the West; pinkish on underparts and pink at the base of the bill more intense on nominate race, but generally not separable in the field. **SIMILAR SPECIES** Similar to Ruddy Ground-Dove (and see that account), but note Common's scaled head and breast, pale base to bill, and unmarked scapulars. **VOICE CALL:** A repeated soft, drawn-out, and ascending *wah-up* double

note given every two to three seconds. **STATUS & DISTRIBUTION** Common resident in open areas along the southern border of the U.S., but declining along the Gulf Coast. Prefers open areas and feeds on the ground. **BREEDING:** Nest is a flimsy collection of twigs on the ground or low in dense shrubs; normally two white eggs; two or three broods each year. **MIGRATION:** Although considered resident, some disperse northward in fall. **VAGRANT:** Casually north to OR, SD, MI, MA, and NS.

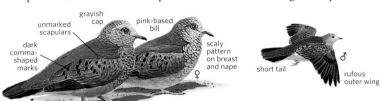

grayish cap

unmarked scapulars

pink-based bill

dark comma-shaped marks

scaly pattern on breast and nape

short tail

rufous outer wing

♀

♂

pinkish breast

RUDDY GROUND-DOVE *Columbina talpacoti*   RUGD ▪ 3

This small, short-tailed dove is widespread throughout the lowlands of Central and much of South America, and is expanding its range northward. It often associates with Inca Doves. Polytypic (4 ssp.; 2 in N.A.). L 6.8" (18 cm) **IDENTIFICATION** Similar in shape to Common Ground-Dove, but slightly larger and sexually dimorphic; wings with bright rufous panels in primaries, and underwing coverts black; tail black with narrow whitish tips on outer two to three rectrices; black spots on wings form vertical lines that extend up onto scapulars. **ADULT MALE:** Head unmarked pale grayish with blue-gray on crown and hind neck; upper parts rufous, most intense on lower back and rump; underparts unmarked rufous; iris reddish brown; orbital ring pale blue-gray; bill gray with black tip; and feet bright red. **ADULT FEMALE:** Head unmarked pale gray-brown; upper parts gray-brown,

and underparts paler and grayer; most have whitish fringes on tertials and greater coverts; soft parts as on male. **JUVENILE:** Similar to female, but males show varying amounts of rufous; black spots on wings obscured or missing. **SIMILAR SPECIES** Both sexes of Ruddy Ground-Dove have black linear markings on scapulars, unlike unmarked scapulars of Common Ground-Dove. Female Ruddy Ground-Doves show no "scaling" on head and breast; no reddish coloration at base of bill; and invariably show white fringes on tertials. Underwing coverts are mostly black. **GEOGRAPHIC VARIATION** The two subspecies found in N.A. are *rufipennis* of eastern Mexico and *eluta* of western Mexico. Adult males of *rufipennis* are a rich rufous above and below, more so than the adult males of *eluta,* and tips of their outer rectrices are rufous instead of whitish. Females and immatures

are probably inseparable in the field. **VOICE CALL:** A monotonous series of evenly pitched disyllabic *ca-whoop* notes given at one-second intervals. **STATUS & DISTRIBUTION** Common to abundant throughout the lowlands of Mexico south to northern Argentina, frequenting open woodlands. The western Mexican population is expanding northward, and a few are found annually, primarily in fall, in southeastern CA, southern AZ, southern NM, and western TX east to Big Bend N.P.; has nested at isolated locations within this area. Casual west to coastal southern CA, southern NV, and southwestern UT. **BREEDING:** Nest similar to that of Common Ground-Dove, but always off the ground; lays two white eggs. **VAGRANT:** In south TX, *rufipennis* is casual.

eluta   *rufipennis*

eluta ♀

grayish cap

dark bill

males with overall rufous coloration; females grayish

black linear markings on scapulars

no scaling on chest

eluta ♂

eluta ♂

short tail

rufous outer wing

NEOTROPICAL FOREST DOVES Genus *Leptotila*

WHITE-TIPPED DOVE *Leptotila verreauxi*   WTDO ■ 2

The White-tipped Dove, a plump, short-tailed terrestrial dove with broad, rounded wings, is found in forested areas of southern Texas, often on the ground. It jerks its tail when alarmed. Polytypic (14 ssp.; 1 in N.A.). L 11.5" (29 cm)

IDENTIFICATION Head gray, palest on face; upperparts unmarked, faintly bronzed olive-brown, and underparts unmarked pale gray; short, square-tipped tail blackish with prominent white tips on outer three rectrices; underwings rufous. **ADULT MALE:** Iridescent green and maroon wash on hind neck; iris yellow; orbital skin red; bill entirely dark; and feet coral red. **ADULT FEMALE:** Like

male, but iridescence on hind neck reduced. **JUVENILE:** Duller; fine buff fringes on scapulars and coverts; lacks iridescence.

SIMILAR SPECIES Most of the doves in this genus are similar, but differ in the amount of white in the tail; any *Leptotila* found away from southern Texas should be identified with care.

GEOGRAPHIC VARIATION Depending on the authority: one species with 14 subspecies or two species with multiple subspecies. The subspecies *angelica* is resident in southern TX.

VOICE CALL: A low-pitched *waa-*

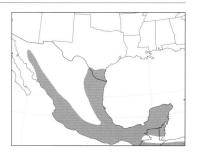

woooo, like the sound produced by blowing across the top of a bottle.

STATUS & DISTRIBUTION Common throughout lowlands of C.A. and S.A.; resident in southern TX and expanding northward; now found to Uvalde and Refugio Counties; casual farther north. **BREEDING:** The nest is a bulky flat of twigs lined with grasses, normally in dense brush close to the ground; typically two creamy-white eggs. **VAGRANT:** Two records from Dry Tortugas, FL (Apr. 2–8, 1995; Apr. 19–May 3, 2003).

POPULATION Appears to adapt well to fragmentation of forests, so is increasing in numbers.

chunky body shape

short, broad slightly rounded tail

pale yellow eye

white forehead and throat

whitish belly

angelica

angelica

white tail tips

QUAIL-DOVES Genus *Geotrygon*

KEY WEST QUAIL-DOVE *Geotrygon chrysia*   KWQD ■ 4

Although named Key West, this most widespread and common of the quail-doves in the West Indies is but a casual stray to Florida. It is rather secretive and is usually found singly, even where it is common, although pairs are also frequently encountered. It feeds on the forest floor, primarily on seeds and fruit. Quail-doves will tend to walk from danger, as opposed to taking flight. The male will perch on low branches, from which it makes its territorial calls. Monotypic. L 12" (31 cm)

IDENTIFICATION Key West Quail-Dove is one of the larger quail-doves. It is bicolored with a prominent white facial stripe; rich chestnut-brown above, and grayish white below. **ADULT MALE:** Crown, nape, and hind neck are washed with iridescent blue-green. A white face is bordered below by a narrow dark chestnut malar stripe. Pale underparts. An orange iris is bordered by red orbital skin. Dark red bill has a brownish tip. Feet are bright coral red. **ADULT FEMALE:**

Duller brown above than adult male, with less iridescence and with a less obvious facial stripe. **JUVENILE:** Duller still than adult female, generally lacking rich chestnut tones, and lacking any hint of iridescence. Scapulars and wing coverts are fringed cinnamon; feet are a much duller red than on adults.

VOICE CALL: A low moaning ventriloquial *ooooo* or *oooowoo*, with the second part accentuated and slightly higher; very similar to the call of White-tipped Dove.

STATUS & DISTRIBUTION Fairly common to uncommon; primarily resident in arid and semiarid woodlands and scrub thickets on the Bahamas, Cuba, and Hispaniola, and locally on Puerto Rico. **BREEDING:** The nest is a fragile platform of twigs lined with dead leaves built in low undergrowth or on the ground; normally lays two creamy-buff eggs. **VAGRANT:**

Possible former resident, prior to about the mid-19th century, on the Florida Keys. Now a casual mid-Oct.–mid-June straggler to the Florida Keys and southern FL, with at least 15 records since 1964, the northernmost being in Palm Beach County. Single birds have been reported in residential areas of southern FL; they frequented roadways in Everglades N.P. in 1979 and Boot Key in 1987.

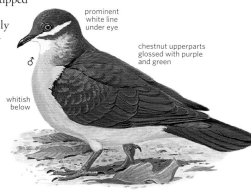

prominent white line under eye

chestnut upperparts glossed with purple and green

♂

whitish below

RUDDY QUAIL-DOVE *Geotrygon montana* RUGD ■ 5

As is typically the case for this genus, the Ruddy Quail-Dove is secretive and quite terrestrial, frequently running rather than flying when alarmed. It is normally encountered on the forest floor in Central and South America. Males readily fly to low branches to vocalize, but otherwise flight is infrequent, except during the courting process. Ruddy Quail-Dove feeds on seeds, fruits, and more so than most congeners, on a regular diet of invertebrates, such as beetles. Polytypic (2 ssp.; nominate in N.A.). L 9.8" (25 cm)

IDENTIFICATION One of the smaller and most widespread of the quail-doves, most often encountered walking on the forest floor. A plump, short-tailed dove, reddish brown above with buff underparts and a pale stripe through face. While it has subtle colors, it gives the general impression of a dark reddish brown, stocky dove. **ADULT MALE:** Reddish brown on upperparts is washed with iridescent purple. Chin, throat, and sides of face are pale buff; cheeks are separated from throat by a broad, reddish brown malar stripe. Iris is yellow-orange;

orbital skin red; bill dark reddish with dusky tip; feet bright coral red. **ADULT FEMALE:** Smaller; darker olive-brown above and below, lacking iridescence; facial stripe less obvious. **JUVENILE:** Coloration like female, but has rufous fringes on feathers.

VOICE CALL: A series of very deep, resonant, monosyllabic coos, *waooo* or *wooo* repeated every three to four seconds, trailing off at the end. In the classic *Birds of the West Indies,* James Bond described the call as "reminiscent of the doleful sound of a fog buoy."

STATUS & DISTRIBUTION Uncommon to locally common; largely resident in lowland humid forests of the West Indies and Mexico southward through C.A. to southern Brazil. **BREEDING:** Ruddy Quail-Dove's nest is a platform of twigs lined with dead leaves, built low in a bush or tree, but sometimes on the ground; normally lays two creamy-buff eggs. **MIGRATION:** Much remains to be learned abut the movements of this species. It is partially migratory or

nomadic, but specific patterns and causes of migration are undiscovered. Numbers vary from year to year, even where the species is common; but the extent and regularity of the bird's movements remain unknown. As an example of the significant travel that can take place, this species has been recorded at sea in the Caribbean. **VAGRANT:** Eight records from the Florida Keys, including the Dry Tortugas, and one at Bentsen-Rio Grande Valley S.P. in TX.

montana ♀

male rich rufous above, female duller

buffy line under eye

♂ *montana*

CUCKOOS, ROADRUNNERS, AND ANIS Family Cuculidae

Almost all of the species in the Cuculidae family are arboreal, inhabiting forests and well-wooded areas. Cuckoos form a remarkably variable family, which is well known for having members that are brood parasites. Various species are known as "rain birds" throughout the world because they are very vocal at the beginning of the rainy season.

Structure Cuckoos have slender bodies and long tails. All species in the family have zygodactyl feet, where the outer toes point backward and the two inner toes point forward. The bill is usually long with a curved culmen. The wing length varies depending on whether the species is a long-distance migrant or a more sedentary species. Terrestrial species—such as the Greater Roadrunner—have long, strong legs.

Behavior Many cuckoos are solitary and are more often heard than seen. Although mostly diurnal, many species call at night. Up to 53 species of cuckoos are brood parasites; their cryptic plumage allows the females to surreptitiously approach the nests of hosts. Black-billed and Yellow-billed Cuckoos are nest builders, but they have been documented as brood parasites. Unlike cuckoos, anis live in noisy social groups and are often cooperative breeders with several females using the same nest and multiple adults feeding the nestlings.

Groove-billed Ani (left) and Smooth-billed Ani (FL, Mar.)

Because of the texture of their plumage, many cuckoos, when wet or in the early morning, dry their bodies by sitting on an open perch with their wings and tail spread, often with their back feathers raised to expose the skin to the sun.

Plumage For most Cuculidae species the plumage is soft and is a brown, gray, or black color, often streaked or barred. A few species exhibit bright green, rufous, or even purple plumage. The tail is generally tipped in a different color, often white. The sexes are similar in plumage in almost all species, with some size dimorphism—females being larger.

Distribution The species of this widespread family occur in tropical and temperate regions worldwide; however, the family reaches its greatest diversity in the Old World. Most species are nonmigratory or short-distance migrants; a few species are long-distance migrants, including the Yellow-billed and Black-billed Cuckoos of North America. Vagrants may turn up far out of range.

Taxonomy The family divides into six subfamilies: Old World cuckoos, coucals, malkohas and cauas, American cuckoos, New World ground-cuckoos, and anis. Some taxonomists suggest that these groups represent distinct families. Worldwide, there are 138 species in 35 genera recognized. In North America, six species regularly occur, while two additional species are rare visitors, primarily to western Alaska.

Conservation Several species in the family rate conservation concern. Most of them occur in tropical forests and on islands; however, the populations of the Smooth-billed Ani in Florida and the Yellow-billed Cuckoo in western North America are also on the list. One species in the family has been extinct since 1850 and two others have not been reported since the early 1900s. Pollution could be a contributing factor—it appears that cuckoos are heavily impacted by pollutants. Studies have shown that chlorinated hydrocarbons build up in their tissues.

OLD WORLD CUCKOOS Genus *Cuculus*

In this genus composed of 10 or 11 species—with most species found in southern Asia and Africa—the most well-known and widespread species is the Common Cuckoo. All species share the same basic plumage of uniform-colored upperparts and paler, often barred, underparts. They have slender bodies and long, graduated tails.

COMMON CUCKOO *Cuculus canorus*   COCU ■ 3

An Old World species, the Common Cuckoo is well known as a brood parasite, affecting a very wide variety of hosts and occurring in numerous habitats ranging from woodlands to farmlands. Hepatic-morph birds are females. Polytypic (4 ssp. worldwide; likely *canorus* in N.A.). L 13" (33 cm)

IDENTIFICATION Common Cuckoo has a slender body, a long rounded tail, and pointed wings. In flight, it has a falconlike appearance; its wings rarely rise above the body. It often perches in the open; when perched, it often droops its wings. It closely resembles Oriental Cuckoo. **ADULT MALE:** Pale gray above and paler below with a white belly narrowly barred with gray. Undertail coverts are white and lightly barred. He has a prominent yellow orbital ring. **GRAY**

MORPH FEMALE: Difficult to separate from adult male, but often has a rusty buff tinge on breast. **HEPATIC-MORPH FEMALE:** Rusty brown above and on breast and heavily barred—black bars are narrower than brown ones. Lower back and rump are a lighter brown and either unmarked or lightly spotted.

GEOGRAPHIC VARIATION Variation in subspecies is slight and possibly clinal. All records in North America likely refer to nominate *canorus*, although many seen in AK have been

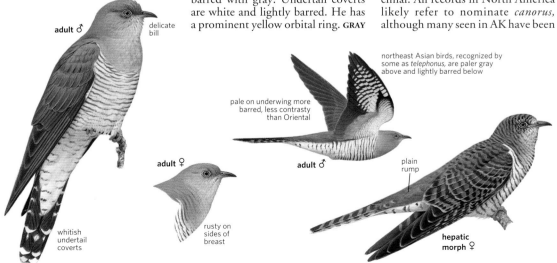

adult ♂
delicate bill

northeast Asian birds, recognized by some as *telephonus*, are paler gray above and lightly barred below

pale on underwing more barred, less contrasty than Oriental

adult ♀

adult ♂

plain rump

whitish undertail coverts

rusty on sides of breast

hepatic morph ♀

paler above with finer barring below than European birds, and perhaps represent *telephonus*, though that subspecies is not recognized by most. Subspecies identification is difficult and is based on variation in plumage coloration.

SIMILAR SPECIES Common Cuckoo is very similar in appearance to Oriental Cuckoo, and the identification of silent birds is difficult. Common Cuckoo is slightly larger with a more delicate bill and is generally slightly lighter in overall color. Oriental Cuckoo has a more contrasty greater covert bar on the underwing and buffy undertail coverts. Hepatic-

morph Common Cuckoo can be identified by its paler unmarked rump. **VOICE** The male's call is a disyllabic *cuc-oo*, with emphasis on the first syllable. The female utters a loud bubbling trill.

STATUS & DISTRIBUTION Very rare spring and summer visitor to central and western Aleutian, St. Lawrence, and Pribilof Islands in Alaska. Casual to mainland. BREEDING: Palearctic, ranging from western Europe to eastern Russia. WINTER: Primarily in southern Africa. VAGRANT: Accidental to Martha's Vineyard, MA (May 3–4, 1981); Watsonville, Santa Cruz

Co., CA (Sept. 28–Oct. 2, 2012); and the Lesser Antilles. **POPULATION** Stable.

ORIENTAL CUCKOO *Cuculus optatus*   ORCU ■ 4

As with other Old World *Cuculus*, the Oriental Cuckoo is a brood parasite. It is encountered in North America only as a casual visitor to western Alaska. As in Common Cuckoo, hepatic-morph birds are females. Polytypic (3 ssp.; *optatus* recorded in N.A.). L 12.5" (32 cm)

IDENTIFICATION Oriental closely resembles the Common Cuckoo. It has a slender body, long rounded tail, and pointed wings. In flight it resembles a small falcon. ADULT MALE: Medium gray above and paler below with a white belly barred with gray. Buffy undertail coverts are generally sparsely barred or unmarked. Prominent unmarked whitish wing linings are visible in flight. Orbital ring is bright yellow. GRAY MORPH FEMALE: Difficult to separate from adult male, but often has a rusty buff tinge on breast. HEPATIC-MORPH FEMALE: Rusty brown above and on breast and heavily barred overall, including rump—black bars are wider than brown bars.

GEOGRAPHIC VARIATION Complicated. Most authorities treat the much smaller resident bird *(C. lepidus)* of Malaysia and Indonesia as its own species, the Sunda Cuckoo. Some authorities, including the AOU, split the northern *C. optatus* as a separate species from the very similar but smaller Himalayan bird *(C. saturatus)*. Both of these species are highly migratory, but it appears that *C. optatus* migrates much farther south and east, as far as Australia.

SIMILAR SPECIES Very similar in appearance to Common Cuckoo and the identification of silent birds is difficult. Oriental Cuckoo is very slightly

smaller, and gray morph birds are generally somewhat darker in overall color with slightly broader barring on belly and buffy undertail coverts. Oriental shows more contrast between the whitish and grayish areas of its underwing than Common Cuckoo. Hepatic-morph Oriental is easily separated from Common Cuckoo based on its broad black barring and barred rump.

VOICE The male's call is a series of hollow notes usually given on the same pitch and in groups of four. The female's call is very similar to the trill of a Common.

STATUS & DISTRIBUTION Casual spring, summer, and fall visitor to western Aleutian, St. Lawrence, Pribilof Islands, and once from mainland AK. BREEDING: Primarily Palearctic,

ranging from northeastern Europe to Korea, Japan, and the Russian Far East. WINTER: Primarily in Indonesia. **POPULATION** Stable.

adult ♂

more contrast between whitish and slate-gray areas on underwing than Common Cuckoo

adult ♂

bill slightly thicker and more curved than Common

hepatic morph ♀

barred rump

ventral barring averages stronger

buffy undertail coverts

nape spot

juveniles of both species are blackish brown with whitish feather fringes above

juvenile

NEW WORLD CUCKOOS Genus *Coccyzus*

Nine species, primarily neotropical, make up this genus. The two most common species in North America are long-distance migrants to South America. Most New World Cuckoos share a basic plumage pattern that consists of brown or gray upperparts with buffy or white underparts. They have slender bodies and long, graduated tails.

MANGROVE CUCKOO *Coccyzus minor*   MACU ■ 2

black mask

adults

thick bill; lower mandible has yellow base

uniform brown wings

buffy underparts

large white spots on black tail

The Mangrove Cuckoo is more often heard than seen as it often perches quietly in a tree. It is widespread in the Caribbean and in coastal habitats from northern Mexico to northern South America. Monotypic. L 12" (31 cm)
IDENTIFICATION It has a slender body, a long tail, and rounded wings. **ADULT:** It has grayish brown upperparts with a grayer crown. Throat and upper breast are white or buffy with remainder of underparts buffy. Long tail is graduated with brown central rectrices tipped in black, remaining tail feathers are black and broadly tipped with white. Bill is strongly curved with a black upper mandible and a black-tipped yellow lower mandible. **IMMATURE:** Black mask is faint to near absent.
GEOGRAPHIC VARIATION Up to 14 subspecies have been described, but the species is now considered monotypic. Previously, the Florida population was placed in *maynardi* (these birds tend to be more richly colored and likely originate from Mexico) and the vagrants to the western Gulf Coast in *continentalis*.
SIMILAR SPECIES Mangrove Cuckoo most closely resembles Yellow-billed Cuckoo, but it is distinguished by its black mask, buffy underparts, and lack of rufous primaries. Its voice is very different as well.
VOICE SONG: A very guttural *gaw gaw gaw*.
STATUS & DISTRIBUTION Uncommon. **BREEDING:** Nests in mangrove swamps and low canopy tropical hardwood forests. **WINTER:** Withdraws from northern portions of its range in FL. **VAGRANT:** Casual, presumably from northern Mexico, along the Gulf Coast from TX to northwestern FL.

BLACK-BILLED CUCKOO *Coccyzus erythropthalmus*   BBCU ■ 1

The Black-billed Cuckoo regularly feeds on caterpillars, and it also greedily consumes tent caterpillars and gypsy moth larvae during outbreaks of those insects. Monotypic. L 12" (31 cm)
IDENTIFICATION Black-billed Cuckoo has a slender body, a long tail, and rounded wings **ADULT:** It has grayish brown upperparts and whitish underparts. Long tail is graduated; it is predominantly brown above, narrowly tipped with white, while undertail is patterned in gray. It has a prominent red orbital ring. Bill is strongly curved and all dark.
IMMATURE: It looks similar to adult, but it has a buffy throat and undertail coverts. This coloration sometimes extends in remainder of underparts. Undertail pattern is muted

red orbital ring

slender, dark bill

slight buff tint to throat

adult

uniform brown wings

buffy orbital ring

adult

juvenile

indistinct tips

small white spots on grayish tail

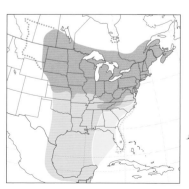

and tips of rectrices are buffy and not as prominent. Orbital ring is buffy as well.
SIMILAR SPECIES Black-billed Cuckoo most closely resembles Yellow-billed Cuckoo, but it is distinguished by its red orbital ring, lack of rufous primaries, different under-

tail pattern, and narrower dark bill.
VOICE SONG: A monotonous *cu-cu-cu* or *cu-cu-cu-cu* phrase.
STATUS & DISTRIBUTION Uncommon. **BREEDING:** Deciduous and mixed forests to open woodlands and brushy habitats. **MIGRATION:** Trans-Gulf migrant, rare in the U.S. after mid-

Oct. **WINTER:** Northern S.A. **VAGRANT:** Casual visitor west to the Pacific coast.
POPULATION Black-billed Cuckoo has shown substantial declines and is considered a species of conservation concern in many areas of the United States.

YELLOW-BILLED CUCKOO *Coccyzus americanus*   YBCU ■ 1

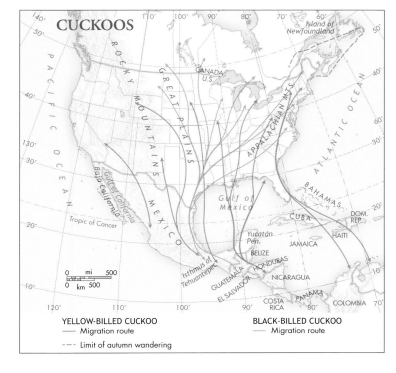

Generally shy and elusive, Yellow-billed Cuckoo is easily overlooked, except for its loud vocalizations. Monotypic. L 12" (31 cm)
IDENTIFICATION It has a slender body and a long tail. **ADULT:** Grayish brown

above, whitish below; crown noticeably grayer. Rounded wings have reddish primaries. Long tail is graduated with brown central rectrices tipped in black; remainder are black and broadly tipped with white. Yellow orbital ring. Bill curved with black culmen; lower mandible is yellow with a black tip. **IMMATURE:** Similar to adult but has buffy undertail coverts. Undertail pattern muted, and tips of rectrices not as prominent. Orbital ring is a dull yellow.
GEOGRAPHIC VARIATION Presently considered monotypic; however, there have been two subspecies described: *americanus* in eastern N.A. and *occidentalis* in the Southwest. Differences between these populations are weak.
SIMILAR SPECIES Yellow-billed Cuckoo most closely resembles Black-billed Cuckoo, but it is distinguished by

yellow orbital ring, rufous primaries, more prominently white-tipped tail, and yellow lower mandible.
VOICE SONG: A rapid staccato *kuk-kuk-kuk* that usually slows and descends into a *kakakowlp-kowlp* ending; sounds hollow and wooden. A loud, braying *coo, coo, coo,* ending slower and softer.
STATUS & DISTRIBUTION Common in eastern N.A., becoming increasingly rare and local in much of the West. **BREEDING:** Open woodlands with dense undergrowth, riparian corridors, and parks. Southwestern populations increasingly limited to riparian corridors. **MIGRATION:** Trans-Gulf migrant as well as southeastward over Caribbean Islands. In spring, the Gulf Coast peak occurs ±1 May; southern Great Lakes ±15 May. Southwestern population arrives primarily in June. In fall, it wanders up eastern seaboard as far north as NF. Southern Great Lakes peak ±20 Aug; Gulf Coast peak ±10 Sept. Rare in the U.S. after Nov. 1. **WINTER:** Casual to accidental along Gulf Coast; these records could pertain to lingering fall migrants. The majority of population winters in South America, as far south as northern Argentina.
POPULATION Populations in the western United States are declining.

ROADRUNNERS Genus *Geococcyx*

Worldwide there are two species in this genus, the Greater and Lesser Roadrunners. These large terrestrial cuckoos are very similar in appearance and are found primarily in arid and semiarid habitats. Roadrunners can run at speeds of up to 15 miles an hour pursuing prey or to escape predators.

GREATER ROADRUNNER *Geococcyx californianus*   GRRO   ▣   1

The Greater Roadrunner is omnivorous, although the majority of its diet includes insects, birds, reptiles and rodents. Monotypic. L 23" (58 cm)
IDENTIFICATION Greater Roadrunner is a large ground-dwelling cuckoo with a bushy crest. It is brown overall and streaked with brown and white. It has short rounded wings with a white crescent in primaries. Its long tail has rectrices edged in white. Note long, heavy bill.
SIMILAR SPECIES Unmistakable. No real contenders for misidentification.
VOICE SONG: A descending series of low coos.
STATUS & DISTRIBUTION Uncommon.
BREEDING: Nests in low woody plants in open scrub habitats, including chaparral, and in open woodlands.

WINTER: Some local movements away from nesting habitat.

crest can be raised or flattened

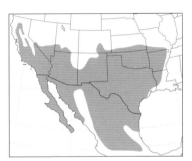

very long, graduated tail with white tips

ANIS Genus *Crotophaga*

The three species in this genus—Great, Groove-billed, and Smooth-billed Anis—exhibit complex social behavior and are cooperative breeders. They share a loose black plumage and a laterally compressed bill.

SMOOTH-BILLED ANI *Crotophaga ani*   SBAN   ▣   3

The Smooth-billed Ani colonized Florida in the late 1930s from the West Indies. The population expanded and the species was locally fairly common until about 1990. Sharp declines followed, and by 2000 only a few individuals remained (e.g., at Ft. Lauderdale); by about 2010 they too had disappeared. Monotypic. L 14.5" (37 cm)
IDENTIFICATION Plumage is entirely black, but with some purplish or greenish iridescent edges to feathers, and it has a shaggy or disheveled appearance. Long tail is frequently held down and wagged. Large, laterally compressed bill often has a high curve to culmen. These are easily separated from Groove-billed; others are more difficult. Flight often looks labored and weak.
SIMILAR SPECIES A vagrant Groove-billed Ani can be very difficult to distinguish from Smooth-billed Ani; it is best identified by voice and close examination of bill shape (see Groove-billed Ani). Grackles are the only other species likely to cause confusion with Smooth-billed Ani.

some show raised base to culmen, but many others do not

bill has no distinct grooves

slightly larger than very similar Groove-billed; most birds best distinguished by voice

Shaggy plumage and large bill are key features to use in identification.
VOICE CALL: A whining, rising *quee-lick*, quite distinct from Groove-billed.
STATUS & DISTRIBUTION Widespread in the neotropics; occurs in brushy or weedy habitats, most often close to water. Now casual from Miami to the Florida Keys; likely these are strays from the Bahamas. **VAGRANT:** Accidental north along Atlantic coast to NC; also LA and OH.

GROOVE-BILLED ANI *Crotophaga sulcirostris*   GBAN ▪ 2

The gregarious Groove-billed Ani is most often found in vocal groups of four to ten birds. It typically roosts with other anis in tight bunches on vines and other protected perches. It (and the Smooth-billed Ani) has a very distinctive flight—weak flaps on rounded wings interspersed with glides. It feeds by gleaning insects off leaves but is often seen chasing grasshoppers on the ground through high grass. Nests may be single to fairly large communal nests shared by up to four pairs. In these large nests, two or three females may brood the eggs simultaneously and the eggs on the bottom rarely hatch. Both males and females defend the territory and care for the young. Throughout most of its range, the Groove-billed Ani is frequently seen around cattle where it feeds on insects stirred up by the livestock. This aspect of its life history is rarely observed in Texas. Polytypic (2 ssp.; nominate in N.A.) L 13.5" (34 cm)

IDENTIFICATION Plumage is entirely black with iridescent purple and green overtones; overall appearance is often shaggy or disheveled. Its long tail appears loosely joined to its body and is often held down. Groove-billed Ani frequently wags its tail when on open perches. Large, laterally compressed bill has a curved culmen. Adults normally have easily observed grooves on upper mandible; on juveniles upper mandible might

be unmarked. Groove-billed's bill shape is similar to Smooth-billed Ani's, but curvature of culmen does not extend above crown. **GEOGRAPHIC VARIATION** The subspecies *pallidula* of southern Baja California is now extinct. **SIMILAR SPECIES** Groove-billed Ani is a rare visitor to Florida and can be difficult to distinguish from the resident, but now nearly extirpated, Smooth-billed Ani. At present, an ani seen in FL is more likely to be a Groove-billed. Both ani species have occurred far out of range. Species identification between the two anis is best accomplished by voice and close examination of bill shape (particularly the bill ridge). Groove-billed is also slightly smaller. Grackles are the only other species likely to cause confusion with Groove-billed Ani. Shaggy plumage and large bill are key features to use in identification. **VOICE CALL:** A liquid *tee-ho*, with the accent on the first syllable. This call is given raucously in quick succession at dawn when groups leave the roost. **FLIGHT CALL:** Soft clucking or chuckling notes. **STATUS & DISTRIBUTION** Widespread in the neotropics from southern TX to northern S.A., and south along the Pacific coast to northern Chile. **BREEDING:** Uncommon to locally common late Apr.–late Sept. Nests in scrub or low-canopied woodlands and in riparian corridors. Occasionally is

long tail

sulcirostris

bill grooves visible at close range

sulcirostris

found in more open habitats. Breeding also occurs in more open scrub, not necessarily near water. **WINTER:** Rare and local as majority of population has withdrawn, presumably into northeastern Mexico. Wintering populations very localized and primarily found along the TX coast and in the lower Rio Grande Valley. **VAGRANT:** Wanders regularly, most often in fall, east along the Gulf Coast to FL. Casual throughout remainder of TX and north to CO and the Midwest. Accidental elsewhere in the eastern U.S., and has occurred as far north as southern ON. Casual to southern AZ, southern NV, and southern CA. **POPULATION** The Groove-billed Ani populations in Texas appear stable. The species is not of conservation concern anywhere in its range.

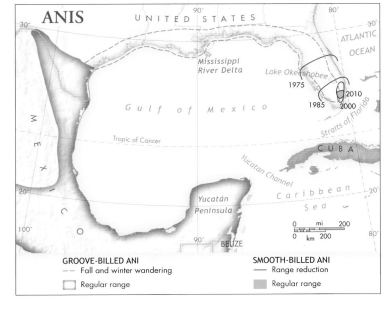

ANIS

GROOVE-BILLED ANI
– – Fall and winter wandering
◻ Regular range

SMOOTH-BILLED ANI
— Range reduction
▨ Regular range

BARN OWLS Family Tytonidae

Barn Owl

Tytonidae owls share many characteristics with owls of the larger Strigidae family (see p. 349).

Structure Tytonids differ from the Strigids by having a heart-shaped facial disk; relatively small eyes; a short, squared tail; and serrated central claws. Their legs are long and feathered, and their toes are bare.

Behavior These owls are essentially nocturnal and sedentary. Their habitats vary, from grasslands and open areas to closed forests. They hunt small mammals and other vertebrates from a perch or in low, quartering flights over open areas; sometimes they hover in place. Voices include shrieks, whistles, hisses, and screeches, unlike the resonant hooting of Strigids.

Plumage Plumage is similar across ages and sexes, but females are generally darker and larger than males. Upperparts range from pale gray and rust through orange-red to sooty blackish, often with sparse contrasting spots. Underparts are often paler, and the facial disk is paler still.

Distribution Tytonidae is a mainly tropical family reaching its greatest species diversity in Australasia, with about 13 species present there. North America's sole representative, the Barn Owl, is the most widely distributed nocturnal raptor, resident on every continent except Antarctica.

Taxonomy The Tytonidae family comprises two genera (*Tyto* and *Philodus*), and about 19 species are recognized.

Conservation Species inhabiting tropical forests are threatened by the logging of large trees, which reduces their forest-dependent food supply as well as the number of large tree cavities available for nesting. The restricted range of island-dwelling species renders them particularly vulnerable to habitat destruction. BirdLife International currently lists four species as vulnerable and two as endangered.

Genus *Tyto*

BARN OWL *Tyto alba* BANO ■ 1

By night, an unearthly shriek may alert a birder to a ghostly white Barn Owl flying overhead. The Barn Owl

underparts vary from whitish to cinnamon-buff; males average paler

looks pale in flight with rounded wings and no dark carpal patches as in Short-eared Owl

dark eyes with whitish heart-shaped face

pratincola ♀

pratincola ♂

long legs

nests and roosts in dark cavities in city and farm buildings, cliffs, and trees. Polytypic (24 ssp.; *pratincola* in N.A.). L 16" (41 cm) WS 42" (107 cm)

IDENTIFICATION Flight features shallow, slow wingbeats, often with long legs dangling. May hover in place while hunting. **ADULT:** A very pale nocturnal owl with a white, heart-shaped facial disk. Dark brown eyes; horn-colored bill. Head and upperparts mottled rusty brown and silvery gray, with fine, sparse black streaks and dots. Underparts vary from cinnamon to white. **MALE:** Palest birds are males. **FEMALE:** Darkest birds are always females; average larger than males.

SIMILAR SPECIES Other than Snowy Owl, this is the palest North American owl. Its flight and wing shape suggest slightly darker Short-eared Owl, but paleness of Barn Owl's plumage is usually evident; also note dark underwing crescents and buffy upperwing crescents on Short-eared.

VOICE Male's territorial song is long, raspy, hissing shriek, *shrrreeee!* Often given many times in sequence, usually in flight

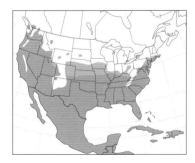

and near the nest. Female's vocalization similar.

STATUS & DISTRIBUTION Rare to fairly common. **YEAR-ROUND:** Occupies low-elevation, open habitats, urban and rural. Density very low in more northerly areas. **WINTER:** Withdraws in winter from colder areas.

POPULATION Has suffered sharp declines in eastern North America. Canada lists it as endangered in the east and of "special concern" in the west. More than a dozen states in the Midwest and Northeast list it as endangered, threatened, or of "special concern."

TYPICAL OWLS Family Strigidae

Great Gray Owl (MN, Jan.)

Owls are chiefly nocturnal predators, ecological counterparts of the diurnal birds of prey but most closely related to nightjars (Caprimulgiformes). Being chiefly nocturnal, they are difficult to detect, identify, and study. In the daytime, a collection of regurgitated pellets on the ground or a noisy mob of songbirds will often point a birder to a roosting or nesting owl. At night, voice is the best means to detect an owl and usually the best way to identify it.

Structure All owls have a relatively large head with immobile eyes that face forward to provide binocular vision and good depth perception. Acute hearing complements owls' keen eyesight. Most have a prominent facial disk formed of stiff feathers whose shape can be altered to help focus sounds. In some genera, ear openings are asymmetrically located in the skull, improving their ability to distinguish the direction and distance to a sound source. Asymmetric external ear structures on many genera provide further discrimination based on the frequency of the emitted sound. Some owl species can detect and capture prey in total darkness, by sound alone. Feather structure contributes to quiet flight. Velvety pile on flight feathers and soft body feathering absorb sound, while comb-like fringes on the leading edge of the outermost primaries reduce the sound of wings cutting the air. Some largely diurnal owls lack these fringes; their flight is not as quiet as that of nocturnal owls. Owls' bills are strongly curved, with a sharp point for tearing prey. They can position their toes two forward and two backward for seizing prey, and their long claws are needle-sharp for clutching and quickly dispatching it. Unlike Tytonids, Strigids generally have circular facial disks, relatively large eyes, and rounded tails; in addition, their central toe's claw is not serrated.

Behavior Most owl species forage by perching and watching for prey. They nest in natural or human-made cavities or in old woodpecker holes; some use old tree nests of raptors, crows, or squirrels. The larger owls prey mainly on mammals and the smallest on insects, with fish and small birds taken by some owls. Owls consume small prey whole but tear apart larger prey before consumption. They periodically regurgitate a dense pellet containing indigestible parts such as bones, chitin, hair, and feathers.

Voice The most commonly heard vocalization is a species-specific territorial song (both sexes vocalize in most species), delivered mainly in the month or two during which male owls establish a territory, locate potential nest sites, and attract a mate. This period of active singing varies among species: It begins about October for the early nesting resident Great Horned Owl and about April for the migrant Flammulated Owl. Owls also utter a variety of hoots, whistles, screams, whines, screeches, barks, or rasps when threatened, alarmed, or begging, for example. Sometimes these cannot be pinned down to a particular species; indeed one of the charms of nighttime owling is hearing a new, unidentifiable sound.

Plumage Owl plumage is similar between sexes, but females are generally larger and sometimes darker than males. Downy young of most species molt directly into adultlike plumage; only a few hold a distinctively colored juvenile plumage for an appreciable period. Owls lack a distinctive breeding plumage, so after first attaining adultlike plumage, an owl's appearance changes little during its lifetime. Most adult Strigids are cryptically patterned in shades of brown, though their facial disk may be quite distinctive. Many have erectile feather tufts at the sides of their crown, called "ear tufts," although they are unrelated to the bird's ears or their hearing; their function is not known for certain.

Distribution Owls occur worldwide in virtually all terrestrial habitats, from the tropics to the Arctic, from below sea level to elevations above 14,000 feet. Most species are mainly resident and sedentary.

Taxonomy Worldwide, about 194 strigid species in about 28 genera are currently recognized. New species are still being described—about ten in the last 20 years—as their voices (which are innate, rather than learned) and other attributes become better known.

Conservation Four island-dwelling strigids are known to have gone extinct since 1600. BirdLife International currently considers seven species critically endangered, eight endangered, and 19 vulnerable; more than half of these are restricted to islands, where habitat loss is their greatest threat.

SCOPS-OWLS Genus *Otus*

These are small to medium-size, mainly nocturnal Old World owls. Most have ear tufts and short, rounded wings. They inhabit forests and semi-open areas with scattered bushes or groups of trees. All have only one song type. About 41 species are currently recognized (one vagrant in N.A.), but more than half are confined to one or a few islands.

ORIENTAL SCOPS-OWL   *Otus sunia*   ORSO ▪ 5

The Oriental Scops-Owl is a small, nocturnal, primarily insectivorous owl of Asia's southeast rim. Its geographically variable vocalizations suggest that perhaps more than one species is involved. Polytypic (9 ssp.; *japonicus* in N.A.). L 7.5" (19 cm) WS 21" (53 cm)
IDENTIFICATION ADULT: Fine dark streaks on head; dark shaft streaks and thin horizontal pencil lines on breast; short ear tufts. Gray-brown, reddish gray, and rufous morphs exist. Yellow eyes; horn green bill with blackish tip. Legs only partially feathered.
SIMILAR SPECIES No other small owl is likely to occur on islands far

offshore mainland Alaska.
VOICE On the breeding grounds, the territorial song of *japonicus* (the vagrant found in N.A.) is a rather low, whistled *tu-tu-tu,* repeated monotonously at short intervals. Probably silent on migration.
STATUS & DISTRIBUTION Uncommon on breeding grounds in Japan. **BREEDING:** Favors deciduous and mixed forest, from Pakistan and India through China to eastern Siberia and the main islands

small size with short ear tufts

three color morphs

rufous morph *japonicus*

of Japan. **MIGRATION:** Four northerly subspecies, including *japonicus,* are largely migratory, wintering mostly in southern China, Thailand, and Malaysia and returning about May. Southerly subspecies are resident. **VAGRANT:** Two N.A. records of rufous morph *japonicus* from the Aleutian Is., AK: Buldir I. (June 5, 1977; a dessicated wing) and Amchitka I. (June 20, 1979).
POPULATION Not globally threatened.

Genus *Psiloscops*

FLAMMULATED OWL   *Psiloscops flammeolus*   FLOW ▪ 2

Once considered rare, the strictly nocturnal, insectivorous Flammulated Owl may actually be the most common owl in its breeding habitat. Its cryptic plumage provides excellent daytime camouflage, but a persistent flashlight-bearing birder tracking a singing owl at night may eventually be rewarded with deep red eye shine returned from a pine cone–size shape perched close to a tree trunk. It nests and roosts in old woodpecker holes or natural tree cavities, sometimes in loose colonies. This New World species has recently been moved to its own monotypic genus. Monotypic. L 6.7" (17 cm) WS 20" (51 cm)
IDENTIFICATION Flight is nervous,

darting, sometimes jerky, interspersed with occasional hovering as it pauses to check for prey. **ADULT:** Small; dark brown eyes; small and often indistinct rounded ear tufts; and variegated rufous and gray plumage. Pale grayish facial disk with variable rufous wash strongest around the eyes. Rufous-tinged creamy spots form bold line on scapulars. Grayish brown bill; feathered legs, bare toes.
GEOGRAPHIC VARIATION Plumage and size vary clinally. Birds in the northwestern part of the range are the most finely marked; those in the Great Basin mountains are grayish and have the coarsest markings; those in the southeast are reddish. Wingspan and weight increase from southeast to northwest, presumably correlated with migration distance.
SIMILAR SPECIES Resembles screech-owls, but it is smaller; it also has dark eyes, shorter ear tufts, and relatively longer, pointier wings.
VOICE Male's advertising song

is a series of single or paired soft, deep, hollow, short hoots repeated every 2–3 seconds, lower pitched than that of a screech-owl (which has a more varied repertoire). Single hoots are often preceded by two grace notes at a lower pitch. Hoots resemble a distant Long-eared Owl's. Female's hoots are higher pitched and more quavering. Ventriloquial; difficult to localize and usually closer than the listener estimates.
STATUS & DISTRIBUTION Uncommon to common during breeding season, but overlooked. Silent and little known the rest of the year. **BREEDING:** Inhabits primarily open montane coniferous forest, especially with ponderosa pine, mixed with oaks or aspen. **MIGRATION:** Highly migratory; winters

reddish type

birds often more reddish in southeastern part of range

short ears are often indistinct

dark eyes

in Central America. North American breeders depart late Sept.–Oct., return beginning late Mar.–late May. Rarely detected during migration.

WINTER: Range, diet, and habits are little known. **VAGRANT:** Accidental in FL, AL, LA, east TX, and 75 miles offshore in the Gulf of Mexico.

POPULATION Canada considers it a species of "special concern" because of its small population and vulnerability to habitat alteration by forest harvest.

SCREECH-OWLS Genus Megascops

Most of these small to medium-size nocturnal owls of the New World have ear tufts and short, rounded wings. They inhabit forests, open woodlands, parks, and arid open and semi-open areas. All have more than one song type and are best identified by voice, combined with habitat and range. This genus (25 sp.; 3 in N.A.) split in 2003 from the genus *Otus*.

EASTERN SCREECH-OWL Megascops asio EASO ■ 1

rufous morph more numerous in Southeast, essentially unknown in West

rufous morph

gray-morph juvenile

ear tufts

yellow eyes

pale greenish bill

strong vertical and horizontal bars on underparts

gray morph

overall pale

western Great Plains
maxwelliae

In eastern wooded suburbs, this small owl is often the most common avian predator. Its whinnying and trilling songs are familiar. It nests in old woodpecker holes or natural tree cavities and readily uses nest boxes. Eastern was formerly classified with Western as a single species; range separation is not yet fully known. Polytypic (5–6 ssp.; all in N.A.). L 8.5" (22 cm) WS 21" (53 cm)

IDENTIFICATION ADULT: Small; yellow eyes; bill yellow-green at base with a paler tip. Ear tufts are prominent if raised; when flattened the bird has a round-headed look. Facial disk is prominently rimmed dark, especially on lower half. Underparts are marked by vertical streaks crossed by widely spaced dark bars that are nearly as wide as streaks. Scapulars have blackish edged white outer webs, forming a line of white spots across shoulder. Feet are proportionally large. Occurs in rufous and gray morphs as well as intermediate brownish plumages; plumages are alike, but female is larger. Markings on underparts are less distinct on rufous morph birds. **JUVENILE:** Similar to adult in coloration, but indistinctly barred light and dark on head, mantle, and underparts; ear tufts not yet fully developed.

GEOGRAPHIC VARIATION Body size and intensity of markings vary clinally: smaller and darker in the south and east, larger and lighter in the north and west. Color morph distribution is more complex. In most areas intermediate brownish birds compose less than 10 percent of the population; but in FL, gray, rufous, and brownish birds are evidently about equally common. Rufous morph becomes more common in the Southeast and outnumbers the gray morph in some areas. Normally only gray morph birds are found on the Great Plains (*maxwelliae*, the palest and most faintly marked) and in southernmost Texas (*mccallii*).

SIMILAR SPECIES Western Screech-Owl's bill is blackish or dark gray at the base, but gray-plumaged individuals are otherwise nearly identical in appearance and habits to gray-morph Eastern Screech-Owls. Where their ranges overlap, the two species are best identified by voice.

VOICE The territorial defense song is a strongly descending and quavering trill up to three seconds long, reminiscent of a horse's whinny (apparently lacking in *mccallii*). Contact song (3–6 secs.) is a single low-pitched quavering trill; it may rise or fall slightly

at the end. Female's voice is slightly higher pitched.

STATUS & DISTRIBUTION Common. Range overlaps that of the Western Screech-Owl in eastern CO, along the Cimarron River, in extreme southwest KS, and in TX east of the Pecos River to near San Angelo. Both species are found at Big Bend N.P. in TX, where the Western is uncommon and the Eastern is rare (no recent records of Eastern); hybrids are known from there and from eastern CO. **YEAR-ROUND:** Resident in a wide variety of tree-dominated habitats: woodlots, forests, river valleys, parks, even suburban gardens below about 4,500 feet. May make local or altitudinal movements in severe winters or during food shortages. **POPULATION** Stable.

WESTERN SCREECH-OWL *Megascops kennicottii* WESO ■ 1

This small, widespread, eared owl of the West inhabits a broad range of semi-open, low-elevation habitats. At dusk it may be seen as it emerges from its roost cavity to perch and look for prey; or it may be heard at the onset of breeding season giving its familiar "bouncing ball" song. Courtship begins in January and February, with male singing near nest. Eggs are laid between late March and late April. It preys on a wide variety of small animals, especially mammals, birds, and invertebrates, and nests in old woodpecker holes or natural tree cavities. It is nocturnal and somewhat crepuscular. Polytypic (8 ssp.; 6 in N.A.). L 8.5" (22 cm) WS 21" (53 cm)

IDENTIFICATION ADULT: Small, with yellow eyes; bill blackish or dark gray at base with pale tip. Ear tufts prominent if raised; when flattened, bird looks round headed. Underparts marked by vertical streaks crossed by much narrower and more closely spaced dark bars. Feet are proportionately large. Sexes alike in plumage; females average slightly larger than males.

GEOGRAPHIC VARIATION Plumage generally monomorphic in a given area: brown or gray-brown in the Northwest, gray in southern deserts. Northwest coastal *kennicottii* more variable; generally dark brownish gray, but a small percentage is reddish. Size increases from south to north and from lowland to higher elevations. Toes feathered in northern populations, bristled in southern deserts.

SIMILAR SPECIES Whiskered is slightly smaller, with proportionately smaller feet, different colored bill, and usually bolder cross barring below. Where ranges overlap, Western is generally found at lower elevations. Eastern's bill is greenish. Best separated from other screech-owls by voice. Flammulated Owl has dark eyes, is much smaller, and generally occupies higher elevations in areas where the species overlap.

VOICE Two songs; both are common. A two-part tremulous whistled trill, first part short and second long, dropping slightly near the end: *dddd-ddddddddr;* also a short sequence of 5–10 hesitating notes, accelerating in "bouncing ball" rhythm, ending in a trill: *pwep pwep pwep pwep pwepwep-wepepepep.* Both sexes sing each song;

dark bill

weaker crossbars than Eastern and Whiskered

Northwest coast *kennicottii*

the female's voice is higher pitched. Agitated barking notes often given in advance of the song.

STATUS & DISTRIBUTION Common. YEAR-ROUND: Inhabits open woodlands, streamside groves, deserts, suburban areas, and parks. Range has expanded eastward along the Arkansas River in CO, along the Cimarron River into extreme southwestern KS, and across the Pecos River in TX, increasing its area of sympatry with Eastern Screech-Owl; slow eastward expansion may also be underway. Hybrids with Eastern are known from CO and from Big Bend N.P., TX.

POPULATION No data, but probably declining slowly as habitat is lost.

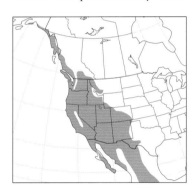

WHISKERED SCREECH-OWL *Megascops trichopsis* WHSO ■ 2

After hiding by day close to a tree trunk or in dense foliage, the nocturnal, mainly insectivorous Whiskered Screech-Owl becomes active at dusk. At the onset of the breeding season, the pair's duets reveal their presence, and imitating their song will often bring them into close view. They nest in natural tree cavities or abandoned woodpecker holes. Polytypic (3 ssp.; *aspersus* in N.A.). L 7.2" (18 cm) WS 18" (46 cm)

IDENTIFICATION ADULT: Small, gray overall with yellow eyes; bill yellowish green at base with pale tip. Facial disk feathers at base of bill have whisker-like extensions. Ear tufts prominent if raised; when flattened, bird looks round headed. Small feet.

SIMILAR SPECIES Very similar to Western Screech-Owl (see above).

VOICE Territorial song a series of 4–8 equally spaced notes, *po po PO po po po,* mostly with emphasis on the third note, falling slightly in pitch at the end. Courtship song a syncopated series of hoots, like Morse code: *pidu po po, pidu po po, pidu po*

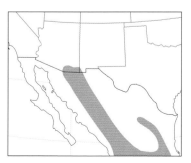

po (pitch same as in Western Screech-Owl), often in a duet with the slightly higher-pitched female.

STATUS & DISTRIBUTION Common. YEAR-ROUND: Inhabits dense oak

dark bill

strong crossbars like Eastern Screech-Owl

aspersus

small feet

and oak-conifer woodlands in the mountains of southeastern AZ and adjacent NM, generally at

higher elevations than the Western Screech-Owl. Found at 4,000–6,000 feet, but mostly around 5,000

feet. **WINTER:** May move to lower elevations.
POPULATION Trends not known.

EAGLE-OWLS Genus *Bubo*

These nocturnal owls, most of which possess prominent ear tufts and powerful talons, are found in Eurasia, Indonesia, Africa, and the Americas. They occur in virtually all habitats except the densest forests and the largest deserts. About 16 species are currently recognized (2 in N.A.). Some authorities include the fish-owls of Asia (genus *Ketupa*) in *Bubo*.

GREAT HORNED OWL *Bubo virginianus*   GHOW   1

Many birders first meet this formidable owl in late winter by finding a female sitting in a large stick nest in a leafless tree. Others are introduced to it as a hulking, eared shape atop a power pole at dusk—perhaps a male bending nearly horizontally as it sings. Primarily a nocturnal perch hunter, Great Horned Owl is a fierce predator that takes a wide variety of prey, but most commonly mammals, up to the size of a large hare. It favors disused tree nests of other large species, such as Red-tailed Hawk, for nesting but also uses cavities in trees or cliffs, deserted buildings, and artificial platforms. It breeds early, with first eggs laid by January in Ohio, later farther north. Young climb onto nearby branches at six to seven weeks and fly well from approximately ten weeks. It often spends its daylight hours dozing in a tree, where the raucous cawing of a chorus of crows may lead one to the bird. Polytypic (approx. 16 ssp.; 9 in N.A.). L 22" (56 cm) WS 54" (137 cm)
IDENTIFICATION In flight, wings are broad and long, pointed toward tip; ear tufts are usually flattened and head is tucked in, producing a blunt profile. Flight is direct; wingbeats are stiff, steady, and mostly below the horizontal. **ADULT:** A very powerful, bulky owl with a broad body. Females larger than males and generally darker. Large head, stout ear tufts, and staring yellow eyes create a catlike appearance. Broad facial disk rimmed with black; whitish superciliary "eyebrows." White foreneck often conspicuous and ruff-like, especially when vocalizing. Upper chest coarsely mottled; rest of underparts crossbarred. Gray bill; densely feathered legs and toes. Plumage and size vary geographically. **JUVENILE:** Downy plumage grayish to buff, with dusky barring. By about October, young birds acquire a complete set of flight feathers and tail feathers that differ in pattern, shape, and wear from those of adults: they have broader and more numerous crossbars, wings with uniformly fresh-

looking rather than variably worn flight feathers, and tail feathers tapered rather than blunt ended.
GEOGRAPHIC VARIATION In general, birds of the eastern subspecies (inc. nominate) are medium size and brownish, with medium pale feet. Birds of the Pacific coast subspecies group are small and dark, with dusky feet. Birds of the interior West subspecies group are large and variably pale, with whitish feet. The palest subspecies, *subarcticus,* is resident across the far north-central portion of the range and has wandered southeast in winter as far as New Jersey.
SIMILAR SPECIES Long-eared Owl is smaller and more slender and weighs much less; it has a dark vertical stripe through its eye and longer, more closely set ear tufts; it also lacks the white throat. Its flight is floppier, and its voice is different. All other large North American owls lack ear tufts. Female and young Snowy Owls somewhat resemble *subarcticus,* but they have a white face and lack ear tufts.
VOICE SONG: Territorial song is a series of three to eight loud, deep hoots in a rhythmic series; the second and third hoots are often short and rapid: commonly *hoo hoo-HOO hoooo hoo;* often longer, *hoo huhuHOO hooooo hoo.* Mostly heard near

dusk and dawn. Male territorial singing begins about November. Duetting commonly begins one to two months before the first egg is laid. Female's voice is higher pitched, closer to a Mourning Dove's; juvenile gives a loud raspy screech.
STATUS & DISTRIBUTION Common and widespread. **YEAR-ROUND:** Resident, sedentary, and territorial within the varied habitats of its breeding range, from forest to city to open desert.
POPULATION Stable.

prominent broad ear tufts on sides of head
color of facial disk varies geographically
pale overall
bulky body shape
barred below
subarcticus

SNOWY OWL *Bubo scandiacus* SNOW ▪ 2

During winter, this charismatic owl might appear far to the south. It will forage at all hours during continuous summer light, mainly from a perch. Lemmings and voles are its main prey. Monotypic. L 23" (58 cm) WS 60" (152 cm)

IDENTIFICATION Flight strong, steady, direct, jerky. Often glides on horizontal wings. **ADULT:** White, large; rounded head; small yellow eyes; blackish bill; heavily feathered legs, feet. Females larger than males. Male's plumage usually broken with narrow, sparse dark bars or spots.

Female's markings larger, darker. Markings less intense as birds age; old males may be pure white. **IMMATURE:** More heavily marked than adult females. **VOICE** Fairly vocal when breeding, otherwise largely silent. Male's song a far-carrying series of two to six deep, low hoots, last often the loudest; higher-pitched female rarely hoots. When disturbed, either sex may utter a repeated *kre* call. High-pitched, drawn-out scream protests intrusion into winter territory.

STATUS & DISTRIBUTION Uncommon to fairly common. **BREEDING:** Open tundra; nests on ground. Nomadic; breeds only where and when prey is abundant. **WINTER:** Retreats from northernmost part of its range. Irruptive; plummeting lemming populations may drive birds as far south as central CA and northern TX, AL, and FL. **VAGRANT:** Accidental to Oahu, HI, and Bermuda.

POPULATION Overall numbers are presumed to have changed little in North America.

round head

immature

color varies from all-white to heavily barred depending on age and sex

---

HAWK OWLS Genus *Surnia*

NORTHERN HAWK OWL *Surnia ulula* NHOW ▪ 2

This largely diurnal Holarctic owl is often seen perched atop a conifer in a mostly open area. Its flight is low and fast (with quick, stiff wingbeats) and highly maneuverable; it occasionally hovers. It stoops onto nearby prey with a smooth, gliding dive off its perch. Nests are found in cavities atop broken trunks, natural tree holes, and old holes of large woodpeckers. Polytypic (3 ssp.; *caparoch* in N.A.). L 16" (41 cm) WS 32" (81 cm)

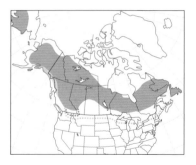

IDENTIFICATION ADULT: Has a long, graduated tail; a falconlike profile; and a black-bordered facial disk. Its underparts are barred brown.
GEOGRAPHIC VARIATION Eurasian nominate subspecies is paler overall than the North American *caparoch;* two supposed AK records of the nominate subspecies may pertain instead to pale *caparoch.*
SIMILAR SPECIES Much smaller Boreal Owl is strictly nocturnal, has short wings and tail, and is streaked below.
VOICE Male's song is a trilling, rolling whistle, *ululululululul . . . , ±12* notes per second, lasting up to 14 seconds, then repeated. Heard mainly at night; reminiscent of the Boreal's song, but longer, higher pitched, sharper.
STATUS & DISTRIBUTION Uncommon. **BREEDING:** Across taiga belt, from the edge of the forested steppe north to timberline. Nomadic; breeds only where and when prey is abundant. **WIN-TER:** Retreats slightly from northern-

most part of mapped range. Irruptions probably tied to vole population cycles, which occasionally send large numbers of birds farther south. Has reached OR, NE, OH, PA, and NJ in winter; accidental on Bermuda.
POPULATION Stable.

black border to facial disk

caparoch

dark barring on underparts

long tail

PYGMY-OWLS Genus *Glaucidium*

These small but aggressive long-tailed owls (approx. 30 sp.; 2 in N.A.) lack ear tufts; many are mainly diurnal. Best identified by voice where ranges overlap, they inhabit deserts, deciduous bottomlands, wooded foothills, and high-elevation coniferous forests; they occur on all temperate continents except Australia. Many species prey mainly on other birds.

NORTHERN PYGMY-OWL *Glaucidium gnoma* NOPO ▪ 2

This aggressive diurnal predator sometimes catches birds larger than itself. Mobbing songbirds may lead a birder to the owl. Chiefly diurnal, it is most active at dawn and dusk. It makes its nests in natural tree cavities and old woodpecker holes. Polytypic (7 ssp.; 5 in N.A.). L 6.7" (17 cm) WS 15" (38 cm)
IDENTIFICATION ADULT: Long tail, dark brown with pale bars; rusty brown or gray-brown upperparts; spotted crown; underparts white with dark streaks; prominent "false eyes" on nape. Females average redder or browner, males grayer. **JUVENILE:** Crown spots indistinct or lacking; other whitish markings indistinct. Flight is undulating, with bursts of quick wingbeats; not as quiet as nocturnal owls.
GEOGRAPHIC VARIATION Birds of interior West subspecies *pinicola* grayest; Pacific birds browner. (See sidebar below.)
SIMILAR SPECIES Other small owls within Northern Pygmy-Owl's geographic and elevation range have short tails, lack nape marks, and have different plumage patterns.
VOICE SONG: A mellow, whistled, *took* or *took-took*, repeated in a well-spaced series. (See sidebar below.) Songs of interior subspecies are lower pitched and faster than

"NORTHERN" group

hoskinsii "CAPE" group

gnoma "MOUNTAIN" group

Northern group type | "false eye" | spotted crown

long tail with pale bars | averages grayer than birds from Pacific region | Mountain group type

those of the coastal subspecies. Also gives a rapid series of *hoo* or *took* notes followed by a single *took*.
STATUS & DISTRIBUTION Uncommon. **YEAR-ROUND:** Dense woodlands in foothills and mountains. In colder areas of its range, descends downslope in winter. **POPULATION** No data on historical changes.

Identification of "Mountain Pygmy-Owl"

Some authorities treat *Glaucidium gnoma gnoma* as a species distinct from Northern Pygmy-Owl, calling it "Mountain Pygmy-Owl." It ranges from the densely wooded foothills and forests of mountains in southeastern Arizona and southwestern NM (Hidalgo Co.), south through Mexico's interior highland forests. In plumage, "Mountain Pygmy-Owl" closely resembles the slightly larger *pinicola*, a subspecies of Northern Pygmy-Owl found in the interior West. Like *pinicola*, "Mountain Pygmy-Owl" is grayer than other Northern Pygmy-Owls, which often have rusty-brown upperparts; it also has a spotted crown; white, dark-streaked underparts; and yellow eyes. In the field, "Mountain Pygmy-Owl" and *pinicola* can be separated reliably only by voice.
"Mountain Pygmy-Owl's"

"Coastal" Northern Pygmy-Owl (CA, Jan.)

"Mountain" Northern Pygmy-Owl (AZ, May)

song is a series of mellow, whistled, and paired *took-took*s, about one pair per second on about the same pitch as *pinicola*'s single *took*s; it often starts hesitantly and then runs into prolonged song, mostly of paired *took-took* notes, but with single notes occasionally thrown in.

Owl vocalizations are inherited, not learned. Within the Northern Pygmy-Owl complex, some authorities elevate *hoskinsii* of southern Baja California—with its slower, paired *took-took* notes—to species status, calling it the "Cape" (or "Baja") Pygmy-Owl. Some also suggest species status for *cobanense*, the "Guatemalan Pygmy-Owl" of Chiapas to Honduras, in which the reddish morph predominates. The song is a fast series of generally paired *took-took* notes, with tripled notes, trills, and single notes thrown in occasionally. ■

FERRUGINOUS PYGMY-OWL *Glaucidium brasilianum* FEPO ▪ 3

Ferruginous Pygmy-Owl is scarce at the northern limit of its range. Still, a birder walking its arid habitat might hear it calling or chance upon it hunting quietly during the day. It roosts in crevices and cavities and nests chiefly in old woodpecker holes or natural tree cavities. Polytypic (approx. 12 ssp.; 2 in N.A.). L 6.7" (17 cm) WS 15" (38 cm)
IDENTIFICATION Not as quiet in flight as nocturnal owls. Flight over longer distances is rather straight, with several rapid wingbeats alternating with glides; like a woodpecker. **ADULT:** Long tail, reddish with dark or dusky bars. Gray-brown upperparts; faintly streaked crown. White throat puffed out and more apparent when bird sings. Yellow eyes; prominent "false eyes" on nape. White underparts streaked reddish brown. **JUVENILE:** Like adults, but crown streaking very indistinct or lacking.
GEOGRAPHIC VARIATION Arizona's *cactorum* is slightly grayer than Texas's *ridgwayi*. Some authorities treat these two subspecies as constituting a separate species, "Ridgway's Pygmy-Owl" (*G. ridgwayi*), and S.A. birds as another species (*G. brasilianum*).
SIMILAR SPECIES The much smaller, nocturnal Elf Owl is short tailed and lacks false eyes. Northern Pygmy-Owl's range overlaps in Arizona, but it inhabits higher elevations, its crown and breast are spotted, and it has fewer, more-whitish tail bands.
VOICE The male's territorial and courtship song is a series of upslurred, high-pitched whistled *pwip!* notes, monotonously delivered at about 2.5 notes per second. Phrases of about ten to more than 100 notes are repeated for minutes at a time, spaced 5–10 seconds apart. Female's voice is higher and wheezier.

streaked crown
"false eye"
cactorum
long tail with rusty bars

STATUS & DISTRIBUTION Rare sedentary resident in southeast AZ; uncommon sedentary resident in south TX. Common throughout most of its range south of the U.S. **YEAR-ROUND:** Inhabits live oak-honey mesquite woodlands, mesquite brush, and riparian areas in TX, where insects are its main prey. In AZ, inhabits riparian woodlands and Sonoran desert scrub (inc. saguaro cactus), hunting mainly reptiles, birds, and small mammals.
POPULATION Much reduced, probably because of loss and fragmentation of habitat. The subspecies *cactorum* was federally listed in Arizona until recently.

Genus *Microthene*

ELF OWL *Microthene whitneyi* ELOW ▪ 2

This is North America's smallest owl. It nests and roosts in old woodpecker holes or other cavities in saguaros and trees. Polytypic (4 ssp.; 2 in N.A.). L 5.8" (15 cm) WS 14.7" (37 cm)
IDENTIFICATION Flight composed of uniformly rapid wingbeats in straight-line hunting strikes; less quiet than other nocturnal owls. Occasionally glides and hovers while feeding. **ADULT:** Tiny; round head; yellow eyes; lacks ear tufts. Wings fairly long, tail very short. Upperparts grayish brown to brown with buff to cinnamon spots on widespread southwest *whitneyi;* face and underparts with substantial cinnamon. **JUVENILE:** Like adult, but head and back markings slightly less distinct.
GEOGRAPHIC VARIATION Weak. Upperparts of *idonea* (south TX) are grayish, with little or no brown; face and underparts have little or no cinnamon.
SIMILAR SPECIES Pygmy-owls are chiefly diurnal, with long tails and "false eyes." The larger screech-owls have ear tufts and streaks below.
VOICE The male's territorial and courtship song is an irregular series of usually 5–7 high-pitched *churp* notes, delivered at ±5 notes per second. Both sexes utter a short, soft, whistle-like contact call, *peeu;* sometimes precedes male's song.
STATUS & DISTRIBUTION Fairly common to common. **BREEDING:** Inhabits desert lowlands, foothills, and canyons, especially among oaks and sycamores. **MIGRATION:** Northern populations migrate to southern Mexico by Oct. and return by Mar., males first. Casual in winter in southernmost TX. **VAGRANT:** Accidental in fall to Los Angeles Co., CA.
POPULATION Numbers reduced where habitat has been destroyed or degraded. Extirpated in southern NV; state-listed as endangered in CA, where nearly extirpated.

yellow eyes
rounded head with no ear tufts
whitneyi
tiny size
very short tail

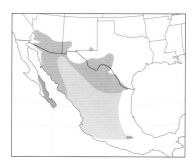

Genus Athene

BURROWING OWL *Athene cunicularia*   BUOW ▪ 1

During daylight, this owl often perches conspicuously at the entrance to its burrow or on a low post. It bobs its head to better gauge an intruder's distance. Primarily nocturnal, it hunts insects, small mammals, and birds from a perch or in flights, where it often hovers. It nests singly or in small colonies; western *hypugaea* is often associated with burrowing mammal colonies, where it usually occupies a disused mammal burrow; *floridana* usually excavates its own burrow. Polytypic (±20 ssp.; 2 in N.A.). L 9.5" (24 cm) WS 23" (58 cm) **IDENTIFICATION** Ground dweller; long legs, unlike all other small owls. **ADULT:** Round head lacking ear tufts. **JUVENILE:** Plain brown upperparts with little or no distinct marking; dark brown chest; pale buff underparts.

GEOGRAPHIC VARIATION Western *hypugaea* pale brown with buff mottling and spotting. Slightly smaller *floridana* is darker brown, with whitish mottling and spotting, and less buffy below. **VOICE** Male's primary song is a soft, repeated *coo-cooooo*, reminiscent of Greater Roadrunner; also gives a chattering series of *chack* notes. **STATUS & DISTRIBUTION** Species' range extends to southern S.A. **BREEDING:** Open country, grasslands, golf courses, airports. **MIGRATION:** Most northerly

populations migrate. **VAGRANT:** Casual in spring and fall to western and southern ON, southern QC, ME, and NC; to AL and northwestern FL on Gulf Coast. Either subspecies might be found along Atlantic coast. **POPULATION** Greatly reduced in much of the northern Great Plains by extermination of prairie dogs, conversion of prairies to cultivation, pesticide use, and habitat destruction. Declines continue. Many states list it as endangered or of "special concern"; endangered in Canada, with fewer than 1,000 pairs thought to remain.

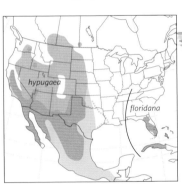

western
hypugaea

heavily spotted
with buffy white
above

round head

juvenile

no dark
barring
below

western
hypugaea

WOOD OWLS *Genera Ciccaba and Strix*

Medium-large to large nocturnal owls with large eyes and large, rounded heads, lacking ear tufts. Their plumage is mostly cryptic; several species have different color morphs. They occupy wooded habitats in much of the world and prey mainly on small mammals and birds. (*Ciccaba* often merged in *Strix*; approx. 20 sp.; three resident, one accidental in N.A.)

MOTTLED OWL *Ciccaba virgata*   MOOW ▪ 5

This medium-size, nocturnal owl inhabits wooded habitats in C.A. and S.A. Polytypic (7 ssp.; likely *tamaulipensis* from northeastern Mexico in N.A.). L 14" (36 cm) WS 33" (84 cm) **IDENTIFICATION** **ADULT:** Round head, no ear tufts; dark brown eyes. Light morph in drier areas (e.g., northern Mex.) has brown facial disk with bold white brows and whiskers; dark brown above with faint brownish barring; mottled dark brown on chest, rest of underparts streaked dark brown. Morph in more humid areas is somewhat darker all over; some individuals much darker. **GEOGRAPHIC VARIATION** Northerly subspecies are smaller, with fine distinct barring above; those in Amazon region

are larger and more reddish brown. **SIMILAR SPECIES** Closely related, larger Barred Owl has prominent barring across upper chest and paler facial disk. **VOICE** Territorial song deep hoots: a single *wh-OWH* and *WOOH,* and longer series, 3–10 hoots, often accelerating and becoming stronger before fading. **STATUS & DISTRIBUTION** **RESIDENT:** *squamulata* from southern Sonora south to Guerrero; *tamaulipensis* from central Nuevo León and Tamaulipas; other ssp. south to northeastern Argentina. **VAGRANT:** One certain record: a likely roadkill found near Bentsen–Rio Grande S.P. (Feb. 23, 1983). An accepted record from Weslaco, TX (July 5–11, 2006) is problematical.

brown facial
disk bordered
by white

round
head
with
dark
eyes

streaked
underparts

SPOTTED OWL *Strix occidentalis*　　SPOW ▪ 2

This gentle-looking owl can be quite confiding when found dozing on a shaded limb during the day, perhaps alongside its mate. Imitating its call at dusk in a mature western forest may summon a mellow, echoing response from a far-away owl, but sometimes a Spotted Owl will fly in silently and unseen and then announce its presence with hair-raising barks. From a perch, it seeks wood rats, flying squirrels, and other small mammals. It nests mainly in tree cavities but also in debris on tree limbs or in old nests of other species. In the southwest, it also uses cliff ledges and caves. Polytypic (3–5 ssp.; 3 in N.A.). L 18" (46 cm) WS 41" (104 cm) **IDENTIFICATION** Strictly nocturnal. Flight is slow and direct, with methodical wingbeats interspersed

color varies from darkest (Northwest) to palest (Southwest)

dark eyes

spotted below, including on breast

with gliding. **ADULT:** Brown overall with round, elliptical, or irregular white spots on its head, back, and underparts; lacks ear tufts. Indistinct concentric circles set in a rounded facial disk emphasize its dark eyes. Sexes alike in plumage; female larger. **GEOGRAPHIC VARIATION** Three subspecies generally recognized in N.A. vary in color (dark to light) and spots (small to large) from north to south. "Northern Spotted Owl" *(caurina)*, which resides from Marin Co., CA, to BC, is the largest. "California Spotted Owl" (nominate) is resident in CA in the Sierra Nevada and from Monterey Co. southward. "Mexican Spotted Owl" *(lucida,* or *huachucae,* according to some authorities) is resident in the Southwest and in Mexico. Some authorities recognize up to two additional subspecies in Mexico. **SIMILAR SPECIES** The slightly larger Barred Owl is barred and streaked below, not spotted. Hybridization has occurred. See Barred Owl. **VOICE** A series of usually four nearly monotonic doglike barks, *hoo! hu-hu hooooh,* is used by either sex to proclaim and defend territory. Female's is higher pitched. Variants include ending with a sharper, louder bark, *hoo! hu-hu ow!,* and renditions of 7–15 notes in a series. Contact call, given mainly by females, is a hollow, upslurred whistle, *cooweeeeip!* Females use a rapid series of 3–7 loud barking notes, *ow!-ow!-ow!-ow!-ow!,* during territorial disputes.

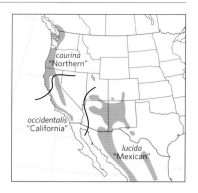

caurina "Northern"

occidentalis "California"

lucida "Mexican"

STATUS & DISTRIBUTION Uncommon resident; decreasing in number due to habitat destruction. Further threatened by predation and hybridization as Barred Owl expands into its range in the Northwest and CA. **YEAR-ROUND:** Inhabits mature coniferous and mixed forests and wooded canyons, usually with multileveled, closed canopies and uneven-aged trees. Some individuals descend to lower elevations in winter. **POPULATION** Relatively widespread within ranges that are presumed to have changed little overall, but numbers probably have declined dramatically within specific habitats because of clear-cutting and even-aged forest management. "Northern Spotted Owl" is listed as threatened in the U.S. and endangered in Canada, where it is estimated that fewer than 100 pairs remain. "Mexican Spotted Owl" is also considered threatened in the U.S.

BARRED OWL *Strix varia*　　BADO ▪ 1

This widespread woodland owl dozes by day on a well-hidden perch but seldom relies on its good camouflage to avoid harm. Instead, it flies away at the least disturbance, seldom tolerating close approach. But where foot traffic is heavy, such as along boardwalks in southern swamps, Barred Owl may sit tight and provide good views at close range. It hunts mainly from a perch but will also hunt on the wing, preying on small mammals, birds, amphibians, reptiles, and invertebrates. It prefers to nest in a natural tree hollow, but it will also use an abandoned stick nest of another species. Polytypic (4 ssp.; 3 in N.A.). L 21" (53 cm) WS 43" (109 cm) **IDENTIFICATION** Large and chunky. Flight is heavy and direct, with slow,

methodical wingbeats; occasionally makes long, direct glides. **ADULT:** Dark brown barring on its ruff-like upper breast; rest of underparts are whitish with bold, elongated dark brown streaks; lacks ear tufts. Central tail feathers expose 3–5 pale bars between their tips and tips of uppertail coverts. Sexes alike in plumage; female larger. Chiefly nocturnal. **JUVENILE:** By about September, young birds acquire a complete set of fresh flight feathers; may be variably worn on adults. Central tail feathers expose 4–6 pale bars. **GEOGRAPHIC VARIATION** Weakly differentiated. Southeastern *georgica* is darker brown than the widespread nominate *varia.* Subspecies *helveola* of southeast Texas is paler.

SIMILAR SPECIES The underparts of the slightly smaller Spotted Owl are spotted overall, not barred and streaked. Hybridization has occurred where

ranges overlap; the hybrids' plumages, voices, and sizes are intermediate between the two species.

VOICE Highly vocal, with a wide range of calls. Much more likely than other owls to be heard in the daytime. Its most common vocalization is a rhythmic series of loud *hoot* or *whoo* notes: *who-cooks-for-you, who-cooks-for-you-all.* Also often heard is a loud, drawn-out *hoo-waaah* that gradually fades away; it is sometimes preceded by an ascending agitated barking. Often a chorus of two or more owls will call back and forth with these and other calls; the female's voice is higher pitched.

STATUS & DISTRIBUTION Common in eastern N.A. Has expanded its range westward in the 20th century along Great Plains river corridors and through Canada's boreal forest to southeastern AK, WA, OR, and northern CA. **YEAR-ROUND:** Resident in mature mixed deciduous and uniform coniferous forests, often in river bottomlands and swamps; also in upland forests. **MIGRATION:** More northerly populations may drift south during late autumn if prey are scarce. **VAGRANT:** Accidental on Bermuda.
POPULATION Stable to increasing in North America.

dark eyes

barred breast

vertical streaks

GREAT GRAY OWL *Strix nebulosa*   GGOW ■ 2

Surprisingly, our largest owl is far from our heaviest: Great Gray Owl's uncommonly dense, fluffy feathering creates an illusion of great bulk. This Holarctic owl is most often found perched low on the edge of a clearing, listening and looking intently for the small rodents on which it preys. It hunts from dusk to just before dawn from a low perch overlooking an open area; during breeding season, it may also hunt in the daytime. It is able to take prey moving unseen beneath snow by plunge-diving feet first into shallow snow or head-first into deeper snow, before thrusting its legs to grasp its target. It favors the abandoned

nests of other birds of prey, but it also uses broken tops of large trees, preferring shaded locations. Polytypic (2 ssp.; nominate in N.A.). L 27" (69 cm) WS 55" (140 cm)
IDENTIFICATION Its flight is heron-like, with slow and deep wingbeats and very little gliding. It often hovers above a suspected prey location before pouncing. **ADULT:** Relatively long tailed, with a disproportionately large head; lacks ear tufts. Pale gray facial disk with (usually) five distinct dark gray concentric circles around each yellow eye, making the eyes seem particularly small. Black-and-white "bow tie" pattern beneath the bill. Plumage largely gray, subtly patterned with combinations of whites, grays, and browns. Upperparts marked with dark and light; underparts boldly streaked over fine barring. Yellowish bill; legs and toes heavily feathered. Sexes alike in plumage; female distinctly larger. **JUVENILE:** Downy plumage is cryptic gray and white. By about August, young birds acquire a complete set of uniformly fresh flight feathers, not variably worn as on adults; terminal bands on flight feathers and tail feathers are whitish when present.
SIMILAR SPECIES Great Gray Owl is unlikely to be mistaken for any other species. The smaller Barred and Spotted Owls have dark eyes, browner plumage overall, and different patterns on their underparts. The similar-size Great Horned Owl has prominent ear tufts and larger yellow eyes; the northern subspecies of

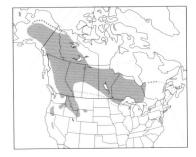

Great Horned *(subarcticus)* is grayest and palest.
VOICE The male's courtship song is a series of 5–10 deep, resonant, muffled *whoo* notes that are longer and higher pitched than the Sooty Grouse's; they gradually drop in frequency and decelerate toward the end of the series. The song may be repeated as many as ten times. The female may answer with a soft, mellow *whoop*.
STATUS & DISTRIBUTION Uncommon. Inhabits boreal forests and wooded bogs in the far north, dense coniferous forests with meadows in mountains farther south. **BREEDING:** Nomadic; breeds where it finds abundant prey and may not breed in years that prey is scarce. **WINTER:** Withdraws downslope or southward, where it is a rare and irregular visitor to the limit of the dashed line on the map. Extreme prey shortages in the north may cause tens or hundreds of owls to winter in areas where normally few are seen. An unprecedented 1,700 Great Gray Owls were recorded in northern MN during the 2004–2005 irruption.
POPULATION Seems to be stable.

huge size

dark rings on facial disk

black and white at bottom of facial disk gives "bow tie" effect

nebulosa

EARED OWLS Genus Asio

Most of these medium-size owls have prominent ear tufts, long wings, well-developed facial disks, bold streaking below, and cryptic plumage. Some are open-country specialists; others occupy wider habitat ranges (e.g., forests, forest edges, shrubby growth, and open country). They occur on all temperate continents except Australia (7 sp.; 3 in N.A.).

LONG-EARED OWL Asio otus   LEOW ▪ 2

long ear tufts that are close together

rufous facial disk

blackish around eyes

overall slender body

heavily streaked and barred below

The Holarctic Long-eared Owl is occasionally seen on its day roost in a dense tree or thicket. If it does not flush immediately, the owl may try to hide, stretching into an unbelievably slim, erect profile, often with a wing cloaking its body. With its facial disk narrowed, its eyes reduced to vertical slits, and its ear tufts raised like little twigs, it can look remarkably like a broken or decayed limb. It hunts small rodents nocturnally, coursing low over the ground, and may occasionally hover before pouncing. It makes a nest in another bird's abandoned stick nest, preferably in a clump of trees rather than in an isolated tree. Polytypic (4 ssp.; 2 in N.A.). L 15" (38 cm) WS 37" (94 cm)

IDENTIFICATION Ear tufts flattened in flight, not visible. (See sidebar opposite.) **ADULT:** Slender; medium size; long, close-set ear tufts. Rusty facial disk; yellow eyes set in vertical dark patches. Upperparts a mix of black, brown, gray, buff, and white. Bold streaks and bars on breast and belly. Central tail feathers expose 5–7 dark bars between their tips and tips of the uppertail coverts. Females average darker, richer buff, more heavily streaked. **JUVENILE:** Tail feathers have more exposed dark bars. By about October, it acquires a set of uniformly fresh flight feathers; adults' secondaries and outer primaries may be variably worn.

GEOGRAPHIC VARIATION Weak and clinal in North America where widespread *wilsonianus* meets the smaller and paler *tuftsi* of western Canada south to northwest Mexico.

SIMILAR SPECIES The larger, much heavier Great Horned Owl has a white throat; its eyes lack dark vertical patches. Short-eared Owl's ear tufts are tiny; its underparts are boldly streaked.

VOICE Generally silent except in breeding season. Male's advertising song is a series of ten to more than 200 low, soft *hooo* notes, evenly spaced about 2–4 seconds apart, each note resembling the cooing note of a Band-tailed Pigeon. The female's answering call is a softer, higher-pitched *whoof-whoof-whoof*. The alarm call is a harsh, barking *ooack ooack ooack.*

STATUS & DISTRIBUTION Rare in Southeast; uncommon elsewhere in N.A.

BREEDING: Densely wooded areas in lowlands and mountains with nearby open foraging areas having fertile, boggy, or arid sandy soil. **MIGRATION:** Those breeding in northern parts of the range with heavy winter snow cover migrate south to areas with more favorable climates; casual on Bermuda, Cuba. **WINTER:** Day-roosting groups of 10–100 or more cluster close together in one or a few neighboring trees.

POPULATION Numbers much reduced in some areas. State-listed as endangered in Illinois, threatened in Iowa, and a species of "special concern" in many other states.

STYGIAN OWL Asio stygius   STOW ▪ 5

close-set ear tufts

dark facial disk

robustus

This is a medium-size, strictly nocturnal forest-dwelling owl. Polytypic (6 ssp.; likely *robustus* in N.A.). L 17" (43 cm) WS 42" (107 cm)

IDENTIFICATION ADULT: Deep chocolate-brown overall, with long, close-set ear tufts. Whitish forehead contrasts with blackish facial disk. Upperparts blackish brown, barred and spotted buff. Underparts dirty buff, with bold dark brown streaks and crossbars. Legs feathered; toes bristled.

SIMILAR SPECIES The closely related Long-eared Owl lacks bold white forehead; rusty facial disk contrasts with brownish plumage; underparts less boldly patterned; legs and toes fully feathered.

VOICE Male's song is a series of deep, emphatic *woof* or *wupf* notes spaced 6–10 seconds apart.

STATUS & DISTRIBUTION Little known. Resident in Mexico (within 200 mi. of U.S. in Sierra Madre Occidental) and the Caribbean locally to northern Argentina. **VAGRANT:** Accidental at Bentsen-Rio Grande S.P., southern TX (Dec. 9, 1994, and Dec. 26, 1996) Subspecies not known, but likely *robustus* (eastern Mex. to C.A.).

POPULATION Considered endangered by the Mexican government.

SHORT-EARED OWL *Asio flammeus* SEOW | 1

There is no more captivating spring-time sight on the western prairies than a male Short-eared Owl in courtship song high overhead and then dropping toward the ground with wing claps that sound like a flag flapping in the wind. It hunts voles and lemmings at any time of the day but is usually crepuscular, coursing low over open areas. Its nest is a shallow ground scrape, usually under a protective cover of grass or low vegetation.

slow, floppy wingbeats
dark primary covert patch on underwing
buffy and blackish wing patches
flammeus
very short ear tufts
blackish around eyes
heavily spotted above
flammeus
streaked below

Polytypic (10 ssp.; 2 in N.A.). L 15" (38 cm) WS 39" (99 cm).
IDENTIFICATION MALE: Holarctic *flammeus* has a boldly streaked tawny breast, paler and more lightly streaked belly. Large, round head; closely spaced ear tufts barely visible; yellow eyes in blackish patches. Upperparts mottled brown and buff, appearing distinctly striped; tawny uppertail coverts. **FEMALE:** Larger, generally darker: upperparts browner; underparts rustier, heavier streaking.
GEOGRAPHIC VARIATION Both sexes of smaller *domingensis* ("Antillean Short-eared Owl") resemble female *flammeus,* but underparts buffier overall with very fine or no streaking on belly; upperparts and uppertail coverts dark brown.
SIMILAR SPECIES Closely resembles Long-eared Owl in flight (see sidebar). Northern Harriers display white uppertail coverts in flight. They compete with Short-eared Owls for prey during daylight hours and sometimes rob them.
VOICE Male's courtship song is a series of 13–16 rapid, deep hoots. Both sexes give hoarse *cheeaw* calls when disturbed at any time of year.
STATUS & DISTRIBUTION Nominate still fairly common in much of the northern part of its N.A. range. Rare visitor

domingensis occurs on the Florida Keys and Dry Tortugas in spring and summer. Most appear to be juveniles, probably dispersing from Cuba, where locally common. **BREEDING:** Tundra, prairie, coastal grasslands, and marshes. **MIGRATION:** Southward to escape areas with complete snow cover. Numerous records far at sea and colonization of distant islands attest to the species' propensity to wander; very rare on Bermuda. **WINTER:** More gregarious; flocks may roost together, occasionally in trees.
POPULATION Has declined in many areas. Canada lists it as a species of "special concern" because of loss of habitat. Listed as endangered, threatened, or "special concern" species by seven northeastern states, where it is all but eliminated as a breeder.

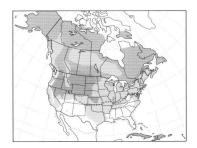

Flight Identification of Long-eared and Short-eared Owls

The Long-eared Owl and the Short-eared Owl appear very similar in flight. They hunt by coursing low over open ground in slow, moth-like flight. Short-eared generally hunts during crepuscular hours, and Long-eared generally hunts at night. Their upper wings have a broad tawny patch at the base of the outer primaries. Their underwings show a distinct carpal mark ("wrist mark"), formed by dark brown to black tips on the outermost under primary coverts. Both marks are usually more prominent on the Short-eared, but differences are not diagnostic. Identification can be made, however, by noting underwing details in flight.

Short-eared Owl's outer four or five primaries are broadly tipped dark brown (with small buffy spots intruding). Slightly nearer the bases of the primaries, a single

dark brown band cuts across the outer primaries, separated from the dark tips by an equally wide buffy band. Two or three narrower, indistinct dark bands cut across the secondaries and the inner primaries near their tips. The overall impression is of a pale underwing with a dark wrist mark, broad dark outer primary tips, and a single dark band in between, close to the dark tips.

Long-eared Owl's outer four or five primaries are crossed by five or more pairs of alternating dark and light gray-brown bands of varying width. The overall impression is of a mainly pale, but patterned, underwing with a dark wrist mark and a series of dark bands extending more than halfway from the primary tips to the wrist mark. ∎

Short-eared Owl

Long-eared Owl

FOREST OWLS Genus *Aegolius*

Small and nocturnal, with large, rounded heads that lack ear tufts, forest owls have well-developed, rounded or square facial disks. They have yellow or orange-yellow eyes and relatively long wings, and they reside in extensive forests, where they prey mainly on small rodents and shrews. One species is Holarctic; the others reside in the Americas (4 sp.; 2 in N.A.).

BOREAL OWL *Aegolius funereus*   BOOW ▪ 2

Males of this striking nocturnal species (known as "Tengmalm's Owl" in the Old World) give their far-carrying song in late winter. It is most often seen around nest boxes. A cavity nester, it mainly uses woodpecker holes. Polytypic (7 ssp.; 2 in N.A.). L 10" (25 cm) WS 23" (58 cm)

IDENTIFICATION ADULT: Resident *richardsoni* has white underparts broadly streaked chocolate brown; whitish facial disk with distinct black border; umber-brown crown densely spotted white; pale bill. Female noticeably larger.

JUVENILE: Chocolate-brown plumage held about June to September.
GEOGRAPHIC VARIATION Eurasian subspecies (e.g., *magnus* of Russian Far East) are paler than the subspecies resident in N.A., *richardsoni*.
SIMILAR SPECIES Northern Saw-whet Owl is smaller; adult has streaked crown, buffy facial disk with no dark border, dark bill, and cinnamon-brown breast streaks; juvenile tawny rust below.
VOICE Male's territorial song is about a two-second-long series of 11–23 low, whistled *toot* notes, repeated after a few seconds' silence; like a winnowing Wilson's Snipe's, but does not fade at the end.
STATUS & DISTRIBUTION Uncommon to rare. **YEAR-ROUND:** Inhabits boreal forest belt, also subalpine mixed conifer-deciduous forests in high Rockies and mountains of Northwest. **MIGRATION:** Adult males usually sedentary, females nomadic; juveniles disperse widely. Casual to northern IL, northern OH, New York City. Prey shortages periodically drive large

numbers south, to winter in areas where normally few are seen. Unprecedented irruption of 2004–2005 saw more than 400 in upper Midwest. One Old World *magnus* reached St. Paul Is., AK (Jan. 1911).
POPULATION No reliable information on trends.

prominent black border to facial disk
pale bill
richardsoni
juvenile *richardsoni*
darker body coloration than juvenile Northern Saw-whet

NORTHERN SAW-WHET OWL *Aegolius acadicus*   NSWO ▪ 2

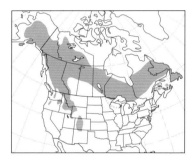

This widespread owl usually tolerates close approach without flushing. Strictly nocturnal, it hunts rodents from a perch and roosts by day in thick vegetation, often near the end of a lower branch. In some regions, often located on day roost. Nests in a woodpecker hole or a nest box. Polytypic (2 ssp.; both in N.A.). L 8" (20 cm) WS 20" (51 cm)
IDENTIFICATION ADULT: Reddish brown

above with crown streaked white; white below with broad reddish streaks; dark bill; reddish facial disk, without dark border. **JUVENILE:** Plumage held about May to September. Dark reddish brown above, tawny rust below.
GEOGRAPHIC VARIATION Distinctive subspecies *brooksi*, endemic to the Queen Charlotte Islands, BC, is much darker and buffier on facial disk and belly than nominate described above; white only on eyebrows.
SIMILAR SPECIES Boreal Owl is larger; adult has a spotted crown, whitish facial disk with black border, pale bill, chocolate-brown breast streaks; juvenile chocolate brown below.
VOICE Male's territorial song is a monotonous series of single, low whistles, about two per second, on a constant pitch. A two- to three-second-long rising nasal whine and a high-pitched *tssst* call are also heard.

no dark border to facial disk
acadicus
dark bill
rufous overall coloration
juvenile *acadicus*

STATUS & DISTRIBUTION Fairly common in breeding range; uncommon to rare over much of winter range. **BREEDING:** Dense coniferous or mixed forests, wooded swamps, tamarack bogs. **MIGRATION:** Some remain year-round on breeding range, but considerable numbers migrate south or downslope in autumn. Casual on Bermuda. **WINTER:** Wide range of habitats; dense vegetation for roosting.
POPULATION Stable, but no data.

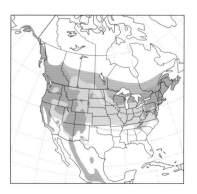

Genus Ninox

BROWN HAWK-OWL *Ninox scutulata*   BRHO ▪ 5

round head with
yellow eyes

japonica

long
banded
tail

Ninox hawk-owls have indistinct facial disks compared with other owls. Polytypic (13+ ssp.; *japonica* in N.A.). L 13" (33 cm)
IDENTIFICATION A medium-size, hawklike brownish owl with a round head, no ear tufts, large golden yellow eyes, and a long, banded tail; underparts heavily streaked with reddish brown. Flies with rapid wingbeats and glides; hunts at dusk.
GEOGRAPHIC VARIATION Some authors have considered the species to be a complex of two or three species, based on differences in vocalizations. The northern group consists

of *japonica* and *florensis*. The two records from N.A. are likely *japonica,* which breeds in Ussuriland, southeastern Manchuria, North Korea, and Japan, migrating in winter south to the Philippines and Borneo.
SIMILAR SPECIES None; in flight might be mistaken for a small falcon or *Accipiter* hawk.
VOICE Unlikely to be heard in North America.
STATUS & DISTRIBUTION Asian species. Northern populations are strongly migratory; southern populations are sedentary. North America's two records are from AK: St. Paul I.

(Aug. 27–Sept. 3, 2007) and Kiska I. (found dead on Aug. 1, 2008; specimen not retained).
POPULATION Apparently stable.

GOATSUCKERS Family Caprimulgidae

Mexican Whip-poor-will (AZ, May)

Caprimulgids occur nearly worldwide and occupy a wide variety of habitats. They collectively are referred to as "goatsuckers," the translation of the family name. According to folklore, nightjars suckled nursing goats during the night. In reality, most species are insectivorous, but because of their nocturnal and secretive habits, strange vocalizations, huge mouth, and, in some species, tendency to forage in areas where livestock occur, this superstition developed.

Structure Goatsuckers are superficially reminiscent of owls because of their large head, large eyes that shine when illuminated at night, cryptic plumage, and largely nocturnal behavior. But instead of capturing and eating prey with sharp bills and talons, goatsuckers are the nocturnal counterpart of swifts and swallows. The bill is tiny, but the gape is exceptionally wide and reveals a cavernous mouth. Insect prey are literally engulfed and swallowed whole rather than being seized and manipulated with the bill. Most nightjars have elongated rictal bristles along the edges of the mouth that presumably help a nocturnal species detect and guide an insect into its mouth; these are not elongated in the more diurnal

nighthawks. Caprimulgids have proportionately long wings; nighthawk wings are more pointed and falconlike, nightjar wings more rounded. In most species the tail is usually fairly long and provides added maneuverability. Short legs and small feet reflect mostly aerial habits, but all can grip perches or walk short distances if necessary. Virtually all species have cryptically colored plumage with complex patterns to provide camouflage in daylight. Most species are small to medium in size.

Behavior Nightjars roost by day on ground, rocks, or branches, usually in dense vegetation, and are most frequently observed during twilight hours or after dark. All species are ground nesters. Loud, distinctive calls during the breeding season usually alert observers to their presence. Nighthawks forage during long sustained flights; nightjars make short sallies from an exposed perch or ground. Most species in North America are highly migratory, but some can survive periods of cold weather by lowering their metabolism and going into torpor—the Common Poorwill, for example.

Distribution North American species occupy a diversity of habitats. All but the Common Pauraque are migratory to some extent, but the most widespread species, the Common Nighthawk, is the champion long-distance migrant, traveling from as far north as northern Ontario to as far south as central Argentina.

Taxonomy Worldwide the family is composed of 92 species in 20 genera; relationships within the family are not well established by phylogenetic analyses.

Conservation Nocturnal habits of most species make study difficult; much remains unknown. Habitat deterioration or destruction and pesticides present the greatest threats, but only two species are listed as critically endangered (Jamaican Pauraque and New Caledonian Owlet-Nightjar; both may now be extinct) and seven species are listed as threatened.

NIGHTHAWKS Genus *Chordeiles*

A New World genus, three of six species occur in North America. Nighthawks are partially diurnal, open-country birds, thus easier to observe and more frequently encountered than other nightjars. All species have relatively long, pointed wings, notched tails, conspicuous pale wing patches, and a patch of puffy white feathers at the bend of the wing when perched.

LESSER NIGHTHAWK   *Chordeiles acutipennis*   LENI ■ 1

Found primarily in arid regions, this species often congregates at water sources morning and evening, rarely active during midday. It is the only breeding nighthawk across most of the extreme southwestern United States lowlands, but limited overlap with the Common Nighthawk presents an identification challenge. Polytypic (7 ssp.; *texensis* in N.A.). L 8–9.2" (20–23 cm)
IDENTIFICATION Generally dark gray to brownish gray above mottled with black, grayish white, or buff; crown and upper back darkest, paler markings more concentrated on upperwing coverts, scapulars, tertials. Secondaries and primaries dark brownish gray; secondaries and basal portions of primaries spotted with buff. Underparts generally buffy, finely barred dark brown; chest and malar area darker and grayer with whitish to buff spotting. Underwing coverts buff mottled with brown. **ADULT MALE:** White throat, subterminal tail band, and primary patch about two-thirds out from bend of wing to wing tip. **ADULT FEMALE:** Buff throat and wing patch; tail band reduced or absent. **JUVENILE:** Upperparts uniformly buffy-gray with fine dark markings and spots.
SIMILAR SPECIES Some Commons very similar in general coloration and size, but have proportionately longer, more slender, more pointed wings, and primary patch is farther from wing tip. Most Commons lack buff spotting on secondaries and primaries. **VOICE CALL:** (Males only) Whistled trill in variable bursts. Flight display a bleating *bao-b-bao-bao*.
STATUS & DISTRIBUTION Common. **BREEDING:** Arid lowland scrub, farmland. **MIGRATION:** In spring, arrives early Mar.–mid-May, peak in Apr.; departs early Aug.–late Oct., peak mid-Aug.–mid-Sept. **WINTER:** Occurs year-round from northwestern and central Mexico south to northern S.A.; rare in extreme southern U.S., including FL. **VAGRANT:** Casual/accidental to northwestern AK, CO, OK, ON, WV, Bermuda.
POPULATION Stable.

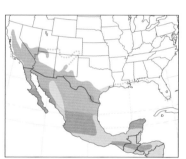

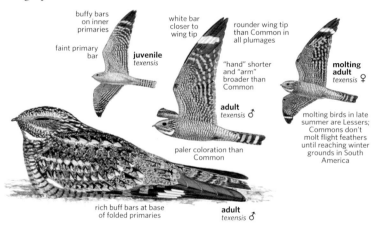

buffy bars on inner primaries

faint primary bar

juvenile *texensis*

white bar closer to wing tip

rounder wing tip than Common in all plumages

"hand" shorter and "arm" broader than Common

molting adult *texensis* ♀

adult *texensis* ♂

molting birds in late summer are Lessers; Commons don't molt flight feathers until reaching winter grounds in South America

paler coloration than Common

rich buff bars at base of folded primaries

adult *texensis* ♂

COMMON NIGHTHAWK   *Chordeiles minor*   CONI ■ 1

This goatsucker performs flight displays and roosts conspicuously. Normally solitary, it sometimes forages or migrates in loose flocks. It is often out hunting or migrating during daylight hours. Polytypic (9 ssp.; 7 in N.A.). L 8.8–9.6" (22–24 cm)
IDENTIFICATION Varies geographically. Upperparts black to paler brownish gray; crown and upper back darkest; paler markings concentrated on upperwing coverts, scapulars, tertials. Underparts barred blackish brown and white or buff, chest and malar area darker, spotted with buff or dull white. **ADULT MALE:**

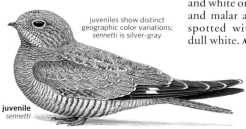

juveniles show distinct geographic color variations; *sennetti* is silver-gray

juvenile *sennetti*

Breeding Range

minor

sennetti

hesperis

henryi

howelli

chapmani

← *aserriensis*

neotropicalis

subspecies uncertain

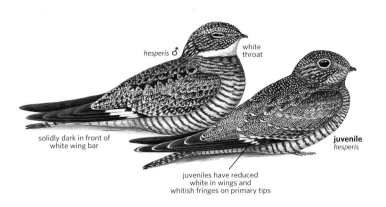

hesperis ♂

white throat

solidly dark in front of white wing bar

juvenile *hesperis*

juveniles have reduced white in wings and whitish fringes on primary tips

White throat patch, subterminal tail band, and primary patch about halfway between bend of wing and wing tip. **ADULT FEMALE:** Throat patch buffy, primary patch smaller, tail band reduced or lacking. **JUVENILE:** Generally paler, more uniform above with finer spotting and vermiculations.

GEOGRAPHIC VARIATION Eastern birds darkest, blackish above, less mottling on back; nominate *minor* (large), *chapmani* (smaller). Great Plains-Great Basin-Southwestern birds paler, grayer; *henryi* (medium size), *howelli* (large), *sennetti* (large), and *aserriensis* (small). Western *hesperis* relatively dark, grayer, large. Geographical color variation is most evident in juvenal plumage: juvenile *minor* and *chapmani* blackish; *sennetti* palest; *hesperis, howelli, aserriensis* intermediate; *henryi* rusty.

SIMILAR SPECIES Position of wing patch, lack of buff spotting on primaries, pointier wing, and darker underwing coverts eliminate Lesser. Separation from Antillean problematic; best told by voice, but Antillean also usually smaller, shorter winged, and buffier on belly and undertail coverts.

VOICE CALL: Nasal *peent* by male in flight; multiple-syllable variation may suggest Antillean. Male courtship dive vibrates primaries, producing "boom."

STATUS & DISTRIBUTION Common. **BREEDING:** Open habitats. **MIGRATION:** In spring, arrives early Apr.–mid-June, peak May (later in Northwest); departs late July–Oct., peak late Aug.–early Sept., when moderate to even large numbers can be consistently seen throughout eastern N.A. (birds initiate migration by late afternoon); stragglers into Nov. **WINTER:** S.A.; casual Gulf Coast. **VAGRANT:** Casual or accidental, mainly fall, HI, northern Canada, U.K. **POPULATION** Declines in much of the East.

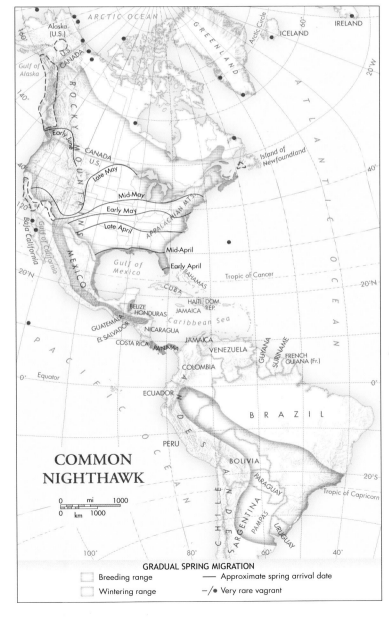

COMMON NIGHTHAWK

mi 1000
km 1000

GRADUAL SPRING MIGRATION

Breeding range — Approximate spring arrival date

Wintering range —/• Very rare vagrant

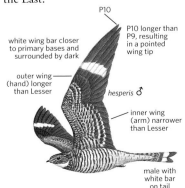

P10

P10 longer than P9, resulting in a pointed wing tip

white wing bar closer to primary bases and surrounded by dark

outer wing (hand) longer than Lesser

hesperis ♂

inner wing (arm) narrower than Lesser

male with white bar on tail

ANTILLEAN NIGHTHAWK *Chordeiles gundlachii*   ANNI ▪ 2

Formerly considered a Caribbean sub-species of the Common Nighthawk, this Florida Keys specialty was elevated to full species status following evidence there that both Commons and Antilleans nested without interbreeding. Polytypic (2 ssp.; *vicinus* in N.A.). L 8.6" (21.5 cm)

IDENTIFICATION Relatively small nighthawk. Upperparts generally blackish, darkest on crown and upper back, varyingly mottled with whitish gray to buff, pale markings more concentrated on upperwing coverts, scapulars, and tertials. Upper breast and malar region blackish spotted with buff or grayish white; lower breast barred dull black and grayish white, belly with pale buff. Undertail coverts buffy. **ADULT MALE:**

White throat, white subterminal tail band, and uniformly dark primaries except for white patch about halfway between bend of the wing and wing tip on five outer primaries. **ADULT FEMALE:** Similar but throat patch buffy, white primary patch smaller, lacks tail band.
SIMILAR SPECIES Voice is only certain identification character. Sight identification of silent, out-of-range individuals is likely impossible. Antillean Nighthawk is virtually identical to southeastern *chapmani* of Common but averages smaller, shorter winged, and buffier on belly and undertail coverts. Some Commons have buffy underparts. Lesser overall very similar and buffy below, but has generally paler upperparts, breast, and primaries; paler, buffier underwing coverts; slightly more rounded wing tips. Antillean and Common have more pointed wing tips. Lesser has primary patch positioned about two-thirds of the way out from bend of wing to wing tip; and buffy spots on primaries and secondaries, usually visible when perched. Female Lesser has buff-and-white or completely buffy primary patch.
VOICE CALL: (Males only) *pity-pit, chitty-chit,* or *killikadik;* also a Common

Nighthawk–like nasal *penk-dik.* During male courtship dive, air rushing through primaries at the terminus of a dive produces hollow, roaring "boom."
STATUS & DISTRIBUTION Uncommon.
BREEDING: Open or semi-open habitats. **MIGRATION:** Highly migratory. Arrives in Apr.; departs Aug.–Sept. **WINTER:** Unknown, presumably S.A. **VAGRANT:** Spring and summer, rare to mainland FL, Dry Tortugas, Virginia Key; accidental LA, NC.
POPULATION Stable; expanded breeding range to FL Keys beginning in the 1940s.

vicinus ♂

slightly shorter wing than Common and pale bar slightly closer to tip

overall similar to Common

vicinus ♂

quite buffy overall, but so are some Commons

NIGHTJARS Genera *Nyctidromus, Phalaenoptilus, Antrostomus,* and *Caprimulgus*

Nyctidromus and *Phalaenoptilus* are monotypic New World genera. *Phalaenoptilus* is characterized by its short tail. All New World goatsuckers formerly placed in *Caprimulgus* are now placed in their own genus, *Antrostomus.* They have cryptic plumage patterns and proportionately long, rounded wings and tails. Only six of 56 species found almost worldwide occur in N.A.

COMMON PAURAQUE *Nyctidromus albicollis*   COPA ▪ 2

This beautifully patterned tropical nightjar is often detected by its golden-red eyeshine along roadsides at night, where it waits to sally after insects. By day it roosts on the ground in thickets, perfectly camouflaged in leaf litter. Polytypic (7 ssp.; *merrilli* in N. A.). L 11.2–12" (28–30 cm)
IDENTIFICATION Upperparts brownish

gray finely vermiculated with black, crown coarsely streaked with black, back and upper wings crisply and complexly mottled with black and buff spots. Auriculars chestnut, stripe below eye (in some birds) whitish. Underparts buffy with narrow dark bars; chest, chin, malar area darker grayish brown. **ADULT MALE:** Throat patch and large patch across outer primaries white, extensive white on third and fourth pairs of tail

feathers. **ADULT FEMALE:** Can be less prominently marked on upperparts, reduced, buffier throat and wing patches, and less white in tail (restricted to feather tips). **JUVENILE:** Similar, but browner, less heavily patterned; lacks throat patch, wing patches smaller and buffier.

rufous-tinged cheek

boldly fringed scapulars

merrilli ♂

rounded wings

merrilli ♂

wing patch; white in males, buffy in females

long tail

a few tail feathers extensively white in males, more restricted white in females

merrilli ♀   *merrilli* ♂

SIMILAR SPECIES Chuck-will's-widow and whip-poor-wills lack white primary patch, have shorter tail and longer wings, and are more uniformly colored.

VOICE CALL: *Whip* or *wheeeeeeer*. SONG: Buzzy whistle, *pur pur perp pur-wheeeeer*.
STATUS & DISTRIBUTION Common. YEAR-ROUND: Woodland clearings

and scrub in southern TX. VAGRANT: Casual in TX north of breeding range to northern Maverick, Bastrop, Grimes, Calhoun Counties.
POPULATION Stable.

COMMON POORWILL   *Phalaenoptilus nuttallii*   COPO ▪ 1

The smallest nightjar is more typically heard giving its forlorn song or detected by its eyeshine at night (typically on roads). Polytypic (5 ssp.; 4 in N.A.). L 7.6–8.4" (19–21 cm)
IDENTIFICATION Typically sits on ground and makes short flights. **ADULT MALE:** Large head; short, rounded tail. Paler (gray) and darker (browner) morphs occur throughout range. Upperparts dark brown to pale grayish brown, mottled pale gray and black.

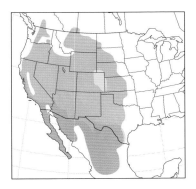

White throat bordered by blackish brown chin, malar area, and chest. All but central tail feathers tipped white. **ADULT FEMALE:** Similar but can have narrower or buffier tail tips.
GEOGRAPHIC VARIATION Subspecies are distinguished by size and dorsal coloration (beware color morphs). Two low-desert subspecies palest: *hueyi* of southeastern CA smallest, *adustus* of southern AZ larger. Widespread northern nominate *nuttallii* and *californicus* of western CA relatively large, darker brown above.
SIMILAR SPECIES Whip-poor-will and Buff-collared Nightjar larger with longer tail, less white on throat.
VOICE SONG: Plaintive whistled *poor will up,* last note heard at close range.
STATUS & DISTRIBUTION Uncommon to locally common. **BREEDING:** Low-middle elevation dry, rocky, open, shrubby habitats. **MIGRATION:** Northern populations apparently migratory, southern populations partially migratory or move to lower elevations in winter. Arrives Feb.–Mar. south, late Apr.–late May north.

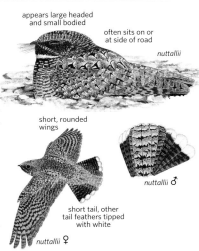

appears large headed and small bodied

often sits on or at side of road

nuttallii

short, rounded wings

nuttallii ♂

short tail, other tail feathers tipped with white

nuttallii ♀

Departs Sept. north, Oct.–Nov. south. **WINTER:** Southern areas of breeding range (CA, AZ, TX) to central Mexico. **VAGRANT:** Accidental southwestern BC, southern MB, ON, MN, MO.
POPULATION Possibly expanding to the northeast.

BUFF-COLLARED NIGHTJAR   *Antrostomus ridgwayi*   BCNI ▪ 3

The Buff-collared Nightjar is primarily a west Mexican species. Polytypic (2 ssp.; nominate in N. A.). L 8.8–9.2" (22–23 cm)
IDENTIFICATION **ADULT:** Upperparts brownish gray with black blotches on crown and scapulars, head and back separated by distinct pale buff collar. Upperwing coverts spotted with buffy-white, primaries brownish black banded with cinnamon-buff.

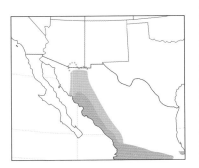

Upper throat dark grayish brown, breast somewhat paler, and belly paler tan barred with dark brown; throat and breast separated by whitish buff to buff foreneck collar. **JUVENILE:** Similar.
SIMILAR SPECIES Mexican Whip-poor-will almost identical but slightly larger, darker, and browner, and lacks distinct pale buff hind neck collar. Common Poorwill smaller, shorter tailed. Lesser Nighthawk has pale primary patches, different proportions.
VOICE CALL: Series of *tuk* notes. SONG: Rapid, accelerating series of 5–6 *cuk* notes, concluding with rapid *cuk-a-CHEE-a,* somewhat reminiscent of the dawn song of Cassin's Kingbird.
STATUS & DISTRIBUTION Rare and irregular in southeastern AZ; once to Upper Guadalupe Canyon, NM. **BREEDING:** Local in desert canyons dominated by mesquite, acacia, and hackberry. **MIGRATION:** In U.S. extreme

dates Apr. 17–Aug. 28; presumably migratory based on lack of winter records, but no specific information. **WINTER:** Western Mexico. **VAGRANT:** Accidental to coastal southern CA (Ventura Co.).
POPULATION Generally stable. The range in the United States was first detected in 1958–1960; this peripheral population has been designated by state wildlife agencies as endangered (NM) and of "special concern" (AZ).

ridgwayi ♂

distinct and complete buff collar

male with less white than whip-poor-wills

ridgwayi

overall paler than Mexican Whip-poor-will

CHUCK-WILL'S-WIDOW   *Antrostomus carolinensis*   CWWI   1

This species is heard much more often than seen. Widespread in open forest and woodlands of the Southeast, this is our largest nightjar. During the day, it roosts on the ground or on a branch. It can be quite startling when flushed at close range. It flies up in a burst of brown, sometimes accompanied by loud grunting notes. Monotypic. L 11.2–12.8" (28–32 cm)

IDENTIFICATION ADULT: Large nightjar, males are larger than females. Large, flat-topped head. Generally dark grayish brown, but considerable individual variation exists, from paler and rustier to darker and browner. Crown and upper back darkest; scapulars palest, with heavy black streaks or spots on these areas. Wings blackish and heavily barred rufous. Tail long, broad, rounded, buffy to rufous, and with black bars and vermiculations. Chin and upper throat brown with fine dark bars. Whitish buff to buff patch on lower throat, contrasting with dark brown to blackish malar area and breast. Belly and undertail coverts pale buff to dark rufous, barred with blackish brown. MALE: Extensive subterminal whitish buff inner webs of outer three pairs of tail feathers. FEMALE: Buffy tips to tail feathers.

SIMILAR SPECIES Eastern Whip-poor-will is smaller, has shorter, narrower, more round-tipped wings, averages grayer overall, has dark throat, and male has different tail pattern.

VOICE CALL: A single or a series of grunting notes, *quaah*. SONG: Whistled *chuck-weo-wid-ow;* the first note is rapid, the *weo* is drawn out, and the song is concluded by a rapid *wid-ow*.

STATUS & DISTRIBUTION Common. BREEDING: Deciduous, evergreen, or mixed woodlands with edges or gaps. MIGRATION: Spring migration mid-Mar.–mid-May, a few to late May. Depart mid-Aug.–Oct., stragglers into Nov. WINTER: Southern FL, northern West Indies, and east-central Mexico south to Colombia; rare in northern FL and along Gulf Coast. VAGRANT: Rare, mainly spring and summer, north of breeding range as far as SD, MN, WI, MI, and southeastern Canada. Casual or accidental west to NM, NV, and coastal northern CA.

POPULATION Stable.

much larger than Eastern Whip-poor-will with big, flat head

gray morph

some are grayer

more restricted white than both whip-poor-will species

rufous-brown throat

rufous morph

most are rufous overall in coloration

lacks white in tail

EASTERN WHIP-POOR-WILL   *Antrostomus vociferus*   EWPW   1

On still summer nights, the song of the "Whip" rings in eastern forests. Daytime encounters with this seldom seen creature are usually by chance when a nesting bird is flushed, but many will permit close study when found roosting. Monotypic. L 8.8–10" (22–25 cm)

IDENTIFICATION ADULT: Relatively rounded wings. Generally grayish brown above with black, pale gray, and buff spotting and vermiculations; thicker black streaks and blotches on crown and scapulars. Wings heavily barred with buff. Blackish brown throat and dark brown upper chest separated by white to buff foreneck collar; rest of underparts buffy with fine dark bars. Male has thick, white tips on outer three pairs of tail feathers; female has smaller, buffy tail tips.

SIMILAR SPECIES Chuck-will's-widow is much larger, with bigger, flatter head, broader and more pointed wings; it averages darker and browner, has buffy brown throat, and male has different tail pattern. Buff-collared Nightjar is slightly smaller, paler, and grayer,

blackish throat and white collar on males

overall darker and less rufous than Chuck-will's-widow

female with buffy tips to outer tail feathers

male with much more white in outer tail than Chuck-will's Widow

with a more prominent pale buff collar across hind neck. Common Poorwill is notably smaller and has a shorter tail. Common Pauraque and all nighthawks have pale primary patch, different tail pattern and shape, and pale throat. See Mexican Whip-poor-will.
VOICE SONG: Melodious, repeated *whip-poor-will,* usually delivered at night. **CALL:** Year-round, a subdued Hermit Thrush-like *tuk* or *quirt,* usually at dusk and dawn.

STATUS & DISTRIBUTION Formerly common but now absent or distinctly uncommon across most of range. **BREEDING:** Dry deciduous or mixed forest with open understory and partly open canopy. **MIGRATION:** Arrives in spring mid-Mar.–mid May, peak Apr., stragglers to late May–early June. Departs late Sept.–Oct., peak mid-Sept.–mid-Oct., stragglers to Nov.–early Dec. **WINTER:** FL and Mexico south to Honduras; rare

elsewhere along southern Atlantic and Gulf coasts. **VAGRANT:** In the East, confirmed records as far out of range as NF and Bermuda; in the West, records from San Diego, CA (Nov. 14, 1970), a territorial singing bird near Santa Cruz, CA (May–June, 2013), and Petersburg, AK (Nov. 22, 1972). **POPULATION** Widespread declines and breeding range contractions have occurred in many parts of the species' range.

MEXICAN WHIP-POOR-WILL   *Antrostomus arizonae*   MWPW   ▪ 1

Until recently considered conspecific with Eastern Whip-poor-will, Mexican Whip-poor-will is a far less familiar bird in the United States, where it is found in rugged and remote country, mostly in the isolated mountain ranges of the Southwest. It roosts mainly on the ground and rarely tolerates a close approach. Polytypic (5 ssp., nominate in N.A.). L 9–10.4" (23–26 cm)
IDENTIFICATION Very similar to slightly smaller Eastern Whip-poor-will; best distinguished by song. Mexican has longer rictal bristles that appear buffy near the base, rather than blackish;

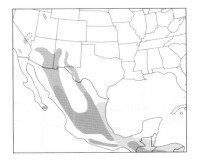

and more black and less white (or, in females, less buff) in outer rectrices.
GEOGRAPHIC VARIATION Nominate *arizonae* is the only subspecies recorded north of Mexico. Subspecies *setosus,* which could reach Texas as a vagrant, is resident from central Tamaulipas to Veracruz. Three other subspecies occur from central Mexico to El Salvador.
VOICE SONG: Similar to Eastern but burrier (more modulated), lower in pitch, and slightly slower, a husky *whirr-p-wiirr.* **CALL:** *Tuk,* similar to Eastern.
STATUS & DISTRIBUTION Locally fairly common. **BREEDING:** In mixed coniferous-oak or dry coniferous woodland with brushy understory, in mountains. **MIGRATION:** Some may winter in or near breeding areas in border areas, but most appear to withdraw southward from the U.S. Arrive in spring Apr.–mid May; departure

longer and browner rictal bristles

male averages less white in outer tail feathers than Eastern Whip-poor-will

arizonae ♂

throat blackish with white collar on males

arizonae ♂

coloration very similar to Eastern Whip-poor-will

white corners on tail hidden when perched

may start by early Aug. **VAGRANT:** Confirmed records as far out of range as northern CA, UT, OR, MT, NE, CO, BC, AB, SK. The easternmost record thus far is from the Nebraska Panhandle; however, a review of specimens in museum collections (especially of late season migrants) could well produce records from the East.
POPULATION In contrast to Eastern Whip-poor-will, Mexican Whip-poor-will appears to have expanded its range in recent decades, with singing males detected in NV, central NM, and southern CA.

GRAY NIGHTJAR   *Caprimulgus indicus*   GRNI   ▪ 5

Also known as "Jungle Nightjar," the Gray Nightjar has been recorded only once in North America. The English name reflects the anticipated split from a mostly sedentary Indian subspecies. Polytypic (5 ssp.; N.A. record pertains to northernmost, migratory *jotaka*). L 11–12.8" (28–32 cm)
IDENTIFICATION Upperparts generally brownish gray, patterned with black, buff, and grayish white spots, streaks, and bars. Flight feathers contrast darker; secondaries and inner primaries banded tawny buff, outer webs of outer primaries spotted with buff. Whitish submoustachial stripe.

Underparts grayish brown, barred with paler gray, buff, or brown. **ADULT MALE:** Large subterminal white patch on outer primaries and white tips to all but central pair of tail feathers. **ADULT FEMALE:** Tawny primary patch, brownish white or brownish buff tail tips.
SIMILAR SPECIES North American goatsuckers either lack primary patches or have different wing, tail, and throat patterns.
VOICE CALL: Loud, rapid series of downslurred *schurks.*
STATUS & DISTRIBUTION

Found in eastern and southern Asia. **BREEDING:** Variety of forest, woodland, and scrub habitats. **MIGRATION:** Northern populations migrate. Arrive presumably Mar.–May; depart Sept.–Nov. **WINTER:** Southern breeding range south to East Indies. **VAGRANT:** Accidental on Buldir I., AK (May 31, 1977).
POPULATION Stable.

♂ *jotaka*

white patch on inner primaries visible in flight

SWIFTS Family Apodidae

White-throated Swift (BC, June)

Superficially swallowlike with their aerial habits, small bills, and long, narrow wings, swifts differ from swallows in their wing structure (the wrist joint is closer to the body), stiff wingbeats (the wings are not swept back against the body), and more rapid flight. Swifts are usually observed in flight because they generally only perch at nests and nocturnal roost sites, which are well hidden in hollows and crevices. Most species forage at high altitudes during the daylight hours and can be hard to observe. Watch for swifts over freshwater, along bluffs and ridgetops, and near roost and nest sites. For species identification, concentrate on overall size, wing and tail shape, dark-and-light patterning on the body, and flight style (wingbeat rate, gliding, etc.).

Structure Swifts are small to medium-size, long-winged aerialists. The largest (up to 6.3 ounces) are 20 times the weight of the smallest. *Apodidae* literally means "without feet," a reference to the swift's tiny (but strong) feet. The "hand" portion of the wing (carpal bones and digits) is quite long relative to the "arm" (humerus and forearm bones); thus the primaries dominate the flight feathers, with the secondaries short and bunched together. There are ten rectrices; spine-like protrusions are formed by the tip of the rachis (a feather's central shaft) in some species and are used as props. The strongly clawed toes assist in clinging to vertical surfaces, the only kind of perching of which true swifts are capable. The bill is small, with a large gape.

Behavior The most aerial of birds, swifts feed throughout the day and sometimes even "roost" on the wing. Some species can attain very high speeds as they cut through the air. The swift's nest is usually a shallow open cup, with materials glued together and attached to a vertical substrate with saliva; some extralimital species build a larger nest

with plant fibers. Their diet consists of aerial arthropods, principally small insects and spiderlings, and they may forage dozens of miles from the nest site. Large numbers of swifts may gather together before dusk around roost sites, often coming together in very vocal, swirling flocks.

Plumage Swifts are generally clad in blackish to gray-brown plumage, usually with contrasting paler and darker regions and sometimes with bold white markings. Most species show short, dense blackish feathering in front of the eyes. Tail shape varies greatly among genera and can also vary with the degree of closure or spreading of the tail: For example, swifts with deep tail forks may hold the tail closed so that it appears as a single point. Sexes are similar, but juveniles are usually distinguished by duller patterning, pale feather fringes on the body, and pale tips to the secondaries and the tertials.

Distribution True swifts are found worldwide in temperate and tropical regions; many species are highly migratory. Being wide-ranging, strong fliers, swifts have a considerable potential for vagrancy: Five of our nine species occur only as vagrants, and the Alpine Swift, an additional European/African species, has occurred twice in the Lesser Antilles.

Taxonomy Swifts comprise 99 species, in 20 genera. These are often placed in three subfamilies: the more primitive Cypseloidinae (now limited to the New World) and the widespread Chaeturinae and Apodinae. Four species breed in N.A.; five others occur as vagrants. The swifts are grouped with hummingbirds in the order Apodiformes. Formerly, four Southeast Asian tree-swifts were placed in their own family (Hemiprocnidae).

Conservation Some swift species have increased with the availability of human-built nest substrates, whereas other species are highly specialized in their nesting habits and have suffered local declines.

CYPSELOIDINE SWIFTS Genera *Cypseloides* and *Streptoprocne*

These are primitive swifts, medium-small to very large in size, with all modern species restricted to the New World. Unlike other swifts, they do not use saliva in their nest construction. Most species nest near waterfalls or other damp, shaded sites, but they can cover huge distances while foraging.

BLACK SWIFT *Cypseloides niger*   BLSW ■ 2

Our largest regularly occurring swift, this species is associated with waterfalls and other damp cliff habitats in western mountains and rugged northwestern coastlines. It is generally rare as a migrant. Polytypic (3 ssp.; *borealis* in N.A.). L 7.3" (19 cm)

IDENTIFICATION A large, blackish swift with long, broad-based wings that show a distinct angle at wrist joint and a rather long, broad, shallowly forked tail. **ADULT:** Sooty black throughout except for frosted whitish chin, forehead, lores, and thin line over eye. In fresh plumage, some thin white fringing is also seen on lower underparts. **JUVENILE:** Resembles adult, but body plumage, coverts, and flight feathers are narrowly fringed with white; this pale scalloping is especially evident on belly and undertail coverts. **FLIGHT:** Wingbeats of Black Swift can

appear relatively languid, but its flight can nevertheless be exceedingly fast; it usually flies with short bursts of wingbeats interspersed with long, twisting glides on slightly bowed wings. Individuals are wide ranging and fly very high, and they are most likely to be observed lower on overcast days.

GEOGRAPHIC VARIATION West Indian nominate is smaller; might reach FL; *costaricensis* breeds from central Mexico to Costa Rica.

SIMILAR SPECIES Vaux's is much smaller, paler, has shorter wings, with rapid wingbeats and little gliding. White-throated can appear all blackish in poor light, but has a thinner tail (often carried in a point), slimmer wings with

a little whitish on and around forehead visible at close range

large and overall blackish color

adult
borealis

soars more than other North American swifts

broader wings and rear of body than White-throated

white scaling visible at close range; retained through fall migration

juvenile
borealis

less of an angle at wrist joint, and a more rapid wingbeat.
VOICE Relatively silent.
STATUS & DISTRIBUTION Generally uncommon; locally much more numerous in coastal BC. **BREEDING:** Locally distributed within its general breeding range, concentrating around steep mountain cliffs with a damp microclimate (often from waterfall spray) and available moss for nest material. Also nests locally in sea cliffs and caves in central CA, northern WA, and BC. **MIGRATION:** Spring migrants can appear after mid-May, migration continues into early June; fall migrants occur mainly.–early Oct., casually into Nov. Generally scarce away from breeding grounds, although concentrations of migrants sometimes noted during fronts and overcast weather. **WINTER:** Wintering range recently discovered to be in western Amazon Basin. **VAGRANT:** Accidental in East.
POPULATION Apparently stable, as most of the nest sites are inaccessible.

WHITE-COLLARED SWIFT *Streptoprocne zonaris*   WCSW ■ 4

The neotropical White-collared Swift has occurred casually but widely in North America. Our largest swift, it has a distinct white collar and soaring flight, which help to identify it. Polytypic (9 ssp.; 2 in N.A.). L 8.5" (22 cm)

IDENTIFICATION Very large (larger and longer-winged than a Purple Martin), with slightly forked tail. Prolonged soaring flight with wings bowed downward. **ADULT:** Blackish throughout, with a bold white collar, broader across chest and narrower on hind neck. **JUVENILE:** Resembles adult, but plumage is sooty with indistinct pale fringes and collar is less distinct.

GEOGRAPHIC VARIATION Most records in N.A. are of the large Mexican subspecies *mexicana,* but a Florida specimen is of the smaller West Indian *pallidifrons.*
SIMILAR SPECIES Black Swift is smaller and lacks a white collar.
STATUS & DISTRIBUTION Vagrant to North America. Widespread from eastern and southern Mexico (north to south Tamaulipas) to northern Argentina. **VAGRANT:** Casual vagrant, recorded widely across North America. The four TX records are from Mar., May, and Dec.; the two FL records are from Sept. and Jan. There are also single records for CA and MI, both in May.

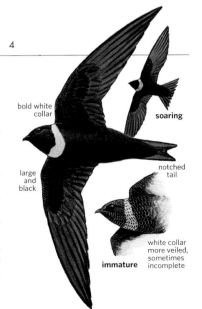

bold white collar

soaring

large and black

notched tail

white collar more veiled, sometimes incomplete

immature

SPINE-TAILED SWIFTS Genera *Chaetura* and *Hirundapus*

The subfamily Chaeturniae includes small to large spine-tailed swifts found in both the New and Old Worlds. It also includes the familiar Chimney Swift, the needletails, and the southeast Asian and Pacific swiftlets, which can echolocate in caves and whose saliva is the source of "bird's nest soup."

CHIMNEY SWIFT   *Chaetura pelagica*   CHSW ■ 1

A small, dark "cigar with wings," this is the common swift of the eastern half of N.A. Its original nest sites (hollow trees, cliffs) have largely been substituted with human-built structures such as chimneys or building shafts. Monotypic. L 5.3" (13 cm)
IDENTIFICATION Small, dark; squared; spine-tipped tail; narrow-based wings often appear markedly pinched at the base during secondary molt in late summer, early fall. **ADULT:** Brownish black overall; paler chin; throat; slightly paler rump. Plumage can appear browner

with wear or appear blacker from contact with chimney soot. **JUVENILE:** Nearly identical to adult, but with whitish tips to the outer webs of the secondaries, tertials. **FLIGHT:** Usually rapid, fairly shallow wingbeats, including quick turns, steep climbs, short glides. V-display of pairs involves long glides with wings raised in a V-pattern and some rocking from side to side.
SIMILAR SPECIES Vaux's Swift is very similar but slightly smaller, paler; differs subtly in shape; has higher-pitched calls.
VOICE CALL: Commonly heard; quick,

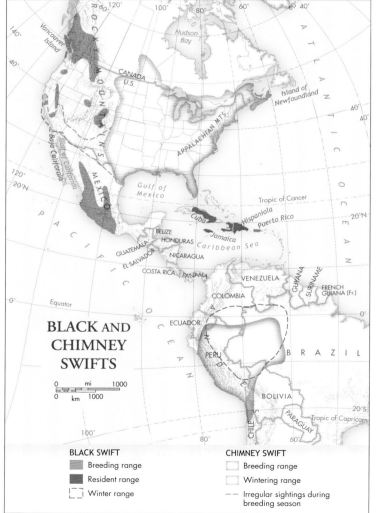

short tail

sooty brown underparts

cigar-shaped body

darkish rump, more uniform with upperparts than Vaux's

soaring

hard chippering notes, sometimes run together into rapid twitter.
STATUS & DISTRIBUTION Common. **BREEDING:** Widespread in variety of habitats; most abundant around towns, cities. Possibly breeds north to NF. Small numbers summer regularly in southern CA (though fewer since 1990s), with breeding documented; possibly also bred in AZ. **MIGRATION:** Migrates in flocks during the day, mainly along the Atlantic coastal plain, Appalachian foothills, and Mississippi River Valley. Large concentrations may appear during inclement weather; hundreds may roost in chimneys. First spring arrivals are in mid-Mar. in southern states; peak arrivals in northernmost breeding areas are late Apr.–mid-May. Most have departed breeding areas by late Sept.–mid-Oct; latest fall migrants occur in early Nov. **WINTER:** Most or all winter in Upper Amazon Basin of S.A.; unrecorded in N.A. in midwinter, but records as late as Dec. **VAGRANT:** Casual away from CA in West, mainly May–Sept.; accidental on Pribilof Is., AK, and in western Europe.
POPULATION Population declines have been noted since the 1980s.

BLACK AND CHIMNEY SWIFTS

0   mi   1000
0   km   1000

BLACK SWIFT
- Breeding range
- Resident range
- Winter range

CHIMNEY SWIFT
- Breeding range
- Wintering range
- – – Irregular sightings during breeding season

VAUX'S SWIFT *Chaetura vauxi*   VASW ■ 1

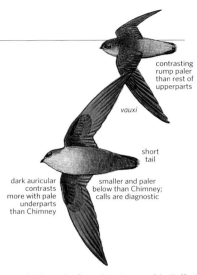

contrasting rump paler than rest of upperparts

vauxi

short tail

dark auricular contrasts more with pale underparts than Chimney

smaller and paler below than Chimney; calls are diagnostic

This is the counterpart of the Chimney Swift in western N.A., occurring almost exclusively west of the Rocky Mountains. Only rarely does it overlap with the Chimney Swift; identifications of out-of-range-birds should be made very carefully, if at all. Vaux's also appears to be the only *Chaetura* to winter north of Mexico, though it is scarce and local at that season. Polytypic (6 ssp.; nominate in N.A.). L 4.8" (12 cm)

IDENTIFICATION Our smallest regularly occurring swift, it is very similar to Chimney Swift in all respects. Distinctions from Chimney are fairly evident on the rare occasions the two species can be directly compared, but lone birds or single species flocks are very difficult to identify. **ADULT:** Brownish overall, paler on throat, breast, and rump. **FLIGHT:** Very rapid, twinkling wingbeats are interrupted by only brief glides; many quick turns and climbs when foraging, as in Chimney. Rocking V-display also similar to Chimney Swift's.

GEOGRAPHIC VARIATION Our birds are nominate *vauxi,* the northernmost subspecies. Five more subspecies are found from Mexico to Venezuela, one of

which (*tamaulipensis* of eastern Mexico) has been collected once in Arizona; it is slightly darker and glossier black than *vauxi*. Dark birds from southern Mexico to Venezuela (*richmondi* group) and small birds on the Yucatán Peninsula (*gaumeri* group) are sometimes treated as separate species.

SIMILAR SPECIES The very similar Chimney Swift shows a gray-brown throat contrasting with darker remaining underparts. In Vaux's, throat is paler, and this pale color extends through breast. Chimney is more uniform above, with a gray-brown rump only slightly paler than back; Vaux's rump and uppertail coverts are paler, showing more contrast to back. Structural characters are subtle but important: Chimney is larger and appears relatively longer winged; its rear body appears slightly longer, thus wing placement appears slightly farther forward. Flight differs subtly, with Chimney gliding more and flapping slightly more slowly.

VOICE High insectlike chipping and twittering notes, often run together into a buzzy trill. Vocalizations resemble those of Chimney, but are considerably higher pitched, more rapid, and more trilled.

STATUS & DISTRIBUTION Common. **BREEDING:** Heavily forested lowland and lower montane areas; old standing snags generally required, but Vaux's will also nest in chimneys. **MIGRATION:** Spring migrants arrive in the Southwest by early Apr., and the northernmost breeding areas are occupied by late May; fall migration is mainly late Aug.–

early Oct. (to late Oct. in south). Fall migrants occur rarely east to UT, eastern AZ, and NM. Huge postbreeding and migrant roosts sometimes noted from WA to southern CA. **WINTER:** Individuals or small flocks sometimes winter in coastal central and southern CA. A few late fall and winter records for LA and FL, including some small flocks, suggest occasional wintering in the Southeast. Most Vaux's winter from central Mexico south through Central America, but details of winter range are poorly known because of the presence of resident subspecies or closely related species.

POPULATION Although Vaux's have become familiar sights in some urban areas—where late summer and migrant roost flocks can be very large—overall numbers have declined with loss of old growth forests in the Northwest.

WHITE-THROATED NEEDLETAIL *Hirundapus caudacutus*   WTNE ■ 5

white throat and lore spot visible at close range

broad wings like a 747, often soars for extended periods

white horseshoe pattern

short wedge-shaped tail

caudacutus

silvery back

pale throat

white tertial tips

This large Asian swift occurs casually on the western Aleutian Islands in spring. Polytypic (2 ssp.; nominate in N.A.). L 8" (20 cm)

IDENTIFICATION Large swift with stubby spine-tipped tail, broad-based wings, and large head. Remarkable high-speed flight consists of rapid and powerful wingbeats; glides on bowed wings. **ADULT:** White throat, dark breast and belly; extensive white undertail coverts and flanks form a large U-shaped patch; white spot on forehead, lores; white on inner webs of tertials. Upperparts show pale brown saddle, contrasting with green-glossed black crown, wings, tail. **JUVENILE:** Similar to adult, but lores grayish, white of

lower underparts scalloped with dark. **GEOGRAPHIC VARIATION** More northerly subspecies *caudacutus* has been recorded in N.A.

SIMILAR SPECIES Fork-tailed has a long notched tail, a white rump, and dark underparts.

VOICE Common call is a soft, rapid insectlike chattering.

STATUS & DISTRIBUTION BREEDING: Nests in hollow trees in eastern Asian forests from southern Siberia to northeastern China, Korea, Sakhalin I., and northern Japan. **WINTER:** Northern populations migrate to New Guinea, eastern Australia. **VAGRANT:** Casual in spring on the outer Aleutian Is., AK (five recs. from late May).

TYPICAL SWIFTS Genera *Apus, Aeronautes,* and *Tachornis*

This widespread group of typical swifts ranges in size from tiny to quite large. Most have forked tails, and many species show white patterning on the rump, throat, or belly. Nests, which are generally made of feathers and plant material, are cemented with saliva to a vertical wall, tree hollow, palm frond, or other protected site.

COMMON SWIFT *Apus apus*   COSW ■ 5

The Common Swift is a familiar and well-studied swift of Eurasia, but it is recorded only casually in North America, with late June records in the far northwestern and northeastern corners of the continent. It is a moderately large and very dark swift, with a paler throat and a strongly forked tail. Polytypic (2 ssp., both likely to have occurred in N.A.). L 6.5" (17 cm)

IDENTIFICATION Dark nearly throughout, with a long and obviously forked tail. In flight the rapid, frenzied wingbeats alternate with long glides. **ADULT:** Plumage is entirely blackish brown, with slight scaled effect to body feathering. Pale throat contrasts to remaining underparts. Upperwings are uniformly dark throughout; on underwings, flight feathers are only slightly paler than wing linings. **JUVENILE:** Similar to adult, but its plumage is blacker and more scaly, with whitish fringes on forehead and pale tips to flight feathers.

GEOGRAPHIC VARIATION Subspecies *pekinensis,* of the eastern half of the breeding range, is slightly paler and

browner, with a more extensive whitish throat patch; Alaska specimen is of this subspecies. Nominate European birds are the likely source of eastern records and reports.

SIMILAR SPECIES See Fork-tailed, which is casual in western Alaska. Compared to Black Swift, Common Swift has a narrower and more deeply forked tail, narrow-based and more pointed wings, and a pale throat patch. Adult Common Swifts have dark foreheads, but pale fringes on forehead of juveniles recall Black Swift. (Note pale throat of juvenile Common.) Black's flight is more leisurely; Common with rapid wingbeats; interspersed with glides. Several related Old World species (e.g., Pallid Swift) are closely similar but not likely to occur in North America.

VOICE A wheezy, screaming *sreeee* or *shreeee;* vocal in pre-roosting gatherings.

STATUS & DISTRIBUTION Vagrant in North America. Common breeder from Europe eastward through northern China, migrating south to winter in the southern half of Africa. **VAGRANT:**

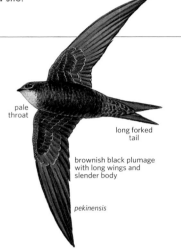

pale throat

long forked tail

brownish black plumage with long wings and slender body

pekinensis

Three late June records: a specimen from the Pribilof Is., AK (June 28, 1950); one photographed there June 28–29, 1986; and one photographed on Miquelon I., off NF (June 23, 1986). Also a single fall record from CA: one photographed near Desert Center, Riverside Co. (Oct. 30, 2013). There have been other likely sightings along the Atlantic coast; recorded in Bermuda.

FORK-TAILED SWIFT *Apus pacificus*   FTSW ■ 4

This is a high-flying, large, fork-tailed Asian swift recorded casually in western Alaska. It is also known as "Pacific Swift." Polytypic (4 ssp.; nominate in N.A.). L 7.8" (20 cm)

IDENTIFICATION This is a large but slender, blackish brown swift with a white rump; it has a long, forked tail, but the tail's deep fork is not always apparent. **ADULT:** Fork-tailed Swift is dusky black above with a white band across rump and contrastingly black uppertail coverts and tail. Underparts appear scaly, with dusky black feathering fringed with whitish feathers; throat is paler. Underwings have dark linings, white-fringed underwing coverts, and somewhat translucent flight feathers. **JUVENILE:** Similar to adult, but its flight feathers are fringed with white.

GEOGRAPHIC VARIATION The nominate is the northernmost breeder (to Russian Far East) and highly migratory, moving as far south as New Zealand in winter. The other three, more southerly

subspecies are smaller and darker; they are short-distance migrants and possibly resident in some areas.

SIMILAR SPECIES No species of swift occurs regularly in western Alaska. Common Swift of Eurasia, recorded in western Alaska, is smaller and lacks the white rump. Common Swift is more uniformly dark, whereas Fork-tailed has a two-toned appearance from below: Its blackish body and wing linings contrast with its silvery flight feathers. The vagrant White-throated Needletail has a short rounded tail and white undertail coverts; it also has pale upperparts without a white rump patch.

VOICE Screaming *sre-eee* resembles calls of the Common Swift, but it is slightly softer and more disyllabic.

STATUS & DISTRIBUTION Breeds in eastern Asia, northeast to the Kamchatka Peninsula; southern Asian populations are sedentary, whereas northern birds winter in southeast Asia and Australia. **VAGRANT:** Casual in AK,

mainly in the western Aleutian and Pribilof Is., but also on St. Lawrence I. and Middleton I. Most records are mid-May–June and Aug.–Sept.

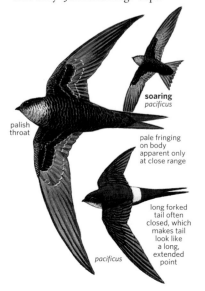

soaring
pacificus

palish throat

pale fringing on body apparent only at close range

long forked tail often closed, which makes tail look like a long, extended point

pacificus

WHITE-THROATED SWIFT *Aeronautes saxatalis*   WTSW   1

The White-throated Swift is a common and characteristic swift of western canyons, cliffs, coastal bluffs, and even urban areas, seemingly equally at home over the remotest desert mountains or around busy urban freeway interchanges. Its bold patterning, streaking flight, and staccato vocalizations attract attention wherever it occurs. Polytypic (2 ssp.; nominate in N.A.). L 6.5" (17 cm)

IDENTIFICATION White-throated Swift is a fairly slender swift with long, scythe-like wings. Its moderately long, notched tail is usually held in a tight double-point but can be widely fanned in maneuvering birds. Bold black-and-white patterning is unique among our regularly occurring swifts. **ADULT:** Blackish to blackish brown on crown, upperparts, flanks, undertail coverts, and flight feathers; paler gray on forehead, lores, and narrow supercilium. White chin, throat, and chest; white continues more narrowly to belly. There is also a large white spot on flanks, extending up toward sides of rump, and distinct white tips to secondaries and tertials.

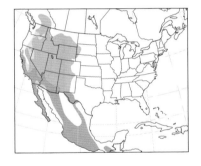

Sexes are similar, but females have a slightly shallower tail fork and less white on secondary and tertial tips. **JUVENILE:** Closely similar to adult, but it appears paler and browner due to indistinct pale fringing on forehead, crown, and undertail coverts. Flight feathers are narrowly edged with white, but white tertial tips are indistinct.

SIMILAR SPECIES Beware that with distant views or poor lighting, White-throated Swifts can often appear all dark. Vaux's and Chimney Swifts have shorter wings and shorter, squared tails; compared with these *Chaetura*, White-throated's long wings appear to be attached farther forward (probably due to longer rear body and tail). See also Black Swift, which is distinguished by its broader, more shallowly forked tail, crook at the wrist joint of the wings, and generally more languid flight with long periods of soaring. In distant birds, the white rump, sides, and throat may suggest a Violet-green Swallow, but White-throated Swift's long, stiff wings, blackish sides and undertail coverts, and long pointed tail make this distinction straightforward.

VOICE Often quite vocal, giving a loud, rapid, shrill *tee-dee, dee, dee, dee* ... or *jee-jee-jee-jee* ... series that drops slightly in pitch.

STATUS & DISTRIBUTION Common. **BREEDING:** In the interior West, the White-throated Swift is mostly found in canyons, river gorges, and other areas of high relief, with records as high as 13,000 feet. It is also found on coastal cliffs and lowlands in southern and central CA and increasingly in

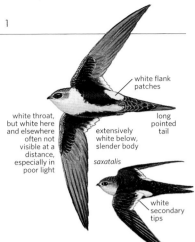

white flank patches

long pointed tail

white throat, but white here and elsewhere often not visible at a distance, especially in poor light

extensively white below, slender body

saxatalis

white secondary tips

urban regions where crevices in buildings, highway overpasses, and other human-built structures provide nest and roost sites. Foraging birds can wander widely over lowland areas, and like most wide-ranging swifts, their local abundance can shift with changes in wind patterns, clouds, and weather fronts. **MIGRATION:** Withdraws from northern part of breeding range, where present mainly Apr.–Sept. **WINTER:** Most birds winter from central CA, southern NV, central AZ, and western TX southward. Numbers often more concentrated in winter when large communal roosts may be used. On colder winter days, activity may be curtailed; some birds may even become torpid. **VAGRANT:** Casual east (mainly in fall) to central and coastal TX and northwest to Vancouver I.; recorded exceptionally in west and central OK (Apr.–May), MO (Nov.), AR (May, Dec.), and MI (Aug.). **POPULATION** Stable.

ANTILLEAN PALM-SWIFT *Tachornis phoenicobia*   ANPS   5

A tiny Caribbean swift with a forked tail, a dark-capped appearance, and a distinctive white rump and belly patch, the Antillean Palm-Swift has been recorded once during summer in Florida. Polytypic (2 ssp.). L 4.3" (11 cm)

IDENTIFICATION Very small size, white rump, shallowly forked tail, and low, batlike flight combine to make this swift distinctive. **ADULT:** Blackish above with broad white rump; white throat and center of belly; blackish breast band, sides, and undertail coverts. **IMMATURE:** Similar to adult, but duller; white areas of underparts buffy. **FLIGHT:** Batlike, with rapid

wingbeats and short glides and twists; generally flies low, among trees and around palms.

GEOGRAPHIC VARIATION The Cuban subspecies *iradii* is likely the source of the FL record.

SIMILAR SPECIES The white rump and longer, slightly forked tail eliminates all dark Chimney Swifts; see Bank Swallow (which lacks a white rump). The much larger White-throated Swift, which lacks a dark band between the white throat and belly, is very unlikely to occur in the Southeast.

VOICE Weak twittering calls.

STATUS & DISTRIBUTION Vagrant. **YEAR-**

ROUND: Greater Antilles, including Cuba. **VAGRANT:** Two were present and photographed at Key West, FL (July 7–Aug. 13, 1972).

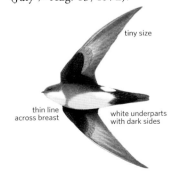

tiny size

thin line across breast

white underparts with dark sides

HUMMINGBIRDS Family Trochilidae

Black-chinned Hummingbird, male (AZ, Apr.)

Hummingbirds, or "hummers" as they are often known, are the smallest of all birds. Solely inhabiting the New World, they are very tiny and have incredible flight powers and brilliant, iridescent colors. These characteristics make field identification problematic, but hummers are readily attracted to sugar-water feeders, where they can be studied. In general, head and tail patterns, bill shape and color, and vocalizations are the best markers. Some birds will remain unidentified. Relatively frequent hybridization among species adds a cautionary dimension.

Structure Hummingbirds have relatively long wings, with ten primaries but only six secondaries; their usually ample tails have ten rectrices; and their feet are tiny. Bills are slender, pointed, and proportionately long, varying from straight to distinctly decurved or arched in profile.

Behavior Flight is fast and acrobatic; the wings beat so fast that they are a blur to the naked eye. Hummingbirds can reverse their primaries while hovering, in effect rotating their wings through 180 degrees and enabling them to fly backward. Aggressive for their size, several species have spectacular dive displays related to the defense of feeding territories and perhaps to courtship. They hover at flowers (or feeders) where they probe for nectar (or sugar water); they dart, spritelike, in pursuit of flying insects, which they sometimes pirate from spider webs. All hummers lay two unmarked white eggs; the male plays no part in nesting. Voices are unmusical and include the simple chip call (given by perched and feeding birds), the warning call, and the flight chase call. Some species are migratory (including most populations in North America).

Plumage North American hummingbirds have dichromatic plumage related to both age and sex; adult males are more brightly colored than females. Juveniles typically resemble adult females, but their upperparts have buff tipping in fresh plumage, and the sides of the upper mandible have tiny scratches, or grooves (adult bills are smooth), that banders use for in-hand aging. Almost all species are metallic green above, but underparts vary greatly in color and pattern. The most common and widespread North American species are often termed the small gorgeted hummingbirds, for the iridescent throat panels, or gorgets, of the adult males, which vary from ruby red and magenta rose to violet and flame orange. Some of the large species have emerald green or royal blue gorgets. Molt occurs mostly on nonbreeding grounds. Thus, immature males of the small gorgeted species migrate south in female-like plumage and return in adultlike plumage. Because flight is integral to their existence, hummingbirds need to molt their wing feathers gradually. Thus, molt of the primaries may require four to five months. The ninth (next-to-outermost) primary is molted last, rather than the straight 1-to-10 sequence typical of most birds. Iridescence is due to the interference of reflected light; a gorget can change in appearance from blackish to flame orange in a split second. Many species show discolored (usually yellow or whitish) throats or crowns from pollen gathered during feeding.

Distribution Hummingbirds achieve their greatest diversity in the Andes of South America, from Colombia to Ecuador. At least one species can be found almost anywhere between southern Alaska and Tierra del Fuego. The birds occur in basically all habitats that support flowering plants; 17 species have nested in North America. Six other species have occurred in North America as rare to accidental visitors from Mexico and the Caribbean.

Taxonomy Species-level and especially genus-level taxonomy needs critical revision. Worldwide, 320+ species are recognized, but much remains to be learned about many taxa in South America. Within Middle and North America, there are 105 to 115 species depending on taxonomy, with 23 species in 15 genera reported from North America.

Conservation Feral predators such as cats can impact local populations. However, long-term data on populations of North America are sparse, because hummingbirds are hard to observe and census. Especially in the West, variations in annual rainfall (and subsequent flower abundance) confound attempts to determine population sizes and trends.

GREEN VIOLETEAR *Colibri thalassinus* GREV ■ 3

This fairly large and overall green hummingbird is a rare visitor to the United States from the pine-oak highlands of Mexico. Polytypic (4 ssp.; nominate in N.A.). L 4.2–4.7" (11–12 cm) Bill 18–22 mm

IDENTIFICATION Green Violetear has a short and slightly decurved bill and a fairly long, broad tail that is slightly notched. Overall dark green color, with distinctive violet-blue auricular and breast patches. **ADULT MALE:** Brighter overall with an extensive, purplish blue chest patch and auricular patches, a golden green crown, a longer, more strongly notched tail. **ADULT FEMALE:** Duller overall than male, with a smaller and less purplish chest patch and smaller auricular patches, a bronzy green crown, and a shorter, less notched tail. **IMMATURE:** Distinguished from adult by its duller upperparts (with fine cinnamon tips when fresh) and dull bluish green

throat and chest with patchy iridescent green and blue feathers.
GEOGRAPHIC VARIATION North American records are of nominate *thalassinus*, distinguished from southern populations (sometimes considered a separate species, "Mountain Violetear," *C. cyanotis*) by its larger size and violet-blue chest patch.
SIMILAR SPECIES Unlikely to be confused with other North American hummingbirds but, as with all extralimital hummers, beware the possibility of hybrids that might resemble violetears, escapes of the southern subspecies, or similar species.
VOICE CALL: Hard short rattles and clipped chips, both given from a perch and in flight. **SONG:** A metallic, mostly disyllabic chipping, often repeated tediously with a jerky cadence from an exposed perch; immatures may give more varied series, including buzzes and rattles.

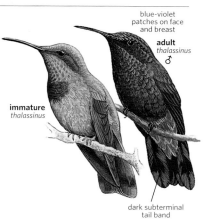

blue-violet patches on face and breast

adult *thalassinus* ♂

immature *thalassinus*

dark subterminal tail band

STATUS & DISTRIBUTION Mex. to S.A. **VAGRANT:** Very rare but annual to TX, mainly in late spring and summer. Casual elsewhere (mainly May–Aug., extremes Apr.–Dec.) in central and eastern N.A., north to MI and ON. Accidental in NM, CA, and AB.

Introduction to Hummingbird Identification

Perhaps more so than with most groups of birds, the field identification of hummingbirds usually requires a good view to confirm what species you are looking at. However, hearing a diagnostic call in a split-second view can convince an experienced observer of a species' identity. And many of the finer points can be appreciated after you have gained a degree of comparative experience— for example, are the primaries relatively tapered or relatively blunt? At first, the frenetic behavior of hummingbirds may seem overwhelming and the differences between species almost esoteric, but with patience and practice, it is possible to identify most birds you see. And many hummingbird species are distinctive, even unmistakable.

The small gorgeted hummers include some of the greatest bird identification challenges in North America, so you're not alone if you have difficulty with them. Still, males are mostly distinctive: Check their gorget color and tail shape details to help with harder identifications. Females and immatures are undeniably problematic, and it can be helpful to start by deciding if your bird is one of the

Anna's Hummingbird, female

Black-chinned Hummingbird, female

gray-and-green species or one of the rufous-and-green species. The latter group shows distinct rufous coloration on the body sides and tail base; the former group lacks rufous tones except as a flank wash on some individuals. In gray-and-green species (the genera *Archilochus* and *Calypte*), head and body proportions, tail length and shape, and voice differences are helpful characters. And details of primary shape are diagnostic: The inner six primaries of *Archilochus* are disproportionately narrow, and the shape of the primary tips is diagnostic within *Archilochus*. Among the rufous-and-green species (genus *Selasphorus*), check tail length relative to wing tip length on perched birds, the brightness and contrast of rufous coloration on the body sides, relative bill length, and tail pattern details. Even after narrowing your choices, some birds will defy specific identification, and this might be true even if an expert examined the birds in the hand (so don't be discouraged). Time and the associated experience you gain will make hummingbirds more manageable, and soon you will be able to discern more and more refined characters. ■

GREEN-BREASTED MANGO   *Anthracothorax prevostii*   GNBM ▪ 4

This striking and ostensibly unmistakable large hummingbird is a casual visitor to the United States from eastern Mexico. Polytypic (4 ssp.; nominate in N.A.). L 4.5–4.8" (11.5–12 cm) Bill 24–30 mm

long curved bill

adult ♀

green vertical stripe on underparts

prevostii

IDENTIFICATION Green-breasted Mango has a thick and arched black bill and a broad tail. This species' size, bill shape, and overall plumage patterns are unlike any other North American hummer. **ADULT MALE & SOME FEMALES:** Plumage is distinctive: deep green overall with a black throat and mostly purple tail. **ADULT FEMALE:** Note blackish median throat stripe, becoming deep green on white underparts and purple tail base. **IMMATURE:** Resembles female but has cinnamon mottling on sides of throat and chest.
SIMILAR SPECIES Unlikely to be confused, but beware the possibility of other mango species wandering from the Caribbean, or escapes of the similar Black-throated Mango (*A. nigricollis;* western Panama to S.A.).
VOICE Not very vocal. **CALL:** A high,

sharp *sip,* as well as fairly hard, ticking chips, and shrill tinny twitters in interactions.
STATUS & DISTRIBUTION Mex. to northern S.A. **VAGRANT:** Casual visitor (mainly in autumn and winter) to coastal lowlands of south TX (12+ records); accidental in fall and winter in GA, NC, and WI.

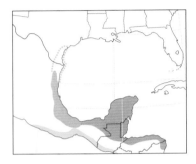

MAGNIFICENT HUMMINGBIRD   *Eugenes fulgens*   MAHU ▪ 2

This large hummer of southwestern mountains often perches prominently on high twigs. Polytypic (2 ssp.; nominate in N.A.). L 4.7–5.3" (12–13.5 cm) Bill 27.5–30 mm
IDENTIFICATION Magnificent has a long black bill and a slightly notched tail. **ADULT MALE:** Distinctive; often looks all dark, since violet crown and green gorget rarely seem to catch the light. **ADULT FEMALE:** Throat and underparts pale gray, mottled green on sides. **IMMATURE MALE:** Resembles female but throat and underparts strongly mottled green, white malar streak more distinct, and central throat usually has a large iridescent-green blotch. **IMMATURE FEMALE:** Resembles immature male but throat lacks iridescent green patches.

SIMILAR SPECIES Note large size, long bill, and face pattern. Plain-capped Starthroat has proportionately longer bill, dark throat patch bordered by whitish malar, and white rump patch. See Blue-throated Hummingbird.
VOICE CALL: A fairly loud, sharp *chik* or *tsik,* given from perch and in flight. Also a higher *piik,* and males, at least, give fairly hard, short rattles, *trrirr* and *trrrr ch-chrr* that may run into prolonged squeaky chattering. Aggression call a squeaky, slightly bubbling, accelerating chatter. **SONG:** A fairly soft, slightly buzzy gurgling warble.
STATUS & DISTRIBUTION Southwest U.S. to Panama. **BREEDING:** Fairly common (Mar.–Oct.) in pine-oak

and oak highlands. **WINTER:** Rare, mainly southeast AZ. **VAGRANT:** Casual west to CA, north (mainly late summer) to UT and WY, exceptionally MI, and eastward (mainly winter) to Gulf states, exceptionally GA and VA.

usually just a white postocular spot; sometimes a short white supercilium

fulgens ♀

limited white on tail

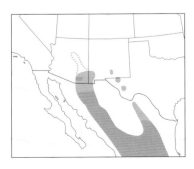

purple crown

large size

green throat

extensive black breast

adult ♂
fulgens ♂

Genus *Heliomaster*

PLAIN-CAPPED STARTHROAT *Heliomaster constantii* PCST 4

This large vagrant from Mexico often makes prolonged flycatching sallies over streams. Like many tropical hummers it shows little age/sex variation. Polytypic (3 ssp.; *pinicola* in N.A.). L 4.7–5" (12–13 cm) Bill 33–37 mm

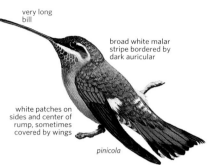

IDENTIFICATION Very long, straight black bill; note throat and face pattern, white rump patch, and Black Phoebe-like call. **ADULT:** Throat patch is dark sooty with iridescent reddish mottling on lower portions (often hard to see). Tail has white tips to all but central pair of rectrices. **IMMATURE:** Resembles adult but throat patch has little or no red, white tail tips average wider. **SIMILAR SPECIES** Compare with female Magnificent and Blue-throated Hummingbirds. **VOICE CALL:** A sharp, fairly loud *peek!,* given from perch and especially in flight. **SONG:** A series of chips interspersed with varied notes. **STATUS & DISTRIBUTION** Mexico to

very long bill

broad white malar stripe bordered by dark auricular

white patches on sides and center of rump, sometimes covered by wings

pinicola

Costa Rica. **VAGRANT:** Casual visitor (May–Nov., mainly late summer) to southern AZ.

Genus *Lampornis*

BLUE-THROATED HUMMINGBIRD *Lampornis clemenciae* BTHH 2

The largest hummingbird in North America, the Blue-throated favors shady understory and edge in watered pine-oak and oak-sycamore canyons. Polytypic (3 ssp.; 2 in N.A.). L 4.8–5.3" (12–13.5 cm) Bill 21.5–26 mm **IDENTIFICATION** Blue-throated has a proportionately short black bill. Its very large tail often looks slightly rounded. Whitish face stripes, bold white corners on tail, and high-pitched call are good field marks. **ADULT MALE:** Blue gorget well defined but can be hard to see. Outer two pairs of rectrices have large white tips. **ADULT FEMALE:** Throat plain dusky gray, white tips to outer three pairs of rectrices; some birds have a

few blue throat feathers. **IMMATURE:** Resembles female but upperparts have buffy-cinnamon edgings in fall, and lower mandible base often pinkish on younger birds. Males have irregular blue gorget patch restricted to central throat. **GEOGRAPHIC VARIATION** Texas subspecies *phasmorus* is slightly greener above and on average shorter billed than *bessophilus* from AZ. **SIMILAR SPECIES** Female and immature Magnificent Hummingbird longer billed with a greenish tail that has only small white corners; more mottled underparts; and a smacking chip call. Magnificent's face pattern typically

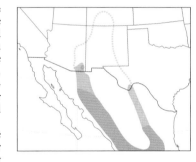

dominated by a bold white postocular spot, but some can be similar to poorly marked Blue-throated. See immature male Broad-billed Hummingbird. **VOICE CALL:** A high, penetrating *siip* or *siik,* given from perch and in flight; less often a fuller *tsiuk.* **SONG:** Apparently a repetition of call, *siip siip siip,* though more complex vocalizations have been reported. **STATUS & DISTRIBUTION** Mexico to southwest U.S. **BREEDING:** Uncommon to fairly common (mainly Apr.–Sept.) in humid mountain forests. **WINTER:** Mexico. Casual in southeastern AZ. **VAGRANT:** Casual north (mainly late summer) to UT and CO, and eastward (mainly winter) to southern LA. Accidental in Tulare Co., CA (Dec. 27, 1977–May 27, 1978; nested and fledged two hybrid young, the male parent was probably an Anna's).

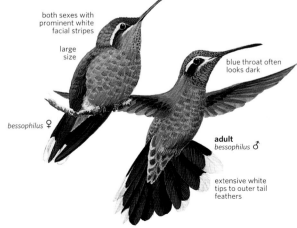

both sexes with prominent white facial stripes

large size

blue throat often looks dark

bessophilus ♀

adult *bessophilus* ♂

extensive white tips to outer tail feathers

Genus *Calliphlox*

BAHAMA WOODSTAR   *Calliphlox evelynae*   BAWO ■ 5

There are few well-documented records of this vagrant from the Bahamas. Polytypic (2 ssp.; nominate in N.A.). L 3.4–3.7" (8.5–9.5 cm) Bill 16–17 mm
IDENTIFICATION Bill slightly arched; fairly long, forked tail, projecting past wing tip on perched birds. **ADULT MALE:** Unmistakable. **ADULT FEMALE:** Outer rectrices have cinnamon bases and tips. **IMMATURE MALE:** Similar to female, but throat usually has some purplish spots; tail slightly longer, more forked.
SIMILAR SPECIES Female easily confused with female or immature *Selasphorus* hummingbirds, which have straighter bills and shorter tails

with bold white tips to outer rectrices. **VOICE CALL:** A high, fairly sharp chipping *tih* or *chi,* often doubled when repeated from perch, *chi chi chi-chi chi.*
STATUS & DISTRIBUTION Endemic to Bahamas. **VAGRANT:** Accidental to south FL (4 recs. 1961–1981; Jan, Apr.–Oct);

extraordinary was an adult male (photographed) at Denver, Lancaster Co., PA (Apr. 20–24, 2013).

Genus *Calothorax*

LUCIFER HUMMINGBIRD   *Calothorax lucifer*   LUHU ■ 2

This small hummer is a summer inhabitant of mountain desert canyons, especially in west Texas. Monotypic. L 3.5–4" (9–10 cm) Bill 19–23 mm
IDENTIFICATION Bill is proportionately long and arched; fairly long, forked tail (usually held closed) projects past wing tip on perched birds. Dusky auricular stripe on buff face of females is diagnostic. **ADULT MALE:** Expansive magenta gorget; longer tail than female. **ADULT FEMALE:** Throat

is pale buff to whitish, rarely with one or a few magenta spots; outer rectrices have rufous bases and bold white tips. **IMMATURE MALE:** Resembles adult female but upperparts in fall are fresher, with fine buffy tips; throat usually has more magenta rose spots. Tail is slightly longer and more forked with narrower outer rectrices.
VOICE CALL: A fairly hard, slightly smacking *chih* or *chi,* at times run

into a rolled *chi-ti.* Varied, rapid-paced chippering in interactions. Male makes a fairly loud wing buzz in displays.
STATUS & DISTRIBUTION Mexico to southwest U.S. **BREEDING:** Uncommon and local (mainly Apr.–Oct.) in arid mountain canyons. **VAGRANT:** Casual visitor away from traditional sites in AZ, NM, and TX.

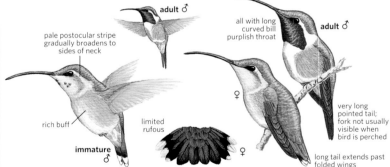

"BLACK-CHINNED HUMMINGBIRDS"   Genus *Archilochus*

Eastern and western counterparts are the Ruby-throated and Black-chinned. Diagnostic of the genus are the relatively narrow inner six primaries. Adult males have shield-shaped gorgets with black chins. Females/immatures have mostly plain underparts with little or no buff wash, no rufous in tail. Primary molt typically starts in fall or later (Sept.–Jan.).

RUBY-THROATED HUMMINGBIRD   *Archilochus colubris*   RTHU ■ 1

The only hummer seen regularly in much of the East. Monotypic. L 3.2–3.7" (8–9 cm) Bill 14–19 mm

IDENTIFICATION Primaries tapered. Tail forked to double-rounded. **ADULT MALE:** Solid ruby red gorget with black chin

and face. **ADULT FEMALE:** Throat whitish. Sides and flanks often washed buff. **IMMATURE MALE:** Resembles adult female but throat usually with ruby spots; tail slightly longer and more forked. Complete molt in winter produces plumage like adult male. **IMMATURE FEMALE:** Resembles adult female but fall plumage fresher; throat lacks red spots.

SIMILAR SPECIES Black-chinned has blunter primaries. Adult male has black throat with violet lower band, shorter tail. Females and immatures similar to Ruby-throated but generally duller green above (especially crown) and dingier below (Ruby-throated bright emerald above and whiter below). Black-chinned tail less forked; often pumps tail strongly while feeding (Ruby-throated usually holds tail fairly still); also note triangular dark patch on lores of female and immature male Ruby-throats. Problem identifications best confirmed by checking details of primary shape. See Costa's and Anna's Hummingbirds.

VOICE CALL: Slightly twangy or nasal chips, *chih* and *tchew,* given in flight and perched. Also varied twittering series. Indistinguishable from Black-chinned, but generally lacks strongly buzzy or sharp, smacking quality of

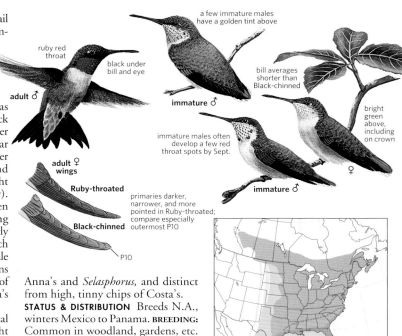

ruby red throat
black under bill and eye
adult ♂
a few immature males have a golden tint above
immature ♂
bill averages shorter than Black-chinned
immature males often develop a few red throat spots by Sept.
bright green above, including on crown
immature ♂
♀

adult ♀ wings
Ruby-throated
Black-chinned
primaries darker, narrower, and more pointed in Ruby-throated; compare especially outermost P10
P10

Anna's and *Selasphorus,* and distinct from high, tinny chips of Costa's.
STATUS & DISTRIBUTION Breeds N.A., winters Mexico to Panama. **BREEDING:** Common in woodland, gardens, etc. **MIGRATION:** Mainly Mar.–May, Aug.–Oct. **WINTER:** Rare (mainly Nov.–Mar.) in the Southeast. **VAGRANT:** Casual in the West and to AK.

BLACK-CHINNED HUMMINGBIRD *Archilochus alexandri*   BCHU   1

The western counterpart of Ruby-throated, Black-chinned regularly pumps its tail. Monotypic. L 3.3–3.8" (8.5–9.5 cm) Bill 16–22 mm
IDENTIFICATION Best marks for all ages are narrow inner primaries and blunt primaries. Double-rounded tail. Age/sex differences as Ruby-throated except as noted. **ADULT MALE:** Black throat with violet-blue lower band. **ADULT FEMALE:** Black-violet spots rare on throat. **IMMATURE MALE:** Throat usually has black-violet spots.
SIMILAR SPECIES Black-chinned is often confused with Anna's and

Costa's, which are chunkier and proportionately bigger headed, shorter billed, and shorter tailed; lack narrow inner primaries of *Archilochus;* and molt wings in summer. Female/immature Anna's slightly larger; underparts more mottled, including undertail coverts (mostly whitish on Black-chinned); throat often with rose-red spots; wags tail infrequently. Female/immature Costa's slightly smaller; face often plainer; wing tips often fall beyond tail tip at rest (shorter than tip on Black-chinned). Anna's call is a

smacking chip, Costa's is a high, tinny *tik,* both distinct from Black-chinned. See Ruby-throated.
VOICE CALL: Indistinguishable from Ruby-throated but distinct from high tinny chips of Costa's; distinctive male wing buzz in flight.
STATUS & DISTRIBUTION Western N.A. to northwest Mexico. **BREEDING:** Common in riparian woodlands, foothills. **MIGRATION:** Mainly Mar.–May, Aug.–Sept. **WINTER:** Mexico. Rare (mainly Oct.–Mar.) in Southeast; casual in Northeast (fall).

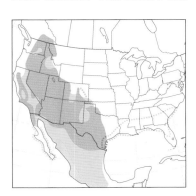

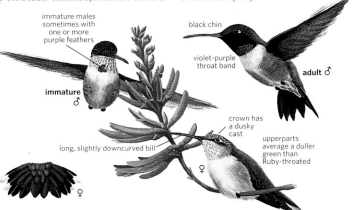

immature males sometimes with one or more purple feathers
immature ♂
black chin
violet-purple throat band
adult ♂
long, slightly downcurved bill
crown has a dusky cast
upperparts average a duller green than Ruby-throated
♀
♀

HELMETED HUMMINGBIRDS Genus *Calypte*

This genus comprises two western species of scrub habitats: Anna's and Costa's. Small, fairly chunky hummers with relatively large heads and short tails; adult males have an iridescent helmet (i.e., gorget and crown), with gorgets elongated at corners. Females and immatures have dingy underparts with no buff wash on the sides, no rufous in tail.

ANNA'S HUMMINGBIRD   *Calypte anna*   ANHU   1

This hummer is a familiar species in West Coast gardens, where it is present year-round. Monotypic. L 3.5–4" (9–10 cm) Bill 16–20 mm **IDENTIFICATION** Tail slightly rounded to double-rounded. **ADULT MALE:** Rose (fresh) to orange-red (worn) gorget and crown. **ADULT FEMALE:** Throat and underparts spotted and mottled dusky to bronzy green, median throat blotched rose red. **IMMATURE MALE:** Resembles adult female but throat and crown usually with more scattered rose spots; white tail tips narrower. Complete summer molt produces plumage like adult male by late fall. **IMMATURE FEMALE:** Resembles adult female but throat often lacks rose spots.

SIMILAR SPECIES Costa's smaller (obvious in direct comparison), and males readily identified (beware occasional hybrids, which look more like male Costa's, sound more like Anna's). Female/immature Costa's proportionately longer billed but shorter tailed, often best told by call: high, tinny pit call and twitters distinct from Anna's. Costa's generally plainer on throat and underparts, without dusky throat spotting. See Black-chinned. **VOICE CALL:** A slightly emphatic to fairly hard *tik* or *tih* and a more smacking *tsik*, in flight and perched. In flight chases, a rapid-paced, slightly buzzy

twittering, *t-chissi-chissi-chissi,* and variations. **SONG:** A high-pitched, wiry to lisping squeaky warble from perch, often prolonged and repeated with pulsating succession. Year-round. **STATUS & DISTRIBUTION** Western N.A. to northern Mexico. **BREEDING:** Common (Dec.–June) in scrub, gardens, etc. **DISPERSAL/MIGRATION:** Some late summer movement upslope to mountains. Local movements complex. **VAGRANT:** Rare north to southeastern AK and casual to southern AK (mainly fall). Casual to the East (fall and winter); accidental NF.

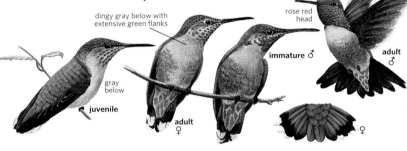

dingy gray below with extensive green flanks

rose red head

immature ♂

adult ♂

gray below

juvenile

adult ♀

♀

COSTA'S HUMMINGBIRD   *Calypte costae*   COHU   1

Costa's is a spectacular small hummer of southwest deserts. Monotypic. L 3–3.4" (7.5–8.5 cm) Bill 16–20 mm **IDENTIFICATION** Tail rounded to slightly double-rounded. **ADULT MALE:** Violet gorget and crown, tail lacks white tips. **ADULT FEMALE:** Throat and underparts dingy white to pale gray, throat sometimes blotched violet. **IMMATURE MALE:** Like adult female but upperparts

fresher February to July, with fine buff tips; auriculars darker; white tail tips narrower. Complete summer molt produces plumage like adult male by early winter. **SIMILAR SPECIES** Anna's larger (obvious in direct comparison); males readily identified (beware occasional hybrids; see Anna's). Female/immature Anna's proportionately shorter billed but longer tailed; often best told by call, a smacking chip and buzzy twitters. Anna's generally more spotted on throat and underparts. See Black-chinned. Problem identifications best confirmed by using primary shape and voice. **VOICE CALL:** A high, slightly tinny, fairly soft *tik* or *ti,* often run into short, slightly liquid or rippling twitters suggestive of Bushtit chatter. **SONG:** A very high-pitched, thin, drawn-out, whining whistle, *tsi ssiiiiiu,* given from perch and,

more loudly, in looping display dives. **STATUS & DISTRIBUTION** Western U.S. to northwest Mexico. **BREEDING:** Fairly common (Feb.–July) in desert washes, dry chaparral. **DISPERSAL/MIGRATION:** Mainly Feb.–May and June–July; local movements complex. **WINTER:** Uncommon to rare and local north to southern NV. **VAGRANT:** Casual north to southern AK and MT, east to KS and TX.

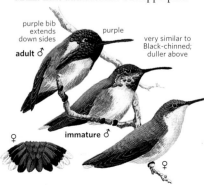

purple bib extends down sides

purple

very similar to Black-chinned; duller above

adult ♂

♀

immature ♂

♀

Genus *Atthis*

BUMBLEBEE HUMMINGBIRD *Atthis heloisa* BUHU ▪ 5

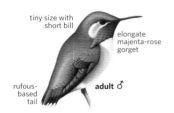

tiny size with short bill

elongate magenta-rose gorget

rufous-based tail

adult ♂

This tiny hummingbird is endemic to montane forests of Mexico, where it feeds inconspicuously with a relatively slow and deliberate, insect-like flight. Some authors place this species and the closely related Wine-throated Hummingbird *(A. ellioti)* of northern Central America in the genus *Selasphorus.* Polytypic (2 ssp.; N.A. recs. unclear).

L 2.7–3" (7–7.5 cm) Bill 11–13 mm
IDENTIFICATION Bill is medium-short, and the rounded to double-rounded tail projects beyond wing tips on perched birds. Both sexes have a femalelike tail with rufous bases and white tips to outer rectrices. **ADULT MALE:** Distinctive, with an elongated magenta-rose gorget. **ADULT FEMALE:** Throat is whitish with lines of bronzy-green flecks, and a whitish forecollar contrasts variably with cinnamon sides. White tail tips are bolder than male. **IMMATURE MALE:** Resembles adult female but throat flecked more heavily with bronzy green, often with rose-pink spots or streaks. **IMMATURE FEMALE:** Resembles adult female but throat has finer dusky spots. Tips of outer tail feathers washed buffy cinnamon when fresh.

SIMILAR SPECIES Female and immature *Selasphorus* hummingbirds are larger with longer bills, more graduated tails, quick flight, and louder and harder calls. Calliope Hummingbird is slightly larger with a short, mostly black tail that falls equal with or shorter than wing tips at rest, and a quicker, darting flight.
VOICE Quiet and inconspicuous except for strong insect-like wing buzz of adult males; at times gives quiet, high chips when feeding. Female wing buzz is soft but stronger and buzzier than Calliope Hummingbird and easily overlooked.
STATUS & DISTRIBUTION Endemic to Mexico. **VAGRANT:** Accidental in southeastern AZ (two enigmatic specimens collected on July 2, 1896).

SELASPHORUS HUMMINGBIRDS Genus *Selasphorus*

In North America this genus comprises seven species, four in western N.A.: Broad-tailed, Rufous, Allen's, and Calliope, and three more in Central America. These small hummers have fairly long, graduated tails that project beyond the wing tips at rest. Males have solid gorgets slightly elongated at the corners; their power-dive displays from heights of 10 to 30 feet are spectacular and species-specific. Females are green-and-rufous, with cinnamon body sides and tail bases.

BROAD-TAILED HUMMINGBIRD *Selasphorus platycercus* BTAH ▪ 1

A common species of western mountains, where the male's diagnostic, cricketlike wing trill is a characteristic sound. Monotypic. L 3.5–4" (9–10 cm) Bill 16–20 mm
IDENTIFICATION Tail weakly graduated. Pale eye ring in all plumages. **ADULT MALE:** Rose red gorget with pale chin and face. Often detected by wing trill. **ADULT FEMALE:** Throat whitish with variable lines of bronzy-green flecks, sometimes one or more rose spots; sides of neck and underparts variably washed cinnamon. **IMMATURE MALE:** Resembles adult female but upperparts fresher in fall, with fine buff tips; throat usually flecked fairly

heavily with bronzy green, often with rose-pink spots; tail averages more rufous at base. Complete molt in winter and spring produces plumage like adult male. **IMMATURE FEMALE:** Resembles adult female, but upperparts fresher in fall, with fine buff tips; tail averages less rufous at base.
SIMILAR SPECIES Female/immature Rufous/Allen's Hummingbirds are slightly smaller and slimmer in build (noticeable in comparison) with more strongly graduated tails that have a more tapered tip. Rufous/Allen's typically have a whiter forecollar contrasting with brighter rufous sides, and lack whitish eye ring often shown by Broad-tailed; their uppertail coverts and tail base have more rufous (adult female Rufous can be all green); and their chip calls are slightly lower pitched. Also see female and immature Calliope Hummingbird.
VOICE Generally higher pitched than Rufous/Allen's. **CALL:** A slightly metallic, sharpish *chip* or *chik,* often doubled, *ch-chip* or *chi-tik.* Adult male's wing trill diagnostic.
STATUS & DISTRIBUTION **BREEDING:** Western U.S. to Guatemala. Common

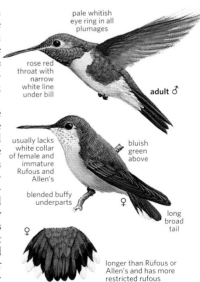

pale whitish eye ring in all plumages

rose red throat with narrow white line under bill

adult ♂

usually lacks white collar of female and immature Rufous and Allen's

bluish green above

blended buffy underparts

♀

♀

long broad tail

longer than Rufous or Allen's and has more restricted rufous

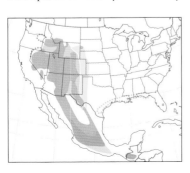

(Apr.–Aug.) in mountains. **MIGRATION:** Mainly Apr.–May, Aug.–Sept. (a few to western Great Plains). **WINTER:** Mainly Mexico. Casual to very rare (mainly Nov.–Apr.) in the Southeast. **VAGRANT:** Casual north to BC and west to coastal southern CA.

RUFOUS HUMMINGBIRD *Selasphorus rufus*   RUHU ■ 1

rufous back

adult ♂

slight golden cast above

white collar

♀

green-flecked ♂

immature ♂

Rufous and Allen's have extensive rufous

This common summer hummingbird of the Northwest is the western species most often found in the East (in fall and winter). Note that all except (most) adult male Rufous are rarely separable in the field from Allen's Hummingbird, so many observations are best termed "Rufous/Allen's." In dive display (also given in migration and by immatures) male climbs to a start point, then dives with a slanted J-form trajectory, typically followed by a short, horizontal fluttering flight before climbing to repeat the dive. Monotypic. L 3.2–3.7" (8–9 cm) Bill 15–19 mm
IDENTIFICATION Adult males often detected by wing buzz, which, like other *Selasphorus,* is produced only in direct flight, not when hovering. **ADULT MALE:** Flame orange gorget; rufous back often has some green spotting, can be solidly green. **ADULT FEMALE:** Throat whitish with lines of bronzy-green flecks strongest at corners, and typically a central splotch of red. **IMMATURE MALE:** Resembles adult female but upperparts fresher in fall, with fine buff

tips; throat usually flecked fairly heavily, and with red spots. Rectrices average narrower and with more rufous at bases, and white tips to outer rectrices narrower. Complete molt in winter produces plumage like adult male. **IMMATURE FEMALE:** Resembles adult female but upperparts fresher in fall, with fine buff tips; rectrices average broader; throat evenly flecked and with no (rarely a few) red spots.
SIMILAR SPECIES Female and immature Allen's Hummingbird safely distinguished only in the hand by narrower outer rectrices relative to age and sex. Adult male Allen's has green back (like very small percentage of Rufous). Male Allen's display dives are U-shaped, not J-shaped, and can be given by immatures in fall and winter. Females of resident southern California subspecies of Allen's have paler and

strongly green-mottled flanks, unlike Rufous. See female and immature Broad-tailed.
VOICE CALL: A fairly hard ticking or clicking *tik* or *chik,* often doubled or trebled, *ch-tik* or *ch-ti-tik.* **ALARM CALL:** A slightly squeaky buzz, *tssiur* or *tsirr,* and squeaky chippering in interactions. Adult male's wing buzz often draws attention; stuttering *ch-ch-ch-ch-chi* at pullout of dive is diagnostic. Immature males make species-specific dives without the sound effects.
STATUS & DISTRIBUTION BREEDING: Northwestern N.A. Common (Mar.–July) in open woodlands and parks. **MIGRATION:** Mainly late Feb.–early May, late June–Sept. Rare (mainly July–Nov.) in the East. **WINTER:** Mexico. Rare (mainly Oct.–Mar.) in the Southeast and casual in southwestern CA.

ALLEN'S HUMMINGBIRD *Selasphorus sasin*   ALHU ■ 1

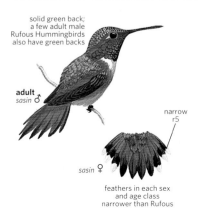

solid green back; a few adult male Rufous Hummingbirds also have green backs

adult *sasin* ♂

narrow r5

sasin ♀

feathers in each sex and age class narrower than Rufous

Virtually endemic to California as a breeding bird, this stunning gem is one of the earliest migrants in North America: Males return from Mexico in January and head south in June! In dive display (also given in migration and by immatures) male climbs to start point and then makes repeated, often fairly low, U-shaped dives followed by a single high climb and steep dive. Polytypic (2 ssp.; both in N.A.). L 3.2–3.5" (8–9 cm) Bill 15–21 mm
IDENTIFICATION Rarely separable in field from Rufous Hummingbird except by breeding range (in which Rufous is a common migrant), and

many birds are best termed "Rufous/Allen's." Age-related variation and plumage sequences are as Rufous Hummingbird (see account) with some exceptions. **ADULT MALE:** Back always green, with rufous uppertail coverts. **ADULT FEMALE:** Uppertail coverts typically rufous (ironically, often green on Rufous). Flanks paler and mottled bronzy green on *sedentarius* subspecies.
GEOGRAPHIC VARIATION Migratory subspecies *sasin* breeds from southwestern Oregon to southern California. It may intergrade with expanding population of slightly larger and

longer-billed resident subspecies *sedentarius*, originally confined to California's Channel Islands, but now breeding on mainland. Males of the two subspecies are identical in the field, but female *sedentarius* has paler rufous and extensively green-mottled sides that create a less-demarcated vest (*sasin* females have brighter rufous, well-demarcated sides).

SIMILAR SPECIES See Rufous Hummingbird. Very fine, almost wirelike outer rectrices of adult male Allen's may be appreciated in good views (e.g., when aggressive males spread their tail at a feeder), but identification best confirmed by photo or in-hand examination. See female and immature Broad-tailed Hummingbird (which rarely overlap in range with Allen's).

VOICE CALL: Not distinguishable in field from Rufous (see account). However, high, drawn-out, shrieky whine at pullout of adult male Allen's power dive is diagnostic (unlike stuttering of Rufous); immature males make species-specific dive displays but without the sound effects.

STATUS & DISTRIBUTION Breeds western N.A., winters Mexico (nominate *sasin*). Subspecies *sedentarius* is a fairly common but local resident in southern CA (north to Ventura Co.) and on the Channel Is. **BREEDING:** Common (Feb.–July) in open woodlands of coastal belt. **MIGRATION:** Mainly mid-Jan.–early Mar., June–Aug., in fall, ranging east to southeastern AZ. **WINTER:** Casual to rare (mainly Aug.–Feb.) in the Southeast. **VAGRANT:** Casual (mainly July–Jan.) in the East.

CALLIOPE HUMMINGBIRD *Selasphorus calliope*   CAHU ▪ 1

adult ♂

purplish red throat streaks

very short bill

overall females and immatures are very similar to Broad-tailed but are smaller, with shorter tail and bill

whitish in supraloral region

♀

♀

shorter than Rufous or Allen's and almost lacks rufous

The smallest breeding bird in North America, the Calliope Hummingbird is a summer resident of western mountain meadows, especially along wooded streams. Monotypic. L 3–3.2" (7.5–8 cm) Bill 12.5–16 mm

IDENTIFICATION Note relatively short and squared tail, which means that Calliope's wing tips often project slightly beyond its tail when perched; bill is medium-short. Adult males are distinctive; females and immatures are buff-and-green and best identified by their small size, and relatively short, mostly black tail. Note whitish lores on many birds. **ADULT MALE:** Elongated gorget of rose stripes rarely looks solid; note white lores and eye ring. **ADULT FEMALE:** Throat is whitish with variable lines of bronzy-green flecks and rarely one or more rose spots. Sides of neck and underparts are washed cinnamon and lack a distinct white "forecollar." Tail is slightly rounded (although it appears squared when closed). **IMMATURE MALE:** Resembles adult female but has fresher upperparts in fall, with fine buff tips. Throat is usually flecked more heavily with bronzy green and often has one or more rose pink spots or streaks; white bill-lip line is often reduced or absent. A complete molt in winter produces an adultlike male plumage. **IMMATURE FEMALE:** Resembles adult female but upperparts are fresher in fall, with fine buff tips, and white bill-lip line is reduced or absent.

SIMILAR SPECIES Note short tail. Other female/immature *Selasphorus* are larger (obvious in direct comparison), with longer and slightly thicker bills, longer and distinctly graduated tails that project beyond wing tips on perched birds. They are more aggressive, often uttering their louder calls and chasing other hummingbirds. However, the female Broad-tailed is remarkably similar in plumage to the Calliope; other than in size and bill length, note Broad-tailed's longer tail, which is mostly green above. Rufous/Allen's Hummingbirds have brighter and more contrasting rufous body sides and more rufous in the tail than Calliope.

VOICE Often fairly quiet and inconspicuous. **CALL:** A relatively soft, high chip, often doubled, *chi* and *chi-ti;* often repeated steadily from perch, less often when feeding. High, slightly buzzier chippering during interactions. Male's wing buzz can attract attention during flight displays.

STATUS & DISTRIBUTION Breeds western N.A., winters Mexico. **BREEDING:** Fairly common (Apr.–Aug.) in mountain meadows and open forest. **MIGRATION:** Mainly late Mar.–early May, July–Sept.; in fall, ranges east to western Great Plains and western TX. **WINTER:** Rare (mainly Oct.–Mar.) in the Southeast. **VAGRANT:** Casual in late fall as far northeast as MN and MA.

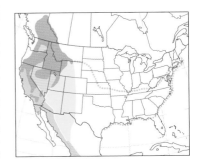

BROAD-BILLED HUMMINGBIRD *Cynanthus latirostris* BBIH ■ 2

This medium-size hummingbird of desert canyons and low mountain woodlands is an inveterate tail wagger. It often attracts attention through its dry chattering calls. Polytypic (5 ssp.; *magicus* in N.A.). L 3.5–4" (9–10 cm) Bill 18.5–23.5 mm
IDENTIFICATION Broad-billed has a cleft tail and a medium-long bill that is reddish basally. **ADULT MALE:** Distinctive, with deep green plumage overall, a blue throat, white undertail coverts, a cleft blue-black tail, and a bright red bill, tipped black. **ADULT FEMALE:** Note broad dark auricular mask offset by a whitish to pale-gray postocular stripe; gray below with limited green spotting on sides. Blackish tail has variable greenish basally and distinct white tips to outer rectrices. Lower mandible is reddish basally. **IMMATURE MALE:** Resembles female but plumage fresh in fall with buff-tipped upperparts. Throat usually has some blue blotching or a solid blue

patch, and older birds are extensively mottled green on underparts. Red on bill base often extends to upper mandible, and tail is more extensively blue-black, with smaller white corners. **IMMATURE FEMALE:** Resembles adult female but plumage fresh in fall, buff-tipped on upperparts; underparts have reduced or no green mottling on sides, and lower mandible base averages paler, more pinkish.
GEOGRAPHIC VARIATION North American breeding birds are of *magicus* in nominate subspecies group, *latirostris*. Two other subspecies groups in Mexico *(lawrencei* and *doubledayi)* may be specifically distinct.
SIMILAR SPECIES Immature males with a blue throat patch might suggest larger and heavier Blue-throated Hummingbird, which has an all-black bill, a very large blue-black tail with bold white corners, and a high-pitched squeak call. A female might be confused with a female White-eared Hummingbird, which is stockier and proportionately shorter billed, with a bold black auricular mask, an even more flagrant white postocular stripe, and a hard chipping call suggesting Anna's Hummingbird. Dullest females could suggest a Black-chinned Hummingbird, but note their reddish

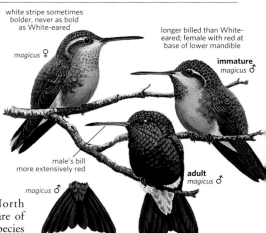

white stripe sometimes bolder, never as bold as White-eared

magicus ♀

longer billed than White-eared; female with red at base of lower mandible

immature *magicus* ♂

male's bill more extensively red

magicus ♂

adult *magicus* ♂

longer, lobed tail more notched

bill base, broad inner primaries, and call.
VOICE CALL: Often gives a dry *cht*, singly or in short series, *ch-ch-cht*, etc., when chattering cadence suggests the common call of Ruby-crowned Kinglet; also a high squeaky chippering in interactions. **SONG:** A high, sharp song, repeated from a perch.
STATUS & DISTRIBUTION Mexico to southwest U.S.; winters mainly in Mexico. **BREEDING:** Fairly common (mainly Mar.–Sept.), in brushy woodland, desert washes, gardens. **WINTER:** Uncommon and local in southern AZ. Very rare in southern CA and TX, mainly in fall and winter. **VAGRANT:** Casual or accidental to OR, ID, CO, and the East.

AMAZILIA HUMMINGBIRDS Genus *Amazilia*

Of about 30 species in this genus, four have occurred north of Mexico: two regular breeders, a casual breeder, and an accidental visitor. The sexes look generally similar. *Amazilia* may represent two genera: a slender-billed group with dark upper mandibles, including Berylline (subgenus *Saucerottia*) and typical *Amazilia*, with broader red bills.

BERYLLINE HUMMINGBIRD *Amazilia beryllina* BEHU ■ 3

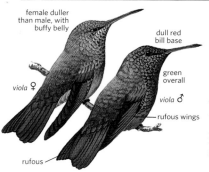

female duller than male, with buffy belly

dull red bill base

green overall

viola ♀

viola ♂

rufous wings

rufous

This common Mexican hummingbird of foothill oak woodlands ranges north on an annual basis. Polytypic (5 ssp.; *viola* in N.A.). L 3.7–4" (9.5–10 cm) Bill 18.5–21 mm
IDENTIFICATION This medium-size hummer has a slender, medium-length bill with red on at least base of lower mandible. Diagnostic bright-rufous wing patch of all plumages noticeable in flight. **ADULT MALE:** Beryl green throat and chest contrast with dusky buff

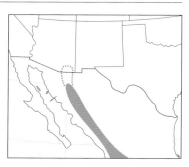

rufous on wing especially visible in flight

♂

belly; crown green; coppery purple tail often looks rufous. **ADULT FEMALE:** Similar to male but duller, with whitish mottled chin, and more grayish belly. **IMMATURE:** Resembles female; dingier below.

SIMILAR SPECIES Buff-bellied Hummingbird (normally no overlap in range) lacks a rufous wing patch.

VOICE CALL: A fairly hard, buzzy *dzirr* or *dzzrit.* **SONG:** Short jerky and squeaky phrases; squeaky warbling.

STATUS & DISTRIBUTION Mexico to Honduras. **SUMMER:** Rare to casual visitor (Apr.–Oct., mainly June–Aug.) in oak zone of southeastern AZ mountains. **VAGRANT:** Casual in summer to southwestern NM and western TX.

BUFF-BELLIED HUMMINGBIRD *Amazilia yucatanensis* BBEH ■ 2

This fairly large hummer is found in south Texas. Polytypic (3 ssp.; *chalconota* in N.A.). L 3.8–4.3" (10–11 cm) Bill 19–22 mm

IDENTIFICATION Medium-length bill; broad, cleft tail; buffy belly; pale eye ring; and mostly rufous tail in all plumages. **ADULT MALE:** Throat and chest iridescent green, contrasting with buffy-cinnamon belly. Mostly rufous tail tipped bronzy green, most broadly on central rectrices. Bright red bill tipped black. **ADULT FEMALE:** Similar to

male but duller, with whitish mottled chin, a shallower tail cleft, and mostly green central rectrices. **IMMATURE:** Like female but throat and chest dingy buff to whitish, mottled iridescent green, belly paler buff, upper mandible mostly blackish, developing red over first year.

SIMILAR SPECIES Berylline Hummingbird (normally no overlap in range) has rufous wings, a darker bill, and a different voice.

VOICE CALL: A clipped to slightly smacking chip, *tik* or *tk,* at times run into a rolled *tirr* or *tsirrr.* In warning, a slightly buzzy *sssir.* In chases, variable and usually fast-paced series of buzzy to lisping calls. **SONG:** A varied arrangement of chips alternating with slurred whistles and wheezy notes.

STATUS & DISTRIBUTION: South TX to Guatemala. **BREEDING:** Fairly common (mainly Apr.–Aug.). **WINTER:** More widespread in TX, rare to casual (mainly Oct.–Mar.) along Gulf Coast to north FL. **VAGRANT:** Casual to AR, coastal GA, and south FL.

pinkish red bill base

chalconota

green breast

buff belly

chalconota

tail extensively rufous

VIOLET-CROWNED HUMMINGBIRD *Amazilia violiceps* VCHU ■ 2

violet crown; duller on juveniles

red-based bill

ellioti

striking snowy white underparts

An unmistakable hummer of western Mexico, with local summer populations in southeastern Arizona and southwestern New Mexico. Polytypic (2 ssp.; *ellioti* in N.A.). L 4–4.5" (10–11.5 cm) Bill 21–24 mm

IDENTIFICATION Bill is medium-long and straightish, and tail is broad and squared to slightly cleft (averaging more so on males). Bright white underparts are unique among North

American hummingbirds. **ADULT:** Sexes similar. Crown and auriculars are violet-blue (rarely appearing turquoise) and tail bronzy greenish to brownish. Bright red bill is tipped black. **IMMATURE:** Resembles adult, but throat and underparts are dingier whitish. Crown is oily bluish (can look dark) with buff feather tips in fresh plumage; upperparts and tail are tipped buff in fresh plumage; and upper mandible is mostly blackish, developing red over the first year.

GEOGRAPHIC VARIATION North American birds are of northern subspecies, *ellioti.* Nominate from central Mexico has a brighter, coppery-bronze tail.

VOICE CALL: A hard chip, *stik* or *tik,* often run into rattled short series; single plaintive chip, repeated from perch, *chieu chieu chieu.*

STATUS & DISTRIBUTION Mexico to southwest U.S., winters Mexico. **BREEDING:** Uncommon (mainly Apr.–Sept.) in arid to semiarid scrub, riparian woodland, and gardens. **WINTER:** Rare in southeastern AZ. **VAGRANT:** Casual northwest to northern CA, east to TX.

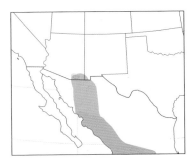

CINNAMON HUMMINGBIRD *Amazilia rutila* CIHU ▪ 5

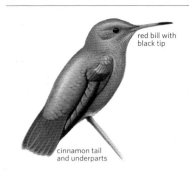

red bill with black tip

cinnamon tail and underparts

This medium-large hummer is a vagrant from the tropical lowlands of west Mexico. Polytypic (4 ssp.; N.A. recs. probably *diluta*). L 4–4.5" (10–11.5 cm) Bill 20.5–23.5 mm **IDENTIFICATION** All plumages solidly cinnamon throat and underparts. **ADULT:** Sexes similar. Black-tipped, bright-red bill. **IMMATURE:** Resembles adult but upperparts tipped cinnamon when fresh; upper mandible mostly blackish, developing red over the first year.

SIMILAR SPECIES Should be unmistakable. See Xantus's Hummingbird. **VOICE CALL:** A hard, clipped tick with a buzzy or rattled quality, *tzk* or *dzk*, can be run into buzzy rattles. In interactions, high squeaks run into excited chatters. **SONG:** A short, varied arrangement of slightly squeaky chips. **STATUS & DISTRIBUTION** Mexico to Costa Rica. **VAGRANT:** Accidental to Patagonia, AZ (July 21–23, 1992) and Teresa, south-central NM (Sept. 18–21, 1993).

Genus *Hylocharis*

WHITE-EARED HUMMINGBIRD *Hylocharis leucotis* WEHU ▪ 3

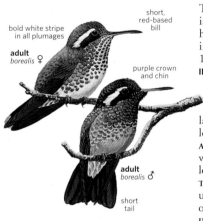

short, red-based bill

bold white stripe in all plumages

adult *borealis* ♀

purple crown and chin

adult *borealis* ♂

short tail

borealis ♂

short, square-ended tail

This medium-size, stocky hummer is common in Mexico's pine-oak highlands. Polytypic (3 ssp.; *borealis* in N.A.). L 3.5–4" (9–10 cm) Bill 15–18.5 mm **IDENTIFICATION** Bill is medium-length and red basally. In all plumages, bold white postocular stripe contrasts with broad blackish auricular mask. **ADULT MALE:** Head often looks black with flagrant white stripe. **ADULT FEMALE:** Throat and underparts whitish, extensively spotted green; lower mandible reddish basally. **IMMATURE MALE:** Resembles female; throat usually has some blue and green. Red on bill extends to upper mandible. **IMMATURE FEMALE:** Resembles adult; throat has sparser and more bronzy spotting. **SIMILAR SPECIES** Unlikely to be confused. See Xantus's Hummingbird and Broad-billed Hummingbird. **VOICE CALL:** A clipped, fairly hard

ticking chip, singly or often doubled or trebled, *ti-ti-tik* or *chi-tik chi-tik*. **SONG:** Rapid, rhythmical chipping interspersed with squeaks and gurgles. **STATUS & DISTRIBUTION** Mexico to Nicaragua. **SUMMER:** Rare and local visitor (mainly Apr.–Oct.) in pine-oak of southeastern AZ mountains. **VAGRANT:** Very rare or casual to southwestern NM and western TX, accidental elsewhere in TX and to CO and MS.

XANTUS'S HUMMINGBIRD *Hylocharis xantusii* XAHU ▪ 5

This unmistakable vagrant shares the congeneric White-eared Hummingbird's bold face pattern. Monotypic. L 3.3–3.8" (8–9 cm) Bill 16–19 mm **IDENTIFICATION** Bill is medium length and red basally. In all plumages a bold white postocular stripe contrasts with a broad blackish auricular mask. **ADULT MALE:** Unmistakable. **ADULT FEMALE:** Throat and underparts cinnamon, lower mandible reddish basally. **IMMATURE:** Resembles adult female but fresh plumage has broadly buff-tipped crown feathers. Males usually have some blue-green throat spots, and red

on bill base extends to upper mandible. **SIMILAR SPECIES** Should not be confused. **VOICE CALL:** Most often gives a low, fairly fast-paced rattled *trrrrr* or *turrrt*. **SONG:** A quiet, rough, gurgling warble interspersed with rattles and high squeaky notes. **STATUS & DISTRIBUTION** Endemic to Baja California, Mexico. **VAGRANT:** Accidental in winter to southern CA: Anza-Borrego S.P. (Dec. 27, 1986) and Ventura (Jan. 30–Mar. 27, 1988) and southwestern BC: Gibsons (Nov. 16, 1997–Sept. 21, 1998).

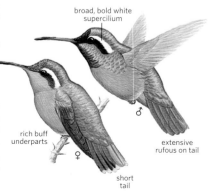

broad, bold white supercilium

rich buff underparts

♀

extensive rufous on tail

short tail

♂

TROGONS Family Trogonidae

Elegant Trogon, male (AZ, Mar.)

Trogons are stunning tropical birds with bright parrotlike colors. Their plumages and rarity in the United States place them among the most sought-after birds.

Structure Their unique profile consists of a long, relatively broad tail; a large, rounded head with large eyes; a small, broad, notched bill; a short neck; and a compact body.

Behavior Trogons sit upright and motionless for extended periods, making them hard to find. Their calls are often the easiest method of locating them, but the calls can carry some distance and have a ventriloquial quality. Flights are usually short; the flash of color gives their presence away. They primarily eat large insects and fruit. Food is taken on the wing, either "flycatching" or, more typically, via sallies where the bird plucks an item from the end of a branch without landing. They are usually seen singly or in small groups, and nest in tree cavities.

Plumage Trogons have soft, dense plumage. Males are brilliantly colored, females more subdued. Males of the two species seen in the U.S. are green and red. The metallic green upperparts occasionally appear greenish blue. Juveniles slowly molt into adultlike plumage over their first year. Adults typically molt in late summer.

Distribution A pantropical family of 44 species, trogons reach their greatest diversity in the Neotropics. Nine species occur in Mexico; two barely reach the U.S.

Taxonomy The 29 neotropical trogon species are placed in four genera, from the large quetzals *(Pharomachrus)* to the largest genus, *Trogon,* which includes all the similar, smaller-size birds.

Conservation Habitat destruction and human development threaten some species. Birder disturbance at well-known breeding locations in Arizona may result in nest failures.

ELEGANT TROGON *Trogon elegans* ELTR ■ 2

This trogon was formerly named the "Coppery-tailed Trogon." Polytypic (4 ssp.; 2 in N.A.). L 12.5" (32 cm) **IDENTIFICATION** Both sexes have a white breast band that borders a red belly; a

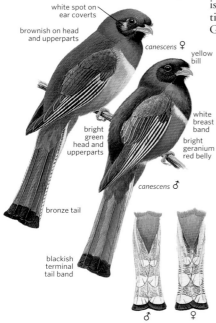

white spot on ear coverts

brownish on head and upperparts

canescens ♀

yellow bill

bright green head and upperparts

white breast band

bright geranium red belly

bronze tail

canescens ♂

blackish terminal tail band

♂   ♀

yellow bill; and a red orbital ring. **MALE:** Bright green head, chest, and back. Very fine barring on underside of tail, each feather with a broad white tip; above, tail gold to greenish copper, with two central rectrices tipped broadly with black. **FEMALE:** Gray-brown where male is green. A broad, white teardrop below and behind eye. Rufous-brown upper tail with a black tip; underside has slightly coarser barring. **JUVENILE:** Like female, but lacking red. Large whitish buffy spots above.

GEOGRAPHIC VARIATION The subspecies *ambiguus* has reached Texas; the male has deeper and more extensive red underparts, the female has a brownish tinge; subspecies *canescens* occurs in Arizona. **SIMILAR SPECIES** The Eared Quetzal is only superficially similar. It is larger and has a thicker body; its bill is black or gray; and it lacks a white breast band and barring on undertail. The Mountain Trogon *(T. mexicanus)*, resident as far north as Chihuahua, Mexico, is a possible vagrant. **VOICE CALL:** Varying croaking or *churr* notes. **SONG:** A series of croaking *co-ah* notes.

STATUS & DISTRIBUTION Fairly common, but local, in southeastern AZ. Rare in southwestern NM. **BREEDING:** Pine-oak woodlands, in association with streamside woodlands, primarily sycamores, mostly at elevations of 4,000–6,500 feet. **MIGRATION:** Routes, duration, and distances largely unknown. In spring, arrives Apr.–early May. Departs in fall by early Nov. **WINTER:** Withdraws from most of AZ, although still annual at that season. **VAGRANT:** Casual to western (Big Bend N.P.) and southern TX.

POPULATION The breeding population fluctuates depending, in part, on drought conditions. Arizona lists the bird as a candidate species on its Threatened Native Wildlife list.

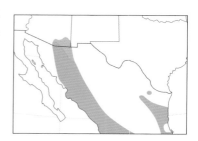

EARED QUETZAL *Euptilotis neoxenus* EAQU ■ 4

This large and showy (but shy) bird from western Mexico is casually seen in our area. Monotypic. L 14" (36 cm) **IDENTIFICATION** Formerly called the "Eared Trogon," for its wispy, postocular plumes. **MALE:** Dark face and crown, almost blackish. Bright green lower throat, upper breast, and most of upperparts; bright red belly. Tail steely blue above, mostly white from

below. **FEMALE:** Like male but head, throat, and most of breast gray; belly paler red. **JUVENILE:** Like female, but duller and with more black at base of underside of tail.

SIMILAR SPECIES Only superficially similar to Elegant Trogon (see that species). **VOICE CALL:** A long upslurred squeal ending in a *chuck* note. And a loud, hard cackling *ka-kak,* sometimes given in flight. **SONG:** A long, quavering series of whistled notes that increase in volume.

STATUS & DISTRIBUTION Casual in mountain woodlands of southeastern AZ; two records in central AZ (Mogollon Rim and Superstition Mts.). Most records from late summer and fall, but some from winter. Its preference for remote canyons might result in underreporting.

slaty gray head and breast on female

dark bill

both sexes lack pale marking on head; "ears" visible only at close range

♂

♀

no breast band

all have a hunchbacked appearance and a disproportionately small head

blue-black tail when viewed from above

unlike Elegant, tail feathers extensively pure white with no fine barring

HOOPOE Family Upupidae

Eurasian Hoopoe (Spain, May)

A unique family composed of a single, flashy and unforgettable species—the Eurasian Hoopoe. **Structure** It has a long, curved bill, a head crest, a rather chunky body, and rounded wings.
Behavior Hoopoes feed primarily on the ground, where they can be unobtrusive. They nest in cavities, either on the ground, in rocks, or in trees. The slow and undulating flight style is reminiscent of a large butterfly. The crest is held back along top of crown; the bird expands it fully when excited or briefly upon landing.
Distribution This Old World species breeds from central Europe and central Asia, through Africa and continental Asia. Northerly populations are migratory; most European breeders winter in sub-Saharan Africa, while central Asian breeders winter in southern Asia.
Taxonomy Most authorities recognize one species with eight subspecies. Some authors elevate the Madagascar and African subspecies to species status.

Genus *Upupa*

EURASIAN HOOPOE *Upupa epops*   EUHO ■ 5

The Eurasian Hoopoe is an Old World vagrant to North America. Polytypic (8 ssp.; nominate in N.A.). L 10.5" (27 cm)
IDENTIFICATION Sexes look similar, but some adult males have a pink tinge to brown head, neck, back, and breast. Crown is a warmer brown, with black tips to crest feathers. Back stripes alternate black and creamy white, and are continuous with black-and-white stripes on wings. **JUVENILE:** It resembles an adult female, but brown plumage is grayer and duller. **FLIGHT:** Very striking. Secondaries and secondary coverts are boldly striped black-and-white, while black primaries have a broad white band near tip and black tail has a white band near base.
SIMILAR SPECIES None.
VOICE CALL: A high-pitched *scheer.* **SONG:** A low resonant *poo-poo-poo*, from which the name hoopoe derives.
STATUS & DISTRIBUTION Accidental, one specimen record of the northern nominate subspecies from the Yukon Delta, western AK (Sept. 2–3, 1975).
POPULATION Most populations are stable, but hunting in southern Europe and Asia is of concern.

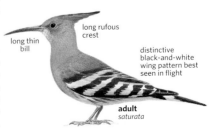

long thin bill

long rufous crest

distinctive black-and-white wing pattern best seen in flight

adult
saturata

KINGFISHERS Family Alcedinidae

Green Kingfisher, female (TX, Nov.)

In general, kingfishers sit on low perches watching for prey below them. Their flight is direct and strong, with rapid wingbeats.

Structure Tremendous variation in size exists within the family. The majority of species are small to medium size. All kingfishers have large heads with long, strong beaks and short legs. Tail length varies from very short and stubby to long with streamers.

Behavior Most species are solitary except during the breeding season. The exception is the large Kookaburra of Australia, which is more social. Kingfishers nest in burrows and usually defend territories all year round. Kingfishers in North America hunt from low perches over water. Many forest species elsewhere hunt primarily for terrestrial vertebrates. Species on the smaller end of the spectrum feed primarily on invertebrates.

Plumage Almost all species are dark above, most commonly blue, green, or brown. Underparts are normally white or rufous, with a few species exhibiting banding or barring. For most species the sexes are similar. Sexual dimorphism is normally expressed with minor differ-

ences, including the pattern of the underparts or changes in throat or tail color.

Distribution Kingfishers are found on all continents except Antarctica. They are also absent from the northernmost reaches of North America and most of Russia and central Asia. The family reaches its greatest diversity in an area encompassing Southeast Asia to Australia. There are three regularly occurring species in North America and one accidental species; two more species occur from Mexico to South America, for a total of six species in the New World.

Taxonomy The family is divided into three subfamilies, only one of which occurs in the New World. Worldwide, there are 91 species in 17 genera recognized. Two N.A. species (Ringed and Belted Kingfishers) formerly in the genus *Ceryle* now reside in the genus *Megaceryle*.

Conservation Many species of kingfisher live in primary forest and are therefore subject to pressures from deforestation. There are 12 species of kingfisher of conservation concern; most are found on islands, and one is considered endangered and two are critically endangered.

Genera Megaceryle and Chloroceryle

RINGED KINGFISHER *Megaceryle torquata*   RIKI ■ 2

The largest of North American kingfishers, the Ringed normally frequents larger rivers but can also be found around ponds and lakes. It typically perches higher than the other, smaller kingfishers found in North America and often flies at surprisingly high altitudes with slow, deep wing beats. Polytypic (3 ssp.; nominate in N.A.). L 16" (41 cm)

IDENTIFICATION Large with a big-headed appearance, short crest and long, heavy bill. **ADULT MALE:** Slate blue above with a prominent white collar and rufous underparts. White underwing coverts easily visible in flight. Massive bill is gray at base, becoming black on distal half. **ADULT FEMALE:** Similar to adult male, but with an obvious slate blue breast band and rufous underwing coverts. **JUVENILE:** Similar

to adults, but male has a narrow slaty band across upper breast that is mixed with cinnamon brown. Slaty breast band of juvenile female is also mixed with cinnamon brown.
SIMILAR SPECIES Distinctive. Smaller Belted Kingfisher has predominantly white underparts.
VOICE CALL: A harsh rattle that is slower and lower than a Belted Kingfisher's. Also a loud single *chack,* somewhat suggestive of a Great-tailed Grackle's; primarily given in flight.
STATUS & DISTRIBUTION Locally com-mon in the Lower Rio Grande Valley, becoming uncommon and local farther north. **BREEDING:** In burrows close to or along rivers. **WINTER:** Wanders from nesting areas during winter, including farther up the Rio Grande and its tributaries. **VAGRANT:** Some regularity to the Texas Hill Country east of current breeding distribution. Accidental to OK and LA.
POPULATION The population in the U.S. appears to be continuing to increase, with a reflected expansion in range.

shaggy crest

huge bill

torquata ♂

most of underparts rufous

overall size much larger than Belted

torquata ♀

rufous belly

BELTED KINGFISHER *Megaceryle alcyon* BEKI 1

The Belted Kingfisher is the most widespread and abundant kingfisher in North America. These birds need clear, still water for fishing, with elevated perches from which to hunt. Monotypic. L 13" (33 cm)
IDENTIFICATION Medium size with a big-headed appearance, prominent shaggy crest, and long, heavy bill. **ADULT MALE:** Slate blue above with a prominent white collar. Underparts white with a single blue breast band. Large bill has a gray base with a black outer half. **ADULT FEMALE:** Similar to adult male, but with an obvious rufous band across upper belly. Rufous also extends down flanks. **JUVENILE:** Very similar to adults in both sexes. Juvenile male has a tawny breast band that is mottled. Juvenile female also has a much reduced rufous belly band.
SIMILAR SPECIES Distinctive. For most of its distribution there is nothing that can cause confusion. In southern TX, the much larger Ringed Kingfisher has rufous underparts.
VOICE CALL: A loud, dry rattle. Belted Kingfishers also make harsh *caar* notes while perched and in flight.
STATUS & DISTRIBUTION Common and conspicuous. **BREEDING:** In burrows close to or along water. **MIGRATION:** A partially migratory species, with the northernmost populations completely leaving the breeding grounds. Belted Kingfisher is resident through much of the U.S., but most of the birds in this area leave to wintering grounds in the extreme southern U.S. and south through C.A. and the Caribbean. **WINTER:** Rare along the Pacific Coast from southeastern AK through BC. Uncommon through the northern and central U.S., becoming common in the southern third. **VAGRANT:** Accidental on the Azores, Iceland and western Europe. **POPULATION** Stable.

shaggy crest

blue and rufous breast bands

single bluish breast band

AMAZON KINGFISHER *Chlloroceryle amazona*   AMKI ▪ 5

Amazon Kingfisher is a widespread, adaptable species found in many aquatic habitats in the neotropics. They prey mostly on crustaceans and fish, patiently watching from a perch over water until prey is spotted, then diving. Amazon Kingfishers tend to hover much less than Belted or Ringed Kingfishers, but they are much more conspicuous than Green Kingfisher. Monotypic. L 11–12" (28–30 cm)

IDENTIFICATION The only medium-size kingfisher with vivid green upperparts in Mexico and most of Central America. Long, heavy blackish bill. **ADULT:** Upperparts and head a deep forest green with an almost oily-looking metallic sheen, broken by broad white collar. Underparts white with a few dark streaks in flanks; male has broad rufous breast band, female a narrower green breast band. **JUVENILE:** Similar to female, with breast band sometimes broken around the midpoint, spotted with buff.

SIMILAR SPECIES Green Kingfisher is much smaller, with a smaller bill and smaller crest and usually forages from a concealed perch. Green-and-rufous Kingfisher *(C. inda)*, a retiring species also smaller than Amazon and found north only to Nicaragua, is easily distinguished by its extensively rufous underparts.

VOICE Often gives a harsh *teck* call. The rarely heard song, given from a treetop, is a whistled *see see see see.*

STATUS & DISTRIBUTION Common resident of lakes, ponds, oxbows, and rivers from northern Mexico south to central Argentina and Uruguay, mostly east of the Andes. The northernmost populations are found in Mexico, north to southern Sinaloa and southern Tamaulipas. **VAGRANT:** Reported several times in Texas before recent, verified records at Laredo, Webb Co. (Jan. 24–Feb. 3, 2010), and San Benito, Cameron Co. (Nov. 9–still present early Dec. 2013).

crested

much larger size than Green Kingfisher with more massive bill

no white on wings

♀

GREEN KINGFISHER *Chlloroceryle americana*   GKIN ▪ 2

Smallest of the North American kingfishers, the Green Kingfisher frequents clear streams and ponds, often perching very close to the surface. The presence of a Green Kingfisher is often betrayed by its nervous calling when approached. Polytypic (5 ssp.; 2 in N.A.). L 8.75" (22 cm)

IDENTIFICATION Small with a big-headed appearance, an inconspicuous crest, and long, fairly heavy bill. **ADULT MALE:** Green above with a prominent white collar. Underparts white with a wide rufous band across chest. Wings are heavily spotted with white. Tail is green with white outer rectrices that are spotted with green. **ADULT FEMALE:** Similar to adult male, but with a green breast band that is mottled with white. There is also green mottling along flanks and across upper belly. Some individuals have a buffy wash across throat and chest. **JUVENILE:** Very similar to adult female but with buffy spotting on the crown and upperparts. **FLIGHT:** Direct and very fast, the bird's white-based outer tail feathers are conspicuous in flight.

GEOGRAPHIC VARIATION Differentiation of the two subspecies that occur in the U.S. is weak. Birds occurring in Texas belong to *hachisukai* and are more heavily spotted with white on lesser wing coverts. Individuals occurring in southeastern Arizona are placed in *septentrionalis* and lack white spotting on lesser coverts.

SIMILAR SPECIES Unmistakable. Both of the other North American kingfishers are much larger and slate blue in color.

VOICE **CALL:** Recalls two pebbles knocked together. Another call is a long series of rapid, but subdued *tick* notes. A squeaky *cheep* is given in flight.

STATUS & DISTRIBUTION Uncommon and often inconspicuous. **BREEDING:** In burrows close to or along rivers. **WINTER:** Susceptible to very cold weather, withdraws from northern portion of its range during severe winters. Rare in AZ, primarily in the San Cruz and San Pedro River drainages. **VAGRANT:** Accidental in TX outside normal range north to the panhandle, and in southwestern NM along the Gila River.

POPULATION Stable, but there are possible declines in Arizona.

white at base of outer tail feathers visible in flight

extensive white spotting on wings

♂

♀

very long bill

rufous breast

green breast band

green upperparts

WOODPECKERS AND ALLIES Family Picidae

Pileated Woodpecker, nestlings (MO, June)

Field identification of woodpeckers is generally straightforward except within a few close species pairs or groups such as the Yellow-bellied Sapsucker complex. These cases are compounded by hybridization, and not all individuals can be identified to species. Twenty-five species occur in North America, two as vagrants and one very likely extinct.

Structure Woodpecker structure and posture render them instantly recognizable (but see creepers and nuthatches). Our species vary from large sparrow to crow size. Adaptations for trunk foraging and excavation include a chisel-like bill, skull and neck muscle adaptations to reduce brain impacts from blows with the bill, stiffened rectrices, and strong claws. Woodpeckers have short legs; their feet have two toes forward and two (rarely one) backward. The extensible tongues are housed in a sling that wraps over the skull to (or nearly to) the nostril.

Behavior Woodpeckers perch along trunks and limbs, moving up in jerky "hitching" movements, using the stiff tail feathers as a prop. The flight of most species is strongly undulating. They hop when on the ground. Foraging often involves drilling into or flaking bark and dead wood for grubs, but the diverse feeding repertoire includes gleaning, lapping ants from the ground, consuming and storing acorns and other mast, excavating seeds from cones, drilling into living plant tissue for sap, and aerial sallies for insects. Most species are quite vocal, but song is replaced by drumming. Woodpeckers excavate nest cavities in trunks or branches, usually in dead wood, that are later used by a variety of birds and other animals.

Plumage Most of our woodpeckers are black and white, often in striking pied or barred patterns. Red adorns the plumage of nearly all our species, but is usually limited to the head (and frequently present only in males).

Distribution Woodpeckers occur on most major continents. They are most diverse in wooded regions, but some species have adapted to arid scrub, grassland, and alpine tundra. Most populations are sedentary; a few species (notably sapsuckers and Northern Flickers) can be strongly migratory, and other species move short distances or are nomadic depending on food availability.

Taxonomy There are 217 species of woodpeckers in 33 genera worldwide. These include the two wrynecks of the subfamily Jynginae, the 29 small piculets (subfamily Picumninae), and roughly 186 "true" woodpeckers (subfamily Picinae). Within the true woodpecker subfamily, several smaller groupings ("tribes") are recognized.

Conservation Some woodpeckers are among our most familiar birds. Almost universally they require standing snags for nest sites. Species requiring tracts of old-growth forests have generally declined, most strikingly the endangered Red-cockaded and the very likely extinct Ivory-billed. Worldwide, BirdLife International lists seven woodpecker species as vulnerable, two as endangered, and three as critically endangered.

Genus *Jynx*

EURASIAN WRYNECK *Jynx torquilla*　　EUWR ■ 5

The Eurasian Wryneck, a vagrant from the Old World, hardly resembles a woodpecker. Patterned in browns and grays, it perches horizontally and has a long, squared tail without stiffened rectrices. Polytypic (4 ssp.; *chinensis* vagrant to N.A.). L 6" (17 cm)
IDENTIFICATION Cryptic patterning, broad gray mantle stripes, buff throat, dark line through eye, sharply pointed bill, and long tail distinctive; sexes similar.

SIMILAR SPECIES Unlike any other woodpecker; at a glance it can suggest a songbird such as a thrush or sparrow.
STATUS & DISTRIBUTION Breeds widely across temperate and boreal Eurasia from Great Britain to northern Japan; highly migratory, wintering in sub-Saharan Africa and from India through southeastern Asia. **VAGRANT:** Two records for western AK: Cape Prince of Wales (Sept. 8, 1945) and

Gambell, St. Lawrence I. (Sept. 2–5, 2003); one found dead in Feb. 2000 in southern IN was probably artificially transported.

dark mask and line on side of back

cryptic pattern in browns and grays

adult

The varied diets of these generalized New World woodpeckers include seeds and fruit; many take flying insects on the wing, and some store acorns and other nuts. Their tails are flat and only moderately stiffened, and the bills are medium to long and very slightly curved. Some species are highly social.

LEWIS'S WOODPECKER *Melanerpes lewis*   LEWO ■ 1

This distinctive large, glossy black woodpecker flies with slow, steady wing beats, recalling a crow in flight. It typically perches openly and makes long and often acrobatic aerial sallies for insects. Most populations are migratory, and large irruptions sometimes occur within the winter range. Monotypic. L 10.5" (27 cm)
IDENTIFICATION Sexes are similar. **ADULT:** Unmistakable glossy greenish black on upperparts and a broad pale gray breast-band that extends around hindneck; belly is pink and face is deep red. **JUVENILE:** Head, face, and foreparts dusky, with no gray collar. Adultlike plumage is attained during winter.
SIMILAR SPECIES Unmistakable. Acorn and Red-headed Woodpeckers show conspicuous white patches on wings and much white on underparts. Distant flying Lewis's strongly suggest small crows.
VOICE Generally silent for a woodpecker. **CALL:** Soft calls, including a series of short, harsh *churr* notes and clicking, squeaky *yick* notes. **DRUM:** Infrequent; a weak roll followed by a few individual taps.
STATUS & DISTRIBUTION Uncommon to fairly common; often gregarious. **BREEDING:** Open arid conifer, oak, and riparian woodlands in interior West; rare in coastal areas. **MIGRATION:** Large diurnal flights are sometimes noted. Arrives in northern interior breeding areas during first half of May; most have departed these areas by mid-Sept. **WINTER:** Oak savannas, orchards, shade trees in towns. Fall and winter movements are irregular, depending upon availability of acorns, conifer seeds; large flights sometimes occur into southern portions of CA, AZ, and NM. Winters irregularly north to WA, BC. **VAGRANT:** Recorded casually through the upper Midwest and Great Plains to southwestern TX; a few records east to NL, New England, mid-Atlantic states and south to the northern tier of Mexican states.
POPULATION Largely eliminated from coastal Northwest due to degradation of pine, oak, and riparian woodlands.

Illustration labels:
- browner head and no collar
- juvenile
- dark red face
- gray collar
- pink belly
- oily green upperparts, no red on face, and little pink on belly
- adults
- slow crowlike wingbeats
- dark wings

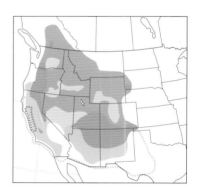

RED-HEADED WOODPECKER *Melanerpes erythrocephalus*   RHWO ■ 1

This flashy, distinctive woodpecker is a familiar sight over much of eastern and central North America but can be surprisingly inconspicuous at times. It occupies a variety of semi-open woodlands. Monotypic. L 9.3" (24 cm)
IDENTIFICATION All ages show white secondaries and a white rump, contrasting with dark remaining upperparts. **ADULT:** Bright red head, neck, and throat contrast with black back and pure white underparts; a narrow ring of black borders red throat. Sexes are similar. **JUVENILE:** Brownish on head and upperparts, with blackish bars through white secondary patch and some brown streaking and scaling below; adultlike plumage is attained gradually over first winter; most first-spring birds retain some black in secondaries.
GEOGRAPHIC VARIATION Although generally considered monotypic, birds west of the Mississippi River Valley *(caurinus)* average larger and sometimes show a tinge of red on the belly.
SIMILAR SPECIES See Red-breasted Sapsucker, which shares the all-red head, but with red extending through the breast, white on the wing coverts (not secondaries), and barred upperparts; note its typical retiring sapsucker habits. Red-bellied Woodpecker commonly co-occurs and has somewhat similar calls, but has a whitish face and throat, barred upperparts, and very different wing pattern.
VOICE CALL: A loud *queark* or *queeah*,

given in breeding season, is harsher and sharper than rolling *churr* of the Red-bellied Woodpecker; also a dry, guttural rattle and, in flight, a harsh *chug*. **DRUM:** A simple or two-part roll, lasting about a second and consisting of 20–25 beats. **STATUS & DISTRIBUTION** Uncommon to fairly common; sometimes perches openly and sallies for insects, but often surprisingly inconspicuous for such a flashy bird. **BREEDING:** Occupies a variety of open woodlands, orchards, and open country with scattered trees. Summers rarely in northeastern UT. **MIGRATION:** Small parties of migrants noted in early fall and late spring. **WINTER:** Withdraws southward from most of the breeding range in the Great Plains and Great Lakes regions. Unrecorded in Mexico. **VAGRANT:** Casual

west to BC, ID, CA, NV, AZ. Scarce vagrant to New England, mainly in fall. **POPULATION** New England breeding populations are nearly gone, and strong declines have been noted in the mid-Atlantic states, some Great Lakes states, FL, and elsewhere.

red head

brownish head

juvenile

barring on secondaries

pure white secondaries

adult

adult

ACORN WOODPECKER *Melanerpes formicivorus*   ACWO   1

white primary patches

female has black forecrown bar ♀

clown head pattern

♂

♂

This conspicuous clown-faced woodpecker of western oak woodlands is remarkable for its social habits, living over much of its range in communal groups of up to four or more breeding males and as many as three breeding females. These groups maintain and protect impressive granaries in which thousands of acorns are stored in holes drilled in tree trunks or utility poles for future consumption; in a study a single tree contained more than

50,000 acorn-storage holes. Acorn Woodpeckers also feed by sallying for flying insects and gleaning trunks, and they often eat ants (as reflected in the species' scientific name). Polytypic (7 ssp.; 2 in N.A.). L 9" (23 cm)

IDENTIFICATION A boldly patterned black-and-white woodpecker with a white patch at base of primaries, a white rump, black chest, streaked black lower breast, and white belly. Head pattern is striking, with a ring of black around base of bill, a red crown patch, a white forecrown narrowly connected to yellow-tinged white throat, and black sides of head setting off a staring white eye. **ADULT:** Iris white. Adult male has white forehead meeting red crown. Adult female is similar, but white forehead is separated from red crown by a black band. **JUVENILE:** Resembles adult but black areas are duller and iris is dark; juveniles of both sexes have a solid red crown like that of adult male.

GEOGRAPHIC VARIATION Pacific coast birds, *bairdi*, have slightly longer and stouter bills than nominate birds of the interior West. There is considerable additional variation in the remaining range south to Colombia, with five additional subspecies.

SIMILAR SPECIES Unmistakable given its group-living habits and loud calls. White-headed Woodpecker has similar white wing patch and black back, but lacks white rump and belly; Lewis's lacks white areas in plumage.

VOICE Acorn is noisy and conspicuous in communal groups, with raucous "Woody Woodpecker" calls. **CALL:**

Loud *wack-a, wack-a* or *ja-cob, ja-cob* series. Also, a scratchy, drawn-out *krrrit or krrrit-kut,* and a high, cawing *urrrk*. **DRUM:** A simple, slow roll of about 10–20 beats.

STATUS & DISTRIBUTION Common. **YEAR-ROUND:** Oak woodlands and mixed oak-conifer or oak-riparian woodlands. Most abundant where several species of oaks occur together. Isolated breeding populations are found on the east side of the Sierra Nevada, CA; on the central Edwards Plateau, TX (virtually extirpated); and possibly in far southwestern CO. **VAGRANT:** Found rarely or casually, primarily in fall and winter, away from woodland habitats along the immediate Pacific coast and in western deserts; accidental north to BC and east to the Great Plains states from ND south to coastal TX.

POPULATION Stable, apart from some local declines resulting from degradation of oak woodlands.

GILA WOODPECKER *Melanerpes uropygialis*   GIWO ■ 1

The Gila is a zebra-backed woodpecker of southwestern desert woodlands, where it can be noisy and conspicuous in tall cactuses such as saguaros. Its U.S. range does not overlap that of the similar Golden-fronted and Red-bellied Woodpeckers. Polytypic (4 ssp.; nominate in N.A.). L 9.3" (24 cm)

IDENTIFICATION Within its range, Gila Woodpecker's barred black-and-white upperparts, pale, grayish tan head and underparts (with a touch of pale yellow on belly), and broken white patch at base of primaries are diagnostic. Rump and uppertail

red crown on male

grayish brown forehead

grayish brown nape

uropygialis ♀

uropygialis ♂

barred central tail feathers

coverts are barred with black, and central tail feathers are white with black bars. **ADULT:** Male Gila Woodpecker has a round red cap on crown, absent in female. **JUVENILE:** Resembles adult but slightly paler and duller.

GEOGRAPHIC VARIATION There are two additional subspecies in Baja California and one other in mainland western Mexico (Sonora); birds from farther south tend to be more heavily barred and darker overall.

SIMILAR SPECIES Gila's range does not overlap with that of Golden-fronted (except at southern end of range in Jalisco, Mexico), but potential vagrants of either species (e.g., in southern NM) would need to be carefully documented. Golden-fronted has extensive solid white on the rump, solid black central rectrices, and yellow-orange nape patch and nasal tufts. The immature female Williamson's Sapsucker is superficially similar to Gila, but pale bars on the back are tan rather than white; chest and sides are barred with blackish; head is darker gray-brown; and bill is shorter. Note also the sapsucker's more retiring behavior. See also Northern and Gilded Flickers, which are larger, show solid white rumps, yellow or red color in wings and tail, and a black crescent on breast.

VOICE CALL: A loud, rolling *churrr* or *whirrrr*, often doubled. Also an insistent, laughing *yip, yip* series. **DRUM:** Infrequent drum is a steady, loud roll.

STATUS & DISTRIBUTION Fairly common. **YEAR-ROUND:** Desert woodlands, including cactus country, mesquite woods, riparian corridors, and lower canyon woodlands; often common in residential areas, date palm groves. Range extends west in CA to the Imperial Valley, east to southwestern NM, and at least formerly barely into southern NV. **DISPERSAL:** There is some movement into wooded foothills in southeastern AZ in fall and winter. **VAGRANT:** Accidental west to eastern San Diego County and on coastal slope in San Bernardino and Los Angeles Cos., and in the Bay Area, CA.

POPULATION Some declines have occurred with clearing of cottonwood-willow riparian associations, as along the lower Colorado River; colonization of the Imperial Valley occurred in the 1930s.

GOLDEN-FRONTED WOODPECKER *Melanerpes aurifrons*   GFWO ■ 1

red crown on male

Closely related to Red-bellied Woodpecker, it largely replaces that species from central and southern Texas south into C.A. Polytypic (12 ssp.; nominate in N.A.). L 9.8" (25 cm)

IDENTIFICATION All show a yellow area on nasal tufts just above bill, as well as a golden yellow to orange nape and hindneck. Rump is extensively pure white, and central rectrices are solidly black. Underparts are largely pale grayish white, with a touch of yellow on lower belly. **ADULT MALE:** Orange-yellow nasal tufts, red crown patch, and mixed red and orange-yellow nape patch. **ADULT FEMALE:** Similar to male, but red crown patch is absent and nasal tufts and nape are purer yellow, less orange. **JUVENILE:** Duller than adults; yellow or orange nasal tufts and nape lacking, but males (and some females) show a small red crown patch.

GEOGRAPHIC VARIATION Nominate *aurifrons* in U.S., but considerable geographic variation, with 11 subspecies in remainder of range. Appearance and vocalizations differ in southern Mexico; birds from southeastern Mexico to Honduras have fine white barring on back, continuous red from crown through nape, and some red on belly (possibly a distinct species).

SIMILAR SPECIES Red-bellied Woodpecker overlaps marginally with Golden-fronted from east-central TX to southwestern OK, with hybridization frequent in the latter area. Red-bellied Woodpecker differs in its extensive red rear crown and nape (extending to bill in males), white bars on central rectrices (solidly black in Golden-fronted), and pink or red tinge on belly (yellow in

gold feathering above bill

aurifrons ♂

aurifrons ♀

gold nape

pure white rump area

dark central tail feathers

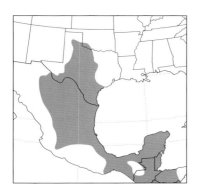

Golden-fronted). Rare individual Red-bellieds can show yellow or orange on crown and nape (see Gila Woodpecker account).

VOICE CALL: A rolling *churr* and cackling *kek-kek,* both slightly louder and raspier than calls of Red-bellied; also a scolding *chuh-chuh-chuh.* **DRUM:** A simple roll, often preceded or followed by single taps.

STATUS & DISTRIBUTION Fairly common. **YEAR-ROUND:** Dry woodlands and brushlands such as oak-juniper savannas and mesquite thickets; also riparian corridors, pecan groves, suburban areas. **VAGRANT:** Wanders casually to northeastern TX, eastern OK, and southeastern NM. Accidental in Cheboygan Co., MI (Nov. 20–Dec. 2, 1974) and FL; the latter record, at least, may pertain to an aberrant Red-bellied, and SC reports of Golden-fronted have conclusively been shown to be abnormal Red-bellieds.

POPULATION Apparently stable. Colonized southwestern OK in the 1950s and now hybridizes there with Red-bellied. Fairly common in the Big Bend region of TX only since the 1970s.

RED-BELLIED WOODPECKER *Melanerpes carolinus*   RBWO   ▄ 1

The Red-bellied Woodpecker is the familiar zebra-backed woodpecker of eastern woodlands and towns. Monotypic (or up to four weakly defined ssp. sometimes recognized). L 9.3" (24 cm)

IDENTIFICATION All Red-bellied Woodpeckers show a black-and-white barred back, white uppertail coverts, grayish white underparts, black chevrons on lower flanks and undertail coverts, and barred central tail feathers. In flight a small white patch shows at base of primaries. **ADULT MALE:** Entire crown, from bill to nape, is red; there is a suffusion of pink or red on center of belly. **ADULT FEMALE:** Red on head is limited to nasal tufts (just above bill) and nape; wash of color on belly is paler, less extensive. In rare individual females, nape and nasal tufts can be yellow-orange instead of red. **JUVENILE:** Resembles adults but duller, with red nasal tuft and nape patches lacking; bill is brownish (black in adults).

SIMILAR SPECIES Compare with Golden-fronted Woodpecker, which has solid black central rectrices, lacks pink or red on belly, and has a different pattern of color on the head.

VOICE In breeding season, Red-bellied gives a rolling *churr;* it also gives also a conversational *chiv chiv;* softer than calls of Golden-fronted Woodpecker. **DRUM:** A simple roll of up to a second, with about 19 beats per second.

STATUS & DISTRIBUTION Common in the Southeast, uncommon to fairly common in the Northeast, Midwest, and Great Plains. **YEAR-ROUND:** Pine and hardwood forests, open woodlands, suburbs and parks. Small populations exist west to southeastern ND, central SD, and northeastern CO. **DISPERSAL:** Not migratory, but at least some individuals in northern range withdraw southward in fall. **VAGRANT:** Wanders casually north to central ON, southern QC, ME, and the maritime provinces of Canada and west to eastern NM; accidental in southeastern WY, ID, NV, and SK.

POPULATION Generally stable. The Red-bellied has expanded its range northward in the Great Lakes region and New England over the last century and is also expanding northwestward in the Great Plains.

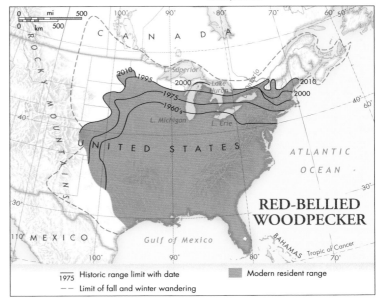

solid red crown and nape on male

red nape

♀

♂

♂

speckled white rump

pink on lower belly

barred central tail feathers

whitish primary patch in Red-bellied, Golden-fronted, and Gila

1975 Historic range limit with date

– – Limit of fall and winter wandering

Modern resident range

RED-BELLIED WOODPECKER

SAPSUCKERS Genus *Sphyrapicus*

The four North American sapsuckers are shy relatives of *Melanerpes* that feed on insects but most characteristically on living plant tissue and sap obtained from rows of drilled holes. All species give distinctive "Morse Code" display-taps; otherwise soft tapping. Sapsuckers are long winged, with strong undulating flight; all species are partially to highly migratory.

YELLOW-BELLIED SAPSUCKER *Sphyrapicus varius*   YBSA   ■ 1

The only sapsucker normally found in the boreal and eastern parts of the continent, this species is our most highly migratory woodpecker. Monotypic (smaller, darker resident birds in southern Appalachians sometimes separated as *appalachiensis*). L 8.5" (22 cm)
IDENTIFICATION Shows less red on head than related Red-naped and Red-breasted, and back is more extensively scalloped with yellow-buff. **ADULT MALE:** Forecrown, chin, and throat red, outlined completely in black; red normally lacking on nape. **ADULT FEMALE:** Similar to male, but chin and throat are entirely white. **JUVENILE:** Head and underparts pale brownish barred with dusky black; upperparts extensively pale buff with dusky barring, becoming white on rump. Unlike Red-breasted and

Red-naped, this juvenal plumage is retained well into winter, with red coloration of adult plumage gradually acquired through fall but black-and-white head and chest pattern not appearing until late winter.
SIMILAR SPECIES See the very similar Red-naped Sapsucker (formerly, along with Red-breasted, considered conspecific with Yellow-bellied).
VOICE This species and Red-breasted and Red-naped Sapsuckers are similar in calls and drums. **CALL:** A nasal *weeah* or *meeww;* on territory a more emphatic *quee-ark.* **DRUM:** A distinctive rhythm of a short roll of several beats, a pause, then two to several brief rolls of 2–3 beats each.
STATUS & DISTRIBUTION Common. **BREEDING:** Deciduous forests, mixed hardwoods and conifers of boreal regions and the Appalachians. **MIGRATION:** Main fall movement is Sept.–Oct.; spring migrants arrive in the Upper Midwest and Northeast during mid-Apr., and the northernmost breeding populations arrive late Apr., early May. **WINTER:** Widespread in the East south of New England and Great Lakes states, south to West Indies and Panama. **VAGRANT:** Rare but regular west to CA in fall and winter, with a

pure white throat

adult ♀

red throat bordered by solid black frame

adult ♂

juvenal plumage usually held well into winter

juvenile

golden buff spots scattered liberally above

few records north to WA. Accidental in Iceland, Britain, and Ireland.
POPULATION Generally stable.

RED-NAPED SAPSUCKER *Sphyrapicus nuchalis*   RNSA   ■ 1

The Rocky Mountain and Great Basin representative of the Yellow-bellied Sapsucker complex, the Red-naped Sapsucker closely resembles the Yellow-bellied Sapsucker, and the two hybridize in southwestern Alberta. Red-naped also hybridizes frequently with Red-breasted from British Columbia south to eastern California; in migration (especially) and winter, these hybrids are noted in small numbers in the Great Basin and southeastern California. Monotypic. L 8.5" (22 cm)
IDENTIFICATION Very similar in all plumages to Yellow-bellied, but with

slightly more red on head. **ADULT MALE:** Red crown is bordered by black, with a small red patch below black nape bar; chin and throat are red, with red color partially invading black malar stripe that outlines throat; breast is black. **ADULT FEMALE:** Similar to male, but chin is white (red on rare individual females), and red on crown and throat is slightly less extensive, more completely bordered by black; red on nape may be nearly or completely absent. **JUVENILE:** Closely resembles juvenile Yellow-bellied, but adultlike face pattern is attained by beginning of October (brown may

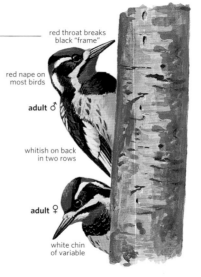

red throat breaks black "frame"

red nape on most birds

adult ♂

whitish on back in two rows

adult ♀

white chin of variable

be retained on breast into midwinter). **SIMILAR SPECIES** Male Yellow-bellied Sapsuckers can rarely show some red on nape. In Red-naped red throat invades or completely covers black border along malar, and pale markings on back are whiter and more restricted. Female Red-napeds with maximal red on chin and throat closely resemble male Yellow-bellied, but usually have white on uppermost chin and a hint of red on nape; note also back pattern differences. Male Red-naped Sapsucker, with maximal red invading the auricular and malar regions, may not be distinguishable from Red-

breasted x Red-naped hybrids; such hybrids usually have only limited black on the breast, auriculars, and sides of crown. **STATUS & DISTRIBUTION** Common. **BREEDING:** Aspen parklands and deciduous groves within open coniferous woodlands or adjacent to montane forest. Breeding range narrowly overlaps that of the Yellow-bellied Sapsucker in AB, with some hybridization. **WINTER:** Riparian and pine-oak woodlands, orchards, and shade trees south to northwestern and north-central Mexico. Winters rarely north on the Pacific coast to WA, and BC. **VAGRANT:** Casual east to KS, NE, OK,

southern TX, and southeastern LA. **POPULATION** Generally stable.

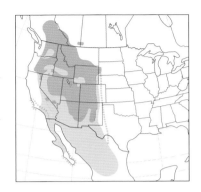

RED-BREASTED SAPSUCKER *Sphyrapicus ruber*   RBSA   1

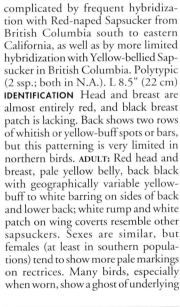

more extensive
and solid red head
than *daggetti*

ruber

daggetti

marks
on back
whitish
and more
extensive
than *ruber*

small
yellow
spots
on back

The Pacific coast representative of the Yellow-bellied Sapsucker complex, the Red-breasted Sapsucker differs from other sapsuckers in its almost entirely red head and breast. Identification is complicated by frequent hybridization with Red-naped Sapsucker from British Columbia south to eastern California, as well as by more limited hybridization with Yellow-bellied Sapsucker in British Columbia. Polytypic (2 ssp.; both in N.A.). L 8.5" (22 cm) **IDENTIFICATION** Head and breast are almost entirely red, and black breast patch is lacking. Back shows two rows of whitish or yellow-buff spots or bars, but this patterning is very limited in northern birds. **ADULT:** Red head and breast, pale yellow belly, back black with geographically variable yellow-buff to white barring on sides of back and lower back; white rump and white patch on wing coverts resemble other sapsuckers. Sexes are similar, but females (at least in southern populations) tend to show more pale markings on rectrices. Many birds, especially when worn, show a ghost of underlying

black in auriculars and chest and black-and-white patterning in malar region; red of worn birds may appear paler and more orange-red. **JUVENILE:** Brown head and extensive brown mottling below; darker than both juvenile Red-naped and Yellow-bellied, with less facial patterning. Adultlike plumage attained by September, though underparts are duller and mottled with brown.
GEOGRAPHIC VARIATION Nominate *ruber* breeds from southern OR northward; *daggetti* occupies the remainder of the range. Nominate birds show deeper and more extensive red on the breast, which is more sharply delineated from the pale belly; more limited white spotting on flight feathers; a deeper yellow wash on belly; and a more extensively black back with limited yellowish buff cross bars. **SIMILAR SPECIES** Compare with Red-naped Sapsucker; extensive hybridization renders many individuals with intermediate head patterns unidentifiable. **STATUS & DISTRIBUTION** Common. **BREEDING:** Moist coniferous forests, mixed oak-conifer riparian woodlands in coastal mountain ranges, usually in lower and wetter habitats than Williamson's. Subspecies *daggetti* is largely limited to montane habitats from about 4,000 to 8,000 feet. Red-breasted Sapsucker frequently hybridizes with Red-naped in the Cascades and eastern Sierra Nevada. **WINTER:** Northernmost populations (e.g., in southeastern AK, BC) withdraw southward (though resident on Queen Charlotte Is.), and higher-

elevation breeders generally withdraw to lower elevations. Winters widely around deciduous trees, orchards, and parks in lowlands adjacent to breeding range from southwestern BC to southern CA and south to northwestern Baja California. Northern nominate birds have been found as far south as San Diego, CA, and southern AZ. **VAGRANT:** Casual east to AZ, NM, central TX, Sonora, and northwest to Kodiak I., AK. **POPULATION** Generally stable.

ruber

daggetti

WILLIAMSON'S SAPSUCKER *Sphyrapicus thyroideus*   WISA   ▪ 1

This sapsucker of western montane conifer and aspen forests is notable for its extreme sexual dimorphism. Polytypic (2 ssp.; both in N.A.). L 9" (23 cm)
IDENTIFICATION Slightly larger than other sapsuckers. **ADULT MALE:** Largely black, with white rump, white postocular and moustachial stripes, large white wing patch, red chin and throat. Belly is bright yellow; flanks scalloped black and white. **ADULT FEMALE:** Head gray-brown; black back has fine pale grayish tan bars; rump white. No white wing patch. Chest black, belly yellow, sides and flanks tan, barred with black. **JUVENILE:** Like respective adults, but duller; male has white chin, female lacks black breast patch.
GEOGRAPHIC VARIATION Nominate *thyroideus* breeds from south-central BC to northern Baja California; more easterly *nataliae* from southeastern BC to AZ, NM, has smaller, narrower bill.
SIMILAR SPECIES Males unmistakable; females suggest flicker, but Williamson's is smaller, shorter billed, lacks red or yellow in flight feathers, has yellow belly. See Gila Woodpecker. Juvenile lacks large white wing covert patch of other juvenile sapsuckers.
VOICE CALL: A strong, slightly harsh *cheeur* or *queeah*. **DRUM:** Short roll followed by several shorter rolls; slower and more regular than other sapsuckers.
STATUS & DISTRIBUTION Fairly common. **BREEDING:** Dry pine, spruce, or fir forests, often where mixed with aspen groves. **MIGRATION:** Most move south in Sept.–Oct. and return Mar.–Apr.; in CA, AZ, NM sedentary or make short altitudinal movements. **WINTER:** Dry conifer or pine-oak woodlands from CA and southwestern U.S. to central Mexico. Casual north to OR, WA, CO; very rare in lowlands of CA, AZ in planted conifers. **VAGRANT:** Rare in fall, winter to western TX, casually to the Edwards Plateau region. Accidental LA, NY.
POPULATION Some declines, especially in Pacific Northwest.

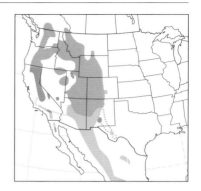

Genus *Picoides*

These woodpeckers are small to medium in size. Most are bark foragers, with chisel-shaped bills and stiffened rectrices. Distributed in the New World, except one "three-toed" species found across Eurasia. Males of most species have red on rear crown, lacking in females and juveniles. Males and juveniles of "three-toed" species have yellow crown patches.

NUTTALL'S WOODPECKER *Picoides nuttallii*   NUWO   ▪ 1

Endemic to oak and mixed woodlands in California and northwestern Baja California, Nuttall's Woodpecker is closely related to the Ladder-backed. Monotypic. L 7.5" (19 cm)
IDENTIFICATION "Ladder-back" pattern; spotted sides; barred flanks; auriculars almost wholly black; uppermost back black. Sexes similar but females lack red patch on nape and hindcrown.
SIMILAR SPECIES Compared with Ladder-backed, Nuttall's shows more black on face; white bars on back are narrower; more extensive black on upper back; white outer tail feathers sparsely spotted rather than barred, purer white below, more cleanly spotted and barred with black; Ladder-backed's underparts are washed with buffy, and markings are finer but often extend across breast as short streaks. Nuttall's nasal tufts are usually white (buffy to dusky in Ladder-backed). Red of crown male is restricted on Nuttall's, unlike Ladder-backed. Calls differ markedly.
VOICE CALL: A short, rolling *prrt* or *pitit* that may be followed by a longer trill, *prrt prrt prrrrrrrrrrrr;* also a loud *kweek kweek kweek* series. **DRUM:** A steady roll

of about 20 taps lasting about a second. **STATUS & DISTRIBUTION** Common. **YEAR-ROUND:** Oak woodlands, mixed oak-conifer and oak-riparian wood-lands, and tall dense chaparral; sea level to about 6,000 feet. Small populations extend onto deserts along riparian corridors. **VAGRANT:** A few wander out onto the western Mojave Desert, CA, casually to Imperial Valley; southwestern OR, western NV. **POPULATION** Generally stable.

LADDER-BACKED WOODPECKER *Picoides scalaris*   LBWO ▪ 1

This common desert woodpecker replaces the closely related Nuttall's in arid regions; the two species are known to hybridize at a few localities in southern California. Polytypic (9 ssp.; *cactophilus* north of Mexico). L 7" (18 cm) **IDENTIFICATION** Barred black-and-white back pattern of Ladder-backed

buffy white face with narrow black frame

buffy white nasal tufts

extensive reddish crown

white bars extend to nape

black crown

♂

♀

buffy cast to underparts with spots on sides and flanks

tail more barred than Nuttall's

Woodpecker extends up to hindneck, with very little solid black on upper back. Underparts are tinged creamy or buffy, with spots on sides, thin bars on flanks, and sparse, short streaks across breast. Outer tail feathers are barred with black. There is as much white as black in face pattern, with lower auriculars being white. Sexes are similar, but male Ladder-backed Woodpecker has extensive red on crown, which is lacking in female.
SIMILAR SPECIES See closely similar Nuttall's Woodpecker.
VOICE CALL: A fairly high, sharp *pik;* suggestive of Downy Woodpecker but louder, sharper, and slightly lower in pitch. Ladder-backed also gives a slightly descending *jee jee jee* series and a louder, slower *kweek kweek kweek.*
DRUM: A simple roll, like Nuttall's, but longer, averaging 1.5 seconds.
STATUS & DISTRIBUTION Common; extensive range south of U.S. to Nicaragua, El Salvador. **YEAR-ROUND:** Dry desert woodlands with yuccas, agaves, cactuses; pinyon-juniper foothills; mesquite woodlands; and riparian corridors. Often common in southwestern towns. Overlaps (and sometimes hybridizes) with Nuttall's very locally on the western edge of the California deserts from Inyo and Kern Counties south to northwestern Baja California. **VAGRANT:** Casual east to vicinity of Houston, TX, and on the Pacific coast near San Diego, CA. **POPULATION** Generally stable, though declines have been noted in TX.

ARIZONA WOODPECKER *Picoides arizonae*   ARWO ▪ 2

This brown-backed species of pine-oak woodlands in the Southwest borderlands was formerly considered conspecific with Strickland's Woodpecker *(P. stricklandi),* a localized species of the high mountains surrounding Mexico City, which is smaller and has white on its back. Polytypic (2 ssp.; nominate in U.S.). L 7.5" (19 cm) **IDENTIFICATION** Solidly brown back is unique among our woodpeckers.

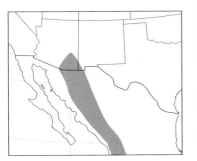

ADULT: Upperparts dark brown, underparts white with heavy brown spotting on breast and barring on flanks, belly. Brown crown, auricular, and malar contrast with large white patch on sides of neck. Male has red nuchal patch, absent in female. **JUVENILE:** Has limited red on crown.
SIMILAR SPECIES Unmistakable.
VOICE CALL: Sharp, high *peeek* call is higher and hoarser than similar call of Hairy Woodpecker. **DRUM:** Rapid roll is like that of Hairy Woodpecker, but longer.
STATUS & DISTRIBUTION Fairly common, but usually wary, inconspicuous. **YEAR-ROUND:** Dry pine-oak woodlands and oak-riparian canyon woodlands from 4,000–7,000 feet. Found east to Peloncillo and Animas Mountains, NM; and southwest to Santa Catalina and Pinaleño Mountains, AZ; a few birds may move into lower foothill oak woodlands in winter.

arizonae ♂

brown upperparts

large spots below

arizonae ♀

DOWNY WOODPECKER   *Picoides pubescens*   DOWO ■ 1

Our smallest woodpecker, the Downy is also among our most widespread and familiar species; it is a confiding bird that often visits feeders. In all respects it suggests a small version of the Hairy Woodpecker, both differing from our other species by the broad white stripe down the back. Polytypic (7 ssp.; all in N.A.). L 6.7" (17 cm)
IDENTIFICATION The small size and often acrobatic foraging on small branches and twigs are distinctive, and

male has red
hindcrown spot

♂

short,
stubby bill

♂

barred outer
tail feathers

♀

Rockies
leucurus ♂

the plumage pattern can be confused only with Hairy. Has hybridized with Nuttall's. **ADULT:** Black crown, auricular and malar; upper back, scapulars and rump black, but a broad white stripe extends down center of back. Underparts unmarked white (to grayish buff in some populations). Outer tail feathers white with limited black spotting; variable white spotting on upperwing coverts and barring on remiges. Male has a small red nuchal patch, lacking in female. **JUVENILE:** As in other pied woodpeckers, both sexes have a pale red patch in center of crown, more extensive in male.
GEOGRAPHIC VARIATION The seven subspecies differ mainly in size (northern birds generally larger), underparts color (white to gray tinged), amount of black in rectrices, and amount of white spotting in wings. Southeastern birds are smaller and slightly grayer below than boreal and northeastern birds. Pacific coast birds have reduced white spotting on the wing coverts and secondaries; such white spotting is most highly developed in birds east of the Rockies. Birds of the Pacific Northwest are tinged gray on the back and gray-buff below.
SIMILAR SPECIES Nearly identical in patterning to Hairy Woodpecker. Downy is much smaller, with a short bill (much shorter than head); outer tail feathers usually show black spots (but these can be lacking, and darkest Hairy subspecies may show a few spots). Pale nasal tufts of Downy are

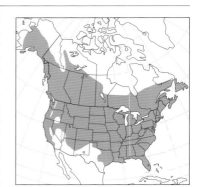

relatively larger than in Hairy. Hairy shows a larger wedge of black from the rear of the malar stripe onto the breast. Note differences in calls.
VOICE CALL: *Pik* call is higher and much softer than Hairy's sharp, ringing *peek*. Commonly gives a distinctive high, slightly descending and accelerating whinny, *kee-kee-kee-kee*.
DRUM: A soft roll, slightly slower than that of Hairy; about 17 beats a second, with drum lasting 0.8–1.5 seconds.
STATUS & DISTRIBUTION Common; uncommon in northern boreal regions.
YEAR-ROUND: Resident in a variety of deciduous woodlands and, more sparsely, in coniferous forests; also found in parks, gardens, and orchards, even in urban regions. Absent from desert Southwest. **DISPERSAL:** Not migratory, but some individuals can disperse long distances. Casual in southern AZ and Queen Charlotte Is., BC.
POPULATION Stable.

HAIRY WOODPECKER   *Picoides villosus*   HAWO ■ 1

Like a large, long-billed version of the Downy Woodpecker, the Hairy is a widespread generalist of a variety of forests and woodlands over most of the continent. Polytypic (about 17 ssp.; 11 north of Mexico). L 9.5" (24 cm)
IDENTIFICATION Plumage pattern is nearly identical to Downy Woodpecker, with long white patch down back, variable white spotting on wing coverts and flight feathers, and mostly unmarked underparts. Outer tail feathers are usually unmarked white. **ADULT:** Male shows red nuchal bar, often divided vertically by black (especially in some eastern populations); red is lacking in female.
GEOGRAPHIC VARIATION Variation is extensive but generally clinal.

Nominate *villosus* is widespread in the East; southeastern birds *(audubonii)* are smaller, buffier (less pure white) below, and with less white on the back. Boreal *septentrionalis*, from interior AK east to QC, is the largest, whitest subspecies. Newfoundland *terranovae* is distinctive, with white back reduced and barred (especially in immatures), some black spotting on outer rectrices, and often with fine black streaking on sides and flanks. In the West, *picoideus* of the Queen Charlotte Is., BC, is most distinctive, with gray-brown underparts; black markings in the white back stripe, sides, and flanks; and strong black bars on outer rectrices. Northwestern *harrisi* and *sitkensis* have gray-brown underparts and face and reduced white on

Rockies
orius ♂

**Maritimes
juvenile** ♂
terranovae ♂

wings. Subspecies *hyloscopus* of CA and northern Baja California is smaller, paler (light gray-buff) below, and whiter on head. Three additional subspecies of the Great Basin and Rocky Mountain regions (*orius, monitcola, leucothorectis*) are moderate to large in size and white to very pale buff below; compared with eastern and boreal birds, they have reduced white on the back and greatly reduced white spotting on the wing coverts. The subspecies *icastus,* ranging from southeastern AZ and southwestern NM south to central Mexico, is similar but smaller. Four Middle American subspecies from eastern and southern Mexico south to western Panama are smaller still (some nearly as small as Downy Woodpecker) and are variably buffy to deep buff-brown on the underparts. There are two additional subspecies on the Bahamas.

SIMILAR SPECIES Downy Woodpecker is similar in pattern but much smaller, with a small, short bill (much smaller than half the length of the head), and black bars on outer tail feathers. See American Three-toed Woodpecker. Note that some populations of Hairy Woodpecker (especially in NF) can show barred backs (especially as juveniles), and some Three-toed

populations have nearly pure white backs.
VOICE CALL: A piercing, sharp *peek* or *pee-ik.* The rattle call ("whinny") is a fast, slightly descending series of these *peek* calls. **DRUM:** Rapid roll of about 25 beats in one second.
STATUS & DISTRIBUTION Fairly common; uncommon to rare in the South and FL. **YEAR-ROUND:** Hairy Woodpecker occupies a wide range of coniferous and deciduous forests from sea level to tree line; such habitats are usually densely wooded, but in some areas are more open and parklike. Moves into burned areas, where most easily located in the Deep South and FL. **DISPERSAL:** Although generally nonmigratory, individuals can disperse long distances. Larger, more northerly birds from Canada regularly occur in the Northeast in fall and winter. Recorded in fall and winter on the southern plains and Pacific coast lowlands well away from breeding habitats.

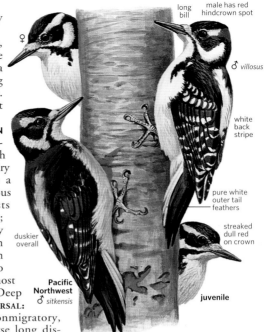

POPULATION Declines that have been noted in many areas are thought to be due to fragmentation of forests, loss of old-growth trees, and nest site competition with European Starlings.

RED-COCKADED WOODPECKER *Picoides borealis*   RCWO ▪ 2

This species has a highly fragmented distribution in southeastern pinewoods, especially old-growth long-leaf pine. Living in family groups, or "clans," its territories can be identified by distinctive nest and roost cavity trees: large living pines with heartwood disease and with abundant resin flowing from holes drilled around the cavity entrance. This oozing pitch protects nests from snakes and other predators. Monotypic (slightly smaller Florida peninsula birds have been separated as *hylonomus*). L 8.5" (22 cm)

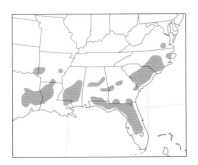

IDENTIFICATION Unique face pattern, with black crown and nape, long black malar, and extensive white auriculars and sides of neck; sides and flanks with short streaks. Adult male's red "cockade" is essentially invisible in the field.
SIMILAR SPECIES Unmarked white face is diagnostic from Downy, Hairy, and Ladder-backed Woodpeckers. Larger Red-bellied Woodpecker shares barred back pattern but has an extensive red nape and lacks black on the head.
VOICE CALL: A raspy *sripp* or *churt* and high-pitched *tsick* or *sklit* are the most frequent and distinctive calls. **DRUM:** Quiet and infrequently heard.
STATUS & DISTRIBUTION Rare and local. **YEAR-ROUND:** Mature lowland woods of long-leaf, loblolly, or other pines with open understory and suitable cavity trees afflicted with heartwood disease. Core remaining populations occur from eastern TX and adjacent OK, AR, LA, and from MS east to FL, the Carolinas and southern VA (still a few). Populations in TN, KY, MD, and MO have disappeared in recent decades. **VAGRANT:** Accidental in northeastern

IL, and twice in south-central OH.
POPULATION Endangered. About 6,000 colonies estimated to exist in 2006, with 2–5 adults per colony. Most colonies are on federal lands, where management strategies have had mixed results, with declines continuing in many areas.

WHITE-HEADED WOODPECKER   *Picoides albolarvatus*   WHWO ▪ 1

white head

♂

♂

white wing patch

♀

long narrow white patch on folded wing

A striking bird of far western pine forests, the White-headed Woodpecker is unique among our species in its white head and solid black body. It often forages on pine cones, extracting seeds; otherwise it mainly flakes away bark on trunks or branches of conifers. Polytypic (2 ssp.; both in N.A.). L 9.3" (24 cm)

IDENTIFICATION A solid black body,

mostly white head, and large white wing patches make this bird unmistakable. **ADULT:** Red nuchal patch, lacking in females. **JUVENILE:** Resembles adult, but both sexes have a pale red wash on crown, and white wing patch is often more interrupted with black spots.

GEOGRAPHIC VARIATION Birds of the southern California mountains, *gravirostris,* are slightly larger billed than the northern nominate subspecies.

SIMILAR SPECIES Acorn Woodpecker, largely black above with white wing patches, may suggest White-headed dorsally, but Acorn's white rump and belly and black in the face simplify identification.

VOICE CALL: Distinct call is a sharp, two or three syllable *pitik, pee-dink,* or *pee-de-dink,* dropping slightly in pitch; 1- and 3-syllable notes are also given. Also a longer series of these notes, a slower *kweek kweek kweek* series, and various softer calls. **DRUM:** A 1–1.5 second roll with about 20 beats a second; probably not distinguishable from drum of Hairy.

STATUS & DISTRIBUTION Fairly common in CA; generally uncommon

from OR and western NV north. **YEAR-ROUND:** Montane coniferous and mixed forests usually dominated by ponderosa or Jeffrey pines, but also sugar pine, white fir, Douglas fir, and incense cedar. **VAGRANT:** Rare in fall and winter at lower elevations adjacent to breeding mountains, and casual in lowlands along CA coast and deserts. Accidental western MT.

POPULATION Generally stable in CA, but many populations, especially from OR north, are decreasing because of logging, even-age stand management, and long-term forest changes.

AMERICAN THREE-TOED WOODPECKER   *Picoides dorsalis*   ATTW ▪ 2

This stocky, large-headed, pied woodpecker has two forward-facing toes and just one rear-pointing toe. Males have a yellow patch on crown. Forages on dead or dying trees, often in burned-over areas. Back pattern varies geographically, from solid white to heavily barred with black. Formerly known simply as "Three-toed Woodpecker," it has been split into our North American species and an Old World species now called the Eurasian Three-toed Woodpecker *(P. tridactylus).* Polytypic (3 ssp.; all in N.A.). L 8.7" (22 cm)

IDENTIFICATION A "three-toed"

fasciatus

bacatus

dorsalis

woodpecker with white patch or barring on otherwise black upperparts; underparts white with black barring on sides and flanks. White spots on primaries, outer secondaries, and tertials. Tail black with white outer rectrices. **ADULT:** Male has yellow crown, becoming spotted with black and white on forecrown; female lacks yellow. **JUVENILE:** Duller than adult; both sexes show some yellow on crown, reduced in female.

GEOGRAPHIC VARIATION Three subspecies differ in back pattern and size. Density of black barring on the back is greatest in eastern *bacatus* of eastern Canada south to New England and the Adirondacks; this is the smallest subspecies. Dorsal barring and size are intermediate in *fasciatus* of boreal regions from AK to SK, and in the Cascades south to southern OR. Rocky Mountain sub-

yellow crown, blended at edges

smaller bill than Black-backed

thin white lines on face

darkest backed subspecies

eastern ♂ *bacatus*

thin white "ladders" on back

eastern ♀ *bacatus*

barred sides and flanks

fairly solid white back, suggestive of Hairy, but note barred sides and flanks

Rocky Mountain ♂ *dorsalis*

fasciatus ♂

species *dorsalis* has a pure white center of back with irregular barring on sides of back; it is also slightly larger than other subspecies.

SIMILAR SPECIES See Black-backed. Can be mistaken for Hairy, especially white-backed Rockies *dorsalis.* Three-toed has heavily barred sides and flanks.

Face pattern of Hairy shows more white on sides of neck and supercilium. Male Three-toed has a yellow crown patch. Beware some juvenile Hairies (e.g., in NF), which show barring on back. **VOICE CALL:** *Pik or kik,* higher than call of Black-backed, less sharp than Hairy's; also a longer rattle. **DRUM:** Deep, resonant drums frequent and variable, range from 11–16 beats a second. **STATUS & DISTRIBUTION** Uncommon. **YEAR-ROUND:** Coniferous forests of spruce and fir, especially where there are large stands of dead trees in burned areas. Has bred south to western SD, MN, Upper Peninsula of MI, and MA. **DISPERSAL:** Small numbers move irruptively to areas adjacent to breeding range in winter, usually after early Oct.; irruptions generally more minor than in Black-backed. **VAGRANT:** Casual south to KS (summer), NE, IA, RI, NJ, DE. **POPULATION** Generally stable, northerly range makes monitoring difficult.

BLACK-BACKED WOODPECKER *Picoides arcticus*   BBWO ■ 2

This is one of two "three-toed" woodpeckers in North America. These are stocky, large-headed pied woodpeckers that have two forward-facing toes and but just one rear-pointing toe (usually held out to the side). Its range overlaps extensively with that of the similar American Three-toed Woodpecker. Both species feed in dead or dying conifers and can be especially prevalent in burned-over forests; they flake away sections of loose bark to obtain insects and their larvae. Worked-over trees can be quite evident. Monotypic. L 9.5" (24 cm) **IDENTIFICATION** A very dark woodpecker with solid black upperparts and heavy black barring below. Outer tail feathers, throat, and central underparts are white; head pattern shows long white submoustachial stripe, broadening at rear; there is at best only a hint of a short white postocular mark. Primaries are spotted with white. **ADULT MALE:** Roundish yellow patch on crown. **ADULT FEMALE:** Similar but crown all black. **JUVENILE:** Black areas duller than in adults; females have a few yellow feathers on crown; males have extensive yellow on crown. **SIMILAR SPECIES** Most similar to darkest eastern subspecies *bacatus* of American Three-toed Woodpecker, which has extensive black barring on back. Black-backed is told by its solid black upperparts, lack of white postocular streak, and bolder white submoustachial stripe; it is also larger overall, with a longer, stouter bill. **VOICE CALL:** A single sharp *pik* or *chik,* lower and sharper than call of American Three-toed. **DRUM:** As in American Three-toed, the deep and resonant drum roll speeds up and trails off slightly toward the end. Black-backed's drum is slightly deeper, faster, and longer than that of American Three-toed. **STATUS & DISTRIBUTION** Uncommon, but can be locally fairly common when responding to insect outbreaks in burnt or otherwise stressed forests. **YEAR-**

sharply delineated yellow crown

black face

solid black upperparts

long, stout bill

barred sides and flanks

solid black crown

♀

♂

ROUND: Conifer forests of spruce, fir, or pine. South of boreal regions and the northern Rockies, the Black-backed is found locally in the Adirondacks of NY, Black Hills of SD and adjacent WY, and the Cascades and Sierra Nevadas from WA to central CA. **DISPERSAL:** Occasional irruptive movements south of regular range into New England, Great Lakes, and Canadian Maritimes, taking advantage of outbreaks of woodboring beetles. Exceptional irruptions have brought birds as far south as IL, OH, PA, NJ; records at the southern periphery of range have decreased in recent decades. **VAGRANT:** Casual in winter south of resident and irruptive range.

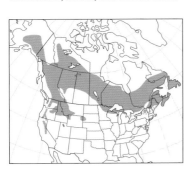

Genus *Dendrocopos*

GREAT SPOTTED WOODPECKER *Dendrocopos major*   GSWO ■ 4

This Eurasian woodpecker is a casual vagrant in western Alaska. Polytypic (14–24 ssp. in Old World; AK specimen *kamtschaticus*). L 9" (23 cm) **IDENTIFICATION** Large size, white auriculars, black back bordered by large white scapular and covert patch, and red undertail coverts. Females lack red nape patch; juveniles have red in crown. **SIMILAR SPECIES** Hairy and Downy Woodpeckers have white backs, black auriculars, and lack red undertail coverts and solid white wing patch. Yellowbellied Sapsucker is patterned below and on back, lacks red undertail coverts, differs in head pattern. See American Three-toed and Black-backed Woodpeckers. **VOICE CALL:** Sharp *kix, kick,* or *chik.* **STATUS & DISTRIBUTION** Resident nearly throughout Eurasia from Great Britain and northwestern Africa east to Japan, Kamchatka; northern populations may migrate irruptively. **VAGRANT:** Casual in AK; at least six recs. from Aleutian Is. (mainly May, Sept.–Oct.) and one rec. in May on Pribilof Is. A male was observed wintering north of Anchorage, AK (fall 2001–Apr. 2002).

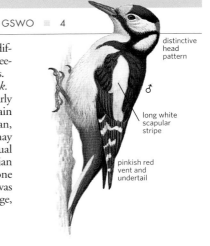

distinctive head pattern

♂

long white scapular stripe

pinkish red vent and undertail

PILEATED WOODPECKER *Dryocopus pileatus* PIWO ■ 1

Our largest woodpecker (other than the likely extinct Ivory-billed), the Pileated is a crow-size, crested woodpecker of forested areas that feeds largely on carpenter ants and beetles extracted from fallen logs, stumps, and living trees. Foraging birds excavate large, rectangular holes in trunks and logs. Polytypic (2–4 ssp.; all in N.A.). L 16" (42 cm)

IDENTIFICATION Mostly black, with a slaty black bill, white chin, white stripe from bill down neck to sides of breast, a white patch at base of primaries, and extensive white on the underwing linings. Flight consists of deep, irregular crowlike wing beats with little or no undulation. **ADULT MALE:** Red crown, crest and malar mark. **ADULT FEMALE:** Forecrown mottled black, malar black.

GEOGRAPHIC VARIATION Differences between subspecies are minor; northern birds average larger.

SIMILAR SPECIES See Ivory-billed Woodpecker; all recent Ivory-billed reports have turned out to be Pileated Woodpeckers or misidentified vocalizations of other species. Crows with aberrant white wing patches can momentarily suggest a Pileated.

VOICE CALL: A long, flickerlike series *kee kee kee kee,* often slightly irregular in cadence. Also, single *wuk* or *cuk* notes. **DRUM:** Loud and resonant, lasting 1–3 seconds with about 15 beats a second; beat rate accelerates slightly, often trails off at very end.

STATUS & DISTRIBUTION Common and widespread in Southeast; uncommon and more localized in the Great Lakes region, boreal areas, and the Pacific coast. **YEAR-ROUND:** Dense coniferous and deciduous forests and woodlots, with suitable presence of large older trees, snags, and downed wood. **VAGRANT:** Wanders casually slightly away from resident range, with documented records in CA from coastal Los Angeles County and the San Joaquin Valley; there are also unsubstantiated reports for east-central AK, CO, UT, northwestern AZ, and southern NM.

POPULATION Generally stable.

shorter, darker bill

♀

black above

♂

white wing patch

♀

extensive white underwing and black secondaries

♂

now likely North America's largest woodpecker

IVORY-BILLED WOODPECKER *Campephilus principalis* IBWO ■ 6

This spectacular bird of old southern bottomland forests is likely extinct. The much publicized announcement of its rediscovery in eastern Arkansas (Apr. 2005) was followed by intensive searches that yielded no certain evidence. None of the purported sightings have been accepted, including those from northern FL. Polytypic (2 ssp.; nominate in U.S., extinct or nearly extinct *bairdii* in Cuba). L 19" (50 cm)

IDENTIFICATION The largest woodpecker in the U.S. and the third largest in the world. Mostly black with a massive creamy white bill, a large crest (male's is red and curves back, female's is black and curves forward), a striking flight pattern with completely white secondaries and white tips to inner primaries, and long white stripes on sides of back. Legs and feet light gray, eyes yellow. Flight is direct with relatively rapid and shallow, continuous wing beats. Juvenile has shorter crest, browner plumage, and brown eyes.

SIMILAR SPECIES Pileated Woodpecker is common in any potential area for Ivory-billed and can easily be mistaken for it, but Pileated is 10 to 20 percent smaller, lacks white secondaries, has more extensive white on underwing linings, lacks white back stripes, and has a white (not black) chin. Bill of

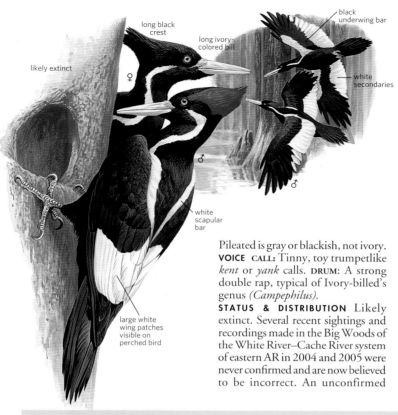

long black crest

long ivory-colored bill

likely extinct

♀

black underwing bar

white secondaries

♂

white scapular bar

large white wing patches visible on perched bird

Pileated is gray or blackish, not ivory. **VOICE CALL:** Tinny, toy trumpetlike *kent* or *yank* calls. **DRUM:** A strong double rap, typical of Ivory-billed's genus *(Campephilus)*. **STATUS & DISTRIBUTION** Likely extinct. Several recent sightings and recordings made in the Big Woods of the White River–Cache River system of eastern AR in 2004 and 2005 were never confirmed and are now believed to be incorrect. An unconfirmed

sighting in 1999 in the Pearl River area of southeastern LA sparked an intensive search, with no subsequent sightings. A recording of calls from eastern TX in 1968 was likely not an Ivory-billed. Ivory-billed Woodpecker has not been reliably documented in the U.S. since Apr. 1944, when a small population in the Singer Tract, near Tallulah, Madison Parish, in northeastern LA lost its habitat to logging and subsequently disappeared. Formerly resident in lowland pine, hardwood, and cypress forests in the southern Mississippi River Valley, Gulf Coast, and southern Atlantic states from NC to FL. Forages by stripping bark from recently dead trees to reach beetle larvae, its primary source of food. **POPULATION** Logging of old-growth bottomland forests throughout the southern U.S. and the control of river flooding led to the decline of Ivory-billed by the mid-1800s and its likely extinction by the 1940s; a tiny Cuban population was well-documented as late as 1948 but not reported since the 1980s.

FLICKERS Genus *Colaptes*

Largely terrestrial, ant-eating woodpeckers of the New World, flickers have long, thin, slightly curved bills; large rounded wings; and flat tails that are only minimally stiffened. Plumage is barred olive to (in our area) brown on the back, spotted below, with colorful shafts on the wings and tail. Our species show a bold white rump and black chest crescent.

GILDED FLICKER *Colaptes chrysoides*   GIFL   2

This "Yellow-shafted" flicker of southwestern desert woodlands was re-split in 1995 from the Northern Flicker complex because of the very limited extent of interbreeding. Polytypic (3 ssp.; *mearnsi* in U.S. and 2 more in Baja California). L 11.5" (29 cm) **IDENTIFICATION** A small flicker with yellow wing flash and yellow tail base; distal half of tail black from below; chest patch deeper and more rectangular than in other flickers, and black spots on lower underparts expand to

bars or crescents. Crown and nape are rich brown, contrasting with gray face. Adult male has a red malar. **SIMILAR SPECIES** Some hybrid "Yellow-shafted" x "Red-shafted" Northern Flickers approach Gilded in looks, having yellow in wings and tail but lacking a red nuchal crescent. Note Gilded Flicker's smaller size, more extensive black on undertail, paler more finely barred back, more extensive chest patch, dark bars or crescents on lower underparts, and rich brown crown. **VOICE CALL:** High descending *klee-yer* and territorial *wick wick wick;* calls higher pitched than Northern Flicker. **DRUM:** Like other flickers. **STATUS & DISTRIBUTION** Fairly common. **YEAR-ROUND:** Desert woodlands, esp. where dominated by saguaro. Rare in limited range in CA and NV. Limited hybridization with Northern Flicker in AZ. **VAGRANT:** Casual to southwestern UT.

POPULATION Presumably stable, though westernmost populations in California have declined.

more cinnamon crown

larger black chest patch

mearnsi ♂

paler back with fainter bars

crescent-shaped black markings below

yellow underwing

♀

NORTHERN FLICKER *Colaptes auratus*   NOFL ▪ 1

This familiar large woodpecker and the closely related Gilded Flicker show flashy color in the wings and a bold white rump in flight. Often feeds while on the ground, where it hops. Polytypic (10 ssp.; 5 in N.A.). L 12.5" (32 cm)

IDENTIFICATION All Northern Flickers show a bold black chest crescent, a white rump, and bright color (salmon-red or yellow) in the shafts and much of the vanes of the flight feathers and on underwing coverts. All Northern Flickers are pale buffy white to rich buff below with black spotting and have brown to gray-brown backs with black barring. **ADULT:** Sexes are similar, but males have a malar mark (red in "Red-shafted," black in "Yellow-shafted") that is lacking in females. "Yellow-shafted" has a gray crown with a red crescent on nape, a tan face and throat, and rich buff underparts with a relatively narrow chest crescent. Flight feathers and underwing linings are golden yellow. "Red-shafted" has a grayish head and throat with pale

brown on forecrown and loral region, lacks red on nape, and has paler buff to creamy underparts with broader chest crescent; flight feathers and underwing linings are salmon pink. Introgressant individuals that combine characters of both groups are widely seen.

GEOGRAPHIC VARIATION The "Red-shafted" and "Yellow-shafted" groups, formerly considered separate species, are easily separated. In an extensive hybrid zone on Great Plains and northern Rocky Mountains from northeastern NM and the Texas Panhandle north to AB, BC, and southeastern AK, a large percentage of individuals encountered show traits of both groups. Intergrades are frequently seen in the area of overlap and throughout the West, but rarely noted in the East. Within the "Red-shafted" group, northwestern *cafer* (southern AK to northwestern CA) is darker than remaining *collaris* group, which includes smaller *nanus* from southwestern TX. There is slight size variation within the "Yellow-shafted" group. Birds of the Central American highlands (*mexicanoides* group) and the Cuba–Grand Cayman region (*chrysocaulosus* group) are distinctive representatives of the "Red-shafted" and "Yellow-shafted" groups, respectively.

SIMILAR SPECIES Gilded Flicker closely resembles Northern Flicker and combines some features of "Yellow-shafted" (yellow wings and tail base) and the "Red-shafted" (head pattern). Gilded Flicker is smaller and shows black on the distal half of the undertail; Northern Flicker undertails are black on about the distal third (note that all flicker tails look mostly black from above). The crown of Gilded is more extensively brown than in "Red-shafted," and the back is paler, more gray-brown, with narrower and more widely spaced black bars (but note that interior western *collaris*

"Red-shafted" are paler backed than northwestern birds). Black crescent on chest of Gilded is thicker and more truncated on sides. Spotting on underparts is broadened into short bars or crescents on the flanks of Gilded; Northern Flickers have round spots throughout underparts.

VOICE **CALL:** A piercing, descending *klee-yer* or *keeew* is given year-round; also a soft, rolling *wirrr* or *whurdle* in flight and a soft, slow *wick-a wick-a wicka* given by interacting birds. On breeding grounds a long, long *wick, wick, wick, wick* series. **DRUM:** A long, simple roll of about 25 beats.

STATUS & DISTRIBUTION Common. **BREEDING:** Widespread in open woodlands, parklands, suburban areas, riparian and montane forests. **MIGRATION:** Northern populations of "Yellow-shafted Flickers" and northern interior "Red-shafted Flickers" are highly migratory. Small flocks of migrants and even large flights are evident from late Sept. through Oct. and in spring in late March and April. **WINTER:** Uncommon to rare north to southern Canada (but common in southwestern BC). **VAGRANT:** "Yellow-shafted" is accidental in western Europe, where the two or three records represent known or possible ship-assisted birds. "Yellow-shafted Flickers" winter regularly to the Pacific coast, though outnumbered there by intergrades. "Red-shafted" is casual east to MB, western MO, eastern TX, AR, LA; possibly also farther east, but most or all may not be pure "Red-shafted."

POPULATION Significant declines have occurred over much of the continent.

all North American flickers have white rumps

"Yellow-shafted" ♀

yellow underwing

red nape and pale brown face

"Yellow-shafted" ♂

male has black "whisker"

brown nape and gray face

male has red "whisker"

rounder spots below than Gilded

"Red-shafted" ♂

pinkish red underwing

"Red-shafted" ♀

CARACARAS AND FALCONS Family Falconidae

American Kestrel, immature male (NJ, Oct.)

The jet fighters of the bird world, falcons are flashy, flamboyant fliers. For centuries captive birds have been trained by humans to hunt, leading to the term *falconry*. Historically, larger falcon species were reserved for royalty, and even today they command extremely high prices. In recent times, falconers have crossbred similar species to achieve optimum size and behavioral characteristics; so escaped falconry birds may challenge birders in field identification. The name "falcon" comes from the term "falcate," meaning sickle-shaped, which describes the bird's wing silhouette.

Structure The typical falcon is a large-headed, dark-eyed raptor with pointed wings and a square-tipped tail. Most are strong fliers and take their prey in the air. Powerful flight muscles give them a thick-chested, stout body. The beak has a notch on the cutting edge of the culmen, which severs the spinal cord of prey. The feet and toes are not designed for killing prey, but for grabbing and holding it immobile. Females are larger than males, dramatically so in larger species. Differing external features are found in the forest-falcon and caracara, such as rounded or square-tipped wings.

Behavior These pointed-winged birds commonly take their prey in flight, either by diving powerfully from above, or by pursuing it from behind and below. Most falcons also hunt from an exposed perch, dropping down to pursue prey from behind. Small birds may be forced to high altitudes; then the falcon dives upon the exhausted prey. Many falcons hunt at dawn and dusk, even by city lights. Caracaras often feed on carrion and will rob food from other raptors. The *Micrastur* forest-falcons are ambush hunters in heavy foliage, much like North American accipiters. Caracaras build stick nests in trees, whereas falcons make scrape nests on a ledge or in a shallow cave. Kestrels favor a cavity and use manmade nest boxes. The nest of another hawk may be used. Northern-nesting species are usually migratory, although the Gyrfalcon does not follow any discernable pattern of passage. Eastern migrants prefer following coastlines, preying on shorebirds and migrant passerines. They often fly in wind conditions unfavorable to other migrating raptors and are not as reluctant to cross open water. Inland migrants soar on fixed wings along mountains' thermal updrafts, like other hawks, but take a direct flight line instead of the swirling "kettles" favored by the buteos. An owl decoy placed atop a long pole at a ridgeline hawk-watch site may attract a migrating falcon to swoop down and harass its historic enemy.

Plumage Except for the kestrels and the Merlin, adult falcons and caracaras of both sexes share similar plumage, while juvenile birds have different plumages than their adults, in some cases only by degree. All are darker above and generally light below. Most falcons have moustache marks on the face.

Distribution Falcons are found worldwide, caracaras and forest-falcons only in the New World. Falcons fly at all altitudes, in open areas throughout North America; caracaras are found in open areas with warm climates.

Taxonomy Recent genetic studies have shown that the Falconidae share a close affinity to the Psittacidae (parrots); both families are now placed just before the order Passeriformes (passerine birds). Within the family Falconidae, there are 37 species of *Falco* worldwide (9 in N.A.), seven forest-falcons (Micrasturinae) and 11 living caracaras (Caracarinae; other sources list them differently).

Conservation Worldwide, four species are endangered, six are listed as near threatened. Pesticides and other toxins in the environment are detrimental to nesting success. The prohibition of DDT and similar chemicals in North America has allowed populations to rebuild. Loss of grassland habitat in the East due to development and reforestation of farmland is a concern.

FOREST-FALCONS Genus *Micrastur*

COLLARED FOREST-FALCON *Micrastur semitorquatus* COFF ■ 5

dark cap and dark point in white face; white collar

light-morph adult *naso*

short wings

long graduated tail

Forest-falcons are tropical New World raptors that look decidedly unlike a falcon. The Collared Forest-Falcon looks more like an accipiter of the northern forests. There is one accepted North America record, a light-morph bird from Bentsen Rio Grande State Park in southern Texas (Jan. 22–Feb. 24, 1994). Polytypic (2 ssp.; 1 in N.A., presumably *naso*). L 20" (52 cm) WS 31" (79 cm)

IDENTIFICATION Dark above and light below; short-winged, long-tailed, and long-legged. It has large eyes for hunting in dense foliage. Black above, with white cheeks, neck collar, and undersides. Cheek has black crescent some call "sideburns." Wings and back dark; tail dark with thin white bars, white tip.

ADULT: Two morphs: a light morph with white or buffy, unmarked underparts and a rare dark morph. **JUVENILE:** Like adult above, but browner. Undersides heavily barred, ground color has buffy wash.

SIMILAR SPECIES Size similar to Cooper's Hawk. Resembles nothing else common to N.A. Sharp-shinned Hawk subspecies *chinogaster* of S. Mexico is dark above, mostly white below, but lacks "sideburns" and is far smaller.

VOICE Quite vocal, especially early and late in the day. Usual call a hollow *how* or *aow,* frequently repeated, plus rapid combinations getting louder and more slurred at the end. Also repeats a whiny *keer keer keer,* which seems to be a "false alarm call" to attract small birds.

CARACARAS Genus *Caracara*

Long-legged raptors of open areas throughout South and Central America. Their structure and behavior are very different from that of a true falcon, appearing more like a vulture or buteo. Eleven species are found in the New World (inc. the extinct Guadalupe Caracara, *C. lutosa,* from Guadalupe I., off Baja California). Only one caracara reaches N.A.

CRESTED CARACARA *Caracara cheriway* CRCA ■ 1

This large, long-legged raptor of the American Southwest is seen standing on bare ground or perched on a telephone pole. Its long, square-tipped wings are unlike a falcon's, as is its long head and neck. It feeds on carrion and small prey. In flight the wingbeats are steady; gliding, the wings are held slightly downward. It soars on thermals with other hawks and vultures. L 23" (58 cm) WS 50" (127 cm)

IDENTIFICATION ADULT: Body, wings, and crown black, contrasting with white face and neck. Upper breast and back are white, finely barred with black. Tail white with dark barring and dark terminal band. Facial skin and cere are orange; legs yellow. **JUVE-NILE:** Similar pattern to adult, but with brownish and buffy coloration. Legs are gray, and facial skin is pinkish. In flight, gives a striking black-and-white look. Whitish tail is tipped in black.

VOICE Generally silent.

STATUS & DISTRIBUTION Rare to common within range. **YEAR-ROUND:** Resident in central FL, along TX coastal prairies, and in a small range in southeast AZ. Range is expanding into central TX. **MIGRATION:** Generally non-migratory. **VAGRANT:** Casual to CA and NM; scattered summer reports north into MN, ON.

POPULATION Stable in FL and AZ, it has been expanding in TX since the 1980s.

blackish cap

slight crest

pink facial skin

adult

barred white chest

black belly

juvenile

streaked chest

conspicuous white wing patches

adults

long white-based tail barred with black

FALCONS Genus *Falco*

Also known as "typical falcons" or "true falcons," there are 38 species described, nine of which appear in North America. Six species are regular nesters here, including the re-introduced Aplomado Falcon. Known for their aggressive nature, aerial acrobatics, and spectacular dives, falcons are often a symbol of sports teams and warplanes. All true falcons typically have pointed wings, a large head, and dark eyes. They occur worldwide, nesting on all continents except Antarctica.

AMERICAN KESTREL *Falco sparverius* AMKE 1

Our smallest and most common falcon, it is usually found in close proximity to open fields, either perched on a snag or telephone wire or hovering in search of prey. The typical falcon-shaped wings are slim and pointed, the tail long and square-tipped. Adult male plumage is easily told from adult females and juveniles of both sexes. All have two bold, dark moustache marks framing white cheeks on the face and have the dark eyes typical of falcons. It hunts insects, small mammals, and reptiles from a perch or on the wing. Will hover above a field on rapidly beating wings, or soar in place in strong winds above a hillside. Flight style is quick and buoyant, almost erratic, with wings usually swept back. Polytypic (17 ssp.; 2 in N.A.). L 10.5" (27 cm) WS 23" (58 cm)

IDENTIFICATION **ADULT MALE:** Head has gray crown, rufous nape with black spot on either side, dark moustaches around white cheeks. Back is bright rufous with black barring on lower back. Tail is patterned with highly variable amounts of black, white, or gray bands. Wings are blue-gray with dark primaries. Underparts are white, washed with cinnamon. **ADULT FEMALE:** Head similar pattern to male, but more brown on crown. Back, wings, and tail are reddish brown with dark barring; subterminal tailband much wider than other bands. Underparts are buffy-white with reddish streaks. **JUVENILE MALE:** Head similar to adult, but less gray and with dark streaks on crown. Back is completely streaked, heavy streaks on breast. **JUVENILE FEMALE:** Very similar to adult female. **FLIGHT:** Light, bouncy flight is usually not direct and purposeful—often with "twitches" or hesitation. Light underwings and generally light body coloration. Males show a row of white dots ("string of pearls") on trailing edges of underwings. Fans tail when hovering.

GEOGRAPHIC VARIATION Widespread nominate *sparverius* is the typical migratory form. Subspecies *paulus,* from South Carolina to Florida, is smaller, the male with less barring on the back and fewer spots on its undersides; essentially nonmigratory.

SIMILAR SPECIES Merlin appears darker in flight due to dark underwings, shorter tail. When perched, looks darker, more heavy-bodied, lacks the two moustaches. Peregrine Falcon is larger, has wider wings, shorter tail, single heavy moustache. Eurasian Kestrel is slightly larger, has one moustache, wedge-shaped tail; males have blue-gray tail and reddish wings.

VOICE Loud, ringing *killy-killy-killy* or *klee-klee-klee* used all year round. Distinctive.

STATUS & DISTRIBUTION Common in open areas, it ranges throughout N.A., including much of Canada and into AK. **BREEDING:** A cavity nester, it uses dead trees, cliffs, occasionally a dirt bank, and even a hollow giant cactus in the Southwest. It will also use man-made nestboxes placed high on trees and telephone poles. Up to five or six young per brood, depending on food availability. **MIGRATION:** Fall migration starts early, in Aug. Northern breeders follow traditional fall migration routes to wintering ranges in the southern U.S. and northern Mexico. Eastern populations use the coastlines more than the inland corridors, and are not reluctant to cross water. Spring migrants are only concentrated along the Great Lakes watch sites. **WINTER:** The majority of birds winter in the southern U.S., often spaced out on every other telephone pole in agricultural areas. A small percentage winter in snow-covered states, the numbers depending on food sources.

POPULATION Overall, numbers are stable. However, increases in the central U.S. are being offset by declines in the Northeast and the West Coast (CA and OR). Eastern populations are thought to be affected by loss of open habitat due to two factors; human development and agricultural abandonment leading to reforestation, with a subsequent increase in Cooper's Hawk predation.

adult ♂ / frequently hovers / adult ♂ / uniformly rufous-brown back and tail / all with two dark facial stripes / adult ♀ / juvenile ♂ / rufous tail / male has bluish gray wings / adult ♂

EURASIAN KESTREL   *Falco tinnunculus*   EUKE   ■   4

Also called "Common Kestrel," this small falcon is from Eurasia. A typical falcon with large head and dark eyes, slim pointed wings and a long tail, it shares many of the characteristics of the American Kestrel, but appears slightly larger and heavier and has only one moustache mark on its face. Undersides on all ages are buffy and streaked. In flight the tail looks slightly wedge-shaped, and from above the wing gives a two-tone appearance. The bird feeds on small prey, including insects, small mammals, and birds, first hovering to spot its quarry, then descending in steps before stooping to the ground. Polytypic (11 ssp.; N.A. records likely of nominate). L 13.5" (34 cm) WS 29" (74 cm)

IDENTIFICATION **ADULT MALE:** Gray head and nape, with a single moustache and paler throat. Reddish back and wing coverts have black streaks, dark primaries. Gray rump and unmarked gray uppertail contrast with wide, dark subterminal tail band and a white tip. Undersides are whitish buffy, with spots on breast, streaked flanks, plain undertail coverts. **ADULT**

FEMALE: Head with similar pattern to male, but more rufous-brown. Back is more heavily barred, rump gray with gray-brown barred tail and wide subterminal band. **JUVENILE:** Similar to female, but usually with buff tips to feathers on upperparts.

SIMILAR SPECIES American Kestrel is smaller, slimmer; has two moustache marks on face. Underparts are usually whiter, tail more rounded. Males have blue-gray wing coverts; red tail; females are less heavily marked.

VOICE A repeated *kee-kee-kee.* Mostly quiet away from breeding grounds.

STATUS & DISTRIBUTION Widespread in Old World, casual to western Aleutians and Bering Sea, accidental to East and West Coasts of U.S., most recently to WA, CA, FL, and MA. **MIGRATION:** Northern European and Asian birds are highly migratory, wintering in southern Europe, the Middle East, and southern Asia **POPULATION** Eurasian Kestrel seems

to be stable across its range, with human persecution being the most common threat. No signs of range expansion to N.A.

two-tone upperwing

juvenile

adult ♂

wedge-shaped tail

pale gray head

adult ♀

single dark moustache

adult ♂

gray tail base

EURASIAN HOBBY   *Falco subbuteo*   EHOB   ■   4

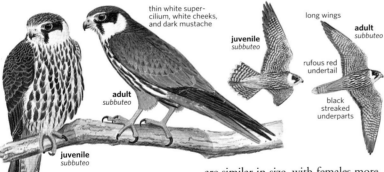

thin white supercilium, white cheeks, and dark mustache

juvenile *subbuteo*

long wings

adult *subbuteo*

rufous red undertail

adult *subbuteo*

black streaked underparts

adult *subbuteo*

juvenile *subbuteo*

Also called "Northern Hobby," and "mini Peregrine" by raptor aficionados, this Old World species is casual in the Bering Sea region and accidental elsewhere. Polytypic (2 ssp.; N.A. records likely of nominate). L 12.3" (31 cm) WS 30" (77 cm)

IDENTIFICATION A medium-size falcon with long thin, pointed wings and a relatively short tail, dark above and whitish below with heavy streaking. A rapid, strong flier, it takes prey on the wing, usually small birds, but also dragonflies and other large insects. Sexes

are similar in size, with females more robust, approximately 5 percent larger, but up to 30 percent heavier. **ADULT:** Dark gray head and upperparts, black moustache sharply contrasting with pure white cheek, short white streak above eye, whitish throat, with whitish underparts heavily streaked. Undertail coverts and legs bright rufous, easier to see when perched. Cere and eye ring yellow. **JUVENILE:** Upperparts are more brownish, rufous tips to new feathers quickly wear off. Head pattern similar to adult, underside ground color may appear more buffy, tail more distinctly light-tipped, undertail coverts and legs

buffy to dull rufous. **FLIGHT:** Long, slim wings and a relatively short tail, it uses its speed more often than its maneuverability to catch prey. At ease in all kinds of wind conditions. Underwings are uniformly grayish with dark markings and darker wingtips. Rufous undertail of adult is visible in good light, but by the time you see that, the identification should already be clinched.

SIMILAR SPECIES American Kestrel is longer-tailed and shorter-winged with a lighter flight; is lighter above, has two moustache marks on face. Merlin is stockier, darker, less heavily streaked underneath. It resembles the far larger Peregrine Falcon in proportions, powerful flight, and dark over light plumage.

VOICE Silent away from breeding grounds.

STATUS & DISTRIBUTION Widespread in the Old World; winters in Africa and Indian subcontinent. **VAGRANT:** Casual in AK, ±20 records mainly in late spring and early summer from the Bering Sea region, inc. near Nome on the Seward Peninsula. Accidental to western WA (Oct.), MA (May), and NF (May).

MERLIN *Falco columbarius*   MERL ▪ 1

rapid, powerful flight with no hovering

adult ♂

♀

columbarius

faint moustache

columbarius ♀

dark brown above

"Black Merlin" ♀

darker than *columbarius* with fainter tail bars and darker head

suckleyi

adult ♂ *columbarius*

dark bluish above

white flank spots

adult ♂ *suckleyi*

♀ *richardsonii*

overall, color of Prairie Falcon

pale bluish gray above

adult ♂ *richardsonii*

A small, dark falcon, most often described as dashing, the Merlin is an active hunter, taking its prey in midair, often after spectacular chases. It primarily feeds on birds, sometimes those larger than itself. Hunting is generally on the wing, but it will hunt from a perch and ambush passing prey. It occasionally soars on spread wings but never hovers. It shows typical falcon features of a dark eye, pointed wings, and a square-tipped tail. Its facial moustache mark is less noticeable than on most falcons. Females are larger than males, and adults have different plumages. Polytypic (9 ssp.; 3 in N.A.). L 12" (31 cm) WS 25" (64 cm)

IDENTIFICATION At a distance, a small, dark falcon either flying or sitting is usually a Merlin. Its wings are wider and tail much shorter than a kestrel's. Usually perches atop a dead snag, tower, or building. When perched, wingtips do not reach tip of tail. The following descriptions apply to the nominate subspecies, known as "Taiga Merlin." **ADULT MALE:** Back and upper wing coverts are blue gray, undersides are whitish streaked with brown with an overall rufous wash to sides and leg feathers. Tail is dark with lighter gray banding. **ADULT FEMALE:** Above, it has a slate brown back and upper wing coverts. Tail bands are buffy on a dark brown tail. Below, buffy underparts are heavily streaked. **JUVENILE:** Apart from size, sexes very similar to each other; plumage like adult female; tail bands buffy. **FLIGHT:** Fast, flickering wingbeats and direct flight set it apart from American Kestrel.

GEOGRAPHIC VARIATION "Taiga Merlin" *(columbarius)* from Alaska to eastern Canada is smallest of the three; averages darker to the east of its range; and is highly migratory, wintering down to the Caribbean, and Central America. "Prairie Merlin" *(richardsoni)* is largest of the three, much paler than "Taiga Merlin," especially adult females, which also have thinner streaks on body. Adult male is paler blue on back than "Taiga." All have wider tail bands than "Taiga." Moustache mark can be very faint. "Black Merlin" *(suckleyi)* is a resident of the Pacific Northwest; generally sedentary, but does wander down to California. About the size of "Taiga Merlin," but much darker overall, with almost no white on face, heavy and broad dark streaking on underparts with contrasting white spots on flanks, and tail bands are often incomplete.

SIMILAR SPECIES American Kestrel is slimmer, has longer tail, less direct flight style. Peregrine Falcon is much larger, has broader base of wings, large moustache mark on head. When perched, its wingtips reach or almost reach tip of tail. Prairie Falcon is similar to adult female and juvenile "Prairie Merlin" in coloration, but his larger, and his larger, more distinct moustache mark on face, and dark axillaries on underwing.

VOICE Rapid *kee-kee-kee* or *klee-klee-klee* heard often around nest. Males are higher pitched than females. On migration, often heard when harassing other raptors, especially Peregrines.

STATUS & DISTRIBUTION Widespread throughout the Holarctic, they are uncommon nesters throughout range. Relatively common on East Coast during fall migration, an uncommon winter bird. **BREEDING:** Uses a mix of habitats in conjunction with open spaces, anything from tundra to coastline, boreal forest to Northwestern rain forest, prairie edges, and parkland. In recent years they have been increasingly found within cities in prairie states and provinces. **MIGRATION:** Merlins follow most traditional migratory routes, but are less likely to soar on

thermals, so are usually underreported on mountain hawk-watches as they cruise by at treetop level. Along coastal routes, primary passage times in fall are mid-Sept. through late Oct., with juveniles often passing first, followed by adults. They are more active migrants during morning and late afternoon, and many hawk watches call 4 p.m. "Merlin Time." **WINTER:** Uncommon in northern states in winter, more likely to be found along coast feeding on shorebirds. "Prairie Merlins" have recently been taking to wintering in Great Plains cities, feeding on small birds and rodents. **POPULATION** Numbers are increasing for "Prairie" and "Taiga" Merlins as they become habituated to cities.

columbarius

suckleyi & wintering columbarius

richardsonii & wintering columbarius

columbarius

RED-FOOTED FALCON   *Falco vespertinus*   RFFA ▪ 5

This medium-size falcon of Eastern Europe and western Asia is larger than the Eurasian Kestrel and Merlin. It feeds mainly on insects, hawking them from midair, often hovering above a

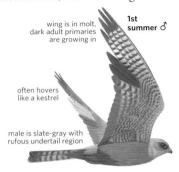

1st summer ♂

wing is in molt, dark adult primaries are growing in

often hovers like a kestrel

male is slate-gray with rufous undertail region

spot in search of prey. On Aug. 7, 2004, the first record of this bird in N.A., a subadult male in heavy flight-feather molt, was found at Katama Airfield, on Martha's Vineyard, MA. Numerous diagnostic photographs were taken, and thousands of birders made the trip to see the bird, which stayed until Aug. 24. The locals christened "Red Socks," and some proclaimed it a "miracle bird" and a harbinger of the October 2004 Boston Red Sox World Series win. Monotypic. L 11" (27 cm) WS 29" (73 cm) **IDENTIFICATION ADULT MALE:** Red Socks was a second-year male, with dark gray head, mostly gray wings and body, and barred flight feathers

and tail. **ADULT FEMALE:** Very different from male, with slate gray back and wings; brown head with black moustache; and white cheek and side of neck. Underparts are buffy with thin streaking. **JUVENILE:** Back and wings browner, with light feather edges. Head pattern similar to adult female. **SIMILAR SPECIES** Mississippi Kite is all gray with pointed wings, but does not hover when foraging. **STATUS & DISTRIBUTION VAGRANT:** A long-distance migrant between Europe and Africa. There are vagrant records to Iceland. Small numbers occur annually to the British Isles, particularly in late spring, and there are at least four records from Iceland.

APLOMADO FALCON   *Falco femoralis*   APFA ▪ 3

This medium-size falcon is a neotropic species whose range formerly just reached into North America. In recent decades captive birds have been released in southernmost Texas. However, at this time, the South Texas Aplomado Falcon is not yet considered established. It is often seen perched atop a single tree or cactus as it scans for prey. Polytypic (5 ssp.; *septentrionalis* in N.A.). L 15–16.5" (38–42 cm) WS 40–48" (102–122 cm) **IDENTIFICATION** A slim, long-winged, long-tailed falcon, tail extending well past wingtips on a perched bird. Adults and juveniles have slightly different plumages. All show a distinctive head pattern of dark eye line and moustache mark, dark crown, and contrasting buffy cheeks and buffy superciliary lines that meet at nape. **ADULT MALE:** Head as described above. Dark gray back and upper wings contrast with buffy chest, black belly band, and rufous belly, undertail coverts, and legs. Flight feath-

juvenile *septentrionalis*

juvenile ♀ *septentrionalis*

bold white supercilium and breast

adult ♂ *septentrionalis*

blackish side patches

buffy streaked breast

long tail

adult ♀ *septentrionalis*

dark belly band and underwing coverts

ers and tail are blackish, with six or more narrow white bands on the tail. Black underparts are often finely barred in white. Eye ring and cere are yellow. **ADULT FEMALE:** Like the male but noticably larger; buffy chest has black streaking, which is usually lacking in males. Females are noticeably larger than males. **JUVENILE:** Shows a similar pattern to adults, but with brownish back and wings, buffy tail bands, buffy underparts, legs, and undertail coverts. Buffy breast is more heavily streaked with black. Cere and eye ring are bluish gray. **FLIGHT:** Shows deep, rapid wingbeats. It is highly maneuverable, able to turn and swoop in pursuit. Can also hover and buoyantly soar. Combination of slim wings and long tail, plus light-dark combination of chest and underparts make it fairly distinctive in flight. **SIMILAR SPECIES** Adult Peregrines have a dark moustache on face and dark back and wings, but are much larger, more stocky, and shorter-tailed. Peregrine and Prairie Falcon lack black underparts.

VOICE Most typically a fast *ki-ki-ki-ki-ki-ki* sound, males higher-pitched than females. **STATUS & DISTRIBUTION** Uncommon, local from Mexico to Argentina. Until the early 20th century likely bred in open country from south TX to AZ. Listed as endangered in 1986. A few wild pairs in southern NM appear to be hanging on, but additional captive-bred birds have been released there in recent years. A recent record from near the Davis Mts., TX, likely originated from a nearby wild population in Chihuahua, Mexico. Reintroduced into southern TX. **POPULATION** Very little data on Mexican populations, U.S. numbers slowly continue to grow.

GYRFALCON *Falco rusticolus*   GYRF ▪ 2

A visitor to the lower 48 states from its breeding grounds in arctic Alaska and Canada, the Gyrfalcon is the largest falcon in the world, capable of taking down flying geese and cranes. It is a powerful flyer, with deceptively slow wingbeats and the longest tail of the large falcons. Gyrfalcons occur in three color morphs—white, gray, and dark (black). Juveniles are different from adults; sexes are similar, but females are larger. The Gyrfalcon is typically seen in the lower 48 states during fall migration and winter; an individual bird may take up winter residence and remain there for a period of months. A bird seen daily around the Boston, MA, harbor was determined to be one that was banded in the area more than five years earlier. In medieval Europe, the Gyrfalcon was highly prized as a falconry bird, as it still is in some Arab countries. Monotypic. L 20–25" (51–64 cm) WS 50–64" (127–163 cm)

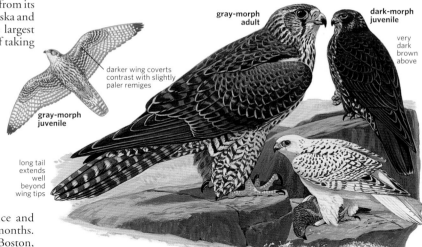

gray-morph adult

dark-morph juvenile

very dark brown above

darker wing coverts contrast with slightly paler remiges

gray-morph juvenile

long tail extends well beyond wing tips

white-morph adult

IDENTIFICATION A large, bulky falcon, with thick pointed wings wide at base, giving a fatter wing silhouette than most large falcons. Females are larger than males. Perched, they sit with an upright posture on cliffs, towers, buildings, and other manmade structures. Wings extend about two-thirds of the way down the tail. **ADULT LIGHT MORPH:** All white with varying amounts of dark markings above, blackish primary tips; dark bands on tail often incomplete. Occasionally shows a faint moustache. **ADULT GRAY MORPH:** Dark gray to gray-brown ground color on back and wings is patterned with lighter gray to blue-gray feather tips, giving a scaled pattern. Tail strongly barred. Head can appear lighter than back, with a dark eye line and paler supercilium,

moustache usually visible. Underparts with barred flanks and spots on breast and belly. **ADULT DARK MORPH:** Blackish brown above with less obvious barring on back and tail; head usually with a "hooded" appearance; moustache usually lost in the overall pattern. Underneath, overall dark with some whitish streaks on breast and barring on flanks, belly, and undertail. **JUVENILE LIGHT MORPH:** Similar overall color to adult, but more heavily marked above with black or brown. Primaries are blackish with complete barring. Tail bands are complete, heavier than adults. **JUVENILE GRAY MORPH:** Darker gray-brown than adult, with back and wing coverts finely edged or spotted in lighter brown. Head color as dark as back, but still showing faint moustache and supercilium. Underparts are heavily streaked with brown, lightest on throat and undertail. **JUVENILE DARK MORPH:** Overall dark brown with heavy streaking underneath, barred undertail coverts. **FLIGHT:** A large falcon with heavy chest, wide pointed wings that are blunt on tips. Long tail is wide-based and tapers down to the tip when folded. Its wingbeats are slow and relatively stiff, yet it flies with great speed. Gray- and dark-morph birds show lighter flight feathers that contrast strongly with dark wing coverts, giving a strong two-tone look. **SIMILAR SPECIES** White morph Gyrfalcon is unmistakable. Albino or pale morph Red-tailed Hawk may look similar, but wing shape and tail differences are easy to spot. Peregrine Falcon is slimmer, shorter-tailed, shows

heavier moustache, usually darker crown. Adult Peregrine is horizontally barred across breast; juveniles show uniformly dark underwings. Peale's Peregrine of the Pacific Northwest resembles a dark morph Gyrfalcon. Prairie Falcon is lighter underneath, has dark axiliaries contrasting with light underwing, and a pale cheek. Northern Goshawk has the same bulk, but differs with blunt rounded wings, wide secondaries, and two-tone upperwings. **VOICE** Usually silent. **STATUS & DISTRIBUTION** An uncommon species, it is a widespread breeder in the Holarctic. **BREEDING:** Nest scrape usually on a cliff, but also uses old Raven nests in a suitable location. **MIGRATION:** Gyrfalcon sightings along traditional hawk migration sites are usually late in the season—Nov.–Dec. There are enough late-Oct. records to make it a species to watch for. **WINTER:** There seem to be no real patterns in wintering birds, but individuals have shown that they will return to the same general area in succeeding years. A favored roost site will be used for the night, with the bird leaving in the morning and arriving in the evening with sufficient fidelity to afford birders a "viewing schedule." Most wintering birds are found above 40° N. **VAGRANT:** Casual to northern CA, OK, northern TX, and east to the mid-Atlantic states. **POPULATION** Circumpolar and out of the range of most habited areas, populations are thought to be stable.

PEREGRINE FALCON *Falco peregrinus* PEFA ■ 1

The Peregrine Falcon is world renowned for its speed, grace, and power in the air. It is distributed across the world, found on every inhabited continent, and on many islands. Formerly called the "Duck Hawk" in North America. It can occur in almost any habitat in North America, but is frequently seen around water—lakes, rivers, or coastal shorelines. Peregrines feed primarily on birds taken in flight, along with the occasional bat or rodent. Its large hind talon (the hallux) is anchored by heavy tendons, allowing it to use the force of its dive to kill its prey on contact. The true velocity of its high-speed dive has been the subject of much debate over the years, with some observers claiming speeds of up to 200 miles an hour. Recently a falconer's Peregrine was outfitted with a skydiver's altimeter and data recorder and trained to chase down a lure released by the skydiver. The bird easily dove to a speed of 247 miles an hour, and the feeling is that it could still go faster! Polytypic (at least 16 ssp.; 3 in N.A.). L 16–20" (41–51 cm) WS 36–44" (91–112 cm)

IDENTIFICATION Peregrine is a raptor with long pointed wings and a medium-length tail. All ages have a dark moustache mark on the face. When sitting wings extend almost to the tip of the tail. They favor an exposed perch, natural or man-made, to scan for prey. The majority of successful kills are made with high-speed dives that overtake from above or behind. Females are larger than males; sexes appear similar. **ADULT:** A dark head and nape, with dark moustache and white cheek. Cere and eye ring are yellow. "Tundra Peregrine" *(tundrius)* shows the thinnest moustache, whitest cheek, and often a light forehead. Some *anatum* have a dark cheek, blending with the moustache. Upperparts are dark gray, underparts are whitish, with clear breast and barring on belly and flanks. Subspecies *anatum* often shows a salmon wash on breast. "Peale's Peregrine" *(pealei)* is without salmon

color and has streaking on breast. **JUVENILE:** Browner above than adults, with brown streaking on buffy underparts. Cere and eye ring are blue-gray. "Tundra" has a light forehead, often a light crown (giving a "blonde" look), and thinnest moustache. Subspecies *anatum* shows a darker head, thicker moustache, thicker streaking on underparts. "Peale's" is overall very dark with a small cheek patch; underparts heavily streaked or almost uniformly dark. **FLIGHT:** Capable of swift level flight, power dives, and soaring on outstretched wings, Peregrine can fly in virtually any wind and weather condition. Typical flight is a smooth, quick, rhythmic, shallow wingbeat. With deeper wingbeats, it gains speed and dives, whereas it executes quick turns after fleeing prey with a fanned tail and choppy wingbeats. When soaring, wing silhouette tapers smoothly from body to tip, like a dipped candle. Underwings are barred on both flight feathers and coverts, giving a uniformly dark appearance.

GEOGRAPHIC VARIATION "Tundra" *(tundrius)* nests on the Alaskan and Canadian tundra, is overall the lightest North American subspecies, and is the longest-distance migrant. "Peale's" *(pealei),* the darkest and most sedentary, is found along the coast of the Pacific Northwest. Continental-breeding *anatum* is intermediate, but variable in features, while movement is on a smaller scale than "Tundra." The release of mixed-subspecies birds as part of the North American reintroduction program has resulted in many eastern birds showing characteristics of "Peale's," selected in hopes that the newly hacked birds will remain in the vicinity of their release points.

adult
anatum

uniform underwing

adult
pealei

juvenile
pealei

heavily spotted breast

heavily streaked underparts

SIMILAR SPECIES Gray- and dark-morph Gyrfalcon is usually larger and darker, with smaller moustache, wider wing bases with two-toned underwing, and a longer tail that tapers from base to tip. Some small, dark male Gyrfalcons can appear similar to dark juvenile female "Peale's." Prairie Falcon is lighter overall, has a thin moustache and dark axiliaries and underwing coverts contrasting with light underwing. When perched, Prairie's wing tips do not reach the tip of the tail.

VOICE A rapid *kak-kak-kak,* not often heard away from the nesting area.

STATUS & DISTRIBUTION The widespread continental form was decimated by the use of pesticides in the 1950s and 1960s, placing it on the Endangered Species List by 1970. A reintroduction program begun in the late 1970s has helped to restore Peregrine numbers throughout. Wintering Peregrines may be found in large cities throughout the southern U.S. Recently taken off the Endangered Species List, the Peregrine is a showpiece for the ability of a species

broad, dark moustachial stripe

juvenile
anatum

long wing tips extend nearly to tail tip

adult
anatum

tundrius

anatum

pealei

mixed stock

to rebound from the effects of pesticides in the environment. Most major cities in the eastern U.S. now host Peregrine nests. However, some Peregrines have been introduced to locations where they did not formerly nest and have had a negative impact on native birds. Populations of "Tundra Peregrines" were never in real danger, but concern continues due to their wintering in countries where DDT use is still allowed. **BREEDING:** Historically nesting in a scrape atop cliffs, they have taken well to man-made structures like tall buildings, bridge beams, and towers, and even use old Osprey and Bald Eagle nests in coastal locations. **MIGRATION:** During the 1970s and early 1980s, a single Peregrine passing by a hawk watch site was cause for celebration. A site in the Florida Keys now records single days with more than 350 Peregrines passing by. Fewer numbers fly the inland ridges, often soaring in with kettles of other raptors to the delight of hawk watchers. Prime fall dates in the East are early to mid-Oct., spring dates in late Apr. Recent surveys of offshore oil rigs in the Gulf of Mexico found that Peregrines pick off exhausted migrants, with any one rig hosting up to a dozen falcons. **WINTER:** Usually found along the coast or around cities, wherever food is available.
POPULATION Slowly increasing in North America.

thinner dark moustache than *anatum*

juvenile *tundrius*

adult *tundrius*

PRAIRIE FALCON *Falco mexicanus*   PRFA ▪ 1

The large, light-colored falcon of the arid western United States, it inhabits foothills, grasslands, and other open country. While it primarily feeds on birds, especially flocking species in winter, it also takes ground squirrels when they are abundant. It takes birds on the wing, usually while flying low to the ground and surprising its prey as they take off. It can also dive down on prey from high above. Monotypic. L 15.5–19.5" (39–50 cm) WS 35–43" (89–109 cm)
IDENTIFICATION A large raptor with pointed wings and a wide, square tail. When perched, wing tips fall noticeably short of tip of tail. Large head, a pale cheek, a pale supercilium, and a thin, dark moustache mark. Sexes appear similar, but females are larger. **ADULT:** Upperparts are brownish with pale barring on wing coverts and back; tail and flight feathers are lightly barred. Underparts are whitish with dark spots on sides and belly. Cere, legs, and eye ring are yellow. **JUVENILE:** Darker brown above than adults with less light barring on wing coverts and back, they appear darker at a distance. Underparts are more buffy and thinly streaked. Cere and eye ring are blue-gray. **FLIGHT:** Wingtips are not as sharply pointed as those of other large falcons, underwings feature light-colored flight feathers and outer wing coverts contrasting with dark axillaries and longer underwing coverts, forming a dark "wingpit" that makes it look broader-winged. Females show more dark on longer underwing coverts than males.
SIMILAR SPECIES Peregrine Falcon shows more contrast between dark above and light below, a thicker, darker moustache, and uniformly darker underwings. Flight style is similar to Prairie, but hunts less often at low altitudes. A light-colored juvenile "Tundra Peregrine" shares the pale supercilium of Prairie Falcon, but lacks Prairie's pale cheek. A pale "Prairie Merlin" has a similar shape and coloration, but is much smaller with faster wingbeats, shows a faint moustache mark, and lacks the dark area of the underwing.
VOICE A loud *kik-kik-kik* around the nest, usually silent at other times.
STATUS & DISTRIBUTION Widespread throughout acceptable habitat from Canada to Mexico, but never abundant. **BREEDING:** Uses a small scrape on a cliff or ledge, high above the ground. It will also use the old nest of another raptor or raven, almost always on a cliff, rarely in a tree or on a building. **MIGRATION:** Movements not well understood. Sightings well away from nesting locations begin by late Aug. Prairie Falcons are noted for their post-breeding dispersal, wandering west to the Pacific and east to the

Great Plains states. **WINTER:** Outside of mapped range, a few are found coastally from WA to CA and into Baja CA, Mex. **VAGRANT:** Records as far east as WI, IL, OH, PA, KY, and AL. The farther away from its traditional range, the greater the suspicion that an individual might be an escaped falconer's bird.
POPULATION Stable, adapting more and more to the spread of cities and western development.

head pattern differs from Peregrine Falcon

pale brown above

adults

blackish axillaries and underwing coverts

LORIES, PARAKEETS, MACAWS, AND PARROTS Family Psittacidae

Monk Parakeets (IL, Feb.)

Throughout tropical and subtropical (rarely temperate) areas of the world live 322 species of these colorful birds. Hundreds of thousands of them were imported into the U.S. during the late 1960s through early 1990s. Numerous species are now at liberty (primarily in CA and FL) as a result of accidental or deliberate releases.

Structure Psittacids are large-headed birds with no visible necks and short legs. Parakeets and macaws are slim with long tails; parrots are chunky, with short tails. Psittacids are primarily arboreal. Their bills are short, thick, and curved, with a powerful, articulated tongue that aids in processing palm nuts and other plant food. In all species, two toes face forward and two backward. Psittacids show the greatest size diversity of all the world's bird families, ranging from 3.7 to 39 inches (8–100 cm).

Behavior Most species are social, gathering in large mixed flocks at nighttime roosts or when foraging; some species are communal breeders. Virtually all nest in cavities in palms, trees, or termite mounds; the Monk Parakeet is the sole exception. Psittacids forage primarily for seeds, nuts, and fruit and often visit bird feeders. Typically noisy in flight, they often grow quiet when feeding or roosting. They can be difficult to locate then, as their plumage blends into the vegetation. Flocks can be identified to group fairly easily—based on shape and flight style—but multiple species are often present.

Plumage Psittacids are primarily green, especially in the New World, but usually show some red, orange, or yellow on the head or in wings or tail. Some species are entirely red, blue, or yellow. Sexes are usually similar, but many parrots take more than two years to reach adult plumage, resulting in confusing immature plumages. Aviculturists have created numerous artificial color morphs that may be seen outside of captivity.

Distribution Widespread. Greatest native diversity is in the tropics; greatest exotic diversity is in southern Florida.

Taxonomy Recent genetic studies have shown that the Psittacidae share a close affinity to the Falconidae (caracaras and falcons); both families are now placed just before the order Passeriformes (passerine birds). Species limits are unknown for many genera.

Conservation Capture for the pet trade during the late 1960s to early 1990s was vast. Unregulated capture and the massive degree of habitat destruction have endangered dozens of species. All psittacids in the U.S. currently are exotic and unprotected. The potential impacts of exotic psittacids in the U.S. on native species, habitats, or agriculture are largely unknown. BirdLife International lists 94 species as threatened and 30 others as near threatened.

Flight Silhouettes of Various Psittacid Genera

The flight patterns of psittacid genera differ.

Brotogeris. These small parakeets have moderately long, pointed tails. In flight several rapid wingbeats are followed by brief closure of bowed wings. Flight is rapid, but seems halting and undulating from wing closures and side-to-side twisting of body.

Psittacula. The Rose-ringed Parakeet is of medium size and has a markedly long, slender tail. It appears relatively small headed, and thus does not seem "front-heavy." Its wingbeats are deeper and more sweeping than those of other parakeets.

Aratinga, Nandayus, and *Myiopsitta.* These medium-size parakeets have long, pointed tails. Bills are moderate to large, giving them a more "front-heavy" look. Flight is rapid and constant; wingbeats are fairly shallow, with wings bowed slightly below the body plane. There is some side-to-side body-twisting.

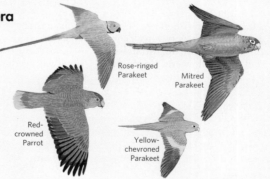

Rose-ringed Parakeet

Mitred Parakeet

Red-crowned Parrot

Yellow-chevroned Parakeet

Amazona. These medium to large parrots seem large headed and markedly "front-heavy" in flight. The tail is squared and moderately short. Wings are bowed down; wingbeats are stiff, continuous, and fast, but flight is slower than in parakeets. ∎

AUSTRALIAN PARAKEETS Genus *Melopsittacus*

This genus consists of a single species, the Budgerigar. One of the world's most abundant psittacids, it numbers perhaps five million individuals. These birds breed communally in cavities, or—as in Florida—in nest boxes, raise multiple broods annually, and feed extensively on grass and other seeds obtained from the ground, and on birdseed in Florida.

BUDGERIGAR *Melopsittacus undulatus*   BUDG ▪ 3

Perhaps the most popular cage bird in the world, the Budgerigar has been bred in captivity since the mid-1800s. A wide variety of artificial color morphs exist, including white, yellow, or blue plumages. Monotypic. L 7" (18 cm)
IDENTIFICATION This is a tiny parakeet—the size of a warbler if the tail is excluded. Sexes are similar. **ADULT:** Head is mostly yellow with black barring on auriculars, hind crown, and nape. Two black spots on each side of throat, with a small purplish patch on malar. Eyes are yellow; bill, legs, and feet grayish. Back and wing coverts are yellow, barred with black. White or yellow wing stripe in flight. Rump green. Tail equal to length of body, with blue central rectrices. Underparts wholly lime

green. **JUVENILE:** Forehead barred, throat unspotted, eyes dark.
SIMILAR SPECIES None; much smaller than any other psittacid found in N.A.
VOICE A series of pleasant high-pitched chittering or chirping; some notes reminiscent of House Sparrows.
STATUS & DISTRIBUTION Exotic in the U.S. Common to abundant in native Australia. **YEAR-ROUND:** Nonmigratory. In U.S. restricted to FL, where a breeding population along the central Gulf Coast since the early 1960s is nearing extirpation. Escapes are possible anywhere.
POPULATION The FL population has declined more than 99 percent since the late 1970s, from perhaps 20,000 individuals to fewer than 100. Nesting competition with House Sparrows

thought to be the primary cause of the decline. Now limited to residential areas at Hernando Beach and Bayonet Point.

escapes may be blue, white, or yellow

blue variant

yellow variant

yellow face

barred above

green underparts

natural coloration

long, pointed tail

Genus *Psittacula*

ROSE-RINGED PARAKEET *Psittacula krameri*   RRPA ▪ EXOTIC

The Rose-ringed Parakeet is the only representative of the North American psittacid fauna native to Africa and India. It is established in a few restricted areas of Florida and California. The species is known as the "Ring-necked Parakeet" in the Old World. Polytypic (4 ssp.; 1–2 in N.A.). L 15.8" (40 cm)
IDENTIFICATION Overall, Rose-ringed Parakeet is yellowish green. It has a slender tail with very long central feathers. A narrow, red orbital surrounds pale yellow eye. Upper mandible is entirely bright red, but lower mandible is mostly or completely black. Remiges are darker green, contrasting with yellowish green coverts. **ADULT MALE:** Chin and throat are black. Black extends backward and encircles the head; back end of collar is tinged with rose pink. Nape is pale azure. **ADULT FEMALE & JUVENILE:** Chin and throat are yellow-green, collar is lacking, and nape is green.
GEOGRAPHIC VARIATION The North American subspecies—*manillensis*, possibly also *borealis*—are both native to the Indian subcontinent. The two

African subspecies *(krameri* and *parvirostris)* have dark-tipped upper mandibles.
SIMILAR SPECIES Green Parakeets and immatures of other birds of the genus *Aratinga* may be mostly or entirely green, but these birds lack the red bill and extremely long central tail feathers of Rose-ringed Parakeet. Collar of the adult male Rose-ringed is diagnostic.
VOICE Highly vocal at roosts. **CALL:** A loud flickerlike *kew* is common, along with various high, shrill notes uncharacteristic of most psittacids.
STATUS & DISTRIBUTION Exotic in the U.S. The most widespread Old World psittacid, Rose-ringed Parakeet, is native to Africa and India. **YEAR-ROUND:** Nonmigratory. In the U.S., Rose-ringed is restricted to CA and FL. Common to abundant in its native range.
POPULATION The CA population is currently increasing. Trends in fl are unknown; some populations are now extirpated. Rose-ringed Parakeets number about 200 in FL, at Fort Myers and Naples. A population

at Bakersfield, CA, numbers about 1,000 birds, with less than 100 at Los Angeles.

red bill

black neck ring edged with rose

♀

adult ♂

adult ♂

very long, thin, pointed tail

ROSY-FACED LOVEBIRD *Agapornis roseicollis* RFLO ■ 2

Lovebirds take their name from mated pairs' habit of resting and roosting in close proximity and probably for their frequent allopreening. Polytypic (2 ssp., probably nominate in N.A.). L 6.2" (16 cm)
IDENTIFICATION Tiny, short-tailed parrot with pale bill. **ADULT:** Pale green overall with cerulean uppertail coverts. Rose-pink face and upper breast, reddish forehead; female duller. **JUVENILE:** Forecrown green, pink of face and throat pale buff-pink, bill duskier than adult. **FLIGHT:** Tiny size and erratic, twisting flight.
SIMILAR SPECIES Among the eight

other species in the genus, Lilian's Lovebird (*A. lilianae*) and Redheaded Lovebird (*A. pullarius*) are most similar to Rosy-faced but have more intense orange or reddish orange colors in face and are not known in the wild in N.A.
VOICE Calls include high-pitched, short *tweet!* and *squeer!* notes; also lower, throatier trills.
STATUS & DISTRIBUTION A small population has become well established around Phoenix, AZ, where some have been found nesting in cacti. Small numbers are also seen in FL and occasionally CA.

POPULATION Severe declines in Angola due to trapping for the cagebird trade.

red band across forehead

pink face and upper breast

also known as Peach-faced Lovebird

NEW WORLD PARROTS Genera *Myiopsitta, Conuropsis, Aratinga, Nandayus, Rhynchopsitta, Brotogeris,* and *Amazona*

This diverse group of about 148 species ranges from northern Mexico to the southern tip of South America. New species are still being discovered in the tropics. Many species with allopatric native ranges occur sympatrically in California and Florida (e.g., the Red-masked Parakeet). As a result, some field marks remain to be worked out. Some psittacids in California and Florida may hybridize, creating additional undescribed plumages, which further complicates identification. Many species not included in this guide may be encountered outside captivity.

MONK PARAKEET *Myiopsitta monachus* MOPA ■ 2

This most widespread and numerous psittacid in North America, the Monk Parakeet survives frigid New England and Illinois winters by feeding on bird-

seed and sheltering in its bulky stick nest. Polytypic (4 ssp.). L 11.5" (29 cm)
IDENTIFICATION Sexes similar. **ADULT:** Green, with forehead, throat, and breast grayish. Bill pinkish orange, legs and feet gray. Remiges and lesser primary coverts blue, remaining coverts green. **IMMATURE:** Forehead tinged with green.
SIMILAR SPECIES Unmistakable.
VOICE Highly vocal, a variety of loud, harsh notes or quieter chattering.
STATUS & DISTRIBUTION Exotic in U.S. and many other countries. Native to lowlands of southern S.A. **YEAR-ROUND:** Nonmigratory. Locally common in

gray forehead

bluish remiges

gray throat and breast

CT and FL Peninsula. Smaller numbers in several other states including IL, NY, OR, and TX.
POPULATION Expanding; many N.A. populations doubling every five to six years. Eradicated in some states (e.g., CA, GA); also destroyed in huge numbers in native range.

CAROLINA PARAKEET *Conuropsis carolinensis* CAPA ■ 6

North America's only native breeding psittacid became extinct about 95 years ago (in wild by 1904). Its stronghold was the great river basins of the Midwest and Deep South, where it favored cottonwoods, cypress swamps, and old-growth bottomlands and fed on a variety of fruits and seeds. Exact reasons for the bird's demise are not well understood. Polytypic (2 ssp.). L 13.5" (34 cm)
IDENTIFICATION Sexes similar. **ADULT:** Green body with yellow head and red-

dish orange face. Bill and orbital ring pale. Yellow and/or orange patches on shoulders, "thighs," and vent. Green coverts and long green tail, yellower below. **IMMATURE:** Entirely green except for orangish patch on forehead.
SIMILAR SPECIES None in former range, early 20th-century reports from FL or GA possibly misidentified exotics.
VOICE Vocal, especially in flight. **CALL:** A loud and harsh *qui* or *qui-i-i-i.*
STATUS & DISTRIBUTION Formerly a

reddish orange face, remainder of head yellow

adult

locally common resident throughout eastern U.S.— recorded north to NY, MI, and MN, and west to central CO.
POPULATION Last one, "Incas," died on Feb. 21, 1918, at the Cincinnati Zoo.

BLUE-CROWNED PARAKEET   *Aratinga acuticaudata*   BCPA ▪ EXOTIC

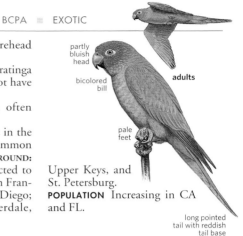

partly bluish head

bicolored bill

pale feet

adults

long pointed tail with reddish tail base

This parakeet is identified by its blue head and pinkish bill. Polytypic (5 ssp.; nominate in N.A.). L 14" (36 cm) **IDENTIFICATION** Mostly green overall, with a long tail. Sexes similar. **ADULT:** Most of head dull blue; may appear green from a distance or in bad light. Prominent white orbital ring, eyes orange (often appear dark), upper mandible pinkish orange, lower mandible dark. Flight feathers dull yellow-green. Reddish inner webs and dull yellow outer webs of tail feathers visible from below. **JUVENILE:** Blue on head restricted to forehead and forecrown. **SIMILAR SPECIES** Other exotic Aratinga parakeets found in N.A. do not have blue heads. **VOICE** A loud *cheeah-cheeah,* often repeated. **STATUS & DISTRIBUTION** Exotic in the U.S. Widely distributed and common in three regions of S.A. **YEAR-ROUND:** Nonmigratory. In U.S. restricted to CA and FL; ±100 in CA at San Francisco, Los Angeles, and San Diego; ±125 in FL, at Fort Lauderdale, Upper Keys, and St. Petersburg. **POPULATION** Increasing in CA and FL.

GREEN PARAKEET   *Aratinga holochlora*   GREP ▪ 2

This Mexican parakeet is most common in the U.S. in Texas. Breeding colonies are found throughout several large and small towns along the Rio Grande. Polytypic (3–4 ssp.; uncertain which occur in N.A.). L 13" (33 cm) **IDENTIFICATION** Green overall, slightly yellower below, with a long tail. Sexes similar. **ADULT:** Head often with scat-

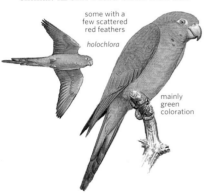

some with a few scattered red feathers

holochlora

mainly green coloration

tered orange feathers. Wide white or beige orbital ring; dull red in some Texas birds. Eyes orange, bill beige, legs and feet gray. May have scattered orange feathers on breast. Underparts paler green. Undersurface of flight feathers pale yellow; coverts yellow-green. **JUVENILE:** Similar to adult; eyes brown. **SIMILAR SPECIES** Juveniles of other *Aratinga* such as White-eyed or Crimson-fronted Parakeet (*A. finschi,* not illustrated) may be encountered in Florida. These have entirely green bodies but reportedly have at least scattered red, orange, or yellow feathers on the underwing coverts. Otherwise, the only entirely green parakeet in the United States. **VOICE** Various chattering calls. In flight, harsh screeches. **STATUS & DISTRIBUTION** Exotic in the U.S. Native to Mexico and C.A., where fairly common. **YEAR-ROUND:** Nonmigratory, but some movement of native birds in response to food supply. In U.S. restricted to FL and TX; ±2,000 in TX, along the Lower Rio Grande Valley. Some also in FL at Fort Lauderdale and Miami, where juveniles of other *Aratinga* species complicate identification. **POPULATION** Increasing in Texas and probably also in Florida.

MITRED PARAKEET   *Aratinga mitrata*   MIPA ▪ EXOTIC

Mitred and Red-masked Parakeets, along with other species not included here, represent a great identification challenge in California and Florida, where several species occur. Mitred Parakeets in Florida roost—and apparently breed—in cavities and chimneys of buildings. Polytypic (2 ssp.; nominate in U.S.). L 15" (38 cm) **IDENTIFICATION** Large, with variable red facial markings. Sexes similar. **ADULT:** Body green. Head has variable amount of red spotting; forehead dusky. Wide, white orbital ring; orange eyes; pale bill; pink legs and feet. Red spotting often on shoulders. Flight feathers yellowish below, green

limited red on head and leading edge of wing

scattered red head feathers

adults
mitrata

long, pointed tail

above—like other *Aratinga.* Underparts often with random red breast feathers. **JUVENILE:** Less red on cheeks and head; eyes brown. **SIMILAR SPECIES** Smaller Red-masked Parakeet usually has more extensive, solid red hood, red shoulders, and gray legs and feet. Juvenile Mitred resembles other *Aratinga* not included here. **VOICE** Strident *scree-ah* and other shrieking calls. **STATUS & DISTRIBUTION** Exotic in the U.S. Native to Andes of Peru, Bolivia, and Argentina, where common. **YEAR-ROUND:** Nonmigratory. In U.S. restricted to CA and FL. **POPULATION** Populations stable or increasing in California and Florida. Some local declines in native range; prior to 1990, exported in great numbers.

RED-MASKED PARAKEET   *Aratinga erythrogenys*   RMPA ■ EXOTIC

Like most other exotic parakeets, but unlike the similar Mitred Parakeet, this species nests singly in natural cavities in oaks and palms. It flocks and roosts with other *Aratinga* species. Monotypic. L 13" (33 cm)
IDENTIFICATION A medium-size parakeet with a red hood and shoulders and a green body. Sexes similar. **ADULT:** Extent of bright red hood variable, may end near mid-bill or extend to throat. Wide white orbital ring; orange eyes; pale bill; gray legs and feet. Wings show red shoulders. Underwings are yellowish with red on lesser coverts. Inner "thighs" red, but inconspicuous. **JUVENILE:** Red on head and underwing coverts much reduced; "thighs" green.

SIMILAR SPECIES Mitred Parakeet larger (conspicuous in mixed flock) has much less red on head (forehead dusky) and crown, and little or no red on shoulders. Mitred has pink legs and feet rather than gray. Confusion possible with immatures of other *Aratinga*.
VOICE Strident *scree-ah* or *skreet* calls, higher pitched than Mitred Parakeet.
STATUS & DISTRIBUTION Exotic in the U.S. Native to western Ecuador and northwestern Peru **YEAR-ROUND:** Nonmigratory. In U.S. restricted to CA, ±300, and FL, perhaps ±200.
POPULATION Populations stable or increasing in California and Florida. Declining and near threatened in limited native range from extensive capture for the pet trade.

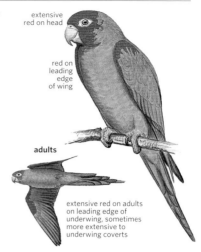

extensive red on head

red on leading edge of wing

adults

extensive red on adults on leading edge of underwing, sometimes more extensive to underwing coverts

WHITE-EYED PARAKEET   *Aratinga leucophthalma*   WEPA ■ EXOTIC

red and yellow underwing primary coverts

scattered red feathers on head and neck

smaller than Mitred

red at bend of wing

adult

South American researchers have recently moved this species out of *Aratinga* and placed it in the genus *Psittacara*. Polytypic (2–4 ssp.). L 12–13" (30–34 cm)
IDENTIFICATION A medium-size parakeet, often seen in flocks. **ADULT:** Plumage mostly green with a white orbital ring and variable red flecking on head. Iris orange. **JUVENILE:** Tail shorter than adult, red flecking in head reduced or absent, underwing pattern muted. Iris brown. **FLIGHT:** Lesser and median primary coverts are red, and greater coverts are yellow in underwing.
GEOGRAPHIC VARIATION Most likely the nominate subspecies is found in

N.A., but this has not been determined.
SIMILAR SPECIES Best told from other *Aratinga* parakeets, particularly Green, Mitred, and Red-masked, by distinct underwing pattern.
VOICE A high, screechy chatter, similar to other *Aratinga* parakeets.
STATUS & DISTRIBUTION Widespread and common over much of northern and central South America, and more recently adapted to urban environments, where it nests in cavities in buildings and under eaves. In FL, where the species has been introduced, it is found in 12 southern counties and has become fairly common in the greater Miami area.

NANDAY PARAKEET   *Nandayus nenday*   NAPA ■ 2

This large parakeet is one of the most successful psittacids in North America. It has bred in sycamore woodlands in California, but prefers palm snags and telephone poles in urban areas in Florida. Formerly known as Black-hooded Parakeet; recently placed in the genus *Aratinga* by South American researchers. Monotypic. L 13.8" (35 cm)
IDENTIFICATION Sexes similar. **ADULT:** Prominent blackish hood with an inconspicuous red or brown border on the hind crown; black bill and dark eyes. Light green body with powder blue breast patch. Bright red "thighs" distinctive. Lower half of upper tail bluish, undersurface of tail blackish.

Orbital ring gray and inconspicuous. **IMMATURE:** Blue breast patch smaller; red "thighs" paler. **FLIGHT:** Primaries and secondaries dark blue above and blackish below, contrast with yellow-green wing linings.
SIMILAR SPECIES None in N.A.
VOICE A harsh *kee-ah*, often doubled, or *chree, chree, chree;* Vocal, especially in flight.
STATUS & DISTRIBUTION Exotic in the U.S. Native to interior of central S.A., where common. **YEAR-ROUND:** Nonmigratory. In U.S. restricted to CA and FL.; ±200 birds in southern CA, mostly coastal Los Angeles, Huntington Beach, and San Gabriel Valley. Established in FL, ±1,000

black head

black remiges

black bill

bluish chest, reduced on immature

red thighs

long, pointed bluish tail

individuals, primarily in the central Gulf Coast.
POPULATION Increasing in Florida and California.

THICK-BILLED PARROT *Rhynchopsitta pachyrhyncha* TBPA ■ 6

This majestic psittacid wandered infrequently from its Mexican haunts to the southwestern U.S. until about 75 years ago. An attempt to introduce a flock into the Chiricauhua Mts. of southeastern Arizona in the 1980s was unsuccessful. Monotypic. L 16.2" (41 cm)
IDENTIFICATION A large green psittacid with a long, pointed tail and large, black bill. Sexes similar. **ADULT:** Red forehead extending in a broad line over the eyes; red shoulders. Underparts wholly green with red "thighs." Orange eyes and dull yellow orbital ring. **IMMATURE:** Bill dusky; red on head limited to forehead; no red on wings. **FLIGHT:** Pattern unmistakable from below. A prominent yellow stripe (greater coverts) is set off by dark green

wing linings and blackish flight feathers. Underside of tail also blackish; leading edge of the wing is red.
SIMILAR SPECIES No other psittacid within former North American range. Distinguished from *Amazona* by dark bill and long, pointed tail. Paler-billed immature similar to smaller, white-billed Mitred or Red-masked Parakeet but orbital ring gray and inconspicuous.
VOICE Screeches, screams, and squawks.
STATUS & DISTRIBUTION Uncommon to rare in old-growth pine forests in the Sierra Madre Occidental of northwestern Mexico. Sighting from NM (2003) was not accepted as a wild bird. **YEAR-ROUND:** Northern populations are migratory. Wanders in response to pine crop success.

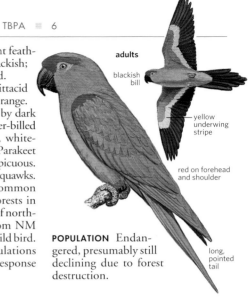

adults

blackish bill

yellow underwing stripe

red on forehead and shoulder

long, pointed tail

POPULATION Endangered, presumably still declining due to forest destruction.

WHITE-WINGED PARAKEET *Brotogeris versicolurus* WWPA ■ 2

In size, *Brotogeris* parakeets fall between Budgerigars and *Aratinga* parakeets. They are mostly green with moderately long, pointed tails. White-winged and Yellow-chevroned parakeets formerly

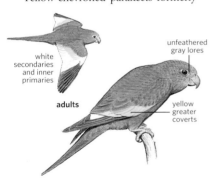

white secondaries and inner primaries

adults

unfeathered gray lores

yellow greater coverts

were considered conspecific under the name "Canary-winged Parakeet," a name still sometimes used for the White-winged Parakeet. Monotypic. L 8.75" (22 cm)
IDENTIFICATION Sexes and ages similar. Perched birds often show little or no white in wings—only yellow of the greater coverts. An inconspicuous gray orbital ring merges with unfeathered gray lores. **FLIGHT:** Large white areas—formed by the white secondaries and inner primaries—conspicuous in flight. Outer primaries bluish.
SIMILAR SPECIES Yellow-chevroned Parakeet very similar. Best field mark on a perched White-winged Parakeet is the unfeathered gray lores and orbital ring.

In flight, white secondaries diagnostic, but note that hybrids with Yellow-chevroned Parakeet apparently occur at San Francisco and Fort Lauderdale.
VOICE Perched birds give various chattering calls. Flight call of *chree* or *chree-chree* richer and slightly lower-pitched than that of the Yellow-chevroned.
STATUS & DISTRIBUTION Exotic in the U.S. Native to the Amazon Basin of northern S.A., where common. **YEAR-ROUND:** Nonmigratory. In U.S. restricted to CA, ±50 at San Francisco and Los Angeles, and to FL, ±200 mainly in Fort Lauderdale, smaller numbers at Miami.
POPULATION U.S. populations have declined considerably since the 1970s.

YELLOW-CHEVRONED PARAKEET *Brotogeris chiriri* YCPA ■ EXOTIC

The Yellow-chevroned was formerly considered conspecific with the White-winged and known as the "Canary-winged Parakeet" (see account above). In Florida both species nest in living date palms by burrowing into the insect debris surrounding the trunks. Polytypic (2 ssp.; apparently nominate only in N.A.). L 8.75" (22 cm)
IDENTIFICATION Sexes and ages similar. Perched birds show a yellow wing bar—similar to White-winged. Body and tail are entirely light green. Green lores are fully feathered, and there is a narrow white orbital ring. Dark eyes; pinkish yellow bill; pink legs and feet.
FLIGHT: Yellow wing bar prominent

in flight. Primaries and secondaries are green.
SIMILAR SPECIES White-winged Parakeet is very similar. Fully feathered lores of Yellow-chevroned Parakeet best field mark when perched. In flight, shows no white in wings.
VOICE Calls are similar to White-winged Parakeet but higher and scratchier.
STATUS & DISTRIBUTION Exotic in the U.S. Native to central S.A., where common. **YEAR-ROUND:** Nonmigratory. In U.S. restricted to CA, ±650 primarily in Los Angeles with a few at San Francisco (where hybrids with White-winged reported) and FL, ±450

mainly at Miami with some at Fort Lauderdale (hybrids also reported).
POPULATION In U.S., population is stable or increasing.

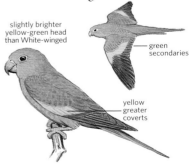

slightly brighter yellow-green head than White-winged

green secondaries

yellow greater coverts

RED-CROWNED PARROT   *Amazona viridigenalis*   RCPA ■ 2

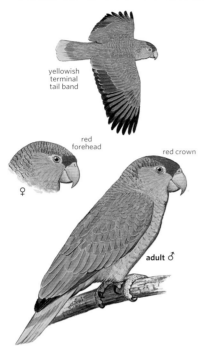

yellowish terminal tail band

red forehead

♀

red crown

adult ♂

The Red-crowned is the most wide-spread and common parrot in North America and the second most successful psittacid overall. It roosts in large, noisy flocks with other species and nests singly in cavities in palms or other trees. Monotypic. L 13" (33 cm)
IDENTIFICATION Chunky green psittacid with variable amount of red on forehead and crown. Pale orbital ring, bill, and cere; yellow eyes. **ADULT MALE:** Extensive red forehead and crown; blue hind crown and nape. **ADULT FEMALE & IMMATURE:** Red on head restricted to forehead; crown bluish. **FLIGHT:** Upper wings have a red patch on secondaries and a dark blue trailing edge. Underwings green. Yellowish band across tip of tail.
SIMILAR SPECIES Similar to Lilac-crowned, which has a burgundy forehead and purplish blue crown and nape, extending farther onto the auriculars. Longer tail of Lilac-crowned Parrot noticeable in flight.
VOICE Loud grating and cawing calls, a distinctive rolling, descending whistle.

STATUS & DISTRIBUTION Exotic in the U.S. Native to northeastern Mexico, where endangered; ±3,000–6,500 remaining. **YEAR-ROUND:** Nonmigratory, but may wander in response to food availability. In U.S. restricted to CA, FL, and TX. Largest numbers in southern CA, ±2,600; ±400 in FL. Hundreds in TX.
POPULATION Expanding in California and Texas; stable or perhaps declining in Florida due to capture for local pet trade. Native population in Mexico endangered and declining.

LILAC-CROWNED PARROT   *Amazona finschi*   LCPA ■ EXOTIC

A close relative of the Red-crowned Parrot, with which it flocks, the Lilac-crowned roosts and rarely hybridizes in California. Monotypic. L 13" (33 cm)
IDENTIFICATION **ADULT:** Sexes similar. Body green, head green with burgundy forehead and purplish blue crown and nape that curves downward behind auriculars. Gray orbital ring, red-orange eyes, dark cere, and pale bill. **IMMATURE:** Eyes brown. **FLIGHT:** Upper wings green with blackish blue primaries and secondary tips; red outer secondaries. Underwings green. Yellowish band across tip of long tail.
SIMILAR SPECIES Very similar to Red-crowned Parrot. The primary differences are the lilac crown and nape versus the red crown (adult

male) and bluish nape. Note also the dark cere and reddish eyes of Lilac-crowned Parrot. In flight, its longer tail is apparent.
VOICE Various loud grating and cawing calls indistinguishable from those of Red-crowned Parrot, but also utters, especially in flight, a distinctive, squeaky, ascending whistled *ker-leek?*
STATUS & DISTRIBUTION Exotic in the U.S. Native to northern Pacific coast of Mexico. **YEAR-ROUND:** Nonmigratory. In U.S. restricted to southern CA, where ±500 occur at Los Angeles and San Diego, nesting in cavities in trees and telephone poles. Escapes seen in southern FL.
POPULATION Increasing in California. Native populations considered near threatened from habitat loss.

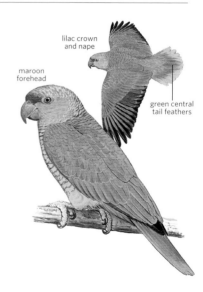

lilac crown and nape

maroon forehead

green central tail feathers

RED-LORED PARROT   *Amazona autumnalis*   RLPA ■ EXOTIC

Established for many years in several southern U.S. states but perhaps overlooked, Red-lored Parrots have adapted readily to urban and suburban areas, where exotic plantings provide abundant food sources.

Polytypic (4 ssp.; nominate presumably in N.A.). L 13" (33 cm)
IDENTIFICATION Bright green body with red frontlet and bluish crown. **ADULT:** Yellow below eye and part of auriculars. **JUVENILE:** Similar to

adult, but little or no yellow in face, less blue in crown. **FLIGHT:** Note yellow tips to tail feathers, dark blue wingtips, and brilliant red outer secondaries.
GEOGRAPHIC VARIATION Populations

in the United States are thought to be of the nominate subspecies, which lacks red on the chin and whose range extends on the Atlantic slope from Mexico to northern Nicaragua. Subspecies *salvini* ranges from Nicaragua to southwestern Colombia and northwestern Venezuela; *lilacina* is found in western Ecuador; and allopatric *diadema* is found only in northwestern Brazil along the lower Río Negro and adjacent northern bank of the Amazon. It is possible that *salvini* (incorporating *lilacina*) and *diadema* could be recognized as distinct species.

SIMILAR SPECIES Adult Red-crowned Parrot has yellowish bill, red nape and crown, no yellow below the eye, and has blue limited to post-ocular area. Juvenile Red-crowned most similar to larger Red-lored but has yellowish bill and lacks blue in crown.
VOICE Abrupt, very loud *kyeik, ack, eck, yoik* calls, similar to other *Amazona* parrots but more shrill than smaller Red-crowned.
STATUS & DISTRIBUTION Fairly common and widespread in wooded lowlands from eastern Mexico to northern South America. Red-lored

yellow on cheek

adult

Parrots have been established since the 1980s in southernmost TX, and smaller numbers have been established in southern FL and southern CA for several decades.
POPULATION Capture for the pet trade and habitat loss have caused some declines in its native range.

YELLOW-HEADED PARROT *Amazona oratrix* YHPA ■ EXOTIC

This species native to the tropics is often considered conspecific with the Yellow-naped (*A. auropalliata*) and Yellow-crowned (*A. ochrocephala*) Parrots under the combined name of "Yellow-crowned Parrot." Polytypic (3 ssp.; *oratrix* in N.A.). L 14.5" (37 cm)
IDENTIFICATION Large size. **ADULT:** Head mostly or entirely yellow, depending on subspecies. White orbital ring, eyes orange, bill pale. Red and yellow markings on shoulder; yellow on "thighs." **IMMATURE:** Head mostly green with yellow crown and face, green wings; dark eyes. **FLIGHT:** Wings green with bluish tips to remiges. Red patch on outer secondaries. Underwings green. Tail green with yellow outer rectrices.
GEOGRAPHIC VARIATION Three subspecies: *oratrix* (head yellow, but nape and breast green), presumably the

subspecies in N.A.; *belizensis* (green nape and breast); *tresmariae* (yellow head and breast and greater amount of yellow at the shoulders). Yellow-naped Parrot breeds in very small numbers in southeastern FL; Yellow-crowned is seen there occasionally.
SIMILAR SPECIES Pale-billed adults are unmistakable. Immature similar to adult Yellow-crowned Parrot (not illustrated) but has dark eyes.
VOICE A variety of shrieks, squawks, and whistles; excellent mimic.
STATUS & DISTRIBUTION Exotic in the U.S. Native from Mexico to northern Honduras. **YEAR-ROUND:** Nonmigratory. In U.S., restricted to CA and FL.
POPULATION Declining in CA and FL; apparently extirpated as a breeding species. Subspecies *oratrix* and *belizensis* endangered due to habitat destruction and capture for the pet trade.

yellow head

adults *oratrix*

pale bill

ORANGE-WINGED PARROT *Amazona amazonica* OWPA ■ EXOTIC

The Orange-winged Parrot is one of several *Amazona* species breeding in small numbers in Florida.

bluish on head and nape

orange base to outer tail feathers

Little is known of its natural history in the United States, but nests in royal palm snags have been found. It feeds on native and exotic fruits, nuts, and flowers. Monotypic. L 12.3" (31 cm)
IDENTIFICATION A small, green Amazona. Sexes similar. **ADULT:** Yellow face and crown divided by azure stripe above and through each eye, extending on to nape. Orange eyes with narrow, purplish orbital ring; pale bill with blackish edges; gray legs and feet. **IMMATURE:** Brown eyes. **FLIGHT:** Upper wings have a small orange-red patch on outer secondaries and a dark blue trailing edge. Underwings green. Tail has a

yellowish band with orange stripes on outer tail feathers.
SIMILAR SPECIES None; yellow and azure head pattern distinctive.
VOICE CALL: A shrill *kee-ik, kee-ik,* as well as various squawks, and whistled notes.
STATUS & DISTRIBUTION Exotic in the U.S. Native to lowlands of northern S.A., where common to abundant. **YEAR-ROUND:** Nonmigratory. In U.S., restricted to FL, where ±100 are found at Fort Lauderdale and Miami.
POPULATION Stable or increasing in Florida. Native birds are heavily trapped for the pet trade and shot for sport.

TYRANT FLYCATCHERS Family Tyrannidae

Gray Kingbird (FL, Apr.)

With some flamboyant exceptions, flycatchers are characterized by shades of brown, yellow, and olive. Several, like many *Empidonax* and *Myiarchus,* present some of the greatest identification challenges in North America.

Structure Tyrant flycatchers tend to perch upright and have a rather broad, flattened bill with a bit of a hook on the tip. They usually have rictal bristles at the base of the bill; the one exception in North America is the Northern Beardless-Tyrannulet. There is considerable variation in structure, and understanding it is a critical part of flycatcher identification. Overall shape, head shape, bill structure, primary projection, and wing formula are particularly important.

Behavior Tyrant flycatchers generally eat insects, but many at least occasionally eat small fruits, usually in migration or on the wintering grounds. Most sit on a perch and wait until they spot an insect, at which point they will go after it. Some species, such as most species in the genus *Contopus,* sally out to grab a flying insect and return to the same or another prominent perch. Others, like the Gray Flycatcher and the Least Flycatcher, often fly from a perch, hover, and strike prey. Vermilion Flycatchers frequently fly from a perch and pounce on or capture prey on the ground. Northern Beardless-Tyrannulets use their short warbler-like bill to glean stationary prey from foliage and branches, more in the manner of a vireo than a typical flycatcher. These differences in foraging style can be very helpful in identifying birds.

Plumage Flycatcher plumages are often studies in subtleties. Males and females are usually identical in plumage. Juveniles and immatures are similar to adults, with subtle differences in color and feather shape. Molt timing differs between some species and provides an excellent way to distinguish some difficult species in fall.

Distribution This is a large, exclusively New World family that reaches its greatest diversity and abundance in the New World tropics: 208 species have been recorded in Ecuador alone. Of some 400 species in 98 genera, 46 in 15 genera are known from North America. Thirty-four species are native breeders; the other 12 are vagrants. Most species in the U.S. and Canada are highly migratory.

Taxonomy Passerines are classified into two suborders: oscines and suboscines. The tyrant flycatchers are the only suboscines that occur north of Mexico. All have simpler syringeal morphology, which results in less impressive vocalizations than other North American passerines. Relationships among some species are poorly understood.

Conservation The most threatened North American flycatcher is the southwestern subspecies of the Willow Flycatcher, which is a federally listed endangered subspecies. *Contopus* may also have a relatively high risk of decline, in part because it has the lowest reproductive rate of all North American passerine genera. Loss of habitat, habitat fragmentation, and habitat degradation pose the biggest threats for most flycatchers.

TYRANNULETS AND ELAENIAS Genera *Camptostoma, Myiopagis,* and *Elaenia*

Birds of this large, diverse group of small- to medium-size flycatchers generally feed on insects, but many at least supplement their diet with fruit. Only the Northern Beardless-Tyrannulet is regularly encountered in N.A.; two others have occurred as vagrants. They are best found by listening for their vocalizations and best identified by differences in facial pattern.

NORTHERN BEARDLESS-TYRANNULET *Camptostoma imberbe*   NOBT ■ 2

Otherwise easily overlooked, this small flycatcher is almost always first detected by its plaintive whistled vocalizations. Its foraging behavior is very different from any other flycatcher regularly found north of Mexico. It hops through foliage, frequently flopping its tail up and down, like a vireo. Polytypic (3 ssp.; 2 in N.A.). L 4.5" (11cm)

IDENTIFICATION Grayish olive above and on breast, fading to dull white or pale yellow below. Short pale eyebrow contrasts with dark eye line. Darker crown often raised, giving a bushy-crested appearance. Darker wings and tail, with two indistinct pale buffy wing bars; very short primary projection. Short blunt-tipped bill has bright pinkish orange base and dark culmen. **JUVENILE:** Similar to adult but with cinnamon-colored wing bars and edging to secondaries.
GEOGRAPHIC VARIATION Widespread nominate subspecies from southern TX averages smaller, with smaller bill, more grayish wash to olive upperparts, and less yellow on underparts than *ridgwayi* (southeastern AZ, southwestern NM, northwestern Mexico).

Differences slight and obscured by individual variation. **SIMILAR SPECIES** Dull but relatively distinctive. Note small crest and tail-dipping behavior. Juvenile Verdins have a stout, sharply pointed bill and are hyperactive. Ruby-crowned

Kinglet has buffy wing bars, very different posture, and behavior and lacks bold eye ring. **VOICE CALL:** An innocuous, whistled *peeuuuu;* also a similar shorter call often followed by high trill: *peeut di-i-i-i-i.* **SONG:** A somewhat variable series of loud, clear, downslurred notes, typically with one or two loudly stressed notes (e.g., *dee dee dee DEE DEE dee*). **STATUS & DISTRIBUTION** Uncommon. **BREEDING:** Often found near streams in sycamore, mesquite, and cottonwood groves. Occurs from south Texas and the Southwest to Costa Rica. **MIGRATION & WINTER:** Most depart northern breeding grounds by Oct., return by Mar.; but small numbers apparently winter in N.A., particularly in south TX. **VAGRANT:** Casual to western TX and

outside mapped range in southern TX. **POPULATION** Unknown. Apparent increase in Pinal County, Arizona, may reflect observer effort.

GREENISH ELAENIA *Myiopagis viridicata*   GREL ▪ 5

This accidental, pewee-size bird usually perches vertically. Polytypic (10 ssp.) L 5.5" (14 cm)

IDENTIFICATION ADULT: Flat head; small, slender blackish bill (often flesh-colored at base below); long tail; very short primary projection. Mostly olive above, head more grayish; short white eyebrow; dark eye stripe. Wings and tail darker, noticeable olive edging, rather bright yellowish edging to secondaries. No wing bars. Olive breast; yellow belly, undertail coverts. **JUVENILE:** Similar but with brownish head and upperparts. **SIMILAR SPECIES** Face pattern recalls

Orange-crowned Warbler, but structure and habits very different. **VOICE CALL:** High and thin, somewhat burry, descending *seei-seeur* or *sleeryip.* **STATUS & DISTRIBUTION** Accidental in U.S. Fairly common to common in resident range. **YEAR-ROUND:** Forest and woodlands, often in more open areas from northern Mexico to northern Argentina. **VAGRANT:** Accidental in Galveston County, TX, May 20–23, 1984. **POPULATION** Unknown.

WHITE-CRESTED ELAENIA *Elaenia albiceps*   WCEL ▪ 5

The White-crested Elaenia is a widespread S.A. flycatcher, one of two dozen tropical and subtropical elaenias that are difficult to identify. Some species are so similar to other elaenias that they can be told in the field only by their calls. For any N.A. elaenia, first make sure it is not an *Empidonax* flycatcher. Records should be substatiated by photos and recordings of its vocalizations. Polytypic (6 ssp.). L 6" (15 cm) **IDENTIFICATION** Structure and bare parts generally similar to North American *Empidonax* species. Upper mandible blackish; lower mandible with pale base. **ADULT:** Sides of crown, face, nape, back, and rump grayish olive. Broad white crown stripe sometimes conspicuous, sometimes partially concealed. Narrow whitish eye ring and lores; two white wing bars. Throat grayish, rest of underparts whitish to grayish. **JUVENILE:** Crown stripe less conspicuous.

SIMILAR SPECIES Small-billed Elaenia *(E. parvirostris)* is most difficult to separate, but it is usually paler on underparts, and with more contrast between olive face and white throat. Small-billed also often has a third pale wing bar and a slightly more prominent eye ring. A difference in wing shape may be helpful: The migratory White-crested subspecies *chilensis*—likely the source of the one confirmed record for N.A.— usually has a longer, more pointed wing tip than Small-billed. **VOICE CALL:** In subspecies *chilensis* note is a *weeo* or *feeo,* different from short *prk* of Small-billed. **SONG:** On breeding grounds, *chilensis* has a short, raspy two-note or three-note phrase. **STATUS & DISTRIBUTION** Common breeder in forest edge, second growth, and scrub in much of South America from southwestern Colombia through the Andes to southern Chile. The

chilensis subspecies breeds in the southern two-thirds of Chile and Argentina and is a long-distance migrant northward to wintering grounds in the Andes of Peru and Bolivia. **VAGRANT:** Accidental to the U.S. The first accepted record was a calling *chilensis* on South Padre I., TX (Feb. 9–10, 2008). A bird at Chicago, IL (Apr. 2012), has been accepted only as "elaenia sp.," pending further evidence. A bird at Santa Rosa I. in northwestern FL (Apr. 1984) was originally identified as a Caribbean Elaenia *(E. martinica)* but was subsequently downgraded to an "elaenia species" and may have been a White-crested.

TUFTED FLYCATCHER   *Mitrephanes phaeocercus*   TUFL ■ 5

This distinctive flycatcher acts like a pewee, perching on conspicuous branches. After sallying, it may return to the same group of favored perches. Polytypic (4 ssp.; presumably 2 in N.A.). L 5" (13 cm)

distinct crest

brownish olive above

cinnamon face and underparts

IDENTIFICATION ADULT: Obvious tufted crest. Bright cinnamon face, underparts; upperparts generally more brownish olive. Brownish olive wings; dull cinnamon wing bars; pale edges to tertials and secondaries. **JUVENILE:** Similar but with pale cinnamon-tipped feathers on upperparts; broader, brighter wing bars. Plumage held briefly.
GEOGRAPHIC VARIATION Texas records presumably *phaeocercus* (eastern Mexico to El Salvador); Arizona record presumably *tenuirostris* (western Mexico).
SIMILAR SPECIES Nearly unmistakable. Buff-breasted is much paler; lacks tuft; has more-contrasting wing bars, blacker wings, and empid-like behavior.
VOICE CALL: A whistled burry *tchurree-tchurree*, sometimes given singly. Also a soft *peek*, similar to Hammond's.
STATUS & DISTRIBUTION Common

to fairly common in resident range. **BREEDING:** Favors somewhat open areas in pine-oak and evergreen woods from Mexico to Bolivia. **WINTER:** Some move to lower elevations; often found in a greater variety of habitats in winter. **VAGRANT:** Five winter and spring records from western TX and western and southeastern AZ.
POPULATION Unknown.

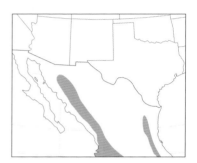

These rather plain, dark olive, medium-size flycatchers (14 sp. in the Americas; three breed north of the Mexico border; one vagrant to FL) have poorly contrasting wing bars and long wings. They sit motionless on fairly high, conspicuous perches, sallying forth to catch insects.

OLIVE-SIDED FLYCATCHER   *Contopus cooperi*   OSFL ■ 1

Olive-sided Flycatchers are usually easy to find, perching conspicuously on the top of a snag or on high dead branches or giving far-carrying vocalizations. Monotypic. L 7.5" (19 cm)
IDENTIFICATION ADULT: Large and proportionately short tailed, with a rather large blocky head. Brownish gray-olive above. Distinctive whitish tufts on side of rump, often hidden by wings but at other times is quite visible. Dull white throat, center of breast, and belly contrast markedly with dark

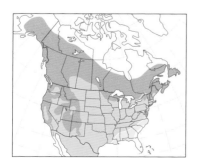

head. Brownish olive sides and flanks appear streaked, giving a heavily vested appearance. Mostly black bill; center and sometimes base of lower mandible are dull orange. All molting occurs on wintering grounds. **JUVENILE:** During fall separated from worn adults by fresh plumage, buff-brown wing bars, and brownish wash to upperparts.
GEOGRAPHIC VARIATION Populations differ subtly; breeders in southern CA average slightly larger with a large bill. See below for geographic variation of vocalizations.
SIMILAR SPECIES Wood-Pewees and Greater Pewees are similar in shape, but their underparts are less strongly patterned; they lack a sharply defined "vest" and are also relatively longer tailed. Greater Pewee also has an orange lower mandible. Eastern Phoebes are sometimes mistaken for this species, but they also lack strong vested appearance, tend to perch lower to the ground, and frequently dip tail downward.

VOICE CALL: A repeated *pep*, often repeated in groups of three or two: *pep-pep-pep pep-pep;* sometimes given singly. **SONG:** In the West far-carrying, distinctive, clear, whistled *QUICK-THREE-BEER;* second note higher pitched; in the East the accent is on the final note.

adult

dark breast sides with diffuse streaking

chunky body shape

mostly dark bill

long wings

short tail

STATUS & DISTRIBUTION Uncommon to fairly common. BREEDING: Boreal forest, subalpine forest, spruce bogs, and mixed conifer or mixed conifer-deciduous forest; prefers relatively open areas, particularly burned areas. More numerous in the West. MIGRATION: In Southwest mostly mid-Apr.–mid-May; in Great Lakes May 10–mid-June. Generally rare migrant in East. WINTER: Mostly in Andes of western S.A.; smaller numbers in southern Mexico and C.A. Casual in winter on coastal slope of southern CA (and outside Andes in S.A.). VAGRANT: Casual on Bering Sea Is. (Pribilofs and St. Lawrence).

POPULATION Olive-sided is listed as near threatened on BirdLife International's IUCN 2004 Red List and is included in the U.S. Fish and Wildlife Service's Bird of Conservation Concern list. Widespread decline detected throughout most of range, including core range in Northwest. Once common in West Virginia, they are now scarce to absent as breeders; extirpated from southern New England, and locally in the West. Threats include

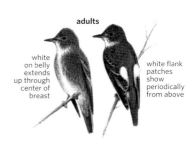

adults

white on belly extends up through center of breast

white flank patches show periodically from above

loss of habitat within breeding and wintering range, fire suppression, and possibly pesticides.

GREATER PEWEE *Contopus pertinax* GRPE ■ 2

While restricted in the U.S. to pine and pine-oak woodlands of Arizona and adjacent New Mexico, this species is not hard to find. It perches conspicuously near the top of a tree, often larger trees than those frequented by Western Wood-Pewees. Its distinctive song carries well. Birds outside of range are often identified by *pip* notes. Polytypic (3 ssp.; *pallidiventris* in N.A.). L 8" (20 cm)

IDENTIFICATION ADULT: Large; usually shows a short spiky, tufted crest. Large bill with a dark upper mandible and a distinctive bright yellowish orange lower mandible. Generally dark olive above and on face with somewhat paler lores. Darker wings and tail, with indistinct wing bars. Paler underparts, especially on throat, which contrasts with grayish breast. Worn summer birds are paler overall, more grayish olive above. Adults molt on breeding grounds. JUVENILE: Similar to adult but with cinnamon wash to upperparts and buffy-cinnamon wing bars. More difficult to separate from adults than other North American *Contopus* due to adult prebasic molt that occurs on breeding grounds.

GEOGRAPHIC VARIATION The subspecies *pallidiventris* occurs in the

slender crest

all-pale lower mandible

blended underparts

spring *pallidiventris*

fall *pallidiventris*

longer tail than Olive-sided

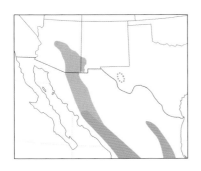

Southwest; it is similar in size to nominate subspecies (central Mexico to Guatemala), but averages slightly paler and grayer. The subspecies *minor* of Central America is smaller.

SIMILAR SPECIES Larger than wood-pewees, with more noticeable tufted crest and larger bill that is bright orange below. (Even the brightest-billed Eastern Wood-Pewees do not have such a bright orange lower mandible.) Note vocalizations. Olive-sided Flycatcher shows a distinctive "vest," is shorter tailed, and lacks a crest. Eastern Phoebe is generally paler with a darker bill, tends to perch lower to the ground, and often dips tail downward. Smaller *Empidonax* have more contrasting wing bars and usually more of an eye ring, and they typically flick their tails.

VOICE CALL: *Puip* or *beek* notes, often repeated in groups (e.g., *puip-puip-*

puip); very similar to Olive-sided Flycatcher's call but averages somewhat higher pitched and softer. SONG: A far-carrying, distinctive, plaintive *ho-SAY ma-RE-ah*. Song often begins with series of repeated introductory *whee-de* or *wee-de-ip* phrases.

STATUS & DISTRIBUTION Uncommon to fairly common. BREEDING: Pine or pine-oak forest; in U.S. often found in riparian areas in steep-sided canyons where pine-oak habitat borders canyon. Ranges south to Nicaragua and El Salvador. MIGRATION: Most return to AZ in mid-Apr. and leave by mid-Sept. Rarely noted in migration. WINTER: Most withdraw from northern portion of range and from higher elevations. VAGRANT: Casual: TX 15+ records, southern and central CA 35+ records (mostly winter); also casual in winter in southern AZ. POPULATION No data.

WESTERN WOOD-PEWEE   *Contopus sordidulus*   WEWP ▪ 1

This species is extremely similar to the Eastern Wood-Pewee and is best identified by range and voice. Vagrants should be identified with great care and preferably documented with recordings of vocalizations, photos, or video. This is the only exclusively western-breeding passerine that winters almost entirely in S.A. Polytypic (5 ssp.; 2 in N.A.). L 6.3" (16 cm)

IDENTIFICATION ADULT: Extremely similar to Eastern Wood-Pewee, with long wings that extend one-third of the way down the tail. Average differences listed below, but plumage somewhat variable. Western Wood-Pewee tends to be slightly darker, browner, and less greenish than Eastern Wood-Pewee, with complete grayish breast band, darker centers to undertail coverts, less pale nape contrast, and duller back. Wing bars, on average, are narrower

and more grayish, contrasting less with wings. Base of lower mandible is primarily dark (usually darker than Eastern Wood-Pewee's) but usually shows some dull orange. Adults molt on wintering grounds, and worn summer birds (and fall birds in N.A.) are essentially identical to Eastern Wood-Pewee in appearance. **JUVENILE:** During fall separated from worn adults by fresh plumage, buff-gray wing bars, and brownish wash to upperparts. Many have more extensive pale coloration on lower mandible than adults (i.e., more like Eastern Wood-Pewee's). On average, wing bars contrast less than on juvenile Eastern Wood-Pewee; also note that lower wing bar is broader and more defined than upper: paler tips on greater coverts are broader and more defined than those on median coverts.

GEOGRAPHIC VARIATION Differences minor and clinal. Compared to widespread *veliei*, coastal breeders from southeastern Alaska through central Oregon *(saturatus)* usually have more of a yellow wash to flanks, a browner breast, and a duskier crown that contrasts more with brownish or olive back. Mexican and Central American subspecies are darker and larger, except for southern Baja California's *peninsulae*, which are paler with a larger bill.

SIMILAR SPECIES Extremely similar to Eastern Wood-Pewee and best separated by range and voice. (Average visual differences compared above.) Most often confused with Willow Flycatcher; note Willow's (and Alder's) relatively short primary projection, smaller size, and tendency to flip tail up. Compare

paler adult / **darker adult**

averages darker plumage with darker lower mandible than Eastern, but identification safely done only by voice

variations
occasionally with all-pale lower mandible

with Greater Pewee, Olive-sided Flycatcher, and Eastern Phoebe.

VOICE CALL: A harsh, slightly descending *peeer;* a short, even, rough *brrt;* and clear descending whistles similar to Eastern's upslurred *pwee-yee,* but on one pitch or descending, not ascending. **SONG:** Has three-note *tswee-tee-teet,* usually mixed with peer notes; heard mostly on breeding grounds.

STATUS & DISTRIBUTION Common. **BREEDING:** Open woodlands. **MIGRATION:** In spring, AZ and CA mid-Apr.–mid-June. From CO to OR, first individuals generally appear in early May, with peak in mid- to late May. In fall, primarily Aug. and Sept. Most have left U.S. by early Oct. **WINTER:** Mostly northwestern S.A. No valid U.S. winter records. **VAGRANT:** Casual in East to IA, MN, WI, IL, IN, LA, MS, FL, MD, MA, ON, and QC, and north to western and northern AK and YK.

POPULATION Breeding Bird Survey trends show widespread declines throughout most of range north of Mexico, but species not classified as threatened, vulnerable, or of "special concern." Causes likely include loss and deterioration of habitat on both breeding and wintering grounds.

EASTERN WOOD-PEWEE   *Contopus virens*   EAWP ▪ 1

This species is extremely similar to the Western Wood-Pewee and is best identified by range and voice. Vagrants should be identified with great care and preferably documented with recordings of vocalizations, photos, or video. Monotypic. L 6.3" (16 cm)

IDENTIFICATION ADULT: Plumage generally dark grayish olive above with dull white throat, darker breast; whitish or pale yellow underparts. Bill has black upper mandible and

dull orange lower mandible, usually with a limited black tip. Long wings extend one-third of the way down the tail. Very similar to Western Wood-Pewee, but spring and early summer adults are usually more olive with less extensive breast band (often produce vested appearance) and a pale smooth gray nape that contrasts slightly. Wing bars are often broader and more contrasty. Adults molt on wintering grounds,

and worn summer birds (and fall birds in N.A.) are essentially identical to Western Wood-Pewee in appearance. **JUVENILE:** During fall separated from worn adults by fresh plumage, buff-gray wing bars, and brownish wash to upperparts. Many have more extensive dark coloration to lower mandible and appear more like Western Wood-Pewee. On average, wing bars stand out more than on Western Wood-Pewee, with

slightly paler nape than Western

usually adults have orange lower mandible

juveniles can have all-dark bill

fresh spring adult

worn late summer adult

long primary projection

juvenile

variations

Lakes); remain later than the Western, regularly into early Oct. **WINTER:** Mostly northern S.A. No valid U.S. winter records. **VAGRANT:** Casual in West to western TX, western OK, western KS, eastern CO, southeastern WY, western NE, western SD, western ND, eastern MT, south-central SK, NM, southern NV, southern AZ, southeastern OR, and CA (12 recs., mostly late spring singing birds). Vagrant records in the West are best confirmed by vocalizations.

POPULATION Breeding Bird Survey shows widespread declines, particularly in central N.A., but species not classified as threatened, vulnerable, or of special concern. Causes for decline unknown.

upper and lower wing bars same color and prominence (unlike Western, which usually has a less noticeable upper wing bar). **SIMILAR SPECIES** Extremely similar to Western Wood-Pewee and best separated by range and voice (see species). Most often confused with Willow and Alder Flycatchers. Note Willow's and Alder's relatively short primary projection (barely reaching beyond base of tail), smaller size, brighter wing bars, and tendency to flip its tail. Wood-pewees also forage from higher prominent perches,

to which they repeatedly return. Compare with Greater Pewee, Olive-sided Flycatcher, and Eastern Phoebe. **VOICE CALL:** A loud, dry *chip plit* and clear, whistled, rising *pawee* notes; often given together: *plit pawee.* **SONG:** A clear, slow plaintive *pee-a-wee;* second note is lower; often alternates with a downslurred *pee-yuu.* **STATUS & DISTRIBUTION** Common. **BREEDING:** Variety of woodland habitats. **MIGRATION:** Primarily circum-Gulf migrant. Most return mid-Apr. (southern TX) to mid.-May (Great

CUBAN PEWEE *Contopus caribaeus* CUPE ■ 5

This Caribbean species is accidental to south Florida. It typically forages from low perches and is relatively tame and, usually, easily approached. Also called the "Crescent-eyed Pewee," it was formerly treated as conspecific with the Hispaniola and Jamaican Pewees, as the "Greater Antillean Pewee." Polytypic (4 ssp.; likely *bahamensis* in N.A.). L 6" (15 cm)
IDENTIFICATION Appears relatively small for a *Contopus,* with upright posture and erectile crest and large

bill. Best told by conspicuous white crescent behind eye. Brownish gray above, darker on head. Noticeably paler underparts, especially belly and undertail coverts. Wing bars relatively weak.
GEOGRAPHIC VARIATION U.S. records likely *bahamensis* of the Bahamas.
SIMILAR SPECIES Wood-Pewees lack bold white crescent behind eye; Eastern and Western Wood-Pewees have longer primary projection. *Empidonax* have more uniform eye ring, prominent wing bars, and smaller bill.
VOICE CALL: Repeated *wheet* or *dee;* also a *vi-vi* similar to call of La Sagra's Flycatcher. **SONG:** A long thin descending whistle, sometimes likened to the sound of a bullet flying through the air: *wheeeooooo.*
STATUS & DISTRIBUTION Common in resident range. **BREEDING:** Pine and other woodlands, forest edge, brushy scrub edges; Bahamas and Cuba. **VAGRANT:** Four southern FL records, three photographed: Boca Raton

distinct crescent-shaped white eye ring

longer tail than Eastern

(Mar. 11–Apr. 4, 1995; another Oct. 29–Nov. 2, 1999), Everglades NP (Sept. 5–26, 2010), and a sight record from Key Largo (Feb. 16, 2001). Several other FL observations either undocumented or not accepted by Florida Ornithological Records Committee.
POPULATION Unknown.

Eleven of the 14 species of empids breed in N.A. These relatively small, drab birds are notoriously difficult to identify. Their plumages are characterized by shades of green, brown, buff, and off-white, with darker wings and tails. Most have wing bars and eye rings. Most species are long-distance migrants. Note that plumages become duller with wear.

YELLOW-BELLIED FLYCATCHER *Empidonax flaviventris* YBFL ■ 1

Even in migration, these birds favor shady forest interiors. They are usually detected by their vocalizations or by

conspicuous circular yellow eye ring

short bill

1st fall

worn fall adult

yellow throat and belly

olive breast

spring

their active foraging with tail and wing flicking. Monotypic. L 5.5" (14 cm)
IDENTIFICATION Appears short tailed, with a large head and rounded crown that appears slightly peaked at back; moderate primary projection. Broad-based bill may appear somewhat large; entirely pale yellow-orange or pink lower mandible. **ADULT:** Mostly olive above; much more yellow below with a fairly extensive olive wash across sides of breast. Olive color from sides of face blends smoothly into more yellow throat. Conspicuous bold eye ring is relatively even, often yellowish, and often slightly thicker behind eye. Wings appear very black with bright white wing bars (less so on worn birds). Worn fall migrants are grayer above and paler below, occasionally nearly grayish white below. Molt occurs on wintering grounds. **FIRST-FALL:** Similar to spring adults but with bold buffy wing bars.
SIMILAR SPECIES Compared to Acadian Flycatcher, note Yellow-bellied's smaller size, shorter primary projection, smaller bill, shorter and narrower tail, olive wash on breast, and molt timing. Very worn fall adults approach Least Flycatcher in appearance, but note Yellow-bellied's more blended throat, uniformly pale lower

mandible, and call. See Pacific-slope Flycatcher for differences with that species; see also Cordilleran Flycatcher.
VOICE CALL: Includes a sharp whistled *chiu,* similar to Acadian's. **SONG:** A hoarse *che-bunk* similar to Least's, but softer, lower, not as snappy.
STATUS & DISTRIBUTION Common. **BREEDING:** Northern bogs, swamps, and other damp coniferous woods. **MIGRATION:** Circum-Gulf migrant. Late in spring, early in fall; e.g., southern Great Lakes mostly May 15–June 5, early Aug.–late Sept. **WINTER:** Eastern Mexico to Panama. **VAGRANT:** Casual mostly in fall to west TX, NM, AZ, NV, and CA.
POPULATION Stable.

ACADIAN FLYCATCHER *Empidonax virescens* ACFL ■ 1

The Acadian has a calm, even languorous look, with often-drooped wings and a minimum of wing and tail flicking. However, its vocalizations

greenish above

spring

are delivered with force and it frequently perches relatively high in trees. Monotypic. L 5.8" (15 cm)
IDENTIFICATION A large empid with long primary projection and a broad tail. There is usually a peak at back of head. Bill is largest of any empid: long and broad based with an almost entirely yellowish lower mandible. **ADULT:** Generally bright olive above and pale below. Usually with pale grayish throat, pale olive wash across upper breast, white lower breast, and faint yellowish belly and undertail coverts. Usually narrow, yellowish, and sharply defined eye ring (but may be faint on some birds). Quite dark wings with prominent wing bars.

Acadians typically appear very worn by mid-summer, almost white below. Molt occurs on breeding grounds. **FIRST-FALL:** Similar to adults but with bold buffy wing bars and more

thin yellowish eye ring

fall

longer, stout bill

long primary projection

worn summer adult

extensive yellow coloration throughout. **SIMILAR SPECIES** Compare with similarly structured Alder and Willow Flycatchers and smaller Yellow-bellied and Least Flycatchers. **VOICE CALL:** *Peek,* similar to first part of its song; louder and sharper than other empid call notes. On breeding grounds gives *pwi-pwi-pwi-pwi-pwi,* flicker like, but softer. **SONG:** An explosive *pee-tsup.* **STATUS & DISTRIBUTION** Common. **BREEDING:** Mature forest, prefers larger tracts. **MIGRATION:** Mostly trans-Gulf migrant. In spring, arrives Gulf Coast in Apr.; southern Great Lakes, May 10–20. In fall, departs quickly after molting, July– late Sept.; rare in U.S. after early Oct. **WINTER:** Mostly Panama to Ecuador. **VAGRANT:** Casual in QC, NB, and NS. Least likely eastern empid in West. Accidental in NM, AZ, SK, and BC. **POPULATION** Stable.

ALDER FLYCATCHER *Empidonax alnorum* ALFL ■ 1

Until the early 1970s, Alder Flycatcher was considered the same species as the Willow Flycatcher. Even today, Alders are often not separable from eastern Willows without hearing their distinctive vocalizations. Many birders refer to unknown birds of this species pair by their old name, the "Traill's Flycatcher." Alders do not do as much tail and wing flicking as some empids. Monotypic. L 5.8" (15 cm) **IDENTIFICATION** A large empid with moderately long primary projection and a fairly broad tail. Moderately long and broad-based bill; almost entirely pinkish lower mandible, sometimes with a dusky area at tip. All molts occur on wintering grounds. **ADULT:** Most have strongly olive-green upperparts. Usually dark head, often with a grayish tone; contrasts with white throat. Paler underparts washed with olive across breast. Usually narrow and well-defined eye ring, but may be almost absent. Quite dark wings with prominent wing bars and tertial edges. Alders usually appear

Identification of an Empid

First, make sure you are really looking at an empid. *Empidonax* are frequently confused with woodpewees and are best separated by structure and behavior. *Empidonax* are active birds that frequently change perches and flick their wings or tail, whereas pewees sit still for long periods of time. While they may spread their wings and tail after landing or to preen, they do not engage in the quick flicks that are characteristic of empids. As can be seen on this Western Wood-Pewee, pewees also have longer wings. To learn *Empidonax,* watch birds that have been identified by song on the breeding grounds. Watch them for long periods of time under a variety of lighting conditions, and look at many different individuals. Pay attention to the exact shape of the bill. Some species (e.g., the Willow) have broad bills, while others (e.g., the Hammond's) have narrow bills. The pattern of the lower mandible is often important: the Hammond's lower mandible is mostly dark; the Least's is mostly pale. Study the wings, particularly the primary projection. Eastern species tend to have brighter tertial edges and wing bars.

Western Wood-Pewee (CA, May)

Hammond's Flycatcher (CA, Apr.)

Dusky Flycatcher (AZ, Feb.)

Willow Flycatcher (CA, May)

Note the Hammond's primary projection is longer than the Least's. Eye ring, throat color, and upperparts coloration are also important. Listen to any vocalizations, particularly call notes. Appearance (e.g., bill color) changes based on age, molt, and season, as well as individual variation. Breeding habitat is also an important clue. Most importantly, remember that some individuals appear intermediate and others are poorly seen. ■

quite worn by mid-summer and during fall migration, with paler upperparts and reduced wing bars and eye ring. **JUVENILE:** Similar to adults but with buffy wing bars. First prebasic molt occurs on wintering grounds. **SIMILAR SPECIES** Extremely similar to Willow Flycatcher; should be identified by voice. On average, Alder has a slightly shorter bill, a more distinct eye ring, and a greener back. It usually has a darker head with a more contrasting white throat than nominate eastern Willow Flycatcher. The western subspecies of Willow Flycatcher appears

less like Alder than do eastern Willows; note western Willow's less well-defined tertial edges, weaker wing bars, shorter primary projection, and less distinct crown spots. Acadian is also very similar, but it has longer primary projection, and its white throat contrasts less with paler face. Compared with wood-pewees, note Alder's relatively short primary projection (barely reaching beyond base of tail), smaller size, brighter wing bars, and tendency to flick tails.

VOICE CALL: A loud *pip* similar to Hammond's Flycatcher but more robust. **SONG:** Harsh and burry *rrree-bee-ah;* last note is almost more of an inflection and may be inaudible at a distance. Sings occasionally on spring migration.
STATUS & DISTRIBUTION Common. **BREEDING:** Damp brushy habitats and wet woodland edges; bogs; birch and alder thickets. **MIGRATION:** Circum-Gulf migrant. Spring is late; rare in south TX before early May; southern Great Lakes peak in late May. Fall is primarily Aug. and Sept. Most

have left U.S. by early Oct. **WINTER:** S.A. **VAGRANT:** Casual to CO, WA, and CA. **POPULATION** Unknown.

1st fall

slight greenish cast on back

short primary projection

thin eye ring

worn fall adult

very dark wings with sharply contrasting tertial edges

strong contrast between face and throat

spring

WILLOW FLYCATCHER *Empidonax traillii* WIFL 1

Willow subspecies in eastern U.S. and Canada (nominate *traillii* and *campestris,* merged by some authorities) are more similar to an Alder Flycatcher than to a "Southwestern" Willow. Separating Willow and Alder usually requires hearing vocalizations. Polytypic (5 ssp.; all in N.A.). L 5.8" (15 cm)

IDENTIFICATION Almost identical in structure to Alder. Bill averages slightly longer in western subspecies; primary projection in western subspecies is shorter than Alder's, but in eastern subspecies equal to Alder's. All molts occur on wintering grounds. **ADULT:** Eastern subspecies' features overlap almost completely with Alder, but upperparts average less strongly olive; head usually paler and contrasts slightly less with white throat; eye ring less prominent. **JUVENILE:** Similar, but buffy wing bars.
GEOGRAPHIC VARIATION Subspecies fairly well

defined but intergrade where ranges meet. Compared to eastern, all western subspecies have shorter primary projection; more blended tertial edges; and duller, buffy, wing bars. (They also differ from Alder in those characteristics.) Northwestern *brewsteri* has a darker brownish head (more like Alder Flycatcher) and upperparts and a brownish breast band. Rocky Mtns. and Great Basin *adastus* is similar to *brewsteri* (farther west and north), but its upperparts and head average paler brownish olive

1st fall *brewsteri*

darker above than *traillii*

extimus similar to but slightly paler than *brewsteri*

spring *extimus*

spring *brewsteri*

long broad-based bill with pale lower mandible

darker face

wing bars and tertial edges less contrasty than *traillii*

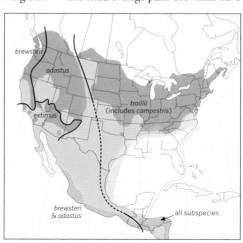

brewsteri

adastus

extimus

traillii (includes *campestris*)

brewsteri & *adastus*

all subspecies

to grayish olive; it also has a more extensive yellow wash to flanks and vent. Southwestern *extimus* (generally in lower elevations near range of overlap with *adastus*) has more grayish brown, contrasting with olive-tinged grayish brown back and wing bars (averages paler than other western subspecies). **SIMILAR SPECIES** See Alder Flycatcher. Also often confused with wood-pewees; note Willow's short primary projection, smaller size, more contrasty plumage, and tendency to flick tail upward. **VOICE CALL:** Thick, rich *whit,* like Least's or Dusky's but more resonant. **SONG:** Harsh, burry *fitz-bew* with variations. Sometimes only a husky, rough

rrrr-up. Often sings on spring migration, sometimes early in fall migration. Songs of the "Southwestern" lower pitched, more drawn out (esp. second phrase). **STATUS & DISTRIBUTION** Fairly common; "Southwestern" uncommon and local. **BREEDING:** Damp brushy habitats, edges of pastures and mountain meadows. **MIGRATION:** Nominate circum-Gulf migrant. Late in spring; rare in south TX before early May; arrives in southern Great Lakes mid-May. Fall primarily Aug., Sept. Most leave east by early Sept., west by early Oct. Casual migrant in Southeast. **WINTER:** Western Mexico to Venezuela; nominate believed to winter farther south (mostly in South America) than western subspecies. **POPULATION** "Southwestern" *extimus* endangered. Species faces widespread decline in North America, largely due to loss and degradation of habitat.

LEAST FLYCATCHER *Empidonax minimus*   LEFL   1

This active and vocal species frequently flicks its wings and flips tail upward. Monotypic. L 5.3" (13 cm) **IDENTIFICATION** Smallest eastern *Empidonax.* Primary projection usually short. Short, broad-based bill with a mostly pale lower mandible, usually with a small dusky area at tip. Appears large headed with short tail. **ADULT:** Bold white eye ring. Grayish brown above with olive wash to back. Underparts pale with gray wash across breast. Wings usually quite dark with contrasting

wing bars. Adult's prebasic molt may begin on breeding grounds but occurs mostly on wintering grounds. A partial prealternate molt begins on wintering grounds but regularly continues into spring migration. **FIRST-FALL:** Similar but with buffy wing bars. First prebasic molt occurs primarily on breeding grounds. **SIMILAR SPECIES** May be confused with almost any other empid. In East, only Willow Flycatcher has a similar *whit* note. Compared to Willow, Alder, and Acadian, Least is smaller, a smaller bill, and shorter primary projection; it also usually has a conspicuous eye ring. It is similar to Dusky, Gray, Hammond's, and Willow Flycatchers in the West. Hammond's often appears similar because both have a relatively large head and a narrow tail; note Hammond's smaller, narrower bill; longer primary projection; and darker underparts, especially on flanks. Dusky and Gray Flycatchers usually appear longer tailed with duller wings; they usually have more extensive dark on underside of mandible. Gray is most easily separated by its tendency to dip its tail downward. **VOICE CALL:** A sharp *whit,* very similar to, but usually sharper and louder than, Dusky's; not as thick as Willow's. Given frequently even in migration. **SONG:** A snappy *che-beck, che-beck,* usually given in a rapid series. Similar to Yellow-bellied's *che-bunk*

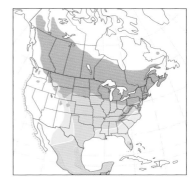

but snappier, higher, more accentuated, and usually quickly repeated. Sings regularly on spring migration. **STATUS & DISTRIBUTION** Common, more local in West. **BREEDING:** Mature deciduous woods, orchards, parks, brushy understory important. **MIGRATION:** Circum-Gulf migrant. Spring to southern TX, mostly mid-Apr.–early May; southern Great Lakes late Apr.–early June, peaking in mid-May. In fall, adults leave very early (e.g., peak at Long Point, ON, mid- to late July). First-fall birds generally early Aug.–mid-Oct., depending on latitude. Rare migrant through most of West; a very few winter in CA. **WINTER:** Mexico to Costa Rica. Rare but regular south of Lake Okeechobee and northwest of Orlando in FL. **VAGRANT:** Casual to AK. Accidental to Iceland. **POPULATION** Stable.

HAMMOND'S FLYCATCHER   *Empidonax hammondii*   HAFL ■ 1

The Hammond's Flycatcher is usually quite active, flicking its wings and tail upward at the same time. The *pip* call note is distinctive, but many Hammond's are relatively quiet in migration, especially in fall. Monotypic. L 5.5" (14 cm)

IDENTIFICATION This relatively small *Empidonax* appears large-headed and short-tailed, and it has long primary projection. Short, very narrow bill, with a mostly dark lower mandible (may be extensively pale on young birds). **ADULT:** Bold white eye ring, usually expands behind eye. Grayish head and throat; grayish olive back; gray or olive wash on breast and sides; yellow-tinged belly. Adult prebasic molt occurs on breeding grounds. Fall birds are much brighter olive above and on sides of breast, much more yellow below. Pre-alternate molt is variable, so some spring birds appear much brighter than others. **FIRST-FALL:**

Similar to adults but with buffy wing bars. First prebasic molt occurs primarily on breeding grounds.

SIMILAR SPECIES Overall, Hammond's is quite distinctive, with long wings, a tiny dark bill and, in fall, very bright plumage. It is most often confused with Dusky and most easily separated by its *pip* call note if vocalizing. Experienced observers will note Dusky's relatively short primary projection and its bill, which is wider, longer, and usually with more pale coloration to base of lower mandible. Gray can be eliminated by these same characteristics and by its tendency to dip its tail, like a phoebe. Least has a larger, more triangular bill that is usually extensively pale; much paler underparts; shorter primary projection; more contrasting dark wings with prominent wing bars and tertial edges; and frequently given *whit* call note. Bright fall birds resemble "Western Flycatchers," but note gray throat and narrow dark-tipped bill.

VOICE CALL: A *pip*, similar to Pygmy Nuthatch's and very different from calls of Dusky, Gray, and Least. **SONG:** First element of song suggests the *che-beck* of Least and particularly *che-bunk* of Yellow-bellied; may be given alone, particularly late in breeding season. Similar to Dusky but hoarser and lower pitched, particularly on second phrase. Phrases tend to be more two-parted. Also lacks the high, clear notes that are typical in Dusky's song.

STATUS & DISTRIBUTION Fairly common. **BREEDING:** Coniferous forest

short, thin-based bill

conspicuous eye ring often almond-shaped

gray throat

olive breast and flanks

overall shape similar to Least

longer primary projection than Dusky

underparts, including flanks, duskier than Least

fall

spring

medium-length, slightly notched tail is edged with gray

usually without shrubby component; Duskies tend to be found in more open habitats at elevations both above and below Hammond's. **MIGRATION:** Spring is mostly early Apr.–early June. Fall is late Aug.–mid-Oct. Throughout most of the West, Hammond's is the most likely empid to be seen in Oct.; regular in CA into late Oct. **WINTER:** Southeastern AZ to western Nicaragua. **VAGRANT:** Mostly in fall. Casual to Great Plains (more frequent than Gray and slightly more frequent than Dusky). Accidental in late fall and winter to eastern N.A.

POPULATION No significant continental change.

GRAY FLYCATCHER   *Empidonax wrightii*   GRFL ■ 1

This empid has the distinctive habit of slowly dipping its tail downward, like a phoebe. Monotypic. L 6" (15 cm)

IDENTIFICATION Gray Flycatcher is a medium-size empid with short primary projection and a fairly long tail. Head often appears proportionately small and rounded. Straight-sided bill is narrow and averages slightly longer than Dusky Flycatcher's. Most Gray's bills have a well-defined dark tip to lower mandible that contrasts with pinkish orange base, but some have an entirely pinkish orange lower mandible. Molting generally occurs on wintering grounds. **ADULT:** Gray above, sometimes with a slight olive tinge in fresh fall plumage (in U.S., mostly seen in southeastern AZ, where Gray

Flycatchers regularly winter). White eye ring is inconspicuous on pale head. Adult prebasic molt occurs on wintering grounds. **JUVENILE:** Similar to adult but more brownish gray overall with buffy wing bars and somewhat fresher plumage in summer and early fall. First prebasic molt usually occurs on wintering grounds but may begin on breeding grounds.

SIMILAR SPECIES This species is very similar to Dusky Flycatcher in terms of plumage and structure. Gray Flycatcher does have a slightly longer, more two-toned bill, but is easily separated from Dusky Flycatcher and all other *Empidonax* by its distinctive phoebe-like tail dipping. Make sure that tail is really dipping downward;

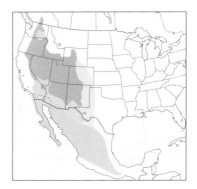

if dipping happens too fast to tell or if it is accompanied by wing flicking, observed bird is not a Gray Flycatcher, because Grays dip their tail rather slowly. Compare with Eastern Phoebe.

VOICE CALL: A loud *wit,* very similar to Dusky Flycatcher's; averaging somewhat stronger. Also very similar to calls of Least and Willow Flycatchers. **SONG:** A vigorous *chi-wip* or *chi-bit,* followed by a liquid *whilp,* trailing off in a gurgle. **STATUS & DISTRIBUTION** Fairly common. **BREEDING:** Dry habitat of the Great Basin, usually in pine or pinyon-juniper. **MIGRATION:** An early spring migrant; the only western empid likely to be seen away from the Southwest or Pacific coast in Apr.; usually arrives on breeding grounds mid-Apr.–mid-May. Fall is primarily mid-Aug.–early Oct. Rare but regular spring and fall migrant on CA coast, where a very few winter. **WINTER:** Southern AZ to central Mexico. **VAGRANT:** Mostly in fall. Casual to Great Plains. Accidental to OH, MA, DE, ON, SK, and AB.

POPULATION The Breeding Bird Survey indicates that this species is experiencing an overall increase.

head is small and rounded

eye ring inconspicuous on pale gray face

bill longer than Hammond's and has extensive pale base with dark tip

short primary projection

winter

spring

long tail, often bobbed down like a phoebe, unlike all other *Empidonax*

DUSKY FLYCATCHER *Empidonax oberholseri* DUFL ▪ 1

This is a relatively sedate species; it occasionally flicks its tail or wings upward. Dusky Flycatchers are generally intermediate between Hammond's and Gray Flycatchers, and many resemble the Least Flycatcher. Monotypic. L 5.8" (15 cm)

IDENTIFICATION A medium-size empid with short primary projection and a long tail. Straight-sided bill is intermediate in length between that of Hammond's and Gray Flycatchers; it is extensively pale at base, fading to a dark tip. Most molting occurs on wintering grounds. **ADULT:** Generally drab. Grayish olive upperparts, with little contrast between head and mantle. More yellowish below, with a pale gray or whitish throat and pale olive wash on upper breast. Usually well-defined white eye ring may not stand out against head. Lores are often paler than on other empids. Adult prebasic molt occurs on wintering grounds. During fall migration, these birds should not be confused with much brighter Hammond's Flycatchers. Fresh late-fall birds appear quite bright with flanks and undertail coverts strongly washed yellow, but in U.S. these birds are usually only seen in southeastern Arizona, where they winter.

SIMILAR SPECIES Structure and plumage similar to Gray Flycatcher; but Grays have distinctive habit of dipping tail downward, like a phoebe. Dusky Flycatcher is also often confused with Hammond's Flycatcher. In addition to noting different molt timing, observe Hammond's longer wings; shorter, thinner, darker bill; and eye ring that expands behind eye. Also note vocalizations. Many Dusky Flycatchers appear similar to Least Flycatchers, particularly worn fall birds. Dusky has a slightly longer and narrower bill; tail usually appears longer; wings are usually not as black and have low contrast wing bars and tertial edges; and eye ring is usually narrower and less pronounced. Least is also paler below.

1st fall

fall birds are dull; compare to Gray

worn fall adult

winter

long tail

short primary projection

mainly yellow below in fresh plumage; throat paler

moderately long but thin-based bill

gives a soft *whit,* very unlike Hammond's

spring

VOICE CALL: A *whit* usually softer than Gray's. Frequently gives a mournful *dew-hic* on breeding grounds, particularly early and late in day. **SONG:** Several phrases consisting of a quick, clear high *sillit;* a rough, upslurred *ggrrreep;* another high *sillit* that may be omitted; and a clear, high *pweet,* which is reminiscent of male Pacific-slope Flycatcher's contact note. This song is often confused with that of Hammond's Flycatcher, but Hammond's song never has high, clear notes characteristic of Dusky's song.

STATUS & DISTRIBUTION Common. **BREEDING:** Open woodlands and brushy mountainsides. **MIGRATION:** Spring is mostly mid-Apr.–early June; most arrive on breeding grounds mid-May. Fall is mostly Aug.–late Sept. **WINTER:** Mostly Mexico (southeastern AZ south to Isthmus of Tehuantepec, Mexico). **VAGRANT:** Casual to Great Plains and AK. Accidental to DE, AL, WI, and NS.

POPULATION The Breeding Bird Survey shows a significant increase in North America.

PACIFIC-SLOPE FLYCATCHER *Empidonax difficilis*   PSFL ▪ 1

This was formerly considered the same species as the Cordilleran Flycatcher, known as the "Western Flycatcher." The two species are extremely similar and all but impossible to identify away from the breeding grounds. "Western Flycatchers" are usually found in shaded areas, even in migration. They tend to be quite active, often flicking the wings and tail. Polytypic (3 ssp.; all in N.A.). L 5.5" (14 cm)

IDENTIFICATION Medium-size empid; fairly short primary projection; tail may appear relatively long. Wide bill; entirely yellow-orange lower mandible. Most molting on wintering grounds. **ADULT:** Brownish green above, yellow below, brownish wash across breast. Broad pale eye ring often broken or very narrow above; expands behind eye. Adults usually appear quite worn by mid-summer and during fall migration. **JUVENILE:** Similar but with buffy wing bars, variably pale underparts. Some quite whitish; may be confused with Least. First prebasic molt on wintering grounds.

GEOGRAPHIC VARIATION Widespread nominate (south to southern CA) has relatively small bill; relatively pale dull olive and yellow plumage with pale lemon wing bars (ochre-buff in juvenile and first-year). Subspecies *insulicola* (endemic to Channel Is.) has been suggested as a separate species: relatively dull; relatively long bill; whitish wing bars (buffy in juvenile and first-year); lower-pitched song; male's position note a rising *tsweep,* unlike more slurred mainland vocalizations. Longer-billed, smaller *cineritius* (vagrant to AZ, breeds in southern Baja California) is dingy whitish in color with lemon tone.

SIMILAR SPECIES Virtually identical to Cordilleran Flycatcher (see species). Pacific-slope's extensive yellow throat and bright orange-yellow or pink lower mandible are different from those of other western *Empidonax.* Yellow-bellied is quite similar, but usual range of overlap (AB) is narrow. Yellow-bellied usually has blacker wings; bolder wing bars and tertial; shorter tail; longer and more evenly spaced primary tips ("Western" shows much less even spacing); more uniform, round eye ring, rarely broken above; and more greenish upperparts. Its head tends to be rounder and not as peaked; calls are most reliable characteristic.

VOICE Very similar to Cordilleran; some vocalizations indistinguishable in the field. **CALL:** Male's position note in mainland populations is a slurred *tseeweep;* male's position note on the Channel Islands is rising *tsweep.* Neither as two-parted as the typical male Cordilleran's position note. **SONG:** A sharp *tsip;* a thin, high slurred *klseeweee;* a loud *PTIK!* Usually contains three separate phrases (usually repeated), all of which are higher pitched and thinner than for other *Empidonax* (except Cordilleran).

STATUS & DISTRIBUTION Common. **BREEDING:** Moist woodlands, coniferous forests, shady canyons. Often builds nests on human-made structures. **MIGRATION:** Thought to move more frequently in lowlands during migration than the Cordilleran. Spring

is early Mar.–mid.-June. Fall is primarily Aug. and Sept., some into Oct. **WINTER:** West Mexico, usually at lower elevations than Cordilleran and closer to the coast. A very few winter in coastal southern CA. **VAGRANT:** Out-of-range birds very similar to Pacific-slope, so identification is extremely difficult and true status as a vagrant is unresolved. Except under exceptional circumstances, vagrants are best considered "Western Flycatchers." Accidental to PA, LA. Some eastern records (e.g., one from AL) identified only as "Western Flycatchers."

POPULATION Little information.

rather long broad-based bill with all-orangish lower mandible

eye ring expands slightly behind eye

usually slight break to eye ring

pale 1st fall *difficilis*

1st fall *difficilis*

some immatures in fall are very dull; can be confused with Least Flycatcher

worn fall adult *difficilis*

olive sides to breast have slight brownish tint

longer tail projection than Yellow-bellied

spring *difficilis*

CORDILLERAN FLYCATCHER *Empidonax occidentalis*   COFL ▪ 1

This was formerly considered the same species as the Cordilleran Flycatcher, known as the "Western Flycatcher." The two species are extremely similar and all but impossible to identify away from the breeding grounds. Polytypic (2 ssp.; *helmayri* in N.A.). L 5.8" (15 cm)

IDENTIFICATION Essentially identical to Pacific-slope Flycatcher, Cordilleran Flycatcher is only readily identifiable by range and by male's two-note *tee-seet* contact call. Average differences in song and other call notes are described below, but these may be matched by Pacific-slope Flycatcher.

Some Cordilleran Flycatchers at least occasionally give contact notes that are indistinguishable from classic position notes of Pacific-slope Flycatcher. In most cases, male Cordilleran Flycatchers will eventually give more classic two-part calls. Recent studies indicate that over western and northwestern

spring

almost identical
to Pacific-slope
in coloration

parts of range, vocalizations are intermediate. On average, the Cordilleran Flycatcher is slightly larger, darker, more green above, and more olive and yellow below, with slightly shorter bill. As with Pacific-slope Flycatcher, most molting takes place on the wintering grounds. First prebasic and prealternate molt averages more extensive in Cordilleran Flycatcher, but this not evident in the field. The only subspecies in N.A. is *helmayri*.
SIMILAR SPECIES Essentially identical to the Pacific-slope Flycatcher. Average

differences presented above, but individual variation makes it impossible to identify any one individual by anything other than range or vocalizations. See the Pacific-slope Flycatcher for separation from other *Empidonax.*
VOICE Extremely similar to Pacific-slope Flycatcher. **CALL:** Cordilleran usually gives a two-note *tee-seet.* Male's position note is the most distinctive difference between Pacific-slope and Cordilleran Flycatchers. A minority of Cordillerans give position notes, at least occasionally, that are extremely similar to those given by Pacific-slope Flycatchers. The whistled *seet* note often seems sharper in Cordilleran than in Pacific-slope. **SONG:** Very similar to Pacific-slope Flycatcher's song, but the first note of the first phrase is higher than the second note.
STATUS & DISTRIBUTION Common. **BREEDING:** Coniferous forests and canyons of the west. Often nests on cabins and other human-made structures. **MIGRATION:** Generally considered rare in lowlands during migration, even within core breeding range. Spring migration is late Apr.–early June; fall is primarily

Aug. and Sept., some into Oct. **WINTER:** Mexico, usually in foothills above 600 meters. **VAGRANT:** Extreme similarity to Pacific-slope makes the identification of out-of-range birds extremely difficult, so true status as a vagrant is unresolved. Thought to be casual on Great Plains, mostly in fall; however many of these records could pertain to the Pacific-slope. Except under exceptional circumstances, vagrants are best considered the "Western Flycatcher." Accidental to LA.
POPULATION Stable.

BUFF-BREASTED FLYCATCHER *Empidonax fulvifrons* BBFL ▪ 2

The Buff-breasted Flycatcher, the most distinctive *Empidonax,* reaches north of the Mexico border, where it is found locally in southeastern Arizona and southwestern New Mexico. Polytypic (6 ssp.; 2 in N.A.). L 5" (13 cm)
IDENTIFICATION Smallest empids in N.A. Short, small bill; appears entirely pale orange from below; fairly long primary projection. Birds molt before leaving for wintering grounds. **ADULT:** Warm brown above with bright cinnamon buff wash to breast; throat paler, but blends into face. Whitish eye ring often somewhat almond shaped; pale wing bars. Worn summer birds paler, more grayish above, usually with at least a hint of buff

across breast. **JUVENILE:** Similar but with duller upperparts, well-defined buffy wing bars. First prebasic molt occurs mostly on breeding grounds.
GEOGRAPHIC VARIATION Subspecies *pygmaeus* (AZ and NM) has relatively dark brownish upperparts, often tinged with gray. Larger nominate (a vagrant to south TX) has pale brownish upperparts, often washed with olive.
SIMILAR SPECIES Distinctive. Even dullest birds usually have a light buff wash to breast. Vagrant Tufted is much richer and deeper cinnamon in color and has a distinctive crest.
VOICE CALL: A soft to sharp *pwic* or *pwit,* sharper and higher than casual *whit* of Dusky. **SONG:** Jerky, with two phrases that are repeated one after the other: *chiky-whew, chee-lick.*
STATUS & DISTRIBUTION Very local in Huachuca and Chiricahua Mountains of AZ. Largely casual in U.S. outside AZ, but recently recorded in Peloncillo Mountains, NM. A small outpost also found at Davis Mountains Preserve, TX, where breeding was detected annually from 2000 until about 2010. Northern limit in eastern Mexico is in mountains near Monterrey (within 100–120 mi. of TX).

BREEDING: Dry pine and oak woodlands, usually near openings with scattered shrubs. **MIGRATION:** Casual in migration. Spring migration in AZ early Apr.–mid-May. Fall departs in Aug.–mid-Sept. **WINTER:** Mexico to central Honduras. **VAGRANT:** Accidental: accepted sight record from El Paso Co., CO (May 1991).
POPULATION Formerly bred north to central Arizona and west-central New Mexico. No information on trends south of the United States.

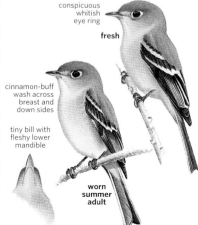

conspicuous whitish eye ring

fresh

cinnamon-buff wash across breast and down sides

tiny bill with fleshy lower mandible

worn summer adult

PHOEBES Genus *Sayornis*

These plump flycatchers have a distinctive trait of dipping their tails downward. Relatively short-distance migrants, they are found in open habitats, often near water or near human structures and sometimes nesting under overhangs on houses or barns or under bridges. The Vermilion Flycatcher is very closely related and shares many traits with the phoebes.

BLACK PHOEBE *Sayornis nigricans*   BLPH ■ 1

A distinctive black-and-white phoebe of the Southwest, this species is almost always found near water. Polytypic (6 ssp.; *semiatra* in N.A.). L 6.8" (17 cm) **IDENTIFICATION** Black overall with white belly and undertail. **JUVENILE:** Plumage similar to adult's, but browner, with two cinnamon wing bars, cin-

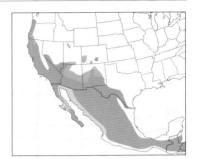

namon tips to feathers on upperparts. **GEOGRAPHIC VARIATION** N.A. *semiatra* (south to western Mexico) has duller and duskier head; birds south of Isthmus of Panama have extensive white in wings. **SIMILAR SPECIES** Distinctive. Has hybridized with Eastern Phoebe (CO); offspring appear intermediate. **VOICE CALL:** Includes a loud *tseew* and a sharp *tsip,* similar to Eastern Phoebe's but sounding more plaintive and whistled. **SONG:** Thin whistled song consists of two different two-syllable phrases: a rising *sa-wee* followed by a falling *sa-sew;* usually strung together one after the other. **STATUS & DISTRIBUTION** Uncommon to common. **BREEDING:** Woodlands, parks, suburbs; almost always near water. **MIGRATION:** Resident over much of range. Breeders return to CO late Mar.–mid-Apr.; depart

early Sept. Fall migrants detected on Farallon Is. (CA) early Sept.–late Nov. **VAGRANT:** Casually appears north and east to northern OR, WA, ID, northern and eastern UT, northern AZ, central TX, OK, and KS. Accidental to FL, southwestern BC, and south-central AK. **POPULATION** Increasing, with range slowly spreading north.

phoebes drop their tail down

mostly blackish except for sharply contrasting white belly

juvenile

EASTERN PHOEBE *Sayornis phoebe*   EAPH ■ 1

The Eastern Phoebe is a rather dull phoebe found in the East and across central Canada. It frequently nests under eaves, bridges, or other overhangs on human-made structures. The Eastern Phoebe is most easily separated from other dull flycatchers by its characteristic habit of dipping its tail in a circular motion. Monotypic. L 7" (18 cm) **IDENTIFICATION** Eastern is brownish gray above; darkest on head, wings, and tail. Its underparts are mostly white, with pale olive wash on sides and breast. Fresh fall adult Easterns

worn summer adult

dark cap and face

fresh fall

pale yellow belly in fresh plumage

are washed with yellow, especially on belly. Molt occurs on breeding grounds. **JUVENILE:** Plumage is briefly held and similar to adult's but browner, with two cinnamon wing bars and cinnamon tips to feathers on upperparts. **SIMILAR SPECIES** Pewees are darker and they have longer wings, but they are most easily separated from phoebes by phoebes' distinctive tail wagging. *Empidonax* flycatchers have eye rings and wing bars, which are absent in Eastern Phoebe. An *Empidonax* flycatcher flicks its tail upward; only Gray Flycatcher dips its tail downward. **VOICE CALL:** Typical call is a sharp *tsip,* similar to Black Phoebe's. **SONG:**

Distinctive, rough whistled song consists of two phrases: *schree-dip* followed by a falling *schree-brrr;* sometimes strung together one after the other. **STATUS & DISTRIBUTION** Common. **BREEDING:** Woodlands, farmlands, parks, and suburbs; often near water. **MIGRATION:** Breeders return to the Midwest mid-Mar.–late Apr. and depart late Sept.–early Oct. **VAGRANT:** Rare in fall and winter to CA; otherwise casual west of the Rocky Mountains and northwestern Great Plains. Accidental to southern YK and northern AK; sight record for England. **POPULATION** Apparently stable.

SAY'S PHOEBE *Sayornis saya*   SAPH ■ 1

This widespread western species is frequently seen perching on bushes, boulders, fences, and utility wires. Polytypic (4 ssp.; 3 in N.A.). L 7.5" (19 cm)
IDENTIFICATION Grayish brown above; darkest on wings and head. Grayish breast contrasts with tawny-cinnamon belly and undertail coverts. Contrasting black tail is particularly obvious in flight. **JUVENILE:** Plumage is briefly held. Similar to adult but browner, with two cinnamon wing bars.

GEOGRAPHIC VARIATION Variation is complicated by individual variation and wear. Compared to more widespread nominate, northwestern breeding *yukonensis* (AK and NT to coastal OR) is smaller billed, with deeper orange underparts and deeper gray upperparts. Resident *quiescens* (deserts of southeastern CA and southwestern AZ) is paler overall.
SIMILAR SPECIES Yellow-bellied kingbirds can appear vaguely similar (esp. Western Kingbird), but all have a heavier bill, dark mask, yellow belly, and olive upperparts; unlike phoebes, they do not dip tail. Female and immature Vermilion Flycatchers have a white throat and white chest with brown streaks; also has pale supercilium contrasting with dark auriculars.
VOICE CALL: Typical call is a thin, plaintive, whistled, slightly downslurred *pee-ee.* **SONG:** A fast whistled *pit-tsear,* often given in flight.
STATUS & DISTRIBUTION Common. **BREEDING:** A variety of open and dry habitats from tundra to desert, usually

with cliffs, canyons, rocky outcroppings, or human-made structures for nesting. **MIGRATION:** Early in spring, late in fall. Bulk of migrants arrive from CO to east of the Sierra Nevada in CA late Mar.–mid-Apr.; southern BC late Mar.–mid-Apr.; and AK early to mid-May. In fall, most depart AB late Aug.–early Sept.; CO and OR late Aug.–Sept. **VAGRANT:** Casual over most of the East (late fall–early winter) with records for every state and province except DE, DC, MD, NH, WV, NU, and PE.
POPULATION Apparently stable.

grayish back

tawny belly

blackish tail

Genus *Pyrocephalus*

VERMILION FLYCATCHER *Pyrocephalus rubinus*   VEFL ■ 1

This stunning bird frequently pumps its tail like a phoebe. Polytypic (13 ssp.; 2 in N.A.). L 6" (15 cm)
IDENTIFICATION Adult male red and brown. Adult female grayish brown above, darker tail; pale throat and supercilium; streaked pale breast; tawny belly and undertail. **JUVENILE:** Similar to adult female but spotted below; belly with yellowish tinge. Immature male acquires red feathers by early winter.
GEOGRAPHIC VARIATION Southwestern *flammeus* has paler brown upperparts with a grayish tinge. Head and breast of adult male average more orange-red in color, often with pale mottling.

Adult male *mexicanus* (TX) has deep, bright red underparts and darker brown upperparts without grayish tinge.
SIMILAR SPECIES Compare an immature with Say's and Eastern Phoebes, noting white throat and streaked breast.
VOICE CALL: A sharp *pseep.* **SONG:** A soft tinkling repeated *pit-a-set, pit-a-see, pit-a-see;* often given in fluttery fight display. Also sings while perched.

STATUS & DISTRIBUTION Common. **BREEDING:** Open woodlands, parks; near water. **MIGRATION:** Returns to northern breeding areas by late Apr.; most depart Sept.–Oct. **VAGRANT:** Casual, mostly in fall, throughout most of lower 48 (fewest along Canadian border and in northeast) and to ON, NS, and QC. Has bred in CO and OK.
POPULATION Apparently stable.

immature ♀ — dark face — streaked breast — pale supercilium — immature ♂ — adult ♀ — pale pinkish belly — adult ♂ — juvenile — spotted breast

These rather big-headed, bushy-crested, relatively slim-bodied, moderately long-tailed flycatchers (22 sp.; four breed in U.S.; two more occur as vagrants) are easy to recognize by virtue of their upright posture, large all-dark bill, grayish brown upperparts, gray chest, yellowish belly, and often rufous-edged primary and tail feathers. Dark brown wings have two pale wing bars; rufous-edged primaries give the folded wing a prominent rusty wing panel. Moving from perch to perch, they sally to foliage (in wooded habitats, primarily sub-canopy) or ground for insects or fruit. Distinctive vocalizations include many daytime calls, which can also be repeated or combined to form male's dawn song. *Myiarchus* are secondary cavity nesters.

Identification of *Myiarchus* Flycatchers

The largest and smallest of these species differ in size substantially, but size overlap among most species and overall similar structure and coloration make identification challenging. Overall, males are slightly larger, but this is not appreciable in the field; the sexes are otherwise identical. *Myiarchus* do not have seasonally variable plumage; but because they have only one complete molt per year (during late summer or fall), their plumage can become considerably duller and paler during the spring and summer from wear and bleaching.

Young birds leave the nest while growing their juvenal plumage. This first plumage is superficially similar to the adult's but typically paler and of a softer, more delicate feather structure. It may include a more extensively rufous tail (incl. rufous on the outer edges of the feathers) and rusty-edged upper-wing and uppertail coverts; secondaries have rusty or buffy edges (adults generally have white or yellow), making the rusty "primary" panel less well defined. In some species, coloration of the secondary edges is an important adult field mark, so individuals must first be correctly aged.

Juvenal plumage is quickly replaced by a more adultlike first basic plumage. Identifications should be made using combinations of characters. Especially important are tail pattern, mouth color, and vocalizations. Determining the extent and pattern of dark brown versus rufous on the tail is vital, particularly for identifying silent individuals in areas of overlap. In adults, the central tail feathers and outer webs of the remaining pairs of feathers are dark brown, so that viewed from above, the closed tail above appears all brown. Depending on species, the inner webs of the outer five pairs have varying amounts of rufous and might have a dark "shaft stripe" of varying width along the feather shaft; the rufous is best studied on a perched bird from below, where the pattern of the outermost pair is visible even when the tail is closed.

Great Crested Flycatcher (TX, Apr.)

Great Crested Flycatcher (TX, June)

Ash-throated Flycatcher (CA, May)

The color of the inside of the mouth (mouth lining) is difficult to see in the field but with patience can sometimes be viewed when a bird calls, yawns, or feeds. *Myiarchus* call frequently during the breeding season, and key vocalizations are diagnostic; migrants and wintering birds call much less frequently. Overall size, bill size and proportions, and subtle differences in plumage coloration are important but much more subjective. Species that vary greatly in size or color will be easily distinguishable from each other, but there is extensive overlap among most species. Distance and lighting are also important with such subtle differences; distribution and breeding habitat can also be helpful in elimination.

In general, vagrants are less likely in spring and summer, so identifications are usually safe based on probability during the breeding season in areas where only one species normally occurs. This approach works best with the Ash-throated and the Great Crested, which are the only species present across much of the west and east, respectively. All *Myiarchus* breeding in North America are migratory and normally winter south of the U.S. border, except for small numbers of Ash-throateds in the extreme Southwest and the Great Cresteds in southern Florida. Vagrants should be identified with extreme caution. A high percentage of vagrant *Myiarchus* are found during fall and winter, usually in coastal regions. Species normally found in an area as breeders or migrants are not necessarily more likely to occur at the wrong season (e.g., the Great Crested Flycatcher is virtually unrecorded in winter in the east outside of FL). The vast majority of vagrant *Myiarchus* are Ash-throateds, the only vagrant species that has occurred in Canada, most of the interior U.S., and along the Atlantic coast north of Florida. Thus, the Ash-throated should be the first option when trying to identify a vagrant or an out-of-season *Myiarchus*. ∎

DUSKY-CAPPED FLYCATCHER *Myiarchus tuberculifer*   DCFL ■ 2

secondaries washed with rufous unlike other North American *Myiarchus*

long bill

moderately yellow belly

olivascens

tail from below

outer tail feathers with almost no rufous

Widespread in the neotropics, our smallest *Myiarchus* barely enters the U.S. Its plaintive call signals its presence in the leafy subcanopy of lower montane forests and woodlands. Range and habitat can overlap with Ash-throated and Brown-crested Flycatchers. Polytypic (13 ssp.; 2 in N.A.). L 6.4–7.3" (16–19 cm)

IDENTIFICATION **ADULT:** Grayish brown crown and cheeks contrast somewhat with more grayish olive back. Outer and middle secondaries edged rufous to rusty-yellow; inner secondaries whitish. Upperwing coverts tipped pale grayish

brown, making wing bars less conspicuous than in other species. Gray throat and breast; yellow belly, with relatively sharp gray-yellow contrast; olive-suffused sides. Black, proportionately long, slender bill; paler base of lower mandible. Orange mouth lining. **JUVENILE:** Similar but duller overall; tail broadly edged rufous on inner webs. **GEOGRAPHIC VARIATION** Two subspecies reach the U.S.: *olivascens,* described above (southeastern AZ, extreme southwestern NM, and northwestern Mexico), and *lawrenceii* (vagrant from eastern Mexico). The latter is slightly larger and darker above and has more extensive rufous inner edges of tail feathers. **SIMILAR SPECIES** Other N.A. *Myiarchus* are larger, have different bill shapes, and more rufous in tail. Ash-throated and much larger billed Brown-crested have flesh-colored mouth lining; calls of both different. **VOICE** **CALL:** Mournful, descending whistled *peeeeuuuu,* sometimes preceded by *whit* note. **DAWN SONG:** A continuous series of mixed *whit* notes, whistles, and trills. **STATUS & DISTRIBUTION** Fairly

common. **BREEDING:** Lower- and middle-elevation montane riparian, oak, or pine-oak woodlands. **MIGRATION:** Migrants seldom detected in lowlands. In spring, exceptionally, arrives late Mar.–early Apr., more typically mid-Apr. In fall, most have departed U.S. by mid-Aug., stragglers recorded to mid-Oct. **WINTER:** Mexico. **VAGRANT:** Subspecies *olivascens* is a rare migrant and breeder in mountains of western TX; very rare but annual in late fall (Nov.) and winter to coastal CA; accidental to OR and in spring to CO; *lawrenceii* is casual in winter in extreme southern TX. **POPULATION** Stable.

ASH-THROATED FLYCATCHER *Myiarchus cinerascens*   ATFL ■ 1

This is the "default" *Myiarchus* throughout much of the West. Monotypic. L 7.6–8.6" (19–22 cm)
IDENTIFICATION **ADULT:** Relatively small, slender; moderately long tailed. Whitish gray throat and pale yellow belly. Brown crown and brownish gray back. Blackish brown wings with two whitish wing bars; rufous edged primaries; secondaries edged white to pale yellow. Outer pairs of tail feathers extensively rufous on inner webs; dark shaft stripe flares at tip so that rufous does not extend to feather tips. Some lack typical tail pattern, and

pattern can vary among feathers. All-dark bill is relatively thin, short to medium length. Mouth lining flesh-color. **JUVENILE:** Duller and paler; browner above; belly more whitish yellow; rufous edged secondaries; tail predominantly rufous with dark shaft stripes on outer webs. Some juvenile middle secondaries or tail feathers can be retained into first basic plumage. **SIMILAR SPECIES** Very similar to Brown-crested Flycatcher, which averages darker gray and brighter yellow below, is larger in all aspects, lacks typical Ash-throated tail pattern, and has different voice. Juvenile Ash-throated tail suggests Great Crested, but size, shape, and plumage should make identification easy. Nutting's best separated by voice, mouth color. **VOICE** **CALL:** Soft *prrrt* (nonbreeders less vocal). **SONG:** *Ka-brick.* **DAWN SONG:** A repeated series of *ha-wheer* and other notes. **STATUS & DISTRIBUTION** Common.

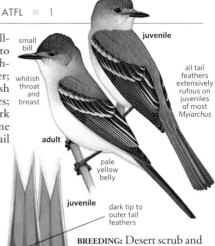

small bill

juvenile

whitish throat and breast

all tail feathers extensively rufous on juveniles of most *Myiarchus*

adult

pale yellow belly

juvenile

dark tip to outer tail feathers

adult

BREEDING: Desert scrub and riparian, oak, or coniferous woodland. **MIGRATION:** In spring, mid-Mar.–mid-May. In fall, late July–Aug., a few to mid-Sept.; stragglers Oct.–Nov. **WINTER:** Extreme southwestern U.S. to Honduras. **VAGRANT:** Rare/casual north to southwestern BC, east to southeastern Canada and to Atlantic and Gulf coasts in the U.S., mainly in fall and winter. **POPULATION** Stable.

NUTTING'S FLYCATCHER *Myiarchus nuttingi*   NUFL ▪ 5

This vagrant from Mexico closely resembles the Ash-throated Flycatcher in size, structure, and coloration. It is also superficially very similar to the larger Brown-crested Flycatcher in overall plumage coloration. Identifications are best confirmed by calls and orange-colored mouth lining. Polytypic (3 ssp.; *inquietus* in U.S.). Based on vocal differences, it has recently been proposed that *flavidor* (Pacific slope from Chiapas to nw. Costa Rica) is a separate species. L 7.2" (18 cm)
IDENTIFICATION ADULT: Relatively small; small billed. Pure gray breast

coloration very similar to slightly larger Ash-throated, but Nutting's has slightly yellower belly; best told by call

slightly more olive above than Ash-throated

inquietus

yellow tint to secondaries when in fresh plumage

dark tip of inner web of outer tail feather not as extensive as Ash-throated

tail from below

and bright yellow belly, with abrupt gray-yellow transition. Outermost secondary edge is rufous; remaining secondaries range (moving inward) from pale rufous or brownish white to white or grayish white. Tail pattern is variable; typically, inner webs of outer five pairs of feathers are extensively rufous with a dark shaft stripe of variable width (as in Ash-throated and Brown-crested), but rufous may or may not extend to the feather tip.
JUVENILE: Essentially identical to Ash-throated; some rusty-edged juvenile middle secondaries can be retained in first basic plumage.
SIMILAR SPECIES Brown-crested (esp. *magister*) and Great Crested Flycatchers are larger and have more extensively rufous tails. Brown-crested has a flesh-colored mouth lining, a dusky cap, and little or no rufous in its tail; and it is smaller and disproportionately longer billed. Ash-throated, which averages slightly larger, has a flesh-colored mouth lining; longer and more pointed wings; paler underparts with a whitish area between whitish gray breast and pale yellow belly; a subtle grayish collar across hindneck; grayer auriculars; and whitish or yellowish white secondary edges. Nutting's Flycatchers that

share typical Ash-throated pattern on outer (sixth) pair of tail feathers (e.g., feather tips dark, no rufous to tip) also have a dark shaft stripe on inner webs of second pair of feathers (lacking in Ash-throated).
VOICE CALL: A sharp, whistled *wheep* (suggesting Great-crested) or *peer*; also a *pip*.
STATUS & DISTRIBUTION Common in normal range. **YEAR-ROUND:** Thorn-scrub and open tropical deciduous forest, from northwestern and central-eastern Mexico south to northwestern Costa Rica. **VAGRANT:** Six records. Casual to AZ (four fall and winter recs.); accidental to southern CA and Big Bend, TX in winter.
POPULATION Presumably stable.

GREAT CRESTED FLYCATCHER *Myiarchus crinitus*   GCFL ▪ 1

This is the "default" *Myiarchus* of eastern N.A. is usually heard more than seen as it forages high in canopy or subcanopy. Monotypic. L 8.5" (21 cm)
IDENTIFICATION Key characters include a broad white stripe on innermost tertial, mostly rufous inner webs of all but central tail feathers, and overall darker plumage. Longer winged than other *Myiarchus*. **ADULT:** Dark gray face and breast (slightly paler throat) contrasts

sharply with bright yellow belly; these colors blend at sides of breast to form olive-green patches. Olive-brown back. Bill rather thick, is black with a pale at base of lower mandible (most). Mouth lining is bright orange-yellow. **JUVENILE:** Similar but has rusty secondary edges (but innermost 2–3 edged white) and wing bars; all tail feathers rufous. Immature brighter, like adult, but wing bars and most secondary edges still rusty.
SIMILAR SPECIES A pale or bleached Great Crested would be superficially similar to a Brown-crested or Ash-throated, but would still exhibit key field marks. See sidebar (page 444).
VOICE CALL: Most familiar is an ascending *whee-eep* but also gives *purr-it* and series of *whit* notes. **DAWN SONG:** A continuously repeated series of modified *whee-eeps*.
STATUS & DISTRIBUTION Common. **BREEDING:** Deciduous or mixed forest. **MIGRATION:** In spring, primarily western circum-Gulf, but some trans-Gulf, mid-Mar.–early June. In fall, both

always with pale base to lower mandible

olive wash to crown and back

steel gray breast

olive-tinged sides

broad white inner tertial edge, thicker at base

lemon yellow belly

rufous slightly more extensive on inner web than Brown-crested

tail from below

circum- and trans-Gulf, mid-July–mid-Oct. **WINTER:** Southeastern Mexico to Colombia and Venezuela; also southern FL. **VAGRANT:** Rare to casual/accidental, mainly in fall, to West Coast, AK, NT, NF, Bermuda, Bahamas, Cuba, and Puerto Rico.
POPULATION Stable.

BROWN-CRESTED FLYCATCHER *Myiarchus tyrannulus*   BCFL ■ 1

This is our largest *Myiarchus*. The western subspecies dwarfs the smaller Dusky-capped; the eastern subspecies is closer in size to the Great Crested and the Ash-throated. Brown-cresteds prefer more mature, undisturbed habitats. Polytypic (7 ssp.;

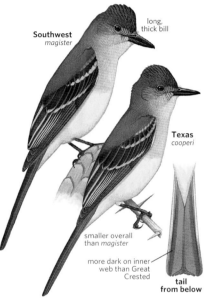

Southwest
magister

long, thick bill

Texas
cooperi

smaller overall than *magister*

more dark on inner web than Great Crested

tail from below

2 in N.A.). L 7.2–9.2" (18–23 cm) **IDENTIFICATION ADULT:** Rather bright yellow belly contrasts abruptly with gray breast. Outer pairs of tail feathers extensively rufous on inner webs; dark shaft stripes and rufous extend to feather tips. Bill proportionately long, heavy, and black. Mouth lining typically flesh-colored. **JUVENILE:** Similar, but secondaries (except whitish inner 2) and wing bars are rusty edged and inner webs of tail feathers are more extensively rufous. **GEOGRAPHIC VARIATION** Larger *magister* breeds in Southwest south through western Mexico; smaller *cooperi* breeds in southern Texas and eastern Mexico south to Honduras. **SIMILAR SPECIES** Great Crested has a darker gray face and breast; broad white edge to innermost tertial; all-rufous inner webs of tail feathers; and paler base to lower mandible. Smaller, paler Ash-throated has whitish transition between gray and yellow on underparts, and entire tail tip (from below) is dark not rufous. Smaller Nutting's and Dusky-capped have orange mouth lining; Dusky-capped has much less rufous in tail. **VOICE CALL:** A sharp *whit*. Breeding,

a rough, descending *burrrk* (or rasp) or *whay-burg*. **DAWN SONG:** Repeated *whit* notes, vibrato whistles, *burrrk* notes, and other complex phrases. **STATUS & DISTRIBUTION** Common. **BREEDING:** Riparian forest, thorn woodland, columnar cactus desert. **MIGRATION:** In spring, arrives in TX exceptionally by mid- to late Mar., more typically early to mid-Apr.; arrives in Southwest late Apr.–early May. In fall, generally departs Aug., rare after mid-Sept. **WINTER:** Mexico to Honduras. **VAGRANT:** Casual, mainly fall-winter, to coastal CA *(magister)* and to coastal TX, LA, and FL *(cooperi,* but one LA specimen rec. of *magister).* **POPULATION** Stable.

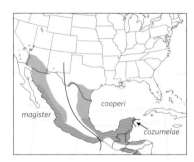

magister

cooperi

cozumelae

LA SAGRA'S FLYCATCHER *Myiarchus sagrae*   LSFL ■ 3

A rare visitor to our area from the northern West Indies, the La Sagra's was formerly considered the same species as the Stolid Flycatcher (Jamaica and Hispaniola). More secretive and sluggish than our other *Myiarchus,* it prefers to forage in dense second growth. At first glance, its coloration and more hunched posture may suggest an Eastern Phoebe. Polytypic (2 ssp.; both in N.A.). L 7.5–8.5" (19–22 cm) **IDENTIFICATION ADULT:** A relatively small *Myiarchus* with a proportionately long, thin, black bill. Pale gray on throat and breast; whitish on belly and undertail coverts, sometimes tinged yellow. Rufous edges to primaries reduced or absent; secondaries thinly edged white. Tail mostly blackish

brown; extent of rufous on inner web of outer pairs of feathers ranges from a very narrow fringe to a fairly broad stripe along inner edge. Pale yellow mouth lining. **JUVENILE:** Secondaries and tail feathers edged with rufous. **GEOGRAPHIC VARIATION** At least some (and presumably most) FL records pertain to *lucaysiensis* of the Bahamas, which is larger, with more rufous in tail. Subspecies *sagrae* (Cuba and Cayman Is.) is slightly smaller, with less rufous in tail, and has occurred once in Alabama. **SIMILAR SPECIES** Other *Myiarchus* are yellower on belly or have more rufous in tail. Size, structure, and tail pattern suggest Dusky-capped Flycatcher (e.g., relatively small overall, proportionately long bill, and reduced amount of rufous in tail). **VOICE CALL:** High-pitched *wheep, whit,* or *wink,* sometimes doubled; also loud *teer teer,* softer *quip quip quip.* **STATUS & DISTRIBUTION** Common in normal range. **YEAR-ROUND:** Pine or mixed woodland, dense scrub and

second growth, mangroves; in FL almost always on or near immediate coast. **VAGRANT:** Casual in southern FL, mainly along extreme southeastern coast and Keys. First recorded in FL in 1982; now about 25 records between late Oct. and mid-May, most during winter. Accidental Orrville, Dallas Co., AL (specimen; Sept. 14, 1963). **POPULATION** Stable.

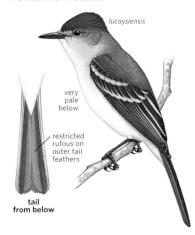

lucaysiensis

very pale below

restricted rufous on outer tail feathers

tail from below

Genera *Pitangus, Myiozetetes, Myiodynastes, Empidonomus,* and *Legatus*

GREAT KISKADEE Pitangus sulphuratus   GKIS ▪ 2

This widespread, spectacular neo-tropical species barely reaches the United States. Its large size, chunky shape, massive black bill, loud voice, and striking plumage of black, white, rufous, and yellow make it unmistakable. Noisy and conspicuous, it employs many foraging tactics (inc. aerial hawking, sallying to foliage, even plunge-diving for aquatic prey); it builds a bulky domed nest with a side entrance, usually situated in the fork of a tree or on a utility pole. Polytypic (10 ssp.; 2 in N.A.). L 9.6" (24 cm)

IDENTIFICATION ADULT: Sexes similar. Black crown surrounded by white; central orange-yellow crown patch (larger in males; not often visible in the field). Black face contrasts with white

supercilium and throat. Rest of underparts bright yellow. Brown back; wings and tail extensively edged rufous. **JUVENILE:** Similar but duller; lacks central crown patch; has rustier brown back and more extensive rufous tail and wing edgings.

GEOGRAPHIC VARIATION Subspecies *texanus* breeds in Texas; vagrants to Arizona and New Mexico reported as *derbianus,* which has a more grayish brown back. A specimen record of S.A. nominate subspecies (CA, 1926) considered unacceptable on origin.

SIMILAR SPECIES This combination of size, shape, and plumage coloration and pattern is unique in North America. Social Flycatcher has superficially similar plumage but is much smaller and less chunky, with a much smaller bill and less contrasting head pattern; it also lacks rufous on tail and wings.

VOICE CALL: Year-round, a loud *crear;* named for breeding season's *KIS-ka-dee;* also a loud, somewhat rising *reeee, chick-weer.* **DAWN SONG:** A repeated series of typical calls, uttered during twilight.

bold white supercilium · *texanus*
dark mask
bright rufous wings and tail
white throat
bright yellow belly
texanus

STATUS & DISTRIBUTION Common. **YEAR-ROUND:** Lowland thorn forest, riparian forest, woodland, desert scrub, even suburban or urban situations; usually near water. Occurs from southern TX and northwestern Mexico to central Argentina. Introduced in Bermuda. **VAGRANT:** Rare beyond normal range in TX. Casual/accidental in southeastern AZ, southern NM, OK, southwestern KS, and southern LA. Some "vagrant" records involve breeding attempts or single birds that remain in an area for months or years and construct a nest.

POPULATION Stable; slowly expanding breeding range northward in TX.

SOCIAL FLYCATCHER Myiozetetes similis   SOFL ▪ 5

small bill
always lacks prominent rufous on wings and tail
adult
primulus

This bird has only recently wandered to extreme southern Texas. Polytypic (7 ssp.; likely *primulus* in N.A.). L 6.7–7.2" (17–18 cm)

IDENTIFICATION ADULT: Medium size; short black bill. Gray crown; white forehead, eyebrows connect indistinctly across nape; concealed reddish orange central patch; rest of face dark gray. Brownish olive back with darker wings. White throat; rest of underparts bright yellow. **JUVENILE:** Similar but duller; no crown patch; rusty-edged wings and tail.

SIMILAR SPECIES See Great Kiskadee for differences.

VOICE CALL: Strident *chee cheechee cheechee cheechee;* also harsh *cree-yooo.*

STATUS & DISTRIBUTION Common, Mexico to northeast Argentina. **YEAR-ROUND:** Open woodland, forest edge, second growth. **VAGRANT:** Accidental; one well-documented record in Bentsen-Rio Grande Valley S.P., TX (Jan. 7–14, 2005). An accepted record at Anzuldaus Co. Park, TX (Mar. 17–Apr. 5, 1990) should be considered problematic at best.

POPULATION Presumably stable.

SULPHUR-BELLIED FLYCATCHER Myiodynastes luteiventris   SBFL ▪ 2

This handsome flycatcher reaches its northern limit in the mountain canyons (with flowing water) of southeastern Arizona. A cavity nester that mainly forages in the canopy, the Sulphur-bellied Flycatcher is often

more easily located by its squeaky call. It usually stays within the canopy, rather than using exposed perches, but will engage in vigorous aerial pursuits of prey. Monotypic. L 8.5" (22 cm)

IDENTIFICATION ADULT: Sexes similar. Moderately large overall with heavy, long, mostly black bill and proportionately short tail. Strikingly patterned head: dark mask and malar stripes contrast with whitish

superciliary, moustachial stripes, and throat; grayish brown crown and back have blackish streaks; yellow concealed central crown patch. Uppertail coverts and tail extensively rufous. Dark brown wings with whitish edgings. Blackish chin, connecting malar stripes. Throat lightly streaked blackish. Rest of underparts pale yellow, heavily streaked blackish on breast and sides. JUVENILE: Similar to adult but duller and paler; upperparts more brownish buff; wing feathers edged with cinnamon-buff; yellow crown patch small or absent.
SIMILAR SPECIES Piratic and Variegated Flycatchers are similar but smaller, with proportionately much

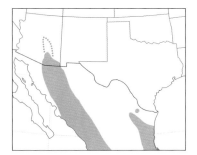

smaller bills, less distinctly streaked and duller yellow underparts, and less rufous in tail. The very similar Streaked Flycatcher, with migratory northernmost (southern Mexico to Honduras) and southernmost (Argentina and southern Bolivia) populations, is a potential vagrant to N.A. Streaked Flycatcher lacks dark chin connection between malar stripes, has a slightly heavier bill with more extensively pale base to lower mandible, and is whiter on belly.
VOICE CALL: Loud *squeez-za* recalls a squeaky toy; also a rasping screech or *weel-yum*. Migrants usually silent. DAWN SONG: Repetitive, warbled *tre-le-re-re* or combinations of other notes. STATUS & DISTRIBUTION Fairly common. BREEDING: Mature sycamore-walnut dominated riparian forest at lower elevations in mountain canyons. Breeds south to Costa Rica. MIGRATION: Seldom detected in AZ lowlands. For spring, arrives in AZ exceptionally by early–mid-May, more typically late May–early June; typically about one month earlier in northeastern Mexico than in northwestern Mexico and AZ (e.g., spring vagrants in TX as early as Apr. 5).

dark mask bordered by broad white lines

rufous tail

dark streaks below with yellowish background color

For fall, breeders depart AZ by early to mid-Sept. Fall vagrants have been recorded into Oct. and early Nov. WINTER: S.A., along eastern base of Andes from Ecuador to northern Bolivia. VAGRANT: Casual in spring and summer to southern TX (inc. breeding recs.), southeastern CO, southern NV, coastal LA, and the Gulf of Mexico off LA. Casual in fall, west to coastal CA and north and east to ON, NB, NF, MA, and along the Gulf Coast in TX, LA, and AL.
POPULATION Stable.

CROWNED SLATY FLYCATCHER *Empidonomus aurantioatrocristatus* CSFL ■ 5

An extraordinary first U.S. record, a Crowned Slaty Flycatcher came to Louisiana in 2008 from at least as far away as Brazil or southern Colombia, which are the usual northern limits of its range. Birds of this species' southern populations migrate northward to the South American tropics for the austral winter. Perhaps this one far overshot its normal distance, as Fork-tailed Flycatchers presumably do when they wander from South America far into the Northern Hemisphere. One other South American species of this bird's genus *Empidonomus,* the Variegated Flycatcher, has several records in the U.S. Polytypic (2 ssp.). L 7" (18 cm)
IDENTIFICATION ADULT: Smoky or brownish grayish overall; crown black with a semiconcealed yellow patch down center, and only slightly crested. Gray supercilium extends from bill to nape; lores and auricular area darker gray. Upperparts may be tinged brownish gray. Wings smoky brown, with narrow pale edges on inner flight feathers and coverts. Tail dull brownish gray. Underparts are

paler brownish gray, slightly paler on belly. Eyes brown, bill, tarsi, and toes black. JUVENILE: Similar to adult but crown dusky (not black) and lacking yellow patch; supercilium paler; tail with narrow rufous margins.
SIMILAR SPECIES Adult's combination of black crown, median yellow stripe, and dark gray face distinguish it from other flycatchers and elaenias. Slaty Elaenia *(Elaenia strepera)* is gray but crown is brown, not black, and a small crown patch is white, not yellow. Juvenile Crowned Slaty Flycatcher might be confused with Piratic Flycatcher or Variegated Flycatcher but lacks these species' whitish submoustachial stripe.
VOICE Usually quiet in nonbreeding season. CALL: High *tzeer.* SONG: On breeding grounds rising *be-bee-beee-beeez,* low whistling *pree-ee-ee-er,* and a series of squeaky notes, or a two-part *tsi-tsitsewt-tsi-tsébidit.*
STATUS & DISTRIBUTION Widely distributed and generally common throughout its range. BREEDING: Deciduous forest, forest edges, and scrub with scattered trees, in interior Brazil, Paraguay, Uruguay, and Argentina.

pale gray supercilium

dark crown with partially concealed yellow center

broad dark eye stripe

aurantioatrocristatus

WINTER: Some populations migrate to spend the austral winter (June–Aug.) as far north as western Brazil, Ecuador, and southeastern Colombia, where they prefer the canopy of humid forests, forest edges, and clearings. VAGRANT: Vagrants have been reported farther north in South America and once to Panama (Dec. 2007). The sole U.S. record was an adult male collected near Johnsons Bayou, Cameron Parish, LA (June 3, 2008).
POPULATION Stable.

VARIEGATED FLYCATCHER   *Empidonomus varius*   VAFL ■ 5

This austral migrant from S.A. occasionally strays to N.A., where it might be confused with the Sulphur-bellied Flycatcher or the Piratic Flycatcher.

varius

larger bill than Piratic with pale base

streaking more distinct than Piratic

rufous on uppertail coverts, tail, and wings

Polytypic (2 ssp.; nominate in N.A.). L 7.25" (18 cm)
IDENTIFICATION ADULT: Relatively small, proportionately long tailed. Black crown and mask; crown encircled by white and with concealed yellow central patch. Dark grayish brown back with subtle pale mottling. Blackish brown upper tail coverts; tail conspicuously edged rufous. Blackish brown wings; coverts and secondaries edged white; primaries faintly edged rufous. Dingy white throat, faintly mottled with gray and bordered by dark malar stripes. Rest of underparts gradually blend to pale yellow on belly and undertail coverts; breast and sides heavily streaked with blackish brown.
JUVENILE: Similar, but duller below, browner above; wings extensively edged rufous; lacks crown patch.
SIMILAR SPECIES Piratic Flycatcher is smaller overall, with a shorter, broader, all-black bill; a shorter tail; more uniform brown back; little rufous in wings or tail; and blurrier breast streaks. Sulphur-bellied and Streaked Flycatchers are larger overall, heavier billed, more distinctly streaked below, and more extensively rufous on tail. Sulphur-bellied Flycatchers are also more yellow below.
VOICE CALL: High, thin *pseee*.
STATUS & DISTRIBUTION VAGRANT: Casual, single fall records from ME, ON, and WA; single spring records from TN and FL.
POPULATION Unknown.

PIRATIC FLYCATCHER   *Legatus leucophaius*   PIFL ■ 4

This neotropical species often chooses a high, conspicuous perch. It bears a superficial resemblance to a miniature Sulphur-bellied Flycatcher. Polytypic (2 ssp.). L 6" (15 cm)
IDENTIFICATION ADULT: Sexes similar. Dark brown crown has semi-concealed yellow central crown patch (larger in male) and is bordered by white eyebrows thinly connecting across nape. Blackish brown face; whitish throat with dark malar streaks. Rest of underparts progressively yellower toward lower belly and undertail coverts; breast and sides have extensive smudgy brown streaks. Dark brown wings and tail; slightly paler back. Wing coverts, secondaries, and tail thinly edged white

to pale yellow. **JUVENILE:** Similar, but lacks crown patch; edges of wing and uppertail coverts tinged rusty.
GEOGRAPHIC VARIATION Both subspecies at least partially migratory. Most U.S. records probably pertain to northern *variegatus* (described above), which breeds from southeastern Mexico to Honduras; nominate, breeding from Nicaragua into South America, is smaller and whiter on central underparts.
SIMILAR SPECIES Variegated is larger, with a proportionately longer, narrower bill; pale-based lower mandible; somewhat more well-defined breast streaks; and conspicuously rufous-edged uppertail coverts and tail. The much larger,

short all-dark bill

variegatus

dark malar streak

darker wings and tail than Variegated

heavier-billed Sulphur-bellied has an extensively rufous tail.
VOICE CALL: Trilled, often repeated, *pi ri ri ri ri ri*.
STATUS & DISTRIBUTION Common.
BREEDING: Open woodland, forest edge, clearings. **WINTER:** S.A. **VAGRANT:** Casual in spring and fall to FL (1), TX (4), and NM (3).
POPULATION Stable.

KINGBIRDS   Genus *Tyrannus*

These large, open-area birds (13 sp.) are conspicuous and noisy. Often seen on exposed perches or sallying to capture flying insects, they will also hover-glean fruit during migration or winter. They may be solitary or in loose flocks. Divided into three basic groups (white-bellied, yellow-bellied, and long-tailed), most are easily identifiable. Only the male sings a dawn song, and only the adults have a colorful, semiconcealed, central crown patch and notched outer primaries. Juveniles have duller, paler, more delicately structured body plumage; and rusty-edged upperwing and uppertail coverts, secondaries, and tail. They molt to more adultlike body plumage by fall but retain juvenile wing and tail feathers.

TROPICAL KINGBIRD   *Tyrannus melancholicus*   TRKI ■ 2

This wide-ranging neotropical species has northern breeding limits in southeastern Arizona and in the lower Rio Grande Valley in Texas. Its deeply notched tail, and relatively heavy bill distinguish it from most other yellow-bellied kingbirds. The Couch's (southern TX and eastern

Mexico) is virtually identical to the Tropical; voice is the best way to distinguish between them. Extralimital records of silent birds (esp. in the east, where Couch's is as likely to occur) are difficult to resolve. Polytypic (3 ssp.; *satrapa* in N.A.). L 9.25" (24 cm)
IDENTIFICATION ADULT: Gray head,

dark mask; orange-red central crown patch usually not visible. Olive back; blackish brown wings and tail; tail conspicuously notched. White throat blends to olive-yellow on upper chest and bright yellow on belly.
GEOGRAPHIC VARIATION Slight differences in coloration and wing, tail, and

bill length; not safely separable in the field. U.S. breeders and, presumably, most nonbreeding records involve northernmost *satrapa*.

SIMILAR SPECIES Couch's has a broader bill that appears shorter, different call, slightly shallower tail notch. On folded wing, Tropical's primary tips visible beyond secondaries appear unevenly spaced (adults). Other yellow-bellied kingbirds differ in smaller bill; less extensively yellow below, darker above; or more square-ended, differently colored tails.

VOICE CALL: High-pitched rapid trill

or twitter. **DAWN SONG:** A repeated combination of 1–2 short *pip* notes followed by a trill.

STATUS & DISTRIBUTION Uncommon in U.S., common elsewhere. **YEAR-ROUND:** Open areas with scattered trees, riparian woodland, forest edge and clearings, even open suburban situations; often near water. **MIGRATION:** Northwesternmost (AZ, Sonora) and southernmost (southern Bolivia and southeastern Brazil to Argentina) populations migratory. In spring, arrives AZ mid-May. In fall, rare in AZ after mid-Sept.; West Coast vagrant recs. mainly late-Sept. into winter. **WINTER:** West-central and east Mexico south; southern TX breeding population present year-round. **VAGRANT:** Rare mainly fall and winter, to West Coast (mainly CA but north to southwestern BC, casually to southeastern AK); less frequently inland in southern CA, western AZ, and TX away from areas of local breeding. Accidental in NM, LA, ME, and Bermuda

(May rec. of nominate subspecies). **POPULATION** Stable or increasing. Slowly expanding breeding range northward; first bred in Arizona in 1938, southern Texas beginning in 1991, and western Texas in 1997.

adults
satrapa

longer, thinner-based bill than Couch's

spring adult ♂
satrapa

notched tail

adult ♂ primary tips
satrapa

uneven spacing of tips

COUCH'S KINGBIRD *Tyrannus couchii* COKI ◼ 2

This southern Texas specialty is virtually identical to the Tropical and is best confirmed by voice in the field. Most Tropical-type kingbirds found in southern Texas, especially away from cities, are mostly Couch's, whereas breeding birds in Arizona and most western vagrants will almost certainly be Tropicals. Monotypic. L 9.25" (24 cm)
IDENTIFICATION ADULT: Gray head, with blackish mask and concealed reddish orange central crown patch. Olive-gray back. Blackish brown wings and tail; tail prominently notched.

broad-based bill

notched tail

even spacing of tips

adult ♂ primary tips

spring adult ♂

yellow more extensive below than Western

White throat blends to olive-yellow breast and bright yellow belly and undertail coverts. **JUVENILE:** Similar but with buff tips to upper-wing and upper-tail coverts.
SIMILAR SPECIES Combination of distinctly notched brownish black tail and olive-yellow breast separate Couch's from other yellow-bellied kingbirds (except Tropical). On adult's folded wing, primary tips visible beyond secondaries are evenly spaced (uneven in Tropical). See Tropical account for other differences.
VOICE CALL: Single or slow series of *kip* notes, suggestive of Western Kingbird; also *queer* or slurred *chi-queer*. **DAWN SONG:** A repeated *tuwit, tuwit, tuwitchew.*
STATUS & DISTRIBUTION Common. **YEAR-ROUND:** Thorn and riparian forest, clearings, forest edge, woodland, overgrown fields with scattered trees, even suburban situations. **MIGRATION:** Partial migrant, with part of northernmost population moving south during winter. In spring, influx of migrants augments the overwintering southern TX population during mid-Mar.–early Apr.

In fall, some of the southern TX population retreats south from late Aug.–mid-Oct., with migratory flocks occasionally seen. **VAGRANT:** Mainly fall and winter but occasionally spring and summer. Rare in TX outside breeding areas. Casual east to southern LA, including breeding record southwest; accidental to west in NM, AZ, CA, to north in AR and MI and east to FL.
POPULATION Stable or increasing. Since the mid 1900s, Couch's has expanded range in TX from immediate vicinity of extreme Lower Rio Grande north to Big Bend (a few), to near southern Edwards Plateau, and to Calhoun County on coast.

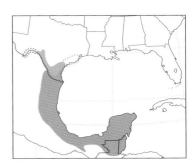

WESTERN KINGBIRD *Tyrannus verticalis* WEKI ■ 1

juvenile

pale gray head
and back

darker wings
contrast with
gray back

adults

worn
fall adult

black tail with
white edge

spring
adult ♂

The Western is the "default" breeding yellow-bellied kingbird across vast areas of the west (esp. arid lowlands); but its distribution and habitat overlap with other yellow-bellied kingbirds (the Cassin's, Thick-billed, Tropical, and Couch's) as well as with the Eastern and Scissor-tailed Flycatchers. From exposed perches on trees, shrubs, or wires, it chases flying insects; it will also take fruit during fall and winter. Monotypic. L 8.75" (22 cm)

IDENTIFICATION ADULT: Pale gray head; darker mask; concealed orange-red central crown patch. Pale grayish olive back. Plain brownish black wings, contrasting paler back. Square-tipped, black tail; white outer webs of outer pair of feathers. White throat subtly blends to pearly gray chest and yel-

low belly. Relatively small, black bill. **JUVENILE:** Duller, paler. **SIMILAR SPECIES** Typical tail pattern unmistakable, but individuals with worn or missing white outer tail feathers might be mistaken for other yellow-bellied species. Cassin's is overall much darker on chest, upperparts; has contrasting white chin, paler wings, gray tail tip. Tropical and Couch's have heavier bills; olive-yellow chests; and dark brown, deeply notched tails. Cassin's, Tropical, and Couch's also have pale-edged upperwing coverts with a more scalloped appearance. Thick-billed has much heavier bill, darker upperparts, yellow central crown patch, paler underparts. (Juvenile's underparts yellower, more Western-like.) Superficially similar to juvenile or immature Scissor-tailed; immature white-bellied kingbirds with tinge of yellow on underparts, *Myiarchus,*

or Say's Phoebe; but key field marks should still be apparent.
VOICE CALL: Single or repeated sharp *kip* notes; also a staccato trill. **DAWN SONG:** A repeated series of *kip* notes and long trills.
STATUS & DISTRIBUTION Common. **BREEDING:** Open country with scattered trees or shrubs; will nest on human-made structures (e.g., utility poles). **MIGRATION:** Mid-Mar.–early June; late July–mid-Sept.; scarce after early Oct, stragglers into Nov. **WINTER:** Central-western Mexico to Costa Rica; also southern FL. **VAGRANT:** In summer, casual/accidental to AK, central and northern Canada; in winter, rare/casual on Pacific coast to central CA, on Gulf and Atlantic coasts, and to Panama, West Indies, and Bermuda.
POPULATION Stable or increasing.

THICK-BILLED KINGBIRD *Tyrannus crassirostris* TBKI ■ 2

First recorded in the U.S. in 1958, this West Mexican species now regularly breeds in southeastern Arizona and extreme southwestern New Mexico. Polytypic (2 ssp., *pompalis* in N.A.). L 9.5" (24 cm)
IDENTIFICATION ADULT: Long, thick-based bill. Blackish brown head; slightly darker mask; concealed yellow central crown patch, paler olive-gray back. Dark brown tail, squared or notched slightly. Pale below with yellowish (brighter in fall) belly.

SIMILAR SPECIES Massive bill, white throat, contrasting dark head, pale yellow to whitish yellow belly make identification easy.
VOICE CALL: Loud, whistled *kiter-re-er;* also a rapid trill similar to Eastern's.
STATUS & DISTRIBUTION Uncommon. **BREEDING:** Canopy of mature sycamore-cottonwood dominated riparian forest.

MIGRATION: In spring, arrives late May–early June. In fall, departs in Sept., rare by Oct. **WINTER:** Western Mexico. **VAGRANT:** Casual to accidental, mainly in fall and winter, in western AZ, CA, southwestern BC, northern TX, Baja California; casual in summer to western TX (inc. breeding records) and CO.
POPULATION Has contracted in southeastern AZ in recent years.

dark crown and
nape with even
darker mask

very thick
bill

1st fall
pompalis

1st fall
pompalis

fresh fall birds
quite yellowish
on belly

dark
brownish
tail

worn
summer
adult
pompalis

whitish
underparts

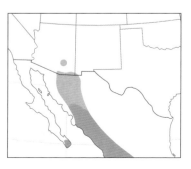

CASSIN'S KINGBIRD *Tyrannus vociferans* CAKI ■ 1

Cassin's Kingbird is fairly widespread at middle elevations in the southwestern and west-central U.S., where it overlaps extensively with Western Kingbird. Monotypic. L 9" (23 cm)
IDENTIFICATION ADULT: Dark gray head and nape (mask less obvious); semi-concealed orange-red central crown patch; dark grayish olive back, brownish wings. Dark gray chest contrasts with white chin, blends to yellow belly and olive flanks. Tail squared or slightly notched; brownish black with an indistinct pale gray terminal

band. Relatively small bill. **JUVENILE:** Similar but wings edged pale cinnamon and pale tail tip less obvious.
SIMILAR SPECIES Western has paler upperparts and chest; much less chin-breast contrast; and darker wings and black tail with white outer web of outer feather pair (but beware Westerns completely lacking outer tail feathers or with worn-off outer webs). Cassin's has pale (but not pure white) outer web. Thick-billed, Tropical, and Couch's have paler or yellower breasts, uniformly brown tails, and heavier bills; Tropical and Couch's also have paler backs and deeply notched tails.

VOICE CALL: Single or repeated, strident *kabeer;* rapidly repeated *ki-dih* or *ki-dear.* **DAWN SONG:** A repeated *rruh rruh rruh-rruh rreahr, rruh ree reeuhr* (possibly confused with Buff-collared Nightjar's).
STATUS & DISTRIBUTION Common. **BREEDING:** Open, mature woodlands, including riparian, oak, and pinyon-juniper. **MIGRATION:** Relatively infrequently detected away from breeding sites. In spring, mid-Mar.–early June. In fall, departure late July–Oct., most in Sept. **WINTER:** Western to central-southern Mexico. Locally resident in coastal southern CA. **VAGRANT:** Casual/accidental to OR, ON, MA, VA, AR, LA, and FL.
POPULATION Stable.

GRAY KINGBIRD *Tyrannus dominicensis* GRAK ■ 2

This mainly West Indian species is a conspicuous summer resident of coastal Florida and, locally, along the eastern Gulf Coast, where it overlaps with the Eastern Kingbird. Monotypic. L 9" (23 cm)
IDENTIFICATION ADULT: Relatively large and large billed. Uniformly gray head and back except for contrasting dark mask and orange to reddish orange central crown patch. Blackish brown upperwings and tail; upperwing coverts edged pale gray to grayish white; and secondaries edged white

to yellowish white. Tail distinctly notched. Underparts white with grayish wash across chest and down sides; lower belly and undertail coverts sometimes tinged pale yellow. **JUVENILE:** Similar but more brownish gray on back, with rusty-edged upperwing and uppertail coverts and tail feathers.
SIMILAR SPECIES Eastern is overall smaller and darker above, has obvious white tail tip, and lacks tail notch. Yellow-tinged immature Gray is superficially similar to Thick-billed, but the latter would be heavier billed, yellower below and darker above, without notched tail. Loggerhead Shrike has a shorter, hooked bill and white on wings and tail.
VOICE CALL: *Pe-cheer-ry;* also a rapid trill similar to Eastern's. **DAWN SONG:** Poorly known; described as "complex chatter."
STATUS & DISTRIBUTION Rare to fairly common. **BREEDING:** Mangroves, open woodland, second growth, forest edge, mainly along immediate coast, including suburban situations. **MIGRATION:** In spring, arrives southern FL mid-Mar., elsewhere along Gulf and

southern Atlantic coasts mid-Apr.–early May; stragglers into June. In fall, departs mid-Sept.–Oct. **WINTER:** Central and southern West Indies, central and eastern Panama, and extreme northern S.A.; rare southern FL, northern West Indies. **VAGRANT:** Casual, mainly in spring, west to LA (recent breeding recs.) and TX, north to Canadian Maritimes; accidental to southwestern BC, WI, IL, KY, MI, ON, and central NY.
POPULATION Stable; expanding breeding range westward along Gulf Coast.

EASTERN KINGBIRD   *Tyrannus tyrannus*   EAKI ■ 1

This small, white-bellied, conspicuous species has the largest distribution of any North American kingbird. Monotypic. L 7.8–9.2" (20–23 cm) **IDENTIFICATION** **ADULT:** Black head blends to slate gray back; central crown patch varies from red to yellow. Dark gray wings; narrow white edgings to upperwing coverts and secondaries. Black tail with conspicuous white terminal band. White underparts; gray patches on sides of breast, paler gray wash across middle of breast. Extensively gray underwing coverts. **JUVENILE:** Generally similar but paler grayish brown upperparts contrast with

blackish mask; white tail tips narrower. **SIMILAR SPECIES** Adults in reasonably fresh plumage are virtually unmistakable. Immatures and worn adults, both of which can have somewhat paler or browner upperparts and reduced white tail tips, can be superficially similar to Gray or Thick-billed but would still be overall smaller and smaller billed. Immature Fork-taileds are also superficially similar but have more head-back contrast, a longer tail, and a whiter center of breast. **VOICE** **CALL:** Single or variety of *zeer, dzeet,* or trilled notes. **DAWN SONG:** A series of complex notes and trills, which are repeated over and over, *t't'tzeer, t't'tzeer, t'tzeetzeetzee.* **STATUS & DISTRIBUTION** Common. **BREEDING:** Open areas in a variety of habitats that have trees or shrubs for nest sites. **MIGRATION:** Diurnal migrant, often observed in loose flocks; at least some trans-Gulf movement. In spring, mid-Mar.–mid-June, peaks mid-Apr.–mid-May; in west mid-May–June. In fall, late July–mid-Oct., peaks mid-Aug.–early Sept.; mostly gone by end of Sept., very rare after early Oct. **WINTER:**

blackish head

juvenile

small bill

white tail tip

S.A., mainly western Amazonia (eastern Ecuador and Peru, western Brazil), but also casually as far south and east as northern Chile, Argentina, Paraguay, eastern Brazil, and Guyana. **VAGRANT:** Rare during migration to Pacific coast, southwestern states, Bermuda, Bahamas, Cuba; casual to AK, southern YT, Hudson Bay, central QC, NF, Greenland. **POPULATION** Generally stable.

LOGGERHEAD KINGBIRD   *Tyrannus caudifasciatus*   LOKI ■ 5

A rare vagrant to the U.S., the Loggerhead Kingbird has an interesting history of disputed identifications in Florida. Only two of at least eight reports since 1971 are now considered valid. The two currently accepted U.S. records did not come until 2007 and 2008 and appear to represent the nominate subspecies from Cuba. Polytypic (7 ssp.). L 9" (23 cm) **IDENTIFICATION** **ADULT:** Blackish crown and face, slight rear crest; rest of upperparts dark gray; white edges on wing coverts; white underparts (underparts

yellowish in subspecies from the Bahamas, *bahamensis*); square tail with narrow white tip (lacking in Puerto Rican and Hispaniolan populations). Loggerhead has a relatively short primary projection compared with similar species. **SIMILAR SPECIES** Gray has a paler crown, with a blackish mask and a much larger bill; tail is notched and lacks Loggerhead's white tip. Eastern is smaller, slimmer, and much smaller-billed and has a wider white terminal tail band. Giant *(T. cubensis),* an endangered Cuban endemic, is rounder headed and has a much thicker bill. **VOICE** **CALL:** Variable, but usually a loud buzzy *tireet,* or *teerrrp* often repeated; mostly silent outside the breeding season. Wings produce a muffled sound in flight. **STATUS & DISTRIBUTION** Endemic to the Caribbean, Loggerhead is a year-round resident of tropical lowland evergreen forest and pine forests up to 6,000-feet elevation. The range includes the northern Bahamas, Cuba, the Greater Antilles, and at least formerly the

dark crown and face

slightly crested

long, moderately thick bill

caudifasciatus

white-edged wing coverts

white below

short primary projection

whitish tail tip

Cayman Is. **VAGRANT:** The two accepted U.S. records are from FL: Key West (Mar. 8–27, 2007) and the Dry Tortugas (Mar. 14–22, 2008), both photographed. Previous reports included birds misidentified as a Gray Kingbird, an Eastern Kingbird, and perhaps a Giant Kingbird, which has historical extralimital records from Great Inagua in the southern Bahamas and from the Caicos Is.

SCISSOR-TAILED FLYCATCHER   *Tyrannus forficatus*   STFL ■ 1

The striking Scissor-tailed Flycatcher is our only regular "long-tailed kingbird." It is not only graceful and beautiful, but also common and easy to observe. Monotypic. L 10–14.8" (25–38 cm)
IDENTIFICATION ADULT: A long, forked tail of male is longer than the female's. Pale gray head and back; blackish wings; extensive white in outer tail. Whitish underparts; pinkish wash on belly. Salmon sides, flanks, underwing coverts; bright red axillaries. **JUVENILE:** Duller yellowish pink on underparts; tail much shorter.
SIMILAR SPECIES Adult is unmistakable. Immature is superficially like the Western Kingbird, but it lacks pure yellow tones on belly and its tail is proportionately longer, narrower, forked, more extensively white.

VOICE CALL: Sharp *bik* or *pup;* also a chatter. **DAWN SONG:** A repeated series of *bik* notes interspersed with *perleep* or *peroo* notes; given when perched or during display flight.
STATUS & DISTRIBUTION Common. **BREEDING:** Open country with scattered trees and shrubs. **MIGRATION:** In spring, arrives mid-Mar.–early Apr. Fall peak Oct., some into Nov. **WINTER:** Southern FL, southern Mexico to central Costa Rica, a few to Panama. Rare in TX, LA; locally regular central south TX and extreme southeastern LA. **VAGRANT:** Rare to casual, to Pacific coast, southeastern AK, southern Canada, Atlantic coast.
POPULATION Range expanded East.

juvenile

juvenile

reddish pink "armpit"

adult ♂

long forked tail with extensive white; female has shorter tail

juvenile

pale gray head and back

orange-buff belly

adult ♂

FORK-TAILED FLYCATCHER   *Tyrannus savana*   FTFL ■ 3

This distinctive, handsome, common neotropical counterpart of the Scissor-tailed Flycatcher is a much sought-after vagrant to North America. Polytypic (4 ssp.; 2 in N.A.). L 14.5" (37 cm)
IDENTIFICATION ADULT: Unmistakable. Overall relatively small bodied compared to other kingbirds, with a proportionately short, thin, narrow bill and a proportionately very long, narrow, forked, mostly black tail. Black head contrasts with pale gray back and white underparts. Yellow central crown patch. **JUVENILE:** Similar but much shorter tail, duller black head; upperwing and uppertail coverts and central tail feathers edged with rufous. Immature similar.
GEOGRAPHIC VARIATION Most of the more than 100 records in N.A. pertain to nominate (illus. and described above), which has on average a darker back and less head-back contrast. Presumably, these records represent "overshooting" northbound austral migrants from southern-most South American breeding populations (occurring here May–July) or "reverse" spring austral migrants (originating from northern S.A. and occurring here late Aug.–early Dec.). Northern *monachus,* occurring from southeastern Mexico to Panama, has been documented at least twice in early winter in southern TX; it has a somewhat paler gray back, and its more distinct white collar separates the crown and back, resulting in more head-back contrast. There are also subspecific differences in the notching pattern of outer primary tips (adults).
SIMILAR SPECIES The structurally similar Scissor-tailed Flycatcher lacks black cap and has salmon-pink on underparts and extensive whitish in tail. The superficially similar Eastern Kingbird is larger and uniformly dark above and has a shorter, broader, white-tipped tail.
VOICE CALL: A sharp *bik* or *plick* and a chattering trill.
STATUS & DISTRIBUTION Common.

immature

shorter tail

black cap

gray back

adult

YEAR-ROUND: Open grassland, scrub, farmland, pine savanna. Generally present year-round from southeastern Mexico to central S.A. Breeds as far south as central Argentina during austral summer; generally absent as a breeder across Amazonia, but widespread during austral winter. Locally migratory or nomadic in Mexico and C.A. **VAGRANT:** Casual/accidental, scattered records north to CA, WA, ID, AB, MN, WI, MI, ON, QC, NB, NS; majority of records from eastern N.A.

very long, forked black tail

BECARDS, TITYRAS, AND ALLIES Family Tityridae

This assemblage of 33 New World tropical and subtropical fruit- and insect-eating species has recently been given family rank now that the genera comprising it have been confirmed through molecular analysis to be each other's closest relatives.

Structure There are few outward characters that unite this family. They are small (purpletufts) to medium-size (tityras) passerines with short to medium somewhat stout bills that are slightly to strongly hooked at the tip. Becards have large, puffy heads and square to graduated tails; the tityras are especially strong billed and have relatively short tails.

Masked Tityra, female (Costa Rica, Dec.)

Behavior Becards and tityras are generally found singly or in pairs, feeding mainly on insects and fruit. Tityras are sluggish, perching openly high in the canopy; they sally for insects and hover for fruit. Becards feed sluggishly at mid-story and canopy levels, with some species joining mixed-species flocks. Songs range from musical whistles in many becards and schiffornis to grunting notes in the tityras. Nests vary greatly, from cavities (tityras) to cup nests (schiffornis, purpletufts) to large, globular structures with side entrances (becards).

Plumage Variable, from solidly black or rufous to strongly patterned. Most becards are sexually dimorphic; immature males resemble females. Tityras, striking pale gray,

black, and white birds, are moderately dimorphic; two of the three species show bare red facial skin and bill bases.

Distribution An exclusively neotropical family. Seven species range as far north as Mexico, and the northernmost species (Rose-throated Becard) formerly reached the U.S. as a regular breeding bird in southeastern Arizona. Two other species that reach northern Mexico have been recorded as vagrants along the U.S. border. The family is most diverse in tropical South America. All are birds of woodlands, river edges, and forests, and many species are primarily found in the canopy.

Taxonomy It has long been known that the becards, tityras, and their allies were part of the vast radiation of tyrant flycatchers and other "sub-oscine" passerine families that are exclusively New World in distribution, but their exact placement has long been debated. Genera now placed in this family were until recently distributed among three families—cotingas, manakins, and tyrant-flycatchers. The becards, all but one in the genus *Pachyramphus*, make up about half the family, with the remainder including three tityras, two mourners *(Laniocera)*, up to seven very similar species in the genus *Schiffornis* (formerly considered manakins), a cotinga-like bird in the genus *Laniisoma*, and three purpletufts *(Iodopleura)*.

Conservation Populations of most species are stable.

TITYRAS Genus *Tityra*

These four cavity-nesting species of the Neotropics are large, chunky, and boldly patterned; they have relatively short tails, heavy hooked bills, and bare reddish facial skin. There is some sexual dimorphism. Tityras are usually observed on exposed perches or in fruiting trees in the canopy of lowland forest.

MASKED TITYRA *Tityra semifasciata*   MATI ■ 5

Common to our south, this unmistakable species has been found only once in the U.S. Polytypic (8 ssp.; *personata* in N.A.). L 9" (23 cm)

IDENTIFICATION MALE: Generally pale gray above (inc. inner secondaries, tertials, and upperwing coverts). Whitish gray below, including underwing coverts. Starkly contrasting black on face, most of wings, and thick subterminal tail band. Bare-skinned facial spectacles and base of heavy bill are pinkish red; tip of bill is black; eye is red. **FEMALE:** Darker and browner on head and back; darker gray tail contrasts less with black subterminal band; lacks black face. **JUVENILE:**

Like female but with subtle darker streaks on crown and back, and whitish edges to inner secondaries, tertials, and upperwing coverts.

SIMILAR SPECIES No other passerine with similar plumage pattern or soft-part colors.

VOICE CALL: A double, nasal grunt, *zzzr zzzrt,* given when perched or in flight; reminiscent of Dickcissel's call.

STATUS & DISTRIBUTION Common. **YEAR-ROUND:** Lowland tropical forest and woodland from northern Mexico (southern Sonora and southern Tamaulipas) to S.A. **VAGRANT:** Accidental, one record, Bentsen-Rio Grande Valley S.P., TX (Feb. 17–Mar. 10, 1990).

pinkish-red bare skin on face and pinkish-red bill base

white with mostly black face

adult ♂ *personata*

BECARDS Genus *Pachyramphus*

Becards comprise 16 species of small- to medium-size birds; they are chunky, slightly crested, and disproportionately large headed and short tailed. Typically found in forest or woodland canopy, they are sluggish, almost vireo-like, and perch with an upright posture and sally short distances with fluttery flight for fruit or insects.

GRAY-COLLARED BECARD *Pachyramphus major* GCBE ■ 5

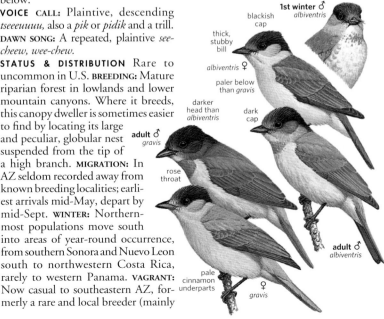

short white
supraloral
stripe

immature ♂
uropygialis

distinctively
marked
graduated
tail

A close but very distinctive relative of the Rose-throated Becard, the Gray-collared Becard of Mexico and Central America has been reported only once in the U.S., an immature male bird in Arizona. Polytypic (5 ssp.). L 6" (15 cm) **IDENTIFICATION** Proportionally large headed, slightly crested, and thick billed. Note white supercilium extending from base of bill to back of eye, and graduated tail with black subterminal marks. ADULT MALE *uropygialis:* Black crown and back, pale gray hind collar and underparts; outer tail tipped with white. ADULT FEMALE: Cinnamon crown bordered black; brown back, buffy underparts, tail tipped with cinnamon. IMMATURE MALE: Intermediate, with cinnamon on back and rump but white markings on wings and tips of outer tail feathers.

In more easterly nominate subspecies, female has a black crown, and pale areas are more buffy cinnamon. **SIMILAR SPECIES** Unlikely to be confused with any other species. Female Rose-throated Becard might be suspected but is larger, often bushy-headed, and has a gray crown with a buffy collar. **VOICE CALL:** A clear titmouse-like *peeu, peeu, peeu.* SONG: A repeated whistled *hoo hoo-dee* or *hoo wee-deet.* **STATUS & DISTRIBUTION** Resident in evergreen, pine-oak, and tropical deciduous forest as high as 8,200 ft elevation, from southeastern Sonora and southern Nuevo León and Tamaulipas, south to north-central Nicaragua. Distinctive pale western Mexican subspecies *uropygialis* has bred at Sahuaripa, Sonora (60 miles

from AZ), indicating this subspecies' potential as most likely to reach the U.S. The single record north of Mexico, at Cave Creek Canyon, Cochise Co., AZ (June 5, 2009), was identified as a second-calendar-year male *uropygialis* attaining adult plumage. **POPULATION** The west Mexican subspecies *(uropygialis)* has a limited range and is uncommon.

ROSE-THROATED BECARD *Pachyramphus aglaiae* RTBE ■ 3

A widespread and numerous tropical lowland species now found only casually in the U.S. Polytypic (8 ssp.; 2 in N.A.). L 7.3" (19 cm) **IDENTIFICATION** ADULT: Relatively small; short, thick, dark bill. MALE: Black crown; remainder of upperparts plain dark gray; light gray underparts with pink throat patch. FEMALE: Slate gray cap; pale rufous hindneck collar; brownish gray back; light buffy underparts. JUVENILE: Similar to female. First-year male more patchy gray and brown; has smaller pink throat patch. **GEOGRAPHIC VARIATION** Subspecies *albiventris* (northwest Mexico; described above) formerly bred north to Arizona; *gravis* (eastern Mexico;

gravis

albiventris

casual to southernmost TX) is larger. Male overall substantially darker, with more extensively black crown and more extensive, darker pink throat patch; female with rufous back, upperwings, and tail; more intensely rufescent below. **VOICE CALL:** Plaintive, descending *tseeeuuuu,* also a *pik* or *pidik* and a trill. DAWN SONG: A repeated, plaintive *see-cheew, wee-chew.* **STATUS & DISTRIBUTION** Rare to uncommon in U.S. BREEDING: Mature riparian forest in lowlands and lower mountain canyons. Where it breeds, this canopy dweller is sometimes easier to find by locating its large and peculiar, globular nest suspended from the tip of a high branch. MIGRATION: In AZ seldom recorded away from known breeding localities; earliest arrivals mid-May, depart by mid-Sept. WINTER: Northernmost populations move south into areas of year-round occurrence, from southern Sonora and Nuevo Leon south to northwestern Costa Rica, rarely to western Panama. VAGRANT: Now casual to southeastern AZ, formerly a rare and local breeder (mainly

near Patagonia); no recent breeding records. Casual in extreme southern TX, mainly in winter (mostly Hidalgo Co.), but a few unsuccessful summer breeding attempts; single records north to Jeff Davis, Kenedy, and Aransas Cos.

1st winter ♂
albiventris

blackish
cap

thick,
stubby
bill

albiventris ♀

paler below
than *gravis*

darker
head than
albiventris

dark
cap

adult ♂
gravis

rose
throat

pale
cinnamon
underparts

♀
gravis

adult ♂
albiventris

SHRIKES Family Laniidae

Loggerhead Shrike (CA, Apr.)

What could be more incongruous than predators that attack mice, bats, birds, and insects, yet sing a soft catbird-like song? Add to this their habit of impaling prey and one gets extraordinarily fascinating birds.

Structure Shrikes' morphology is designed for carnivorous, raptor-like capturing and killing: broad wings and a long tail combine for speed and maneuverability; large jaws and powerful, hooked bills enable them to break the necks of their prey and carry them away. The legs and feet of shrikes are much weaker than those of true raptors and are of little use in hunting, although they occasionally do grab and carry small prey with their feet.

Behavior From an elevated perch, a shrike may watch attentively over long periods for prey. It strikes in a smooth movement, swooping down, attacking quickly, and carrying its quarry to another perch. Impaling prey on sharp objects may serve various purposes, including storage for a later meal, holding the food to tear it apart, and to mark territory with a ring of kills.

Plumage The Northern and Loggerhead Shrikes (residents of North America) are patterned distinctively in black, gray, and white with black masks. Both are gray above and paler below, with black wings marked by white patches and long, black, white-edged tails.

Distribution The 30 species (26 in the genus *Lanius*) now comprising this family are distributed mainly in the Northern Hemisphere, although the ranges of some extend well south in Africa and Asia.

Taxonomy Shrikes are so different from other birds that their placement within the passerine order has been unsettled. Previously placed between waxwings and starlings, they were moved more recently to a position between flycatchers and vireos.

Conservation Many species around the globe are in trouble, some seriously. Suggested causes include loss of habitat, elimination of prey by pesticides, warming climate, and trapping for food in Southeast Asia. One species, the Mountain Shrike from the Philippines, is listed as near threatened. In the United States and Canada, two Loggerhead subspecies are listed as endangered and one as threatened.

BROWN SHRIKE *Lanius cristatus*   BRSH ■ 4

This small Asian vagrant usually perches on the side of a bush or tree, often less conspicuously than our North American shrikes. It feeds primarily on insects but will also take small birds. Polytypic (4 ssp.; likely nominate in N.A.). L 7" (18 cm)

IDENTIFICATION No white patch on wings. **ADULT MALE:** Warm brown upperparts; brighter rump and upper-tail coverts; buffy underparts; narrow black mask; white supercilium. **ADULT FEMALE:** Mask less distinct; sides, flanks barred pale brown. **JUVENILE:** Mask limited to brown ear coverts; upperparts, sides, flanks faintly barred.

GEOGRAPHIC VARIATION Nominate most likely to occur in N.A.

SIMILAR SPECIES Juvenile Northern (only brownish shrike in N.A.) is larger, longer billed, distinctly barred across underparts. Three other Old World shrikes (*L. tigrinus, L. collurio, L. isabellinus*), though not recorded in N.A.,

female has duller head pattern and some barring on sides and flanks

juvenile cristatus

adult ♀ cristatus

white supercilium above dark mask

overall brown above

stubby bill

buffy-brown wash on sides and flanks

adult ♂ cristatus

tail uniform with back

should be separated when identifying this species.

VOICE CALL: Harsh *chacks* and *churucks*. **SONG:** Soft warbles interspersed with sharp notes.

STATUS & DISTRIBUTION Common in Asia. **BREEDING:** Forest edges and open areas with bushes from central Siberia to Kamchatka and Anadyr (Russian Far East); also China and Japan. **WINTER:**

Similar habitats in India, Southeast Asia, and East Indies. **VAGRANT:** Casual in N.A. (6+ recs. in AK; 4 in coastal northern CA, including 2 of wintering birds; 1 in NS). Of 7+ fall records, most were juveniles. Accidental in Europe

POPULATION Declining. Causes unclear, although habitat loss and trapping for food may account for declines in Southeast Asia.

LOGGERHEAD SHRIKE *Lanius ludovicianus*   LOSH ▪ 1

adult

black wings
with white
primary patch

black extends
across forehead

darker
gray
back

stubby
black bill

juvenile

This lively little predator occurs in many habitats, from remote deserts to suburban areas, where it perches on trees, shrubs, poles, fences, and utility wires. It captures small rodents and birds but favors large insects, typically swooping down smoothly, cruising low, and then flying up abruptly to another perch. Polytypic (6 ssp.; 6 in N.A.). L 9" (23 cm)

IDENTIFICATION ADULT: Medium-gray

upperparts; stubby bill; wide black mask extends thinly across forehead; black wings. **JUVENILE:** Paler gray with slight brownish tint; faintly barred overall.

GEOGRAPHIC VARIATION Subspecies differ slightly in bill shape and overall coloration; *mearnsi* (San Clemente Is., CA) has darkest gray upperparts.

SIMILAR SPECIES Northern Shrike is larger, paler, and larger billed; its mask does not extend across forehead.

VOICE CALL: Harsh *kee, kaak,* and *chek* sounds. **SONG:** Repeated chirps, squeaks, warbles, buzzes, and chips.

STATUS & DISTRIBUTION YEAR-ROUND: Open country with scattered trees and shrubs, desert scrub, grasslands, farms, parks. **MIGRATION & WINTER:** Northernmost birds move into southerly portions of range.

POPULATION Declining rapidly in most regions, especially in the East, where *migrans* is now rare to very rare in eastern Midwest and Northeast. In U.S., *mearnsi* is listed as endangered; in Canada, *migrans* is endangered and *excubitorides* is threatened.

NORTHERN SHRIKE *Lanius excubitor*   NSHR ▪ 1

As a distance, this hunter may appear more kestel-like, but it perches more horizontally and ceaselessly bobs its tail. It is best known during winter, when it perches on bare treetops or shrubs. It is merciless in pursuing prey. The pursuits (sometimes largely aerial) at times go on for minutes. Polytypic (7 ssp.; 1 in N.A.). L 10" (25 cm)

IDENTIFICATION ADULT: Pale gray upperparts; long, heavy, sharply hooked bill; black, narrow mask, tapering on lores; white forehead; black wings with white patch across base of primaries; grayish white underparts; long, black, white-edged tail. **FIRST-WINTER:** Slight brownish tint, grayish as season progresses; mask

indistinct; underparts barred pale brown. **JUVENILE:** Brownish overall; dark brown patch behind eye; underparts conspicuously barred.

GEOGRAPHIC VARIATION Six Eurasian subspecies (often known as Great Gray Shrike) and one generally recognized in North America *(borealis)* vary in size, plumage coloration. Populations breeding from Iberia and North Africa to China and India have been split as Southern Gray Shrike *(L. meridionalis)*.

SIMILAR SPECIES Loggerhead is smaller, smaller billed, darker on back; its mask extends thinly across forehead.

VOICE CALL: Loud *keek, shak,* other sharp sounds. **SONG:** Soft, catbird-

like trills, twitters, whistles, warbles, mews, squeaks, and harsh notes.

STATUS & DISTRIBUTION Uncommon. **BREEDING:** On taiga and edge of tundra. **MIGRATION:** Regularly moves short distances south of breeding range; irregularly invades in varying numbers the northern U.S. in late fall and winter. **WINTER:** Rare but regular in northern AZ, NM, and TX. Has occurred to southern CA, and to TN and NC. Casual in Bermuda.

POPULATION The trend in N.A. is unknown because of remote breeding range, but declines in Europe.

gray
forehead

paler gray
back

faint mask with
white eye ring

longer
bill with
distinct
hook

brownish on head
and upperparts

barred

immature

juvenile

VIREOS Family Vireonidae

Black-capped Vireo, male (TX, Apr.)

Vireos are known more for their vocal repertoire than for having fancy plumage. Fairly small birds, they have green or gray plumage that vaguely resembles that of warblers. A few vireos are colorful, but most blend with their environs—thickets, dense brush, and trees—or have a preference for skulking. As a result, vireos are more often heard than seen.

Structure Vireos are heavily built for a small songbird; some species are almost stocky. Vireo bills are rather thick and blunt, with a hook at the end of the upper mandible. Their legs are thick and strong, usually blue-gray tones.

Behavior The movements of vireos are deliberate. Upon landing, a vireo is likely to hold its position for a few seconds or longer. It frequently cocks and twists its head, scanning its surroundings for food or potential predators. Some species move along branches with tail cocked in the air. Food is mostly insects on the summering grounds, although fruit and seeds are taken; in winter many species eat primarily fruits and seeds. Vireos glean prey from the underside of leaves and limbs and while hovering.

They sometimes use their strong legs to hold prey while they eat. Vireos can sing continuously. Some are great mimics; since their songs are learned in the months after hatching, many develop notes from nearby breeding birds (e.g., White-eyed Vireos interjecting notes from Acadian Flycatcher and Western Scrub-Jay). Some species sing for hours, their songs appreciated more for their variation and length than for their melody.

Plumage The sexes look generally alike with differences readily apparent only in the Black-capped Vireo. Some species have more subtle differences between the sexes. Plumages are usually similar all year, although most species are brighter in fresh fall feathers after molting (which occurs on the breeding grounds). Vireos are typically gray or green above, white or yellowish below. Keys to identification are: the presence or absence of wing bars, distinctiveness of wing bars; presence of an eye ring and spectacles, presence of supercilia, and presence and boldness of adjacent eye lines or lateral crown stripes.

Distribution Widely distributed across the United States and Canada; nearly all vireos are migratory. Some species undertake short migrations; others move from Canada to the Amazon Basin.

Taxonomy Fifty-two species of vireos, a New World family, have been described, but vireo taxonomy is evolving. Work is ongoing to determine if some species might involve multiple species (e.g., Hutton's, and particularly Warbling). The North American species are all currently placed in the genus *Vireo*. While superficially similar to warblers, vireos are most closely related to shrikes (Laniidae).

Conservation Many vireo species are susceptible to Brown-headed Cowbird parasitism; local programs to prevent such parasitism have resulted in strong population rebounds. Habitat changes or losses impact several species. BirdLife International lists three species as threatened and two others as near threatened.

Genus *Vireo*

THICK-BILLED VIREO   *Vireo crassirostris*   TBVI   ▪ 4

This Caribbean species is a casual visitor to southeastern Florida from the Bahamas. Polytypic (5 ssp.; likely nominate in N.A.). L 5.5"(14 cm)

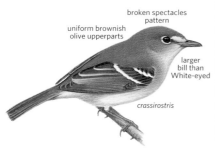

broken spectacles pattern

uniform brownish olive upperparts

larger bill than White-eyed

crassirostris

IDENTIFICATION Bigger than White-eyed Vireo, with larger, slightly stouter and grayer bill. Thick-billed is olive-brown above, occasionally showing some grayish olive on sides of neck and auriculars; breast is also tinged with brownish. Yellow eye ring is broken above eye, and yellow supraloral mark is particularly thick. There are two bold whitish wing bars. Underparts are dull and rather uniform pale olive-brown, with whitish undertail coverts.
SIMILAR SPECIES Many reports from south Florida are probably misidentified White-eyed Vireos. Note Thick-billed Vireo's broken eye ring,

pale olive-brown underparts, and particularly lack of gray on nape or contrasting clear yellow on flanks. Iris is darker than on adult White-eyed, but similar to darker-eyed, young White-eyed Vireos.
VOICE CALL: Slow and harsh *sheh*, or *chit* notes. **SONG:** Similar to that of White-eyed, but harsher.
STATUS & DISTRIBUTION Common resident in the Bahamas in mangroves and more commonly thickets. Uncommon and restricted to northern cays in Cuba. Casual year-round visitor to southeastern FL, accidental in Florida Keys.

WHITE-EYED VIREO *Vireo griseus* WEVI ■ 1

Explosive song from a thicket or tangle of vines usually announces the White-eyed. Polytypic (6 ssp.; 3 in N.A.). L 5" (13 cm)

IDENTIFICATION ADULT: Bold face pattern with yellow spectacles and white iris. Grayish olive above, gray neck and two whitish wing bars. Whitish below, pale yellow sides and flanks. **IMMATURE:** Gray or brown iris through fall.
GEOGRAPHIC VARIATION Compared to widespread nominate, *maynardi* (Florida Keys) is grayer above, with less yellow below and a larger bill. Subspecies

micrus (southern TX) is smaller than *maynardi* but similar in color.
SIMILAR SPECIES Can be confused in worn plumage with eastern subspecies of Bell's, which has a dark eye, fainter wing bars, and different face pattern. See also Thick-billed Vireo.
VOICE CALL: A raspy *sheh-sheh,* often repeated, suggestive of a House Wren. **SONG:** Loud, often explosive, five- to seven-note phrase; usually begins and ends with a sharp *chick.* Great mimic.
STATUS & DISTRIBUTION Common. **BREEDING:** Secondary deciduous scrub, wood margins. **MIGRATION:** Two southern subspecies largely sedentary.

Northernmost breeders of nominate subspecies migratory; southern populations probably resident. Arrives southern Great Lakes late Apr.–mid-May. Most fall migrants gone by early Oct., stragglers to early Nov. **WINTER:** Southeastern U.S., Mexico, and Caribbean. Attempted wintering recorded as far north as southern ON into Dec. **VAGRANT:** Rare to the Maritimes and NF, primarily in the fall; rare but annual to CA (mostly spring), otherwise casual to the West.
POPULATION Some declines in the western part of its range.

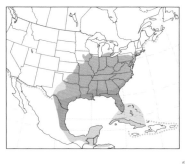

white iris, yellow spectacles

pale gray nape

yellow sides and flanks

thin white wing bars

Florida Keys *maynardi*

BLACK-CAPPED VIREO *Vireo atricapilla* BCVI ■ 2

This endangered, highly sought-after species is arguably our most stunning vireo. Only in the Black-capped do clear differences exist between the plumages of male and female vireos. Its restricted range and its penchant for thick scrub make viewing this species difficult. Small and very active. Monotypic. L 4.5" (11 cm)

IDENTIFICATION Both sexes have prominent white spectacles on which eye ring is broken at top. Other field marks include two yellowish wing bars, a greenish back, and reddish eyes. **MALE:** Glossy black cap. **FEMALE:** Slaty gray cap; back and flanks average paler. **IMMATURE:** Cap gray, with

males molting in some black during first year. Eyes brown. Females are more buffy below.
SIMILAR SPECIES Unmistakable, more likely to be confused with a Ruby-crowned Kinglet than any other vireo.
VOICE CALL: *Tsidik,* recalling Ruby-crowned Kinglet, and *zhree,* like Bewick's Wren. **SONG:** A hurried, restless *which-er-chee, chur-ee;* two- or three-note phrases, repeated with variations. Can be crisp and emphatic, or wren-like chatter. Whisper songs are softer, less warbled. Persistent vocalizer, singing through the day Mar. through Aug.
STATUS & DISTRIBUTION Endangered. **BREEDING:** Scrubby deciduous vegetation, usually oaks, in rocky hill country. **MIGRATION:** Spring migration

occurs late Mar. in TX (earliest 13 Mar.), mid- to late Apr. in OK. Fall migration occurs Aug.–Sept.; most birds depart by early Sept. **WINTER:** Pacific slope of Mexico. No U.S. records. **VAGRANT:** Eastern NE, east-central NM, and southern ON.
POPULATION The species was eliminated from KS in the 1930s and now is nearly gone from OK. Major factors include habitat destruction or deterioration and brood parasitism by Brown-headed Cowbird. Overgrazing removes key vegetation and attracts cowbirds; fire suppression adversely impacts the species.

slaty cap

adult ♀

adult ♂

black cap, white spectacles

immature ♀

BELL'S VIREO *Vireo bellii* BEVI ■ 1

Bell's is usually located by its distinctive song, often given from inside dense low and mid-level thickets. This rather plain, short-winged vireo frequently bobs its long tail. Polytypic (4 ssp.; all in N.A.). L 4.7" (12 cm)

IDENTIFICATION Plumage coloration varies with subspecies. All have two ill-defined white wing bars (lower one more prominent), indistinct white spectacles that are broken in front and back, and dark lores.

GEOGRAPHIC VARIATION Four subspecies become progressively greener above and yellower below from west to east. Endangered West Coast *pusillus* is grayish above, whitish below, with a trace of green or yellow in fresh fall birds. Nominate eastern subspecies is greenish above, yellowish below. Southwestern *arizonae* and *medius* are intermediate.

SIMILAR SPECIES Larger Gray Vireo has a longer tail, poorly defined wing bars, and an eye ring without a hint of spectacles.

VOICE CALL: A somewhat nasal, wrenlike *chee.* **SONG:** Many birds alternate—*cheedle-ee, cheedle-ew;* notes may be sharp or slurred, often delivered in couplets, ending with ascending or descending notes.

STATUS & DISTRIBUTION Uncommon to locally common. **BREEDING:** Moist woodlands, bottomlands, mesquite, and, in Midwest, shrubby areas on prairies. **MIGRATION:** Seldom seen on migration. First spring arrivals to southern breeding range in late Mar., May to northern breeding locales. Fall migrations occur Aug.–Sept. **WINTER:** Not well known, but primarily Mexico; scattered records for southern tier of states. **VAGRANT:** Casual north to OR and in Midwest to ON and along Gulf and Atlantic coasts north to NH.

POPULATION Declines in eastern TX, southern CA *(pusillus),* and AZ due to habitat loss and cowbird parasitism.

Eastern *bellii* is much more olive and yellow than *pusillus;* Southwestern *medius* and *arizonae* are intermediate

overall grayish color · one rather faint wing bar · rather long tail · subdued head pattern

West Coast *pusillus*

GRAY VIREO *Vireo vicinior* GRVI ■ 2

The Gray Vireo, drab as it may be, is a highly sought-after species. It is an easy bird to miss due to its choice of rather warm, out-of-the-way habitats and its penchant for hiding in undergrowth. It is best located by its persistent vocalizations. Gray is an active forager and constantly flicks and whips its tail in a manner reminiscent of a gnatcatcher. Monotypic. L 5.5" (14 cm)

IDENTIFICATION Seemingly featureless, upperparts are gray and underparts are whitish. Gray has a thin white eye ring and variably pale lores. Wings are brownish gray, with wing bars when fresh, the lower one more prominent. Long tail is gray and edged white. Bill is short and thick, even for a vireo. In fresh fall plumage, Gray Vireo can show slightest hint of green to rump and uppertail coverts and a slight yellowish wash along flanks.

SIMILAR SPECIES Plumbeous Vireo is similar, especially when worn, but is more boldly marked, has a longer primary projection, and a shorter tail. Compared to the West Coast subspecies of Bell's *(pusillus),* the larger Gray has a slightly bolder eye ring, heavier bill, and different vocalizations.

VOICE CALL: Shrill, descending whistled notes, sometimes delivered in flight. **SONG:** A series of *chu-wee chu-weet* notes, faster and sweeter than Plumbeous.

STATUS & DISTRIBUTION Fairly common, but local. **BREEDING:** Semiarid foothills and mountains with a variety of thorn scrub, junipers, or oaks. **MIGRATION:** Almost never seen. In spring, arrive CA and TX mid–late Mar.; early May in CO. Some fall migrants depart by late Aug., most by early Sept. **WINTER:** Mostly northwestern Mexico, local in western TX, southern AZ, and a few have recently been discovered in southern CA (Anza-Borrego S.P.). **VAGRANT:** Casual along southern CA coast and offshore islands, one remarkable specimen record for WI (Oct.).

POPULATION TX population expanding over past 30 years; CA's fragmented and reduced in size.

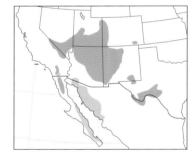

whitish eye ring but no spectacled effect · long tail, which is flipped about · faint wing bar · short primary projection

YELLOW-THROATED VIREO *Vireo flavifrons*   YTVI ■ 1

This is a large, colorful vireo and a strong, though slow-paced, singer. It moves sluggishly, which combined with its camouflaged coloration, can make it difficult to locate high in the leaves of the tall shade trees it favors. Yellow-throated often cocks its head as it surveys its surroundings or methodically searches for insects. Monotypic. L 5.5" (14 cm)
IDENTIFICATION Bright yellow spectacles, throat, and breast of this vireo are distinctive. Its wings are dark gray, with two bold, white wing bars. Crown and back are olive, rather bright, contrasting with a gray rump. Immature plumage is similar to that of adult but

paler yellow, sometimes with a slightly buffy throat.
SIMILAR SPECIES Unlike other vireos, but compare with Pine Warbler, with which it is confused, particularly in winter. Pine Warbler has a greenish yellow rump, streaked sides, thinner bill, and less complete and distinct spectacles. Vocalizations of the two are different—Pine Warblers often give a high, thin note when moving between branches. Yellow-breasted Chat, a particularly bulky warbler with a bill more suited to a vireo, has white spectacles and lacks wing bars.
VOICE CALL: Includes a rapid series of harsh *cheh* notes, similar to those of "Solitary" Vireo complex. **SONG:** Slow repetition of *de-a-ree, three-eight;* burry, low-pitched two- or three-note phrases separated by long pauses: It often gives a whisper song, which is more warbled and less burry.
STATUS & DISTRIBUTION Fairly common. **BREEDING:** Deciduous and mixed deciduous-coniferous habitats. **MIGRATION:** Long-distance, trans-Gulf migrant. Early spring

yellow spectacles / thick bill / yellow throat and breast / white wing bars / gray rump / white belly

migrant, with arrivals in southern states mid–late Mar., mid Apr. farther north, early May in Great Lakes. Fall migrations Aug.–Oct. (migrants noted as early as late July). Latest records are mid-Oct. in northern and middle latitudes, early Nov. in south. **WINTER:** Tropical lowlands of C.A., Bahamas, and Caribbean to northern S.A. More scarce in U.S. than the numerous reports would suggest. Most reports are misidentified Pine Warblers or Yellow-breasted Chats. Rare in southernmost FL, casual in southern CA, southern TX. **VAGRANT:** Very rare in the West, more in spring than fall.
POPULATION Apparently stable, with some local fluctuations.

PLUMBEOUS VIREO *Vireo plumbeus*   PLVI ■ 1

Formerly part of the "Solitary" Vireo complex—along with Blue-headed and Cassin's—Plumbeous is the grayer, Rocky Mountain species. Appearing large and bulky, the bird moves through trees in a deliberate manner foraging for insects and sings frequently, even in nonbreeding seasons. Polytypic (4 ssp.; nominate in N.A.). L 5.5" (14 cm)
IDENTIFICATION Nominate described. A rather colorless species, but with a strong pattern composed of bold white spectacles and wing bars contrasting with lead gray upperparts. Generally

white below, except on sides of breast, which are gray, sometimes tinged olive. Outer tail feathers are broadly edged white, visible from below; tail looks long on this species. In fresh fall plumage, yellow, if present on underparts, is restricted to flanks, and rump might show a greenish tint.
SIMILAR SPECIES Larger and bigger billed than Cassin's, with sharper head and throat contrast and gray upper parts. Avoid confusion with worn spring Cassin's, which can be dull gray above and seemingly lack green or yellow. Compare worn summer Plumbeous to Gray. Gray has a complete eye ring, but lacks supraloral mark that completes the spectacles, has a shorter bill, and waves its longer tail like a gnatcatcher. Grays also have short wings and a shorter primary projection.
VOICE CALL: *Cheh* notes similar to Blue-headed. **SONG:** Burry notes almost indistinguishable from Cassin's, but usually starts with clear notes; more burry sounding than Blue-headed.

STATUS & DISTRIBUTION Fairly common. **BREEDING:** Montane forests of pine, and oak-juniper, locally in deciduous woodlands. **MIGRATION:** Late spring migrant, primarily May. Late fall migrant, primarily late Sept.–mid-Oct., with stragglers to Nov. **WINTER:** Primarily Mexico. Rare in southern coastal CA, south-central AZ, casual southern NM, TX. **VAGRANT:** Spring records from ND, MA, and ON. Fall records from NJ, LA, AB, and NS.
POPULATION Recently found breeding in northwest NV and southeastern OR.

white spectacles / gray above with two white wing bars / long primary projection / slaty gray sides / shorter tail than Gray / *plumbeus*

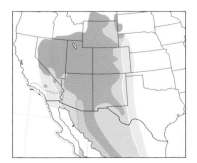

CASSIN'S VIREO   *Vireo cassinii*   CAVI ▪ 1

Cassin's Vireo is the western counterpart of—and formerly considered conspecific with—the Blue-headed Vireo. Cassin's, very similar to the Blue-headed Vireo in both plumage and behavior, is routinely heard singing on the breeding grounds, often throughout the day. The movements and behavior of the Cassin's are like those of the Blue-headed. Occasionally, the Cassin's flicks its wings like a kinglet. Polytypic (2 ssp.; nominate in N.A.). L 5.3" (13 cm).
IDENTIFICATION Olive-gray head of Cassin's contrasts slightly with its greenish back. There is a strong pattern

of white spectacles and wing bars (the latter can be yellowish) contrasting with dark upperparts. Flanks are heavily washed with olive-yellow; throat and breast are whitish, typically dingy. Tail is dark blackish brown above, with olive or gray edges. From below, outer tail feathers are edged with white when fresh. Sexes generally look alike, although plumage of males is slightly brighter than that of females. Plumage of immature Cassin's vireos is duller, especially that of females, which can have entirely green heads and lack any white in tail.
SIMILAR SPECIES Cassin's Vireo is similar to Blue-headed (see sidebar below). Compare the all-green head of the immature female Cassin's to that of Hutton's Vireo. Note the difference in eye ring, subtle difference to underparts, and vocalizations.
VOICE CALL: Call of Cassin's consists of scold notes similar to those of Blue-headed. **SONG:** Jerky two- to four-note song almost identical to the song of Plumbeous; much more burry than Blue-headed. Also very similar to the song of Yellow-throated Vireo.
STATUS & DISTRIBUTION Common. **BREEDING:** Coniferous and mixed forest.

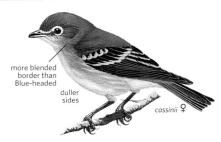

more blended border than Blue-headed

duller sides

cassinii ♀

MIGRATION: Early spring migrations typically occur in late Mar., peak in late Apr. west of the Sierra Nevada in CA, a little later east of the mountains; early May near Canada east of the Cascades. Fall migrations are more protracted: Aug. (northern populations) to Sept. is the bulk of passage; lingerers through Oct., particularly on the coast; a few through Nov. Migrants are more likely throughout the southwestern and the western Great Plains in fall than in spring. **WINTER:** Primarily Mexico. Rare in coastal CA, casual in interior CA, southern AZ, NM, and western TX. **VAGRANT:** Cassin's is casual north to AK; accidental in the East.
POPULATION Recent increases have been noted in some western populations of Cassin's Vireo.

Cassin's Vireo versus the Blue-headed Vireo

Field identification of these vireos is complex; some birds must be left unidentified. Two main problems exist. First, the Blue-headed in the westernmost part of its range is said to look and sound a bit more like the Cassin's. Second, with both species, adult males are the boldest and brightest, immature females the dullest. While the Blue-headed is more colorful—richer greens, brighter yellows, and more—these characters are not reliable for specific identification. Focus on the following key marks:
1. If the head has a distinct blue cast, it is a Blue-headed. Even bold Cassin's have a lead-gray to olive-gray head. If the head is green, lacking any gray, it is an immature female Cassin's. Any other head color can be matched by both species.
2. While all "Solitary" Vireos are whitish below, many Blue-headed appear brighter white, whereas most Cassin's appear dingy.

Cassin's Vireo, male (CA, Apr.)

Blue-headed Vireo, female/immature

3. All birds show contrast, but on the Blue-headed the transition from dark head to white throat is crisp, particularly noticeable at the lower edge of the auriculars. In the Cassin's these transitions are more blended. This mark takes experience to assess; the differences can be minor.
4. Both species can show white in the tail, but the Blue-headed has more, and all ages and sexes have white. In adult males, the entire outer web of the outermost tail feather is white (with fairly extensive white on the penultimate rectrix), with white wrapping around the tip and forming a broad edge on the inner web. Adult male Cassin's have white; it often covers much of the outer web, but is not as extensive. White is less evident in other age/sex classes.
A combination of marks, studied in good light, is necessary to tell these vireos, as is the ability to leave some unidentified to species. ■

BLUE-HEADED VIREO *Vireo solitarius* BHVI ■ 1

The Blue-headed Vireo is typically found in late spring or summer by its song. It forages at mid-level, yet can be difficult to find among the leaves. In summer it primarily eats insects and appears quite inquisitive, frequently cocking its head as it slowly forages in the branches of a tree. Polytypic (2 ssp.; both in N.A.). L 5.3" (13 cm)

IDENTIFICATION Bright blue-gray to gray hood clearly contrasts with a bright olive back and white spectacles and throat. Wing bars and tertials are yellow-tinged, and dark secondaries have greenish yellow edges. Tail is dark above, with greenish edges, and white is easily seen in outer tail from below.

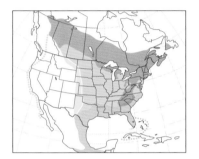

Underparts are clean white with bright yellow (sometimes mixed with green) on sides and flanks. Sexes mostly look alike, but some variation, with adult males being most colorful. **IMMATURE:** Like adult, but some females are duller. **GEOGRAPHIC VARIATION** Larger Appalachian *alticola* has more slaty back; only flanks are yellow.

SIMILAR SPECIES Identification is complex; compare to Plumbeous Vireo and especially Cassin's (see sidebar opposite).

VOICE CALL: A nasal *cha-cha-cha-cha*, given as a single *cha* or repeated; a typical vireo scold. **SONG:** Short, clear notes with various intervals, similar to Red-eyed, but slower. Blue-headed Vireo can sing the song of a Yellow-throated Vireo (with which it has hybridized) and is a good mimic (e.g., singing a White-eyed Vireo song or giving a Yellow-bellied Flycatcher call).

STATUS & DISTRIBUTION Common. **BREEDING:** Mixed woodlands. **MIGRATION:** Short-distance *(alticola)* to medium-distance migrant. Earliest vireo to move north in spring, to mid-Atlantic and southern Midwest by mid-Apr.; higher elevations or latitudes not until

white spectacles, bluish gray head

sharp contrast between auricular and throat

yellow sides with some olive

solitarius ♂

May. Latest vireo to depart in fall, more protracted migration in fall. Most migration starts mid to late Sept. away from Appalachians. Not expected in southern states until Oct. Midwest peak late Sept.–early Oct.; some remain into Nov. **WINTER:** Winters in southern states south to northern C.A. Rare north of mapped range. **VAGRANT:** Annual to the West, primarily CA in fall (most records late Sept.–Oct.), casual there in winter and spring. Casual in other western states. **POPULATION** Overall increases in recent years, with some local declines due to habitat destruction.

HUTTON'S VIREO *Vireo huttoni* HUVI ■ 1

This active vireo often flicks its wings in a kingletlike manner. Its thicker bill is the first clue that this is not a kinglet. Quite vocal, Hutton's is often heard before seen. Outside the breeding season, Hutton's vireos form mixed-species flocks. Polytypic (11 ssp.; 7 in N.A.). L 4.7" (12 cm)

IDENTIFICATION Similar to Ruby-crowned Kinglet. Hutton's is greenish to olive gray above, with an eye ring broken above eye; pale lores. Two whitish wing bars; greenish yellow

huttoni group

stephensi group

mexicanus

vulcani

edges of secondaries and primaries connect to lower wing bar.

GEOGRAPHIC VARIATION Seven subspecies in North America are divided into Pacific group and Interior (or Stephen's) group. Pacific group birds are smaller and greener. Interior group birds are larger, paler, and grayer. More than one species might be involved. See map for separate (allopatric) ranges.

SIMILAR SPECIES Most likely to be confused with Ruby-crowned Kinglet; Hutton's has a thicker bill and paler lores and lacks a dark area below its lower wing bar. Immature female Cassin's Vireo can be greenish, but has prominent white spectacles and whitish throat.

VOICE CALL: Low *chit* and raspy *rheee*, often followed by nasal, descending *rheee-ee-ee-ee*. **SONG:** Pacific Group birds give a repeated or mixed rising *zu-wee* and descending *zoe zoo*. Interior Group, a harsher *tchurr-ree*.

STATUS & DISTRIBUTION Uncommon to fairly

common. **BREEDING:** Mixed evergreens and woodlands, particularly live oaks. **YEAR-ROUND:** Largely resident, some seasonal dispersal to lower elevations, as early as July in CA. **VAGRANT:** Casual in southeastern CA and southwestern AZ; accidental in western NV. **POPULATION** Recent expansion to the Edwards Plateau in TX.

thicker bill than Ruby-crowned Kinglet

eye ring broken above eye, unlike Cassin's Vireo

overall dull

pale supraloral spot

worn summer

pale edges on primaries connect to lower wing bar

overall bright

interior subspecies average more grayish than coastal subspecies and fresh plumaged fall birds are greener than worn summer ones

fresh fall

WARBLING VIREO *Vireo gilvus*   WAVI ▪ 1

An unmarked vireo, the Warbling is most often located by its song. The birds tend to work mid and top parts of broad, leafy trees. Polytypic (5 ssp.; 3 in N.A.). L 5.5" (14 cm)
IDENTIFICATION Among the dullest of

Eastern *gilvus* distinctly larger

pale lores

spring *gilvus*

fall *gilvus*

fall *swainsoni*

yellow brightest on sides

paler primary coverts

smaller than *gilvus*; comparably sized to Philadelphia

vireos, lacking wing bars and spectacles; gray with brownish or greenish tones to upper parts. Face pattern is ill-defined, with a dusky postocular stripe and pale lores; white eyebrow lacks a dark upper border. Underparts are typically whitish. **FALL:** In fresh plumage, greener above with yellow wash on flanks and undertail coverts.
GEOGRAPHIC VARIATION Two western subspecies, especially *swainsoni*, are smaller than the nominate eastern subspecies, have a slighter bill, and tend to be more olive above. Eastern and western birds likely represent two species; they breed almost sympatrically in a wide band from southern AB to CO Front Range. Two additional subspecies in Mexico, fit into the western group.
SIMILAR SPECIES Fall birds greenish above, often with extensive yellow below, and can be confused with Philadelphia. On Warbling Vireo, yellow restricted to sides and flanks; it lacks dark lores of Philadelphia.
VOICE CALL: A nasal *eahh* mobbing call; typical vireo. Note commonly uttered in flight. **SONG:** Eastern *gilvus* song is delivered in long, melodious, warbling phrases. Song of western *swainsoni* similar but less musical, higher tones.

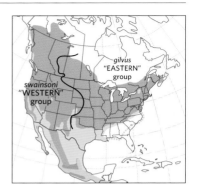

gilvus "EASTERN" group

swainsoni "WESTERN" group

STATUS & DISTRIBUTION Common. **BREEDING:** Deciduous woodlands, primarily riparian areas. **MIGRATION:** Western birds have prolonged spring migration (early Mar.–late May). Eastern subspecies a circum-Gulf migrant, rare on eastern Gulf Coast; arrives mid-Apr. in TX, by early May to Great Lakes. Peak fall migration in northern U.S. late Aug.–mid-Sept. Mostly gone from U.S. by mid-Oct.; stragglers at southern areas to Nov., rarely Dec. **WINTER:** Mostly Mexico and C.A. Casual to southern CA, southern AZ, and southern LA. **VAGRANT:** Western AK.
POPULATION Overall stable.

PHILADELPHIA VIREO *Vireo philadelphicus*   PHVI ▪ 1

This smallish green vireo is usually seen singly amidst tall, leafy trees. Its rather bright coloration may suggest a warbler. Monotypic. L 5.3" (13 cm)
IDENTIFICATION **FACE PATTERN:** Dull white eyebrow and dark eye line. Green above, with contrasting grayish cap and variably yellow below; palest on belly.

dark eye line extends through lores

fall birds often quite bright yellow

fall

shorter tail

dark primary coverts

spring

yellow of equal intensity extends across breast

No wing bars or spectacles. **FALL:** Fresh plumage, usually brighter green above and brighter yellow below.
SIMILAR SPECIES Warbling Vireo is very similar. Note Philadelphia's dark eye line extending through lores, darker cap, darker primary coverts, and yellow continuing to center of throat and breast. Some spring Philadelphias can be very dull with detectable yellow only with the best of views. Tennessee Warbler, superficially similar, has a thinner bill, white undertail coverts (usually), and usually utters high-pitched flight notes while it forages.
VOICE CALL: An *ehhh* sometimes given though often silent. But may vocalize in response to "pishing." **SONG:** Very closely resembles that of Red-eyed Vireo but averages slightly slower, thinner, and higher-pitched.
STATUS & DISTRIBUTION Uncommon. **BREEDING:** Most northerly breeding vireo; open woodlands, streamside willows and alders. **MIGRATION:** In spring first arrivals usually mid-Apr. (TX), and early May (southern Great Lakes);

peak late Apr.–early May, and mid-May, respectively. Broad fall migratory path; more numerous on East Coast and in Southeast than in spring. Movement mid-Sept.–mid-Oct.; arrivals in southern states not expected until late Sept. Lingering birds casual to Nov. **WINTER:** Southern C.A. Accidental in southern CA. **VAGRANT:** Casual to rare (CA, >150 records) in the West. Casual to the Azores; two Oct. records for Europe.

RED-EYED VIREO *Vireo olivaceus*   REVI ■ 1

breeding
olivaceus

black lateral
crown stripe

dark line
through eye

1st fall
olivaceus

olive above, white below

pale yellow
lower flanks
and undertail
coverts

A large vireo, the Red-eyed Vireo is one of the most common songbirds in eastern woodlands. It moves sluggishly through the canopy of broadleaf forests, making it hard to detect, and often picks food by hover-gleaning. It sings incessantly, often throughout the day. Polytypic (10 ssp.; nominate in N.A.). L 6" (15 cm)

IDENTIFICATION Bold face pattern with white eyebrow, bordered above and below with black. Ruby red iris of adult visible at close range. Gray to blue-gray crown contrasts with olive back and darker wings and tail. Lacks wing bars. **FALL:** Flanks and undertail coverts usually washed olive or olive-yellow. **IMMATURE:** Brown iris; often extensive yellow or olive-yellow wash

on undertail coverts and flanks, which may extend up to the bend of the wing.

GEOGRAPHIC VARIATION The *chivi* subspecies group of South America has been considered a separate species. **SIMILAR SPECIES** Resembles Black-whiskered Vireo, but Red-eyed has a bold, black lateral crown stripe above its white eyebrow; more green above, less brown; red eye (adult); and will not show the diagnostic black whisker. Yellow-green Vireo can be similar. **VOICE CALL:** A whining, down-slurred *myahh.* **SONG:** A deliberate *cheer-o-wit, cher-ee, chit-a wit, de-o;* a persistent singer of a variable series of robinlike short phrases **STATUS & DISTRIBUTION** Common. **BREEDING:** Woodlands. **MIGRATION:**

Long-distance migrant. First spring arrivals on Gulf Coast by late Mar., late Apr. in East/Midwest; peaks during Apr. on Gulf Coast, mid-May in East/Midwest. Migration continues into June farther north. Fall migration peaks late Aug. through Sept., most depart by early Oct., some linger to Nov. Southern peak is early Sept.–early Oct. **WINTER:** Winters in northern S.A. No documented winter records for N.A.—reports at this season suspect. **VAGRANT:** Rare but annual across Southwest; a few annually along CA coast. Very rare on the Azores and 75+ records for Europe, primarily late Sept.–mid-Oct.

POPULATION Stable. Expanded into OR, UT, and then Newfoundland in the mid-20th century.

BLACK-WHISKERED VIREO *Vireo altiloquus*   BWVI ■ 2

This Caribbean counterpart of the Red-eyed Vireo is a summer visitor along the Florida coast. Territorial birds are vocal throughout the day, but are heard much more often than seen. Polytypic (6 ssp.; 2 in N.A.). L 6.3" (16 cm)

IDENTIFICATION Dark malar stripe can be quite evident at times, but often hard to see, owing to lighting, angle, and wear. White eyebrow is

prominent, bordered below by black eye line and above by gray crown and, sometimes, a faint black border. Bill is large, and eye is amber. Lacks wing bars and is whitish below, with variable, pale yellow wash on sides and flanks. **IMMATURE:** Duller than adult, with buffier underparts.

GEOGRAPHIC VARIATION Breeding subspecies is *barbatulus* (FL, Cuba, Haiti); *altiloquus* (the rest of the Greater Antilles) is a vagrant with an indistinct, brownish gray eyebrow and even larger bill. **SIMILAR SPECIES** Most similar to Red-eyed Vireo. Compare also to Yellow-green Vireo and Yucatan Vireo. **VOICE CALL:** A thin, unmusical *mew.* **SONG:** Deliberate 1- to 4-note phrases (most typically 2 to 3 notes); notes are loud and clear, separated

less evident dark
lateral crown stripe

black whisker

longer
bill than
Red-eyed

barbatulus

by a distinct pause, less varied and more emphatic than Red-eyed.

STATUS & DISTRIBUTION Common. **BREEDING:** Coastal mangrove swamps and, less commonly, to a few miles inland. **MIGRATION:** Spring migrants arrive late Mar.–mid-Apr. Harder to assess fall migrants as males not as vocal. Most depart by early Sept. **WINTER:** Mostly Amazonia, but limits of subspecies' winter ranges poorly known. **VAGRANT:** Subspecies *barbatulus* is casual along the Gulf Coast from eastern TX (about 20 records, mostly early Apr.–late May; twice in fall) to northwestern FL.

Accidental up Atlantic coast to VA. A few specimen records have been identified as nominate *altiloquus* (FL, LA).

POPULATION Most FL populations are stable; the Tampa Bay population has been adversely affected by cowbird parasitism.

YELLOW-GREEN VIREO *Vireo flavoviridis*   YGVI ▪ 3

The Yellow-green Vireo is primarily a Central American and Mexican species sighted along the U.S. border, particularly in southern Texas. Its behavior and general appearance are similar to Red-eyed Vireo's, including the ability to blend into the leaves through which it forages. When agitated, it often cocks its tail and raises its crown feathers. Yellow-green Vireo was re-split from Red-eyed Vireo in 1987. Polytypic (3 ssp.; nominate in N.A.). L 6" (15 cm)
IDENTIFICATION Typically yellowish green above, with rather bold, yellow edgings to flight feathers; lacks wing bars or spectacles. Head pattern is similar to that of Red-eyed but more blended—a gray crown blends into back, and pale supercilium is bordered by rather indistinct lines. Dark line above supercilium is often very faint or absent. Underparts are whitish, with a yellow wash to sides and undertail coverts. Yellow continues up sides of breast

and blends into ear coverts. Iris is ruby red, and bill is long, often with pale coloration on upper mandible as well as lower. **FALL:** Brightest in fresh plumage, with more extensive yellow ventrally. **IMMATURE:** Similar to adult but iris is brown.
GEOGRAPHIC VARIATION Subspecies are weakly differentiated; nominate in N.A.
SIMILAR SPECIES Similar to Red-eyed, but bill longer. Strong yellowish wash on flanks extends to sides of face, while yellow of Red-eyed is usually restricted to flanks. Yellow-green has a more diffuse face pattern; dark borders to supercilium reduced or absent. Bill has pale tones (pink, blue) versus black bill of Red-eyed. Upperparts are more yellowish green, with brighter edges to remiges. Black-whiskered is much duller, with less green above, substantially less yellow below, and a longer, dark bill. Compare to Warbling and Philadelphia Vireos.

VOICE CALL: A soft, dry *rieh;* chatter often repeated. **SONG:** A rapid but hesitant series of notes suggestive of a House Sparrow.
STATUS & DISTRIBUTION Rare to casual. **BREEDING:** Very rare, but probably annual breeder in southern TX (Rio Grande Valley). Main breeding range is Mexico south to Panama, primarily utilizing lowland forests and forest edges. **VAGRANT:** Casual in spring on the upper Gulf Coast (mid-Apr.–May) east to FL. Very rare, but now recorded annually from coastal CA (late Sept.–Oct.; one July rec.); a few interior southern CA records. Casual to southern AZ and southern NM in summer.
POPULATION Stable, but current studies needed. Reliance on forest edge, in part, protects populations from effects of deforestation.

duller gray cap than Red-eyed with less-evident lateral crown stripe

yellowish olive above

often cocks tail

flavoviridis

large bill

yellow sides and flanks, some yellowish into face

YUCATAN VIREO *Vireo magister*   YUVI ▪ 5

A large brown vireo from the Yucatán Peninsula, it is accidental in the United States. Monotypic. L 6" (15 cm)
IDENTIFICATION Perhaps the most noticeable features are brown plumage and large, heavy bill. A broad pale supercilium contrasts with a dark eyeline. Throat and underparts are dull whitish, sometimes tinged buff; sides and flanks are washed with grayish brown. No spectacles or wing bars. Short primary projection, as one would expect with a generally sedentary species.
SIMILAR SPECIES Like a large-billed

Red-eyed that lacks olive tones. Its shorter wings and overall structure give it an ungainly look—very different from more streamlined Red-eyed. Yucatan's bill is much larger, and it lacks dark lateral crown stripe and has a brownish (not gray) crown.
VOICE CALL: Sharp, nasal *benk* notes, often strung together in a series; also a softer, dry chatter. **SONG:** Rich phrases, delivered hesitantly.
STATUS & DISTRIBUTION Fairly common within home range. **YEAR-ROUND:** Resident on Yucatán Peninsula and some nearby islands; also Grand Cayman. **VAGRANT:** Accidental in TX—

1 remarkable record from the upper TX coast, near Crystal Beach (Apr. 28–May 27, 1984).
POPULATION Current data needed; no indications of concern.

large heavy bill

short primary projection

grayish brown sides and flanks

CROWS AND JAYS Family Corvidae

Northwestern Crow (AK, Mar.)

Identification of many species is remarkably easy, but crows and ravens present some of the greatest identification challenges in North America. These problems are often complicated by distant views of birds in flight. Where two or more species overlap, birders should focus on bill size and structure, tail shape, and overall proportions. Vocalizations are often the best way of separating similar species. Because everyone hears vocalizations differently, it may be helpful to transcribe your own depictions of vocalizations, noting different calls at different times of year (e.g., juvenile begging calls) and between different populations. Much remains to be learned about regional variation in many species of corvids (e.g., American Crow, Gray Jay), and detailed observations and photographs will aid in our understanding.

Structure Corvids vary considerably in size, but all species have strong legs and feet and a straight bill. Males are generally larger than females. Size differences may occasionally be helpful in sexing paired breeders but are otherwise of little help in sexing birds in the field.

Behavior Corvids are omnivorous, and many species store seeds, nuts, and other foods for consumption during winter months. Some corvids gather in large flocks, particularly during the nonbreeding season. Several species are cooperative breeders, with helpers that assist with nest rearing. Corvids tend to have a diverse array of vocalizations. Most species give a few vocalizations most frequently, but do not be surprised to hear calls not described in this guide. Corvids can also mimic sounds. While jays are most known for hawk imitations, several species give very soft songs into which they incorporate a variety of mimicked sounds.

Plumage Coloration varies considerably, from the brightly colored Green Jay to the grayish coloration of the Clark's Nutcracker and the Gray Jay. Many North American jays are blue, and all crows and ravens are entirely or mostly black. Most corvids appear very similar throughout the year. Sexes are identical. Juvenile plumage is held only briefly and is generally duller but similar to that of the adult. Juveniles and immatures of some species have pale coloration to the bill that they may hold for one or more years. All corvids undergo a single annual (prebasic) molt after breeding. The first prebasic molt is usually partial; adult prebasic molts are complete. In some species first-year birds may be aged by looking for differences in wear in the wing coverts. Crows and ravens retain juvenal flight feathers for one year. By spring immatures are often quite worn, and retained brownish flight feathers can contrast markedly with fresher, replaced black wing coverts.

Distribution Corvids are found on all continents except Antarctica. In North America these birds range from the high Arctic to the Sonoran desert and utilize nearly every habitat in between. Migration and dispersal is diurnal, but most corvids are nonmigratory or engage in limited movements. Some species stage occasional irruptions, during which time individuals can be far from their normal range. Migration and dispersal tendencies may vary between different populations of the same species.

Taxonomy Corvid taxonomy continues to undergo revision. Worldwide there are between 115 and 120 species in the family Corvidae. There are 20 species recognized in North America, but a strong case could be made for splitting the Western Scrub-Jay and placing the Northwestern Crow with the American Crow. Corvids are believed to have arisen from an Australian ancestor that gave rise to a large group of some 650 species, which includes such diverse Old World groupings as the birds-of-paradise, Old World orioles, wood-swallows, and paradise flycatchers. The only other families from this assemblage that are found in North America are the vireos and the shrikes.

Conservation Some species are quite adaptable and do well near humans. Habitat destruction and fragmentation pose the most serious threats to this family, with additional mortality caused by pesticides and traps. Populations of many corvids declined dramatically with the spread of the West Nile virus.

GRAY JAY *Perisoreus canadensis* GRAJ ▪ 1

Bold and cunning, Gray Jays usually travel in pairs or family groups, but they often stay in cover and remain fairly quiet. They feed on the ground and in trees. While many corvids store food, the Gray Jay and its Asian counterpart, the Siberian Jay *(P. infaustus),* are the only ones known to produce saliva that allows them to "glue" food together for storage. They sometimes visit feeders. Polytypic (8–11 ssp.; all in N.A.). L 11.5" (29 cm)

IDENTIFICATION This is a fluffy, long-tailed jay with a small bill and no crest. **ADULT:** Grayish above, paler below, usually with variable dark markings on back of head and pale tip to tail. **JUVENILE:** Sooty-gray overall, with a faint white moustachial streak. They have molted by early fall, after which they closely resemble adults. **FLIGHT:** Typically a burst of wingbeats, followed by slow unsteady glides.

GEOGRAPHIC VARIATION A highly variable species, but variation is somewhat clinal and many intermediate individuals occur where subspecies meet. Eight to eleven subspecies now generally recognized. Nominate *canadensis* (boreal forest from NT and northern AB to NF and PEI) is intermediate in overall coloration, with a white collar and forehead and medium-gray hindcrown and nape; upper parts only slightly paler than hindcrown nape, with prominent dark shaft streaks; wings are

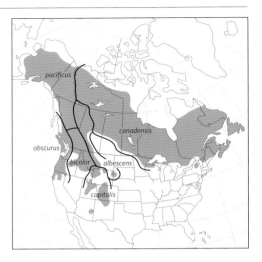

moderately edged with white. Subspecies *nigricapillus* (NL, QC, NS) is darker overall with more extensive black on crown. Subspecies *capitalis* (central and southern Rockies from eastern ID and WY to AZ and NM) has an extensive pale crown that makes head appear mostly white; upperparts considerably paler than *canadensis;* wings edged with frosty white. Subspecies *bicolor* (western MT to southeastern BC and northeastern OR) is intermediate between *capitalis* and *canadensis,* with mostly whitish crown and frosty white wing edges of *capitali*s but with darker upperparts that have dark shaft streaks. Subspecies *albescens* (east side of Rockies from southeastern YT and east-central BC to the Black Hills east to central MB and northwestern MN) is also intermediate between *capitalis* and *canadensis,* but darker coloration on nape and hindcrown is more extensive than on *bicolor.* Subspecies *obscurus* (Pacific coast from southwestern BC to northwestern CA) has dark brownish gray upperparts (with white shaft streaks), with extensive brownish gray on crown and nape and paler underparts than other subspecies. Subspecies *griseus* (southwestern BC to northern CA) is slightly larger than *obscurus* and is paler gray. Subspecies *pacificus* (AK to YT and northwestern BC) similar to *obscurus* with brownish gray upperparts, but underparts heavily washed brownish gray.

SIMILAR SPECIES Clark's Nutcracker has longer bill, black-and-white wings and tail, lacks dark coloration on head. Juveniles distinctive; usually seen with adults.

VOICE Relatively quiet. **CALL:** A whistled two-part *wheeoo* and a low *chuck.* Also a screeching *jaaay.*

STATUS & DISTRIBUTION Common. **BREEDING:** Takes place in late winter in northern and mountain coniferous forests; less commonly in mixed forest. **DISPERSAL:** This species only infrequently strays south in the East, and to lower altitudes in the West; accidental to northeastern IA, MA, and southern NY.

POPULATION Birds on the south edge of range may be declining. Vulnerable to traps for smaller fur-bearing mammals.

sooty overall

juvenile

pale tail tips

pale tail tips

boreal adult
canadensis

dark nape

whitish crown

paler on head and back than nominate subspecies

pale tail tips

southern Rockies adult
capitalis

dark on nape extends to crown

small bill

pale gray underparts for all subspecies

Northwest adult
obscurus

TROPICAL JAYS Genera *Psilorhinus* and *Cyanocorax*

These two genera include 17 species of tropical jays, only two of which are found north of Mexico—the Green Jay and the Brown Jay. The Brown Jay was recently moved out of *Cyanocorax* into its own monotypic genus. Those species that have been studied have social breeding systems.

BROWN JAY *Psilorhinus morio*   BRJA ■ 3

The range of this neotropical species barely reached southern Texas. Brown Jays usually travel in boisterous flocks of 5–10 birds. In Texas it was most readily found at feeding stations. Polytypic (3 ssp.; *palliatus* in N.A.). L 16.5" (42 cm)

IDENTIFICATION A very large jay with long broad tail. Dark sooty-brown overall except for dirty whitish belly and undertail coverts. Juvenile has a yellow bill, yellow legs, and narrow yellow eye ring, all turning black by second winter. Transitional birds have bills that are blotched with yellow and black. **FLIGHT:** Slow and unsteady, with heavy wingbeats interspersed with short glides.

GEOGRAPHIC VARIATION All birds north of Mexico belong to *palliatus,* characterized by uniformly brown tail and whitish belly with sooty wash. More southerly subspecies are polymorphic.

SIMILAR SPECIES Nearly unmistakable; crows and ravens uniformly black.

VOICE Noisy; less varied than most other jays. **CALL:** A loud, displacing *KAAH KAAH* or *KYEEAH,* similar to Red-shouldered Hawk's but higher and more offensive; usually given endlessly. Also produces a less voluminous popping or hiccuping sound.

STATUS & DISTRIBUTION Discovered in TX in the early 1970s, it was uncommon and local in woodlands mainly below Falcon Dam. By 2000 numbers had markedly decreased and it disappeared about 2011; now casual, at best.

thin yellow orbital ring
yellow bill
juvenile *palliatus*
heavy black bill
pale belly
very large size
adult *palliatus*

GREEN JAY *Cyanocorax yncas*   GREJ ■ 2

The range of this neotropical species extends into southern Texas. While brightly colored, Green Jays blend in well with woodlands, and they are best found by listening for the characteristic vocalizations. The Green Jay is a regular visitor to feeding stations throughout the Lower Rio Grande Valley. Polytypic (11–14 ssp.; *glaucescens* in N.A.). L 10.5" (27 cm)

IDENTIFICATION Unmistakable long-tailed jay with blue crown, complex black-and-blue face pattern, and black breast. Upperparts are bright green; underparts are paler and tinged with yellow; bright yellow on undertail coverts and outer tail feathers. Juveniles are duller with brownish olive head and throat, dull green upperparts, and paler yellow underparts. **FLIGHT:** Quick wingbeats interspersed with short glides; generally prefers to fly in forest interior.

GEOGRAPHIC VARIATION All birds north of Mexico belong to northern *glaucescens.* Eleven to fourteen subspecies in the New World. The five Andean subspecies are sometimes considered a separate species, Inca Jay (*C. yncas*); if so, the scientific name of Middle and North America birds would become *C. luxuosus.* Interestingly, the species is absent from Nicaragua to Panama.

SIMILAR SPECIES Unmistakable.

VOICE An array of chatters, clicks, mews, rattles, squeaks, and raspy notes are frequently given, although generally quiet when feeding and less vocal during the breeding season. **CALL:** A series of four or five harsh electric calls, *jenk jenk jenk jenk,* and a dry scolding *cheh-chech.*

STATUS & DISTRIBUTION Locally common resident in brushy areas and streamside growth of the Lower Rio Grande Valley as far west as Laredo and north to Live Oak County and sporadically north to San Antonio. Found in suburbs (e.g., McAllen, Brownsville), where habitat is suitable.

POPULATION Status is unknown.

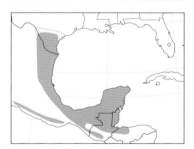

glaucescens
blue-and-black head
body largely greenish
yellow outer tail feathers

Genus *Gymnorhinus*

PINYON JAY *Gymnorhinus cyanocephalus*   PIJA ■ 1

Finding this social species is usually feast or famine. Listen for the Pinyon Jay's far-carrying calls and scan for roving flocks of a few individuals to several hundred birds. Pinyon Jays nest in loose colonies beginning in winter. Considered monotypic, but three ssp. sometimes recognized. L 10.5" (27 cm)

IDENTIFICATION A short-tailed blue jay with a long spikelike bill. **ADULT:** Blue overall with white streaks on throat. **JUVENILE:** Ashy gray underparts without much blue coloration. **FLIGHT:** Direct with rapid wingbeats, unlike scrub-jay's undulating flight.

GEOGRAPHIC VARIATION Subspecies

variation has been described as follows: Birds from northeastern portion of the range, *cyanocephalus,* have a shorter, more decurved bill, with pale plumage; western *cassini* (from north-central OR and southern ID to southern NV and central NM) average darker with straighter bill; and *rostratus,* in southwest portion of range, has a longer and wider bill and intermediate plumage coloration.

SIMILAR SPECIES Scrub-jays and Mexican Jays have paler underparts that contrast with bluer upperparts and do not occur in large flocks; shorter, thick bills; and much longer tails. The smaller Mountain Bluebird is often found in similarly large flocks but has a much shorter bill, thrush shape, and white undertail coverts.

VOICE Suggestive of Gambel's or California Quail, or nasal like a nuthatch. **CALL:** Most commonly heard call, frequently given in flight, is a soft *hwaau* given repeatedly, or a single *hwauu'hau.*

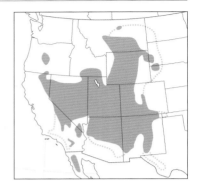

STATUS & DISTRIBUTION Common. **RESIDENT:** Pinyon-juniper woodlands of interior mountains and high plateaus; also yellow pine woodlands. **DISPERSAL:** Generally nonmigratory. When cone crop fails, may irrupt outside of normal range. **VAGRANT:** Casual: chiefly in fall to the Plains states, southwestern deserts, and coastal CA; irregular to western TX. Accidental to southern WA, western IA, southwestern SK, northwestern Mexico.

POPULATION Major declines may have occurred with loss of pinyon-juniper woodlands in the mid-20th century. Not threatened or endangered.

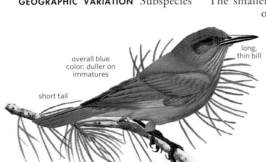

overall blue color; duller on immatures

short tail

long, thin bill

CRESTED JAYS Genus *Cyanocitta*

Two species of relatively small northern jays have conspicuous crests. The upperparts are blue, and the wings and tail are barred with black. Unlike other North American jays, Blue and Steller's Jays use mud to build their nests, usually placing the mud between the large sticks on the outside of the nest and fine rootlets or similar material that line the nest.

STELLER'S JAY *Cyanocitta stelleri*   STJA ■ 1

The Steller's Jay is a characteristic species of coniferous and mixed forests of western North America. It is a bold and aggressive species frequently found scavenging in campgrounds, picnic areas, and feeding stations in the West. It is less gregarious than most other jays. The bird's flight is strong and steady, with wings rarely flexed above horizontal. Polytypic (13–16 ssp.; 8 in N.A.). L 11.5" (29 cm)

IDENTIFICATION A nearly unmistakable dark blue, black-crested jay with variable white or blue markings on head. Wings and tail are a vivid blue, with fine black barring. Head, including crest; back; and throat are blackish. Juveniles are washed

with brownish or grayish to upperparts and are duller below.

GEOGRAPHIC VARIATION Extensive among the 13–16 subspecies from Alaska to Nicaragua, but more limited and clinal among seven of eight subspecies north of Mexico, all of which have blue streaks on forehead. The most distinctive subspecies in North America is *macrolopha* of central and southern Rockies to northern MX, which has a long crest, paler back, and white streaks on forehead and over eye. It is also strongly

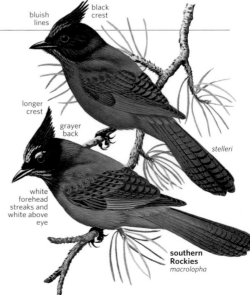

black crest

bluish lines

longer crest

grayer back

stelleri

white forehead streaks and white above eye

southern Rockies *macrolopha*

genetically distinct from the other N.A. subspecies, which are shorter crested and differ primarily in size, head patterning, and overall coloration. Some, including nominate *stelleri* (Pacific coast from AK to southwestern BC) are darker backed. The largest subspecies, *carlottae,* from the Queen Charlotte Islands off British Columbia, is almost entirely black above. Another group of subspecies from central Mexico south to Nicaragua have short blue crests.

SIMILAR SPECIES Nearly unmistakable. Crest, shorter tail, and lack of white in body separate Steller's Jays from scrub-jays. Blue Jay, our other crested jay, is paler blue, has white in wings, tail, and face, as well as whitish underparts.

VOICE Vocal with a diverse array of squacks, rattles, harsh screams. **CALL:** A piercing *sheck sheck sheck* and a descending harsh *shhhhhkk.* Steller's Jay frequently mimics other species, particularly raptors, and also incorporates calls of squirrels and household animals, such as dogs and chickens.

STATUS & DISTRIBUTION Common. **BREEDING:** A variety of coniferous and mixed coniferous forests. **DISPERSAL:** Generally resident, but irruptions casually occur in fall and winter in both the blue-fronted subspecies group and in *macrolopha,* which has reached casually southeastern CA along the lower Colorado River and the mountains of the east Mojave. During such irruptions, individuals can appear to lower elevations of the Great Basin, the Great Plains, southern CA, and southwestern deserts. Accidental east to northeastern IL, eastern NE, eastern KS, and central TX.

POPULATION Apparently stable.

BLUE JAY *Cyanocitta cristata*   BLJA ▪ 1

The Blue Jay is a familiar and widespread bird throughout the East. Generally loved or hated, it has acquired a Jekyll-and-Hyde reputation. It is a frequent visitor to backyard feeding stations, where its raucous and rambunctious behavior is well known, but this adaptable jay is equally at home stealthily moving through the forest, plundering other birds' nests, searching for nuts, or quietly raising its young. Polytypic (4 ssp.; all in N.A.). L 11" (28 cm)

IDENTIFICATION Blue is a nearly unmistakable crested jay with black barring and white patches on blue wings and tail; its underparts are paler with a dark necklace. Juveniles are more grayish above with gray lores and more limited white markings on wings.

GEOGRAPHIC VARIATION Minor, obscured by broad overlap. Four subspecies are generally recognized. Northern *bromia* average largest and brightest. Southeastern *cristata* generally show a subtle violaceous wash; *semplei* from south FL is smallest. Western *cyanotephra* are generally duller and paler blue.

SIMILAR SPECIES Nearly unmistakable. Occasionally hybridizes with Steller's Jay (e.g., in the vicinity of the Front Range of Colorado); hybrids appear intermediate between the two parent species.

VOICE Vocal with a diverse array of vocalizations. **CALL:** A piercing *jay jay jay;* a musical *yo-ghurt;* frequently imitates raptors, particularly Red-shouldered Hawk. Actively migrating birds are typically silent.

STATUS & DISTRIBUTION Common. **BREEDING:** A variety of mixed forests, woodlands, suburbs, and parks. **MIGRATION:** Diurnal migrant. Northern populations move south in varying numbers from year to year. During migration, large loose flocks can fill the sky, particularly along the shores of the Great Lakes and other northern locations known for raptor concentrations. The spring peak is during May. In fall, first detected away from breeding grounds as early

as July (typically earliest in big flight years); peaks in Great Lakes late Sept.–mid-Oct. **WINTER:** Distribution generally similar, but departs from northernmost breeding locales. **VAGRANT:** Casual chiefly in fall and winter west of the Rockies, recorded most frequently in the Northwest. Casual to northern CA and accidental southern CA and AZ. Accidental to Bermuda.

POPULATION Numbers are increasing in parts of the West. Domestic cats are likely the most significant human-caused source of mortality.

SCRUB-JAYS Genus *Aphelocoma*

Of the six species in *Aphelocoma* jays, four are found north of Mexico. Generally medium or small in size, they have relatively long tails, are predominantly blue above and gray below (or uniformly blue), and lack a crest. Species limits (esp. with the Western) are poorly understood. Some species are social breeders.

FLORIDA SCRUB-JAY *Aphelocoma coerulescens*   FLSJ ■ 2

The Florida Scrub-Jay is the only bird endemic to Florida, where it has been the subject of intense study since the late 1960s. These birds are cooperative breeders and are usually encountered in small family groups. While they are often hidden in scrub, flock members share sentinel duty in which individuals perch up on exposed perches and scan for predators. Many Florida Scrub-Jays are extremely tame and are well known for taking food from the hand. Monotypic. L 11" (28 cm)

IDENTIFICATION Similar to other scrub-jays, with blue head, wings, and tail, and pale underparts. But this species is slightly longer tailed, with a shorter and broader bill, and has blue auriculars. Forehead is whitish, blending into eyebrow; back is distinctly paler gray-brown, grayer than other scrub-jays; and slightly grayish belly and flanks are faintly streaked. Juvenile is duller with a sooty grayish or brownish wash on head and back.

SIMILAR SPECIES Only scrub-jay found in Florida. Blue Jay has black-and-white wings and tail, and is usually crested.

VOICE CALL: Raspy and hoarse calls are vaguely reminiscent of Western Scrub-Jay, but are lower and harsher.

STATUS & DISTRIBUTION Fairly common, but local. Resident in scrub and scrubby flat woods of FL. Optimal habitat is produced by fire, consisting of scrub, mainly oak, about ten feet high with small openings. Also found along roads and vacant lots near overgrown scrub habitat. **MIGRATION:** Nonmigratory. Most travel only a few kilometers during their entire life, making it extremely unlikely for an individual to show up out of range. **POPULATION** Declined by some 90 percent during the 20th century due to habitat loss and fire suppression.

whitish forehead

grayish back

long tail

ISLAND SCRUB-JAY *Aphelocoma insularis*   ISSJ ■ 2

The Island Scrub-Jay is restricted to Santa Cruz Island in the Channel Islands south of Santa Barbara, California. Its behavior is quite similar to that of the Western Scrub-Jay, and like that species, pairs actively defend year-round territories. Island Scrub-Jays forage at all levels of foliage and forage on the ground more regularly than the Western Scrub-Jay. As with many island endemics, this species has a more varied diet than the similar mainland representative. It regularly caches acorns, lizards, and even deer mice. Monotypic (formerly considered conspecific with Western and Florida Scrub-Jays). L 12.5" (31 cm)

IDENTIFICATION ADULT: Very similar to coastal populations of Western Scrub-Jay but 15 percent larger, 40 percent heavier, and with a heavy bill that is up to 40 percent larger than largest-billed mainland scrub-jays. Island Scrub-Jay is darker blue above, with a darker brown back, blacker auriculars, and bluish undertail coverts. **JUVENILE:** Duller, with a sooty-brown wash and little blue coloration.

SIMILAR SPECIES Only scrub-jay found on Santa Cruz Island; no scrub-jays have been found on any of the other Channel Islands.

VOICE CALL: Similar to calls of Pacific populations of Western Scrub-Jays, slightly louder and harsher.

STATUS & DISTRIBUTION Locally

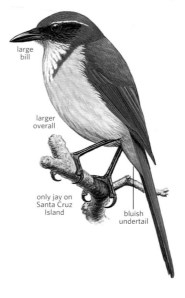

large bill

larger overall

only jay on Santa Cruz Island

bluish undertail

common on Santa Cruz I., where it is found in a variety of habitats; optimal habitat consists of oak chaparral. Approximately half of the adults on Santa Cruz I. are nonbreeding "floaters" that are generally found in marginal habitat, including pine

and riparian scrub. **MIGRATION:** Nonmigratory; unrecorded away from Santa Cruz I.
POPULATION Approximately 12,500 individuals were recorded on Santa Cruz Island in 1997, a large number for the size of the island. Population

trends unknown, but anecdotal evidence indicates some declines, perhaps as a result of West Nile virus, though this disease has not yet been documented on Santa Cruz I. Much remains to be learned about this restricted species.

WESTERN SCRUB-JAY *Aphelocoma californica*   WESJ   1

Plumage and behavior differ greatly between interior and coastal populations. Coastal populations Western Scrub-Jays are confiding, tame, and easily seen. Interior populations are more secretive and often are seen darting from bush to bush or are simply heard giving their harsh calls. Pairs of all North American subspecies hold territories year-round. Polytypic (14 ssp.; 8 in N.A.). L 11" (28 cm)
IDENTIFICATION This is a long-tailed jay with a small bill and no crest. **ADULT:** Generally bluish above, gray below with a contrastingly paler throat and upper breast, and a variable bluish band on chest. **JUVENILE:** Much grayer overall, showing very little blue on head. **FLIGHT:** Usually undulating with quick deep wingbeats.
GEOGRAPHIC VARIATION The subspecies of Western Scrub-Jays form three well-defined subspecies groups that may represent separate species: The coastal *californica* group is darker blue with a well-marked blue breast and a distinct brown patch on the back, and has white flanks. The interior *woodhouseii* group is a duller blue, with blended upperparts, with grayish-

brown flanks, and lacks a distinct breast band; *texana* from central TX is richer blue. The *sumichrasti* group is confined to southern Mexico.
VOICE CALL: Gives a variety of harsh calls, most frequently a raspy *shreeep,* often repeated in a short series of shorter notes: *shuenk shuenk shuenk shuenk.* Interior vocalizations are higher pitched and more two-syllabled — an upslurred *jrr-eee.*
STATUS & DISTRIBUTION Common. **BREEDING:** Scrubby and brushy habitats, particularly with oak, pinyon, and juniper; also found in gardens, orchards, and riparian woodlands. **DISPERSAL:** Largely resident, but prone to occasional irruptions, particularly

members of the interior *woodhouseii* group. Irruptions often coincide with movements of Pinyon Jay, Steller's Jay, and Clark's Nutcracker. **VAGRANT:** Casual to southwestern BC, eastern WA, southeastern CA, central KS. Accidental to southern MB, northeastern IL, and northwestern IN.
POPULATION The *californica* group is spreading northward in the Northwest.

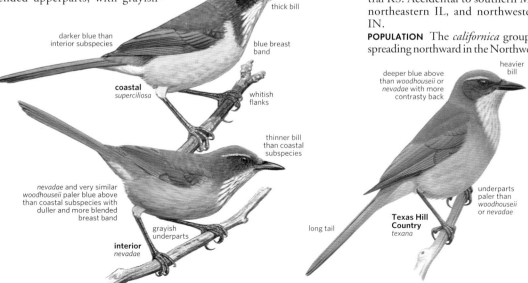

MEXICAN JAY *Aphelocoma wollweberi*   MEJA ■ 2

This species travels in large, raucous flocks and has a cooperative breeding system like the Florida Scrub-Jay; it regularly visits bird feeders. Polytypic (5 ssp.; 2 in N.A.). L 11.5" (29 cm)
IDENTIFICATION A relatively thick-bodied, broad-tailed, broad-winged jay with a large bill and no crest. **ADULT:** Generally bluish on face and above, with a slight grayish cast to back and a brownish patch on center of back. Underparts pale gray. **JUVENILE:** Much grayer overall, very little blue on head. Juvenile *arizonae* has a yellowish bill that gradually turns

black by third year. All ages of *couchii* have black bills.
GEOGRAPHIC VARIATION Three subspecies groups, two in N.A.: western *wollweberi* group includes *arizonae* (reaches AZ and NM) larger with dull pale grayish blue upperparts and a uniformly gray throat and breast; eastern *couchii* group (represented only by *couchii*, found north to Chisos Mts., TX) with blue head and whitish throat that contrasts with pale grayish underparts. Split in 2011 from Transvolcanic Jay (*A. ultramarina*) of central MX.
SIMILAR SPECIES May be confused with scrub-jays. Western Scrub-Jay best distinguished by its slimmer body, thinner tail, contrasting dark cheeks, white eyebrow, whitish throat offset by at least a

faint blue breast band, and gray back. Western Scrub-Jays tend to flock less.
VOICE Relatively limited. **CALL:** Most common is a rising *week*, often repeated in a short series; calls generally less raucous than scrub-jay's, but may be heard more frequently as they travel in family groups. Texas birds similar, but less musical, somewhat more like Western Scrub-Jay. Texas birds also give a mechanical rattle.
STATUS & DISTRIBUTION Common. **RESIDENT:** Pine-oak canyons of southwestern mountains. **DISPERSAL:** Largely resident; does not wander like Western Scrub-Jay. **VAGRANT/ACCIDENTAL:** extreme western TX; record from south-central KS based on lost specimen.
POPULATION Not threatened.

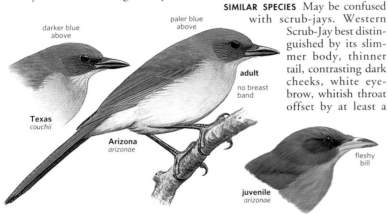

darker blue above

paler blue above

Texas
couchii

Arizona
arizonae

adult

no breast band

juvenile
arizonae

fleshy bill

NUTCRACKERS Genus *Nucifraga*

CLARK'S NUTCRACKER *Nucifraga columbiana*   CLNU ■ 1

While generally found far from most human habitations, nutcrackers are very adaptable and are commonly seen at scenic overlooks and picnic grounds in the West. During summer and fall, nutcrackers store thousands of pine cone seeds, which they eat during winter and feed to their offspring during the very early nesting season (most have laid eggs by March or early April). Monotypic. L 12" (31 cm)
IDENTIFICATION A short-tailed, long-winged corvid with a long bill. Head and body pale grayish overall with boldly contrasting black-and-white wings and tail. **ADULT:** Paler whitish area on throat, forecrown, and around eyes. **JUVENILE:** More uniform face and generally washed with brown. **FLIGHT:** Direct, with deep crowlike wingbeats.
SIMILAR SPECIES Gray Jay has much shorter bill and more uniform grayish wings and tail, without boldly contrasting black-and-white wings.

VOICE Quite varied; most are nasal and harsh, but some with clicks and cackles. **CALL:** Commonly gives a very nasal, grating, drawn-out *shra-a-a-a;* a nasal *whaah* similar to that of Pinyon Jay; and a slow rattle sometimes likened to the croaking of a frog.
STATUS & DISTRIBUTION Locally common. **BREEDING:** Prefers forests dominated by at least one species

of large-seeded pine. In summer, often found in higher coniferous forests near timberline. **MIGRATION & DISPERSAL:** Generally resident, but some populations regularly move to lower elevations in fall. In late spring, following breeding, most birds move upslope to higher

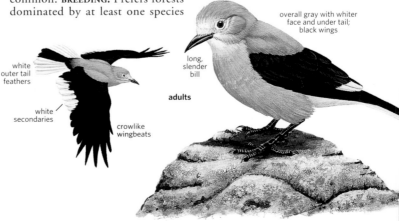

white outer tail feathers

white secondaries

crowlike wingbeats

adults

long, slender bill

overall gray with whiter face and under tail; black wings

elevations. Nutcrackers irrupt out of core range and into nearby deserts and lowland areas of the West following major cone crop failures. Most individuals are seen in late Sept. and Oct., but may appear as early as late July or early Aug. Major irruptions generally occur every 10–20 years and are thought to follow two or more years of good cone production immediately followed by a failure of all seed sources.

VAGRANT: During irruptions casual to the Plains states, southeastern AK, southern YK, and coastal CA, including offshore islands. **ACCIDENTAL:** As far east as western ON, central MB, northeastern MN, northern IL, AR, and LA, and north to the central NT. Most extralimital records are Aug.–Nov.
POPULATION Breeding Bird Survey suggests populations generally increased from 1966 to 1996.

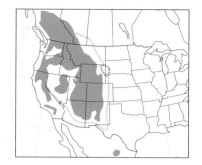

MAGPIES Genus *Pica*

Two members occur in N.A.; at least one other occurs in the Old World. That species was split from the Black-billed Magpie. All species are patterned in black and white and have long tails. They build stick nests and are usually found in open habitats with scattered trees. Less social than some species, many magpies still form larger flocks during fall.

BLACK-BILLED MAGPIE *Pica hudsonia*   BBMA ▪ 1

With their large size, bold pied plumage, and fondness for open areas, Black-billed Magpies are easily seen. They hop and walk on the ground with a swaggering and confident gait. Monotypic. L 19" (48 cm)
IDENTIFICATION An unusually long-tailed black-and-white corvid with a black bill. In good light, black on wings and tail shine with iridescent green, blue, and violet. Juvenile has milky grayish-colored iris and fleshy pinkish gape. Upperparts are washed dull brownish, and belly is more cream-colored. Immatures have narrower and more pointed outer tail feathers. **FLIGHT:** Relatively slow, with steady, rowing wingbeats, but easily able to quickly change direction in flight. Magpies usually swoop up or down to perch.
GEOGRAPHIC VARIATION While monotypic, birds in the south average smaller and tend to show bare dark grayish skin below eye (similar in shape to the yellow skin on Yellow-billed Magpie).

SIMILAR SPECIES Black bill and range distinguish this species from Yellow-billed Magpie; larger size of Black-billed Magpie sometimes appreciable in the field. Until recently, Black-billed Magpie was lumped with the widespread Magpie (*Pica pica*) of Eurasia and northwestern Africa, but all evidence suggests a much closer relationship with Yellow-billed.
VOICE Quite varied, but most vocalizations are rather harsh. All are very similar to that of Yellow-billed Magpie. They are strikingly faster and lower pitched than Old World populations of magpie. **CALL:** Frequently gives a whining, rising *mea;* sometimes these are more drawn out and questioning: *meeaaah.* Also a quickly repeated *shek-shek-shek* with each phrase repeated 3–5 times (not as harsh or piercing as analogous vocalization of Steller's Jay).
STATUS & DISTRIBUTION Common. Resident of open woodlands and thickets in rangeland and foothills; nests along watercourses and other areas with trees and shrubs, but foraging birds use very open areas. **MIGRATION & DISPERSAL:** Generally considered to be nonmigratory, but varies regionally and by year. Dispersing flocks form as early as July and typically consist of a few to a hundred birds; occasionally forms flocks of several hundred. Movements may be upslope, downslope, or in any direc-

large white wing patch

black-and-white coloration

long tail

tion. Banding recoveries have shown atypical movements of more than 300 miles. Most birds are thought to return near where they hatched to breed. **VAGRANT:** Casual towards Pacific coast in Pacific Northwest (fewer records in recent decade), western WI, IA, and northern TX, mostly in fall. In summer found north as far as northern AK, northeastern NT. Other sightings occur casually in the East and may pertain to escaped cage birds, particularly away from the Great Lakes.
POPULATION Declined throughout the Great Plains with the slaughter of bison and targeted eradication. Adapting and now found in many suburban areas.

YELLOW-BILLED MAGPIE *Pica nuttalli* YBMA ■ 2

This highly-prized California endemic is quite gregarious; roosting and feeding in flocks, it usually nests in small colonies. The range does not overlap with that of the Black-billed Magpie. Monotypic. L 16.5" (42 cm)
IDENTIFICATION Similar to Black-billed Magpie but with yellow bill and yellow skin below eye. The extent of yellow is variable, sometimes fully encircling eye, sometimes confined to below eye. Differences may be related to age, state of molt, individual variation, or some combination thereof. Smaller than Black-billed. **JUVENILE:** Milky grayish iris with brownish wash, most noticeable on belly. **IMMATURE:** Differs as in Black-billed Magpie. Also note that inner tail feathers are sometimes replaced and contrast with dull retained outers. **FLIGHT:** Very simi-

lar to that of Black-billed Magpie; wingbeats are somewhat faster due to Yellow-billed Magpie's smaller size.
SIMILAR SPECIES Yellow bill and yellowish skin below eye should easily distinguish this species from Black-billed, which is also larger. Beware of Black-billed Magpies carrying twigs, food items, or other pale or yellowish objects—such sightings have resulted in many erroneous reports of Yellow-billed Magpie.
VOICE Very similar to that of Black-billed Magpie, perhaps higher pitched and clearer.
STATUS & DISTRIBUTION Common, but distribution somewhat clumped. **BREEDING:** Prefers oaks, particularly more open oak savanna; also orchards and parks. Found in rangelands and foothills of central and northern

Central Valley, CA, and coast range valleys south to Santa Barbara County.
MIGRATION & DISPERSAL: In late summer forms flocks that may wander short distances; these almost always remain within the breeding range. **VAGRANT:** Casual north almost to OR. Yellow-billed Magpies are not prone to wandering, and observations away from the breeding range may pertain to escapes.
POPULATION West Nile virus has severely impacted this species. Loss of habitat poses another significant risk. Some populations have been locally extirpated (e.g., extirpated in 1981 from Pacific Grove, Monterey Co.; gone by 1900 from western Los Angeles Co. and Ventura Co.), and others have become fragmented due to habitat loss. Pesticides targeted at small mammals and traps pose additional threats.

variable yellow skin around eye

yellow bill

CROWS AND RAVENS Genus *Corvus*

Crows and ravens form the largest group of corvids. Generally considered intelligent and adaptable, *Corvus* includes some of the best studied and most well-known species. Of the 40 species worldwide, seven are found in the continental U.S. and Canada. Crows and ravens are large, with mostly black plumages, sometimes with grayish or whitish markings on the head or neck. Ravens are larger than crows with a bigger bill and deeper vocalizations. Crow and raven identification presents some of the greatest identification challenges, complicated by variation within the species. Focus on structural differences (particularly the bill and nasal bristles) and vocalizations, and beware of regional differences.

EURASIAN JACKDAW *Corvus monedula* EUJA ■ 4

The Eurasian Jackdaw is a widespread Eurasian and north African species that is strictly a vagrant to North America. During the 1980s, it staged a small invasion into the Northeast, with singles and small groups found in several northeastern states and provinces. Some (or all) of these these birds may have been assisted by ships. Bold and inquisitive. Polytypic (4 ssp.). L 13" (33 cm)
IDENTIFICATION A small, mostly black crow with pale gray nape wrapping

around sides of head. **ADULT:** Eerie pale gray iris. **JUVENILE:** Similar to adult but with dark iris and darker nape and side of head.
GEOGRAPHIC VARIATION No information on subspecies that appeared in North America; presumably *monedula* of Scandinavia or *spermologus* of western and central Europe.
SIMILAR SPECIES Other crows and ravens are larger without pale area to nape.

gray nape and face with pale gray eye

small size

VOICE CALL: A sharp, abrupt *chjek,* typically repeated several times.
STATUS & DISTRIBUTION Widespread Old World species from British Isles and northwestern Africa east to Siberia and Kashmir; generally sedentary. Within native range found in a variety of open habitats. **VAGRANT:** Casual to Iceland and aboard ships in the mid-North Atlantic; two records for Greenland. Other records from Faroe Is., Canary Is., and Japan.

First recorded in N.A. at Nantucket, MA, in late Nov. 1982; this bird was joined by another in July 1984; at least one remained until 1986. During the mid- and late 1980s found at several locations in the northeast, including a flock of 52 near Port-Cartier, QC; others at Miquelon, NS, ON, ME, RI, CT, and PA. A pair at the Federal Penitentiary at Lewisburg, PA, from May 1985 through at least 1991 nested or

attempted to nest several times. The CT, RI, and PA records were ultimately rejected by state records committees, and it is generally thought that many (or all) North American records may pertain to ship-assisted birds.
POPULATION Abundant and increasing throughout most of Old World range. Vagrants to QC were poisoned or shot by the Quebec Fish and Game Department.

AMERICAN CROW *Corvus brachyrhynchos* AMCR 1

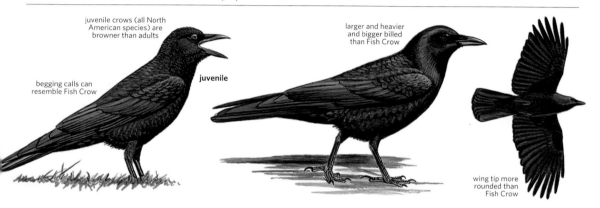

juvenile crows (all North American species) are browner than adults

begging calls can resemble Fish Crow

juvenile

larger and heavier and bigger billed than Fish Crow

wing tip more rounded than Fish Crow

The American Crow is the "default" crow across most of N.A. It overlaps broadly with the Common Raven, and to a lesser extent with the Chihuahuan Raven, Fish Crow, and Northwestern Crow. Study of vocalizations, bill structure and size, tail shape, and overall structure of this species will greatly aid in the identification of other crows and ravens. Regional variation in size of the American Crow poses challenges, particularly in the Northwest. Polytypic (4 ssp.; all in N.A.). L 17.5" (45 cm)
IDENTIFICATION Largest crow in North America, with uniformly black plumage and fan-shaped tail. Bill is larger than other American crows, but distinctly smaller than either raven. On rare occasions individuals show white patches in wings. **JUVENILE:** Brownish cast to feathers; grayish eye, and fleshy gape (quickly darkening after fledging). **IMMATURE:** Tends to show worn brownish wings that contrast with fresher black wing coverts. **FLIGHT:** Steady, with low rowing wingbeats. Does not soar.
GEOGRAPHIC VARIATION Four poorly defined subspecies generally recognized. While variation is largely clinal, differences between extremes

are apparent. Northern *brachyrhynchos* and eastern and southern *palus* are essentially inseparable. Florida Peninsula *pascuus* has relatively long bill, long tarsus, and large feet. Also differs in behavior, never forming flocks; not found in urban areas and has more extensive vocal repertoire. It has been suggested that smaller western subspecies *hesperis* is more closely related to Northwestern Crow than to other subspecies of American Crow—the entire relationship between American and Northwestern Crow remains unclear.
SIMILAR SPECIES Compare with very similar Fish Crow and nearly identical Northwestern Crow (both most easily separated by voice); Common and Chihuahuan Ravens are much larger with much heavier bills and wedge-shaped or rounded tails.
VOICE CALL: Adult's familiar *caw* generally well known. Voice of *hesperis* generally lower pitched than other subspecies. Juvenile's begging call is higher pitched, nasal, and resembles the call of Fish Crow.
STATUS & DISTRIBUTION Common to abundant. **BREEDING:** A variety of habitats, particularly open areas with scattered trees. **MIGRATION & DISPERSAL:**

Diurnal migrant. In spring, arrive mid-Feb.–late Apr. Fall migration generally more protracted than in spring. Most depart north-central BC and AB by late Sept; peak in Great Lakes early Oct.–mid-Nov. Uncommon to rare migrant and winter visitor in deserts of the West. **WINTER:** Throughout much of the lower 48. **VAGRANT:** Casual to Imperial Valley, CA, southwestern AZ, southwestern TX, northwestern Sonora, MX.
POPULATION Expanded with clearing of forests and planting of woodlots in prairies. Many populations experienced dramatic declines with the spread of West Nile virus early this century. Nevertheless long-term populations generally stable.

NORTHWESTERN CROW *Corvus caurinus* NOCR ▪ 1

This small crow of the Pacific Northwest is virtually identical to *hesperis* American Crows. Field identification within the suspected range of overlap in WA is probably impossible. Many authorities consider the Northwestern Crow a subspecies of the American Crow. Monotypic. L 16" (41 cm)

IDENTIFICATION Virtually identical to American Crow. Northwestern averages smaller, with quicker wing flaps than most American Crows. Unfortunately, *hesperis* American Crows found in the Pacific Northwest are essentially identical. A tendency for nasal bristles to be placed along sides of bill on American Crow versus more on top of culmen in Northwestern Crow has been suggested, but there is considerable overlap, and this is unreliable. Voice often reported to be the most useful characteristic in separating this species from American Crows, but *hesperis* American Crows are so similar as to render this nearly impossible to use.

SIMILAR SPECIES Over most of range, likely to be confused only with Common Raven, which is much larger, with large heavy bill, and wedge-shaped tail. Separated from American Crow in most areas by range.

VOICE CALL: Distinctly lower and more nasal than most populations of American Crow; *hesperis* American Crows also have lower and more nasal calls similar to Northwestern Crows.

STATUS & DISTRIBUTION Resident of coastal areas and islands in Pacific Northwest. Common from south-coastal AK

(Kodiak I.) to southern BC. Once a coastal resident in WA south through Puget Sound to Grays Harbor. Deforestation in late 1800s allowed American Crows to invade southern portions of this region by the early 1900s. Puget Sound population now generally considered to be small *hesperis* American Crows or hybrids. Within WA, birds thought most likely to be phenotypicly pure Northwestern Crows are limited to western Olympic Peninsula and San Juan Islands.

MIGRATION & DISPERSAL: Generally sedentary with local movements within breeding range.

POPULATION Stable or increasing.

nearly identical to American Crow

TAMAULIPAS CROW *Corvus imparatus* TACR ▪ 3

The range of this species barely extended into south Texas near Brownsville; it is now extirpated. It does not overlap with any other crow. It was formerly considered the same species as the Sinaloa Crow *(C. sinaloae)* of northwestern Mexico, then called the Mexican Crow. Monotypic. L 14.5" (37 cm)

IDENTIFICATION A small, very glossy crow with a short, small bill. Recalls Fish Crow, but glossier; no range overlap. **JUVENILE:** Duller than adult, but still glossier than adults of other North American crows.

SIMILAR SPECIES Only crow regularly found in southern Texas. Chihuahuan Raven is common in the same habitat, but is larger, stockier, and has a much larger and heavier bill. May be confused with male Great-tailed Grackle, particularly molting individuals that can have short tails. Still readily separated by crow's larger bill, stockier build, and thicker legs and dark eyes. Also note distinctive vocalizations.

VOICE CALL: Rough, nasal, froglike croaking *ahrrr,* typically repeated several times and sometimes doubled, *rah-rahk.* Juvenile's begging call reportedly similar to that of Fish Crow.

STATUS & DISTRIBUTION Common to fairly common in northeastern Mexico to northern Veracruz. Invaded the Brownsville area in 1968, and was common through the 1970s and 1980s (hundreds). Found principally in winter but a few summered and successfully nested there. The species declined greatly in the 1990s and was rare by 2000. No sightings since about 2010. It was best known in TX from the Brownsville Dump.

POPULATION Presumably stable in Mexico.

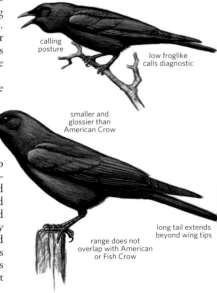

calling posture

low froglike calls diagnostic

smaller and glossier than American Crow

long tail extends beyond wing tips

range does not overlap with American or Fish Crow

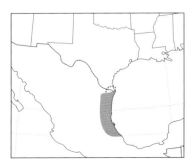

FISH CROW *Corvus ossifragus* FICR ▪ 1

This glossy, gregarious corvid largely replaces the American Crow in coastal and tidewater regions of the Southeast. Its range is expanding up the Mississippi River Valley, its tributaries, and elsewhere away from the coast. While subtle structural differences exist, the most reliable way to identify this species is by call. Monotypic. L 15.5" (39 cm)
IDENTIFICATION Very similar to American Crow but with disproportionately longer tail, longer wings, shorter legs, and smaller feet. Fish Crows have a bluish violet or greenish gloss over most of wings and body (less so on lower underparts) that is more extensive than in American. Bill is usually more slender, but female crows have smaller bills and small-billed female Americans may have bill that appears very similar to Fish's. **JUVENILE:** Brownish cast to feathers; grayish eye, fleshy gape (quickly darkening after fledging). **IMMATURE:** Tends to show worn brownish wings that contrast with fresher black wing coverts. **FLIGHT:** Similar to American Crow but with slightly stiffer and quicker wingbeats. When gliding, wings often appear swept back at tips, creating a less-even trailing edge to wing than American Crow. Small-headed and long-tailed appearance is often most

noticeable in flight. Apt to hover and soar, unlike American Crow.
SIMILAR SPECIES American Crow averages somewhat larger with proportionately shorter wings and tail, but such differences difficult to use in the field; best identified by call. Note that some juvenile begging calls of American Crow resemble Fish Crow vocalizations. Closely related to Tamaulipas Crow, but does not overlap in range; Tamaulipas glossier, smaller, with proportionately even longer tail. Rarely overlaps with Common Raven, but raven much larger with very large bill and wedge-shaped tail.
VOICE Possibly more limited repertoire than American Crow. **CALL:** Typically a high nasal *ca-hah*, second note lower; has a similar inflection and emphasis to American's sometimes casual dismissal of something with an audible *uh-uhh*. Also gives low, short *awwr*. Begging calls of Fish Crow similar in quality to American's begging calls, but are shorter and cut off abruptly.
STATUS & DISTRIBUTION Common. Resident Usually near fresh and salt water lakes, rivers, beaches, marshes, or estuaries. **MIGRATION & DISPERSAL:** In many regions (e.g., MD, TN, central NY), spring migration/return of

breeders appears to peak in mid-Mar.–Apr. Fall migration more protracted: AR late Aug.–early Nov.; departs central NY Sept.–Oct.; MD mid- or late Sept.–mid- to late Dec. (peaks Oct. 20–Dec. 10). **VAGRANT:** Casual: MI, southern ON, southern FL peninsula and Keys, NS, and Bahamas. **WINTER:** Becomes more localized throughout range as species gather in large flocks. Because birds become less vocal in winter, the northern limit of wintering birds is somewhat uncertain. Withdrawal from northern portions of range, particularly in the interior.
POPULATION Increasing in abundance, and populations spreading northward. First recorded OK in 1954; KY 1959; MO 1964; ME 1978; KS 1984; IN 1988; VT 1998.

calling posture

smaller than American Crow with slight purplish gloss

bill averages smaller and more slender than American Crow

shorter legs than American Crow

wings more pointed than American Crow and flies with more rapid flaps

CHIHUAHUAN RAVEN *Corvus cryptoleucus* CHRA ▪ 1

Separating the smaller Chihuahuan Raven from the Common Raven is exceptionally difficult, if relying on subtle differences in size, bill and tail shape, and vocalizations. White bases to the neck feathers are the most reliable field mark, but these are usually hidden. Chihuahuan Ravens

are very gregarious, particularly during the winter, when they often form large flocks (occasionally up to several thousand birds). Where trees are lacking, they will nest on utility poles, windmills, and abandoned buildings or under bridges. Monotypic. L 19.5" (50 cm)

IDENTIFICATION Very similar to Common Raven. Major difference is white (not gray) bases to neck feathers. Bill is slightly shorter; tail is slightly less wedge shaped and shorter; nasal bristles on top of bill extend farther out onto bill; throat feathers are less thick and shaggy. **JUVENILE:** Brownish cast to feathers; grayish eye, fleshy gape (quickly darkening after fledging). **IMMATURE:** Tends to show worn brownish wings that contrast with fresher black wing coverts. **FLIGHT:** Very similar to Common Raven, but less wedge-shaped tail often most easily seen on soaring birds.

GEOGRAPHIC VARIATION Birds of central and southern TX average smaller, but more study required to determine if slight differences warrant subspecies status.

SIMILAR SPECIES Chihuahuan Raven is extremely similar to Common Raven. Most reliable difference is white-based feathers to neck, often visible only under windy conditions. Additional clues include shorter bill with longer nasal bristles; less "shaggy" throat feathering; smaller

white on neck shows when wind ruffles feathers

nasal bristles extend well over half the length of the culmen

shorter, more rounded tail than Common Raven

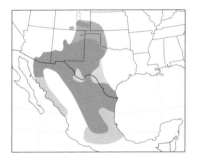

size; shorter, less wedge-shaped tail; and, on average, slightly higher-pitched vocalizations. All these characteristics are subjective, and where range overlaps some birds should probably be left unidentified, particularly given distant or brief views. Heavier bill and wedge-shaped tail distinguish Chihuahuan Raven from crows.

VOICE Less varied than that of Common Raven. **CALL:** Drawn-out croak is usually higher pitched and more crowlike than most Common Raven vocalizations. Common Ravens may give Chihuahuan Ravenlike vocalizations. Listening to a bird (or a flock of birds) for a period of time is

preferable in order to hear full variety of calls given.

STATUS & DISTRIBUTION Uncommon to common. Resident of desert areas and dry open or shrubby grasslands, usually with scattered trees or shrubs. **MIGRATION & DISPERSAL:** Generally considered to be nonmigratory, but forms large roaming flocks in winter. Birds in northern portion of range are thought to withdraw, but this may vary between years. In some years, local movements through much of the range.

POPULATION Declined in eastern Colorado, western KS during the late 1800s and early 1900s; extirpated from southeastern WY. Now strictly casual in NE.

COMMON RAVEN *Corvus corax* CORA ■ 1

In northern or desert areas, this is one of the first species to greet the dawn as they patrol the roads for overnight roadkill. This large raven found throughout much of the Northern Hemisphere is more frequently found singly, in pairs or small groups, but in many regions is sometimes found in foraging or roosting flocks of several hundred, even several thousand, birds. Polytypic (11 ssp.; 4 in N.A.). L 24" (61 cm)

IDENTIFICATION Largest corvid in

the Americas, with uniformly glossy black plumage, long, heavy bill, and long wedge-shaped tail. Bases of neck feathers are gray. Nasal bristles on top of bill cover basal third to half of bill. Throat is covered by thick and shaggy

feathers. **JUVENILE:** Brownish cast to feathers; grayish eye, and fleshy gape (quickly darkening after fledging). **IMMATURE:** Tends to show worn brownish wings that contrast with fresher black wing coverts. **FLIGHT:**

larger than Chihuahuan but beware of smaller and smaller-billed Common Ravens from California

longer bill than Chihuahuan and nasal bristles extend about half way out culmen

long, wedge-shaped tail

shaggy throat feathers

large size, larger than Red-tailed Hawk; frequently soars

Wingbeats shallower than crows. Frequently soars; pairs frequently engage in a variety of aerial acrobatics, sometimes even turning upside down. Glossy black plumage is often most apparent in flight, when ravens often appears "greasy," as if covered with oil.
GEOGRAPHIC VARIATION While subspecies variation is largely clinal, differences between extremes sometimes apparent. Northern and Eastern *principalis* large with long bill of medium depth. Residents of southwestern Alaska to Russian Far East *(kamtschaticus)* are largest, with broadest and longest bill. Western *sinuatus* and particularly southwestern *clarionensis* smaller, with smaller bill, shorter wings and tail.
SIMILAR SPECIES See Chihuahuan Raven, which can be extremely similar. Crows are much smaller with much smaller bills and fan-shaped tails.
VOICE Extremely varied, with local dialects and individual specific calls reported. **CALL:** Common call is a low, drawn-out croak *kraaah;* also

a deep, nasal and hollow *brooonk.* Juvenile begging calls are relatively high-pitched, but there is much individual variation. Calls can be similar to Chihuahuan Raven.
STATUS & DISTRIBUTION Generally common, but more local on southern periphery of range in East. **BREEDING:** Diverse array of habitats. Tends to prefer hilly or mountainous areas, but found on tundra, prairies, grasslands, towns, cities, isolated farmsteads, forests, even Arctic ice floes. **MIGRATION & DISPERSAL:** Generally considered sedentary, but poorly understood. Regular spring passage noted along Front Range of Colorado late Jan.– late Mar. **VAGRANT:** Casual chiefly in winter to Great Plains, southern Great Lakes, and lower elevations of Atlantic coast states.
POPULATION Declined greatly in the 19th and early 20th centuries due to loss of habitat, shooting, poisoning, and disappearance of bison on the Great Plains; extirpated from Alabama, North Dakota, South Dakota, and the southern

Great Lakes. Populations are now expanding into some of their former territory in parts of the East (e.g., a few are now regular again in eastern Ohio), Great Lakes, and northern Plains. Listed as endangered in Tennessee and Kentucky. Shooting, trapping, and habitat degradation continue to pose threats for this species, but it is becoming more tolerant of humans; birds are often found in cities and towns in the West.

LARKS Family Alaudidae

Horned Lark (UT, June)

Larks—small, generally cryptically colored terrestrial passerines—number some 92 species worldwide. They occur on all continents except on Antarctica.
Structure Bill shape varies from long and decurved to thick and conical. The hind toe is usually long and straight. The tertials are unusually long, often completely covering primaries on folded wing; nine primaries are visible (the tenth is vestigial or reduced).
Plumage Larks are generally dull-colored, in browns,

rufous, buff, black, and white, with many species streaky or with nondescript uniform plumage. Most species are cryptically colored; some species are more boldly marked.
Behavior These ground-dwelling birds walk or run rather than hop. Most give complex songs in display flight.
Distribution Larks reach their greatest diversity in Africa, where 80 percent of species occur, and lowest in South America, where they are represented by an isolated breeding population of Horned Lark *(E. a. peregrina)* near Bogotá, Colombia. Two species occur in North America: the Horned Lark (naturally) and the Sky Lark (introduced and also as a vagrant). The Horned Lark is the most widely distributed species in the family, with 42 subspecies occurring across North and South America, Eurasia, and North Africa.
Taxonomy Relationships remain unclear. Historically the alaudids are thought to be the most primitive of the oscine passerines despite their complex songs. Some DNA evidence places the larks near the Old World sparrows, while recent genetic studies place them closer to the Old World warblers, bulbuls, and swallows.
Conservation Lark habitat in general is threatened by human agricultural activities, as well as by urbanization. Many species have very restricted ranges and habitat requirements.

Genus *Alauda*

SKY LARK   *Alauda arvensis*   SKLA ■ 3

This Eurasian species raises a slight crest when agitated. Polytypic (11 ssp.; 2 in N.A.). L 7.2" (18 cm)

IDENTIFICATION Brown with pointed bill. Upperparts heavily streaked, pale supercilium; buffy white underparts with necklace of fine dark streaks on breast and throat. Narrow white trailing edge on secondaries and inner primaries and white outer tail feathers visible in flight. **ADULT:** Sexes similar. **JUVENILE:** Less distinctly marked than adult. Upperparts spotted and speckled, throat less streaked. Distinctive buff fringes on wing coverts.

GEOGRAPHIC VARIATION Western European nominate *arvensis* is described above. The northeastern Asian *pekinensis* is more richly colored on breast, darker and more heavily streaked above, and has longer wings.

SIMILAR SPECIES Pipits have thinner bills; sparrows, longspurs, and buntings have thicker bills. They all lack white trailing edge on wing.

VOICE FLIGHT CALL: A liquid *chirrup* with buzzy overtones. **SONG:** A very long (2–4+ min.) series of buzzy, trilling, churring notes, given from the ground or in aerial display flight.

STATUS & DISTRIBUTION Locally introduced population *(arvensis)* on Vancouver I., BC, sedentary. Asian subspecies *(pekinensis)* rare in spring to Aleutians and Pribilofs (casual in summer, has bred), casual in spring north to St. Lawrence I. **BREEDING:**

Open habitats with short grasses and low herbs in BC; alpine tundra on Pribilofs. **MIGRATION:** Breeding birds on Pribilofs migratory, AK records primarily May–early June. **WINTER:** Sedentary population in BC uses wide variety of open habitats. **VAGRANT:** Record from CA likely *pekinensis;* records from WA and Queen Charlotte Is. may have been *pekinensis.* Nominate *arvensis* introduced on Hawaiian Is. Casual to Leeward Is. *(pekinensis).*

POPULATION Introduced population in BC has declined from about 1,000 in the 1950s to 100 in 1995, attributed to urbanization.

arvensis

white trailing edge to secondaries

short crest can be raised

fresh *arvensis*

worn *arvensis*

pekinensis

darker, richer, and more heavily streaked than *arvensis*

juvenile *arvensis*

Genus *Eremophila*

HORNED LARK   *Eremophila alpestris*   HOLA ■ 1

The head pattern is distinctive in all subspecies: black "horns" and white or yellowish face and throat with broad black stripe under eye. The Horned Lark usually nests on bare ground, including plowed fields, and

lays two to five eggs (mid-Mar.–mid-July). Polytypic (42 ssp.; 21 in N.A.). L 6.8–7.8" (17–20 cm)

IDENTIFICATION Nape, back, rump, and wing brown streaked with dusky to black. Black bib. Tail black with two outer tail feathers edged light gray on outer web. Breast and belly yellow to white, breast patch black. Head boldly marked with black lores, cheek, and ear tufts; white to yellow eyebrow, ear coverts, and throat. Short, stout bill. Black legs. **ADULT:** No seasonal variation. Sexes similar, but males generally larger and darker. Female duller on head and face, cheek patch; breast band smaller and gray instead of black; back may be more striped; "horns" smaller and rarely erected. **JUVENILE:** Lacks black facial and breast markings of adult, making dark eye more conspicuous on brownish head. Back spotted with whitish to gray, giving a mottled appearance. Underparts generally creamy-buff with streaking on sides and breast. Tail as adult. Bill and legs paler yellowish to flesh in younger birds.

GEOGRAPHIC VARIATION In North America, 21 subspecies are recognized, with most intergrading. Migratory subspecies have longer wings, and birds in hotter environments have longer legs. Subspecies are mainly distinguished by plumage. Back color matches the color of the soil in local habitat; face and throat varies from yellow in Northeast to whitish or white in prairie and desert subspecies. Nominate *alpestris* of the Northeast is the largest and darkest of five eastern subspecies, and has a yellow throat. White-throated "Prairie" subspecies *(praticola)* is widespread in

Lapland Longspur for comparison

vagrant to Alaska ♂ *flava*

northwestern Canada, Alaska ♂ *arcticola*

central Arctic coast breeder ♂ *hoyti*

widespread in East in winter

winter ♂

"Northern Horned Lark" *alpestris*

alpestris

northeastern California ♂ *sierrae*

California Central Valley ♂ *rubea*

black tail with white outer tail feathers

coastal Northwest ♂ *strigata*

upperparts spotted with white

juvenile *ammophila*

faintly streaked breast

juvenile often mistaken for Sprague's Pipit

alpestris

small "horns"

praticola breeder over much of Midwest and East

"Prairie Horned Lark" *praticola*

winter ♂ *ammophila*

duller than male

southwestern deserts ♂ *ammophila*

Gulf Coast *giraudi*

except juveniles, all have dark breast band

♀ *ammophila*

the upper Midwest. Ten subspecies, generally pale, are western in distribution; one of the palest is Great Plains subspecies *enthymia*. Another western group consists of six small subspecies that are largely rufous on upperparts. Restricted to northeastern California, subspecies *sierrae* is very yellow below and russet above. Confined to the Channel Islands, California, *insularis* is rather streaked above and below. The ruddy-colored *rubea* is found in central California.

SIMILAR SPECIES A juvenile Horned is frequently mistaken for a Sprague's Pipit. Note thicker bill, longer tail and wings, and pale spotted upperparts of Horned, and paler buffy face and head of Sprague's.

VOICE CALL: Includes a high-pitched *tsee-titi,* given from ground or in flight. Flight call resembles American Pipit's, but softer. **SONG:** Typically, two or three thick introductory *chit* notes, followed by a high-pitched, rapid, jumbled series of tinkling notes rising slightly in pitch, given from ground or in display flight.

STATUS & DISTRIBUTION Common and widespread in open country. Movements complex; some subspecies resident and others highly migratory. **BREEDING:** Open, barren country, avoiding forests. Prefers barren ground and shorter grasses, deserts, brushy flats, and alpine habitat. **MIGRATION:** In spring, diurnal migrant in flocks.

Over much of Midwest, winter birds are largely nominate *alpestris,* but *praticola* returns to more northern breeding areas in the East by early Feb. in southern ON to early Apr. in MB. Timing is complex in much of the West, including altitudinal movements by alpine subspecies. Arrive in AK generally by mid-Apr. In fall, peak movements in the East late Oct.–Nov. Begin departing from AK in Aug. In Southwest, move as early as mid-June–Sept. **WINTER:** Occur in similar habitat to breeding grounds, including beaches, sand dunes, and airports. **VAGRANT:** Casual to southern FL; one Palearctic subspecies (*flava*) casual to west, southwest, and south-central AK. **POPULATION** Slight increases in the Southeast. Central Prairie Provinces showed significant declines from 1968–1975. Declines in many western states.

SWALLOWS Family Hirundinidae

Cliff Swallows (CA, Apr.)

This family consists of a group of accomplished aerial-foraging songbirds. The terms "swallow" and "martin" are used fairly interchangeably, with the square-tailed species generally being referred to as martins and the fork-tailed species as swallows. In the Old World, the Bank Swallow is called the Sand Martin.

Structure Hirundinidae is a very homogeneous family of birds, all with very similar body structures, and generally unlike any other passerines. Swallows have long pointed wings with ten primaries (the tenth extremely reduced); small compressed bills with a wide gape, sometimes with rictal bristles; short legs; and square to deeply forked tails. Feathers of the lores are directed forward, which shades the eyes, a useful adaptation for aerial insectivores. The structure of the syrinx (vocal apparatus) is well differentiated from other passerine families, though for the most part, swallows are unremarkable singers.

Plumage Swallows and martins have variable plumage on the whole, but with much consistency within genera. The birds often are iridescent blue or green above, and dark, white, or rufous below. Many species show pale rumps or dark breast bands, and a few Old World species are striped below. Sexual dimorphism is weak, with the New World martins (genus *Progne*) a notable exception.

Behavior All species are insectivorous, though one species (Tree Swallow) is able to feed on wax myrtle berries in winter, allowing it to winter farther north than most species. Swallows feed on the wing, and all are very accomplished fliers. Many temperate species undergo long-distance migrations, some among the longest of any passerines. Many species, including all those breeding in North America, are social in the breeding season, some nesting in colonies and sometimes with other swallow species. Some species are less gregarious or solitary. Nests are built either of mud—sometimes reinforced with vegetation and attached to vertical faces of natural and man-made structures—or within cavities excavated by other species. The Bank Swallow is the only cavity-nesting species that excavates its own nest. The rare Blue Swallow *(Hirundo atrocaerulea),* from southern Africa, is the only species that nests below ground, using potholes or aardvark burrows in open grassland.

Distribution 84 species in 20 genera are found worldwide, except in Antarctica and the high Arctic, with the greatest diversity in Africa (29 breeding sp.), and secondarily in Central and South America (19 sp.). Eight species breed in North America and seven additional species have been recorded as vagrants.

Taxonomy Swallows are well differentiated from other passerine families, and within the family there is little variation. Two species of river-martin, one occurring in West Africa and the other (possibly extinct) occurring in Thailand, form the subfamily Pseudochelidoninae. All other swallow species are placed in the subfamily Hirundininae. Although swallows were formerly considered to be most closely related to tyrant flycatchers (Tyrannidae), recent DNA evidence suggests that they are actually more closely related to Old World warblers (Sylviidae), babblers (Timaliidae), white-eyes (Zosteropidae), chickadees (Paridae), and long-tailed tits (Aegithalidae).

Conservation Many species, including all North American breeding species, have adapted well to the influence of humans on their environment. In fact, some species now rarely nest in natural situations, preferring to nest on manmade structures. A few species have very restricted ranges and are considered threatened or endangered.

MARTINS Genus *Progne*

These largest swallows, with broad-based wings (adapted for sustained soaring), moderately forked tails, and strongly decurved upper mandibles, are the only swallows of N.A. showing obvious sexual dimorphism. Taxonomy of this genus is unsettled, with at least eight species described. All members of *Progne* are called martins but are not directly related to martins of the Old World.

PURPLE MARTIN *Progne subis* PUMA ▪ 1

The Purple Martin is the largest swallow in North America. Males are all dark, glossy blue-black; females and immatures duller above and grayish below. This is an extremely popular and well-known bird due to its willingness to nest in structures provided by humans. Polytypic (3 ssp.; all in N.A.). L 7.5" (19 cm)

IDENTIFICATION ADULT MALE: All glossy blue-black above and below, wings and tail dusky black, distinctly notched tail. In hand, small concealed white tuft on sides of rump and sides of body are visible. **ADULT FEMALE:** Duller above than

1st spring ♂

eastern ♀
subis

eastern ♀
subis

darker overall than western subspecies

dark purple overall

adult ♂

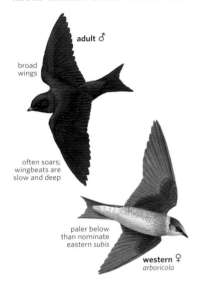

adult ♂

broad wings

often soars; wingbeats are slow and deep

paler below than nominate eastern *subis*

western ♀
arboricola

male, with more scattered patches of blue-black above. Grayish collar on hind neck. Throat, breast, and flanks dusky brown, paler on center of belly. Undertail coverts grayish with dusky centers. **IMMATURE:** Similar to adult female; young males show some blue-black on head and underparts, dark shaft streaks on ventral feathers, and sometimes a less distinct collar on hind neck; females, paler below and browner above, and lack dusky centers on undertail coverts. **FLIGHT:** Somewhat starlinglike shape. Graceful, liquid wingbeats interspersed with gliding and soaring.

GEOGRAPHIC VARIATION Females of western (*arboricola*) and desert (*hesperia*) subspecies with whiter underparts and forehead. Immature males tend

to look more like females of eastern nominate subspecies *(subis)*.

SIMILAR SPECIES Very similar to, sometimes indistinguishable from, other martins except Brown-chested. Female Purple Martin is the only species with contrasting gray collar on hind neck and pale forehead. Mottled undertail coverts.

VOICE CALL: Most frequently gives a *chur* call in many situations. When alarmed or excited, gives a *zwrack* or *zweet* call. **SONG:** Usually a series of chortles, gurgles, and slightly harsher croaking phrases. Also gives a churring, chortling "dawn song" around potential nest sites upon arriving on the breeding grounds in early spring.

STATUS & DISTRIBUTION Fairly common but a local and declining summer resident. **BREEDING:** In the East, colonially almost exclusively in

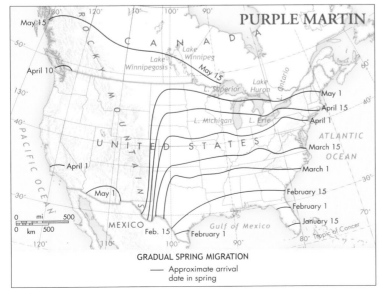

PURPLE MARTIN

May 15

April 10

April 1

May 1

Feb. 15

Lake Winnipeg

Lake Winnipegosis

May 15

L. Superior

Lake Huron

L. Michigan

L. Erie

February 1

Gulf of Mexico

CANADA

UNITED STATES

MEXICO

ATLANTIC OCEAN

May 1

April 15

April 1

March 15

March 1

February 15

February 1

January 15

Tropic of Cancer

0 mi 500
0 km 500

GRADUAL SPRING MIGRATION
— Approximate arrival date in spring

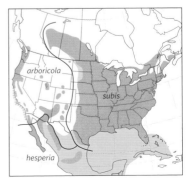

arboricola

subis

hesperia

artificial sites near human habitations. In the West, frequently solitarily, more often in natural cavities in forested areas, and in saguaro cactus in desert Southwest. **NEST:** In cavity excavated by another species, or in artificial structures; three to six eggs (late Mar.–late May). **MIGRATION:** In spring, arrives as early as mid-Jan. in TX, FL, and Gulf Coast; early Mar. in VA and KS, mid-Apr. in southern Canada, May

in AZ, early May in MT. During fall, in the East, very large aggregations of thousands of birds form locally in late summer. Passage peaks late July–Sept., beginning as early as late May, with stragglers until early Oct. In Southwest, scarce Aug.–late Sept. **WINTER:** South American lowlands east of the Andes south to northern Argentina (rarely) and southern Brazil. **VAGRANT:** Casual to NV and

AK. Accidental in Bermuda and U.K. **POPULATION** Causes of long-term declines unknown. Competes for nest cavities with the introduced European Starling and House Sparrow. In the West, logging has reduced availability of natural nest cavities. Increased availability of human-provided nest sites has had a positive effect on populations. Sharp declines in southern CA.

CUBAN MARTIN *Progne cryptoleuca*   CUMA ■ 5

This vagrant from the Caribbean has been confirmed only once in North America. Monotypic. L 7.5" (19 cm) **IDENTIFICATION** **ADULT MALE:** Like Purple Martin. **ADULT FEMALE:** Similar to Purple Martin, with darker upperparts, duskier brown throat, breast, and flanks contrasting with unmarked white belly and undertail; lacks paler grayish collar. **IMMATURE:** Similar to adult female. **SIMILAR SPECIES** Male indistinguishable from Purple Martin in field; in hand shows some concealed white feathers on belly, longer more deeply

forked tail. See above for female. May be conspecific with Caribbean Martin *(dominicensis)* and Sinaloa Martin *(sinaloae),* which show clean, white underparts in adult. **VOICE** **CALL:** Gurgling; a high-pitched *twick-twick.* **SONG:** A strong, melodious warble close to that of Purple Martin. **STATUS & DISTRIBUTION** A local and breeding endemic on Cuba and Isle of Pines. Non-breeding (Sept.–Feb.) range unknown, presumably S.A. **VAGRANT:** Key West, FL (imm. male specimen; May 9, 1895).

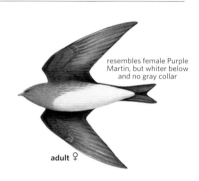

resembles female Purple Martin, but whiter below and no gray collar

adult ♀

GRAY-BREASTED MARTIN *Progne chalybea*   GYBM ■ 5

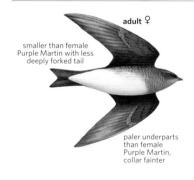

adult ♀

smaller than female Purple Martin with less deeply forked tail

paler underparts than female Purple Martin, collar fainter

This swallow has occurred in N.A. twice, both in TX. Monotypic. L 7" (18 cm) **IDENTIFICATION** **ADULT MALE:** Steel blue or purple glossed upperparts. Face, breast, sides, and flanks gray-brown or sooty gray; chin and throat paler. Belly and undertail coverts pure white. **ADULT FEMALE:** Duller than adult male with paler chin and throat; whiter below, browner forehead. **IMMATURE:** Duller than adult female with little gloss above.

SIMILAR SPECIES Female Purple Martin has distinct gray collar; female Southern Martin has mostly or entirely dark underparts; female Cuban Martin has more contrasting white belly. **VOICE** **CALL:** In contact, a *cheur.* Alarm and aggressive calls include *zwat, zurr,* or *krack.* **SONG:** A rich, liquid gurgling. **STATUS & DISTRIBUTION** Lowland resident, southern Mexico to northern Argentina. **VAGRANT:** Two TX specimens, Starr Co., 25 April 1980 and Hidalgo Co., 18 May 1889.

SOUTHERN MARTIN *Progne elegans*   SOMA ■ 5

This species of S.A. has occurred once in N.A. Monotypic. L 7" (18 cm) **IDENTIFICATION** **ADULT MALE:** Upperparts and underparts glossy dark, violet blue. Wings and tail black with slight bluish green gloss. Tail slightly forked. **ADULT FEMALE:** Smaller than male. Sooty black above, glossed blue on back; dusky brown below, often with pale edgings giving a scaly look. Uppertail coverts sometimes with narrow pale terminal margins. Most uniformly dark below. **IMMATURE:** Similar to adult female.

SIMILAR SPECIES Male slightly smaller than Purple Martin, with a slightly longer and more forked tail. In hand, lacks concealed white patch on sides and flanks. Female darker below than any other female martin. **VOICE** **CALL:** A harsh contact call and a high-pitched alarm call. **SONG:** A short warbling. **STATUS & DISTRIBUTION** Breeds in southern S.A.; migrates north in Austral winter to western Amazon Basin. **VAGRANT:** one specimen in Key West, FL (Aug. 14, 1890).

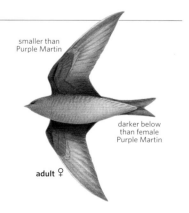

smaller than Purple Martin

darker below than female Purple Martin

adult ♀

BROWN-CHESTED MARTIN *Progne tapera* BCMA ■ 5

adult
fusca

distinct breast band with
downward point like
smaller Bank Swallow

This species suggests an oversize Bank Swallow. Polytypic (2 ssp.; *fusca* and possibly nominate in N.A.). L 6.5" (16 cm) **IDENTIFICATION ADULT:** Subspecies *fusca* described. Upperparts brown. Dark brown band on upper breast; with pear-shaped dusky marks from center of breast to upper belly; sides and flanks brownish. The more northerly breeding subspecies *tapera* is slightly smaller and overall a paler brown and lacks dark extension of breast band. **SIMILAR SPECIES** Bank Swallow is smaller.

VOICE CALL: Gives a *chu-chu-chip* call when in flocks. **STATUS & DISTRIBUTION** S.A. from Colombia to northern Argentina. Southern breeding subspecies *(fusca)* is an Austral migrant to northern S.A. **VAGRANT:** Casual (mostly *fusca*) fewer than ten records to eastern N.A., recorded MA to FL. One photographed at Patagonia Lake S.P., AZ (Feb. 3, 2006) is sole record for the West. One on July 1, 2006, at New London, CT, was thought to be the nominate.

Genus *Tachycineta*

These swallows are mainly iridescent blue or green above and white below. The genus contains nine species, all occurring in the New World. Two species breed in North America, and two species have been recorded as vagrants. The tails of *Tachycineta* are shallowly notched to deeply forked.

TREE SWALLOW *Tachycineta bicolor* TRES ■ 1

Familiar denizens across North America, Tree Swallows nest in abandoned cavities in dead trees or nest boxes provided for them by admiring humans. These birds are hardier than other swallows and can feed on seeds and berries during colder months. Monotypic. L 5.8" (15 cm)

adult

1st spring ♀

adult

dark
brownish
above

white tertial tips

juvenile

fall adult

spring adult

dark blue
upperparts

IDENTIFICATION SPRING ADULT: Upperparts iridescent greenish blue. Wings and tail dusky blackish, underparts entirely clean white. Lores black, ear coverts blue-black. Tail slightly forked. A few females retain brown back of immature into first spring, possibly longer. **FALL ADULT:** Upperparts sometimes appear more greenish than in spring. Tertials and secondaries edged pale grayish. **IMMATURE:** Upperparts gray-brown, with pale grayish edges on tertials and secondaries. Underparts clean white with indistinct dusky brown wash across breast, faintest in center. **FLIGHT:** Small but chunky swallow with a shallowly forked tail and broad-based triangular wings. **SIMILAR SPECIES** Smaller Violet-green Swallow has white of cheek extending behind and above eye, and has white patches on sides of rump. Tree Swallow flies with slower wing beats and glides more than Violet-green. Bank Swallow similar to immature, but smaller and with distinct, clean-cut brown breast band, different face pattern (white behind ear), and paler brown rump. **VOICE SONG:** An extended series of variable chirping notes—*chrit, pleet, euree, cheet, chrit, pleet.* **STATUS & DISTRIBUTION** Common. **BREEDING:** Open areas,

usually near water, including marshes, fields, and swamps. **NEST:** Brings a few pieces of vegetation to nest cavity and often includes feathers; two to eight eggs (May–July). **MIGRATION:** In spring, arrives earlier in north than other swallows, usually mid-Mar.–early Apr. Departs later than other species, peaking late Sept. in MO and VA, lingering until late Nov. as far north as the Great Lakes. **WINTER:** Southern U.S., south through C.A. and the Caribbean. Open areas near water and nearby woodlands. **VAGRANT:** Casual in Bermuda and the Azores; accidental in Guyana, near Trinidad, Greenland, and England. **POPULATION** The breeding range is expanding southward; the population is increasing in eastern and central United States.

MANGROVE SWALLOW   *Tachycineta albilinea*   MANS ■ 5

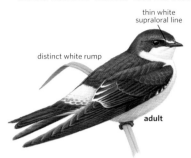

thin white
supraloral line

distinct white rump

adult

A very small swallow, the Mangrove Swallow is never found far from water, frequently in coastal areas. Monotypic. L 5.2" (13 cm)
IDENTIFICATION ADULT: Crown and back iridescent greenish, auriculars and lores black. Narrow white line above lores usually meets on forehead. Rump white, uppertail coverts dark. Wings and slightly forked tail dusky blackish, underwing coverts clean white. Tertial and inner secondary edges white. Underparts white, extending onto side of neck as a partial collar, sometimes with a dark diffuse wedge on sides of upper breast. **JUVENILE:** Gray-brown where adults are greenish blue. **FLIGHT:** Small size evident, with short, broad wings and short tail. Rapid, shallow wingbeats. **SIMILAR SPECIES** Tree Swallow is noticeably larger, lacks white rump and supraloral. Common House-Martin is larger, bluish above, lacks white supraloral, and has a more deeply forked tail. White-rumped *(T. leucorrhoa)* has not been recorded from N.A., but is larger, bluish above, and has narrower white edges on tertials.
VOICE SONG: A series of *chrit or chriet* notes mixed with buzzier chirps.
STATUS & DISTRIBUTION Common non-migratory resident of lowlands, especially near water, from northwestern and northeastern Mexico and C.A. **VAGRANT:** One adult in Viera Wetlands, Viera, Brevard Co., FL (Nov. 18–25, 2002).

VIOLET-GREEN SWALLOW   *Tachycineta thalassina*   VGSW ■ 1

The western counterpart of the Tree Swallow, the Violet-green Swallow has narrower wings and a shorter tail. This species frequently uses nest boxes. Polytypic (2 ssp., nominate in N.A.). L 5.3" (13 cm)
IDENTIFICATION ADULT MALE: Upperparts of male are dark velvet green, more bronze on crown, becoming purple on rump and uppertail coverts. Underparts are white, extending onto cheek, to behind and above eye. Lores are dusky. White of flanks extends onto sides of rump, forming two white patches from above in flight. Wings and short, slightly forked tail are black. Underwing coverts grayish. **ADULT FEMALE:** Somewhat duller than male, browner on head, face and ear coverts. Throat is washed slightly with ashy brown. **JUVENILE:** Similar to adults, but green and purple are replaced with brownish. White on cheek and behind eye is much less extensive. Slight pale brownish wash on breast. Tertials and inner secondaries are very narrowly edged pale grayish. **FLIGHT:** Longer, narrower wings and shorter tail than Tree Swallow; Violet-green has a more fluttering flight style.
SIMILAR SPECIES Tree Swallow is larger, entirely greenish blue above, and lacks white around back of eye and on sides of rump. The larger White-throated Swift has narrower pointed wings and black-and-white underparts. Vagrants reaching the Bering Sea region should be carefully compared to the all-white-rumped, barely smaller Common House-Martin.
VOICE CALL: Various short notes, including a twitter, and *chee-chee* notes. Alarm call is a *zwrack* similar to that of Purple Martin.
STATUS & DISTRIBUTION Common. **BREEDING:** Occurs in a variety of habitats, including open areas in montane coniferous and deciduous forests, coastal regions, and even in higher elevation desert areas. **NEST:** It brings a variety of grasses and stems, as well as feathers, to an abandoned cavity in trees, rock crevices, dirt

banks, or columnar cactuses; 4–6 eggs (Apr.–May). **MIGRATION:** In spring, arrives in southern AZ and southern CA by late Feb., AK by early May. Departs AK by mid-Aug., lingering into mid-Oct. in BC and WA. In the Southwest departs late Sept.–late Oct. Overall, a late fall migrant; early fall migrants reported from desert regions are usually misidentified Tree Swallows. **WINTER:** Tidal flats to interior mountains. A few birds in southern CA, but most from Mexico south to Guatemala, El Salvador, and Honduras. **VAGRANT:** Casual in western AK, including Bering Sea islands, and east of the Rockies, with most sightings in the East Sept.–Nov.
POPULATION There is no documented effect of human activity on numbers; the introduction of the House Sparrow and European Starling may have reduced populations in urban areas of southern Canada.

adults

small size with fluttery wingbeats

♂

juvenile

♂

white up sides of rump

white above eye

adult ♂

mainly green upperparts and snowy white underparts

BAHAMA SWALLOW *Tachycineta cyaneoviridis*   BAHS ▪ 4

Endemic to the northern Bahamas, this species has one of the most restricted ranges of any swallow species in the world. Monotypic. L 5.8" (15 cm)
IDENTIFICATION ADULT MALE: Iridescent green upperparts, dusky lores. Underparts, including lower cheek to below eye and underwing coverts, white. Wings and deeply forked tail dusky bluish black above. Margins of outer tail feathers sometimes whitish. **ADULT FEMALE:** Slightly duller

than male, with some slaty brown on ear coverts and breast washed with pale, sooty brown. Shorter tail. **JUVENILE:** Duller than adults, upperparts brownish with greenish mainly on mantle and wing coverts. Underparts white washed with sooty brown on breast. Shorter, less forked tail. **FLIGHT:** Resembles Barn Swallow in shape, but Tree Swallow in plumage.
SIMILAR SPECIES Tree Swallow has a much less forked tail; Barn Swallow is blue above and buffy and white below; Violet-green Swallow has large white spots on sides of rump.
VOICE CALL: A metallic *chep* or *chi-chep.*
STATUS & DISTRIBUTION Uncommon to fairly common. **BREEDING:** Pine forests in the northern Bahamas (Grand Bahama, Abaco, Andros). Uses abandoned cavities in trees and sometimes on buildings; two to four eggs (Apr.–July). **WINTER:** Some

winter on the breeding grounds, with some moving south to the southern Bahamas and rarely to eastern Cuba (Nov.–Mar.). **VAGRANT:** Casual in southern FL (9+ records, but none since 1992).
POPULATION Species is vulnerable due to small geographic range, and world population estimated at approximately 2,500 pairs.

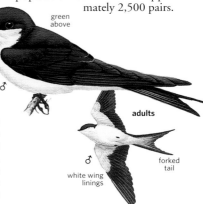

green above

adults

♂

forked tail

♂

white wing linings

NORTHERN ROUGH-WINGED SWALLOW *Stelgidopteryx serripennis*   NRWS ▪ 1

Named for the inconspicuous serrations on the first primary feather on the wing, this swallow is unique among birds of N.A. Polytypic (6 ssp.; 2 in N.A.). L 5" (13 cm)
IDENTIFICATION ADULT: Upperparts brown, with rump slightly paler; throat and breast pale brownish; underparts

long wings; flies with slow and floppy wingbeats

juvenile

cinnamon wing bars

dusky wash on throat and breast

dirty white. Square tail. **JUVENILE:** Similar to adults, bright cinnamon-edged tertials and wing coverts. **FLIGHT:** This long-winged species flies with slow floppy wingbeats.
GEOGRAPHIC VARIATION Southwestern subspecies *(psammochrous)* is paler than nominate. Some treat *ridgwayi,* mainly from the Yucatán and which has different vocalizations, as a separate species.
SIMILAR SPECIES Smaller Bank Swallow has clean white throat that wraps around rear of auriculars, a distinct brown breast band, and a paler brown back and rump that contrasts with darker wings. Immature Tree Swallow has white throat and dusky on breast usually not meeting in center.
VOICE CALL: Series of loose, low-pitched, upwardly inflected *brrt* notes, a buzzy *jrrr-jrrr-jrrr-jrrr, or a* higher pitched *brzzzzzt.*
STATUS & DISTRIBUTION Fairly common over a wide altitudinal range. **BREEDING:** Open areas; often near rocky gorges, road cuts, gravel pits, or exposed sand banks. **NEST:** Singly or in small colonies in natural burrows or crevices, and on man-

made structures; 4–8 eggs (May–June). **MIGRATION:** In eastern N.A. arrives in early Mar. in FL to mid-Apr. in MI. In West, arrives in mid-Feb. in southern CA to mid-Apr. in OR and WA. In eastern N.A. fall migration prolonged, late July–early Oct. A few are trans-Gulf migrants. In West, mainly in Aug. with some into Sept. **WINTER:** Lowlands and foothills from Mex. to C.A., with small numbers in the Caribbean. Some winter in U.S. along Gulf Coast from TX to FL Panhandle and in southeastern CA and southern AZ. **VAGRANT:** Rare in YT and southeastern AK; casual northern AK.
POPULATION Stable.

Genus *Riparia*

BANK SWALLOW *Riparia riparia* BANS ▪ 1

Our smallest swallow, the Bank is one of the few passerines with nearly world-wide distribution. Polytypic (2 ssp.; nominate in N.A.). L 4.8" (12 cm)

IDENTIFICATION ADULT: Pale gray-ish brown upperparts contrast with darker wings, underparts clean white with distinct brown breast band. Slightly forked tail. **JUVENILE:** Similar to adult. Wing coverts and tertials narrowly edged buff. **FLIGHT:** Sweeps wings backward with quick flicking wingbeats. Narrow, pointed wings, slim body. Flight feathers contrast-ingly dark from above.

SIMILAR SPECIES Northern Rough-winged Swallow is larger, has a dusky throat gradually blending into dirty white underparts. Immature Tree Swallow is larger, shows an incom-plete breast band.

VOICE CALL: Most frequently heard is a series of buzzy *trrrt* notes, not too different from the buzzing sound of high-tension lines.

STATUS & DISTRIBUTION Locally com-mon. **BREEDING:** Lowland areas along coastlines, rivers, lakes, and other wetlands; also in gravel quarries and road cuts. Small to large colonies in burrows in sand banks or cliffs; one to nine eggs (Apr.–June). **MIGRA-TION:** Departs S.A. in Feb., through U.S. from early Mar. in South to late May in North. During fall migration, often in large flocks. Among the earli-est to depart in the East, with most birds departing the Great Lakes by early Sept.; late July to early Nov. in FL; peak in late July and Aug. **WINTER:** Most often in wetlands areas, nearly throughout S.A. as far south as central

Chile and northern Argentina, Pacific slope of southern Mexico, rarely in eastern Panama and in the Caribbean. Rare in southern CA, southern TX.

POPULATION Road building and quar-ries have changed breeding distribu-tion in many areas. Erosion negatively affects breeding colonies.

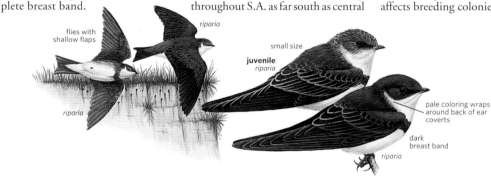

flies with
shallow flaps

riparia

riparia

small size

juvenile
riparia

pale coloring wraps
around back of ear
coverts

dark
breast band

riparia

Genus *Hirundo*

BARN SWALLOW *Hirundo rustica* BARS ▪ 1

This most widely distributed and abun-dant swallow in the world is familiar to birders and nonbirders. Polytypic (6–8 ssp.; 3 in N.A.). L 6.8" (17 cm)

IDENTIFICATION ADULT MALE: Deep iri-descent blue crown, back, rump, and wing coverts; deeply forked tail with large white spots; rich buff to rufous forehead and underparts; iridescent blue patches on sides of breast, some-times with very narrow connection in center. Wings and tail black. **ADULT FEMALE:** Similar to male but with paler underparts and less deeply forked tail. **IMMATURE:** Duller above than adults, with shorter but still deeply forked tail, buffy throat and forehead, whit-ish underparts. **FLIGHT:** Very pointy wings, deeply forked tail, and low zig-zagging flight are distinctive. **HYBRID:**

Rarely reported. Cliff and Cave Swal-lows in N.A., and Common House-Martin in Old World.

GEOGRAPHIC VARIATION Subspecies *erythrogaster* breeds in N.A. Eurasian subspecies *(rustica)* is accidental; it has clean white underparts with broader dark breast band. East Asian sub-species *(gutturalis)* is casual; similar to *rustica* but smaller, usually with incomplete blue breast band.

SIMILAR SPECIES No other swallow in North America shows such a deeply forked tail. Shorter-tailed juveniles in flight may suggest Tree Swallow but will always show partial dark breast band and buff throat, white spots on tail.

VOICE CALL: In flight repeats a high-pitched, slightly squeaky *chee-jit.*

SONG: A long series of squeaky war-bling phrases, interspersed with a nasal grating rattle.

STATUS & DISTRIBUTION Common. **BREEDING:** In various habitats in

lowlands and foothills with nearby open areas and water. Former natural nesting locales, including caves and cliff faces, have now mostly been abandoned in favor of a variety of manmade structures, including barns. A cup-shaped bowl made entirely of mud and the bird's saliva; three to seven eggs (May–June). MIGRATION: In spring, arrives in extreme southern U.S. late Jan.–early Feb., peaking in mid-May in northeastern U.S. Early arrival in southeastern AK in mid-May. Departs northeastern U.S. as early as mid-July, peaking late Aug.–early Sept.; late Sept.–early Oct. in southern CA. Main routes through C.A. and through Caribbean, but also a trans-Gulf migrant. WINTER: Often in fields and marshes mainly in lowlands; rarely in southern U.S.; more frequently now in southern CA. Uncommon from Mex. south through C.A. Most common throughout S.A. as far as central Chile and northern Argentina; breeding noted in Buenos Aires. VAGRANT:

North American subspecies casual or accidental in western and northern AK, and HI, southern Greenland, Tierra del Fuego, Falkland Is. Eurasian subspecies *(rustica)* accidental in AK (specimens from Pt. Barrow and St. Lawrence I.), NT, NU, and southern Labrador. East Asian subspecies *(gutturalis)* casual Aleutians and Bering Sea islands, AK; accidental mainland AK and Queen Charlotte Is., BC. POPULATION Common worldwide. In North America, breeding range has been expanded by using manmade structures for nesting.

Eurasian
rustica

complete
breast band

white
underparts

paler below
than adult

dark
erythrogaster

long, forked
tail with
white at base

juvenile
erythrogaster

shorter
tail

erythrogaster

bluish above

COMMON HOUSE-MARTIN *Delichon urbicum*   COHM ◼ 4

This Old World species is casual in migration in western Alaska and on St. Pierre and Miquelon. Polytypic (3 ssp.; 1, possibly 2, in N.A.). L 5" (13 cm)

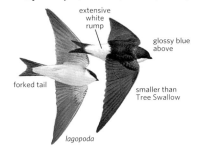

extensive
white
rump

glossy blue
above

forked tail

smaller than
Tree Swallow

lagopoda

IDENTIFICATION ADULT MALE: Deep blue upperparts, white rump and underparts, forked tail, whitish underwing coverts. ADULT FEMALE: Similar to male; grayer underparts. JUVENILE: Duller upperparts, less deeply forked tail.
GEOGRAPHIC VARIATION Western European subspecies *(urbica)* shows dark uppertail coverts below white rump, forked tail. East Asian subspecies *(lagopoda)* shows white uppertail coverts continuous with white rump, slightly less deeply forked tail.
SIMILAR SPECIES Larger Tree and comparably sized Violet-green Swallows lack

a solidly white rump. Most similar is the Asian House-Martin *(D. dasypus)*, not recorded from N.A., which shows duller upperparts, gray-brown underwing coverts, pale gray wash on throat and undertail coverts, less deeply forked tail. VOICE CALL: A scratchy *prrit*. SONG: A soft twittering.
STATUS & DISTRIBUTION BREEDING: Western Europe to Russian Far East. VAGRANT: AK (one specimen of *lagopoda* from Nome, about ten other records from Bering Sea islands), St. Pierre and Miquelon Is. (1 sight record, likely *urbica*).

Two of the three species in this New World genus are found in North America. Four Old World species are sometimes included in this genus. Characterized by square tails, some rufous on the head, orangish or chestnut rumps, and white-striped dark backs, all build mud nests. This genus is sometimes merged with the more widespread *Hirundo*.

CLIFF SWALLOW *Petrochelidon pyrrhonota*   CLSW ◼ 1

Look for their large nesting colonies on cliffs as well as on buildings and under bridges. Polytypic (5 ssp.; all in N.A.). L 5.5" (14 cm)
IDENTIFICATION ADULT: Square

tail; orangish buff rump; dark cap extending below eye; chestnut cheeks, sides of neck, and throat; bluish black on lower throat; whitish to buffy to chestnut forehead

patch (feathering immediately above bill darkish). Underparts, including flanks, whitish. JUVENILE: Similar to adult, but entire head usually dark brownish black, sometimes with small pale grayish, whitish, or rusty forehead patch. Rump paler buff. Chin and upper throat pattern quite variable, some mixed with white, full black, gray, or cinnamon. **FLIGHT:** Short triangular wings and square tail. Whitish underparts and paler underwings contrast with dark head and throat. **HYBRID:** Very rare; hybrids with Barn Swallow and Tree Swallow (once) known. Cave Swallow hybrids possible but not confirmed.

GEOGRAPHIC VARIATION Considerable intergradation between five recognized subspecies. Northern nominate (inc. much of the East), *ganieri* (East, west of Appalachians from central TN south), *hypopolia* (NT to central CA), *tachina* (mainly southern CA), and *minima* (Southwest). They differ in overall darkness, including color of forehead and

throat. Most distinctive is Southwest *minima* with chestnut forehead.
SIMILAR SPECIES Cave Swallow (see sidebar below).
VOICE CALL: A single low-pitched churr, and a higher keer when disturbed. **SONG:** A subdued squeaky twittering given in flight and near or from within nest.
STATUS & DISTRIBUTION Locally common. **BREEDING:** Various habitats, including grasslands, towns, open forest, and river edges wherever there are cliff faces or escarpments for nesting. Small to large colonies, on cliff faces and on man-made structures (sometimes with Barn Swallows), rarely at cave entrances. A gourd-shaped structure built entirely of mud and saliva; one to six eggs (Apr.–June). **MIGRATION:** Always via C.A. Departs winter range in early Feb., a few by mid-Jan. to

CA. Arrives in IL in early Apr., and AK in mid-May. Departs after nestlings fledge, sometimes as early as late June. Peak is Aug.–early Sept., earlier in Southwest (July–early Aug.). A few may linger to early Nov. in East. **WINTER:** Grasslands, agricultural areas,

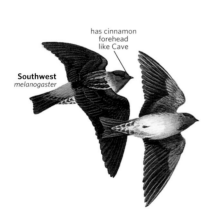

Strong weather systems, characterized by prolonged strong west or southwest winds, sometimes bring Cave Swallows to the Atlantic and Great Lakes states late October through November. Cliff Swallows have typically left for their South American winter areas by then, but occasionally lingerers are recorded, presenting an identification challenge. The subspecies of Cave Swallow most likely occurring in the East during and following these weather events is thought to be the southwestern *pallida*, not a bird from the geographically closer *fulva* West Indies group.

Adults of the eastern subspecies of Cliff Swallow (*pyrrhonota*) show a bright cream-colored forehead, unlike the Cave's chestnut forehead. Cliffs can be distinguished by their dark heads, including the throat, contrasting

Cave Swallow (NJ, Nov.)

Cliff Swallow (CA)

with whitish underparts. The Cave has a pale buff throat contrasting little with the breast.

Juvenile Cliffs typically have dark heads, but some have a few paler markings on the throat. Juvenile Caves are similar to adults; a small dark cap is relatively easy to see against pale buff cheeks and throat, even in flight. ■

near towns, and in marshes. S.A., from southern Brazil south to south-central Argentina. **VAGRANT:** Casual to Bering

Sea islands, AK (spring), and the Azores. Accidental to Wrangel I., Russian Far East, southern Greenland, and U.K.

POPULATION Has expanded its range into the Great Plains and eastern N.A. in the past 150 years.

CAVE SWALLOW *Petrochelidon fulva*   CASW ■ 1

This small, square-tailed, buff-rumped swallow has greatly expanded its breeding range by utilizing man-made structures. Polytypic (6 ssp.; three in N.A.). L 5.5" (14 cm)

IDENTIFICATION ADULT: Square tail and orangish buff to russet-chestnut rump. Has a small blackish cap extending to top of eye, which gives a somewhat masked look; broad pale wrap-around buff collar (further isolates black cap) cheeks, and throat; chestnut forehead patch. Wings, tail, and uppertail coverts blackish; back is black with narrow white stripes. Underparts are whitish with pale buff flanks. **JUVENILE:** Similar to adult, but with less extensive forehead patch, variably paler chin and throat, browner back lacking distinct white stripes, and duller flanks. **FLIGHT:** Has a small black cap with pale buff collar and face visible at some distance. Glides frequently. **HYBRID:** Rarely with Barn Swallow, suspected with Cliff Swallow.

GEOGRAPHIC VARIATION Subtle. Subspecies breeding in Florida and Cuba (*cavicola*) is slightly smaller, shows tawny-buff flanks, richer rufous-buff rump; subspecies breeding in southwestern U.S. (*pallida*) is slightly larger, shows grayish buff flanks, variable orange-buff rump. Slightly smaller *fulva* from Hispaniola is not separable in the field from *cavicola*.

Cliff and Cave Swallows both have fairly square-ended tails and buffy rumps

cinnamon forehead

buffy rump

rump slightly more cinnamon than Cliff

juvenile

pale collar contrasts with dark cap

adult
pallida

Southwest
pallida

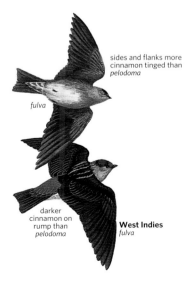

sides and flanks more cinnamon tinged than *pelodoma*

fulva

darker cinnamon on rump than *pelodoma*

West Indies
fulva

SIMILAR SPECIES Cliff Swallow, especially southwestern subspecies (see sidebar p. 494).

VOICE CALL: Various chattering notes, including a short, clear *weet* or *cheweet,* ascending or descending in pitch. **SONG:** A complex squeaky twittering, slightly higher pitched and more prolonged than that of Cliff Swallow.

STATUS & DISTRIBUTION Locally common, but patchy in distribution. Migration little known, and wintering grounds of most U.S. breeding birds not fully known. **BREEDING:** Natural or man-made structure often near water. Has only recently expanded northward into the U.S.; first recorded nesting in TX in 1910, NM in 1930, and FL in 1987. Continues to expand in TX, but still fairly restricted in southern FL and NM. In colonies in twilight zone of caves, sinkholes, culverts, and under bridges (sometimes with Barn and Cliff Swallows), often high up and inaccessible in caves. A crescent-shaped half bowl built entirely out of mud and saliva, sometimes with bat guano; three to five eggs (Apr.–Aug.). **MIGRATION:** Movements and routes little known. In spring, arrives mid-Feb.–early Mar. in NM. Departs from NM late Oct.–mid-Nov. Routes largely unknown, but has been reported from western Mexico, Curaçao, Costa Rica, and Panama (sight records). **WINTER:** FL birds may winter in Caribbean. Winter range of Cave Swallows nesting in NM unknown. Texas birds move south

to unknown areas, but since the 1980s hundreds have overwintered in southern TX, often along streams, rivers, or lakes. **VAGRANT:** Recent annual late fall invasions on eastern seaboard and in Great Lakes (see sidebar p. 438). A few spring records there as well. Casual to accidental in spring or summer in AZ, CA (also in winter), central Great Plains, MS, AL—all or nearly all records thought to refer to pallida from the Southwest. Two spring specimens from Sable I., NS (others seen) have been ascribed by various authorities as either *cavicola* or *fulva* from the West Indies.

POPULATION Cave Swallows are potentially vulnerable to disturbance at nesting colonies, although populations appear stable. The species is expanding its breeding range to north and east using man-made bridges and culverts, most dramatically in TX since the mid-1980s. Cave Swallows are vulnerable to temperature extremes and adverse weather conditions in NM and TX.

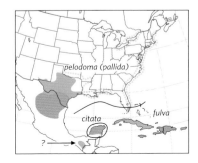

pelodoma (pallida)

citata

fulva

?

CHICKADEES AND TITMICE Family Paridae

Bridled Titmouse (AZ, May)

These lively sprites top many people's lists of favorite birds. The North American Paridae divides neatly into two groups: brightly patterned chickadees (7 sp.) and mostly drab titmice (5 sp.). They inhabit tundra, high mountains, desert scrub, old-growth forests, small woodland patches, urban parks, and suburban backyards. Nearly all are sedentary. The northernmost species can endure frigid winters, migrating only in intermittent irruptions south or moving short distances down-elevation for the winter when food is scarce. Among the most intensively studied North American birds, parids have provided immense knowledge and understanding of physiological adaptation, hybridization, vocalization, genetic variations and relationships, socialization, and many other aspects of avian behavior.

Structure Parids are small birds, either round-headed (chickadees) or prominently crested (titmice). Their relatively short, rounded wings and long tails are aerodynamically suited for agility in crowded habitats. Their short, sharply pointed bills are designed to grasp tiny arthropods and crack small seeds.

Behavior Acrobatic feeders, parids glean high and low, flitting from leaf to leaf and branch to branch in noisy chatter, nearly always easy to see and hear. Their vocal repertoires are exceedingly complex. Most species sing in clear whistles, and all have sputtering variations of *tsick-a-dee* or *chirr-chirr,* plus a seemingly endless array of gargles, twitters, and *zeets*—each sound used in a particular behavioral context. Parids eagerly visit bird feeders. All are cavity-nesters, and many use nest boxes.

Plumage The chickadee's standard features include a dark cap, white or partly gray cheeks, and a black bib; their body colors are various shades of gray or brown.

Titmice are typically dull, unpatterned gray or brown, brightened only by rusty flanks in several species and by a brightly patterned face in one. The sexes in most species look alike, and juveniles differ little from adults.

Distribution The Paridae includes more than 54 species around the globe, with 12 occurring in North America. At distributional extremes, Black-capped and Boreal Chickadees span the continent, while only the northernmost Mexican Chickadees and Bridled Titmice and the easternmost Gray-headed Chickadees reach N.A. The family includes habitat generalists such as the Carolina Chickadee, at home in remote southeastern river bottomlands and crowded urban parks, as well as species with strong habitat preferences such as the Oak Titmouse, which relies heavily on the woodlands of its name.

Taxonomy The classification of North American chickadees has long been stable, but titmice have undergone considerable flux. The AOU divided the Plain Titmouse into Oak and Juniper species in 1997. Black-crested and Tufted Titmice, traditionally classified as separate species, were merged in 1976 and then separated again in 2002. The Black-capped Chickadee is the most variable of our parids, with at least nine subspecies recognized in three geographic groups. The least geographically variable parid in N.A. is the Tufted Titmouse, which is monotypic.

Conservation Loss of habitat may be a potential threat to a few species, especially for the Oak Titmouse in California, where its favored oak woodlands are being destroyed for suburbs and agriculture. BirdLife International lists one species as threatened and three others as near threatened, none in North America.

CHICKADEES Genus *Poecile*

Worldwide this Northern Hemisphere genus consists of 15 species, named chickadees in the New World and tits in the Old World, all of them formerly classified in the genus *Parus.* Seven of these species occur in North America, but the Mexican Chickadee and the Gray-headed Chickadee breed here only at the limits of their ranges.

MOUNTAIN CHICKADEE *Poecile gambeli*   MOCH ▪ 1

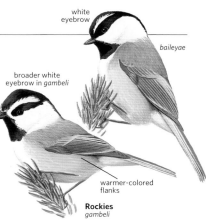

white eyebrow

baileyae

broader white eyebrow in *gambeli*

warmer-colored flanks

Rockies *gambeli*

This chickadee can be found above 12,000 feet, sometimes to timberline and beyond. It spends a lot of time in the postbreeding season caching conifer seeds for the winter. It forages higher in tall trees than most chickadee species, and it often spends the nonbreeding season foraging in mixed-species flocks. Polytypic (4 ssp.; 3 in N.A.). L 5.3" (13 cm)

IDENTIFICATION Black cap and bib, and white supercilium, which consists of white-tipped feathers, is unique among chickadees. White cheeks; grayish, brownish, or olive upperparts, depending upon the subspecies; greater wing coverts, secondaries, and tertials indistinctly edged with pale gray.

GEOGRAPHIC VARIATION Nominate *gambeli* in the Rocky Mountains is tinged with buff or brownish olive on back, has buffy sides and flanks, and

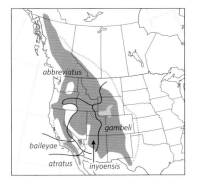

abbreviatus

gambeli

baileyae

atratus *inyoensis*

has a prominent white supercilium. California's *baileyae* is entirely grayish on back, sides, and flanks, sometimes with a slight olive tinge. It has a less distinct supercilium than *gambeli*. The other N.A. subspecies (*inyoensis*) has various features intermediate between these two. Recently, species recognition was proposed for the Rocky Mountain *gambeli* group (inc. *inyoensis*) and the Pacific *baileyae* group, based on genetic, morphological, and vocal differences. Somewhat intermediate *inyoensis* found east of the Sierra Nevada complicates the issue.

SIMILAR SPECIES When supercilium is less conspicuous, as in worn *baileyae,* it could be confused with Black-capped. Wing edgings offer a helpful distinction between the two: pale gray and inconspicuous on Mountain, prominently white on Black-capped. Mountain Chickadee's white eyebrow and smaller bib should distinguish it from Mexican Chickadee.

VOICE CALL: A hoarse *chick-dzee-dzee-dzee,* as well as a variety of buzzes and chips. **SONG:** Typically a three- or four-note descending whistle, *fee-bee-bay,* or a *fee-ee-bee-bee* on one pitch, with many local dialects. The *fee-bee-bay* version has been compared to the melody of "Three Blind Mice;" another song (*baileyae* group) sounds like "cheese-bur-ger." Some of its vocalizations are very similar to those of Black-capped, and a good look may

be necessary to ascertain the species.

STATUS & DISTRIBUTION Common. **BREEDING:** Primarily found in montane coniferous forests and mixed woodlands, but locally in coastal foothill canyons, pinyon-juniper and desert riparian woodlands. Sometimes use nest boxes. **WINTER:** Some descend to lower elevations in foothills, riparian woodlands in valleys, and suburban areas, where it sometimes visits feeders. More or less regular to coastal CA; more rarely to southern AZ (away from Santa Catalina Mts.). **VAGRANT:** Casual, mostly in winter, east to SK, ND, SD, KS, NE, OK, and TX Panhandle.

POPULATION Breeding bird surveys show a long-term decline in many parts of Mountain Chickadee's range, but the causes have not been established.

MEXICAN CHICKADEE *Poecile sclateri*   MECH ▪ 2

This Mexican species barely reaches the U.S. It calls frequently, and in the postbreeding season it forages among mixed-species flocks of as many as

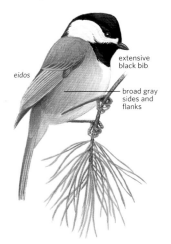

eidos

extensive black bib

broad gray sides and flanks

a hundred birds. Polytypic (4 ssp.; only *eidos* occurs north of Mexico). L 5" (13 cm)

IDENTIFICATION Black crown; black bib, largest of any chickadee species, extends well down onto breast; white cheeks; dark gray upperparts, sides, and flanks.

SIMILAR SPECIES Only chickadee within its limited U.S. range.

VOICE CALL: Husky, nasal buzzes, such as *zhree;* various *tsip* or *chit* calls. **SONG:** A rapid, slightly buzzy warble, *chee-lee, chee-lee, chee-lee;* no clear whistled song.

STATUS & DISTRIBUTION Uncommon to fairly common, but local, at high elevations. **YEAR-ROUND:** Mainly in montane coniferous forests to at least 9,000 feet at its northern range limits—the "sky islands" in the

Chiricahua Mountains in southeast AZ and the Animas and Peloncillo (scarce) Mountains in southwestern NM. Some birds move to lower elevation (±6,000 ft.) woodlands in the postbreeding season.

POPULATION No trend data are available.

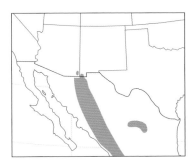

CAROLINA CHICKADEE *Poecile carolinensis*   CACH ▪ 1

The Carolina Chickadee is quite at home in cities and towns, readily using nest boxes and bird feeders. In fall and winter, the Carolina Chickadee forages in mixed flocks with nuthatches, woodpeckers, warblers, and other woodland species. The only chickadee in the Southeast, Carolina is smaller and duller than most other chickadee species. Polytypic (4 ssp.; all in N.A.). L 4.8" (12 cm)

IDENTIFICATION Cap and bib are black; cheeks are white, tinged at rear with pale gray when fresh; back and rump are gray, sometimes with an olive wash; greater coverts are gray without pale edgings; secondaries and tertials are indistinctly edged in dull white or pale gray; flanks are pale grayish, tinged buffy when fresh in fall.

GEOGRAPHIC VARIATION Four subspecies: largest and palest on upperparts in western portions of the range *(atricapilloides);* smallest and darkest in Florida *(carolinensis);* the other two subspecies from south-central U.S. *(agilis)* and east-central U.S. *(extimis)* are intermediate. Differences are weak and clinal where ranges meet. Subspecies are not field-identifiable.

SIMILAR SPECIES Absence of white wing edgings on greater coverts and grayish wash on flanks, even when slightly tinged with buff, help to separate it from Black-capped. Carolina interbreeds extensively in some areas with Black-capped along a belt from Kansas to New Jersey; hybrid offspring are not easily identifiable (see sidebar opposite). Paler sides and flanks distinguish it from Mountain Chickadee when the usually

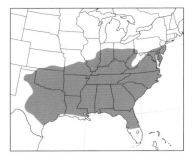

conspicuous white supercilium of the latter species is not visible.

VOICE CALL: A fast and high-pitched *chick-a-dee-dee-dee.* Compare to Black-capped's call. **SONG:** A 4-note whistle, *fee-bee fee-bay,* the last note lowest in pitch. A Carolina may learn Black-capped's two- or three-note song in areas where their ranges overlap, so do not identify by song alone at the contact zone.

STATUS & DISTRIBUTION Common. **YEAR-ROUND:** Open deciduous forests, woodland clearings and edges, suburbs, and urban parks; in the Appalachians it prefers lower elevations than Black-capped, but it has been recorded regularly as high as 6,000 feet. **VAGRANT:** Wanders casually short distances north; single records in MI and ON are extraordinary.

POPULATION Apparently declining in some regions in recent decades, particularly in the Gulf Coast states, but its range has expanded northward along much of the contact zone with Black-capped, particularly in OH and PA.

atricapilloides

slightly shorter tail than Black-capped

grayish tinge to rear cheek

fresh fall *extimis*

slightly duller flanks than Black-capped

extimis

BLACK-CAPPED CHICKADEE *Poecile atricapillus*   BCCH ▪ 1

The most widespread, numerous, and geographically variable chickadee, this bird brightens winter days at bird feeders and eagerly takes advantage of nest boxes. It is curious, with little or no fear of humans, and it is famous for willingly, after a little "training," taking seeds and nuts from the hand. During fall postbreeding movements, its noisy little parties usually contain titmice, nuthatches, woodpeckers, and other species. The calls of a flock of Black-cappeds in the fall often signal the presence of migrant warblers and vireos. Polytypic (9 ssp.; all in N.A.). L 5.3" (13 cm)

IDENTIFICATION Black cap; white cheek; black bib; gray upperparts; greater coverts, secondaries, and tertials edged conspicuously white in fall and winter (less distinct in summer when white fringes are worn off); sides and flanks buffy or pinkish when fresh, fading to pale buff by summer when worn; outer tail feathers edged prominently white.

GEOGRAPHIC VARIATION Northern and eastern birds include *turneri, fortuitus, septentrionalis, bartletti,* nominate *atricapillus,* and *practicus.* Northwestern *occidentalis* is small and dark-backed, with narrow white wing edgings and flanks heavily washed with buffy tan. Two subspecies from the interior West, *garrinus* and especially *nevadensis,* are large and pale-backed

with broad white wing edgings, and pale buffy sides and flanks.

SIMILAR SPECIES Although Mountain Chickadee is Black-capped's closest relative, Carolina Chickadee likely will

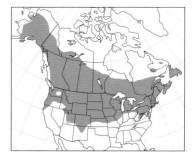

cause the most trouble, especially in late spring and summer when Black-capped's white wing edgings are worn. Where their ranges overlap, the two species resemble each other and they also hybridize, intergrading characters (see sidebar below). Otherwise, the combination of bright white cheek, especially in fresh plumage, pure

darkest subspecies

occidentalis

nevadensis

palest subspecies

black cap and bib, gray upperparts, bright wing edgings, and pinkish sides and flanks should distinguish Black-capped from other species.
VOICE CALL: *Chick-a-dee-dee-dee,* lower and slower than Carolina. **SONG:** A clear, whistled two-noted *fee-bee* or three-noted *fee-bee-ee,* the first note higher in pitch. In the Pacific Northwest some birds sing *fee-fee-fee* with no change in pitch.
STATUS & DISTRIBUTION Common. **YEAR-ROUND:** Deciduous and mixed woodlands, clearings, suburbs, and urban parks. Occurs in the Appalachians at higher elevations than Carolina. **FALL & WINTER:** Makes irregular irruptions south, usually not far into Carolina's range, but irregular to southeastern MO, eastern KY, eastern VA, MD, casually as far south as northernmost AZ, OK, and TX.
POPULATION Stable or increasing in nearly all regions, though its range may be contracting in areas where Carolina is expanding.

worn summer *atricapillus*

uniform white cheeks

prominent white-edged secondaries

fresh fall *atricapillus*

warm colored wash on flanks

Separating Black-capped and Carolina Chickadees

The differences between these two species are usually straightforward, except where the two species overlap and interbreed. Keep in mind the following:

1. Greater coverts. Brightly edged with white on the Black-capped, but inconspicuously bordered in dull whitish or gray on the Carolina.

2. Fringes of secondaries, tertials, and outer retrices. Prominently white on the Black-capped, but indistinctly dull whitish or grayish white on the Carolina.

3. Flanks. Pinkish tinged contrasting with pale gray breast and belly on the Black-capped, but dull grayish buff (and entirely grayish when worn) on the Carolina.

4. Wear of feather tips. The Black-capped's wing and tail edgings less bright in spring and summer, but still more conspicuous than those of the Carolina.

5. Black-capped has a purer white cheek; feathers on rear of cheek of Carolina are tipped with gray when fresh in fall and winter.

However, at the contact zone of the two species, a belt from Kansas to New Jersey, the hybrids

Black-capped Chickadee, fresh

Carolina Chickadee, fresh

or products of hybrid ancestry show a befuddling combination of Black-capped and Carolina characteristics. Intermediacy in plumage features is also evident in the Appalachian Mountains, where the Black-capped and Carolina are found at higher and lower elevations respectively. Vocalizations are not particularly helpful here either. A bird that looks like one species may learn to sing the other's song or, bilingually, the songs of both species. Some sing aberrant songs that are not exactly representative of either species. The *chick-a-dee* calls (typically higher-pitched and faster for the Carolina) may be more reliable at the zone, but even these do not always exactly match the normal sounds or are confusingly intermediate. Other features frequently mentioned in field guides—for example, differences in the extent of white on the cheek and relative sharpness of the bib's lower edge—are little or no help as well.

Where these two species occur together, it is probably best to leave many of them unidentified. ■

CHESTNUT-BACKED CHICKADEE *Poecile rufescens*   CBCH ▪ 1

rich chestnut back, sides, and flanks

pure white cheeks

rufescens

chestnut back

gray sides

coastal central California
barlowi

The Chestnut-backed Chickadee is curious about humans, and spends the postbreeding and winter seasons foraging noisily in mixed-species flocks. Unlike other parids (except for the Mountain Chickadee), it forages high in the canopies of tall conifers. This species is the smallest chickadee in North America. One of the three "brown-backed" species (along with Boreal and Gray-headed), its nominate subspecies is the most richly colored parid. Polytypic

(3 ssp.; all in N.A.). L 4.8" (12 cm)
IDENTIFICATION Chestnut color of back is unique, but it varies in intensity. Cap is dark brown, shading to black at its lower edge from bill through eye; cheek is white; bib is black; back and rump are rufous; greater wing coverts are edged white; breast and belly are whitish; sides and flanks are bright rufous or dull brown, depending on subspecies; tail is brownish gray.
GEOGRAPHIC VARIATION The widely distributed *rufescens* is the most colorful subspecies, with a rich chestnut back, bright white edgings on greater wing coverts, and extensively bright rufous flanks. The other two subspecies are restricted to coastal CA; they differ notably on underparts. In Marin County, *neglectus* has reduced, pale chestnut flanks that do not contrast conspicuously with whitish breast and belly; *barlowi,* found from the San Francisco Bay Area to Santa Barbara County, has grayish flanks, tinged with olive brown. The sparse and local populations at mid-elevations on the west slope of the Sierra south to northern Madera County (typically found in well-vegetated, moist canyons with tall, mature conifers) are the nominate subspecies.
SIMILAR SPECIES Boreal Chickadee also has a dark brown cap and rich brown sides and flanks, but it has predominantly grayish cheeks, lacks a rufous tint on back and rump, and shows no white wing edgings.

VOICE CALL: A hoarse, high-pitched, rapid *sik-zee-zee* or just *zee-zee;* also a characteristic sharp *chek-chek.* SONG: No whistled song is known.
STATUS & DISTRIBUTION Common. YEAR-ROUND: Coniferous forests, especially of Douglas fir, and mixed and deciduous woodlands. DISPERSAL: Postbreeding movements have been noted to higher elevations in BC and to lower elevations in OR. VAGRANT: Wanders irregularly inland as far as southwestern AB and casually south of its usual range to Ventura County, CA.
POPULATION Its range has expanded southward and eastward in recent decades from humid coastal regions to the drier eastern San Francisco Bay Area *(barlowi);* it can now be found on the CA coast to Goleta, Santa Barbara County. Numbers appear to be stable in the northern portion of the range, but recent declines in the interior expansion area have raised conservation concerns.

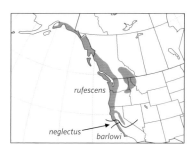

rufescens

neglectus   *barlowi*

BOREAL CHICKADEE *Poecile hudsonicus*   BOCH ▪ 1

This species is one of North America's largest chickadees and, like the Gray-headed Chickadee, it survives winters in the far north by hoarding large supplies of food. Unlike the Gray-headed, it sometimes makes long-distance irruptions southward

(less so in recent decades), apparently in autumns when food is scarce, and to the delight of birders in central and mid-Atlantic states. Boreal was sometimes called the Brown-capped Chickadee in older literature—an apt name, although the rest of the upperparts are brownish as well. Many observers comment on its tameness around humans in much of the year, but those who seek Boreal during the breeding season know that it is exceedingly quiet and inconspicuous. Polytypic (4–5 ssp.; all in N.A.). L 5.5" (14 cm)
IDENTIFICATION Largely gray cheek is unique among chickadee species. Cap is dull brown; bib is black; back and rump are brownish in

eastern portions of the range and grayer in western regions; wings are plain brownish with no pale edges on coverts or secondaries; sides and flanks are extensively washed with dull chestnut, brown with a reddish tinge, or tawny, depending upon subspecies.
GEOGRAPHIC VARIATION The four generally recognized subspecies are: *hudsonicus* (widespread from AK to eastern Canada; inc. *stoneyi* from northern AK and northwestern Canada); *columbianus* (southern AK, western Canada); *littoralis* (southeastern Canada, northeastern U.S.); and *fanleyi* (south-central Canada). At their extremes the plumage differences are notable, but the characters

juvenile
hudsonicus

grayish rear cheek

brownish cap

worn summer *hudsonicus*

fresh fall *hudsonicus*

pinkish buff flanks

plainer, more uniform greater coverts

vary clinally across the range, and subspecies are not field-identifiable. **SIMILAR SPECIES** Brown cap, plain wings, and dusky auriculars distinguish it easily from Black-capped Chickadee. Dull wings and darker face also separate it from Chestnut-backed and Gray-headed Chickadees. Relatively grayish northwestern Boreal could bring to mind Gray-headed, but wings and face differ enough to identify either species. Gray cheek is much darker and more conspicuous than pale grayish area at rear of a Carolina Chickadee's cheek. **VOICE CALL:** A wheezy, nasal, drawled *tseek-day-day;* also a characteristic single, high *see* or *dee*. **SONG:** A clear trill. No whistled song is known. It is usually less vocal than Black-capped with which it shares much of its range.

STATUS & DISTRIBUTION Fairly common. **YEAR-ROUND:** Prefers spruce and balsam fir forests, ranging north to the tree line on the taiga; to a lesser extent mixed woodlands. Its range overlaps widely with the ranges of the Gray-headed in AK and Chestnut-backed in BC, WA, ID, and MT, but no hybridization is known. **MIGRATION:** Regular short-distance movements southward have been reported within MN, SK, and MB. **WINTER:** Move slightly south of breeding range and have occurred casually as far south as IA, IL, IN, OH, WV, PA, VA, MD, DE, and NJ—but few, if any, records in recent years. **POPULATION** Trend data are unreliable because populations have not been surveyed in much of the range. Numbers increase greatly over short periods of years during spruce budworm outbreaks, then decline, making judgments about trends difficult. Numbers are considered stable in most regions but surveys suggest a gradual long-term decline in recent decades, particularly along the southern edge of the range in the northern tier of the U.S. Extensive logging in Canadian boreal forests poses a potentially serious threat.

GRAY-HEADED CHICKADEE *Poecile cinctus* GHCH ■ 3

Known in Eurasia as Siberian Tit, this species is perhaps our hardest-to-see breeding bird and remains little known. It remains on its home territory through the Arctic winter, surviving on hoarded food supplies. The bird has a small head and long tail. Polytypic (4 ssp.; 1 in N.A.). L 5.5" (14 cm) **IDENTIFICATION** Brownish gray cap; white cheek including auriculars; black bib, frayed at corners; brown back and rump with grayish tinge; greater coverts, secondaries, and tertials prominently edged white; buffy to cinnamon sides and flanks; whitish breast and belly. Ragged bib edges are usually a good field mark. **GEOGRAPHIC VARIATION** Three subspecies in northern Eurasia; *lathami*, in North America. Differences are weak and clinal, although *lathami* averages paler and grayer overall, with slightly darker cinnamon flanks. Vocal and genetic studies of *lathami* would be useful to determine relationships to Old World subspecies. **SIMILAR SPECIES** Brownish tint of its cap and back differentiate it from Black-capped. Grayish northwestern Boreal are possibly confusing, but white auriculars and bright wing edgings of Gray-headed are distinctive, as are the vocalizations. **VOICE CALL:** A loud *deer deer deer* or *chi-urr chi-urr*. Apparently no clear, whistled song. **STATUS & DISTRIBUTION** Rare. **YEAR-ROUND:** Coniferous forests in Europe and Asia; mainly willows and stunted spruces along waterways in Brooks

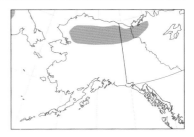

Range, east through the northern YT to Mackenzie in northwest Canada. Distribution in N.A. is not well known; for instance, there are only a half dozen records in YT since 1926, when 13 were collected on the Old Crow River. Traditional locations in AK include Kelly Bar at the confluence of the Kelly and Noatak Rivers north of Kotzebue and the Canning River area in Arctic N.W.R. **VAGRANT:** A few winter records for Fairbanks (at feeders) are the only valid vagrant records. No valid records for Alaska Range or Denali N.P. **POPULATION** The species is declining in much of its extensive Eurasian range.

pure white cheeks

white edgings to greater coverts and secondaries

longish tail

slightly paler flanks than Boreal

lathami

TITMICE Genus *Baeolophus*

All five species in this genus are New World titmice formerly classified in the genus *Parus*. Three breed almost entirely in North America, and two have substantial distributions in Mexico. Revisions to the taxonomy recently split the Plain Titmouse into Juniper and Oak species and separated the Black-crested Titmouse from the Tufted Titmouse.

BRIDLED TITMOUSE *Baeolophus wollweberi*   BRTI ■ 2

Slightly smaller than other titmice, the Bridled sometimes behaves chickadee-like and at other times forages more slowly, more deliberately, and less noisily than other parids. Bridleds commune in substantial flocks, first as family groups and then, through fall and winter, in larger numbers of two dozen or more of their own species. They also form the nucleus of mixed-species flocks, often including

distinct crest with black-and-white head pattern

phillipsi

Mexican Chickadees and sometimes Juniper Titmice, which forage together. Polytypic (4 ssp.; 2 in N.A.). L 5.3" (13 cm)

IDENTIFICATION Bridled's strikingly patterned face is an unmistakable field mark. **ADULT:** A black and white (the "bridle") face; a tall, dusty crest; a black throat, appearing as a small bib. Gray upperparts sometimes have an olive tint; underparts are paler, washed with dull olive on flanks; belly sometimes has a yellow tint in fresh plumage. **JUVENILE:** crest is shorter; face pattern is duller; bib is gray.

GEOGRAPHIC VARIATION Subspecies *phillipsi* (AZ and southwestern NM, south of the Gila River) is slightly paler than *vandevenderi* (AZ and southwestern NM, north of the Gila River).

SIMILAR SPECIES Facial pattern and black throat separate it from all other titmice. Black bib might suggest a chickadee, when seen from below, but with a better look, crest and face will easily eliminate that possibility.

VOICE CALL: A rapid, high-pitched *tsicka-dee-dee* or a harsh *tzee-tzee-tzee-tzee,* similar to Juniper. **SONG:** A rapid series of clear, whistled, identical *peet-peet* notes.

STATUS & DISTRIBUTION Common.

YEAR-ROUND: Oak, juniper, and sycamore woodlands in mountains of southeastern AZ and southwestern NM, up to 7,000 feet. The main portion of its range is in Mexico. **DISPERSAL:** Flocks make local movements to riparian areas at lower elevations in winter. **VAGRANT:** Accidental in winter to Bill Williams Delta, western AZ (Feb. 17–Mar. 20, 1977). Reports from TX not well documented.

POPULATION The trend is uncertain because of insufficient monitoring, particularly in Mexico, but Bridled's limited breeding range and the potential loss of its preferred habitats by logging and conversion to agriculture pose concern. The loss of oak woodlands has extirpated the species from some locations in Mexico.

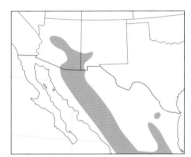

TUFTED TITMOUSE *Baeolophus bicolor*   TUTI ■ 1

The active and noisy Tufted Titmouse, North America's most widespread titmouse, is remarkably uniform morphologically, genetically, vocally, and behaviorally throughout its range. Besides gleaning trees and shrubs for arthropods, it spends more time on the ground searching leaf litter than do chickadees and most other titmouse species. Tufted Titmouse does not usually associate with mixed-species flocks; after the breeding season it spends a lot of time in small foraging parties that typically consist of parents and their offspring. Tufted frequents well-vegetated

juvenile

tall crest

dark forehead

pinkish flank patch

adult

urban and suburban areas, willingly uses nest boxes, and regularly visits bird feeders. Monotypic. L 6.3" (16 cm)

IDENTIFICATION ADULT: Gray crest; black forehead; gray upperparts; pale gray or whitish breast and belly; flanks prominently washed in rust or orange. **JUVENILE:** Gray forehead, rather than black; slightly paler crest than in adult. **HYBRID:** Tufted and Black-crested Titmice interbreed in a north-south 15- to 25-mile-wide belt extending from southern OK to coastal TX, producing offspring with variably inter-

mediate dusky crests and grayish foreheads that make identification near impossible.

SIMILAR SPECIES The only other titmice likely found in its range is Black-crested Titmouse, which has a pale forehead. Juniper Titmouse is much plainer, with a shorter crest. It also lacks constrastingly dark forehead and does not have brightly washed pinkish flanks.

VOICE CALL: A harsh, scolding *zhee zhee zhee.* **SONG:** A loud, whistled *peto peto peto* or *wheedle wheedle wheedle,* often repeated monotonously.

STATUS & DISTRIBUTION Common.

YEAR-ROUND: Deciduous woodlands, suburbs, urban parks, wherever else trees are large enough to provide nest holes. **DISPERSAL:** Wanders casually north of its usual limits in fall and winter; records from SD, MN, QC, and NB.

POPULATION This species is increasing in nearly every part of its range. Expansion northward has been under way for at least a century, and the movement continues through northern New England and southern Canada. Its extreme sedentary nature probably accounts for the slow progression.

BLACK-CRESTED TITMOUSE *Baeolophus atricristatus*   BCTI ■ 2

Except for its black crest, which no other titmouse has, and forehead, the Black-crested Titmouse could be a mirror image of the larger Tufted Titmouse in behavior and appearance. Throughout the summer, fall, and winter, this lively, noisy, and continually active bird forages in small family groups, searching trees and shrubs for insects, spiders, and their eggs. Confiding and curious, it shows little fear of humans and does not hesitate to use nest boxes. Polytypic (3 ssp.; all in N.A.). L 5.8" (15 cm)

IDENTIFICATION ADULT: Black crown and crest; dull white forehead; medium to dark gray upperparts, sometimes with a slight olive tinge; pale gray underparts with a cinnamon wash on the flanks. **JUVENILE:** Gray crest, sometimes washed with brown; dusky forehead. **HYBRID:** Black-crested and Tufted Titmice interbreed freely in a

north-south 15- to 25-mile-wide belt extending from southern OK, southwards, just east of Balcones Escarpment to the central TX coast. Here hybrids outnumber pure birds. Offspring show variably intermediate shades of dusky crests and gray or brownish foreheads, making it near impossible to assign them to a species. Most hybrids have a chestnut patch on forehead.

GEOGRAPHIC VARIATION Three subspecies in N.A.: *castaneifrons* (southwestern OK to south TX); *paloduro* (north and Trans-Pecos TX); and nominate (south TX and northeastern MX). They differ slightly in size and plumage coloration; they are not field-identifiable.

SIMILAR SPECIES Juveniles with gray crests may be difficult to distinguish from Tufted or Juniper Titmouse. Look for juvenile Black-crested's whitish forehead and crown darker than upperparts.

VOICE CALL: A harsh, scolding *jree jree jree,* sharper and more nasal than Tufted. **SONG:** A series of slurred *chew chew chew* or *pe-chee-chee-chee* phrases, faster than the Tufted's similar vocalizations.

STATUS & DISTRIBUTION Common. **YEAR-ROUND:** Black-crested inhabits oak and other deciduous woodlands

juvenile

black crest

white fore-head

adult

along watercourses, open tracts with scattered trees, arid scrubby habitats, and urban and suburban areas. **VAGRANT:** Casual to Guadalupe Mountains, TX and west TX Panhandle. It has not yet been recorded in NM, even though the northern part of hte Guadalupe Mts. are in NM.

POPULATION No trend data exist, but fragmentation of oak-woodland habitats has raised concern. The Black-crested's return to full species status in 2002 has brought renewed attention to monitoring its numbers and its potential conservation needs.

OAK TITMOUSE *Baeolophus inornatus*   OATI ■ 1

Once classified with the Juniper Titmouse as a subspecies of the aptly named "Plain Titmouse," the Oak Titmouse is easily separable from all

other timice species except the Juniper. Its lack of field marks is, ironically, an important field mark. Polytypic (4 ssp.; 3 in N.A.). L 5" (13 cm)

IDENTIFICATION The bird is a drab brownish or grayish brown color overall, but is paler below. It has a short crest.

GEOGRAPHIC VARIATION Four subspecies, one restricted to Baja California; variation is weak and clinal. The northernmost subspecies, *inornatus,* from Oregon to south-central California, has medium brownish gray or olive-brown upperparts, pale gray underparts with pale brown tinge on flanks, and a short bill. Subspecies *affabilis* found in southwestern California is notably larger, has darker gray-brown or olive-brown upperparts, flanks washed in dusky brown, and a longer bill; *mohavensis,* restricted to the Little San Bernardino Mountains in San Bernardino and Riverside Counties, is smaller, paler, and grayer than *affabilis.*

SIMILAR SPECIES The closely related Juniper Titmouse is slightly larger, paler, and grayer, but similarly plain. The two are best separated by range and vocalizations; they are sedentary and are sympatric only in a small area of northern California (Modoc Plateau).

VOICE CALL: A hoarse *tsick-a-deer.* **SONG:** A series of clear, whistled sets of alternating high and low notes, such as *peter peter peter* or

worn
affabilis

short grayish crest

slight brownish tint above

fresh
affabilis

teedle-ee teedle-ee, with many variations.

STATUS & DISTRIBUTION Common. **YEAR-ROUND:** Primarily oak and pine-oak woodlands on the Pacific slope. **VAGRANT:** Casually reported east of its usual limits in CA, including once near the south end of the Salton Sea, Imperial Co.

POPULATION Declining in California because of losses in its preferred oak habitat, primarily by encroaching suburban and agricultural development, although disease could become a new threat with the appearance of Sudden Oak Death Syndrome.

JUNIPER TITMOUSE *Baeolophus ridgwayi* JUTI ■ 1

The Juniper Titmouse travels through woodlands in small family parties, does not habitually forage in mixed-species flocks, and will visit bird feeders. Formerly classified as a subspecies of "Plain Titmouse," the Juniper has a much larger range, though occupies it more sparsely than its close relative, the Oak Titmouse; its total population has been estimated at less than half that of the latter species. Polytypic (2 ssp.; both in N.A.). L 5.3" (13 cm)

IDENTIFICATION Sexes and ages similar. Drab medium-gray upperparts; grayish white underparts; short crest.

GEOGRAPHIC VARIATION Despite its wide distribution, Juniper shows little geographic variation. Two

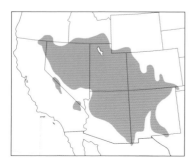

different subspecies are recognized: nominate and the more westerly zaleptus—they are not identifiable in the field.

SIMILAR SPECIES Juniper's utter lack of field marks sets it apart from all other titmice except for Oak, which is slightly smaller, darker, and brownish tinted. To the eye, differences appear slight, but genetic, morphological, ecological, and vocal characters set the species apart. They are best distinguished from each other by range and vocalizations; fortunately, they are mostly sedentary, coming into contact only at the periphery of their ranges in a small area on the Modoc Plateau in northern California. Research at the contact zone indicates that there is little or no interbreeding and that physiological adaptations to temperature and precipitation differ significantly between the species.

VOICE CALL: A hoarse, rapid *tsick-dee,* faster than Oak, more like Bridled Titmouse. **SONG:** A variety of long, rolling trills on a uniform pitch and 3-part groups of *wheed-leah, wheed-leah, wheed-leah.* Male has a large repertoire of song types.

STATUS & DISTRIBUTION Uncommon.

almost identical to Oak Titmouse, but slightly grayer and narrower billed; best told by calls and song in narrow range of overlap

YEAR-ROUND: Primarily juniper and pinyon-juniper woodlands in the intermountain region and southward. It can be found at elevations up to 7,000–8,000 feet in AZ, NM, and CA.

POPULATION Its relatively small numbers, low density, limited habitat preferences, and recent declines reported in the southern Rockies and the Colorado Plateau region have raised conservation concerns.

PENDULINE TITS AND VERDINS Family Remizidae

Verdin (NM, Nov.)

These are very small birds that generally resemble chickadees. All species have small, sharply pointed bills. Some species, like the Verdin, are rather solitary, while others are quite gregarious and nest in loose colonies. Nests of most species are a tightly woven bag or ball with a tubular side entrance.

Structure Most species are small-bodied, with a short tail. Although small, the Verdin of North America and the Penduline Tit of Eurasia, are notably larger and longer tailed than other family members. The bill is short and sharply pointed. The legs are long and slender.

Behavior They are acrobatic, easily moving through dense foliage and even hanging upside down while foraging. They often hold larger prey items under a foot while tearing it apart with the bill.

Plumage Most of the species in this family have nondescript plumage. They primarily have gray, olive, or yellow upperparts with pale, or white, underparts. One member of the family, the Penduline Tit, has rather ornate plumage.

Distribution The family reaches its greatest diversity in Africa. Singles species are found in North America and Europe; two are present in Asia.

Taxonomy Worldwide, there are ten species in five genera. The Remizidae family is sometimes classified as a subfamily of the Paridae. A diverse assemblage of species and further research may result in one or more members moved to other families. The Penduline Tit is sometimes treated as a complex of up to four species.

Conservation No member of the family is considered of conservation concern, although two are considered rare, and many aspects of their biology are poorly known.

Genus *Auriparus*

VERDIN *Auriparus flaviceps*   VERD ▪ 1

An acrobatic species, the Verdin is generally solitary away from nest sites. It is most easily detected by its surprisingly loud calls. Nest is an intricately woven ball of spiderwebs and small twigs. The male often builds several structures during the nesting season; both sexes roost in these year-round. Polytypic (6 ssp.; 2 in N.A.). L 4.5" (11 cm)

IDENTIFICATION Very small body size, medium-length tail. Sexes similar. **ADULT:** Dull gray overall, darker on upperparts, with a yellow face and chestnut shoulder. Small bill sharply pointed bill with a straight culmen. **JUVENILE:** Similar to adult, but paler gray and lacking yellow face and chestnut shoulder making them appear very plain. Juveniles also have pale base to bill.

GEOGRAPHIC VARIATION Variation in two N.A. subspecies is weak and clinal. Western birds *(acaciarum)* tend to have more brownish upperparts than those from southern New Mexico to central TX *(ornatus)*.

SIMILAR SPECIES Yellow face and chestnut shoulder of adults distinguish them from other chickadee-like birds. Juvenile similar to Lucy's Warbler (particularly female) but can be distinguished by its slightly heavier bill with a pale base. Bushtit has a longer tail and smaller bill with a curved culmen. Gnatcatchers have longer, black tails and prominent eye rings.

VOICE CALL: A clear *tschep* and rapid *chip* notes. **SONG:** A plaintive three-note whistle, *tee tyew too,* with the second note higher.

STATUS & DISTRIBUTION Common; uncommon in UT and rare in OK and TX Panhandle. **YEAR-ROUND:** Resident in desert scrub and other brushy habitats, including mesquite woodlands. Formerly resident in coastal southern San Diego Co., CA. **VAGRANT:** Otherwise casual in southwestern CA north to coastal southern Santa Barbara Co.

POPULATION Stable, although very little data concerning this species.

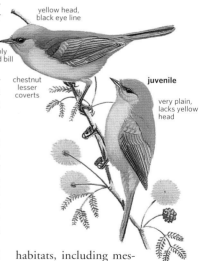

yellow head, black eye line

sharply pointed bill

chestnut lesser coverts

juvenile

very plain, lacks yellow head

LONG-TAILED TITS AND BUSHTITS Family Aegithalidae

This family is characterized by very small birds with short wings and long tails. The small body size combined with rather loose contour feathers gives these birds the look of tiny fluff balls.

Structure Small bodied; long tailed. All species have small black, conical bills; legs generally long and slender.

Behavior Sociable birds that live in small family groups during breeding season and congregate in larger flocks in nonbreeding season. Typically thought to be monogamous, but cooperative breeders and nest "helpers" recorded.

Plumage In general, species in this family are gray or brown above and white below. Most have a black or brown mask, often combined with a black bib. Others

Bushtit, male (WA, Apr.)

are fairly uniform in color, such as the Bushtit. Sexes are similar in plumage, although juveniles are often distinctive.

Distribution This family's greatest diversity is in the Himalaya and western China; single species in Europe and N.A.

Taxonomy Sometimes included as a subfamily of Paridae; however, they differ in several important aspects, and DNA analysis confirms the distinctiveness of the family. The Bushtit is the only New World representative and is put in its own genus. Worldwide there are 11 species in four genera.

Conservation No member of the family is considered of conservation concern, but some Asian species are not well known.

Genus *Psaltriparus*

BUSHTIT *Psaltriparus minimus*   BUSH ◼ 1

black ear coverts

**"Black-eared Bushtit"
juvenile ♂**
plumbeus

gray crown

interior ♂
plumbeus

pinkish buff face

female with pale eye

interior ♀
plumbeus

long tail

grayish face

coastal ♂
minimus

brown crown

This acrobatic species is most often encountered as a small- to medium-size flock moves through woodland. Its presence is usually announced by the almost constant calling between members of the groups as they move from tree to tree. Nest is an intricately woven hanging structure. Polytypic (10 ssp.; 5 in N.A.). L 4.5" (11 cm)

IDENTIFICATION Very small body size and long tail distinguish this species from other chickadee-like birds. **ADULT:** Pale gray overall, darker on upperparts, with very small bill. Interior birds have brown ear patch with a gray crown; coastal birds have a brown cap. Adult females have pale eyes. **JUVENILE:** Similar to adult except in Chisos Mts., TX, where juvenile males can

show a black ear patch, as can some adult males. Juvenile females develop pale eyes a few weeks after fledging.

GEOGRAPHIC VARIATION Three well defined subspecies groups: "Coastal" *minimus* group with five ssp.; three in N.A.: *saturatus* (Pacific Northwest), *minimus* (Pacific coast), *californicus* (inland from coast). "Interior" *plumbeus* group with one subspecies *plumbeus* (east of the Sierra Nevada). "Black-eared" *melanotis* group with four subspecies; *dimorphicus* in the Chisos Mts., TX. The "Coastal" and "Interior" groups meet near Lone Pine, Inyo Co., CA.

SIMILAR SPECIES Immature Verdin more uniform in overall body plumage and has a shorter tail. Gnatcatchers have longer, black tails, prominent eye ring.

VOICE Sharp twittering *tsip* or *tseet*.

STATUS & DISTRIBUTION Common. **YEAR-ROUND:** Resident in a variety of woodland, scrub, and residential habitats; south to Guatemala. **DISPERSAL:** Some movement to lower elevation in northern part of range. **VAGRANT:** Very rare ("Interior" group) to western Great Plains and "Coastal" group to Salton Sea region, CA.

POPULATION Populations appear to be stable in most of the United States.

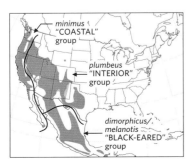

minimus
"COASTAL"
group

plumbeus
"INTERIOR"
group

*dimorphicus/
melanotis*
"BLACK-EARED"
group

NUTHATCHES Family Sittidae

Pygmy Nuthatch (CO, Feb.)

Nuthatches are familiar acrobatic birds, with one or more species occurring throughout much of the U.S. and Canada. All but the Brown-headed Nuthatch are commonly seen at bird feeding stations. When not nesting, nuthatches frequently gather in mixed-species flocks that include chickadees, creepers, kinglets, and warblers. Experienced observers listen for nuthatch calls to try to detect other species. Identification is straightforward, except when a Brown-headed or Pygmy Nuthatch wanders far from its normal range.

Structure Nuthatches are small, chunky, and compact birds, with a short tail. They are well adapted to creeping about at all angles on tree trunks and branches with legs that are relatively short and with strong toes and claws. Their bills are rather long, straight, and pointed; their wings are short.

Behavior Nuthatches are usually seen on the surfaces of tree trunks and branches, climbing in any direction in pursuit of invertebrates and seeds. Unlike woodpeckers and creepers, they do not use their tail for support and may go down trees head first. The smaller species often forage among the outermost needles on branches. Occasionally they drop to the ground to feed. To open a seed, a nuthatch wedges it into a crevice and pounds on the seed with its bill. This behavior, known as *hacking,* likely gave rise to the family name. During fall and winter, nuthatches store nuts, seeds, and invertebrates in caches. The Brown-headed Nuthatch is one of few bird species known to use a tool: Individuals have been seen holding bark in their bill and using it to flake off another piece of bark. All nuthatches nest in cavities, mostly tree holes and cavities. Some Brown-headed and Pygmy Nuthatches are cooperative breeders. Most are vocal year-round, with their loud whistles, trills, and calls often belying their presence even before they are revealed by their active foraging style.

Plumage Nuthatches are blue-gray above, typically with black or brown markings on the head. The underparts are pale, sometimes washed with buff or rufous coloration. The sexes are similar or identical in all nuthatches; with good views, White-breasted and Red-breasted Nuthatches can be sexed in the field. Plumages are similar throughout the year with a prealternate (early spring) molt that is limited or absent. All species have a partial first pre-basic (fall) molt. Under ideal field conditions, aging is possible by looking for worn, brownish colored wing coverts that contrast with the blue-gray mantle on first-year birds. By late spring the worn appearance of adults makes aging very challenging, even in the hand.

Distribution Nuthatches are widely distributed throughout the Northern Hemisphere. They reach their greatest diversity in south Asia, where 15 species occur. Most nuthatches are nonmigratory, but in North America even the generally sedentary species are occasionally seen far from their normal range.

Taxonomy This is a relatively small family with roughly 24 species worldwide, four of which occur in North America. All but one species (the Wallcreeper, in its own subfamily Ticshodromadinae) belong to the genus *Sitta.* Species limits among the nuthatches continue to be debated. A recent study suggests that the subspecies of the Brown-headed Nuthatch found on Grand Bahama Island may be a separate species: the Bahama Nuthatch *(Sitta insularis).* Work continues to determine whether the White-breasted may involve multiple species.

Conservation Habitat loss and degradation largely caused by shifts in cultivation, grazing, and timber harvesting pose the biggest threats to this family. Worldwide, the White-browed Nuthatch (western Myanmar) and the Algerian Nuthatch (northern Algeria) are considered endangered by BirdLife International; the Giant and Beautiful Nuthatches (Southeast Asia) are vulnerable.

RED-BREASTED NUTHATCH *Sitta canadensis* RBNU ■ 1

The distinct nasal calls usually announce the presence of the Red-breasted Nuthatch. During the breeding season, the Red-breasted is usually found in forests dominated by firs and spruces; during migration and winter, it is found in a variety of habitats. When not breeding, the Red-breasted can be seen in small flocks with other nuthatches, chickadees, kinglets, and Brown Creepers. Monotypic. L 4.5" (11 cm)

IDENTIFICATION Small, with a prominent supercilium that contrasts with darker crown and eye line. **ADULT MALE:** Bold face pattern: Inky black crown and nape; prominent white supercilium extending from sides of forehead to sides of nape and separating

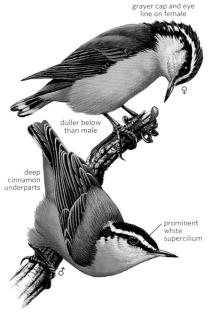

grayer cap and eye line on female

♀

duller below than male

deep cinnamon underparts

prominent white supercilium

♂

crown from very broad, black eye line. Upperparts otherwise a deep blue-gray. Wings are edged blue-gray. Whitish chin and throat blend into buff breast and rich buff belly, flanks, and undertail coverts. **ADULT FEMALE:** Similar to male, but black on head is paler, more lead-colored, often contrasting with blacker nape. Wings are edged dull gray with a brownish or olive tinge. Upperparts are duller gray; underparts are less richly colored. **IMMATURE:** Usually duller than adults of the same sex. Wings coverts, primaries, and secondaries are uniformly brownish gray without edging visible in adults. Many individuals not safely aged under normal field conditions, particularly by late spring when even bright adults have become quite worn. **FLIGHT:** Undulating and resembling a short-tailed woodpecker; white diagonal subterminal band on tail sometimes visible in flight.

SIMILAR SPECIES This is the only North American nuthatch with a broad white supercilium and contrasting broad dark eye line. Chinese Nuthatch *(S. villosa)* is found as close as southern Ussuriland (Rus.) but is unlikely to occur as a vagrant to North America. A vagrant Chinese Nuthatch would be differentiated by its ill-defined eye stripe, duller underparts, and lack of a diagonal white subterminal band on tail.

VOICE CALL: A nasal *yank* that is variable but typically repeated. Classically described as similar to a toy tin horn. Short versions are sometimes given in flight. Calls are higher and more nasal than White-breasted Nuthatch's. **SONG:** A rapid, repeated series of *ehn ehn ehn* notes; reminiscent of calls.

STATUS & DISTRIBUTION Common to abundant. **BREEDING:** Northern and subalpine conifers, particularly spruces and firs. Occasionally breeds south of mapped breeding range, usually in conifer plantations or residential neighborhoods with conifers. **MIGRATION:** Irruptive; often moving in two- to three-year cycles but variable. Northernmost migrate annually; southernmost are generally resident. First detected away from breeding grounds as early as July (typically earliest in big flight years); peaks in Great Lakes Sept.–mid-Oct. Spring migration less pronounced, but migrants are seen through May in much of the lower 48 states. **WINTER:** Highest densities typically occur along the U.S.-Canada border (e.g., Northeast; MI-WI border; south-central BC and northwestern WA). **VAGRANT:** Casual in mainland north Mexico, north Baja California, and west AK; Bermuda (4 recs.); Iceland (1 rec.); England (1 rec.).

POPULATION Bird Breeding Survey shows significant increase throughout the breeding range. In the East, resident range is expanding southward.

PYGMY NUTHATCH *Sitta pygmaea* PYNU ■ 1

The endearing Pygmy Nuthatch is easily found, thanks to its unremitting vocalizations. It is one of only a few cooperative-breeding passerines: A third of breeding pairs are assisted by one to three male helpers, usually relatives. Pairs roost together, and juveniles roost with parents. In winter, congregations of a dozen or more birds may roost together: There are reports of more than 150 individuals roosting in a single tree. Pygmies usually forage well out on branches. They regularly descend to the ground

to feed. Polytypic (6 ssp.; 3 in N.A.). L 4.3" (11 cm)

IDENTIFICATION Tiny; found in long-needled pine forests in the West. **ADULT:** Dusky-olive crown; small pale nape spot. Lores, eye line contrast darker. Rest of upperparts blue-gray. Underparts pale, variably washed with buff on flanks. Sexes identical. **IMMATURE:** Average duller. Brownish gray wing coverts, secondaries, and primaries, contrast with blue-gray mantle. Aging difficult in the field, particularly by spring. **FLIGHT:** Undulating and

resembling that of a miniature short-tailed woodpecker; white diagonal subterminal band on tail is at times visible in flight.

GEOGRAPHIC VARIATION Three subspecies are found north of the U.S.-Mexico border (three confined to Mex.). Widespread *melanotis* has blackish lores and eye line that contrast noticeably with olive crown. Nominate *pygmaea* (coastal central CA) is smaller, with dusky lores and eye line that contrast only slightly with crown. Flanks extensively washed buff.

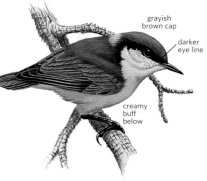

grayish brown cap

darker eye line

creamy buff below

Subspecies *leuconucha* (montane San Diego County) is larger, with uniformly gray-olive crown, lores, and postocular; relatively large pale spot on nape; flanks only lightly washed buffy gray.

SIMILAR SPECIES Best separated from White-breasted and Red-breasted by face pattern; from Brown-headed by range. Also note Pygmy's more grayish olive crown, smaller pale nape spot,

and buffier undertail coverts. Pygmy also has shorter primary projection (difficult to see on these active birds). Note vocalizations.

VOICE CALL: Rapid, clear, high-pitched notes usually given in a series of three or more notes: *bip-bip-bip* or *kit-kit-kit*. Calls vary in pitch and intensity, but are given almost incessantly. Calls of nominate *pygmaea* are very rapid. **FLIGHT CALL:** Similar but softer: *imp imp*. **SONG:** A rapid, high-pitched sequence of two-note phrases: *ki-dee, ki-dee, ki-dee,* similar to alarm calls.

STATUS & DISTRIBUTION Common. **BREEDING:** Long-needled pine forests of the West. Range essentially mirrors those of Ponderosa pine, Jeffrey pine, and similar species. **DISPERSAL:** Casually to very rarely disperses into lowlands during nonbreeding season (mostly late July–mid.-Dec.). In recent

years, nearly annual in southeastern CO. **VAGRANT:** Casual in southwestern BC, western KS, western OK, central MT. Accidental in eastern ND and MN (same bird), IA, eastern KS, and eastern TX.

POPULATION It is a species of special concern in ID, WY, and CO, where it is an indicator species of healthy Ponderosa pine forests. Threats include logging, fire suppression, and grazing.

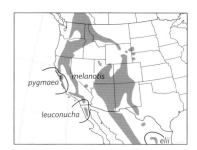

pygmaea
melanotis
leuconucha
elii

BROWN-HEADED NUTHATCH *Sitta pusilla*   BHNU ▪ 1

This energetic, vocal species is more common and easier to find than the other two southeastern pine specialties (Red-cockaded Woodpecker and Bachman's Sparrow). Brown-headed Nuthatches spend most of their time foraging well out on branches and on treetops, frequently hanging upside down searching for insects and spiders, or feeding on cones. This species is perhaps better known for using tools than any other bird in N.A: An individual will take a bark scale, hold it in its bill, and use it as a wedge to pry off other bark scales as it searches for insects and spiders. Apparently, this behavior is more common in years of poor cone crops. Polytypic (3 ssp.; 2 in N.A.). L 4.3" (11 cm)

IDENTIFICATION Tiny; found in pine forests in the Southeast. **ADULT:** Warm brown to chestnut crown and upper portion of auriculars; relatively large whitish nape spot. Underparts pale, washed with dull buff on flanks. Sexes identical. **IMMATURE:** On average

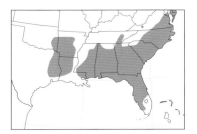

slightly duller. **FLIGHT:** Flight style similar to Pygmy's.

GEOGRAPHIC VARIATION Weak and clinal within the U.S. The subspecies *caniceps* (found in southern and central FL) may be smaller than nominate birds, with a larger bill and paler gray head; many doubt the validity of this subspecies. Recent work suggests that highly endangered birds *(insularis)* found on Grand Bahama Island constitute a separate species.

SIMILAR SPECIES Easily separated from White-breasted and Red-breasted Nuthatches by face pattern and size; best separated from Pygmy by range. Also note Brown-headed's more brownish crown, larger pale nape spot, and less buff undertail coverts. Brown-headed has shorter primary projection, and its subterminal tail band is grayer and more limited, but this is difficult to see on these active birds. Note vocalizations, all of which tend to be lower pitched and less pure than Pygmy's.

VOICE CALL: A repeated two-part note like the squeak of a rubber duck, *KEW-deh*. At times, up to ten or more notes may follow the first note. Feeding flocks also give a variety of twittering, chirping and talky *bit bit bit* calls. Flight calls similar. Differences between songs and calls poorly understood.

STATUS & DISTRIBUTION Fairly

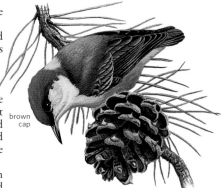

brown cap

common. **BREEDING:** Pine forests of the Southeast, particularly the loblolly-shortleaf in the upper Coastal Plain and the longleaf-slash association in the lower Coastal Plain. Favors mature open woodlands with dead wood (e.g., burn areas) but regularly found in young and medium-age pine stands. **DISPERSAL:** Very limited postbreeding dispersal. **VAGRANT:** Casual to KY (recently nested in western part of state). Accidental southeastern WI, northern IN, northern IL, eastern KY, eastern OH, PA, and southern NJ.

POPULATION Declining throughout range; threatened by clear cutting and fire suppression, which degrade or destroy habitat. The species likely benefits from conservation efforts focused on Red-cockaded Woodpecker, which occupies same range and habitat but is much more sensitive.

WHITE-BREASTED NUTHATCH *Sitta carolinensis*   WBNU ▪ 1

Less gregarious than other nuthatches. Commonly seen at bird feeders. Polytypic (11 ssp.; 6 in N.A.). L 5.8" (15 cm)
IDENTIFICATION The species has a black crown and nape that contrast with a white face and breast. **ADULT MALE:** It has a uniformly black crown and nape. Upperparts are blue-gray. **FEMALE:** Paler crown often contrasts with a blacker nape; in the Southeast, head pattern is more similar to the male's.
GEOGRAPHIC VARIATION See sidebar.
SIMILAR SPECIES The large size and

white face with an isolated dark eye is distinctive.
VOICE CALL: Varies between subspecies; see sidebar below. **SONG:** "Eastern" group: series of repeated whistles on one pitch. "Pacific" group: similar but higher pitched. "Interior West" group: Distinctly different, more staggered in delivery.
STATUS & DISTRIBUTION Fairly common. **BREEDING:** A variety of deciduous and mixed-forest habitats, generally preferring relatively open woods. **DIS-PERSAL & MIGRATION:** Poorly known. Does not undertake large-scale irruptions (unlike Red-breasted), but "Interior West" birds regularly move onto the Great Plains, and some "Eastern" birds disperse west. Found in southeastern CO (where it does not breed) early Aug.–early May. **VAGRANT:** Casual to deserts of southeastern CA, mostly fall (mostly *aculeata*). Accidental ("Interior West") to Ventura Co., CA in winter. Also casual (ssp. undetermined) to Vancouver I., BC; and offshore CA (Santa Cruz Is. and the Farallones);

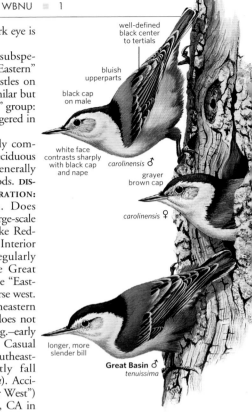

well-defined black center to tertials

bluish upperparts

black cap on male

white face contrasts sharply with black cap and nape

carolinensis ♂

grayer brown cap

carolinensis ♀

longer, more slender bill

Great Basin ♂
tenuissima

Sable I., NS; and one remarkable fall record for Bermuda.
POPULATION Generally stable, possible declines in the Southeast.

Variation in the White-breasted Nuthatch

There are nine subspecies of the White-breasted Nuthatch in Mexico and North America, five of which occur north of the U.S.-Mexico border. All of the subspecies can be placed into three groups on the basis of plumage characteristics, genetics, and calls. Songs appear to differ between the subspecies groups as well, but more study is needed. The three subspecies groups may constitute three separate species, but for now that question remains under review.

The "Eastern" (or *carolinensis*) group consists of nominate *carolinensis* and *atkinsi* (recognized by some) from FL. This is the palest subspecies group. The centers of tertials and wing coverts are sharply defined, deep black, and contrast distinctly with blue-gray upperparts. Crown stripe is broader, bill is thicker and shorter, and underparts are whiter than in either western group. Their nasal calls—*yank, yank*—are the classic, slow, low-pitched calls associated with most published descriptions of this species.

The "Interior West" (or *lagunae*) group consists of Rockies *nelsoni* and more westerly *tenuissima*, and five additional subspecies in Mexico. This group includes the

birds from the Cape District of southern Baja California *(lagunae)* on the basis of their calls and genetics. It is the darkest subspecies group with much darker upperparts and more extensive dark coloration to flanks, particularly when compared to "Eastern" group birds. Within this group there is a clinal variation with the darkest birds occurring in Mexico, where only throat and sides of head are bright white. Their very short, rapid succession of high-pitched nasal calls—*nyeh-nyeh-nyeh-nyeh*—have a laughing quality about them, and are very different from the calls of "Eastern" or "Pacific" group birds.

The "Pacific" (or *aculeata*) group consists of *aculeata* and one other subspecies from northern Baja California. This group is somewhat intermediate in plumage between the "Interior West" and "Eastern" groups but more closely resembles "Interior West" birds. They are slightly paler overall, particularly on mantle and flanks, and paler centers of tertials and wing coverts show little contrast with the rest of upperparts. Their calls—*eeerh, eeerh*—are higher-pitched, longer, and harsher than the "Eastern" group. ∎

CREEPERS Family Certhiidae

Brown Creeper (OH, Feb.)

The Brown Creeper is the only certhiid found in the Americas.

Structure Creepers have short legs, long stiff tail feathers, long curved claws and toes, and relatively long and dense feathers.

Behavior Creepers are almost always found on trees, using their stiff tail as a prop. When foraging, they typically begin near the base of the tree and jerkily spiral upward, sometimes onto larger branches, until near the top, then fly to the base of the next trunk. In constant motion, they glean and probe for arthropods and insects. Creepers are vocal, but their high-pitched calls and songs may be difficult to hear. The nests of all but the Spotted Creeper are placed behind loose bark.

Plumage The upperparts are finely patterned with a mixture of white, rufous, brown, black, and buff markings. The underparts are pale. Unlike other passerines, tree creepers retain the central rectrices until all others have been replaced. The first prebasic molt is incomplete, but all tail feathers are replaced. There is no prealternate molt.

Distribution Found in forested regions of the Northern Hemisphere, creepers reach their greatest diversity in the Himalaya.

Taxonomy There are eight species in two genera. Some DNA comparisons suggest that creepers are part of a much larger family that includes nuthatches, chickadees, wrens, gnatcatchers, and gnatwrens. The unusual Spotted Creeper *(Salpornis spilonotus)*, placed in its own subfamily, has a discontinuous range in Africa and India.

Conservation Mature and old-growth forests, the species' preferred habitat, face increasing threats from logging and other forms of habitat destruction, deterioration, and fragmentation.

Genus *Certhia*

BROWN CREEPER *Certhia americana*   BRCR ▪ 1

A generally solitary species, it sometimes joins mixed-species flocks in migration and winter. Polytypic (15 ssp.; 10 in N.A.). L 5.3" (13 cm)

IDENTIFICATION Sexes alike. Distinctive shape and foraging behavior. Cryptic upperparts streaked brown, buff, and black; underparts pale with warm wash on flanks; rump buff, tawny, or rufous. **FLIGHT:** Broad, pale wing stripe at base of flight feathers prominent in flight.

GEOGRAPHIC VARIATION Subspecies north of the U.S.-Mexico border divide into three groups: Western birds

(alascensis, occidentalis, stewarti, phillipsi, zelotes, montana, leucosticta) small, dark, and long billed. Eastern birds *(americana, nigrescens)* larger, generally paler, and shorter billed. Mexican *albescens* more closely related to Middle American subspecies and darker with white spotting that contrasts more; its range extends into southeast AZ and southwest NM. Three color morphs (reddish, brown, and gray) in several populations complicate identification. **VOICE** Very high-pitched. **CALL:** A soft, sibilant *seee,* usually buzzier and doubled in western birds, *tseeesee.* **SONG:** Variable, but consists of several notes, *seee seeedsee sideeu.* Eastern songs more complex, quavering, and usually end on a high note; western songs more rhythmic and often end on a low note; *albescens* also differs.

STATUS & DISTRIBUTION **BREEDING:** Coniferous, mixed, or swampy forests. Wider habitat variety during migration and winter. **MIGRATION:** Peak late Mar.–mid-Apr. and Oct.–Nov. in Midwest/Northeast.

POPULATION Populations thought to have declined throughout much of N.A. with the felling of old-growth forests.

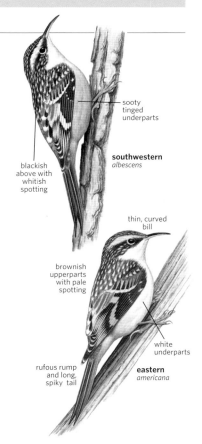

sooty tinged underparts

southwestern *albescens*

blackish above with whitish spotting

thin, curved bill

brownish upperparts with pale spotting

white underparts

rufous rump and long, spiky tail

eastern *americana*

WRENS Family Troglodytidae

Rock Wren (CA, Apr.)

North America's wrens are diverse (11 species in N.A.; 76 worldwide) yet distinctive; members of Troglodytidae are rarely confused with species in other families. Our wrens are brown, and most have slender, decurved bills. Tail length and structure are variable; in many postures, however, wrens exhibit a characteristic cocked tail. With good views, sight identification is straightforward. Wrens are often found in dense cover, though, and can be fidgety. Song and microhabitat preferences are important for identification.

Structure In many situations, wrens strike a rotund, potbellied pose. When foraging or evading detection, they appear to flatten or prostrate themselves. The tail varies from very short to fairly long, but it is distinctive—typically cocked up and/or flipped about. Wrens have medium to very long, decurved, usually slender bills. Species in North America range in size from small to medium.

Behavior Wrens are often observed creeping about substrates close to or on the ground. Specialized microhabitat preferences are good cues to identification (e.g., Cactus Wrens within thorny shrubs, Rock Wrens about talus slopes). Songs are complex, variable, and distinctive, with several species (e.g., Winter and Canyon Wrens) among our finest vocalists. Birds of some species sing through the night (e.g., Marsh Wren) and/or during the winter (e.g., Carolina Wren). Several species (e.g., Marsh Wren) build conspicuous "dummy nests" that are abandoned after construction.

Plumage All wrens of North America are largely brownish, darker above and paler below. Some species tend toward grayish or reddish tones, but these differences are usually of limited relevance to species-level identification.

Important marks to note include streaking and spotting on upperparts, barring and shading on underparts and tail, and the supercilium (strong, weak, or nearly absent). Below the species level, coloration is important in assigning individuals to different subspecies.

Distribution The wrens' center of diversity is Middle America, with more than 30 species in Mexico. All of the lower 48 host multiple wren species, but several do not range north to Canada, and only the Pacific Wren is regularly found in Alaska. The "Eurasian Wren" is the only troglodytid that occurs in the Old World (recent research, however, indicates that the "Eurasian Wren" comprises multiple species). Migratory strategies range from sedentary (e.g., Cactus Wren), to intermediate-distance migration (e.g., Sedge Wren). Long-term range shifts appear to be under way for several species.

Taxonomy The Troglodytidae are regarded by most authorities as a valid, monophylous taxon. Relationships among and within lower-order taxa are less clear-cut, however. For example, the genera *Salpinctes* and *Catherpes* are sometimes lumped, and the taxonomic affinities of *Troglodytes* are unclear. At the level of currently recognized species, future splits may be in the offing for House and Marsh Wrens.

Conservation Most wren species in North America appear to be stable or increasing. Below the species level, population-change patterns are complex, as with the Bewick's Wren (eastern populations are in severe decline; several western populations are increasing and expanding). Habitat loss threatens local populations of Marsh and Sedge Wrens. Climate change may be implicated in the Carolina Wren's ongoing range expansion.

Genus *Salpinctes*

ROCK WREN   *Salpinctes obsoletus*   ROWR ■ 1

The well-named Rock Wren inhabits rocky and pebbly habitats of all sorts: from scree fields above timberline, to talus slopes and bajadas, to washes and road cuts in open desert. The nest site is indicated by a peculiar trail of pebbles. Polytypic (10 ssp.; 2 in N.A.). L 6" (15 cm)
IDENTIFICATION Usually seen creeping about ground, but sometimes perches conspicuously and bobs up and down in exaggerated, jerky movements. **ADULT:** One molt a year; sexes similar. Brown and buff above. Tail

pattern in flight distinctive, with broad buffy terminal band, broken at center. Underparts paler; throat and breast gray-white; belly pale buff. Upperparts spotted, with white; underparts with faint dark streaks. Fresh birds (early fall) more contrastingly patterned than worn birds (midsummer). **JUVENILE:** Similar to adult, but less contrast.
GEOGRAPHIC VARIATION Nominate subspecies is widespread; *pulverius* is resident on San Nicolas and San Clemente Is., CA.
SIMILAR SPECIES Structure and behavior generally wrenlike, but rarely confused with other species. Canyon Wren's habitat superficially similar, but tends more toward sheer, rock faces.
VOICE Varied, but most vocalizations diagnostic. **CALL:** an emphatic *ch'-pweee,* with overall buzzy quality; softer twittering audible from nearby. **SONG:** *Bweer bweer bweer, chiss chiss chiss, swee swee swee,* a repetitive complex of jangling series, suggesting a distant, weak-voiced Northern Mockingbird.
STATUS & DISTRIBUTION Fairly common. Range extends south to

Costa Rica. **BREEDING:** Arid country with sparse vegetation; usually in the immediate vicinity of loose, pebbly substrates. **MIGRATION:** Withdraws from northern portion of range; altitudinal withdrawal occurs throughout range. **WINTER:** Relative winter ranges of latitudinal versus altitudinal migrants poorly known. **VAGRANT:** Casual well east of core range (mostly fall and winter). **POPULATION** Stable.

buffy tail tips

cinnamon rump

faintly streaked breast

Genus *Catherpes*

CANYON WREN   *Catherpes mexicanus*   CANW ■ 1

The clear, sweet notes of the Canyon Wren are a characteristic sound of rimrocks and other mountainous country in western North America. The species can be difficult to see, but a good view reveals distinctive behavior, morphology, and plumage. Polytypic (8 ssp.; 4 in N.A., differences weak and obscured by individual variation). L 5.9" (15 cm)
IDENTIFICATION Bill extremely long. Flattened profile perched and in flight. Rarely seen away from rock outcroppings or canyon walls; usually seen creeping or probing about crevices and rock faces. Behavior mouselike.

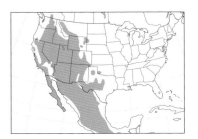

ADULT: One molt a year; sexes similar. White breast contrasts with reddish brown belly. Tail, wings, and back mainly reddish brown; head grayish. With close view, note spotting and barring on tail, wings, back, and belly. **JUVENILE:** Similar to adult, but with less spotting and barring.
SIMILAR SPECIES General structure and behavior call to mind other wrens, but plumage and song distinctive. Microhabitat preference often separates Canyon from similar species. Can overlap with Rock Wrens, but these typically found in more pebbly and sandy environments (e.g., talus slopes, quarries).
VOICE Vocalizations distinctive. **CALL:** shrill *beet,* given singly or repeatedly; year-round. **SONG:** Series of approximately 15 clear, whistled notes, descending in pitch and slowing somewhat; usually ends with fewer than five rasping notes, audible at close range. Audibility of song varies with terrain and orientation of observer;

songs given fairly infrequently (interval less than one minute).
STATUS & DISTRIBUTION Fairly common. South to Mexico. **BREEDING:** Canyons, rock faces, outcroppings. **MIGRATION:** Largely sedentary. **WINTER:** As breeding, with some wandering to lower elevations. **VAGRANT:** Casually to 100+ miles from area of regular occurrence.
POPULATION Populations presumed to be secure, but little study; recreational climbing cited as possible threat.

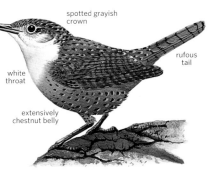

spotted grayish crown

very long bill

white throat

rufous tail

extensively chestnut belly

HOUSE WREN   *Troglodytes aedon*   HOWR ▪ 1

Dull in appearance but notable for its effervescent song, the House Wren is a common summer inhabitant of scrublands and woodland edges throughout much of North America. Variation in plumage and call notes is extensive. Polytypic (32 ssp.; 2 in N.A.). L 4.7" (12 cm)

IDENTIFICATION A small wren with medium-length bill and tail, it responds readily to pishing. Overall jizz is of a plain, typical wren. **ADULT:** One molt a year; sexes similar. All populations brownish above, paler gray-brown or gray below. Supercilium, often indistinct, is paler than crown and auriculars. **JUVENILE:** Variable, but many differ from adults in having warmer buff and rufous tones, along with indistinct scalloping on throat and breast.

GEOGRAPHIC VARIATION Two widespread, poorly differentiated subspecies in N.A.: Western *parkmanii* and eastern *aedon*. Eastern birds average more rufescent, western birds grayer. Breeding birds in the mountains of southeastern AZ, classified with *parkmanii*, have a slightly buffier throat and breast and a bolder supercilium;

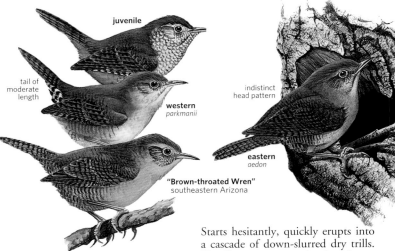

juvenile

tail of moderate length

western *parkmanii*

indistinct head pattern

eastern *aedon*

"Brown-throated Wren" southeastern Arizona

cahooni and closely related *brunneicollis* of northern MX are more richly colored; treated as a separate species until 1973 ("Brown-throated Wren"). Subspecies *martinicensis* (Martinque I.) is extinct; *guadeloupensis* (Guadeloupe I.) may be extinct.

SIMILAR SPECIES Main point of confusion are Winter and Pacific Wrens; both are smaller, shorter-tailed, more heavily barred on flanks and crissum, and usually darker; Pacific Wren has the richest coloration. House Wren is further distinguished by differences in song, call, timing of migration, and microhabitat preferences.

VOICE Most vocalizations have a "dry" quality. **CALL:** Most given singly or in stuttering series. A scratchy, frequently heard call is similar to that of Common Yellowthroat. Another note, which is raspy, resembles that of Blue-gray Gnatcatcher. **SONG:** Loud, bubbly.

Starts hesitantly, quickly erupts into a cascade of down-slurred dry trills. Usually easy to recognize, but due to considerable variation may be confused with certain non-troglodytids (e.g., Rufous-crowned Sparrow).

STATUS & DISTRIBUTION Common. **BREEDING:** Most vegetated habitats, except for dense forests, open grassland, marshland, and desert. Favored microhabitats include clearings, edges, residential neighborhoods. **MIGRATION:** Most birds in N.A. migratory. House Wrens migrate later in the spring and earlier in the fall than Winter and Pacific Wrens. **WINTER:** Generally as breeding, but with greater tendency for dense cover. Not a hardy species; stragglers north of core wintering range uncommon. **VAGRANT:** Sometimes noted to offshore or peninsular locales from which otherwise absent.

POPULATION Long-term population increases documented in many areas. Species is tolerant of humans and readily accepts nest boxes.

PACIFIC WREN   *Troglodytes pacificus*   PAWR ▪ 1

The Pacific Wren was long classified as a subspecies of a single worldwide species, the Winter Wren (called simply the Wren in Europe), all lumped under the scientific name *T. troglodytes*. In 2010 the American Ornithologists' Union separated western North American populations as a distinct species. Observers were long aware of notable differences in Pacific populations' plumage and vocalizations. Additionally, substantial

"genetic distance" in mitochondrial DNA indicated an absence of interbreeding between eastern and western populations at their contact zone in Alberta, Canada. Polytypic. (7 ssp.; all in N.A.). L 3.5–4.5" (9–11 cm)

IDENTIFICATION Sexes similar. **ADULT:** Upperparts dark rufous; face dark brown with buff supercilium; tail banded black and rufous; throat and breast bright buff or grayish depending on subspecies; dark barring on

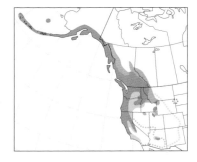

wings, tail, flanks, and belly; relatively long bill. JUVENILE: Similar to adult but with less distinct barring on belly, and sometimes with faint breast-scalloping.
GEOGRAPHIC VARIATION Differences among subspecies are complex and subtle, primarily based on size, plumage coloration, and pattern of dark markings. "Alaskan Island" group is largest, with longest bill, and uniformly pale brownish to grayish; populations in Pacific Northwest and western Canada are most rufescent; interior populations in western ID–WY are grayer. Validity of classifying some populations as subspecies has been questioned.
SIMILAR SPECIES Winter Wren is paler, duller, with less richly rufous and buff coloration; bill relatively short; throat whitish. House Wren is larger, longer tailed, paler, especially on throat and breast; barring does not

extend forward of legs; supercilium is less distinct.
VOICE Vocalizations diagnostic between Pacific and Winter Wren. SONG: A long, loud, complex, and beautiful series of trills, similar to Winter Wren but faster, slightly higher pitched, somewhat harsher, and less melodic. CALL: a *chimp*, often doubled, with a quality reminiscent of Wilson's Warbler. Winter Wren's call is more like a *kelp* than a *chimp*.
STATUS & DISTRIBUTION Fairly common in its habitat. Nests primarily in old-growth coniferous and mixed forests with dense understory and often near water. BREEDING: Primarily Pacific coast from AK to central CA, and in scattered interior regions. WINTER: Within breeding range and south to southern CA, AZ, rarely NM. VAGRANT: Accidental to northern AK.
POPULATION Stable in much of range,

but declining severely in the northwestern rain forest, particularly in BC.

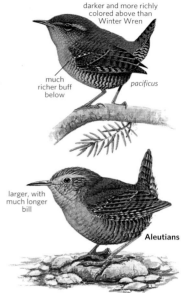

darker and more richly colored above than Winter Wren

much richer buff below

pacificus

larger, with much longer bill

Aleutians

WINTER WREN *Troglodytes hiemalis*   WIWR   1

Tiny, nervous, and unforthcoming, the Winter Wren breeds mainly in cool, shady forests and tangles at northern latitudes and higher elevations. Its long, loud, complex, and striking song is frequently rated among the best among all North American birds. Polytypic (2 ssp.). L 3.5" (9 cm)
IDENTIFICATION Diminutive in all respects. Small-bodied, stub-tailed; bill short for a wren. Hides among dense shrubbery, tangled roots, etc. Surprisingly hard to glimpse, even when close. Note constant twitching and jerking motions, with tiny tail held straight up. Sexes similar. ADULT:

overall coloration duller and paler than Pacific Wren; vocalizations diagnostic

hiemalis

All populations brown to dark brown, pale brownish to grayish underparts and prominent barring on belly, flanks, and crissum; whitish throat. JUVENILE: Similar to adult, but with less distinct barring on belly.
GEOGRAPHIC VARIATION Widespread nominate *hiemalis* has dark upperparts with rufous tinge, breast washed pale brownish; Appalachian *pullus* has less rufescent upperparts and moderate brownish wash on breast, decidedly darker and less rufescent, underparts lighter brown with vermiculations on abdomen and flanks heavier; bill smaller and more slender.
SIMILAR SPECIES Pacific Wren is brighter and darker overall, more richly rufescent, and with dark barring; throat and breast rich buff. House Wren is larger, longer tailed, paler, especially on throat and breast; barring does not extend forward of legs; supercilium is less distinct.
VOICE Vocalizations are diagnostic between Pacific and Winter Wren. SONG: A long, loud, complex, and

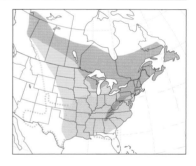

beautiful series of trills, similar to Pacific Wren but slower, slightly lower pitched, and more melodic. CALL: A *kelp*, unlike Pacific's *chimp*.
STATUS & DISTRIBUTION BREEDING: Fairly common in coniferous forest habitat across Canada, northern U.S., south in Appalachians to GA. WINTER: Winters south to Gulf Coast states, and in Southeast from VA to GA, rarely to northwestern FL; casual farther south in FL, to Southwest and CA.
POPULATION Stable overall.

Genus *Cistothorus*

MARSH WREN *Cistothorus palustris*   MAWR   1

The rambunctious and highly vocal Marsh Wren may be heard in cattail marshes throughout much of North

America all through the summer, at any time. Like most wrens, the Marsh is a recluse, but its densities are often

high in favored breeding areas, and patient watching usually results in a good view. The ovoid dummy nests of the species are conspicuous. Vocal and plumage variation is extensive and often field-discernible, and population status is locally and regionally complex. Polytypic (15 ssp.; 14 in N.A.). L 5" (13 cm)

IDENTIFICATION Usually stays close to cover in dense cattails, even when singing. Glimpses tend to be brief, often of flying birds. In flight note rounded wings and tail, overall reddish brown coloration. Marsh Wrens are highly social and aggressive (both inter- and intraspecifically), and careful study may result in interesting behavioral observations. **ADULT:** Sexes similar. Most subspecies have a conspicuous supercilium sharply set off against dark brown crown and gray-brown auriculars; black back streaked with prominent white stripes; wings, tail, rump, and belly variably chestnut. **JUVENILE:** More smudgy-looking, less contrastingly marked than adult, with few if any white streaks on back.

GEOGRAPHIC VARIATION Extensive. Two distinct subspecies groups, "Eastern" and "Western" divide on the central Great Plains. These groups are separated mainly by song (see below). The "Eastern" *palustris* group is comprised of six subspecies, most of which are darker and more richly colored, typified by widespread *dissaeptus* (illustrated). The subspecies *griseus* of the south Atlantic coast is decidedly gray, the dullest Marsh Wren subspecies. The

"Western" *paludicola* group, composed of eight subspecies in N.A. (one more in MX), is basically similar, but colors more muted and contrast somewhat weaker, especially on crown.

SIMILAR SPECIES See Sedge Wren.

VOICE Geographic variation in songs extensive. **CALL:** Common call note a soft *chuck,* without the sharp, grating quality of the call notes of many other wren species. At close range, softer stuttering notes, more like those of other wren species, may be heard. **SONG:** All songs have a liquid or gurgling sound, more so than the dry, stuttering song of Sedge Wren. Liquid quality is more pronounced in eastern birds than in western birds, which also have a much larger repertoire. General pattern is similar to that of Sedge or House Wrens, with several well-spaced introductory notes followed by a more complex series: *Chik chik chu-u-u-u-u-u-u-rrr.*

STATUS & DISTRIBUTION Locally common. **BREEDING:** All populations favor dense cattail marshes. Eastern birds most strongly concentrated in estuarine marshes fairly far inland, where conditions are less brackish. Eastern birds also occur in large cattail marshes in association with large inland lakes and marshes. Western birds occur in similar situations, but are also frequently noted as nesters in very small cattail marshes (e.g., on golf courses), usually eschewed by eastern birds. **MIGRATION:** Complex. Some populations highly migratory, others partially migratory,

others nonmigratory. Migration in general poorly detected, owing to reclusive behavior of migrants. Migratory populations usually back on breeding grounds by mid-May. Withdrawal to wintering grounds begins by midsummer; most migrants are detected by early fall in the West, but not until Oct. in the East. Even on migration, most individuals are found in cattails and similar vegetation. **WINTER:** Cattail marshes generally favored, but a broader array of habitats is accepted. Dense cover almost always required. **VAGRANT:** Accidental in fall to southern AK (ssp. unknown).

POPULATION General pattern is of increasing numbers in western populations vs. decreasing numbers in the East, possibly reflecting western birds' tendency to be more generalized in habitat selection. Destruction or alteration of wetlands, especially in the East, has caused local declines and extinctions.

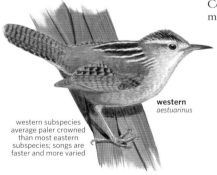

western subspecies average paler crowned than most eastern subspecies; songs are faster and more varied

western
aestuarinus

resident in coastal marshes from South Carolina to east-central Florida

a very gray subspecies

south Atlantic coast
griseus

more solid crown, more prominent white stripes on back, more prominent whitish supercilium, and longer bill than Sedge Wren

most eastern subspecies are, on average, a little darker and more richly colored than western subspecies

eastern
dissaeptus

SEDGE WREN *Cistothorus platensis* SEWR ◼ 1

Few North American birds show as much distributional complexity as the Sedge Wren. Nesting begins in the northern portion of the range in late spring and shifts southward by midsummer. Although a weak flier, it is the only wren of North America. that withdraws completely from the breeding range in winter. Sedge Wrens are seldom detected on

migration, but vagrants are recorded far from main routes. Highly disjunct and largely nonmigratory populations occur south to islands off southern South America. Polytypic (18 ssp.; only stellaris in N.A.). L 4.5" (11 cm)

IDENTIFICATION Small. Bill short for a wren (species was formerly known as "Short-billed Marsh Wren"). In winter and migration can be difficult to

glimpse, with most individuals staying under cover. Where locally common on breeding grounds easy to observe due to constant singing and frequent interactions. **ADULT:** Two molts a year (early spring, late summer); sexes similar. Buffy overall, subtly but extensively streaked and spotted. Note fine white streaks across brown crown, fairly coarse black-and-white steaks on buffy-

faint white
streaks on back

streaked crown
with indistinct
supercilium

shorter bill
than Marsh
Wren

buff extends
across breast

orange back. Supercilium faint but discernible; buff-tinged. Unmarked underparts warm buff, brighter on breast and throat. Tail and wings broad, rounded; extensively but weakly barred with brown and buff. JUVENILE: More muted overall than adult; streaking on upperparts more subdued, and buffy tones on underparts weaker.
GEOGRAPHIC VARIATION One subspecies, *stellaris,* in N.A.; variation discrete and extensive (17 additional ssp.) farther south to southern S.A.
SIMILAR SPECIES Marsh Wren often treated as point of confusion, perhaps

more because of taxonomic and nomenclatural affinities (both species were formerly known as "marsh wrens") than because of actual field-based characters. Marsh Wren differs in many respects, including darker colors and more contrast overall, solid brown crown, conspicuous white supercilium, and black back streaked with white. The two species rarely overlap on breeding grounds, with Marsh in dense palustrine or estuarine cattail marshes, Sedge Wren in tall grasses in damp uplands.
VOICE Most vocalizations sharp, chattering. CALL: Sharp *chap,* like introductory notes in song. Close-up, soft sputtering series, as given by other wrens. SONG: Several sharp notes followed by a short series of faster notes: *chip chip chip ch'ch'ch'ch'ch'ch'ch.*
STATUS & DISTRIBUTION Complex. Locally common, but also scarce in or absent from many areas of seemingly suitable habitat. BREEDING: Loosely colonial in tall grasses; wet soil preferred, but extensive standing water a deterrent. Breeding in northern portion of range (upper Midwest, western Great Lakes) occurs earlier than

breeding farther south (central Great Plains).
MIGRATION: Begins northward movement by early May. Many individuals retreat from northern breeding grounds in early summer to staging or breeding grounds farther south. Fall migration drawn out; late summer to mid-fall; bulk of records in Oct. WINTER: As breeding, but more general, with some individuals in wet marshes, others in drier scrub. VAGRANT: Annually to hundreds of miles from area of regular occurrence; casual to West Coast. Most vagrants noted in fall.
POPULATION Overall stable but hard to monitor given erratic and opportunistic movements.

Genus *Thryothorus*

CAROLINA WREN *Thryothorus ludovicianus* CARW ■ 1

The adaptable and highly vocal Carolina Wren is a familiar inhabitant of gardens and woodlands in the Southeast. Climate-related range shifts in the species are well documented. Polytypic (10 ssp.; 6 in N.A.). L 5.5" (14 cm)
IDENTIFICATION Active, inquisitive. In most of range, the most brightly colored wren in its habitat. ADULT: One molt a

bold white
supercilium

rufous-brown
upperparts

extensively
rich buffy
underparts

year; sexes similar. Upperparts bright reddish brown; breast and belly warm buffy-orange; throat whitish. White supercilium; long, conspicuous. JUVENILE: Overall tones duller.
GEOGRAPHIC VARIATION Six subspecies north of Mexico; four others in Mexico and Central America. Populations fairly homogeneous north of about 32° N. More heterogeneous farther south, but field identification to subspecies is difficult.
SIMILAR SPECIES Where ranges overlap, Bewick's Wren may present confusion. Bewick's has colder colors, thinner bill, and longer, expressive white-cornered tail. Songs different.
VOICE Loud and frequent. CALL: Varied. A hollow, liquid *dihlip.* Another call sounds like a stick being run across a wire-mesh fence. SONG: Rich and repetitious. Most songs consist of short, repeated phrases; song may start or end with single notes—*chip mediator mediator mediator meep.* Primitive antiphonal singing is sometimes heard: One bird begins with characteristic song; mate finishes with low rattle.

STATUS & DISTRIBUTION Common. BREEDING: Dense vegetation, frequently near human habitation. MIGRATION: Largely sedentary. WINTER: As breeding. VAGRANT: Very rare to casual to MN, SD, CO, NM, and AZ. These vagrants may be better thought of as vanguards in range expansion.
POPULATION Northward range expansion is fairly sustained; westward expansion is erratic. Sudden range contractions follow harsh winters, but expansion resumes eventually. Tolerant of humans; projected beneficiary of global warming.

Genus *Thryomanes*

BEWICK'S WREN *Thryomanes bewickii* BEWR ■ 1

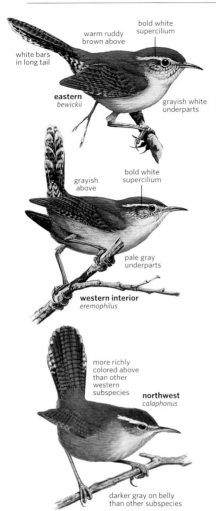

white bars
in long tail

warm ruddy
brown above

bold white
supercilium

eastern
bewickii

grayish white
underparts

grayish
above

bold white
supercilium

pale gray
underparts

western interior
eremophilus

more richly
colored above
than other
western
subspecies

northwest
calaphonus

darker gray on belly
than other subspecies

Vocal and plumage variation in the Bewick's Wren is extensive, but all adults have a very long tail, tipped in white and flicked side to side. Eastern populations have disappeared from much of former range and are in continuing decline. Polytypic (15 ssp.; 11 in N.A.). L 5.1" (13 cm)

IDENTIFICATION ADULT: Sexes similar. Plumage variable, but all subspecies gray-brown to rufous-brown above, gray-white below, with long, pale supercilium. White-tipped tail has black bands.

GEOGRAPHIC VARIATION Complex, extensive. "Pacific Coastal" *spilurus* group comprises nine subspecies, six in N.A.; as a group brownish above, pale gray below; *calaphonus* from the

Northwest is the most richly colored above and darkest below. Subspecies *leucophrys* from San Clemente I. and *brevicaudus* from Guadalupe I. are extinct. "Interior West" *eremophilus* group comprises five subspecies, four in N.A.; widespread *eremophilus* is dull and grayish above, subspecies farther east are more warmly colored above. "Eastern" *bewickii* group (inc. *altus*) is rufescent brown above and whitish below.

SIMILAR SPECIES In most of range distinctive tail and supercilium prevent confusion with other wrens. See Carolina Wren.

VOICE Varied. In many places in the West, Bewick's produces the local "mystery song." **CALL:** Variable, but many notes with raspy or buzzy quality, some quite loud. **SONG:** Most songs combine 1–5 short, breathy, buzzy, notes (sometimes run together in a short warble) with a longer, often loose, trill. Single buzzy or nasal notes may be introduced. Overall rhythm and tone remind many observers of Song Sparrow, but elements in Bewick's song are usually thinner, buzzier. In the "Interior West" songs are simpler than those from the far West or *bewickii*.

STATUS & DISTRIBUTION Western populations fairly common. Eastern *bewickii* has sharply declined in recent decades, reduced to remnant breeding populations centered in the Ozarks, a few

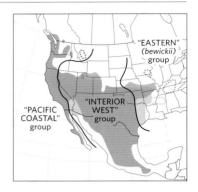

"EASTERN"
(*bewickii*)
group

"PACIFIC
COASTAL"
group

"INTERIOR
WEST"
group

up the Tennessee Valley. **BREEDING:** Shrubby vegetation always a requirement. Western individuals frequent arid juniper foothills, lush riparian corridors, residential districts, etc. Eastern *bewickii* inhabits open woodlands. **MIGRATION:** Western populations largely sedentary. Eastern populations migratory, with general pattern of dispersal south and west following breeding. **WINTER:** As breeding, with generally warmer and lower-elevation component. **VAGRANT:** Extralimital records difficult to assess. Apparent vagrants in the East may be remnants, natural migrants, or actual vagrants.

POPULATION Subspecies *bewickii* once ranged northeast to NY and ON. Competition with House Wren thought a major factor in the Bewick's decline.

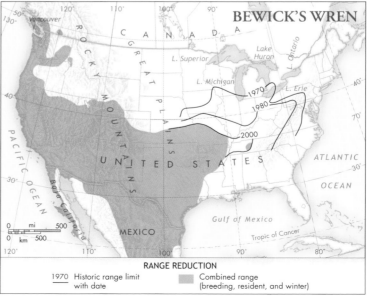

BEWICK'S WREN

RANGE REDUCTION

1970 Historic range limit
with date

Combined range
(breeding, resident, and winter)

CACTUS WREN   *Campylorhynchus brunneicapillus*   CACW ▪ 1

Calling to mind a small thrasher, the oversize Cactus Wren is a distinctive inhabitant of the thorn-scrub country of the desert Southwest. A good clue to this wren's presence is its characteristic dummy nest, placed conspicuously but inaccessibly amid chollas, acacias, etc. Polytypic (7 ssp.; 3 in N.A.). L 8.7" (22 cm)

IDENTIFICATION Large-proportioned, with bulky build, broad tail, relatively thick bill. **ADULT:** One molt a year; sexes similar. Plumage spotted and streaked all over; overall tones brown, black, and white. Note variable dark spotting on undersides and conspicuous white supercilium. In flight rounded wings and tail are brown overall but heavily spotted and barred with white. **JUVENILE:** Similar to adult, but buffier overall, with spotting and barring more muted.

GEOGRAPHIC VARIATION Widespread *anthonyi* and *guttatus* in TX. "San Diego" Cactus Wren *(sandiegensis),* resident in coastal southern CA, is distinguished from other subspecies by different pattern on underparts—black spotting is more uniform, and background color is more extensively white.

SIMILAR SPECIES Other wrens smaller and usually in other habitats, so confusion unlikely. Similar in certain respects to Sage and Bendire's Thrashers.

VOICE Most vocalizations low-pitched and grating. Species highly vocal, often singing through the hottest times of the day. **CALL:** Low, growling, clucking sounds, sometimes given in a loose series, *chut chut chut.* **SONG:** A distinctive, unmusical series of pulsing, chugging notes, reminiscent of an old car trying to start—*churr churr churr.*

STATUS & DISTRIBUTION Locally common in southwestern U.S. and northern Mex. **YEAR-ROUND:** Towns, washes, and open desert with thorn scrub. **VAGRANT:** Very few records of wanderers.

POPULATION "San Diego" subspecies threatened by habitat loss; species tolerant of humans. Local distribution limited by availability of nest sites.

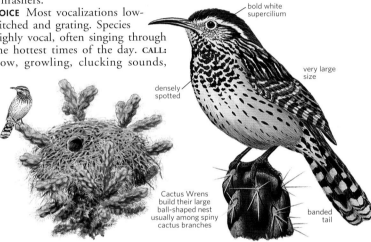

bold white supercilium

very large size

densely spotted

banded tail

Cactus Wrens build their large ball-shaped nest usually among spiny cactus branches

SINALOA WREN   *Thryophilus sinaloa*   SIWR ▪ 5

This nonmigratory Mexican species, resident as far north as northern Sonora, has been discovered four times north of Mexico, all in southeastern Arizona. Polytypic (3 ssp.; likely the duller, northern *cinereus* in N.A.). L 5.25" (13 cm)

IDENTIFICATION A skulker. Shaped like a Carolina Wren, with a long, thin bill and a short, often upraised tail. Sexes similar. **ADULT:** White face with dark eye stripe and white supercilium; sides of neck streaked black and white. Upperparts brown, wings and tail brown with black bars. Underparts with pale gray breast, pale buff lower breast and belly; undertail coverts barred black and white (hence alternate English name "Bar-vented Wren"). **JUVENILE:** Duller overall, with pale rusty underparts.

SIMILAR SPECIES Western interior

Bewick's is gray above and pale grayish below. Carolina is much brighter, with extensively rich buffy underparts and deep rusty brown upperparts. Happy Wren *(T. felix),* another west Mexican endemic and potential stray into U.S., has black eye line, more boldly striped face, and entirely buff underparts including breast.

VOICE SONG: Loud, rich phrases, often with rapid repetition of notes, such as *weet-weet—weet, chuee-chuee-chuee.* **CALL:** Wide variety of notes, rough, buzzy, sometimes chattering.

STATUS & DISTRIBUTION West Mexican endemic, resident in tropical deciduous forest, gallery forest, mangrove forest, and thickets in thornscrub from northern Sonora to south to western Oaxaca. **VAGRANT:** Four AZ records: Patagonia-Sonoita Creek Preserve (Aug. 25,

2008–fall 2009); Huachuca Canyon (Apr. 14–18, 2009); Huachuca Canyon (Sept. 2–Oct. 4, 2013); Tubac (Sept. 11–21, 2013).

POPULATION Stable; has expanded northward in Sinaloa, breeding at least to Santo Domingo 70 mi south of AZ.

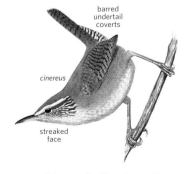

barred undertail coverts

cinereus

streaked face

GNATCATCHERS AND GNATWRENS Family Polioptilidae

Blue-gray Gnatcatcher, male (ON, May)

Small and intensely active, the gnatcatchers can present an identification challenge in the southwestern border regions where two or more species may occur together, but only the familiar Blue-gray Gnatcatcher is found over most of North America. This small New World family also includes three Neotropical species known as gnatwrens.

Structure This is a generally uniform group of small (4–5 in) insectivorous birds. The bills are very thin, moderately short in gnatcatchers but long in the gnatwrens. Gnatcatchers are all closely similar in structure, being slender with long, expressive tails; the two *Microbates* gnatwrens have shorter tails. All have wings that are relatively short and rounded.

Behavior Active insectivorous birds that flit actively through twigs and foliage, often cocking and waving the long tail. They forage by gleaning, but also by hover-gleaning or making short aerial sallies. They are typically found in pairs. They build small, soft deep cup-shaped nests. Both sexes incubate for about 14 days.

Plumage Generally gray to brownish in color, often with black markings on the crown or elsewhere on the head, and in most species white in the outer tail feathers. The sexes and age groups are generally similar,

but breeding season males in temperate zone species acquire black head markings.

Distribution Exclusively New World; gnatcatchers are most diverse in middle and tropical South America, occurring mainly in lowland forest, thornscrub, and arid scrub; one species (Blue-gray Gnatcatcher) breeds north through much of the United States and even southernmost Canada. The gnatwrens are found in tropical lowland forest understory and borders from southeastern Mexico to Brazil and Bolivia. Most species are sedentary, but most Blue-gray Gnatcatcher populations are migratory.

Taxonomy Long linked taxonomically to the ill-defined "Old World Warbler" family Sylviidae, the 13 gnatcatchers (genus *Polioptila*) and three gnatwrens (*Microbates* and *Ramphocaenus*) are, in fact, more closely related to the predominantly New World wrens (Troglodytidae) and creepers (Certhiidae). There may well be more cryptic species within *Polioptila*. Four species occur in North America north of Mexico (one being marginal along the Arizona/Sonora border).

Conservation Some range-restricted species are considered rare and in some cases threatened; in North America the California Gnatcatcher is threatened within its U.S. range by extensive urban development of its scrub habitats.

Genus *Polioptila*

BLUE-GRAY GNATCATCHER *Polioptila caerulea* BGGN ▪ 1

This species is quite active. Polytypic (7 ssp.; 2 in N.A.). L 4.3" (11 cm) **IDENTIFICATION** Long tailed, with outer tail feathers almost entirely white (tail from below looks white).

Bill is thin and pale gray. **BREEDING MALE:** Blue-gray above, including most of head and back. Crown has a black line at forecrown that extends along sides of crown; white eye ring contrasts

with gray face. Wings brownish gray; tertials blackish, edged white. Underparts entirely white. **NONBREEDING MALE:** Black on crown absent, resulting in grayish crown. **FEMALE:** Like nonbreeding male but grayer above. **GEOGRAPHIC VARIATION** Nominate *caerulea* more extensively white tail; western *obscura* has a black base to outer rectrices that extend beyond undertail coverts. Western male slightly less blue on back, with black forehead mark that is thicker, and less like a supraloral line found in nominate *caerulea*. Western females are dingier above.

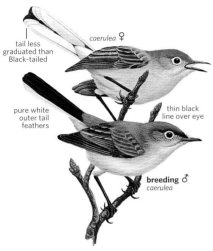

tail less graduated than Black-tailed

caerulea ♀

pure white outer tail feathers

thin black line over eye

breeding ♂
caerulea

SIMILAR SPECIES Most confusion is likely to occur with Black-tailed, and to a lesser extent, California Gnatcatcher. These species have different calls; California is also darker below. The best feature is tail pattern. Blue-gray is almost entirely white on outer rectrices; Black-tailed and California have mostly black outer rectrices with white tips or edges. Tail shape also differs. In Black-tailed in CA (and Black-capped) the outer tail feathers are short and progressively increase in length, giving a graduated tail shape with an intricate pattern of white tips. In Blue-gray, outer tail feathers are much longer, so on closed tail, only outers are visible. Be aware that in late summer gnatcatchers molt their tails. Blue-grays will look mostly dark from below when their outer rectrices are dropped. See Black-capped Gnatcatcher. **VOICE CALL:** A querulous *pwee* or various mewing calls. Western birds have lower, harsher notes, more like wrens; Easterns' common call slightly more wiry, thin. **SONG:** Thin, wiry notes; lower and harsher in western populations. **STATUS & DISTRIBUTION** Common.

BREEDING: Various woodlands. **MIGRATION:** Earliest migrants reach Great Lakes in late Mar., typically in early Apr. Peak late Apr. through early May, with stragglers to later in the month. Fall migration starts as early as late June or July, in southern states. Farther north migration starts mid-Aug., with peak mid-Sept. Small numbers seen into Oct. **VAGRANT:** Annual in fall in small numbers to Atlantic Canada, Aug.–Nov. Casual in spring to Atlantic Canada; to BC and Ottawa in spring and fall. **WINTER:** Southern U.S., south to Guatemala and Honduras. **POPULATION** Northward expansion in northeastern U.S. and southeastern Canada occurred in the 20th century.

CALIFORNIA GNATCATCHER *Polioptila californica* CAGN 2

A southern California and Baja California endemic, this gnatcatcher is restricted to coastal chaparral in our area. Formerly considered conspecific with the Black-tailed Gnatcatcher, it was split based on morphological and vocal differences. Polytypic (3 ssp.; nominate in U.S., other two in Baja California). L 4.3" (11 cm) **IDENTIFICATION** A dark, rather dingy gnatcatcher with a dark tail. Black outer rectrices are edged white, including tip, but white does not extend noticeably inward from tip. From below, tail looks dark. Crown and upperparts dark gray, occasionally with blue tones. Wings

decidedly brownish. Underparts are more grayish than whitish, and flanks have buffy wash. **BREEDING MALE:** Crown is black to include slightly below eye. Face and lores are paler; tertials are edged grayish. **NONBREEDING MALE:** Crown is gray, with black reduced to a streak over eye. As greater wing coverts wear, they appear browner. **FEMALE:** Like nonbreeding male, but plumage browner, with brownish or grayish brown edges to tertials; flanks strongly washed with buff; lacks any black on crown. **SIMILAR SPECIES** Similar to Black-tailed, but with darker, dingier gray below. Tail of California is edged white, as opposed to white tips of outer tail feathers of Black-tailed, and California has a less distinct eye ring. Ranges do not overlap in U.S. **VOICE CALL:** A rising and falling *zeeer,* rather kittenlike. **SONG:** A series of *jzer* or *zew* notes. **STATUS & DISTRIBUTION** Threatened. **YEAR-ROUND:** Local resident in sage

more restricted white on outer tail feathers

californica ♀

black cap like Black-tailed

breeding ♂
californica

overall darker above and below than Black-tailed

scrub of southwest CA, north to southern Ventura Co. **POPULATION** The northern nominate subspecies, which extends into northern Baja California to 30° N, is threatened by habitat destruction; only a few thousand pairs believed to remain in CA. Population of southern subspecies appears stable. The nominate is federally listed as threatened.

BLACK-TAILED GNATCATCHER   *Polioptila melanura*   BTGN   ■   1

The Black-tailed is an inhabitant of desert thornscrub, partial to washes. Although small, this feisty songbird will aggressively defend its nest against larger predators. Formerly considered conspecific with the California Gnatcatcher. Polytypic (3 ssp.; 2 in N.A.). L 4" (10 cm)

IDENTIFICATION Large white terminal spots on graduated tail feathers; short dark bill. **BREEDING MALE:** Glossy black cap extends past eye, contrasting with thin, white eye ring. Tertial edges usually whitish. **NONBREEDING MALE:** Like breeding male but dark cap is replaced by thin dark line over eye. **FEMALE:** Like nonbreeding, but lacks dark line over eye; back washed with brown.

GEOGRAPHIC VARIATION Three subspecies: One *(curtata)* is from Tiburon I. in the Gulf of California. In the U.S., *lucida,* which breeds in the Sonoran and Mojave deserts (CA and AZ), has less white in tail, a noticeably paler gray base to mandible, and, in female, a back with a rather brownish wash. Nominate *melanura,* which breeds in the Chihuahuan desert (NM and TX), has more white in tail, a rather uniform dark bill with only slight paling at base, and a lighter brown wash on back of female.

SIMILAR SPECIES Most likely confused with Black-capped Gnatcatcher as birders search for the latter species in southeastern Arizona. Most easily distinguished by white tips to outer rectrices of Black-tailed, compared to nearly all-white outers of Black-capped; tail of Black-capped is also more graduated. Also note on breeding males the more extensive black cap coming below the eye and longer bill and a more faint eye ring, usually limited to a crescent below eye. In nonbreeding Black-capped,

note pale face including auriculars, whereas in nonbreeding Black-tailed the face is rather more uniform with the crown. See also Blue-gray Gnatcatcher entry for separation from that species.

VOICE CALL: Rasping *cheeh* and hissing *ssheh;* in general, its vocalizations are rather wrenlike. **SONG:** A rapid series of *jee* notes.

STATUS & DISTRIBUTION Fairly common. **YEAR-ROUND:** Resident in variety of arid habitats in the Southwest U.S., as well as in northern and central Mexico. A few isolated populations exist in CA, notable in the Panamint, Inyo Co., and near Mojave, Kern Co. This species moves very little, if at all. A specimen record from San Antonio, TX, is one of few records that might come from outside its known range. Any suspected Black-tailed Gnatcatcher outside its known range should be thoroughly documented.

strongly graduated tail with white tips to outer tail feathers

breeding ♂
black cap comes just to eye; in winter, male has gray cap with short black line above eye

♀

BLACK-CAPPED GNATCATCHER   *Polioptila nigriceps*   BCGN   ■   3

This western Mexican species is rare in southeastern Arizona. Polytypic (2 ssp.; *restricta* in N.A.). L 4.3" (11 cm)

IDENTIFICATION Long bill is usually dark. Tail is strongly graduated, with outer rectrices mostly white. **BREEDING MALE:** Black cap and lores; black extends below eye. Head and back blue-gray with indistinct white eye ring; face rather pale. Underparts whitish, with grayish wash on breast. **NONBREEDING MALE:** Black crown absent; dark line over eye, or dark flecks. **FEMALE:** Like nonbreeding, but no black on crown.

SIMILAR SPECIES Separated from Blue-gray Gnatcatcher by more graduated outer tail feathers, longer bill, and, in breeding males, more extensive black in crown. From Black-tailed Gnatcatcher, in all plumages, outer rectrices are mostly white and more graduated; in breeding plumage, dark crown extends below eye, and eye ring, if present, is less distinct. In nonbreeding, note the different call and tail pattern of Black-capped, as well as its obviously longer bill.

VOICE CALL: A rising, then falling *mee-ur,* somewhat like California Gnatcatcher; also suggestive of Bewick Wren's call. **SONG:** A jerky warble.

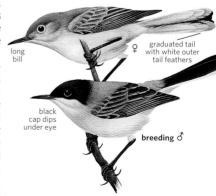

long bill

♀
graduated tail with white outer tail feathers

black cap dips under eye

breeding ♂

STATUS & DISTRIBUTION Western Mexican species is very rare (primarily resident) of southeastern AZ. Recently discovered in Guadalupe Canyon in both AZ and NM.
POPULATION Stable.

DIPPERS Family Cinclidae

American Dipper (UT, June)

Structure Portly overall, appearing bobtailed with strong legs and feet. The bill is short, compressed, and sturdy. Wings are short and rounded.

Behavior Highly distinctive foraging techniques. Often seen wading into rushing water, submerging head while looking for food—mostly aquatic larvae. Also swims on the surface for short distances, peering under water, while paddling with feet. Using its wings, a dipper can submerge completely and maneuver near stream bottoms by grasping the substrate—remaining under water for up to 30 seconds. When not in water, it stands on rocks and boulders, pumps its body up and down in highly stereotyped motion, and flashes white-feathered eyelids.

Plumage Dense and waterproof. All species exhibit significant brown or gray, some with white and/or chestnut. The American Dipper is one of the plainer species.

Distribution Found in large montane watershed complexes in the Americas, Europe, N. Africa, and Asia.

Taxonomy Worldwide, five closely related species, all placed in a single genus. A close relationship with the wrens (Troglodytidae) has been posited, but some researchers have argued instead for a closer alliance with the mimic-thrushes (Mimidae).

Conservation Most dipper populations are naturally isolated and hence susceptible to local extinctions from habitat alteration and water pollution.

Few passerines are more specialized than the highly aquatic dippers. They spend their entire lives in the immediate vicinity of, and often actually in, rushing streams and rivers. Accordingly, details of their physiology and behavior are unique. There is one North American species, which is widespread in the West. Its absence from the Appalachians is reckoned by some experts as a biogeographic mystery.

Genus *Cinclus*

AMERICAN DIPPER *Cinclus mexicanus*   AMDI ▪ 1

One of the most charismatic species of the American West, the American Dipper is at home along rushing streams and rivers, from timberline to sea level. Although its attire is drab gray and brown, its behavior, physiology, and vocalizations are remarkable. It readily jumps into streams of rushing water and swims underwater, procuring its prey. Whitewashed rocks in streams are a likely sign of their presence. Polytypic (5 ssp.; *unicolor* in N.A., 4 others from Mex. to Panama). L 7.5" (19 cm)

IDENTIFICATION Plump and monochromatic. Frequently bobs. Legs and feet pale; wings short; tail very short. **ADULT:** Slate gray with somewhat browner head; dark bill. One molt per year, in late summer. **JUVENILE:** May retain distinctive plumage into early winter; note paler tones overall, especially throat; distinctive white scalloping to contour and wing feathers; yellowish bill. **FLIGHT:** Alcid-like: low over the water, buzzy.

SIMILAR SPECIES Unmistakeable.

VOICE CALL: Loud, sharp *bzeet,* given frequently by flushing birds; easily heard. On landing, may give fast series of these notes. **SONG:** Amazingly long (exceptionally to five mins.) series of repeated, varied,

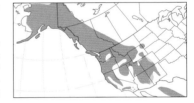

will swim under water to obtain food

overall paler than adult

juvenile

flesh legs

unmistakable— chunky and sooty gray overall

thrasher-like phrases. Song often ventriloqual; nearby birds may sound faint, far away.

STATUS & DISTRIBUTION Fairly common. AK to Panama. **BREEDING:** Near rushing streams. Year-round resident where streams remain ice-free. **MIGRATION:** High-elevation breeders move to nearby ice-free streams at lower elevations; some farther. Juveniles generally move to lower elevations than adults. **VAGRANT:** Casual to TX; accidental to MN.

POPULATION Overall population is stable but naturally fragmented; some local populations are at risk of extinction.

BULBULS Family Pycnonotidae

Red-whiskered Bulbul

Bulbuls are native to tropical areas of the Old World. Popular as cage birds, several escaped species have established naturalized populations in many regions. In the United States, bulbuls are associated with parks and neighborhoods that are landscaped with exotic plants and trees that provide fruit and nectar year-round.

Structure Bulbuls are slim with long tails; some have small crests.

Plumage Commonly brown, olive, or gray, with red, yellow, or black markings. Many species have bold head patterns and brightly colored undertail coverts. Juvenile plumage generally resembles that of adult.

Behavior Highly gregarious during the nonbreeding season, most bulbuls forage and roost in large flocks. Primarily frugivorous, but also feed on insects, which are gleaned from foliage or captured in midair by flycatching or hawking. Many pairs of Red-whiskered Bulbuls in Florida are accompanied by a third adult, but cooperative breeding not documented.

Distribution Their native range extends from the western Palearctic to the Orient and to the Philippines. The Red-whiskered is found in North America as a result of escaped cage birds; another species is also naturalized in Hawaii.

Taxonomy A large family with 119 species in 26 genera. One newly discovered species, the Bare-faced Bulbul (*Pycnonotus hualon*) was just described from Laos in 2009.

Conservation The United States has banned further importation of Red-whiskered Bulbul because of their potential threat to crops. Most bulbuls are common in their native ranges, but more than a dozen species are threatened, some critically.

Genus *Pycnonotus*

RED-WHISKERED BULBUL *Pycnonotus jocosus*   RWBU ■ 2

The Red-whiskered Bulbul is the sole representative of this Old World family in North America, although the Red-vented Bulbul (*P. cafer*) occurs in Hawaii. Bulbuls are popular cage birds, and U.S. populations formed from escapees. Polytypic (8 ssp.; *emeria* in FL; CA ssp. unknown). L 7" (18 cm)

IDENTIFICATION A small, slender songbird with a conspicuous crest and long tail, brown above and white below. Sexes similar. **ADULT:** Black forehead and crest, remainder of upperparts brown. Black eyes, bill, legs, and feet. Small red "whiskers" just below and behind eyes. Narrow black line separates white

cheek from white throat and malar. White breast and belly, with blackish "spur" on sides of breast forming incomplete band. Flanks and vent washed with buff; red undertail coverts. Brown tail, broadly tipped with white on all but central rectrices. **JUVENILE:** Brown crown and crest, red "whiskers" lacking, buffy undertail coverts.

SIMILAR SPECIES Few other crested birds occur within narrow U.S. range. May be confused with Phainopeplas in California, which are wholly black or grayish, with white patches in outer primaries. Bulbuls in Florida were originally confused with Tufted Titmice, which are casual to the Miami area.

VOICE CALL: *Kink-a-jou* or *chip-pet-ti-grew; peet* given at roost or when alarmed. SONG: Rolling musical whistle *chee-purdee, chee-purdee-purdee.*

STATUS & DISTRIBUTION Exotic in the U.S.; importation now banned. Common in native range from India to China and to the northern Malay Peninsula; ssp. found in FL native from India to southwestern Thailand.

long black crest

small red auricular spot

juvenile

black breast mark

red undertail coverts

white tail tips

YEAR-ROUND: Non-migratory. In U.S., restricted to CA (<100 in the San Gabriel Valley) and FL (perhaps a few hundred individuals at Kendall and Pinecrest, southwest of Miami). Florida birds nest Feb.–July in shrubs, palms, or small trees, usually those not native to the U.S.

POPULATION In FL, the range is stable or expanding slightly; in CA, eradication efforts during the 1970s–1980s and increasing urbanization have reduced the population.

KINGLETS Family Regulidae

Golden-crowned Kinglet, female (NY, Oct.)

Structure Kinglets have rounded bodies, medium-length tails, relatively short wings, and small, thin bills.

Plumage Grayish or olive with bright olive-edged flight feathers, two white wing bars, concealed colorful crown patches. Like the Golden-crowned Kinglet, most species worldwide have boldly marked heads, particularly lateral black crown stripes. On the other hand, the Ruby-crowned Kinglet is much more subdued and can confused with vireos.

Behavior Frequently observed in feeding flocks in migration, in low shrubbery or canopy. Insectivorous and in winter feed on insects in buds and under bark, allowing them to winter farther north than other insectivores. When foraging, they tend to quiver or flick their wings, often hovering below leaves to glean insects.

Distribution Six species worldwide: two in North America, one in Eurasia, one in Europe to North Africa, one endemic to the Canary Islands, and one endemic to Taiwan.

Taxonomy All species are in a single genus, *Regulus,* and were formerly classified with the Old World warblers (Sylviidae); recent genetic studies suggest they are not closely related.

Conservation Generally stable.

The two North American kinglet species are tiny forest songbirds that are rather easily identified as they are among the smallest of our songbirds. Breeding in the upper canopy of boreal forest and near spruce bogs, kinglets are generally common in migration and in winter, and are usually encountered in small flocks, often mixing with other songbirds, including chickadeees and, during migration, warblers.

Genus *Regulus*

GOLDEN-CROWNED KINGLET *Regulus satrapa*   GCKI   1

A tiny, thin-billed, wing-flicking insectivore, it has a conspicuously striped head. Polytypic (5 ssp.; 3 in N.A.). L 4" (10 cm)
IDENTIFICATION ADULT MALE: Dull grayish olive above, paler whitish below, head boldly marked with white supercilium, blackish lores and eye line. Yellow crown bordered broadly with black. Orange in center of yellow crown most visible during display or when bird is agitated. Black at base of secondaries, contrasting with white wing bar. **ADULT FEMALE:** Similar to male, no orange in crown. **IMMATURE:** Somewhat pointier tail feathers than adult; a few males may lack orange in crown.
GEOGRAPHIC VARIATION Western subspecies *(apache* and *olivaceus)* slightly

smaller, somewhat brighter, with longer white supercilium, and longer bill than widespread eastern satrapa. Within the West, more westerly subspecies *apache* is larger and paler, than *olivaceus.*
SIMILAR SPECIES Ruby-crowned Kinglet, readily distinguished by head pattern and call.
VOICE CALL: When flocking, a very high, sibilant jingling, *tsii tsii tsii.* Also, a quiet, high single note, *tsit,* and a thin sibilant *seee* similar to call of Brown Creeper. **SONG:** An extended version of the call, becoming louder and chattering toward the end; *tsii tsii tsii tsii tiii djit djit djit djit.*
STATUS & DISTRIBUTION Common. **BREEDING:** Mainly boreal forests, a few in mixed or deciduous forests and conifer plantations. Also breeds in central Mexico and Guatemala. **SPRING MIGRATION:** Spring and fall migration difficult to detect in some areas with resident populations. Winter residents depart Gulf States before Apr. Peaks across continent late Mar.–late Apr. **FALL MIGRATION:** Begins mid-Sept. over much of range, peaking in East in Oct. and early

orange-red median crown stripe, bordered by yellow
whitish underparts
yellow median crown stripe
bold whitish supercilium bordered by black on both sides

Nov. **WINTER:** Many remain in breeding range through winter. Primarily south of Canada and north of Mexico in a wide variety of habitats. Numbers in southern CA and the Southwest (always small) vary from year to year.
POPULATION Breeding range increasing in the East and the Midwest due to plantings of spruce and pine. Adversely affected by logging.

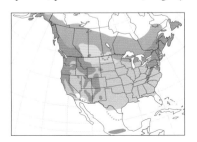

RUBY-CROWNED KINGLET *Regulus calendula*   RCKI ■ 1

This tiny, thin-billed, wing-flicking insectivore is grayish olive with a conspicuous broken white eye ring. Polytypic (3 ssp.; 2 in N.A.). 4.2" (11cm)
IDENTIFICATION ADULT MALE: Bright greenish gray above, paler below; broad, teardrop-shaped white eye ring, slightly broken at the top (especially) and bottom; lores olive. Ruby crown patch visible only during display, or when bird is agitated. Black at base of secondaries, contrasting with white lower wing bar. **ADULT FEMALE:** Similar to male, no ruby crown patch. **IMMATURE:** Somewhat pointier tail feathers than adult, a few males may have orange, yellow, or olive crown patch.
GEOGRAPHIC VARIATION Northwestern subspecies *grinnelli* (breeding coastal southeastern AK and BC) is slightly darker and more brownish than widespread nominate *calendula*. Subspecies *obscurus* on Guadalupe I., MX is extinct.
SIMILAR SPECIES Compare with Hutton's Vireo.
VOICE CALL: Most frequently heard is a husky two-syllable *ji-dit*. **SONG:** Often heard in migration; begins with 2–3 very high-pitched notes, abruptly changing to a rich, and surprisingly loud warble: *tsii tsii tsii chew chew chew teedleet teedleet teedleet.*
STATUS & DISTRIBUTION Common. **BREEDING:** Breeds in boreal spruce-fir forests, preferably near water and especially in black spruce bogs. Breeding begins immediately when females arrive in early May. **SPRING MIGRATION:** As late as early May in Mexico. Mar. to early May, in southern U.S. Early Apr. to late May, peaking late Apr. to early May in central and northeastern U.S. **FALL MIGRATION:** Begins mid-Sept.; peaks late Sept. to mid-Oct.; continuing through mid-Nov. Arrives as early as late Sept. in FL and Mex. **WINTER:** Not as hardy as Golden-crowned, and winters farther south in a broad range of habitats. Primarily southern and western U.S. through Mex. to Guatemala. **VAGRANT:** Yucatán Peninsula of Mex., Bahamas, western Cuba, Jamaica (sight rec.), Greenland (2 recs.) and Iceland (2 recs.). Many records from ships off Atlantic coast and single recs. for Ireland and the Azores.
POPULATION Breeding areas in the western United States may be adversely affected by logging and wildfire.

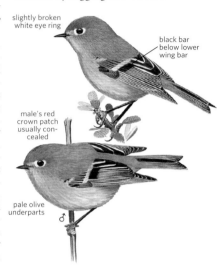

slightly broken white eye ring

black bar below lower wing bar

male's red crown patch usually concealed

pale olive underparts

LEAF WARBLERS Family Phylloscopidae

Arctic Warbler (AK, June)

This is a uniform group of small insectivorous woodland species found across Eurasia as well as in Africa and Indonesia; they are only marginal in North America. Seventy-three species are currently recognized in two genera, *Phylloscopus* (62) and *Seicercus* (11). Identification in both genera is very difficult between some species; vocalizations (song and contact notes) are often the best characters.

Structure These small insectivores have short, fine bills and slender legs. The wings are moderate in length to rather long and pointed in the more migratory species.
Behavior Leaf warblers are very active foliage gleaners that also take insect prey on short aerial sallies; *Phylloscopus* warblers are mainly arboreal, while *Seicercus* warblers mainly inhabit undergrowth. Songs vary from simple to monotonous trills; rattles; or high, thin whistles.
Plumage *Phylloscopus* is very conservative morphologically, all species being olive-gray, greenish or brownish above and paler below, often with a yellow, buff, or pale olive wash; most show a dark transocular line and pale supercilium, and some species show one or two thin wing bars, lateral crown stripes and a pale central crown stripe, and/or a pale rump patch. The *Seicercus* warblers of Southeast Asia are mostly brighter yellow; most show a bold, pale eye ring; many show tawny or chestnut head markings; and some have bold wingbars.
Distribution Leaf warblers breed in Africa, Eurasia, and the Indonesian and Philippine archipelagos, with one species occurring east to New Guinea; they are most diverse in the Palearctic, especially at higher elevations

in eastern Asia and the Himalaya region. One Asian species (Arctic Warbler) breeds east across the Bering Strait to central Alaska. Most temperate-zone Palearctic species are strongly migratory, wintering in Africa, India, and Southeast Asia. Six species (Dusky, Wood, Willow, and Yellow-browed Warblers, Common Chiffchaff, and Pallas's Leaf-Warbler) have been recorded as vagrants in western and northern Alaska; Dusky and Arctic Warblers have also been recorded in fall south to California; and there is a Baja California record of Yellow-browed Warbler.

Taxonomy The leaf warblers were formerly considered part of the "Old World Warbler" family Sylviidae, but recent molecular data have led taxonomists to split that huge and ill-defined family into many smaller families. Within the Phylloscopidae there has been an explosion of recently recognized species, particularly in eastern Asia and the Himalaya region; this follows intensive study by ornithologists incorporating vocal and molecular data after traditional morphological studies had failed to appreciate the diversity within the family. The Arctic Warbler may soon be split into three species, at least two of which occur in North America.

Conservation Most species appear to have stable populations, but some range-restricted Asiatic species are considered vulnerable or near-threatened.

Genus Phylloscopus

WILLOW WARBLER *Phylloscopus trochilus* WILW ■ 5

The Willow Warbler is a vagrant from Eurasia. It occasionally wags its tail, usually immediately after landing. Polytypic (3 ssp.; presumably *yakutensis*, but no specimen). L 4.5" (11 cm)

IDENTIFICATION Uniform green to brownish olive above, without wing bars. A narrow, pale supercilium is offset by a thin, dark eye line. Below, whitish; some birds have pale yellow wash on throat and breast. Legs are typically pale.

GEOGRAPHIC VARIATION The subspecies *yakutensis* (breeds eastern Siberia and Russian Far East), can have very little green in plumage and might have dark legs. Nominate *trochilus* and *acredula* from Europe, greener above and more yellow below.

SIMILAR SPECIES Most similar to Common Chiffchaff. The mainly eastern subspecies, *tristis,* is browner than nominate Chiffchaff, and looks quite similar to easterly *yakutensis* of Willow Warbler. The best character is the longer primary extension of Willow, which is at least 75 percent of the length of the exposed tertials, compared to shorter projection of Chiffchaff, which is approximately 50 percent. Otherwise, auriculars of Chiffchaff are uniformly dark, whereas they are slightly paler on Willow. Legs are usually paler in Willow, but eastern birds can have dusky legs. Arctic Warbler is longer winged, greener overall, with a bolder supercilium, a pale wing bar, and a larger bill. Tennessee Warbler has a different shape, more pointed bill, and contrastingly white undertail coverts.

VOICE CALL: A disyllabic *hoo-eet.*

STATUS & DISTRIBUTION Breeds in Eurasia, winters in Africa. In recent years recorded nearly annually from St. Lawrence I., AK in early fall (10+ recs.); also recorded from Pribilofs. AK sightings have been of rather bright birds (dorsally quite greenish) unlike descriptions for *yakutensis.*

overall plain with no wing bars

St. Lawrence Island birds have been more olive above, yellowish below

small bill

adult
yakutensis

long primary projection

COMMON CHIFFCHAFF *Phylloscopus collybita* COMC ■ 5

Common Chiffchaff is a widespread and common summer resident throughout much of Eurasia. Polytypic (6 ssp.; *tristis* in N.A.). L 4.5" (11 cm)

IDENTIFICATION Active, medium-size leaf warbler with a fine dark bill, dark legs, and somewhat short primary projection (equal to about 50 percent of the length of the exposed tertials); four outer primaries (P5–P8) are emarginated (three in Willow Warbler). Subspecies *tristis* has a gray-brown crown and mantle (lacking green tinge). Moderately prominent supercilium is whitish, as is the breast (lacking strong yellowish tones). Ear coverts often tinged rusty. Wing and tail feathers often narrowly edged yellow-green. Fall immatures may show a short, pale wing bar.

GEOGRAPHIC VARIATION Eastern subspecies *tristis* ("Siberian Chiffchaff") has white to pale buff underparts and brownish upperparts that separate it from most subspecies found in Europe and Asia Minor.

SIMILAR SPECIES Eastern *tristis* is similar to *yakutensis* Willow Warbler; see that species for details on separation.

VOICE CALL: For ssp. *tristis,* typically a plaintive, flat *peep;* less often *swee-oo* similar to European subspecies.

STATUS & DISTRIBUTION "Siberian Chiffchaff" breeds in coniferous taiga woodlands from the Urals east to the Kolyma River; winters chiefly from the Middle East to India; casual migrant to Japan. **VAGRANT:** Two records for N.A., both at Gambell, St. Lawrence I., AK (Jun. 6–7, 2012, and Sept. 22–23, 2013). This subspecies is also a regular stray, mainly fall, to western Europe.

POPULATION Stable; least concern.

gray-brown crown and mantle

plain wings

"Siberian Chiffchaff"
adult *tristis*

whitish buff underparts

short primary projection past tertials

dark legs

WOOD WARBLER *Phylloscopus sibilatrix* WOWA ▪ 5

The Wood Warbler is a vagrant from Eurasia. Monotypic. L 5" (13 cm)
IDENTIFICATION Perhaps the most

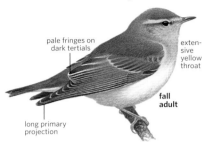

pale fringes on dark tertials

extensive yellow throat

fall adult

long primary projection

colorful of the leaf warblers, Wood Warbler has bright green upperparts and no wing bars. Feather edges to wing coverts are usually yellow, as are dark-centered tertials. A bold yellow supercilium is offset by a dark eye line. Below, throat and upper breast are a clear lemon yellow; remainder of underparts are cleanly white. Note very long primary projection. Both bill and legs are pale, usually pinkish.
SIMILAR SPECIES Not similar to the other potential *Phylloscopus,* as its throat and breast are distinctly yellow, contrasting with otherwise

white underparts. Arctic Warbler, the only likely congener, usually shows a wing bar and lacks yellow throat and breast of Wood Warbler.
VOICE CALL: A plaintive *tew.*
STATUS & DISTRIBUTION This species breeds in Eurasia, but no further east than central Russia; winters in Africa. Two records from AK: Oct. 9, 1978, on Shemya Island (Aleutians) and Oct. 7, 2004, on St. Paul Island. A few fall records from Japan suggest a weak pattern of dispersal east of the breeding range during the fall.

DUSKY WARBLER *Phylloscopus fuscatus* DUWA ▪ 4

A vagrant from Asia, this brown *Phylloscopus* spends more time close to, or on, the ground, but will feed in trees on occasion. It works its way through low bushes or on the ground, flicking its wings and calling regularly. Polytypic (3 ssp.; nominate in N.A.). L 5.5" (14 cm)
IDENTIFICATION Upperparts of this warbler are dusky brown and rather uniform; this species lacks wing bars. Bill is short, thin, and dark; legs are dark. Dusky Warbler exhibits a subtle, but distinct face pattern: A dark eye

line includes lores and contrasts with a pale supercilium that is invariably dingy whitish in front of eye. There is a faint but noticeable white eye ring. Underparts are dingy, darkest across breast where it might be dusky, but otherwise creamy with buffy brown wash on flanks and undertail coverts. Unlike that of most *Phylloscopus,* tail is slightly rounded.
SIMILAR SPECIES Arctic Warbler, the most likely *Phylloscopus* in our area, differs in several ways from Dusky. Arctic is green, has a wing bar on greater coverts, has a thicker bill that is usually pale on lower mandible, has pale mottling in auriculars and has a different call. Dusky is most similar to Radde's Warbler *(P. schwarzi),* which has not been recorded in our area. Radde's has a thicker bill that is noticeably pale on lower mandible. Radde's often has yellow or olive tones, which Dusky lacks,

and undertail coverts of Radde's are particularly contrasting and yellowish buff. Supercilium in front of Radde's eye is more muted and yellowish. Finally, the calls of the two birds differ.
VOICE CALL: A hard *tschik,* recalling the *chip* of Lincoln's Sparrow.
STATUS & DISTRIBUTION Asian species, casual on islands off western AK, and in fall, off south coastal AK and in CA; there are two fall records from Baja California.

long pale supercilium

small fine bill

brownish upperparts

PALLAS'S LEAF-WARBLER *Phylloscopus proregulus* PALW ▪ 5

In plumage, structure, and behavior, this tiny denizen of taiga forest of eastern Asia bears remarkable resemblance to kinglets (family Regulidae). Monotypic (formerly 5 ssp.). L 3.5" (9–10 cm)

yellow median crown stripe

yellow rump

tiny bill

fall immature

IDENTIFICATION A tiny, kinglet-size bird, most reminiscent in North American context of Golden-crowned Kinglet, with plump body and short tail. Overall, patterned contrastingly in green, brown, yellow, and white. Boldly striped head distinctive, with a bright yellow median crown stripe, dark greenish brown lateral crown stripe, yellow supercilium, blackish eye line; has whitish tertial tips and wing bars. Lemon yellow rump, especially conspicuous during flight, when the bird habitually flits and hover-gleans around vegetation, much as kinglets do.

GEOGRAPHIC VARIATION Most authorities now treat Pallas's as monotypic. Current treatment of leaf-warblers in the complex that has included Pallas's elevates many former subspecies to full species.
SIMILAR SPECIES Most likely to be confused with Chinese Leaf-Warbler or other taxa recently split from the Pallas's complex (Gansu, Lemon-rumped, Sichuan), but all have much less vivid tones overall, with the whitish supercilium. Yellow-browed Warbler and Hume's Leaf-Warbler *(P. humei),* the latter unrecorded

in N.A., are also similar, but both lack the yellow rump and vivid yellow supercilium and median crown stripe.

VOICE CALL: a soft, nasal, rising *chuee*. SONG: A long, ringing series of trills and clear whistles.

STATUS & DISTRIBUTION Breeds in coniferous and mixed taiga woodlands from in northern Mongolia and southern Siberia east to Sakhalin I. and northeastern China. Winters primarily in southeastern China and northern Vietnam, but hundreds or more now also winter regularly in Europe and Great Britain. One photographed at Gambell, St. Lawrence I., AK (Sept. 25–26, 2006), represents the only N.A. record.

POPULATION Considered widespread and common in its large range.

YELLOW-BROWED WARBLER *Phylloscopus inornatus*   YBWA ■ 4

This bird is a vagrant from Asia. Monotypic (most authorities now recognize the polytypic Hume's Warbler, *P. humei*, as a separate species). L 4.5" (11 cm)

IDENTIFICATION A small leaf warbler, greenish olive above, with an obvious supercilium and wing bars. Bill is small and thin. Greater coverts have a large wing bar; median covert bar is smaller and less distinct, particularly when worn. Tertials are dark, with pale edges. There is a dark patch at base of secondaries, like that shown by Ruby-crowned Kinglet.

SIMILAR SPECIES Boldly marked fall birds are distinguished by bold wing bars, unlike other expected *Phylloscopus*. Arctic Warbler is larger, with a thicker bill, and has one thinner bar on greater wing coverts.

VOICE CALL: A high, ascending *swee-eet*.

STATUS & DISTRIBUTION Breeds in Asia and winters in southern Asia (mostly Southeast Asia), with a well-established pattern of vagrancy to Europe and the Middle East. Casual to western Aleutians and Bering Sea islands (fewer than 10 recs.).

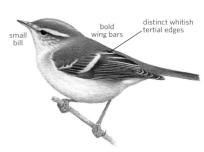

small bill — bold wing bars — distinct whitish tertial edges

Accidental in winter to Baja California Sur, Mexico.

ARCTIC WARBLER *Phylloscopus borealis*   ARWA ■ 2

A species of Eurasia, the Arctic Warbler breeds in Alaska. Recent studies have revealed that the Arctic Warbler may in fact comprise three separate cryptic species. These studies are based on disctinct molecular and vocal differences. In North America, the breeding subspecies *kennicotti* is closely allied to the widespread (in Eurasia) *borealis*, but migrants from the Aleutians and perhaps elsewhere will belong to another species. Polytypic (4 ssp.; 2 in N.A.). L 5" (13 cm)

IDENTIFICATION Arctic Warbler's long, yellowish white supercilium often curves upward behind eye, and contrasts with a dark eye line. Auriculars are mottled, not uniform. Upperparts are olive, and greater coverts have a thin, whitish wing bar. There is occasionally an indistinct wing bar on median coverts; long primary extension. Underparts are whitish, with some brownish olive on sides and flanks; they might be more yellowish in fall. Bill is rather large and extensively pale; legs and feet are straw-colored.

GEOGRAPHIC VARIATION Alaskan breeders are small-billed *kennicotti*, closely allied to nominate *borealis*. Specimens of migrants from western Aleutians are either *xanthrodryas* (breeding in Japan on Honshu, Shikoku, and Kyushu) or more likely *examinandus* (breeding in southern Kamchatka north to at least 56°, Sakhalin, the Kuril Is., and part of Hokkaido); both are bigger billed and more yellow below than *kennicotti*. Full species status has been proposed for *examinandus* and *xanthrodryas* (vocalizations, including songs and call notes, strongly differ). Under this arrangement, *kennicotti* would remain a subspecies of *P. borealis*.

SIMILAR SPECIES More olive than Dusky Warbler, with a larger bill and at least one wing bar. Willow Warbler lacks a wing bar and has a thinner bill.

VOICE CALL: A buzzy *dzik*. SONG: A long, loud series of toneless, buzzy notes.

STATUS & DISTRIBUTION Eurasian species breeding east to AK. Fairly common in western and central

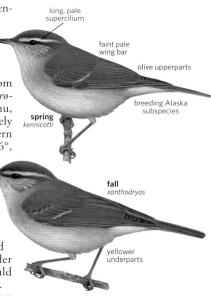

long, pale supercilium — faint pale wing bar — olive upperparts — breeding Alaska subspecies

spring *kennicotti*

larger bill — fall *xanthrodryas* — yellower underparts

AK; nonbreeding subspecies have occurred as migrants on western Aleutians (not *kennicotti*). There are five fall records for CA and an additional record for Baja California. The subspecies of these records has not been determined, but several were banded, so mensural data is likely available.

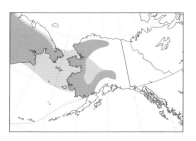

SYLVIID WARBLERS Family Sylviidae

Wrentit (CA, Oct.)

The "sylviids" had long been a huge and dominant assemblage of small "warbler-like" insectivorous birds found throughout the Old World. The family, as formerly constituted, was the legacy of taxonomies based on superficial morphological characters, many of which have been proved to be less than reliable indicators of evolutionary relatedness. There is no current consensus on the composition of this family, but it is either dominated by or virtually restricted to the 25+ warblers of the genus *Sylvia,* one of which has been recorded as a vagrant in western Alaska. Various other genera that had long been considered babblers in the family Timaliidae may also belong in the Sylviidae, including the enigmatic Wrentit of western North America (which is considered a sylviid by the American Ornithologists' Union).

Structure Sylviids are slender-billed and medium- to long-tailed warblers. The Asian parrotbills, sometimes placed within the Sylviidae, have remarkably thick bills with a strongly arched culmen, and long, graduated tails.

Behavior Foliage-gleaning; food is primarily arthropods, but most species take small fruits at least part of the year and may even be important seed-dispersal agents. Songs range from rich warbling with fluty tones to scratchier jumbles in *Sylvia.* Open cup nests.

Plumage *Sylvia* are primarily gray above and vary from whitish to gray or vinaceous below; many species have black crowns or other black markings. In many species the iris may be red, yellow, or white, and some have bright red orbital rings.

Distribution The genus *Sylvia* (including *Parisoma*) is found in Europe, Africa, and western Asia; most diverse in the Mediterranean region. Most winter in southern Europe or Africa. The easternmost Eurasian species, the Lesser Whitethroat, has been recorded once in western Alaska. Most species occupy scrub habitats and broadleaf woodlands. Other "babblers" currently placed in the Sylviidae by various authorities are primarily Asian.

Taxonomy The definition and taxonomic composition of the Sylviidae is definitely a "work in progress," and various genera are likely to be transferred into or out of the family in future years. The enigmatic Donacobius of tropical South America seems closely related to the Sylviidae, but uncertainties abound, and it is now usually given family rank. As many as 70 species in some 17 genera are considered sylviids by recent authors; many of these were previously placed in with the babblers (Timaliidae) or in smaller families (e.g., Paradoxornithidae for the 20 species of parrotbills). The sole New World breeding species, the Wrentit, was formerly afforded unique family status (Chamaeidae), then merged into the babbler family Timaliidae. The timaliids, ultimately, proved to be polyphyletic, with some of its former members, including the Wrentit, transferred to the Sylviidae.

Conservation No threatened species.

Genus *Sylvia*

LESSER WHITETHROAT *Sylvia curruca*   LEWH ■ 5

The Lesser Whitethroat, a Eurasian species, is an accidental vagrant to Alaska. Polytypic (7–9 ssp.). L 5.5" (14 cm)

IDENTIFICATION Slightly longer tailed than the leaf warblers in *Phylloscopus,* with white in outer rectrices. Crown is gray with a dark mask from base of bill to auriculars. Wings and back are warm brown, with darker centers to tertials, but otherwise unmarked. Tail is dark from above, and square. Below, throat and undertail coverts are white; flanks and sides have a tan-brown wash.

GEOGRAPHIC VARIATION Up to nine subspecies, but an uncertain number of species because some authorities have split the "northern" group from the "desert" group. The northern subspecies, *blythi* and *curruca,* have more well-defined face masks and more contrasting brown wings compared to other subspecies from farther south.

SIMILAR SPECIES While identification within the whitethroat complex is difficult, there are no challenges in separating this species from North American birds.

VOICE CALL: A sharp *tik,* often repeated, and various chattering notes.

STATUS & DISTRIBUTION Breeds in Eurasia and winters from Africa to south-west Asia. One record: Sept. 8–9, 2002, from St. Lawrence I., thought to pertain to the Northern complex. A few fall records from Japan and Korea suggest a weak pattern of vagrancy.

gray crown

dark mask

1st fall
blythi

Genus *Chamaea*

WRENTIT *Chamaea fasciata*   WREN ▪ 1

Wrentits are curiously long-tailed and skulking denizens of dense, non-forested chaparral on the West Coast. Their taxonomic placement has always been a conundrum; in the recent past they were placed with babblers (Timaliidae), now with the sylviid warblers. Polytypic (5 ssp.; all in N.A.). L 6.5" (17 cm)

IDENTIFICATION Recognized by distinctive round, fluffy body and long, round-tipped tail usually cocked at an angle. Is usually heard rather than seen. Whitish eye; short and stout bill; and lightly streaked buffy or cinnamon underparts are distinctive. Muted plumage is alike in all ages and sexes.

GEOGRAPHIC VARIATION Subspecies not clearly differentiated: Birds from northern, wetter areas darker, more cinnamon below; subspecies from southern dry sites grayish above, dull buffy-grey below.

SIMILAR SPECIES Unmistakable.

VOICE CALL: Wooden-sounding, rattled *churrrrrrr,* or longer *krrrrrrrrrrr.* **SONG:** Male song a *pit-pit-pit-pit-pit-pit-trrrrrrrr,* which has a bouncing-ball quality as notes speed up into the trill. Female song similar, but lacks the trill, and the *pit* notes are spaced evenly.

STATUS & DISTRIBUTION Common. **YEAR-ROUND:** Found in dense chaparral or Huckleberry-Salal thickets from sea level to 6,000 feet. Totally resident within range. The Columbia River marks the northern end of their range; no records for WA. **POPULATION** Stable.

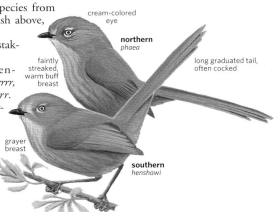

REED WARBLERS Family Acrocephalidae

Sedge Warbler

Formerly in the Sylviidae, these are medium- to large-size, rather plain-looking and difficult to identify warblers that inhabit reed beds and other marshes, with a few species in dry woodlands and brushlands. As currently constituted, this family includes 62 species in 5 genera, of which *Acrocephalus* (42 species) dominates. One species, the Sedge Warbler, has been recorded as a vagrant in western Alaska; another is resident in the leeward Hawaiian Islands.

Structure Moderately large "warblers" with rather long and strong bills, sloping foreheads, and rounded (*Acrocephalus*) or squared (*Iduna, Hippolais*) tails.

Behavior Most species skulk within reed beds and other dense habitats, being most detectable when singing. Songs consist mostly of rapid jumbled chatters and churring notes, often incorporating mimicry and in many species given at night.

Plumage Generally brownish to grayish above and paler gray or whitish below, washed with buff, yellow, or brown. A few species of *Acrocephalus* have strong head patterns and dorsal streaking. African "yellow warblers" (now in the genus *Iduna*) are bright yellow below and relatively short-winged and long-tailed.

Distribution They are found widely in the Old World, from western Eurasia and Africa (plus Madagascar) east to the Australasian region and on some oceanic islands of the Indian and south and central Pacific Ocean. Eurasian species are mostly migratory, many wintering in Africa.

Taxonomy Once again molecular work has helped sort out the relationships among the "Old World warblers," confirming the monophyly of the genera comprising this family. The African genus *Chloropeta* is now merged into *Iduna* (which itself was split off from the Eurasian genus *Hippolais*).

Conservation Many species are decreasing through habitat loss and degradation. Several oceanic island species are considered endangered. Europe's Aquatic Warbler has long been of conservation concern and is regarded as Europe's most endangered songbird.

SEDGE WARBLER *Acrocephalus schoenobaenus*   SEWA ■ 5

bold supercilium and dark eye line

faint streaks on back

1st fall

Sedge Warbler is an Old World warbler of the genus *Acrocephalus,* an infamously difficult group of birds to identify, but Sedge Warbler is one of the most distinctive and readily identified of the *Acrocephalus*—though its retiring habits can make it difficult to study. Monotypic. L 4.5–5" (11.5–13 cm)

IDENTIFICATION Wedge-shaped tail and long primary projection quite distinctive. Buffy olive brown above, with darker brown streaks in mantle; rump and uppertail coverts a contrasting rich rusty brown, most readily observed in flying bird. Head distinctly patterned with a long, bold supercilium; dark eye stripe; brownish auriculars; and dark-streaked crown (often showing indistinct brownish median stripe). **ADULT:** Whitish below, with buff tones in breast and flanks. **IMMATURE:** Shows light streaking or stippling on breast.

SIMILAR SPECIES The most similar species, Moustached Warbler *(A. melanopogon),* occurs east to Central Asia and is unrecorded in N.A.; it is readily distinguished from Sedge by shorter primary projection, whiter and differently shaped supercilium, more richly colored upperwings, and darker crown. The very rare, little-known, highly migratory and likely endangered larger Streaked Reed Warbler *(A. sorghophilus)* of eastern Asia (breeds northeast China, possibly north to Amurland, Russian Far East) is more diffusely patterned on the back, but is otherwise quite similarly plumaged.

VOICE CALL: A dry *chirr* or *errrrr.* **SONG:** a long, loud rollicking set of trills, whistles, and staccato notes, often included mimicked calls of other species.

STATUS & DISTRIBUTION BREEDING: Dense marsh vegetation from northwestern Europe to western Siberia and northwestern China. **WINTER:** Sub-Saharan Africa. **VAGRANT:** Only N.A. record comes from Gambell, St. Lawrence I., AK (Sept. 30, 2007). The only other record of any reed warbler (genus *Acrocephalus*) in N.A. also comes from Gambell (Sept. 9, 2010), a bird initially identified as a Blyth's Reed Warbler *(A. dumetorum)* but lacking enough documentation to make a positive identification beyond the genus.

POPULATION Some population declines over much of this species' large range.

GRASSBIRDS Family Megaluridae

Middendorff's Grasshopper-Warbler (Japan, June)

This Old World family includes small to medium-size, mostly skulking birds that often stay hidden near the ground in grassy thickets. The two species recorded in North America are casual migrants to the westernmost Aleutians and other islands in the Bering Sea region. The family includes 56 species in 10 genera.

Structure Grassbirds are rather long- and stout-billed and long-legged as insectivorous "warblers" go; they have strongly graduated tails and dense, long undertail coverts. They range from about 4.5 to 9 in.

Behavior These are retiring, and in many cases—especially for species within the genus *Locustella*—extremely secretive birds that skulk well within vegetation and are reluctant to fly. The two Australasian songlarks, part of genus *Megalurus,* inhabit more open areas and are considered beautiful songsters (with elaborate flight songs). The songs of most *Locustella* are mechanical and repetitious, often insect-like in quality.

Plumage Nearly all species are basically nondescript, clad in shades of brown above and usually somewhat paler below; some species are streaked above, and a few have ventral streaking or diffuse spotting (often with strongly marked undertail coverts).

Distribution Grassbirds are found from Africa and Eurasia east to Australasia. They inhabit brushy and scrubby areas, forest interiors, swamps, marshes, and tall grasses. The two species recorded in Alaska have the most northeasterly Asian breeding ranges of any megalurid species (to the Kamchatka Peninsula); one of these (Lanceolated Warbler) has reached California once.

Taxonomy Like so many small insectivorous birds of the Old World, the grassbirds were formerly placed in the now-splintered "mega-family" Sylviidae. Asiatic species of *"Bradypterus"* are now placed in the genus *Locustella,* so the former genus (11 species) is now restricted to Africa. The most species-rich genus is *Locustella* (24 species); many of these species are migratory, some highly migratory. Most authorities now consider the valid name of this family to be Locustellidae.

Conservation Populations of most species are stable; the subspecies of Fernbird *(Megalurus punctatus rufescens)* from the Chatham Is. off New Zealand is extinct.

MIDDENDORFF'S GRASSHOPPER-WARBLER *Locustella ochotensis*   MIGW ■ 4

If not singing from an exposed perch, this species, a vagrant from Asia, skulks in grasses and can be difficult to see. Monotypic (formerly conspecific with *L. pleskei* and *L. certhiola*). L 6" (15 cm)
IDENTIFICATION A large, chunky, brown warbler with a relatively thick bill. Brown upperparts look more mottled than streaked; rump and uppertail coverts are lighter and warmer brown. A whitish supercilium is rather bold and offset by a dark eye line. Underparts are whitish, with buff flanks and sides of breasts, with streaks at sides of breast. Graduated tail is brown, with

a dark subterminal mark and a whitish tip. As plumage wears, underparts become whitish and lose their streaks. **FALL:** Plumage more colorful when fresh, with much of underparts washed yellowish buff; supercilium buffy. **FLIGHT:** A frustrating species that is often flushed from underfoot, but otherwise not seen well as it dives back into grass. In flight this is a large warbler, with an obviously warmer rump and tail; tail looks wedge-shaped and has a pale tip.
SIMILAR SPECIES The other congener, Lanceolated Warbler, is smaller, more obviously streaked above and below,

and lacks pale tip to tail. There are no North American species that should be confused with Middendorff's, although Pallas's Grasshopper-Warbler, *L. certhiola,* from farther west in Asia and an annual fall vagrant to the U.K., is very similar though more strongly streaked on the back.
VOICE CALL: An insectlike *trat-at-at-at.*
STATUS & DISTRIBUTION Asian breeder, winters primarily in the Philippines and vicinity. Casual to westernmost Aleutians mostly fall; also recorded (mainly fall) in Bering Sea region from Nunivak I., Pribilofs, and St. Lawrence I.

LANCEOLATED WARBLER *Locustella lanceolata*   LANW ■ 5

Lanceolated Warbler is a vagrant from Asia. Quite secretive, it runs and walks on the ground. Like other *Locustella,* it can be extraordinarily difficult to see if not singing. Its movements on the ground are certainly mouselike. One clue to their presence are agitated chiffing or chacking notes when alarmed. Watch also for flicking wings. On its winter ground in southeast Asia, often found in quite dry situations. Singing birds perch up on grasses; otherwise unlikely to be seen much off the ground. Monotypic. L 4.5" (11 cm)
IDENTIFICATION A midsize warbler

with black streaks on a brown crown, mantle, and rump. Wings are brownish and lack wing bars; tertials are dark, with pale brown edges. There is a pale, thin supercilium on an otherwise streaked face. Below, breast, flanks, and undertail coverts are streaked; ground color might be buff, or wears to whitish. Throat is usually white and generally unmarked.
SIMILAR SPECIES Somewhat similar to the larger Middendorff's Grasshopper-Warbler, but that species lacks extensive streaking above, including on crown, and is streaked primarily on sides of breast. Also, Middendorff's

has a dark eye line, contrasting with a bolder, pale supercilium, and is larger. Finally, tail feathers of Middendorff's have pale tips and a dark subterminal patch. Otherwise, plumage of a Lanceolated Warbler recalls a sparrow or pipit; bill is too thin for a sparrow, and pipits have larger bills and are structurally different, with longer tails that have white.
VOICE CALL: A distinctive, metallic *rink-tink-tink,* delivered infrequently; also an explosive *pwit* and excited rapid series of *chack* of *chiff* notes when disturbed. **SONG:** A thin, insectlike reeling sound, like line moving through a fishing reel.
STATUS & DISTRIBUTION Up to 25 occurred on the Aleutian island of Attu during the spring and summer of 1984. Two more were found at Attu on June 2–6, 2000; up to four recorded (with nesting) from Buldir I. in summer 2007. Accidental to St. Lawrence I. (Sept. 2013) and to Farallones Is., CA (Sept. 11–12, 1995). This is an annual fall vagrant to the U.K.

OLD WORLD FLYCATCHERS Family Muscicapidae

Bluethroat, male (AK, June)

The core of this family is a suite of upright perching birds that engage in sallying flights like New World tyrant flycatchers, although these families are not closely related. Of the 15 species recorded from North America, records of most species are from the western Aleutians or other Bering Sea islands in Alaska in late spring, a few in summer and fall. Two species, the Northern Wheatear and Bluethroat, breed in arctic N.A.

Structure Old World Flycatchers have rather flattened bills with a slight hook at the tip and well-developed rictal bristles at the base. Their tails are short and either square or slightly graduated, the wings in most of the migratory species are long, and the legs and feet are short and relatively weak. Many of the chats are ground-inhabiting, with long legs; most have expressive tails (varying from rather short to long), which can be cocked, shivered, or flicked open and shut. All have ten primaries.

Behavior Like tyrant flycatchers, Old World flycatchers sally out after insects and return, often to the same perch. Most species are arboreal to one degree or another. Some use canopies of trees or exposed snags at the tops of trees for hunting perches (much as in some *Contopus* tyrant flycatchers), while others forage lower in the undergrowth or from the ground. Many species engage in wing and tail flicking; others sit still after returning to a perch. They are solitary overall, and their songs tend to be not overly complex. Call notes can be important for identification but are infrequently heard away from breeding grounds in *Muscicapa* (more frequent in *Ficedula*). Wheatears and other chats are ground- and rock-loving; many are conspicuous in open habitats while others skulk within underbrush. One group, the Asiatic forktails, forage along mountain streams.

Plumage In some genera the plumage is subdued, and all post-juvenal plumages look alike, while in others (e.g., *Ficedula*) there is strong sexual and sometimes seasonal and age variation. Chats vary from plain to strikingly colored; many show conspicuous tail markings.

Distribution Widespread in the Old World, throughout Eurasia and Africa. Most European and northern Asian species are strongly migratory.

Taxonomy Molecular studies show that many groups formerly in the thrush family (Turdidae) are closer to Old World flycatchers.

Conservation BirdLife International currently recognizes 26 species as near-threatened, 17 as vulnerable, and 14 as endangered.

Genus *Muscicapa*

GRAY-STREAKED FLYCATCHER *Muscicapa griseisticta*   GSFL ■ 4

spots form distinct streaks on white underparts

spring

very long primary projection

1st fall

All *Muscicapa* (23 sp.) are dull, and plumages except juvenal and first fall, are alike. Formerly called the Gray-spotted Flycatcher, the Gray-streaked is the most regularly occurring Old World flycatcher to North America (e.g., 27 recorded from Attu I. on June 2, 1999).Monotypic. L 6" (15 cm)

IDENTIFICATION ADULT: Moderately large and very long winged, wing tips projecting nearly or actually to tail tip. Distinct stripes on sides of throat and well defined dark ventral streaks against a white background color; grayish brown above with somewhat ill-defined eye ring and darker streaks visible on forehead. Undertail coverts are always whitish and unmarked. JUVENILE: Palely spotted with dark bars above.

SIMILAR SPECIES Pattern of markings below distinctive from Dark-sided Flycatcher (see that entry), though note that species is variable in this regard. Markings below are more distinctive than in Spotted Flycatcher and note unmarked white throat and much longer wings. Crown of Spotted more extensively streaked.

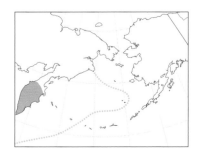

VOICE CALL: A thin *seet* and a fairly loud *speet-teet-teet* have been described. SONG: Perhaps not described. STATUS & DISTRIBUTION Uncommon in a variety of woodland habitats.

BREEDING: Southeastern Siberia to Sakhalin and southern Kamchatka. WINTER: From Philippines and northern Borneo to western New Guinea. MIGRANT: Eastern China, rarely Vietnam.

VAGRANT: To western Aleutians and more recently small numbers from St. Paul in the Pribilof Is. Almost all records are for late spring (late May to mid-June); three Sept. recs. in 2004.

ASIAN BROWN FLYCATCHER *Muscicapa dauurica*   ABFL ■ 5

This accidental vagrant to western Alaska is vaguely suggestive of a Least Flycatcher in appearance. Unlike Spotted and Dark-sided Flycatcher adults, Asian Browns have a complete post-breeding molt on the breeding grounds rather than after reaching the winter grounds. This bird tends to perch somewhat more within the canopy of woodlands than other flycatchers. Polytypic (3–4 ssp.; nominate in N.A.). L 5.3" (13 cm)

distinct whitish eye ring

spring *dauurica*

unstreaked underparts

pinkish-based bill

1st fall *dauurica*

shorter primary projection

IDENTIFICATION ADULT: Grayish brown above with distinct white eye ring and supraloral line. Whitish below with grayish wash on sides of breast and, more rarely, diffuse streaks. Primary projection is relatively short, with wing tips falling to only base of tail. Distinct and sharply separated pale base to lower mandible on bill. JUVENILE: Distinct white spots on upper parts, bolder white markings on wing, and some fine dark scaling on breast. GEOGRAPHIC VARIATION Complex. The one Alaskan specimen belongs to the most migratory, northerly *dauurica*. Three to four other subspecies in Southeast Asia, one of which *(williamsoni)* is sometimes considered specifically distinct. SIMILAR SPECIES This species is much less marked ventrally than the other *Muscicapa* flycatchers recorded from Alaska, especially Dark-sided or Gray-streaked. Also wing tips project far less down the tail in this species. Note, too, the distinctly pale-based lower mandible and distinct white eye ring and supraloral line. The other *Muscicapa* are darker billed, and only Dark-sided has an eye ring that approaches the distinctness of Asian Brown.

VOICE CALL: Includes a series of high, thin, sharp notes, and a short, thin, and high-pitched *tzi*. SONG: Similar to Dark-sided but shorter and less pleasant, consisting of short trills, squawky notes and mixed with two- to three-note whistled phrases. STATUS & DISTRIBUTION Common and widespread in Asia. BREEDING: Nests in rather open mixed woodlands. Nominate subspecies breeds from southern Siberia and northern Mongolia to northeast China, Sakhalin, and southern Kuril Islands south to the Himalaya. WINTER: Found in a variety of open woodlands, including parks and gardens, from India, southern China, and throughout mainland Southeast Asia and the Greater and Lesser Sundaes. Other races are less migratory and found in mainland Southeast Asia and northeast Borneo. VAGRANT: Three Alaska records, a specimen from Attu I. (May 25, 1985), one photographed at Gambell, St. Lawrence I. (June 9, 1994), and one photographed on St. Paul I. (Sept. 6–7, 2013). POPULATION Stable as far as is known.

SPOTTED FLYCATCHER *Muscicapa striata*   SPFL ■ 5

A familiar bird of Europe, this species has been recorded only once in North America. It tends to perch at mid-levels in vegetation, but is usually found on prominent perches. Polytypic (7 ssp.). L 6" (15 cm) IDENTIFICATION ADULT: Grayish brown above including head and face, and indistinct whitish eye ring. Fine dark streaking on crown and forecrown. Below, lacks distinct dark malar and pale submustachial (faint dark markings are present) stripes on sides of throat, but does show faint blurry streaks against a pale buffy background on throat and across chest. Bill is mainly dark. Although appearing long winged, wing tips do not extend more than halfway down tail. JUVENILE: Has

prominent buff spotting above and dark mottling below. Fall immature has more prominent edges to tertials and wing feathers and buff tips to inner greater coverts, which form an ill-defined wing bar. GEOGRAPHIC VARIATION Moderate, but differences are largely clinal. Seven subspecies are recognized, the closest to Alaska being the overall rather pale *neumanni*. SIMILAR SPECIES Gray-streaked Flycatcher is much more prominently streaked below against a whiter background color, has streaking on crown restricted to forehead, and has wing tips that extend to about the tip of the tail. Asian Brown Flycatcher is cleaner underneath, has a more prominent whitish eye ring

and supraloral line, and has a more extensive pale base to lower mandible. Dark-sided Flycatcher is overall darker; it has darker sides and flanks with inverted whitish stripe up the midsection, a more prominent whitish eye ring and supraloral line, and

fine streaking on crown

faint eye ring

instinct streaks on breast

adult

has longer primary tip projection. **VOICE CALL:** Includes a thin and squeaky *zeeee*, a sharp *chick*, and a doubled note. **SONG:** Weak, high-pitched, somewhat squeaky and disjointed, often with long intervals between phrases.

STATUS & DISTRIBUTION Breeds in a variety of rather open woodland environments from western Europe north to Scandinavia and south to northwest Africa, east to about Lake Baikal in central Siberia, and south to the Middle East and north and

west Pakistan. **WINTER:** Africa south of the Sahara. **VAGRANT:** One photo-documented record from Gambell, St. Lawrence I. (Sept. 14, 2002) is surprising given the distance from the region of closest regular occurrence in eastern Siberia.

DARK-SIDED FLYCATCHER *Muscicapa sibirica*   DSFL ■ 4

This Asian species, known previously as the Siberian Flycatcher, is not as regular to western Alaska as the Gray-streaked Flycatcher. In Asia it nearly always perches on conspicuous bare branches, often high in the canopy. Polytypic (4 ssp.; nominate in N.A.). L 5.3" (13 cm)
IDENTIFICATION ADULT: Grayish brown above with dark head and rather conspicuous white eye ring; whitish

spring
sibirica

brownish sides and flanks

long primary projection

1st fall
sibirica

below with brownish wash on sides and flanks and some diffuse to fairly obvious streaks on center of breast; whitish half-collar and rather pale buffy-brown supraloral spot; long primary projection that extends mostly or actually to tip of tail. **JUVENILE:** Darker above with pale buff spots/streaks. More streaking ventrally and with buff tips and edges to secondary coverts and tertials.
GEOGRAPHIC VARIATION The Alaska specimens refer to the most migratory subspecies, nominate *sibirica*, which breeds in coniferous woodlands from central Siberia to Kamchatka, Japan, and northeast China and winters in Southeast Asia, including western Indonesia and the Philippines. Three other subspecies—all shorter-distance migrants—breed in the Himalayan region; some winter south to the Malay Peninsula.
SIMILAR SPECIES Like the slightly larger and bigger-billed Gray-streaked, Dark-sided is quite long winged, though usually is slightly shorter. It can be told from Gray-streaked by more diffuse breast streaking against a more brownish rather

than white background color. Note also brownish cast to supraloral spot and white half-collar. Head is more uniformly dark, with no darker evident forehead streaks and eye ring contrasts more. Though diagnostic on Dark-sided, dark markings on undertail are often concealed. Stripes on side of face (malar and submustachial) are not as distinct as in Gray-streaked. Asian Brown Flycatcher has more distinctly bicolored lower mandible, is much paler ventrally, has a bolder eye ring, and has much shorter wings.
VOICE CALL: Very high, short and rapid metallic trills and a single high down-slurred note. **SONG:** A weak and subdued series of high thin notes followed by musical trills and whistles.
STATUS & DISTRIBUTION Uncommon to common. **VAGRANT:** Casual in spring (one mid-Sept. record) to the western Aleutians and to St. Paul Island in the Pribilof Islands, Alaska, mainly from late May to mid-June. Accidental Pt. Barrow, AK, in spring.
POPULATION Stable as far as is known.

Genus *Luscinia*

RUFOUS-TAILED ROBIN *Luscinia sibilans*   RTRO ■ 5

To a North American observer, the Rufous-tailed Robin recalls a miniature Hermit Thrush. It is generally a secretive species, skulking in wet forested gullies in Asia. Also known as "Swinhoe's Robin," the species was first described by 19th-century English naturalist Robert Swinhoe, who described many other Asian vertebrates. Monotypic. L 5.3" (13 cm)
IDENTIFICATION A small but rather plump thrushlike bird, typically found on the ground, but also perches in trees; pale pinkish legs appear disproportionately long for the size. When foraging on the ground this

species stands erect and has a habit of "shivering" rear of body and tail, presumably to startle insect prey. Plain brown upperparts contrast with rufous uppertail coverts and tail and rusty wings. Underparts dingy cream color, brownish olive on flanks, with dusky scaling or scalloping mostly on throat, breast, and flanks. This pattern fainter in immature or worn plumages. Buff eye ring and supraloral area; white of throat incurs a bit below dusky auricular.
SIMILAR SPECIES Few similar species. Other species in genus *Luscinia* should be considered, though their

differences will be apparent with good views. Female and young Siberian Blue Robins are less richly colored above, lack rufous tail, and often show bluish rump, while Red-flanked Bluetails would show unmarked underparts,

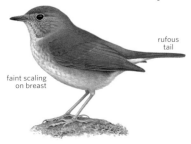

rufous tail

faint scaling on breast

bluish tail, and at least some orange in flank. Nominate Hermit Thrush has similar plumage but is much larger and has proportionately shorter legs, with speckled rather than scaly underparts. **VOICE SONG:** A loud, vibrato series of varied downward trills, given in phrases. **CALL:** A subtle *tuk-tuk*.

STATUS & DISTRIBUTION BREEDING: in well-wooded forest with heavy undergrowth, often near streams, mostly in the taiga zone, from Yenisey east to the Kamchatka Peninsula, south to Sakhalin I. and northeastern China. **WINTER:** From southeastern China through

Indochina; a few west to Thailand. **VAGRANT:** Four N.A. records, all from Alaska: singles at Attu I. (June 4, 2000, and June 8–9, 2008); and singles at St. Paul I. (June 4, 2008, and Sept. 6–7, 2012). **POPULATION** Considered fairly common within its large range.

SIBERIAN RUBYTHROAT *Luscinia calliope*   SIRU ▪ 3

Despite the male's distinctive ruby throat, this species of thrush can be quite difficult to see well, due to its skulking habits and its preference for dense, shrubby vegetation. Up to three subspecies recognized by some, but often regarded as monotypic. L 6" (15 cm)
IDENTIFICATION MALE: Unmistakable. Its crown, nape, and back are grayish brown, with rump, upper-tail coverts, and tail being somewhat rustier. Bright white supercilium; malar stripe bordered above and below by black; and chin and throat are a distinctive opalescent, with a black-spotted lower border. Chest and upper sides are gray, lower sides and flanks are washed cinnamon, and belly and vent area are white. Bill is somewhat long and black, and legs are dark yellow. **FEMALE:** Similar in pattern, but browner than male and lacks black on face, with streaked auriculars and a white throat, some throats with varying amounts of red. Immature females have no red or pink on throat.

SIMILAR SPECIES Immature Bluethroat and Red-flanked Bluetail are potentially confusing. Bluethroat is distinguished by a distinct dark necklace and rufous tail base, and Bluetail has distinct orange sides.
VOICE CALL: A short, hard *chak;* a muffled, creaking *arrr;* and a whistling *ee-lyu.* **FLIGHT NOTE:** Unknown. **SONG:** A variable chatty and calm warbling, with a variety of high whistles and harder, lower notes, and often mimicry of other bird species.
STATUS & DISTRIBUTION Uncommon. **BREEDING:** In dense thickets with or without overstory, from Kamchatka west to the Ural Mountains; winters in southeast Asia. **VAGRANT:** Accidental

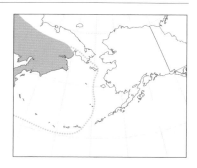

in Europe. In western Aleutians, rare in spring; rare to casual in fall. Casual spring and fall elsewhere in western Alaska (nearly annual on Pribilof Is.). One winter record from ON.

ruby red throat bordered by black and white stripes

uniform-colored tail

adult ♂

1st fall ♂

pink-throated adult ♀

shorter, fainter supercilium than Bluethroat

♀

SIBERIAN BLUE ROBIN *Luscinia cyane*   SBRO ▪ 5

The habitual, rapid tail-quivering behavior of this skulking, terrestrial species is distinctive. Polytypic (2 ssp.; likely *bochaiensis* in N.A.). L 5.5" (14 cm)
IDENTIFICATION MALE: Deep blue above; white underparts. **FEMALE:** Brownish olive above, buffy-white throat, and buff breast and flanks with darker scaling on throat and sides; some have blue in tail base. **IMMATURE:** Cinnamon tips to outer greater coverts. Immature males resemble adult females, but have dull blue uppertail coverts and usually have blue in the scapulars and some wing coverts. Immature females lack blue in tail.

SIMILAR SPECIES Vaguely resembles male Black-throated Blue Warbler. Female's olive-brown upperparts and bicolored bill might suggest a small *Catharus* thrush, but note shorter primary projection. See also Red-flanked Bluetail.
VOICE CALL: A hard *tuck* or *dak.* **FLIGHT NOTE:** Unknown. **SONG:** Loud, rapid, explosive *try try try* and *tjuree-tiu-tiu-tiu-tiu,* usually introduced by fine, spaced *sit* notes.
STATUS & DISTRIBUTION Uncommon coniferous-forest breeder, central Asia east through Kamchatka; winters southern China through S.E. Asia to Borneo. **VAGRANT:** Accidental on Attu I.

AK (May 21, 1985), and St. Lawrence I. (Oct. 2–5, 2012); also accidental to Europe. One disputed report from YT. **POPULATION** Declining due to habitat loss.

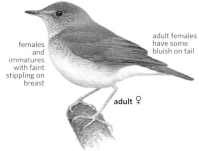

females and immatures with faint stippling on breast

adult females have some bluish on tail

adult ♀

BLUETHROAT *Luscinia svecica*   BLUE ■ 2

This species is a skulker, though singing males can be conspicuous. Polytypic (10 ssp.; nominate in N.A.). L 5.5" (14 cm)
IDENTIFICATION Runs along the ground, usually with tail cocked. **MALE:** Distinctive, though fall through winter extent of blue on throat is reduced by wide, white fringes to those feathers. **FEMALE:** Medium brown upperparts; distinct whitish supercilia bordered above by blackish lateral crown stripe; white malar stripes; black lateral throat stripes; and white throat. Black necklace crosses otherwise whitish underparts, though sides and flanks are washed buff. Bill black with yellowish base. **JUVENILE:** Darker than adult females, with

cinnamon tipping to upperparts' feathers; some older immatures distinguished by thin cinnamon tips to upperwing coverts and, in males in late spring and summer, white admixture in blue throat patch. **FLIGHT:** In all plumages, rufous patches at base of tail are distinctive and easily seen in flight.
SIMILAR SPECIES Similar to Siberian Rubythroat in structure, in skulking, and in terrestrial habits.
VOICE CALL: Common call a dry *tchak;* also a whistled *hiit* and a hoarse *bzru.* **SONG:** Strong, clear, and varied, usually beginning with metallic *ting* notes, speeding up and becoming a jumble of notes, often including imitations of other birds.

STATUS & DISTRIBUTION Fairly common breeder from western Europe east through Russian Far East to northern AK and northern YT. **BREEDING:** Nests in willow riparian in U.S. and Canada; also in other soggy habitats in Old World. **MIGRATION:** Medium- to long-distance migrant. Spring: Arrives in western AK ±25 May. Fall: Most depart AK in Aug., but small numbers are still moving in Bering Strait area through early Sept. Winters primarily from northeastern Africa east through southeastern Asia. **VAGRANT:** Casual to Pribilofs and western Aleutian Is. Accidental in fall to San Clemente I., CA (Sept. 14–18, 2008).

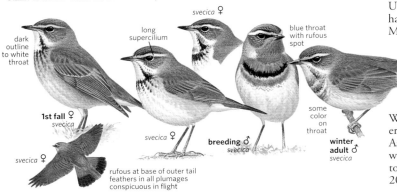

dark outline to white throat

long supercilium

svecica ♀

blue throat with rufous spot

some color on throat

1st fall ♀ *svecica*

svecica ♀

breeding ♂ *svecica*

winter adult ♂ *svecica*

svecica ♀

rufous at base of outer tail feathers in all plumages conspicuous in flight

---

Genus *Tarsiger*

RED-FLANKED BLUETAIL *Tarsiger cyanurus*   RFBL ■ 4

This is one of many Siberian species that has occurred in North America, primarily in the western Aleutians. Some split the two more southerly races as a separate species, Himalayan Bluetail *(L. rufilata),* the adult male of which has a pale blue forehead and supraloral stripe. Polytypic (3 ssp.; nominate in N.A.). L 5.5" (14 cm)
IDENTIFICATION Adult male distinctive.

Female similar in pattern, but head, back, and wings brownish gray; supraloral stripes indistinct and gray; well-defined, thin white eye ring, broken slightly in front by black eye lines. White chin and throat contrast strongly with brownish chest. Most first-year males not readily separable from females.
SIMILAR SPECIES Siberian Blue Robin

somewhat similar, but lacks orange sides and flanks and white supraloral stripes and has pinkish gray legs. Females have brown tails. Compare with male Eastern Bluebird.
VOICE CALL: A whistled short *hweet* or *veet,* often repeated, and a hard, throaty *keck* or *trak,* usually doubled. **FLIGHT NOTE:** Unknown. **SONG:** Fast, clear, melancholy verse, *itru-churr-tre-tre-tru-truur,* repeated at length.
STATUS & DISTRIBUTION Common breeder in taiga and other coniferous forests in northern Asia, west through northwestern Russia to Finland; winters S.E. Asia. **VAGRANT:** Casual spring and fall in western Aleutians, Pribilofs, and St. Lawrence I. (fall); single fall records from Farallones Is. and San Clemente I., CA; and one from Vancouver, BC, in winter. Rare in Europe west of Finland.

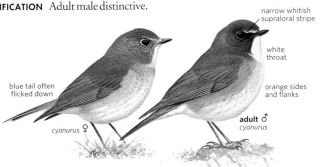

narrow whitish supraloral stripe

white throat

blue tail often flicked down

orange sides and flanks

cyanurus ♀

adult ♂ *cyanurus*

Genus *Ficedula*

This is a large genus of about (taxonomy dependent) 29 species, three of which have been recorded in North America. There is marked sexual, age, and in some cases seasonal dimorphism. Species in *Ficedula* often perch with their wings drooped. They nest in holes or cavities in trees or in nest boxes.

NARCISSUS FLYCATCHER   *Ficedula narcissina*   NAFL ■ 5

Adult males are one of the most stunning members of the Old World Flycatchers. Females and immatures are more subdued. The species has occurred only twice in North America, both events involved males on Attu Island, Alaska. Polytypic (3 ssp.; nominate in N.A.). L 5.3" (13 cm)
IDENTIFICATION *(narcissina)* **ADULT MALE:** Unmistakable with black head and upperparts, except for bold orange supercilum and rump and conspicuous white wing patch. Ventrally extensively orange. **FEMALE:** Brownish above with olive tinge on lower back and rump; whitish below with some mottling on sides of throat and breast and a grayish brown wash on flanks; sometimes with pale yellow wash on belly.
IMMATURE MALE: Resembles female through winter but much more like adult males by spring, except for browner wings and patches of brown elsewhere on upperparts.
GEOGRAPHIC VARIATION Three subspecies. Subspecies *elisae*, with an isolated breeding range in northeast China and wintering primarily on the Malay Peninsula, is often split and recognized as its own species, Green-backed (or Chinese) Flycatcher. Adult

males lack supercilium and are green dorsally; females are duller and lack wing and rump patch, but are otherwise olive above with a contrasting rufous rump and tail and yellowish below. Another green-backed subspecies, *owstoni,* is resident on the southern Ryukyu Islands, Japan, is also sometimes split as Ryukyu Flycatcher.
SIMILAR SPECIES Yellow-rumped Flycatcher *(F. zanthopygia)* a possible stray to western Alaska; adult male resembles Narcissus but has an extended white spur from wing patch and white supercilium. Duller female has wing patch, spur and a yellow rump.

VOICE CALL: *Tink-tink.* Song: A warble with repeated three-syllable notes.
STATUS & DISTRIBUTION BREEDING: Nominate *narcissina* breeds from Sakhalin, the Kuril Islands, and Japan in deciduous, mixed, or coniferous woods with dense undergrowth; winters in forests of the Philippines, sparingly to Borneo. **VAGRANT:** Accidental, two specimen records, both from Attu Island, western Aleutians (May 20–21, 1989, and May 21, 1994).
POPULATION Nominate *narcissina* stable as far as known. Subspecies *owstoni* has suffered severe declines; *elisae* no doubt vulnerable, but little data.

spring ♀

1st spring ♂

♀

olive above

unmistakable black and yellow-orange with white wing patch

adult ♂

reddish-fringed tail

MUGIMAKI FLYCATCHER   *Ficedula mugimaki*   MUFL ■ 5

A long-winged flycatcher that feeds from mid- to upper levels in the canopy of forests in eastern Asia; recorded only once in N.A. Monotypic. L 5.3" (13 cm)
IDENTIFICATION ADULT MALE: Blackish head and upperparts with bold, but short supercilium, white wing patch. Below, extensively orange from chin to belly. **FEMALE:** Grayish brown above, orangish on throat and breast, two thin wing bars. **IMMATURE MALE:** Overall closer to female; usually shows some indication of a supercilium.
VOICE CALL: A rattled harsh *trrrt.*

SONG: A fast and musical warble.
SIMILAR SPECIES No confusion species in N.A.; Rufous-chested *(F. dumetoria),* resident of lowland evergreen forests in southeast Asia, is plumaged similarly.
STATUS & DISTRIBUTION Overall uncommon. **BREEDING:** Central and eastern Siberia and northeastern China in mature mixed forest. **WINTER:** In forests of Southeast Asia and east to Sulawasi. **VAGRANT:** Accidental. An accepted record from Shemya I., western Aleutians (May 24, 1985,

supported only by marginal photos); accidental in U.K.
POPULATION Stable as far as is known.

1st year ♂

extensive orange on underparts in all plumages

long primary projection

TAIGA FLYCATCHER   *Ficedula albicilla*   TAFL ■ 4

This Eurasian Flycatcher has a distinctive black tail, which is often flicked rapidly upward. White

patches on outer tail feathers are best seen in flight. It often drops to catch prey on the ground and forages

mostly from low ot mid levels in the vegetation. Monotypic. L 5.3" (13 cm)

IDENTIFICATION SUMMER ADULT MALE: Distinctive red throat; upperparts brownish, face and chest ash-grayish; whitish eye ring; underparts paler with buffy wash on sides and flanks. Black uppertail coverts and tail with white lateral patches, not usually visible on folded tail. FEMALE & WINTER MALE: Plumage similar but red throat lacking and face and chest less grayish. Fall immatures show faint pale wing bars and tertial edges. The red throat of the breeding-plumages male is acquired as early as Feb.

SIMILAR SPECIES Summer males with red throat are unmistakable; distinctive on all birds is black tail with white patches, which is rapidly raised. Taiga Flycatcher (also known as "Red-throated Flycatcher") was formerly treated as a subspecies of Red-breasted Flycatcher (*F. parva*), which breeds farther west in central and eastern Europe to southwest Siberia and Iran; it winters to Pakistan and northern India. On Red-breasted adult males, red extends to chest; females and immature males very similar, but in all Taiga Flycatchers the longest uppertail coverts are black (brown in Red-breasted), breast is grayer, and bill is more uniformly dark.

VOICE CALL: Given year-round, a brief and fast series of dry rattled notes, *trrrrr*; a high-pitched *tzee*. SONG: An Old World bunting-like song.

STATUS & DISTRIBUTION Common in Asia in open woodland, breeding from eastern European Russia east to Mongolia and Kamchatka and wintering from eastern India and through mainland Southeast Asia. VAGRANT: Casual to western Alaska (mainly spring); 20+ records from western Aleutians; also recorded from Pribilof Is. and St. Lawrence I.; casual to northwest Europe. Accidental to northern CA, along Putah

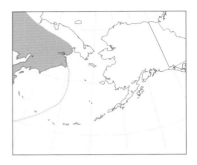

Creek, in Solano and Yolo Counties (Oct., 25, 2006).

POPULATION Stable as far as is known.

reddish throat

breeding adult ♂

spring ♀

grayish wash on breast

1st fall

black uppertail coverts and tail with white oval patches, best viewed in flight, in all plumages

♀

Genus *Phoenicurus*

COMMON REDSTART *Phoenicurus phoenicurus*   CORE ▪ 5

On this familiar European species, the reddish tail is most readily apparent in flight. Polytypic (2 ssp.; likely nominate in N.A.). L 5.5" (14 cm)

IDENTIFICATION A slim, thrushlike bird with alert demeanor and upright stance; frequently quivers tail, especially just after perching. ADULT: Male in breeding plumage: upperparts a soft gray, face and throat black set off from gray by white supraloral; breast and flanks a rich chestnut orange; tail rich rufous orange except for brownish gray central rectrices. Adult males in fall, in fresh plumage, have their colors muted. FEMALE: Tail like male's, but lacks head pattern and vivid colors; breast buffy, upperparts mouse brown. IMMATURE: Like female, but young males begin to show traces of adult colors and patterns.

GEOGRAPHIC VARIATION The range of nominate *phoenicurus* extends from Europe to central Siberia and northern

Mongolia, while subspecies *samamisicus* ranges from the southern Balkans and Greece eastward to Iran, Turkmenistan, and Uzbekistan.

SIMILAR SPECIES Any *Phoenicurus* out of range, such as in N.A., should be studied carefully to rule out congeners such as Daurian (*P. aureorus*), which has base of secondaries and tertials white, and the eastern subspecies of Black (*P. ochruros rufiventris*). *Luscinia* species such as Bluethroat and Rufous-tailed Robin should also be considered, as many have rufous in tail and could be confused with a redstart.

VOICE CALL: A sweet, upslurred, whistled *wheet*. SONG: A variable series of run-together notes, both sweet and grating, often beginning with a standard repeated double-note but finishing differently.

STATUS & DISTRIBUTION BREEDING: From Europe (and sparingly

northwestern Africa) eastward to Siberia, south to Mongolia and Central Asia. WINTER: Nominate subspecies winters in Africa, mostly south of the Sahara, while *samamisicus* winters in northeastern Africa and the Arabian Peninsula. VAGRANT: The only N.A. record is of a single male at St. Paul I., AK (Sept. 8–10, 2013).

diffuse supercilium

immature ♂ *phoenicurus*

rusty orange breast and flanks

rusty orange tail with dark central tail feathers

WHEATEARS Genus *Oenanthe*

NORTHERN WHEATEAR *Oenanthe oenanthe* NOWH ▪ 2

North American–breeding Northern Wheatears are among the longest-distance migrant passerines in the world. Polytypic (4 ssp.; 2 in N.A.). L 5.8" (15 cm)

IDENTIFICATION In nominate *oenanthe*, male has upperparts gray with contrasting white rump; white underparts with buff tinge to chest. Wide black eye line expands into auriculars. Wings blackish; tail pattern—black with large white basal corners— unique in North America. Female similar in pattern, but less contrasty and with strong buffy-brown cast. Males of *leucorhoa* similar, but a bit larger with richer buff on chest extending down sides to flanks; females browner. Adult males in fall and winter similar to females. **JUVENILE:**

Distinctive; upperparts brownish with buff or cinnamon tipping.
GEOGRAPHIC VARIATION The nominate subspecies *oenanthe* is widespread breeder in Eurasia and breeds in Alaska and northwesternmost Canada; *leucorhoa* breeds in Greenland, Nunavut, and northern Newfoundland. Both subspecies winter in sub-Saharan Africa.
SIMILAR SPECIES Other wheatears are similar, particularly Isabelline *(O. isabellina)*.
VOICE CALL: A tongue-clicking *chack* and a short whistle, *wheet*. **SONG:** Scratchy warbling mixed with call notes and imitations of other birds; often given in flight.
STATUS & DISTRIBUTION Common breeder across high-latitude Northern Hemisphere, through north-central Canada, wintering in sub-Saharan

Africa. **VAGRANT:** Very rare along East Coast, primarily in fall; casual to accidental elsewhere south of North American breeding range; one wintered in LA and another in south TX. Determination to subspecies in fall is problematic.

STONECHATS Genus *Saxicola*

STONECHAT *Saxicola torquatus* STON ▪ 4

All North American records are attributable to the *maura* group of subspecies, split by some as "Siberian Stonechat." Polytypic (23 ssp.; specimens from North America refer to easternmost *stejnegeri*). L 5.25" (13 cm)
IDENTIFICATION Small size, short tail, thin bill, white patch on inner wing, contrastingly pale rump, and upright posture are unmistakable. **MALE:** Black head, back, wings, and tail; wide white ear surround; white belly, sides, and rump; and orange wash on chest. **FEMALE:** Pale brown head with indistinct paler supercilium, brown eye line, whitish throat, buffy-orange wash on chest and rump, and dark-centered back feathers. **IMMATURE:** Similar to adult female, but males have traces of black in face.

VOICE CALL: A shrill, sharp whistle; a throaty, clicking *vist trak-trak*.
STATUS & DISTRIBUTION Common. **BREEDING:** In Old World. "Siberian Stonechat" breeds in open scrubland and grassland with scattered shrubs in Asia and northeasternmost Europe; winters from Japan south to southeastern Asia and west to northeast Africa. **VAGRANT:** Casual to various AK locations, but interestingly not the Aleutians; one fall record from NB and another for San Clemente I., CA.

THRUSHES Family Turdidae

Hermit Thrush (OH, Oct.)

The thrush family houses some very distinctive species, with which few others can be confused, including some tough identification challenges. Chief among the latter group are the brown thrushes in the genus *Catharus,* though female bluebirds also cause their share of consternation. Relatively close and unobstructed views are required to identify the thrushes, as is strong experience in the variation, both individual and regional/subspecific, in plumages of the various species. Two North American members of *Turdus* are called robins, though they are not related to the original recipients of that name (members of the genus *Erithacus*). Two other species recently had their English names changed from "robin" to "thrush" (White-throated and Clay-colored Thrushes).

Structure Thrushes are generally plumper, more compact than other similar-size birds, but generally have quite long wings; in fact, thrushes average among the longest-winged passerines relative to body size. There is relatively little variation in bill shape among the family, with all North American species possessing short- to medium-length and thin beaks. Many species are terrestrial and thus have long legs.

Plumage Although most thrushes are monomorphic in plumage (males and females appear similar), the family includes nearly the whole range of plumage variation from monomorphic to dimorphic. Some *Turdus* species exhibit sexual dimorphism, although in the American Robin this is confounded by individual and subspecific variation. The bluebirds are fairly dimorphic, with distinct male and female plumages. The forest-dwelling *Catharus* thrushes exhibit subtle-to not-so-subtle subspecific differences in plumage coloration. All North American thrushes undergo a single annual molt after breeding; pre-alternate molts

are absent or, at best, limited. The family is characterized by the typically spotted plumage of juveniles, a trait that few other families match. Upon completing their post-juvenal molt, young birds achieve the essentially adult plumage.

Behavior Flight style is strong and fairly uniform across the family, with quick, flicking wingbeats followed by very short glides with the wings held closed. This style produces a fairly level flight, unlike the flight of most other passerines, which tends to undulate. Some *Turdus* (e.g., the American Robin) forage in a very ploverlike manner, with short runs and stops when they search the ground for invertebrate prey. American Robins are purported to be able to hear earthworms (a staple for the species) under the surface of the ground, which may explain their common behavior of cocking their heads sideways—to better listen (or watch) for prey.

Distribution The thrushes make up one of the most widespread bird families, with the largest radiations in the Old World. In fact, six of the 26 species dealt with here are primarily birds of the Palearctic; another eight are neotropical in origin. Half of the North American thrushes have wide distributions across the continent: Four are primarily western species and two are eastern specialties. All species of thrush breeding in North America are at least partly migratory. Most of the North American species are medium-distance migrants; four species of brown thrush regularly occur south of Mexico, and two of these winter solely in South America.

Taxonomy The Old World chats (seven recorded from N.A.) recently were moved to the Old World Flycatcher family (Muscicapidae, see p. 534) as a result of molecular genetic studies.

Conservation Of the North American breeding species, the Wood Thrush (and perhaps other brown thrushes) is threatened by habitat loss and fragmentation; it is particularly affected by the subsequent higher rates of nest predation and parasitism in forest fragments. BirdLife International classifies nine species of thrush as vulnerable (including Bicknell's Thrush); five as endangered, one of which, the Oloma-o *(Mydastes lanaiensis)* is likely extinct; four as critically endangered, including the Puaiohi *(Mydastes palmeri)* from Kauai; and four as extinct: Bonin Thrush *(Zoothera terrestris),* Kamao *(Myadestes myadestinus),* Amaui *(M. woahensis),* and Grand Cayman Thrush *(Turdus ravidus).*

BLUEBIRDS Genus *Sialia*

This is a strictly New World genus composed of three species. All are frugivorous (plucking small fruits from perches or while hovering) and insectivorous (sallying to the ground for prey). The Mountain Bluebird is noted for its hovering ability. Eastern and Mountain bluebirds hybridize; an apparent Eastern x Western Bluebird has been found in Colorado.

EASTERN BLUEBIRD *Sialia sialis*   EABL   1

sialis ♀

white belly and undertail coverts

The Eastern Bluebird is a species familiar to millions in eastern North America. Polytypic (8 ssp.; 3 in N.A.). L 7" (18 cm)
IDENTIFICATION MALE: Bright blue above, with orange throat, ear surround, chest, sides, and flanks. **FEMALE:** Differs in that upperparts are less blue (often grayish); has partial whitish eye ring, and whitish throat bordered by brown lateral throat stripes. **JUVENILE:** Very similar to Western Bluebird, but with tertials fringed cinnamon; immatures discernable with duller upperparts and browner primary coverts. Flight as in other bluebirds; pale wing stripe less obvious than in Western Bluebird.

GEOGRAPHIC VARIATION Three subspecies in North America; compared with widespread and somewhat migratory, short-billed nominate *sialis,* resident *fulva* of southeastern Arizona and likely southwestern NM (rare) slightly larger and paler; and southern Florida *grata* with longer bill.
SIMILAR SPECIES Male Western Bluebird with all-blue head, including chin and throat, and is generally a deeper blue with gray vent and undertail. Male Western usually has some rufous on upperparts, particularly on scapulars. See female bluebird sidebar below.
VOICE CALL: Musical, typically two-noted *too-lee.* This call is also given in flight. **SONG:** Mellow series of warbled phrases; varied.
STATUS & DISTRIBUTION Common in the eastern three-fifths of the lower 48 states and in southern Canada; uncommon and local in southeastern AZ. **BREEDING:** Nests in open woodland, second-growth habitats, and along the edges of fields and pastures, placing nest in cavity; readily accepts nest boxes. **MIGRATION:** Short- to medium-distance migrant. Spring: arrives Great Lakes ±25 Feb; southern SK ±1 April. Departs northernmost range during Oct. Usually migrates in flocks. **WINTER:** almost always in flocks, often mixed with yellow-rumped, pine, and palm warblers and/or dark-eyed juncos; mainly central and southern U.S., but to south-central CO, central NM, and northeastern Mexico. **VAGRANT:** Casual west to AB and UT; recent local colonization in western CO.
POPULATION Nest boxes have apparently helped reverse a decline.

juveniles of all bluebirds are spotted

rufous in both sexes wraps around sides of neck and includes throat

juvenile *sialis*

sialis ♂

overall paler than nominate *sialis*

southwestern ♂ *fulva*

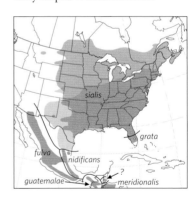

sialis

grata

fulva   *nidificans*

guatemalae   *meridionalis*

WESTERN BLUEBIRD *Sialia mexicana*   WEBL   1

This species typically prefers more wooded breeding habitats than does the Mountain Bluebird, though co-occurs widely with it. Polytypic (6 ssp.; 3 in N.A.). L 7" (18 cm)
IDENTIFICATION Western Bluebird is shorter-winged and shorter-tailed than Mountain Bluebird, producing similar wing/tail ratio, and different primary projection/tertial length ratio. **MALE:** Rich blue on head (including chin and throat), wings, and tail. **FEMALE:** Grayer head and back; paler orange underneath; at least partial whitish eye ring. **JUVENILE:** Similar to Eastern Bluebird, being strongly spotted dark on underparts, whitish on upperparts and wing coverts, but with tertials fringed grayish; immatures distinguishable by browner primary coverts, duller upperparts. Flight is level and typical of thrushes. Migrates diurnally in flocks, occasionally with Mountain Bluebirds, from which they can be distinguished by more obvious pale underwing stripe (created by darker wing linings and flight feathers) and, with experience, by shorter, rounder wings.
GEOGRAPHIC VARIATION The widespread subspecies *bairdi* (breeds west to central UT, southern AZ) is larger and with more extensive chestnut on upperparts than more westerly *occidentalis; jacoti* of southeastern NM and trans-Pecos TX is smallish, with extensive dark chestnut on upperparts.
SIMILAR SPECIES Unlike plumage of Eastern Bluebird, male's head entirely blue; females confused with other

bluebird species (see sidebar below).
VOICE CALL: Similar to other bluebirds, but a more single-noted and harder *few,* though some calls are fairly strongly two-noted; this call also given in flight.

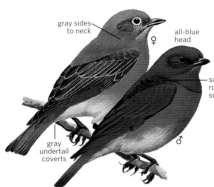

gray sides to neck

all-blue head

♀

some dark rufous scapulars

♂

gray undertail coverts

SONG: Consists of a series of call notes and is primarily heard at dawn.
STATUS & DISTRIBUTION Uncommon to fairly common at mid- and low elevations in western lower 48, extending north into Canada in BC. **BREEDING:** Nests in parklike, low-elevation pine and mixed forests, rich riparian bottomlands, and oak savanna. **MIGRATION:** Short-distance migrant that rarely strays far from nesting areas. Spring arrival central CO ±15 Mar.; eastern WA ±10 April. Fall: Most depart southern BC ±31 Oct.; northern CO ±30 Sept. **WINTER:** Primarily in southwestern U.S. and northern Mexico, but range to southern WA on Pacific slope; northern extent of wintering variable, dependent on food (juniper or

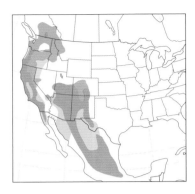

mistletoe berries). **VAGRANT:** Casual east to ND, western KS, and eastern TX.
POPULATION Populations depressed by alteration of pine habitats due to fire suppression.

Identifying Female Bluebirds

Many features provide clues to the identification of female bluebirds, and identification should be based on more than one. Except for the Mountain Bluebird, vagrancy in these species is quite rare, so range is a good first clue.

The Mountain Bluebird is usually the species most easily ruled in or out, as much about it is distinctive. The Mountain Bluebird has a distinctive structure, being longer-winged, longer-tailed, and longer-legged than the other two bluebirds; and its penchant for hovering while foraging is a further excellent identification feature. The female Mountain Bluebird's white eye ring is obvious on its otherwise bland face; both the Eastern and Western exhibit partial whitish eye rings (behind the eye). Eastern and Western Bluebirds have darkish lateral throat stripes, with the Eastern's being darker than those on the Western (which can be quite vague) and contrasting more with the Eastern's white throat; the Western's throat is gray. The Eastern Bluebird has a distinctive orange ear surround. All three species can show gray backs, but the Western Bluebird often has at least a slight orange wash, particularly on the scapulars; the Eastern Bluebird can be quite blue-backed. The Mountain Bluebird has a more slender bill than do the other species.

The Mountain Bluebird typically has gray underparts, but a substantial minority have varying amounts of an orange tint on the chest and, occasionally, the upper sides, particularly when fresh in fall; this color does not extend to the flanks, which are a cold gray-brown and contrast sharply with the white undertail coverts. The Western Bluebird is usually orange on the chest, sides, and flanks; some duller females can nearly be matched by bright Mountains. The Eastern Bluebird is always orange on the chest, sides, and flanks, typically more so, even, than the Western. The bellies and vent areas on both the Mountain and the Eastern Bluebird are bright white, whereas the Western Bluebird is off-white to gray in this area, occasionally exhibiting some dark feather centers to the undertail coverts.

Though the differences in underwing pattern may be difficult to discern, this feature can be quite useful, particularly if the Eastern Bluebird can be excluded on other grounds. The Western Bluebird has the darkest underwing of the three species; thus it has, contrastingly, the brightest pale wing stripe. The Eastern Bluebird's wing stripe shows less contrast with other plumage than the Western's. Again, the Mountain Bluebird is relatively distinctive here, having the palest underwing with the least contrasting wing stripe. ■

Eastern Bluebird, female (OH, Mar.)

Western Bluebird, female (CA, Apr.)

Mountain Bluebird, female (NM, Oct.)

MOUNTAIN BLUEBIRD   *Sialia currucoides*   MOBL ■ 1

The Mountain Bluebird often hovers above open country in a wide variety of western habitats. Monotypic. L 7.25" (18.5 cm)
IDENTIFICATION Long-winged aspect, particularly very long primary projection of Mountain Bluebird, sets this species apart in all plumages. **ADULT:** male's almost entirely sky-blue plumage (tipped with some brownish in fall when fresh) is unique. Females are extensively gray underneath, though some, in fresh plumage, have orangish cast to chest and upper sides. **JUVENILE:** Variably spotted white above, though nearly lacking in some, and vaguely spotted dark below. **FLIGHT:** Level and typically thrush-like. A flocking diurnal migrant, female has translucent flight feathers; both

sexes have long, pointed wings. **SIMILAR SPECIES** Females, particularly orange-chested individuals, can be mistaken for other bluebird species (see sidebar opposite), and grayer females for Townsend's Solitaire (see that species account). **VOICE CALL :** A somewhat soft bluebird whistle, less two-noted than that of Eastern and softer, less single-noted than that of Western. This call is also commonly given in flight. **SONG:** A series of notes similar to that of the call note, but more variable.
STATUS & DISTRIBUTION Common and widespread in western N.A. **BREEDING:** Nests in huge elevational and latitudinal range (grasslands through forests to alpine tundra), requiring nesting cavities or niches in proximity to open country for foraging; readily takes to bluebird boxes. **MIGRATION:** Short- to long-distance migrant. Arrivals or migrants apparent in spring starting ±15 Feb. in southerly areas, northern extreme of breeding range occupied ±1 May. Fall: Departs northernmost breeding areas ±30 Sept.; more southerly areas in Oct., though quite variable. **WINTER:** Mainly lower

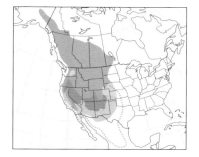

48 states and northern Mexico, but extent and local occurrence of wintering highly variable and dependent on food resources (e.g., in CO, particularly juniper berries). Regularly found in flocks in agricultural fields, where hovering foraging behavior put to good use. **VAGRANT:** Numerous fall and winter records from east of the species' normal range, to the Atlantic coastal states and Maritime provinces; found alone and in flocks of Eastern Bluebirds.
POPULATION Populations of Mountain Bluebird have risen concomitant with a great increase in the provision of nest boxes in the species' range.

SOLITAIRES   Genus *Myadestes*

BROWN-BACKED SOLITAIRE   *Myadestes occidentalis*   BBSO ■ 5

This Mexican and Middle American thrush, a close relative of Townsend's Solitaire and with a similar high-mountain habitat preference, has occurred at least twice in Arizona. A popular cage bird in Mexico; occurrences in the U.S. are questioned by some. Polytypic (5 ssp.). L 8" (21 cm)
IDENTIFICATION ADULT: Gray and rufous overall. Head gray with darker lores and white eye crescents; upperparts brownish olive, wings slightly darker and rufous; underparts gray; tail dark grayish to blackish with mostly white outer rectrices. **JUVENILE:** Wings and tail as adult, but head and body whitish buff, scalloped dark brown.
SIMILAR SPECIES Townsend's Solitaire is similar in shape but with bold and complete white eye ring, and dark gray wings with conspicuous buff wing patches. Compare also Slate-colored Solitaire *(M. unicolor)* of

southern Mex. and Middle America, which is slaty gray overall.
VOICE CALL: Loud, upslurred *wheeoo.* **SONG:** Amazing series of initial hesitant call notes, accelerating into a long rocking garble of flutelike notes, sung year-round.
STATUS & DISTRIBUTION Resident in pine-oak and montane evergreen forests to at least 11,000 ft on western and eastern slopes from northern Sonora and southern Nuevo León south to central Honduras. Range possibly advancing northward in Sonora; currently found north to Sierra Huachinera about 80 miles from AZ. **VAGRANT:** Two accepted U.S. records, both in southeastern AZ: Madera Canyon in the Santa Rita Mountains (Oct. 4, 1996) was initially rejected based on doubt about its origin, then accepted in 2011 as valid; Miller Canyon in the Huachuca Mountains (July

16, 2009) and presumably the same individual 2.7 miles away at Ramsey Canyon (July 18–Aug. 1, 2009).

TOWNSEND'S SOLITAIRE   *Myadestes townsendi*   TOSO ▪ 1

The voice of the Townsend's Solitaire is often the first indication of its presence, and it is a characteristic sound in western coniferous forests. Polytypic (2 ssp.; nominate in N.A.). L 8.5" (21.5 cm)

IDENTIFICATION A long and slender, somewhat aberrant thrush, Townsend's Solitaire often perches conspicuously on the highest available perch. In some wintering situations, individuals hold territory with extensive supply of small fruits (e.g., juniper berries) defended by call note and, rare in North American birds, song. **ADULT:** Monomorphic. Entire head and body medium gray, though underparts can be slightly paler. Thin, but crisp, white eye ring; short, black bill; and short, black legs are distinctive. Central third of long tail is bordered by black and with outermost rectrix on each side mostly white. Wings are distinctive, being primarily gray with whitish fringes to dark tertials and greater coverts with striking buffy-orange patches (but which are duller in some individuals). **JUVENILE:** Juveniles are heavily spotted throughout body and wing coverts, with a less obvious wing pattern. Older immatures are virtually identical to adults, except that occasional birds retain a few juvenal wing coverts, as is typical in thrushes; at least through early spring, most should have more worn flight feathers. **FLIGHT:** It is really in flight that Townsend's Solitaire is most distinctive, with floppier flight than is typical of thrushes, and a bold, buffy-orange wing stripe, more obvious from below than from above. Tail is often flicked open and closed in flight, with the bird resembling a long, thin junco.

SIMILAR SPECIES The most common confusion is with duller female Mountain Bluebirds, but they exhibit blue at least in their tail and wings. They have shorter tails and relatively longer wings and lack the solitaire's bold wing pattern. That wing pattern is surprisingly similar to that of Varied Thrush, but nothing else about the species is. Brown-backed Solitaire *(M. occidentalis),* a Mexican species of accidental occurrence in southeastern Arizona, has a brown back and wings, a white malar stripe, and a long, absolutely distinctive song.

VOICE CALL: A single *thnn,* similar to the toot of a pygmy-owl, usually repeated at intervals. **FLIGHT NOTE:** Unknown. **SONG:** A finch-like warble, at times quite long and disjointed, with occasional call notes interspersed and with no distinct pattern.

STATUS & DISTRIBUTION Common. **BREEDING:** Nests in a wide variety of open conifer-dominated habitats, often with rock outcrops and/or cliffs on which the birds place their nests. **MIGRATION:** Short- to long-distance migrant, with a strong downslope aspect. Spring: Most birds have departed low-elevation winter areas by ±1 May, but some are still there in mid-May. Fall: Arrive at low elevations ±1 Sep., but general arrival there about ±25 Sept. **WINTER:** Throughout lower 48 breeding range, even at very high elevations (to 11,000 ft in CO), and in lowlands adjacent to foothills and mountains; abandons all Canadian breeding range except for extreme southmost areas; winters to central Mexico. **VAGRANT:** Very rare, but regular fall and winter vagrant to the East, especially to the northern tier of states and southeastern Canada.

juvenile

boldly spotted plumage

overall gray color with prominent white eye ring

prominent buffy wing stripe

buffy wing patches

mostly white outer tail feathers

BROWN THRUSHES   Genus *Catharus*

This terrestrial New World genus consists of 12 species; the five breeding north of Mexico have similar plumage patterns and retiring behaviors. Identification is complicated by subspecific variation. Boreal species exhibit a wide, pale wing stripe in flight. Two species of the neotropical radiation of this genus have occurred accidentally in N.A.

ORANGE-BILLED NIGHTINGALE-THRUSH *Catharus aurantiirostris*   OBNT   ▪ 5

This typically shy denizen of moist, high-elevation forests in Central and South America has found its way north of Mexico at least twice. Polytypic (up to 14 poorly differentiated ssp. in the Neotropics; 1 in N.A., presumably *clarus*). L 6.5" (16.5 cm)
IDENTIFICATION Orange-brown above with brighter uppertail coverts, wings, and tail. Underparts generally pale gray and whitish, with chest and flanks darkest. Bright orange bill, legs, and eye rings provide obvious field marks. In flight this species lacks pale wing stripe typical of boreal genus *Catharus*.
SIMILAR SPECIES Compared with Orange-billed, Black-headed Night-

ingale-Thrush is darker, more olive, on upperparts with distinctly blackish crown and face. Russet Nightingale-Thrush *(occidentalis)* is less bright above, and has vague darkish spotting below and a black bill.
VOICE CALL: A nasal, Gray Catbird-like *meeer*. **FLIGHT NOTE:** Unknown. **SONG:** A jerky, short, scratchy though loud warble of varied phrases.
STATUS & DISTRIBUTION Fairly common inhabitant of montane understory. **VAGRANT:** This elevational migrant has occurred twice in spring in the Lower Rio Grande Valley of TX (April 8, 1996, at Laguna Atascosa N.W.R.; May 28, 2004, at Edinburg).

A territorial singing male was in Spearfish Canyon, Black Hills, SD (July 10–Aug. 19, 2010).

orange bill, orbital ring, and legs

orange-brown upperparts, pale gray below

BLACK-HEADED NIGHTINGALE-THRUSH *Catharus mexicanus*   BHNT   ▪ 5

A recent record in south Texas provided for this montane Middle American species' addition to the North American avifauna. Polytypic (4 ssp.; 1 in N.A., presumably nominate *mexicanus*). L 6.5" (16.5 cm)
IDENTIFICATION Orange eye ring, bill, and legs are only bright parts of otherwise dark bird. Crown and face blackish with nape a bit paler; rest of upperparts and wings olive. Chin, throat, and vent area whitish; rest of underparts gray. A skulker.
SIMILAR SPECIES Orange-billed

Nightingale-Thrush is similar in pattern and bright orange soft parts.
VOICE CALL: Buzzy or petulant *mew, rreahr*. **FLIGHT NOTE:** Unknown. **SONG:** Similar to Orange-billed Nightingale-Thrush, but more melodious; 6–8 thin, high-pitched, flutey whistles and trills; some phrases repeated.
STATUS & DISTRIBUTION Fairly common in humid montane forest; some suggestion of altitudinal migration. **VAGRANT:** Accidental, one record in lower Rio Grande Valley, at Pharr, Hidalgo Co., TX (May 28–Oct. 29, 2004).

blackish crown and face

orange orbital ring, bill, and legs

VEERY *Catharus fuscescens*   VEER   ▪ 1

Among a retiring genus, the Veery is shy and relatively poorly known. Polytypic (4–5 ssp.; all in N.A.). L 7.3" (18 cm)
IDENTIFICATION For a *Catharus*, generally rather distinctive; less spotted below, brighter reddish above, and lacking eye rings. **ADULT:** Depending

on subspecies, upperparts entirely dull reddish brown to bright rufous-brown, though small minority of western birds nearly gray-brown. Large, non-contrasting, pale grayish loral area and brown to reddish lateral throat stripes fairly weak; throat whitish to buffy. Chest washed buff, some

pinkish buff, and vaguely to distinctly spotted with dull reddish brown on upper chest; any spotting on lower chest is grayish. Belly white, sides and flanks pale gray. Bill horn-colored with dark culmen. **JUVENILE:** Spotted whitish above, dark below; some older immatures distinguishable by

How to Identify a *Catharus* Thrush

When viewing a *Catharus* thrush, one should discern as many features as circumstances allow and, particularly for a poorly seen individual, be willing to let it go unidentified. First consider the time of year: All but the Hermit Thrush are exceedingly rare north of Mexico from November through March. The Hermit also has a distinctive tail-lifting behavior that, if observed, can help to eliminate other contenders. If the bird is in profile, concentrate on the precise facial pattern (inc.

the presence or absence of an eye ring), the color of the lateral throat stripes, the color of the flanks, the primary projection, and the color of the tail. From the front, consider the distinctness of the pale loral area, the color of the lateral throat stripes, and the color and distinctness of chest spotting. From behind, look for back/tail color contrast and note primary projection. Individuals in shadow can look quite different from those in dappled or strong light. ▪

presence of retained juvenal wing coverts with buffy tips.

GEOGRAPHIC VARIATION Subspecies vary primarily in upperparts brightness and color and distinctness of chest spotting; subspecific identification in field problematic due to slight differences and individual variation; western *salicicola* has dullest upperparts (brown with reddish tinge) and distinct brownish spots; central Canadian and upper Midwest *levyi*

face pattern like Gray-cheeked with faint partial eye ring around rear of eye

western
salicicola

faintly spotted chest

eastern
fuscescens

gray flanks

(medium-dull reddish brown) and eastern Canadian *fuliginosus* (bright reddish brown) with distinct reddish brown spots; northeastern U.S. *fuscescens* medium-bright reddish brown with indistinct reddish spots; southern Appalachian *pulichorum* dark reddish brown with moderately distinct brownish spots.

SIMILAR SPECIES Other *Catharus* more spotted underneath, less reddish above. See "russet-backed" Swainson's Thrush.

VOICE CALL: Abrupt, rough and descending veer; also a slow *wee-u* and a harsh chuckle; alarm call a sharp, low *wuck*. **FLIGHT NOTE:** Low *veer*. **SONG:** Flutelike, slow, somewhat mournful, downward-spiraling *veeerr veeerr veeerr*.

STATUS & DISTRIBUTION Fairly common, but western birds restricted by limited suitable riparian habitat. **BREEDING:** Most nest in deciduous and/ or mixed forest but *salicicola* is primarily a shrubby willow riparian inhabitant. **MIGRATION:** Long-distance migrant;

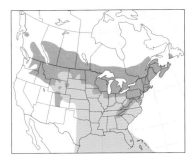

eastern birds trans-Gulf migrants, though many aspects of migration unknown. In spring, peaks on Gulf Coast ±25 Apr., southern Great Lakes ±15 May; casual-to-rare migrant in West, arrival in montane CO breeding areas ±25 May. In fall, Upper Midwest and Northeast peak ±10 Sept., though late migrants linger in northeastern U.S. into Oct. Winters solely at southern periphery, and a bit south, of Amazon Basin. **VAGRANT:** Casual in CA, mainly in fall; accidental in Europe.

POPULATION Forest fragmentation may negatively impact populations.

BICKNELL'S THRUSH *Catharus bicknelli* BITH ■ 2

Bicknell's Thrush is basically a slightly smaller version of the Gray-cheeked. Monotypic. L 6.8" (17 cm)
IDENTIFICATION Nearly indistinguishable from Gray-cheeked, away from breeding areas. **ADULT:** Upperparts brown, usually with a warmer cast; tail usually distinctly reddish. Compared to Gray-cheeked, wings are color of upperparts, except primaries fringed rufous; wings shorter; primary projection about equal to length of longest tertial. Pale loral area not contrasty; partial eye rings grayish; auriculars plain grayish brown. Black lateral throat stripes contrast with auriculars and white throat. Chest spotting blackish, usually in a wash of buff; flanks olive-gray. Unmarked belly and

vent white. Mandible mostly dull to medium yellow with dark tip; maxilla all dark. **JUVENILE:** Spotted buff above, blackish below; white throat. Some older immatures distinguishable by retained juvenal wing coverts with buffy tips.
SIMILAR SPECIES Gray-cheeked very similar, but larger, longer-winged, primaries edged with duller brown, and less rufous in tail. Unless in hand or diagnostic vocalizations are heard (and preferably recorded), migrants are best left unidentified, especially of out of range.
VOICE CALL: Similar to notes in song, single *wee-ooo;* also a sharper *shrip*. **FLIGHT NOTE:** Sharp, buzzy *peeez* or *cree-e-e* is higher, less slurred than that of Gray-cheeked. **SONG:** Similar to Gray-cheeked's, more even in pitch, beginning with low *chuck* notes; high-pitched phrases rise slightly toward end and terminate, after pause, with *shre-e-e.*
STATUS & DISTRIBUTION Uncommon to rare and local. **BREEDING:** Mostly in dense stunted spruce and/or balsam fir forests at tops of northeastern mountains; lower in eastern QC. **MIGRATION:** Medium-distance migrant, generally along Atlantic coastal plain; a Sept.

specimen from northwestern OH. Typically arrives on breeding grounds ±25 May, slightly earlier in Adirondacks, slightly later in QC. Departs New England mid–late Sept.; probably peaks southern NJ late Sept.–early Oct. **WINTER:** Range of elevations, primarily on Hispaniola, but mostly in montane forests. **VAGRANT:** Accidental to Bermuda, Cuba.
POPULATION Introduced exotic pests are killing trees with its breeding habitat. Deforestation is a major concern on its range-restricted winter grounds in the West Indies, mainly in the Dominican Republic.

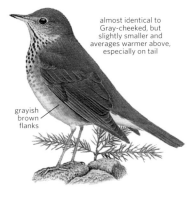

almost identical to Gray-cheeked, but slightly smaller and averages warmer above, especially on tail

grayish brown flanks

GRAY-CHEEKED THRUSH *Catharus minimus* GCTH ■ 1

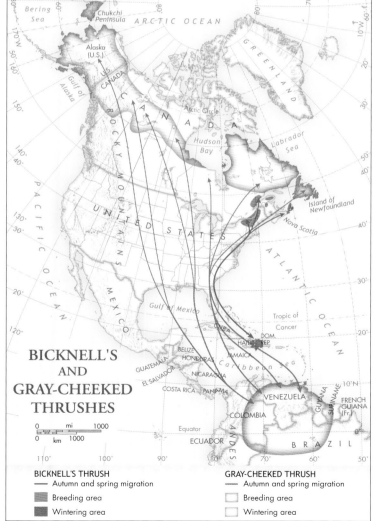

faint partial eye ring
around rear of eye

grayish
lores

1st fall
aliciae

aliciae

grayish
brown
flanks

minimus

Gray-cheeked Thrushes that breed in the Russian Far East and winter in South America are among the longest-distance passerine migrants. Polytypic (2 ssp.; both in N.A.). L 7.3" (18 cm)

IDENTIFICATION A cold gray-brown, long-winged *Catharus*. **ADULT:** Upperparts brownish olive; uppertail coverts and tail with slight reddish cast. Face quite plain with vague pale gray loral area and partial eye ring (particularly behind eye). Thin, buff to gray malar stripe and white throat contrast strongly with black lateral throat stripes. Chest spotting black and rounded, extending above and below cream wash on breast. Sides and flanks washed brown (with gray or olive cast). Belly and vent unmarked white. Bill yellow, with dark culmen and tip; yellow usually extending beyond nostrils and contrasting with gray lores and white throat. **JUVENILE:** Spotted buff above, blackish below; some older immatures distinguishable by presence of retained juvenal wing coverts with buffy tips. **GEOGRAPHIC VARIATION** Subspecific identification in field problematic due to slight differences and individual variation. Widespread *aliciae* described. Newfoundland *minimus* upperparts and flanks grayish olive; breast lightly washed cream; and bill darker and duller, with yellow usually not extending beyond nostrils.

SIMILAR SPECIES Colder-colored than other *Catharus;* quite similar to Bicknell's (see that entry). **VOICE CALL:** Variable; downward-slurred *wee-ah,* higher, more nasal than that of Veery; also a thin *pweep*. **FLIGHT NOTE:** *Che-errr,* which is similar to that of regular call note, but more nasal, stressing highest pitch in middle; ending more abruptly than that of Bicknell's Thrush. **SONG:** A Veery-like descending spiral, but thinner, higher, more nasal, and with a stutter. **STATUS & DISTRIBUTION** Uncommon. **BREEDING:** Nests in soggy areas, usually conifer bogs or willow or alder thickets; also in stunted spruce at high elevation. **MIGRATION:** Long-distance, trans-Gulf migrant. Spring: Gulf Coast peak ±1 May; arrival in western AK ±1 June, when some still migrating through LA. Fall: peak in eastern U.S. ±1 Oct. Winters at lower elevations in northwestern SA (south to eastern Ecuador). **VAGRANT:** Casual to accidental in West; annual in spring in eastern CO; casual to northern Europe (>45 to U.K.), Azores; accidental to Surinam. **POPULATION** As far as known, population stable, but bears watching.

BICKNELL'S AND GRAY-CHEEKED THRUSHES

mi 0 — 1000
km 0 — 1000

BICKNELL'S THRUSH
— Autumn and spring migration
▨ Breeding area
▨ Wintering area

GRAY-CHEEKED THRUSH
— Autumn and spring migration
☐ Breeding area
☐ Wintering area

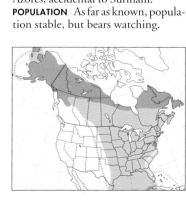

SWAINSON'S THRUSH *Catharus ustulatus* SWTH ■ 1

The most common migrant *Catharus* in much of N.A. Polytypic (6 ssp.: all in N.A.). L 7" (18 cm)

IDENTIFICATION Typically bold, buffy eye ring and supraloral patch of most Swainson's Thrushes distinctive. Is divided into two subspecies groups, "Olive-backed" *swainsoni* group and "Russet-backed" *ustulatus* group. **ADULT:** "Olive-backed" group described. Upperparts olive-brown. Chest spots dark brown and round and usually entirely within buffy wash. Sides and flanks olive-gray. Tail similar in color to upperparts. **JUVENILE:** Spotted buff above, dark gray-brown below, with white throat.

GEOGRAPHIC VARIATION Six subspecies, some of questionable validity; subspecific identification in field prob-

lematic beyond subspecies group. "Olive-backed" group: Widespread *swainsoni* described; *appalachiensis* upperparts slightly darker and same color as uppertail coverts; *incanus* of western AK to northern BC and north-central AB has upperparts paler and grayer, underparts whiter, and chest spotting nearly black. "Russet-backed" group: Northwestern coastal *ustulatus* (southeastern AK to northwestern CA) and *phillipsi* (Queen Charlotte Is.) upperparts brown, tinged reddish, with breast washed brownish with indistinct spots, and flanks warm brown; and *oedicus* of rest of CA is duller above; eye rings of western subspecies usually thinner. The two groups have different breeding grounds (though they approach each other rather closely), migration routes, and winter grounds. No intergrade specimens are known, the two groups may well be separate species.

SIMILAR SPECIES "Russet-backed" subspecies is regularly confused with western *salicicola* Veery, but note differences in color of breast spotting, flank color, and facial pattern.

VOICE CALL: Russet-backed gives a liquid *dwip*, Olive-backed a sharper *quirt;* also a rough, nasal chatter introduced by call note. **FLIGHT NOTE:** Clear, spring peeper-like *queep*. **SONG:** Flutelike, similar to Veery in spiraling pattern, but spiraling upward.

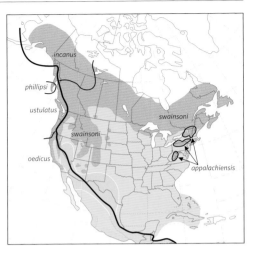

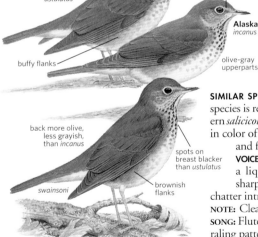

buffy eye ring and supraloral line

upperparts with a russet cast, like many Veeries

Pacific coast
ustulatus

buffy eye ring and supraloral line

Alaska
incanus

olive-gray upperparts

buffy flanks

back more olive, less grayish, than *incanus*

spots on breast blacker than *ustulatus*

brownish flanks

swainsoni

STATUS & DISTRIBUTION Common, but in West restricted by limited suitable habitat. **BREEDING:** Mostly in deciduous and/or mixed forest; birds in western U.S. primarily nest in riparian habitat with combination of shrubby willow understory and deciduous or coniferous overstory. **MIGRATION:** "Olive-backed" birds are trans-Gulf migrants; "russet-backed" birds migrate through the CA deserts (spring) and along Pacific coast. Peaks on Gulf Coast ±25 Apr., southern Great Lakes ±15 May, and CA deserts 10–25 May; arrival in montane WY breeding areas ±25 May; spring migration continues into early June as far south as KS and southeastern CA. In fall, most migrate Sept.–mid-Oct., arrival on CO plains ±25 Aug; both groups are largely absent as migrants over much of the Southwest. "Russet-backed" birds winter from western Mexico to northwest Costa Rica; "olive-backed" birds winter in S.A. **VAGRANT:** Casual to accidental to northern AK and NT; casual to Europe.

HERMIT THRUSH *Catharus guttatus* HETH ■ 1

The Hermit Thrush is the only *Catharus* expected in winter in the U.S. Polytypic (9–13 ssp.; all in N.A.). L 6.8" (17 cm)
IDENTIFICATION Slow tail lift is distinctive in the genus. **ADULT:** In eastern *faxoni*, upperparts brown with moderate rufous wash. Eye ring whitish, occasionally not complete in front. Loral area darker than on other *Catharus* and not contrasting. Tail distinctly reddish, contrasting with upperparts. Boldly spotted below. Sides and flanks washed

buffy-brown; brownish gray in other races. Bill dark with pinkish base to mandible. **JUVENILE:** Spotted whitish to buff above, blackish below.
GEOGRAPHIC VARIATION Three subspecies groups. Subspecific identification in field problematic due to slight differences and individual variation, but separation to group more likely. "Lowland Pacific" *guttatus* group includes *guttatus* (coastal southern AK to western BC), *nanus* (southern Alaska islands),

verecundus (Queen Charlotte Is.), *vaccinius* (coastal southwestern BC and northwestern WA), *jewetti* (northwestern WA to northwestern CA), *slevini* (interior south-central WA to west-central CA), *munroi* (central BC and western AB to northern MT), and *oromelus* (interior southern BC east to northwestern MT and south to northeastern CA). All are small and most are dark brownish, with little to no reddish coloration on back, and with richly colored tails.

"Interior Western Montane" *auduboni* group includes *sequoiensis* (Sierra Nevada of CA), *polionotus* (eastern CA east to northwestern UT and AZ), and *auduboni* (southeastern WA east to southern MT, south to southern AZ and NM and western TX). All with pale, grayish brown backs and duller, less contrasting tails; the last subspecies having distinctive buff undertail coverts. "Northern" *faxoni* group includes *faxoni,* see description p. 550 (southern NT south to southern AB, east to NF and MD) and *euborius* (central AK and northern BC); birds in this group are richly colored above with bright reddish tails, and have tawny brown flanks (flank colored duller and variable in *euborius*).

SIMILAR SPECIES Some similarity to Gray-cheeked, Bicknell's, and Swainson's Thrushes (see entries). **VOICE CALL:** Soft, low blackbirdlike *chuck*; also a rising, whiny catbirdlike *wheeee*. **FLIGHT NOTE:** Clear, somewhat complaining *peew*. **SONG:** Flutelike; begins with long, clear whistle followed by series of rather clear phrases; successive songs often alternate pitch direction of song, rising then falling. **STATUS & DISTRIBUTION** Common. **BREEDING:** Typically in conifer-dominated forests, usually in areas of relatively little undergrowth; also in deciduous forests. **MIGRATION:** Medium- to long-distance migrant, with western montane breeders wintering farthest south (to southern Guatemala). Early migrant in spring, with southern Great Lakes peak ±25 Apr.; arrival in farthest reaches of breeding range ±10 May, though with stragglers still in West in early June. Fall: Western montane birds begin migration in early Sept., but *faxoni* and *guttatus* group peak in East ±15 Oct. Western subspecies winter in coastal states from WA south to southern Guatemala; *faxoni* group winters

in Southeast (from southern NJ south), a few to southern Midwest, south to northeastern Mexico. **VAGRANT:** Regular to Bering Sea islands in migration and to Bermuda fall through spring; accidental north to northern AK and Canada, Greenland, Greater Antilles, and Europe.

all juvenile *Catharus* have spotted plumage

juvenile *faxoni*

faxoni

thin eye ring

warm brown upperparts with contrasting rufous tail

brownish buff flanks

grayish upperparts

auduboni

smaller, darker, and with deeper rufous tail than *auduboni*

guttatus

grayish flanks

pale rufous tail

Genus *Hylocichla*

WOOD THRUSH   *Hylocichla mustelina*   WOTH   ■ 1

This is one of the finest songsters living in our eastern deciduous forests. It occupies a monotypic genus. Monotypic. L 7.8" (20 cm) **IDENTIFICATION** A chunky, well-marked brown thrush of eastern N.A. **ADULT:** Orange-brown upperparts, much brighter on rear crown and nape. Obvious white eye ring barely broken by dark gray eye line.

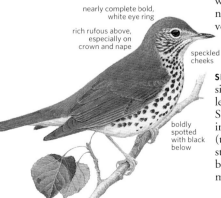

nearly complete bold, white eye ring

rich rufous above, especially on crown and nape

speckled cheeks

boldly spotted with black below

Distinctive black-and-white striped auriculars and spotting on throat forming streaks. Underparts white with large and distinct, oval black spots extending onto lower belly and flanks, the latter washed warm brown. Bill grayish pink with darker culmen. **JUVENILE:** Spotted whitish above, dark below; some older immatures distinguishable by flanks possibly less warm, and through fall by thin cinnamon tips to inner coverts forming very slight, partial wing bar. **FLIGHT:** Wide, pale wing stripe as in boreal *Catharus,* but wing base wider with pronounced secondary bulge. **SIMILAR SPECIES** *Catharus* thrushes similar, but all smaller, less potbellied, less heavily spotted; most not as bright. Strong white eye ring not matched in *Catharus,* nor is auricular pattern (though Gray-cheeked has vaguely-streaked auriculars). Brown and Long-billed Thrashers also similar, but with much longer tail, bill; yellowish eyes.

VOICE CALL: A rolling *popopopo* and a rapid, staccato *pit pit pit*. **FLIGHT NOTE:** Sharp, nasal *jeeen*. **SONG:** Beautiful, flutelike *eee-o-lay,* with last note accented and highest in pitch; this song usually introduced with quiet *po* notes, not audible at long distance, and ending with a buzzy or trilled whistle. **STATUS & DISTRIBUTION** Fairly common but declining in East. **BREEDING:**

Nests in deciduous forest. **MIGRATION:** Medium- to long-distance trans-Gulf migrant. Spring peaks Gulf Coast ±15 Apr.; southern Great Lakes ±15 May. In fall, some move early (late Aug.), but most still present near breeding grounds into Sept.; Gulf Coast peak in first half of Oct. Winters from eastern Mexico east and south to Panama. **VAGRANT:** Casual to accidental migrant in West and north to SK. Accidental in western Mexico, northern S.A., Iceland, U.K., and Azores.

POPULATION Forest fragmentation and resultant cowbird parasitism negatively impacts populations in parts of breeding and wintering ranges.

ROBINS Genus Turdus

Of the huge genus Turdus (as many as 66 species), North America north of southern Texas supports just a single breeding species, American Robin. However, five Old World species have occurred in N.A. Another three Mexican/neotropical species have occurred in the U.S., with the Clay-colored Thrush recently adding south Texas to its breeding range.

EYEBROWED THRUSH Turdus obscurus   EYTH ■ 3

The Eyebrowed Thrush is a rare, but nearly annual, spring visitor to western Alaska. Monotypic. L 9" (22.5 cm)
IDENTIFICATION Superficially similar to American Robin, but all plumages distinct. Head and throat of males are largely slaty gray with strong white supercilia, eye arcs, and chin, and black eye line; those of females are similar, but paler and browner with strong white malar stripe. Back, wings, and tail are brown, the latter without white tail corners. Sides and flanks are orange-rufous; belly and undertail coverts unmarked white. Bill yellowish with variably dusky tip. Legs yellowish brown to flesh.
SIMILAR SPECIES Duller female American Robins somewhat similar, but easily identified by leg and upperparts colorations and precise head pattern.
VOICE CALL: Variable; high, thin, penetrating *dzee*, a doubled *chack* or *tuck*, and a single *tchup*.
FLIGHT NOTE: High *tseee*. **SONG:** Clear, whistled, mournful series of phrases of two or three syllables followed by a pause, and then a lower-pitched twittering; somewhat imitative of other birds.

white eyebrow
olive upperparts
♀
browner lower throat
dark gray throat
extensive white belly
♂

STATUS & DISTRIBUTION Rare spring migrant in western Aleutians, casual in fall; casual on Bering Sea islands; accidental elsewhere in western AK and north to Barrow. In some springs, groundings of migrants widespread in this area (e.g., 180 in May 1998). One in northeastern Kern Co., CA (May 28, 2001) was presumably of an individual that wintered in the New World.

DUSKY THRUSH Turdus naumanni   DUTH ■ 4

This striking Asian thrush occurs very rarely in Alaska. Polytypic (2 ssp.; all documented AK records are of *eunomus*). L 9.5" (24 cm)
IDENTIFICATION Male blackish above; bright white supercilium. Underparts blackish with variable extent of white fringing; wings mostly rufous. Female similar to male, but upperparts brown rather than black; face less white; wings brown. Whitish chest crescent on both sexes variable in extent, but distinctive.
GEOGRAPHIC VARIATION Subspecies *eunomus* described above; "Naumann's Thrush" (*naumanni*, more southerly breeding) is rufous on head, where *eunomus* is white, and on belly, where *eunomus* is black; wings brown in both sexes. Apparent intergrades are frequent.
SIMILAR SPECIES Varied Thrush vaguely similar, but head and wing patterns distinctive.
VOICE CALL: Variable; series of *shack* notes; a chattering

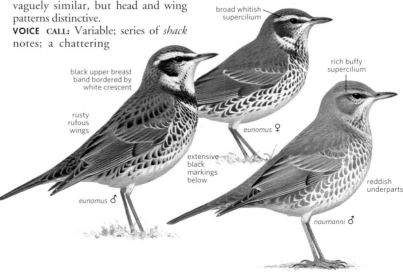

broad whitish supercilium
black upper breast band bordered by white crescent
rusty rufous wings
rich buffy supercilium
eunomus ♀
extensive black markings below
reddish underparts
eunomus ♂
naumanni ♂

kwaawag; kwet-kwet; European Starling–like *spir.* **FLIGHT NOTE:** Thin *geeeh* or *shrree,* often repeated, or thin, high *huuit.* **SONG:** Flutey and melodious series of phrases, often ending in faint trill or *twitter, tryuuu-tvee—tryu,* with accent in first phrase.

STATUS & DISTRIBUTION Common breeder in Siberia and the Russian Far East; winters primarily in eastern China, but also in Japan; a few in some winters as far south to northern Thailand. **VAGRANT:** Very rare migrant in western Aleutians, casual elsewhere.

Recorded at St. Lawrence I. and twice in YT; has also been recorded to southeastern AK and coastal BC. None of the few sight records of "Naumann's Thrush" from the Aleutians and Pribilofs have been substantiated with specimens or photographs.

EURASIAN BLACKBIRD   *Turdus merula*   EUBL ▣ 5

This all-dark *Turdus* is distinctive in North America, but similar to other blackish species in Middle and South America. Polytypic (16 ssp.; 1 in N.A., presumably nominate *merula).* L 10.5" (26.5 cm)
IDENTIFICATION Overall behavior and habitat choices are very similar to those of American Robin. **MALE:** All black; orange-yellow eye ring and bill; dark legs. **FEMALE:** Browner; pale throat and dark-streaked chest; eye ring and bill duller.

SIMILAR SPECIES Unlike all other thrushes in northeastern North America where occurrence is most likely.
VOICE CALL: Variable; a deep *pok* and a hard *chack-ack-ack.* **FLIGHT NOTE:** Quiet, rolling *srrri.* **SONG:** Similar to that of American Robin.
STATUS & DISTRIBUTION
VAGRANT: Twelve records from Greenland and one fully accepted record from NF (a specimen

found dead Nov. 16, 1994). A few other Canadian records of uncertain origin.

male all-black with orange-yellow orbital ring and bill

merula ♂

REDWING   *Turdus iliacus*   REDW ▣ 4

A relatively small, short-tailed Old World *Turdus,* the Redwing has distinctive streaked underparts and reddish orange flanks and wing linings. The last feature gave the species its name. Polytypic (2 ssp.: both ssp. may have reached N.A., but no specimens). L 8.3" (21 cm)
IDENTIFICATION Upperparts from forehead to tail are medium grayish brown, with a strong face pattern of long white supercilium, black eye line, white-streaked auriculars, white malar stripe, black lateral throat stripe, and white chin and throat. Wings are colored same as back; tail is blackish. Chest and upper belly are white with extensive dark brown to blackish streaking. Streaking of chest sides extends into reddish orange lower sides and flanks. lower belly is unmarked white; outer undertail coverts have dark centers.
GEOGRAPHIC VARIATION There are two poorly marked subspecies, with *coburni,* a breeder in Iceland, the Faroe Islands, and, recently, southern Greenland, being slightly larger and slightly darker in all plumage aspects. Which subspecies accounts for northeastern North American records is uncertain (though quite possibly *coburni),* but recent records from WA and southern AK probably refer to nominate *iliacus.*
SIMILAR SPECIES Redwing is unlikely to be confused with any other thrush in

North America but a molting juvenile American Robin; however, distinctive face pattern and lack of other juvenile traits (such as white upperparts spotting) should easily rule out that option. Fieldfare is vaguely similar, but distribution of orange underneath is quite different, as is head pattern. Streaked underparts and strong black-and-white head pattern might cause confusion with one of the tropical streaked flycatchers, but behavior and underparts pattern are quite different. In flight, Redwing appears short-tailed; not reddish orange wing linings.
VOICE CALL: Variable; a nasal *gack;* a harsh *zeeh;* an abrupt *chup,* sometimes extended to *chidik.* **FLIGHT NOTE:** A long, somewhat hoarse *stoooof* or a thin, high *seeeh.* **SONG:** Quite variable across individuals, with tone and structure being more important than particular phrases; short phrases and squeaky twitters interspersed with long (up to 6 seconds) silent intervals.
STATUS & DISTRIBUTION
BREEDING: From Russian Far East west to Iceland, with recent colonization of a small area in southern Greenland. Redwing winters primarily in southern Europe, northwest Africa, and southwest Asia. **VAGRANT:** The species is a casual visitor, primarily in late fall

and winter, to the rest of Greenland and to NF, and is accidental farther south in northeastern N.A. (to Long Island, NY, and eastern PA). Recent winter records from Olympia, WA, and Seward, AK, were the first in western N.A.

distinct whitish buff supercilium

rusty red wing linings

streaked underparts

rusty red flanks

immature

FIELDFARE *Turdus pilaris*   FIEL ■ 4

The Fieldfare is a distinctive Eurasian breeder of casual occurrence in northeastern North America, where most occurrences follow large-scale, hard-weather movements out of northwestern Europe. Monotypic. L 10" (25 cm)

IDENTIFICATION Head is primarily medium gray with a black loral area; buffy-white to orange chin; throat with fine blackish streaking; and dark lateral throat stripe connected to black lower corner of auriculars. Bill varies from yellow to orange with a black tip of variable extent.

flashy white wing linings

gray head

purplish brown back

buffy breast with bold arrow-shaped spots

gray rump

immature

Upper back, scapulars, and wing coverts are purplish brown; flight feathers are blackish. Lower back and rump are pale gray and contrast strongly with rest of upperparts; this feature is particularly noticeable in flight. Chest and sides are buffy to orange with blackish feather centers forming arrowhead-shaped spots, lower ones typically larger. Belly and undertail coverts are white, the latter with dark centers or shaft streaks. Longish tail is black with vague gray corners. Legs are gray. Primary projection is about equal to length of longest tertial. In flight bird's contrastingly white wing linings are obvious. Birds in first basic are a bit dingier, less contrasty, with a thin whitish lower wing bar and more extensive black on bill.

SIMILAR SPECIES Vaguely similar to American Robin, but brown upper back and wings, pale gray lower back and rump, and white wing linings of Fieldfare are distinctive.

VOICE CALL: Variable; *chack,* similar to Dusky Thrush, is often doubled. **FLIGHT NOTE:** Thin, nasal *tseee* or *weeet.* **SONG:** Series of chattery or warbling notes with squeaky chuckles interspersed; without flutelike quality typical of genus. **FLIGHT SONG:** Chattering, more drawn out, and faster.

STATUS & DISTRIBUTION Common breeder west from the western Russian Far East through northern Europe and locally in southern Greenland. Winters primarily in southern Europe and southwestern Asia, but the species' fall migration has a strong facultative aspect, and many individuals continue only as far south as necessary to avoid snow. **VAGRANT:** Casual in late fall and winter in northeastern N.A.; also casual in western and northern AK in late spring/early summer. Accidental to ON, MN, and BC.

SONG THRUSH *Turdus philomelos*   SOTH ■ 5

Among Europe's most cherished songbirds, Song Thrush has been an inspiration to musicians and writers throughout its range but has been a special muse for English poets, from Chaucer to Shakespeare, Wordsworth to Hughes. Its scientific name, *philomelos,* means "lover of song." Polytypic (3 ssp.). L 8–9.25" (20–23 cm)

IDENTIFICATION Larger and more husky of build than *Catharus* thrushes but smaller than American Robin. Upperparts warm brown. Underparts whitish with ochre-buff wash on breast and flanks, liberally spotted with dark arrowhead-shaped spots except on lower belly. Narrow eye ring, pale submoustachial bordered by dark malar stripe, and auriculars somewhat mottled, most strongly at edges. **JUVENILE:** Slightly paler above than adult, with buff-tipped upperwing coverts in fresh plumage. In flight,

shows reddish buff underwing coverts. **SIMILAR SPECIES** In N.A. context, Song Thrush might suggest Swainson's Thrush but is larger and more heavily and extensively marked in face and underparts.

GEOGRAPHIC VARIATION Migratory nominate subspecies is found through most of the species' large Eurasian range. The dark subspecies *hebridensis,* restricted to Isle of Skye and Outer Hebrides, Scotland, is sedentary, while warm-toned *clarkei,* breeding in the rest of the British Isles and in western Europe (France, Belgium, and the Netherlands), is a partial migrant. **VOICE CALL:** a sharp *zip* and longer *seep.* **SONG:** Typically given at dusk, a bold, very long series of repeated notes in phrases, some ethereal sounding, others grating, incorporating imitations of other species and even man-made sounds.

STATUS & DISTRIBUTION **BREEDING:** From the British Isles and most of northern and central Europe across the Ukraine nearly to Lake Baikal. **WINTER:** Northernmost birds from Scandinavia and Russia migrate farthest south, into the Mediterranean and Middle East, whereas birds from the western portions of the range are

dark-bordered auricular

plain brown upperparts

extensively patterned below with arrow-shaped marks

fall immature

sedentary or partly migratory, some wintering near breeding grounds or on the Iberian Peninsula. **VAGRANT:** The only N.A. record comes from Saint-Fulgence, eastern QC (Nov. 11–17, 2006). An annual vagrant in small numbers to Iceland, chiefly in fall; one record from Greenland.

POPULATION Song Thrush populations have declined by 50–70 percent in some areas of England and Europe since the 1970s.

CLAY-COLORED THRUSH *Turdus grayi*   CCTH   3

Formerly just a wintering resident in southernmost Texas, this widespread tropical thrush now breeds there in small numbers. It is somewhat more shy than American Robin. Polytypic (8 ssp. recognized, differences are minor; *tamaulipensis* in N.A.). L 9" (23 cm)
IDENTIFICATION A typical *Turdus*, it is particularly common in Middle America and essentially replaces American Robin there. **ADULT:** Lacks eye ring. Upperparts medium brown, underparts tan, with throat vaguely streaked brown and pale buff. Eyes are reddish orange, bill is greenish yellow with a dark base,

and variably colored legs range from dull pink through greenish brown to gray. **JUVENILE:** Similar to adult but with pale buff shaft streaks on back feathers, scapulars, and median coverts, and with buff tips to median and greater coverts.
SIMILAR SPECIES See White-throated Thrush.
VOICE CALL: Variable; a distinctive cat-like and rising *jerereee* (similar to a call of Long-billed Thrasher) and some typical thrush notes, including doubled *tock.* **FLIGHT NOTE:** High, thin *siii,* weaker than American Robin. **SONG:** Clear, whistled phrases recalling that of American Robin, but with a much wider variety of phrases.
STATUS & DISTRIBUTION Abundant in neotropics; uncommon to rare in U.S. Most frequently seen at the various reserves in the lower Rio Grande, but also seen in nearby well-wooded suburban yards, particularly in winter. **BREEDING:** Nests in areas of dense thickets, streamside brush, and

woodlands. **WINTER:** Small breeding population is supplemented in winter (numbers vary from year to year) with birds from northeast Mexico. Found from northeast and northwest Mexico south to northern Equador. **VAGRANT:** Has occurred as far north up the Rio Grande as San Ygnacio, TX; accidental to Big Bend N.P.; also accidental to Huntsville, TX, and northern NM.
POPULATION Very common in Middle America. No concern.

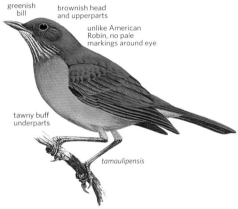

greenish bill

brownish head and upperparts

unlike American Robin, no pale markings around eye

tawny buff underparts

tamaulipensis

WHITE-THROATED THRUSH *Turdus assimilis*   WTTH   4

This retiring tropical thrush is widespread in Middle America. It is very similar looking and closely related to the White-necked Thrush (*T. albicollis*), a widespread South American thrush found east of the Andes. They are lumped by some authorities. In North America, White-throated is a casual winter visitor to southernmost Texas. Polytypic (14 ssp.; 1 in N.A., presumably *suttoni*) L 9.5" (24 cm)
IDENTIFICATION White-throated is a darkish *Turdus* of moist montane forests

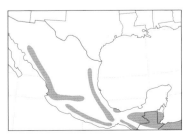

of Mexico, where it occupies middle and canopy levels. This species can be difficult to see well due to its shy nature. Upperparts are dark brown; underparts dark tan, with throat strongly streaked blackish and white, and undertail coverts somewhat paler. A distinctive white band on upper chest requires a frontal view. Bold yellow-orange eye ring surrounds dark eyes; bill is greenish brown, and legs are dull pinkish brown.
SIMILAR SPECIES White-throated Thrush is unlikely to be confused with anything other than Clay-colored. Note upperpart color and presence or absence of eye ring and bib color.
VOICE CALL: Quite variable; short, guttural *ep* or *unk;* nasal *rreuh;* whistled *peeyuu.* **FLIGHT NOTE:** Unknown. **SONG:** Similar to that of American Robin, but with wider variety of phrases, and sometimes repeating phrases like a mimid.
STATUS & DISTRIBUTION VAGRANT:

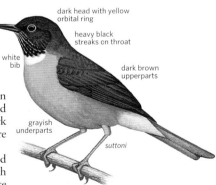

dark head with yellow orbital ring

heavy black streaks on throat

white bib

dark brown upperparts

grayish underparts

suttoni

White-Throated Thrush has occurred during four winters in lower Rio Grande Valley of TX (8–10 recs. documented in winter 2004–2005). Another subspecies, *assimilis,* from western Mexico, is found north to southeast Sonoroa and might reach the Southwest. It is paler, more of a grayish brown above, and has a grayish, not yellow, orbital eye ring.

RUFOUS-BACKED ROBIN *Turdus rufopalliatus* RBRO ■ 3

The Rufous-backed is a Mexican robin with a distinctively orange bill. "Grayson's Thrush" (*graysoni*, breeds on Tres Marias Is., off western Mex.) is regarded by some authorities as a distinct species; it could possibly occur in the United States. Polytypic (2–3 ssp.; *rufopalliatus* in N.A.). L 9.3" (24 cm)

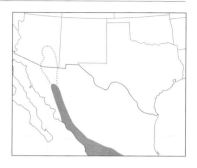

IDENTIFICATION Superficially similar to American Robin, but all plumages distinct. Head and upper back are largely gray with variable extent (due to age?) of orangish brown on crown; lacks eye ring. Lower back and wing coverts bright orangy-rufous, contrasting with gray rump and tail. Black and white throat streaks extend onto upper chest; rest of underparts orange, except for white central belly and undertail coverts. Legs are orangish pink, eyes are orange, and bill is dull to bright orange.

SIMILAR SPECIES Duller individuals might be mistaken for American Robin, but bill and leg colors, plumage pattern, and lack of eyering distinctive.

VOICE CALL: Variable; a typical *Turdus* throaty trebled *chok,* a plaintive,

drawn-out, descending whistle, *peeeuuuuu.* **FLIGHT NOTE:** High *sseep.* **SONG:** A leisurely, clear, low-pitched, warbled series of phrases with a repetitive pattern, recalling that of American Robin.

STATUS & DISTRIBUTION Rare but nearly annual fall and winter visitor to the southwestern U.S., primarily southern AZ, but with records scattered from southern TX to coastal southern CA.

gray head with no white around eye

rufous back and wing coverts

long streaks on throat

white extends up to lower breast in a point

rufopalliatus

AMERICAN ROBIN *Turdus migratorius* AMRO ■ 1

This species' often confiding nature, distinctive plumage, pleasing song, and acceptance of human-dominated habitats make it one of the most beloved of N.A. birds. Polytypic (7 ssp.; 5 in N.A.). L 10" (25 cm)

IDENTIFICATION A distinctive, pot-bellied bird. Forages on lawns and other areas of short vegetation for earthworms and other invertebrates in a run-and-stop pattern typical of terrestrial thrushes. **ADULT:** Depending on sex and subspecies, head, with white eye arcs, varies from jet black to gray, with white supercilia and throat, blackish lores and lateral throat stripe. Underparts vary, often in tandem with head color, from deep, rich reddish maroon to gray-scalloped, peachy orange. Males tend to be darker, females grayer, but overlap makes determining sex of many problematic. Throat streaked black and white; vent and undertail coverts white. Upperparts medium gray; tail blackish, with white corners. Bill color yellow with variable, season-dependent, black tip. Legs dark. **JUVENILE:** Spotted dark on underparts; whitish on upperparts and wing coverts. Older immatures not distinguishable from adults; small percentage retain a few juvenal wing coverts or other feathers. **FLIGHT:** Quick, flicking wingbeats followed by short, closed-wing glides. Wing linings

migratorius

white spot

juvenile *migratorius*

heavily spotted underparts

bold white broken eye ring

yellow bill

brick red underparts

migratorius ♀

migratorius ♂

color of underparts; remiges blackish. **GEOGRAPHIC VARIATION** Nominate *migratorius* is found in the north and much of the East; it is described above. North Pacific coastal *caurinus* and widespread western *propinquus* (larger, paler) with white tail corners small or lacking; Canadian maritime *nigrideus* dark brownish to blackish above, underparts deep rufous, medium-size tail corners; southeast U.S. *achrusterus* smaller, upperparts browner, smaller white tail corners. The subspecies *(confinis)* from the Laguna Mts., in Baja California Sur's Cape District is often recognized as a separate species, "San Lucas Robin."

SIMILAR SPECIES Duller females possibly mistaken for Eyebrowed Thrush. Juveniles possibly confused with spotted thrushes. **VOICE CALL:** Variable; low, mellow single *pup;* doubled or trebled *chok* or *tut;* shriller and sharper *kli ki ki ki ki;* high and descending, harsh *sheerr.* **FLIGHT NOTE:** Very high, trilled, descending *sreeel.* **SONG:** Clear, whistled phrases of two or three syllables *cheerily cheery cheerily cheery,* with pauses; lacks the burry quality of many tanagers; *Pheucticus* grosbeaks typically have different tempo. **STATUS & DISTRIBUTION** Common and widespread. **BREEDING:** Wide variety

of wooded or shrubby habitats with open areas. **MIGRATION:** Short to medium-distance migrant. Departs northerly winter-only areas by ±10 Apr.; arrival northern Great Lakes ±20 Mar.; central AK ±1 May. Strong facultative aspect (particularly in East) in fall, so variable timing; departs southern Canada ±20 Oct. **WINTER:** Mainly lower 48 and Mexico; also southernmost ON and BC, Bahamas (rare), northern Guatemala. **VAGRANT:** Widely in Europe; casual to Jamaica and Hispaniola; accidental Azores. **POPULATION** Strong adaptability and widespread distribution suggest little concern.

RED-LEGGED THRUSH *Turdus plumbeus*   RLTH ■ 5

This sedentary Caribbean species has been recorded just once from Florida (found as close to Florida as Grand Bahama). Polytypic (6 ssp.; *plumbeus* in N.A.). L 10.5" (24 cm)
IDENTIFICATION Characters in widespread Caribbean regions differ extensively. Florida bird's white chin and black throat indicated nominate *plumbeus* subspecies of northern Bahamas. **ADULT:** Upperparts gray. Underparts from almost entirely gray with black throat and white chin, gray with reddish brown abdomen, to white throat with black stripes and whitish abdomen. **JUVENILE:** Similar to adult, but duller.
SIMILAR SPECIES Typically thrush shaped, but uniformly dark slate gray

body, bright red legs and feet, and red eye ring are unlikely to be confused with any other species.
VOICE CALL: A soft *week week.* **SONG:** Usually two or three phrases repeated, such as *chirra, chirra, weeup, wheet,* interspersed with thin *wheet* or *weep* notes—much slower, lower pitched, and less musical than American Robin.
STATUS & DISTRIBUTION Widespread resident of West Indies in northern Bahamas, Cuba, Hispaniola, Puerto Rico, Dominica, and Cayman Is. Habitats include tropical deciduous and montane evergreen forests, scrub, thick undergrowth, yards, shade coffee plantations—usually seen on the ground in leaf litter. **VAGRANT:** The U.S. bird was

photographed at Maritime Hammock Sanctuary, Melbourne Beach, FL (May 31, 2010).
POPULATION Stable.

mostly black throat with white chin

red orbital ring and legs

plumbeus

Genus *Ixoreus*

VARIED THRUSH *Ixoreus naevius*   VATH ■ 1

This species' ethereal song is a distinctive aspect of wet northwestern forests. Polytypic (4 ssp.; all in N.A.). L 9.5" (24 cm)
IDENTIFICATION Orange legs and dark bill with yellow mandible base. **MALE:** Blue-gray above, orange below, with broad orange supercilia, black auriculars and chest band, complicated pattern of orange-on-black wings, and gray scalloping on flanks and lower belly. **FEMALE:** Similar but upperparts brown; auriculars mostly dark gray; thinner, indistinct chest band gray; wings brown. **JUVENILE:** Similar to adult females, but chest heavily scalloped with dark gray-brown; central belly white. Older immatures similar

juvenile *meruloides*

orange wing bars and markings on wing

meruloides ♀

orange supercilium

meruloides ♂

black breast band

to adults, though males less blue above, females browner; all have browner tails. **FLIGHT:** Similar to American Robin, but orange-and-black wing linings and bold orange wing stripe. **GEOGRAPHIC VARIATION** Subspecies differentiation based on female plumages; northern *meruloides* paler; north Pacific coastal *naevius* and *carlottae* upperparts darker, tawny tinged, underparts orange; southern interior *godfreii* paler, upperparts reddish tinged. **SIMILAR SPECIES** American Robin has no wing pattern, lacks orange throat and supercilium and flank pattern of Varied Thrush. In flight Townsend's Solitaire's wing pattern quite similar, but species is thinner, with long, thin, white-edged tail. See Siberian Accentor. **VOICE CALL:** A low *tschook* similar to Hermit Thrush, but harder; a high *kipf*; a thin, mournful whistle,

woooeee. **FLIGHT NOTE:** Short, humming whistle. **SONG:** Series of long, eerie whistles of one pitch, with successive notes at different pitch and long inter-note intervals. **STATUS & DISTRIBUTION** Common. **BREEDING:** Nests in moist, typically conifer-dominated, habitats in Northwest; in tall willow riparian north of tree line. **MIGRATION:** Short- to medium-distance migrant with some only moving altitudinally. Spring: Departs southern winter areas ±15 Mar., though some still there early May; arrives western AK ±30 Apr. Fall: Departs northern breeding areas ±15 Sept.; first arrivals in southern wintering areas ±10 Oct. **WINTER:** Coastal AK to southern CA (numbers vary considerable from year to year in CA) and parts of northern Rockies; rare to casual south and east of northern Rockies. **VAGRANT:** Subspecies *meruloides*

rare, but regular fall and winter vagrant to East, particularly northern tier of states and southern Canada. Casual to Bering Sea islands; accidental to Iceland and U.K. **POPULATION** Logging in breeding range may be negatively impacting birds.

Genus *Ridgwayia*

AZTEC THRUSH   *Ridgwayia pinicola*   AZTH ▪ 4

The Aztec Thrush is a Mexican visitor to the U.S. Southwest. Although by nature shy and inconspicuous, it is boldly patterned and distinctive. Monotypic. L 9.3" (24 cm)
IDENTIFICATION ADULT MALE: Blackish hood, browner on crown, face, and back; blacker on throat and chest, with blackish rump and tail. White uppertail coverts form a U-shaped band; tail tip is white. A dark brown vent strap separates white belly from white undertail coverts. Wings have an intricate pattern of black, white, and dark brown, with obvious white tertial and secondary tips. **ADULT FEMALE:** It is similarly patterned to male, but all dark colors are paler and it has a more obvious streaked aspect to head and chest; tips of tertials and secondaries are gray. **JUVENILE:** Back, scapulars, and chest are streaked with buff; belly is creamy-colored and

scaled with brown; and a dark eye line contrasts with a whitish, streaked supercilium. Some older immatures possibly distinguished by presence of retained juvenal wing coverts. **FLIGHT:** Bold pattern is distinctive; whitish wing stripe contrasts with blackish secondaries with white tips. **SIMILAR SPECIES** Distinctive, but compare with juvenile Spotted Towhee. **VOICE CALL:** Harsh, slightly burry and whining *wheeerr* and a *whining*, slightly metallic *whein*. **FLIGHT NOTE:** Unknown. **SONG:** Probably a louder and steadily repeated variation of call note. **STATUS & DISTRIBUTION** Uncommon to rare. **BREEDING:** Pine forests in western

and southern Mexico. **MIGRATION:** Northernmost breeders may be partially migratory, perhaps accounting for pattern of occurrence in U.S. Southwest. Winters primarily in breeding habitat and range, often joining mixed-species flocks of other frugivores (*Turdus* thrushes, Gray Silky-flycatchers). **VAGRANT:** Casual to southeastern AZ sky islands (mostly late summer and early fall, occasionally in small numbers) and Sierra Madre Oriental of Mexico (late fall, early winter). There are a handful of records from western and southern TX (fall, winter, and spring).

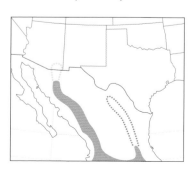

juvenile

spotted plumage; note white wing pattern

blackish head, breast, and back

♂

extensive white on wing

white belly

white rump and tail tip

♀

browner breast than male

MOCKINGBIRDS AND THRASHERS Family Mimidae

Long-billed Thrasher (TX, Apr.)

Thrashers and mockingbirds, along with the Gray Catbird, constitute the Mimidae found in North America. Most of the species are known for their long, varied songs of repeated phrases; some, the Northern Mockingbird in particular, are well-known mimics of other birds. In general, the family reaches its highest diversity in the desert Southwest, where several species of thrasher coexist. They live in a variety of habitats, from wet thickets to desert washes and chaparral hillsides. Most are large, brown, ground-dwelling birds with long, decurved bills.

Structure Generally large for passerines, these birds have long tails and long, decurved bills used to probe for food items in leaf litter and holes in the ground. Their legs are rather long, with long toes used for running. Several species have distinctive yellow or orange eyes.

Behavior Most of the species are well adapted to their desert environment. They tend to be quite secretive, remaining well hidden during the heat of the day, but they often climb to the top of trees to sing. Members of the Mimidae family generally feed on insects and seeds, although several species take advantage of seasonal fruit production and feed on berries and cactus fruit, mainly in the fall and winter. Most are very territorial, defending both summer and winter territories. Some, particularly thrashers, form pairs that last multiple years. Most species are resident within their range. One species, the Gray Catbird, is a neotropical migrant, whereas some others (the Sage, Bendire's, and Brown Thrashers) move south seasonally and winter mainly in the southern United States. Some species (e.g., the Curve-billed Thrasher), although nonmigratory, disperse after the breeding season and have been detected well away from their breeding range.

Plumage Most of the thrasher species are brown, with varying amounts of streaking or spotting on the underparts. Some are concolor and characterized by their rusty crissums; others are heavily streaked underneath. The sexes are similarly plumaged in all species.

Distribution Mimidae comprises about 34 species in 12 genera worldwide. Ten species in four genera breed in North America. Two additional species are casual or accidental north of Mexico: the Blue Mockingbird, a vagrant from Mexico, and the Bahama Mockingbird, a vagrant from the West Indies.

Taxonomy The majority of the species are monotypic and show little, if any, geographical variation. Some, such as the Le Conte's and the Long-billed Thrashers, have additional subspecies father south in Mexico. The Curve-billed Thrasher can be divided into two distinct subspecies groups (possibly distinct species) that may overlap in southeastern Arizona.

Conservation Several of the thrashers, in particular those species with local ranges (e.g., Le Conte's, California, Long-billed), are somewhat threatened by loss of natural habitat to urbanization and increased agriculture. The Northern Mockingbird, the Gray Catbird, and, to a lesser extent, the Bendire's Thrasher actually benefit from increase in disturbed habitats. Outside of the U.S. the Cozumel Thrasher *(Toxostoma guttatum)* from Isla Cozumel, Mex., has not been seen for over a decade and may be extinct.

BLUE MOCKINGBIRD *Melanotis caerulescens*   BLMO ▪ 5

This fancy Mexican endemic, an casual stray in the Southwest, is normally secretive, remaining hidden low or on the ground under dense thickets in thorn forest. It is usually detected first by hearing one of its loud, varied calls. Polytypic (2 ssp.; likely *caerulescens* in N.A.). L 10" (25 cm)

IDENTIFICATION Larger than the Northern Mockingbird; more thrasherlike in shape and behavior. Dark slaty blue overall (appears blackish in poor light); black mask. **ADULT:** Appears to have paler blue streaking on head, throat, upper breast. Bill slightly longer than a mockingbird's, with a slight decurvature. Red eye.

IMMATURE: Grayer blue overall with brownish tinge to primary edges, less noticeable streaking; eye not as red. **SIMILAR SPECIES** No other bird in N.A. is entirely blue with a black mask. **VOICE CALL:** Quite variable. Typical calls include a *chooo* or *chee-ooo*. Also a loud *wee-cheep* or *choo-leep*, a low *chuck*, and a sharp *pli-tick*. **SONG:** Mockingbird-like rich series of repeated phrases. **STATUS & DISTRIBUTION** Endemic to Mexico. Casual in southeastern AZ and southernmost TX; origin questioned on records from coastal CA (a wintering adult in Long Beach) and NM. The AZ records all involve wintering birds. Known to move altitudinally in fall and winter.

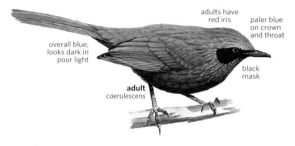

overall blue, looks dark in poor light

adults have red iris

paler blue on crown and throat

black mask

adult *caerulescens*

GRAY CATBIRD Genus *Dumetella*

The only member of this monotypic genus is the Gray Catbird, a neotropical migrant that is smaller and shorter billed than other members of the Mimidae family. With a uniform gray body, a black cap, and a rufous crissum, the Gray Catbird is quite unlike other thrashers.

GRAY CATBIRD *Dumetella carolinensis*   GRCA ▪ 1

This very distinctive mimid generally remains hidden in the understory of dense thickets in eastern woodlands and residential areas. It often cocks its longish, black tail, and it is usually detected by its harsh, downslurred *mew* call, reminiscent of a cat's *meow*. A chunky, medium-size bird, it is larger than *Catharus* thrushes yet smaller than other thrashers. No other North American bird has a uniform dark gray plumage. Monotypic. L 8.5" (22 cm)
IDENTIFICATION Sexes similar. Body entirely dark gray, with black cap, black tail, and chestnut undertail coverts.
SIMILAR SPECIES Plumage unique. Its

mimicking song vaguely resembles songs of other thrashers or the American Dipper, with which it overlaps in the western portion of its range. *Mew* calls can be confused with calls of the Hermit Thrush or the Spotted and Green-tailed Towhees.
VOICE CALL: A nasal, catlike, downslurred *mew*. Also a *quirt* note and a rapid chatter alarm when startled. **SONG:** A variable mixture of melodious, nasal, and squeaky notes, interspersed with catlike *mew* notes. Some individuals are excellent mimics. Normally sings and calls from inside dense thickets.
STATUS & DISTRIBUTION Common,

but secretive. **BREEDING:** Nests in dense thickets along edge of mixed woodland. In the West, nests along willow- and alder-lined montane streams. **MIGRATION:** Nocturnal migrant. Trans-Gulf and Caribbean migrant. Sometimes abundant during migratory fallouts along the Gulf Coast of TX and LA and in southern FL. Spring peak in TX mid-April–early May. **WINTER:** Mainly southeastern U.S., Mexico, northern C.A., and Caribbean islands. **VAGRANT:** Casual during migration and winter in the Southwest and along Pacific coast.
POPULATION Western birds limited by loss of riparian habitats.

blackish cap

dark chestnut undertail coverts

short, slender dark bill

overall steel gray

THRASHERS Genus *Toxostoma*

Seven species of *Toxostoma* thrasher are found in the United States (4 more endemic to Mexico). Generally large, brown, ground-dwelling birds of arid regions, most have long, decurved bills used for foraging in ground litter or holes. They often run with their long tail cocked in the air. Known for complex songs, some species mimic other species' songs.

BROWN THRASHER *Toxostoma rufum*   BRTH ▪ 1

The widespread thrasher of eastern N.A., this is a generally secretive bird of dense thickets and hedgerows that frequently sings from open exposed perches at the top of trees. Polytypic (2 ssp.; both in N.A.). L 11.5" (29 cm) **IDENTIFICATION** Upperparts entirely bright rufous; underparts white to buffy-white, especially on flanks; extensive black streaking. Wing coverts with black subterminal bar and white tips, forming two wing bars. Bill long and slender with little decurvature. Yellow eye. **GEOGRAPHIC VARIATION** Western

longicauda larger, paler, with less extensive streaking than eastern nominate. **SIMILAR SPECIES** Most similar to the Long-billed Thrasher of southern TX, which is more grayish above and has a longer, more decurved bill; redder eye; shorter primary projection. **VOICE CALL:** A low *churr* and a loud, smacking *spuck,* somewhat resembling the call note of a "Red" Fox Sparrow. **SONG:** A long series of varied melodic phrases, each phrase often repeated two or three times. Rarely mimics other bird species.

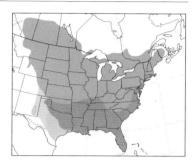

STATUS & DISTRIBUTION BREEDING: Uncommon to dense thickets throughout the eastern U.S. **MIGRATION:** Birds from the northern portion of the breeding population migrate south in the fall, augmenting resident populations in the South. **VAGRANT:** Rarely wanders west to AZ and CA. Casual or accidental to AK, BC, YT, NT, and NF. Casual in winter in northern Mexico. **POPULATION** Declines have been noted in the Northeast, probably as a result of habitat loss.

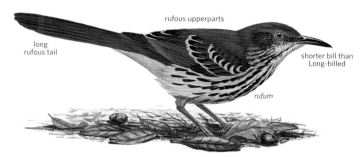

rufous upperparts

long rufous tail

shorter bill than Long-billed

rufum

LONG-BILLED THRASHER *Toxostoma longirostre*   LBTH ▪ 2

Superficially similar to the Brown Thrasher, the Long-billed replaces the Brown in south Texas and northeast Mexico. It is found in dense thickets and in the understory of remaining natural woodland habitat in south Texas, particularly in the lower Rio Grande Valley. Habits similar to Brown Thrasher's, particularly in feeding behavior and singing from exposed perches. Polytypic (2 ssp.; *sennetti* in N.A.). L 11.5" (29 cm) **IDENTIFICATION** Sexes similar. Coloration of upperparts grayish brown;

very black streaking below; face quite gray, contrasting with browner head and back; bill black, long, and distinctly decurved; yellow-orange eye. **SIMILAR SPECIES** Most similar to Brown Thrasher, with which it barely overlaps in winter. (See account above.) **VOICE CALL:** Most common call is similar to the Brown Thrasher's loud smacking *spuck;* also known to give a mellow *kleak* and a loud whistle *cheeooep.* **SONG:** Similar to the Brown's long series of melodious phrases, yet the individual phrases are not duplicated as often.

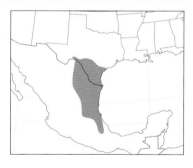

Song differs from the overlapping Curve-billed Thrasher's by being slower and more musical. **STATUS & DISTRIBUTION** Resident throughout breeding range in southern TX and northern Mexico. **VAGRANT:** Few extralimital records, but recorded casually in western TX; accidental in NM and CO. **POPULATION** Loss of most of the native brushland to agriculture in the lower Rio Grande Valley has likely had a serious impact on overall population.

shorter primary projection than Brown

grayish brown upperparts

long curved bill

gray face

sennetti

CURVE-BILLED THRASHER *Toxostoma curvirostre*   CBTH ■ 1

The Curve-billed Thrasher is the common thrasher of the rich, cactus-laden Sonoran Desert and mesquite brushlands of the Chihuahuan Desert and southern Texas. Polytypic (6–7 ssp.; 2 groups in N.A., thought by some to be two separate species). L 11" (28 cm)

IDENTIFICATION Sexes similar. A largish, pale brown thrasher with uniform brown upperparts and round, somewhat blurry spots on the underparts. Wings have noticeable whitish wing bars, particularly in eastern birds. Tail has pale tips, the extent of which depends on subspecies. Bill is relatively long, black, and distinctly decurved. Eye is distinctly orange-yellow. **JUVENILE:** Recently fledged birds have less distinct spotting than do adults, and their bills are significantly shorter and less decurved.

GEOGRAPHIC VARIATION Two groups: The more easterly *curvirostre* group represented in U.S. by *oberholseri* (includes *celsum*) is found from east side of the Chiricahuas, AZ, to southern TX. It has clearer spotting below, more distinct white wing bars, and more extensive white tips to the tail feathers.

juvenile
palmeri

thick, curved bill

striking orange iris

iris may average more golden yellow than *oberholseri*

whitish wing bars

spots less obvious below than *oberholseri*

round spots below

oberholseri

palmeri

smaller, less contrasty tail tips and wing bars than *oberholseri*

white tail tips

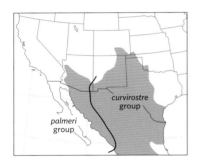

The more westerly *palmeri* group; *palmeri* in U.S. in southern AZ. The group has less distinct breast spots and less conspicuous white tips to the tail feathers. Calls differ. The degree of contact between the two groups (west or south side of Chiricahuas) remains unstudied.

SIMILAR SPECIES Adults distinctive; note different habitat and calls compared with Bendire's Thrasher (see sidebar below). Juvenile Curve-billed easily confused with Bendire's, but Bendire's usually retains at least some fine dark streaking on the underparts, largely lacking on juvenile Curve-billed.

VOICE CALL: Very distinctive loud *whit-wheet* or *whit-wheet-whit* in palmeri group; in *curvirostre* group the call notes are all on one pitch. **SONG:** Long and elaborate, consisting of low trills and warbles, seldom

repeating phrases. Quite different from Bendire's.

STATUS & DISTRIBUTION Common resident in desert habitats, particularly those rich in cholla and other cactuses. Particularly common in suburban neighborhoods that retain natural desert vegetation. Also found in mesquite-dominated desert washes. **VAGRANT:** Extralimital records mostly pertaining to *palmeri* from CA, NV, ID, and various states in the Midwest. **POPULATION** Common but experiencing habitat loss through urban development and increased agriculture in southern AZ and southern TX.

curvirostre group

palmeri group

Curve-billed and Bendire's Thrashers

In southeastern AZ and southwestern NM, Curve-billed and Bendire's Thrashers overlap in range and habitat. Although the Bendire's is very locally distributed, it can often be found with the more widespread Curve-billed. Adults in fresh plumage are more easily distinguished than worn adults. The larger Curve-billed has a longer, decurved bill; heavier, more blurry spotting on the underparts; and a distinctive, frequent two-note *whit-wheet* call *(palmeri)* or *whit-whit (oberholseri)*. The slightly smaller Bendire's has a relatively short, mostly straight bill (sometimes with

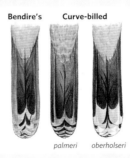

Bendire's   Curve-billed

palmeri   *oberholseri*

slight decurvature) and finer, more distinct arrow-shaped spotting on the underparts. Its call is a seldom-heard low *chuck*. Juveniles and worn adults pose a greater identification challenge: The juvenile Curve-billed has a shorter, straighter bill and, typically, a pale area along the gape. Both birds when worn have virtually no obvious spotting on the underparts, making structural differences and call more important. White tips to the tail are much more reduced in the Bendire's in comparison with the Curve-billed's *oberholseri* but are quite similar to the tail pattern of *palmeri*. ■

BENDIRE'S THRASHER   *Toxostoma bendirei*   BETH ▪ 2

Although locally distributed in the desert Southwest, Bendire's Thrasher is easiest to find during the late winter and early spring, when singing activity is at its peak. It is normally somewhat secretive, with its pale sandy brown coloration blending in nicely with the sparse desert environment it prefers (Feb.–May). Males sit up on shrubs, fences, power poles, and

spots on breast are triangular when fresh

shorter bill with pale base to lower mandible

fresh

worn

overall color a warmer brown than Curve-billed

roofs to sing, making them more conspicuous. Monotypic. L 9.8" (25 cm) **IDENTIFICATION** Sexes similar. Easily confused with similar Curve-billed Thrasher, but slightly smaller. Overall plumage light brown above, slightly paler below; distinct arrow-shaped spots across the breast; buffy flanks and undertail coverts; relatively short, straight (or slightly decurved) bill, usually with pale at the base; pale yellow eye. Tail with narrow pale tips to the feathers. **SUMMER ADULT:** Late spring and summer adults in worn plumage show little or no spotting. **SIMILAR SPECIES** See sidebar opposite. **VOICE CALL:** A seldom heard low *chuck*. **SONG:** Different from Curve-billed Thrasher's. A long series of warbling, melodic phrases, some with a harsh quality. Phrases often repeated two or three times each. **STATUS & DISTRIBUTION** Uncommon and local in sparse brushy desert. **BREEDING:** Most common in the yucca-dominated Chihuahuan Desert

and in the Sulphur Springs Valley in southeastern AZ. Rarer and more difficult to find at higher elevations across northern AZ, southernmost NV, and southern UT. A few nest in southeastern (Inyo Co. south to Riverside Co.) CA. **WINTER:** Birds from the northern portion of the range migrate to low-elevation desert in southern AZ and northern Sonora. Some birds move into suburban habitats. **VAGRANT:** Very rare visitor in fall and winter to coastal CA. **POPULATION** The greatest threat to the Bendire's is the destruction of habitat due to intense urbanization in AZ.

CALIFORNIA THRASHER   *Toxostoma redivivum*   CATH ▪ 2

This is the characteristic thrasher of the chaparral foothills along the California coast and the Sierra Nevada, and it is the only large, uniform thrasher within its limited range. With habits similar to the Crissal Thrasher's, it is often detected by call or song, since it remains well hidden in dense vegetation on hillsides, except when it runs on the ground between thickets with its tail cocked in the air. In early spring, adult males sing from exposed perches at the tops of shrubs. Polytypic (2 ssp.; both in N.A.). L 12" (30 cm) **IDENTIFICATION** Sexes similar. Dark overall, with long, black, decurved bill, which is heavier than the bills of other thrashers. Tawny buff belly, flanks, and undertail coverts; paler than the Crissal's chestnut crissum. Dark brown breast; pale throat and eyebrow contrast with dark, patterned face; dark eye.

very long tail

tawny undertail coverts

GEOGRAPHIC VARIATION The northern *sonomae* and southern California nominate populations differ slightly. **SIMILAR SPECIES** Only large, uniform brown thrasher within this range. Similar in coloration to Crissal Thrasher, but note non-overlapping ranges. Vagrant Curve-billed and Bendire's Thrashers potentially overlap, but note California's more uniform coloration and heavy, blacker bill. California Thrasher does overlap with California Towhee, which is chunky and similarly colored, but note the thrasher's much larger size and strongly decurved bill.

white throat

long, strongly downcurved bill

overall brown coloration

VOICE CALL: A low, flat *chuck* and *chur-erp*. **SONG:** Loud and sustained song made up of mostly guttural phrases, many repeated once or twice. Known to mimic other bird species and sounds. **STATUS & DISTRIBUTION** Fairly common resident on the chaparral hillsides and other dense brushy areas throughout most of coastal CA, the foothills of the Sierra Nevada, and northern Baja California. **VAGRANT:** Very sedentary, but has wandered casually north to southern OR. **POPULATION** The species has experienced losses in parts of its range due to development of preferred chaparral habitat.

LE CONTE'S THRASHER *Toxostoma lecontei* LCTH ■ 2

This attractive, pale thrasher—with a coloration matching the sparse, sandy desert environment it inhabits—is generally found in sparser desert than are other thrashers. The Le Conte's behavior is similar to that of the Crissal, with which it overlaps: It is secretive but is often seen running between shrubs with its tail cocked in the air; it sings and calls from exposed perches but otherwise stays hidden. Although much paler than Crissal, it too has unspotted underparts. Polytypic (3 ssp.; 2 in N.A.). L 11" (28 cm)

IDENTIFICATION Sexes similar. Generally pale grayish brown in coloration, with light tawny undertail coverts, thin black malar stripe, black lores, and a long, black decurved bill. Long, black tail contrasts with pale body. Dark eye.

GEOGRAPHIC VARIATION Nominate subspecies occupies much of range. Isolated *macmillanorun* from western San Joaquin Valley and adjacent Carrizo Plains, CA, is slightly darker. A third subspecies *(arenicola)* is found in west coastal Baja California.

SIMILAR SPECIES Le Conte's is unlike any other North American thrasher, but care should be taken where Crissal overlaps in range. Le Conte's is much paler overall, but lighting conditions can make it look darker. Note its contrasting dark eye and lores; its tawny undertail coverts are lighter than Crissal's. It overlaps with Curve-billed and Bendire's in southern Arizona but is more scarce and paler overall.

VOICE CALL: Includes an ascending, whistled *tweeep* or *suuweep* and a shorter Crissal-like call. SONG: Very similar to Crissal's, being loud, melodious, and consisting of numerous phrases, sometimes repeated. Little or no mimicry of other species. Heard mostly at dawn and dusk.

STATUS & DISTRIBUTION Uncommon and local resident in sparse, sandy-soiled desert of southwestern AZ, southern CA, southern NV, and extreme southwestern UT. Also found in northeastern Sonora, Mexico, and Baja California. Very seldom wanders away from resident range.

POPULATION Populations generally declining due to urbanization and conversion of habitat into agricultural lands.

paler overall than Crissal

thin, long, strongly downcurved bill

tawny undertail coverts

very long tail

CRISSAL THRASHER *Toxostoma crissale* CRTH ■ 2

This normally secretive but distinctive thrasher sometimes sings quietly from exposed dead branches that stick up from the tops of shrubs in dense thickets. Found in a variety of habitats, from brushy desert to juniper hillsides at higher elevation, it is very difficult to find when not singing. Sometimes it can be seen running on the ground with tail cocked, similar to other desert thrashers. Crissal Thrasher feeds by probing its long bill into ground litter or holes. Along with the California and the Le Conte's, it forms a complex of uniformly colored thrashers with long, black, decurved bills and chestnut crissums. Polytypic (3 ssp.; 2 in N.A.). L 11.5" (29 cm)

IDENTIFICATION Sexes similar. Uniformly dark grayish brown; often shows a paler white throat, distinct black malar stripe, and chestnut undertail coverts. Worn individuals in late spring and early summer often lack the clear white throat, and the black malar stripe can appear quite faded. Yellow eye. Often detected by first hearing call or song.

GEOGRAPHIC VARIATION Differences minor; eastern nominate birds are slightly darker than westerly coloradense.

SIMILAR SPECIES Crissal Thrasher is larger, darker, and more uniform than Curve-billed and Bendire's Thrashers, with which it overlaps in desert washes in southern Arizona; all three can be found at the same locations. More similar in appearance to California Thrasher, but note non-overlapping range. Overlaps with Le Conte's (sometimes found together at the same location), but note Le Conte's much paler overall color, smaller size, less-patterned malar and head, and lighter tawny undertail coverts. Note also different calls.

VOICE CALL: A loud repeated *chideery*

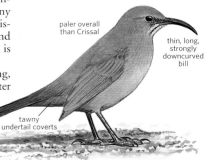

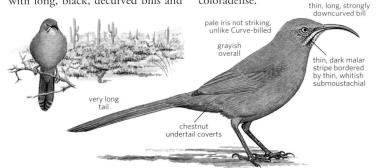

very long tail

pale iris not striking, unlike Curve-billed

grayish overall

thin, long, strongly downcurved bill

thin, dark malar stripe bordered by thin, whitish submoustachial

chestnut undertail coverts

or *churry-churry-churry* or *toit-toit-toit*. **SONG:** Varied long series of musical phrases, its cadence more leisurely and less harsh than the Curve-billed's. **STATUS & DISTRIBUTION** Uncommon resident in the understory of dense

mesquite and willows along streams; also ranging up to lower montane slopes at higher elevations. Very secretive, more so than other desert thrashers. Not known to wander like the Curve-billed does.

POPULATION Local population threats due to conversion of habitat into agriculture and urbanization. Not as adaptable to such threats as either the Curve-billed or the Bendire's.

Genus *Oreoscoptes*

SAGE THRASHER *Oreoscoptes montanus* SATH 1

This distinctive, small thrasher—characteristic of the open sagebrush plains of the montane West—is often seen singing atop sage bushes or along fence lines. It flies low from bush to bush and often runs on the ground with its tail cocked. On breeding grounds, the Sage is the only thrasher with heavily streaked underparts and a short, relatively straight bill. The Sage Thrasher, which is smaller than other thrashers, is migratory, vacating

breeding grounds during the winter. In migration and winter it favors open brushy areas, but can also be found feeding in fruiting trees, particularly junipers and Russian olives. Monotypic. L 8.5" (22 cm)
IDENTIFICATION Sexes similar. **ADULT:** Generally gray-brown above, white below with heavy black streaking and a salmon-buff wash to the flanks. Wings with thin, distinct white wing bars. Tail long with white corners. Obvious black malar stripe, pale yellow iris, and short, relatively straight bill. **LATE SUMMER ADULT:** Can show virtually no streaking on the underparts when in very worn plumage. **JUVENILE:** Back and head streaked. Streaking on underparts reduced and not as black, especially on flanks. Eye darker.
SIMILAR SPECIES The Brown and Long-tailed Thrashers also have heavily streaked underparts, but they differ by having more rufescent upperparts and longer, more decurved bills. Worn adults can lack streaking and look very similar to the Bendire's Thrasher, but the Sage has a more contrasting head pattern and retains suggestion of white wing bars. Note difference in calls and song. (See juvenile Mockingbird.)
VOICE CALL: Gives several calls, including a *chuck* and a high-pitched

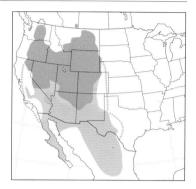

churr. **SONG:** Long series of warbled phrases.
STATUS & DISTRIBUTION Fairly common **BREEDING:** Restricted to specialized sagebrush habitat, mainly within and west of the Rocky Mountains. **MIGRATION:** In spring a very early migrant; in fall, most individuals leave breeding grounds by mid-Oct. Flocks occasionally detected during migration. **WINTER:** Mainly found in desert scrub habitat in the Southwest and northern MX. **VAGRANT:** Disperses widely. Regular vagrant to coastal CA, casual east of breeding range, with numerous fall and winter records from many eastern states and provinces east to the Atlantic seaboard.
POPULATION No threat known.

small size · short bill · streaked underparts · hitish il tips · worn

---

MOCKINGBIRDS Genus *Mimus*

Two members of this genus occur in North America; seven others are neotropical. Generally medium-size with long tails and relatively straight bills, they have gray or brown coloration, with varying amounts of white in the wings and tail. Mockingbirds, widely known as great songsters that mimic other species, are territorial and often sing at night.

NORTHERN MOCKINGBIRD *Mimus polyglottos* NOMO 1

This very common, conspicuous mimid of the southern United States is known for its loud, mimicking song, often heard during spring and summer nights in suburban neighborhoods. Both sexes aggressively defend nesting and feeding territories. They flash

their white outer tail feathers and white wing patches conspicuously during courtship and territorial displays. Seen often on wires and fences in towns, the Northern often feeds on berries during the winter. Monotypic. L 10" (25 cm)
IDENTIFICATION Sexes similar. **ADULT:**

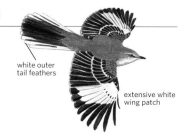

white outer tail feathers · extensive white wing patch

About the size of an American Robin, but thinner and longer tailed. Upperparts gray, unstreaked; underparts grayish white, unstreaked; long black tail has white outer tail feathers; conspicuous white wing bars; white patch at the base of primaries contrasts with blacker wings. Black line through a yellow eye. Bill relatively short and straight. JUVENILE: Underparts can be heavily spotted; upperparts with pale edging give back

caption: juvenile

and head a streaked appearance; black line through eye less distinct; eye darker. SIMILAR SPECIES The Loggerhead Shrike is similarly colored, but note distinct shape differences, particularly in the bill; shrike lacks white wing bars and has more extensive black mask. (Also see Sage Thrasher.) A long-staying Tropical Mockingbird (*M. gilvus*) from near Sabine Pass, TX, was not accepted on origin; it is resident as close as the Yucatán. VOICE CALL: A loud, sharp *check*. SONG: Long, complex song consisting of a mixture of original and imitative phrases, each repeated several times. Excellent mimic of other bird species. Often sings at night. STATUS & DISTRIBUTION Common and conspicuous. BREEDING: Nests in a variety of habitats,

including suburban neighborhoods. MIGRATION: Birds in the northern portion of range and at higher elevations migrate south during fall and winter. Birds in the southern portion of range are resident. VAGRANT: Birds are found casually north of mapped range to AK; accidental to Europe. POPULATION Range is expanding as a result of urbanization.

BAHAMA MOCKINGBIRD *Mimus gundlachii* BAMO 4

This Caribbean species occurs as a very rare vagrant to the south Florida mainland, the Florida Keys, and the Dry Tortugas. Generally more secretive than the Northern Mockingbird, the Bahama Mockingbird remains hidden in dense thickets. It behaves and looks more like a thrasher than a mockingbird, often running between thickets. It is also browner and more streaked above and below than the Northern Mockingbird. Polytypic (2 ssp.; nominate in N.A.). L 11" (28 cm) IDENTIFICATION Sexes similar. ADULT: Distinctive black streaks on a brown

caption: gundlachii

caption: distinct streaks on lower sides, flanks, and undertail coverts

caption: lacks white wing patch of Northern Mockingbird

caption: darker tail than Northern Mockingbird

caption: whitish tail tips

back extend up neck onto head; black streaking found on lower belly, flanks, and undertail coverts. Wings have two relatively narrow white wing bars but no white at base of primaries. Ttail has white tips to outer feathers. Head has obvious black malar stripe and white eyebrow. Bill is longer and slightly more curved than Northern Mockingbird's. IMMATURE: Less patterned than adult, with less distinct streaking on back and flanks. Dark malar is less pronounced. SIMILAR SPECIES Immature Northern Mockingbird has spotting or streaks below, but this is more confined to breast, compared with the Bahama's streaked flanks. All plumages of the

Bahama lack the white flash in primaries; all plumages also have white in tail confined to the tips of the outer feathers. VOICE CALL: A loud *chack* similar to the Northern Mockingbird's. SONG: More thrasher-like, with a complex mixture of phrases. Not known to mimic. STATUS & DISTRIBUTION The nominate subspecies is a rather uncommon resident in the Bahamas and on cays off northern Cuba. The subspecies *hillii* is a local resident on Jamaica. VAGRANT: Very rare visitor, primarily in spring, to parks with dense thickets along the coast of southern Florida, with most records from between West Palm Beach and Key West. Also recorded from the Dry Tortugas in spring. POPULATION Potential loss of breeding habitat due to development, and loss of habitat after hurricanes, may impact this species in the Bahamas.

STARLINGS Family Sturnidae

European Starling

With their glossy plumages and sociable ways, it is not surprising that starlings and mynas have been popular with bird fanciers for generations. The approximately 114 species of Sturnidae are native to the Old World, but several species have been introduced across the globe. The four species treated here have had varying degrees of success in North America: The widespread European Starling is probably the most successful exotic, and the adaptable Common Myna is flourishing in south Florida; the Hill Myna is declining in Florida, and the Crested Myna has been extirpated.

Structure Most starlings are of medium size and medium build, with fairly thick, slightly decurved bills and sturdy feet. Mynas are on average larger, chunkier bodied, thicker billed, and shorter tailed than many starlings, and many species have ornate plumes or bare-part projections on the head and face.

Behavior Many species are gregarious, breeding colonially and gathering in immense numbers during the nonbreeding season. Social behavior is often advanced. Vocalizations are complex; many notes are coarse and unmusical, but several species are capable of a large array of sounds. Starlings and mynas adapt well to human-altered and human-inhabited landscapes, and introductions have succeeded in many regions.

Plumage Starlings vary in plumage, but many are largely black and iridescent, with patches of bright oranges, reds, and yellows on the feathers. Bare parts are frequently brightly colored. The species treated here are primarily black and iridescent with limited areas of white and bright yellow patches on some bare parts.

Distribution Ancestrally, starlings and mynas, which tend to be cavity nesters, were associated with woodlands in Africa, Southeast Asia, and their regional island complexes. For millennia, the family has been increasingly cosmopolitan in its range and generalist in its habitats. In North America, the European Starling is widely established; other established species are local.

Taxonomy Most authorities treat the Sturnidae family as a valid monophylous taxon. Where the Sturnidae family fall among the passerines, however, is unclear: Recent evidence points to affiliations with the Turdidae, Muscicapidae, and Mimidae. Within the Sturnidae, there may be a major break between the African and Asian taxa. The polyphyletic mynas belong to the Asian group, as does the European Starling.

Conservation Worldwide, the Sturnidae run the gamut from abundant nuisances to species in danger of extinction. Habitat loss and capture for the cage-bird trade have been factors in the decline of several species. BirdLife International lists four species as vulnerable, one as endangered, and three as critically endangered; five starlings with small island ranges have gone extinct. Established or escaped populations in North America are sometimes targeted for control.

Genus *Sturnus*

EUROPEAN STARLING *Sturnus vulgaris* EUST ■ 1

Widespread and abundant in much of North America, the introduced European Starling is arguably and problematically the most successful bird on the continent. Often characterized as bold, this bird is actually fairly wary and can be difficult to approach. Polytypic (13 ssp.; nominate in N.A.). L 8.7" (22 cm)

IDENTIFICATION Stocky and short tailed, often seen strutting about lawns and parking lots. Flight profile distinctive: Wings look triangular in flight and can appear somewhat translucent. **ADULT:** One molt per year, but fresh fall adults look very different from summer birds. On freshly molted birds, black plumage has white spots all over; by winter, spots start to disappear from wear and then they actually break off; and by spring, the birds are glossy black all over, with strong suffusions of iridescent pinks, greens, and ambers. Slender and pointed bill usually gray in fall and yellow by winter. **MALE:** By spring has blue-based bill. **FEMALE:** By spring has pink-based bill; paler eyes. **JUVENILE:** Distinctive; dark gray-brown feathering all over. Birds

short tail

vulgaris

triangular wings

plumage heavily spotted with whitish

fall
vulgaris

slender dark bill

short tail

juvenile *vulgaris*

dull, blurred streaks on belly

overall grayish brown

yellow bill

glossy plumage

winter *vulgaris*

breeding ♂
vulgaris

begin a complete molt into adult-like plumage soon after fledging, and briefly exhibit a striking mosaic of juvenal and adult feathers.

GEOGRAPHIC VARIATION A specimen from Shemya I., western Aleutians, was nominate *vulgaris*, thus from N.A.; Asian *poltaratskyi* is casual to Japan.

SIMILAR SPECIES Structure is distinctive, but sometimes confused with unrelated blackbirds, which often occur with starlings in large flocks. Blackbirds more slender bodied, with longer tails and less pointed wings. Flight profile more like a waxwing's or a meadowlark's than blackbird's.

VOICE Highly varied. **CALL:** Commonly heard calls include drawn-out, hissing *sssssheeeer* and whistled *wheeeeoooo*. **SONG:** Elaborate, lengthy (>1 min. long), with complex rattling and whirring elements, and overall wheezy quality; call notes may be incorporated into song. Imitates other species, especially those with whistled notes (e.g., Killdeer, Eastern Wood-Pewee).

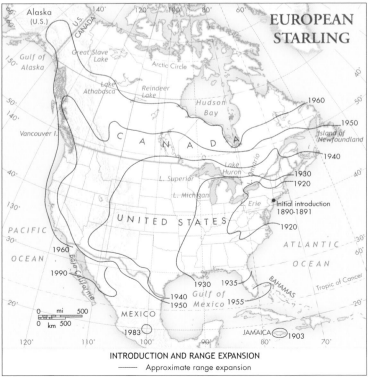

EUROPEAN STARLING

Alaska (U.S.)

Gulf of Alaska

Great Slave Lake

Arctic Circle

Lake Athabasca

Reindeer Lake

Hudson Bay

1960

1950

Island of Newfoundland

Vancouver I.

C A N A D A

1940

L. Superior

Lake Huron

1930
1920

L. Michigan

Erie   Initial introduction 1890-1891

1920

PACIFIC

ATLANTIC

OCEAN

OCEAN

1960

1990

Baja California

UNITED STATES

1930 1935

1940 Gulf of
1950 Mexico 1955

BAHAMAS

Tropic of Cancer

MEXICO

1983

JAMAICA 1903

0   mi   500
0   km   500

INTRODUCTION AND RANGE EXPANSION
——— Approximate range expansion

STATUS & DISTRIBUTION Old World species. Abundant, still expanding range in the Americas. **BREEDING:** Needs natural or artificial cavities. Often evicts native species from nest holes. **MIGRATION:** Withdraws in winter from northern portion of range. **WINTER:** Gregarious, with largest concentrations around cities, feedlots. **POPULATION** Successfully introduced in Central Park, New York, 1890–91; across continent by late 1940s. Population currently exceeds 200 million. A specimen taken on Sept. 12, 2012, at Shemya I., AK, appears to be *vulgaris,* thus originating from N.A., rather than *poltaratskyi,* which has reached Japan.

Genus *Acridotheres*

COMMON MYNA *Acridotheres tristis*   COMY ■ 2

This Asian native is well established in south Florida. It is a popular cage bird (the "House Myna"), and escapes may be seen anywhere. Polytypic (2 ssp.). L 9.8" (25 cm)
IDENTIFICATION Chunky; often in loose interspecific or intraspecific flocks. **ADULT:** One molt per year. Molt schedule variable; Florida data lacking. Brown body; black head with yellow-orange bare patch around eye; yellow-orange bill; dull yellow-orange feet; white patches at base of primaries (conspicuous in flight), tail tip, and undertail coverts. **JUVENILE:** Duller overall.
GEOGRAPHIC VARIATION Populations in N.A. presumably of widespread nominate subspecies.
SIMILAR SPECIES Resembles other *Acridotheres* mynas. Outside of N.A., hybridizes with the Crested and other *Acridotheres* mynas, especially in introduced settings.
VOICE Varied, but most vocalizations raucous; accomplished mimic. **CALL:** Loud, slurred whistles; short, grating notes. **SONG:** Long series of short notes, given by both sexes, with liquid, rolling syllables. Call notes often incorporated into song.
STATUS & DISTRIBUTION First detected in FL in 1983; now common in southern FL. **YEAR-ROUND:** Warehouses, malls, fast-food restaurants, etc.; nests primarily in artificial cavities, especially signage. **VAGRANT:** Escapes possible anywhere.
POPULATION FL population has increased considerably; perhaps thousands of pairs currently breeding.

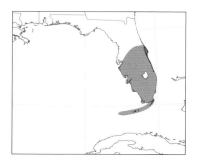

yellow skin around eye
yellow bill
blackish head
brown above and below
white wing patch
tristis
white undertail coverts
white tail tip

Genus *Gracula*

HILL MYNA *Gracula religiosa*   HIMY ■ EXOTIC

Even when judged by the overall standards of the myna clan's vocal excellence, the Hill is notable for its superior abilities as a songster and mimic. An established population in south Florida is waning. Polytypic (10 ssp.; *intermedia* in N.A. from India, Thailand, Indochina, and southern China). L 10.6" (27 cm)
IDENTIFICATION A large and chunky myna. **ADULT:** One molt per year; sexes similar. Black overall; white primary bases, shows as a bold white patch in flight. Warty yellow nuchal and suborbital bare patches conspicuous. Dark eyes, orange bill, yellow-orange feet. **JUVENILE:** Plumage browner, duller; bare parts less colorful.
SIMILAR SPECIES Chunkier, marginally larger than *Acridotheres* mynas. Wings broader, more rounded. Tail tip not white, as in *Acridotheres*. Hill Myna is usually found in trees; *Acridotheres* mynas are found on the ground.
VOICE Varied; many vocalizations, especially of birds in captivity, have "human" qualities. **CALL:** Tremendously varied, and with marked variations among subspecies in its native range in southern Asia. Most are of short duration, some extremely loud. **SONG:** Complex; short to long series of chuckles and whistles. Song elements on average mellower than those of *Acridotheres* mynas.
STATUS & DISTRIBUTION A native of Asian forests from India to the Philippines. Uncommon around Miami, FL. **YEAR-ROUND:** Arboreal; nests primarily in natural cavities. **VAGRANT:** Escapes from captivity (not true vagrants) possible anywhere.
POPULATION Remnant introduced population in Miami area currently numbers 25 to 100 breeding pairs.

yellow wattle
thick orange bill
white wing patch
intermedia

ACCENTORS Family Prunellidae

Siberian Accentor (ID, Feb.)

The accentors are an Old World family, found from Spain, across Eurasia, to the Russian Far East and Japan, with one species, the Yemen Accentor, a montane endemic to that country. The greatest diversity of species is found in the mountains of central Asia. Some species are migratory; others are resident.

Structure Accentors are small to medium-size passerines, similar in size and shape to sparrows although accentors have thin, warbler-like bills. Some species are stocky. The tail is of medium length and typically notched.

Plumage Accentors are brownish above, often with rufous tones, and streaked on the upperparts. Some species are largely grayish below while others are buffy or cinnamon below, some showing contrasting white or black throats. Many have a contrasting face pattern with a noticeable paler or warm-colored supercilium.

Behavior These ground foragers specialize on insects, but take seed in winter. Most are little known, except the familiar Dunnock of Europe and the largest member of the family, the Alpine Accentor, a widespread species across Eurasia.

Distribution Restricted to the Palearctic, from the British Isles to Japan, the family is confined to temperate latitudes as far south as North Africa and southern Asia. Several species are restricted to alpine or highland areas. One species has been introduced to South Africa and New Zealand.

Taxonomy There are 13 accentors, all in the genus *Prunella*. Thrushes may be closely related to accentors, but recent genetic studies have not adequately determined the placement of the accentors. Evidence places the accentors in the Passeridae, a group that includes finches and sparrows, tanagers, blackbirds, sunbirds, weavers, pipits and other taxa. Therefore, accentors appear to be distantly related to the thrushes, which are in an entirely different group.

Genus *Prunella*

SIBERIAN ACCENTOR *Prunella montanella*   SIAC ■ 4

The Siberian Accentor is a much-hoped-for vagrant in Alaska, due to its subtle but striking appearance and its being the one member of its family that visits N.A. Polytypic (2 ssp; *badia* in N.A.). L 5.5" (14 cm)

IDENTIFICATION A sparrow-like bird that forages on the ground, with hunched posture. Unlike a sparrow, it shows a thin warbler-like bill. The structure is otherwise like a sparrow, with a medium-length, notched tail. **ADULT:** Rufous brown above with diffuse streaking, contrasting with marked face and tawny-to-cinnamon underparts. Lacks bold wing bars. Thin, buffy lower wingbar is well formed. Head is dark, with a bold and contrasting tawny supercilium, as well as a diffuse tawny spot on rear ear coverts, gray central crown stripe, and gray nape. Tawny throat contrasts with dark auriculars. Flanks are streaked rufous brown. Belly and vent are white. Legs are pink. Bill is dark, sometimes showing a paler horn base to lower mandible. **IMMATURE:** Much like adult, duller in plumage with less grey on nape; some retain dark speckling on breast from juvenile plumage.

SIMILAR SPECIES Siberian Accentor's bold, dark-and-tawny head pattern is distinctive for a small, ground-dwelling songbird. Rustic Bunting, another Asian vagrant, shows a similar rusty brown plumage and bold face pattern. However, the bunting has a conical bill, whitish supercilium, and whitish throat. Similarly,

the tawny-colored Smith's Longspur has a bold head pattern, but has a conical bill, white supercilium, and white face patch.

VOICE CALL: High-pitched *tsee-ree-seee*, with the quality of Brown Creeper calls. **SONG:** A squeaky, high-pitched warble, like Winter Wren, although slower and less complex.

STATUS & DISTRIBUTION Breeds in Asia; winters in central and eastern China and the Koreas. **VAGRANT:** Casual but nearly annual in recent years in fall from St. Lawrence I.; recorded elsewhere on Bering Sea Islands and western Aleutians, AK. Winter and early spring records from interior (near Fairbanks) and southeastern AK, BC, AB, MT, ID, and WA suggest that a few regularly winter in northwestern N.A. Most winter records have been at bird feeders. Vagrant in northern Europe. In eastern Asia, this is an early spring migrant.

dark crown with gray center
bold, rich buff supercilium
gray nape
rufous streaks on back
thin bill
dark cheek with a few buffy spots
rich buff below
badia

WAGTAILS AND PIPITS Family Motacillidae

Eastern Yellow Wagtail (AK, May)

Slender and highly migratory passerines from the Old World, wagtails and pipits have long tails and characteristic white outer tail feathers. Wagtails are typically black and white or yellow, while pipits are brown and streaked, blending more into their environment. Both share the distinctive behavior of walking on the ground and wagging their tails. They are generally found on or near the ground, usually in tundra habitats or near streams and rivers. Most species in North America are mainly found in the Arctic of western Alaska; five species (two wagtails and three pipits) breed in North America, while five Eurasian species occasionally occur as migrants and vagrants.

Structure Pipits are generally small to medium-size, slender birds with long tails, except for the Sprague's Pipit, which has a relatively short tail. Their bills are also long and slender, and somewhat pointed. The pipits adapted long legs and toes, unusual for passerines, for walking on the ground. Wagtails are shaped similarly to pipits, but they have much longer tails.

Behavior Wagtails and pipits feed on the ground, walking and searching for insects. Usually both pipits and wagtails forage at the edge of streams and pools of water, often continuously wagging their tails as they walk, as well as bobbing their heads and necks like a chicken. Some species, the Sprague's Pipit in particular, rarely if ever wag their tails. During the spring and summer, wagtails sing from exposed perches on shrubs, small trees, roofs, and even rusting machinery. Pipits have elaborate flight displays, but sometimes sing from the ground. Some pipits remain in flight singing for up to 20 minutes. Wagtails have a distinctive undulating flight, while pipits fly more directly, but both call often in flight; their species-specific calls aid in identification. The Sprague's Pipit, when flushed, rises gradually and then circles around before it plunges back to Earth, breaks its fall immediately above the ground, and disappears into the grass. Mostly diurnal migrants, wagtails and pipits use rivers and coasts as navigation aids (except the Sprague's, which migrates from the Great Plains to grasslands in the southern United States and northern Mexico). The American Pipit forms large single-species flocks in winter.

Plumage Wagtails are sexually dimorphic with several age, seasonal, and sex-related plumages—generally combinations of black, white, and yellow, with long tails and white outer tail feathers. Telling identification features include the extent of white in the wing, the color of the back, and the pattern of black on the underparts. Pipits are not sexually dimorphic, but some do have different winter plumages. Most are brown with plain or streaked backs, varying amounts of streaking on the underparts, and white outer tail feathers. Important features to notice include the degree of streaking on the back and underparts, the presence of buff on the underparts, and leg coloration.

Distribution North American wagtails breed almost exclusively in coastal arctic Alaska; but are regularly reported on offshore islands in the Bering Sea during migration; and some have been seen along the West Coast, casually inland, in fall and winter. Pipits generally restrict themselves to arctic regions as well, breeding on rocky tundra; however, the American Pipit also breeds above tree line in the Rocky Mountains and Pacific states, and the Sprague's Pipit only breeds on the short-grass prairies of the Great Plains.

Taxonomy Worldwide there are at least 64 species in five genera. Recent genetic studies have influenced the determination of the species limit for some wagtails (White and Eastern Yellow) and pipits (American).

Conservation The motacillids are not threatened in North America; however, overgrazing and the introduction of non-native grasses are limiting the breeding and wintering grounds for the Sprague's Pipit. BirdLife International lists three species as vulnerable and two as endangered.

WAGTAILS Genus *Motacilla*

Motacilla is mainly an Old World genus, with two breeding and two vagrant species in North America, predominantly in the Arctic. Highly migratory, they winter in Asia, Europe, and Africa. They have distinctive calls and an undulating flight pattern. The immatures present the most trouble in terms of species identification.

EASTERN YELLOW WAGTAIL *Motacilla tschutschensis*   EYWA ▪ 2

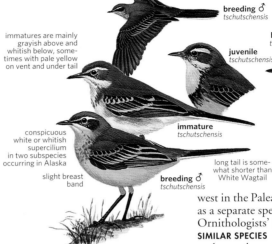

breeding ♂
tschutschensis

breeding ♀
tschutschensis

juvenile
tschutschensis

immatures are mainly grayish above and whitish below, sometimes with pale yellow on vent and under tail

breeding ♂
simillima

immature
tschutschensis

averages brighter than nominate subspecies with less evident breast band

conspicuous white or whitish supercilium in two subspecies occurring in Alaska

long tail is somewhat shorter than White Wagtail

slight breast band

breeding ♂
tschutschensis

Formerly considered a subspecies of the Yellow Wagtail, the Eastern Yellow Wagtail is one of the more characteristic arctic passerines of western Alaska. Birders commonly see the bird singing from the tops of willows or feeding in the short, grassy tundra. Eastern Yellow walks on the ground, where it searches for insects. Polytypic (3–5 ssp.; 1–2 in N.A.). L 6.5" (17 cm)

IDENTIFICATION Eastern Yellow Wagtail has a slender body with relatively long legs. **BREEDING MALE:** Quite distinctive. It generally is bright yellow underneath and green above with a gray crown and has a narrow white supercilium contrasting with dark gray cheeks. It has a relatively long black tail with white outer tail feathers; its wings have pale yellow edging to coverts and flight feathers, forming indistinct wing bars. **FEMALE:** She is slightly duller than male and not as yellow underneath and is less dark on cheek and less gray on crown. **JUVENILE:** More pipitlike. Lacks yellow and is a dull grayish brown above, with narrow white supercilium, and grayish white below, with varying amounts of speckling on sides of breast and buffy flanks. It also has narrow

white wing bars and a white throat with varying amount of dark malar. **FIRST WINTER:** Similar to juvenile, but it has less dark in malar region.

GEOGRAPHIC VARIATION Two subspecies have been identified from North America. The Alaska breeding nominate *tschutschensis* has a speckled breast band. The subspecies *simillima*, regular on the Aleutian and Pribilof Islands, averages brighter yellow underneath and greener above, and typically lacks speckling on sides of breast. Some authorities do not recognize this subspecies and merge it into *tschutschensis*. Two other Asian subspecies are distinctive: *macronyx* with a dark face and *taivana* with a bold yellow eyebrow. These East Asian subspecies are genetically distinct and differ vocally from the more numerous subspecies. The birds found to the

west in the Palearctic are now treated as a separate species by the American Ornithologists' Union.

SIMILAR SPECIES Adults appear similar to the much rarer (in N.A.) Gray Wagtail female, but Gray Wagtail actually has blacker wings and a longer tail. An immature can be confused with an immature White Wagtail, but Eastern Yellow Wagtail is browner, has less distinct wing bars, and usually lacks dark on sides of breast. See also Citrine.

VOICE CALL: A loud *tsweep*. **SONG:** A series of high-pitched *tszee tszee tszee* notes.

STATUS & DISTRIBUTION Common, but has a restricted range in N.A. **BREEDING:** Nests in willows in tundra along the Bering Sea in western and northern AK. **MIGRATION:** Trans-Beringian migrant; regular on islands in Bering Sea. Often seen migrating in large numbers in late summer along the Bering Sea. **WINTER:** Mainly S.E. Asia and Philippines. **VAGRANT:** Casual in early fall along West Coast of mainland U.S., with more than a dozen records from coastal CA.

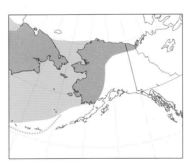

CITRINE WAGTAIL *Motacilla citreola*   CIWA ▪ 5

A rare vagrant to North America, the Citrine Wagtail is typically associated with water. It often mixes with Yellow and Eastern Yellow Wagtails on winter grounds in Southeast Asia and Europe. Polytypic (3 ssp.). L 6.5" (17 cm)

IDENTIFICATION Resembles Eastern Yellow Wagtail in size and general

proportions. **BREEDING MALE:** Unmistakable; brightly yellow on head and underparts with a gray back and unique black nape. Wings have bold white tips and edges to coverts and tertials. **FEMALE & WINTER MALE:** Note hollow, grayish brown cheek completely surrounded by yellow; grayish back and crown; and white wing

bars. **FIRST-WINTER:** Gray and white; resembles female, but lacks yellow in its plumage. Note diagnostic head pattern described above.

SIMILAR SPECIES Other than the breeding male, Citrine Wagtails resemble Eastern Yellow Wagtails, but note Citrine's much broader and encircling supercilium, more prominent

dark ear coverts with pale center, completely surrounded by pale yellow

bold white wing bars and tertial edges

winter adult ♂

white wing bars, gray upperparts, and pale lores.

VOICE CALL: Very similar to the loud *tsweep* call of Eastern Yellow Wagtail. **SONG:** Often given from atop a willow bush; resembles song of White Wagtail.

STATUS & DISTRIBUTION Locally common Eurasian species. Breeds in wet areas with low vegetation, open bogs, soggy riverbanks, and meadows. **WINTER:** Mainly in southern Asia; accidental in N.A. **VAGRANT:** One winter record from Starkville, MS (Jan. 31–Feb. 1, 1992) and an immature from Comox, Vancouver I., BC, was discovered on Nov. 14, 2012, and remained through the winter.

WHITE WAGTAIL *Motacilla alba*   WHWA ■ 3

Formerly considered two species (White and Black-backed Wagtails), the White Wagtail generally prefers more disturbed habitats than the Eastern Yellow Wagtail and favors rocky breakwaters, around abandoned, rusting machinery, garbage dumps, or buildings. It sings from a high exposed perch and walks on ground when feeding. Spotted by its distinctive call notes and undulating flight. Polytypic (9–11 ssp.; 3 in N.A.). L 7.3" (18 cm)

IDENTIFICATION Sexually dimorphic with complicated age differences. Adults distinct, overall gray, black throat, black on hind crown, white in wings; immatures more problematic. Tail black with white outer feathers. Separation of two subspecies complicated due to hybridization. **BREEDING MALE:** In *ocularis*, a gray back, black hind crown and nape, white forehead, black throat extending to bill, and a thin black line through eye. Wing coverts entirely white. In *lugens*, a black back, nape, and hind crown; black extending farther forward on crown. Black throat with white chin. Wider black line through eye. Entire wing, including flight feathers, mostly white. **BREEDING FEMALE:** In *lugens*, some gray in back. **WINTER FEMALE:** In *ocularis*, a gray back and coverts have less white, forming more distinct wing bars. In *lugens*, some black on its scapulars and mostly white wing coverts (patch). **IMMATURE:** Variable. Generally uniform gray above, whiter below, with varying amounts of black on bib. Separation to subspecies should be done very carefully or preferably not at all.

GEOGRAPHIC VARIATION The subspecies *ocularis* is the regular breeder on AK mainland and St. Lawrence I.; *lugens* is a casual summer visitor to the Aleutians and St. Lawrence I. These two subspecies are the only ones with a black post-ocular line; the other subspecies are white-faced. They hybridize to an unknown extent where they come into contact in northern Kamchatka; no detailed study has been done from this region. Nine other subspecies occur across Europe and North Africa. Nominate European *alba*, breeding as close as Iceland and Greenland, has a white face.

SIMILAR SPECIES Juvenile looks similar to Eastern Yellow juvenile, but White is grayer and has more distinct wing bars. Note different calls.

VOICE CALL: A distinctive 2-note *chizzik* given in flight; also a *chee-whee* given from the ground or an exposed perch. The vocalizations are identical between subspecies.

STATUS & DISTRIBUTION Rare in western AK. **BREEDING:** Nests sparsely in villages on Seward Peninsula and St. Lawrence I.; most breeding birds are *ocularis*, but mixed pairs with *lugens* or hybrids have been encountered. **WINTER:** Mainly eastern Asia. **VAGRANT:** Both *lugens* and *ocularis* are casual along the West Coast, mainly in fall and winter. Accidental in Sonora, Baja California, the Southwest, and the East. Records involve both *lugens* and *ocularis*, but some were not identified to subspecies. European *alba* is accidental on the East Coast.

long tail

all wagtails fly with undulating flight and call often

long tail

breeding adult ♂
lugens

adults show extensive white in wing

black back

breeding adult ♂
lugens

breeding
ocularis

pearly gray back

both subspecies have a black eye line

black bib extends to chin

chin usually white

breeding adult ♀
lugens

winter adult ♂
lugens

identification of immatures to subspecies is problematic

immature
lugens

European subspecies has white face with no dark eyeline

breeding ♂
alba

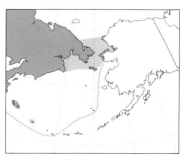

GRAY WAGTAIL *Motacilla cinerea*   GRAW ◼ 4

This Eurasian wagtail is a casual spring migrant to western Alaska. Almost always associated with water, the long-tailed Gray Wagtail typically walks on the ground, often at the edge of a stream or river. Like the White Wagtail, it is often detected in flight by a distinct two-note call and its undulating flight pattern. Polytypic (3–6 ssp.; *robusta* in N.A.). L 7.8" (20 cm)
IDENTIFICATION ADULT MALE: Unmistakable. It is quite distinctive, with yellow below, gray above, a narrow white supercilium, an extensive black throat, and black wings. Wings have white tertial edging, and a white base to secondaries forms a wing stripe that is visible in flight only. It also has pale flesh-colored legs. **FEMALE:** It does not have a black throat and is duller than male, lacking much of the yellow on underparts; it is brightest yellow on undertail coverts. The most distinct

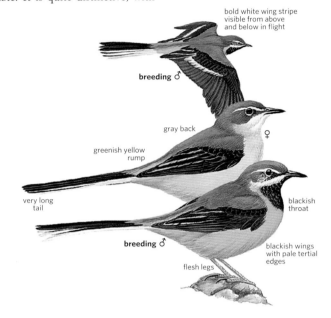

bold white wing stripe visible from above and below in flight

breeding ♂

gray back

greenish yellow rump

very long tail

breeding ♂

flesh legs

blackish throat

blackish wings with pale tertial edges

♀

field marks are its very long tail, longer than other wagtail species, and lack of wing bars. **JUVENILE:** Similar to female, but browner overall, buffy across breast, and has two buffyish wing bars. It already has bright yellow undertail coverts.
GEOGRAPHIC VARIATION Much of the geographic variation is clinal, notably the tail length, which decreases in Eurasia from west to east, thus East Asian *robusta* has a slightly shorter tail. It also averages slightly deeper yellow below. Some authorities merge *robusta* into nominate *cinerea* and recognize only three subspecies.
SIMILAR SPECIES The combination of yellow underparts and black throat make the male unique. The female, however, looks similar to Eastern Yellow Wagtail, but lacks wing bars and has grayer upperparts, a much longer tail, and duller underparts that contrast with brighter yellow undertail coverts. Gray Wagtail also has flesh-colored legs, whereas Eastern Yellow's are black.
VOICE CALL: A two-note metallic *chink chink,* somewhat similar to White's call, and very different from the single note call of Yellow Wagtail.
STATUS & DISTRIBUTION Eurasia. **MIGRATION:** Regular spring migrant to the western Aleutian Is., otherwise a casual late spring stray to islands in the Bering Sea (Pribilof Is. and St. Lawrence I.). **VAGRANT:** Accidental to CA and BC in the fall.

PIPITS Genus *Anthus*

Three breeding and three vagrant species occur in North America. They winter in grasslands and agricultural fields, where their cryptic plumage protects them. The species exhibit little or no sexual dimorphism, except for Red-throated Pipit. Some species are secretive. The majority of N.A. species are gregarious, especially in migration and winter. All are highly migratory.

TREE PIPIT *Anthus trivialis*   TRPI ◼ 5

similar to Olive-backed Pipit, but browner above with more distinct back streaks

fine flank streaks

Also known as the "Brown Tree Pipit," this mainly western Palearctic species is an accidental visitor to North America. It will perch up in trees, but it also walks on ground. It is more secretive than American Pipit. Polytypic (2 ssp.; nominate in N.A.). L 6" (15 cm)
IDENTIFICATION Sexes similar. Overall brown, boldly streaked blackish above, with a necklace of fine streaking across breast with underlying yellowish wash and faint streaking down flanks. Median coverts with pale tips, forming one indistinct wing bar. Face pattern diffuse with short, narrow, pale eye line. Note unstreaked rump and uppertail coverts. Pale flesh-colored legs. In fall, fresher with rich buff wash across breast.
SIMILAR SPECIES Olive-backed has a less streaked back and a bolder face pattern with a dark spot in the rear of the ear coverts. Pechora and non-adult Red-throateds have bold white streaks on back and bold black streaks down flanks; in Red-throateds, background color of the underparts is uniform throughout.

VOICE CALL: A thin, buzzy *teez*, similar to Olive-backed and very different from American Pipit.
STATUS & DISTRIBUTION Common in Eurasia, where it breeds in a variety of forested habitats. **VAGRANT:** Three records in N.A. of the nominate subspecies from the northern Bering Sea region: two records from Gambell, St. Lawrence I. (June 6, 1995, and Sept. 21–27, 2002) and one specimen record from Wales at the end of the Seward Peninsula (June 23, 1972). The latter record was of a singing bird in display flight.

OLIVE-BACKED PIPIT *Anthus hodgsoni* OBPI ■ 3

This Eurasian pipit is a rare migrant to North America, where it mainly appears in western Alaska and treeless areas. The Olive-backed is somewhat more secretive than the American Pipit. It constantly pumps its tail when it walks on the ground. Polytypic (2 ssp.; *yunnanensis* in N.A.). L 6" (15 cm)
IDENTIFICATION Sexes look similar. Olive-backed is a boldly patterned pipit with a relatively plain, faintly streaked grayish olive back and head; rich buff breast that contrasts with white belly; boldly streaked black breast, sides, and flanks; and a broad, dark malar stripe. Uniquely patterned face has a supercilium that is buff in front of eye and white behind; in addition, there is a blackish spot at rear of auriculars that is bordered above by a white spot. Bright pinkish legs. Freshly molted birds are buffier below, more olive above, and more boldly patterned.
GEOGRAPHIC VARIATION The subspecies *yunnanensis* breeds in northern Asia and winters in Southeast Asia and is only faintly streaked on back. The more sedentary subspecies *hodgsoni* from the Himalaya and central and southern Japan is more strongly streaked on back. In Hokkaido the two subspecies intergrade.
SIMILAR SPECIES No other pipit has the same combination of white underparts, relatively plain olive back, rich buff wash on streaked breast, and unique face pattern.
VOICE CALL: A high, thin, buzzy *tseee*, similar to Tree Pipit, but shorter and less descending than Red-throated Pipit call.
STATUS & DISTRIBUTION The subspecies *yunnanensis* breeds across Russia and winters in Southeast Asia. **MIGRATION:** Rare, but regular migrant to the western Aleutians (e.g., Attu). **VAGRANT:** Casual stray in both spring and fall to islands in the Bering Sea (Pribilofs and St. Lawrence). Accidental in fall to CA and Baja California, and in spring to NV (specimen, May 16, 1967, from ten miles south of Reno was the first N.A. record!).

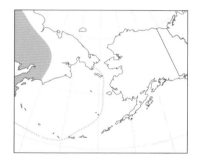

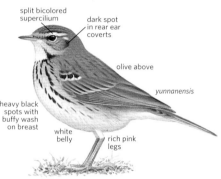

split bicolored supercilium
dark spot in rear ear coverts
olive above
yunnanensis
heavy black spots with buffy wash on breast
white belly
rich pink legs

PECHORA PIPIT *Anthus gustavi* PEPI ■ 4

back feathers with black centers and rufous edges
streaks through nape
white "braces"
distinct primary projection, unlike other pipits
buffy wash on breast
white belly
gustavi

This mainly Asian pipit is quite secretive and usually solitary. It remains hidden in grass and is often silent when flushed but call is a good field character. Polytypic (2–3 ssp.; *stejnegeri* in N.A.). L 5.5" (14 cm)
IDENTIFICATION Sexes similar. Primary tips project beyond tertials, unlike on other pipits. Rich rufous-brown above with heavy streaks, contrasting white "braces" on sides of back; streaked nape; rich buffy yellow wash across breast and flanks is heavily streaked black and contrasts with white belly; two distinct white wing bars; diffuse face pattern; rump streaked; pinkish legs.
GEOGRAPHIC VARIATION Nominate subspecies occupies most of range. Aleutian specimens were identified as stejnegeri that breeds only on Commander Is. Recent authorities have merged that subspecies into nominate *gustavi*. Another, more southerly subspecies *menzbieri* breeding in northeast China and southeasternmost Russian Far East is smaller and duller, and has a markedly different song; winter grounds unknown, but has been collected in Philippines.
SIMILAR SPECIES First-winter or adult female Red-throated Pipit, which sometimes lacks red coloration, looks similar. See Red-throated Pipit.
VOICE CALL: A hard *pwit* or *pit*. Very different from other pipits.
STATUS & DISTRIBUTION Local breeder across much of subarctic Russia, including Commander Is.; winters mainly in Philippines, also northern Borneo and Wallacea. **VAGRANT:** Casual spring migrant in western Aleutians. Almost annual in recent years (fall) at Gambell, St. Lawrence I.

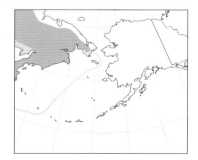

RED-THROATED PIPIT Anthus cervinus   RTPI ■ 3

The Red-throated male is one of the more recognizable pipits in North America. Secretive, it hides in the grassy tundra, but it can be conspicuous while feeding in the open at the edge of small pools. It often calls when flushed, helping in the identification of females and immatures. In spring, the male sings from atop boulders or in a display flight. Polytypic? (2 ssp,; *cervinus* in N.A. birds, but most authorities regard the species as monotypic). L 6" (15 cm)

IDENTIFICATION BREEDING MALE: Unmistakable. It has a unique bright tawny-red head, throat, and breast, which varies in intensity; rest of underparts are buffy white with streaking along sides of breast and flanks. **ADULT FEMALE:** It shows little or no red (if present it's mainly confined to throat and upper breast) and has heavier black streaking

on breast and sides. Back is heavily streaked, sometimes quite buffy, other times white. Crown is finely streaked, but nape is unpatterned. It also has noticeable whitish wing bars, little or no primary projection, and pale legs. **IMMATURE:** It shows no red anywhere. Streaking (sometimes very white) above is heavy, and more so on breast and sides. **SIMILAR SPECIES** Adults in bright plumage do not look like any other pipit. Immatures and dull females can be confused with much rarer Pechora Pipit, but note Red-throated's uniformly colored underparts, unpatterned nape,

unbroken eye ring, and lack of primary projection. The two species also have very different calls. **VOICE CALL:** Given in flight; a high, piercing *tseee* that descends at the end. **SONG:** A long, varied series of chirps, whistles, and buzzy notes, given from the ground or from a display flight. **STATUS & DISTRIBUTION** Common on Russian arctic tundra. **BREEDING:** Has bred in short, rocky tundra in western AK along Bering Sea. **MIGRATION:** Regular spring and fall migrant to islands in Bering Sea (mainly St. Lawrence I.); spring peak in early June; fall peak in late Aug. **VAGRANT:** Rare fall migrant along CA coast, mainly in Oct.; casual inland to southeastern CA and AZ.

breeding ♀

breeding adult females average more restricted red throat and more streaks on breast, but there is overlap

extensive brick red throat

pale nape

prominently streaked back

breeding ♂

no primary extension past tertials

broad dark malar

heavily spotted and streaked below with uniform underlying color

immature

SPRAGUE'S PIPIT Anthus spragueii   SPPI ■ 2

In winter, the Sprague's Pipit is one of the more secretive grassland birds from the Southwest to eastern Texas, yet in summer its song is a characteristic sound of the western prairies, when displaying males give continuous varied song from high altitude. Female and wintering individuals are difficult to spot on the ground. When flushed, the bird calls alarmingly, rises

to a great height while circling, plummets straight down with folded wings, opens its wings just before impact, and then disappears into the tall grass. However, it is usually reluctant to flush, preferring to walk away rather than fly. Unlike other pipits, Sprague's does not bob its tail. Monotypic. L 6.5" (17 cm)

IDENTIFICATION Sexes similar. It is stocky and rather short tailed for a pipit, heavily streaked on back and crown, rather buffy brown underneath, with fine streaking across breast, pale legs and base of mandible, and extensive white outer tail feathers. Very plain buffy face with large, dark eyes serves as the best field mark. **JUVENILE:** It is more scaled on back and has bolder, white wing bars. **SIMILAR SPECIES** The plumage of immature

Red-throated Pipit looks similar, but note plainer face and lack of streaking on flanks of Sprague's; their flight call and behavior when flushed are also different. A juvenile Horned Lark (see p. 484) may be confused with a Sprague's, but it has a different tail pattern, calls, and behavior. **VOICE CALL:** A loud squeaky *squeet,* often given two or three times in succession. **SONG:** Given in flight; a series

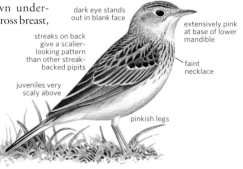

dark eye stands out in blank face

extensively pink at base of lower mandible

streaks on back give a scalier-looking pattern than other streak-backed pipits

faint necklace

juveniles very scaly above

pinkish legs

of descending *tzee* and *tzee-a* notes. **STATUS & DISTRIBUTION** Uncommon. **BREEDING:** Nests in grassy fields in open prairie. **WINTER:** Solitary. Found in tall patches of ungrazed grass in the southern U.S. to south Mexico. **VAGRANT:** Very rare in fall and winter to CA; accidental in the East.

POPULATION Overgrazing and the introduction of non-native grasses are negatively affecting populations on both summer and winter grounds.

AMERICAN PIPIT *Anthus rubescens*   AMPI   ■   1

Formerly treated as a Holarctic species, the "Water Pipit." North American and Asian subspecies were later recognized as a full species, the American Pipit (known as "Buff-bellied Pipit" in Europe). This ground-dwelling species of high latitude and high elevation rocky tundra is widespread across North America's Arctic and mountaintops during summer, and it can be abundant in winter in agricultural fields, forming huge single species flocks across the southern United States. During migration in winter, often seen in wet fields, sewage ponds, and other open mesic habitats. Forages out in the open, often in flocks, frequently bobbing its tail. Polytypic (3–5 ssp.; all occurring in N.A., can be divided into 3 ssp. groups). L 6.5" (17 cm)

IDENTIFICATION Sexes similar. Plumage highly variable, ranging from clear buffy underparts with no streaks, to whitish underparts with heavy streaks, depending upon subspecies and time of year. Upperparts generally gray to brown, unstreaked. Tail with white outer tail feathers. **ADULT BREEDING:** Subspecies *alticola*: Pale gray above, unstreaked tawny-buff underparts, buffy auriculars with buffy supercilium, dark legs. Subspecies *rubescens* (including *pacificus*): Darker gray above, buffy below, streaks across breast that extend faintly down flanks, gray auriculars with buffy supercilium, and dark legs. Subspecies *japonicus*: Similar to *rubescens,* but with blacker streaking, a browner, slightly streaked back, and pale legs. **ADULT WINTER:** Subspecies *alticola*: Pale overall with some faint streaking across breast, grayer auriculars, and buffy supercilium more prominent on grayer face. Subspecies *rubescens*: Grayish brown above with faint streaking, whitish below, buffier on flanks, heavy dark streaking across breast (variable, but always more than *alticola*), decidedly white supercilium, and dark legs (usually). Subspecies *japonicus*: Boldly patterned. More streaking on upperparts and thick black streaks on a white background on underparts, prominent white wing bars, and pinkish legs (sometimes pale on N.A. subspecies too).

GEOGRAPHIC VARIATION N.A. subspecies divide into three groups: the nominate *rubescens* group, breeding on N.A. tundra (*rubescens,* slightly paler *pacificus,* and *geophilus;* the latter two not recognized by some authorities); the Rocky Mountain group *(alticola);* and the Asian group (*japonicus,* which occurs as a rare migrant, mainly in western AK). **SIMILAR SPECIES** All other N.A. pipits are browner and more streaked on upperparts, except Olive-backed. **VOICE CALL:** Given in flight; a sharp *pip-it.* Call of *japonicus* higher pitched, often single noted. **SONG:** A rapid series of *chee* or *cheedle* notes, typically given in flight display, lasting 15–20 seconds. **STATUS & DISTRIBUTION** Common. **BREEDING:** Widespread and scattered during the breeding season in rocky tundra across the Arctic *(rubescens),* high mountaintops in the Rocky Mountains and CA *(alticola),* and the Pacific Northwest *(pacificus).* **WINTER:** Abundant and widespread in agricultural areas and open country from southern U.S. south through Mexico. Rocky Mountain birds winter mainly in Mexico. Subspecies *rubescens* is more widespread and overlaps with *alticola.* **MIGRATION:** Found virtually anywhere in open country or by water across the U.S.; casual to the Azores. Asian *japonicus* regular spring (early June) and fall (late Aug.–Sept.) visitor to islands in the Bering Sea, casual to very rare (in fall) south to CA. **POPULATION** Winter birds are susceptible to the overuse of pesticides in agricultural areas.

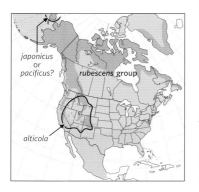

Labels on illustrations: dark malar stripe broadens at bottom; rather plain back; streaked underparts; leg color varies from dark to showing some fleshy tones; winter *rubescens*; more heavily and darkly streaked below; white wing bars; pink legs; winter *japonicus*; richly buff and unstreaked underparts; slender bill; breeding *alticola*; necklace; breeding *rubescens*

japonicus or pacificus?; rubescens group; alticola

WAXWINGS Family Bombycillidae

Bohemian Waxwing (MN, Jan.)

Finding waxwings is often feast or famine, particularly near the periphery of a species' range. They may be abundant one year, gone the next. Even during a single season, several different flocks may visit the same location, making it appear as though one flock is present all winter. Flocks are often quite tame and allow observers to approach closely, particularly when they are in a state of inebriation after consuming fermented fruits. Once waxwings are found, identifying them is straightforward: Differences in the wings, tail, and underparts between species are easily seen. Experienced birders often check flocks carefully in winter—stray waxwings often are found in flocks of the "other" waxwing species.

Structure Waxwings are relatively chunky birds with a pointed crest, which is conspicuous when the birds are perched. They have fairly long, pointed, and triangular wings and a short, square, or slightly rounded tail with very long undertail coverts. In flight this gives them an appearance similar to European Starlings. When perched, their distinctive silhouette makes them unlikely to be confused with other species. The short, broad bill is well adapted for grasping and gulping down large berries.

Behavior No other bird family in North America is more addicted to fruit than the Bombycillidae. From fall through early spring waxwings feed almost exclusively on sugary fruits; however, they will also consume developing fruits (flowers) and insects in springtime. Their fondness for fruit has led to their proclivity to form flocks. Except when nesting, waxwings are almost always seen in flocks. In winter flocks number in the hundreds, often even in the thousands. During these times they frequently forage in fruiting trees, sometimes with American Robins, bluebirds, and Pine Grosbeaks. Even during the breeding season, nesting pairs are often found in close proximity to each other, near an abundant food source. Lacking defended territories, waxwings evolved an unusual characteristic among passerines—they do not sing.

Plumage Waxwing plumages are characterized by a subtle yet elegant blend of soft and silky smooth browns and grays. Waxwings have dark throats and masks. The wings vary from the simple white edging on the inner edge of the tertials on the Cedar Waxwing, to the more elaborate patterns found in the Bohemian and Japanese Waxwings. Bombycillids have a modified basic molt strategy with a partial first prebasic molt where few wing coverts are replaced. While sexually dimorphic, differences between the sexes are subtle and may be challenging to see under most field conditions. Waxwings are named for the waxy droplets at the end of their secondaries, which are unique to the family. Synthesized from carotenoid pigments, these droplets form as extensions of the rachis that project beyond the feather veins. These may be lacking, or at least reduced on young birds, and most developed in adult males. The link between diet and feather color has been well studied in the Cedar. Typically, the Cedar only synthesizes yellow carotenoid from its diet, resulting in the classic yellow tail tip. In much of eastern North America, however, the introduction of exotic honeysuckle—with the red carotenoid pigment rhodoxanthin—has led to the Cedar molting in tail feathers with an orangish tip; the exact color depends upon how much honeysuckle the Cedar eats during the molt.

Distribution Waxwings are found only in the Northern Hemisphere. The Bohemian Waxwing is Holarctic, while the Japanese Waxwing inhabits eastern Asia, and the Cedar Waxwing resides primarily in North America.

Taxonomy There are only three waxwing species; all three belong to the genus *Bombycilla*. Waxwings are closely related to silky-flycatchers (Ptilogonatidae); some taxonomists merge the two families. Waxwings differ from silky-flycatchers in having pointed wings, a minute tenth primary, and waxy droplets at the tips of the secondaries. They lack the well-developed rictal bristles found in silky-flycatchers. The Palmchat *(Dulus dominicus)* of Hispaniola is also closely related. DNA-DNA hybridization suggests that bombycillids are closely related to dippers, Old World flycatchers, thrushes, and starlings.

Conservation Since waxwings congregate at berries near human habitations, they face a cornucopia of dangers posed by residential environments, particularly collisions with windows and cars and cat predation. The Japanese Waxwing is classified by BirdLife International as near threatened due to habitat loss and degradation. The populations of both North American species are thought to be increasing or stable.

BOHEMIAN WAXWING   *Bombycilla garrulus*   BOWA   2

The Bohemian Waxwing often winters in flocks of several hundred birds. In summer, it travels in pairs or small flocks and frequently perches on tops of black spruces. Stray individuals to the south are almost always found with Cedar Waxwings. Polytypic (2–3 ssp.; 2 in N.A.). L 8.3" (21 cm)
IDENTIFICATION Starling-size; sleek crest; grayish overall with face washed in chestnut; rufous undertail coverts; tip of tail yellow. White primary tips make

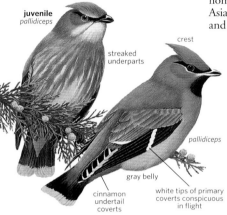

juvenile
pallidiceps

crest

streaked
underparts

pallidiceps

gray belly

cinnamon
undertail
coverts

white tips of primary
coverts conspicuous
in flight

wing appear notched with white. Sexes similar; male averages slightly larger throat patch, broader yellow tip to tail, and more extensive waxy tips to secondaries. First-winter birds lack white tips to primaries and waxy tips reduced. **JUVENILE:** Streaky below with white throat (June–Oct.). **FLIGHT:** White bases to primary coverts appear as contrasting white band in flight.
GEOGRAPHIC VARIATION Weak. Subspecies *centralasiae* (often merged with nominate *garrulus*) is a vagrant from Asia to western Aleutians (specimens) and Pribilof Islands. Paler than *pallidiceps* of N.A. with little contrast between forehead and rest of head; often darker on undertail coverts.
SIMILAR SPECIES See Cedar Waxwing. In flight, European Starlings appear similar in shape and flight style.
VOICE CALL: Commonly a high, sharply trilled *zeeee*, lower and distinctly more trilled than Cedar Waxwing's (more rattlelike).
STATUS & DISTRIBUTION Uncommon to irregularly common. **BREEDING:** Open coniferous or

mixed woodlands. **MIGRATION:** In fall departs interior AK in Sept. First arrivals in SK late Sept.; huge flocks arrive Dec. First arrivals in northern U.S. mid-Oct.–early Nov. In spring, most leave lower 48 by mid-Mar., but some into Apr. **WINTER:** Largest concentrations occur across southern BC and AB and south-central SK. Irregular to Northeast, usually in small numbers; annual in ME, Maritimes, NF. **VAGRANT:** Rare and irregular to dashed line on map. Casual to southern CA, northwestern AZ, northern TX, northwestern AR, southern NJ, southern MD.
POPULATION Stable.

CEDAR WAXWING   *Bombycilla cedrorum*   CEDW   1

The Cedar Waxwing is easily found in open habitat where there are berries. It times its nesting to coincide with summer berry production, putting it among the latest of North American birds to nest. It is highly gregarious; flocks of hundreds, occasionally thousands, are encountered during migration and winter. Now regarded as monotypic. L 7.3" (18 cm)
IDENTIFICATION Smaller than Bohemian Waxwing, with pale yellow belly and whitish undertail coverts. Tip of tail

usually yellow, broadest in adult males, narrowest in immature females. Some birds (esp. immatures) have an orange tail tip, a result of consuming non-native honeysuckle fruit during molt. **JUVENILE:** Streaky below with white chin and bold malar stripe (June–Nov.).
SIMILAR SPECIES Bohemian Waxwing is similar but larger, grayer, has rufous undertail coverts, white bar on primary coverts and chestnut wash on face. A juvenile Cedar can be separated from a Bohemian by its lack of white wing patches, and lack of any rufous on undertail coverts.
VOICE CALL: Commonly a high trilled *zeeeee*, higher and less trilled than Bohemian. Also a long, high, pure *seeeee;* and a shorter descending *sweeew,* longer than analogous call of Bohemian. Does not sing.
STATUS & DISTRIBUTION Common. **BREEDING:** Open woodlands and old fields; rare but regular breeder in southeastern AK and southern YT. **MIGRATION:** Arrival in Great Lakes region is not usually until May and departure Sept.–

Oct. **WINTER:** Irregular. Southern U.S. to C.A., rarely to Panama. Generally rare in the West Indies; most numerous on Cuba, where occasionally fairly common. **VAGRANT:** Casual to central AK; accidental to Iceland, U.K., and Azores.
POPULATION Increasing, likely due in part to spread of exotic fruiting plants.

juvenile

streaked
underparts

crest

yellowish
belly

yellow
tail tip

white
undertail
coverts

SILKY-FLYCATCHERS Family Ptiliogonatidae

Phainopepla, male (AZ, May)

Members of this small Middle American family are often found in close proximity to fruit, although insects may also form a large percentage of silky-flycatchers' diet. The family's common name describes their soft, sleek plumage and agility in catching insects on the wing. They are not at all closely related to flycatchers, which has led some people to call them by the simpler name "Silkies." Identification of all silky-flycatcher species is straightforward, with differences in coloration, structure, and patterning obvious.

Structure Silky-flycatchers are slender thrush-size birds with long tails and small bills. Short crests are standard. Their rounded wings help them make quick sallies from branches to catch insects.

Behavior Silky-flycatchers are seen in small flocks, pairs, or alone. They typically perch near the tops or edges of shrubs and trees. Most call frequently, making them easy to see. During the breeding season, they may form loose colonies when there is an abundance of fruit; during these times adults defend nests from other members of the same species but usually do not defend food supplies. They construct shallow cup nests placed in crotches of trees and shrubs.

Plumage Ptilogonatids have plumages characterized by black, gray, or pale brown, often with yellow or white highlights. The plumages are generally smooth and blended, without well-defined lines or distinct markings of color. Even juveniles lack spots, bars, streaks, or other distinctive markings. All species are sexually dimorphic, with males' plumage brighter or exhibiting more contrast. In Phainopepla there is a supplemental plumage (before the first prebasic molt). The first prebasic molt varies from partial to complete. There is no prealternate molt.

Distribution Silky-flycatchers are found from the southwestern United States to western Panama. Aside from the desert-loving Phainopepla, they primarily inhabit mountain ranges. Movements in Middle America require more study, but since these species feed heavily on fruit, it is likely all species stage at least local movements.

Taxonomy Ptilogonatidae breaks down into three genera with four species. The family is very closely related to waxwings and sometimes merged with that family. Silky-flycatchers differ from waxwings in having rounded wings and prominent rictal bristles; the tenth primary is also larger and the ninth primary is smaller. The monotypic genus *Phainoptila* (Black-and-yellow Silky-flycatcher found in the highlands of Costa Rica and western Panama) appears very different—much more thrushlike; however, unlike thrushes, the juvenal plumage is plain and unspeckled and the short tarsi are covered with broad transverse scales—both characteristics of silky- flycatchers.

Conservation Species populations are apparently stable. Habitat destruction and degradation pose the biggest threat to silky-flycatchers. Habitat fragmentation may not pose as great a threat as it does for many families since all silky-flycatchers are found to some degree in edge habitats. Some species are caught and kept in captivity.

SILKY-FLYCATCHERS Genus Ptiliogonys

GRAY SILKY-FLYCATCHER Ptiliogonys cinereus   GRSF ◼ 5

This Middle American species is an accidental visitor to Texas and perhaps elsewhere. The Gray Silky-flycatcher feeds primarily on berries of mistletoe and other plants, as well as insects; it often perches conspicuously atop trees. Polytypic (2–4 ssp.). L 7.5" (20cm)

IDENTIFICATION Structurally similar to Phainopepla, but with a bushy crest that is not as wispy. **ADULT MALE:** Generally blue-gray overall, with orangey-yellow flanks and bright yellow undertail coverts. Wings and tail dark; tail has a broad white base to outer tail feathers. Dark lores contrast with white eyering and paler forecrown. **ADULT FEMALE:** Similar to male, but coloration more subdued, with more brownish tones. Yellow often confined to undertail coverts. **JUVENILE:** Similar to adults, but duller overall. **GEOGRAPHIC VARIATION** There is some discussion as to how much variation truly exists, as there are subtle differences in age and sex that have likely not been fully resolved. Records for the U.S. are *otofuscus* from northwestern Mexico or nominate *cinereus* from eastern and central Mexico.

SIMILAR SPECIES This species is unmistakable; note its structure, upright posture, bushy crest, eye ring, yellow at least on undertail coverts, and white patches in the long tail.

VOICE CALL: Gives varied, chattering notes; flight calls and some calls given when perched may suggest call notes of *Piranga* tanagers.

STATUS & DISTRIBUTION Accidental in TX: Laguna Atascosa N.W.R. (Oct. 21– Nov. 11, 1985) and El Paso (Jan.12– Mar. 5, 1995). Records from coastal southern CA controversial; natural occurrence questioned. **BREEDING:** Pine-oak and pine-evergreen forest in mountains of Mexico and Guatemala. **MOVEMENTS:** Largely resident, but occasionally nomadic; moves in large flocks, may move locally downslope in winter; possibly withdraws from northwestern Mexico. Potential for extralimital records pertaining to escapes complicates understanding natural distribution.

gray crest

yellow undertail coverts

white based tail

adult ♂

Genus *Phainopepla*

PHAINOPEPLA *Phainopepla nitens* PHAI ■ 1

Phainopepla—Greek for "shining robe"—is the only ptilogonatid regularly found north of the U.S.–Mexico border. The Phainopepla is often solitary, but it may gather to feed on seasonally abundant crops. It feeds heavily on mistletoe berries in winter; its diet includes more insects at other seasons. It vigorously defends feeding territories, and excrement may pile several inches high under territorial perches. It perches upright, usually in the open on tops and sides of trees and shrubs. Polytypic (2 ssp.; both in N.A.). L 7.7" (20 cm)

IDENTIFICATION Slender thrush-size bird with a shaggy crest and long tail. Red eyes (both sexes). Males substantially

larger than females. **MALE:** Shiny black; white wing patches very conspicuous in flight. **FEMALE:** Grayish; white edging on all wing feathers (less so on primaries and secondaries). White wing patches absent. **JUVENILE:** Similar to adult female but with buffy wing bars and brownish upperparts; dark eye. Immature male typically begins to acquire black feathers in fall, producing black patches. **FLIGHT:** Fluttery but direct; bold wing patches revealed on adult male.

SIMILAR SPECIES Generally unmistakable. Northern Mockingbird has white wing patches, but a very different shape and horizontal posture. Mockingbirds are paler gray with darker wings and tail (tail with white corners).

GEOGRAPHIC VARIATION Smaller *lepida* (U.S. Southwest to Baja California and northwest Mexico) and larger *nitens* (southwest TX to central Mexico) differ in wing and tail length; they are not separable in the field.

VOICE CALL: A distinctive, querulous, low-pitched,

whistled *wurp?* May imitate other species (e.g., Red-tailed Hawk and Northern Flicker). **SONG:** A brief warble that includes a whistled *wheedle-ah;* infrequently heard.

STATUS & DISTRIBUTION Common. **BREEDING:** Late winter–early spring in mesquite brushlands; in summer moves into cooler, wetter habitats and raises a second brood. Highest densities occur in riparian woodlands and dense mesquite thickets. **MIGRATION:** Short-distance migrant. Begins spring migration late Mar.–early Apr.; arrives northern CA early to mid-Apr. Fall return occurs Aug.–Nov. **WINTER:** Mostly lowlands from Sonoran Desert to central Mexico. **VAGRANT:** Casual in southern OR, CO, KS, south-central TX. Accidental to eastern N.A. with records for ON, WI, RI, and MA.

POPULATION Apparently stable; potential threats include continued destruction of mesquite forests and riparian woodlands in the Southwest.

conspicuous crest

adults with red iris; brownish iris of juvenile retained well into fall

glossy black color

♂

females and juveniles overall grayish

♀

♂

prominent white wing patches visible in flight; pale gray in female

OLIVE WARBLER Family Peucedramidae

Olive Warbler, female (AZ, Apr.)

Long considered an aberrant wood-warbler, the Olive Warbler was recently determined to be a distant relative, and placed in its own family with uncertain relationships.

Structure Olive Warblers are similar to *Setophaga* wood-warblers but have a slightly longer, thin bill, a slightly decurved culmen, a distinctly notched tail, and ten primaries.

Plumage The adult male's tawny orange head coloration is unlike any wood-warbler, and a white patch at the base of the primaries is shared with only the Black-throated Blue Warbler.

Behavior Vocalizations consist of loud whistled notes (the females also sing). They flick their wings like kinglets. Olive Warblers allow their nestlings to soil the nests with droppings, a behavior unknown in wood-warblers.

Distribution Occurs in montane coniferous and mixed pine-oak forest from southeastern Arizona and southwestern New Mexico to extreme western Nicaragua. The northern subspecies is partially migratory, whereas the southern subspecies are sedentary.

Taxonomy On the basis of anatomy, Olive Warblers may be related to Old World warblers, but some behavioral differences and recent genetic evidence contradict this and suggest a closer relationship to the finches. More study will resolve the taxonomic position. Olive Warblers are generally considered to be more primitive than wood-warblers.

Conservation Populations likely affected by logging.

Genus *Peucedramus*

OLIVE WARBLER Peucedramus taeniatus   OLWA ■ 2

A distinctive warblerlike resident of southwestern pine forests, with a loud song and kingletlike wing flicking. Polytypic (5 ssp.; *arizonae* in N.A.). L 5.2" (13 cm)

IDENTIFICATION ADULT MALE: Plumage acquired by second fall. Tawny orange head and upper breast with distinct black ear patch. Gray back. Two broad white wing bars and a white spot at base of primaries. Tail distinctly notched, with large white spots. **ADULT FEMALE:** Crown, nape,

and back grayish. Ear patch mottled gray and black, surrounded by yellowish. Throat and breast yellowish. **IMMATURE MALE:** Similar to adult male but duller, showing more mottled black and gray cheek patch, more yellowish head, including crown and upper breast. **IMMATURE FEMALE:** Like adult female but head duller and back more brownish.

GEOGRAPHIC VARIATION Northern subspecies *arizonae* in N.A. is larger, paler, and greener-backed than the four southern subspecies.

SIMILAR SPECIES Immatures are similar to an immature female Hermit Warbler, which lacks white at base of primaries and has an all yellow face, indistinctly notched tail, and different bill shape, calls, and songs.

VOICE Very distinct. **CALL:** A soft, whistled *phew*, similar to Western Bluebird. Also a hard *pit*. **SONG:** A loud *peeta peeta peeta peeta*, similar to Tufted Titmouse.

STATUS & DISTRIBUTION Fairly common in mountains of southeastern AZ and southwestern NM. **BREEDING:** Higher elevation pine and fir forests. **NEST:** High in conifer, far from trunk, 3–4 eggs (May–June). **MIGRATION:** Arrive in AZ early Apr. Departure schedule unknown. **WINTER:** Most move south of AZ; very few remain in breeding range, sometimes at slightly lower elevations. **VAGRANT:** West TX and portions of NM outside breeding range.

adult ♂
arizonae

tawny head

adult ♀
arizonae

1st spring ♂
arizonae

long thin bill

pale patch at base of primaries

1st fall
arizonae

pale center to dark mask

LONGSPURS AND SNOW BUNTINGS Family Calcariidae

Snow Bunting (AB, Oct.)

Recent molecular work using mitochondrial and nuclear DNA has shown that this small group of sparrow-like tundra and grassland species is well differentiated genetically from the New World sparrows and Old World buntings of the family Emberizidae and best placed in its own family. Six species are recognized, all of which are found in North America.

Structure All are medium-small (6–7 in) sparrow-like, conical-billed, seed-eating birds with relatively long wings and short legs.

Behavior Ground-dwelling seedeaters, longspurs and snow buntings walk in a shuffling gait; they are strong fliers, rising well into the air with an undulating flight and dropping steeply back to the ground. Territorial songs are given from the top of a shrub and also, in most species, in flight. They are gregarious in the nonbreeding season and prefer open country, often flocking with Horned Larks and often flushing in groups to avoid predators.

Plumage Snow Buntings are largely white with black markings in the breeding season, and washed with tan on head and upperparts in winter; they always flash much white in the wings and tail. Longspurs are generally brownish with conspicuous white markings in the tail; breeding males are marked with black on the head and variably on the underparts, and in most species show areas of rich buff or chestnut in the plumage. Females, winter males, and immatures are duller.

Distribution The family is limited to the Arctic and north temperate regions of North America and Eurasia, from tundra habitats to short-grass prairies and other open habitats. Two species (Snow Bunting and Lapland Longspur) occur across the northern Holarctic Region; McKay's Bunting is restricted to Bering Sea islands, and the remaining three longspur species breed only in North America. All species are migratory but generally winter within subarctic and north temperate regions.

Taxonomy Longspurs and snow buntings have traditionally been placed within the large family Emberizidae, and at least superficially seem to fit well within that family (some authorities had even recommended merging the longspur genus *Calcarius* into the Old World bunting genus *Emberiza*). Molecular data, however, indicate that the calcariids split off from the other groups of "nine-primaried oscines" quite early and deserve family rank of their own. Among the longspurs, McCown's has been treated with the other three species in the genus *Calcarius* but is now, once again, placed in its own genus *Rhynchophanes*. McKay's Bunting has been regarded by some as a well-marked subspecies of the Snow Bunting.

Conservation Populations are generally stable, although the two prairie longspurs (Chestnut-collared and especially McCown's) have fairly restricted ranges on the Great Plains and have suffered from habitat degradation of native grassland to agriculture. The McKay's Bunting breeds almost exclusively on two small islands in the Bering Sea and must be at risk, even though still common within its tiny range. The two most widespread species (Snow Bunting and Lapland Longspur) remain abundant.

LONGSPURS Genera *Calcarius* and *Rhynchophanes*

Four longspur species (3 endemic to N.A.; 1 widespread, circumpolar) occur in open areas and use trees or bushes only for song perches. In winter, they form large flocks, often with paler Horned Larks. They often call when flushed and while in their strong, undulating flight. Based on genetic evidence, McCown's was restored to its own monotypic genus.

LAPLAND LONGSPUR *Calcarius lapponicus*   LALO   1

A flocking species, in migration and winter, they often join Horned Larks and Snow Buntings. Look for Lapland's darker overall coloring and smaller size. It may flock with other longspurs as well. Polytypic (5 ssp.; 3 in N.A.). L 6.3" (16 cm)

IDENTIFICATION Outer two tail feathers on each side of tail are partly white, partly dark; reddish indented edges on greater coverts and on tertials. Very

long primary projection past longest tertial. **BREEDING ADULT MALE:** Black, well-outlined head and breast; a broad white or buffy stripe extends back from eye and down to sides of breast; reddish brown nape. **BREEDING ADULT FEMALE:** Duller version of male. **WINTER:** Bold, dark triangle outlining plain buffy ear patch; dark streaks (female) or patch (male) on upper breast; dark streaks on side; broad buffy eyebrow and buffier underparts; belly and undertail are usually white. **JUVENILE:** Buffy and heavily streaked above and on breast and sides.

GEOGRAPHIC VARIATION Five

extensive black head bordered by white
breeding ♂

chestnut nape

breeding ♀

dark chest band
winter ♂

well-defined dark border to auriculars

very long primary projection

restricted white in outer tail feathers

dark lateral crown stripes surround paler center to crown

juvenile

rufous-edged greater coverts and tertials

whitish belly

winter ♀

buffy fall ♀

subspecies show weak variation in coloration. Eastern *subcalcaratus* breeds from north-central Northwest Territories to Labrador and winters west to Texas; the remainder of the North American range is occupied by *alascensis*, which is comparatively much paler. Asian *coloratus* is dark and bright compared to *alascensis*, and it has occurred as a rare or casual breeder on islands in Alaska's Bering Sea.

SIMILAR SPECIES See sidebar (p. 585).

VOICE CALL: A musical *tee-lee-o* or *tee-dle*. **FLIGHT NOTE:** A dry rattle distinctively mixed with whistled *tew* notes; other longspurs do not include the *tew* notes with their rattle, but Snow Bunting does; Lapland's rattle is more distinct. **SONG:** Rapid warbling, frequently given in short flights; only on breeding grounds.

STATUS & DISTRIBUTION Common. Circumpolar distribution. **BREEDING:** High arctic tundra. **WINTER:** Fields, grain stubble, airports, beaches. **MIGRATION:** Spring late Feb.–early Apr.; fall early Sept.–late Nov, peaking late Oct.–early Nov. in most of the U.S. **POPULATION** Stable.

CHESTNUT-COLLARED LONGSPUR *Calcarius ornatus*   CCLO   1

The fairly secretive Chestnut-collared Longspur favors denser grass in migration and in winter than does either the McCown's Longspur or the Lapland Longspur. Flocks of Chestnut-collared Longspurs flush from

underfoot and can vanish again into ankle-high grass. Chestnut-collared mixes less frequently with other longspurs, and it very rarely wanders into the open (unlike the Lapland Longspur and the McCown's Longspur).

Its *kittle* call, given upon flushing, is the easiest way to identify it. Monotypic. L 6" (15 cm)

IDENTIFICATION White tail is marked with a blackish triangle. Primary projection past longest tertial is short and primary tips barely extend to base of tail. Fall and winter birds have grayish, not pinkish, bills. **BREEDING ADULT MALE:** Black-and-white head,

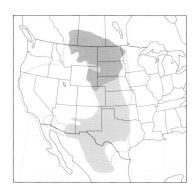

buffy face, and black underparts are distinctive; a few have chestnut on underparts. Whitish lower belly and undertail coverts. Upperparts are black, buff, and brown, with chestnut collar, whitish wing bars. **WINTER MALE:** Paler; feathers are edged in buff and brown, largely obscuring black underparts. Male has small white patch on shoulder, often hidden; compare to Smith's Longspur. **BREEDING ADULT FEMALE:** Usually shows some chestnut on nape. **WINTER FEMALE:** Like breeding female, but paler. **JUVENILE:** Pale feather fringes give upperparts a scaled look.

SIMILAR SPECIES Juvenile is best distinguished from juvenile McCown's Longspur by its tail pattern and bill shape. See sidebar below.

Winter Longspurs

Winter longspurs present one of the greater challenges among North American emberizids, in part because the shifting flocks can be quite furtive on the ground, making viewing difficult. Details of plumage, shape, and tail pattern are useful, but identification is always aided by a consideration of call notes, habitat, and behavior. Lapland Longspur most resembles Smith's Longspur; Chestnut-collared is most like McCown's. The bird's tail pattern provides a good starting point and is best seen in flight, especially as the tail is flared upon landing. Lapland and Smith's Longspurs have white outer tail feathers (whiter in the Smith's). Chestnut-collared and McCown's have largely white tails with differing patterns of black: a black triangle in Chestnut-collared and a black inverted "T" in McCown's. On the ground, Smith's and Lapland Longspurs have a long primary extension, while McCown's and Chestnut-collared have a comparatively short primary extension past the longest tertial. When on the ground, Smith's is best told from Lapland by its wing panel, which is a contrasting chestnut color in Lapland and noncontrasting buffy in Smith's. Also note Smith's smaller, slimmer bill; finer, sparser streaking below; often broken rear border to the facial frame; and white (adult male) or white-edged median coverts. All Smith's are extensively buffy below, but this can be matched, rarely, in Lapland.

McCown's Longspur is large and pale, with a large pinkish bill and distinctive dull face pattern (recalling female House Sparrow); sometimes its chestnut median coverts are shown. Chestnut-collared Longspur is shaped more like a Lapland or Smith's but tends to be plainer gray-brown, is more evenly colored overall with a less distinct face pattern, and has a grayish rather than a pinkish bill. All four species' flight calls are distinctive when learned and invaluable to identification. Focus on Lapland's dry rattle, interspersed with tew or jit notes; Smith's slower, clicking rattle; McCown's softer, more abrupt rattle, interspersed with a distinctive pink note; and Chestnut-collared's unique kittle call. Habitat, while not diagnostic, can be an important clue. McCown's prefers the most open country (heavily grazed grasslands, plowed fields, flat dirt areas). Lapland similarly prefers open areas that are mostly lacking in vegetation, but it may be slightly more regular in short grass areas (e.g., airports) than McCown's. Both Chestnut-collared and Smith's are found in short grass areas, typically in ankle-high grass. Chestnut-collared is often found in areas that have some patches of bare ground interspersed; Smith's is usually in denser, more complete areas of grass. The two denser-grass species are also more furtive and often flush from almost underfoot; Lapland and McCown's are more likely to flush at a distance. ∎

McCown's Longspur, winter

Lapland Longspur, winter

Smith's Longspur, winter

Chestnut-collared Longspur, winter

VOICE FLIGHT NOTE: Distinctive two-syllable *kittle,* repeated one or more times. Also gives a soft, high-pitched rattle and a short *buzz* call. **SONG:** Pleasant rapid warble, given in song flight or from a low perch; heard only on breeding grounds.
STATUS & DISTRIBUTION Fairly common. Winters south to northern Mexico. **BREEDING:** Nests in moist upland prairies and typically prefers moister areas with taller, lusher grass than does McCown's Longspur.

veiled black breast and belly

winter ♂

small darkish bill

faint streaks on breast

winter ♀

dark triangle on white tail

MIGRATION: Spring migration mid-Mar.–mid-Apr.; fall migration Sept.–mid-Nov. Migration is primarily through western and central Great Plains. Male Chestnut-collared Longspurs may wander during mid-summer. Rare in CA in migration and winter. **VAGRANT:** Casual during migration to eastern North America and the Pacific Northwest. Accidental in winter and mid-summer to the East Coast.
POPULATION Like McCown's Longspur, Chestnut-collared Longspur has suffered due to the destruction of native prairies by overgrazing and the conversion to large-scale agriculture. Formerly, Chestnut-collared Longspur bred in Kansas and was more widespread throughout its current range.

SMITH'S LONGSPUR *Calcarius pictus*   SMLO ▪ 2

Habitat provides a good initial clue for this species, as it is very faithful in winter to areas of ankle-high grass or alfalfa, where it can be extremely secretive, with whole flocks virtually invisible until they are flushed. This behavior is similar to that of the Chestnut-collared Longspur, but it is generally unlike that of Lapland and McCown's, which usually prefer more open areas and behave less secretively. Smith's sometimes flocks with Laplands; identification is often easiest by strong buff underparts and by call, since all longspurs call when flushed and when in flight. Monotypic. L 6.3" (16 cm)

IDENTIFICATION Outer two tail feathers on each side of tail are almost entirely white. Bill thinner than on other longspurs. Note long primary projection past longest tertial, a bit shorter than Lapland's, but much longer than Chestnut-collared's or McCown's; shows rusty edges to greater coverts and tertials. **BREEDING ADULT MALE:** Bold black-and-white head pattern; rich buff nape and underparts; white patch on shoulder, often obscured. **BREEDING ADULT FEMALE:** Duller, crown streaked, chin paler. Dusky ear patch bordered by pale buff eyebrow; pale area on side of neck often breaks through dark rear edge of ear patch. Underparts pale buff with thin reddish brown streaks on breast and sides. Much less white on lesser coverts than males. **WINTER & IMMATURE:** Like female.

distinctive black-and-white head pattern

fine streaks on buffy underparts

breeding ♂

long primary projection, but shorter than Lapland

breeding ♀

more white in outer tail feathers than Lapland

winter ♂

white lesser coverts often best noted in flight

SIMILAR SPECIES See sidebar (p. 585).
VOICE CALL: Short, nasal *tseu.* **FLIGHT NOTE:** Dry, ticking rattle, harder and sharper than the call of Lapland and McCown's. **SONG:** Rapid, melodious warbles, ending with a vigorous *wee-chew.* Heard in spring migration and on the breeding grounds; delivered only from the ground or a perch.
STATUS & DISTRIBUTION Uncommon. **BREEDING:** Open tundra and damp, tussocky meadows. **WINTER:** Open, ankle-high grassy areas. **MIGRATION:** Spring late Mar.–late May in U.S.; arrives on breeding grounds late May–early June; fall early Sept.–mid-Nov. Typically migrates earlier in fall and later in spring than Lapland. Regular spring migrant in the Midwest, east to western IN; rare migrant in western Great Lakes. **VAGRANT:** Casual to both coasts, mainly in migration (mostly fall).
POPULATION Stable.

MCCOWN'S LONGSPUR *Rhynchophanes mccownii*   MCLO   2

This species generally favors more barren country than do other longspurs (except the Lapland) and flocks most often with Horned Larks. Look for McCown's chunkier, shorter-tailed shape and slightly darker plumage, its mostly white tail, its much thicker bill, and its undulating flight. On breeding grounds it is easily separated from Chestnut-collared by its unique display flight. Monotypic. L 6" (15 cm)

IDENTIFICATION McCown's white tail is marked by a dark inverted T-shape. Note also stouter and thicker-based bill than found on other longspurs. Its primary projection is slightly longer than Chestnut-collared's; in a perched bird, wing extends almost to tip of short tail. **BREEDING ADULT MALE:** Black crown, black malar stripe, black crescent on breast; gray sides. Upperparts streaked with buff and brown, with gray nape and rump; chestnut median coverts form contrasting

crescent. **BREEDING ADULT FEMALE:** Streaked crown; may lack black on breast and show less chestnut on wing. **WINTER ADULT:** Large pinkish bill with a dark tip; feathers edged with buff and brown. Winter adult female paler than female Chestnut-collared, with fewer streaks on underparts and a broader buffy eyebrow. Some winter males have gray on rump and variable blackish on breast; retain chestnut median coverts. **JUVENILE:** Streaked below; pale fringes on feathers give upperparts a scaled look; paler overall than juvenile Chestnut-collared.

SIMILAR SPECIES See Chestnut-collared Longspur for identification of juveniles. See also sidebar p. 585.

VOICE FLIGHT NOTE: A dry rattle, a little softer and more abrupt than Lapland's. A unique *pink* note is especially useful for identification; it may recall a soft Bobolink or Purple Finch flight note. **SONG:** Heard only on breeding grounds; a series of exuberant warbles

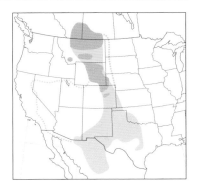

and twitters, generally given in a distinctive song flight, unlike that of Chestnut-collared. McCown's Longspur rises to a considerable height and then delivers its song as it floats slowly back to the ground with its wings held in a sharp dihedral. The appearance recalls a falling leaf or a floating butterfly.

STATUS & DISTRIBUTION Uncommon to fairly common, but range has shrunk significantly since the 19th century. Winters south to northern Mexico. **BREEDING:** Dry shortgrass prairies. **WINTER:** Dry shortgrass prairies and fields, also plowed fields, airports, and dry lake beds. Very rare visitor to interior CA and NV. **MIGRATION:** Spring migration early Mar.–mid-May; fall migration mid-Sept.–late Nov. Migrates primarily through western Great Plains. **VAGRANT:** Casual in coastal CA and OR; accidental to the East Coast.

POPULATION Breeding range drastically reduced since the 1800s; for example, the species formerly bred in Oklahoma, South Dakota, and western Minnesota. Contraction in range and overall decrease in population is due to land management practices that have greatly reduced shortgrass prairie. Conversion to large-scale agriculture is especially to blame.

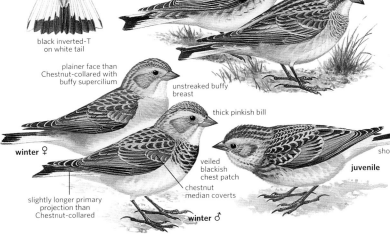

black crown

black chest patch

breeding ♀

breeding ♂

black inverted-T on white tail

plainer face than Chestnut-collared with buffy supercilium

unstreaked buffy breast

thick pinkish bill

winter ♀

short tail

veiled blackish chest patch

chestnut median coverts

juvenile

slightly longer primary projection than Chestnut-collared

winter ♂

Genus *Plectrophenax*

SNOW BUNTING *Plectrophenax nivalis*   SNBU   1

The Snow Bunting's breeding and flocking behaviors are similar to the Lapland Longspur's. Polytypic (4 ssp.; 2 in N.A.). L 6.8" (17 cm)

IDENTIFICATION Long black-and-white wings; breeding plumage acquired

by end of spring, entirely by wear. Bill black (summer) or orange-yellow (winter). Males usually show more white (esp. wings). **JUVENILE:** Grayish, streaked; buffy eye ring. First-winter plumage browner than adult's.

GEOGRAPHIC VARIATION Larger, whiter, bigger-billed *townsendi* is resident on Aleutians, Pribilofs, and Shumagin I. (also Commander Is., Russia); nominate occurs throughout rest of N.A. range.

SIMILAR SPECIES See McKay's. Sometimes confused with albinistic sparrows. Note plumage, behavior.
VOICE CALL: Often given in flight. Sharp, whistled *tew;* short buzz; and musical rattle or twitter. Rattle and *tew* notes like Lapland's (rattle softer, *tew* clearer). **SONG:** Loud, high-pitched musical warbling, only on breeding grounds.

STATUS & DISTRIBUTION Fairly common. **BREEDING:** Tundra, rocky shores, talus slopes. **WINTER:** Shores, weedy fields, grain stubble, plowed fields, roadsides. **MIGRATION:** Fall late Oct.–early Dec.; spring early Feb.–late Mar. **VAGRANT:** Very rare to northern CA and FL; casual to southern CA, AZ, TX, and Bahamas.
POPULATION Stable.

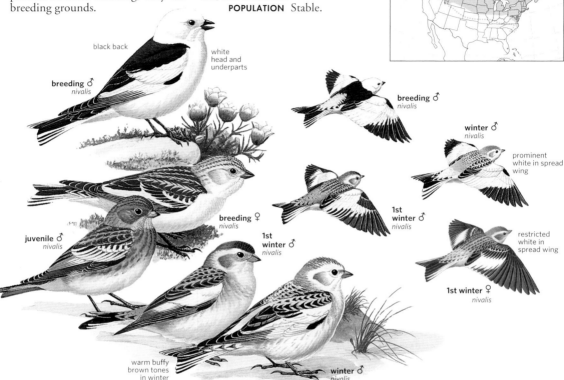

black back
white head and underparts
breeding ♂
nivalis

breeding ♂
nivalis

winter ♂
nivalis
prominent white in spread wing

juvenile ♂
nivalis

breeding ♀
nivalis

1st winter ♂
nivalis

1st winter ♂
nivalis

restricted white in spread wing

1st winter ♀
nivalis

warm buffy brown tones in winter

winter ♂
nivalis

MCKAY'S BUNTING *Plectrophenax hyperboreus* MKBU ▪ 2

This is one of the rarest breeding birds in North America. Monotypic. L 6.8" (17 cm)
IDENTIFICATION Like Snow, but comparable plumage more white in tail, primaries. **BREEDING:** Male mostly white, including back (black in Snow), with less black on wings and tail. Female like Snow, but white panel on greater coverts. **WINTER:** Edged with rust or tawny brown; male whiter overall than Snow. **JUVENILE:** Like juvenile Snow.

more faintly marked crown than female Snow Bunting
back paler overall than female Snow Bunting
white greater coverts
reduced black in wings and tail as compared to Snow Bunting
white back
breeding ♀
breeding ♂
paler overall than Snow Bunting
breeding ♂
winter ♂

SIMILAR SPECIES Snow Bunting is very similar; beware of hybrids, which appear to be frequent. Male McKay's more extensively white; black limited to feather tips. Female McKay's whiter in tail, wing coverts, and primaries than male Snow. Juveniles very similar.
VOICE Calls and song identical to Snow's.
STATUS & DISTRIBUTION Uncommon. McKay's is known to breed regularly only on Hall and St. Matthew Is. in the Bering Sea. A few often present in summer on the Pribilofs, and it has bred there; sometimes present in migration on St. Lawrence I. Rare to uncommon in winter along Bering Sea coast of AK from Kotzebue south and to Cold Bay; casual in winter southward, in interior of AK and on Aleutians. **VAGRANT:** Accidental south along Pacific coast to BC (2 rec.), WA (3 recs.), and OR (2 recs.).
POPULATION Stable, but vulnerable, given its restricted range on two small islands.

WOOD-WARBLERS Family Parulidae

Yellow-rumped Warbler (NJ, Oct.)

Wood-warblers include some of our most colorful birds. Most members of this exclusively New World family are neotropical migrants. Most are fairly closely related, and a number of hybrids—even between genera—are known.
Structure Most are small songbirds with variable, colorful plumage in spring, sometimes more muted in fall, and nine visible primaries. Some show rictal bristles, an adaptation to an insectivorous diet; others lack them.
Plumage Most species molt into a breeding plumage, more colorful than their winter or nonbreeding plumage. Yellow, green, and olive figure prominently in many plumages and prominent wing bars are found on some species. Many species show extensive white on the outer tail feathers, which may appear as white spots.
Behavior All species in North America are generally insectivorous, but some feed on fruit or nectar in winter. Most species build a cup nest, placed on the ground or low to high in a shrub or tree; only two species nest in cavities. Many species have two song types: an often-accented, primary song used for defending territory and an often-unaccented, alternate song used in a variety of situations (e.g., near the nest when paired with a female). Songs are quite varied. Many eastern species undertake trans-Gulf migrations in spring and fall.

Distribution About 114 species breed from Alaska to northern Argentina. Fifty species breed annually in North America (not inc. Bachman's Warbler which is very likely extinct), with the majority of those in the East; six additional species have been recorded as vagrants.
Taxonomy About half of the world's wood-warblers are in two genera: 34 species in *Setophaga* and 24 species in *Basileuterus*. A number of species, many in N.A., are classified in their own genus, and their closest relatives within the family are uncertain. Recent molecular work radically changed the taxonomy of wood-warblers. The American Redstart *(Setophaga)* was found to be firmly embedded within genus *Dendroica,* and by the Rule of Priority (which taxon was named first) all *Dendroica* become *Setophaga.* Additional significant changes include: moving most species out of the genera *Vermivora* and *Oporornis* and a completely new linear sequence. The two waterthrushes were found to be very different from the Ovenbird, a fact that ornithologist Kenneth Parkes discovered decades earlier when he discovered that their juvenal plumages were totally different. The new waterthrush genus, *Parkesia,* honors his insights. The Yellow-breasted Chat is perhaps the most aberrant family member; and an unusual tropical species, the Wrenthrush *(Zeledonia coronata),* is most likely also a member of the wood-warbler family.
Conservation Wood-warblers have experienced significant population declines in recent decades. Most North American wood-warblers migrate to the tropics, which puts them at risk not only through potential habitat destruction on both breeding and wintering grounds, but on their migration corridors as well. In particular, the hazardous, difficult trans-Gulf migration often results in high mortality. Some species have restricted and specialized breeding ranges. Brown-headed Cowbird parasitism affects these rare species most dramatically, but as forests become fragmented many species become more vulnerable. Worldwide, BirdLife International lists nine species as near-threatened, seven as vulnerable, six as endangered, and two as critically endangered (Semper's Warbler, *Leucopeza semperi,* and the likely extinct Bachman's Warbler).

Formerly this genus included the two waterthrushes; now it contains only the Ovenbird and is basal (most primitive) to the wood-warbler family.

OVENBIRD *Seiurus aurocapilla*   OVEN   1

rufous crown with blackish lateral crown stripes

olive above

bold white eye ring

extensively spotted below

A familiar voice from the forest, the Ovenbird spends much all of its time walking on the ground or low branches often with its tail cocked. Its domed, well-hidden nest, with its side entrance, is built on the ground and encloses three to six eggs in June. Polytypic (3 ssp.; all in N.A.). L 6" (15 cm)

IDENTIFICATION **ADULT:** Olive upperparts, black lateral crown stripes, bold white eye ring, hard-to-see orange crown patch. Underparts white; bold black streaks on breast, sides. **JUVENILE:** Very different from adult; brown

above with indistinct blackish streaking and mottling; buffy below with faint markings; head pattern subdued. **GEOGRAPHIC VARIATION** Slight variation among three subspecies, usually not evident in the field. Western *cinereus* is grayer and paler on upperparts than eastern nominate; *furvior* (breeds NF) is slightly darker.

SIMILAR SPECIES Northern and Louisiana Waterthrushes are slimmer, brown backed; lack white eye ring. Waterthrushes habitually bob their tails.

VOICE **CALL:** A loud, sharp *tsick,* often in a rapid series when alarmed. **FLIGHT CALL:** Thin, high *seee.* **SONG:** Primary song a loud, ringing *cher-tee cher-tee cher-tee* or *tea-cher tea-cher tea-cher* in a rising crescendo. Also an elaborate flight song.

STATUS & DISTRIBUTION Common in mature forests, rare in the West. Migrates on a broad front in the East. **BREEDING:** Deciduous or mixed forests dominated by deciduous trees. **MIGRATION:** Arrives on the Gulf Coast late Mar.; in much of the South early Apr., peaks there mid-Apr.–early May; in the upper Midwest peaks

mid-May. Departs as early as late July, earliest arrivals in the South early Aug.; peaks in much of the East in mid-Sept., stragglers into Dec. (sometimes in northern states). **WINTER:** Primary and second growth forests from southern TX and coastal NC to southern FL south through C.A. and the Caribbean. **VAGRANT:** Rare but regular migrant in the West. Accidental in Ecuador, Greenland, and Great Britain.

POPULATION Recent significant declines likely due to forest fragmentation on breeding grounds. Trends on wintering grounds unclear.

Both species in this genus have loud, ringing songs. They have streaked breasts and are terrestrial, walking, not hopping, around wet areas. These two species were formerly placed in the genus *Seiurus* with Ovenbird.

LOUISIANA WATERTHRUSH   *Parkesia motacilla*   LOWA   1

Larger and longer-billed than the Northern Waterthrush, Louisiana has a distinctive ringing song. Between April and June, it will nest in a hollow of a fallen tree's root system or a streamside bank and lay three to

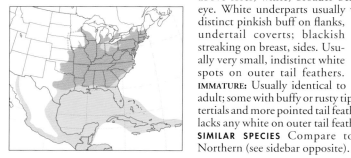

six eggs. Monotypic. L 6" (15 cm)

IDENTIFICATION Bobs tail up and down, with some side-to-side motion. Large bill, bright pink legs. **ADULT:** Olive-brown crown, back, wings, tail. Two-tone supercilium: pale buff above lores; white, broader behind eye. White underparts usually with distinct pinkish buff on flanks, undertail coverts; blackish streaking on breast, sides. Usually very small, indistinct white spots on outer tail feathers. **IMMATURE:** Usually identical to adult; some with buffy or rusty tips on tertials and more pointed tail feathers; lacks any white on outer tail feathers. **SIMILAR SPECIES** Compare to Northern (see sidebar opposite).

VOICE **CALL:** Loud, rich *chik* or *chich,* less ringing or metallic than Northern's. **FLIGHT CALL:** High *zeet.* **SONG:** Loud, rollicking primary song with clear slurred whistles ending in sputtering notes, like Swainson's and

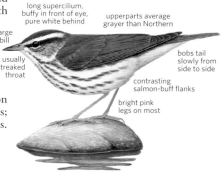

long supercilium, buffy in front of eye, pure white behind

upperparts average grayer than Northern

large bill

usually unstreaked throat

bobs tail slowly from side to side

contrasting salmon-buff flanks

bright pink legs on most

Yellow-throated Warblers': *seeeu seeeu seeeu seewit seewit ch-wit it-chu.*
STATUS & DISTRIBUTION Uncommon. **BREEDING:** Along streams in hilly deciduous forest, in cypress swamps and bottomland forest. **MIGRATION:** Arrives early, typically Gulf Coast in mid-Mar., peaks late Mar.–mid-Apr., reaches Great Lakes mid- to late-Apr.

Migration generally complete by mid-May. Departs early, mostly by mid- to late July; migrates through eastern N.A. during early to mid-Aug., a very few into early Sept., casually to early Oct., exceptional later. **WINTER:** Rivers and streams in hilly or mountainous areas from coastal northern Mexico through C.A. and Caribbean

to extreme northwestern S.A. Very rare southern FL, AZ. **VAGRANT:** Casually overshoots in spring migration to ND, QC, NS. Casual in West (inc. CA, NM, AZ, NV, CO, Baja California); six recs. from Venezuela.
POPULATION Range is expanding in the Northeast; populations stable or slightly declining elsewhere.

NORTHERN WATERTHRUSH *Parkesia noveboracensis*   NOWA ▪ 1

The waterthrushes are appropriately named, as they are almost always found near water, and on or near the ground, whether in migration or when breeding in northern forests. Often located by its loud song and calls, the Northern lays one to five eggs in a hollow of a fallen tree's root system or a streamside bank between May and June. The species is now regarded as monotypic. L 5.8" (15 cm)
IDENTIFICATION Vigorously bobs tail up and down. Dull pink legs. **ADULT:** Olive-brown crown, back, wings, and tail; yellowish, buffy, or white supercilium (even width throughout); underparts whitish tinged with yellowish, or mainly whitish, with blackish streaks on throat, breast, and sides. Usually very small, indistinct white spots on outer tail feathers. **IMMATURE:** Usually identical to adult; some with buffy or rusty tips on

tertials or more pointed tail feathers, lacking any white on outer tail feathers.
GEOGRAPHIC VARIATION Slight, clinal variation not well correlated with the three formerly recognized subspecies. Yellowest, in Northeast, but much variation throughout range.
SIMILAR SPECIES Compare to Louisiana Waterthrush (see sidebar below).

VOICE CALL: Sharp, ringing, metallic *chink.* **FLIGHT CALL:** Buzzy *zeet.* **SONG:** Primary song is loud and ringing, rather staccato: *twit twit twit sweet sweet sweet chew chew chew.*
STATUS & DISTRIBUTION Common; trans-Gulf, circum-Gulf, trans-Caribbean migrant; migrates on a broad front through the East. **BREEDING:**

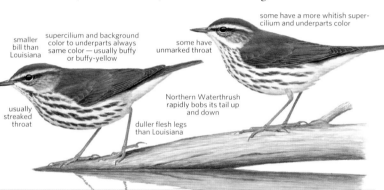

some have a more whitish supercilium and underparts color

smaller bill than Louisiana

supercilium and background color to underparts always same color — usually buffy or buffy-yellow

some have unmarked throat

Northern Waterthrush rapidly bobs its tail up and down

usually streaked throat

duller flesh legs than Louisiana

Identification of Waterthrushes

The two species of waterthrush—Northern and Louisiana—are very similar in appearance, habits, and habitat. They can occur together during migration and in the overlap zone of their breeding grounds. Use the following points in combination to separate them.

Singing birds are easily identified by voice. See their species accounts for descriptions of their songs and calls.

Get a good look at head patterns. A Northern's supercilium is uniform in width and color (usually yellow, sometimes white). A Louisiana has a subtly bicolored supercilium, buffy above the lores and white behind, that broadens noticeably.

Compare the waterthrushes' throat pattern as well. Both species show a narrow malar streak. The Northern also shows small streaks

Northern Waterthrush (NJ, Sept.)

Louisiana Waterthrush (CA, Sept.)

on the throat, whereas the Louisiana usually shows a clean white throat or a very few streaks at most.

Underparts differ noticeably, too. The Northern most often shows yellowish uniformly colored underparts, though sometimes it may be white, with bolder darker streaking. The Louisiana shows white underparts, often with pinkish-buff flanks, and less-distinct streaking.

Note bill size. The Louisiana has a distinctly longer and heavier bill than the Northern.

Leg color is also helpful in identification. Generally, the Louisiana's legs are a brighter pink than the Northern's.

Finally, consider the bird's behavior. The Louisiana's tail bobbing is more exaggerated, but slower and more circular than the Northern's rapid up-and-down bobbing. ■

Wooded areas with slow-moving water (e.g., swamps, bogs, margins of lakes). **MIGRATION:** In spring, a slightly more westerly route: arrives on Gulf Coast early Apr., peaks late Apr.–early May; arrives in the Great Lakes and Northeast late Apr., peaks mid-May. In the West, smaller numbers late Apr.–early June. In fall, a slightly more easterly route: Departs breeding grounds in July, main arrival early Aug., peaks Sept. in northern U.S., few linger into Oct. **WINTER:** Mangroves, rain forest,

second growth from coastal northern Mexico through C.A., Caribbean, and northern S.A. to northeastern Peru and northern Brazil. A few overwinter in southern FL, along Gulf Coast. **VAGRANT:** Casual in late spring to northern slope of AK, Wrangel I., Chukotski Peninsula; also Greenland, U.K., France, and the Azores. **POPULATION** Breeding areas appear stable and secure; wintering habitat (esp. mangroves) threatened with development.

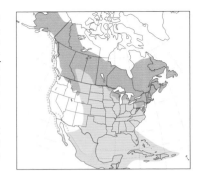

Genus *Helmitheros*

WORM-EATING WARBLER *Helmitheros vermivorum*   WEWA ◾ 1

This inconspicuous warbler feeds mostly on caterpillars by probing into suspended dead leaves. Its nest is found at the base of a sapling on a hillside or ravine, usually under dead leaves; it lays four to six eggs (May–June). Monotypic. L 5.3" (13 cm)
IDENTIFICATION Large bill; flesh-colored legs; short tail. **ADULT:** Rich buffy head with bold black lateral crown and eye stripes, buffy underparts, grayish olive upperparts.

olive-brown back

boldly striped head

dead leaf specialist

pumpkin buff color on head and underparts

IMMATURE: Nearly identical to adult, with duller head stripes and rusty tips on tertials, greater and median wing coverts, which often wear off by fall.
SIMILAR SPECIES Swainson's Warbler lacks bold head stripes and rich buff coloration.
VOICE CALL: Includes a soft *chip*.
FLIGHT CALL: A doubled *zeet-zeet*, also given at other times. **SONG:** A simple, dry, high-pitched trill similar to Chipping Sparrow's, but drier and less staccato, and to Pine Warbler's, but less musical. Note habitat of singer.
STATUS & DISTRIBUTION Fairly common but often inconspicuous; migrates across a broad front. **BREEDING:** Large tracts where deciduous and mixed forests overlap with moderate to steep slopes. **MIGRATION:** In spring, arrives in FL and Gulf Coast in last half of Mar. continuing to early May; arrives in Midwest mid-Apr.–early-May. In fall, inconspicuous departure from breeding grounds mid-July–Aug. in

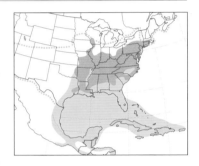

northern areas, through Sept. in southern areas. **WINTER:** Forest and scrub habitats of eastern Mexico south to Panama and the northern Caribbean (inc. the Greater Antilles). **VAGRANT:** Rare to casual in West, including more than 100+ records from CA, 35+ from AZ, and 30+ from NM. One record from Venezuela.
POPULATION Probably stable; however, it is difficult to census populations accurately. The species is sensitive to forest fragmentation.

Genus *Vermivora*

This genus comprises a total of three species. They can be recognized by their smaller size, shorter tails, and thin, pointed bills. These species have tail spots and lack rictal bristles. Their vocalizations consist primarily of a series of buzzes.

GOLDEN-WINGED WARBLER *Vermivora chrysoptera*   GWWA ◾ 2

Boldly patterned with black, white, gray, and yellow, the Golden-winged Warbler is an active and acrobatic forager. It makes its nest on or near the ground at the base of a shrub or other leafy plants, laying four to five eggs (May–June). Monotypic. L 4.8" (12 cm)
IDENTIFICATION ADULT MALE: Gray upperparts with a bright yellow cap. Broad wing bars appear as a single

large yellow patch. White underparts. Black cheek and throat contrast with white malar and supercilium. Large white spots on outer tail feathers. **ADULT FEMALE:** Similar to male, but upperparts more olive-gray, cheek and throat gray, duller yellow cap blending in with gray nape; two yellow wing bars usually distinct, not forming a large yellow patch on wing. **IMMATURE:** Duller than adult. Male more olive

above, with duller yellow cap; black throat and cheek often slightly fringed with buff. Female more olive above with duller yellow cap contrasting little with grayish olive nape, paler gray throat and cheek patch. **HYBRID:** See Blue-winged Warbler.
SIMILAR SPECIES Throat pattern and foraging behavior suggest Black-capped Chickadee. Black-throated Gray Warbler is similar when seen

from below, but it shows distinct black streaking on sides.
VOICE CALL: Most often heard call is a gentle *tsip,* similar to Field Sparrow's

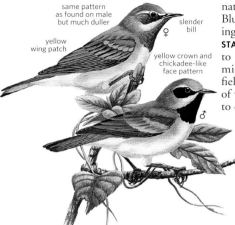

same pattern as found on male but much duller
♀
slender bill
yellow wing patch
yellow crown and chickadee-like face pattern
♂

and identical to Blue-winged Warbler's.
FLIGHT CALL: A high, slightly buzzy *tzii,* often doubled. **SONG:** Primary song is a high-pitched, buzzy *zeee bee bee bee,* with a higher-pitched first note. Alternate song is indistinguishable from Blue-winged Warbler's, with a stuttering first note and a flatter second note.
STATUS & DISTRIBUTION Uncommon to rare and declining; a trans-Gulf migrant. **BREEDING:** Wet shrubby fields, marshes, and bogs on edge of woodlands, most often restricted to early successional habitat. **MIGRATION:** In spring, peaks in early May in MI and PA, to late May in northern WI. In fall, often departs breeding grounds undetected during Aug., with latest dates in early to mid-Oct. **WINTER:** Woodlands and forest borders from southern Mexico to Panama. Casual in northern

Colombia, Venezuela, and Caribbean. **VAGRANT:** Casual in West, recorded from most western states and provinces; about 75 records from CA. Accidental to UK (winter) and the Azores. **POPULATION** A species of special concern. Population declining as Blue-winged moves into its range.

BLUE-WINGED WARBLER *Vermivora cyanoptera* BWWA ■ 1

A bright golden yellow bird with a distinct buzzy song, the Blue-winged Warbler breeds in brushy fields. It nests on or near the ground at the base of a shrub or in a clump of grasses, laying 4–5 eggs (May–June). Monotypic. L 4.8" (12 cm)
IDENTIFICATION ADULT MALE: Bright yellow forehead, crown, and underparts contrast with greenish yellow nape and back; white undertail coverts. Bluish gray wings with two distinct white wing bars. Black lores with short black post-ocular line behind eye. Large white spots on outer tail feathers. **ADULT FEMALE:** Duller than male, with yellow forehead blending into yellowish olive crown, nape, and back; whitish undertail coverts. Wing bars less distinct, sometimes tinged yellow. Lores and eye line slightly duller. **IMMATURE:** Duller, similar to adult

female. **HYBRIDS:** Blue-winged x Golden-winged Warbler hybrids are well documented, and two types have been named; "Brewster's" is more frequent, as all first-generation hybrids and some later-generation hybrids result in this type. A few backcrosses result in the rarer "Lawrence's." Some hybrids do not resemble these two types and combine characters of both parents. Hybrids usually sing the songs of one species or the other.
SIMILAR SPECIES Yellow Warbler and Prothonotary Warbler do not have black eye lines. Pine Warbler is duller and also lacks an eye line, and its underparts are duller with diffuse olive streaks.
VOICE CALL: Most often heard call is a gentle *tsip,* identical to Golden-winged Warbler's and similar to Field Sparrow's. **FLIGHT CALL:** A high, slightly buzzy *tzii,* often doubled. **SONG:** Primary song is typically a high-pitched, buzzy, inhaled-exhaled *beeee-bzzzz.* Alternate song is similar, but with a stuttering first note and a flatter second note: *be-ee-ee-ee-bttttt.*
STATUS & DISTRIBUTION Locally common in brushy meadows and second-growth woodland edges; a

black eye line
♀
slender bill
♂
white undertail coverts
two white wing bars on grayish wings

"Brewster's Warbler" first-generation hybrids and backcrosses exhibiting genetic dominance have Blue-winged head pattern

Blue-winged x Golden-winged hybrids
♀
♂
♂

"Lawrence's Warbler" rarer recessive hybrids from backcrosses have Golden-winged head pattern

trans-Gulf migrant. **BREEDING:** Overgrown old fields and brushy swamps, usually in early to mid-succession. **MIGRATION:** In spring, early Apr.–early May in TX; late Apr.–mid-May in OH, late Apr.–late May in MN. In fall, usually departs breeding grounds undetected. As early as late July in NJ, peaking in late Aug., a few to late Sept.

Occasional Nov. recs. in Southeast, a few farther north. **WINTER:** Humid evergreen and semideciduous forest and edge from southern Mexico to Costa Rica, rarely south to Panama and West Indies. **VAGRANT:** Casual in West, including SK, AB, WY, CO (approx. 40 recs.), NM (approx. 12 recs.), AZ (9 recs.), NV, WA, OR,

and CA (30+ recs.). Rare north to Atlantic provinces in fall; accidental in Iceland, Azores.
POPULATION Expanding northward in Great Lakes into range of Golden-winged Warbler, but declining due to reforestation of open habitats and urban sprawl in northeastern United States.

BACHMAN'S WARBLER *Vermivora bachmanii*   BAWA ■ 6

This likely extinct warbler was a denizen of open areas in southern swamps. Monotypic. L 4.8" (12 cm)

gray crown; breast gray or yellowish

adult ♀

adult ♂

large black throat patch extending down through breast; black forecrown

IDENTIFICATION SPRING MALE: Thin, pointed, slightly decurved bill. Yellow forehead, supercilium, eye ring, chin, and shoulders. Black crown and bib, gray nape. White spots on outer tail feathers. **SPRING FEMALE:** Duller than male with no black. Grayish on crown, throat, and breast; yellow forehead. **IMMATURE:** Male shows less black on crown and bib than adult male; smaller tail spots. Female duller than spring female.
SIMILAR SPECIES Hooded has yellow undertail coverts and larger white tail spots; bill not decurved. Immature female Bachman's is similar to several species, notably nominate subspecies of Orange-crowned, but distinguished by pointed, decurved bill; complete white eye ring; and whitish undertail coverts. Female Common Yellowthroat and immature Yellow Warbler have also

been confused with female Bachman's. **VOICE CALL:** A buzzy *zip* or *zeep* given by both sexes. Little known. **SONG:** A rapid series of buzzy notes, similar in quality to alternate song of Blue-winged and Golden-winged, delivered rapidly and on one pitch.
STATUS & DISTRIBUTION Probably extinct; last confirmed rec. was spring 1962, near Charleston, SC. Formerly a local breeder from southernmost MO to SC. Most records were migrants from mainland FL, Key West, and southeastern LA. **BREEDING:** Probably open canebrakes in swamps and bottomlands. **WINTER:** Cuba and Isle of Pines in semideciduous forest, forested wetlands, and forested urban open space. **VAGRANT:** Casual to NC and VA.
POPULATION Degradation of breeding habitat is the most likely cause of decline and probable extinction.

Genus *Mniotilta*

BLACK-AND-WHITE WARBLER *Mniotilta varia*   BAWW ■ 1

This distinctive black-and-white striped warbler, which creeps along horizontal branches while foraging, is more like a nuthatch than most other wood-warblers. Between April and June, it nests on the ground against

a shrub, tree, stump, rock, or log and produces four to six eggs. Monotypic. L 5.3" (13 cm)
IDENTIFICATION Long, slightly decurved bill; long hind claw grips tree trunks. **SPRING MALE:** Striped

black and white overall, with black cheek and throat; black crown with white median stripe. First-spring males show much white, or all white, on chin and throat. **SPRING FEMALE:** Similar to male, but with grayish buff cheek, narrow black line behind eye, white chin and throat, narrower streaks on sides. **FALL ADULT:** Male with chin and throat mottled with white (or nearly entirely white), smaller black cheek patch. Female very similar to spring, but sometimes with pale buffy wash on underparts. **IMMATURE:** Male similar to adult male, but with clear white cheek; female dullest, with richer buff on flanks and undertail coverts; paler, blurry side streaks; buffy cheek patch.
SIMILAR SPECIES Spring male Black-polls show white cheeks; Black-throated Grays lack white median crown stripe, have solid gray back,

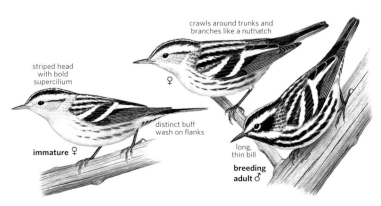

crawls around trunks and branches like a nuthatch

striped head with bold supercilium

♀

distinct buff wash on flanks

immature ♀

long, thin bill

breeding adult ♂

small yellow spot in front of eye. Yellow-throateds and Pines forage in a similar manner but show much yellow. Brown Creepers creep vertically and are striped with brown, not black.
VOICE CALL: A dull *chip* or *tik*. FLIGHT CALL: A doubled *seet-seet*. SONG: Primary song a long, slightly variable, thin, rhythmic *weesee weesee weesee weesee weesee weesee,* sometimes described as sounding like a squeaky wheel.
STATUS & DISTRIBUTION Common and widely distributed; a mid- to long-distance migrant, generally along a broad front. BREEDING: A variety of habitats, including deciduous and mixed woodlands, both mature and

second growth. MIGRATION: In spring, arrives earlier than many warbler species, reaching central FL and TX in early Mar., PA and NJ by mid-Apr., New England by early May. Peaks in Great Lakes in early May. Departs breeding areas in late July; main migration from late Aug. through late Sept.; stragglers into Nov. WINTER: A wide variety of habitats, including mature forest, mangroves, and open areas, from coastal NC to FL and southern TX through C.A. and Caribbean to northern S.A., south to Ecuador (rare) and Peru (casual). VAGRANT: Rare but regular migrant through much of the West, also noted rarely in winter. Casual to WA, AK,

and in fall to Iceland, Faroes, U.K., Ireland, and the Azores.
POPULATION Appears stable, but may be negatively affected by severe forest fragmentation.

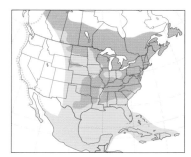

Genus Protonotaria

PROTHONOTARY WARBLER *Protonotaria citrea*   PROW ▪ 1

long, stout spiky bill

white undertail coverts

adult ♂

brilliant yellow head and underparts

contrasting gray wings

♀

A denizen of wooded swamps, this is the only eastern wood-warbler that nests in natural and artificial cavities (3–7 eggs, Apr.–June). Monotypic. L 5.5" (14 cm)
IDENTIFICATION ADULT MALE: Bright golden yellow head, underparts; beady black eye; long black bill; greenish back; bluish gray wings, tail; white undertail coverts; large white tail spots; short tail. Flesh-colored lower mandible in fall.
ADULT FEMALE: Similar to male, but greenish olive wash on rear of crown,

nape; smaller white spots on tail. IMMATURE: Male like adult male but with some stippling on crown. Female with extensive olive on crown, nape, cheek.
SIMILAR SPECIES Blue-winged shows black eye line, whitish wing bars. The smaller Yellow shows pale edges to wing coverts and tertials and has yellow tail spots and a smaller bill.
VOICE CALL: Loud, dry *chip,* like Hooded's. FLIGHT CALL: Loud *seeep.* SONG: Simple, loud, ringing series of notes: *sweet sweet sweet sweet sweet sweet.*
STATUS & DISTRIBUTION Fairly common, often local; mainly a medium-distance trans-Gulf migrant. Endangered in Canada. BREEDING: Larger wooded areas, mainly swamps and along rivers, always over water. MIGRATION: Arrives Gulf Coast mid-Mar., NJ mid-Apr., MN mid-May.

Departs northern breeding areas Aug.; most by early Sept.; stragglers into Oct., rarely Nov.
WINTER: Mangrove swamps, dry tropical forest, southern Mexico through C.A. (scarce north of Costa Rica) and Caribbean to extreme northern S.A. Casual southern FL, southern TX, and CA. VAGRANT: Rare to casual in western states.
POPULATION Declining or stable. Sensitive to habitat fragmentation on breeding and wintering grounds.

Genus *Limnothlypis*

SWAINSON'S WARBLER   *Limnothlypis swainsonii*   SWWA ■ 2

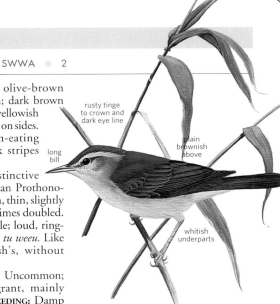

rusty tinge to crown and dark eye line

plain brownish above

long bill

whitish underparts

This skulking warbler is very difficult to observe. It usually forages by shuffling through leaf litter (quivering its tail). Its well-hidden nest, in dense ground vegetation, hosts two to five eggs (Apr.–May). Monotypic. L 5.5" (14 cm)
IDENTIFICATION Large; heavy bodied; short tail; flat forehead, long pointed bill with flesh-colored lower mandible and pinkinsh legs. **ADULT:** Rufous-brown crown; duller olive-brown above; pale supercilium; dark brown eye line; whitish or pale yellowish underparts, grayish tinge on sides.
SIMILAR SPECIES Worm-eating has conspicuous black stripes on head.
VOICE CALL NOTE: Distinctive *chip;* louder, sweeter than Prothonotary's. **FLIGHT CALL:** High, thin, slightly buzzy *swee* notes, sometimes doubled. **SONG:** Somewhat variable; loud, ringing *whee whee whee wee tu weeu.* Like Louisiana Waterthrush's, without sputtering notes.
STATUS & DISTRIBUTION Uncommon; medium-distance migrant, mainly across eastern Gulf. **BREEDING:** Damp bottomland hardwoods, canebrakes, and rhododendron thickets. **MIGRATION:** Arrives late Mar.–late Apr. Departs between mid-Sept. and mid-Oct., a few as early as mid-Aug. or as late as mid-Nov. **WINTER:** Swamps and river floodplain forests in Yucatan, Belize,

Guatemala, and western Caribbean.
VAGRANT: Overshoots in spring, where recorded casually or accidentally north to WI, MI, OH, NJ, NY, MA, ON, NS; accidentally west to KS, NE, CO, AZ, and NM.
POPULATION Probably stable.

Genus *Oreothlypis*

This genus comprises a total of eight species—seven of which occur in N.A. Six of the seven species were formerly placed in *Vermivora* and Crescent-chested was in the now-defunct genus *Parula*. All are plain-winged, lack tail spots, and have trilled songs.

COLIMA WARBLER   *Oreothlypis crissalis*   GWWA ■ 2

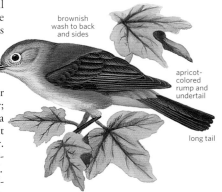

brownish wash to back and sides

apricot-colored rump and undertail

long tail

The Colima Warbler breeds only in the Chisos Mountains of west Texas and adjacent mountain ranges in Mexico. It nests on the ground in grasses, often those shaded by overhanging vegetation, and it produces three to four eggs between May and June. Monotypic. L 5.8" (15 cm)
IDENTIFICATION This is the largest and longest-tailed species in the genus *Oreothlypis;* it only occasionally wags its tail. **ADULT:** Grayish head, brownish back, and yellowish olive rump. Olive-brown sides and flanks. Undertail coverts are tawny buff. Bold white eye ring. Partially concealed rufous crown patch. **JUVENILE:** Similar to the adult but lacks rufous crown patch and has pale buff wing bars.
SIMILAR SPECIES Virginia's Warbler is smaller, shorter tailed, and grayer; it has no brown coloring. It has a distinct yellow patch on breast, and it bobs its tail, unlike Colima Warbler. It has likely hybridized with Virginia's in the high David Mts., TX.
VOICE CALL: A loud, sharp *plisk,* similar to calls of Nashville, Virginia's, and Lucy's Warblers. **SONG:** A simple musical trill in which final two notes are downslurred, similar to song of Orange-crowned Warbler.
STATUS & DISTRIBUTION Fairly common but very local. **BREEDING:** Montane areas of oak, pinyon, and juniper. Essentially unknown as a migrant. Arrives on breeding grounds as early as mid-Mar., with all birds

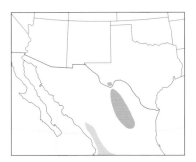

arriving on territory by early May. All have departed the Chisos Mountains by mid-Sept. **WINTER:** Brushy areas of humid montane forests in southwestern Mexico, including Colima, hence its English name.
POPULATION Chisos Mts., TX, population varies, with between 40 and 80 breeding pairs on censuses from 1960s to 1990s (fewer in drought years).

LUCY'S WARBLER *Oreothlypis luciae* LUWA ■ 1

A small gray bird of the arid southwest deserts, the Lucy's Warbler is one of only two species of cavity-nesting warblers. It makes its nest in a natural cavity, including crevices, under loose bark, or in an abandoned woodpecker hole. It produces three to seven eggs between April and June. Monotypic. L 4.3" (11 cm)

IDENTIFICATION This very small warbler flicks its short tail. **ADULT MALE:** Pale gray upperparts, slightly darker wings and tail, white underparts. Dark eye stands out on pale gray face, which shows white lores and an indistinct white eye ring. Chestnut crown patch and rump are both often concealed. Outer tail feather shows a small white

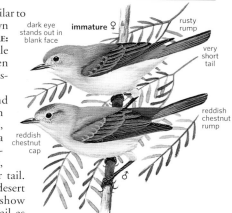

dark eye stands out in blank face

immature ♀

rusty rump

very short tail

reddish chestnut rump

reddish chestnut cap

♂

spot near tip. **ADULT FEMALE:** Similar to male, but paler face, smaller crown patch, and paler rump. **IMMATURE:** Similar to adult female, with pale buff cast on underparts. When fresh, immatures have two indistinct pale buffy wing bars.

SIMILAR SPECIES Virginia's and Colima Warblers show yellowish on rump and undertail coverts, not chestnut. Bell's Vireo has a thicker bill and wing bars. Juvenile Verdin has a much sharper, thicker-based bill and a longer tail. Females of Yellow Warbler's desert subspecies (*sonorana*) always show some greenish on wings and tail as well as pale yellowish spots on tail. **VOICE CALL:** A sharp *chink*, similar to the call of Virginia's Warbler. **FLIGHT CALL:** A weak *tsit*. **SONG:** A loud, lively, sweet song somewhat similar to Yellow Warbler's: *tee-tee-tee-tee-tee-sweet-sweet-sweet.*

STATUS & DISTRIBUTION Fairly common. **BREEDING:** Dense lowland riparian mesquite, cottonwood, and willow woodlands, mainly in the Sonoran Desert. **MIGRATION:** Earlier in spring than most migrants; arrives in southern AZ in early Mar. In fall, departs breeding

grounds early, sometimes beginning as early as late July. Most gone from AZ by early to mid-Aug., with latest in early Oct. Rare on CA coast late Aug.–late Nov. **WINTER:** Thorn forest and riparian scrub in western Mexico. **VAGRANT:** Casual in southwestern CO. Accidental in ID, OR, LA, and MA. **POPULATION** Nests in unusually high densities for a noncolonial species. Declines due to the loss of riparian habitat from water projects, the cutting of mesquite trees, and increase in thickets of introduced tamarisks.

VIRGINIA'S WARBLER *Oreothlypis virginiae* VIWA ■ 1

A tail-bobbing gray, white, and yellow warbler of our southwestern mountains, the Virginia's makes an open-cup nest of grasses on or near ground in a clump of vegetation. It lays three to five eggs between May and June. Monotypic. L 4.8" (12 cm)

IDENTIFICATION SPRING MALE: Gray upperparts with yellowish green rump and uppertail coverts. Prominent white eye ring. Whitish underparts with a yellow patch on breast and yellow undertail coverts. Extensive rufous crown patch often concealed. **SPRING FEMALE:**

Like male, with less extensive yellow on breast and smaller rufous crown patch. **FALL ADULT:** Browner upperparts, buffy-tinged underparts. Yellow breast patch mixed with gray. Rufous crown patch obscured with gray. **IMMATURE:** Brownish gray upperparts, buffy underparts, yellow breast patch usually small or absent. Rufous crown patch very limited or absent. **JUVENILE:** Similar to immature, but with two pale buff wing bars. **SIMILAR SPECIES** Western subspecies of Nashville Warbler also bobs its shorter tail, but it shows entirely yellow underparts, and its upperparts are greenish, not gray. Colima Warbler is larger and browner, it lacks yellow on breast, and its undertail coverts are not yellow. **VOICE CALL:** A loud, sharp *chink*. **FLIGHT CALL:** A very high, clear *seet*. **SONG:** Similar to song of Nashville Warbler's western subspecies, but less structured and on 1 pitch: *s-weet, s-weet, s-weet, sweet-sweet-sweet.*

STATUS & DISTRIBUTION Fairly common. **BREEDING:** Scrubby areas in steep-sloped pinyon-juniper and oak

woodlands. **MIGRATION:** In spring, later than most other warblers, arriving in West late Apr.—mid-May, late Apr. in CO, and early May in WY. In fall, generally early Aug.—early Oct. in AZ and NM. **WINTER:** Arid to semiarid scrub in highlands in southwest Mexico. Casual in southern TX and coastal southern CA. **VAGRANT:** During migration it has

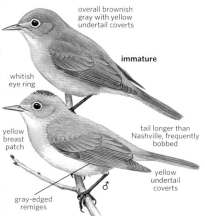

overall brownish gray with yellow undertail coverts

immature

whitish eye ring

yellow breast patch

gray-edged remiges

tail longer than Nashville, frequently bobbed

yellow undertail coverts

♂

been recorded casually north to OR and coastal CA. Casual to accidental in southeastern TX, southwestern LA, ON, NS, Labrador, MI, NJ, NB, IL, MO, Belize, northern Guatemala, and Grand Bahama Is.

POPULATION Apparently stable on breeding grounds, but number of vagrants to coastal CA has declined.

TENNESSEE WARBLER *Oreothlypis peregrina* TEWA ▪ 1

This rather plain but common spruce-budworm specialist nests on the ground at base of small shrub or tree (3–8 eggs, June). Monotypic. L 4.8" (12 cm)

IDENTIFICATION SPRING MALE: Bright olive green upperparts, contrasting gray crown and nape, distinct white supercilium and underparts, distinct blackish eye stripe. Faint whitish wing bar. **SPRING FEMALE:** Duller than male, less contrasting crown and back, tinged yellow below. **FALL ADULT:** Duller than spring adult, usually some pale yellowish below, less contrasting nape and back. **IMMATURE:** Bright olive-green upperparts. Less distinct supercilium, buffy to pale yellowish; dusky eye line. Pale to bright yellowish underparts. Bright white undertail

slender, straight spike-like bill

bright green back, faint wing bars

fall

fall

breeding ♀

bright green upperparts

whitish undertail coverts on most, sometimes pale yellow

gray cap and dark eye line

white belly

short tail

breeding ♂

coverts, sometimes tinged pale yellow.
SIMILAR SPECIES Similar to Orange-crowned, but with straighter bill and shorter tail; is plain below and nearly always has white undertail coverts. Philadelphia and Warbling Vireos have thicker bills with hooked tips.
VOICE CALL: Sharp *tsit*. **FLIGHT CALL:** Thin, clear *see*. **SONG:** Loud, staccato series of *chip* notes, increasing speed in two or three distinct steps.
STATUS & DISTRIBUTION Fairly common; mostly a trans-Gulf migrant. **BREEDING:** Mainly mixed and coniferous forest and bogs in boreal zone. Arrives early Apr. on Gulf Coast. Bulk of migration in Great Lakes during mid- to late May. Adults depart

breeding grounds very early, often in July. Peaks mid-Sept.–early Oct., a few to late Oct. **WINTER:** Southern Mexico to Panama and northwestern S.A. Rarely in coastal CA. **VAGRANT:** Rare migrant in CA, mostly casual to rare elsewhere in the West. Rare in AK (some breed), casual to southwestern Greenland, Iceland, UK, and the Azores.
POPULATION Often common.

ORANGE-CROWNED WARBLER *Oreothlypis celata* OCWA ▪ 1

This rather yellowish green warbler, named for its least conspicuous character, forages lower than

immature ♀ celata

grayish head

slender bill slightly downcurved

grayish underparts

split eye ring

longer tail than Tennessee

even drab birds have yellowish undertail coverts

blurry streaks on breast

celata ♂

many species. It lays four to five eggs (Mar.–May in west, June in east) in its nest on or near the ground. Polytypic (4 ssp; all in N.A.). L 5" (13 cm)
IDENTIFICATION ADULT MALE: Dusky olive green upperparts, grayer on crown and nape. Whitish or yellowish narrow broken eye ring, indistinct dusky eye line. Greenish yellow underparts with indistinct blurry streaks. Undertail coverts always brighter yellow than belly. **ADULT FEMALE:** Duller and grayer than male. **IMMATURE:** Duller, similar to adult female.

GEOGRAPHIC VARIATION Northern nominate *celata* is dullest, West Coast *lutescens* is brightest yellow, Rocky Mountains and Great Basin *orestera* is intermediate, and coastal southern California *sordida* (mainly Channel Is.) is darkest green.
SIMILAR SPECIES Compare with

brighter overall and more of a lemon yellow below

lutescens ♂

Tennessee Warbler. In West, Yellow and Wilson's Warblers are similar to *lutescens,* but both show plain faces with prominent dark eyes.

VOICE CALL: Most frequent call a very distinctive, hard *stick* or *tik.* **FLIGHT CALL:** A high, thin *seet.* **SONG:** A high-pitched loose trill becoming louder and faster in the middle, weaker and slower at the end; faster in *lutescens.*

STATUS & DISTRIBUTION Common in West, uncommon in East. **BREEDING:** Brushy deciduous thickets and second growth, from boreal forest in East to a great variety of habitats in West. In spring, northern subspecies takes a more westerly route up the Mississippi River Valley, mid-Apr.–late May. Western *lutescens* moves shorter distances, earlier peaking in late Mar.–

early Apr. In fall, northern subspecies move later than other warblers, peaking in Oct. in much of northern U.S. Western subspecies moves earlier, mid-Aug.–early Oct. in AZ and CA. **WINTER:** Generally in thickets and shrubby areas from southeast U.S. and CA, through Mexico to Guatemala. Rarely lingers in northern U.S. **VAGRANT:** Rare to casual on Bahamas, Cuba, Jamaica, Cayman Is.

POPULATION Reasonably stable, especially in West, with no significant threats identified.

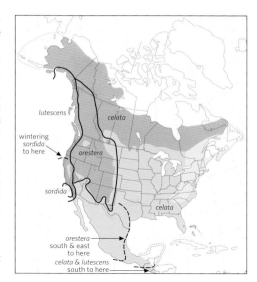

NASHVILLE WARBLER *Oreothlypis ruficapilla*   NAWA   ▪   1

The Nashville is a small olive-and-yellow warbler with a broad white eye ring and a gray head. The breeding ranges of the eastern and western subspecies do not overlap. In May–June, it lays four to five eggs in a nest on or near the ground, in thickets or bases of shrubs or small trees. Polytypic (2 ssp; all in N.A.). L 4.8" (12 cm)

IDENTIFICATION ADULT MALE: Gray head contrasts with olive back, wings, and tail; bold white eye ring; yellow underparts. Small white area around vent. Rufous crown patch generally hidden. **ADULT FEMALE:** Duller than male, with more olive-gray head. Smaller rufous crown patch. **IMMATURE:** Duller than adult female, with more brownish olive upperparts. Whitish throat.

GEOGRAPHIC VARIATION Compared

to eastern nominate, western *ridgwayi* is brighter yellow-green on rump and brighter yellow on breast; has more white on vent; and often wags its longer tail.

SIMILAR SPECIES Connecticut is much larger, with a gray hood, including throat and upper breast. Virginia's has less yellow below and is gray above. Nashville is grayer above and always has olive-edged flight feathers.

VOICE CALL: A distinctive flat *tink*; sharper in *ridgwayi.* **FLIGHT CALL:** A high, thin *tsip* or *seet.* **SONG:** A two-part, loose series of sweet notes; first part bounces, second part is slightly faster: *see-bit, see-bit, see-bit, see see see see see.* Western subspecies has a similar but sweeter series of *see-bit* notes with less pattern.

STATUS & DISTRIBUTION Common. **BREEDING:** Open deciduous or mixed forests, including spruce bogs in east. **MIGRATION:** Eastern birds are circum-Gulf migrants. Spring peak late Apr.–early May in much of

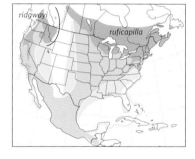

eastern N.A.; much of fall migration late Aug.–late Oct. Western birds migrate through Southwest and along coast, most in Apr. in West; in fall, somewhat earlier than the easterns. **WINTER:** Wooded areas in highlands and coastal areas, coastal CA (rare) through Mexico to Guatemala. **VAGRANT:** Very rare in fall and winter in Caribbean. Casual to AK. Accidental in Greenland.

POPULATION Stable.

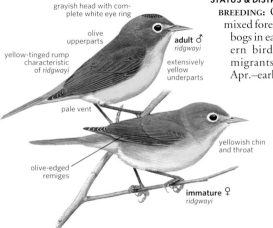

grayish head with complete white eye ring

olive upperparts

yellow-tinged rump characteristic of *ridgwayi*

adult ♂ *ridgwayi*

extensively yellow underparts

pale vent

olive-edged remiges

yellowish chin and throat

immature ♀ *ridgwayi*

adult ♂ *ruficapilla*

shorter tail than *ridgwayi*

rump duller and more olive than *ridgwayi*

CRESCENT-CHESTED WARBLER *Oreothlypis superciliosa*   CSWA   4

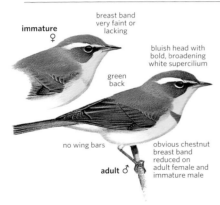

immature ♀ — breast band very faint or lacking

bluish head with bold, broadening white supercilium

green back

no wing bars

obvious chestnut breast band reduced on adult female and immature male

adult ♂

This distinctive tropical montane species occurs casually north of Mexico and was formerly placed in the genus *Parula*. Polytypic (5 ssp.; 2 poss. in N.A.). L 4.3" (11 cm)
IDENTIFICATION Blackish upper mandible; blackish lower mandible with flesh-colored base. **ADULT MALE:** Gray head with broad white supercilium and small white arc below eye; green back; yellow throat and breast with distinct crescent-shaped chestnut band on breast; whitish belly and undertail coverts. Gray wings and tail. **ADULT FEMALE:** Smaller chestnut breast band. **IMMATURE:** Breast band is much reduced or lacking.
GEOGRAPHIC VARIATION Very slight variation. Northernmost and palest subspecies *(sodalis)* likely has occurred in Arizona, while slightly darker, brighter subspecies *(mexicana)* may have occurred in Big Bend, Texas (if sight record is correct).
SIMILAR SPECIES Both Northern and Tropical Parulas lack white supercilium and have white wing bars. Rufous-capped Warbler is rufous on crown and lacks chestnut breast band.
VOICE CALL: A soft *sik*, similar to but softer than Orange-crowned's. **FLIGHT CALL:** A high, thin *sip*. **SONG:** A short, flat buzz, reminiscent of a "Bronx cheer": *t-t-t-t-t-t-t-t-t.*
STATUS & DISTRIBUTION Uncommon in montane pine-oak woodlands of Mexico to Nicaragua; casual vagrant to N.A. in spring, fall, and winter. **BREEDING:** Not in N.A. Montane pine-oak and cloud forests from Mexico to Nicaragua. **VAGRANT:** At least four confirmed records for AZ and two unconfirmed records for TX.

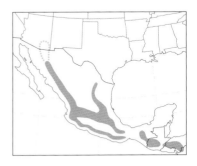

Genus *Geothlypis*

Most, but not all, of this group of skulking wood-warblers prefer marsh habitats. Twelve species; five in N.A. Vocalizations are rhythmic and chanting. This genus was expanded in 2011 to include the Kentucky, Mourning, and MacGillivray's Warblers (all formerly in *Oporornis*). The latter two species fit well genetically into *Geothlypis*, but Kentucky less so.

KENTUCKY WARBLER *Geothlypis formosa*   KEWA   1

This skulking warbler is more often heard than seen. In May or June, it lays three to six eggs in a nest at the base of a small shrub. Monotypic. L 5.3" (13 cm)
IDENTIFICATION Chunky, short tail, pale pink legs. **ADULT MALE:** Olive-green upperparts, wings; black cap; yellow supercilium forming spectacles; black lores, triangular cheek patch; bright yellow underparts. **ADULT FEMALE:** Similar to male, but less extensive black on crown, cheek. **IMMATURE:** Male similar to adult female. Female dullest; suggestion of face pattern; black mostly replaced by dark olive.
SIMILAR SPECIES Common Yellowthroats do not show yellow spectacles.
VOICE CALL: Low, distinctive *chup*, similar to Hermit Thrush's. **FLIGHT CALL:** Loud, buzzy *zeep*. **SONG:** Somewhat variable series of rich rolling notes, somewhat similar to Carolina Wren's: *churree churree churree churree.*
STATUS & DISTRIBUTION Fairly common, primarily a trans-Gulf migrant. **BREEDING:** Dense understory in deciduous or mixed woodlands, often near streams. **MIGRATION:** Arrives Gulf Coast late Mar.–early Apr., northern extreme of range early May. In fall, some take a more easterly route than in spring. Departs early Aug., peaks late Aug.–early Sept., stragglers to late Sept. and (exceptional) early Oct. **WINTER:** Forested lowlands, second growth from northeastern Mexico to Panama and extreme northeastern Colombia and northwestern Venezuela. **VAGRANT:** Overshoots north in spring. Rare to casual in West. Casual in winter to FL, TX, and CA.
POPULATION Generally stable; declining in some areas. Probably affected negatively by tropical deforestation.

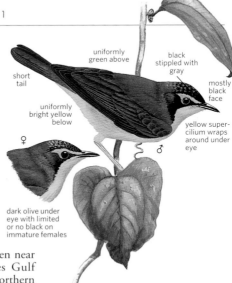

uniformly green above

black stippled with gray

short tail

mostly black face

uniformly bright yellow below

♀

yellow supercilium wraps around under eye

♂

dark olive under eye with limited or no black on immature females

COMMON YELLOWTHROAT *Geothlypis trichas* COYE ▦ 1

This common skulking species nests on or near the ground in thick clumps of sedges and grasses, often in wetlands (1–6 eggs, Apr.–July). Polytypic (14 ssp.; 11 in N.A.). L 5" (13 cm)
IDENTIFICATION Short wings, long tail; flesh-pink legs. Nominate described. **ADULT MALE:** Broad black face mask with grayish border. Bright yellow throat, undertail coverts; paler belly. **ADULT FEMALE:** Face brownish olive; yellowish throat. **IMMATURE:** Somewhat like adult female; many males show some black on face.
GEOGRAPHIC VARIATION Mask of eastern adult males usually bordered by gray; in West, such as widespread

occidentalis, mask bordered by whitish. Nominate found in much of East; in *chryseola* (western TX, southern AZ) underparts often entirely yellow, mask border tinged with yellow. Great Plains *campicola* and more northerly *yukonicola* (if recognized) have more restricted yellow throats, white underparts.
SIMILAR SPECIES Kentucky has yellow spectacles. Larger immature Mourning and MacGillivray's are larger, shorter tailed, with all-yellow underparts.
VOICE **CALL:** Includes husky *tschep;* rapid chatter. **FLIGHT CALL:** Buzzy *dzip.* **SONG:** A somewhat variable, loud, rolling *wichity wichity wichity wichity wich.*
STATUS & DISTRIBUTION **MIGRATION:**

Peaks late Apr.–early May and Sept. in East. In West, extended spring peak; fall Sept.–mid-Oct. Stragglers late Oct. through much of range. **WINTER:** Dense vegetation, usually in or near wetlands; south to extreme northwestern S.A. A few to Great Lakes. **VAGRANT:** Casual to Azores; accidental Greenland, U.K. **POPULATION** Stable.

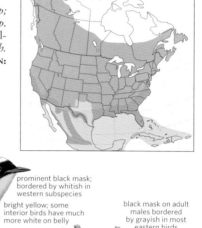

narrow whitish eye ring

yellowish under tail

yellowish tinge on neck

color below varies

occidentalis ♀

adult ♂ southwestern *chryseola*

underparts nearly solid yellow

prominent black mask; bordered by whitish in western subspecies

bright yellow; some interior birds have much more white on belly

black mask on adult males bordered by grayish in most eastern birds

long tail

immature male has some black in face

adult ♂ *occidentalis*

adult ♂ *trichas*

immature ♂ *occidentalis*

GRAY-CROWNED YELLOWTHROAT *Geothlypis poliocephala* GCYE ▦ 4

An aberrant *Geothlypis.* Polytypic (7 ssp.; *ralphi* in N.A.). L 5.5" (14 cm)
IDENTIFICATION Moderately thick, bicolored bill, curved culmen; long tail often cocked and waved. **ADULT**

MALE: Entirely gray crown; black lores extend below eye; broken, white eye ring; all yellow below, brownish on flanks. **ADULT FEMALE:** Similar to male, but olive-gray crown, slate-gray lores. **IMMATURE:** Similar to adult female.
SIMILAR SPECIES Common Yellow-throat has thinner all-dark bill, lacks strongly curved culmen. Larger, thicker-billed Yellow-breasted Chat has white spectacles. A possible Gray-crowned x Common Yellowthroat hybrid was photographed in Rio Grande Valley.
VOICE **CALL:** Includes a distinctive, nasal *cheed-l-eet.* **SONG:** Primary song is a rich,

thick bill with pinkish base

ralphi ♂

long tail

ralphi ♀

gray head with broken white eye ring and dark lores

scratchy warble like song of a *Passerina* bunting or Blue Grosbeak.
STATUS & DISTRIBUTION Historically resident (through 1910) but now casual in extreme southern TX. At least six records in the last 25 years.

MACGILLIVRAY'S WARBLER *Geothlypis tolmiei*   MGWA   1

A western version of the Mourning Warbler, the MacGillivray's Warbler is a close relative, and the two hybridize frequently where their ranges meet in Alberta. MacGillivray's breeds mainly in montane areas, where it makes a nest for 3–5 eggs on or slightly above the ground in dense shrubs or dense undergrowth between May and July. Polytypic (2 ssp.; both in N.A.). L 5.3" (13 cm)

IDENTIFICATION **ADULT MALE:** Blue-gray

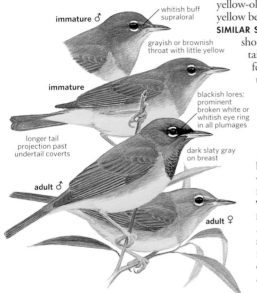

immature ♂

whitish buff supraloral

grayish or brownish throat with little yellow

immature

blackish lores; prominent broken white or whitish eye ring in all plumages

longer tail projection past undertail coverts

adult ♂

dark slaty gray on breast

adult ♀

head and throat form a hood; black lores; black and gray mottling on upper breast do not form a solid patch. Bold white crescents above and below eye. Yellow below. **ADULT FEMALE:** Similar to male, but with paler gray hood, paler throat. **IMMATURE:** Similar to adult female, but with more brownish olive hood; throat grayish, rarely yellow tinged.

GEOGRAPHIC VARIATION The two subspecies are not field-identifiable. Nominate from Pacific coast is more yellow-olive on upperparts and deeper yellow below than interior *monticola*.

SIMILAR SPECIES Mourning Warbler is shorter tailed, has longer undertail coverts; song and calls different. Adult male Mourning usually lacks black lores and usually shows an extensive black breast patch; white eye arcs are absent or very narrow. Immature Mourning Warblers show a yellowish throat and an incomplete breast band; white around eye, when present, is thinner, often forms a nearly complete ring. **VOICE** **CALL:** A loud, sharp *tsik*. **FLIGHT CALL:** A penetrating *tseep*. **SONG:** Primary two-part song, with a somewhat burry first part and a variable (higher- or lower-pitched) second part: *churry churry churry tree tree tree*

or *sweet sweet sweet sweet peachy peachy*. **STATUS & DISTRIBUTION** Common; widespread migrant in the West. **BREEDING:** Open areas of montane mixed and coniferous forests, usually near rivers and streams. **MIGRATION:** In spring, arrives in southern AZ and southern CA in early Apr., peaking mid-May over much of the West. In fall, departs breeding areas during Aug., peaking late Aug.–mid-Sept., with some to mid-Oct. **WINTER:** Densely vegetated habitats, from northwestern Mexico to central Panama. Casually in southern CA, AZ, and TX. **VAGRANT:** Casual in East, recorded from ON, MN, IL, IN, TN, MO, GA, FL, MD, NJ, CT, and MA. **POPULATION** Stable population, probably benefits from logging activities, which create second-growth.

MOURNING WARBLER *Geothlypis philadelphia*   MOWA   1

This skulking warbler arrives on its breeding grounds later than most warblers. Some can be very difficult to separate from the similar MacGillivray's and are best told by call. Built in June or July, its well-concealed nest is on or near the ground in tangles and shrubs (2–5 eggs). Monotypic. L 5.3" (13 cm) **IDENTIFICATION** Chunky, short tailed. Flesh-colored legs, mostly pinkish bill somewhat dusky on upper mandible. Plumage somewhat variable. **ADULT MALE:** Blue-gray head and throat form a hood; black bib on upper breast sometimes extends onto throat; lores sometimes black; very rarely with thin, broken eye ring. Olive-green back, wings, and tail. Bright yellow lower breast, belly, and undertail coverts. **ADULT FEMALE:** Similar to male, but with paler gray hood and lacking black bib. Shows nearly complete, very thin eye ring in fall. **IMMATURE:** Similar to

adult female, but with more olive-gray hood and incomplete olive-gray bib; yellowish throat, sometimes quite pale. Shows a thin, broken eye ring (thicker on some); some males may show limited black in bib. **SIMILAR SPECIES** Adult male MacGillivray's has different song and calls, shows black lores and slaty mottling on throat; all ages and sexes also show distinct broadly broken eye arcs, above and below eyes, and longer tail projection. Connecticuts show distinct, complete eye rings and shorter tail projection; they walk instead of hop. Immature and female Common Yellowthroats are longer tailed, lack gray on crown, have a smaller all-dark

immature males often with some black stippling on lower throat

immature ♂

dull yellow supraloral

short tail projection past undertail coverts

immature

yellow breaks through breast; extends into throat on most immatures

black chest patch

adult ♂

adult ♀

bill, and show no yellow on belly. **VOICE CALL:** A distinctive, sharp, scratchy *chit* or *jip.* **FLIGHT CALL:** A thin, sharp *seep.* **SONG:** Variable, most often two-part, with first part louder and slightly burry, second part faster and lower: *curry churry churry chorry chorry.* Also, a one-part song, somewhat similar to some songs of Kentucky Warbler. **STATUS & DISTRIBUTION** Fairly common; circum-Gulf migrant. **BREEDING:** Clearings in disturbed second growth, mixed forests. **MIGRATION:** In spring, a few arrive in southern TX as early as late Apr., peaking in Upper Midwest

in late May–early June. Departs breeding grounds in early Aug., peaking late Aug.–early Sept., casually to late Oct., accidentally later. **WINTER:** Dense thickets, overgrown fields, shrublands, and other semi-open areas, often near water; southern Nicaragua to Ecuador. **VAGRANT:** Very rare in much of Southeast. In West mainly from CA, where rare: 120+ recs., most from fall, but including two exceptional coastal winter records. Casual elsewhere in the West. Accidental in fall to AK. Two recs. from Baja California, three specimens from Greenland, **POPULATION** Stable; may be one of few

species to benefit from human activities, especially from logging activities that create the openings it prefers.

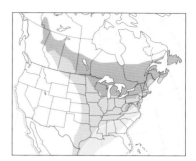

Genus *Oporornis*

This genus now contains only the Connecticut Warbler. The other three members (Kentucky, Mourning, and MacGillivray's) were moved into *Geothlypis.* Unlike those species the Connecticut has a distinctive walking gait. Surprisingly, its closest relative may be Semper's Warbler *(Leucopeza semperi),* a species from St. Lucia (Lesser Antilles) that is likely extinct.

CONNECTICUT WARBLER   *Oporornis agilis*   CONW ■ 2

This secretive, skulking, very terrestrial denizen of northern spruce bogs is one of the most challenging wood-warblers for birders to see. Not surprisingly, its June-built nest is well concealed on or near the ground in thickets or at the base of a shrub (3–5 eggs). Monotypic. L 5.8" (15 cm) **IDENTIFICATION** Large and chunky; short tail projection and long primary extension; habitually walks on or near ground. Pinkish legs; bill mostly pinkish with dusky brown culmen and tip. **SPRING MALE:** Gray head, throat, and upper breast form a distinct hood, paler on throat; prominent complete white eye ring, occasionally broken at rear; olive upperparts, pale yellow underparts. **SPRING FEMALE:** Similar to male, but with duller hood tinged brownish olive. **FALL ADULT:** Similar to spring, but washed with olive on head. **IMMATURE:** Similar to fall female, some duller; extensively washed with olive-brown on head;

throat buffy with brownish olive chest band; buff-tinged eye ring. **SIMILAR SPECIES** MacGillivray's and Mourning Warblers are smaller with longer tail projection; show dark lores with broken or absent eye rings; and hop instead of walk. Nashville Warbler is much smaller and shows yellow on throat and breast. **VOICE CALL:** A loud, nasal *chimp,* rarely heard. **FLIGHT CALL:** A buzzy *zeet,* also given when perched. **SONG:** Variable primary song loud, staccato, and more emphatic toward end (similar in quality to Northern Waterthrush's): *chuppa-cheepa chuppa-cheepa chuppa-cheep.* **STATUS & DISTRIBUTION** Uncommon; long-distance migrant. **BREEDING:** Boreal forest, spruce-tamarack bogs, muskeg, poplar woodlands, and deciduous forest. **MIGRATION:** In spring, through Caribbean to FL, then northwest to breeding grounds; later than most other warblers. Arrives in FL early May–late

May, peaks in Great Lakes in late May, stragglers into early June. More easterly route in fall, begins in late Aug., peaks in last half of Sept., stragglers to late Oct. on southern Atlantic coast. Most fly nonstop across the Atlantic from southern New England or the mid-Atlantic to S.A.; 75 were found on Bermuda (Sept. 25, 1987) after the passage of Hurricane Emily. **WINTER:** Most poorly known of all wood-warblers; woodlands, forest edge, and dense shrubby second growth. A few recorded south of Amazon in Pantanal of Brazil to Bolivia. **VAGRANT:** Casual in spring east of Appalachians and to LA, TX, and CA. Very rare to accidental in West, in fall most recs. along Pacific coast (100+ recs. from CA). Rare to casual in Atlantic Canada. Casual to northern Baja California, Honduras, Costa Rica, and western Panama. **POPULATION** Probably stable.

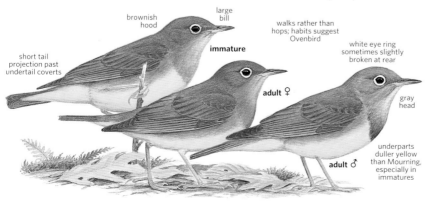

short tail projection past undertail coverts

brownish hood

large bill

immature

walks rather than hops; habits suggest Ovenbird

adult ♀

white eye ring sometimes slightly broken at rear

gray head

adult ♂

underparts duller yellow than Mourning, especially in immatures

Most of these species were formerly placed in the now-defunct genus *Dendroica*. This newly expanded genus includes both Northern and Tropical Parulas and the Hooded Warbler, for a total of 25 species in N.A. and nine other species resident from the West Indies for a total of 34 species. Most species have pale wing bars or patches and tail spots, and shorter and thicker bills than warblers in *Oreothlypis* and *Vermivora*. Many species have distinctive songs, but these songs have notable individual variation.

HOODED WARBLER *Setophaga citrina* HOWA ■ 1

The Hooded is often seen only as a flash in open deciduous woodland. Its nest is placed low in a patch of shrub and holds two to five eggs (May–June). Monotypic. L 5.2" (13 cm) **IDENTIFICATION** Opens and closes tail, revealing extensive white on outer tail feathers; large bill, eye. **ADULT MALE:** Black hood, throat; contrasting bright yellow forehead, cheeks, dark lores; olive-green upperparts; bright yellow underparts. **ADULT FEMALE:** Like male, but variable head pattern: most show only narrow black border; a few show nearly complete black hood. First-spring similar to immature. **IMMATURE:** Male similar to adult; hood feathers extensively tipped yellow and olive. Female dullest; no black on throat, crown;

indistinct yellow superciliary, throat; olive cheeks.
SIMILAR SPECIES Smaller, immature female Wilson's Warbler has smaller bill and eye; lacks dark lores, white tail spots. Yellow Warbler has yellow tail spots.
VOICE CALL: Loud *chink*. **SONG:** Clear and whistled, with emphatic ending: *ta-wee ta-wee ta-wee ta-wee tee-too.*
STATUS & DISTRIBUTION Fairly common; primarily trans-Gulf migrant. **BREEDING:** Mixed hardwood forests in north; cypress-gum swamps in south, preferring larger woodlots. **MIGRATION:**

Arrives Gulf Coast mid-Mar.; Midwest late Apr.; northern range early May. Slightly more easterly route in fall: earliest arrive FL mid-July; latest depart early Sept., small numbers in South into early Oct., casually into Nov. **WINTER:** Males lowland mature forest; females scrub, secondary forest, disturbed habitats. C.A.; rarely southern FL, northern Colombia, Venezuela. **VAGRANT:** Rare spring and fall to much of West, especially CA (has bred), AZ. Accidental in winter to southern states, inc. SC, CA, TX; two records from U.K.
POPULATION Stable. May be sensitive to forest fragmentation.

rapidly opens and closes tail

adult ♂

black hood

all Hoodeds have dusky lores

adult ♀

older females have some black to extensive black in hood and necklace

extensive white in outer tail feathers

immature ♀

large dark eye and blank face

AMERICAN REDSTART *Setophaga ruticilla* AMRE ■ 1

This bird is conspicuous in lower levels of vegetation as it frequently fans its tail, showing large tail spots, and as it sallies out to catch insects. It lays one to five eggs in its nest, built against the main trunk of a tree or woody shrub or, sometimes, on a horizontal branch away from the trunk (May–June). Monotypic. L 5.3" (13 cm) **IDENTIFICATION** Flat, flycatcher-like bill with prominent rictal bristles. **SPRING MALE:** Largely black with conspicuous orange patches at sides of breast, on bases of wing feathers (forming broad wing stripe), and extensively at bases of outer tail feathers. White belly and undertail coverts. First-spring male sometimes nearly

identical to spring female; most often with irregular black blotches on face, head, back, and underparts. **SPRING FEMALE:** Male's orange patches replaced with yellow (orange in some older females), light gray head, olive green back, white underparts. **FALL ADULT:** Male very similar to spring, but with narrow buffy edges on black body feathers. Female nearly identical to spring; more brownish back contrasts less with gray head. **IMMATURE:** Similar to adult female, often with narrower wing stripe. Sexes often indistinguishable; some males show orange-yellow patch on sides of breast.
SIMILAR SPECIES Slate-throated and

Painted Redstarts similar in behavior, but only show white in tail.
VOICE CALL: Thin *chip*, like Yellow

Warbler's. **FLIGHT CALL:** Penetrating, clear *seep*. **SONG:** Significantly variable, high-pitched series, sometimes with downslurred, slightly burry final note: *zee zee zee zee zweeah.* Some versions similar to songs of other warbler species. **STATUS & DISTRIBUTION** Common to locally abundant; migrates along a broad front. **BREEDING:** Wide variety of open wooded habitats. Usually arrives Gulf Coast first half of Apr., peaks mid-May through much of northern U.S. Departs breeding grounds July, peaks Upper Midwest late Aug.–mid-Sept., some into early Oct. **WINTER:** Forest, woodland, lower and middle elevations from southern FL and southern Mexico through Caribbean to northern S.A. south to Peru. **VAGRANT:** Casual or accidental in Iceland, U.K., France, and the Azores. **POPULATION** Slight declines, significantly in northern New England; increases in CT, WI, and QC.

gray head with white eye ring

often spreads and holds open tail

adult ♂

♀

extensive yellow base to outer tail feathers

dark lores

1st spring ♂

1st-spring male usually with some blackish spotting on throat or breast

KIRTLAND'S WARBLER *Setophaga kirtlandii* KIWA 2

The rarest warbler in North America, the large Kirtland's forages lower than most warblers and persistently pumps its tail when foraging. Its nest on ground under lowest branches of jack pines, where it mingles with ground vegetation; here it will deposit three to six eggs in June. Monotypic. L 5.8" (15 cm)
IDENTIFICATION ADULT MALE: Bluish gray upperparts, with black lores and black below eye; fine black streaks on crown, bolder on back. Yellow underparts with bold black streaks on sides, white undertail coverts. Broken whitish eye ring, indistinct wing bars. **ADULT FEMALE:** Similar to male, but duller, no black on face, less bold streaks on sides. Upperparts tinged brownish gray. **IMMATURE:** Male similar to adult female. Female more brownish above.
SIMILAR SPECIES Female and immature Magnolias are smaller; show green on back, yellow rump, more prominent wing bars, white band at base of tail, and do not pump tail. Smaller Prairie wags its tail but has a yellow face pattern (except immature female).
VOICE CALL: A low, forceful *chip,* similar to Prairie Warbler. **FLIGHT CALL:** A thin, high *zeet.* **SONG:** A loud, deliberate, rich, throaty *chew chew chee chee wee wee,* suggesting Northern Waterthrush.
STATUS & DISTRIBUTION Endangered, but locally fairly common within its limited range and breeding habitat. Rarely observed in migration or on wintering grounds. **BREEDING:** Extensive, dense stands of young jack pine (7–15 yrs. old) on sandy soil in central MI with far fewer on Upper Peninsula; a small population of about 50 now breeds in WI, and a few singing males have been noted in southern ON. **MIGRATION:** Migrants rarely noted, but a few reported from Lake Erie region in early to mid-May; also recorded FL and historically coastal GA. Arrives on breeding grounds around May 12 (earliest May 3). Some leave as early as late Aug., some linger as late as early Oct. Fewer reports of fall migrants along route than in spring; some recent records for central Appalachians; casually to northwestern OH, northern IN, and northeastern IL. **WINTER:** Dense scrubby undergrowth mainly in central Bahamas; rarely observed. Recently found in moderate numbers on Eleuthera. **VAGRANT:** Exceptional records for MO and ME.
POPULATION Endangered. Extensive management of breeding habitat (through burning or clear-cuts and replanting of jack pines, as well as removal of parasitic Brown-headed Cowbirds) has allowed the population

to increase from a low of 167 singing males in 1974 to 2,090 singing males in 2012. In addition to continually removing cowbirds, managing larger tracts of breeding habitat has proven to be more effective in increasing population than managing smaller tracts.

female paler yellow below, spotted across breast

grayish lores

adult ♀

immature ♀

black lores with white eye ring broken on sides

adult ♂

bobs tail like Prairie and Palm Warblers

CAPE MAY WARBLER *Setophaga tigrina* CMWA ■ 1

The Cape May is one of the most aggressive wood-warblers. Tiger-striped and chestnut-cheeked spring males are distinctive, but drab immature females can be challenging. Their nests can be found near the top of a spruce or fir near the trunk, with four to nine eggs (June). Monotypic. L 5" (13 cm)

IDENTIFICATION Slightly decurved bill is most finely pointed in genus. Short tail, pale patch on side of neck behind darker ear coverts; indistinct in immatures. **SPRING MALE:** Sides of neck yellow with chestnut cheek patch, olive above with black streaks and a yellow rump. Yellow underparts with extensive black streaking; whitish undertail coverts, white wing patch. **SPRING FEMALE:** Similar to male, but duller with paler yellow on face and underparts; indistinct wing bars. **FALL ADULT MALE:** Less chestnut in cheek; duller crown and back. **IMMATURE:** Male duller than adult male, lacking chestnut in cheeks, slightly reduced white in wing. Female streaked below; very dull grayish sometimes with some yellow on face; underparts (sometimes tinged yellow) with faint streaks. Note greenish edges on flight feathers.

SIMILAR SPECIES Immature female is similar to "Myrtle" Yellow-rumped and Palm Warblers, which are larger and browner above, with heavier bills and longer tails; they lack yellowish patch on sides of neck. Dull "Myrtle" has a contrasting bright yellow rump. Tail-pumping Palm has less streaking below and yellow undertail coverts.

VOICE CALL: A high, sharp *tsip*. **FLIGHT CALL:** A soft buzzy *zeet*. **SONG:** Primary song a high-pitched penetrating *seet-seet-seet-seet-seet*. Alternate song lower pitched with several two-syllable notes: *seetee seetee seetee seetee seetee;* similar to Bay-breasted Warbler's.

STATUS & DISTRIBUTION BREEDING: Coniferous forest and bogs with spruce. Favors areas with spruce budworm infestations. **MIGRATION:** In spring, more westerly route. Rare on western Gulf Coast. Arrives on FL coast in late Mar., peaks in late Apr. In Midwest, arrives early May, peaks in mid-May. In fall, peaks in Great Lakes and Northeast in mid-Sept., mid-Atlantic in late Sept., FL in late Sept.–mid-Oct. **WINTER:** In a variety of habitats, mainly in Caribbean, where it feeds substantially on nectar using its specialized semitubular tongue. **VAGRANT:** To West, mostly late Sept.–early Oct. In CA, 200+ recs. Also recorded in OR, WA, and AK; two records from Scotland.

POPULATION Variable depending on spruce budworm outbreaks.

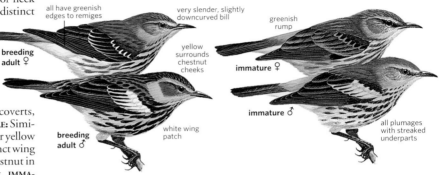

all have greenish edges to remiges

very slender, slightly downcurved bill

greenish rump

breeding adult ♀

yellow surrounds chestnut cheeks

immature ♀

breeding adult ♂

white wing patch

immature ♂

all plumages with streaked underparts

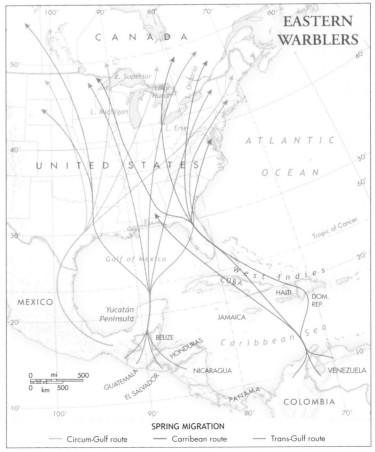

EASTERN WARBLERS

SPRING MIGRATION
— Circum-Gulf route   — Carribean route   — Trans-Gulf route

YELLOW WARBLER   *Setophaga petechia*   YWAR ■ 1

This very familiar and widespread warbler is often associated with willows. It builds its nest in an upright fork of a bush, sapling, or tree and lays four to five eggs (May–June). When the nest is parasitized by Brown-headed Cowbird, Yellow Warbler is known to build a new nest on top of the old one. Polytypic (43 ssp.; ±10 in N.A.). L 5" (13 cm)
IDENTIFICATION Only wood-warbler with yellow tail spots. **SPRING MALE:** Entirely yellow; crown sometimes tinged chestnut. Conspicuous black eye on yellow face. Wing feathers broadly edged with yellow. Chestnut streaks on breast and sides. **SPRING FEMALE:** Similar to male, but duller yellow on head and underparts. Breast streaking indistinct or absent. **IMMATURE:** Duller and generally lacking chestnut breast streaks; crown and face more olive-yellow, as on back. Sometimes very olive or grayish overall.
GEOGRAPHIC VARIATION The 43 subspecies are often arranged into three main groups based on plumage characteristics and geographic distribution: "Northern" *(aestiva)*, "Golden" *(petechia)*, and "Mangrove" *(erithachorides)* groups. "Northern" group consists of nine subspecies, eight breeding across N.A., they tend to be darker in the north and paler in the south; southwestern subspecies *(sonorana)* is most distinctive, showing minimal chestnut breast streaking. "Golden" group consists of 18 subspecies, mostly in Caribbean. They are represented by a single subspecies *(gundlachi)* in N.A., which breeds in mangroves in south FL and Cuba; it shows an extensively olive crown and shorter wings. "Mangrove" group (16 ssp. found coastally from Mexico to Galápagos) is represented by two to three subspecies in the U.S., of which *oraria* has recently been found breeding in coastal southern TX; all adult males in this group except *aureola* (from Galápagos and Isla de Coco) have chestnut heads.
SIMILAR SPECIES Duller immatures of northwestern

(rubiginosa), Alaskan *(banksi),* and south Florida *(gundlachi)* subspecies can be very grayish yellow and can be confused with Orange-crowned or Wilson's. Yellow Warbler can be recognized by its plain, unmarked face with a narrow yellow eye ring and by its yellow tail spots.
VOICE CALL: A husky, downslurred *tchip* or a thinner *tsip*. **FLIGHT CALL:** A buzzy *zeet*. **SONG:** Primary song somewhat variable between individuals, usually most similar to *sweet sweet sweet sweeter than sweet*. Alternate songs are longer and more complex, lacking accented upslurred or downslurred ending; sometimes similar to songs of Chestnut-sided Warbler or American Redstart.
STATUS & DISTRIBUTION Common in second-growth and shrubby areas; a long-distance migrant to C.A., S.A., and southern Caribbean. **BREEDING:** Wet deciduous thickets and early successional habitats, usually those dominated by willows. **MIGRATION:** In East, mostly circum-Gulf or along western Gulf, arriving in southern U.S. by mid-Apr. and in Great Lakes in late Apr.–early May. In West, variable depending on subspecies, but generally by late Mar. in southern AZ, late Apr. in OR and WA, and late May in AK. In East, one of the earliest departing warblers in fall, with birds leaving Great Lakes beginning in mid-July and rarely later than mid-Sept.

(the more northerly eastern *amnicola* departs later) Small numbers along the Gulf Coast into mid-Oct. Later in West, beginning in late July, peaking in late Aug. and early Sept., and extending into mid-Oct. **WINTER:** A variety of wooded and scrubby habitats from northern Mexico to northern S.A. to central Peru and northern Brazil. Rare in southern CA. **VAGRANT:** "Mangrove" group is accidental to southern CA, southern AZ, and Rockport, TX. "Northern" group birds are casual in fall in Europe: recorded from Greenland (two recs.), Iceland, and Britain (about five recs.).
POPULATION Widespread and common, population appears stable.

dark eye stands out
in blank face

immature ♀
aestiva

pale
tertial edges

adult ♂
aestiva

immature ♀
gundlachi

whitish
underparts

red streaks
on under-
parts

immature ♀
amnicola

overall duller and
darker than *aestiva*

faint
streaks

short tail with
yellow tail spots

♀ *aestiva*

immature
rubiginosa

"Golden" adult ♂
gundlachi

gundlachi found in
Cuba and south Florida

adult ♂
sonorana

sonorana found in
Southwest

"Mangrove" adult ♂
oraria

oraria found in eastern Mexico
and coastal south Texas

immatures of northern
subspecies are duller and
darker and are washed
with olive or brownish

CERULEAN WARBLER *Setophaga cerulea*   CERW ■ 2

This beautiful sky-blue warbler sings, forages, and nests higher in the tree-tops than most other species. It lays two to five eggs in a nest built on the limb of a tall deciduous tree (mid-story or canopy) in June. Monotypic. L 4.8" (12 cm)

IDENTIFICATION Long wings, short tail. **ADULT MALE:** Cerulean upper-parts, brightest on crown; black streaks on back; white underparts, narrow dark blue band across lower throat; bold dark blue streaks on sides. Whitish supercilium and reduced breast band in first-spring male. Two white wing bars. **ADULT FEMALE:** Blu-ish green crown, back, and rump; unstreaked back; whitish underparts variably washed with yellow; gray eye line, broad and long yellowish supercilium, white arc under eye; diffuse greenish streaks on sides, no breast band. Two white wing bars. **IMMATURE:** Similar to adult female. Male bluish gray above, streaked back,

more prominent side streaks. Female dullest; greenish upperparts with little or no bluish tint; more yellow on underparts (esp. throat).

SIMILAR SPECIES Immature female Blackburnians are most similar to immature Ceruleans, but show dark auricular patch with broad pale super-cilium connected to yellow patch on sides of neck and more olive-gray upperparts with pale streaking on sides of back. Immature Black-throated Grays are similar, but never bluish or greenish.

VOICE CALL: A slurred *chip*. **FLIGHT CALL:** A buzzy *zzee*. **SONG:** A variable, buzzy, accelerating, rising series of notes, the last note highest: *zhee zhee zhee zizizizizi zzzziiiii*. Similar to some songs of Northern Parula, but the rising buzzy note at end is distinctive.

STATUS & DISTRIBUTION Uncommon and declining; trans-Gulf migrant. **BREEDING:** Mature deciduous woods with open understory. **MIGRATION:** In spring, almost never in large numbers. Arrives on Gulf Coast early to mid-Apr.; arrives in Great Lakes early May,

with latest migrants through end of May. In fall, departs breeding grounds very early and sometimes arrives on wintering grounds as early as Aug. Some linger on breeding grounds into Sept., rarely to Oct. in South. **WINTER:** Canopy and forest borders at middle and lower elevations on eastern Andean slopes from Colombia and Venezuela south to Peru, rarely to Bolivia. **VAGRANT:** Casual in West: recorded from CO, NM, AZ, NV, and CA (13+ recs.), as well as Baja California. Also casual in New England and Atlantic Canada, Bermuda, Bahamas, and Greater Antilles. Accidental in Iceland.

POPULATION Formerly common and widespread, but with steep declines recently. Some expansion of range into northeastern U.S. Identified causes of declines include deforestation, fragmentation, management of forests for even-aged stands, and loss of some key tree species including oaks, elms, chestnuts, and sycamores.

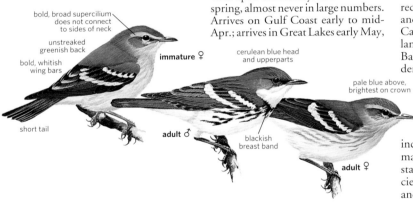

bold, broad supercilium does not connect to sides of neck

unstreaked greenish back

bold, whitish wing bars

immature ♀

short tail

cerulean blue head and upperparts

adult ♂

blackish breast band

pale blue above, brightest on crown

adult ♀

NORTHERN PARULA *Setophaga americana*   NOPA ■ 1

A small, short-tailed warbler, the Northern Parula inhabits mossy environments and is often found singing high in the treetops. It usually builds its nest in bunches of epiphytes (*Usnea* lichen in north, *Tillandsia* in south) on the end of a branch, ranging from low to high in a tree; it lays between three and five eggs from April to July. Monotypic. L 4.5" (11 cm)

IDENTIFICATION Bicolored bill: black upper mandible and yellow lower mandible. **SPRING MALE:** Blue-gray upperparts, including head and sides of throat, with greenish upper back;

two broad, white wing bars; white eye ring broken at front and rear; black lores. Yellow throat and breast with reddish and black bands across breast; white belly and undertail coverts. Large white spots on outer tail feath-ers. **SPRING FEMALE:** Similar to male, but duller blue-gray above; breast bands are much reduced or absent. **IMMATURE:** Similar to spring female, but duller with more greenish upper-parts and smaller white tail spots. **HYBRID:** "Sutton's" Warbler is a very rare hybrid of Northern Parula with Yellow-throated Warbler. Its appear-ance is similar to Yellow-throated

Warbler, but it has Northern Parula's green back patch, no white ear patch, and fewer side streaks.

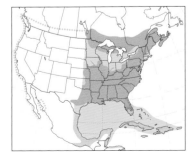

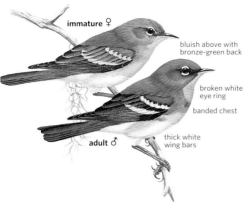

immature ♀

bluish above with
bronze-green back

broken white
eye ring

banded chest

adult ♂

thick white
wing bars

SIMILAR SPECIES Tropical Parula has a black face mask, but it lacks black or chestnut bands on breast; it also lacks white eye ring. The dullest immature female Magnolia Warblers show a complete eye ring, more extensive yellow coloring on underparts, and

white tail spots that form a band across center of tail. **VOICE CALL:** A sharp *tsip*. **FLIGHT CALL:** A high, weak, descending *tsif*. **SONG:** Primary song a rising trill with a terminal note that drops sharply. The song of western populations has a less emphatic, upslurred ending. Alternate song is more complex and wheezy: *b-zee-b-zee-b-zeee-zee-zee-zee-up*. **STATUS & DISTRIBUTION** Common breeding bird, typically departing the region entirely in winter. Eastern birds migrate through the Caribbean; western birds migrate through TX and Mexico. **BREEDING:** Moist deciduous, coniferous, or mixed woodlands in north; hardwood bottomlands along rivers and swamps. Almost always associated with

epiphytic growth. **MIGRATION:** Arrives widely in southern U.S. late Mar.–early Apr., in northeastern U.S. in mid-May. Rare over much of the West. In fall, it departs breeding grounds in Aug., with migration peaking Sept.–mid-Oct. Rare in West (scarcer than in spring). **WINTER:** In a wide variety of habitats, preferring undisturbed areas to disturbed areas. In Caribbean, most of Mexico south to El Salvador. Rare from Nicaragua to Panama. Casual in LA, NM, and SC, and a few in south TX and CA. **VAGRANT:** Casual or accidental in AK (sight rec. on Middleton I.), Greenland, Iceland, U.K. (more than 15 recs.), France, and the Azores. **POPULATION** Due to habitat requirements, populations are locally variable. Long-term trends are stable, but recent shorter-term trends show declines.

TROPICAL PARULA *Setophaga pitiayumi*   TRPA   3

This close relative of the Northern Parula—with which it is sometimes considered conspecific—mainly reaches North America in south Texas. From April to May, the Tropical Parula lays two to five eggs in a mass of *Tillandsia* on the outer branches of trees at low to moderate heights. Polytypic (14 ssp.; 1 probably 2 in N.A.). L 4.5" (11 cm) **IDENTIFICATION** Black upper mandible, yellow lower mandible. **SPRING MALE:** Bright blue-gray upperparts, including head, with greenish upper back; two broad white wing bars, small black face mask. Yellow throat and belly, orange-yellow breast, white undertail coverts. Large square white spots on outer tail feathers. **SPRING FEMALE:** Similar to male, but duller above. Black face mask is lacking or much duller; breast shows little or no orange. **IMMATURE:** Similar to spring female, but more greenish above and duller yellow below. **GEOGRAPHIC VARIATION** The east Mexican subspecies *(nigrilora)* breeds north to south Texas. The west Mexican subspecies *(pulchra)* is the probable vagrant to Arizona. It differs from the eastern subspecies by its larger size, longer tail, more white on greater wing covert tips, and more cinnamon on flanks. South American subspecies differ more subtly from each other. **SIMILAR SPECIES** Northern Parula shows a broken white eye ring, and

blue-gray coloring extends to sides of throat. It has a white belly and less black on face (lores only). Males show a breast band of black and chestnut. Tropical Parula may hybridize with Northern Parula. **VOICE CALL:** A thin slurred *chip,* similar to Northern Parula's call note. **SONG:** The primary song is a rising trill that ends abruptly with a separate buzzy, downslurred terminal note. This song is unlike that of Northern Parula's eastern populations, but it is similar to the song of Northern Parula's western populations. The alternate song is quite variable and more complex, but it always has a buzzy quality. **STATUS & DISTRIBUTION** Uncommon in south TX, where it is partially migratory. **BREEDING:** In N.A., mainly in live-oak woodlands with abundant epiphytes, especially Spanish moss. **MIGRATION:** In spring, they arrive in south TX in mid-Mar. In fall, they depart south TX by Sept. **WINTER:** A few overwinter in the southern part of the breeding range (south TX). Some postbreeding dispersal to the north and east happens in fall and winter. Winter range of TX breeding birds uncertain. **VAGRANT:** West and central TX, southeastern AZ, MS, CO, and Baja California.

POPULATION The southern Texas population declined significantly after a devastating freeze in the winter of 1951. Habitat destruction may have hindered Tropical Parula's recovery. The species is common in most of its neotropical range.

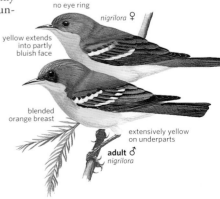

no eye ring

nigrilora ♀

yellow extends
into partly
bluish face

blended
orange breast

extensively yellow
on underparts

adult ♂
nigrilora

MAGNOLIA WARBLER   *Setophaga magnolia*   MAWA   ▪ 1

Easily identified in all plumages by its unique black-and-white tail pattern, this boldly patterned warbler is very common. It nests low in dense conifer in June and lays three to five eggs. Monotypic. L 5" (13 cm)

IDENTIFICATION SPRING MALE: Gray crown with black mask and broad white line behind eye; black back and uppertail coverts; yellow rump. Yellow underparts with broad black streaks on breast and sides. White belly and undertail coverts. Broad white wing bars usually merged into a single patch. Black tail with broad white band near base. SPRING FEMALE: Variable; duller than male, with narrower wing bars and breast streaks, grayish mask, some greenish on back, less purely gray crown. FALL ADULT: Narrow white eye ring on grayish cheek. Male lacks black on head, shows greenish and black on back, has narrower wing bars and breast streaks; black uppertail coverts narrowly edged grayish. Female duller, with more greenish on back, less breast streaking, narrower wing bars. Uppertail coverts broadly edged grayish. IMMATURE: Similar to fall adult female, but duller, sometimes nearly lacking streaks on breast and sides (esp. female). Green back sometimes with narrow black streaks.

SIMILAR SPECIES Immature Prairie and Kirtland's Warblers are superficially similar, but Magnolia shows a unique tail pattern and complete white eye ring on gray face, and it does not wag its tail. In particular, some first-spring female Magnolias are frequently confused with the larger Kirtland's Warbler, but note Kirtland's nearly continuous tail bobbing (Magnolia does not wag its tail) and very different tail pattern.

VOICE CALL: A unique nasal *enk,* most often heard in fall migration and on wintering grounds. FLIGHT CALL: A high, buzzy *zee.* SONG: Primary song is a variable short, musical *weeta weeta wit-chew,* accented on the final notes. Alternate song is an unaccented *sing sweet.*

STATUS & DISTRIBUTION Fairly common to common breeding bird; very common in fall migration. Mainly a trans-Gulf migrant. BREEDING: Dense young coniferous or mixed woodland, usually fairly low to ground. MIGRATION: In spring, arrives on Gulf Coast in mid-Apr., peaks in northern U.S. in late May, arrives on breeding grounds in late May–early June. In fall, departs breeding grounds mid-Aug., peaks through Sept., and lingers as late as late Oct. (sometimes into Dec. and Jan. in LA and FL). WINTER: Shrubby second growth, wooded, and agricultural areas in Caribbean and

from southern Mexico to Costa Rica. VAGRANT: Rare spring and fall migrant to Pacific states. Casual to the Azores. Accidental in Barbados, Trinidad, Tobago, Venezuela, northern Colombia; two records from UK.

POPULATION Stable or slightly increasing. Uses second growth. Parasitism by Brown-headed Cowbird may be an increasing problem.

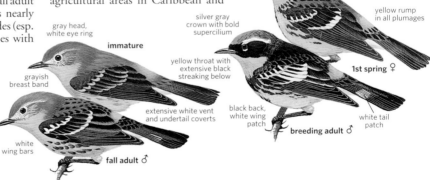

gray head, white eye ring

immature

silver gray crown with bold supercilium

yellow rump in all plumages

yellow throat with extensive black streaking below

grayish breast band

1st spring ♀

extensive white vent and undertail coverts

black back, white wing patch

breeding adult ♂

white tail patch

white wing bars

fall adult ♂

CHESTNUT-SIDED WARBLER   *Setophaga pensylvanica*   CSWA   ▪ 1

Commonly seen foraging low in small trees and shrubs with drooped wings and cocked tail, the Chestnut-sided Warbler lays three to five eggs in a nest among the shrubby understory, near the ground, between May and July. Monotypic. L 5" (13 cm)

IDENTIFICATION Distinct white tail spots in all plumages. SPRING MALE: Bright yellow crown; black lores, eye line, and whisker; white cheek and underparts; extensive chestnut on sides. Wings with two pale yellow wing bars. SPRING FEMALE: Similar to male, with greener crown, duller upperparts, less black on face; the blackish in malar region is often broken and does not meet the chestnut on the sides; and less chestnut on the sides (first-spring males usually with a shorter chestnut strip, not reaching the flanks). FALL ADULT: Distinctive; very different from spring adult. Bright lime green upperparts. Pale grayish cheek, throat, and upper breast. Whitish underparts. Conspicuous white eye ring. Less chestnut on sides than in spring plumage, especially in female. IMMATURE: Similar to fall adult, but all females and some males completely lack chestnut on sides.

SIMILAR SPECIES Fall adult and immature Chestnut-sided Warblers superficially resemble Bay-breasted Warbler and Blackpoll Warbler, but neither of

lime green upperparts

yellowish wing bars

gray face with white eye ring

grayish white underparts

immature

these show bright green upperparts or a distinct white eye ring. **VOICE CALL:** A loud, sweet *chip,* like Yellow Warbler's call note, but not in a series. **FLIGHT CALL:** A very burry, slightly musical *breeet.* **SONG:** Primary song is similar in quality to Yellow Warbler's, but its phrasing is different,

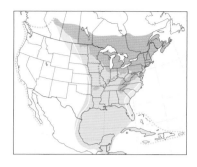

with variations similar to *please, please, pleased to meetcha,* the last note with a distinct drop. Alternate song is rather nondescript and is more similar to Yellow Warbler's song and to some songs of American Redstart, with variations similar to *wee-weewee-wee-chi-tee-wee.* **STATUS & DISTRIBUTION** Fairly common breeder in early successional second growth. Medium-distance migrant to C.A. **BREEDING:** Northern hardwood and mixed woodland. **MIGRATION:** In spring, arrives in southern U.S. in early Apr. and in MN by mid-May. Common migrant along western Gulf Coast, rare and irregular in Caribbean. In fall, departs breeding grounds in Aug.; more easterly migration continues without a clear peak through Sept. and occasionally through mid-Oct. **WINTER:** Southern Mexico through Panama; most

numerous in Costa Rica in a variety of forested and second-growth habitats. **VAGRANT:** Rare in the West; has wintered in AZ and CA. Accidental in AK. Casual in northern S.A. Accidental in Greenland and Scotland.

POPULATION Chestnut-sided Warbler has benefited from deforestation, which has opened up breeding habitat. Population declines since the 1960s are possibly related to urbanization as well as to reforestation and maturation of habitat.

breeding adult ♀

usually cocks tail

greenish crown

yellow crown

chestnut sides

breeding adult ♂

BLACKBURNIAN WARBLER *Setophaga fusca*   BLBW   ■   1

The fiery orange throat and head markings of the adult male Blackburnian Warbler are unique among wood-warblers. In June, Blackburnian constructs a nest high in a conifer, on limbs well out from the trunk, and produces three to five eggs. Monotypic. L 5" (13 cm) **IDENTIFICATION** In all plumages, Blackburnian shows a dark cheek patch; a broad, pale supercilium connected to pale sides of neck; pale stripes on sides of back (braces). **SPRING MALE:** Black triangular cheek patch is surrounded by fiery

orange, most conspicuous on throat as bird is often viewed from below. Black crown with orange patch; black back with conspicuous pale yellowish stripes on each side of back (braces). Underparts pale yellow, shading to white on undertail coverts. Large white wing patch. Extensive white on outer tail feathers. **SPRING FEMALE:** Similar to male, but duller, with grayish cheek patch, crown, and side streaks; orange areas on head less intense and yellower; back striped brownish and olive with less prominent braces; less white in wing, forming two distinct wing bars. **FALL ADULT:** Male with duller orange coloring; its black areas are veiled with olive edges; and it has less white in wing, forming two wing bars. Female has

yellower head markings than in spring; also has more olive upperparts. **IMMATURE:** Male is similar to a fall female, but with a black eye line above grayish cheek patch; yellower throat and head markings; stronger back streaking; and blacker side streaks. Female immature has dullest plumage, with throat and supercilium colored pale yellow to buffy (sometimes almost whitish) and indistinct streaks on sides.

SIMILAR SPECIES An immature female Cerulean Warbler shows plainer greenish upperparts, a less conspicuous ear patch, and a shorter tail; its supercilium is not joined to pale areas on sides of neck. Both Bay-breasted Warbler and

breeding ♀

bold white wing patch

fiery orange throat

adult males have buffy belly

breeding adult ♂

all Blackburnians have pale mantle lines

fall adult ♂

dark triangular auricular patch

bold, broad supercilium connects to pale sides of neck

immature ♀

Blackpoll Warbler lack pale stripes on back and show an inconspicuous supercilium.

VOICE CALL: A rich *tsip*. **FLIGHT CALL:** A buzzy *zzee*. **SONG:** Primary song is a very high-pitched, ascending series of notes, ending with an almost inaudible trill: *see-see-see-see-ti-ti-ti-siiii*. Alternate song begins at a high pitch and concludes with a lower-pitched ending: *tsee-tsee-tsiii-chi-chi*.

STATUS & DISTRIBUTION A common boreal forest breeder. A long-distance, mainly trans-Gulf migrant to S.A.

BREEDING: Mature coniferous or mixed forests; also deciduous forests in southern U.S. **MIGRATION:** In spring, it takes a more westerly route, arriving on Gulf Coast in early Apr. and peaking in ON in mid-May. In fall, it takes a more easterly route, departing breeding grounds early and arriving in the Great Lakes by early Aug., where it peaks in mid-Aug. It peaks in the south during Sept. and on through early Oct. Stragglers can be found into Nov. and Dec. **WINTER:** Most numerous in the montane forests of northern Andes,

but also in Amazonia and northern S.A. A few winter in Costa Rica to Panama. **VAGRANT:** Recorded rarely or casually in all western states; more than 500 records from CA. Accidental in Greenland, Iceland, and U.K.

POPULATION Breeding populations are apparently stable, though the Appalachian breeding population is threatened by the loss of hemlock trees killed by introduced insect pests. The loss of preferred mature evergreen forests on wintering grounds is also a potential cause of declines.

BAY-BREASTED WARBLER *Setophaga castanea*   BBWA   ▪ 1

Bay-breasted's fall plumage presents identification challenges. It makes a nest for four to seven eggs on a horizontal branch of dense conifer (June–July). Monotypic. L 5.5" (14 cm)

IDENTIFICATION SPRING MALE: Black face; chestnut crown, throat, sides; cream neck patch; two broad white wing bars. **SPRING FEMALE:** Duller than male. Cheek mottled blackish; paler, less extensive chestnut on crown, throat, sides; buffy split eye ring. **FALL ADULT:** Sexes similar, female slightly duller. Crown, nape, and back yellowish olive green, with indistinct black streaks. Pale olive face, indistinct pale supercilium; buff throat, breast, belly, undertail coverts; some or little chestnut on flanks; dark legs, feet. **IMMATURE:** Similar to dullest fall adult female, but

less conspicuous streaking on back; clear whitish buff underparts; much reduced (male) or absent (female) chestnut on flanks.

SIMILAR SPECIES In fall, Pine lacks streaks on back; yellow of throat extends behind auriculars; wing bars contrast less; longer tail extends well beyond undertail coverts; primary projection beyond tertials shorter. Blackpoll more similar (see sidebar opposite).

VOICE CALL: Loud, slurred *chip*. **FLIGHT CALL:** Buzzy *zeet*. **SONG:** Variable series of very high-pitched lisping notes: *see-see-swee-see-see-swee-swee-see*.

STATUS & DISTRIBUTION Fairly common, variable depending on spruce budworm outbreaks; mainly a trans-Gulf migrant, rarely through Caribbean. **BREEDING:** Boreal forest, mainly in mature dense,

spruce-fir forests. **MIGRATION:** Arrives Gulf Coast mid- to late Apr., peaks Great Lakes last half of May. In fall, more easterly than in spring, departs late July; peaks eastern N.A. late Aug.–mid-Sept.; stragglers to late Oct., later on Gulf Coast. **WINTER:** Forest edges, second growth from Costa Rica through Panama to northwestern Colombia, northern Venezuela. Feeds mainly on fruit. **VAGRANT:** Casual to accidental to nearly all western states (rare in CA); sight rec. for central AK. Casual in Labrador, Greenland; accidental in U.K.

POPULATION Spraying for spruce budworm has reduced the breeding populations.

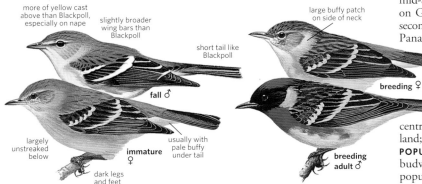

more of yellow cast above than Blackpoll, especially on nape

slightly broader wing bars than Blackpoll

short tail like Blackpoll

fall ♂

largely unstreaked below

dark legs and feet

immature ♀

usually with pale buffy under tail

large buffy patch on side of neck

breeding ♀

breeding adult ♂

BLACKPOLL WARBLER *Setophaga striata*   BLPW   ▪ 1

The black-and-white males are easily identified in spring, but in fall much duller plumages make identification challenging. In June, Blackpolls place three to five eggs in their nest, which is built against the trunk of a tree, low to

the ground. Monotypic. L 5.5" (14 cm)

IDENTIFICATION SPRING MALE: White cheek; black cap and malar stripe; black-streaked olive-gray back; bold black streaks on sides of white underparts. Two bold white wing bars,

orange-yellow legs and feet, yellowish lower mandible. **SPRING FEMALE:** Much individual variation. Olive-gray upperparts, including cheeks, with dark streaks on back; crown sometimes heavily streaked. Face with dark eye

line, dark streaking on malar. Bold or indistinct side streaks. Some more yellowish olive above and below, resembling fall birds. **FALL ADULT:** Upperparts dull olive with black streaks on back; underparts whitish or yellowish with narrow streaks on sides; white undertail coverts. Face pattern similar to spring female's. Pale or dark legs, always with yellow on soles of feet. **IMMATURE:** Similar to fall adult, but usually greener above and yellower below with less streaking on sides. Sexes usually indistinguishable.

SIMILAR SPECIES In fall, Pine Warbler lacks streaks on underparts and usually also on upperparts; it always has a much longer tail and shorter primary projection. In fall, Bay-breasted is very similar (see sidebar below). Spring male is superficially similar to Black-and-white Warbler, which has a dark cheek, a white median crown streak, and a different foraging style.

VOICE CALL: A loud, sharp *chip*. **FLIGHT CALL:** A loud, sharp, buzzy *zeet*. **SONG:** A series of very high-pitched staccato notes, inaudible to some, usually louder in the middle: *tsit tsit tsit tsit tsit tsit tsit.*

STATUS & DISTRIBUTION Common; undertakes extremely long nonstop overwater migrations. **BREEDING:** Boreal spruce forest and spruce-alder-willow thickets. **MIGRATION:** In spring, takes a more westerly route, through the Caribbean as far west as coastal TX; later than most other wood-warblers. Departs wintering areas in Apr. and arrives on breeding grounds mid-May–early June. Peaks in Midwest and mid-Atlantic states mid- to late May, with a few into early June. In fall, a more easterly route, including an overwater flight from northeastern coastal U.S. to northern S.A.; a few south through eastern Caribbean. Peak numbers through northeastern U.S. from mid-Sept. to early Oct., with some as late as Oct. and Nov. **WINTER:** A variety of wooded habitats in northern S.A. east of Andes south to northern Bolivia. **VAGRANT:** Rare migrant in West, primarily in spring away from Pacific coast. Rare but regular fall migrant on Pacific coast. Very rare in Costa Rica and Panama. Casual in fall to Greenland, Iceland, U.K., islands off France, and the Azores. **POPULATION** Stable.

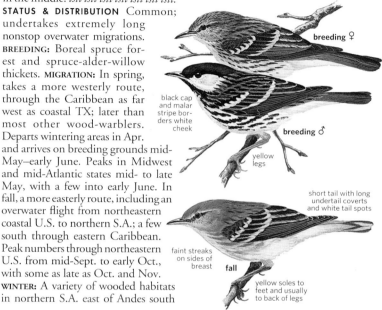

breeding ♀

black cap and malar stripe borders white cheek

breeding ♂

yellow legs

short tail with long undertail coverts and white tail spots

faint streaks on sides of breast

fall

yellow soles to feet and usually to back of legs

Separation of Bay-breasted and Blackpoll Warblers in Fall

These two species are perhaps the most difficult of the "confusing fall warblers." The Pine Warbler is often very similar too, but it can usually be distinguished by its different face pattern, shorter wings, and much longer tail projection. Adult Bay-breasteds and Blackpolls appear considerably different in fall than in spring but will often show enough similar plumage to be identifiable. Dull, fall immatures create the greatest confusion.

Both species show a similar face pattern, with a dark eye line and a pale broken eye ring. This tends to be better defined in the Blackpoll.

Both species have greenish olive backs with blackish streaks. Streaks tend to be more conspicuous on the Blackpoll and less so on the Bay-breasted, but this is somewhat variable. The Bay-breasted has a brighter yellowish olive on the crown, nape, and back. When noting wing bars, look for the Bay-breasted's often broader white wing bars.

Bay-breasted Warbler, immature (GA)

Blackpoll Warbler, immature (NY)

The Bay-breasted tends to be buffier from throat to undertail coverts, often buffiest on vent; it also has whitish undertail coverts and, sometimes, a little chestnut (or richer buff) on the flanks. The Blackpoll tends to be yellowish with contrastingly white (or occasionally yellow) undertail coverts. Unlike the Bay-breasted, it also usually shows narrow but conspicuous dark streaks on the sides of the breast and flanks.

Both species have white tail spots and long undertail coverts; however, the Blackpoll's undertail coverts tend to be longer and whiter, whereas the Bay-breasted's are slightly shorter and pale buff, but can be nearly whitish.

Foot color is also important. Although immatures of both species can show dark legs, the soles of the Bay-breasted's feet are bluish gray whereas the Blackpoll's are yellowish. Sometimes this yellowish color occurs narrowly up the rear of the tarsus. ∎

PINE WARBLER   *Setophaga pinus*   PIWA   ◼ 1

One of the most appropriately named warblers (prefers pines), the Pine Warbler has a drab plumage that makes identification challenging away from breeding areas. It often forages on the ground in fall and winter with Yellow-rumped Warblers and Chipping Sparrows. It builds a nest high in a pine on a horizontal limb, often at the end of a branch, and lays three to five eggs. Polytypic (4 ssp.; 2 in N.A.). L 5.5" (14 cm)

IDENTIFICATION ADULT MALE: Olive green upperparts. Yellow throat, breast, and belly, extending to rear of olive cheeks; white lower belly and undertail coverts. Dull olive to blackish indistinct streaks on sides of breast. Indistinct broken yellow eye ring and supercilium. Two white wing bars; large white tail spots on outer tail feathers; tail extends well past the undertail coverts. **ADULT FEMALE:** Similar to male, but paler yellow. **IMMATURE:** Duller than adult; female quite brownish above and whitish below.

GEOGRAPHIC VARIATION Widespread N.A. *pinus* and *florida* from south FL not field identifiable. Two others are resident in West Indies (*achrustera* from the Bahamas and *chrysoleuca* from Hispaniola).

SIMILAR SPECIES Yellow-throated Vireo is larger and more sluggish and has a thicker hooked bill, conspicuous yellow spectacles. In fall, Bay-breasteds and Blackpolls are similar. Pine lacks streaking on upperparts; has yellow or pale extending up behind contrasting auriculars, longer tail, and shorter primary projection. Wing bars contrast less.

VOICE CALL: A slurred *tsup,* similar to call of Yellow-throated and Grace's Warblers. **SONG:** A musical trill, most similar to Chipping Sparrow's or Dark-eyed Junco's song, but usually softer, more musical, and shorter, varying in speed. Occasionally two-parted songs, with the second part being faster and higher pitched.

Also similar to songs of Worm-eating and Orange-crowned.

STATUS & DISTRIBUTION Common, occurring in N.A. year-round, with northern populations migratory. **BREEDING:** A broad range of pine habitats. **MIGRATION:** One of the earliest spring migrant warblers in many areas: begins northward movement in late Feb., arriving by mid-Apr. in southern Great Lakes, mid-Apr. in New England, and late Apr.–early May in northernmost breeding areas. One of the latest fall migrant warblers: departs northernmost breeding areas as early as late Aug. but peaks late Sept.–mid-Oct., rarely into Nov. and with stragglers into Dec. and Jan. **WINTER:** Pine forests in southeastern U.S., rarely to northern Caribbean. **VAGRANT:** Rare, mainly in fall, in Atlantic Canada. Rare or casual in northern Great Plains and Prairie Provinces. Very rare in West, with most records (60+) from coastal southern CA in fall and winter. Casual in southern Caribbean in winter, accidental in fall in Greenland. **POPULATION** Stable or increasing.

immature ♀
pinus

brownish above

pale wraps around auriculars

whitish below

long tail projection past undertail coverts

immature ♂
pinus

unstreaked back in all plumages, unlike Blackpoll and Bay-breasted

adult ♀
pinus

dark auriculars contrast sharply with throat

adult ♂
pinus

white lower belly and undertail coverts

BLACK-THROATED BLUE WARBLER   *Setophaga caerulescens*   BTBW   ◼ 1

The sexes of this species are strikingly different in plumage and easy to identify. Females lay two to five eggs in a June-built nest found in a fork of low shrub or sapling. Polytypic (2 ssp.; both in N.A.). L 5.3" (13 cm)

IDENTIFICATION ADULT MALE: Dark blue crown, back, shoulders. Black face, throat, flanks. White underparts. Large white wing patch at base of primaries. **ADULT FEMALE:** Greenish gray upperparts, buffy underparts. Dusky ear coverts, whitish stripe above eye, narrow white crescent below eye. Small white wing spot at base of primaries. **IMMATURE:** Similar to adult; pale spot at base of primaries smaller or absent. Male green-tinged above; black throat and breast feathers tipped with grayish white. Female duller above, buffy yellow

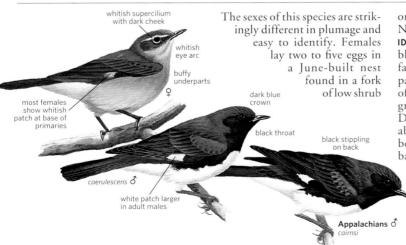

whitish supercilium with dark cheek

whitish eye arc

buffy underparts

♀

most females show whitish patch at base of primaries

dark blue crown

black throat

black stippling on back

caerulescens ♂

white patch larger in adult males

Appalachians ♂
cairnsi

below; eye stripe indistinct, yellowish. **GEOGRAPHIC VARIATION** Males, particularly adults, of the southern Appalachians *cairnsi* tend to have more black streaking on upperparts. **SIMILAR SPECIES** Female's head pattern resembles that of female and immature Yellow-rumped ("Myrtle") Warblers, but it lacks wing bars or yellow rump. **VOICE CALL:** Popping *tuk*. **FLIGHT CALL:** Distinctive, prolonged *tseet*. **SONG:** Primary song a slow series of buzzy notes, rising at the end: *zhee zhee zhee zeeee;* sometimes faster (may resemble the Cerulean Warbler's). Alternate song shorter: *zree zree zhrurrr*.

STATUS & DISTRIBUTION Fairly common. Migrates mainly along the eastern seaboard. **BREEDING:** Large tracts of deciduous and mixed woodland. **MIGRATION:** Arrives in Southeast mid-Apr., peaks in Great Lakes mid-May. Departs breeding grounds mid-Aug; peaks somewhat later than other warblers (late Sept.–early Oct.) in Great Lakes; occasionally lingers to Nov. **WINTER:** Forested areas of Greater Antilles; some Bahamas, a few in Yucatan Peninsula, Belize. **VAGRANT:** Very rare to the West in fall; casual to the Azores; accidental to Iceland.

POPULATION Stable or increasing over most of its range; declining at southern edge.

PALM WARBLER *Setophaga palmarum*   PAWA   ▪ 1

The most terrestrial *Setophaga,* the Palm wags its tail vigorously. Found in palms only in some areas of its winter range, it nests in sphagnum on the ground under a short conifer (4–5 eggs, May–June). Polytypic (2 ssp.; both in N.A.). (L 5.5" (14 cm)

IDENTIFICATION SPRING MALE: Chestnut crown, olive-tinged and lightly streaked grayish brown back, yellow-olive rump. Yellow supercilium and submoustachial stripe, distinct dark eye line, narrow crescent below eye whitish. Bright yellow throat and undertail coverts contrast with whitish belly; thin dark chestnut malar streak and narrow streaks on breast and sides. **SPRING FEMALE:** Very similar to male; often not distinguishable. Tends to have less

chestnut in crown, duller and paler yellow areas. **FALL ADULT:** Chestnut in crown reduced or lacking; dull whitish supercilium, submoustachial, and throat; less distinct breast and side streaks. Bright yellow undertail coverts (typically the only yellow at this season). **IMMATURE:** Very similar, often indistinguishable from adult. More pointed tail feathers.

GEOGRAPHIC VARIATION The more widespread "Western" Palm (nominate, described above) breeds from the Hudson Bay region west. The slightly larger "Yellow" or "Eastern" Palm *(hypochrysea)* breeds from eastern QC to Atlantic Canada and northern New England, and shows entirely yellow underparts (chin to undertail coverts) with little or no contrast; yellowish narrow crescent

below eye; yellow-green tinged upperparts; and broader, brighter chestnut breast and side streaks. "Yellow" is duller in fall but still distinguishable; a "Western" can show yellow on belly, but always with contrastingly brighter undertail coverts and whitish

supercilium. Intermediates can occur. **SIMILAR SPECIES** Prairie and Kirtland's share tail-wagging habit but lack contrasting bright yellow undertail coverts. Similarly, female Cape May and fall Yellow-rumped lack bright yellow undertail coverts and do not wag their tails. **VOICE CALL:** A distinct, sharp *chick*. **FLIGHT CALL:** A high, light *seet* or *see-seet*. **SONG:** A somewhat buzzy or gravelly series of notes, uttered somewhat unenthusiastically and often more forcefully in the middle: *zwee zwee zwee zwee zwee zwee zwee.* **STATUS & DISTRIBUTION** Common, more easily observed in winter.

BREEDING: Bogs in open coniferous boreal forest. **MIGRATION:** In spring, earlier than most other wood-warblers: "Western" migrates through Mississippi Valley to Canada, typically arrives in Upper Midwest mid- to late Apr., peaking in late Apr.–early May; "Yellow" migrates northeast along the Atlantic coast, arrives by mid-Apr. (NS) to early May (NF). In fall, later than most other wood-warblers: "Western" takes a more easterly route than in spring, mainly arriving mid-Sept., peaking late Sept.–early Oct. through much of the East; "Yellow" is very rare west of the Appalachians, migrating slightly later. **WINTER:** A

variety of woodland, second-growth, thickets, and open areas. "Yellow" primarily along Gulf Coast (LA to FL); nominate throughout southeast (TX to NC, rarely farther north), Caribbean, and eastern coastal Mexico to Honduras. **VAGRANT:** Recorded in almost every western state and province. Rare annually in fall to coastal CA, sometimes in small numbers. Casual in Costa Rica, Panama, northwestern Colombia, western Venezuela. One England rec.: a bird found dead on shoreline (1976). **POPULATION** Generally stable or increasing. Breeding grounds difficult to census, better data from wintering grounds.

YELLOW-RUMPED WARBLER *Setophaga coronata*   YRWA ▪ 1

These are probably the best known and most frequently encountered wood-warblers. Although variable, all Yellow-rumped Warblers possess a bright yellow rump, which is shared with only two other species. Yellow-rumped's unique ability to digest the waxes in bayberries allows it to winter farther north than other warblers. Built in June on the fork of a horizontal conifer branch near the trunk at low to moderate height, the Yellow-rumped Warbler's nest contains three to six eggs. Polytypic (4–6 ssp.; 3–5 in N.A.). L 5.5" (14 cm)

IDENTIFICATION SPRING MALE: Crown and back blue-gray streaked with black. Yellow crown patch, distinct rump patch, and patches at sides of breast. White or yellow throat. Black streaks on upper breast and side. White wing bars. White spots

in outer tail feathers. **SPRING FEMALE:** Similar to male, but brownish above with smaller tail spots. **FALL ADULT:** Similar to spring adult, but generally browner above in both sexes with less black on breast. **IMMATURE:** Similar to spring female; some immature females very dull with indistinct streaking and much reduced yellow on sides of breast.

GEOGRAPHIC VARIATION Four to six subspecies are placed into two groups that were formerly considered full species until 1973: "Myrtle" and "Audubon's." They hybridize extensively in

portions of BC and AB. "Myrtle" (nominate over most of range and *hooveri,* not recognized by many, in

Subspecies *coronata,* "Myrtle Warbler"

coronata & *auduboni* intergrade zone

"Audubon's Warbler"

more extensively black overall

Southwest breeding ♂

most have pale yellow in rounded throat patch

yellow crown patch

fall ♀

yellow throat

breeding ♂
auduboni

"Myrtle Warbler"
coronata

breeding ♀

angled whitish throat

yellow patches on sides of breast

breeding ♂

browner above than fall "Audubon's"

yellow rump

distinct whitish supercilium

whitish patch sharply angled on sides of throat

fall ♀

Northwest) shows a white throat, black lores and ear coverts, a white line above lores and eye, two distinct white wing bars, a broken white eye ring, and large white spots on outer three tail feathers. "Audubon's" Warbler (*auduboni* over most of range, similar *memorabilis*, not recognized by many, of the Rockies and Great Basin, larger and darker *nigrifrons* of northwestern Mexico, and largest and darkest *goldmani* of Chiapas and Guatemala) shows a yellow throat, bluish gray sides of head (including ear coverts), two broad white wing bars often forming a distinct patch, a broken white eye ring, and white spots on outer four or five tail feathers. Some birds breeding in the southwestern mountains are intermediate between *auduboni* and darker-faced *nigrifrons*. Telling "Myrtle" from "Audubon's" is more difficult in winter. See sidebar.
SIMILAR SPECIES Compare to Magnolia and Cape May Warblers, which have yellow rumps but also yellow underparts. Palm Warbler shows yellow undertail coverts.
VOICE CALL: "Myrtle" gives a loud, husky, flat *chek;* "Audubon's" gives a loud and richer *chep*. **FLIGHT CALL:** A high, clear *sip*. **SONG:** A variable,

loosely structured trill, sometimes with two parts—the first higher pitched and the second lower and trailing off at the end: *chee chee chee chee wee wee wee we.* Louder and richer on breeding grounds than in migration. "Audubon's" song is similar, but it is simpler and weaker.
STATUS & DISTRIBUTION Common breeder in coniferous woodlands; very common short- to medium-distance migrant to central U.S. south to Caribbean and central Panama. **BREEDING:** Northern boreal and mixed forest, and montane coniferous woodland. **MIGRATION:** In spring, generally arrives earlier than other warblers, returning to northern breeding areas by late Apr. In fall, generally migrates later than other warblers, peaking in northern portions of nonbreeding range in late Sept.–mid-Oct. **WINTER:** A wide variety of habitats from central U.S. south to Gulf Coast; small numbers

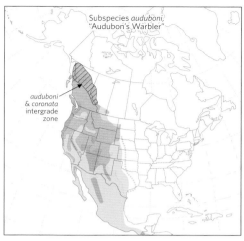

Subspecies *auduboni*, "Audubon's Warbler"

auduboni & *coronata* intergrade zone

irregularly through Carribean and south to C.A. and western Panama. **VAGRANT:** "Audubon's" Warbler is casual in eastern North America; a spring Attu I. record may represent a ship assist. "Myrtle" Warbler is rare on the Azores; casual or accidental to Baffin I., Greenland, Iceland (7 recs.), Ireland, Great Britain (22 recs.), Madeira, the European mainland, and Russian Far East.
POPULATION Breeding and wintering populations appear stable.

Separating "Myrtle" and "Audubon's" in Winter

In winter, adult and immature birds of the two groups that compose the Yellow-rumped Warbler—"Myrtle" and "Audubon's"—become duller and more difficult to distinguish. In the West, "Myrtle" and "Audubon's" winter together in many areas, although "Myrtles" prefer wetter areas (e.g., riparian woodland along streams or ponds). In the East, "Audubon's" is a possibility almost anywhere as a vagrant. The following points will help identify these subspecies groups in winter:

"Audubon's Warbler" (CO, Oct.)

1. Overall color. "Myrtle" Warbler is browner than "Audubon's."
2. Throat color. "Myrtle always shows a whitish throat, whereas "Audubon's" often shows a pale yellow throat. In a few dull immature female "Audubon's," however, the throat is sometimes pale buff or whitish.
3. Throat pattern. In "Myrtle," the whitish throat extends part way around the rear of the ear coverts in a small curved point. In "Audubon's," the yellowish or pale buff throat does not extend behind the ear coverts, and the line between the cheek and the throat is relatively straight.
4. Face pattern. "Myrtle" shows darker ear coverts, a

faint white supercilium, and a small white crescent below the eye. "Audubon's" more uniform brown head shows a broken white eye ring (eye crescents).
5. Tail spots. This characteristic can also be useful in separating these birds. "Myrtle" will usually show white spots on only the outer two or three tail feathers, with the outer being distinctly larger than the inner. "Audubon's," however, will show more uniformly sized white spots on the outer four or five tail feathers.
6. Call notes. "Myrtle" gives a flatter, softer *chek* or *chup,* and "Audubon's" gives a richer *chep* or *chip*.
The photograph above shows a dull brownish "Audubon's" in the fall with essentially no yellow visible; the individual is probably an immature female. Although the throat appears whitish at first glance (perhaps suggesting "Myrtle"), the color is really pale buff and, importantly, it does not curl up behind the ear coverts. The face pattern is actually the best clue to this bird's identity. Note the broken eye ring and the lack of a whitish supercilium, both good field marks for "Audubon's." ∎

YELLOW-THROATED WARBLER *Setophaga dominica*   YTWA ▪ 1

The foraging behavior of this long-billed warbler, creeping along branches, is unusual among species in the genus *Setophaga*. It lays its eggs (3–5) in its nest high in canopy, often in clumps of Spanish moss or pine needles, between April and June. Polytypic (1–3 ssp.; all in N.A.). L 5.5" (14 cm)

IDENTIFICATION ADULT MALE: Blackish forehead; gray crown and upperparts; yellow throat and breast; black triangular ear patch; bold black streaks on sides; white supercilium, crescent below eye, and patch at sides of neck. **ADULT FEMALE:** Similar to male, but with less black on forehead. **IMMATURE:** Similar to adult, but with little black on forehead. Brownish tint above, ear patch duller. Belly and undertail coverts washed with buff.

GEOGRAPHIC VARIATION Western subspecies *(albilora)* shows an entirely white supercilium (rarely yellow), including above lores; more extensive black on forehead; often a small white spot at base of chin. Eastern subspecies

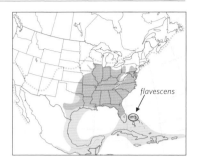

flavescens

(dominica) shows yellow above lores (rarely white) merging to white at front of eye; rarely shows a small white spot at base of chin. Subspecies in FL panhandle and adjacent AL *(stoddardi)* with longer, narrower bill not field distinguishable from *dominica* population on Delmarva Peninsula. Some believe that the species should be treated as monotypic. The now-endemic Bahama Warbler *(S. flavescens),* was regarded until recently as a subspecies of Yellow-throated Warbler. It is resident on Grand Bahama and Abaco in the northern Bahamas *and* not recorded in North America. It is distinctive and shows a noticeably longer bill with decurved culmen and straight lower mandible, little black on forehead, brownish gray upperparts, much narrower supercilium with yellow to rear of eye, much smaller white spot on side of neck, narrow streaks on whitish undertail coverts, and yellow extending to belly.

SIMILAR SPECIES Grace's broad supercilium is more extensively yellow; it has much less black on cheek and face and a different foraging style. Immature male Blackburnians show streaked back and lack white in supercilium and on sides of neck.

VOICE CALL: A high, soft *chip,* similar to Pine's, identical to Grace's. **FLIGHT CALL:** A clear, high *see.* **SONG:** A somewhat variable series of clear, ringing, downslurred notes, rising and weaker

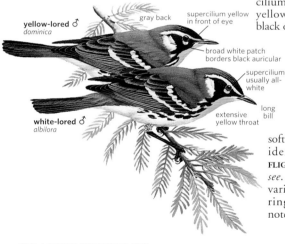

yellow-lored ♂
dominica

gray back

supercilium yellow in front of eye

broad white patch borders black auricular

supercilium usually all-white

extensive yellow throat

long bill

white-lored ♂
albilora

at the end: *tee-ew tee-ew tew tew tew tew wi.* Lacks an alternate song.

STATUS & DISTRIBUTION Common woodland species, partial migrant. **BREEDING:** In South prefers cypress swamps and live-oak stands, especially those with large amounts of Spanish moss. In North, prefers bottomland with large sycamores or dry upland pine-oak forests. **MIGRATION:** Spring migration very early. Migrants reach southern breeding grounds by mid-Mar. and northern breeding grounds mid- to late Apr. Spring overshoots seen Great Lakes *(albilora)* and to NY and CT *(dominica?).* Departs breeding grounds mid-Aug.–late Sept. in northernmost areas. A few remain well north into Dec. and Jan. **WINTER:** Swamps and more open areas in southeastern U.S. and semiopen woodlands, city parks, and gardens in Caribbean and northeastern Mexico to Costa Rica. **VAGRANT:** Rare in spring migration north of breeding range; very rare to casual in spring and fall in western states. Accidental from the Azores. **POPULATION** Stable, with breeding range expanding northward. In northwestern FL, subspecies *stoddardi* is rare and declining.

PRAIRIE WARBLER *Setophaga discolor*   PRAW ▪ 1

This small tail-wagging warbler, often found in low scrub, has a distinctive, buzzy song. It builds its nest low to moderately high in the middle of a small tree clump (over water in coastal FL) and lays three to five eggs (Apr.–June). Polytypic (2 ssp.; both in N.A.). L 4.8" (12 cm)

IDENTIFICATION Distinctive facial pattern: yellow supercilium, dark line through eye, broad pale crescent below eye, dark lower border to cheek. Wags tail. **ADULT MALE:** Black and yellow face pattern; olive upperparts, usually with

reddish spots on back; yellow underparts, paler on undertail coverts; bold black streaks on sides; distinct black spot on side of lower throat; yellowish wing bars. **ADULT FEMALE:** Similar to male, but dark olive and yellow face pattern, sometimes a little black on lower cheek. Less distinct reddish spots on back and black streaks on sides. **IMMATURE:** Duller than adult. Male with very limited or no black on face, narrow side streaks. Female dullest, with indistinct face pattern showing no black, and yellow replaced by pale

whitish, very indistinct side streaks. **GEOGRAPHIC VARIATION** Nominate (most of range) and sedentary *paludicola* (coastal mangroves of FL) are not field-separable. **SIMILAR SPECIES** Pine is larger, shows no black on head or reddish on back, does not wag tail as habitually as

Prairie. Immature Magnolia shows yellow rump, white band across base of tail, complete white eye ring, and does not wag its tail. **VOICE CALL:** A smacking *tsip,* or *tchick,* similar to Palm's and especially Kirtland's. **FLIGHT CALL:** A thin *seep.* **SONG:** Primary song a rapid or slower series of buzzy notes evenly ascending in pitch: *zee zee zee zee zee zee zee zee zee.*

adult ♀
reddish streaks on back
adult ♂
yellow subocular bordered by black
pale chin
whitish subocular bordered by curved slate gray moustachial stripe
faint yellowish wing bars
yellow underparts, including undertail coverts, with streaks on sides and flanks
immature ♀

STATUS & DISTRIBUTION Fairly common, declining in some areas. **BREEDING:** Not on prairie. Nominate in a variety of shrubby old fields, dunes, pine barrens, early successional habitats. **MIGRATION:** Arrives on Gulf Coast of FL mid-Mar.; arrives mid-Apr. to Ohio River Valley, by mid-May in Great Lakes. In fall, reaches FL and Bahamas mid- to late July. Northern breeders depart early Aug.–early Oct., stragglers through Dec. **WINTER:** Second growth and forest edge, mangroves, and gardens mainly in Caribbean and most of FL, rarely to coastal TX and NC. A few from coastal Yucatán to El Salvador. **VAGRANT:** Rare but regular to CA and Atlantic Canada in fall; casual to accidental through much of West in spring and fall (inc. southeastern AK); accidental to the Azores. **POPULATION** Declining in parts of breeding range.

GRACE'S WARBLER *Setophaga graciae*   GRWA ☐ 1

This short-billed montane pine specialist of the Southwest is one of the least-known North American passerines. In the upper half of a conifer, well away from the trunk, it makes a nest for its three to four eggs between May and June. Polytypic (4 ssp.; nominate in N.A.). L 5" (13 cm) **IDENTIFICATION ADULT MALE:** Gray upperparts with black streaks on crown and back. Yellow throat and breast, white belly and undertail coverts. Bold black streaks on sides of breast and belly, yellow supercilium becoming white behind eye, small yellow crescent below eye, black lores and moustachial area. Two white wing bars, large white spots on outer tail feathers. **ADULT FEMALE:** Similar to male, but duller, with brownish gray back; finer streaks on crown, back, and sides; gray lores and moustachial

area. **IMMATURE:** Male similar to adult female, with upperparts unstreaked and more brownish, duller yellow underparts, little black on crown and face, finer streaks on sides, buff wash on flanks. Female duller than immature male, lacking black on crown and face; more buff on flanks and belly. **GEOGRAPHIC VARIATION** Slight and clinal, with back color brownest in northernmost subspecies, *graciae* (breeding in N.A.), to more blue-gray in three Central American subspecies; yellow of throat paler in north and deep orange-yellow in south. **SIMILAR SPECIES** Yellow-throated is larger with a bolder face pattern and different foraging style. "Audubon's" Yellow-rumped shows yellow rump and dark cheek patch. **VOICE CALL:** A soft, slurred *chip,* identical to Yellow-throated's. **FLIGHT CALL:** A very high, thin *sip.* **SONG:** An accelerating series of *chip* notes, most often two-parted: *chew chew chew chew chew chew chee chee chee chee;* similar to Virginia's and "Audubon's" Yellow-rumped's. **STATUS & DISTRIBUTION** Uncommon to fairly common, short-distance migrant. **BREEDING:** Mainly montane open, parklike forests of tall pines. **MIGRATION:** Arrives on breeding

♀
short yellow supercilium turns white behind eye
yellow chin and throat
♂

grounds Apr.–early May. Departs breeding grounds as late as late Sept. **WINTER:** Pine forest to pine-oak woodland, from western Mexico to Nicaragua. **VAGRANT:** Casual on the central and especially southern CA coast (40 recs.). One recent record from IL is the only record from eastern N.A. **POPULATION** Range is increasing northward, but numbers may have declined in some areas.

BLACK-THROATED GRAY WARBLER *Setophaga nigrescens* BTGW ■ 1

This boldly patterned warbler forms a superspecies group with the Golden-cheeked, Townsend's, Black-throated Green, and Hermit Warblers, sharing call notes and long tails with extensive white in the outer tail feathers. It nests on a horizontal branch (low to moderate height) in a deciduous or coniferous tree and produces three to five eggs (May–June). Polytypic (2 ssp.; both in N.A.). L 5" (13 cm)

IDENTIFICATION ADULT MALE: Black-and-white head with broad white supercilium and malar; tiny yellow spot on lores. Gray back with black streaks, two white wing bars. White underparts. **ADULT FEMALE:** Similar to

male, but grayer head, white chin, black throat patch mixed with white. **IMMATURE:** Similar to adult female; male with more restricted black on upper throat and crown; female with brownish gray on upperparts, buffy white underparts (chin to tail), with narrow streaks restricted to sides.

GEOGRAPHIC VARIATION Southern *halseii* slightly larger and paler, with more white in tail and heavier side streaks than northern nominate subspecies.

SIMILAR SPECIES Townsend's has a green back and yellow underparts. Spring male Blackpoll has all-white cheek. Black-and-white has a striped back, different foraging style. Immature male Ceruleans show bluish on upperparts, yellow on breast.

VOICE CALL: A dull *tip*. **FLIGHT CALL:** A high, clear *see*. **SONG:** Primary song is a series of two-syllable buzzy notes, the second syllable louder and higher pitched, the final note falling: *buzz see buzz see buzz see buzz see buzz see wueeo*. Alternate song is longer, more complex, lacking downward-inflected ending.

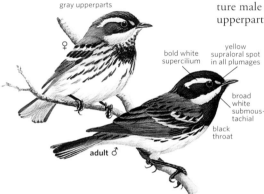

gray upperparts

♀

bold white supercilium

yellow supraloral spot in all plumages

broad white submoustachial

black throat

adult ♂

STATUS & DISTRIBUTION Fairly common. Short- to medium-distance migrant. **BREEDING:** Open mixed or coniferous woodland with brushy undergrowth. **MIGRATION:** CA arrivals peak mid- to late Apr. in south, mid-Apr.–mid-May in north. In fall, migration in OR mid-Aug.–mid-Sept., rarely late Oct. In southern CA, good movement typically occurs Oct. 14–22. **WINTER:** A variety of forest, scrub, and thickets in Mexico from Baja California Sur to central Oaxaca. A few in central CA. Rare in LA, southern FL, southern AZ, southern TX. **VAGRANT:** Casual to accidental in migration east to NS, MA, NJ; north to MT. **POPULATION** Population is stable or slightly increasing.

TOWNSEND'S WARBLER *Setophaga townsendi* TOWA ■ 1

This boldly patterned warbler of the Pacific Northwest gleans insects on its breeding grounds; on its wintering grounds, it will also exploit honeydew excreted by sap-sucking insects. Its well-concealed nest, host to three to five eggs, may be found at various heights in a conifer, from May to June. Monotypic. L 5" (13 cm)

IDENTIFICATION Dark auriculars surrounded by yellow, yellow on breast, extensive white on outer tail feathers, two white wing bars. **ADULT MALE:** Black cheek surrounded by yellow,

with small yellow crescent below eye; black crown. Olive-green back streaked black. Black chin, throat, and broad side streaks. Yellow breast and sides; white belly and undertail coverts. **ADULT FEMALE:** Duller than male, with olive-green crown and auriculars; yellow chin and throat with limited black; narrower side streaks. **IMMATURE:** Male very similar to adult female. Female much duller, with no streaking on crown and back, no black on chin and throat, very narrow side streaks. **HYBRID:** Townsend's frequently hybridizes with Hermit where their ranges meet in WA and northern OR. Most hybrids closely resemble Hermit, especially in face, which is often entirely yellow. Crown varies from black to yellow and breast from yellow to white; flank streaks are extensive or absent; and back is olive green or gray. Most individuals tend to show a Hermit-faced pattern

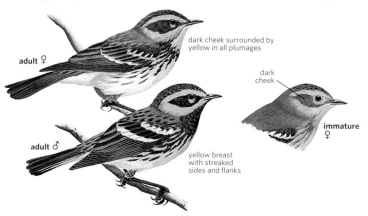

adult ♀

dark cheek surrounded by yellow in all plumages

dark cheek

immature ♀

adult ♂

yellow breast with streaked sides and flanks

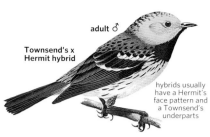

adult ♂

Townsend's x
Hermit hybrid

hybrids usually
have a Hermit's
face pattern and
a Townsend's
underparts

with yellow on the underparts more closely matching Townsend's. Less frequently, hybrids show Townsend's head pattern; a green or gray back with black streaks; and a breast with little or no yellow.

SIMILAR SPECIES Black-throated Green is superficially similar to adult females and immatures, but it lacks dark auricular and yellow on throat and breast. Immature female Black-burnian shows less green and more streaking on back, including pale streaks on sides of back and a trace of a black forehead stripe.

VOICE CALL: A high, sharp *tsik*. **FLIGHT CALL:** A high, thin *see*. **SONG:** Primary song a variable, buzzy *weazy weazy weazy dzeee*. Alternate song a buzzy *zi-zi-zi-zi-zi-zi, zwee zwee*.

STATUS & DISTRIBUTION Fairly common in coniferous forests of the Pacific Northwest. A medium- to long-distance migrant. **BREEDING:** Tall coniferous and mixed woodlands, preferring mature or old-growth forest. **MIGRATION:** Migrants from Central America arrive in southern CA in mid-Apr., peaking late Apr.–mid-May. Arrives as early as late Apr. in AK, with most arriving in May. Departs AK by early Aug., lingering into early Oct. Peaks late Aug.–early Sept. in OR; peaks mid-Aug.–mid-Oct. in CA. **WINTER:** A variety of habitats in coastal Pacific Northwest south to central CA, fewer to southern CA. Also rare in Central Valley, low western Sierra, and the Southwest; also winters commonly in montane forests from northwestern

Mexico to Costa Rica. **VAGRANT:** More than 90 records from east of Great Plains, from about half of the eastern states. Casual to accidental in Bermuda, Bahamas, northern AK, western Aleutians (to Shemya I.), and other Bering Sea islands.

POPULATION May be expanding northward in Alaska. Stable to increasing in rest of range.

HERMIT WARBLER *Setophaga occidentalis*   HEWA   1

grayish
upperparts

yellow face

adult ♀

black
throat

adult ♂

white underparts
with no or limited
flank streaks

immature
with olive
cast to
back

yellow face
with little dark
in auriculars

immature
♀

This striking golden-headed warbler, named for its secretive behavior, is often difficult to see high in its conifer haunts. It is a sister species (closest relative) to the Townsend's Warbler and frequently hybridizes with it where the two ranges overlap in the Northwest. It appears

that Hermit Warblers are dominated by Townsend's. (See "Population.") Between the bottom and middle of a tall conifer, it nests on a branch away from the trunk and produces four to five eggs between May and June. Monotypic. L 5.5" (14 cm)

IDENTIFICATION ADULT MALE: Yellow face, crown; black chin, throat, upper breast, lower nape. Gray back with black streaks. White underparts. Two white wing bars. Extensive white on outer tail feathers. **ADULT FEMALE:** Duller than male, with lower nape, small olive green throat patch, dull white to light gray sides, smaller tail spots; back without prominent streaks. **IMMATURE:** Male with olive on crown and ear coverts; chin and throat extensively mottled with olive green, yellow, and black. Dull whitish or grayish underparts. Female dullest, with extensive olive on crown and ear coverts; no black on back, nape, or

throat; yellow face, throat, and narrow eye ring. Underparts tinged buffy.

SIMILAR SPECIES Female and immature Olive Warblers are similar to immature female, but show bright white patch at base of primaries and have much different voice. Black-throated Green shows yellowish on underparts. See Townsend's for hybrids with Hermit.

VOICE CALL: A flat *tip*, similar to Black-throated Green's. **FLIGHT CALL:** A high, clear *sip*. **SONG:** Primary song a variable *weezy weezy weezy weezy zee* (last note is highest in pitch). Alternate song is variable and more complex: *che che che che cheeo ze ze ze ze ze ze seet*.

STATUS & DISTRIBUTION Fairly common in mountain forests; a medium- to long-distance migrant. **BREEDING:** Montane coniferous forest. **MIGRATION:** Arrives in southern AZ and southern CA in mid-Apr., peaking late Apr.–mid-May. Departs breeding grounds early, some by mid-July, peaking mid-Aug.–early Sept., with a few to mid-Oct. **WINTER:** Montane pine-oak and other forest from northwestern Mexico to Nicaragua. Also rarely in live oak and coniferous woodland in coastal CA. **VAGRANT:** Casual to accidental central and eastern U.S. and Canada. Accidental to Panama.

POPULATION Stable, though competition and hybridization with Townsend's may be reducing its range.

GOLDEN-CHEEKED WARBLER *Setophaga chrysoparia*   GCWA ▪ 2

This aptly named warbler breeds only in central Texas and lays three to four eggs in a nest near the trunk of a 20- to 30-foot-tall Ashe Juniper, about two-thirds of the way up (Mar.–Apr.). Monotypic. L 5.5" (14 cm)

dark eye line

immature ♀

unmarked, bright yellow cheek in all plumages

black back

adult ♀

all plumages with white vent

adult ♂

IDENTIFICATION Clear yellow cheeks, black line from eye to bill. **ADULT MALE:** Bright golden yellow sides of head, black line from bill to nape. Black crown, back, chin, throat, breast, broad side streaks. White belly, undertail coverts; two white wing bars. **ADULT FEMALE:** Duller than male; duller yellow face, black eye line narrower. Black-streaked olive-green upperparts. Chin, throat mixed with yellow. Narrower streaks on sides. **IMMATURE:** Male similar to adult female. Female with white chin, throat; very restricted blackish patches on sides of breast; narrower side streaks. **SIMILAR SPECIES** Black-throated Greens lack black crown, eye line, and back. Immature females do not show dark line through eye; have less extensive yellow face, pale yellow belly and undertail coverts. Hermit Warbler is plainer, grayer. **VOICE CALL:** A *tsip,* similar to Black-throated Green's. **FLIGHT CALL:** A high, thin *see.* **SONG:** A variable, short buzzy *dzeee dzweeee dzeezy see.* Alternate song simpler: *zee zee zee zee see.* **STATUS & DISTRIBUTION** Endangered; a medium-distance migrant. **BREEDING:**

Mature juniper-oak woodlands of central TX. **MIGRATION:** Arrives breeding grounds early to mid-Mar.; departs as early as late June, but typically present to early Aug., stragglers until mid-Aug. **WINTER:** Montane pine-oak forests from southern Mexico to Nicaragua. **VAGRANT:** Casual to accidental on upper TX coast, NM, CA, FL, and Virgin Is. (sight record). **POPULATION** Estimated 4,800–16,000 pairs in 1990, generally declining prior to that time. Prime habitat is relatively unfragmented old-growth juniper-oak woodlands, which have sometimes proven contentious to protect.

BLACK-THROATED GREEN WARBLER *Setophaga virens*   BTNW ▪ 1

A familiar bird of mixed and deciduous woodlands, the Black-throated Green Warbler is often detected by its buzzy, cheery song. It usually makes its nest low to moderately high in a conifer—away from the trunk—and nurtures three to five eggs (May–June). Polytypic (1–2 ssp., likely monotypic). L 5" (13 cm) **IDENTIFICATION** Yellow face, plain green back, yellow sides to vent.

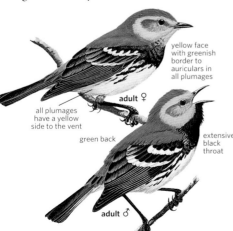

yellow face with greenish border to auriculars in all plumages

adult ♀

all plumages have a yellow side to the vent

green back

extensive black throat

adult ♂

greenish cheek

immature ♀

ADULT MALE: Yellow sides of head with olive-yellow auriculars; greenish olive crown, back, and rump. Black chin, throat, breast, and broad side streaks. Two white wing bars, whitish to pale yellowish belly and undertail coverts. **ADULT FEMALE:** Similar to male, but chin and upper throat mottled with (or entirely) whitish or pale yellowish; narrower side streaks. **IMMATURE:** Belly and undertail coverts more strongly tinged with yellow. Male similar to adult female. Female with whitish chin and throat (no black); very narrow side streaks. **GEOGRAPHIC VARIATION** Smaller overall size and bill length of isolated southeastern coastal subspecies *waynei* not detectable in the field and not recognized by some authorities.

SIMILAR SPECIES Golden-cheeked always shows a dark line from bill through eye to dark nape. Some hybrids of Townsend's and Hermit can resemble Black-throated Green Warbler. **VOICE CALL:** A soft, flat *tsip.* **FLIGHT CALL:** A high, sweet *see.* **SONG:** Primary song is variable with a whistled, buzzy quality; the last note is the highest: *zee-zee-zee-zoo-zee.* Alternate song a variable, more deliberate *zoo zee zoo zoo zee.* **STATUS & DISTRIBUTION** Common; primarily a trans- and circum-Gulf

migrant, some move through FL to Caribbean. **BREEDING:** A wide variety of habitats: mainly coniferous and mixed forest, but sometimes deciduous. Subspecies *waynei* breeds in cypress swamps. **MIGRATION:** Arrives in Gulf states in late Mar., in Midwest by late Apr., peaking in mid-May. Departs breeding grounds as early as late July, peaking mid- to late Sept. into late Oct.; stragglers widely recorded in North into Nov. **WINTER:** Mature montane forests from northeastern Mexico to Panama; wide variety of habitats in Caribbean. Small numbers winter in southern FL, southern TX. Rare in northern Colombia and western Venezuela. **VAGRANT:** Multiple recs. for most western states, most in fall and many (300+) in CA. Casual in coastal BC, southeastern AK, and NT. Casual to the Azores; accidental in Greenland, Iceland, and Germany (specimen from 1858).
POPULATION Numerous and relatively stable, but some local trends indicate extended, slight declines.

Genus *Basileuterus*

With 25 species, this mainly tropical genus—the second largest in the wood-warbler family—reaches its greatest diversity in South America. Only three species have been recorded in North America as vagrants. Most species are olive to gray above, yellow to whitish below, with some patterning on the crown or face.

FAN-TAILED WARBLER   *Basileuterus lachrymosus*   FTWA   4

This unique, large warbler is similar in behavior to redstarts but is more secretive. The Fan-tailed is often seen near the ground or walking. Polytypic (3 ssp.; *tephra* in N.A.). L 5.8" (15 cm)
IDENTIFICATION Long, graduated white-tipped tail held open, swung up

and down and side to side; long pink legs. **ADULT:** Distinct head pattern: blackish head with broken white eye ring, white supraloral spot, and yellow crown patch; gray upperparts, yellow underparts with tawny-orange wash on breast. **IMMATURE:** Face slightly paler.
SIMILAR SPECIES Yellow-breasted Chat has less yellow on underparts and lacks yellow crown patch.
VOICE CALL: A distinctive, high, thin *tsew* or a penetrating *schree.* **SONG:** Primary song a series of sweet notes ending in a sharp upslur or downslur: *che che che a-wee wee che-cheer.*
STATUS & DISTRIBUTION Breeds in moist, shady, steep-walled ravines from Pacific slope of northwestern Mexico south to Nicaragua, with a disjunct population in coastal east-central Mexico. **VAGRANT:** Casual spring and fall to southeastern AZ (8 recs.); one lingered into late June. Accidental to east-central NM in spring and to west TX (Big Bend) in fall.

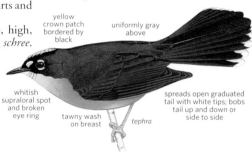

yellow crown patch bordered by black

uniformly gray above

whitish supraloral spot and broken eye ring

tawny wash on breast

spreads open graduated tail with white tips; bobs tail up and down or side to side

tephra

RUFOUS-CAPPED WARBLER   *Basileuterus rufifrons*   RCWA   3

Inhabiting dense brush in montane areas, Rufous-capped has occurred north of Mexico casually and has even attempted to breed. Polytypic (8 ssp.; 2 in N.A.). L 5.3" (13 cm)
IDENTIFICATION Short, thick bill. It cocks and waves its long tail like a wren or gnatcatcher. **ADULT:** Distinct head

pattern with rufous on crown and cheeks, broad white supercilium, dark lores; grayish olive upperparts; bright yellow throat and breast, white belly, grayish buff sides and flanks.
GEOGRAPHIC VARIATION Subspecies likely recorded from TX *(jouyi)* is generally darker and more richly colored than that recorded from AZ *(caudatus)*.
VOICE CALL: A hard *tik,* sometimes doubled or in a series. **SONG:** Primary song is a rapid, variable series of *chip* notes, chirps, and trills, usually changing in pitch and pace.
STATUS & DISTRIBUTION From northern Mexico south to Panama and extreme northwestern S.A. **VAGRANT:** Casual from spring to fall to western and central TX (20+ recs.) and southeastern AZ (15+ recs., inc. one attempted nesting in 1977). Most recent AZ records have come from French Joe Canyon in the Whetstone Mts. and Florida Canyon in the Santa Rita Mts.; at both locations they are presently resident. Most recent TX records have been from the southern Edwards Plateau. Accidental Guadalupe Canyon.

long tail is often cocked

bright rufous crown and auriculars; bold white supercilium

olive-brown above

bright yellow throat and white submoustachial

GOLDEN-CROWNED WARBLER   *Basileuterus culicivorus*   GCRW ■ 4

The most widespread member of the genus, the Golden-crowned occurs only casually north of Mexico. Polytypic (13 ssp.; *brasherii* in N.A.). L 5" (13 cm)
IDENTIFICATION ADULT: Olive-gray upperparts; bright yellow underparts. Distinctive head pattern: yellow central crown patch, broad black lateral stripes, grayish olive auriculars, dull yellowish olive supercilium, gray loral spot, narrow broken yellow eye ring. **IMMATURE:** Very similar to adult; some show slightly less distinct head pattern. **SIMILAR SPECIES** Orange-crowned and Wilson's are superficially similar, lack distinct crown stripes. Worm-eating is buffy, not yellow; has a larger bill.
VOICE CALL: Often repeated, slightly buzzy *tuck.* **SONG:** Several clear whistled notes, ending with a distinct upslur: *see-whew-whew-wee-see?*
STATUS & DISTRIBUTION Disjunct populations from west-central and southeastern Mexico through much of C.A., northern S.A., the Guianas, and much

of eastern Brazil to northern Argentina and Uruguay. **VAGRANT:** Casual, mainly in winter from Nov.–Mar. (one rec. Oct.; one late Apr. rec.), to extreme southern TX (20+ recs.). Accidental in spring to southern coastal TX and east-central NM.

greenish supercilium

yellowish green median crown stripe and blackish lateral crown stripe

plain grayish olive above with yellowish underparts

broken eye ring

brasherii

Genus *Cardellina*

Formerly, the genus was monotypic with only the Red-faced Warbler, but it has been expanded to include the beautiful Red and Pink-headed Wablers (formerly in *Ergatius*) and Wilson's and Canada Warblers. The latter two and the Hooded Warbler were formerly in the now-defunct *Wilsonia*. All have long tails that are frequently "flipped."

WILSON'S WARBLER   *Cardellina pusilla*   WIWA ■ 1

This small warbler feeds very actively, often fly-catching. Wilson's lays eggs (2–7) in a nest on or near the ground at the base of a shrub or small tree (Apr.–June). Polytypic (3 ssp.; all in N.A.). L 4.8" (12 cm)
IDENTIFICATION Flips long tail up and down and in a circular motion. Nominate described and illustrated. **ADULT MALE:** Olive-green upperparts (inc. wings, tail, cheeks); solid shiny black cap; bright yellow forehead, lores, eye ring, broad superciliary, entire underparts. **ADULT FEMALE:** Similar to male, but smaller black cap, usually nearly absent or restricted to front half of crown (often mottled with olive). **IMMATURE:** Similar to adult; male with extensive olive mottling on black cap, female usually with olive crown, forehead; lacks black cap.
GEOGRAPHIC VARIATION Nominate breeds through much of boreal Canada; *pileolata* (AK, Rocky Mountains) has brighter yellow forehead and underparts. Pacific coast *chryseola* is brightest overall: yellowish wash on cheeks, orange-tinged forehead.
SIMILAR SPECIES Larger immature female Hoodeds have larger bill, dusky lores, white in tail. Shorter-tailed Orange-crowneds are drabber with diffuse breast streaks, dusky eye line, more pointed bill. Immature Yellow Warbler shows yellow tail spots, pale-edged wing feathers.
VOICE CALL: A somewhat nasal *timp.* **FLIGHT CALL:** A sharp, slurred *tsip.* **SONG:** Primary song short and chattering, dropping in pitch toward end: *chi chi chi chi chi chi chet chet.* Songs of *pileolata* and *chryseola* louder and faster.
STATUS & DISTRIBUTION Common; medium- to long-distance migrant, more common in West. **BREEDING:** Wet situations with dense ground cover, low shrubs. **MIGRATION:** Eastern birds are circum-Gulf migrants, arrive in southern TX late Apr., in Great Lakes late May. Pacific coast birds arrive in southern AZ as early as late Feb., but migration extends to late May (early spring migrants are nearly all *chryseola*); arrives in southwestern BC late Apr. In East, earliest fall migrants mid-Aug.,

olive face

adult ♀
pusilla

black cap

dark eye stands out in plain face

pusilla ♂

auriculars contrast less with supercilium than *pusilla* and *pileolata*

gold forehead

yellow wash on back

chryseola

brighter yellow below

pileolata

pusilla

chryseola

peak late Aug.–mid-Sept., stragglers casually into Nov. Pacific coast birds depart mid-July. WINTER: Wide variety of habitats; northwestern Mexico to central Panama, rarely to extreme southeastern TX, southern LA, coastal southern CA. VAGRANT: Casual in winter to Colombia, northern Bahamas, Cuba, Jamaica. Accidental in Greenland, U.K. POPULATION Declining (esp. in the West), possibly due to cowbird parasitism and loss of riparian habitats.

CANADA WARBLER *Cardellina canadensis* CAWA ■ 1

Sometimes called the "Necklaced" Warbler, in June, the Canada Warbler builds a well-concealed nest for two to six eggs on or near the ground. Monotypic. L 5.3" (13 cm) IDENTIFICATION Active; often cocks tail, flicks wings. Pink legs. ADULT MALE: Plain blue-gray upperparts (inc. wings, tail); black forehead, patch below eye; distinct complete white or yellow eye ring, yellow lores, forming spectacles; bright yellow underparts; rows of black spots across breast form a necklace; white undertail coverts. Black markings reduced in first-spring males. ADULT FEMALE: Duller than male, lacks black on face; less distinct dark necklace. IMMATURE: Male similar to spring female; more olive-gray above. Female duller; very indistinct necklace formed by grayish olive spots. SIMILAR SPECIES Kirtland's and Magnolia Warblers both have white in tail. VOICE CALL: Sharp *tchup* or *tik.* FLIGHT CALL: High *zzee.* SONG: Primary song variable, staccato series of jumbled notes; often begins with loud *chip* note, then short pause: *chip . . . chupety swee-ditchety chip.* STATUS & DISTRIBUTION Common; mainly a circum-Gulf migrant. BREEDING: Cool, moist mixed forests with dense understory. MIGRATION: Late in spring; arrives southern TX after mid-Apr., Upper Midwest mid-May, continues into early June. Peaks in mid- to late Aug., continues into mid-Sept. Casual mid-Oct.–early Nov. WINTER: Dense montane undergrowth; from northern S.A. in Andes to central Peru. VAGRANT: Very rare to casual in West; casual in Caribbean and to Greenland; accidental Iceland. POPULATION Declining, likely from forest succession and loss of forested wetlands.

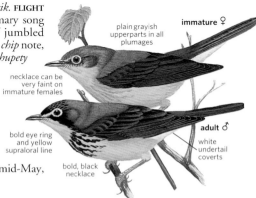
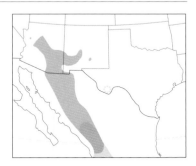

RED-FACED WARBLER *Cardellina rubrifrons* RFWA ■ 2

This strikingly plumaged southwestern mountain species is easily identified. In May–June, it constructs a nest in a depression on the ground, often on a slope at the base of woody plants, and often with an overhang that helps conceal and protect the four to six eggs within. Monotypic. L 5.5" (14 cm) IDENTIFICATION Small bill; head appears peaked. Flips long tail around in a similar manner to a *Wilsonia.* ADULT MALE: Bright red face, upper breast, and sides of neck; black crown and cheeks; white nape spot, rump, and underparts; a single white wing bar on median coverts; gray upperparts. ADULT FEMALE: Nearly identical to male, but with duller, more orange-red face. IMMATURE: Similar to adult female, but with more brownish gray upperparts. Some males nearly as bright as adult males. SIMILAR SPECIES Painted Redstart is the only other warbler with red in plumage (but not on face), and it shows large white patches in wings and tail. Chickadees are similarly gray above, and Red-faced Warbler is sometimes similarly acrobatic, but once the red face is seen there can be no confusion. Also note that chickadees do not show white nape spots or white rumps. VOICE CALL: A sharp *chup* or *tchip,* suggesting Black-throated Gray. SONG: Primary song a series of assorted thinner notes with an emphatic ending: *wi tsi-wi tsi-wi si-wi-si-whichu.* STATUS & DISTRIBUTION Fairly common; medium-distance migrant. BREEDING: Montane mixed and deciduous woodland. MIGRATION: Not often observed in spring migration; arrives late Apr.–late May. Departs as early as Aug., some into Sept. WINTER: Humid montane forest, pine-oak forest, riparian woodland from northwestern Mexico south to eastern Honduras. VAGRANT: Casual in TX (mostly Big Bend N.P.) and CA (all but one from southern CA); accidental in CO, WY, LA, and GA. POPULATION Difficult to determine; possibly declining slightly. May be vulnerable to logging.

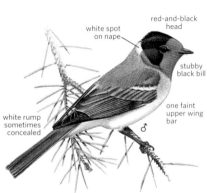

Genus Myioborus

When it was first described, the Painted Redstart was thought to be closely related to the American Redstart, which was previously named mainly for its red tail patches. All 12 members of this mainly tropical genus show white tail patches, thus they are sometimes called whitestarts, and only two species show any red in the plumage at all.

PAINTED REDSTART *Myioborus pictus*   PARE ■ 2

With its bold black, red, and white coloration and its conspicuous wing- and tail-fanning behavior while creeping along branches, the Painted Redstart is unlikely to be confused with any other warbler species. It makes its nest on an embankment, often near water—well hidden under rocks, grasses, or roots—and lays three to seven eggs in May. Polytypic (2 ssp.; nominate in N.A.). L 5.8" (15 cm)

IDENTIFICATION ADULT: Mainly black with large white patches in wing, white on outer three tail feathers, and a small white crescent below eye. Red (female's red slightly paler) lower breast and belly; undertail coverts mixed slate and white. **JUVENILE:** Retains juvenal plumage later than most warblers, June–August. Similar to adult, but without red. Sooty lower underparts; undertail coverts mottled grayish and white. Red belly is acquired in late summer or early fall.

SIMILAR SPECIES Slate-throated Redstart lacks white wing patches and eye crescent, shows less white in tail, and has slaty (not black) upperparts.

VOICE CALL: A unique, scratchy, whistled *sheu* or a richer *cheree,* similar to calls of Pine Siskin. **SONG:** Primary song a somewhat variable series of rich, two-part syllables, ending with one or two inflected notes: *weeta weeta weeta wee.*

STATUS & DISTRIBUTION Common; a short-distance migrant. **BREEDING:** Pine-oak woodlands in foothills and mountains. **MIGRATION:** Arrives on breeding grounds mid- to late Mar.

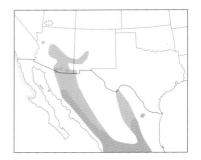

white crescent under eye   *pictus*

bold white wing patch

red belly

white outer tail feathers

juvenile *pictus*

sooty breast

Departs breeding grounds in Sept., with some lingering into mid-Oct. **WINTER:** Pine-oak woodlands, sometimes at lower elevations than when breeding; from northwestern Mexico south to northern Nicaragua, where it mixes with sedentary subspecies. Rare in winter in breeding range. **VAGRANT:** Rare in fall and winter in southern CA (more than 100 recs.). Casual in northern CA, Baja California, southwestern AZ, southwestern UT, southwestern CO, and southern TX. Accidental in, BC, MT, WI, MI, southern ON, OH, NY, MA, LA, MS, AL, and GA. **POPULATION** Stable.

SLATE-THROATED REDSTART *Myioborus miniatus*   STRE ■ 4

An easily identified, flashy warbler, the Slate-throated Redstart occurs casually north of Mexico. Polytypic (12 ssp.; nominate in N.A.). L 6" (15 cm)
IDENTIFICATION Slate-throated frequently fans its long, graduated tail. **ADULT MALE:** Slate-gray upperparts and wings; a blackish face, throat, and sides; a dark chestnut crown patch; a black tail with large white spots on outer tail feathers; red breast and belly; undertail coverts mottled white and slate gray. Central and South American subspecies show orange or yellow breast and belly. **ADULT FEMALE:** Nearly identical to adult male, but with duller red underparts; a slate-gray face, throat, and sides; and smaller chestnut crown patch. **JUVENILE:** Similar to adult, but lacks red underparts; paler slaty coloring on breast and belly; undertail coverts mottled with cinnamon-brown.

SIMILAR SPECIES See Painted Redstart.
VOICE CALL: A single, high *tsip,* similar to call of Chipping Sparrow. **SONG:** Primary song a variable series of *sweet s-wee* notes, often accelerating toward the end of the series. Although the pattern can suggest Painted Redstart, it is higher pitched and thinner, with slurred single notes rather than

chestnut crown patch

plain dark gray wings

no white under eye

more graduated tail than Painted with white only in tips

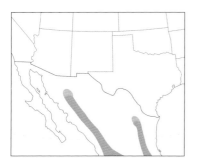

doubled ones. Some versions recall the songs of Yellow-rumped or Magnolia Warblers.

STATUS & DISTRIBUTION Breeds in montane coniferous and mixed forest from northwestern and central Mexico south to northern S.A., south in Andes to central Bolivia. **VAGRANT:** Casual to accidental in spring (mostly Apr.–May) to southern AZ (7 recs.), southeastern NM (1 rec.), and western and southern TX (more than five recs.).

Genus *Icteria*

Icteria, the most aberrant genus in the wood-warbler family, comprises only a single species—the Yellow-breasted Chat—that is characterized by its large size, exceptionally thick bill, long tail, and unique voice. While perhaps not a wood-warbler, it is unclear what it is.

YELLOW-BREASTED CHAT   *Icteria virens*   YBCH   1

The Yellow-breasted Chat is heard more often than it is seen: It is skulky and secretive in its brushy and often impenetrable habitat. In keeping with its secretive nature, Yellow-breasted builds a well-concealed nest near to the ground in a dense thicket or shrub; here it lays three to five eggs from May to June. Polytypic (3 ssp.; 2 in N.A.). L 7.5" (19 cm)

IDENTIFICATION Very large—the largest of all wood-warblers (although its classification as a wood-warbler is debated)—with a thick bill and a long tail. Nominate subspecies described. **ADULT MALE:** Bright yellow throat and breast sharply contrast with white belly and undertail coverts and with pale grayish flanks. Grayish olive head has white spectacles, black lores, and a very narrow white submoustachial stripe. Upperparts, wings, and tail are all olive green. Bill is black, with some gray at base of lower mandible. **ADULT FEMALE:** Very similar to male, but head is more olive and lores are duller. It also shows yellowish pink at base of lower mandible and pale buffy on flanks. **IMMATURE:** Similar to, but slightly duller than, adult female.

GEOGRAPHIC VARIATION Subtle. Western subspecies *(auricollis)* is longer tailed (sometimes known as "Long-tailed Chat"). It upperparts are more grayish olive; it has a broader white submoustachial stripe; and it shows richer orange-yellow on breast (though this can be affected by diet) than eastern nominate, *virens,* described above.

SIMILAR SPECIES Unmistakable if seen well, but usually skulks.

VOICE CALL: Yellow-breasted Chat—though not often heard unless giving its loud song—gives a variety of calls, including a harsh *chough,* a nasal *air,* and a sharp *cuk-cuk-cuk.* **SONG:** It has an extensive repertoire, consisting of a series of irregularly spaced scolds, chuckles, mews, rattles, and other unmusical sounds. It often incorporates harsh *sheh sheh sheh sheh* calls and a higher-pitched *tu-tu-tu* series. Its songs are suggestive of mockingbirds or thrashers, especially given its tendency to mimic other birds. Sings in flight in spectacular display flight with tail tucked and deep, slow-motion wing beats.

STATUS & DISTRIBUTION Uncommon; a medium-distance migrant, nominate *virens* is both a trans-Gulf and circum-Gulf migrant. **BREEDING:** Low, dense vegetation with open canopy, including shrubby habitat in wetland areas and early second growth. In East *(virens),* favors second-growth; in West *(auricollis),* favors riparian areas. **MIGRATION:** During spring migration in the East, it arrives on the Gulf Coast in mid-Apr., reaching the Midwest by early May. In the West, it arrives

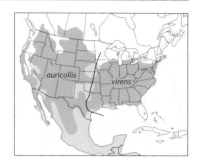

in CA and southern AZ in mid-Apr., reaching northernmost breeding areas in Canada by late May. Its departure from the breeding grounds in fall is difficult to detect once singing stops in late summer. It moves mainly late Aug.–late Sept., with stragglers remaining into late Oct. **WINTER:** Shrub-steppe with dense, low cover of woody vegetation from northern Mexico south to western Panama. Rarely in southern U.S. and casually farther north. **VAGRANT:** Casual in spring migration north of breeding range; regular in fall to Atlantic Canada; rare in fall to Bermuda; casual in fall and winter to Bahamas; very rare in Cuba and possibly Grand Cayman Is. Three specimens from Greenland. **POPULATION** Reports show precipitous declines throughout much of the Yellow-breasted Chat's range. There has been a near total withdrawal from southern New England. Populations are more stable in much of the West, but those in California and some adjacent areas are greatly reduced. Reasons for the declines are unclear.

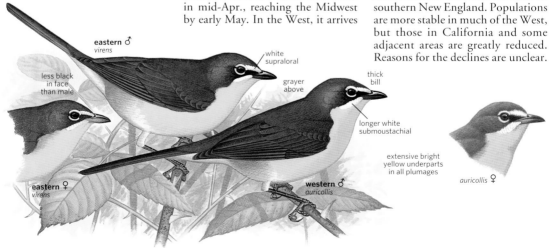

eastern ♂ *virens* · white supraloral · less black in face than male · grayer above · thick bill · longer white submoustachial · extensive bright yellow underparts in all plumages · eastern ♀ *virens* · western ♂ *auricollis* · *auricollis* ♀

FAMILY UNCERTAIN *Incertae Sedis*

Bananaquit (Bahamas, Mar.)

Formerly, the exclusively neotropical family Coerebidae included the honeycreepers, conebills, dacnises, and flower-piercers, which were assigned to other families (mainly Thraupidae, tanagers). The Bananaquit was not reassigned, and it remained the sole member of Coerebidae until the AOU removed that family from taxonomic status in 2005.

Structure The Bananaquit has a long, thin, decurved bill. Its tongue is specialized for feeding on nectar.

Plumage Variable, with a tendency toward dark hues.

Many subspecies have a gray throat, and in the southern Caribbean some are almost entirely black. The amount of yellow on the underparts is also variable.

Behavior Though fairly social, the birds can be territorial especially at food sources, which include nectar, fruits, and berries. Sometimes they pierce the bases of larger flowers to gain access to nectar, and they often hang upside down when feeding. Their diet also includes small insects. They build untidy globular nests and use empty nests for roosting at night.

Distribution The Bananaquit is generally found in lowlands throughout the Caribbean except in Cuba, from southern Mexico south through Central America, and in South America south to northern Peru and southeastern Brazil. It is absent from a large area of the Amazon basin.

Taxonomy Genetic data have shown that the genus *Coereba* is related closely to various other species and does not warrant classification in a separate family. As of 2013, precise relationships of the genus were not clear; placement within another family was left undetermined pending further research. The formal term for this status is the Latin *incertae sedis,* meaning that the taxonomic position is uncertain.

Conservation The Bananaquit is common throughout its range and is well adapted to the presence of humans.

Genus *Coereba*

BANANAQUIT   *Coereba flaveola*   BANA ■ 4

The Bananaquit is a small, short-tailed, warbler-like, black-and-white bird with a yellow breast and rump, and a thin, distinctly decurved bill. It feeds on nectar, often piercing through the bases of flowers. (41 ssp.; *bahamensis* in N.A.). L 4.5" (11 cm)

IDENTIFICATION ADULT: Head black with long white supercilium and throat. Back, wings, and tail black, with small white tips on outer tail feathers and white patch at base of primaries. Breast yellow. Belly and undertail coverts white. Red gape corners. **JUVENILE:** Duller than adult, with yellowish supercilium and little or no yellow on dull white underparts. Has duller rump patch and duller gape.

SIMILAR SPECIES Adults are unlike any other North American bird. Juveniles may superficially resemble a wood-warbler (Parulidae), but are readily distinguished by their decurved bill.

VOICE CALL: An unmusical *tsip.* Young birds give a *chit, chit, chit* similar to warblers. **SONG:** Several ticks followed by rapid clicking in Bahamas. Other subspecies give a thin, high-pitched, tumbling trill.

STATUS & DISTRIBUTION Common in second growth and near human settlements. Resident throughout the Caribbean, except Cuba, from southern

bold white supercilium

juvenile
bahamensis

yellowish rump

curved bill with reddish gape

yellow breast

white patch at base of primaries

adult
bahamensis

white tail spots

Mexico through Panama, and South American lowlands to southern Brazil. **BREEDING:** Related to rainfall, peaking in Feb.–Apr. in Bahamas. An untidy, globular nest in the middle to top of shrub or small tree on outer branch (2–4 eggs). **VAGRANT:** Casual in southern FL (40+ recs.), mainly Jan.–Mar.

POPULATION Bananaquit is very common within its extensive range, and the bird seems to do well in gardens with abundant flowers.

TANAGERS Family Thraupidae

Western Spindalis, male *zena* (FL, May)

Ranging from intensely multicolored to relatively dull brown or gray in plumage, tanagers are common and remarkably diverse through the New World tropics. Work by taxonomists in recent years has obliterated any former definitions of what constituted a "tanager." Many birders are understandably confused that all North American birds called tanagers are now placed in the cardinal and grosbeak family (Cardinalidae). Only one species in the tanager family—as currently defined by the American Ornithologists' Union (A.O.U.) and followed in this volume—occasionally wanders to southern Florida from the West Indies, the Western Spindalis. Three North American species currently placed in the Emberizidae are probably better considered tanagers, the White-collared Seedeater and Yellow-faced and Black-faced Grassquits (pp. 632–633), as is the Bananaquit (previous page). These taxonomic changes and others can be expected in the near future.

Structure Part of the huge radiation of nine-primaried oscines, tanagers are small to moderately large songbirds exhibiting great variations in bill shape including relatively large-billed fruit-eaters, stout-billed seed eaters, and thin-billed insect and nectar feeders. They have a slender to slightly stocky build and medium-length wings and tails.

Behavior Foraging behavior is as diverse as the morphologies shown by the members of this family; frugivory (fruit eating) is common, but there are insect gleaners, nectar feeders, "flower piercers," and even aerial salliers. Most species are mainly found in pairs or travel in small flocks; many join mixed-species flocks. Tanagers are not gifted vocalists, with most species giving thin chips, twitters, squeaks, or scolds; a few groups have more pleasant warbles or high whistles.

Plumage Many tanagers sport brilliant plumage, as in the almost psychedelic genus *Tangara.* In many species the sexes are similar, but in others the females are considerably duller. Future taxonomic changes will probably move many plainly marked genera—with plumages of grays and browns to yellows, and including many streaked, sparrow-like birds—into the tanager family.

Distribution Tanagers are found through much of the New World, from Mexico and the West Indies south to southernmost South America. Most species are sedentary or undertake short altitudinal movements, but the family includes some austral and intratropical migrants. Three enigmatic taxa are endemic to south Atlantic Ocean islands, and the famous "Darwin's Finches" (now usually placed in the Thraupidae) occur on the Galápagos Is. and Cocos I. off the northwest coast of South America.

Taxonomy Tanagers are perhaps most closely related to the New World sparrows (family Emberizidae). Recent taxonomic upheavals have drastically changed the composition of this family, with many genera transferred to other families (Cardinalidae, Fringillidae). Many more genera that have been proposed for transfer into the Thraupidae (mainly from the Emberizidae) await action by the A.O.U.'s Committee on Classification and Nomenclature. As constituted by most authorities (including the South American Classification Committee of the A.O.U.), there are 383 species in more than 100 genera.

Conservation Many species have very restricted ranges and are rare within those ranges, such as the Cherry-throated Tanager of Brazil, the Azure-rumped Tanager of Chiapas and Guatemala, and the chat-tanagers of Hispaniola. Others suffer from habitat degradation and/or exploitation for the pet trade.

Genus Spindalis

Generally distributed in the Bahamas and West Indies, the four species in this genus were previously considered one species—the "Stripe-headed Tanager." Small tanagers with relatively short, stout bills. The male's plumage is a combination of black, white, and rich tawny; the female is very dull grayish green with a white spot in the wings.

WESTERN SPINDALIS *Spindalis zena*   WESP ■ 3

Formerly named the Stripe-headed Tanager, this fancy West Indian tanager is a rare visitor from the Bahamas to coastal southeastern Florida and the Florida Keys. Typically found in fruiting trees, where it feeds sluggishly, the Western Spindalis sometimes sits motionless for long periods. Males have a striking plumage: a mixture of tawny brown, black, and white—unlike any other North American bird. Females are totally different and much duller. Western Spindalis is smaller than other North American tanagers and has a very small bill. Polytypic (5 ssp.; 3 in N.A.). L 6.8" (17 cm)

IDENTIFICATION MALE: Plumage unmistakable. Most striking are black-and-white stripes on head. Its wings are black with entirely white greater coverts forming a large white wing patch; most of flight feathers have a white edging. Underparts are tawny with gray flanks; lower belly and undertail coverts are white, and back is black (or green in some subspecies), contrasting with tawny rump. Tail is black with white outer tail feathers. A white undersurface

to tail is visible when viewed from underneath. In flight, Western Spindalis shows large white patches at base of both primaries and secondaries. FEMALE: Much duller and completely different from male, female's plumage is a drab grayish olive with darker auriculars, paler throat and malar. Wing coverts' narrow white edging and primaries' white base together form a white tick visible on folded wing, similar to that found on female Black-throated Blue Warbler.

GEOGRAPHIC VARIATION Black-backed birds from the central and southern Bahamas *(zena)* make up the bulk of the Florida reports. Green-backed birds from the northern Bahamas *(townsendi)* have been reported. Intergrades between the two Bahamian subspecies are possible. Also, there is a recent record from Key West of an adult male green-backed Cuban subspecies *pretrei*. Two additional subspecies are found on Grand Cayman I. *(salvini)* and Cozumel I. *(benedicti)*.

SIMILAR SPECIES Males are unlikely to be confused with other species. Females, however, present more of a problem. They look most similar to female Black-throated Blue Warblers, which also show a white spot on folded wing.

Note that Western Spindalis has a stout gray tanager-like bill and white edging to wing coverts. It is also distinguished from female Black-throated Blue Warbler by its very different behavior and larger size.

VOICE CALL: A thin, high *tseee,* which is given singly or in a short series. SONG: Seldom heard in the U.S. In the Bahamas, a series of high thin notes ending in buzzy phrases.

STATUS & DISTRIBUTION A common resident in the Bahamas and northern West Indies. A rare visitor to parks in coastal southern FL (casual on the FL Gulf Coast) and the FL Keys; particularly found in fruiting trees. Most of the reports from spring, fall, and winter. There is one breeding record.

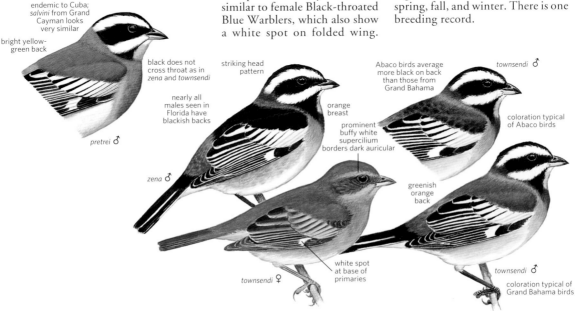

endemic to Cuba; *salvini* from Grand Cayman looks very similar

bright yellow-green back

pretrei ♂

black does not cross throat as in *zena* and *townsendi*

nearly all males seen in Florida have blackish backs

zena ♂

striking head pattern

orange breast

prominent buffy white supercilium borders dark auricular

Abaco birds average more black on back than those from Grand Bahama

townsendi ♂

coloration typical of Abaco birds

greenish orange back

townsendi ♀

white spot at base of primaries

townsendi ♂

coloration typical of Grand Bahama birds

EMBERIZIDS Family Emberizidae

Baird's Sparrow (ND, June)

In North America, the Emberizidae family is best represented by a bewildering array of sparrows, most of which are cryptically patterned with various shades of brown and are affectionately known to many birders as the quintessential LBJs: "Little Brown Jobs." While sparrow identification is one of the greater challenges for the beginning birder, familiarity with distinctive shapes of the genera, as well as with the habits and habitats of sparrows themselves, simplifies the process.

Structure Emberizids are fairly small perching birds, ranging in size from very small seedeaters and grassquits to the much larger towhees. Their bills are conical, their wings are typically short and rounded, and their tails are often long.

Behavior Behavior is often quite skulking, particularly so in some species. The exception is during the breeding season, when males sing from elevated perches or perform skylarking flight songs. The songs can be faint and unremarkable, but more often they are quite complex and beautiful. Most species have several calls, which may include a simple chip and a high, thin lisping note. The latter note is often given as the flight note by nocturnal migrants (e.g., the White-throated Sparrow) but is used more frequently as a contact call when on the ground.

Plumage Typically, the emberizid plumage is a dull collection of brown, gray, black, and white, but some species in North America have areas of bright yellow feathering or greenish color on the back and wings. In many species the sexes are similar, but a few species (e.g., grassquits, seedeaters, and certain towhees) show moderate to pronounced sexual dimorphism.

Distribution Emberizids occur worldwide, except in Australasia. They are most diverse in the New World, with just 42 Old World species, most of which are in the genus

Emberiza. Although New World diversity is centered in the tropics, some 60 species are known north of Mexico. Preferred habitats are often open country, including fields, deserts, gardens, marshes, riparian areas, and wood edges; a few species prefer the forest interior. Most North American species are short-distance migrants, but some species inhabiting the southern United States are resident. Nests are typically an open cup that is placed on the ground or low in shrubbery. A typical clutch would include several eggs (2–5), which are often unmarked and whitish or pale blue, though occasionally they are patterned. The winter diet is primarily composed of seeds and other vegetable matter; in summer that diet is heavily supplemented by insects.

Taxonomy Emberizidae is currently the second largest passerine family in the world (second only to the tyrant flycatcher family Tyrannidae), comprising some 321 species in 71 genera. Recent molecular work has suggested that a transfer of many emberizid genera into the tanager family (Thraupidae) is called for; these groups include the *Sporophila* seedeaters, the *Tiaris* grassquits, and a great many other West Indian and Neotropical genera (as well as the iconic "Darwin's Finches" of the Galápagos Islands). Furthermore, our American sparrows appear to be well-differentiated genetically from the Old World buntings (*Emberiza* and related genera), and many taxonomists now propose the erection of a new family, the Passerellidae, for the New World sparrows.

Conservation These small birds are often fairly catholic in their habitat requirements, so most species have stable populations. However, certain species are more specialized in their habitat requirements, and some populations are severely depressed, endangered, or extirpated; habitat loss is almost always to blame.

SEEDEATERS AND GRASSQUITS Genera *Sporophila* and *Tiaris*

These neotropical genera include 34 *Sporophila* species and five *Tiaris* species. Many authorities believe that they should be placed in family Thraupidae (tanagers). All are small, with strong sexual dimorphism. *Sporophila* seedeaters have a distinctive "Roman nose" bill and a fairly long, rounded tail; *Tiaris* grassquits have a short, conical bill and a short, squared tail.

WHITE-COLLARED SEEDEATER   *Sporophila torqueola*   WCSE ■ 3

During spring, males sing their cheery song from high perches. Females and nonbreeders can be secretive as they forage near the ground. Loosely colonial as nesters, White-collareds may form small flocks in the nonbreeding season. Polytypic (4 ssp.; *sharpei* in N.A.). L 4.5" (11 cm) **IDENTIFICATION** Size, bill shape, and rounded tail are distinctive. **ADULT MALE:** Black cap and wings, white crescent below eye, an incomplete buffy collar, white throat and breast washed with buff, narrow white wing bars, and white patch at base of primaries. **IMMATURE MALE:** Paler, browner, less distinctly marked. **FEMALE:** Brown overall; darker on wings with paler buffy color on breast; narrow wing bars.

GEOGRAPHIC VARIATION Texas breeders are relatively indistinctly marked *sharpei*. Escapes in California are apparently the more boldly marked, buff-breasted western Mexican subspecies *torqueola* or *atriceps*.

SIMILAR SPECIES Females can be confused with *Passerina* buntings or Lesser Goldfinch; note the seedeater's smaller size, more arched bill, more rounded tail, different coloration, and distinctive calls. **VOICE CALL:** Distinct, high *wink;* loud, whistled *chew;* and husky, rising *che.* **SONG:** A variable, clear-toned *sweet sweet sweet sweet cheer cheer cheer* that is pitched high, then low; recalls song of American Goldfinch.

STATUS & DISTRIBUTION Mexico south to west Panama. **YEAR-ROUND:** Uncommon and local along Rio Grande in Starr, Zapata, and Webb counties, where it prefers stands of cane. Formerly occurred farther east along Rio Grande, but now casual there. **EXOTIC:** Rare but regular in urban southern CA, southeastern AZ, and southern TX; no established populations. **POPULATION** Formerly more widespread, it lost ground in the U.S. due to habitat loss and possibly pesticide use. Small extant U.S. population is stable but vulnerable.

Illustration labels: sharpei ♀; tiny size; thin, pale wing bars; short, stubby bill with strongly curved culmen; buffy below; 1st winter ♂ *sharpei*; blackish head; pale buffy white collar; distinctly patterned wings; adult ♂ *sharpei*

YELLOW-FACED GRASSQUIT   *Tiaris olivaceus*   YFGR ■ 4

This bird feeds on grass seeds in vacant lots and weedy edges. Polytypic (5 sp.; 2 in N.A.). L 4.3" (11 cm) **IDENTIFICATION** Very small; short tail; small, conical bill; straight culmen. **ADULT MALE:** Black head, breast, and upper belly; golden yellow eyebrow, throat, and crescent below eye; olive above. **ADULT FEMALE & IMMATURE MALE:** Traces of same head pattern; olive above, lacking black below. **GEOGRAPHIC VARIATION** Subspecies *pusillus,* which has a more extensive black breast, occurs from Mexico to S.A. and accounts for Texas records; Florida records pertain to *olivaceus* of

Cuba, Jamaica, and the Cayman Is., which has more restricted black on face and underparts. **SIMILAR SPECIES** Male distinctive; female nondescript with suggestion of the male face pattern. An escaped cagebird species, the Cuban Grassquit *(T. camorus),* is occasionally seen in southern FL. The species is endemic to Cuba and declining. It has a yellow frame around a

Illustration labels: West Indies *olivaceus*; black bib; ♀; adult ♂; bright yellow supercilium and throat; extensive black breast; mainland *pusillus*; adult ♂

black (male) or chestnut (female) face.
VOICE CALL: A high-pitched *sik* or *tsi*.
SONG: Thin, insectlike trills.

STATUS & DISTRIBUTION Common. Resident. **RANGE:** Middle and South America, Caribbean islands. **VAGRANT:**

Casual in southern FL and south TX (fewer than five recs. from each region). **POPULATION** Stable.

BLACK-FACED GRASSQUIT   *Tiaris bicolor*   BFGR   4

Behavior like the Yellow-faced Grass-quit's. Polytypic (8 ssp.; *bicolor* in N.A.). L 4.5" (11 cm)
IDENTIFICATION Very small; short tail; a small, conical bill and a straight cul-men. **ADULT MALE:** Black face and upper breast, dark olive elsewhere; black bill.

FEMALE: Extremely nondescript, pale gray below, gray-olive above; pale horn bill. **IMMATURE:** Like female.
GEOGRAPHIC VARIATION Multiple subspecies in Caribbean and north-ern S.A.; FL records pertain to Bahamas/Cuba subspecies, *bicolor.*
SIMILAR SPECIES Male unmistak-able; compare with Yellow-faced Grassquit. Sparrows and *Passerina* buntings are larger and are more patterned about the wings, breast, and face.
VOICE CALL: A lisping *tst.* **SONG:** A buzzing *tik-zeee.*
STATUS & DISTRIBUTION Com-mon. Resident. **RANGE:** West Indies (largely absent from Cuba). **VAGRANT:** Casual in southeastern FL (eight scattered records).

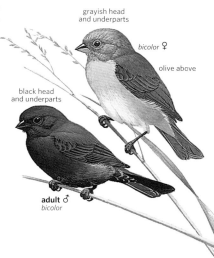

grayish head and underparts

bicolor ♀

olive above

black head and underparts

adult ♂
bicolor

Genus *Arremonops*

Of the four species in this genus, only the Olive Sparrow reaches the U.S.; the other species occur in Middle and South America. All are greenish on the back with head stripes; they tend to be fairly secretive, preferring dense brush where they forage on the ground. They are nonmigratory, typically occurring in pairs (sexes similar), and never form flocks.

OLIVE SPARROW   *Arremonops rufivirgatus*   OLSP   2

This fairly secretive sparrow stays close to dense cover, and it rapidly darts for the undergrowth when it is startled. Typically quite vocal, the Olive Sparrow can be heard call-ing often and singing at any season. Polytypic (9 ssp.; nominate in N.A.). L 6.3" (16 cm)
IDENTIFICATION Olive's plumage is a dull olive above; underparts are an unmarked pale gray. A brown stripe is found on each side of Olive's crown.
JUVENILE: Buffier than adult, with pale wing bars. Neck and breast are both faintly streaked.
GEOGRAPHIC VARIATION Of about nine subspecies in Middle America, only *rufivirgatus* occurs in the United States.
SIMILAR SPECIES Green-tailed Towhee is larger, and it has a distinct rufous cap. It also has a brighter greenish coloring on back, wings, and tail, as well as a white throat.
VOICE CALL: A dry *chip,* given singly or repeated in a rapid series when

agitated. Also a buzzy *speeee.* **SONG:** An accelerating series of dry *chip* notes.
STATUS & DISTRIBUTION Common. Non-migratory. Southern TX to Costa Rica. **YEAR-ROUND:** Dense undergrowth, brushy areas, mes-quite thickets, live oak. **VAGRANT:** Occurs casually slightly north of mapped range.
POPULATION Although still common in its native habitat, Olive Sparrow has suffered significantly in the U.S. as mesquite thorn forest has given way to agriculture and residential

uses, especially in the Lower Rio Grande Valley.

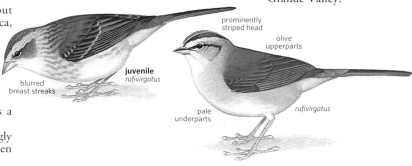

prominently striped head

olive upperparts

juvenile
rufivirgatus

blurred breast streaks

pale underparts

rufivirgatus

TOWHEES Genera *Pipilo* and *Melozone*

The six species of North American towhees were all formerly placed in *Pipilo*. Recently three species—Canyon, California, and Abert's—were moved to *Melozone*, which includes three species of ground sparrows from Middle America. Towhees are large sparrows with long, rounded tails. They feed on the ground, scratching with both feet at once.

SPOTTED TOWHEE *Pipilo maculatus* SPTO ▮ 1

The Spotted and Eastern Towhees (formerly combined as the Rufous-sided Towhee) have a narrow hybrid zone in the central Great Plains. In general, a female Spotted differs less from a male than in Eastern. Polytypic (20 ssp.; 9 in N.A.). L 7.5" (19 cm)

IDENTIFICATION MALE: Plumage like Eastern Towhee's, except for white tips on median, greater coverts forming two white wing bars, variable white spotting on back and scapulars, lack of rectangular white patch at primary bases. **FEMALE:** Similar to male Spotted, but with a slate-gray hood (variable by ssp.). **JUVENILE:** Like juvenile Eastern, but lacks white primary patch.

GEOGRAPHIC VARIATION Nine N.A. subspecies show weak to moderate variation. "Interior" group birds have extensive white spotting above, prominent white tail corners; "Coastal" group birds dark overall with white back spotting, reduced tail corners (variable). Subspecies *oregonus* is darkest; white increases southward to *megalonyx*. Within the "Interior" group, *arcticus* shows the most white spotting and most extensive white corners to tail; other subspecies have less white, but are similar to one another. Head color of females varies geographically from brownish to blackish.

SIMILAR SPECIES Distinguished from Eastern Towhee by white spotting on back, white wing bars, lack of patch at bases of primaries, and call. Hybrids occur in the Great Plains and winter to the south; often combine characteristics of both parents.

VOICE Song and calls also show great geographical variation. **CALL:** A descending and raspy mewing in *montanus;* an upslurred, questioning *queee* in *arcticus* and coastal subspecies. All subspecies also give a high, thin lisping *szeeueet* that drops in middle (poss. flight note), like Eastern's call, and various *chip*'s when agitated. **SONG:** "Interior" group gives introductory notes, then a

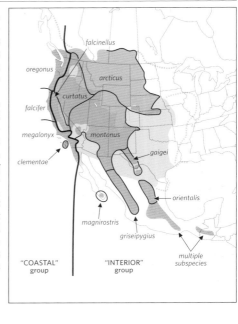

falcinellus

oregonus

arcticus

curtatus

falcifer

megalonyx

montanus

gaigei

clementae

orientalis

magnirostris

griseipygius

multiple subspecies

"COASTAL" group   "INTERIOR" group

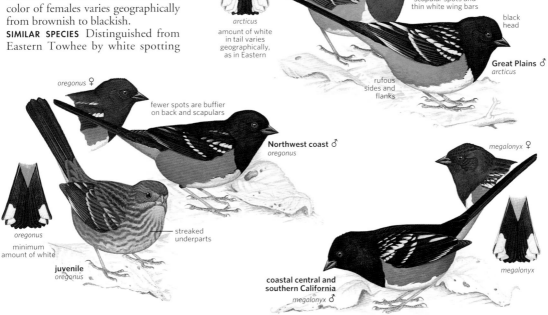

extensive white

arcticus ♀

slaty-brown head

red iris

white back and scapular spots and thin white wing bars

black head

arcticus
amount of white in tail varies geographically, as in Eastern

Great Plains ♂
arcticus

rufous sides and flanks

oregonus ♀

fewer spots are buffier on back and scapulars

Northwest coast ♂
oregonus

megalonyx ♀

oregonus
minimum amount of white

streaked underparts

juvenile
oregonus

coastal central and southern California
megalonyx ♂

megalonyx

montanus

Rockies/
Great Basin ♂
montanus

in most
subspecies
of Spotted,
females have
blackish
head with
grayish cast

montanus ♀

trill. "Pacific" birds sing a simple trill of variable speed.

STATUS & DISTRIBUTION Common. Some populations are largely resident; others are migratory. The most migratory subspecies is *arcticus.* Resident south to Guatemala. Subspecies are *oregonus* (OR to BC), *falcifer* (coastal northwest CA), *megalonyx* (coastal central to southern CA), *clementae* (certain Channel Islands); *arcticus* (Great Plains), *montanus* (Rocky Mountains),

falcinellus (south-central CA to OR), *curtatus* (primarily in Sierra Nevada), and *gaigei* (resident in mountains of southeastern NM and western TX). **MIGRATION:** Fall primarily Sept.–Oct.; spring Mar.–early May. **VAGRANT:** Subspecies *arcticus* is casual to East.
POPULATION Stable in most areas. Subspecies *clementae* is extirpated from San Clemente I., one of California's Channel Is., due to overgrazing by introduced goats; persists on Santa Catalina and Santa Rosa Is. Another island subspecies *(consobrinus)* from Guadalupe I., off Baja California, is extinct.

EASTERN TOWHEE   Pipilo erythrophthalmus   EATO ▪ 1

The Eastern behaves similarly to the Spotted Towhee. Polytypic (4 ssp.; all in N.A.) L 7.5" (19 cm)
IDENTIFICATION Conspicuous white corners on tail and white patch at base of primaries. **MALE:** Black upperparts, hood; rufous sides, white underparts. **FEMALE:** Black areas replaced by brown. **JUVENILE:** Brownish streaks below.
GEOGRAPHIC VARIATION Four subspecies show weak to moderate variation. Bird's overall size and extent of white in its wings and tail decline from northern part of range to the Gulf Coast; bill, leg, and foot sizes increase. Large nominate subspecies (breeds in North) has red irides, most extensive white in tail. Smaller *alleni* of Florida paler and duller, with straw-colored irides. Intermediate southern subspecies

canaster (west) and *rileyi* (east) have variably orange to straw-colored irides.
SIMILAR SPECIES See Spotted Towhee.
VOICE CALL: Emphatic, upslurred *chewink;* in *alleni,* a clearer, even-pitched or upslurred *swee.* Also a high-pitched *szeeueet,* dropping in middle (poss. flight note). Various *chip*'s when agitated. **SONG:** Loud ringing *drink your tea,* sometimes with additional notes at beginning or shortened to *drink tea.*
STATUS & DISTRIBUTION Fairly common. **BREEDING:** Partial to second growth with dense shrubs and extensive leaf litter, coastal scrub or sand dune ridges, and mature southern pinelands. **MIGRATION:** Resident, except for partially migratory nominate subspecies. Migration primarily Oct. and Mar. **VAGRANT:** Casual to

CO and NM. Accidental to AZ, ID, and Europe.
POPULATION Recent declines, especially in North, are due to urbanization. Southern populations more stable.

juvenile
erythrophthalmus

erythrophthalmus ♀

milk chocolate
upperparts and head

yellowish
white iris

Florida ♂
alleni

black head
and back

limited
white

alleni

extensive
white

streaked
underparts

white base to
primaries in all
plumages

rufous
sides

erythrophthalmus ♂

erythrophthalmus

GREEN-TAILED TOWHEE   Pipilo chlorurus   GTTO ▪ 1

The ground-loving Green-tailed Towhee emerges infrequently in the open to feed. In breeding season, males sing from exposed

perches. Monotypic. L 7.3" (19 cm)
IDENTIFICATION Olive upperparts and tail; gray head and underparts, fading to whitish belly. Distinct head pattern

with obvious reddish crown, white loral spot, distinct white throat, dark moustachial stripe, and white malar stripe. **JUVENILE:** Two faint wing bars, streaked plumage overall, olive-tinged upperparts. Lacks reddish crown. **SIMILAR SPECIES** Chipping Sparrow lacks green back and tail, is drastically smaller and more arboreal, and has a notched tail. See Olive Sparrow. **VOICE CALL:** Catlike *mew*, like Spotted Towhee's but clearer and more

two-parted. Also a thin, high *tseeee* (poss. flight note) and varied *chip's* when excited. **SONG:** Whistled notes beginning with *weet-chur* and ending in a raspy trill. **STATUS & DISTRIBUTION** Fairly common. Breeds south to northern Baja California. **BREEDING:** Dense brush and chaparral on mountainsides, high

plateaus, and sage steppes. **MIGRATION:** Spring late Mar.–mid-May, most mid-Apr.–early May. Fall July–Oct., peaking Sept.–early Oct. Uncommon migrant generally, rare along West Coast. **WINTER:** South to central Mexico; favors brushy draws and desert thickets. **VAGRANT:** Casual in fall and winter throughout the East. **POPULATION** Has probably declined as sagebrush steppes have been converted to agricultural and grazing land.

olive wings and tail

dull rufous crown

white supraloral and submoustachial

white throat
gray chest

spring

fall

extensively streaked below

juvenile

CANYON TOWHEE *Melozone fusca*   CANT   1

As its name implies, this species is common on shallow, rocky canyon slopes and rimrock in the Southwest. It is similar to the California Towhee, and the two were formerly considered the same species—the Brown Towhee. Polytypic (10 ssp.; 3 in N.A.). L 8" (20 cm)

IDENTIFICATION Plumage is pale gray-brown, fading to whitish on belly, with cinnamon-buff undertail coverts. Rufous-brown cap, buffy eye ring, buffy throat framed by necklace of black streaks typically forming black spot at base of throat. **JUVENILE:** Lacks rufous crown, has narrow buff wing-

bars, and is faintly streaked below. **GEOGRAPHIC VARIATION** Three N.A. subspecies show weak and clinal variation in measurements, overall coloration, and prominence of rufous cap. Two small, dark subspecies inhabit central and southwestern Texas *(texanus)* and the Sonoran and Chihuahuan Deserts of Arizona, New Mexico, and western Texas *(mesoleucus);* the more westerly subspecies has a much stronger rufous cap. Northern *mesatus* is large and pale, with a brown cap that is tinged rufous. **SIMILAR SPECIES** California Towhee has never been known to overlap in range, even as a vagrant. Compared to California Towhee, Canyon is paler, grayish rather than brown; it has a shorter tail and more contrast in reddish crown, giving a capped appearance. Crown is sometimes raised as a short crest. Canyon has a larger whitish belly patch with a diffuse dark spot at its junction with breast, a

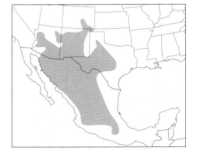

paler throat bordered by finer streaks, lores same color as cheek, and a distinct buffy eye ring. Songs and calls are also very distinctive. See Abert's Towhee. **VOICE CALL:** Shrill *chee-yep* or *chedup*. **SONG:** More musical, less metallic, than California's; opens with a call note, followed by sweet slurred notes. Also gives duet of lisping and squealing notes, like California Towhee. **STATUS & DISTRIBUTION** Common. Resident; no regular movements. **YEAR-ROUND:** Arid, hilly country; desert canyons. **VAGRANT:** Casual even a short distance out of range to southeastern UT and southwestern KS. **POPULATION** Stable.

pale rufous crown

buffy throat

mesoleucus

whitish belly with fairly conspicuous breast spot

CALIFORNIA TOWHEE *Melozone crissalis*   CALT   1

A widespread, common denizen of chaparral through most of coastal California, this species often occurs

in pairs year-round, like the Canyon and the Abert's Towhees. California Towhee and the similar Canyon

were formerly treated as conspecific, though they have never been known to overlap in range, even as vagrants.

Polytypic (8 ssp.; 6 in N.A.). L 9" (23 cm)
IDENTIFICATION Brownish overall. Buff throat bordered by a distinct broken ring of dark brown spots; no dark spot on breast, unlike Canyon. Lores same color as throat, contrast with cheek; warm cinnamon undertail coverts. **JUVENILE:** Faint cinnamon wing bars; faint streaking below.
GEOGRAPHIC VARIATION Six subspecies in U.S. show weak and clinal variation in size and overall coloration. Generally, size decreases from north to south, with the three inland subspecies *(bullatus, carolae, eremophilus)* averaging larger than the three coastal subspecies *(petulans, crissalis, senicula).* Coloration is generally darker to the north and paler to the south, but is fairly dark in *senicula* of coastal southern California.
SIMILAR SPECIES See Canyon and Abert's Towhees.
VOICE CALL: Sharp metallic *chink* notes; also gives some thin, lispy notes and an excited, squealing series of notes, often delivered as a duet by a pair. **SONG:** Accelerating *chink* notes with stutters in the middle.
STATUS & DISTRIBUTION Common. Resident; no known movements. **YEAR-ROUND:** Chaparral, coastal scrub, riparian thickets, parks, and gardens.
POPULATION Stable, except for the federally threatened subspecies *eremophilus*, which is limited to Inyo County, CA.

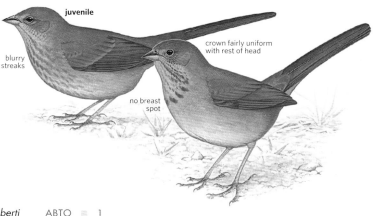

juvenile

blurry streaks

no breast spot

crown fairly uniform with rest of head

ABERT'S TOWHEE *Melozone aberti* ABTO 1

This towhee of the Southwest's low deserts can be fairly secretive inside thickets but is locally quite common, especially along the Colorado River and at the Salton Sea. It is similar to the Canyon and California Towhees, but its range does not overlap with the California Towhee, and the Canyon uses different habitats where it overlaps with Abert's. Polytypic (2 ssp.; both in N.A.). L 9.5" (24 cm)
IDENTIFICATION A pale sandy, gray-brown, overall; black face (particularly lores); warm brown upperparts, paler underparts; with cinnamon undertail coverts; tail contrastingly darker.
GEOGRAPHIC VARIATION The two subspecies are weakly defined and probably not separable in the field. The nominate occurs in southeastern Arizona, extending barely into New Mexico. Subspecies *dumeticolus* occupies the rest of the range, including western Arizona, southeastern California (and northwestern Mexico) to southern Nevada and southwestern Utah. Compared to the nominate, it is paler, with a faint reddish tinge to upperparts and underparts.
SIMILAR SPECIES Canyon and California Towhees are similar, but only the former overlaps in range. Prominent black face, lack of streaking on throat, and lack of a contrasting cap easily separate Abert's from both species. Note also habitat differences: Canyon is found on slopes and hills; Abert's prefers moister, lower-lying areas, including mesquite thickets and riparian scrub.
VOICE CALL: Piping or shrill *eeek*. Also high *seeep;* various *chip*'s when agitated. **SONG:** An accelerating series of *peek* notes, often ending in a jumble; frequently sings in a duet.
STATUS & DISTRIBUTION Common. Resident; no known movements. Occurs south only to northern Sonora and Baja California. Inhabits desert woodlands, mesquite thickets, riparian growth, orchards, suburban yards. Found typically at lower altitudes than the similar Canyon.
POPULATION Abert's Towhee persists throughout its range, but it has declined rangewide due to the severe degradation of riparian habitat caused by the overuse of water in the Southwest, by grazing, and by the increase of invasive exotic plants. These impacts have been especially marked along the Colorado River.

pale bill

overall warm brown coloration

blackish around bill

The Generic Approach to Sparrow Identification

Identifying sparrows uses somewhat different skills from those needed for the typical approach to bird identification. To identify a sparrow, consider first its shape, then its habitat and habits, and finally its field marks. Each genus has a distinctive silhouette; note especially tail length and shape. For example, *Ammodramus* has a flat head, big bill, and short tail, whereas *Spizella* is a round-headed, small-billed bird with a long, notched tail. Habitat is also useful: *Ammodramus* favors short-grass fields and marshes; *Melospiza* is found in brushy or weedy fields or tall grass. A hard-to-flush, skulking sparrow is likely an *Ammodramus* or *Aimophila*. Certain others (e.g.,

Le Conte's Sparrow, genus *Ammodramus*

Chipping Sparrow, genus *Spizella*

Spizella, Melospiza, Zonotrichia) are often conspicuous. Plumage characters to note include its appearance from below (streaky or plain), the tail pattern, and the exact face pattern (eye ring, malar stripes, and lateral or median crown stripes). ∎

Genera *Aimophila* and *Peucaea*

During the breeding season, territorial males sing from exposed perches. At other seasons they are secretive as they forage on the ground in dense grass. Individuals do not flock, but rather travel alone or in pairs. Upon flushing, they usually fly directly away, often pumping their long rounded tails like *Melospiza* sparrows. They fly low over the vegetation before diving back into cover; sometimes they will perch in a bush. Other than contact notes, they are silent outside the breeding season.

RUFOUS-CROWNED SPARROW *Aimophila ruficeps*   RCSP ∎ 1

Habitat is one of the best clues for this species: It is closely tied to dry, rocky slopes. The Rufous-crowned Sparrow feeds on the ground and might flush away from an observer into a bush or cactus, but it often sits up on rocks or bushes to survey the area. Its distinctive song and calls aid detection. This species, like most other *Aimophila,* is almost always found singly or in pairs and does not occur in flocks (though family groups may forage together in late spring and summer). It is considerably less

secretive, even when not singing, than other *Aimophila*. Polytypic (18 ssp.; 6 in N.A.). L 6" (15 cm). **IDENTIFICATION** Gray head with dark reddish crown, distinct whitish eye ring, rufous line extending back from eye, and single black malar stripe on each side of face. Gray-brown above, with reddish streaks; gray-brown below; tail long, rounded. **JUVENILE:** Buffier above, with streaked breast and crown; may show two pale wing bars.

GEOGRAPHIC VARIATION Eighteen subspecies show moderate variation in size and coloration. Six U.S. subspecies fall into two groups: the small, warm-toned, "Pacific coast" group and the large, pale "southwestern" group. Birds of the "Pacific coast" group are small, with reddish upperparts; they include the

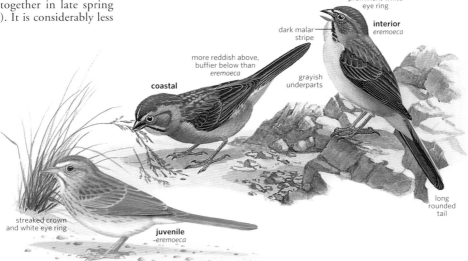
dull rufous crown; prominent white eye ring

interior
eremoeca

dark malar stripe

more reddish above, buffier below than *eremoeca*

coastal

grayish underparts

long rounded tail

streaked crown and white eye ring

juvenile
eremoeca

northern *ruficeps,* which is smaller and somewhat warmer in color than the *canescens* of southern California; the subspecies *obscura* is similar to *canescens,* but is limited to southern California's Channel Islands. The "southwestern" group includes the pale gray *eremoeca,* which is found over much of the range's eastern interior, the widespread southwestern subspecies *scottii,* which is pale and reddish, and the slightly darker *rupicola* from southwestern AZ.

SIMILAR SPECIES Similar to the smaller Rufous-winged, but Rufous-crowned Sparrow has a larger bill, just one whisker mark, a contrasting white malar, and a bolder eye ring; it also lacks Rufous-winged's rufous shoulder and pale mandible. Chipping Sparrow and other *Spizella* are superficially similar, but shape and behavior of Rufous-crowned are distinctive.

VOICE Song and calls are extremely helpful for detecting and identifying this species. **CALL:** A distinctive, sharp, *dear,* often given in a series; also drawn-out *seep* notes (poss. flight note). **SONG:** A rolling, bubbly, series of rapid *chip* notes; can sound very similar to House Wren's song, but note Rufous-crowned's shorter duration and more explosive quality.

STATUS & DISTRIBUTION Fairly common. Occurs south to southern Mexico. **YEAR-ROUND:** Strongly prefers rocky hillsides and steep grassy slopes with areas of open ground or bare rock. Largely resident, with some limited local movements downslope. **VAGRANT:** Very rare or casual to central or southern Great Plains; accidental to northeastern Kern Co., CA and, remarkably, to WI.

POPULATION Fire suppression and development have degraded the scrub habitats that this species prefers in Texas and, especially, in southern California *(canescens).* Some southern California populations occur in highly imperiled coastal sage-scrub habitat. Populations in Arkansas and eastern Oklahoma are small and isolated.

RUFOUS-WINGED SPARROW *Peucaea carpalis* RWSP ▪ 2

light rufous crown with fine black streaks

broad pale gray eye-brow with black eye line

distinct blackish malar and moustachial stripes

bicolored bill

rufous lesser coverts usually hidden

white underparts

all *Peucaea* sparrows have large bills and long, rounded tails

all *Peucaea* sparrows but Rufous-winged tend to be secretive, except for singing territorial males

juvenile

pattern suggestive of adult but duller and streaked below

In the small section of southeast Arizona where this species occurs, it is uncommon to fairly common. It sings from exposed perches, but even when it is not singing, it tends to be more conspicuous than other *Peucaea.* It often flies to the tops of bushes and fence lines when flushed. Polytypic (2 ssp.; nominate in N.A.). L 5.8" (15 cm)

IDENTIFICATION ADULT: Pale gray head with reddish eye line and cap; faint gray median crown stripe. Distinctive double whisker is the result of both black moustachial and malar stripes on face. Two-toned bill, with pale mandible. Gray-brown back, streaked with black; whitish underparts; two whitish wing bars. Reddish lesser

coverts distinctive but usually difficult to see as they are covered by the scapulars. Grayish white underparts, without streaking. **JUVENILE:** Less-distinct facial stripes; buffier wing bars; darker bill; lightly streaked, whitish breast and sides; plumage can be seen as late as Nov.

GEOGRAPHIC VARIATION Nominate *carpalis* in the U.S. is larger than the Mexican subspecies *cohaerens.*

SIMILAR SPECIES Small bill and body, long tail, and rufous cap could recall Chipping Sparrow (or other *Spizella*), but note that tail is rounded rather than notched. Color of eye line also helps eliminate Chipping, as does double whisker mark. Rufous-winged is not found is flocks, unlike

Chipping and the other *Spizella.* See also Rufous-crowned Sparrow, another non-flocking species.

VOICE CALL: Distinctive, sharp, high *seep.* **SONG:** Two primary song types: an accelerating series of sweet *chip* notes lasting several seconds, and a shorter song consisting of 2–4 *chip* notes followed by fast trill, *tink tink tidleeeee.*

STATUS & DISTRIBUTION Uncommon to fairly common, but local. Almost no migration or vagrancy known, but recently discovered in Guadalupe Canyon, NM. Occurs south to central Sinaloa in western Mexico. **YEAR-ROUND:** Rather level areas of desert grassland mixed with brush, shrubs, and cactus; often occurs along washes.

POPULATION Heavy grazing has impacted habitat of this species, and it is absent from many areas that it historically occupied. It is considered an indicator species for healthy grassland.

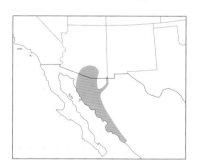

BOTTERI'S SPARROW *Peucaea botterii* BOSP ■ 2

Behavior like Cassin's Sparrow, but does not skylark. Polytypic (9 ssp.; 2–3 in N.A.). L 6" (15 cm)

IDENTIFICATION Large and plain with large bill; fairly flat forehead; long, rounded, dusky-brown tail. Gray upperparts streaked with dull black, rust, or brown; underparts unstreaked; whitish throat and belly; grayish buff breast and sides. **JUVENILE:** Buffy belly, broadly streaked breast, narrowly streaked sides.

GEOGRAPHIC VARIATION Nine subspecies show marked variation in measurements and upperparts coloration. The subspecies *arizonae*, breeding in southeastern Arizona (and extreme southwestern New Mexico) and northern Mexico, is more reddish above and buff on breast; *texana*, of

extreme southern Texas and northeastern Mexico, is slightly grayer.

SIMILAR SPECIES Best identified by voice. See Cassin's and Bachman's Sparrows.

VOICE CALL: Variable high *tsip* notes; also a high, piercing *seep* (poss. flight call). **SONG:** Several accelerating high *tsip* or *che-lik* notes, followed by two short trills and a longer series of notes accelerating into a trill, like a bouncing ball. A secondary song type (more prevalent in late summer) omits accelerating series. May perform perch-to-perch or perch-to-ground song flights, but does not skylark like Cassin's does.

STATUS & DISTRIBUTION Uncommon. Occurs south to Nicaragua. **BREEDING:** Fairly tall grasslands and prairies, often shrubby; also open grassy woodlands. Subspecies *arizonae* closely tied to tall, dense grasslands; *texana* found in coastal prairies. **MIGRATION:** Migrants

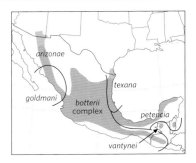

return late Mar. *(texana)* or early May *(arizonae)*, depart by Oct. **VAGRANT:** Accidental (has bred; likely *mexicana* ssp. from nearby north-central Mexico on probability) in west TX.

POPULATION Both subspecies found in the United States have declined significantly because of degradation of their grassland habitat as a result of grazing, development, and the conversion to agriculture.

head pattern more poorly defined than Bachman's

redder above than *texana*

arizonae

streaked upperparts

slightly larger bill than Cassin's

buffy, unstreaked rear flanks

texana

streaked below

juvenile *texana*

CASSIN'S SPARROW *Peucaea cassinii* CASP ■ 1

Its melodic song and skylarking display flight are distinctive. When not singing it is extremely secretive,

whitish eye ring

dark "anchors" on upperpart feathers

whitish tertial fringes

faint flank streaks

overall appears like a large Brewer's

streaked below

juvenile

flushing only when it is almost trod upon. When flushed it behaves like Botteri's and Bachman's (also in *Peucaea*), flying low away from the observer and rarely shows itself again. Monotypic. L 6" (15 cm)

IDENTIFICATION A large, drab, grayish sparrow, with a large bill and fairly flat forehead. Long, rounded tail is dark gray-brown; outer tail feathers have indistinct white tips, less apparent with worn. Gray upperparts are streaked with dull black, brown, and variable amount of rust (a few are decidely rusty above); underparts are grayish white, usually with a few short streaks on lower flanks. **JUVENILE:** Streaked below; paler overall than juvenile Botteri's Sparrow.

SIMILAR SPECIES Best

identified by voice. Botteri's Sparrow is generally browner, and it lacks dark flank streaks and barred tail with pale tail corners. Cassin's Sparrow has distinctive black crescents on its uppertail coverts and scapulars (where Botteri's Sparrow

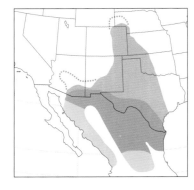

is streaked), its wings are more patterned, and its white-fringed tertials have black centers. Botteri's does not have a song flight.
VOICE CALL: High *stit* given when excited; also *psyit* call (poss. flight call). **SONG:** Often given in brief fluttery song flight, in which the bird rises to a height and then floats down. Song typically begins with a soft double whistle, a loud, sweet trill, a low whistle, and a final, slightly higher note; alternate versions include a series of *chip*'s ending in a trill or warbles. Also gives a trill of *pit* notes.
STATUS & DISTRIBUTION Fairly common; variable abundance in response to rainfall. Occurs south to central Mexico. **BREEDING:** Breeds in grasslands with scattered shrubs, cactus, yucca, and mesquite. **MIGRATION:** Returns to breeding grounds in early Apr. or later, departs Sept.–early Oct. **VAGRANT:** Casual to the Midwest,

whitish tail corners

East, CA (numbers were found in the east Mojave, San Bernadino Co., in spring 1978 after heavy winter rainfall), and NV.
POPULATION Loss of grasslands has caused declines rangewide.

BACHMAN'S SPARROW *Peucaea aestivalis*   BACS ■ 2

This highly secretive species has been known to crawl down gopher holes to escape pursuit. Outside the breeding season, Bachman's and Cassin's Sparrows behave similarly, but in spring and summer Bachman's sings from high perches, often an open pine branch, and it does not perform song flights. Polytypic (3 ssp.; all in N.A.). L 6" (15 cm)
IDENTIFICATION This large sparrow has a large bill, a fairly flat forehead, and a long, rounded tail. It is gray above, heavily streaked with chestnut or dark brown; sides of head are buff or gray; belly whitish. **JUVENILE:** Distinct eye ring; streaked throat, breast, and sides, often into first winter.
GEOGRAPHIC VARIATION Three subspecies show moderate color variation. Subspecies *bachmani* breeds in northeastern portions of the range and is very similar to the slightly richer reddish *illinoensis* from the western portions of breeding range. Southeastern *aestivalis* (SC to FL) is darkest and grayest.
SIMILAR SPECIES Botteri's Sparrow does not overlap in range and differs in voice. It also has richer coloration and more-patterned wings. Vagrant

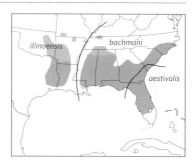

illinoensis   *bachmani*

aestivalis

Cassin's and Bachman's might overlap; note Cassin's white tail corners and drabber plumage. Swamp Sparrow is similar in color but of entirely different shape, with a grayer face and darker back. Note voice differences.
VOICE CALL: High *tsit*. **SONG:** One clear, whistled introductory note, followed by a trill or warble on a different pitch. Pitch and speed of trill highly variable. The song is ventriloquial, and singing bird can be hard to locate.
STATUS & DISTRIBUTION Uncommon. Apparently extended its range north in the early to mid-1900s (probably because farm abandonment and logging created vast areas of early successional forest), occurring north to PA, MD, OH, IN, IL, and south ON; it has since greatly contracted its range. **YEAR-ROUND:** Inhabits dry, open, grassy woods, especially pines, and scrub palmetto. Also occupies regenerating pine forests that are 7–20 feet high. **MIGRATION:** Northern populations are migratory; timing is poorly known, and it is very rarely detected in migration, no doubt due to its secretive nature. Returns to northern breeding grounds in Apr. or May; fall migration occurs early Aug.–mid-Oct. **VAGRANT:** Accidental to KS, MI, NJ, NY, and throughout its former range.
POPULATION Considered near threatened by BirdLife International. Formerly common in open, old-growth pine forests, but that habitat has been entirely lost to logging. Bachman's Sparrows currently persist in young pine stands (less than 15 years old) and what few older stands (more than 70 years old) remain in the Southeast. Fire suppression and cutting of pine stands favor shrubby undergrowth rather than the grass that Bachman's prefers.

brighter rufous above than nominate subspecies

illinoensis

buffy breast, sides, and flanks

prominent head and back streaks

aestivalis

streaked below

juvenile
aestivalis

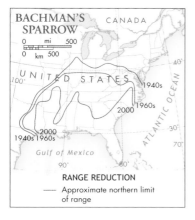

BACHMAN'S SPARROW

CANADA

0   mi   500
0   km   500

UNITED STATES

1940s
1960s
2000
2000
1940s 1960s

Gulf of Mexico

ATLANTIC OCEAN

RANGE REDUCTION
— Approximate northern limit of range

Genus *Spizella*

All seven species in this genus occur north of Mexico. They are all distinctive in that they are fairly small, slender sparrows with small heads and small bills. Their proportionately long tails are distinctly notched. They typically feed in flocks on or near the ground, and they fly immediately up to a tree or bush if disturbed. Their flight is light, buoyant, and undulating, with frequent delivery of their flight call. Their songs vary from clear whistled phrases to dry buzzes and rattles.

CHIPPING SPARROW *Spizella passerina* CHSP ▪ 1

Usually conspicuous, they may form sizable flocks during the nonbreeding seasons, often mixing with juncos, Lark or Clay-colored Sparrows; Pine, Yellow-rumped or Palm Warblers; or bluebirds. They frequently give their distinctive flight note upon flushing, usually flying up to a tree or other elevated perch to survey the intrusion. Polytypic (5 ssp.; 2 in N.A.). L 5.5" (14 cm)

IDENTIFICATION In all plumages dark eye stripe extends through lores to base of bill, gray nape and cheek, gray unstreaked rump, two white wing bars, and lack of a prominent malar stripe. **BREEDING ADULT:** Bright chestnut crown, distinct white eyebrow. **WINTER ADULT:** Browner cheek, streaked crown with some rufous color. **FIRST-WINTER:** Similar to winter adult, but brownish crown; buff-tinged breast and sides. **JUVENILE:** Underparts prominently streaked; crown usually lacks rufous; may show slightly streaked rump. Plumage often held into October, especially in western subspecies. **GEOGRAPHIC VARIATION** Five subspecies show moderate variation in color

and measurements. Nominate eastern subspecies is small and fairly dark, with rich rufous upperparts. Western subspecies *arizonae* is larger and paler (breeds from Great Plains west).
SIMILAR SPECIES Clay-colored and Brewer's Sparrows differ from winter and immature Chippings by their pale lores, prominent malar and submoustachial stripes (particularly Clay-colored), and brownish rumps, though the rump is often concealed by the folded wings; lack chestnut on cap. Brewer's has a streaked nape and rump, and a duller face pattern with a more distinct eye ring; Clay-colored is typically warmer buff-brown on breast, especially in fall, and has a broader, pale supercilium and a more strongly contrasting gray nape.
VOICE CALL: High *tsip;* sometimes a rapid twitter when excited. **FLIGHT NOTE:** High, sharp *tseet;* sharper at beginning than in Brewer's or Clay-colored. **SONG:** Rapid trill of dry *chip* notes, all one pitch; speed can vary considerably.

STATUS & DISTRIBUTION Common. Occurs south to Nicaragua. **BREEDING:** Lawns, parks, gardens, woodland edges, pine-oak forests. **MIGRATION:** Spring mid-Mar.–mid-May; fall late July–early Nov., peaking Sept.–late Oct. Rare in winter north of mapped range. **VAGRANT:** Casual to western AK. **POPULATION** Stable. Has largely benefited from human activities, including the clearing of forests and creation of open, grassy parks.

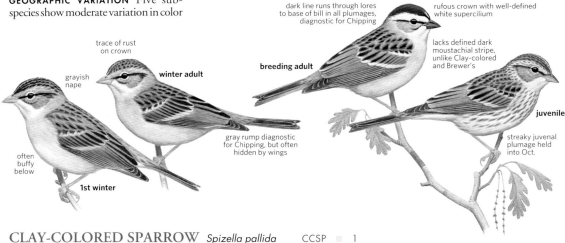

trace of rust on crown

grayish nape

winter adult

dark line runs through lores to base of bill in all plumages, diagnostic for Chipping

breeding adult

rufous crown with well-defined white supercilium

lacks defined dark moustachial stripe, unlike Clay-colored and Brewer's

juvenile

gray rump diagnostic for Chipping, but often hidden by wings

streaky juvenal plumage held into Oct.

often buffy below

1st winter

CLAY-COLORED SPARROW *Spizella pallida* CCSP ▪ 1

The Clay-colored's behavior is similar to the Chipping's. It may flock with that species or with Field Sparrows. Monotypic. L 5.5" (14 cm)
IDENTIFICATION Brown crown with black streaks and a distinct buffy white or whitish central stripe.

Broad, whitish eyebrow; pale lores; brown cheek outlined by dark postocular and moustachial stripes; conspicuous pale submoustachial stripe. Gray nape; buffy brown back and scapulars, with dark streaks; rump not streaked and color does not

contrast with back as in Chipping. Adult in fall and winter is buffier overall. **IMMATURE:** Much buffier; gray nape and pale stripe on sides of throat stand out more. **JUVENILE:** Breast and sides streaked.
SIMILAR SPECIES Identification of

fall *Spizella* can be quite challenging and should focus on details of head pattern. Brewer's Sparrow is most similar, but it is duller overall and less distinctly marked. Its face pattern is indistinct, lacking prominent white supercilium, whitish moustachial, and pale central crown stripe. Unlike Clay-colored, nape of Brewer's is streaked and does not contrast with head and breast. Clay-colored Sparrow usually shows a buffy wash across breast, especially in fall, while Brewer's is paler on breast. Winter and immature Chipping are also similar but tend to be darker on back, with chestnut tones rather than buff and tan. Chipping Sparrow's face is much more strongly marked and set off by a prominent dark eye line that extends through lores. Chipping Sparrow typically does not have Clay-colored's prominent whitish supercilium and moustachial, and its usually grayish breast is unlike Clay-colored's buff breast.

VOICE CALL: High *tsik,* given repeatedly when excited. **FLIGHT NOTE:** A thin *sip,* like Brewer's Sparrow. **SONG:** A series of three to four long, insectlike buzzes; can recall the song of Golden-winged Warbler.

STATUS & DISTRIBUTION Fairly common. Winters south to southern Mex. **BREEDING:** Fairly common in brushy fields, groves, streamside thickets. **WINTER:** Primarily in Mexico, uncommonly in southern TX, rarely in southern FL and AZ. Casual in winter to both coasts. **MIGRATION:** Primarily through interior mid-Apr.–mid-May and late Aug.–late Oct. Rare in fall, casual in winter and spring to both coasts. **VAGRANT:** Casual in fall to AK, YT, and NF (also in winter).

POPULATION Stable.

bold head pattern, including strong median crown stripe and buffy cast to plumage

buffy auricular

breeding

all have pale lores, like Brewer's, but unlike Chipping

grayish nape contrasts with rest of buffy plumage

dark moustachial and broad, pale submoustachial stripes

streaking on underparts

juvenile

quite buffy below

fall

BREWER'S SPARROW *Spizella breweri* BRSP ■ 1

This is a sparrow of the dry Great Basin desert, intermontane valleys, and mountain meadows. In winter and migration it is usually found in dry, sparse desert scrub; the similar Clay-colored Sparrow more typically found in lusher, grassier habitats, though there is much overlap. Brewer's Sparrow may form flocks or mix with other sparrows in migration and winter. Polytypic (2 ssp.; both in N.A.). L 5.5" (14 cm)

IDENTIFICATION Brown crown with fine black streaks. Distinct whitish eye ring, grayish white eyebrow; overall indistinct face pattern. Pale brown ear patch with darker borders, pale lores, dark malar stripe. Buffy brown, streaked upperparts; buffy brown rump may be lightly streaked. Immature, fall adults, and winter adults are somewhat buffy below. **JUVENILE:** Buffier overall, lightly *(breweri)* to prominently *(taverneri)* streaked on breast and sides.

GEOGRAPHIC VARIATION Two subspecies show moderate variation in measurements and overall color and pattern. The widespread nominate occupies the breeding range across most of the continent, where it nests primarily in sage steppe habitat. The other subspecies, *taverneri,* known as "Timberline" Sparrow, has a largely disjunct breeding range and is thought by some to represent a different species. It breeds in the subalpine zone of the Canadian Rockies from northwestern Montana to east-central Alaska and is thought to winter in north-central Mexico. It has a slightly different song, a larger bill, heavier black streaking on nape and upperparts, a stronger face pattern, and a darker gray breast

rather distinct white eye ring and pale lores

head pattern like Clay-colored but more muted and without contrasting pale median crown stripe

breweri overall a sandy color, less rich than Clay-colored

pale brown rump

juvenile *breweri*

streaking on underparts, often rather faint

breeding *breweri*

taverneri

breweri

that contrasts more with the belly. Juveniles are heavily streaked with blackish below. Despite these average differences, reliable field identification may not be possible (except by breeding range and habitat) due to variation within *breweri*.

SIMILAR SPECIES Clay-colored Sparrow is most similar. Note Clay-colored's unmarked gray nape, which contrasts with face and breast, and more

prominent facial pattern with a well-defined median crown stripe, whitish moustachial, and bold supercilium.

VOICE CALL: High *tsik,* given repeatedly when excited. **FLIGHT NOTE:** A thin *sip,* like Clay-colored Sparrow's. **SONG:** A series of varied, bubbling notes and buzzy trills at different pitches; entire song is often very long in duration.

STATUS & DISTRIBUTION Common. Winters south to central Mexico. **BREEDING:** Sage steppe habitats across the West, especially in extensive

stands of big sagebrush *(Artemisia tridentata).* **WINTER:** Found in dry desert scrub, including saltbush and creosote; also weedy roadsides. **MIGRATION:** Spring Mar.–mid-May, peaking mid-Apr.–mid-May; fall early Aug.–early Oct., peaking in Sept. **VAGRANT:** Casual to Great Lakes. Accidental in fall to LA, MA, and NS; in spring to IL and ON; and in winter to IL.

POPULATION Breeding Bird Survey data show a gradual decline, presumably due to the loss of sage steppe habitats to grazing, agriculture, and the invasion of exotic plants.

larger than *breweri* with markings stronger and darker, more like Clay-colored

grayish auricular

"Timberline" *taverneri*

breeding

juvenile

streaks below heavier and darker than juvenile *breweri*

AMERICAN TREE SPARROW *Spizella arborea* ATSP 1

One of the hardiest sparrows, this is the only one likely to winter in much of the far northern U.S. and southern Canada, where the Dark-eyed Junco can also be found. At that season it is frequently mistaken for the Chipping Sparrow, but the two rarely overlap in winter. The American Tree Sparrow often occurs in flocks of up to 50 birds. In habitat and behavior, they are much like Field Sparrows, but American Tree Sparrows are more frequent at bird feeders. Polytypic (2 ssp.; both in N.A.). L 6.3" (16 cm)

IDENTIFICATION Gray head and nape crowned with rufous; rufous stripe behind eye; gray throat and breast, with dark central spot; rufous patches on sides of breast; sides of flanks rich buff and unmarked. Back and scapulars streaked with black and rufous. Outer tail feathers thinly edged in white on outer webs. Grayish white underparts with buffy sides. **WINTER:** More buffy; rufous color on crown sometimes forms a central stripe. **JUVENILE:** Streaked on head and underparts.

GEOGRAPHIC VARIATION Two subspecies show weak variation in measurements and overall coloration. The small, dark nominate subspecies breeds eastward from the eastern Northwest Territories and winters eastward from the central Great Plains. The western *ochracea* is larger and paler, and it winters from the central Great Plains west.

largest *Spizella*

breeding *arborea*

bicolored bill

rufous patch at breast sides and buffy flanks

winter *ochracea*

breast spot

juvenile *ochracea*

SIMILAR SPECIES See Field Sparrow. The smaller Chipping Sparrow rarely overlaps in range (except in certain areas in migration); it has a distinct dark eye line in all plumages.

VOICE CALL: Sharp, high, bell-like *tink;* sometimes with a more lispy quality (poss. flight note). Flocks also give a musical *teedle-eet.* **SONG:** Usually begins with several clear notes followed by a variable, rapid warble.

STATUS & DISTRIBUTION Fairly common. Uncommon to rare west of Rockies. **BREEDING:** Breeds along edge of tundra, in open areas with scattered trees, brush. **WINTER:** Weedy fields, marshes, groves of small trees. **MIGRATION:** One of the late-fall and early-spring migrants. Fall migration in U.S. typically mid- or late Oct.–late Nov.; spring migrants depart

mid-Mar.–early Apr.; accidental in U.S. south of Canada after late May (late Apr. in midlatitudes). **VAGRANT:** Casual to southern CA, central TX, and the Gulf Coast.

POPULATION Largely stable.

FIELD SPARROW *Spizella pusilla*   FISP ▪ 1

In migration and winter, the Field Sparrow forms small pure flocks (5–50 birds); it may also mix with other sparrows. It usually gives its flight call when flushed, typically employing a bounding flight to fly up ahead and drop back in the grass or fly into a hedgerow. Flocks will often gather in the same tree or bush to investigate a disturbance. Polytypic (2 ssp.; both in N.A.). L 5.8" (15 cm)
IDENTIFICATION Entirely pink bill is distinctive. Gray face with reddish crown, distinct white eye ring, and indistinct reddish eye line. Back is streaked except on gray-brown rump. Rich buffy-orange unstreaked breast and sides;

grayish white belly; pink legs. **JUVENILE:** Streaked below; buffy wing bars.
GEOGRAPHIC VARIATION Two subspecies show well-marked variation in measurements and overall coloration. Nominate breeds roughly east from eastern Dakotas and eastern Texas; *arenacea* breeds to west. The longer-tailed *arenacea* is larger, paler, and grayer; *pusilla* is especially rufous on auriculars, buffier below, and richer above.
SIMILAR SPECIES Larger American Tree has a two-toned bill and a black central breast spot. Breeding-plumaged Chipping shares rufous cap but has a dark eye line, is plain gray below without buff on breast or flanks, has a darker bill, and lacks bold eye ring. See also Worthen's Sparrow.
VOICE CALL: A high,

sharp *chip,* similar to call of Orange-crowned Warbler. **FLIGHT NOTE:** High, loud *tseees.* **SONG:** A series of clear, plaintive whistles accelerating into a trill.
STATUS & DISTRIBUTION Fairly common. Winters south to extreme northeast Mexico. **MIGRATION:** Spring mid-Mar.–early May, peaking mid-Apr.; fall, mainly mid- to late Oct. Rare migrant in east CO and Maritimes. **BREEDING:** Open, brushy woodlands, power-line cuts, overgrown fields. **WINTER:** Prefers open fields with tall grass, often near hedgerows. **VAGRANT:** Casual to NF. Casual to accidental west of mapped range as far as CA. **POPULATION** Stable.

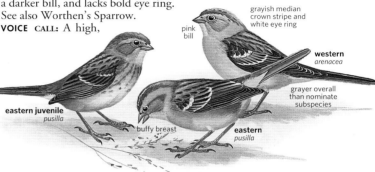

eastern juvenile *pusilla*

buffy breast

eastern *pusilla*

pink bill

grayish median crown stripe and white eye ring

western *arenacea*

grayer overall than nominate subspecies

BLACK-CHINNED SPARROW *Spizella atrogularis*   BCSP ▪ 1

This attractive sparrow is typical of chaparral and brushy slopes, where its distinctive song may be heard. Migrants are very rarely detected. In winter they may be found singly, in loose association with other sparrows, or in small flocks. Polytypic (4 ssp.; 3 in N.A.). L 5.8" (15 cm)
IDENTIFICATION Medium gray overall; rusty back and scapulars, with black streaks; bright pink bill. **BREEDING MALE:** Black lores and chin; whitish gray lower

belly. Long tail all dark. **FEMALE:** Black chin reduced or absent. **WINTER:** Lacks black chin. **JUVENILE:** Like winter, but faint streaks below.
GEOGRAPHIC VARIATION The four subspecies show slight variation in overall coloration and size. Subspecies *cana* breeds from south-central CA to northern Baja California; slightly darker *caurina,* breeds in central CA (Marin to San Benito Cos.); larger *evura* breeds east of the Sierra Nevada to AZ, NM, and western TX; and nominate *atrogularis* breeds in north-central Mexico.
SIMILAR SPECIES Medium gray underparts and rufous back of Black-chinned are distinctive; other *Spizella* are paler on underparts. Black-throated Sparrow is sometimes confused with Black-chinned Sparrow, due mostly to their similar names.
VOICE CALL: A high *tsik* is given when agitated; very similar to other *Spizella.* **SONG:** Plaintive song begins slowly and accelerates rapidly, like a bouncing ball. Begins with slow *sweet . . . sweet . . . sweet,* continuing in a rapid trill; somewhat like the song

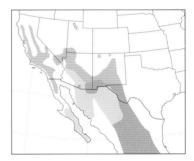

of Field Sparrow, but accelerates much more rapidly and ends with a faster trill. **FLIGHT NOTE:** High, thin *seep.*
STATUS & DISTRIBUTION Uncommon to fairly common. **BREEDING:** Inhabits brushy arid slopes in foothills and mountains. **MIGRATION:** Rarely detected in migration; spring arrival seems to peak in early Apr.; most birds seem to be gone from breeding grounds by late Aug., with a very few individuals noted in Sept. **VAGRANT:** Casual to central TX and accidental to southern OR.
POPULATION Mostly stable in CA from development.

breeding ♂

pink bill

dark chin

breeding ♀

streaked rufous back

medium gray head and underparts

juvenile

WORTHEN'S SPARROW *Spizella wortheni*   WOSP   5

The type specimen of this poorly known Mexican species was taken in the United States, but it has not been found there since. In many respects it

solid rusty crown

blackish legs and feet

is similar to the Field Sparrow. Monotypic. L 5.5" (14 cm)

IDENTIFICATION Bright pink bill, solid rufous cap, grayish face and underparts, and dark legs separate this species from Field Sparrow.

SIMILAR SPECIES Field Sparrow is similar but has pinkish legs, a less-solid but more extensive rufous cap (extends to forehead), and a different song.

VOICE CALL: High, thin *tssip* (poss. flight note); probably also high *chip*'s when excited. **SONG:** Dry, chipping trill suggesting Chipping Sparrow's song.

STATUS & DISTRIBUTION Very rare and local. Breeds on the Central

Mexican Plateau, primarily in Coahuila and particularly Nuevo Léon. Currently it is best known from one huge black-tailed prairie dog colony. **BREEDING:** Breeds in shrubby deserts, overgrown fields, grassy woodland edge. **VAGRANT:** One record from Silver City, NM (June 16, 1884); could have been a member of an isolated population that was quickly extirpated by overgrazing.

POPULATION Considered endangered by BirdLife International. Its population may be as small as several hundred birds. Decline is presumably a result of overgrazing, which degrades its breeding habitat.

Genus *Pooecetes*

VESPER SPARROW *Pooecetes gramineus*   VESP   1

The Vesper Sparrow (monotypic genus) is a bird of open country often found feeding along roadsides or in open areas within the grassland; it avoids thick brush. It often associates with Savannah, Brewer's, or Lark Sparrows. Polytypic (3 ssp.; all in N.A.). L 6.3" (16 cm)

IDENTIFICATION Large; pale underparts marked with fine and distinct but short streaks; prominent white eye ring; dark ear patch bordered in white along lower and rear edges; supercilium indistinct. Chestnut lesser coverts distinctive but often

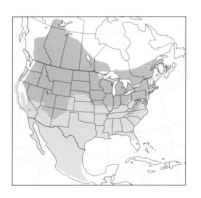

concealed. Distinctive white outer tail feathers best seen in flight.

GEOGRAPHIC VARIATION Three subspecies show moderate differences in measurements and overall coloration. Eastern nominate subspecies is smaller and slightly darker overall than widespread *confinis*. Small, dark *affinis* breeds in coastal dunes of western WA and OR.

SIMILAR SPECIES The shorter-tailed Savannah Sparrow is smaller, lacks eye ring and white outer tail feathers.

VOICE CALL: High *chip*'s when excited. **FLIGHT NOTE:** High, rising *pseeet;* calls less often in flight than does Savannah. **SONG:** Rich and melodious, two to three long, slurred notes followed by two higher notes, then a series of short, descending trills.

STATUS & DISTRIBUTION Fairly common (uncommon in East). Winters south to southern Mexico. Breeds in dry grasslands, farmlands, open forest clearings, sagebrush. **WINTER:** Dry grasslands, grassy margins of agricultural fields, roadsides. Prefers shorter grass with open areas interspersed.

pale wraps around dark-bordered auricular

bold white eye ring; no strong eye line as in Savannah

streaking on breast less extensive than smaller Savannah

eastern *gramineus*

western *confinis*

chestnut coverts usually concealed

gramineus

white outer tail feathers

paler, grayer than nominate subspecies

MIGRATION: Spring late Mar.–early May, peaking in mid-Apr.; fall early Sept.– mid-Nov., peaking late Sept.– late Oct; fall migration is later in the East. **VAGRANT:** Casual to NF.

POPULATION Moderate declines throughout its range are probably due to changes in farming practices and decline in quality of sage steppe and grassland habitats in the West, especially *affinis*.

Genus *Chondestes*

LARK SPARROW *Chondestes grammacus*   LASP   1

This sparrow (monotypic genus) feeds on bare ground. Few other sparrows are likely to fly high overhead during

daylight. They call both when flushing and when flying overhead. Gregarious in nonbreeding season. Polytypic

(2 ssp.; both in N.A.). L 6.5" (17 cm)
IDENTIFICATION Largest open-country sparrow. Long, rounded tail with

prominent white corners. **ADULT:** Distinctive harlequin face pattern of black, white, and chestnut. Bright white underparts marked only with dark central breast spot. **JUVENILE:** Duller face; fine, black streaks on breast, sides, and crown lost gradually through fall. **GEOGRAPHIC VARIATION** Two subspecies show weak variation. Eastern *grammacus* darker overall with wider black back streaks than western *strigatus*.
SIMILAR SPECIES None.

VOICE CALL: Sharp *tsik,* often a rapid series and frequently delivered in flight. High *chip*'s when excited. **SONG:** Begins with two loud, clear notes, followed by a series of rich, melodious notes and trills and unmusical buzzes. Sings one of the longest sparrow songs. **STATUS & DISTRIBUTION** Fairly common. Primarily west of the Mississippi;

once bred as far east as NY and western MD. Rare migrant throughout the East, mainly in fall. Prairies, roadsides, farms, open woodlands, mesas. **VAGRANT:** Accidental to AK, northern Canada, and Europe.
POPULATION Stable, though eastern populations are declining and it is extirpated from some former breeding areas in the East.

distinctive head pattern · *grammacus* · dark breast spot · whitish at base of primaries · confiding behavior · extensive white in tail · **juvenile** *grammacus* · subdued head pattern · streaked below

---

Genus *Amphispiza*

Two species are currently recognized in this genus—the medium-size Black-throated Sparrow and the larger Five-striped Sparrow. Although they have striped faces and distinctive throat patterns, they strongly differ in behavior and all vocalizations.

FIVE-STRIPED SPARROW · *Amphispiza quinquestriata* · FSSP · 3

Extremely local in southernmost Arizona. The first U.S. breeding was confirmed only in 1969. The small populations appear to be somewhat cyclical or irruptive; only a few locations are occupied consistently. During breeding season territorial males can be found singing from the tops of bushes in the few steep canyons it occupies. Previously placed in the genus *Aimophila*; the move to *Amphispiza* is questioned by some. Polytypic (2 ssp.; *septentrionalis* in N.A.). L 6" (15 cm).

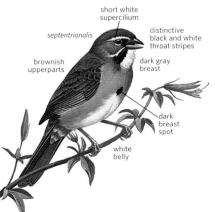

septentrionalis · short white supercilium · distinctive black and white throat stripes · brownish upperparts · dark gray breast · dark breast spot · white belly

IDENTIFICATION Dark brown above; gray head, breast, and sides. Dark central breast spot. **JUVENILE:** Overall dull, with reduced head pattern; lacks streaks found on juveniles of other sparrows.
GEOGRAPHIC VARIATION Two subspecies show slight variation in color; nominate subspecies is restricted to Mexico.
SIMILAR SPECIES Distinctive, but could be confused with juvenile Black-throated Sparrow. On Five-striped, note gray sides, flanks, and breast with central spot; wide dark malar bordered by white throat; and thin white submoustachial.
VOICE Best located by song. **CALL:** Gruff *churp;* also higher *pip* and *seet* notes when excited. **SONG:** Slow series of short, high, varied phrases, with chipping and trilled qualities; typical phrase is a *churp* followed by a short, variable trill; fairly long pauses between phrases.
STATUS & DISTRIBUTION Uncommon. West Mexican species; locally west of the Sierra Madre from Jalisco north to southeastern AZ, with seasonal movements in certain areas; U.S. range

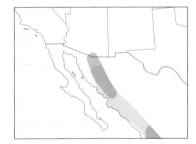

restricted to a few canyons in extreme southeastern AZ (e.g., California Gulch, Sycamore Canyon; sometimes in canyons on south side of Santa Rita Mts.). **BREEDING:** Highly specialized habitat in the U.S.: favors tall, dense shrubs on rocky, steep hillsides and canyon slopes. **NONBREEDING:** Status outside of breeding season poorly known, perhaps present, but highly secretive or some (most?) may withdraw to Mexico. Most records late Apr.–late Sept.
POPULATION The population in the U.S. is small and therefore vulnerable, but it is not threatened by human activity.

BLACK-THROATED SPARROW *Amphispiza bilineata*   BTSP ▪ 1

Perhaps our most beautiful and confiding sparrow. Polytypic (9 ssp.; 3 in N.A.). L 5.5" (14 cm)
IDENTIFICATION ADULT: Black lores and triangular black patch on throat and breast contrast with white eyebrow, submoustachial stripe, and underparts.

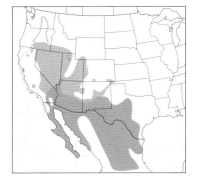

Upperparts plain brownish gray. **JUVE-NILE:** Plumage often held well into fall. No black on throat; breast and back finely streaked. Bold white eyebrow.
GEOGRAPHIC VARIATION Nine subspecies (3 in U.S.) show moderate but clinal variation in overall color and tail pattern. Nominate, breeding in central and southern TX, is smallest, with darker back and more extensive white in tail. Larger *opuntia* (breeds west to southeastern NM and southeast CO) and *deserticola* (breeds in remaining western portion of range) are paler with reduced white in tail.
SIMILAR SPECIES Juvenile like Sagebrush and Bell's; note bold eyebrow.
VOICE CALL: Faint, high, metallic *tink;* also gives series of soft, metallic tinkling notes, especially when flushed. **SONG:** Two clear notes followed by a trill.
STATUS & DISTRIBUTION Fairly common

bold white supercilium and submoustachial stripe

deserticola

large black throat patch

bold white supercilium

white throat

juvenile *deserticola*

in a range of desert habitats. Resident south to central Mexico. **VAGRANT:** Rare to casual migrant in coastal CA. Casual to eastern N.A. in fall and winter.
POPULATION Stable.

Genus *Artemisiospiza*

Formerly these two species (previously treated as one, the Sage Sparrow) were placed in the genus *Amphispiza* with Black-throated Sparrow. However recent genetic studies indicated that they belong in their own genus, which is also supported by distinct behavioral differences.

SAGEBRUSH SPARROW *Artemisiospiza nevadensis*   SAGS ▪ 1

The Sagebrush Sparrow is found in the sagelands in the intermountain West. It spends much of its time on the ground, running between shrubs with its tail raised in the air. When flushed, it flies with irregular wingbeats, and then seems to hit the ground running. When perching, it may seem agitated, delivering a tinkling call and twitching its tail nervously. Monotypic. L 5.8" (15 cm)
IDENTIFICATION Males larger than females. **ADULT:** Grayish brown head; white eye ring; white supraloral area, not extending past eye; broad white submoustachial stripe; weak malar stripe; pale brown, narrowly streaked back; whitish below with large dark brown central breast spot; buffy wash with

distinct streaks on flanks; conspicuous white edge outer web of outer tail feather and inner web at tip. **JUVENILE:** Duller, paler, more streaked, face pattern muted but still with pale supraloral spot.
SIMILAR SPECIES Smaller Bell's of coastal subspecies *belli* is much darker and largely unstreaked on back (faintly streaked on scapulars); has duller pale edges to outer tail feathers. Interior Bell's *(canescens)* is paler, more like Sagebrush, but has very faintly streaked back and prominent dark malar stripe; tail pattern is intermediate. See juvenile Black-throated Sparrow.
VOICE CALL: Twittering series of thin,

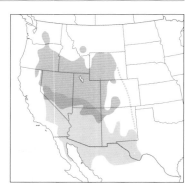

junco-like notes. **SONG:** Strongly defined, pulsating pattern of rising and falling phrases, seemingly in a minor key.
STATUS & DISTRIBUTION BREEDING: Primarily in sagebrush and saltbush desert scrub in Great Basin; arrives late Mar. **WINTER:** Found across the Southwest in arid plains with sparse bushes. **VAGRANT:** Casual to Plains states (e.g., SD, KS, NE) and southern BC; accidental to KY and NS (photos possibly indicate *canescens* Bell's).
POPULATION Declining because of rangewide loss and degradation of sage steppe habitat.

short whitish superciliary

faint back streaks

runs on the ground with tail up

juvenile

BELL'S SPARROW *Artemisiospiza belli* BESP ▪ 1

Bell's Sparrow is a dark-headed, dark-backed resident in California's coastal chaparral (favors chamise) and inland desert scrub. It is difficult to see (esp. *belli*) when not singing atop a bush. For over a century, Bell's Sparrow and Sagebrush Sparrow were confusingly classified and reclassified as either one species or two. In 2013 the AOU separated them into two species based on differences in mtDNA, morphology, ecology, and limited gene flow at the contact zone in eastern California. Polytypic (4 ssp., 3 in N.A.). L 5.4" (14 cm)

IDENTIFICATION Males larger than females. **ADULT** (*belli* subspecies): Dark gray head, conspicuous white eye ring, white supraloral spot, broad white submoustachial stripe; dark malar stripe; unstreaked brownish back. Large black central breast spot; buffy wash and distinct streaks on flanks; plain, dark tail with no white on outer feathers. **JUVENILE:** Similar to Sagebrush Sparrow but slightly darker.

GEOGRAPHIC VARIATION Nominate *belli* of coastal CA, with dark brown back and bold black malar; *canescens* of San Joaquin Valley and Mojave Desert with gray-brown back and dark malar, and indistinct thin streaks on back; *clementae* of San Clemente I., CA, very similar to *belli*. Though *canescens* is intermediate between *belli* and Sagebrush, it is genetically allied to *belli*.

SIMILAR SPECIES Larger Sagebrush Sparrow is much paler, streaked on back, weaker malar stripe, paler brownish breast spot, white outer tail feathers. See juvenile Black-throated Sparrow.

VOICE CALL: High, metallic, thin junco-like notes, delivered as a single note or

as a twitter. Vocalizations differ between Bell's and Sagebrush, but more study is needed. **SONG:** A jumbled series of rising and falling phrases.

STATUS & DISTRIBUTION BREEDING: Year-round resident in dense chaparral and sage scrub in coastal CA; inland in saltbush desert scrub and along western slopes of the central Sierra Nevada, and south to central Baja California. **WINTER:** Resident within breeding range, but many *canescens* winter from Salton Sea area to western AZ.

POPULATION Declining because of habitat loss for building development and conversion to agriculture; *clementeae* is listed as Threatened under the U.S. Endangered Species Act, due to predation by introduced pigs, goats, feral cats, rats, and other grazing animals.

plain back

central breast spot

streaks on flanks

coastal *belli*

canescens

intermediate bewteen Sagebrush and *belli* Bell's

Genus *Calamospiza*

LARK BUNTING *Calamospiza melanocorys* LARB ▪ 1

A characteristic bird of fence lines and open prairie during summer, this species tends to form large flocks during the nonbreeding season. Flocks of other sparrows are much looser: Birds flush individually even if the integrity of the flock is maintained. With the Lark Bunting, the entire flock seems to flush as one. In flight, it looks short and round-winged. Monotypic. L 7" (18 cm)

IDENTIFICATION Stocky, thick-necked and barrel chested for a sparrow, with a fairly short, squared-off tail and a thick, blue-gray bill. Males may migrate south in breeding plumage, which they often retain through August. **BREEDING MALE:** Entirely black with white wing patches; males in molt are a patchwork of brown and black. **FEMALE:** Brown above; pale below with regular sharp, brown streaking; buffy sides; brown primaries. **WINTER MALE:** Like female, but with black primaries, face, and throat. **IMMATURE:** Like female; males usually have some black on face, chin, or lores.

SIMILAR SPECIES Female and

early spring ♂

short tail with white terminal spots

all with thick, bluish bill

on winter males, black surrounds bill

streaking below darker than on females

winter ♂

indistinct face pattern

all with pale wing patches, buffy white in females

breeding ♂

immature similar to some sparrows and *Haemorhous* finches. Note diagnostic white or buffy-white wing patch (sometimes concealed), blue-gray bill, and its shape and behavior. Most finches do not spend as much time on ground.
VOICE FLIGHT NOTE: Distinctive, soft *hoo-ee.* **SONG:** A varied series of rich whistles and trills.
STATUS & DISTRIBUTION Common.

Winters south to central Mexico. **BREEDING:** Dry plains and prairies, especially in sagebrush. Annual shifts depending on precipitation. **WINTER:** Open grasslands, roadsides, dry, weedy fields. **MIGRATION:** Spring mid-Apr.–mid-May; fall early Aug.–mid-Oct. **VAGRANT:** Rare to West Coast. Casual to Pacific Northwest and to the East. **POPULATION** Apparently stable.

Genus *Passerculus*

SAVANNAH SPARROW   *Passerculus sandwichensis*   SAVS   1

The Savannah Sparrow is distinguished by its small size and behavior. Among the most common sparrows of open country, Savannahs tend to be flighty: When flushed they usually fly up and away from the observer in an undulating or stair-step flight that is distinctive. Occasionally, they fly low over the grass more like an *Ammodramus.* When flushed, they will often assume an elevated perch to survey the intrusion; at other times they drop back to the grass. Their distinctive flight note is given frequently upon flushing. Polytypic (21 ssp.; 14 in N.A.). L 5.5" (14 cm)
IDENTIFICATION A fairly small sparrow with a short, square tail. Heavily streaked below; face pattern well marked with a supercilium, strong eye line, dark malar stripe, and pale moustachial; no eye ring. Most birds have distinctive yellow supraloral. Plumage otherwise highly variable by subspecies.
GEOGRAPHIC VARIATION The 14 subspecies found north of Mexico show marked variation in body and bill size and in overall color. They are divided into four distinct, field-identifiable groups. The "Typical" group (8 ssp.)

is fairly small and small billed. Eastern birds are generally more richly colored (darker in the north), while western birds are generally paler and grayer; palest is the western interior breeding *nevadensis* and Arctic subspecies. Second, distinctive *princeps,* known as "Ipswich Sparrow," breeds on Sable Island, Nova Scotia, and winters on East Coast beaches; it can be similar to *nevadensis,* but note its larger size, larger bill, and even paler plumage. Third, "Pacific" group birds are like "Typical" Savannahs in size and bill shape, but they are quite dark in plumage; they are increasingly darker from north to south, with the darkest being *beldingi* (illustrated) of southern California's salt marshes. Finally, "Large-billed" Sparrow is a very distinctive complex involving several subspecies that breed in Mexican salt marshes; only *rostratus* reaches the U.S. Its bill is very large compared to other Savannahs, being especially thick at base. It is pale brownish gray overall, and its underparts are marked with broad, blurry, reddish brown streaks.
SIMILAR SPECIES See Vesper Sparrow. Savannah Sparrow is similar to Baird's Sparrow, but it differs in structure, lacks

double moustachial stripes, and is not extensively orangish about head. The larger, browner Song Sparrow has a longer, rounded tail, lacks yellow above eye, and differs in behavior and habitat.
VOICE All subspecies similar in voice, except for "Large-billed." **CALL:** High *tip.* **FLIGHT NOTE:** A thin, descending *tseew;* in "Large-billed" a more metallic *zink.* **SONG:** Two or three *chip* notes, followed by a long, buzzy trill and a final *tip* note. Compare to that of Grasshopper Sparrow. "Large-billed" has short, high introductory notes, followed by about three rich, buzzy *dzeee*'s; it usually does not sing in the U.S.

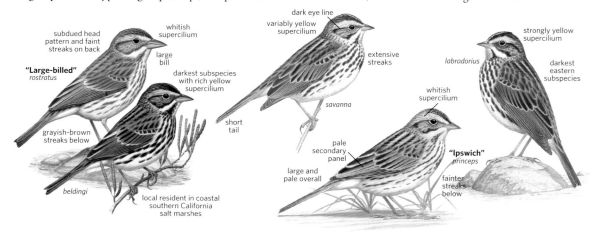

STATUS & DISTRIBUTION Common. Occurs south to Guatemala. May be found in a variety of open habitats, usually with short grass, including grasslands, airports, agricultural fields, salt marshes, and beaches. "Large-billed" occurs exclusively in saline flats and shorelines of the Salton Sea and, more rarely, coastal southern CA (formerly to central CA, where now casual); "Ipswich" breeds on Sable I., NS, and winters on East Coast beaches to northeastern FL. **MIGRATION:** Timing varies, but generally in spring Mar.– mid-May, peaking in Apr.; and in fall mid-Aug.–early Nov., peaking in mid-Oct. "Large-billed" occurs in the U.S. primarily from mid-July–Feb.; it has not yet been found breeding at the Salton Sea. **VAGRANT:** Europe and Asia. **POPULATION** Most populations are stable. "Pacific" birds (e.g., *beldingi*) have declined drastically due to destruction of CA salt marshes. "Large-billed" has declined significantly due to destruction of marshes in the Colorado River Delta (Gulf of California). "Ipswich" is vulnerable due to its limited breeding area on a small island.

Genus *Ammodramus*

The nine members of this genus (7 in N.A., two in S.A. sometimes placed in their own genus *Myospiza*) are notoriously difficult to observe, except during breeding season, when males sing from exposed perches. At other seasons, they forage on ground in and are usually seen only when flushed. They are more difficult to flush than are other sparrows, often flushing at one's feet. Individuals typically do not flock. Birds are silent when flushed (unlike Savannah Sparrow) and their short flights are low to the ground and weak, ending with a sudden drop into the grass. All have comparatively flat heads; large bills; fairly short, sometimes spiky, tails; and wheezy or whispering songs. Several species migrate in juvenal plumage.

GRASSHOPPER SPARROW *Ammodramus savannarum*   GRSP   1

Males sing conspicuously from elevated perches, but otherwise the Grasshopper is fairly secretive: it does not forage in the open, though it is more inclined to perch in the open than other *Ammodramus*. Polytypic (12 ssp.; 4 in N.A.). L 5" (13 cm)
IDENTIFICATION Small, chunky; short tail, large bill, flat head. Buffy breast, sides; usually without obvious streaking. Pale, whitish central stripe on dark crown; white eye ring; often yellow-orange spot in front of eye. Buffier below in fall, never as bright as Le Conte's. **JUVENILE:** Brown streaks on breast, sides.
GEOGRAPHIC VARIATION The four North American subspecies vary in dorsal color from Florida's dark *floridanus* to reddish *ammolegus* of southeastern Arizona and southwestern New Mexico. Eastern *pratensis* is slightly richer colored overall than western *perpallidus,* ranging east through the Great Plains.
SIMILAR SPECIES Lacks Le Conte's broad buffy-orange eyebrow, blue-gray ear patch, heavily streaked breast and sides. Beware juvenile Grasshopper Sparrows with streaked breasts. Compare also with female Orange Bishop and Savannah Sparrow.
VOICE CALL: High, sharp *tsik*, often doubled; sometimes more excited, higher *tik*, often in rapid succession, near nest. **FLIGHT NOTE:** High, lisping *tseee*, sometimes from perch or from inside cover. **SONG:** One to three high *chip* notes followed by a brief, grasshopper-like buzz; also sings a rambling series of squeaky and buzzy notes, often in flight.

richly colored supraloral spot; rest of supercilium gray

large bill

buffy breast

full eye ring and strong median crown stripe

supraloral spot paler

extensive fine streaking below with underlying buffy wash

juvenile *perpallidus*

short, spiky tail as in other *Ammodramus*

fresh fall *perpallidus*

buffiest subspecies overall and the most rufous one above, although a few *perpallidus* and *pratensis* are rufous above too

a few reddish streaks on sides of breast

ammolegus

floridanus

much more blackish above, down through uppertail coverts

STATUS & DISTRIBUTION Uncommon to common. **BREEDING:** Pastures, grasslands, old fields, airports, palmetto scrub. **MIGRATION:** Reaches breeding grounds in southern CA late Mar.; midlatitudes of U.S. mid-Apr.; fall peaking in Oct. **VAGRANT:** Very rare to BC, Atlatic provinces.
POPULATION Subspecies *floridanus* is federally endangered. Declines due to conversion of native grassland, suppression of grassland fires. Other declines most pronounced along East Coast.

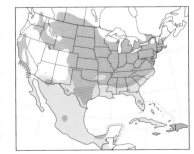

BAIRD'S SPARROW *Ammodramus bairdii*   BAIS ■ 2

One of the rarer sparrows, this species is poorly known due to behavior and identification difficulties; migration and wintering areas incompletely known. The Baird's is secretive, especially away from breeding grounds. It looks more like Savannah than its congeners,

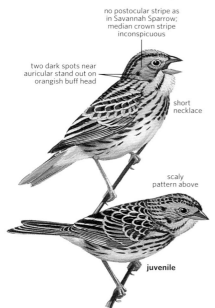

no postocular stripe as in Savannah Sparrow; median crown stripe inconspicuous

two dark spots near auricular stand out on orangish buff head

short necklace

scaly pattern above

juvenile

to separate the two a good view of the different head patterns is necessary. It typically flushes only at close range, silently flies short distances, and drops back into cover. This differs from the behavior of some Savannahs, which are more apt to flush at a distance, call in flight, rise to significant heights in flight, and perch in the open upon landing. Monotypic. L 5.5" (14 cm)

IDENTIFICATION Orange-tinted head (which is duller on worn summer birds); note especially two isolated dark spots behind ear patch, no postocular line. Double whisker (malar and moustachial) lines. Wide-spaced, short dark streaks on breast form a distinct necklace; chestnut on scapulars. **JUVENILE:** Head paler, creamier; central crown stripe finely streaked; white fringes give scaly appearance to upperparts; underparts more extensively streaked.

SIMILAR SPECIES See Savannah and Henslow's Sparrows.

VOICE CALL: Excited *tink* or *tsip* notes near nest. **FLIGHT NOTE:** High, thin *tseee*. **SONG:** Two or three high, thin notes followed by a single warbled note and a low trill.

STATUS & DISTRIBUTION Uncommon and local. **BREEDING:** Rolling short-

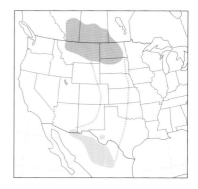

grass prairies with scattered shrubs or weeds. **WINTER:** Large open grasslands, with clumps of grass interspersed with bare areas. **MIGRATION:** Rarely seen, through Great Plains (west KS to west MN, also north NM). Spring migration primarily in early May; in fall primarily Sept., a very few to mid-Oct. Rare but probably overlooked in western TX and CO. **VAGRANT:** Casual breeder east to west MN; accidental in summer to WI. Accidental in fall to East (NY and MD) and CA (5 recs.).

POPULATION Significant declines throughout breeding range due to conversion of native prairie to agriculture, which is unsuitable for this species.

HENSLOW'S SPARROW *Ammodramus henslowii*   HESP ■ 2

On breeding grounds males deliver their songs from small shrubs or tall grass; at other times the Henslow's is very secretive, preferring to feed on or near the ground in fairly dense grass. After being flushed several times it may perch in the open for a few minutes before dropping back into cover. Polytypic (2 ssp.; both in N.A.). L 5" (13 cm)

IDENTIFICATION Large flat olive head; large gray bill; malar stripe; moustachial stripe; white eye ring; distinct streaking across pale breast, extending down flanks; dark chestnut wings and back; narrow pale fringes to back feathers when fresh. **JUVENILE:** Paler, yellower, less streaking below. Compare to adult Grasshopper Sparrow.

GEOGRAPHIC VARIATION Two weakly differentiated subspecies are not field separable: *susurrans* in East; smaller-billed, paler nominate in Midwest.

SIMILAR SPECIES Most similar to Baird's, but greenish head, nape, and most of central crown stripe; wings extensively dark chestnut. Also confused with Grasshopper; note Henslow's greenish head and central crown stripe; malar and moustachial stripes; dark chestnut wings and back; breast streaking. Beware juvenile Grasshopper with streaked breast.

VOICE CALL: Sharp *tsik.* **FLIGHT NOTE:** High *tseee,* like Grasshopper's. **SONG:** Distinctive, short *tse-lick.*

pea-soup green head with blackish lateral crown and postocular stripes

large bill

rich dark chestnut on upperparts

necklace

juvenile

largely unstreaked breast

STATUS & DISTRIBUTION Uncommon, local. **BREEDING:** Wet shrubby fields, weedy meadows, reclaimed strip mines. **WINTER:** Weedy old fields or wet second growth, also understory of pine woods. Winters casually north to IL, IN; accidental to New England.

MIGRATION: Poorly known; arrives at more southerly breeding areas by mid-Apr. **VAGRANT:** Accidental west of mapped range to ND, CO (two sight records), and NM. Casual to New England, Maritimes, and coastal mid-Atlantic.

POPULATION Formerly bred on mid-Atlantic coastal plain and in Northeast; now largely extirpated. Signifcant long-term declines due to loss of native grassland habitat. Recent management efforts have led to recovery in some areas.

LE CONTE'S SPARROW *Ammodramus leconteii*   LCSP ▪ 1

Fairly common but secretive, the Le Conte's Sparrow scurries through matted grasses like a mouse. Compared to Grasshopper, it prefers lower-lying, often wet fields and marsh edges. When flushed, it flies away from the observer with a characteristically weak flight, flaring its narrow tail feathers upon landing. Monotypic. L 5" (13 cm)
IDENTIFICATION White central crown stripe, becoming orange on forehead; chestnut streaks on nape; and straw-colored "cornrows" on back distinguish Le Conte's from sharp-tailed sparrows. Bright, broad, buffy orange

eyebrow; grayish ear patch; thinner bill; and orange-buff breast and sides distinguish it from Grasshopper Sparrow. Dark streaks on sides of breast and flanks. **JUVENILE:** Buffy; tawny crown stripe; breast heavily streaked. Some migrate in juvenal plumage, which can be retained through late Oct.
SIMILAR SPECIES Nelson's and Saltmarsh Sparrows are most similar but have grayish central crown stripe; unstreaked gray nape; distinct white lines on back; larger bill; and more extensive gray face. Juveniles more similar, but Le Conte's is paler, sandier, and streaked on nape. See Grasshopper Sparrow account.
VOICE CALL: Sharp *tsit.* **FLIGHT NOTE:** High *tseeet,* like Grasshopper Sparrow's. **SONG:** Short, high, insectlike buzz: *tsit-tshzzzzzzz.*
STATUS & DISTRIBUTION Fairly common. **BREEDING:** Wet grassy fields, bogs, marsh edges. **WINTER:** High marsh and marsh edges, low-lying grassy fields, especially those with broomsedge or switchgrass components. **MIGRATION:** Commonly through Great Plains east to Mississippi River Valley; uncommon to rare east to Ohio River Valley.

median crown stripe orangish on forehead, then white
orange-buff supercilium
purplish pink nape streaks
broad buff back streaks
necklace and fine streaking on sides and flanks
juvenile
some juveniles migrate south in juvenal plumage

Mid-Sept.–mid-Nov.; spring Apr.–mid-May. **VAGRANT:** Casual to CA (about 35 recs.) and NM, and accidental to BC, WA, OR, and AZ. Occurs casually in migration along East Coast, primarily in fall; rare to casual in winter north to NJ and west to southeastern CO and western TX.
POPULATION Apparently stable.

NELSON'S SPARROW *Ammodramus nelsoni*   NESP ▪ 1

Secretive and hard to find away from the breeding grounds, Nelson's Sparrow (formerly "Nelson's Sharp-tailed Sparrow") has warm, soft plumage colors that make it distinctive once it is found. Its behavior is like Le Conte's Sparrow. Sometimes it responds to pishing, especially when forced to higher ground during high tide. It has a somewhat smaller bill and rounder head than Saltmarsh Sparrow, though this is less pronounced in the Atlantic coast subspecies *(subvirgatus)*. Polytypic (3 ssp.; all in N.A.). L 4.8" (12 cm)
IDENTIFICATION Orange-buff triangle on indistinctly marked face; gray median crown stripe and nape; streaked buffy breast contrasts sharply with white belly; back marked with

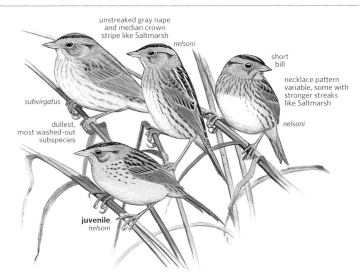

unstreaked gray nape and median crown stripe like Saltmarsh
nelsoni
short bill
necklace pattern variable, some with stronger streaks like Saltmarsh
nelsoni
subvirgatus
dullest, most washed-out subspecies
juvenile *nelsoni*

black and white stripes; wings tinged with rich rufous. Narrow, sharply pointed rectrices. JUVENILE: Fainter median crown stripe; duller nape; variably thicker eye line; less contrast above; lacks streaking across breast.

GEOGRAPHIC VARIATION Three subspecies with disjunct breeding ranges. Interior *nelsoni* is brightest, with rich orange on face and breast; distinct streaking on breast and flank; distinct white streaks on darker back. Subspecies *subvirgatus* of the Maritimes and coastal ME is much duller overall (recalling Seaside Sparrow); has diffuse streaking below; grayer underparts; longer bill. In *alter,* from James and Hudson Bays, brightness is intermediate, streaks blurred, still, it is doubtfully separable from *nelsoni* in the field.

SIMILAR SPECIES See Le Conte's for field characters. Note also that Le Conte's winters across much of the southern U.S., unlike Nelson's, and thus any orange-headed sparrow seen outside the narrow migration timeframes for Nelson's is very likely a Le Conte's.

VOICE CALL: Sharp *tsik.* FLIGHT NOTE: A high, lisping *tsiis,* shorter and less strident than Grasshopper's; regularly delivered from ground in cover. SONG: A wheezy *p-tshhhhhhh-uk,* ending on a lower note; flight display unlike Saltmarsh's, with song delivered at zenith of a short vertical flight, followed by a slow descent.

STATUS & DISTRIBUTION Fairly common. BREEDING: Fresh marshes, wet meadows, bogs *(nelsoni);* salt marshes of James and Hudson Bays *(alterus);* upper reaches of coastal salt marshes and coastal cranberry bogs *(subvirgatus).* WINTER: Coastal salt marshes, in same habitats as the Saltmarsh Sparrow, though may also occur in fresher upper reaches of marshes where Saltmarsh does not occur. Accidental in winter in the interior; very rare but annual at large tidal estuaries in winter in coastal CA. MIGRATION: Spring migration takes place during mid-May–late May; fall migration late Sept.–late Oct. Migration of *nelsoni*

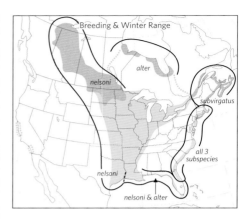

primarily through interior broadly through the Mississippi River Valley; rare inland migrant eastward. VAGRANT: Subspecies *nelsoni* is casual in migration (mainly fall) to CA and CO. Accidental to AZ, NM, WY, and WA.

POPULATION The diking and draining of salt marshes in the Maritimes and southern ME has reduced available breeding habitat and numbers of *subvirgatus.* Interior populations are probably stable.

SALTMARSH SPARROW *Ammodramus caudacutus* SALS ■ 1

This species and Nelson's Sparrow were formerly classified as a single species, "Sharp-tailed Sparrow," and more recently had their English names shortened by dropping "sharp-tailed" from them. The Saltmarsh Sparrow is restricted to salt and brackish marshes and is virtually unknown in freshwater. It hybridizes with Nelson's Sparrow *(subvirgatus)* where ranges overlap in southern Maine and New Hampshire. Its behavior is similar to Nelson's, though it may be less secretive. It has a flat head and a fairly long, pointed bill. Its narrow tail is composed of narrow, sharply pointed rectrices. Polytypic (2 ssp.; both in N.A.). L 5" (13 cm)

IDENTIFICATION Orange face with extensive gray cheek; long dark postocular line; narrow dark malar stripe sets off white throat; gray nape and central crown stripe; reddish brown back with strong white stripes; indistinct buffy or whitish breast marked by extensive dark streaks across breast and along flanks. JUVENILE: Like juvenile Nelson's, but cheek, crown, nape, and back are darker; streaks below more widespread and distinct.

GEOGRAPHIC VARIATION Two subspecies differ subtly; not safely field separable. Nominate breeds from New Jersey to southern ME; *diversus* breeds from NJ to northern NC and is smaller-billed and duller on average. Winter ranges overlap.

SIMILAR SPECIES Similar to Nelson's, but longer bill and flatter head; eyebrow streaked with black behind eye; orange-buff face triangle contrasts strongly with paler, very distinctly streaked underparts; dark markings around eye and head more

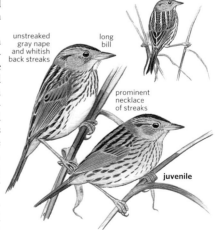

unstreaked gray nape and whitish back streaks

long bill

prominent necklace of streaks

juvenile

sharply defined. See also Seaside (esp. juveniles).

VOICE CALL: Like Nelson's Sparrow. FLIGHT NOTE: Like Nelson's; regularly delivered from ground in cover. SONG: Softer, more complex than Nelson's. Flight display unlike Nelson's: song delivered in rapid, low flight over grass. Note Nelson's song flight rises to an apex with a descent back to the ground.

STATUS & DISTRIBUTION Fairly

common. **BREEDING:** Breeds in extensive coastal salt marshes (esp. those with a large component of saltmeadow cordgrass, which forms short-grass pastures within drier portions of the marsh). **WINTER:** Less tied to saltmeadow cordgrass; frequently forages along dikes, ditches, pool edges, often in stands of taller smooth cordgrass or even shrubs during high tides. **MIGRATION:** Returns to breeding grounds between late Apr. and mid-May; most depart by early Nov.; stragglers often linger north of wintering areas into Dec. **VAGRANT:** Accidental to PA (2 recs.), NS (1 rec.). **POPULATION** Diking and draining of salt marshes along Atlantic coast has depressed numbers, but populations are probably stable. Considered near threatened by BirdLife International.

SEASIDE SPARROW *Ammodramus maritimus*   SESP ■ 1

This conspicuous *Ammodramus* can be seen taking long, low flights over marshes. Most subspecies are dull except for the yellow supraloral patch. Their large size in relation to other marsh sparrows is alone a good field mark. Males sing from exposed perches or deliver flight songs (flying up from perch, then fluttering down as song is delivered). In winter, it is readily drawn into view by pishing. Fairly chunky, it has a flat head; long spikelike bill; and short, pointy tail. Polytypic (9 ssp.; all in N.A.; *nigrescens* extinct). L 6" (15 cm)

IDENTIFICATION Distinct yellow supraloral patch; dark malar stripe; distinct whitish throat; broad, pale submoustachial. Breast varies (gray, whitish, or buffy), some streaking. Upperparts grayish with variable brown or olive coloration; variable reddish cast to wings; back variably streaked. **JUVENILE:** Warmer, paler, buffier than adult, especially on face and breast; breast streaking more distinct; may lack distinct yellow supraloral.
GEOGRAPHIC VARIATION The Seaside varies widely in overall color; forms four distinct groups: "Atlantic coast" *maritimus* (breeds from VA north, winters from NJ to FL) and *macgillivraii* (breeds NC to northeast FL) are dull grayish olive above, smoky gray on the breast with diffuse streaking. "Gulf Coast" (*sennetti* of southeast TX, *fisheri* from TX to northwestern FL, *peninsulae* of western FL) has a buffier breast and face, more distinct breast and back streaking. "Cape Sable," the very localized *mirabilis,* inhabits a small area in the Everglades, with an olive cast to the back and nape, more distinct breast and back streaking. "Dusky" *(nigrescens),* the darkest, became extinct in June 1987. ("Cape Sable" and "Dusky" were recognized as separate species prior to 1973.)
SIMILAR SPECIES Juvenile regularly confused with Saltmarsh Sparrow. Bigger Seaside has larger bill; longer, less spiky tail; flimsy, fresh juvenal feathering.
VOICE CALL: Loud *tsup,* sometimes doubled. Rapid, high *tik* notes when excited on breeding grounds. **SONG:** Wheezy *tup zheee-errr,* like Red-winged Blackbird's, but softer, buzzier. Secondary song more complex, stuttering.
STATUS & DISTRIBUTION Fairly common. Grassy tidal marshes; favors taller grass. **MIGRATION:** Resident (excl. *maritimus*); returns to breeding grounds between mid-Apr. and early May; most departed by mid-Oct.; stragglers often linger north of wintering areas into Nov., Dec. **VAGRANT:** Accidental inland to PA, NC, and TX. Casual to the Maritimes.
POPULATION Stable. Imperiled in New England and FL; habitat limited, populations small. Only 4,000 to 6,000 "Cape Sable" birds remain. "Dusky" and population along St. John's River (sometimes given as a separate subspecies,

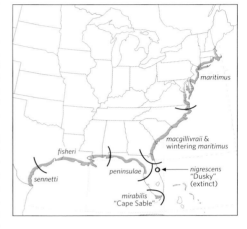

pelonotus) are extinct; caused by diking of salt marshes and spraying of insecticides.

all with yellow supraloral spot

long bill

whitish throat and submoustachial

maritimus

fisheri

buffy breast, especially in fresh plumage

blackish overall

slightly greenish above

nigrescens extinct

mirabilis

juvenile *maritimus*

FOX SPARROW *Passerella iliaca*   FOSP ◼ 1

The Fox Sparrow feeds on the ground, usually in dense cover. Like the towhee, it habitually scratches with both feet moving together. Pishing may bring Fox into the open, where it will chip aggressively; often it approaches cautiously from dense cover. Often visits feeders. Polytypic (16 ssp.; all in N.A.). L 7" (18 cm)

IDENTIFICATION This large, chunky sparrow has a medium-length tail, a large head, and a thick-based bill. Its plumage is highly variable by subspecies. Generally, it is dark above, with pale underparts that are heavily marked by dark triangular spots, which coalesce to form a central breast spot. Most subspecies have a reddish rump and tail; reddish coloring in wings.

GEOGRAPHIC VARIATION See sidebar opposite.

SIMILAR SPECIES "Red" Fox Sparrow is occasionally confused with Hermit Thrush, but note "Red" Fox's bill shape, streaked flanks, streaked back, and different behavior. Lincoln's and Song Sparrows are both smaller, with thinner bills, sharper breast streaking and proportionately longer tails; and they are as likely to be found in grassy or weedy areas than in wooded thickets.

VOICE CALL: Large-billed Pacific subspecies give a sharp *chink* call, like California Towhee's; others give a *tschup* note, similar to but louder than Lincoln's Sparrow's, and similar to but softer than Brown Thrasher's. **FLIGHT NOTE:** High, rising *seeep,* given commonly on ground and in thickets. **SONG:** Sweet, melodic, warble composed of seven or more phrases: for example, *too-weet-wiew too-weet tuck-soo-weet-wiew.* Sweeter in the northern reddish subspecies; includes harsher or buzzy trills in other subspecies. Frequently sings on the winter grounds.

STATUS & DISTRIBUTION Common. **BREEDING:** Dense willow and alder thickets ("Red"); montane willow and alder thickets ("Slate-colored") or coastal forests and thickets ("Sooty"); montane thickets and chaparral ("Thick-billed"). **WINTER:** Undergrowth and dense thickets in coniferous or mixed woodlands, chaparral. **MIGRATION:** Fall migration late Sept.–late Nov.; Spring migration early Mar.–late Apr. "Red" group migrates several weeks earlier in the spring and later in the fall than western groups do.

POPULATION Stable.

Breeding Range

unalaschcensis

insularis
sinuosa
annectens
townsendi
zaboria   "RED" group
iliaca
"SOOTY" group *fuliginosa*   *altivagans*
olivacea
megarhyncha   *schistacea*
brevicauda
"THICK-BILLED" group   "SLATE-COLORED" group
stephensi
swarthi
canescens

Winter Range

"SLATE-COLORED" group

"SOOTY" group

"RED" group

"SOOTY," "SLATE-COLORED" and "THICK-BILLED" groups

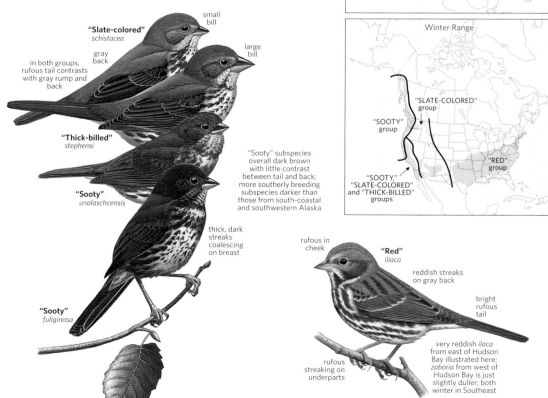

"Slate-colored" *schistacea*

small bill

large bill

gray back

in both groups, rufous tail contrasts with gray rump and back

"Thick-billed" *stephensi*

"Sooty" *unalaschcensis*

"Sooty" subspecies overall dark brown with little contrast between tail and back; more southerly breeding subspecies darker than those from south-coastal and southwestern Alaska

thick, dark streaks coalescing on breast

"Sooty" *fuliginosa*

rufous in cheek

"Red" *iliaca*

reddish streaks on gray back

bright rufous tail

rufous streaking on underparts

very reddish *iliaca* from east of Hudson Bay illustrated here; *zaboria* from west of Hudson Bay is just slightly duller; both winter in Southeast

All three species are of medium size, with fairly long, rounded tails. Favoring brushy areas along streams, pond edges, bogs, swales, or marshes, they may feed on the ground or in rank grasses. Their tails are usually pumped in flight, flipped upon landing, and cocked when perched. Secretive but inquisitive, they are responsive to pishing. All are quite vocal.

SONG SPARROW *Melospiza melodia*   SOSP ▪ 1

Probably the most widespread sparrow, Song Sparrows commonly visit feeders and are responsive to pishing. In winter they may form small flocks, often with other sparrows, including Lincoln's, Swamp, or *Zonotrichia*. Polytypic (39 ssp.; 29 in N.A.; 10 restricted to Mexico; recent work will substantially reduce the number of ssp.). L 4.8–6.8" (13–17 cm)

IDENTIFICATION Distinctive long, rounded tail, often flipped in flight and when landing. Broad, grayish eyebrow; broad. Upperparts usually streaked. Underparts whitish; streaking on sides, breast, often converge in central spot. **JUVENILE:** Buffier; fine streaking.
GEOGRAPHIC VARIATION Extensive, marked variation in measurements,

overall color. Identification to individual subspecies often not possible in the field, though birds are often separable to one of the five geographic groups. Perhaps most distinctive is "Alaska Island" group: four subspecies resident on the Aleutian Island chain and east to Kodiak Island, all of which are very large (close to size of Fox Sparrow),

Fox Sparrow Subspecies Groups

Fox Sparrows comprise four fairly distinct subspecies groups that differ by consistent plumage traits, range, and voice. Some authorities consider these subspecies groups to be separate species. Identification to exact subspecies is rarely possible, given clinal variation, extreme similarity of certain subspecies, and the occurrence of intergrade populations, but they can usually be identified to subspecies group. Key features are overall upperparts coloration; color and extent of spotting below; presence of back streaking; bill size; and call note.

"Red" Group: Subspecies *iliaca* and the slightly duller, darker *zaboria* make up "Red" Fox Sparrow group. They have bright reddish brown on tail, rump, wings, and face; back is gray with reddish brown streaks; underparts are white, except for a pale yellow-buff wash on throat and upper breast, and have contrasting, blurry reddish brown streaks; also narrow white wing bars. Reddish breast spotting well defined on white underparts. The face pattern of "Red" Fox Sparrows is striking, with bright reddish brown crown and ear coverts prominent on otherwise pale gray face. Juveniles of these subspecies have less distinct breast spotting and a brownish gray crown and back with no streaking. These birds breed in the boreal forests and taiga of the far north from the Seward Peninsula, AK *(zaboria)*, to NF *(iliaca)* and both subspecies winter in the southeastern U.S.

"Sooty" Group: Birds in "Sooty" group are similar in pattern those in "Red" group but lack the wing bars, streaked back, and warm reddish brown tones of the "Red." Color within the "Sooty" group varies from a cold blackish brown *(fuliginosa, townsendi,* inc. *chilcatensis* which is not recognized by many) to a somewhat paler dark brown (e.g., *sinuosa, annectens, insularis,* and the palest *unalaschensis)*. Breast of "Sooty" group birds is so densely spotted that its central splotch extends across entire breast and its flanks are more mottled than streaked. Base coloration of underparts of a juvenile "Sooty" is a warm buff. Some in "Sooty" group show gray on head like "Slate-colored" birds, and are tinged with brown on wings, rump, and tail. Breast spotting is heavy and dense, often forming a checkerboard-like

pattern. These birds breed along the Pacific coast from the Alaska Peninsula to northwestern WA and they winter south to southern CA.

"Slate-colored" Group: "Slate-colored" has a medium gray head, nape, and unstreaked back, with contrasting dull red-brown wings, rump, and tail. Subspecies differ subtly, from *olivacea,* with a reddish cast to the head and back; to *schistacea* and *swarthi,* with more gray-brown on the head and back; to intermediate *canescens,* with less brown on the back and a larger bill (tending toward the "Thick-billed"). Markings below, while generally like those of "Red," are a much darker brown and are more sharply defined. "Slate-colored" and "Thick-billed" groups can sometimes show very faint whitish wingbars. A problematic subspecies of northwestern Canada, *altivagans,* is intermediate between the "Slate-colored" and "Red" groups but lacks the rich reddish tones of "Red" group birds. It has only a very faintly streaked gray back. "Slate-colored" birds breed in many of the Great Basin ranges and the northern and central Rockies; and winter in CA, with a very few in AZ and NM.

"Thick-billed" Group: Birds in this group have consistently large, very thick-based bills; plumage is similar to "Slate-colored." This subspecies group typically has a gray mandible (as does "Slate-colored"), but its bill can be yellowish or orangish (as in "Red" or "Sooty" birds); bill color may vary seasonally. Bill size varies from the smaller-billed *megarhyncha* (includes *fulva* and *monoensis,* from OR to northern CA) to *brevicauda* (northwestern CA) and the southern *stephensi* (central and southern CA mountains and northern Baja California), which have especially large bills. Breast spotting is most distinct and blackest in the "Thick-billed" group.

Voice: "Red" and "Slate-colored" subspecies give a *tchewp* call note; this is slightly less downslurred in "Sooty," which has a smacking *tschup;* "Thick-billed" birds and some *canescens* "Slate-coloreds" have a distinctive *chink,* similar to California Towhee's. Song differences are subtle and complex, but "Red" Fox Sparrows birds tend to sing sweeter, clearer notes; others use more buzzes and trills. ∎

large-billed, and generally sooty-colored. No other subspecies groups are as large in overall size; the subspecies from the western Aleutians (e.g., *maxima*) are largest. "Pacific Northwest" group (inc. *morphna*) is large and dark sooty or dark rusty colored; streaking on breast and flanks often has a dusky background. "California" group (inc. *heermanni* and those resident on the Channel Islands) are small and dark, with distinctly marked rich reddish wings, very gray faces; sharp blackish streaking below contrasts markedly with upperparts. "Southwestern" birds *(fallax* and *saltonis)* are quite pale reddish brown with well-defined breast streaking, contrasting little with the back and wings. Typical of "Eastern" group is the nominate, which is medium-size and fairly brown-backed with moderate contrast in breast streaking.

SIMILAR SPECIES See Lincoln's, Swamp, Savannah, Fox, and Vesper.

VOICE **CALL:** A nasal, hollow *chimp;* also high *chip*'s when excited. **FLIGHT NOTE:** A clear, rising *seeet.* **SONG:** Three or four short clear notes followed by a buzzy *tow-wee,* then a trill.

STATUS & DISTRIBUTION Common. Winters south to northern Mexico; isolated resident population in central Mexico. **BREEDING:** Brushy areas, especially dense streamside thickets, marsh edges. **WINTER:** Migratory populations tend to winter primarily in brushy areas, rank weedy fields, swampy woods; often with Swamp and Lincoln's. In the West, Song Sparrow is more closely tied to ponds or streams with lush growth. **MIGRATION:** Most populations are migratory; Pacific and southwestern populations generally resident. Spring early Mar.–late Apr., peaking late Mar.; fall mid-Sept.–mid-Nov., peaking late Oct. **VAGRANT:** Europe.

POPULATION Stable in most portions of range. The subspecies resident on Santa Barbara I., *graminea*, bacame extinct due to overgrazing. Degradation of salt marsh habitats around the San Francisco Bay and desert riparian habitats in the Southwest has also negatively affected certain populations.

melodia
many subspecies have blurry streaks below often with central spot on breast
juvenile
heermanni
buffy wash with blurry streaks on breast and sides
compare carefully to Lincoln's Sparrow
thicker streaks on breast and sides than Lincoln's
California subspecies are very dark
heermanni
all Melospiza have long, rounded tails
northwestern subspecies are dark and ruddy
whitish submoustachial
morphna
Alaskan subspecies, especially those on Aleutians, are very large
maxima
fallax
smallest and palest subspecies

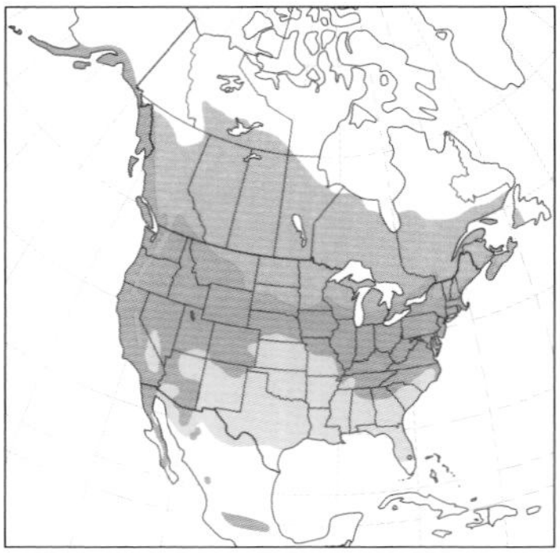

LINCOLN'S SPARROW *Melospiza lincolnii* LISP 1

This sparrow stays in or close to dense cover, flushing away from the observer with its short tail flipped on landing. It often raises its slight crest when disturbed. Polytypic (3 ssp.; all in N.A.). L 5.8" (15 cm)

IDENTIFICATION Buffy wash and fine streaks on breast and sides contrast with whitish, unstreaked belly. Note broad gray eyebrow, whitish chin, and eye ring. Briefly held juvenal plumage is paler overall than juvenile Swamp Sparrow's. Distinguished from juvenile Song Sparrow by it s shorter tail, slimmer bill, and thinner malar stripe, often broken.

GEOGRAPHIC VARIATION The three subspecies show weak, clinal variation in measurements and general coloration. The nominate breeds in the northern portion of the breeding range (AK to the Maritimes), *alticola* breeds in the montane west, and *gracilis* breeds in the Pacific Northwest. Generally, *alticola* is largest and *gracilis* is smallest and darkest.

often appears slightly crested when agitated
broad gray supercilium
well-outlined buffy submoustachial stripe
finer streaking than Song against buffy breast
juvenile
similar to adult but overall, pattern a little more muted and more streaked

SIMILAR SPECIES Song is larger with a longer, rounded tail; it lacks an eye ring, and streaking on underparts looks as if it were painted with a broad brush. Lincoln's has pencil-thin streaks, is grayer on face (esp. supercilium), and has a buffy wash to breast that Song lacks, but beware a juvenile Song with a buffy breast and finer breast streaking. Similar-size Savannah is found more often in open fields and often flies high when flushed. (Lincoln's almost never does.) Savannah also lacks gray face, rich reddish brown back and wings, and buffy wash on breast. See also the sometimes streak-chested juvenile Swamp Sparrow. VOICE CALL: A hardy flat *tschup,* repeated in a series as an alarm call. FLIGHT NOTE: Sharp, buzzy *zeee,* similar to that of Swamp Sparrow but sharper; given commonly on ground and in thickets. SONG: A rich, loud, series of rapid bubbling notes, ending with a trill. STATUS & DISTRIBUTION Common in West, uncommon in East. Winters south to Central America. BREEDING: Bogs, mountain meadows, wet thickets. WINTER: Thickets, overgrown fields, dense brush, cutover woods. MIGRATION: Spring migration late Mar.–mid-May; fall migration Sept.–early

Nov. VAGRANT: Casual on the Azores. POPULATION Stable.

SWAMP SPARROW *Melospiza georgiana*   SWSP 1

Sometimes abundant, Swamp Sparrows do not form a cohesive flock in winter; in rank brush or tall grass they may cluster with Song and Lincoln's. They can be brought into view with other sparrows by pishing. They deliver a loud, metallic call note when excited and at dawn and dusk, when marshes and brushlands are filled with their chorus of *chip* notes. A Swamp often flies away, straight and low, when flushed, and may flip its long tail (esp. when landing). It typically does not call in flight, but often calls upon landing. Polytypic (3 ssp.; all in N.A.). L 5.8" (15 cm)
IDENTIFICATION Gray face; white throat; rich rufous wings contrast with upperparts; reddish brown back streaked with black. BREEDING ADULT: Reddish crown, gray breast, whitish belly. WINTER ADULT & IMMATURE: Buffier overall, especially on flanks; duller crown streaked, divided by gray central stripe. JUVENILE: Briefly held plumage usually even buffier; darker overall than the juvenile Lincoln's or Song; redder wings and tail.

GEOGRAPHIC VARIATION Three subspecies show moderate variation. The nominate and *ericrypta* divide the breeding range roughly at the Canadian border (nominate south, *ericrypta* north). Both small-billed and warm; nominate slightly warmer flanks and cap, grayer back. The local *nigrescens* (breeds in brackish marshes from NJ to MD) has poorly known winter range, longer bill, darker back, black streaking in the crown, no buff tones to the flanks.
SIMILAR SPECIES The Song and Lincoln's are similar but heavily streaked below. Juvenile and first-winter Swamp can be streaked lightly, but never as prominently as the Lincoln's or Song. Unlike the Lincoln's, the Swamp lacks an eye ring; also has a somewhat longer tail and richer red-brown wings, with gray chest.
VOICE CALL: Hard, metallic *chip,* somewhat similar to that of Eastern or Black Phoebe. FLIGHT NOTE: A prolonged, buzzy *zeee,* similar to Lincoln's, but softer. SONG: A slow, single-pitched musical trill. Some deliver a faster trill.

STATUS & DISTRIBUTION Common. Winters south to central Mexico. BREEDING: Nests in dense, tall vegetation in marshes, bogs, alder swales, wet hay fields. WINTER: Tall grass marshes, brushy fields, wet open woodland, esp. cutover forest. Rare in migration and winter in the West. MIGRATION: Spring mid-Mar.–early May, peaking mid-Apr.; fall mid-Sept.–mid-Nov., peaking mid-Oct. VAGRANT: Casual to AK, Europe.
POPULATION Most populations stable; mid-Atlantic subspecies *nigrescens* sharply declined recently (reasons poorly known).

Genus *Zonotrichia*

This genus comprises five species (four breed in Canada and the northern U.S.; one in the highlands of C.A. and throughout S.A.). Adults are distinctive, with strong face patterns; immatures are more indistinctly marked. Songs are loud whistled phrases, often simple but melodic. They typically feed in flocks along dense cover, darting for safety at once when danger approaches.

WHITE-THROATED SPARROW   *Zonotrichia albicollis*   WTSP ▪ 1

This woodland sparrow of the East frequents feeders or woodland edges. Monotypic. L 6.8" (17 cm)
IDENTIFICATION Two distinct color color morphs. White-striped morph has white eyebrow (yellow in front of eye) and black lateral crown stripes; eyebrow stripe is buffier in winter. Tan-striped morph has tan eyebrow (duller yellow in front of eye) and dark brown lateral crown stripes. Both morphs have conspicuously outlined white throat (with thin malar stripe in tan-striped morphs) and mostly dark bill. Upperparts rusty brown; underparts grayish, usually with diffuse streaking on breast of tan-striped morph. **FIRST-WINTER:** Browner and duller than adult; poorly defined white throat patch bordered by dark malar stripe; usually streaked on breast and sides. **JUVENILE:** Pale grayish eyebrow and throat; breast and sides heavily streaked.
SIMILAR SPECIES See White-crowned.
VOICE CALL: A sharp *pink,* like White-crowned, but higher; also a drawn-out, lisping *tseep.* **SONG:** Thin whistle, generally one to two single notes, followed by three to four long notes, sometimes tripled: *Pure sweet Canada Canada Canada.* Often sings in winter.
STATUS & DISTRIBUTION Common. **BREEDING:** Clearings in deciduous or evergreen forests, bogs. **WINTER:** Common in woodland undergrowth, brush, gardens. A very few winter south to northern Mexico. **MIGRATION:** Spring late Mar.–mid-May, peaking mid-Apr.; fall mid-Sept.–early Nov., peaking mid-Oct. Rare but regular in migration and winter in the West. **VAGRANT:** Rare to southeastern AK; casual to northern AK and Europe. **POPULATION** Stable.

yellow supraloral spot

dark bill

tan and dark brown head stripes

tan-striped morph

white-striped morph

juvenile

white throat

richly colored back

WHITE-CROWNED SPARROW   *Zonotrichia leucophrys*   WCSP ▪ 1

One of the most common winter sparrows, the White-crowned Sparrow typically occurs in flocks, which may involve more than 100 birds. They feed in short grass or open areas adjacent to woodlands, hedgerows, or brush piles. They may pop out in response to pishing and raise their crown feathers when agitated. Polytypic (5 ssp.; all in N.A.). L 7" (18 cm)
IDENTIFICATION For bill color, see sidebar; whitish throat; brownish upperparts; mostly pale gray underparts. **ADULT:** Black-and-white striped crown. **IMMATURE:** Tan and brownish head stripes. **JUVENILE:** Brown and buff head; streaked underparts.
GEOGRAPHIC VARIATION See sidebar opposite.
SIMILAR SPECIES White-throated Sparrow has a dark bill, prominent white throat, and yellow supraloral. Immature similar to immature Golden-crowned, but note the latter's dark gray bill, indistinct head pattern, and dingy underparts.
VOICE CALL: Loud, metallic *pink.* Flocking birds give a husky chatter. **FLIGHT NOTE:** Sharp *tseep.* **SONG:** One or more thin, whistled notes followed by a variable series of notes.
STATUS & DISTRIBUTION Common in

brown and light tan head stripes

immature
leucophyrs

lower white stripe normally cut off just in front of eye

pinkish bill

adult
leucophyrs

Breeding Range

gambelii & leucophrys intergrade zone

gambelii & oriantha intergrade zone

gambelii

leucophrys

pugetensis

oriantha

nuttalli

the West, uncommon in East, rare in the Southeast. Winters south to central Mexico. **BREEDING:** Clumps of bushes or stunted trees on taiga and tundra; coastal scrub, chaparral for "Pacific" group birds. **WINTER:** Hedgerows, desert scrub, brushy areas, wood edges, and feeders. **MIGRATION:** Spring mid-

Mar.–late May, peaking early to mid-Apr., later in the Great Lakes Region (nominate) and in the West *(oriantha);* fall early Sept.–mid-Nov., peaking mid-Oct. **VAGRANT:** Casual to Europe, Azores, and Asia. **POPULATION** Stable.

Winter Range

pugetensis & gambelii

gambelii & leucophrys

nuttalli, pugetensis, & a few gambelii

gambelii

leucophrys

gambelii & oriantha

like *leucophrys,* but lateral crown stripe more extensive and darker

immature *oriantha*

bill darker red

adult like *leucophrys,* but slightly more blackish in supraloral, usually dark reddish bill, and slightly paler gray below

adult *oriantha*

yellow bill

head stripes duller

immature *pugetensis*

breeds in mountains of West; most winter in Mexico

like *oriantha,* but note head pattern and bill color

supraloral area whitish

yellow-orange bill

wintering White-crowned over much of West

juveniles of all subspecies streaked below

juvenile *gambelii*

immature *gambelii*

adult *gambelii*

blackish brown streaks down back with pale brown edges

many have dark malar streak

adult *pugetensis*

buffy wash on sides and flanks

White-crowned Sparrow Subspecies

The White-crowned Sparrow is divided into three well-defined subspecies groups: "dark-lored" (Rocky Mountain *oriantha* and eastern *leucophrys*), "Pacific" (northern *pugetensis* and southern *nuttalli*), and "Gambel's" (western *gambelii*). The supraloral area gives the best first clue: dark in "dark-lored," pale in "Pacific" and "Gambel's." Bill yellow in "Pacific," orange in "Gambel's," and pink in "dark-lored." "Pacific" birds often show a prominent malar stripe, rare in others. Primary extension is comparatively short in "Pacific," longest in "Gambel's," and intermediate to longish in "dark-lored." "Pacific" lacks gray striping on its back, showing only dark brown and tan; gray striping is fairly prominent on back of "Gambel's" and "dark-lored." "Pacific" also has more brownish coloration to its breast and flanks. First-winter birds show a distinctive difference that can be very hard to spot in the field: first-winter "Pacific" has a small patch of yellow feathers right at the bend of its wing; on "dark-lored" and "Gambel's," this patch is white.

Songs differ somewhat: rising and falling *zuuuu zeee jeee jeee zee* for "Gambel's"; clearer and more rapid with different introductory note and more trills (esp. last two notes) for "Pacific"; song of "dark-lored" is inconsistent between the two subspecies. Within the resident subspecies *nuttalli* there are many different song dialects, even from nearby locations. Call notes are flatter in "Pacific;"

oriantha is sharp, almost like a Blue Grosbeak's call. Another behavioral distinction is that "Pacific" birds place their nests in low bushes, while "dark-lored" and "Gambel's" birds place it directly on leaf litter.

Subspecies *leucophrys* of "dark-lored" group breeds mainly in the Canadian tundra east of Hudson Bay. In the same group, *oriantha* breeds in the Rocky Mountains and in the higher elevations of the Sierra and Cascade, and very locally in southern California. Subspecies *pugetensis* and *nuttalli* of "Pacific" group reside along Pacific coast from extreme southeastern Alaska to southern California. Subspecies *gambelii* breeds from western Alaska to Hudson Bay. "Pacific" *pugetensis* is rare to uncommon in Central Valley and rare south of Los Angeles Co. on the coast. Inland in southern California it is casual (accidental in southern Nevada). "Gambel's" is rare to casual in the East; "dark-lored" is casual during migration in coastal California. Intergradation commonly occurs between *leucophrys* and *gambelii* in the Hudson Bay region. Despite its very similar appearance to *leucophrys, oriantha* may actually be more distantly related to *leucophrys* than *gambelii* is to *leucophrys*. The subspecies *pugetensis* has recently been found breeding in Cascade mountain passes north of Seattle alongside *gambelii;* the two subspecies appear to be acting as separate species. ∎

HARRIS'S SPARROW *Zonotrichia querula* HASP ■ 1

The behavior of this sparrow is much like that of the White-throated. Regularly occurring in medium-size flocks, it mixes sometimes with the White-throated, White-crowned, or other sparrows. Monotypic. L 7.5" (19 cm)

IDENTIFICATION Large sparrow, larger than other *Zonotrichia*. Bright pink bill; black crown, face, and bib variable by age and plumage; all show post-ocular spot. **BREEDING ADULT:** Extensive black from crown to throat; gray face. **WINTER ADULT:** Buffy cheeks; throat all black or with white flecks or partial white band. **IMMATURE:** Resembles winter adult with less black; white throat bordered by dark malar stripe. **SIMILAR SPECIES** Only other sparrows with black face are otherwise

dissimilar Black-throated and Black-chinned. Male House Sparrow (not a true sparrow; see p. 720) has same black throat, but is smaller, has bolder wing bars and a chestnut crown, does not have bright pink bill, and behaves totally unlike an emberizid sparrow. **VOICE CALL:** Loud *wink*. Flocking birds give a husky chatter. **FLIGHT NOTE:** A drawn-out *tseep*. **SONG:** Series of long, clear wavering whistles, often beginning with two notes on one pitch followed by two notes on another pitch. **STATUS & DISTRIBUTION** Uncommon to fairly common. **BREEDING:** Stunted boreal forest. **WINTER:** Open woodlands, brushlands, hedgerows. **MIGRATION:** Spring late Mar.–late May;

fall early Oct.–early Nov. **VAGRANT:** Rare to casual east and west of narrow range. **POPULATION** Stable.

black crown and bib

large pink bill

breeding

winter adults

dark postocular spot in all plumages

dark malar stripe

dark chest patch

brownish flank streaking

immature

GOLDEN-CROWNED SPARROW *Zonotrichia atricapilla* GCSP ■ 1

The adults of this Pacific species are distinctive, but the immatures are the plainest-faced *Zonotrichia*. The bird prefers denser brush than White-crowned, and its behavior is similar to White-throated. Monotypic. L 7" (18 cm)

IDENTIFICATION Brownish back streaked with dark brown; breast, sides, and flanks grayish brown. Bill dusky above, pale below. **ADULT:** Yellow patch tops black crown; less distinct in winter. **IMMATURE:** Yellow on forehead more restricted and duller, and bordered by brown rather than blackish lateral

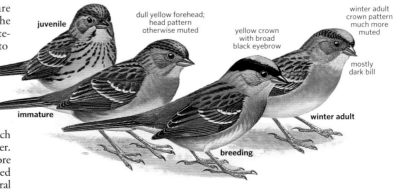

juvenile

dull yellow forehead; head pattern otherwise muted

immature

yellow crown with broad black eyebrow

breeding

winter adult crown pattern much more muted

mostly dark bill

winter adult

crown stripes. JUVENILE: Dark streaks on breast and sides. Plumage briefly held. SIMILAR SPECIES White-crowned Sparrows have distinct head striping at all ages; Golden-crowned has a plainer, brown face, with almost no contrast, darker underparts, and dusky bill. Yellow crown is usually present as a trace of color on forecrown and above eyes. VOICE CALL: A flat *tsick*. Flocking birds give a husky chatter. FLIGHT NOTE: A soft *tseep*. SONG: A series of three or more plaintive, whistled notes, often with

each on a descending note: *oh dear me.* STATUS & DISTRIBUTION Fairly common. Winters south to northern Baja California. BREEDING: Stunted boreal bogs and in open country near tree line, especially in willows and alders. WINTER: Dense woodlands, tangles, brush, chaparral. MIGRATION: Spring migration mid-Mar.–mid-May, peaking mid-Apr; fall migration mid-Sept.–early Nov., peaking mid-Oct. VAGRANT: Casual to the East. POPULATION Stable.

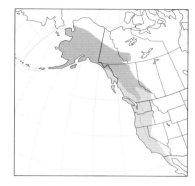

JUNCOS Genus *Junco*

This genus includes three species (or, according to some authors, four or more), two of which occur north of Mexico. These birds are shades of gray above and white below, with pink bills and prominent white outer tail feathers. Their behavior (often feeding on the ground singly or in groups; flying up suddenly to a tree) may recall *Spizella* sparrows.

DARK-EYED JUNCO *Junco hyemalis* DEJU 1

Dark-eyed Juncos are unique sparrows that nest on or near the ground in forests. In winter, they typically form flocks and often associate with other species, including Chipping Sparrows, Pine and Palm Warblers (in the southeastern U.S.), and bluebirds. Polytypic (14–15 ssp.; 12 in N.A.). The subspecies *insularis* from Guadalupe I., Mexico should perhaps be regarded as a separate species and is not included in the subspecies groups below. L 6.3" (16 cm)
IDENTIFICATION A fairly lean sparrow with a long notched tail and a small pinkish or horn-colored bill (bicolored in *dorsalis*). Two prominent white outer tail feathers in most subspecies; three outermost in "Whitewinged." Most subspecies have a gray or brown head and breast sharply set off from a white belly, but are otherwise highly variable. (See Geographic Variation.) MALE: Typically darker with sharper markings. FEMALE: Typically browner with more indistinct markings. JUVENILE: Heavily streaked,

often with a trace of adult pattern. GEOGRAPHIC VARIATION The 12 subspecies in N.A. show marked variation and fall into five major groups: "Slate-colored" (2 subtle ssp., and *cismontanus*), "Oregon" (5 subtle ssp. in N.A., 2 additional ssp. in Baja California Norte), "White-winged" (1 ssp.), "Pink-sided" (1 ssp.), and "Gray-headed" (2 distinctive ssp.) The groups have at times been considered separate species. "Slate-colored" is the most widespread and only form found regularly in the East. It breeds throughout the species' range east of the Rockies and in the northern region; it winters mainly in the East and is uncommon to rare in

the West. "Oregon Junco" breeds in the West Coast states north to southern Alaska and east to central Nevada and western Montana; it winters

Breeding Range

"Oregon" ♀
shufeldti

grayish hood

buffy rufous sides

blackish hood

"Oregon" ♂
shufeldti

all subspecies have white outer tail feathers

"Slate-colored" ♀
hyemalis

"Slate-colored"
hyemalis

some have faint white tips to wing coverts, similar to "White-winged"

females browner above

juvenile

all juvenile juncos are streaked

throughout the West and Great Plains and is casual to the East. "White-winged Junco" is the most local, breeding exclusively in the Black Hills region and wintering along the eastern edge of the Rockies; it is casual to accidental in western Texas, Arizona, and southern California. "Pink-sided" breeds in the northern Rockies, centered in Yellowstone N.P. and ranging from northern Utah to the Cypress Hills of extreme southeastern Alberta and southwestern Saskatchewan; it winters in the southern Rockies, Southwest, and western Great Plains, rarely to the West Coast, and is accidental to the East. "Gray-headed" is the subspecies of the southern Rockies, breeding through much of Nevada, Utah, and Colorado south to central Arizona and western Texas; it winters in the southwest and southern Rockies states and is rare to the West Coast and accidental to the East. The male "Slate-colored Junco" has a white belly contrasting sharply with a dark gray hood and upperparts, usually with very little contrast between the hood and back; immatures can have some brown wash on the back and crown. In female, the amount of brown on head and at center of back varies; it's more extensive in immatures. "Slate-colored Junco" comprises two subspecies: widespread nominate and larger, bluer-billed *carolinensis,* which is resident in the Appalachians from Pennsylvania to northern Georgia. An additional subspecies, *cismontanus,* is often grouped with "Slate-colored." It breeds from the Yukon to central British Columbia and Alberta and may winter throughout

the West; it is casual to the East. In coloration, *cismontanus* is intermediate between "Slate-colored" and "Oregon," with males showing a blackish hood that contrasts with a usually grayish back (occasionally with some brown). Females and immatures are very similar to the "Oregon" Juncos, but are less distinctly hooded. Male "Oregon" Junco has a slaty to blackish hood, contrasting sharply with its rufous-brown to buffy-brown back and sides; female has duller hood color. Of the five "Oregon" subspecies, the more southerly subspecies are paler. Distinctive "White-winged Junco," *aikeni,* is mostly pale gray above, usually with two thin white wing bars; it is also larger and bigger-billed, with more white on its tail. It is most similar to "Slate-colored" (which can rarely have narrow wingbars) but is larger and paler, with contrasting blackish lores and more extensive white in tail. "Pink-sided" Junco, *mearnsi,* has broad, bright pinkish cinnamon sides, a blue-gray hood, a poorly defined reddish brown back and wings that do not contrast markedly with flanks, and blackish lores. Females duller, but retain basic pattern; they can resemble "Oregon" females closely. In "Gray-headed" Junco," pale gray head and dark lores resemble head pattern of "Pink-sided," but flanks are gray rather than pinkish, and back is marked by a very well-defined patch of reddish hue that does not extend to wings and that contrasts sharply with rest of body. A distinctive subspecies, *dorsalis,* is sometimes known as "Red-backed Junco" and is resident from northwestern Arizona through New Mexico to the Guadalupe Mountains of western Texas. It differs from the more widespread, migratory, northerly breeding

caniceps in having an even paler throat and a larger, bicolored bill that is black above and bluish below. Intergrades between some subspecies are frequent. Common intergrades are: "Pink-sided" x "Oregon" and "Pink-sided" x "Gray-headed." Subspecies *cismontanus* may be a broad intergrade population of "Oregon" x "Slate-colored Juncos." Identification to subspecies group thus requires caution to eliminate the possibility of an intergrade; for intergrades, look for intermediate characteristics: For example, a darker, more contrasting hood on a "Pink-sided" indicates the influence of "Oregon" genes; reduced pink sides and a well-defined reddish back on a "Pink-sided" indicate "Gray-headed" parentage.

SIMILAR SPECIES See Yellow-eyed Junco.

VOICE Songs and calls among subspecies are generally similar, but songs and calls of "Gray-headed" *dorsalis* are more suggestive of Yellow-eyed Junco. **CALL:** Sharp *dit.* **FLIGHT NOTE:** A rapid twittering. **SONG:** A musical trill on one pitch; often heard in winter.

STATUS & DISTRIBUTION Common. Breeds south to northern Baja California; winters south to northern Mexico. **BREEDING:** Breeds in coniferous or mixed woodlands. **WINTER:** Found in a wide variety of habitats, Dark-eyed Junco tends to avoid areas of denser brush; it especially favors feeders, parks, and open forest without an understory. **MIGRATION:** Withdraws from wintering areas during Apr., typically early–mid-Apr. Fall arrivals first appear in late Sept., peaking in late Oct. **VAGRANT:** Southern FL and Europe. **POPULATION** Stable.

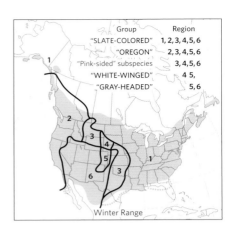

Group	Region
"SLATE-COLORED"	1, 2, 3, 4, 5, 6
"OREGON"	2, 3, 4, 5, 6
"Pink-sided" subspecies	3, 4, 5, 6
"WHITE-WINGED"	4, 5,
"GRAY-HEADED"	5, 6

Winter Range

larger and paler gray than "Slate-colored"

"White-winged" *aikeni*

♂

faint white tips to wing coverts create wing bars

bluish gray hood and dark lores

♂

pale gray head and underparts

pale bill

dark lores

broad pinkish buff sides and flanks

"Pink-sided" *mearnsi*

"Gray-headed"

rufous back

caniceps

mostly dark upper mandible

paler throat than *caniceps*

dorsalis

sometimes called "Red-backed Junco"

pinkish buff meets across lower breast on some

YELLOW-EYED JUNCO   *Junco phaeonotus*   YEJU   ■   2

This distinctive bird, with piercing yellow eyes, replaces the Dark-eyed Junco in certain mountain ranges adjacent to the Mexican border. It behaves much like Dark-eyed, but it rarely forms large flocks. Polytypic (5 ssp.; *palliatus* in N.A.). L 6.3" (16 cm)
IDENTIFICATION Bright yellow eyes, black lores. Bicolored bill with dark maxilla, pale mandible. Pale gray above; bright rufous back; rufous-edged greater coverts and tertials; pale gray underparts. **JUVENILE:** Similar to juvenile "Gray-headed" Dark-eyed Junco, but with rufous wing panel like adult; eye is brown, becoming pale

before changing to yellow of adult.
GEOGRAPHIC VARIATION Five subspecies show marked variation; only *palliatus* occurs north of Mexico.
SIMILAR SPECIES "Gray-headed" group of Dark-eyed Junco is most similar. Yellow-eyed differs not only in eye color, but also in rich rufous on greater coverts and tertials. (Beware "Gray-headed" x "Pink-sided" intergrades, which may show dull rufous on wings.) Note also that only *dorsalis* ("Gray-headed" group) has bicolored bill.
VOICE CALL: Sharp *dit,* softer than Dark-eyed Junco. **FLIGHT NOTE:**

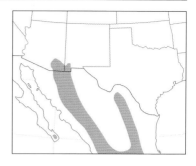

A high, thin *seep,* similar to call of Chipping Sparrow, lower than call of Dark-eyed Junco; also rapid twitter as in Dark-eyed. **SONG:** A variable series of clear, thin whistles and trills.
STATUS & DISTRIBUTION Fairly common. Breeds in mountains south to Guatemala. In the U.S., found only in southeastern AZ and southwestern NM (Peloncillo and Animas Mts., Hidalgo Co.); recently a few discovered near Silver City. Coniferous and pine-oak slopes, generally above 6,000 feet. Generally resident, but some move to lower altitude in winter. **VAGRANT:** Casual to western TX.
POPULATION Stable.

striking yellow eye

rufous back

bicolored bill

palliatus

pale throat

rufous greater coverts

juvenile
palliatus

Genus *Emberiza*

The 39 species in this Old World genus make it the largest in the Emberizidae. *Emberiza* buntings are typically small to medium-size and are sexually dimorphic—males often boldly marked and quite colorful. Most have white outer tail feathers. There are often gregarious in migration and winter and sometimes form mixed-species flocks. They feed on the ground, much like juncos, and fly up to a tree when disturbed. Nine *Emberiza* buntings are vagrants to North America, though the Rustic Bunting is annual on the westernmost Aleutians. Details of plumage pattern, as well as call notes, are important for identification.

PINE BUNTING   *Emberiza leucocephalos*   PIBU   ■   5

Two fall records of the Pine Bunting from Attu I., and one from St. Paul I., AK, are the sum of its North American occurrences. Polytypic

dusky chestnut head with white cheek patch

red-brown flank streaks and rump

fall adult ♂
leucocephalos

(2 ssp.; nominate in N.A.). L 6.5" (17 cm)
IDENTIFICATION A largish, fairly long-tailed bunting with a medium bill. **MALE:** Distinctive white cheek is unique. In breeding plumage, male is rusty about face and throat, and it has a white median crown stripe. In winter plumage, rust is replaced by gray-brown and median stripe is lost. **FEMALE:** Rusty supercilium; underparts are whitish and finely streaked; white cheek is less prominent than on male; rusty rump. **IMMATURE:** Like female, but often duller and lacking white cheek patch entirely, though it has a white spot (at least) at rear auriculars, similar to that found on Rustic Bunting.

GEOGRAPHIC VARIATION Pine Bunting comprises two subspecies, but only *leucocephalos* has been found in North America.
SIMILAR SPECIES White cheek patch is distinctive on adults. Immatures could be confused with either Rustic Bunting or Little Bunting. Note, however, Pine Bunting's weak malar; its lack of both a strong eye ring and a crest; and its dull or whitish cheek, which is never chestnut like that found on Little Bunting. **FEMALE & IMMATURE:** Yellow-breasted Bunting is also similar to Pine Bunting; however, Yellow-breasted always shows some tint of yellow on face or underparts.
VOICE CALL: Abrupt *spit* or *tic,* often

doubled, *spi-tit;* also a hoarse *jeeit.* **SONG:** Long and variable, repetitious warbling phrases. **STATUS & DISTRIBUTION** Pine Bunting breeds across northern Russia; winters south to central-south Asia. **VAGRANT:** Accidental to Attu Island, AK, where there are two records: Nov. 18–19, 1985, and Oct. 6, 1993; and St. Paul Island, Pribilofs (Oct. 2, 2012). Casual in western Europe.

LITTLE BUNTING   *Emberiza pusilla*   LIBU   ■   4

First recorded in Alaska in 1970, this species had only three records between 1970 and 1990. Since then it has been found repeatedly in fall at Gambell (St. Lawrence I.), and it may prove annual with continued coverage. Monotypic. L 5" (13 cm)
IDENTIFICATION Small, with short legs, a short tail, and a small triangular bill;

fall birds, especially immatures, are duller
small size
immature
chestnut median crown stripe and auricular with broad dark lateral crown stripes
prominent whitish eye ring
straight culmen
breeding ♂

bold creamy white eye ring, chestnut ear patch, and two thin pale wing bars. Whitish and heavily streaked underparts; white outer tail feathers. **BREEDING ADULT:** Chestnut crown stripe bordered by black stripes; many males have chestnut on chin. **IMMATURE & WINTER ADULT:** Chestnut crown, tipped and streaked with buff and black.
SIMILAR SPECIES Female Rustic may be confused with Little Bunting. Note Rustic Bunting's larger size, heavier bill with pink lower mandible, diffuse rusty streaking below, and lack of eye ring.
VOICE CALL: Sharp *tsick.* **SONG:** Short, variable series of rising and falling notes.

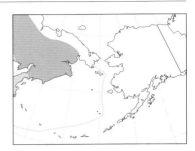

STATUS & DISTRIBUTION Common. Breeds from Scandinavia to Russian Far East; winters from Nepal and India to S.E. Asia. Rare in western Europe. **VAGRANT:** Very rare to St. Lawrence I. in fall (late Aug.–early Oct.); accidental elsewhere in western AK (3+ recs.), OR (Jan. 28, 2013), CA (3 recs., Sept.–Nov.), and Baja California Sur.

RUSTIC BUNTING   *Emberiza rustica*   RUBU   ■   3

A regular migrant in extreme western Alaska, this attractive bunting is also the most regular *Emberiza* south of Alaska. Regarded by most as monotypic. L 5.8" (15 cm)
IDENTIFICATION Has a slight crest, whitish nape spot, and a prominent pale line extending back from eye; white outer tail feathers. **BREEDING MALE:** Black head; prominent white supercilium; upperparts bright chestnut, with buff and blackish streaks on back; white underparts, with chestnut breast band and streaks on sides. **FEMALE:** Brownish head pattern, with pale spot at rear of ear patch. **IMMATURE**

& WINTER MALE: Similar to female. **SIMILAR SPECIES** Lapland Longspur is similar to Rustic Bunting, but its behavior is different (e.g., longspur is unlikely to perch in a bush). Among other plumage differences, Lapland lacks pale ear spot, crested appearance, and rufous streaking below. Rustic differs from female and immature Reed Bunting and from Pallas's Bunting in its crested appearance; its stronger face pattern, pale median crown stripe, and small pale nape spot; extensive rufous streaking on breast and flanks; and its more prominent wing bars. See also Little Bunting.

pale spot at rear of crown and in ear coverts in all plumages
straight culmen
chestnut markings below
♀
often raises slight crest
breeding ♂
black-and-white head pattern
bright chestnut breast band and streaks down sides and flanks

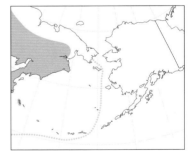

VOICE CALL: Hard, sharp *jit* or *tsip,* lower and sharper than Little Bunting's call. **SONG:** A soft bubbling warble.
STATUS & DISTRIBUTION Common. Eurasian species; breeds from northern Scandinavia east to Kamchatka and northern Sakhalin I. (Russian Far East). It winters primarily in eastern China, Korea, and Japan. Rare in western Europe. **MIGRATION:** Uncommon spring migrant on western Aleutians, rare in fall; very rare on other islands in Bering Sea. **VAGRANT:** Casual in fall and winter on West Coast from southern AK south to CA.

GRAY BUNTING   *Emberiza variabilis*   GRBU ■ 5

This is a large bunting with a heavy, pinkish bill. It shows no white in its tail. Regarded by most as monotypic. L 6.8" (17 cm)

IDENTIFICATION Bill is pinkish at base. **BREEDING MALE:** Gray overall, prominently streaked with blackish on back and wings. **WINTER & IMMATURE MALE:** Broadly edged with buff on upperparts, paler below. Immature plumage is largely held through first spring.

ADULT FEMALE: Patterned like male but brown instead of gray; chestnut rump and tail is conspicuous in flight.

SIMILAR SPECIES Male Gray could be confused with "Slate-colored" Junco, but it has a gray belly and lacks white in tail. Female Gray is like other *Emberiza,* only large, with a thick bill and a distinctive rusty rump and tail, without white outer tail feathers.

VOICE CALL: Sharp *zhii*. **SONG:** Warbling series of loud notes.

STATUS & DISTRIBUTION Breeds in East Asia; winters primarily in Japan. **VAGRANT:** Three records from western Aleutian Islands, AK: May 18, 1977; May 27, 2005 (Shemya Is.); and May 19, 1980 (Attu I.).

distinctive head pattern with gray on sides of nape

large size

♀

rufous-brown rump, often best seen in flight

even by first spring plumage, intermediate between male and female

mostly pale bill in all plumages

overall dark gray with black back streaks

♂

immature ♂

lacks white in tail in all plumages, unusual in *Emberiza* buntings

YELLOW-THROATED BUNTING   *Emberiza elegans*   YTBU ■ 5

This East Asian species has been recorded only once from North America, in western Alaska. Polytypic (2 ssp.; probably *elegans* in N.A.). L 6" (15 cm)

IDENTIFICATION Adults prominently crested. **MALE:** Black auriculars; supercilium white at front, yellow at rear; prominent black crest over yellow crown; yellow throat; brown, streaked back; whitish breast with brownish streaking. **FEMALE & IMMATURE:** Duller version of male; auricular brownish.

GEOGRAPHIC VARIATION Nominate *elegans* (includes *ticehursti*) breeds in southeastern Russian Far East, northeast China, Korea, and Tsushima I., Japan. Winters western Japan, Korea, and eastern China. Darker *elengantula,* with strong blackish ventral streaking, is largely a resident in central China.

VOICE CALL: Sharp *tzick*. **SONG:** Long series of sweet rollicking phrases.

STATUS & DISTRIBUTION Breeds in eastern Asia; winters in coastal eastern Asia. **VAGRANT:** One record: Attu I., AK (May 25, 1998).

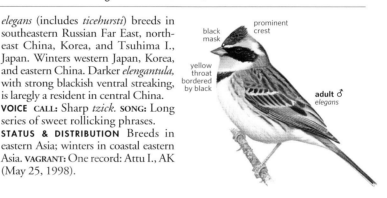

black mask

prominent crest

yellow throat bordered by black

adult ♂
elegans

YELLOW-BREASTED BUNTING   *Emberiza aureola*   YBSB ■ 5

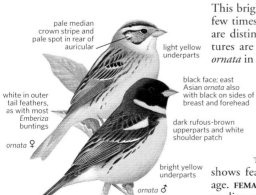

pale median crown stripe and pale spot in rear of auricular

light yellow underparts

white in outer tail feathers, as with most *Emberiza* buntings

ornata ♀

black face; east Asian *ornata* also with black on sides of breast and forehead

dark rufous-brown upperparts and white shoulder patch

bright yellow underparts

ornata ♂

This bright bunting has appeared a few times in western Alaska. Males are distinctive; females and immatures are subtle. Polytypic (2 ssp.; *ornata* in N.A.). L 6" (15 cm)

IDENTIFICATION White outer tail feathers. **BREEDING MALE:** Rufous-brown upperparts, black face, bright yellow underparts, white patch on lesser and median wing coverts. **WINTER ADULT MALE:** Usually shows features of breeding plumage. **FEMALE:** Striking head pattern: median crown stripe; ear patch with dark border and pale spot in rear; yellowish underparts with sparse streaking; unmarked belly; brown lateral crown stripes with internal black streaking; whitish median crown stripe. **IMMATURE:** Like female, often paler yellow.

SIMILAR SPECIES Pine, Reed, and Pallas's lack yellow wash, striking face.

VOICE CALL: Sharp *tzip,* similar to Rustic's. **FLIGHT NOTE:** Flat *stuck*.

STATUS & DISTRIBUTION Breeds Scandinavia to Russian Far East; winters mainland S.E. Asia. Rare in western Europe. **VAGRANT:** Casual in AK (five records): St. Lawrence I. has two (late June, Sept.), Buldir I. has one (late June), and Attu I. has two (both late May).

POPULATION Recent sharp declines due to illegal trapping for food at migration and, in particular, wintering sites in southern China and Southeast Asia.

YELLOW-BROWED BUNTING   *Emberiza chrysophrys*   YBWB ■ 5

Among the Eurasian buntings (family Emberizidae), Yellow-browed has one of the smallest breeding ranges, limited to a rather small area in eastern Russia. Its scientific name, *chrysophrys*, is derived from the Greek words for "golden-browed," and in fact the golden portion of the supercilium, strongly recalling White-throated Sparrow, distinguishes this species from others in the genus. Monotypic. L 5–5.5" (13–14 cm)

slight crest
broad yellowish supercilium
pale spot in rear auricular
fall

IDENTIFICATION A relatively large-headed and crested bunting, similar in size to Rustic Bunting and slightly larger than Little Bunting. Stout bill with dusky maxilla, pink mandible. Legs pale pink. Head pattern quite striking in all plumages: narrow white median crown stripe, dark lateral crown stripe, supercilium yellow mostly (becoming whitish behind eye), dark auriculars with pale spot at rear corner, white malar, and blackish submalar line. Head pattern most striking in breeding adult male, in which auriculars are black. Underparts mostly whitish, with flanks and breast washed buff and with fine streaks. Rump and uppertail coverts tinged reddish brown in all plumages. Crown feathers often raised to a crest. **SIMILAR SPECIES** In female and immature plumages, care should be taken to exclude Tristram's Bunting *(E. tristrami),* which lacks yellow in supercilium, and the much less strikingly patterned Yellow-breasted Bunting *(E. aureola),* which is washed with buffy yellow in face and underparts and lacks dorsal rufous tones of Yellow-browed. **VOICE CALL:** A short, sharp *tsick.* **SONG:** A short, variable series of sweet and more modulated notes; often starts with a soft, long note. **STATUS & DISTRIBUTION BREEDING:** Mixed forest edge in southeastern Siberia and Russian Far East. **WINTER:** Central and eastern China. **VAGRANT:** One at Gambell, St. Lawrence I., AK (Sept. 15, 2007), represents the only N.A. record. **POPULATION** Uncommon but apparently stable in its limited range.

PALLAS'S BUNTING   *Emberiza pallasi*   PALB ■ 5

A vagrant from Asia, the Pallas's Buntings often flicks its tail when perched. Polytypic (4 ssp.; *polaris* in N.A.). L 5" (13 cm)
IDENTIFICATION White outer tail feathers. Like Reed, but smaller; shorter tail; smaller, two-toned bill with pink base (black in breeding male) and straight culmen; grayish lesser coverts (diagnostic but often concealed). **BREEDING MALE:** Like breeding male Reed, but lacks all rust. **FEMALE & IMMATURE:** Like female Reed, but mostly unmarked below, lacks median crown stripe, has less distinct eyebrow and lateral crown stripe, browner rump.

VOICE CALL: Loud *cheeep,* recalling House Sparrow's, very unlike Reed Bunting's call. **STATUS & DISTRIBUTION** Pallas's Bunting breeds from north-central Siberia east to the Chukotskiy Peninsula, Russian Far East; also breeds Mongolia; winters primarily in coastal eastern Asia. **VAGRANT:** Three spring and three fall (all in Sept. from Gambell, St. Lawrence I.) AK records; a specimen at Barrow (June 11, 1968) was *polaris.* Accidental in British Isles.

smaller and smaller billed than Reed with straight, not curved, culmen
pale lower mandible
no median crown stripe
polaris ♀
pale wing bars
striking head pattern
breeding ♂ *polaris*
immatures often have breast streaking in fall
juvenile *polaris*

REED BUNTING   *Emberiza schoeniclus*   REBU ■ 4

Like the Pallas's Bunting, the Reed Bunting is a casual visitor to western Alaska. Polytypic (18 ssp.; *pyrrhulina* in N.A.). L 6" (15cm)
IDENTIFICATION Reed has solid chestnut lesser wing coverts. Note also its gray bill with curved culmen; cinnamon wing bars; dark lateral crown stripes; and paler median crown stripe. **BREEDING MALE:** Black head, throat; broad white submoustachial stripe; white nape; upperparts streaked black and rust; gray rump. White underparts with thin reddish streaks alongsides and flanks. **FEMALE:** Pale buffy rump, broad buffy-white eyebrow.

IMMATURE: Like female, but more buff. **GEOGRAPHIC VARIATION** Fifteen subspecies; U.S. records are east Asian *pyrrhulina.* **SIMILAR SPECIES** See Pallas's and Rustic. **VOICE CALL:** A falling *seeoo.* **FLIGHT NOTE:** A hoarse *brzee.* **STATUS & DISTRIBUTION** Common. Widespread Eurasian species. **VAGRANT:** Casual on westernmost Aleutians (about 10 recs.) in late spring; accidental in fall to Gambell, St. Lawrence I. (Aug. 28–30, 2002).

paler brown median crown stripe
curved culmen
dark bill
striking head pattern
contrasting rufous lesser coverts, often hidden
extensive rufous on wing
pyrrhulina ♀
breeding ♂ *pyrrhulina*
fall ♂ *pyrrhulina*
black on head veiled on adult male Reed and Pallas's in fresh fall plumage

CARDINALS AND ALLIES Family Cardinalidae

Dickcissel, male (OH, July)

In North America, the Cardinalidae is represented by some of the region's brightest and most striking species, such as the Northern Cardinal; Indigo and Painted Buntings; and Scarlet, Western, and Summer Tanagers. Members of the family have stout to very thick bills. Most are strongly sexually dimorphic, with males typically uniquely colored and unmistakable, but females, and in some cases immature males, being duller and more challenging to identify. Important features used to identify similar females of closely related species include bill size and shape, overall color, and presence or absence of wing bars. All breeding species except the Northern Cardinal are nocturnal neotropical migrants. The Rose-breasted Grosbeak and Indigo Bunting are among the more common species seen during migratory fallouts along the Gulf Coast, and Western Tanagers are familiar migrants through the West.

Structure Considered part of the nine-primaried oscines, members of the Cardinalidae family are variable in size, with cardinals, grosbeaks, and *Piranga* tanagers relatively large for passerines and buntings relatively small (about the size of a sparrow). Most have distinctly shaped conical bills adapted for eating seeds and fruit. Some species have disproportionately large bills (Yellow Grosbeak). Species in the genus *Cardinalis* are characterized by a long, pointed crest.

Behavior They mainly feed on a combination of insects (breeding season), seeds, and fruit (fall and winter). Grosbeaks in particular switch to almost exclusively fruit beginning in late summer. Some are conspicuous singers during the breeding season (Northern Cardinal and Blue Grosbeak); others sing from inside dense vegetation (Painted Bunting). Songs are mostly similar across the family: a series of somewhat robin-like phrases, often paired; the Dickcissel's harsher song is an outlier. Calls usually consist of sharp single notes, often metallic in quality and difficult to differentiate within a genus. Members of the genus *Passerina* (buntings and Blue Grosbeak) are known for tail-twitching behavior. Except for the Northern Cardinal, all are medium- to long-distance nocturnal migrants, wintering from Mexico into northern South America.

Plumage Bright male plumages make these some of the most distinctive and recognizable species in North America. Females of many species are equally famous for their dull brown or green plumages and difficult identifications. Males attain adult plumage by their second winter, normally with distinct first-summer plumage. Immature males typically resemble adult females and are well known to establish territories and sing. Winter adult males usually have brown edging to body feathers, obscuring bright plumage.

Distribution The family occurs from Canada south through tropical South America. Members of the genus *Cardinalis* are generally resident, while all other North American cardinalids vacate their breeding grounds and migrate south for winter. Cardinalids are generally more prevalent south of the United States, with no fewer than 30 additional species found exclusively in Central and South America. Fourteen species regularly breed in North America, and four others occur as vagrants from Mexico. The Crimson-collared Grosbeak and Blue Bunting are casual in the Rio Grande Valley, while the Yellow Grosbeak is casual in southeastern Arizona. The Flame-colored Tanager is very rare but regular in southeastern Arizona and has bred. Species that breed in N.A. are found in a variety of habitats, from deciduous forest canopies to oak-conifer woodlands, arid brushy hillsides, and weedy meadows. Several species are associated with riparian woodlands.

Taxonomy The Cardinalidae is closely allied with the Emberizidae, Thraupidae, and other nine-primaried songbird families; the family includes 47 species in 14 genera. The Dickcissel, in the genus *Spiza*, has sometimes been aligned with the icterids, but genetic studies place it with the cardinalids; the family also includes the three *Granatellus* chats (Mexico to South America) formerly considered wood-warblers. Closely related species in the genus *Pheucticus* (grosbeaks) hybridize when they come in contact, as do Indigo and Lazuli Buntings in the genus *Passerina*.

Conservation As most species are found in a variety of disturbed and semidisturbed habitats, they are generally considered common and not threatened at present. Pesticide use in the neotropics may adversely affect Dickcissel populations on the wintering grounds. BirdLife International considers four neotropical species as near-threatened. In the East, the Scarlet Tanager has had its population decline because of forest fragmentation and negative effects from cowbird parasitism. Western populations of the Summer Tanager are declining because of loss of riparian habitats. The Painted Bunting has declined greatly in parts of its range. In addition to habitat issues, many neotropical species have suffered from exploitation for the cage-bird trade.

TANAGERS Genus *Piranga*

After recent genetic studies the *Piranga* tanagers were moved out of family Thraupidae (Tanagers) and placed in family Cardinalidae. The nine species, exclusively found in the Americas, include four regular breeding and one casual visitor in the ABA area. Males are mostly bright red in plumage; females are duller, greenish yellow. North American species are neotropical migrants.

SUMMER TANAGER *Piranga rubra* SUTA ■ 1

The brightly plumaged male Summer Tanager is one of the more spectacular breeding birds of North America. The uniform blood-red feathers seen against a bright green background are quite a sight. Summer feeds mainly on fruit, except during breeding season. Rather large and often sluggish, it usually sits still for long periods of time. Quite vocal, Summer is often detected by its distinctive call. The bill is generally bulky and long and ranges from gray to pale horn in color. The head often shows a slight crested appearance. Polytypic (2 ssp.; both in N.A.). L 7.8" (20 cm)
IDENTIFICATION Sexually dimorphic. **BREEDING MALE:** Adult male is unmistakable—all bright red—and achieves its brightest plumage by end of its second calendar year. **BREEDING FEMALE:** Generally mustard yellow-orange, with greener wings and upperparts. Some may have dull red mixed in plumage or may lack orange tones, resembling quite closely immature female Scarlet Tanagers. **FIRST-SPRING MALE:** Mixture of red and greenish yellow, sometimes blotchy, sometimes with entirely red head and breast. **FIRST-WINTER FEMALE:** Can lack all mustard tones and look very similar to a Scarlet Tanager.
GEOGRAPHIC VARIATION Adult male eastern *rubra* is deeper red than adult male western *cooperi*. Both sexes of *cooperi* are larger, paler overall, and have larger and paler bills than *rubra*.
SIMILAR SPECIES Adult males differ from Scarlet Tanager by lacking black wings. Male Hepatic Tanager is more brick red in color and has a grayer bill,

grayer flanks, and a dark grayish cheek patch. Also note difference in call notes. Females lacking all orange tones more problematic and can be confused with either female Scarlet or Hepatic Tanagers. Note yellow undersurface of tail in Summer, lacking in Scarlet. Summer also has a longer tail and longer bill than Scarlet. Female Hepatics less uniform underneath, with brighter yellow throats and crowns contrasting with grayer underparts and back.
VOICE CALL: A distinctive *pi-tuck* or *pi-ti-tuck* or *ki-ti-tuck,* sometimes extended to several notes. **SONG:** An American Robin–like series of warbling phrases. Easily confused with Rose-breasted's or Black-headed Grosbeak's. **FLIGHT NOTE:** A soft, wheezy *verree.* **JUVENILE:** Begging note similar to Black-headed Grosbeak's *veeooo.*

arrive mid-Apr.–early May. **MIGRATION:** Nominate *rubra* is mainly a trans-Gulf migrant, common from along upper TX coast to coastal FL. Regular at the Dry Tortugas. **WINTER:** Mainly southern Mexico through C.A., uncommonly to northern S.A. Rare in winter in southern U.S. from FL to CA. Accidental in U.K.
POPULATION Western population threatened by loss of riparian habitats.

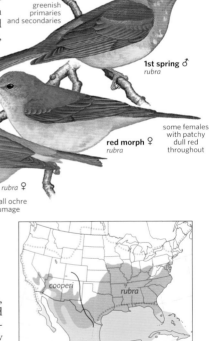

red head and patches elsewhere

greenish primaries and secondaries

1st spring ♂
rubra

overall rosy red

large bill

some females with patchy dull red throughout

red morph ♀
rubra

rubra ♀

overall ochre plumage

adult ♂
rubra

cooperi

rubra

STATUS & DISTRIBUTION In the East, common in the canopy of oak and pine-oak woodland. In the Southwest, common in cottonwood-willow habitats along permanent streams and rivers. **BREEDING:** Nesting birds

HEPATIC TANAGER *Piranga flava* HETA ■ 2

Fairly common throughout the pine and pine-oak forests of the Southwest, the male Hepatic Tanager is often seen singing from exposed perches and is normally found in different habitat and at higher elevation than the similar, brighter Summer Tanager.

Mainly insectivorous during summer, Hepatic is known to eat fruit during fall and winter. Polytypic (15 ssp.; 2 in N.A.). (15 ssp.). L 8" (20 cm)
IDENTIFICATION MALE: Entirely dark brick-red body, mixed with gray on back and flanks. Auriculars are gray,

and bill is heavy and dark gray. **FEMALE:** Somewhat variable. Mostly greenish, with strong gray overtones to back, head, and flanks, contrasting with a bright yellow to yellow-orange throat and forehead. Bill similar to male's. **IMMATURE MALE:** Like female. **JUVENILE:**

Similar to female, but more streaked above.

GEOGRAPHIC VARIATION In the United States, the subspecies *hepatica,* from southeastern California and southern Arizona, tends to be larger and duller than *dextra,* from southeastern CO to western TX, with more extensive gray on auriculars and flanks.

SIMILAR SPECIES Males similar to the brighter red Summer, which lacks

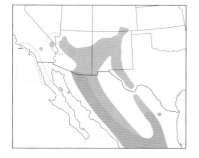

gray auriculars and grayish wash on back and flanks and typically has paler bill. Female Hepatic's grayer overall plumage, contrasting with its bright yellow throat, is quite different from uniform yellow-orange plumage of female Summer. Female Western can be as gray as Hepatic, but the larger Hepatic lacks wing bars. Note very different calls from other *Piranga.*

VOICE CALL: A single low sharp *chuck.* **SONG:** Robinlike and clearer than Western's; more similar to the song of Black-headed Grosbeak, with which it overlaps.

STATUS & DISTRIBUTION Fairly common in the Southwest's mixed coniferous forests (5,000–8,000 feet), extending south through much of

S.A. **MIGRATION:** Rather rare in lowlands, mainly in riparian areas, along streams. **WINTER:** Rare but regular winter resident on oak hillsides, mainly in the Patagonia-Nogales area of southeastern AZ. **VAGRANT:** Rare to coastal CA; accidental to IL, WY, and LA.

POPULATION Populations in Texas and Arizona are susceptible to long-term drought.

SCARLET TANAGER *Piranga olivacea* SCTA ■ 1

The breeding male Scarlet Tanager is one of the easier North American birds to identify. Often seen in small flocks during migration, Scarlet sings on the breeding grounds and feeds high in the canopy. It moves sluggishly and can be difficult to spot. Monotypic. L 7" (18 cm)

IDENTIFICATION Sexually dimorphic. Both sexes have whitish wing lining. **BREEDING MALE:** Unmistakable, brilliant red all over, with black wings and tail. Bill, somewhat short and stubby, is thick at base. **BREEDING FEMALE:** Females are entirely yellow-green, with yellower throat and sides, dark wings and tail, a thin eye ring, and wing coverts with greenish edging. Some adult females show weak wing bars. **WINTER ADULT:** Entirely greenish-yellow, but retains black scapulars, wings, and tail. Late summer birds can be blotchy red. **IMMATURE MALE:**

Resembles adult female but tends to be brighter yellow and has black scapulars and wing coverts. **IMMATURE FEMALE:** More problematic. Entirely greenish yellow with grayer wings and tail. **SIMILAR SPECIES** Immature females are similar to some Summers that lack orange tones. Note Scarlet's structural differences and gray undersurface to tail, which is yellowish in Summer. An adult female Western typically shows more distinct wing bars, a grayer back contrasting with yellower rump, and a longer, paler bill.

VOICE CALL: A hoarse *chip* or *chip-burr,* unlike other tanagers. **SONG:** Robinlike but raspy. Very similar to Western Tanager's, a *querit queer querry querit queer.* **FLIGHT NOTE:** A whistled *puwi.*

STATUS & DISTRIBUTION Commonly nests in deciduous forests in the eastern half of N.A. **BREEDING:** Arrives late Apr.–mid-May. **MIGRATION:** Trans-Gulf migrant. **WINTER:** Mainly in Amazonia and the foothills of the Andes in S.A. **VAGRANT:** Very rare in southern CA, mainly in Oct. and Nov; casual elsewhere in

West, U.K., and the Azores. Accidental to AK.

POPULATION Sensitive to forest fragmentation and parasitism by Brown-headed Cowbirds.

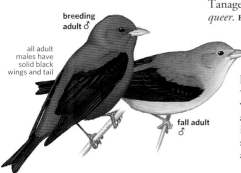

WESTERN TANAGER *Piranga ludoviciana*   WETA ▪ 1

1st fall ♂

gray "saddle"

gray morph ♀

larger bill than Scarlet

winter adult ♂

white wing bars and tertial edges

some very dull below

♀

reddish head

breeding adult ♂

black "saddle" and yellow rump

The striking black-and-yellow Western Tanager, with its bright red head, is one of the more characteristic summer species of western pine forests. Although brightly plumaged, it can be quite inconspicuous when feeding on insects high in the canopy or singing for long periods without moving. During migration, the Western Tanager often feeds on fruit and is more conspicuous. Western is the only regular North American tanager with wing bars. Note its yellow underwing linings in flight. Its bill is small for a *Piranga,* larger in size than Scarlet's, but smaller than other species' bills. Monotypic. L 7.3" (19 cm)
IDENTIFICATION Sexually dimorphic. **BREEDING MALE:** Plumage is unmistakable. Bright yellow underparts, yellow rump, black back, and conspicuous wing bars contrast with black wings (yellow upper bar, lower white bar) and bright red head. **BREEDING FEMALE:** Duller, mostly greenish-yellow below, grayish back and wings, with two thinner, pale wing bars. Underparts variable; some have brighter yellow belly and flanks, while others are quite gray. **NONBREEDING ADULT MALE:** Similar to breeding male, but loses red head. Duller red confined in varying amounts to forehead and chin. Black in plumage not as crisp, with some greenish edging to back. **FIRST-FALL MALE:** Generally yellow-green, yellower below with a yellow upper wing bar, and white lower bar. Bill noticeably pale. **FIRST-FALL FEMALE:** Can be very dull grayish yellow, still with two thin whitish wing bars.
SIMILAR SPECIES Adult males unlikely to be confused with other tanagers. In southeastern Arizona, beware of

confusing Western with Flame-colored Tanager, which has two white wing bars, a striped flame-and-black back, white tips to tertials, and white-tipped tail feathers. Western has hybridized with Flame-colored; offspring has a mixture of Western and Flame-colored characteristics. Female Flame-colored has a streaked back, larger bill, and white tips to both its tertials and tail. Also note that some female Scarlets show pale wing bars, but female Westerns usually show a yellow upper bar. Also, some worn Westerns show little or no wing bars, but when compared to Scarlet, Western has a grayer back, a longer tail, and a larger bill.
VOICE CALL: A *pit-er-ick,* with a noticeably rising inflection. Very different from both Summer's and Scarlet's, but indistinguishable from Flame-colored's. **SONG:** Similar in tone and pattern to both Scarlet's and Flame-colored's. **FLIGHT NOTE:** A whistled *howee* or *weet.*
STATUS & DISTRIBUTION A common bird of western coniferous forests, although it breeds in deciduous riparian habitats as far north as southeast

AK. **MIGRATION:** A common migrant throughout the lowlands of the West in spring (mid-Apr.–early June) and fall (mid-July–early Oct.). **WINTER:** Found mainly from central Mexico south (rarely) to Costa Rica and Panama. Uncommon to rare in winter in southern coastal CA. **VAGRANT:** Very rare wanderer, mainly in fall, to eastern U.S. **POPULATION** Not threatened.

FLAME-COLORED TANAGER *Piranga bidentata*   FCTA ▪ 3

head and below bright yellow with a tinge of orange

1st spring ♂
bidentata

A montane resident from Mexico to western Panama, the Flame-colored Tanager has become a rather regular summer visitor to the mountain canyons of southeastern Arizona, with several recent nesting records. It behaves much like other *Piranga* tanagers, mainly singing and feeding high up in cottonwoods and sycamores, with virtually all records from between 5,000 and 7,000 feet

in elevation. Males are very striking in appearance. In Arizona, the Flame-colored Tanager occasionally forms mixed pairs with the similar-sounding Western Tanager. Its alternative name is Stripe-backed Tanager. Polytypic (4 ssp.; probably 2 in N.A.). L 7.3" (19 cm)
IDENTIFICATION Sexually dimorphic. **ADULT MALE:** Males are quite variable. Some have bright, flame orange-red

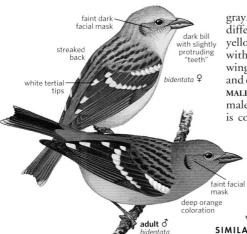

faint dark facial mask

streaked back

dark bill with slightly protruding "teeth"

white tertial tips

bidentata ♀

faint facial mask

deep orange coloration

adult ♂
bidentata

heads and underparts, whereas others are duller. Characteristic feature of male is orange-red back streaked with black. Rump is also usually orange-red with black streaks. Flame-colored has a distinctive darker cheek patch, normally with a blackish rear border. Its wings are blackish, contrasting with two distinct white wing bars and white tips to all tertials. Flanks are grayer, and blackish tail has white tips to outer feathers, visible from underneath. Bill of Flame-colored is rather heavy and

gray. **ADULT FEMALE:** Female is much different in color, being greenish yellow, but is similarly patterned with a streaked back, two white wing bars, and white tips to tertials and outer tail feathers. **FIRST-SPRING MALE:** Yellower than typical adult male, its reddish orange coloring is confined to forehead and face. **IMMATURE FEMALE:** Back has less-distinct streaking. **GEOGRAPHIC VARIATION** Males of the western Mexican subspecies *bidentata* are more orange than the eastern subspecies *sanguinolenta*, which tends to be redder.
SIMILAR SPECIES Flame-colored Tanager overlaps with the similar-sounding Western Tanager in the mountains of Arizona, where the species have been known to form mixed pairs. Hybrids are well documented, with some resembling male Flame-coloreds, though they usually have some noticeable Western feature, such as a yellow upper wing bar, a yellow unstreaked rump, or intermediate white spotting on the tertials or tail tips. Hybrid females undoubtedly occur, but apparently they are more difficult to distinguish.

VOICE CALL: A rising *pit-er-ick,* virtually indistinguishable from the rising call of Western; also gives a huskier, low-pitched *prreck.* **SONG:** Similar to Western Tanager's.
STATUS & DISTRIBUTION Rare to casual spring and summer visitor to montane canyons in southeastern AZ. **BREEDING:** Has nested several times in AZ. All nests have been in well-wooded mountain canyons in the Santa Rita, Huachuca, and Chiricahua Mountains of southeastern AZ, usually where sycamores occur. **VAGRANT:** Casual in the Chisos and Davis Mts. of western TX and in the Lower Rio Grande Valley.

Genus *Rhodothraupis*

CRIMSON-COLLARED GROSBEAK   *Rhodothraupis celaeno*   CCGR   ▪   4

This uniquely colored grosbeak is endemic to northeastern Mexico but wanders casually north to southern Texas, mainly in fall and winter. The winter of 2004–2005 was exceptional, with at least 15 individuals found.

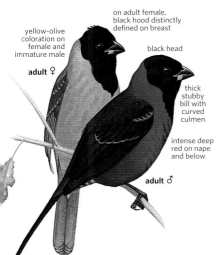

yellow-olive coloration on female and immature male

adult ♀

on adult female, black hood distinctly defined on breast

black head

thick stubby bill with curved culmen

intense deep red on nape and below

adult ♂

This species often skulks on or near the ground and sometimes raises and lowers its rear crown feathers. It prefers to eat greens leaves, but also feeds on fruit in trees (e.g., mulberries). No other North American bird has the color combination of an adult male Crimson-collared Grosbeak. Monotypic. L 8.5" (22 cm)
IDENTIFICATION Sexually dimorphic; thick, stubby bill with a curved culmen. **ADULT MALE:** Pinkish red underparts and collar contrast with the black hood and bib. It is mostly blackish above, with two thin pinkish wing bars. **ADULT FEMALE:** Females have a similar pattern, with pinkish red replaced by greenish yellow. **IMMATURE MALE:** Like immature female. Can show pinkish red blotches. **IMMATURE FEMALE:** Like adult female, but less black in hood and bib.
SIMILAR SPECIES Males unmistakable. Female superficially similar in body, head, and bib color to the adult male Audubon's Oriole, but note the oriole's

different shape, particularly bill length and shape, and its white wing bars.
VOICE CALL: A penetrating, rising and falling *seeiyu.* **SONG:** A varied warble.
STATUS & DISTRIBUTION Endemic to northeastern Mexico. Casual fall and winter visitor to southern TX, mainly to well-vegetated parks and refuges (also residential yards in towns) in the Lower Rio Grande Valley. Some individuals remain until spring.

CARDINALS Genus *Cardinalis*

The Northern Cardinal and Pyrrhuloxia are characterized by long, pointed crests and stout, conical bills. They are mainly resident within their home range. Males are brightly plumaged; females and immatures are duller. Cardinals feed mainly on seeds and fruit and commonly come to seed feeders.

NORTHERN CARDINAL *Cardinalis cardinalis* NOCA ▪ 1

The bright red male Northern Cardinal, with its conspicuous crest, is one of the most recognizable birds in N.A. It is found abundantly through virtually all of the eastern United States in a variety of habitats, including suburban gardens. Although the cardinal can be secretive and remain hidden in thickets, males usually sing from exposed perches. The species is commonly attracted to feeders and open areas with birdseed. Its thick, reddish cone-shaped bill is specialized for cracking seeds. Polytypic (19 ssp.; 5 in N.A.). L 8.8" (22 cm)

IDENTIFICATION Sexually dimorphic. **ADULT MALE:** Plumage unmistakable. Males are uniquely colored, with a bright red body, a black face, and an obvious, pointed crest. **ADULT FEMALE:** Females are similarly shaped, but are buffy brown in coloration, with a reddish tinge on wings, tail, and crest. **JUVENILE MALE:** Similar to adult female, but it is generally browner overall and has a bill with less reddish coloration. **JUVENILE FEMALE:** Lacks reddish tones in wings and tail.

GEOGRAPHIC VARIATION Five subspecies described north of Mexican border. Size and coloration varies clinally from east to west, with the eastern birds *(cardinalis)* being smaller, shorter crested, duller red, and slightly more black on face than the western birds *(superbus)*. The other three N.A. subspecies, *canicaudus, floridanus,* and *magnirostris,* are intermediate.

SIMILAR SPECIES In the East, Northern Cardinal is not really confused with any other species. Note that the adult male Summer Tanager is also bright red, but it lacks both crest and black face. In the Southwest, Northern Cardinal overlaps with the very similarly shaped Pyrrhuloxia, but note their color differences. Female and immature Pyrrhuloxias are very similar to female and immature cardinals. Pyrrhuloxia has a noticeably yellow bill that has a distinct downward curve to culmen, whereas the cardinal has a distinctly straighter culmen as well as a pointier, reddish bill. Plumage of the female Pyrrhuloxia is grayer with very little red.

VOICE **CALL:** A sharp, somewhat metallic *chip.* **SONG:** Variable. A liquid, whistling *cue cue cue,* or *cheer cheer cheer,* or *purty purty purty.* Both sexes sing virtually all year, though females sing less frequently than the males do.

STATUS & DISTRIBUTION Very common. **BREEDING:** Throughout the East, found in a variety of habitats, including woodland edges, swamps, streamside thickets, and suburban gardens. In the West, restricted mainly to southwestern TX, southern NM, and southern AZ, where it is common in mesquite-dominated habitats, usually near water. Very rare and likely extirpated along the Colorado River in southern CA. Generally nonmigratory.

POPULATION Cardinals generally expanded their range northward in the 20th century.

long crest

black surrounds red bill

straight culmen

cardinalis ♂

dark bill

juvenile ♂
cardinalis

cardinalis ♀

overall buffy brown and dull red

longer crest than eastern birds

less black around red bill than eastern birds

male brighter red than eastern birds

southwestern ♂
superbus

PYRRHULOXIA *Cardinalis sinuatus* PYRR ▪ 1

This bird's coloration and bill distinguish it from the cardinal. It often sings from exposed perches, particularly power lines. It also frequents birdseed feeders. Polytypic (3 ssp.; 2 in N.A.). L 8.8" (22 cm)

IDENTIFICATION Sexually dimorphic. **ADULT MALE:** Unmistakable. Plumage consists of a pearly gray body and blood-red face, bib, center to breast and belly, and tip of crest. Thick, yellow bill and curved culmen. Red-edged primaries and tail feathers. **ADULT FEMALE:** Duller, lacks most of red coloration. **JUVENILE:** Lacks red tones in wings and tail. Bill not as yellow as adult's. **GEOGRAPHIC VARIATION** Nominate subspecies *sinuatus* breeds from south Texas to southern NM; *fulvescens* breeds in southern AZ (underparts of

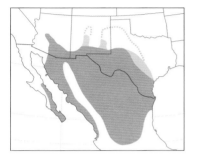

adult male are paler orange-red rather than deep red). **SIMILAR SPECIES** Females and juveniles easily confused with female and juvenile Northern Cardinal. Distinguished by more curved culmen, yellower bill, fewer red tones, and spikier crest. **VOICE CALL:** A *chink*, like cardinal's, but decidedly more metallic. **SONG:** Reminiscent of, but thinner and shorter than, cardinal's liquid whistles. **STATUS & DISTRIBUTION** Fairly common. **BREEDING:** Thorny brush and mesquite thickets in Southwest lowland desert; more arid environments than cardinal. Often in brushy borders near houses in desert ranchland. **WINTER:** More widespread, north from

breeding areas in AZ, NM, and TX. **VAGRANT:** Casual, mainly in fall, winter, and spring to southern CA, as well as the Great Plains. Accidental to ON. **POPULATION** Still common, but potentially threatened by loss of natural desert habitats in the Southwest.

fulvescens ♀
yellowish bill with strongly curved culmen
grayish overall with pale buffy breast
fulvescens ♂
overall gray with patchy bright red

GROSBEAKS Genus *Pheucticus*

Two species breed in N.A.; one is a vagrant from Mexico. These large passerines feature brightly plumaged males and duller females and immatures. Males attain breeding plumage after two years. With generally large, conical bills, they feed on insects during the breeding season and fruit during fall and winter. They are known to frequent seed feeders.

YELLOW GROSBEAK   *Pheucticus chrysopeplus*   YEGR ■ 4

A resident of the thorn forest and riparian areas north through western Mexico, the Yellow Grosbeak is usually quite conspicuous, owing to its large size and bright yellow coloration. Monotypic. L 9.3" (24 cm) **IDENTIFICATION ADULT MALE:** Bright yellow with black-and-white wings: white median coverts, white-tipped greater coverts, and white-based primaries form a white patch on folded wing. White tips to tertials. Black tail with white inner webs. **ADULT FEMALE:** Duller yellow and more streaking on

overall bright yellow, but female and immatures duller
huge bill
adult ♂
black wings with extensive white
adult ♀

upperparts. Wings and tail browner. **IMMATURE MALE:** Like female, with a yellower head. It achieves adult male plumage by its second winter. **SIMILAR SPECIES** Evening Grosbeak has a massive bill and white in wing but is more gold in color and usually travels in flocks. Black-headed Grosbeak female is more buffy underneath and has a more patterned head and a smaller bill.

VOICE CALL: A *Pheucticus*-type *eek*, like Black-headed's. **SONG:** Rich and warbled, similar to Black-headed's. **STATUS & DISTRIBUTION** Primarily a Mexican species. Casual late spring and early summer to riparian areas in the lowlands and canyons of southeastern AZ. There is a winter record from Albuquerque, NM. Records from IA and southeastern CA are questioned on origin.

ROSE-BREASTED GROSBEAK *Pheucticus ludovicianus*   RBGR ■ 1

breeding adult ♂

winter adult ♂

rose red breast and wing linings

breeding adult ♂

rich buff chest, often with a few pink feathers; reddish pink wing linings

1st fall ♂

pale bill

1st spring ♂

strong streaking across breast

♀

The striking Rose-breasted Grosbeak is a common bird of wooded habitats across much of eastern and midwestern North America. Singing from the canopy of a deciduous forest, even a brightly colored male can be difficult to locate. Monotypic. L 8" (20 cm) **IDENTIFICATION** Sexually dimorphic. Takes more than a year to acquire adult plumage. **ADULT MALE:** Black head, back, wings, and tail contrast with gleaming white underparts and rump. Bright rosy pink patch on breast. Rose-red wing linings. Large, pinkish bill. **ADULT FEMALE:** Mainly brown above with streaks, paler below with extensive dark streaking. Yellow wing linings. **WINTER MALE:** Molts into winter plumage before migrating. Brown-edged head and upperparts, barred rump, and dark streaks on sides and flanks. Wings as in adult. **FIRST-FALL MALE:** Buffy wash with fine streaks across breast, pink wing linings, usually a few

pink feathers on sides of breast. **FIRST-SPRING MALE:** Similar to adult male, but browner, particularly wings and tail. **SIMILAR SPECIES** Adult males unmistakable. Plumage of females and first-fall males is very similar to plumage of female Black-headed Grosbeak (see sidebar below). **VOICE CALL:** A sharp *eek*, squeakier than Black-headed's. **SONG:** A robinlike series of warbled phrases, but shorter. Songs

are similar to those of Black-headed Grosbeak. **STATUS & DISTRIBUTION** Common. **BREEDING:** Nests commonly in deciduous forest habitats. **WINTER:** Mainly Mexico, C.A., and rarely Cuba; casual in southern U.S., including coastal CA. **MIGRATION:** Peak spring migratory period in eastern U.S. mid-Apr.–mid-May; peak fall migratory period Sept.–mid-Oct. Often seen in flocks during migratory fallouts. **VAGRANT:** Regular during late spring and fall in the Southwest. Casual, mainly in Oct., to U.K., Europe, and the Azores. **POPULATION** Possible decline due to forest fragmentation.

Identification of Female *Pheucticus* Grosbeaks

One of the more difficult identification challenges among the Cardinalidae involves telling the female and immature *Pheucticus* grosbeaks apart. Although the breeding ranges of the Rose-breasted and the Black-headed Grosbeaks rarely overlap, both species have a tendency to wander, particularly during the fall, often including females and immatures. Characteristics to focus on are the extent of streaking and amount of buff on the underparts and the coloration of the bill.

Determining the age and sex of the grosbeak is usually a good first step in identifying it to species. The female Rose-breasted is generally whitish underneath and more heavily streaked. In addition, the streaks are made bolder by the near or sometimes complete absence of any buff coloration on the breast, nape, and chin. The female Black-headed in comparison tends to be quite buffy underneath with little or no

streaking. The very worn female Black-headed can be quite white underneath, but it will still show little or no streaking. The female Rose-breasted usually has a pale, pinkish bill, whereas the Black-headed's bill is bicolored: dark gray above, light gray below. Females of both species have yellow underwing linings, although the color is brighter yellow in Black-headed.

The first-fall male Rose-breasted has a buff wash across the breast and can look very similar to the female Black-headed. Typically, however, the first-fall male Rose-breasted has more streaking on the center of its breast, which the female Black-headed lacks altogether, and it usually has some pink feathers visible on the breast. The immature male Rose-breasted has pink wing linings (visible in flight), compared to yellow wing linings on the female Black-headed. Also beware of hybrids, particularly in the Great Plains, where the two species sometimes

BLACK-HEADED GROSBEAK *Pheucticus melanocephalus*   BHGR   1

The western counterpart to the East's Rose-breasted Grosbeak; the two occasionally hybridize where they come into contact on western Great Plains. Found in a variety of habitats, from open coniferous forests to montane riparian areas. Neotropical migrant. Adult males are striking; females and immatures are duller. Bill is large and conical, typically bicolored; wing linings are yellow in both sexes. Polytypic (2 ssp.; 2 in N.A.; subspecies are poorly differentiated; coastal birds, *maculatus,* have shorter wings, smaller bills, and a tawnier supercilium than nominate interior birds). L 8.3" (21 cm)
IDENTIFICATION ADULT MALE: Cinnamon-colored body including collar, with nearly black head, mostly black back, black wings. Center to lower belly yellow. **ADULT WINTER:** Similar to

adult male, but head less black, supercilium and crown strip cinnamon, and streaks on back more prominent. **FIRST-SUMMER MALE:** Like adult winter male, but wings less black. **FIRST-WINTER MALE:** Like female, but underparts more cinnamon without streaking, and underside of tail gray. **ADULT FEMALE:** Mainly dull brown, with streaked back, buffy-white nape and supercilium. Underparts variable, from buffy to almost white, with varying amount of streaking on sides of breast and flanks.
SIMILAR SPECIES Adult male unmistakable. Female and first-winter male is easily confused with the female and first-winter male Rose-breasted. Note different amount of streaking on underparts and color of bill (see sidebar below).
VOICE CALL: A sharp *eek,* similar to the Rose-breasted's, but decidedly less squeaky. **SONG:** Robinlike series of

warbled phrases, like Rose-breasted's. Virtually indistinguishable from the Hepatic Tanager's song, where the two species overlap.
STATUS & DISTRIBUTION Common throughout entire West. **BREEDING:** Nests in a variety of habitats, from mixed coniferous forest to montane riparian. Arrives on breeding ground from early Apr. (in south) to mid-May (in north). **MIGRATION:** Migrates in spring singly or in smaller groups, unlike Rose-breasted, which sometimes migrates in flocks. In fall, not uncommon to see several in a fruiting tree. Adults begin migrating south by mid-July. Juveniles migrate later, Aug.–mid-Sept., a few into early Oct. **WINTER:** Mainly Mexico. **VAGRANT:** Casual or accidental wanderer, mainly in fall and winter, to virtually all eastern states and provinces. Spring overshoots known from AK and NT.
POPULATION No known threats.

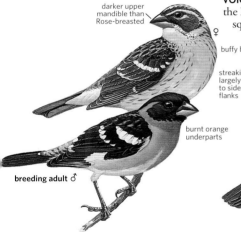

darker upper mandible than Rose-breasted

buffy breast

streaking largely limited to sides and flanks

burnt orange underparts

breeding adult ♂

rich orange-buff underparts

1st fall ♂

Black-headed Grosbeak, female

Rose-breasted Grosbeak, female

Rose-breasted Grosbeak, first-fall male

come together and interbreed. A hybrid can show characteristics passed down from each of its parents, such as a bright yellow underwing lining coupled with a pale, pinkish bill.

Voice also plays a role in identification. Although the calls of both species are similar, the Rose-breasted's is a noticeably higher-pitched, squeakier *eek.* ∎

Genus *Cyanocompsa*

BLUE BUNTING *Cyanocompsa parellina*   BLBU ■ 4

This Middle American endemic is a rare, irregular visitor to southern Texas. Rather secretive. Most records are from feeders in the Lower Rio Grande Valley. Polytypic (4 ssp.; records from U.S. likely of the nominate group of three ssp.). L 5.5" (14 cm)

IDENTIFICATION Sexually dimorphic. **ADULT MALE:** Blackish blue body with brighter blue crown, cheeks, shoulder, and rump. Lacks wing bars. **IMMATURE MALE:** Like adult male, but with brownish cast to wings. **FEMALE:** Uniformly rich, buffy brown, lacking streaks or wing bars. Stout dark bill with curved culmen.

SIMILAR SPECIES Female Varied Bunting, also plainly colored, has less rusty brown and is smaller billed, with a bluish tint to primaries and tail.

VOICE CALL: A metallic *chink,* similar to the Hooded Warbler's. Western Mexico *indigotica,* in which female is overall paler and grayer, gives a very different call note (perhaps a species-level difference?). **SONG:** A series of high, clear warbled phrases, usually beginning with two separate notes.

STATUS & DISTRIBUTION Casual to rare, mainly in winter, in brushy areas in the Rio Grande Valley of southern TX. **VAGRANT:** Accidental to upper TX coast and southwestern LA in winter.

overall a rich cinnamon-brown

large dark bill

deep, dark blue overall with paler blue highlights

adult ♂

♀

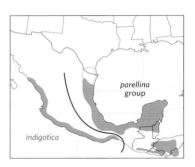

parellina group

indigotica

BUNTINGS Genus *Passerina*

These birds are relatively small-bodied with small, conical bills. Adult males are very brightly plumaged; females are very dull brown or green, some with wing bars, some without. All have the characteristic behavior of twitching the tail while perched. They feed mainly on insects during breeding season, seed during fall and winter. They are nocturnal neotropical migrants.

BLUE GROSBEAK *Passerina caerulea*   BLGR ■ 1

Closely related to the other *Passerina* buntings, the larger Blue Grosbeak is a bird primarily of the southern United States. Found in overgrown fields, along brushy roadsides, or in riparian habitats, it usually stays low in the brush but is often seen singing from an exposed perch. Like buntings, it has the distinctive behavior of twitching and spreading its tail. It feeds mainly on seeds. Polytypic (7 ssp.; 3 in N.A.; eastern N.A. birds slightly smaller and larger-billed than western N.A. birds). L 6.8" (17 cm)

IDENTIFICATION Highly sexually dimorphic. **ADULT MALE:** Large male is deep blue and has bright chestnut

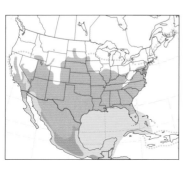

thick bill

caerulea ♀

deep buffy color with rusty buff wing bars

immature *caerulea*

chestnut wing bars

variable amount of blue on head

breeding adult ♂ *caerulea*

1st spring ♂ *caerulea*

frequently twitches tail

wing bars. Also note black in face and chin and indistinct blackish streaking to upperparts. **ADULT FEMALE:** Females very different: pale grayish brown body overall. Buffy brown median coverts, lighter lower wing bar, indistinct streaking on back, some blue tinge to scapulars and tail. Note large, massive bill. **FIRST-SUMMER MALE:** A mixture of male and female plumage, favoring female with a mostly blue head, some blue patches on breast, and more blue tinge to primaries and tail. Chestnut in wings more extensive than in female. Some are very female-like with just a touch of blue on forehead.

FIRST-WINTER MALE & FEMALE: Similar. Overall richer rufous brown than adult female, with chestnut wing bars. No streaking on underparts. **SIMILAR SPECIES** Males similar to male Indigo Bunting, but note larger size, larger bill, and chestnut wing bars. Females more problematic. Very similar to female Indigo Buntings, but more rufescent in fall and winter. Wider and more rufescent upper wing bar. Indigo Bunting always with faint to moderate streaking on underparts (variable) and much smaller bill. Calls and song very different. **VOICE CALL:** A loud, sharp *chink*.

SONG: A series of rich, rising and falling warbles. **STATUS & DISTRIBUTION** Common. **BREEDING:** Found in a variety of brushy habitats, often near water, all across the southern U.S. **MIGRATION:** Arrives on the southern breeding grounds by early to mid-Apr., northern range by mid-May. Eastern population trans-Gulf migrants. **WINTER:** Both eastern and western populations winter in Mexico and C.A. south to Panama. Rare in Cuba and Bahamas. **VAGRANT:** Rare in fall north to New England and Atlantic provinces. Accidental as far north as southeastern AK.

INDIGO BUNTING *Passerina cyanea*   INBU   ■   1

The male Indigo Bunting is commonly seen as a breeding species and at migration hot spots. In the spring, the Indigo may be present in large flocks, particularly during migratory fallouts, and is often seen in brushy habitat or along weedy margins of fields and roads, where it sits up and twitches its tail. It sometimes hybridizes with the Lazuli Bunting. Monotypic. L 5.5" (14 cm) **IDENTIFICATION** Highly sexually dimorphic. **SUMMER ADULT MALE:** Plumage unmistakable, entirely bright blue. **WINTER ADULT MALE:** Blue obscured by brown and buff edging; mottled brown and blue early in winter. **SUMMER FEMALE:** Dull brown, usually with two faint wing bars and indistinct streaking on underparts. Whitish throat, small conical bill with straight culmen, relatively long primary projection. **FIRST-SUMMER MALE:** Patchy dark blue and brown. **IMMATURE & WINTER FEMALE:** More rufescent overall than breeding female, with blurry streaking on breast and flanks.

SIMILAR SPECIES Males smaller than male Blue Grosbeak, but lack chestnut wing bars, black on face, and dark streaks on back; also has smaller bill. Females are similar to other female *Passerina* buntings (see sidebar p. 680). **VOICE CALL:** A dry, metallic *pik.* **SONG:** A series of sweet, varied phrases, usually paired. **STATUS & DISTRIBUTION** Common. **BREEDING:** Found in brushy borders to mainly deciduous woodland throughout eastern U.S. In the Southwest, mainly found in riparian habitats. **WINTER:** Mainly Mexico through C.A., rarely to northern S.A. Also on Caribbean islands. Rare along Gulf Coast and southern FL. **MIGRATION:** Mainly nocturnal. Arrives on breeding

grounds mid-Apr.–early June. In fall, mainly mid-Sept.–mid-Oct. **VAGRANT:** Rare to Pacific states and Atlantic provinces; casual on the Azores. **POPULATION** Western birds may be limited by decrease in riparian habitats.

faint wing bars

all females have blurry streaks below

breeding adult ♀

deep blue color

breeding adult ♂

1st spring ♂

variable patchy blue and brown plumage

mostly brown but with some veiled dark blue showing

winter adult ♂

fall females and immatures a richer cinnamon-brown color than breeding females

fall ♀

LAZULI BUNTING *Passerina amoena*   LAZB ■ 1

A western counterpart to the Indigo Bunting, the Lazuli Bunting is found in a variety of habitats, and its blue, white, and rich buff plumage makes it one of the more attractive songbirds of the West. It is quite similar to other *Passerina* buntings in behavior: frequents brushy borders to fields and roads, seen sometimes in single-species flocks, and often twitches and spreads its tail while perched. Occasionally it hybridizes with the Indigo Bunting. Monotypic. L 5.5" (14 cm)

IDENTIFICATION Highly sexually dimorphic. Achieves adult plumage by second winter. **ADULT MALE:** Distinctive male is bright turquoise blue above and on throat with a rich cinnamon-buff wash across breast; white on belly with a thick white upper wing bar. **WINTER MALE:** As in Indigo Bunting, blue plumage obscured by buffy-brown edging to feathers, with blue visible on head and throat. **ADULT FEMALE:** Very different, duller, resembling female Indigo Bunting. Note drab grayish brown coloration with buffy wash across breast, two narrow white wing bars, and lack of streaking on underparts. **WINTER FEMALE:** Like breeding female but with warmer buff wash on breast, grayer throat, and buffier wing bars. **FIRST-SUMMER MALE:** Like adult male with some brown feathers intermixed with blue.

JUVENILE: Like female, but with distinct fine streaking across breast. **SIMILAR SPECIES** Male unmistakable. Females confused with female Indigo and Varied Buntings and with female Blue Grosbeak. Note overall grayer plumage, warm buff wash across breast, lack of streaking on underparts, which is present on all female Indigos (except in juvenile, which has finer, less blurry streaks), and more distinct narrow white wing bars. **VOICE CALL:** A dry, metallic *pik,* similar to the Indigo's. **SONG:** A series of varied phrases, sometimes paired, but faster and less strident than Indigo's. **STATUS & DISTRIBUTION** Fairly common. **BREEDING:** Found in open deciduous or mixed woodland and in chaparral, particularly along streams and rivers. **MIGRATION:** Congregates during migration in the Southwest (late July–Nov.), where molting occurs, before continuing migration to Mexico. **WINTER:** Mainly western slope of Mexico. Rare in southeastern AZ. **VAGRANT:** Casual, mainly in spring and late fall north and east of breeding range.

winter ♀

rich and extensively buffy below but unstreaked, unlike all female Indigos, which have blurry streaks

juvenile

fine dark pinstripes below, the common plumage seen in late Aug. and Sept. in much of West

buffy orange breast band

breeding adult ♀

strong white wing bars

rich powder or lazuli blue on head

white patch on median coverts

breeding adult ♂

buffy breast contrasts with white belly

Identification of Female *Passerina* Buntings

Although male *Passerina* buntings are some of the easier species to identify in North America, the females present several challenges. There are five species in this genus in North America, many of which overlap distributionally: the Indigo, Lazuli, Varied, and Painted Buntings and the Blue Grosbeak. Key characteristics include overall coloration, especially of the underparts; presence or absence of streaking below; presence or absence of wing bars; length of primary extension; and bill shape. Be aware that in some species, one-year-old males look very much like females, but they sing. Also, fresh birds in fall are usually warmer brown in color.

The Indigo is probably the most common and widespread. Females are drab brown and whiter on the throat, and always have at least some streaking on the underparts. The wings typically show faint wing bars. In the fall, females are brighter cinnamon. The Lazuli female can be similar, but always has a warmer buff wash across the breast with no streaking, (but note that juvenile Lazuli has fine but distinct streaks across the breast) more prominent wing bars, and a grayer, less contrasting coloration to the throat. While the female Indigo has only a hint of blue coloration to its tail feathers, on the female Lazuli a drab grayish blue will often extend from the tail onto the rump, especially during spring and summer. The Varied is generally much drabber, with more uniform underparts, no streaking, and very faint wing bars. Varied Bunting has a distinctively curved culmen and very short primary extension. Fresh fall birds can be quite cinnamon in color. Painted Bunting is generally easier to identify, as it is always green on the upperparts and dull yellowish below. Immature females and very worn adults can lose much of their coloration, but still retain at least some green color to the back. The female

VARIED BUNTING *Passerina versicolor*   VABU ▪ 2

One of the more colorful of the buntings, this Southwest specialty appears uniformly dark in poor light. Found in more arid habitats than other buntings, it is often seen in desert washes or on cactus-laden hillsides of canyons. Similar in behavior to other *Passerina* buntings and an inhabitant of dense mesquite thickets, the Varied Bunting is usually difficult to see, but it often sings from exposed perches. Males are stunning in good light; females are extraordinarily dull. Polytypic (3 ssp.; 2 in N.A.). L 5.5" (14 cm)

IDENTIFICATION Highly sexually dimorphic. SUMMER MALE: Quite colorful, but colors difficult to see. Striking are bright red nape, deep blue head and rump, and dark purplish body. Bill is rather small, with distinctive curved culmen. SUMMER FEMALE: Extremely dull and featureless: almost entirely grayish brown, with very faint suggestion of wing bars, but no obvious pale edging to tertials. Note curved culmen. WINTER MALE: As in other *Passerina* buntings, its color is obscured by brownish edging to many feathers. WINTER FEMALE: Similar to summer female, but slightly more rufescent overall. FIRST-SPRING MALE: Like adult female, but with varying amounts of purple on forehead; often sings like adult male and defends breeding territory.

GEOGRAPHIC VARIATION Nominate subspecies *versicolor* from southeastern New Mexico and southern Texas has duller red nape, purple throat with reddish tinge, and pale blue rump. Subspecies *pulchra* (including *dickeyae*) from southeastern Arizona (and Baja) has brighter red nape, purple throat with no red tones, and

purplish blue rump. Females grayer in Arizona *pulchra*.

SIMILAR SPECIES Males unmistakable in good light. Females more of a challenge. Lack of distinguishing characteristics actually an identifying feature of female Varied, but note distinctive curved culmen. Female Lazuli has distinct pale wing bars and a buff wash across breast. Female Indigo is

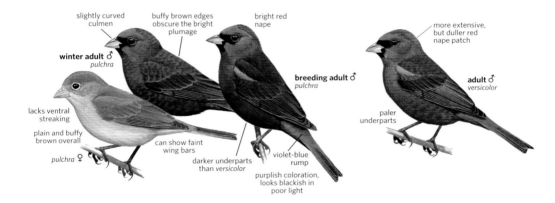

purplish coloration,
looks blackish in
poor light

Indigo Bunting, female (TX, Apr.)

Lazuli Bunting, female (CA)

Varied Bunting, female (AZ)

Blue Grosbeak is considerably larger than other *Passerina* buntings and has prominent buffy wing bars, a larger bill, and even some blue coloration on its shoulders. Although not a *Passerina,* the female Blue Bunting can be similar in size and overall appearance, but is a rich cinnamon reddish brown, and has plain wings and a thick, dark bill with a distinct curved culmen.

Hybridization can further complicate matters, resulting in variable intermediate plumages and sometimes making exact identification almost impossible. ▪

darker, always with some streaking on underparts. First-year female Painted Buntings can be very drab but are still greenish above.

VOICE CALL: A loud, rich *chip,* similar to other buntings. **SONG:** A series of rich, sweet, unrepeated phrases, very similar to Painted Bunting's and hard to differentiate in areas of overlap in Texas.

STATUS & DISTRIBUTION Common, but local. **BREEDING:** In summer, inhabits dense, thorny thickets in desert washes, canyon hillsides, and sometimes along streams and rivers. Mostly found at lower elevations. **MIGRATION:** Most of the North American breeding population migrates south in winter. Population in extreme south TX may be resident. TX breeding population arrives mid-Apr.; AZ population very late arriving, mostly late May–early June (very rare before mid-May). **WINTER:** Western Mexico, mainly Sonora to Oaxaca. Casual in southwestern TX. **VAGRANT:** Accidental to CA (late fall and winter) and ON (spring).

PAINTED BUNTING *Passerina ciris*   PABU ■ 1

Found across the South in two discrete populations, the adult male Painted Bunting, with its incongruous combination of red, blue, and green, can be considered North America's gaudiest songbird. Despite its bright plumage, it can be amazingly difficult to see, often singing from inside dense thickets, but at other times sings from telephone lines. Females look very different, lacking the bright colors, but except for young juveniles are quite green, almost recalling a miniature female Scarlet Tanager. The Painted regularly comes to seed feeders during the nonbreeding season. Polytypic (2 ssp.; 2 in N.A.). L 5.5" (14 cm)

IDENTIFICATION Sexually dimorphic. **ADULT MALE:** Unmistakable. Bright red underparts and rump, lime-green back, dark wings with green edging, and dark blue head (except for red chin and center to throat), red eye ring, dark tail. **ADULT FEMALE:** Very different, lacking any bright coloration. Upperparts lime green, dull yellowish below. Greenish edging on wings and tail. **FIRST-WINTER MALE:** Resembles female. **FIRST-SUMMER MALE:** Also like female, but some show scattered blue feathers on head or red on breast. Achieves adult plumage by second winter. **FIRST-FALL FEMALE:** Trickiest plumage to identify. Very drab, grayish, lacking much of the adult female's green tones. Upperparts are greener than plain underparts. Subsequent molt during the fall renders a greener plumage.

GEOGRAPHIC VARIATION Although the variation in plumage between the eastern nominate subspecies *ciris* and the western *pallidior* is slight, migratory and molt patterns differ. Eastern birds molt on the breeding grounds and then winter in FL and Caribbean, while western birds molt at migratory staging grounds or on the winter grounds in Mexico and Central America.

SIMILAR SPECIES The adult male is unlike any other North American bird. Green upperparts of the immature male and all female plumages separate Painted from all other buntings. The plain and always unstreaked female and immature Varied Bunting can look similar, but they are browner, lacking any green tones above.

VOICE CALL: A loud *chip,* sweeter than other *Passerina* buntings' calls. **SONG:** A rapid series of varied phrases, thinner and sweeter than Lazuli Bunting's, but surprisingly similar to Varied's.

STATUS & DISTRIBUTION Locally common. **BREEDING:** Often secretive in brushy thickets and woodland borders, often along streams and rivers, along the southeast coast (NC, SC, GA, FL) and much of the south-central U.S., north to western TN

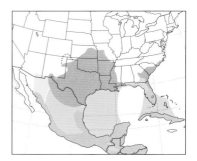

and southwest IL, west to western TX and southeastern NM. **WINTER:** Mainly Mexico south to Costa Rica and Panama; also southern FL, Bahamas, and Cuba. **MIGRATION:** Spring migration relatively early (early Apr.–early May). Fall migrations differ by population; eastern birds mainly late Sept.–Oct. and western birds late July–early Oct. Western birds have interrupted migration where they molt out of breeding plumage, mainly in Sonora, Mexico, but rarely to southeastern AZ. **VAGRANT:** Rare to casual, primarily in spring and fall north and west of mapped breeding range. This species is kept in captivity, so vagrant records are sometimes debated in regard to origin, esp. those of adult males.

POPULATION The Atlantic coast breeding population is limited and affected by loss of habitat.

bright green upperparts similar overall in coloration to larger female Scarlet Tanager

yellow eye ring

♀

adult ♂

unmistakable

overall plain and grayish, usually with some greenish on lower back

juvenile

Genus *Spiza*

DICKCISSEL *Spiza americana*   DICK ■ 1

The Dickcissel, a sometimes abundant migrant and breeding bird of the central United States, gets its name from the verbal interpretation of the male's song. Usually found in open prairie or weedy agricultural fields, its numbers and breeding distribution vary from year to year. Males often sing while sitting up on a tall weed, shrub, or wire. In migration, the Dickcissel is usually detected when flying over, giving a distinctive, flat call. Large flocks congregate during migration and on the winter grounds. Sometimes the Dickcissel is found with House Sparrows at feeders in the East. Monotypic. L 6.3" (16 cm)

IDENTIFICATION Sexually dimorphic. **ADULT MALE:** Distinctively patterned; grayish brown and sparrow-like. Underparts with a black bib surrounding white chin, yellow breast, whitish lower belly and undertail coverts. Grayish back, boldly streaked black, and gray rump, unstreaked. Wings with bright chestnut shoulder patch. Gray nape. Complex face pattern: gray auriculars; yellow eyebrow above and in front of eye, white behind eye; yellow malar, widening to white neck mark. Bill rather long and conical. **WINTER ADULT MALE:** Similar to breeding male, but browner overall and with a less-distinct black bib. **ADULT FEMALE:** Similar to male but duller and browner. Varying amount of black speckling instead of black bib. Eyebrow duller, buffy behind eye. **IMMATURE:** Has fine streaking on breast and flanks and bolder streaking on back; immature male has relatively broad, rusty tips to median coverts, breast with some yellow, lacks bib; immature female is very dull with narrow, buff tips to median coverts.

SIMILAR SPECIES Immatures, which lack chestnut shoulder and have no yellow on underparts, can resemble female or juvenile House Sparrows. Note distinct white upper wing bar and lack of any streaking on underparts of House Sparrow. Note also House Sparrow's shorter bill and Dickcissel's much longer primary projection.

VOICE CALL: A flat *bzrrrrt*, often given in flight. **SONG:** A variable *dick dick di'cissel.*

STATUS & DISTRIBUTION Common, sometimes locally abundant. Numbers fluctuate annually outside core breeding range. **BREEDING:** Nests in open, weedy meadows, in agricultural fields, and in prairie habitats. **WINTER:** Mainly the Llanos in Venezuela; very rare in northeastern U.S. and Atlantic provinces in winter, usually at feeders, and in the southern portion of breeding range. **MIGRATION:** Mainly a nocturnal migrant, but large flocks are seen migrating during the day. Flocks coalesce, sometimes reaching numbers in the thousands. Spring period mid-Apr.–mid-May; fall period mid-Aug.–mid-Oct. **VAGRANT:** Rare, mainly in fall to East and West Coasts; casual on the Azores.

POPULATION Large-scale conversion of native grasslands to agriculture in the main breeding areas has likely had a negative impact on population size.

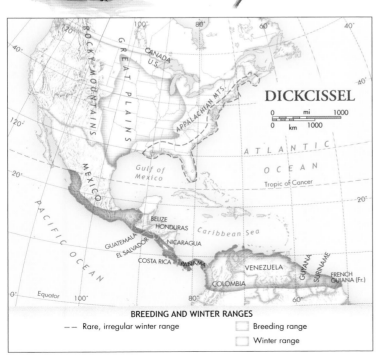

breeding ♀

winter adult ♂

chestnut lesser and median coverts

pale supercilium

large bill

longish wings

black bib and yellow breast

immature ♂

breeding ♂

broad, pale yellow submoustachial

chestnut median coverts

immature ♀

DICKCISSEL

0   mi   1000

0   km   1000

ROCKY MOUNTAINS

GREAT PLAINS

CANADA
U.S.

APPALACHIAN MTS.

ATLANTIC OCEAN

MEXICO

Gulf of Mexico

Tropic of Cancer

PACIFIC OCEAN

BELIZE
HONDURAS
GUATEMALA
EL SALVADOR
NICARAGUA
COSTA RICA
PANAMA

Caribbean Sea

VENEZUELA

COLOMBIA

GUYANA
SURINAME
FRENCH GUIANA (Fr.)

Equator

BREEDING AND WINTER RANGES

-- Rare, irregular winter range

Breeding range

Winter range

BLACKBIRDS Family Icteridae

Altamira Oriole (TX, Feb.)

Contrary to popular thought, blackbirds are not all black. In fact, the family name, Icteridae, refers to their yellow color. So really these are the "yellow birds."

Structure Blackbirds are large and sturdy songbirds. They all have strong bills, ranging from short and finchlike, to sharp, to stout and heavy. The culmen is flattened and straight. Blackbirds have strong legs, particularly the species which feed terrestrially. Overall, blackbirds are stocky, but some orioles are slim, while the meadowlarks are rotund and stocky. Tail length in most species is medium, extremes being the short tails of the meadowlarks and the long and keel-shaped tails of the grackles. In the tropics, other groups, such as the caciques and oropendolas, are found, and the latter are among the largest neotropical passerines.

Plumage Generally blackbirds show some black, yellow, orange, or red on the plumage. Females of some species are brownish and streaked. Many of the blackbirds and grackles have largely or fully black plumages, often with blue or green iridescence. Flash colors may be present as red on epaulet (shoulder) or yellow on head. The black species often show bright yellow eyes. The meadowlarks' upperparts are streaked and patterned for adequate camouflage; however, their underparts are bright yellow with a distinctive black V on the breast. Orioles are the brightest blackbirds: Their plumages are a mix of yellow, orange, or chestnut with black. Several species have black hoods, while others have a black face mask and black bib, contrasting with a yellow or orange head and underparts. Female orioles vary depending on

the species: Some are dull and yellowish, lacking the strong patterns of the adult male, while others are like adult males in plumage. Young males may look similar to females. The tropical blackbirds include some very showy species, but the presence of bright yellow, orange, or red flash colors is typical.

Behavior All blackbirds share special musculature that allows them to open their bill with great force, allowing them to insert the bill tips in the ground or a crevice and then open the bill with strength, creating an opening for them from which to extract food. This "gaping" is characteristic of blackbirds and shared by the starling family. Other common behaviors include "bill tilting," in which an individual will point its bill up toward the sky. This aggressive signal often is given to a nearby individual during feeding and in territorial disputes. During territorial singing, blackbirds make the flash colors obvious, flaring red epaulets, exposing yellow breasts, or twisting yellow heads. In display, the grackles deeply keel the tail, ruffle the plumage, and drop the wings. Blackbirds have interesting mating systems. Orioles tend to be largely monogamous; some of the resident southern species defend territories year-round, with a great deal of female involvement. More migratory species show duller female plumages, and most territorial defense is performed by the male. Larger grackles and marsh-nesting blackbirds have polygamous (many mates) unions. Usually a male defends a large, high-quality habitat and several females settle in his territory and mate with him. Among the grackles, the females congregate in a breeding colony and the largest and most aggressive male defends the "harem." Defending a harem is a behavior nearly unknown in birds. Cowbirds lay their eggs in the nests of other species, doing away with parental care altogether. Many territorial species are highly social during the nonbreeding season, forming large mixed-species flocks. The Tricolored Blackbird is also highly social and colonial during the summer. It is the only strictly colonial land bird extant in North America.

Distribution Restricted to the New World, blackbirds successfully exploit habitats from grasslands and urban areas to forests and shrublands from Alaska to Tierra del Fuego. The most migratory species breed in North America, with the Bobolink showing the longest and most impressive migration.

Taxonomy There are nearly 100 species of blackbirds in five main family lineages. North American blackbirds fit into three of these groups: the orioles, the blackbird-grackle group, and the meadowlark and allies. Interestingly, the Bobolink and Yellow-headed Blackbird are part of the meadowlark allies!

Conservation The Tricolored Blackbird and Rusty Blackbird are of conservation concern, as their populations have drastically declined in the last few decades. Eleven species outside of North America are listed as threatened or near threatened.

BOBOLINK *Dolichonyx oryzivorus*   BOBO ■ 1

The breeding-plumaged Bobolink is a charismatic, attractive bird with a black-and-white plumage. Monotypic. L 7" (18 cm)
IDENTIFICATION Small with a sparrow-like bill, very long wings, and pointed tail feathers. **SUMMER MALE:** Black (feathers tipped buff when fresh), with buffy nape and large white patches on scapulars and lower back to uppertail coverts. **SUMMER FEMALE:** Streaked and sparrow-like, with pink-based bill and pinkish legs. Dark crown with crisp white median stripe, buffy face, a short but obvious postocular stripe broadens above ear. Above brown, with yellowish buff streaking. Crisp streaking restricted to breast sides and flanks. **FALL:** Similar to summer female, but warmer yellowish below.
SIMILAR SPECIES Adult male Bobolink is distinctive. Male Lark Bunting is black with white on wings, but not on nape or body. Females and fall plumage are superficially sparrow-like, but note Bobolink's larger size and bright

pink legs. Also, streaking is crisp and restricted to flanks, while head shows a dark crown with a buff central stripe, a short postocular stripe, and pale lores. Le Conte's Sparrows are much smaller, streaked on breast, white on belly, brighter orange-buff on face, and short winged.
VOICE CALL & FLIGHT NOTE: A loud and sharp *pink*. **SONG:** A euphoric, complex bubbling and gurgling that gives the bird its name, *bob-o-link bob-o-link blink blank blink*. Usually the song is given during an aerial display.
STATUS & DISTRIBUTION Common. **BREEDING:** Old fields. **MIGRATION:** Arrive in FL in mid-Apr., and in the Northeast early May, after breeding congregates in marshes to molt, then heads south Aug.–late Sept., lingerers to Oct. Mostly a migrant through FL, fewer in Gulf Coast states. In fall, most appear to take a overwater route

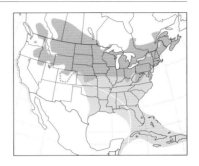

to S.A., hence it is very common on Bermuda in fall (exceptionally, after Hurricane Emily on Sept. 25, 1987, 10,000+ were counted there). **WINTER:** In south-central S.A. **VAGRANT:** Rare, primarily in fall on West Coast. Casual to the Azores, Europe, and the Galapagos Is.; accidental AK.
POPULATION Surveys suggest a general decline in numbers.

buffy nape
black face and underparts
buffy edges to feathers when fresh
early spring ♂
breeding ♂
white rump
dark pink bill
strong head pattern
rich yellow-buff overall
strong back streaks
very long primary projection
fall
breeding ♀
spiky tail tips

BLACKBIRDS Genus *Agelaius*

The five species belonging to this genus are restricted to North America, Central America, and the Caribbean. A group of similar South American species were previously included. Largely black plumages; bright red, tawny, or yellow "epaulets" or shoulder patches; and unmusical and screechy songs characterize the males.

TAWNY-SHOULDERED BLACKBIRD *Agelaius humeralis*   TSBL ■ 5

The Tawny-shouldered is an accidental vagrant from the West Indies. Polytypic (2 ssp.; likely *scopulus* from nearby Cuba; nominate from Hispaniola). L 8" (20 cm)
IDENTIFICATION Like a small and slim Red-winged Blackbird, but plumage entirely black with tawny lesser coverts, median coverts tawny with yellowish tips. Bill slim and sharply pointed.
SIMILAR SPECIES Red-winged Blackbird is larger and bulkier, with a red

epaulet and a wider yellow border. Tawny-shouldered is more arboreal and has a different song.
VOICE CALL: A *chuck* note typical of blackbirds. **SONG:** A muffled buzzy drawn-out *zwaaaaaaaa*, lasting just over a second in length.
STATUS & DISTRIBUTION Accidental. **YEAR-ROUND:** Arboreal, preferring open woods, edge, and parklike settings. **VAGRANT:** Two visited the Key West Lighthouse (Feb. 27, 1936).

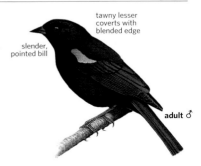

tawny lesser coverts with blended edge
slender, pointed bill
adult ♂

RED-WINGED BLACKBIRD *Agelaius phoeniceus* RWBL ▪ 1

This species is one of the most widely distributed, abundant, well-known, and well-named birds in N.A. Polytypic (22 ssp.; 14 in N.A.). L 8.7" (22 cm)
IDENTIFICATION A medium-size passerine with a sharply pointed black bill. **SUMMER MALE:** Black, including soft parts, with bright red shoulder patch or "epaulet," bordered by yellow in most subspecies. **WINTER MALE:** Feathers finely tipped with brown. **FEMALE:** Well streaked throughout, with whitish supercilium. Brown and streaked above, densely streaked below. Peachy wash on chin and throat. Dull reddish edges to lesser coverts. **IMMATURE MALE:** As winter male, but feathers more broadly edged with brown. Trace of supercilium; epaulet shows black spotting.
GEOGRAPHIC VARIATION Most subspecies poorly defined, but "California Bicolored" group *(californicus, mailliardorum)* characterized by dark females appearing solidly blackish brown on lower breast to vent; greatly reduced supercilium. Males show black median coverts, so there is no yellow border.
SIMILAR SPECIES Male Tricolored has a thinner, pointed bill, a white epaulet border, colder gray-buff upperpart markings

in winter, and a different voice. See sidebar to separate Tricolored females.
VOICE CALL: *Chuk.* **SONG:** A hoarse, gurgling *konk-la-ree;* variable.
STATUS & DISTRIBUTION Abundant. **YEAR-ROUND:** Open or semi-open habitats, closely associated with farmland. **BREEDING:** Usually in cattail marsh, but also in moist open, shrubby habitats. **MIGRATION:** Southern populations resident; northern ones migratory, arrive in Northeast late Feb. and leave by Nov. **VAGRANT:** Casual to Arctic AK, Greenland, and Iceland.

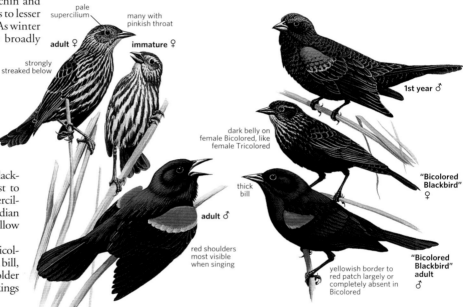

pale supercilium
many with pinkish throat
adult ♀
immature ♀
strongly streaked below
1st year ♂
dark belly on female Bicolored, like female Tricolored
thick bill
adult ♂
red shoulders most visible when singing
"Bicolored Blackbird" ♀
"Bicolored Blackbird" adult ♂
yellowish border to red patch largely or completely absent in Bicolored

Identification of Female Red-winged and Tricolored Blackbirds

The separation of female Red-winged and Tricolored Blackbirds in California, where their ranges overlap, is one of the toughest field identification challenges for birdwatchers. Birders tend to ignore the females and concentrate on the males, which is a valid way to identify the species, but really not that useful when one comes across a lone female or a vagrant. In order to make an adequate separation one needs to look at the bill shape, wing structure, and tones and colors on the upperparts in fresh plumage.

Compared to the strongly dark-streaked female Redwinged Blackbirds from most of North America, the females of many California Red-winged subspecies are largely dark, and when worn can look largely solidly blackish above and unstreaked and solidly blackish on the belly and vent, with dark streaks on a pale background restricted to the throat and upperbreast. This plumage coloration is generally the same pattern shown by female Tricolored Blackbirds.

Structurally, these species do differ somewhat. Tricolored Blackbirds show thinner-based and longer bills than most California Red-winged Blackbirds, but note that the subspecies breeding in the Kern River Valley *(aciculatus)* is characterized by its long and slender bill. The Tricolored has a more pointed wing shape, which can be looked for on the perched bird. The outer primary, P9, is shorter than P6 on the Red-winged but longer on the Tricolored. In addition, on the folded wing, the distance between the longest primary (P8) and the next (P7) is noticeable on a Tricolored but minimal in a Red-winged.

In fresh plumage (during fall and early winter), a Red-winged Blackbird is edged with rufous, golden, and warm buffy edges on the upperparts, while these same areas are cold gray-buff on a Tricolored Blackbird. Often on a Red-winged there are two obviously paler, more yellowish lines of streaks on the back,

TRICOLORED BLACKBIRD *Agelaius tricolor*   TRBL ■ 2

The Tricolored is very similar in appearance, but not in behavior, to the Red-winged Blackbird. The great majority is restricted to California. Monotypic. L 8.7" (22 cm)

IDENTIFICATION This medium-size passerine with a long, sharply pointed bill is seldom alone, breeding in colonies and wintering in medium- to large-size flocks. **SUMMER MALE:** Black with slight blue iridescence, including soft parts, with dark red shoulder patch or epaulet. Epaulet is bordered broadly with white. **WINTER MALE:** As summer male, but shows gray-buff feather tips throughout body, and epaulet border is creamy white. **FEMALE:** Dark, but streaked, most strongly on throat and breast; streaked breast contrasts with solid dark belly; poorly developed supercilium. Dull reddish edges to lesser coverts. **IMMATURE MALE:** As winter male, but more heavily edged and fringed and often showing buff supercilium.

Epaulet shows black spotting.
SIMILAR SPECIES Red-winged Blackbird males have thicker-based bills and a yellow or no epaulet border.
VOICE CALL: *Kuk,* lower in pitch than similar Red-winged call. **SONG:** Distinctive; a nasal, drawn-out *guuuaaaak* or *ker-gwuuuuaaaa,* lasting 1–1.5 seconds.
STATUS & DISTRIBUTION Fairly common. **YEAR-ROUND:** Open or semi-open habitats, closely associated with farmland. **BREEDING:** Highly colonial, traditionally in large CA marshes, more recently in dense thickets of introduced Himalayan blackberry. **MIGRATION:** After breeding, many move toward the coast, starting in mid-July. **WINTER:** Closely associated with rangeland, dairy operations. **VAGRANT:** Previously vagrant to WA, OR, and NV, but has now bred in isolated small colonies in those states. Casual to southeastern CA.
POPULATION Currently the population

dark belly
♀

is estimated at 230,000 birds. Once a single colony in Glenn Co., CA, held more than 200,000 birds, and many colonies had more than 100,000. Habitat change and colonial nesting make it a vulnerable species. A survey detected a potential 37 percent decrease in numbers between 1994 and 1997!

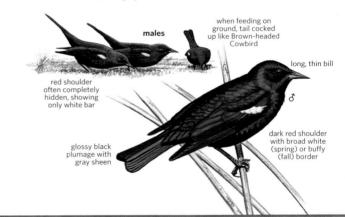

males

when feeding on ground, tail cocked up like Brown-headed Cowbird

long, thin bill

♂

red shoulder often completely hidden, showing only white bar

glossy black plumage with gray sheen

dark red shoulder with broad white (spring) or buffy (fall) border

Tricolored Blackbird, female (CA, May)

Red-winged Blackbird, female (IL)

Red-winged Blackbird, *aciculatus,* female (CA)

like suspenders. The white tips to the median coverts (upper wing-bar) are broader and more noticeable on a Tricolored. Finally, some adult female Red-wingeds show a peachy or pinkish wash to the throat, absent in Tricoloreds.

Worn summer females, which by this point lack the paler feather tips on the upperparts, may be unidentifiable if not studied closely, although a Tricolored usually shows a stronger white bar on the median coverts. ■

YELLOW-HEADED BLACKBIRD *Xanthocephalus xanthocephalus* YHBL ■ 1

The Yellow-headed Blackbird is a beautiful marsh-dwelling blackbird of the prairies and the West. Monotypic. L 9.5" (24 cm)

IDENTIFICATION Large and bulky; males noticeably larger than females. **ADULT MALE:** Black with a bright yellow head (darker in winter), accented by a triangular black mask. White primary coverts create obvious wing patch in flight. Yellow around vent, difficult to see in the field. Bill, eyes, and legs blackish. **ADULT FEMALE:** Brownish, with yellow supercilium and breast. Whitish throat with dark lateral stripes and yellow malar stripes above them. **IMMATURE MALE:** Like female, but substantially larger, more extensive yellow

on head and neck, black lores; noticeable white on primary coverts, although not a fully formed patch. **JUVENILE:** Cinnamon head, contrasting with whitish throat and brownish body. Two whitish wing bars and cinnamon fringes to tertials.

SIMILAR SPECIES No other bird in North America looks like male Yellow-headed Blackbird. Duller females are identified by being brown, unstreaked, and having a contrasting yellow breast.

VOICE CALL: A rich, liquid *check*. **SONG:** Two song types, both with a mechanical, unpleasant, or at least unusual, sound that renders them unmistakable. The primary song, *kuk, koh-koh-koh ... waaaaaaaa,* lasts four seconds; the final nasal scraping sound is separated from the introductory notes, accompanied by an asymmetrical display where neck and head are bent to one side. The second song, a croaking *kuuk-ku, whaaa-kaaaa,*

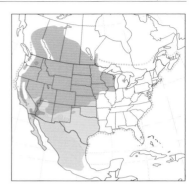

deep yellow throat and breast

deep yellow head

immature ♂

white wing patch on primary coverts

spring adult ♂

spring adult ♂

immature ♂

lasts two seconds, accompanied by a symmetrical display.

STATUS & DISTRIBUTION Common. **BREEDING:** Deepwater wetlands of cattail, rushes, or *Phragmites;* it forages in adjacent grasslands or farmland. **MIGRATION:** First southbound birds in July, but most move Aug.–late Sept., in spring males arrive one–two weeks before females, arriving mid-Apr.– mid-May. A diurnal migrant, usually moving in single-sex flocks. **WINTER:** A few birds in agricultural areas in border states from CA to southern TX. Most in Mexico. Females winter farther south than males. **VAGRANT:** Rare in East away from mapped range. Casual in AK, NT, and Bermuda. Accidental in Europe.

POPULATION There was a general increase in population during the 1970s. Local droughts can greatly alter their numbers.

BLACKBIRDS Genus *Euphagus*

The two species in *Euphagus* are closely related to the grackles and the neotropical genus *Dives*. They resemble grackles in having ruff-out displays that accompany the song: the tail is cocked, the wings are drooped, the body is ruffled, and often the pale eyes are prominent. Unlike the grackles, *Euphagus* have standard-shaped tails.

RUSTY BLACKBIRD *Euphagus carolinus* RUBL ■ 1

The Rusty is a blackbird of swampy forests, adorned with rusty in fall and winter. Polytypic (2 ssp.; both in N.A.). L 9" (23 cm)

IDENTIFICATION A slim blackbird with a slender, pointed bill. **SUMMER MALE:** All black. Black bill and legs; bright yellow eyes. **WINTER MALE:** Black with wide cinnamon, buff, or warm edges. In fall and early winter edging wide

and bold, breast and back appear largely rusty. Buff supercilium and malar. Coverts and tertials tipped rusty; rump blackish. As winter progresses edging wears, revealing black plumage. **SUMMER FEMALE:** Blackish gray with darker wings, tail, and lateral throat stripes. Bill black, legs black, and eyes yellow. **WINTER FEMALE:** Widely edged rusty and buff. Similar to winter male, but paler

and even more rusty, and rump grayish. **GEOGRAPHIC VARIATION** Subspecies *nigrans,* which breeds in the Maritimes and Newfoundland, is not field separable from the widespread nominate subspecies.

SIMILAR SPECIES Brewer's Blackbird is similar, but it has a shorter bill and curved culmen. In winter, wide rusty edging on Rusty's plumage is distinctive, although immature male Brewer's can show some dull buff on breast and upperparts. Brewer's never show rusty edges to tertials. Male Rusty in summer blackish, lacking strong gloss of Brewer's; its iridescence is dull green and even throughout body. Summer female Rusty more grayish than Brewer's, and shows yellow eye. A small percentage of Brewer's females also show yellow eyes, but female Brewer's more brownish. Common Grackle has a long, graduated tail often held in a deeply keeled shape. Common Grackles show more complex and brighter iridescence patterns than Rusty Blackbird. **VOICE CALL:** A *chuck,* not as deep as that of a grackle. **SONG:** A squeaky, sweet, rising *kush-a-lee* or *chuck-la-weeeee.* A secondary song begins with two or three musical notes followed by a harsher long note. Females sing a weaker version. **STATUS & DISTRIBUTION** Fairly common. **BREEDING:** Bogs in

boreal forest. **MIGRATION:** Diurnal migrant; arrives southern Canada by late Mar., southbound late Sept.–Nov. **WINTER:** Wet open woodlands, or fields near wetlands. **VAGRANT:** Very rare to casual to Pacific region from BC to CA, also casual or accidental to Pribilofs Is., St. Lawrence I., Russian Far East, and Bermuda.

POPULATION Numbers have declined precipitously since 1960, with some sources estimating a 90 percent drop between the 1960s and 1990s.

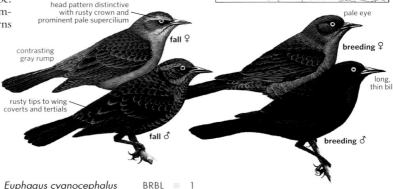

head pattern distinctive with rusty crown and prominent pale supercilium

fall ♀

contrasting gray rump

rusty tips to wing coverts and tertials

fall ♂

pale eye

breeding ♀

long, thin bill

breeding ♂

BREWER'S BLACKBIRD   *Euphagus cyanocephalus*   BRBL ■ 1

A common and widespread ground-dwelling icterid of the West, the Brewer's in some ways replaces the Common Grackle ecologically. Monotypic. L 9" (23 cm). **IDENTIFICATION** A slim blackbird, slightly stouter build than Rusty. **MALE:** Basically black with bright yellow eyes. Strongly iridescent, with a bright blue or purplish blue sheen on head, while body is greenish. **FEMALE:** Dull brownish gray and unstreaked; dark eyes. **IMMATURE MALE:** Many young males show some buffy feather tips on breast and warmer buff tipping on back. **SIMILAR SPECIES** A summer Rusty Blackbird is similar, but it has a more slender pointed bill. Male Brewer's is more strongly glossy, showing blue on head and green on body. In winter some young male Brewer's show buff tipping on breast, head, and upperparts, but tertials are nearly always entirely black. Female Brewer's is darker and browner, than female Rusty, which has grayish color, particularly the rump. Female Brewer's typically show a dark eye, although a few show yellowish eyes. Common Grackle has a long, graduated tail often held in a deeply keeled shape. Common Grackles show more complex and brighter iridescence patterns than Brewer's,

with a characteristic abrupt break in color from head iridescence to that of body. **VOICE CALL:** A *chak,* or *chuk,* similar to that of Rusty Blackbird. When alarmed, it gives a whistled *teeeuuuu* or *sweee.* **SONG:** Often gives a faint and unappealing raspy *schlee* or *schrrup* during the ruff-out display; both sexes sing and display. **STATUS & DISTRIBUTION** Common. **YEAR-ROUND:** Varied habitats, inc. urban areas, golf courses, agricultural lands, open shrubby areas, forest clear-cuts, and riparian forest edges. Requires open ground for foraging and some dense vegetation or edge for nesting. Farther east, where sympatric with Common Grackle, it takes more open sites than the grackle. **MIGRATION:**

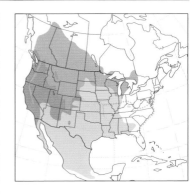

Poorly understood, easternmost populations more migratory. Extent of movements correlated with snow cover. Wintering groups somewhat nomadic. **VAGRANT:** Casual to central and northern AK, NT, and northeastern VA. **POPULATION** Slightly declining.

buffy tipping to head, back, and much of upperparts

wings uniformly dark

immature ♂

♂

bill slightly thicker than Rusty

dark eye

♀

colored plumage with more gloss than Rusty, visible in good light when close

MEADOWLARKS Genus *Sturnella*

The meadowlarks are open-country blackbirds, with a distinctive rotund and short-tailed starling-like shape. They fly with an odd flight style, using shallow fluttery wingbeats. The two North American species are yellow below, with a black V on the breast; most of the seven species in this genus are red breasted.

WESTERN MEADOWLARK   *Sturnella neglecta*   WEME   ▪ 1

The song of the Western Meadowlark is emblematic of the West. Polytypic (2 ssp.; both in N.A.). L 9.5" (24 cm) **IDENTIFICATION** Rotund, stocky, medium-size blackbird with a long bill, short tail, strong legs, and pointed tail feathers. **SUMMER ADULT:** Cryptically patterned above; yellow below with a black V on breast. Crown brown with white median crown stripe, dark post-ocular stripe, yellow supralores. Yellow on throat invades malar area. Gray-buff flanks streaked brown, vent and under-tail coverts whitish. Fresh birds appear scaly due to complete pale fringing of feathers, but worn individuals look pale streaked as pale tip wears. Coverts pale brown with separate thin dark bars that remain separate to center of shaft. Similarly, central tail feathers are pale brown with discrete narrow dark brown bars. White on outer three tail feathers, a small brown strip remains on outer corner of outer two rectrices, but next one in (R4) largely dark with only a white wedge on inner vane. Bill gray with darker culmen and tip, legs dull pink, eyes dark. **WINTER ADULT:** Pale tips cloud black V on breast. Slightly more buffy yellow underparts; more scaly looking upperparts. **JUVENILE:** Similar to winter adult, but duller face pattern, paler yellow below, and breast streaked in a V, not solid.

GEOGRAPHIC VARIATION Subspecies *confluenta* of the Pacific Northwest is darker than the nominate, and it shows dark bars on tail feathers and coverts that widen at center of each feather and join up with adjacent dark bars, like the Eastern Meadowlark.
SIMILAR SPECIES Eastern Meadowlark is extremely similar. The south-western form of Eastern Meadowlark (known as "Lilian's") is even more similar to the Western than the more widespread eastern subspecies due to its pale plumage. To separate these look-alikes one needs to concentrate on the voice, extent of yellow of throat, plumage patterns, and tail pattern. Some vocalizations are diag-nostic, such as the blackbird-like call of Western. The two-parted song is lower in frequency and lacks the ascending whistles of Eastern (includ-ing "Lilian's") song; however, the song is learned, and in rare cases the meadowlarks can learn each other's songs—this is not the case for the call. Western shows more yellow on throat; it extends to malar area, and this can be surprisingly easy to see in a scope view. Western is generally paler than Eastern, but similar to "Lilian's," showing a pale gray-brown overall color, rather than the warmer, more

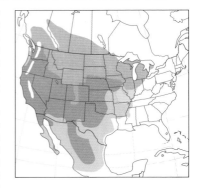

saturated brown of Eastern. Western shows pale gray-buff flanks, like "Lil-ian's," and Eastern has darker, mid-tone buff flanks with stronger streaks. Wing coverts show up as a grayish brown panel with narrow dark bars on Western, while on Eastern they are warm brown to cinnamon brown, with wider dark bars. Western shows largely white outer two tail feathers, while Eastern shows largely white outer three tail feathers.
VOICE CALL: A low *chupp* or *chuck*. Females give a dry rattle, males a slower rolling note. **FLIGHT NOTE:** A sweet whistled *weeet*. **SONG:** Males have melodious and flute-like song lasting approximately 1.5 seconds. Two phrases, starting with several clear whistles, and a terminal phrase which is more gurgled, bubbling and complex, *tuuu-weet-tooo-twleedlooo*.
STATUS & DISTRIBUTION Common. **BREEDING:** Dry grasslands, agricul-tural areas. **MIGRATION:** Diurnal migrant; north-ern populations migra-tory, southern ones resident. Eastern breed-ers are also easternmost in winter. Spring arrival dependent on snow melt, usually Mar.–Apr., fall movements peak Sept.–Oct. **WINTER:** Dry grassy sites. **VAGRANT:** Casual in AK, NT, and Hudson and James Bays. Casual to East Coast from NS to GA. **POPULATION** Slowly declining.

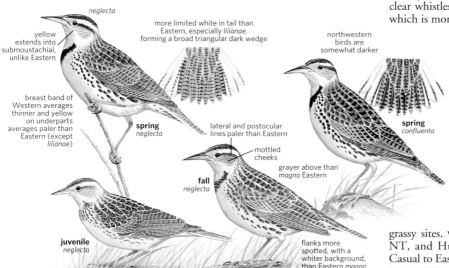

yellow extends into submoustachial, unlike Eastern

neglecta

more limited white in tail than Eastern, especially *lilianae*, forming a broad triangular dark wedge

northwestern birds are somewhat darker

breast band of Western averages thinner and yellow on underparts averages paler than Eastern (except *lilianae*)

spring *neglecta*

lateral and postocular lines paler than Eastern

mottled cheeks

grayer above than *magna* Eastern

spring *confluenta*

fall *neglecta*

juvenile *neglecta*

flanks more spotted, with a whiter background, than Eastern *magna*

EASTERN MEADOWLARK *Sturnella magna* EAME ■ 1

The sweet, whistled song betrays the presence of this ground-loving blackbird. Polytypic (16 ssp.; 4 in N.A.). L 9.5" (24 cm)

IDENTIFICATION Shaped and plumaged like Western Meadowlark but with longer legs. **SUMMER ADULT:** Cryptically patterned above; intense yellow below with bold black V on breast. Warm buff flanks crisply streaked blackish. Coverts warm brown with dark bars that widen and meet adjacent dark bars at the feather shaft. Similarly, central tail feathers show confluent dark bars along shaft. Outer three tail feathers largely or entirely white. **JUVENILE:** Similar to winter adult, but paler yellow below and breast V streaked.

GEOGRAPHIC VARIATION Sixteen subspecies recognized, four in North America. The most distinct, and perhaps a good species, is *lilianae*, "Lilian's." Found in the desert Southwest, it is smaller, has longer wings and legs, and is generally paler than typical Easterns. It shows pale gray-brown plumage, like a Western Meadowlark, and separate and narrow bars on tail and greater coverts. It has extensive white on tail, with outer three rectrices entirely white, and the next in with substantial white. Although the calls are the same as for the Eastern, the song of "Lilian's" is slightly more complex and lower in pitch. The subspecies *hippoprepis* from Cuba has very different vocalizations, more like Western (song), and likely is a separate species.

SIMILAR SPECIES Western Meadowlark is very similar; see that account. The call of Eastern is diagnostic; the higher-pitched chatter is unlike the dry rattle of a Western. Eastern lacks yellow on the malar and is generally darker than a Western, showing a saturated brown overall color. Eastern shows largely white outer three tail feathers; white is even more extensive on "Lilian's" Meadowlark. "Lilian's" shows the pale plumage and discrete, separate barring as in Western, but it lacks streaking on the pale cheek, thus showing a great deal of contrast with the dark eye line and crown, and whitish supercilium and cheeks.

VOICE CALL: A buzzy *dzert;* also a chatter given by both sexes, higher pitched than rattle of Western Meadowlark. **FLIGHT NOTE:** A sweet whistled *weeet.* **SONG:** Three to five or more loud, sliding, descending whistles lasting approximately 1.5 seconds, *tsweee-tsweee-tsweeeooo.*

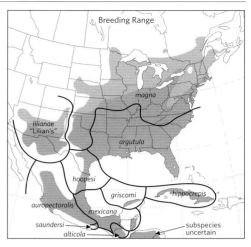

Breeding Range

magna
lilianae "Lilian's"
argutula
hoopesi
auropectoralis
griscomi
hippocrepis
mexicana
saundersi
alticola
subspecies uncertain

STATUS & DISTRIBUTION Common. **BREEDING:** Grasslands and old field habitats; where sympatric with Western, takes moister grassland and shrubby edge habitats. "Lilian's" in desert grassland. **MIGRATION:** Diurnal migrant; northern birds migratory, southern ones resident. Spring arrival dependent on snow melt, usually Mar.–Apr., fall Oct. **WINTER:** Farmland, grasslands, and rangelands. **VAGRANT:** Casual to NF, ND, CO, southwestern AZ, MB, and Bermuda.

POPULATION General declines have been detected from the 1960s to the 1990s due to habitat loss.

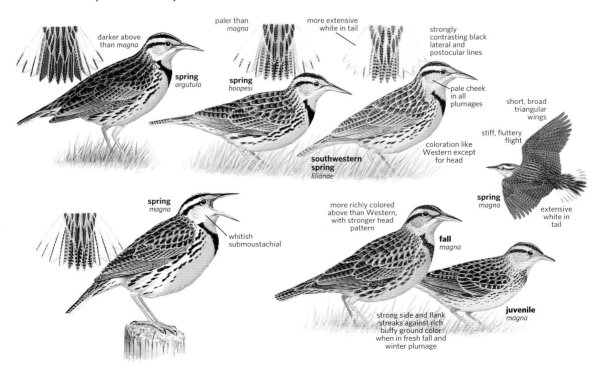

darker above than *magna*

spring *argutula*

paler than *magna*

spring *hoopesi*

more extensive white in tail

strongly contrasting black lateral and postocular lines

pale cheek in all plumages

short, broad triangular wings

stiff, fluttery flight

coloration like Western except for head

southwestern spring *lilianae*

spring *magna*

whitish submoustachial

spring *magna*

extensive white in tail

more richly colored above than Western, with stronger head pattern

fall *magna*

juvenile *magna*

strong side and flank streaks against rich buffy ground color when in fresh fall and winter plumage

GRACKLES Genus *Quiscalus*

There are seven grackle species worldwide, one of which is now extinct—the Slender-billed Grackle (*Q. palustris*) of central Mexico was last recorded 1910. The males have a glossy black plumage and often have yellowish eyes. During breeding displays, the strongly graduated tail is held deeply keeled, giving a V-shaped cross section.

BOAT-TAILED GRACKLE   *Quiscalus major*   BTGR ▪ 1

Inhabits coastal marshes of the Atlantic and Gulf Coasts. Polytypic (4 ssp.; all in N.A.). L 15–16.5" (38–42 cm)
IDENTIFICATION A large grackle with a moderate-size bill, a rounded crown. **ADULT MALE:** Entirely black with obvious blue iridescence, becoming violet on head. Eyes yellow to dull yellowish or even honey brown; bill and legs black. Long, deeply keeled tail. **ADULT FEMALE:** Smaller than male. Brown above; warm tawny on head and below, darker on belly and vent. Eyes usually dark. Does not hold tail in deep keel.
GEOGRAPHIC VARIATION The four subspecies vary primarily in eye color, some having yellow eyes, others brown. They are not field identifiable other than by breeding range.
SIMILAR SPECIES The very similar Great-tailed Grackle was once treated conspecific. Boat-tailed shows a steeper forehead, rounder crown, and smaller bill; this is more obvious on males than on females. Male often shows dark eyes. Display differs where wings are flipped high over the back for Boat-tailed. Female is warmer colored than a Great-tailed and usually has dark eyes; dark lateral throat stripes absent or indistinct.
VOICE CALL: A low *clak* or a *kle-teet*. **SONG:** Distinct from Great-tailed. A continuous, long, harsh trilling song interspersed with other notes, *jeeb-jeeb-jeeb tireeet chrr chrr chrr chrr tireet tireet tireet tireet.* Wing-flipping display accompanies lower *chrr* notes.
STATUS & DISTRIBUTION Common. **YEAR-ROUND:** Coastal marshes, but various open habitats in FL. **VAGRANT:** Casual in New England.
POPULATION Stable, has been expanding its range since the 1890s.

juvenile

1st fall ♂

distinctive rounded head

♀

cinnamon-buff underparts

where ranges overlap, note dark eye of *major* Boat-tailed, unlike yellow eye of all Great-tailed

blue-green gloss overall

western Gulf Coast adult ♂

long, keel-shaped tail

Subspecies of Great-tailed Grackle

Eight Great-tailed Grackle subspecies are recognized, but only three are found in North America. These northern subspecies are *prosopidicola,* found in the east of the Great-tailed's range west to central Texas; *monsoni,* found from central Arizona east to western Texas; and *nelsoni,* found in California and western Arizona. All three subspecies of the Great-tailed are spreading northward in the United States. For the most part, there is little information regarding which subspecies have spread to which areas, therefore the range descriptions given above are tentative. And some intergradation may be occurring now that these subspecies are coming widely into contact.

The males of these subspecies are similar, differing mainly in size, with *prosopidicola* and *monsoni* being large subspecies while *nelsoni* is noticeably smaller. With regards to plumage, a male *monsoni* shows on average more of a purplish gloss, but this is variable. Differences in plumage are much more marked in females. In general, *monsoni* females are darkest below, *nelsoni* palest, and *prosopidicola* intermediate, although

Great-tailed Grackle, female (AZ, Feb.)

GREAT-TAILED GRACKLE *Quiscalus mexicanus*   GTGR 1

This huge blackbird is hard to ignore due to its boisterous nature. Polytypic (8 ssp.; 3 in N.A.). L 15–18" (38–46 cm)
IDENTIFICATION Long, deeply keeled tail. Large, thick bill, with nearly straight culmen. Flat crown; shallow forehead. **ADULT MALE:** Entirely black with obvious violet-blue iridescence. Eyes yellow; bill and legs black. **ADULT FEMALE:** Smaller than male. No keeled tail. Brown above with dull iridescence on wings and tail; buffy on head and below, becoming darker brown on belly and vent. Dark lateral throat stripes usually obvious.

Eyes yellow. **IMMATURE MALE:** Smaller than the adult male, with shorter tail, dull iridescence, browner wings, and frequently dark eyes. **JUVENILE:** Like female, but paler and shows diffuse streaking below.
GEOGRAPHIC VARIATION See sidebar below.
SIMILAR SPECIES The very similar Boat-tailed Grackle overlaps with Great-tailed Grackle in southwestern coastal Louisiana and eastern Texas. See Boat-tailed Grackle.
VOICE CALL: A low *chut;* males may give a louder clack. Eastern males give a striking ascending whistle *twooo-eeeeeeee!* **SONG:** The eastern bird sings a four-part song beginning with harsh notes similar to the breaking of twigs, then a soft

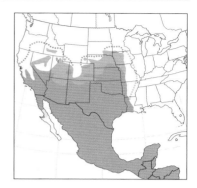

juvenile

streaked underparts

buffy supercilium

♀

cinnamon-buff underparts

long, rather thin bill

♂

purple gloss with no head and body contrast as in Common Grackle

western ♀

nelsoni is smaller and paler than other subspecies

very long, keel-shaped tail

undulating *chewechewe,* and then twig-breaking notes and finally several loud two-syllable *cha-wee* calls, *crrrk crrrk chewechewe crrk cha-wee cha-weewlii.* Subspecies *nelsoni* sings a repeated series of notes, ending in a more accented note *chk-chk-chk-chap-chap-chap-chap-CHWEEE,* often interspersed with various other repeated notes.
STATUS & DISTRIBUTION Abundant. **YEAR-ROUND:** Open habitats with dispersed trees, from agricultural to urban. **MIGRATION:** Not well understood; more are wintering farther north now. **VAGRANT:** Casual to the north of its range, from BC east to NS.
POPULATION The species has experienced a great range and population increase in the United States, showing a 3.7 percent annual increase from 1966 to 1998.

closer to *monsoni.* In fact, *nelsoni* females may be pale grayish below with a nearly white throat; this coloration is strikingly different from the buff to warm brown underparts of *prosopidicola* and *monsoni.* The pale plumage combined with the small size sets *nelsoni* well apart from *prosopidicola* and *monsoni.*

Historically, the mountains of central Mexico divided the general population of Great-tailed Grackles into an eastern and central group and a western group. The "Western" Great-tailed Grackles—from *nelsoni* in the north, to coastal forms in

Great-tailed Grackle, female (TX, Nov.)

west Mexico south to Guerrero—are small, they have a noticeably different song than the more eastern populations, and there are genetic differences. However, with the opening of more grackle-friendly habitats throughout the area due to agricultural and urban development, this previously isolated population has come into contact with eastern Great-tailed Grackles. In Arizona intergradation between *nelsoni* and *monsoni* appears to be common, and birds that have *monsoni* mitochondrial DNA are by measurements small and like *nelsoni.* ■

COMMON GRACKLE *Quiscalus quiscula* COGR ▪ 1

The Common Grackle is a common and often urban blackbird of eastern North America. Polytypic (3 ssp.; all in N.A.). L 12.6" (32 cm) **IDENTIFICATION** A large blackbird with strong legs and a long, graduated tail that is held in a deep keeled shape during the breeding season. The iridescence of head is different from that of body, and changes abruptly. **ADULT MALE:** Entirely black with noticeable iridescence in good light. Widespread form (see Geographic Variation) shows bronze gloss to body, blue head, and purple or blue iridescence on wings and tail. Eyes are bright yellow, while legs and bill are black. **ADULT FEMALE:** Smaller and duller than male and does not hold tail in deep keel shape; in widespread *versicolor* bluish head still sharply contrasts with brown body. **JUVENILE:** Brown, with dark eyes and faintly streaked on breast. **GEOGRAPHIC VARIATION** "Bronzed Grackle" *(versicolor)*, found northwest of the Appalachians, has bronze iridescence on body, a blue head, and purplish tail and wings. "Purple Grackle" *(stonei)*, found southeast of the Appalachians, has a purplish body and head, with a

blue or greenish glossed tail. "Florida Grackle" *(quiscula)*, ranging from Florida to southern Louisiana and South Carolina, has a greenish iridescence on its back. **SIMILAR SPECIES** Brewer's and Rusty Blackbirds lack long, graduated tail of Common Grackle, and they never hold it in a keeled shape. Boat-tailed and Great-tailed Grackles are much larger, with even more striking tails. Common Grackles show a clear, abrupt division between the gloss color of head and body. **VOICE CALL:** A loud and deep *chuck*. **SONG:** A mechanical, squeaky *readle-eak*. Both sexes sing. **STATUS & DISTRIBUTION** Abundant. **YEAR-ROUND:** Open and edge habitats, urban areas, agricultural lands, golf courses, swamps, and marshes. **MIGRATION:** Diurnal migrant; southern populations resident. Arrive at breeding areas mid-Feb.–mid-Mar. and early Apr. in

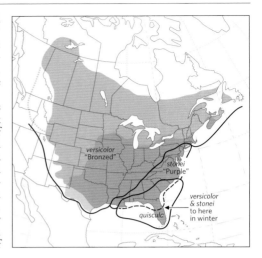

northernmost sites. Begins southward movements as early as late Aug., peaking Oct.–early Nov. **VAGRANT:** Very rare or casual in Pacific states, BC, and Bermuda. Casual to far north, AK, YT, NT, and Churchill, MB. **POPULATION** Historical population increases, but significant declines in East in the last 30 years; range expanding in Northwest.

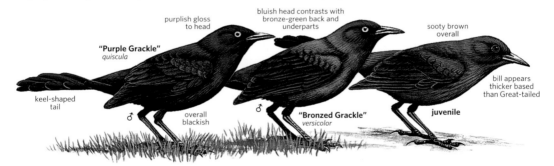

"Purple Grackle" *quiscula*
purplish gloss to head
bluish head contrasts with bronze-green back and underparts
sooty brown overall
keel-shaped tail
♂
overall blackish
♂
"Bronzed Grackle" *versicolor*
juvenile
bill appears thicker based than Great-tailed

COWBIRDS Genus *Molothrus*

The true cowbirds, of which there are five species, are all obligate brood parasites, laying their eggs in the nests of other species. Parental care is performed entirely by the hosts. Unlike many cuckoos, cowbirds are generalists, using various host species. They have stocky bodies and short tails. Male plumage shows glossy black; females are duller.

SHINY COWBIRD *Molothrus bonariensis* SHCO ▪ 3

Shiny Cowbird is a recent invader to the Southeast, from South America via the Caribbean. It is one of the few brood parasites in North America. Polytypic (7 ssp.; *minimus* in N.A.). L 7.5" (19 cm) **IDENTIFICATION** A smallish, stocky blackbird very similar in shape to Brown-headed Cowbird, but longer billed and perhaps with a longer tail and slimmer body. **ADULT MALE:** Entirely

black with violet-blue iridescence on head and anterior part of body, becoming less violet and more pure blue toward posterior parts of body. Eyes dark; legs and bill black. **ADULT FEMALE:** Dull brownish throughout, with paler supercilium and darker wings and tail. Legs and bill black. **JUVENILE:** Similar to female, but obscurely streaked below. **SIMILAR SPECIES** Male Shiny's strong

gloss and violet head separate it from Brown-headed Cowbird. Male is similar in color to a Brewer's Blackbird, but Shiny is smaller and stockier, and has a dark eye and thicker bill. A female Shiny is extremely similar to a Brown-headed. Shiny is generally darker, lacking white throat, but with a stronger face pattern with a noticeably paler supercilium, a longer and slimmer body, and a longer, shiny black bill. A female Brown-headed shows a paler bill with a horn or yellowish base to lower mandible. On closed wings, the secondaries do not show obvious pale fringes on the Shiny. Finally, Shiny has shorter, more rounded wings; the outermost primary (P9) is equal in length to P7 or P6 but noticeably shorter than P8, the second outermost primary.

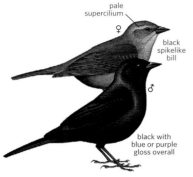

pale supercilium

♀

black spikelike bill

♂

black with blue or purple gloss overall

VOICE CALL: A soft *chup;* females give a chatter. **SONG:** Primary song strange sounding, liquid and bubbling, lasting 2–3 seconds. It begins with several purring bubbly notes and then a screechy series of high-pitched notes, *blurr-glurr-glurr-pt-tcheeeEEE.* Males also give a flight whistle, which may be considered a secondary song rather than a call. It is a more complicated and long series of short whistles, and it shows a great deal of geographic variation. Whistle is given both in flight and while perched. **STATUS & DISTRIBUTION** Rare in FL. **BREEDING:** Open and edge habitats; agricultural areas for foraging and forest edge habitats for finding host species. **MIGRATION:** Not studied, but new arrivals from farther south appear Mar.–Apr. **WINTER:** Open sites, agricultural areas, feeders. **VAGRANT:** Casual north of FL, records from NB, ME, OK, TX (2 recs.), LA (several),

AL (many), MS (several), TN (1 rec.), VA (1 rec.), NC (several), SC (several), GA (several). Most records away from FL have occurred in the spring. **POPULATION** A recent arrival to North America, first detected in Florida in 1985. After a flurry of observations in the late 1980s and early 1990s, sightings have decreased, particularly away from south FL.

BRONZED COWBIRD *Molothrus aeneus* BROC ▪ 1

The largest of our cowbirds, the Bronzed has a rather sinister appearance thanks to the male's typically hunchbacked look and bloodred eye. It gives an odd hovering display that is unique. Polytypic (4 ssp.; 2 in N.A.). L 8.7" (22 cm)
IDENTIFICATION A thick-set cowbird with a large and deep black bill. Males in particular may look proportionately small headed, especially when ruffling the nape feathers in display. Legs black. **ADULT MALE:** Entirely black with bronzed body iridescence, becoming blue-green on wings and tail. Eyes bright red. **ADULT FEMALE:** Varies geographically, see below. More widespread eastern subspecies

has a black female plumage, lacking strong gloss, with browner wings and tail. Eyes red. **JUVENILE:** Similar to female but dark brown.
GEOGRAPHIC VARIATION The more widespread *aeneus,* found from south-central TX eastward, and *loyei* from NM and west TX to southeastern CA. Males are similar, but females black in *aeneus* and grayish brown in *loyei.*
SIMILAR SPECIES Bronzed is larger than other cowbirds, and adults show bright red eyes and thick bills. Juveniles could be mistaken for female Brown-headed Cowbirds, but note their larger size, larger bulk, and thick bill.
VOICE CALL: A rasping *chuck.* Females give a rattle. **SONG:** A series of odd squeaky gurgles, *gluup-gleeeep-gluup-bloooop.* Flight whistle highly geographically variable, given in flight and while perched. About four seconds long, it

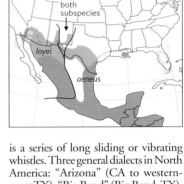

both subspecies

loyei

aeneus

is a series of long sliding or vibrating whistles. Three general dialects in North America: "Arizona" (CA to westernmost TX), "Big Bend" (Big Bend, TX), and "South Texas" (east of Big Bend). **STATUS & DISTRIBUTION** Uncommon to fairly common. **YEAR-ROUND:** Open shrubland, forest edge, and agricultural areas (esp. those associated with livestock). **BREEDING:** Brood parasite, specializes on sparrows and orioles. **MIGRATION:** Not well known. Spring movements in TX in Mar., southbound movements in Sept. Small but increasing numbers winter in FL. **WINTER:** In flocks, particularly in agricultural areas. **VAGRANT:** Casual to coastal southern CA and southern NV. Accidental in NS, MO, and MD. **POPULATION** The range and population of this species began a marked expansion in the 1950s, but currently the population appears stable. It has recently spread to FL and Gulf states.

ruff often raised, giving thick-necked look

red eye

thick bill

highly iridescent plumage

loyei ♂

grayish plumage, unlike female *aeneus*

red eye

loyei ♀

less scaly above with larger bill than juvenile Brown-headed

juvenile *loyei*

blackish plumage

Texas ♀ *aeneus*

BROWN-HEADED COWBIRD *Molothrus ater* BHCO ■ 1

The most widespread brood parasite in North America. Polytypic (3 ssp.; all in N.A.). L 7.5" (19 cm)

IDENTIFICATION A smallish, compact and stocky blackbird with a short and thick-based bill, almost finchlike. Foraging birds are commonly seen on ground with tail cocked. **ADULT MALE:** Glossy black body, with a greenish iridescence, contrasts with a brown head. Eyes are dark; bill and legs are black. **ADULT FEMALE:** Face has a beady-eyed look due to dark eyes; lores are pale; underparts are obscurely streaked. **JUVENILE:** Like female, but is more strongly streaked below, and upperparts are scaly-looking due to pale feather fringes.

GEOGRAPHIC VARIATION Three subspecies, differing mainly in size and darkness of females, are not field identifiable.

SIMILAR SPECIES Male's glossy black body and brown head diagnostic. Female extremely similar to Shiny Cowbird female. However, Brown-headed is paler overall, showing a whitish throat and a pale face with a beady-eyed look. Shiny has a more marked dark eye line and paler supercilium, giving it a more striking face pattern. Compared to a Shiny, Brown-headed is more compact, with a shorter tail and bill. Brown-headed's dark bill shows a pale or horn base to lower mandible; on Shiny, bill is shiny black. On closed wings, secondaries show obvious pale fringes on Brown-headed. Streaked juvenile is easily confused with more strongly and darkly streaked female Red-winged Blackbird; note bill shape differences.

VOICE CALL: A soft *kek*. Females give a distinctive dry chatter, while males may give a single modulated whistle, particularly just after taking off. **SONG:** Primary song is a series of liquid, purring, gurgles followed by a high whistle. Song of Brown-headed has the highest frequency range of any species in North America. Males also give a flight whistle, which is a geographically variable series of two to five whistles, often

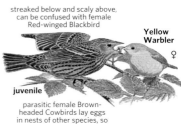

streaked below and scaly above, can be confused with female Red-winged Blackbird

Yellow Warbler ♀

juvenile

parasitic female Brown-headed Cowbirds lay eggs in nests of other species, so young raised by other parents

frequency modulated. Flight whistle is given both in flight and while perched. Primary songs appear to be hardwired, while flight whistles are learned; this accounts for why "dialects" are found in the latter but not in the former.

STATUS & DISTRIBUTION Common. **BREEDING:** Open and edge habitats. **MIGRATION:** Northbound mid-Mar.–mid-Apr., southbound late July–Oct. **POPULATION** Increased its range and population greatly during 1800s, when eastern forests were cleared. Brown-headed was restricted to Great Plains, where the buffalo herds were, and spread east and west from there. More recently, numbers have declined.

immature ♂ in molt

young males seen in this transitional plumage in late summer and early fall can be confusing

conical bill, smaller than Bronzed Cowbird, distinctive for all Brown-headed Cowbirds

brown head contrasts with black body

♂

Brown-headed Cowbirds often feed with tail cocked up

♀

short tail

ORIOLES Genus *Icterus*

The most colorful icterids, largely orange, yellow, or chestnut, orioles are generally slim and long tailed with sharply pointed bills. They weave a characteristic nest that looks like a hanging basket. Plumage patterns are plastic, and species with very similar patterns (e.g., Hooded and Altamira, Baltimore and Orchard) are quite distantly related. In addition, sexual dichromatism is heavily influenced by the role females have in territorial defense and migratory tendency. More migratory orioles show a greater degree of difference in plumage, while in tropical, resident species the female may be as brightly plumaged as the male. Immature male plumage is retained for at least a year.

ORCHARD ORIOLE *Icterus spurius* OROR ■ 1

This smallest oriole is common in the East and Midwest. Polytypic (3 ssp.; 2 in N.A.). L 7" (18 cm)

IDENTIFICATION Bill slightly downcurved. **ADULT MALE:** Black hood and back; chestnut below and on rump. Wings black with chestnut shoulder, white lower wing bar, and white edging to flight feathers. Tail entirely black. **ADULT FEMALE:** Olive above; bright yellow below. Two crisp white wing bars and white edging to flight feathers.

IMMATURE MALE: Similar to female, but by first spring shows a neat black bib and lores, often some chestnut spotting on face or especially on breast.

GEOGRAPHIC VARIATION Nominate is breeding subspecies in N.A. In north-

eastern Mexico, the endemic breeder *fuertesi* has occurred three times in Cameron Co., TX. In adult males, ochre replaces standard chestnut coloration. It is sometimes recognized as a full species, "Fuertes's Oriole" or "Ochre Oriole."
SIMILAR SPECIES Immature and female Orchards yellow below, not orange or orange-yellow as in Baltimore. Female

and immature Hoodeds similar to Orchard, although larger Hooded has longer, more graduated tail, with more downcurved bill (beware of short-billed juvenile Hooded). An eastern Hooded is more orange than an Orchard; western Hooded similarly colored. Orchard's *chuck* call deeper and huskier than a similar call rarely given by young Hoodeds; *wheet* call of Hooded is not given by Orchard.
VOICE CALL: A sharp *chuck,* often in a series.
SONG: A musical, springy, and rapid warbled song interspersed with raspy notes.
STATUS & DISTRIBUTION Fairly common. **BREEDING:** Open woodlands, urban parks, and riparian

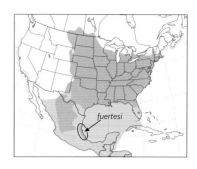

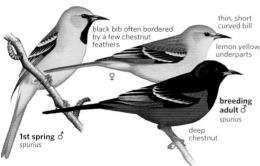

black bib often bordered by a few chestnut feathers

thin, short curved bill

lemon yellow underparts

breeding adult ♂
spurius

♀

1st spring ♂
spurius

deep chestnut

woodlands particularly in the west of range. **MIGRATION:** Trans-Gulf migrant in spring with arrival about Apr. 1, moves south as early as mid-July, but most head south in Aug. **WINTER:** From Mexico to northern S.A. **VAGRANT:** Rare west to CA, AZ, and Maritimes. Casual to OR and Bermuda; accidental to southeastern AK.

HOODED ORIOLE *Icterus cucullatus*   HOOR ▪ 1

This slim oriole prefers palms. Polytypic (5 ssp.; 3 in N.A.). L 8" (20 cm)
IDENTIFICATION Long, graduated tail. Thin, noticeably downcurved bill. **ADULT MALE:** Black bib, face, and back contrasting with orange or yellow-orange head and underparts. Black wings, with black shoulders, two white wing bars, white fringes on flight feathers. **ADULT FEMALE:** Olive above; yellowish or dull orange below. **IMMATURE MALE:** Like female, but by spring shows a neat black bib and lores.
GEOGRAPHIC VARIATION Five subspecies in two groups: *cucullatus* group, includes *sennetti,* from south TX, with more orange, shorter bill, more black on forehead. Similar *cucullatus* from

along the Rio Grande (Big Bend and a bit downriver) more deeply orange; adult males almost orange-red around bib. The *nelsoni* group includes *nelsoni* found from NM to CA, is more yellow, with a longer, more downcurved bill; males have a more restricted bib.
SIMILAR SPECIES Male is similar in pattern to Altamira, but slimmer, with a slenderer bill, and a black shoulder and white upper wing bar. Also see Orchard.
VOICE CALL: A whistled *wheet* and a short chatter. Also a *chut,* from juveniles, like Orchard. **SONG:** A quick

breeding adult ♂
sennetti

orange tint

♀ *sennetti*

black bib

scaly back

winter adult ♂
nelsoni

breeding adult ♂
nelsoni

longer, more curved bill than Orchard

coloration like smaller Orchard

nelsoni ♀

black bib on *cucullatus* and *sennetti* more extensive than *nelsoni,* extending to forehead

deep reddish orange around bib

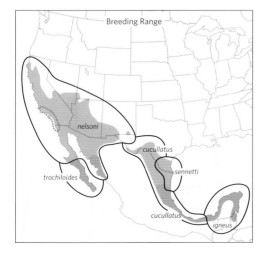

Breeding Range

nelsoni

cucullatus

trochiloides

sennetti

cucullatus

igneus

breeding adult ♂
cucullatus

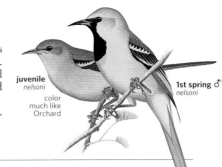

and abrupt series of springy, nasal, or whiny notes, lacking sweet whistled sounds of other orioles.
STATUS & DISTRIBUTION BREEDING: Open areas with scattered trees, riparian areas, and suburban and park settings.

MIGRATION: Arrives during Mar., departs in Aug. **WINTER:** Mainly in Mexico. **VAGRANT:** Casual to BC and WA, annual in OR; accidental in ON, KY, and southern YT.
POPULATION Declines in southern TX.

BLACK-VENTED ORIOLE *Icterus wagleri*   BVOR ■ 5

The striking Black-vented Oriole is a vagrant from Mexico. Polytypic (2 ssp.; presumably northern *casteneopectus* in N.A.). L 8.7" (22 cm)
IDENTIFICATION Slim and long-billed. **ADULT:** Entirely black above, including hood down to mid-breast. Below bright orange-yellow with black crissum. Black breast separated from yellow underparts by a narrow chestnut area. Black wings with bright yellow shoulders; tail entirely black. **IMMATURE:** Variable. Olive above, obscurely streaked on back, and yellow-orange below; greenish shoulder patch. Black on face and breast varies from black lores and chin with scattered black feathers on throat to black face and extensive black bib with irregular border and scattered black feathers on head.

SIMILAR SPECIES Absence of wing bars and white edges to wing feathers separate it from all adult North American orioles. Black vent is diagnostic. In addition peachy-yellow underparts are not matched by any of similarly patterned orioles. Immatures may be confused with Hooded Oriole; however, Black-vented lacks wing bars, has a messy black bib (when present), and is streaked on back.
VOICE CALL: A nasal *nyeh;* also a mechanical chatter. **SONG:** A series of nasal notes and squeaky whistles.

STATUS & DISTRIBUTION Casual vagrant. **YEAR-ROUND:** Open forest and edge from northern Mexico to Nicaragua. **VAGRANT:** Casual to western (2 recs., including one that spent three summers at Big Bend, N.P. from 1968–1970) and southern TX (2–3 recs.) and near Patagonia, AZ (Apr. 18, 1991).

variable black on throat

orange shoulder bar

extensive black hood bordered by chestnut feathers on breast

adult
casteneopectus

persimmon wash to orange below

extensive black under tail

1st spring
casteneopectus

very long black tail held in a point

STREAK-BACKED ORIOLE *Icterus pustulatus*   STBO ■ 4

This Mexican oriole has a stippled, streaked back. Polytypic (10 ssp.; *microstictus* in N.A.). L 8.3" (21 cm)
IDENTIFICATION This oriole has a thick-based, straight, pointed bill. **ADULT MALE:** Body is bright orange; head is reddish orange. Restricted black is found on lores and narrow bib; entirely black tail. Back is stippled with blackish streaks, and black wings are densely edged with white, forming a nearly solid white panel on closed secondaries and primaries. There are bold white wing bars. **FEMALE & IMMATURE:** Resemble adult male; however,

1st fall ♀
microstictus

all have thick-based, straight bills

adult ♀
microstictus

black to blackish broken streaks on back typical of northern subspecies, *microstictus*

deep orange head on adult male

extensive white on wings, including patch at base of primaries

microstictus ♂

body plumage is duller yellowish, becoming more orange around face; back and tail are greenish. Wings have less extensive white edging than on male.
SIMILAR SPECIES Male's streaked back, extensive white on wings, and reddish orange head together are diagnostic. Female Streak-backed is similar to an immature male Bullock's Oriole, but her streaked back and white on wings also identify Streak-backed. Bullock's also shows a largely blue-gray bill with black culmen, while entire upper mandible of Streak-backed is black.
VOICE CALL: A low *wrank,* also a sweet

chuwit like a House Finch, a dry chatter. **SONG:** Melodious whistled song similar to that of a Bullock's Oriole, but with a stop and start pattern.
STATUS & DISTRIBUTION Vagrant, has bred in southeastern AZ. Found from northern Mexico to Costa Rica, on Pacific slope. **YEAR-ROUND:** Open woodlands, forest edge. **VAGRANT:** Casual mainly in fall to southern CA and southern AZ; accidental to OR, UT, NM, eastern TX, and WI.

SPOT-BREASTED ORIOLE *Icterus pectoralis* SBRO ■ 2

The Central American Spot-breasted was introduced to the Miami, Florida, area. Polytypic (4 ssp.; nominate in N.A.). L 9.5" (24 cm)
IDENTIFICATION A large oriole with a sturdy bill. **ADULT:** Bright orange body with a black back. A small face mask and narrow bib are black; rows of black spots immediately below bib are distinctive. Black wings show an orange shoulder, a white spot at base of folded primaries, and a white wedge on folded tertials. Tail is entirely black. **IMMATURE:** Less intensely orange than adult, with olive-green back and tail, as well as duller pattern on wings. **JUVENILE:** Duller than immature and lacking black lores and bib; bill may show pinkish base to lower mandible.
SIMILAR SPECIES Spotted breast is diagnostic; no other oriole with a large white wedge on folded tertials is found in North America.
VOICE CALL: A nasal *nyeh*, also a sharp *whip* and a short chatter. **SONG:** A lengthy, repetitive yet pleasing set of warbled whistles, some of which are delivered

juvenile

immature

white patch on tertials and at base of primaries

orange shoulder patch

spots on sides of breast

adult

slowly and clearly. One song style more repetitive, another more variable; often songs start with a note repeated twice. Female's song less complex than that of male.
STATUS & DISTRIBUTION Uncommon in south FL; native to C.A. **YEAR-ROUND:** Parks and urban habitats in FL; open shrubby woodlands in native range.
POPULATION Spot-breasted was first found nesting in Florida in 1949. The population has oscillated since the introduction but is now on a decline. Its range in south FL has correspondingly grown and shrunk.

ALTAMIRA ORIOLE *Icterus gularis* ALOR ■ 2

A Lower Rio Grande specialty, this oriole builds a long hanging nest. Polytypic (6 ssp.; *tamaulipensis* in N.A.). L 10" (25 cm)
IDENTIFICATION Largest and stockiest oriole in North America; in addition, bill is very thick at base. **ADULT:** Bright orange body, with deeper orange face and bold black mask and narrow bib. Back black, contrasting with orange lower back, rump, and uppertail coverts.
Wing black, with an orange shoulder, a well-marked white lower wing bar. (On primaries, most of white restricted to a bold white patch at their base.) Tail entirely black on male; often a variable amount of olive on female tail. **IMMATURE:** Similar to adult, but duller orange and back and tail olive. Wings blackish, lacking orange shoulder and white edging greatly restricted. **JUVENILE:** Duller still, with buffy wing bars and no black on lores or bib.
GEOGRAPHIC VARIATION The six subspecies vary in minor ways in size and saturation of color. Only *tamaulipensis* is found in Texas.
SIMILAR SPECIES Hooded is similar to an adult

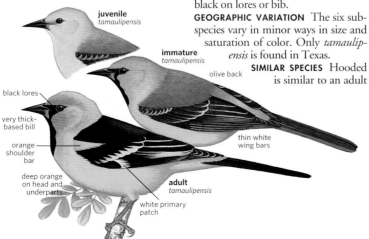

juvenile
tamaulipensis

immature
tamaulipensis
olive back

black lores

very thick-based bill

orange shoulder bar

deep orange on head and underparts

thin white wing bars

adult
tamaulipensis

white primary patch

Altamira in pattern, but Altamira is much thicker billed. Altamira also shows an orange shoulder patch. Immatures are also separable by size and bill, but note thinner bib on Altamira.
VOICE CALL: Contact call is a nasal *ike*, or *yehnk*, often repeated. **SONG:** A series of loud musical whistles, often interspersed with harsher notes. Songs are repeated several times before switching to a different song type. Song is easily imitated by a human whistler.
STATUS & DISTRIBUTION Uncommon non-migratory resident. **YEAR-ROUND:** Open woodlands, particularly riparian willows. Often places nest in a mimosa.
POPULATION Threatened in TX.

BULLOCK'S ORIOLE *Icterus bullockii*   BUOR ■ 1

The widespread and common oriole of the West. Usually regarded as monotypic. L 8.7" (22 cm)

IDENTIFICATION Long wings; relatively short tail; straight, sharply pointed largely blue-gray bill. **ADULT MALE:** Black eye line, crown, nape, and back. Bright orange supercilium. Bright orange on underparts and rump. Very narrow black bib. Wings black with very extensive white wing patch on coverts. Black tail with an orange base to outer rectrices. **FEMALE:** Orange to orange-yellow on head and breast, showing a ghost pattern of male face pattern, with darker eye line and brighter yellowish supercilium. Back gray, pale whitish gray belly. **IMMATURE MALE:** Like female, but by spring shows black lores and bib as well as brighter orange breast.

JUVENILE: Much duller than female, bill pink or orange-pink at base. Wings duller, brownish black with buffier and less well-developed wing bars.

SIMILAR SPECIES Orange supercilium, black eye line, and solid white wing patch are diagnostic for adult male Bullock's. Female and immature can be confused with a dull Baltimore Oriole (see sidebar below). Female and immature Hooded Orioles are entirely yellow below, slimmer, and longer tailed and have a thin, downcurved bill. Hybrids with Baltimore Orioles show features intermediate between the species.

VOICE CALL: A rapid series of *cha* notes on one pitch, given by both sexes; also a sweet but faint *kleek*, or *pheew*. **SONG:** A musical, lively series of whistles ending in a sweeter note: *kip, kit-tick, kit-tick, whew, wheet*. In comparison to Baltimore's songs, songs are shorter, not as melodic, and a lot less variable.

STATUS & DISTRIBUTION Common. **BREEDING:** Mainly open woodlands

and riparian areas, esp. fond of cottonwoods. **MIGRATION:** Spring arrival in south Mar.–Apr., crosses into Canada by early to mid-May. Adult males southbound beginning early July, females and immatures late July–Aug., a few into Sept. **WINTER:** Most retreat to Mexico, but a few in coastal urban habitats in southern CA. **VAGRANT:** Casual to the East, particularly in fall and winter.

POPULATION Gradual declines.

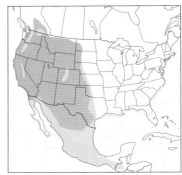

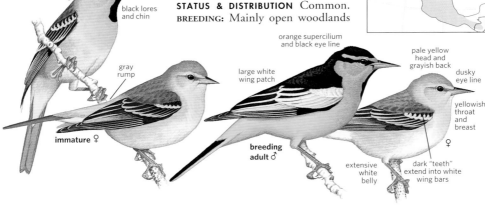

1st spring ♂

black lores and chin

gray rump

immature ♀

large white wing patch

orange supercilium and black eye line

breeding adult ♂

extensive white belly

pale yellow head and grayish back

dusky eye line

yellowish throat and breast

♀

dark "teeth" extend into white wing bars

Baltimore Oriole versus Bullock's Oriole in Winter

Two identification issues exist when attempting to separate a Baltimore Oriole from a Bullock's Oriole in winter. First, these species hybridize commonly in a narrow zone in the Great Plains; and second, a dull immature Baltimore is similar in plumage and pattern to a female Bullock's. These issues may be related, as it is impossible to know if some of the Baltimores showing Bullock's-like plumages may in fact be hybrids. Adult male hybrids are easy to identify as they show a mix of features intermediate between the species, but female and immature hybrids may not always be readily identifiable. Fortunately the area where hybridization is common is a narrow zone, and hybrids are a small

Baltimore Oriole, immature (NJ, Aug.)

proportion of the population of these two oriole species.

The tricky identification issue lies in separating the duller fall/winter Baltimore Orioles from the Bullock's Orioles. These dull Baltimores are grayish above and yellowish on the throat and breast, with a grayish lower breast and belly as in the Bullock's. However, in comparison to the Bullock's, these dull Baltimores tend to have more extensive yellow on the underparts and often show an orange tone. The yellow of the throat and upper breast blends into the grayer belly, usually on the Bullock's this change is more abrupt.

The Bullock's Oriole tends to have a face pattern that mirrors that of the adult male Baltimore, with

BALTIMORE ORIOLE *Icterus galbula*   BAOR ▪ 1

The common oriole throughout much of the East. Monotypic. L 8.7" (22 cm)
IDENTIFICATION Long wings; relatively short tail; straight, sharply pointed bill. **ADULT MALE:** Orange, with black hood and back. Wings black with orange shoulder, white lower wing bar. Tail black with black-based orange outer rectrices. **FEMALE:** Variable. Some like male, but usually lack solid black head and have greenish orange tail. Typical female orange below, with brownish orange face and back, spotted or blotched dark on back and crown. Wings blackish with two white wing bars (upper one wider). **IMMATURE MALE:** Variable. Like female, but no black on face, more extensively

orange below, bill often with pinkish or orange tone. **IMMATURE FEMALE:** Much duller, often with a paler, grayish white belly. **JUVENILE:** Duller than immature, olive back. Pink tone to bill.
SIMILAR SPECIES Dull immature Baltimore is very similar to Bullock's Oriole (see sidebar below). A female Orchard Oriole is yellow below and greenish above, slimmer, and smaller; an immature male Orchard has a crisp black bib, often bordered with a few chestnut feathers.
VOICE CALL: A whistled *hew-li* and a dry chatter. **SONG:** A series of musical, sweet whistles; quite variable.
STATUS & DISTRIBUTION Common. **BREEDING:** Deciduous forest, forest-edge parkland, riparian forest. **MIGRATION:** Northbound along Gulf Coast, but many are trans-Gulf migrants. Arrive Gulf Coast by early Apr., and Canada by early May. Southbound late July–early Aug., peaking late

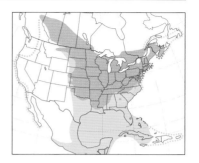

Aug.–early Sept. in north, and mid-Sept.–mid-Oct. on Gulf Coast. Often a fairly common migrant on Bermuda (some winter there). **WINTER:** From southern Mexico to northern S.A. in moist forest and shade coffee plantations. A small number in the U.S. South and FL, largely in urban settings. **VAGRANT:** Rare in NF and CA; casual to Pacific Northwest and western Europe. **POPULATION** Slight declines.

all with thick wing bars and orangish below

fall immatures

yellow-ochre rump

fall immature ♂

strongly contrasting wing bars

black hood

maximum black spring adult ♀

1st spring ♀

breeding adult ♂

---

extensive yellow on the auriculars and supercilium that contrasts with a darker eye line. These areas are a darker olive-gray, contrasting with the yellow throat, on the Baltimore Oriole; the separation of a darker face from yellow malar and throat can be quite abrupt. The Bullock's does not show an abrupt change in color here, the yellow of the face being continuous with that of the throat and malar.
     On the upperparts, the Baltimore Oriole usually shows darker centers to the mantle feathers; those of the Bullock's Oriole are nearly or entirely unmarked gray. As well, the rump of the Baltimore Oriole has a yellowish wash, while the Bullock's Oriole shows a grayish rump. The undertail coverts of

Bullock's Oriole, immature (CA, Sept.)

a dull Baltimore is usually (possibly always) yellow, whereas they are often gray, sometimes yellow, on a Bullock's. Gray undertail coverts and rump are good identifying features for a Bullock's, as opposed to a dull Baltimore.
     The wing bars on the Baltimore Oriole are separate, while they are often connected by pale greater covert edges in the Bullock's Oriole. Additionally, the Bullock's tends to show an upper wing bar with intruding dark "teeth" on each white median covert, although this feature is variable and can be matched by some Baltimores. The calls of these species are similar, but the Baltimore has a drier, more stuttering rattle. ▪

AUDUBON'S ORIOLE *Icterus graduacauda*   AUOR ■ 2

The secretive Audubon's Oriole stays low in the understory. Polytypic (4 ssp.; *audubonii* in TX). L 9.5" (24 cm)
IDENTIFICATION This bird is a thick-set, large oriole with a moderately thick, straight bill. Basal half of lower mandible is blue-gray; legs are also blue-gray. **ADULT:** Black hood contrasts with greenish yellow back and lemon yellow underparts. Black wings have a yellow shoulder, a white lower wing bar, and white fringes on tertials and on secondaries, but fringes do not reach to base of secondaries, creating a

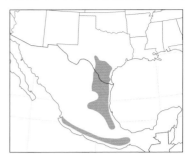

dark bar there. Tail is black. Female shows a duller, more greenish back. **JUVENILE:** Lacks black hood of adult and is greener above; wings dull black-ish, and tail greenish.
SIMILAR SPECIES This oriole is the only one with an isolated black hood. All our other hooded orioles have black backs. A juvenile may be confused with a juvenile Altamira Oriole, but the Audubon's is yellowish, not orange, and shows a duller, grayish head color. A juvenile Hooded Oriole is also similar, but smaller, much slimmer, and slimmer billed, with clear buffy white to white wing bars.
VOICE CALL: A nasal *nyyyee;* and a high-frequency buzz. **SONG:** A long song of melancholy, tentative, and slow whistles. The whistles sound flat and

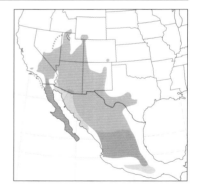

in the same pitch. Both sexes sing.
STATUS & DISTRIBUTION Uncommon; a nonmigratory resident. **YEAR-ROUND:** A variety of wooded habitats, but they all share dense low and midstories. Found in mesquite thickets to riparian thickets. **DISPERSAL:** May wander north of regular breeding range in winter. **VAGRANT:** Accidental southern IN.
POPULATION Appears to be stable, but neither the range or population in TX is extensive or large.

labels on illustration: juvenile *audubonii*; olive-yellow back; black head; adult ♂ *audubonii*; yellow

SCOTT'S ORIOLE *Icterus parisorum*   SCOR ■ 1

Scott's Oriole is North America's only desert specialist oriole. Monotypic. L 9" (23 cm)
IDENTIFICATION Note long, straight, sharply pointed bill. **ADULT MALE:** Black head, back, and breast, sharply separated from lemon yellow underparts. Wings black with yellow shoulder. Tail black with yellow bases to outer rectrices. **ADULT FEMALE:** Olive above, with obscure streaking on back; few with black on throat. Below dull olive-yellow. **IMMATURE MALE:** Similar to female, and some not separable, but back more densely marked and underparts brighter yellow. Typically black bib and face, and variable black spotting on crown and sides of head. Immatures of both sexes hold their plumage into winter and show no traces of black on throat or face.
SIMILAR SPECIES Audubon's and Scott's Orioles are the only yellow orioles in N.A. A black back, two white wing bars, and yellow on tail identify Scott's. A Scott's female is much more olive and generally muddier colored than other orioles; its obscurely streaked back may suggest a female Streak-

backed Oriole, but Scott's lacks orange on face and black mask and bib, and has a thinner-based bill.
VOICE CALL: A harsh *chuck* or *shack* and a scolding *cheh-cheh.* **SONG:** A fluty warbled set of whistles; the low pitch and richness of notes resemble the song of a Western Meadowlark at times. Females sing a softer and weaker song than the males.
STATUS & DISTRIBUTION Common. **BREEDING:** Desert, particularly at interface of low desert to higher elevation. Often in yucca or agave as well as juniper. **MIGRATION:** In spring arrive late Mar.–early Apr., leave in fall late July–mid-Sept. **WINTER:** Retreats mainly to Mexico. **VAGRANT:** Casual

north of range in Great Basin, MN, and to LA. Accidental to KY, WI, and ON.

labels on illustration: 1st spring ♂; yellow shoulder; adult ♂; long, straight bill; stippled dusky black above; immature ♀; bright lemon yellow underparts; olive-yellow below, duller than female Hooded

FRINGILLINE AND CARDUELINE FINCHES AND ALLIES Family Fringillidae

Pine Siskin (CA, Oct.)

Fringillid songbirds range in size from four to eight inches and have characteristic conical-shaped bills used for feeding on seeds. Some are dainty with tiny bills, while others are chunky with large bills, but nearly all have short, notched tails. Most species have undulating flights with distinctive flight call notes. Several species occur in boreal or montane habitats and are widespread, while others stay in very restricted ranges, mainly in western North America. Some fringillids are arboreal and prefer coniferous forests, while others are terrestrial and live in tundra habitats or above tree line. Many are attracted to seed feeders in the winter.

Structure All species share the conical or finchlike bill, although its size varies considerably—from tiny in the Hoary Redpoll to massive in the Hawfinch. The size and shape of the bill greatly aids birders in separating closely related species, as with redpolls and *Haemorhous* finches. In addition, the crossbills have uniquely shaped bills with crossed tips, specialized for extracting seeds from pinecones. Fringillids usually have deeply notched tails that are typically short for songbirds; their wings, however, vary in shape. Many fringillids are long-distance migrants, and thus evolved elongated primaries, but some less migratory species (e.g., the House Finch and the Lesser Goldfinch) have shorter, rounder wings.

Behavior All North American fringillids are at least partially migratory in winter, vacating breeding areas in search of fluctuating food supply. Many fringillids are very erratic in their dispersal and are often irruptive—being present in areas in very large numbers some years, only to be absent other years; redpolls and crossbills are famous for this behavior. The rosy-finches and a few others vacate higher elevations during the winter in search of food. Most species feed on seeds, finding seed feeders very attractive. All species share a distinctive undulating flight, but each can be recognized in flight by its species-specific flight call. In most species, the song is not as important as group contact calls; the Evening Grosbeak, however, rarely, if ever, sings.

Plumage All species are highly sexually dimorphic, the males being brightly colored, as in goldfinches and *Haemorhous* finches, and females duller, browner, and often streaked. Some species, such as Brambling and Pine Grosbeak, take longer than one calendar year to achieve adult breeding plumage, usually molting into it during the second winter. Some species—American Goldfinch for instance—have a distinct winter plumage, while most have a distinct first-winter plumage duller than adult breeding birds. Many species have combinations of wing bars and white in the tail, both more conspicuous in flight.

Distribution The Fringillidae family consists of about 168 species in 42 genera worldwide. In North America, there are 24 species in 11 genera, of which 16 breed and eight occur as vagrants. Several species generally are found in coniferous forests across the boreal zone in Canada and northern United States, as well as in pine and spruce-fir forests in the mountainous west. A number of species (e.g., goldfinches and Pine Siskin) are widespread during the breeding season, but then form gregarious single-species flocks in the winter that can be found virtually anywhere. On the other, hand the Brown-capped Rosy-Finch has a very restricted breeding range and limited dispersal.

Taxonomy Fringillidae members are considered aligned with other 9-primaried oscines and are most closely related to the Emberizidae and Cardinalidae families. Taxonomists sometimes further divide the family into two subfamilies, the Fringillinae, which includes the Brambling and the Common Chaffinch, and Carduelinae, which includes all other North American finches. Our knowledge of taxonomic relationships within the family is an evolving process. For instance, recent studies of the Red Crossbill suggest that as many as nine species may be involved, based on differences in bill size, distribution, and flight call notes.

Conservation Certain species that breed in coniferous forests may be adversely affected by clear-cutting and thinning of old-growth forest. Some species (e.g., House Finch) are expanding their range, while others (e.g., Purple Finch) have ranges that are contracting. BirdLife International lists eight species as threatened and four more as near threatened.

OLD WORLD FINCHES Genus *Fringilla*

Of this highly migratory Old World genus, only the Brambling and the Common Chaffinch occur in North America as rare migrants or vagrants. Medium-size birds, the males are generally colorful, while the females appear drabber and browner. They usually feed on the ground.

COMMON CHAFFINCH *Fringilla coelebs*   COCH ■ 4

brown lateral crown stripes

blue-gray crown and nape

♀

♂

pinkish below

extensive white on wings in all plumages

This Palearctic finch, one of the more common songbirds in Europe, is a casual visitor to North America's Northeast. Found on a variety of breeding ground habitats, the Common Chaffinch is highly migratory, making it an excellent candidate for vagrancy. It often feeds on seeds on the ground. It has an undulating flight. The white markings on its wings and tail are very conspicuous.

Polytypic (18 ssp., nominate from mainland Europe or *gengleri* from the British Isles most likely to appear in N.A.). L 6" (15 cm)
IDENTIFICATION Common Chaffinch is highly sexually dimorphic. **MALE:** A pale gray crown and nape surround a brown face. Back is brown; underneath is pinkish. Very conspicuously marked wings are black with white lesser coverts; each has a white wing bar and a white base to primaries. Flight feathers have noticeable pale edges. Tail is dark with white outer feathers. Conical bill is somewhat long. **FALL MALE:** He resembles the summer male, but many feathers are edged in brown, which wears off by late winter. **FEMALE & JUVENILE:** Both birds are much duller than the male—mostly brown, darker above

and buffier below. A conspicuous gray area surrounds a brown cheek. Wing markings appear as in male, but coloration is slightly browner overall.
SIMILAR SPECIES No other species in North America resembles a male Chaffinch. Winter female American Goldfinch is somewhat similar to female Chaffinch, but Chaffinch is noticeably larger and uniformly buffier brown underneath and has a much larger bill and a different distribution of white in wings.
VOICE **CALL:** A metallic *pink-pink* or *hweet.*
STATUS & DISTRIBUTION Casual or accidental vagrant from Eurasia to northeastern N.A., mainly in fall and winter. Accepted records from NF, NS, ME, and MA. Additional reports from the western U.S., as well as several other states in the East, are usually considered to be escaped cage birds.

BRAMBLING *Fringilla montifringilla*   BRAM ■ 3

The Brambling, a highly migratory Eurasian finch, is a regular vagrant to western Alaska, sometimes occurring in small flocks. The male is strikingly patterned in black, white, and rufous. It has an undulating flight, and its characteristic white rump is best observed in flight. Monotypic. L 6.3" (16 cm)
IDENTIFICATION Brambling is sexually dimorphic. **BREEDING MALE:** In spring, adult male has a black head and back; bright tawny-orange throat, breast, and shoulders; and white lower belly with dark spotting on flanks. Tail is black and deeply notched. Wings are black with white lesser wing coverts, a

buffy lower wing bar, and white bases to primaries that form a small white patch on folded wing. Rump is white, but it is often covered by folded wings. Conical bill is relatively small. **WINTER MALE:** Similar to breeding male, but black plumage is overlaid by a brown edging that wears off throughout winter, revealing striking plumage underneath. **FEMALE:** Browner overall, head is grayish brown, with a grayer face and lateral black stripes on sides of nape. Below, a duller rufous. Rump is white.
SIMILAR SPECIES No other species has the same color combination and all-white rump. Female Brambling could be somewhat suggestive of a Harris's Sparrow, but note Brambling's white rump and lack of black on breast.
VOICE **CALL:** A nasal *check-check-check;* often given in flight. Also a nasal *zwee.*
STATUS & DISTRIBUTION Fairly common but irregular migrant to islands in the central and western Aleutians. Rare on islands in the Bering Sea (St. Lawrence and

Pribilofs), more regular in late spring (early Jun.), when small groups ocasionally form. Casual in fall and winter to Canada and northern U.S., and western states south to central CA and NV.

glossy black head and back

orange breast and shoulders

breeding ♂

pale-fringed feathers with blackish bases

white rump in all plumages

fall ♂

gray face bordered by black

pale nape spot

♀

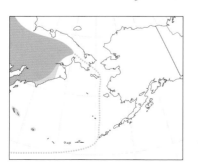

Characterized by their bright rose-pink bellies and wings, there are three North American members of this genus and one vagrant species from Asia. They are usually found in the Arctic or above tree line in the West's high mountains, preferring the tundra often seen at the edge of snowfields. During winter they disperse to lower elevations in flocks and often gather at foothill feeders.

GRAY-CROWNED ROSY-FINCH   *Leucosticte tephrocotis*   GCRF ■ 1

The Gray-crowned Rosy-Finch is the most widespread of the rosy-finches. Found in alpine and rocky coastal areas (AK), it frequents steep rocky cliffs during the breeding season. Often seen at the edge of snowfields. Gray-crowneds frequently form mixed flocks with other rosy-finch species during winter at higher elevations in interior West (visits feeders). Polytypic (6–8 ssp.; all in N.A.). L 5.3–8.3" (14–21 cm).
IDENTIFICATION In general, Gray-crowned is a chunky ground-dwelling brown finch with bright rosy-pink tinting its lower belly, rump, lesser wing coverts, and edging to wing feathers. Both sexes have blackish bills in summer and yellowish bills in winter. All subspecies have a distinctive black forecrown and pale gray hindcrown. **MALE:** Belly and wings are a brighter rosy-pink. **FEMALE:** Overall, she is a paler brown than male; there's also less pink on underparts and a lighter pink edging to wing feathers. **JUVENILE:** It is entirely grayish brown with pale edging to wing coverts and flight feathers, and a darker forecrown. Bill is darker and not as yellow as winter bird's.
GEOGRAPHIC VARIATION Six subspecies are currently recognized. The migratory gray-headed *littoralis* is found in the interior West. Two largely resident subspecies—*umbrina* from the Pribilofs, St. Matthew and Hall Is., and slightly paler *griseonucha* of Aleutians, western AK Peninsula, Shumagin and Semidi Is.; also Commander Is. (Russian Far East)—share the same extensive gray

face of *littoralis* but are much larger. Migratory *tephrocotis* and similar-appearing *wallowa* and *dawsoni* have a narrower gray headband, the lower face being brown, not gray.
SIMILAR SPECIES Unmistakable gray-cheeked Gray-crowned does not overlap with other rosy-finches during breeding season. Head pattern of Gray-crowned subspecies with brown cheeks and Black look similar, but Black is always much blacker or a charcoal gray, lacking brownish tones. Female Brown-capped could also easily be mistaken for an immature female Gray-crowned.
VOICE CALL: A high, chirping *chew,* sometimes given in repetition and often given in flight. **SONG:** A series of descending *chew* notes; sometimes given in flight display. It is seldom heard.
STATUS & DISTRIBUTION Fairly common. **BREEDING:** Resident in the Aleutian and Commander Is. *(griseonucha)* and Pribilofs, St. Matthew, and Hall Is. *(umbrina)* Islands, where it frequents coastal cliffs and rocky tundra. Migratory *littoralis* mainly found in coastal mountains from AK south to WA, OR, and northern CA. Interior subspecies *tephrocotis* group mainly found above tree line from the Brooks Range of northern AK south through the northern Rocky Mts. to ID; also northeastern OR *(wallowa)* and in the Sierra Nevada

and White Mts., CA *(dawsoni).* **WINTER:** Both migratory subspecies (*littoralis* and *tephrocotis*) winter south of breeding range, mostly at higher elevations throughout the interior west, south to CO, UT, northern NM, and northern CA. The winter distribution is somewhat erratic, with species invading farther south some years in search of food. **VAGRANT:** Casual in northern AZ (*tephrocotis* and *littoralis*) and various states and provinces east of regular breeding and wintering ranges. Accidental east to ON, ME, and QC.

Breeding Range

tephrocotis
umbrina
griseonucha
tephrocotis may winter east to here; accidental farther east
littoralis "Hepburn's"
tephrocotis
wallowa
various subspecies
dawsoni
littoralis winters east to here; casual farther east

dark sooty overall with little pink

juvenile *tephrocotis* group

extensive gray face

"Hepburn's Rosy-Finch" winter ♂ *littoralis*

yellow bill on all fall and winter rosy-finches, becoming black by spring

gray head band

extensive gray face, like *littoralis*

Pribilof, St. Matthew, and Hall Islands *umbrina* and Aleutian and Commander Islands (Russian Far East) *griseonucha* are much larger than *littoralis*

brownish edges above, compare to female Black Rosy-Finch

winter ♂ *tephrocotis*

Pribilofs winter ♂ *umbrina*

winter ♀ *tephrocotis*

like male but with less pink

BLACK ROSY-FINCH *Leucosticte atrata*   BLRF ■ 2

The Black Rosy-Finch is restricted to the central Rocky Mountains in the interior west. It frequents rocky tundra above tree line, often feeding on the ground at the edge of snowfields. During winter, it will join with other rosy-finches to form large mixed-species flocks. It often visits established seed feeders. Monotypic. L 6" (15 cm)
IDENTIFICATION Black Rosy-Finch is much darker than similarly patterned *tephrocotis* Gray-crowned Rosy-Finch. Both sexes have blackish bills in summer and yellowish bills in winter. **MALE:** Almost entirely black body has some grayish edging to breast and back feathers. Pale silvery gray of hind crown extends down to level of the eye, forming a gray headband. A pink blush extensively covers lower belly and upper tail coverts. Obvious pink coverts form a pink wing patch on each of blackish wings. **BREEDING FEMALE:** She is patterned similar to male, but she is more of a charcoal gray color than black. Whitish lower belly has little or no pink. Pink wing coverts are much lighter than male's. **JUVENILE:** Body is uniformly gray with two buffy-cinnamon wing bars and a pale eye ring. Bill is pale.
SIMILAR SPECIES The male Black Rosy-Finch is much blacker than any other rosy-finches. The female is also dark, but there are gray tones, too, though no brown, as in Gray-crowned.
VOICE CALL: A high chirping *chew* is like other rosy-finch calls.

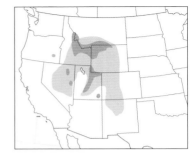

STATUS & DISTRIBUTION Fairly common within its limited range. **BREEDING:** It is found in high-elevation rocky tundra, usually above tree line in the central Rocky Mountains from southern MT and ID to southeastern OR, NV, and northern UT, where it occurs on top of isolated high mountains. **WINTER:** It forms large flocks that move down to lower elevations and south to northern NM. Casual to northern AZ and eastern CA.

silver-gray edges above on fresh plumaged females and males

dark sooty gray overall with reduced pink

winter ♀

stunning — black overall with contrasting rose and silver-gray head band

breeding ♂

BROWN-CAPPED ROSY-FINCH *Leucosticte australis*   BCRF ■ 2

The breeding range of Brown-capped Rosy-Finch is restricted to Colorado and southern Wyoming. Brown-capped can be told from other rosy-finches by the lack of gray on the hind crown or nape. It behaves much like other rosy-finches, breeding above tree line in rocky tundra and often feeding on the ground at the edge of snow patches. Like the "interior" Gray-crowned and Black Rosy-Finches, the Brown-capped migrates to lower elevations during the winter and frequents seed feeders. But unlike those species, it is completely restricted to the Rocky Mountain Region, with no substantiated records of vagrants. Monotypic. L 6" (15 cm)

IDENTIFICATION Unlike other rosy-finches, most Brown-cappeds lack gray on head. Both sexes have blackish bills in summer and yellowish bills in winter. **MALE:** Body almost entirely rich brown with a darker, almost blackish crown. Lower belly is bright reddish pink, while wing coverts and rump are extensively edged in pink. **BREEDING FEMALE:** Much drabber than male, with a uniformly darker brown body and little or no pink on belly, rump, or wing coverts. **IMMATURE FEMALE:** Distinguished by a pale grayish brown body, more blended head pattern, light pink wing coverts, and very little pink on belly.

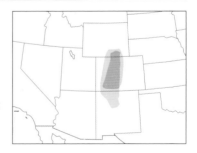

SIMILAR SPECIES Brown-capped is generally browner and paler than any plumage of Black Rosy-Finch. Some individuals have varying amounts of gray on hind crown, making field identification from an immature *tephrocotis*-group Gray-crowned difficult or impossible.
VOICE CALL: A high, chirping *chew* is similar to other rosy-finch calls.
STATUS & DISTRIBUTION Fairly common in proper habitat. **BREEDING:** Like other rosy-finches, it

dark sooty cap

some worn 1st-spring and summer females from *tephrocotis* group of Gray-crowned look very similar

Brown-capped unrecorded away from Rockies region

winter ♀

rich brown back and breast with no gray head band

extensive rosy-pink on wings and below

breeding ♂

prefers high-elevation rocky tundra, but it has a more limited breeding range, which is restricted to above tree line in the eastern Rocky Mountains of CO and southern WY. **WINTER:** Moves to lower elevations and joins mixed-species flocks with other rosy-finches. In winter, regular south to the Sandia Mts., northern NM. It has not been recorded farther west in AZ or CA, like the other two rosy-finch species. Often visits seed feeders.

ASIAN ROSY-FINCH Leucosticte arctoa ASRF ▪ 5

The Asian Rosy-Finch inhabits some of the least hospitable, and accessible, parts of Central and northeastern Asia, nesting well above tree line in most of its remote range. Polytypic (5 ssp.; 1 in N.A.). L 6.3" (16 cm)

IDENTIFICATION Overall appears somewhat dark or dirty, with very dark gray or blackish underparts, brownish upperparts. **ADULT** *(brunneonucha):* Males show blackish crown, face, and underparts, with rose pink stippling on flanks; dark gray back with brownish streaks contrasts with caramel-colored nape and postocular stripe in adult male plumage, postocular stripe tending toward chestnut in some individuals. Upperwing and uppertail coverts with rich pinkish magenta. Females show almost no pink tones but share general tonal pattern of adult males. Colors are most intense and well defined at onset of breeding season. **IMMATURE:** Similar to adult female, but tones even duller and less contrasting, but even first-winter birds show some chamois tone in nape (rather than in face, as in some Gray-crowned) and a mostly grayish face and throat.
GEOGRAPHIC VARIATION There are five named subspecies, mostly allopatric as breeders, and poorly known. They differ substantially in plumage. Eastern *brunneonucha*, which breeds east to Kamchatka and the Kuril Is., is believed to be the subspecies involved in the AK record.
SIMILAR SPECIES Asian is only likely to appear as a vagrant in the range of *griseonucha* or *umbrina* Gray-crowned, which are notably larger, bulkier birds. Adult Asians are fairly distinctive, appearing to N.A. birders perhaps like a "Brown-capped Rosy-Finch above, a Black Rosy-Finch below" in general terms, but Asian Rosy-Finch's colors and their distribution in the plumage are not shared by any other species: in adults, the contrast between blackish face and brown nape is distinctive.
VOICE CALL: A brief *pyut* or soft *tyew*, similar to other rosy-finches. **SONG:** Similar to other rosy-finches, a series of twittering notes, delivered from the ground.
STATUS & DISTRIBUTION **BREEDING:** On treeless montane plateaus, rocky tundra, moraines near snowfields and glaciers, scree slopes, cliffs, and rocky shorelines from western Siberia through the Kamchatka Peninsula, and south to Kazakhstan, Mongolia, and northeastern China. **WINTER:** Movements may vary among subspecies, but most appear to winter downslope from breeding areas. Eastern *brunneonucha* is the most migratory subspecies. Often fairly common in coastal areas of northern Japan in winter, where readily observed at feeding stations. **VAGRANT:** The only N.A. record is from Adak I., AK (Dec. 30, 2011).

blackish forecrown and face

warm brown nape

rose-pink on flanks and wings

winter ♂
brunneonucha

PINE GROSBEAK Pinicola enucleator PIGR ▪ 1

The large, plump Pine Grosbeak is one of the more characteristic species of the boreal forest. Typically found in singles or in pairs, the Pine Grosbeak frequently forms small groups during the nonbreeding season. Migratory. Polytypic (9–10 ssp.; 7–8 in N.A.). L 9" (23 cm)
IDENTIFICATION A large, and somewhat chunky, long-tailed finch with a stubby, curved bill. **ADULT MALE:** Mostly rich pinkish-red body with dark wings and two white wing bars. Depending on subspecies, there are varying amounts of gray on underparts, particularly on flanks and belly, as well as varying amounts of dark centers to back feathers. **ADULT FEMALE & IMMATURE MALE:** They are mostly gray overall with two distinct white wing bars and a variably yellowish olive wash to head, back, and rump. **VARIANT:** Some females and immature males are quite russet on head and rump.
GEOGRAPHIC VARIATION Of the N.A. subspecies, *californica* from Sierra Nevada, CA, is most sedentary. Other N.A. subspecies are northern, boreal *leucura* (includes *alascensis*); coastal Pacific *flammula; carlottae* from the Queen Charlotte Is. and Vancouver I.; western

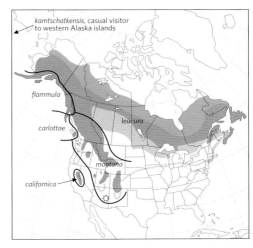

kamtschatkensis, casual visitor to western Alaska islands

flammula

carlottae

leucura

montana

californica

interior *montana;* and *eschatosa* from southeastern Canada. Another sub-species, *kamtschatkensis* from Asia, has been collected on the western Aleutians and the Pribilofs, AK. These subspecies differ in size and tail length, bill shape, and vocalizations and may represent two or more species.

VOICE CALL: Geographically quite variable. Includes a whistled *pui pui pui* or *chii-vli*. SONG: A short musical warble.

STATUS & DISTRIBUTION Fairly common in coniferous forest. WINTER: May move into deciduous woods and orchards. In the East, *leucura* is irruptive in fall and winter, and is casual south to IL, OH, and MD in fall and winter, casual in

Bermuda; in the West, *montana* is irruptive and appears casually south to southeastern OR, eastern CA, AZ (formerly breed in White Mts.), and the Great Plains.

stubby black bill with curved culmen

yellow-olive head and back; body otherwise gray

leucura ♀

white wing bars

deep pinkish red

russet variant *leucura*

adult ♂ *leucura*

Genus *Pyrrhula*

EURASIAN BULLFINCH   *Pyrrhula pyrrhula*   EURB ■ 4

The bright pink male of this striking Eurasian species is one of the more recognizable species to occur in North America. Sometimes occurs in small flocks. Polytypic (12 ssp.; *cassinii* in N.A.). L 6.5" (17 cm)

IDENTIFICATION A chunky finch with a very short, stubby bill. MALE: Intense reddish pink underparts and cheeks contrast with black crown and chin. Gray back contrasts with black wings (one prominent white wing bar) and tail and distinct white rump and undertail coverts. FEMALE: Similarly patterned, but brown where the male is pink. JUVENILE:

Resembles female, but has a brown cap.

SIMILAR SPECIES No other North American bird has this color combination.

VOICE CALL: A soft, piping *pheew.*

STATUS & DISTRIBUTION Eurasian species. Casual in spring (late May–early June) to the western Aleutians (one fall rec.) and the Bering Sea islands (one fall rec.). Casual in winter to AK mainland.

bright reddish pink color characteristic of northeast Asian *cassinii*

cassinii ♂

gray nape and back

white rump in all plumages

cassinii ♀

"RED" FINCHES   Genera *Carpodacus* and *Haemorhous*

In North America, three breeding species (formerly in *Carpodacus,* now *Haemorhous*) and one Eurasian vagrant (remains in *Carpodacus*) make up these highly sexually dimorphic genera. The males are bright red or pink; the females are brown and streaked. To help separate the similar-looking species, focus on the bill size and shape, the degree of streaking on flanks and undertail coverts, and distinctive calls.

COMMON ROSEFINCH   *Carpodacus erythrinus*   CORO ■ 4

bright red head and rump and extensive red below often with buff tinge

adult ♂ *grebnitskii*

known widely in Old World as Scarlet Rosefinch

indistinct face pattern

grebnitskii ♀

long primary projection

immature male like female until just over a year of age

olive-tinged above with pale wing bars

This Eurasian species is a very rare migrant in western Alaska, where other "red" finch species are less likely to occur. Virtually all reports are from the Aleutians and Bering Sea islands. It is typically found in singles, but it has occurred in small groups. It is usually seen feeding on the ground. It is more likely to be found during the spring, when some of the birds have been

bright adult males. Polytypic (at least 5 ssp. recognized; *grebnitskii* in N.A.). L 5.8" (15 cm)

IDENTIFICATION Note the long primary projection past the tertials. MALE: Head, breast, back, and rump are a pinkish red. Back is marked with indistinct dark streaking, while flanks and lower underparts have little or no diffuse streaking. It lacks distinct eyebrows. Bill is finchlike, but it has a strongly curved culmen. Wings have two indistinct pinkish wing bars. FEMALE & IMMATURE: Both are a very drab brown with diffuse streaking above and below, except on paler throat. **SIMILAR SPECIES** This species is by far the most likely red finch to occur in the Aleutians and Bering Sea islands, although Purple Finch has been documented from St. Lawrence I. The male differs from Purple Finch in having a more uniform face and no distinct eyebrows. In addition, Purple's culmen is decidedly straight, making it appear more pointed. House Finch has

browner cheeks and back, and more prominent streaking on a paler belly. Common Rosefinch female looks similar to female House, but she is noticeably drabber, with more diffuse streaking overall. Throat of female House is usually streaked, whereas it is plain in Common Rosefinch. Note again the distinctly curved culmen in Common Rosefinch, although House female also has curved culmen.

VOICE CALL: A soft, nasal *djuee.*
STATUS & DISTRIBUTION Common Rosefinch is a casual spring migrant (late May–early June) on the western Aleutians (Attu, Shemya, and Buldir Is.) and islands in the Bering Sea (e.g., St. Paul and St. Lawrence). It is rarer in the fall, when sightings mostly pertain to females or immatures. Casual east in Aleutians to Adak; accidental (fall to Farallones, CA).

HOUSE FINCH *Haemorhous mexicanus* HOFI ▪ 1

The attractive House Finch is one of the more common and recognizable species throughout the United States. Originally a "western" species of semiarid environments, House Finch was introduced in the East in the 1940s; it has now expanded its range and spread to virtually every state, as well as a multitude of habitats. It has become very common in suburban areas and is easily attracted in large numbers to seed feeders. Polytypic (12 ssp.; 2 in N.A.). L 6" (15 cm)
IDENTIFICATION House Finch is a relatively small *Haemorhous* finch with a longish, slightly notched tail, short wings, and a distinctly small bill with a curved culmen. **MALE:** Breast, rump, and front of head are typically red, but color can vary to orange or occasionally

yellow. Red breast is clearly demarcated from a whitish belly with dark streaks. Top of crown and auriculars are brown. Back is brown and noticeably streaked. Wings have two pale indistinct wing bars each. **FEMALE:** She is much drabber, lacking all-red coloration of male. Brown body has distinct, blurry streaking above and below. She lacks distinct pale eyebrow found on male.
GEOGRAPHIC VARIATION There is clinal variation, as well as individual variation, and effects of diet on plumage coloration complicate the separation of different subspecies. Subspecies *frontalis,* found nearly throughout North America, sports a generally more orange-red to yellow (more rarely) breast and has less distinct streaking on belly. The subspecies *clementis* from San Clemente I., CA, and the Islas Coronados, Mexico, is brighter red with bolder streaking on a whiter belly.
SIMILAR SPECIES Identifying the male and female House Finch from other *Haemorhous* finches requires care. The male House differs from the male Purple Finch not only by having a smaller, more curved bill, but also by lacking a distinct eyebrow, having a brown cap and auricular patch, and being heavily streaked on belly. Told from male Cassin's Finch by brown cap and eyebrow and curved bill. Other telltale differences between the species include the Cassin's pink cheek and pinkish tone on its back, and on female and immature Cassin's, the much finer and crisp streaks on its belly. The male Common Rosefinch is more rose-pink overall and lacks

distinct streaking on its belly. The female finches are more problematic. The female House Finch has a very plain face, unlike Purple and Cassin's, which both show distinct eyebrows. The female House Finch also tends to have browner underparts than the two as well, with blurry streaks below. Also note House's smaller, more curved bill. The female Common Rosefinch looks similar, but she is drabber, with less distinct streaking below.
VOICE CALL: Most commonly a whistled *wheat.* **SONG:** Lively and high-pitched, consisting of varied three-note phrases that usually end in a nasal *wheeer.*
STATUS & DISTRIBUTION A very common, often abundant resident throughout much of the U.S., extending north into much of extreme southern Canada and south into Mexico. Both western and introduced eastern populations appear to be spreading. **MIGRATION:** Some northern populations appear to be migratory, moving south in winter.
POPULATION The human modification of natural habitats, particularly the increase of seed feeders throughout the East, greatly benefits the House Finch populations. Only natural island populations appear to be threatened.

males can be yellow to orange

variant ♂ *frontalis*

broad reddish eyebrow

red throat and breast

brownish streaks on flanks

typical ♂ *frontalis*

square-ended tail

curved culmen

muted brown back

very indistinct face pattern

frontalis ♀

PURPLE FINCH *Haemorhous purpureus* PUFI ■ 1

short thick bill with slightly curved culmen

adult ♂ *purpureus*

dark auricular and distinct white facial streaks

distinct dark streaks above

white belly and flanks with no brownish, as in *californicus*

Pacific region adult ♂ *californicus*

distinct and rather broad streaks below

overall a deep rose-red color

unlike *purpureus*, has a brownish tinge to back, flanks, and belly

pale facial stripes less obvious

♀ *purpureus*

browner, less contrasty streaking below against a buffier background color

Pacific region ♀ *californicus*

less obvious streaking above against a more olive background color

This migratory rose red (not purple) finch is fairly common throughout across much of North America's boreal forest and south through the Pacific states. Generally found in less disturbed habitats than House Finch. Polytypic (2 ssp.; both in N.A.). L 6" (15 cm)
IDENTIFICATION A rather chunky *Haemorhous* finch with a short-ish, strongly notched tail; *purpureus* described. **MALE:** Body mostly rose red, brightest on head and rump. Pink eyebrow contrasts with darker cheek. Back brownish with noticeable streaks and pinkish ground color. Below extensively streaked. Undertail coverts white. Bill rather large, with a straight culmen. **FEMALE:** Underparts whitish, heavy dark brown streaking that does not extend to white undertail coverts. Head boldly patterned with whitish eyebrow and submoustachial stripe that contrast with a dark brown cheek and malar stripe. Crown and back have pale streaks. Immature males maintain female-type plumage for over a year.
GEOGRAPHIC VARIATION Two distinct subspecies. Nominate resides in boreal forests of much of N.A.; male has longer wings and is brighter overall, female has a bolder head pattern and whiter underparts. Pacific region's *californicus* male is less bright and has a brownish wash on its back and sides; female is much buffier underneath with a less bold face pattern and more blurred back streaking. These two subspecies

may represent two distinct species.
SIMILAR SPECIES Males and females most similar to Cassin's, which do not overlap with each other in East, but do in West. Male Cassin's is lighter pink, particularly on underparts and eyebrow. See sidebar below for females.
VOICE CALL: A musical *chur-lee,* and a sharp *pit* (*californicus;* softer and more musical in nominate) given in flight. **SONG:** A rich warbling; shorter than Cassin's, lower pitched and more strident than House. Songs of nominate subspecies longer and more complex.
STATUS & DISTRIBUTION Fairly common. **BREEDING:** Open coniferous forests and mixed woodland in East

purpureus

californicus

and North, and montane coniferous forest and oak canyons in West. **WINTER:** Eastern birds migrate south to lower latitudes, sometimes irrupting with major invasions south to the southern U.S.; casual to Bermuda. Western birds primarily move to lower elevations. Rare throughout much of the interior West (both subspecies).

Identification of Female Purple and Cassin's Finches

One of the more challenging identification problems in the West involves separating adult female Purple and Cassin's Finches. The females (and immature males) of both species are basically brown with coarse dark streaking on whitish underparts. Both species also have similar face patterns with a pale eyebrow, a pale submoustachial streak, and a darkish malar stripe. Further complicating the identification, the Purple Finch has two distinct subspecies that differ just enough to be problematic. Visible differences between the three are subtle, and perhaps best used in combination with each other; however, note that each has different call notes. The Purple Finch gives a quick, musical *chur-lee* when perched and a sharp *pit* in flight. The calls of the eastern *purpureus* race are clearer and higher-pitched. The Cassin's call when perched can be similar to the Purple's, but in flight it gives a distinctive high-

pitched *kee-up* or, the longer version, *tee-dee-yip*.

The female Eastern Purple Finch (*purpureus*), overall the bolder patterned of the species, has a very bold face pattern, with a broad eyebrow, a whiter submoustachial stripe, and a prominent brown malar streak. The underparts are white with very dark streaking and a slight buffy wash on the flanks. The white, unstreaked undertail coverts are a critical field mark. The bill is relatively long, with a slightly curved culmen.

The face pattern on the western Purple Finch (*californicus*) resembles the face pattern on the Cassin's; the former, however, appears more washed-out. The underparts on *californicus* look more like those on a House Finch, with very blurry streaks. Again note that the undertail coverts are white and unstreaked. The bill has a slightly curved culmen.

The Cassin's Finch is slightly larger than either

CASSIN'S FINCH *Haemorhous cassinii*   CAFI   ■   1

A species of the montane West is slightly larger and longer winged than the similar Purple Finch, which it occasionally overlaps with during winter. It is often seen in small flocks, mainly in coniferous forest, but it often appears in the interior lowlands, particularly in spring. Monotypic. L 6.3" (16 cm)

IDENTIFICATION Highly sexually dimorphic with adult males pink and females and immatures brown. Generally lighter pink than other *Haemorhous* finches, with distinctive fine streaking on undertail coverts. **MALE:** A bright pinkish-red crown contrasts sharply with a brown streaked nape. Back is heavily streaked and washed pink. Fairly wide eyebrow and submoustachial stripe are both pale pink. Light pink throat and breast blends into white on lower belly. Varying amounts of fine black streaking cover flanks and undertail coverts. Bill is longer and more pointed than other *Haemorhous* finches. Immature males for over a year resemble females. **FEMALE:** Upperparts are brown and streaked, while underparts are white with fine, crisp streaking, which is heaviest on breast and flanks. Undertail coverts are also finely streaked. Rather diffuse face pattern has a noticeably pale eyebrow and submoustachial stripe.
SIMILAR SPECIES Male is most similar to male Purple Finch, but there is only limited overlap in range during the breeding season. Cassin's eyebrow and streaking on back tend to be wider and frostier; it usually has fine streaking on

flanks and undertail coverts as well. Primary projection is noticeably longer in Cassin's, as is bill. Note different flight calls. See sidebar below for separating female Cassin's and Purples. Cassin's and House overlap more, with Cassin's typically found in coniferous forest and House in lowlands, but they might overlap in winter when Cassin's populations irrupt to lowlands. Note Cassin's pink eyebrow, finer black streaking on flanks, frostier upperparts, longer primary projection, and longer bill with a straight culmen.
VOICE CALL: In flight gives a dry *kee-up* or *tee-dee-yip*. **SONG:** A lively, varying warble, longer and more complex than Purple or House Finches'.
STATUS & DISTRIBUTION Fairly common in montane coniferous forests. **BREEDING:** Found throughout much of the Rocky Mountains, west into the Cascades in WA and OR, and the Sierra Nevada and southern mountain ranges in CA. **WINTER:** Unpredictable. Often stays in breeding range, but periodically drops to lower elevations and/or moves well south. Winters as far south as the mountains of central Mexico. Much more common in the lowlands of the interior West than Purple Finch, which is rare there. **VAGRANT:** Casual to eastern CO, NE, and KS; northern TX; AK; and CA coast.

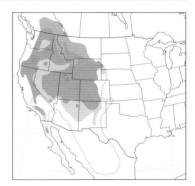

contrasting crimson-red crown

wide brown malar streak

adult ♂

pale pink throat and breast with white belly

pale eyebrow, but face pattern less distinct than Purple

larger bill than Purple with straight culmen

cinnamon tinge to auriculars

adult ♀

long primary projection, longer than Purple

notched tail as in Purple

all plumages have streaked undertail coverts, unlike most Purples

immature male plumage like female until about one year of age

"Eastern" Purple Finch, female (NJ, Jan.)   "Western" Purple Finch, female (CA, Jul.)   Cassin's Finch, female (CA, Feb.)

Purple. It has very white underparts with finer black streaking, particularly on the flanks, and the streaking extends to the undertail coverts. The face pattern, similar to that of *californicus,* often shows a pale eye

ring. The long bill has a straight culmen. At close range, note the Cassin's longer, less evenly spaced primary extensions and the more distinct white edging to its flight and tail feathers. ■

CROSSBILLS Genus *Loxia*

Found in coniferous forests, crossbills use their highly specialized bills with crossed tips to extract seeds from pinecones. Highly nomadic, the species disperse great distances in search of food. The taxonomy is complicated; in North America, only two species are recognized, but variable populations with different bill sizes and calls suggest more.

RED CROSSBILL   *Loxia curvirostra*   RECR   ■   1

Because they specialize in pinecones, the Red Crossbill is found almost exclusively in coniferous forests across the boreal zone of Canada and in the montane west. Dependent upon the local pinecone crop, it moves around both seasonally and annually. An invasive and irruptive species, it often becomes abundant in an area only to remain a few weeks or months, then vanish when the cones are depleted. Polytypic (19 ssp.; 7 in N.A.). L 5.5–7.8" (14–20 cm)

IDENTIFICATION Highly sexually dimorphic. Chunky, medium-size bird with a shortish notched tail, large head, and a stout, uniquely crossed bill. **MALE:** It typically is entirely brick red with darker, unmarked wings (some show narrow whitish wing bars). Back usually has some dark scaling, while undertail coverts are whitish with black chevrons. **FEMALE:** Entire body is a dull yellowish olive, while unmarked wings are dark and throat gray. May show some patches of red on body. **JUVENILE:** Mostly brown, but paler below with heavy dark streaking. Varying narrow white wing bars (usually with upper bar thinner than lower). **FIRST-YEAR MALE:** Resembles female, but is more orange.

GEOGRAPHIC VARIATION Very complicated. Red Crossbill subspecies are referred to as "types," differentiated mainly by range, overall size, bill size, and most important, flight call notes. As many as nine different types have been identified so far in N.A. (potentially another 13 in Europe and Asia). Several potential separate species may be involved. The identification of different types is very difficult in the field. Given that many of the types can turn up in virtually any part of Red's range and that different types sometimes flock together, a certain identification to type is very problematic—the best way to document records is to record their sounds. Only broad extremes will be covered here (for more information refer to technical literature). The smallest and small-billed crossbills have been put into *minor* (type 3); they generally are found farther north (into AK), but they are very widespread. The largest Red is the *stricklandi* (type 6), with a relatively huge bill; it is restricted to southeastern Arizona. Two populations that have very restricted ranges include *percna* (type 8) found in Newfoundland and an unnamed subspecies (type 9) in southern Idaho. The remaining types (1, 2, 4, 5, and 7) are generally put into *pusilla*. They are medium-size birds

with medium-size bills; some are restricted to the West, others are more widespread.

SIMILAR SPECIES White-winged Crossbill is the only other crossbill in N.A., and in all plumages it has a distinctive patterned wing with all-white lesser coverts (forming a wider upper wing bar) and a prominent lower wing bar. A male White-winged is generally pinker than the brick-red male Red. Use wing patterns to separate females and juveniles. Some Reds have narrow wing bars, but note different pattern. Flight notes are also different.

VOICE CALL: Generally a *kip-kip* or *chip-chip* usually given in double notes, but sometimes difficult to discern individual calls as multiple individuals call simultaneously. Given in flight and vary between types. Identification

variant ♂

some with faint whitish wing bars

juvenile

female typically has olive body with darker wings and gray throat

typical ♀

northern
♀ *minor*

overall size and bill size, especially, varies significantly and geographically

southwestern
♂ *stricklandi*

overall reddish coloration

typical ♂

of different calls to different types not safely done in the field. SONG: Begins with several two-note phrases, ending in a warbling trill; often given from the top of a conifer.

STATUS & DISTRIBUTION Fairly common. BREEDING: Irregular breeder anywhere there are pine or spruce-fir forests, as far south as GA, and north to AK and NF. Different subspecies appear to form single-species flocks and may have non-overlapping breeding ranges, but more study is needed. MIGRATION: Wander greatly in search of food. Known to "invade" the lowlands across much of N.A. well away the breeding grounds; in those years often found in ornamental pines in parks and cemeteries. Casual to Bering Sea islands, AK; and in winter to the Southeast and Bermuda.

POPULATION Logging in old-growth coniferous forests may affect food supplies; trees generally produce cones only once they reach 60 years old. In Newfoundland, the population is also threatened by introduced red squirrels, which outcompete the Red Crossbills for pinecones, and loss of breeding habitat.

WHITE-WINGED CROSSBILL *Loxia leucoptera*   WWCR ■ 2

The White-winged Crossbill is one of two species in North America with a distinctive finchlike bill that crosses at the tip; its cousin, the Red Crossbill, is the other. White-winged inhabits the northern boreal forest from Newfoundland to Alaska; highly nomadic, its populations move around both seasonally and annually because of fluctuating spruce cone crops. It forms large single-species flocks during the nonbreeding season. It is also an irruptive migrant, but it is generally found farther north than Red. Polytypic (2 ssp.; nominate in N.A.). L 6.5" (17 cm)

IDENTIFICATION Boldly patterned black wings, with very contrasting white lesser coverts (often covered by scapulars), bold white wing bar, and white tips to tertials, identify this species in all plumages. ADULT MALE: Pink body. Grayish flanks and undertail coverts. Black scapulars and lores. Dark on rear portion of auricular forms a crescent on side of head. WINTER MALE: Paler pink overall. FEMALE: Brownish olive body, with very indistinct streaking on underparts. Pale yellow rump. Dark lores. Auricular patch more distinct. IMMATURE MALE: Like adult male, but yellow with patches of orange. JUVENILE: Brown with heavy streaks all over. Wings have thinner white wing bars than adult.

GEOGRAPHIC VARIATION Old World *fasciata* breeds in the boreal forest and differs vocally from nominate *leucoptera* and like represents a distinct species. A former subspecies from the island of Hispaniola is now recognized as a full species, Hispaniolan Crossbill (*L. megaplaga*).

SIMILAR SPECIES Red Crossbill is the only other North American species with a bill distinctly crossed at tip; however, White-winged male is pinker and all plumages have a bolder wing pattern. Know that some Reds show narrow white wing bars. Pine Grosbeak, which has white wing bars and is red overall, lacks crossed bill and is a third again larger.

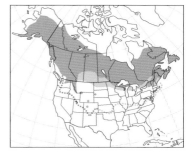

VOICE CALL: A rapid series of dry *chet* notes and a double or triple *chik-chik;* given in flight. Generally softer than a Red's call. SONG: A series of mechanical trills and whistles, usually rising in pitch; given from a treetop or in flight display.

STATUS & DISTRIBUTION Fairly common, but erratic. BREEDING: Frequents northern boreal forests, but irregular and nomadic. Present at locations one year, absent the next. Timing and success dependent upon spruce cone production. WINTER: Generally south of normal breeding areas, but inconsistent year to year. Regular in the northern states. VAGRANT: Casual to northern CA, NM, TX, FL, and Bermuda.

POPULATION As with Red Crossbill, logging practices in coniferous forests may significantly affect populations; it takes nearly 60 years for mature forests to produce cones.

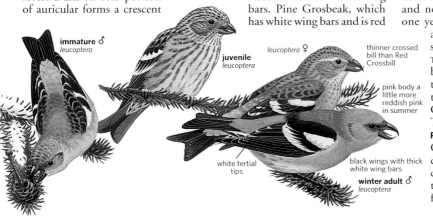

immature ♂
leucoptera

juvenile
leucoptera

leucoptera ♀

thinner crossed bill than Red Crossbill

pink body a little more reddish pink in summer

white tertial tips

black wings with thick white wing bars

winter adult ♂
leucoptera

REDPOLLS Genus *Acanthis*

COMMON REDPOLL *Acanthis flammea*   CORE ■ 1

Common and Hoary Redpolls are two closely related finches of the boreal forest and Arctic tundra scrub. The Common is the more widespread species of the two, usually inhabiting subarctic forest during the summer and frequenting seed feeders in southern Canada and northern United States during the winter, when they form large flocks. Adults have characteristic red cap or "poll." Polytypic (2 ssp.; both in N.A.). L 5.3" (13 cm)

IDENTIFICATION Common is generally a relatively small, streaked finch with a small, pointed bill; short, deeply notched tail; two white wing bars; black chin; red cap; and varying amounts of red underneath. **BREEDING MALE:** Cap is bright red. Upperparts are brown with distinct streaking. bright rosy red of throat and breast extends onto cheeks. white flanks and undertail coverts have fine black streaking; paler rump has distinctive streaking. **BREEDING FEMALE:** Lacks red breast of male and has variable amounts of streaking underneath, usually confined to sides. **WINTER**

MALE: Duller. Buffy wash on sides and rump. **WINTER FEMALE:** Also buffier on sides. **IMMATURE:** First-year birds resemble an adult female, but they tend to be buffier. **JUVENILE:** Brown and streaked, it acquires red cap in late summer molt.

GEOGRAPHIC VARIATION Two breeding subspecies in N.A. The small-billed and smaller *flammea* has less coarse streaking and is widespread across Canada to Alaska; the large-billed and larger *rostrata* has coarser streaking underneath and is found on Baffin Island and Greenland. Both overlap during winter, but the clinal variation makes identification problematic.

SIMILAR SPECIES Great care is needed to separate Common from the very similar-looking Hoary. The breeding adult male Hoary is a very frosty white above, and white below with a very pale pink blush on breast. Females

and immatures are much more difficult; rely on the differences in bill size and shape, the presence or absence of streaking on rump, the quality of the streaking on flanks and undertail coverts, and to a lesser degree, location (see sidebar below). The juvenile Common can resemble a juvenile Pine Siskin, but it lacks yellow in the wing. The extent of interbreeding between Common and Hoary Red-polls is unknown.

VOICE CALL: When perched, gives a *sweee-eet;* flight call a dry rattling *jid-jid-jid-jid.* **SONG:** A lengthy series of trills and twittering rattles.

STATUS & DISTRIBUTION Common. **BREEDING:** Found in the subarctic forests and tundra across northern Canada and much of AK. Subspecies *rostrata* breeds in tundra scrub, where it overlaps with the Hoary. **WINTER:** Forms large flocks. Irruptive migrant south through much of Canada to northern U.S. Generally winters farther south than Hoary. Rare and irregular to Bermuda. **VAGRANT:** Casual or accidental anywhere in southern U.S.

juvenile
flammea

breeding ♀
flammea

breeding ♂
flammea

extensive pinkish red breast

winter ♀
flammea

faint flank streaks

streaked undertail coverts

winter ♀
rostrata
larger and darker than *flammea* with more extensive streaking below

breeds in Iceland, Greenland and Baffin Island; a few in winter in East in redpoll invasion years

winter ♂
flammea

Redpoll Identification

The task of identifying Common Redpolls from Hoary Redpolls ranges from very easy to virtually impossible. Usually, a single field character is not enough to nail down an identification—use a suite of characters. Critical field marks include the overall color of the bird, the color of the streaking on the upperparts, the color of the pink or red on the underparts, the amount and quality of the streaking on the flanks, and whether or not the streaking extends onto the undertail coverts. Pay particular attention to whether or not the rump is streaked. Also focus on the length of the bill. The songs and calls of these two birds are virtually indistinguishable in the field, so not all individuals are identifiable. Also keep in mind that

Common Redpoll, female (NH, Feb.)

these two species often mix together into the same foraging flock.

Identifying adult males in breeding plumage is fairly straightforward. The Common is generally much darker overall than the Hoary, with extensive rosy-red on much of the underparts and very dark streaking on the back. The bold streaking on its flanks extends to the undertail coverts. The Hoary, however, is very pale overall and has very frosty streaking on the back, faint streaking on the flanks, and no streaking on the undertail coverts. The pink on its breast is more of a pale blush. In general, the main character difference between the two species is the color or pattern on the rump: The Common's is darker and more heavily streaked, the Hoary's is whiter

HOARY REDPOLL *Acanthis hornemanni* HORE ■ 2

winter ♀
exilipes

slightly
smaller bill
than Common

very
pale pink

pale rump

faint flank
streaks

winter ♂
exilipes

larger and
overall paler
than *exilipes*

paler upperparts
than Common

winter ♂
hornemanni

The Hoary Redpoll is very similar to its sister species, the Common Redpoll, but it is generally found farther north in tundra habitats during the breeding season. The adults have the characteristic red cap or "poll" and are generally white or frosty above and below. During the nonbreeding season, Hoary forms large flocks and winters mainly in Canada, though it sometimes mixes with Common Redpolls at seed feeders in the northern United States. Polytypic (2 ssp.; both in N.A.). L 5.5" (14 cm)

IDENTIFICATION Hoary is a relatively small finch with a short, notched tail and a very short, stubby bill. Like Common, it sports a distinctive red cap in all plumages except juvenile. **BREEDING MALE:** Very frosty white with contrasting red cap and pale pinkish blush on breast. Wings with rather bold wing bars and white edging on secondaries. Rump white, usually unstreaked. Flanks with fine

black streaking, undertail coverts usually clean white, or with a hint of streaking. Bill very short. **BREEDING FEMALE:** Generally pale overall. Typically white below with fine black streaking on sides and flanks. Rump white with little or no streaking. Back streaking quite white. **WINTER ADULT:** Buffier overall on sides. **IMMATURE:** Streaking more prominent on sides. **JUVENILE:** Lacks red cap. Pale brown overall with streaking above and below.

GEOGRAPHIC VARIATION Two subspecies breed in N.A. Nominate *hornemanni,* found in Baffin Island and Greenland, is larger, larger billed, and overall paler than the more widespread *exilipes.*

SIMILAR SPECIES Hoary Redpoll is most confused with the very similar Common Redpoll. Separating adult males is fairly easy; the whiter (frostier) Hoary has a pale pinkish blush on its breast, while Common's breast is rose red. Females and immatures present more of a problem. Main

characteristics used to separate the two include size and shape of bill, presence or absence of streaking on rump, degree of streaking on flanks and undertail coverts, and to a lesser extent, location (see sidebar below). The juvenile Hoary could be confused with a juvenile Pine Siskin, but Pine Siskin has yellow in the wings.

VOICE CALL & SONG: Very similar to Common Redpoll, essentially indistinguishable in the field.

STATUS & DISTRIBUTION Common in tundra scrub. **BREEDING:** Found across the high Arctic, generally above tree line, or in willows and alders, across northern Canada and northern AK. Generally (at least for *exilipes*) found farther north in nonforested habitats as compared to Common Redpoll. **WINTER:** Forms large flocks that migrate south across much of Canada, rarely into extreme northern U.S. Casual into northern tier states, usually associated with Common Redpolls coming to seed feeders.

and unstreaked. In another character difference, the Common has a longer, more pronounced bill, while the Hoary has a tiny, insignificant bill. This lends the face of the Hoary an even more "pushed in" look than that of the Common, whose longer bill stands out more distinctly from the dark patches surrounding it.

Separating the females and immatures of the species can be quite challenging. In general, the female Hoary is paler and has less coarse streaking on the sides; whiter (typically) unstreaked undertail coverts; bolder, often broader, wing bars; paler streaking on the back; whiter (usually) unstreaked rump, and shorter, pug bill. The female and immature Hoary also tend to show a greater amount of white edging to

Hoary Redpoll, female (NH, Feb.)

their flight feathers, which is evident on their greater coverts in flight.

Both species of Redpoll have distinct subspecies that reside in Greenland, but sometimes wander to eastern Canada in winter. The Greenland subspecies of Common Redpoll is darker than that of the mainland, with a larger patch of black on the throat and a greater amount of brown overall. The Greenland subspecies of Hoary Redpoll is paler overall than its continental counterpart, with less streaking on the sides, a frostier back, and very little pink on the breast of the breeding male. These birds are easier to differentiate between species, but may prove problematic when narrowing down to subspecies. ■

EVENING GROSBEAK AND HAWFINCH Genus *Coccothraustes*

Both species in this finch genus are large with very short tails, striking black-and-white wing patterns, and very large bills. The Evening Grosbeak breeds in North America, whereas the Hawfinch occurs as a vagrant from Asia. Both form flocks during the nonbreeding season that feed in fruiting trees.

EVENING GROSBEAK Coccothraustes vespertinus   EVGR ■ 1

This noisy finch forms large, gregarious flocks during winter, when it often frequents feeders. It is found during summer mainly in coniferous forests across boreal Canada and in the Rocky Mountains; its winter movements are both erratic and irruptive. Polytypic (3 ssp.; 2 in N.A.). L 8" (20 cm)
IDENTIFICATION A large, stocky, finch with very short tail and heavy bill. **MALE:** Body is a rich golden brown, becoming darker olive-brown on head and black on crown. Both forehead and eyebrows are bright golden yellow. White secondaries and tertials are distinctive. Note large bill. **FEMALE:** Body is grayish brown above, buffier on underparts, collar, and rump. Throat whitish with distinctive dark malar stripe. Blackish wings with white patch at base of primaries; somewhat darker secondaries. Large white spots at end of tail. Wing and tail pattern distinctive in flight. **JUVENILE:** Male like adult male, only body a duller uniform brown and eyebrows are a dull yellow.
GEOGRAPHIC VARIATION The three described subspecies in N.A. eastern subspecies *vespertinus* tend to have shorter bills and broader yellow eyebrows than southwestern *montanus*. Western *brooksi* is also long billed, and includes populations with three different flight calls See "Voice."
SIMILAR SPECIES See Hawfinch.
VOICE CALL: Flight calls vary

geographically (five types from five regions have been described) and individual birds give only one type of call. These vocalizations likely foster assortative mating and future research may reveal that they are actually separate species with little or no interbreeding. Eastern birds (type 3) give a ringing *clee-ip* or *peer;* there are four other call types in the West, best learned from audio recordings or by studying sonograms.
STATUS & DISTRIBUTION BREEDING: Coniferous forest and mixed woods; mainly in mountains in West. **WINTER:** Sometimes common at lower elevations and south of breeding range, but in the East southern invasions have been much more infrequent in recent decades. Rare to casual in southern states. **VAGRANT:**

Casual in spring north to AK. Rare and irregular to Bermuda. Accidental to U.K. and Norway.

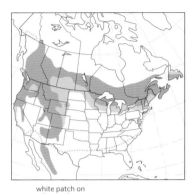

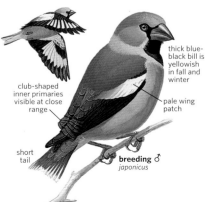

white patch on inner wing

yellow forehead and eyebrow

vespertinus ♂

large bill

breeding ♂ *vespertinus*

vespertinus ♀

white primary patch

juvenile ♂ *vespertinus*

short tail with white tail spots

female with gray on head and back; buffy below

breeding ♀ *vespertinus*

HAWFINCH Coccothraustes coccothraustes   HAWF ■ 4

This Eurasian species related to the Evening Grosbeak is a rare migrant

to the outer Aleutians, and has been recorded from islands in the Bering Sea. Typically it is much more wary than Evening Grosbeak, with many reports of fly-by individuals. It sometimes occurs in small groups, but it tends to be elusive. When on the ground, it walks with a parrotlike waddle. Polytypic (6 ssp.; *japonicus* in N.A.). L 7" (18 cm)
IDENTIFICATION A large, stocky, short-tailed finch with a massive bill. **BREEDING MALE:** Body is yellowish brown above and pinkish brown below, while head is a more golden brown. Throat

and lores are black; nape and collar are gray. The black wings have white secondaries and bases to the primaries

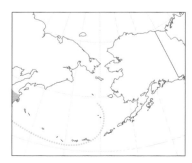

thick blue-black bill is yellowish in fall and winter

club-shaped inner primaries visible at close range

pale wing patch

short tail

breeding ♂ *japonicus*

that form a conspicuous white band on the folded wing and make a striking black-and-white pattern in flight. Tail feathers have broad white tips. Bill is blue-black in the spring and turns yellowish in the fall. FEMALE:

Overall, duller than the male and has grayer wing patches. SIMILAR SPECIES No other finch looks like a Hawfinch, VOICE CALL: A loud, explosive *ptik*. STATUS & DISTRIBUTION A rare but

regular spring migrant in the outer Aleutians (e.g., Attu), with a maximum annual count of 30+ in 1998. Elsewhere a casual spring and fall visitor to western AK islands in Bering Sea.

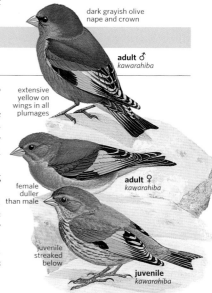

dark grayish olive nape and crown

adult ♂
kawarahiba

extensive yellow on wings in all plumages

female duller than male

adult ♀
kawarahiba

juvenile streaked below

juvenile
kawarahiba

Genus *Chloris*

ORIENTAL GREENFINCH *Chloris sinica*   ORGR ■ 4

A casual spring migrant in the outer Aleutians, this Asian finch sometimes occurs in small groups. No other finch in N.A. is brown with large yellow wing patches. Polytypic (5 ssp.; 1 in N.A.). L 6" (15 cm) IDENTIFICATION Small, stocky brown finch with a deeply notched tail; short, stubby bill; and a large yellow wing patch in primaries. ADULT MALE: Uniformly olive-brown, with greener face and rump; dark gray nape and crown; white patch on tertials and tips of primaries visible on folded wing. Yellow undertail coverts and base of tail feathers. ADULT FEMALE: Duller, lacks gray

crown and nape. JUVENILE: Similar to female, but finely streaked below. SIMILAR SPECIES Female and juvenile Oriental Greenfinch are vaguely similar to smaller Pine Siskin, but yellow in wing is much broader and back in unstreaked. VOICE CALL: Includes a twittering rattle and a nasal *zweee*. STATUS & DISTRIBUTION Asian species; southern Russia to Japan. VAGRANT: Casual (ssp. *kawarahiba*) in outer Aleutians; one Bering Sea island rec. (St. Paul). A wintering bird at Arcata, CA (Dec. 4, 1986–Apr. 3, 1987), was judged to be questionable of origin.

SISKINS AND GOLDFINCHES Genus *Spinus*

EURASIAN SISKIN *Spinus spinus*   EUSI ■ 5

Eurasian counterpart of Pine Siskin. Monotypic. L 4.8" (12 cm) IDENTIFICATION Shaped like a Pine Siskin, but looks like a cross between

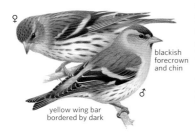

♀

blackish forecrown and chin

♂

yellow wing bar bordered by dark

a siskin and a goldfinch. MALE: Fairly olive above and heavily streaked. Black forecrown and chin. Yellow throat and breast. Prominent black streaking on flanks. Olive auriculars surrounded by yellow. Two bold yellow wing bars; extensive yellow edging on flight feathers and at base of tail. Yellow rump. FEMALE: No black on crown or chin. Yellow on underparts restricted to sides of breast, wash on face, eyebrow, and rump. JUVENILE: Brown streaked above, whitish below, with heavy black streaking.

SIMILAR SPECIES Males unlike other carduelids in N.A. Female and especially juvenile very similar to female or juvenile Pine Siskin, but note blacker wing coverts on Eurasian. VOICE CALL: A two-note whistled *ti-lu* or a short rattle. STATUS & DISTRIBUTION One certain record of a male from Attu I., Aleutians (May 22, 1993). Casual in northeast N.A., with records from ON, ME, MA, NJ, and off NF (St. Pierre & Miquelon), but origin of some (most?) questioned.

PINE SISKIN *Spinus pinus*   PISI ■ 1

A widespread and conspicuous breeding species of coniferous forest across the boreal zone of Canada and northern United States, as well as in mountainous areas of the west, the Pine Siskin is an irregular and less predictable winter visitor virtually anywhere in the U.S. It forms large flocks during the nonbreeding season and is commonly attracted to seed feeders. Polytypic (3 sp.; nominate in N.A.). L 5" (13 cm) IDENTIFICATION It is entirely brown

and streaked and has prominent yellow in the wing, a short deeply notched tail, long wings, and a longish, pointed bill. MALE: Brown and streaked above, below whitish with coarse dark streaking. Two prominent wing bars, lower one extensively yellow. Distinct yellow edging to flight feathers and tail, conspicuous in flight and on folded wing. Some males very yellowish with reduced streaking. FEMALE: Similar to male, but yellow in wings and tail greatly reduced. JUVENILE: Is quite

buffy yellow, but fades by late summer. **SIMILAR SPECIES** It is the only carduelid that is entirely brown and streaked. All *Haemorhous* finches are considerably larger with thicker bills and lack yellow flash in the wing characteristic of siskins. **VOICE CALL:** Most commonly a buzzy, rising *zreeeeee*; also gives a harsh, descending *chee* in flight. **SONG:** A lengthy jumble of trills and whistles

similar to that of American Goldfinch, but huskier. **STATUS & DISTRIBUTION** Common and gregarious. **BREEDING:** Found in coniferous forests of the north and mountainous west. **WINTER:** Range erratic from year to year, likely due to fluctuating food supply; can be found virtually anywhere. Often associates with goldfinches. **VAGRANT:** Rare and irregular to Bermuda; casual to Bering Sea islands, AK.

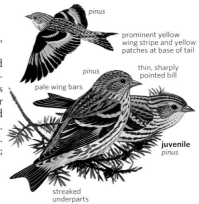

pinus

prominent yellow wing stripe and yellow patches at base of tail

pinus

thin, sharply pointed bill

pale wing bars

juvenile
pinus

streaked underparts

LESSER GOLDFINCH *Spinus psaltria*   LEGO   ■   1

This very common carduelid finch of the Southwest breeds in a variety of habitats at different elevations. It is often seen in small flocks feeding along brushy roadsides, particularly where thistle grows. It is often detected by its distinctive flight calls. Polytypic (5 ssp.; 2 in N.A.). L 4.5" (11 cm)
IDENTIFICATION A relatively small, sexually dimorphic finch. **BREEDING MALE:** Differs depending on subspecies. In "green-

backed" *hesperophila,* bright yellow underparts and black cap contrast with an olive-green back. Note white at base of primaries forms a white "tick" on folded wing. In male "black-backed" *psaltria* (*mexicanus* of some authors), bright yellow underparts contrast with entirely black upperparts. **FEMALE:** More uniform yellow-green, often yellower below and greener above, but less contrasting. Note distinctive white at base of primaries. Some females very dull and drab, mostly gray with some yellowish green wash to body; wings duller, but still with white "tick." **IMMATURE MALE:** It lacks full black cap of adult, but it still has black on forehead (*hesperophila*) or has some black intermixed with green on back and crown (*psaltria*).
GEOGRAPHIC VARIATION Green-backed *hesperophila* is more widespread in Southwest, with black-backed *psaltria* breeding from CO to southern TX. Intergradation occurs clinally between TX and CO.
SIMILAR SPECIES Males are very distinct from other goldfinch species. Female American is larger and has white undertail coverts, a wide buffy lower wing bar with very little white at base for flight feathers, and a pale pinkish

bill. A very pale female Lesser can show whitish undertail coverts, but it has a different wing pattern than all American Goldfinches and is always greener than female Lawrence's.
VOICE CALL: Includes a plaintive, kitten-like *tee-yee*. **SONG:** Very complex jumble of musical phrases, often mimicking other species.
STATUS & DISTRIBUTION Very common; spreading north and east. **BREEDING:** A variety of habitats at different elevations from arid lowlands to high pine forests, often found near water. **WINTER:** Northern and high elevation populations migrate to southern U.S. and Mexico, augmenting resident populations there. **VAGRANT:** Casual north and east of mapped range in Great Plains. Accidental in the East.

hesperophila ♀

smaller than American

pale yellow underparts

pale ♀
hesperophila

black-backed adult ♂
psaltria

green-backed adult ♂
hesperophila

blackish cap contrasts with green back

white patch at base of primaries

immature ♂
hesperophila

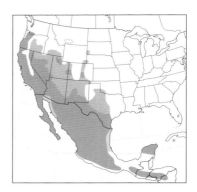

LAWRENCE'S GOLDFINCH *Spinus lawrencei*   LAGO   ■   2

The subtle combination of gray, yellow, and black makes the male Lawrence's Goldfinch one of the more striking carduelids in North America. Largely restricted during the breeding season to the foothills and montane valleys in California, it has a unique irregular dispersal pattern some winters to the southwestern United States. Often seen in small groups, it frequents

brushy areas, preferring those found along riparian corridors. Like other goldfinches, it favors feeding on thistle plants. Monotypic. L 4.8" (12 cm)
IDENTIFICATION BREEDING MALE: Very sharp plumage, acquired through wear. Mostly gray body; lighter gray cheeks; black front of crown, face, and chin; and yellow breast. Black wings with extensive yellow forming two broad wing

bars and yellow edging to flight feathers. Yellow rump. Distinct circular white tail spots. **BREEDING FEMALE:** Much duller than male. Grayish body, browner above. Reduced amount of yellow on breast. Two prominent yellowish wing bars, with yellow edging to primaries. **WINTER MALE:** Like breeding male, but much browner above and duller below. **WINTER FEMALE:** Like breeding female, but much browner. **JUVENILE:** Streaked; unlike other goldfinches.

SIMILAR SPECIES Male unique. Female could be confused with very dull Lesser female, which is still greenish; Lawrence's has whiter undertail coverts, more prominent wing bars, and different tail pattern. Note very different call notes among goldfinch species.

VOICE CALL: A very distinctive, rather soft bell-like *tink-ul.* **SONG:** A series of jumbled musical twittering, often interjecting *tink* notes, and mimicking other species like Lesser.

STATUS & DISTRIBUTION Rather uncommon and local. **BREEDING:** Prefers drier interior foothills and montane valleys of CA and northern Baja. Some populations resident, while others highly migratory. Has bred casually in central AZ. Breeding areas not consistent from year to year. **WINTER:** Irregular fall and winter movement to the southwestern U.S., occasionally wintering in moderate numbers in southeastern AZ and northern Mexico, casual to southern NM and west TX. AZ wintering birds begin arriving by mid-Oct. Some years the wintering grounds remain largely unknown. **VAGRANT:** Casual to northern AZ and NM, and to central OR.

POPULATION Stable but little studied.

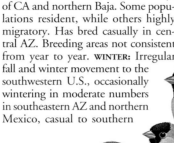

brownish wash on back in fresh fall plumage in both sexes, wears to more grayish by spring

winter ♀

winter ♂

distinctive white oval tail spots like a *Setophaga* wood-warbler

breeding ♂

prominent yellow on wings in all plumages

yellow breast

juvenile

AMERICAN GOLDFINCH *Spinus tristis* AMGO ◼ 1

The brightly colored male American Goldfinch is unmistakable. It regularly visits seed feeders. It is often very gregarious, especially during the nonbreeding season, when it flocks to roadsides and brushy fields to feed on thistle and sunflowers. It is often heard in flight. Polytypic (4 ssp.; all in N.A.). L 5" (13 cm)

IDENTIFICATION A relatively large carduelid. **BREEDING MALE:** Unmistakable. Body entirely bright lemon yellow with white undertail coverts. Jet black cap. Black wings with yellow lesser coverts; two white wing bars. Pink bill. **BREEDING FEMALE:** Very different from male. Underparts very yellow with white undertail coverts, olive green above. **WINTER MALE:** Cinnamon brown above and on breast and flanks, with white lower belly and undertail coverts, yellowish wash on throat and face, and muted black on forehead. Wings more boldly patterned. Yellow lesser coverts. Wide, whitish lower wing bar. Bill darker than in breeding season. **WINTER FEMALE:** Mostly drab gray body with black wings and two bold buffy wing bars. White undertail coverts. Dark bill. **IMMATURE MALE:** Black on forehead reduced or lacking. Lesser coverts duller. **JUVENILE:** Resembles adult female. Unstreaked.

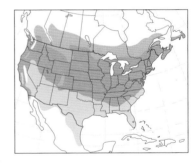

GEOGRAPHIC VARIATION Differences among subspecies rather slight, but breeding adult males *salicamans* (CA) do not achieve bright yellow coloration. **VOICE CALL:** A *per-chik-o-ree* or *ti-di-di-di,* given mainly in flight; distinctly different from other goldfinches. **SONG:** A long series of musical phrases. Not known to mimic other species.

STATUS & DISTRIBUTION Common throughout much of U.S. and southern Canada. **BREEDING:** A variety of habitats, from weedy fields to open second growth woodland, and along riparian corridors, particularly in West. **VAGRANT:** Casual to AK.

breeding ♀
tristis

black wings and tail; white under tail

breeding ♂
tristis

bright yellow body

black forehead and forecrown

yellow shoulder; brownish back

winter adult ♂
tristis

whitish undertail coverts

prominent wing bars

winter ♀
tristis

juvenile
tristis

OLD WORLD SPARROWS Family Passeridae

House Sparrow, female and male (IL, Oct.)

With about 34 species in three genera, Old World sparrows are widespread in Africa and the Palearctic. Two species have been successfully introduced to North America, with different results: The House Sparrow is ubiquitous to about 60° N, whereas the Eurasian Tree Sparrow is more restricted in range. Our two introduced species are found in small flocks in human-dominated landscapes.

Structure Old World sparrows are small, with medium to slender builds. Most species have short legs and short, thick bills. Proportions of the wings and tail are unremarkable in all species.

Behavior Most Old World sparrows are seen on or near the ground, where feeding typically occurs. Many species are gregarious, and some are abundant. Several passerid species have flourished in human-influenced environments, and many are tame and approachable.

Plumage Old World sparrows are mainly clad in browns, grays, and russets. In many species, there are marked differences between juveniles and adults, as well as between the sexes. Seasonal variation can be pronounced, too.

Distribution The genus *Passer* (to which both N.A. species belong) contains the most species, the most widespread, and the most diverse in its habitat preferences. Two other genera, not introduced to North America, are more specialized: Widespread *Petronia* (rock sparrows) is restricted mainly to warm, arid climes; *Montifringilla* (snowfinches), centered around the Tibetan Plateau, occurs in barren habitats at high elevations.

Taxonomy Old World sparrows are not closely related to New World sparrows in the family Emberizidae. Instead, their closest alliance is with the family Ploceidae, in which they were formerly placed. Taxonomists differ in the division of species between the two families.

Conservation Especially in the genus *Passer,* many species enjoy commensal relationships with humans. The House Sparrow, however, long regarded as a textbook example of a beneficiary of human activity, is currently in sharp decline in parts of its native range.

Genus *Passer*

HOUSE SPARROW *Passer domesticus* HOSP ▪ 1

The cheery and sociable House Sparrow is more closely associated with humans than any other widely established North American exotic. Introduced to New York City in 1851, the species today flourishes in both large cities and remote agricultural outposts—just so long as there is some trace of human influence. It aggressively defends nest cavities, possibly to the detriment of native species. It is more gregarious in winter. Polytypic (12 ssp.; nominate in N.A.). L 6.3" (16 cm)

IDENTIFICATION Tame; gregarious. Flight more direct, often higher, than native sparrows. Bill thick, conical; legs short; stocky build. One molt per year, but seasonal variation pronounced. **ADULT MALE:** Worn (breeding) male contrastingly marked; throat and breast black, areas behind eye and on nape are chestnut. On a freshly molted bird, blackish and reddish regions are obscured by gray feather tips. Bill black in summer; yellowish base to lower mandible in winter. **ADULT FEMALE:** Mainly gray-tan. Buffy eye stripe; gray-brown crown and auriculars. Bill more yellowish than male's; tip and culmen are dusky. **JUVENILE:** Variable. Plain overall. Resembles adult female.

GEOGRAPHIC VARIATION North American population (nominate) exhibits extensive geographic variation, with clinal variation: larger birds with shorter appendages in colder climes, darker plumages in more humid environments.

gray crown

blackish bill

buffy eyebrow

pale bill

domesticus ♀

plain underparts; dark streaked back

chestnut-brown nape

black bib

fall ♂
domesticus

white wing bar

head pattern more subdued

bill paler

breeding ♂
domesticus

SIMILAR SPECIES Males distinctive; plainer females and juveniles present a combination of structural and plumage characters that separate them from native sparrows. A female Orange Bishop may be confused with House Sparrow.

VOICE All vocalizations simple. **CALL:** Varied, but three notes are prevalent: throaty *jigga,* usually given by agitated birds; soft *chirv,* often heard in flight; honest-to-goodness *chirp,* given in various settings. **SONG:** Short series of pleasant *chirp* notes.

STATUS & DISTRIBUTION Locally abundant. **YEAR-ROUND:** Cities, farms, and other human-transformed environments. **VAGRANT:** Several recs. for western Canada and AK, including from St. Lawrence I.; they probably originate from nearby Chukotka (Russian Far East) where they have been introduced (not wild birds).

POPULATION Some Palearctic populations are in sharp decline, possibly due to changing land use patterns.

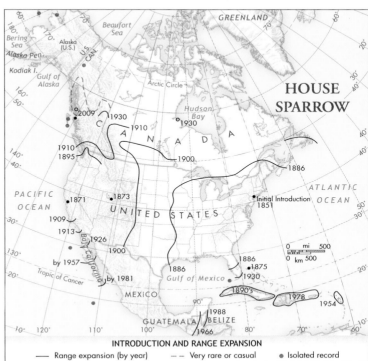

HOUSE SPARROW

INTRODUCTION AND RANGE EXPANSION

—— Range expansion (by year)　— — Very rare or casual　● Isolated record

EURASIAN TREE SPARROW *Passer montanus* ETSP ■ 2

juvenile
montanus

chestnut-brown cap and bold blackish spot on white cheeks

whitish collar

adult
montanus

Like the House Sparrow, the Eurasian Tree Sparrow was deliberately introduced to North America (St. Louis, 1870). Unlike the House Sparrow, its range has not expanded greatly. Polytypic (9 ssp.; nominate in N.A.). L 5.9" (15 cm)

IDENTIFICATION Usually in small flocks. **ADULT:** One molt per year. Sexes similar. Key mark is white cheek with large black spot. Crown and nape chestnut; neck collar white; throat black. On worn plumage (spring,

early summer), russet and chestnut regions browner, duller. **JUVENILE:** Resembles adult; buffier, duller overall; bill straw yellow.

GEOGRAPHIC VARIATION None in North America. Our subspecies is nominate *montanus;* there are eight other subspecies in the Old World.

SIMILAR SPECIES White neck collar, black cheek spot, and chestnut-brown across entire crown separate slightly smaller Eurasian Tree Sparrow from House Sparrow.

VOICE Varied; all notes simple. **CALL:** Monosyllabic utterances, sharper and more metallic than House Sparrow's, typically without the pleasing, liquid qualities: *chet, kip,* etc. **SONG:** Series of sharp monosyllabic notes, interspersed with more liquid, disyllabic notes.

STATUS & DISTRIBUTION Locally common. Occupies an area of about 10,000 sq. mi., mainly in west-central IL; range extends to west of St. Louis, MO, and north to southeastern IA. **YEAR-ROUND:** Near human habitation. **VAGRANT:** Generally northward. Out-of-range individuals difficult to establish as vagrants; some may be at vanguard of range expansion, others may be escapes. Presumed vagrants noted MN, WI, ON, MB. Presumed escapes or ship-assisted birds have been observed along West Coast.

POPULATION The range in N.A. is expanding slowly, generally northward.

WEAVERS Family Ploceidae

Orange Bishop, breeding male (AZ, July)

Known for their elaborate, suspended nests woven from grass and other plant fibers, the weavers are a large family of finchlike birds, primarily of the African tropics. Their colorful plumage makes them popular captives, resulting in naturalized populations in many regions of the world.

Structure They are small to medium size with conical bills; most have relatively short tails, short, rounded wings, and short but strong legs. Breeding male widowbirds *(Euplectes)* and whydahs *(Vidua)* grow stunning tails that are three to four times their body length.

Plumage Breeding males are brightly colored—red, orange, or yellow bodies, often with contrasting black faces or masks—in many species. Nonbreeding males resemble females, which resemble sparrows.

Behavior Many species are highly colonial, foraging and roosting in large flocks; some are colonial nesters. Weavers are largely sedentary, making only local or seasonal movements, but some species are migratory. They feed primarily on seeds and grain, although some species also take insects, fruit, or nectar.

Distribution They are found primarily in Africa, but some species are also in Eurasia, Australia, and Oceania.

Taxonomy Authorities vary on taxonomy (±118–124 sp. in ±16–17 genera in at least four subfamilies). The passerids and the estrildid finches are the ploceids' closest kin.

Conservation BirdLife International lists one species as critical, six as endangered, seven as vulnerable, and three as near threatened. Some are serious crop pests in their native ranges and are killed in large numbers; populations of others have been reduced by capture for the pet trade.

BISHOPS Genus *Euplectes*

This genus comprises nine species known as bishops and eight as widowbirds. Breeding male bishops are brightly colored with short tails, while breeding male widowbirds are less showy but grow spectacular tails that are shed after breeding. All species are endemic to Africa, but many occur elsewhere as a result of escaped or released captives.

ORANGE BISHOP Euplectes franciscanus   ORBI ▪ EXOTIC

The Orange Bishop is the only weaver found in numbers in North America. Polytypic (2 ssp.; presumably nominate in N.A.). L 4" (10 cm)

IDENTIFICATION Small finches with stout bills and short tails that are often flicked open. **BREEDING ADULT MALE:** Orange upperparts with black crown and face. Black eyes and bill; pink legs and feet. Black wings and tail with brown feather edges. Black breast and belly; orange undertail coverts obscure undertail. **FEMALE, IMMATURE,**

& NONBREEDING MALE: Sparrowlike. Head buffy with pale supercilium and streaked crown and nape. Black eyes prominent on pale face; pink bill, legs, and feet. Brown upperparts heavily streaked with black. Breast and sides of neck buffy with dark streaking.

SIMILAR SPECIES Breeding male Red Bishop *(E. orix)*, occasionally seen in N.A., is very similar but it has a black throat and chin. Females and nonbreeding males resemble sparrows (esp. Grasshopper Sparrow), but note stout, wholly pink bill, short tail, and pattern of black streaking on upperparts.

VOICE CALL: Includes a sharp *tsip* and a mechanical *tsik-tsik-tsk.* **SONG:** High-pitched and buzzy.

STATUS & DISTRIBUTION Exotic in U.S. Native to sub-Saharan Africa. **YEAR-ROUND:** Nonmigratory. In southern CA (mostly Los Angeles and Orange Cos.), ±500 found in flood control basins and channeled rivers. About 15 birds found in Phoenix, AZ.

POPULATION Declines in CA.

displaying ♂
franciscanus

breeding ♂
franciscanus

unmistakable, but other species of African bishops are similarly colored

dark eye stands out on blank face

thick, pinkish bill

streaked back

franciscanus ♀

short, blunt tail often flicked open

ESTRILDID FINCHES Family Estrildidae

Nutmeg Mannikin (CA, Feb.)

A large family (± 140 sp. in 28 genera), Estrildids are found in tropical regions of the Old World, primarily in the African tropics. They are typically open-country birds. Several species are common in aviculture, and some have established exotic populations in many regions of the world, including North America. **Structure** They range from tiny to small, and have pointed tails, short, rounded wings, and proportionately large, conical bills.

Plumage Most estrildids are brightly colored, while most *Lonchura* are brown, black, or white. They exhibit no seasonal variation. The bill is often brightly colored, azure in many *Lonchura*. Some species show marked sexual dimorphism; in others, the sexes look similar.

Behavior Many species are highly colonial outside the breeding season, but most breed solitarily. Estrildids feed chiefly on the ground or in low vegetation, primarily on seeds and grain with some fruit or insects taken.

Distribution They are found in Africa, Asia, Australia, and on South Pacific islands.

Taxonomy The taxonomy is confusing and undergoing extensive revision as a result of DNA studies. Generally, there are three main groups: the waxbills of the African tropics; the grassfinches and allies of Australasia and the western Pacific; and the mannikins/munias and allies of Africa and Australasia. Estrildids are closely related to weavers (Ploceidae) and sparrows (Passeridae).

Conservation Most estrildids are common to abundant in their native ranges, however, BirdLife International does list three species as critical and eight as vulnerable.

Genus *Lonchura*

The members (±41 sp.) of this genus are known by several names: mannikin (generally those from Africa or New Guinea), munia (generally those from Asia), silverbill, or sparrow. All are small, pointed-tailed granivorous finches, usually found in flocks. Many naturalized populations are found around the world, including one in North America.

NUTMEG MANNIKIN *Lonchura punctulata*   NUMA ■ EXOTIC

The Nutmeg Mannikin the only munia found in numbers in N.A., is largely restricted to southern California, where flocks feed on grass seeds in weedy river channels and residential areas. It is also known as the "Scaly-breasted Munia" and in the cage trade as "Spice Finch." Polytypic (12 ssp.; apparently only nominate in N.A.). L 4.5" (11 cm). **IDENTIFICATION ADULT:** Sexes similar. Rufous face and breast; remainder of head and upperparts brown. Extensive black scaling on otherwise white lower breast to undertail. Yellow-orange uppertail coverts. **JUVENILE:** Upperparts wholly light brown. Auriculars and underparts peach, with paler throat and undertail coverts. Dark eyes; gray bill.

SIMILAR SPECIES A juvenile is indistinguishable from other *Lonchura* species that occur in North America, such as Chestnut Munia *(L. atricapilla)* or Tricolored Munia *(L. malacca)*. Told from an immature bunting by smaller size and thinner tail.

VOICE CALL: *Beee* or *ki-bee*, repeated frequently. **SONG:** High pitched and quiet; nearly inaudible.

STATUS & DISTRIBUTION Exotic in the U.S. Native from India to China and much of the Orient. Birds in CA from Indian subcontinent; provenance of birds in FL unknown. **YEAR-ROUND:** Nonmigratory, but movements of more than 20 miles noted in CA based on banding recoveries. In U.S., restricted to CA (thousands; primarily Los Angeles and Orange Cos., but locally in small numbers to San Francisco and to San Diego) and FL (a few; Pensacola and Miami). **POPULATION** Increasing in CA.

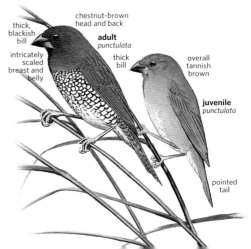

chestnut-brown head and back

thick, blackish bill

adult *punctulata*

intricately scaled breast and belly

thick bill

overall tannish brown

juvenile *punctulata*

pointed tail

ADDITIONAL READING

Bird books offer essential reference and guidance. Following is a short list of recommended sources, arranged by general interest and family. The National Geographic *Field Guide to the Birds of North America*, 6th edition, is an essential, portable field guide for birders of every stripe. CDs of bird songs and calls are available from a variety of commercial sources. Apps for tablets and smart phones are available from most bird book publishers, including National Geographic.

Nationwide, status and distribution information is supplied quarterly by the journal *North American Birds,* published by the American Birding Association—every serious birder should subscribe. Numerous state and regional guides (including breeding bird atlases) are available and are very useful in detailing local status and distribution. To guide readers to the best local birding areas, site guides exist for all regions of the country. Popular birding magazines and ornithological journals round out the printed record.

Electronically, websites and birding forums are the preferred source for keeping track of current sightings and rare bird information. As a starting point, we include a few of the main Internet portals below.

GENERAL

American Ornithologists' Union [AOU]. *Check-list of North American Birds,* 7th ed. AOU, 1998. Standard taxonomic authority for North American birds, annual supplements available online <checklist.aou.org>

Baicich, Paul J., and Colin O. Harrison. *Nest, Eggs, and Nestlings of North American Birds,* 2nd ed., 2005. Princeton University Press, 2009. Descriptions and illustrations of over 670 breeding species.

Brazil, Mark. *Birds of East Asia: China, Taiwan, Korea, Japan, and Russia.* Princeton University Press, 2009. Best illustrated field guide to the region.

Clements, James F. *The Clements Checklist of Birds of the World*, 6th ed. Cornell University Press, 2007. The checklist endorsed by the ABA and updated online by Cornell University.

Cramp, Stanley, and K. E. L. Simmons, eds. *The Birds of the Western Palearctic,* Vols. 1–9. Oxford University Press, 1977–1994. Standard reference to European birds; updated version available as CD-ROM.

del Hoyo, Josep, Andrew Elliot, Jordi Sargatal, and David Christie, eds. *Handbook of the Birds of the World.* Volumes 1–16. Lynx Ediciones, 1992–2011. Monumental and comprehensive treatment of all bird species; Special Volume (2013) includes newly described species and a global index. Available online (access fee).

Dickinson, Edward C. and J. V. Remsen *The Howard and Moore Complete Checklist of the Birds of the World,* 4th ed., vol. 1. Aves Press, 2013. Up-to-date checklist with subspecies and ranges for all non-passerines*;* vol. 2 in press; in the interim, the 3rd ed. (2003) covers the passerines.

Howell, Steve N. G., Ian Lewington, and Will Russell. *Rare Birds of North America.* Princeton University Press, 2014. In-depth information on North American rarities with extensive illustrations.

Howell, Steve N. G., and Sophie Webb. *A Guide to the Birds of Mexico and Northern Central America.* Oxford University Press, 1995. Most complete guide to Mexican birds.

Mullarney, Killian, Lars Svensson, Dan Zetterström, and Peter J. Grant. *The Complete Guide to the Birds of Europe.* Princeton University Press, 2000. Exceptional field guide to European birds.

Poole, A., and F. Gill, eds. *The Birds of North America.* The Academy of Natural Sciences and The American Ornithologists' Union, 1992–2002. Accounts of all North American breeding birds; includes Hawaii. Published individually by species; available online (access fee).

Pranty, Bill, et al. *ABA Checklist: Birds of the Continental United States and Canada,* 7th ed. American Birding

Assoc., 2008. With detailed accounts of occurrences of rare species; updates available online.

Pyle, Peter, et al. *Identification Guide to North American Birds, Parts I and II.* Slate Creek Press, 1997 and 2008. In-depth, banders' guide to North American birds; extensive treatment of subspecies and molt.

Raffaele, Herbert, et al. *A Guide to the Birds of the West Indies.* Princeton, 1998. Illustrated field guide to the region.

Ridgely, Robert S., and Guy Tudor. *Birds of South America: The Oscine Passerines,* vol. 1 and *The Suboscine Passerines,* vol. 2. University of Texas Press, 1989 and 1994. All the passerine species that occur in South America, with selected species illustrated.

WEBSITES

American Bird Conservancy: abcbirds.org
American Birding Association: aba.org
Birding News Portal: birding.aba.org
Birds of North America: bna.birds.cornell.edu (fee based)
Cornell Laboratory of Ornithology Bird Pages:
  allaboutbirds.org
Handbook of the Birds of the World Alive: hbw.com
  (fee based)
ID Frontiers: birding.aba.org/maillist/IDF
Neotropical Birds: neotropical.birds.cornell.edu/portal/home

DUCKS, GEESE, AND SWANS

Behrens, Ken, and Cameron Cox. *Seawatching: Eastern Waterbirds in Flight.* Houghton Mifflin, 2013.

Madge, Steve, and Hillary Burn. *Waterfowl: An Identification Guide to the Ducks, Geese, and Swans of the World.* Houghton Mifflin Co., 1988.

CURASSOWS AND GUANS

Delacour, Jean, and Dean Amadon. *Curassows and Related Birds.* Lynx Ediciones, 2004.

NEW WORLD QUAIL / PARTRIDGES, GROUSE, AND TURKEYS

Madge, Steve, and Phil McGowan. *Pheasants, Partridges, & Grouse.* Princeton University Press, 2002.

GREBES

Ogilvie, Malcolm, and Chris Rose. *Grebes of the World.* Bruce Coleman, 2002.

Konter, André. *Grebes of Our World: Visiting All Species on Five Continents.* Lynx Ediciones, 2001.

ALBATROSSES / SHEARWATERS AND PETRELS / STORM-PETRELS

Brooke, Michael. *Albatrosses and Petrels of the World.* Oxford University Press, 2004.

Flood, Bob, and Ashley Fisher. *North Atlantic Seabirds: Petrels.* Pelagic Birds & Birding Multimedia Identification Guides, 2013.

Howell, Steve N. G. *Petrels, Albatrosses, & Storm-Petrels of North America.* Princeton University Press, 2012.

Robb, Magnus, Killian Mullarney, and the Sound Approach. *Petrels Night and Day.* The Sound Approach, 2008.

Stallcup, Rich. *Ocean Birds of the Nearshore Pacific.* Point Reyes Bird Observatory, 1990.

BOOBIES AND GANNETS / CORMORANTS / DARTERS / PELICANS

Johnsgard, Paul A. *Cormorants, Darters, and Pelicans of the World.* Smithsonian Institution Press, 1993.

Nelson, J. Bryan. *The Sulidae: Gannets and Boobies.* Oxford University Press, 1978.

HERONS, BITTERNS, AND ALLIES

Hancock, James, and James Kushlan. *The Herons Handbook.* Harper & Row, 1984.

HAWKS, KITES, EAGLES, AND ALLIES / CARACARAS AND FALCONS

Dunne, Pete, David Sibley, and Clay Sutton *Hawks in Flight,* 2nd ed. Houghton Mifflin Co., 2012.

ADDITIONAL READING

Ferguson-Lees, James, and David A. Christie. *Raptors of the World*. Houghton Mifflin, 2001.

Liguori, Jerry. *Hawks from Every Angle: How to Identify Raptors in Flight*. Princeton University Press, 2005.

Wheeler, Brian K. *Raptors of Eastern North America* and *Raptors of Western North America*. Princeton University Press, 2003.

RAILS, GALLINULES, AND COOTS

Taylor, Barry. *Rails: A Guide to the Rails, Crakes, Gallinules and Coots of the World*. Yale University Press, 1998.

STILTS AND AVOCETS / OYSTERCATCHERS / LAPWINGS AND PLOVERS / JACANAS / SANDPIPERS, PHALAROPES, AND ALLIES

Chandler, Richard. *Shorebirds of North America, Europe, and Asia*. Princeton University Press, 2009.

Hayman, Peter, John Marchant, and Tony Prater. *Shorebirds: An Identification Guide to the Waders of the World*. Croom Helm Ltd., 1986.

O'Brien, Michael, Richard Crossley, and Kevin Karlson. *The Shorebird Guide*. Houghton Mifflin, 2006.

Paulson, Dennis. *Shorebirds of North America: The Photographic Guide*. Princeton University Press, 2005.

SKUAS AND JAEGERS / AUKS, MURRES, AND PUFFINS / GULLS, TERNS, AND SKIMMERS

Gaston, Anthony J., and Ian L. Jones. *The Auks*. Oxford University Press, 1998.

Grant, P. J. *Gulls: A Guide to Identification*, 2nd ed. Buteo Books, 1986.

Howell, Steve N. J. and Jon Dunn. *Gulls of the Americas*. Houghton Mifflin, 2007.

Olsen, Klaus Malling, and Hans Larsson. *Terns of Europe and North America*. Princeton University Press, 1995.

Olsen, Klaus Malling, and Hans Larsson. *Gulls of North America, Europe and Asia*. Princeton University Press, 2004.

PIGEONS AND DOVES

Gibbs, David, Eustance Barnes, and John Cox. *Pigeons and Doves*. Yale University Press, 2001.

CUCKOOS, ROADRUNNERS, AND ANIS

Erritzoe, Johannes, et al. *Cuckoos of the World*. Christopher Helm, 2012.

OWLS

Duncan, James R. *Owls of the World: Their Lives, Behavior and Survival*. Firefly Books, 2003.

König, Claus, Friedhelm Weick, and Jan-Hendrik Becking. *Owls: A Guide to the Owls of the World*, 2nd ed. Yale University Press, 2008.

Voous, Karel H. *Owls of the Northern Hemisphere*. The MIT Press, 1988.

GOATSUCKERS

Cleere, Nigel. *Nightjars of the World*. Princeton University Press, 2010.

SWIFTS

Chantler, Phil, and Gerald Driessens. *Swifts: A Guide to the Swifts and Treeswifts of the World*, 2nd ed. Yale University Press, 2000.

HUMMINGBIRDS

Howell, Steve N. G. *Hummingbirds of North America: The Photographic Guide*. Academic Press, 2002.

Williamson, Sheri L. *Hummingbirds of North America*. Houghton Mifflin Co., 2001.

KINGFISHERS

Fry, C. Hilary, Kathie Fry, and Alan Harris. *Kingfishers, Bee-eaters and Rollers*. Princeton University Press, 1992.

WOODPECKERS AND ALLIES

Short, Lester L. *Woodpeckers of the World*. Delaware Museum of Natural History, 1982.

Winkler, Hans, David A. Christie, and David Nurney. *Woodpeckers: A Guide to the Woodpeckers of the World*. Houghton Mifflin Co., 1995.

LORIES, PARAKEETS, MACAWS, AND PARROTS

Forshaw, Joseph M., and Frank Knight. *Parrots of the World*. Princeton University Press, 2010.

Juniper, Tony, and Mike Parr. *Parrots: A Guide to Parrots of the World*. Yale University Press, 1988.

SHRIKES

Harris, Tony. *Shrikes and Bush-Shrikes*. Princeton University Press, 2000.

Lefranc, Norbert. *Shrikes: A Guide to the Shrikes of the World*. Yale University Press, 1997.

CROWS AND JAYS

Madge, Steve, and Hilary Burn. *Crows & Jays*. Princeton University Press, 1994.

SWALLOWS

Turner, Angela, and Chris Rose. *Swallows & Martins: An Identification Guide and Handbook*. Houghton Mifflin Co., 1989.

CHICKADEES AND TITMICE / NUTHATCHES / CREEPERS

Harrap, Simon, and David Quinn. *Chickadees, Tits, Nuthatches & Treecreepers*. Princeton University Press, 1996.

WRENS / DIPPERS / MOCKINGBIRDS AND THRASHERS

Brewer, David. *Wrens, Dippers, and Thrashers*. Yale University Press, 2001.

OLD WORLD WARBLERS

Kennerley, Peter and David Pearson. *Reed and Bush Warblers*. Christopher Helm, 2010.

THRUSHES

Clement, Peter. *Thrushes*. Princeton University Press, 2000.

STARLINGS

Feare, Chris, and Adrian Craig. *Starlings and Mynas*. Princeton University Press, 1999.

WAGTAILS AND PIPITS

Alström, Per, and Krister Mild. *Pipits and Wagtails of Europe, Asia and North America*. Princeton University Press, 2003.

WOOD-WARBLERS

Curson, Jon, David Quinn, and David Beadle. *Warblers of the Americas: An Identification Guide*. Houghton Mifflin Co., 1994.

Dunn, Jon L., and Kimball L. Garrett. *A Field Guide to Warblers of North America*. Houghton Mifflin Co., 1997.

Stephenson, Tom, and Scott Whittle *The Warbler Guide*. Princeton University Press, 2013.

TANAGERS

Isler, Morton L., and Phyllis R. Isler. *The Tanagers*. AC Black, 1999.

EMBERIZIDS

Beadle, David and James Rising. *Sparrows of the United States and Canada: The Photographic Guide*. Academic Press, 2002.

Byers, Clive, Jon Curson, and Urban Olsson. *Sparrows and Buntings: A Guide to Sparrows and Buntings of North America and the World*. Houghton Mifflin Co., 1995.

BLACKBIRDS

Jaramillo, Alvaro, and Peter Burke. *New World Blackbirds: The Icterids*. Princeton University Press, 1999.

FRINGILLINE AND CARDUELINE FINCHES AND ALLIES / OLD WORLD SPARROWS

Clement, Peter. *Finches and Sparrows: An Identification Guide*. Princeton University Press, 1993.

ILLUSTRATIONS CREDITS

ART CREDITS

The following artists contributed the art for this book: Jonathan Alderfer, David Beadle, Peter Burke, Marc R. Hanson, Cynthia J. House, H. Jon Janosik, David L. Malick, Killian Mullarney, Michael O'Brien, John P. O'Neill, Kent Pendleton, Diane Pierce, John C. Pitcher, H. Douglas Pratt, David Quinn, Chuck Ripper, N. John Schmitt, Thomas R. Schultz, Daniel S. Smith, Patricia A. Topper, and Sophie Webb.

Back cover images: Red Phalarope by Killian Mullarney; Eastern Screech-Owl by Donald L. Malick; Red-tailed Hawk by N. John Schmitt; Yellow Warbler by Thomas R. Schultz; Baltimore Oriole by Peter Burke; 5, D. Pratt; 8 (UP), D. Pierce; 8 (CTR), D. Beadle; 8 (LOLE), J. Schmitt; 8 (LOCTR & LORT), T. Schultz; 9 (UPLE), J. Alderfer; 9 (UPRT), C. House; 9 (CTR), T. Schultz; 9 (LO), J. Alderfer; 11, C. House; 12, K. Mullarney; 13, K. Mullarney; 14, T. Schultz; 15 (UP), D. Quinn; 15 (LO), J. Alderfer; 16, C. House; 17, C. House; 18, J. Alderfer; 19, C. House; 20 (leucopareia and minima), C. House; 20 (taverneri, hutchinsii, and flight), J. Schmitt; 21, C. House; 21 (parvipes), J. Schmitt; 22, C. House; 23 (UP), C. House; 23 (UP-heads), K. Mullarney; 23 (LO), J. Schmitt; 24, C. House; 25 (UP & LO), C. House; 25 (UP-flight), J. Schmitt; 26-28, C. House; 28 (LO-head), J. Schmitt; 29, C. House; 30 (UP), C. House; 30 (UP-fulgivula), T. Schultz; 30 (LO), J. Alderfer; 31, C. House; 32 (UP & LO), C. House; 32 (UP-female), J. Schmitt; 33, C. House; 34, K. Mullarney; 35 (UP & LO), C. House; 35 (UP-female), J. Schmitt; 36-42 (UP), C. House; 42 (UP-flight), J. Alderfer; 42 (LO), J. Alderfer; 43, J. Alderfer; 44 (UP & LO), C. House; 44 (UP-adult males), J. Alderfer; 45, C. House; 46 (UP), C. House; 46 (UP-fusca), K. Mullarney; 46 (UP-stejnegeri), J. Alderfer; 46 (LO), K. Mullarney; 47–49 (UP), C. House; 49 (UP-hybrid adult male), J. Alderfer; 49 (LO), C. House; 50, C. House; 50 (LO-merganser), J. Alderfer; 51-53, C. House; 54 (UP), J. Schmitt; 54 (UPCTR, LOCTR, & LO), C. House; 55, J. Schmitt; 56–58 (UP), K. Pendleton; 58 (LO), J. Schmitt; 59 (UP-ridgwayi), K. Pendleton; 59 (UP-floridanus and taylori), J. Schmitt; 59 (LO), K. Pendleton; 61 (UP), K. Pendleton; 61 (UP-flying), J. Schmitt; 61 (LO)–69 (UP), K. Pendleton; 69 (UP-displaying male), J. Alderfer; 69 (LO), J. Alderfer; 70, K. Pendleton; 70 (displaying male), J. Alderfer; 71, K. Pendleton; 72-75, D. Quinn; 76, J. Alderfer; 77 (UP), J. Alderfer; 77 (LO), J. Janosik; 78, J. Alderfer; 79, J. Janosik; 79 (winter adult), J. Alderfer; 80, J. Janosik; 80 (winter adult), J. Alderfer; 81, D. Pierce; 81 (flight), J. Alderfer; 83-89, J. Alderfer; 90 (UP), D. Beadle; 90 (LO), J. Alderfer; 91 (UP), M. O'Brien; 91 (LO)–93, J. Alderfer; 94-95, M. O'Brien; 96-100, J. Alderfer; 101 (UP), M. Hanson; 101 (LO), J. Alderfer; 102 (UP), J. Alderfer; 102 (LO), M. Hanson; 102 (LO-head), J. Alderfer; 103, J. Alderfer; 103 (with dark underwing), M. Hanson; 104, J. Alderfer; 105 (UP), J. Alderfer; 105 (LO), M. Hanson; 105 (LO-small flight and swimming), J. Alderfer; 106, M. Hanson; 108-115, J. Alderfer; 116, J. Janosik; 116 (dorotheae), J. Alderfer; 117-118, J. Janosik; 119-120, D. Pierce; 121 (UP), J. Janosik; 121 (CTR & LO), T. Schultz; 123-131, J. Alderfer; 132-135, J. Janosik; 135 (californicus), J. Alderfer; 137 (UP), P. Burke; 137 (LO), D. Quinn; 138 (UP), P. Burke; 138 (LO), J. Alderfer; 139, D. Pierce; 140 (UP), J. Alderfer; 140 (LO), D. Pierce; 140 (LO-modesta), D. Quinn; 141-142, D. Quinn; 143, D. Pierce; 143 (flight), T. Schultz; 144, D. Pierce; 144 (flight), T. Schultz; 145 (UP), D. Pierce; 145 (LO), T. Schultz; 146, D. Pierce; 146 (coromandus and flight), T. Schultz; 147 (UP), D. Quinn; 147 (LO), D. Pierce; 148-149, P. Burke; 150, D. Pierce; 151 (UP), D. Pierce; 151 (LO), P. Burke; 152, P. Burke; 154-156, D. Malick; 158 (UP), K. Pendleton; 158 (UP-perched juvenile, flying adult female, flying pale juvenile), J. Schmitt; 158 (LO), D. Malick; 159 (UP), D. Malick; 159 (LO), K. Pendleton; 160 (UP), J. Schmitt; 160 (LO), D. Malick; 161, D. Malick; 161 (3rd year), J. Schmitt; 162, J. Schmitt; 163, D. Malick; 164, J. Schmitt; 165, D. Malick; 165 (flying juvenile), J. Schmitt; 166 (UP), D. Malick; 166 (UP-flying juvenile), J. Schmitt; 167 (UP), D. Malick; 167 (UP-flying juvenile), J. Schmitt; 168 (UP), D.

Beadle; 168 (LO), J. Schmitt; 168 (LO-perched juvenile), D. Malick; 169-174, J. Schmitt; 174 (perched), D. Malick; 175, J. Schmitt; 175 (perched), D. Malick; 176-177, D. Malick; 177 (flight), J. Schmitt; 178-179, J. Schmitt; 180, D. Malick; 180 (perched juvenile and flying dark-morph adult), J. Schmitt; 181, D. Malick; 183, M. Hanson; 184–187 (UP), T. Schultz; 187 (LO), M. Hanson; 188 (UP), T. Schultz; 188 (CTR & LO), D. Beadle; 189 (UP), D. Quinn; 189 (LO), M. Hanson; 190 (UP), M. Hanson; 190 (LO), D. Quinn; 191, M. Hanson; 192, J. Alderfer; 193, M. Hanson; 195-196, D. Pierce; 197, D. Beadle; 198, D. Quinn; 199, J. Janosik; 200, D. Quinn; 201, J. Janosik; 201 (UP-frazari), T. Schultz; 203 (UP), K. Mullarney; 203 (LO), J. Alderfer; 204-206, J. Alderfer; 207 (UP), K. Mullarney; 207 (LO), D. Quinn; 208 (UP), D. Quinn; 208 (LO), J. Pitcher; 208 (LO-flight), D. Smith; 209, J. Pitcher; 209 (LO-flight), D. Smith; 210, J. Pitcher; 210 (flight), D. Smith; 211, J. Pitcher; 211 (flight), D. Smith; 212, K. Mullarney; 213 (UP), K. Mullarney; 213 (UP-dorsal flight), D. Smith; 213 (LO), K. Mullarney; 213 (LO-flight), D. Smith; 214, J. Janosik; 216-220, J. Pitcher; 220 (flight), D. Smith; 221 (UP), J. Pitcher; 221 (UP-flight), D. Smith; 221 (LO), J. Pitcher; 221 (LO-flight), D. Smith; 222, M. O'Brien; 223, J. Pitcher; 223 (flight), D. Smith; 224 (UP), D. Quinn; 224 (LO), J. Pitcher; 225, D. Quinn; 226 (UP), K. Mullarney; 226 (UP-flight), D. Smith; 226 (LO), J. Schmitt; 227 (UP), J. Schmitt; 227 (LO), D. Smith; 228, D. Smith; 229 (UP), J. Schmitt; 229 (UP-flight), D. Smith; 229 (LO), J. Schmitt; 229 (LO-flight), D. Smith; 230 (UP), D. Quinn; 230 (CTR), J. Schmitt; 230 (CTR-flight & LO), D. Smith; 231 (UP), D. Smith; 231 (LO), J. Alderfer; 231 (LO-flight), D. Smith; 232-233, J. Alderfer; 233 (LO-flight), D. Smith; 234, J. Pitcher; 235 (UP), J. Pitcher; 235 (UP-flight), J. Alderfer; 235 (LO), T. Schultz; 236 (UP), T. Schultz; 236 (UP-flight), D. Smith; 236 (LO), J. Pitcher; 237 (UP), J. Pitcher; 237 (LO), K. Mullarney; 237 (LO-flight), D. Smith; 239, J. Pitcher; 239 (LO-flight), D. Smith; 240–242 (UP), J. Pitcher; 242 (LO), T. Schultz; 242 (LO-flight), D. Smith; 243-244, J. Pitcher; 244 (breeding female), T. Schultz; 245 (UP), J. Pitcher; 245 (UP-flight), D. Smith; 245 (LO), J. Pitcher; 245 (LO-flight), D. Smith; 246 (UP), J. Alderfer; 246 (UP-flight), D. Smith; 246 (LO), T. Schultz; 246 (LO-flight), D. Smith; 247, T. Schultz; 247 (flight), D. Smith; 248, J. Pitcher; 248 (LO-flight), D. Smith; 249, J. Pitcher; 249 (flight), D. Smith; 250, K. Mullarney; 250 (UP-flight), D. Smith; 250 (LO), K. Mullarney; 250 (LO-flight), D. Smith; 251, K. Mullarney; 251 (flight), D. Smith; 252–256 (UP), J. Alderfer; 256 (LO), D. Quinn; 257 (UP), D. Quinn; 257 (LO), J. Alderfer; 258-259, K. Mullarney; 260, D. Quinn; 262-267, T. Schultz; 269, C. Ripper; 269 (winter flight), J. Alderfer; 270, C. Ripper; 270 (flight), J. Alderfer; 271, J. Alderfer; 272 (UP), C. Ripper; 272 (LO), J. Alderfer; 273, C. Ripper; 273 (mandtii), J. Alderfer; 274 (UP), C. Ripper; 274 (LO), J. Alderfer; 275-279, C. Ripper; 280–283 (UP), J. Alderfer; 283 (LO), C. Ripper; 284 (UP), C. Ripper; 284 (LO), J. Alderfer; 285, J. Alderfer; 287 (UP), T. Schultz; 287 (LO), J. Alderfer; 288-292, T. Schultz; 293 (UP), D. Quinn; 293 (LO), T. Schultz; 294–319 (UP), T. Schultz; 319 (LO), D. Quinn; 320-328, T. Schultz; 330, D. Pratt; 331 (UP), D. Beadle; 331 (LO)–333 (UP), D. Pratt; 333 (CTR), J. Schmitt and J. Alderfer; 333 (LO), D. Quinn; 334-337, J. Schmitt and J. Alderfer; 337 (UP-female), J. Alderfer; 338 (UP), J. Alderfer; 338-341, D. Malick; 342-343, D. Quinn; 344-347, D. Pratt; 348, D. Malick; 350 (UP), T. Schultz; 350 (LO), D. Malick; 351–357 (UP), D. Malick; 357 (LO), T. Schultz; 358-362, D. Malick; 363, D. Quinn; 364 (UP), J. Schmitt; 364 (LO), T. Schultz; 365, T. Schultz; 366 (UP), J. Schmitt; 366 (LO), C. Ripper; 367, J. Schmitt; 367 (spread tails), C. Ripper; 368 J. Schmitt; 368 (LO-spread tail), C. Ripper; 369 (UP), T. Schultz; 369 (UP-spread tail), S. Webb; 369 (LO), D. Beadle; 371–375 (UP), D. Beadle; 375 (LO), D. Beadle; 377, S. Webb; 378 (UP), S. Webb; 378 (LO), D. Pratt; 379, D. Pratt; 380 (UP), D. Pratt; 380 (LO), S. Webb; 381-382, D. Pratt; 383 (UP), D. Beadle; 383 (LO), D. Pratt; 384-387, D. Pratt; 388 (UP), D. Beadle; 388 (CTR), D. Pratt; 388 (LO), S. Webb; 390, J. O'Neill; 391, D. Quinn; 393, D. Malick; 394 (UP), J. Alderfer; 394 (LO), D. Malick; 394 (LO-flight), J. Schmitt; 395, D.

ILLUSTRATIONS CREDITS

Quinn; 396-406, D. Malick; 406 (LO-bacatus), T. Schultz; 407 (UP), D. Malick; 407 (LO), D. Quinn; 408-410, D. Malick; 412 (UP), T. Schultz; 412 (LO), D. Malick; 413, D. Malick; 413 (dorsal flight), J. Schmitt; 414 (UP), D. Malick; 414 (UP-flight), J. Schmitt; 414 (LO), J. Schmitt; 415, D Malick; 415 (suckleyi and flight), J. Schmitt; 416 (UP), T. Schultz; 416 (LO), D. Malick; 416 (LO-flight), J. Schmitt; 417, D. Malick; 417 (flight), K. Pendleton; 418, D. Malick; 418 (flight), K. Pendleton; 419, D. Malick; 419 (LO-flight), K. Pendleton; 420-427, J. Schmitt; 429 (UP & CTR), D. Beadle; 429 (LO), J. Schmitt; 430-441, D. Beadle; 442-443, D. Pratt; 445-447, P. Burke; 448 (UP), D. Pratt; 448 (LO), J. Alderfer; 449, T. Schultz; 450, P. Burke; 451, J. Alderfer; 452 (UP), J. Alderfer; 452 (LO), D. Beadle; 452 (LO-front view), J. Alderfer; 453 (UP), J. Alderfer; 453 (LO), D. Beadle; 454 (UP), D. Pratt; 454 (LO), D. Beadle; 455, D. Pratt; 455 (LO-immature), J. Schmitt; 456, J. Alderfer; 457 (UP), T. Schultz; 457 (LO), J. Alderfer; 458, D. Quinn; 459, D. Pratt; 459 (UP-flight), J. Schmitt; 460, D. Beadle; 461, D. Pratt; 462 (UP), D. Pratt; 462 (LO), P. Burke; 463–465 (UP), D. Pratt; 465 (LO), J. Schmitt; 466, D. Beadle; 467, D. Pratt; 468 (UP), D. Pratt; 468 (LO), D. Beadle; 470–474 (UP), D. Pratt; 474 (LO), J. Schmitt; 475, J. Schmitt; 475 (texana), T. Schultz; 476 (UP), J. Schmitt; 476 (UP-juvenile), D. Pratt; 476 (LO)–478, D. Pratt; 479-482 (UP), J. Schmitt; 482 (UP-flight), J. Alderfer; 482 (LO), J. Schmitt; 482 (LO-flight and calling), J. Alderfer; 484-485, D. Beadle; 487, D. Pratt; 487 (western female), T. Schultz; 488, D. Beadle; 489 (UP), D. Beadle; 489 (LO), D. Pratt; 490 (UP), J. Alderfer; 490 (LO)–493 (UP), D. Pratt; 493 (LO), D. Quinn; 494-495, D. Pratt; 495 (pelodoma flight), D. Beadle; 497-504, M. O'Brien; 505-506, J. O'Neill; 508-510, D. Pratt; 511, T. Schultz; 513-515 (UP), D. Pratt; 515 (UP-pacificus), D. Beadle; 515 (LO), J. Schmitt; 516, J. Schmitt; 517 (UP), J. Schmitt; 517 (LO), D. Pratt; 518, D. Pratt; 518 (calaphonus), J. Schmitt; 519 (UP), D. Pratt; 519 (UP-nest & LO), J. Schmitt; 521-524, D. Pratt; 525-526, J. Schmitt; 527 (UP), D. Quinn; 527 (LO), J. Alderfer; 528-530, D. Quinn; 531, M. O'Brien; 532-540 (UP), D. Quinn; 540 (LO), J. Alderfer; 541, D. Quinn; 543-545 (UP), D. Pratt; 545 (LO), T. Schultz; 546, D. Pratt; 547, D. Beadle; 548-551, T. Schultz; 552-554, D. Quinn; 555 (UP), D. Pratt; 555 (LO), P. Burke; 556, D. Pratt; 557 (UP), T. Schultz; 557 (LO), D, Pratt; 558, D. Pratt; 560 (UP), J. Alderfer; 560 (LO), D. Pratt; 561, D. Pratt; 562-565, J. Schmitt; 566-568, D. Pratt; 569 (UP), J. Alderfer; 569 (LO), D. Pratt; 570, D. Quinn; 572, D. Pratt; 573 (UP), D. Quinn; 573 (LO), D. Pratt; 574 (UP), D. Pratt; 574 (LO)–577, D. Quinn; 579, D. Pratt; 581 (UP), J. Alderfer; 581 (LO), D. Pratt; 582, D. Pratt; 584-588, D. Pierce; 588 (UP-1st winter female flight), J. Alderfer; 590 (UP), D. Pratt; 590 (LO), T. Schultz; 591, T. Schultz; 592–594 (UP), D. Pratt; 594 (LO), T. Schultz; 595–597 (UP), D. Pratt; 597 (LO), T. Schultz; 598, D. Pratt; 599, T. Schultz; 600 (UP), D. Beadle; 600 (LO), D. Pratt; 601 (UP), T. Schultz; 601 (LO), P. Burke; 602-603, T. Schultz; 604-606, D. Pratt; 607, T. Schultz; 608-612, D. Pratt; 612 (fall male), D. Beadle; 613-614, D. Pratt; 615, T. Schultz; 616, D. Pratt; 616 (fall female coronata), T. Schultz; 618-622, D. Pratt; 623-624, P. Burke; 624 (LO), T. Schultz; 625 (UP), D. Pratt; 625 (LO), D. Beadle; 626, D. Pratt; 627, P. Burke; 627 (female auricollis), T. Schultz; 630, T. Schultz; 632 (UP), D. Pratt; 632 (UP-adult male & LO), P. Burke; 633 (UP), D. Pratt; 633 (LO)–637, P. Burke; 638, D. Pierce; 638 (juvenile), T. Schultz; 639–644 (UP), T. Schultz; 644 (LO), D. Pierce; 645, D. Pierce; 646, D. Beadle; 647-649, D. Pierce; 649 (UP-canescens), D. Beadle; 650 (UP), D. Pierce; 650 (LO), T. Schultz; 651, T. Schultz; 652, D. Pierce; 653-654, J. Schmitt; 655-656, D. Pierce; 656 (iliaca), T. Schultz; 658 (UP), D. Pierce; 658 (UP-juvenile & LO), T. Schultz; 659, D. Pierce; 660 (UP), D. Pierce; 660 (LO), T. Schultz; 661, T. Schultz; 662-663, D. Pierce; 663 (flight), J. Schmitt; 664, D. Pierce; 664 (mearnsi), D. Beadle; 665 (UP), D. Pierce; 665 (LO)–668, D. Quinn; 670-673, P. Burke; 674, D. Pierce; 674 (superbus), D. Beadle; 675-678, D. Pierce; 679-682, T. Schultz; 683, D. Pierce; 685 (UP), T. Schultz; 685 (LO), D. Beadle; 686, D. Pratt; 687, D. Pratt; 687 (males), J. Schmitt; 688-689, D. Pratt; 690-691, T. Schultz; 692-694, D. Pratt; 695 (UP), P. Burke; 695 (LO), D. Beadle; 695 (LO-aeneus), D. Pratt; 696, J. Schmitt; 697-702, P. Burke; 704 (UP), D. Quinn; 704 (LO), D. Pierce; 705, J. Schmitt; 705 (umbrina), D. Pierce; 705 (littoralis & male tephrocotis), D. Beadle; 706, J. Schmitt; 706 (LO-male), D. Pierce; 707, J. Schmitt; 708 (UP & CTR), D. Pierce; 708 (LO), T. Schultz; 709-711, T. Schultz; 712-714, D. Pierce; 714 (rostrata), J. Schmitt; 715-717 (UP), D. Pierce; 717 (LO), D. Quinn; 718-719, D. Pierce; 720-723, J. Schmitt.

PHOTO CREDITS

Most of the photographs in this book were contributed by the following photographers: Richard Crossley, Rob Curtis/The Early Birder, Mike Danzenbaker, Kevin T. Karlson, Maslowski Wildlife Productions, Robert Royse, Larry Sansone, Brian E. Small, and Tom Vezo.

Cover, Phoo Chan/National Geographic Your Shot; 2-3, Jim Zipp; 10, R. Crossley; 15, Michael S. Peters; 24 (LE), R. Curtis; 24 (CTR), L. Sansone; 24 (RT), T. Vezo; 31 (UP), B. Small; 31 (LO), L. Sansone; 38, L. Sansone; 39, B. Small; 55, B. Small; 56, T. Vezo; 60, James Cumming/NG Your Shot; 72, K. Karlson; 76, T. Vezo; 80, K. Karlson; 82, M. Danzenbaker; 88, M. Danzenbaker; 103 (UP), L. Sansone; 103 (LO), M. Danzenbaker; 107, R. Royse; 116, Tom J. Ulrich; 119, K. Karlson; 120, Judd Patterson; 122, Glenn Bartley; 127, R. Crossley; 132, K. Karlson; 133, L. Sansone; 136, K. Karlson; 149, K. Karlson; 153, R. Royse; 156, Darran Rickards; 157, T. Vezo; 166, K. Karlson; 167, L. Sansone; 174, Brian K. Wheeler; 175, Brian K. Wheeler; 182, T. Vezo; 185 (UP), B. Small; 185 (LO), R. Curtis; 192, Chris Jimenez; 193, Troy Lim/NG Your Shot; 194, T. Vezo; 197, Christopher L. Wood; 198, Maslowski; 200, T. Vezo; 202, T. Vezo; 210, R. Crossley; 211, R. Crossley; 214, B. Small; 215, R. Crossley; 222 (LE), T. Vezo; 222 (RT), B. Small; 238 (LE), R. Crossley; 238 (RT), L. Sansone; 243, R. Crossley; 260, M. Danzenbaker; 261, B. Small; 264, Ran Schols; 265, B. Small; 268, Art Wolfe/artwolfe.com; 286, R. Crossley; 301, R. Crossley; 305 (UP), L. Sansone; 305 (LO), M. Danzenbaker; 326 (UP), B. Small; 326 (LO), Monte M. Taylor; 329, Maslowski; 341, Tom J. Ulrich; 348, Taxi/Getty Images; 349, B. Small; 355, B. Small; 361 (LE), Jeff Nadler; 361 (RT), Steve Metz; 363, Glenn Bartley; 370, Glenn Bartley; 376, B. Small; 377, B. Small; 389, Dorothy E Harris/NG Your Shot; 391, Axel Hilger/NG Your Shot; 392, B. Small; 395, Danny Brown/NG Your Shot; 411, R. Crossley; 420, R. Curtis; 428, R. Curtis; 435, B. Small; 444, B. Small; 456, Chris Jimenez; 458, Maslowski; 460, B. Small; 464, B. Small; 469, B. Small; 483, B. Small; 486, M. Danzenbaker; 494 (LE), M. Danzenbaker; 494 (RT), R. Crossley; 496, Maslowski; 499, K. Karlson; 505, R. Royse; 506, Roger van Gelder; 507, Maslowski; 511, Maslowski; 512, B. Small; 520, Matthew Studebaker; 523, Alan Fuchs; 524, Bhalchandra Pujari/Dreamstime.com; 525, K. Karlson; 526, Matthew Studebaker; 530, B. Small; 531, Dobyter/Dreamstime.com; 532, M. Danzenbaker; 534, Mehrdad Sadat; 542, Maslowski; 544 (LE & RT), R. Royse; 544 (CTR), Bob Steele; 559, T. Vezo; 567, Verastuchelova/Dreamstime.com; 570, L. Sansone; 571, Maslowski 578, B. Small; 580, Maslowski; 582, R. Royse; 583, Brian Wolitski/NG Your Shot; 585 (LE), Howard Cummings; 585 (CTR LE & CTR RT), B. Small; 585 (RT), L. Sansone; 589, R. Crossley; 591 (UP), K. Karlson; 591 (LO), B. Small; 613 (UP), Giff Beaton; 613 (LO), T. Vezo; 617, L. Sansone; 628, K. Karlson; 629, Reinhard Geisler; 631, B. Small; 638, B. Small; 669, B. Small; 677 (LE), R. Royse; 677 (CTR & RT), B. Small; 681, B. Small; 684, M. Danzenbaker; 687 (LE), B. Small; 687 (CTR), T. Vezo; 687 (RT), Bob Steele; 692, B. Small; 693, K. Karlson; 700, K. Karlson; 701, B. Small; 703, Bob Steele; 711 (LE), K. Karlson; 711 (CTR & RT), B. Small; 714, Garth McElroy; 715, Garth McElroy; 720, R. Curtis; 722 (UP), Tom Gatz/USFWS; 723, Kimball L. Garrett.

INDEX

INDEX

INDEX

INDEX

INDEX

INDEX

INDEX

INDEX

INDEX

INDEX

INDEX

INDEX

INDEX

CONTRIBUTORS

The following authors and other contributors formed the talented team that wrote and edited the first edition of *Complete Birds of North America* (2006). They include many of North America's most well-known ornithologists, birders, and birding authors. Under the direction of editor Jonathan Alderfer and associate editor Jon L. Dunn, the first-edition team created a state-of-the-art handbook on the identification of North American birds. The superlative work of map consultant Paul Lehman complemented the written text, and a team of illustrators produced numerous new illustrations (Illustrations Credits appear on pages 726–727).

Since 2006, new identification information has come to light, additional bird species have been recorded in North America, status and distribution has changed for many species, and genetic and observational studies (augmented by the increasingly sophisticated analysis of vocalizations) have had a profound effect on bird taxonomy—this second edition incorporates as much of that new information as would fit in an additional 72 pages. The editorial team of Jonathan Alderfer and Jon L. Dunn working with mapping expert Paul Lehman remains unchanged. Writers Edward S. Brinkley, Kimball L. Garrett, and Paul Hess contributed new species and family accounts. Artists Jonathan Alderfer, N. John Schmitt, and Thomas R. Schultz created additional illustrations.

Found below are short biographies, listed alphabetically, of all editorial and writing contributors. Since new material by other authors and the editors has been woven into the second edition text, the original author credits are no longer accurate and have been dropped from the family accounts and table of contents. However, at the end of each biography found below is a list of the bird families that each contributor originally wrote (or co-wrote) for the first edition or the editing work they did.

Jonathan Alderfer is a widely published author and field guide illustrator. One of the nation's foremost birding artists, he is well known in the birding community for his expertise as a field ornithologist and for his knowledge of North American birds. With Jon Dunn he co-authored the sixth edition of *National Geographic Field Guide to the Birds of North America* and was a contributing artist. More of his artwork can be viewed at www.jonathanalderfer.com. He lives in Washington, D.C. *Editor (both editions).*

Jessie H. Barry works at the Cornell Lab of Ornithology, where she is the project leader for Merlin. Merlin Bird ID is an app designed to help people identify North America's most common birds. Jessie has a passion for bird ID and sharing it with others. She is a member of the Cornell Lab's Team Sapsucker and holds the national record for most species of birds seen in 24 hours, 294 species. *Ducks, Geese, and Swans.*

Edward S. (Ned) Brinkley is a lifelong birder whose interests include migration, seabird identification and distribution, and the effects of weather and climate on birds. He has guided birding tours around the world, and since 2001, he has served as editor of *North American Birds,* a quarterly journal of ornithological record, and has published widely in the field, including *The National Wildlife Federation Field Guide to North American Birds* (2007). He lives in Cape Charles, Virginia. *Additional Species Accounts (2nd edition).*

Steven W. Cardiff is the bird collection manager at Louisiana State University Museum of Natural Science (since 1984) and also serves as chair of the Louisiana Bird Records Committee and as Arkansas-Louisiana regional editor for *North American Birds.* Among his many interests: status and distribution of Louisiana birds, Gulf of Mexico seabirds, identification/plumages/molt of large gulls, and

natural history of the Yellow Rail and *Myiarchus* flycatchers (he co-authored the Ash-throated and Brown-crested Flycatcher *Birds of North America* accounts with Donna L. Dittmann). Steve is a lifelong birder who grew up in southern California, moved to Baton Rouge, Louisiana, to attend LSU in 1979, and has lived in St. Gabriel, Louisiana, since 1990. *Rails, Gallinules, and Coots; Nighthawks and Nightjars; Tyrant Flycatchers.*

Allen T. Chartier picked up a pair of binoculars and a field guide at the age of 11, and has been hooked on birds ever since. His strong interests in participating in scientific projects, writing, and teaching have led him beyond birding, to become a bird bander specializing in hummingbirds, and a freelance environmental educator. He has served on the Michigan Bird Records Committee and the editorial board for Michigan's state bird journal, *Michigan Birds and Natural History,* for many years. Among the many birding publications he has authored or co-authored are *A Birder's Guide to Michigan* and *Michigan's Second Breeding Bird Atlas.* He lives in Inkster, Michigan. *Larks; Swallows; Kinglets; Olive Warbler; Wood-Warblers; Bananaquits.*

Cameron D. Cox currently works as a birding and photography tour leader for Tropical Birding, where he shares his knowledge of and passion for birds and birding. In 2013, he co-authored the Peterson *Reference Guide to Seawatching: Eastern Waterbirds in Flight,* a book inspired by his years observing waterbirds in Cape May, New Jersey. His expertise at challenging bird identifications also served him well as the former photo quiz editor for *Birding,* the flagship magazine of the American Birding Association. *Ducks, Geese, and Swans.*

Donna L. Dittmann is a former member of the American Birding Association (ABA) Checklist Committee and has been secretary of the Louisiana Bird Records Committee since 1988. Avian interests range from genetics to identification and molt, and she has written a variety of technical and popular articles. She is co-founder of the Yellow Rails and Rice Festival. More information can be found at her website: snowyegretenterprises.com. She works at LSU Museum of Natural Science where she is the genetic resources collections manager, permits coordinator, and museum preparator. Donna is a native of San Francisco, California; moved to Louisiana in 1984; and has lived in St. Gabriel, Louisiana, since 1990. *Rails, Gallinules, and Coots; Nighthawks and Nightjars; Tyrant Flycatchers.*

Jon L. Dunn is an expert on identification and distribution of North American birds. Since 1983 he has served as the chief consultant and lead author for all six editions of *National Geographic Field Guide to the Birds of North America.* Among the other books he has published are *Field Guide to the Warblers of North America* and *Birds of Southern California,* both co-authored with Kimball L. Garrett. He has served as a longtime member of the California Bird Records Committee and is currently on the American Ornithologists' Union Committee on Classification and Nomenclature and the American Birding Association Checklist Committee. He is a senior tour leader for WINGS birding tours. He lives in Bishop, California. *Old World Flycatchers; Associate Editor (both editions).*

Ted Floyd is the editor of *Birding,* the flagship publication of the American Birding Association. Floyd has served on the boards of directors of Colorado Field Ornithologists and Western Field Ornithologists, and is a frequent speaker at bird festivals and ornithological society meetings. He is the author of four bird books and more than 100 articles and technical papers on birds and other aspects of natural history. Floyd has a special interest in bird vocalizations,

CONTRIBUTORS

especially nocturnal songs and calls. He lives in Lafayette, Colorado. *Wrens; Dippers; Starlings; Old World Sparrows.*

Kimball L. Garrett is a lifelong resident of the Los Angeles area, where he has served as Ornithology Collections Manager at the Natural History Museum of Los Angeles County since 1982. He is past-president of Western Field Ornithologists, a longtime member of the California Bird Records Committee, and a current member of the ABA Checklist Committee. Among the many birding books and articles he has published are several co-authored with this guide's two principal editors. *Tropicbirds; Boobies and Gannets; Pelicans; Cormorants; Darters; Frigatebirds; Swifts; Woodpeckers and Allies; Additional Family Accounts (2nd edition).*

Matthew T. Heindel has been birding for over 30 years with a focus on identification, migration, and status and distribution. His articles have appeared in *Birding, Birder's Journal, Western Birds, Audubon Field Notes,* and *North American Birds,* and he was editor of the California Christmas Bird Counts for many years. He served several terms on the California Bird Records Committee, including multiple stints as the chairman and vice-chairman. He currently lives near San Antonio, Texas. *Thick-knees; Lapwings and Plovers; Oystercatchers; Stilts and Avocets; Jacanas; Sandpipers, Phalaropes, and Allies; Pratincoles; Trogons; Hoopoes; Vireos; Old World Warblers and Gnatcatchers.*

Paul Hess, a retired newspaper editor, has co-authored or edited nearly two dozen books on birds and natural history for various publishers including National Geographic. He writes a "News and Notes" column for *Birding,* the American Birding Association's magazine; has contributed many articles to the state journal *Pennsylvania Birds;* and has chaired the Pennsylvania Ornithological Records Committee. He has received state and regional awards for outstanding contributions to Pennsylvania ornithology and bird conservation. He lives in Natrona Heights, Pennsylvania. *Shrikes; Chickadees and Titmice; Editing (1st edition); Additional Species Accounts (2nd edition).*

Steve N. G. Howell is an international bird tour leader with WINGS, a research associate at the California Academy of Sciences, and a widely published author. His more recent books include the widely acclaimed photographic guide *Petrels, Albatrosses, and Storm-Petrels of North America* (2012, Princeton University Press). Beyond birds, Steve's interests include chocolate, flyingfish, and tequila. *Loons; Grebes; Cranes; Skuas, Gulls, Terns, and Skimmers; Hummingbirds.*

Marshall Iliff grew up in Annapolis, Maryland, and has been birding for 25 years. He has worked as a tour guide for Victor Emanuel Nature Tours and has served on three state Records Committees. His long interest in helping birders to collect better data led to his current position as eBird project leader, along with Brian Sullivan and Chris Wood. eBird (www.ebird.org) is a free website where all birders can keep their personal records, track their lists, and see what others are seeing. The open-access database has become invaluable to the science and conservation communities and continues to grow, and it is poised to truly revolutionize how birders contribute to the understanding and protection of the birds they love. *Emberizids.*

Alvaro Jaramillo was born in Chile but began birding in Toronto, Canada. He was trained in ecology and evolution with a particular interest in bird behavior. He is the author of *Birds of Chile* and *New World Blackbirds: The Icterids.* Alvaro writes the "Identify Yourself" column in *Bird Watcher's Digest.* He co-authored the Emberizids chapter for the *Handbook of the Birds of the World* and is writing

a photo guide to the birds and wildlife of Patagonia. He runs Alvaro's Adventures, a birding tour company that also emphasizes natural history and culture while traveling. Alvaro lives with his family in Half Moon Bay, California. *Ducks, Geese, and Swans; Skuas, Gulls, Terns, and Skimmers; Babblers; Accentors; Blackbirds.*

Ian L. Jones has studied seabird behavior, ecology, and conservation since 1982. He received his M.Sc. in zoology researching Ancient Murrelets. His Ph.D. focused on Least Auklets in the Pribilof Islands, Alaska, and his postdocs concerned Crested Auklets at Buldir, Aleutian Islands, Alaska, and founding a seabird demography project at Triangle Island, British Columbia. Ian has led seabird research and conservation projects in Newfoundland and the northwestern Hawaiian Islands. His main focus is auklets (*Aethia* spp.) in the Aleutians sector of the Alaska Maritime National Wildlife Refuge, pursuing conservation projects, especially the impact of introduced Norway rats. Movement and migration of auklets outside the breeding season is currently (2014) his major focus. Ian has been a professor of biology at Memorial University in St. John's, Newfoundland, since 1995 and with collaborators has published more than 100 scientific papers. *Auks, Murres, and Puffins.*

Paul Lehman has birded extensively throughout North America since he was young and has written many papers dealing with the continent's bird distribution and field identification. He has been the principal range-map maker for many of the popular field guides used today, including multiple editions of the *National Geographic Society Field Guide to the Birds of North America.* He is a past editor of *Birding,* the American Birding Association's magazine, and a former geography instructor with special interests in meteorology, biogeography, and the effects of weather on bird migration and vagrancy. He currently lives in San Diego, California, and previously resided in Cape May, New Jersey; Santa Barbara, California; and the New York City area. *Maps (both editions); Editing (1st edition).*

Tony Leukering has published extensively on his primary birding interests: migration, identification, and distribution. He is the online photo-quiz editor for the American Birding Association; past photo-quiz editor for the Colorado Field Ornithologists (CFO); primary author of *In the Scope,* a column for *Colorado Birds* that covers identification issues from a Colorado perspective; and past chair of the Colorado Bird Records Committee. In 2008, he was awarded the CFO's Ron Ryder Award, given for scholarly contributions to Colorado field ornithology. Tony has been a working ornithologist for 30 years, primarily at bird observatories scattered from coast to coast; he currently lives in Largo, Florida. *Thrushes.*

Mark W. Lockwood is the conservation biologist for the Trans-Pecos region in the State Parks Division of the Texas Parks and Wildlife Department. He has worked extensively with endangered bird species in Texas as well as having a lifelong interest in the status and distribution of birds in the state. Mark has been a member of the Texas Bird Records Committee since 1995 and has been the Texas regional editor for *North American Birds* since 2000. He is the author or co-author of five books, including *Birds of the Texas Hill Country* and the *TOS Handbook of Texas Birds.* Mark lives in Alpine, Texas. *Curassows and Guans; New World Quail; Cuckoos, Roadrunners, and Anis; Kingfishers; Penduline Tits and Verdins; Long-tailed Tits and Bushtits.*

Guy McCaskie is a semi-retired engineer with a lifelong interest in birds. He is a past-president of Western Field

CONTRIBUTORS

Ornithologists, a member of the California Bird Records Committee since 1970, and a co-author of the southern California regional report for North American Birds and its predecessors since 1963. Guy is one of the co-authors of *Birds of the Salton Sea* (2003, University of California Press) and is keenly interested in that region, birding it far more than anyone else, since the early 1960s. He formerly lived in Great Britain but is a longtime resident of Imperial Beach, California. *Pigeons and Doves.*

Bill Pranty has birded extensively in Florida for more than 35 years. His studies emphasize the documentation of Florida's diverse avifauna, with a focus on its exotic species. His research has added three species to the *ABA Checklist*—Purple Swamphen, Nanday Parakeet, and Common Myna—all of them exotics from Florida. Pranty is chairman of the ABA Checklist Committee and a technical reviewer for and frequent contributor to *Birding* magazine. In addition to dozens of peer-reviewed papers, Pranty is the author or co-author of five books, among these the latest book on Florida ornithology. He lives in Bayonet Point, Florida. *Bitterns, Herons, and Allies; Ibises and Spoonbills; Storks; Flamingos; Limpkins; Parakeets, Macaws, and Parrots; Bulbuls; Weavers; Estrildid Finches.*

David E. Quady began birding while in graduate school, as something to do while camping. But soon camping became simply something to do to facilitate birding. His fondness for owls began with a close, pre-dawn encounter with a vocalizing Northern Spotted Owl and has grown to the point that he now sometimes wonders why people bother to go birding in the daytime. He has birded widely in North America and abroad, leads field trips and teaches owl classes for Golden Gate Audubon Society, and is vice president of Western Field Ornithologists. He lives in Berkeley, California. *Barn Owls; Typical Owls.*

Kurt Radamaker has been interested in birds and nature most of his life. He grew up in southern California, where he started birding at the age of 8 and by 15 had completed Cornell Laboratory's Seminars in Ornithology. He worked as an ornithology teacher for the University of La Verne in La Verne California, and has authored or co-authored a handful of books on birds and contributed articles for a number of journals including *Western Birds* and *Birding*. He has a keen interest in birding in Mexico and was editor of the Mexican journal *The Euphonia* for five years. He is a member of the Arizona Bird Committee and president of Arizona Field Ornithologists. He resides in Cave Creek, Arizona. *Bitterns, Herons, and Allies; Ibises and Spoonbills.*

Gary H. Rosenberg has been a bird-watcher since he was a young boy, when his father introduced him to bird-watching. He studied biology at Arizona State University and then ornithology at Louisiana State University, where he earned a master's degree in zoology, studying birds and specializing on Amazonia river islands in Peru. Since 1986, Gary has been a professional birding tour leader, first working for WINGS and then starting his own company—Avian Journeys. He has led hundreds of birding tours in the neotropics and across North America. Gary has published numerous scientific and popular papers and has been part of the discovery of two birds new to science. He is the co-editor of the Arizona reports for *North American Birds* and has served as secretary of the Arizona Bird Committee since 1987. He lives in Tucson, Arizona. *Mockingbirds and Thrashers; Wagtails and Pipits; Tanagers; Cardinals, Saltators, and Allies; Fringilline and Cardueline Finches.*

Brian Sullivan has conducted fieldwork on birds for the past 20 years. Birding travels, photography, and field projects have taken him across North America, to Central and South America, Antarctica, and the Arctic. He has written and consulted on various books and popular and scientific literature on North American birds and is a co-author of *The Crossley ID Guide: Raptors* and the forthcoming *Princeton Guide to North American Birds*. He is currently project leader for eBird and photographic editor for the Birds of North America Online at the Cornell Laboratory of Ornithology. He also served as photographic editor for the American Birding Association's journal *North American Birds* from 2005 to 2013. *Albatrosses; Shearwaters and Petrels; Storm-Petrels.*

Clay Taylor started birding in Connecticut in 1975. Viewing and photographing raptors became his prime passion while attending college in Rochester, New York, near the Braddock Bay hawk watch site. As a birder, tour leader, and hawk bander in the 1980s he started the hawk-banding project at Braddock Bay that developed into Braddock Bay Raptor Research. He joined Swarovski Optik in 1999 and is now the naturalist market manager, finding time to visit many of the top hawk migration sites during his travels. His Corpus Christi, Texas, yard list has 23 species of raptors seen from the property. *New World Vultures; Hawks, Kites, Eagles, and Allies; Caracaras and Falcons.*

Chris Wood has worked at the Cornell Lab of Ornithology for the past ten years as eBird project leader. During this time eBird has grown into one of the largest citizen science projects in the world, with observations from every country, more than 150,000 users, and gathering some five million records per month. He is widely recognized as a leading authority on bird identification and distribution and has written and consulted on various books and popular and scientific literature on birds. For fun Chris leads bird-watching trips for the tour company WINGS. These activities are all built around Chris's primary interest—developing ways to connect birders with scientists and the conservation community to better understand and conserve birds and their ecosystems. *Partridges, Grouse, and Turkeys; Tyrant Flycatchers; Jays and Crows; Nuthatches; Creepers; Waxwings; Silky-flycatchers.*

NATIONAL GEOGRAPHIC COMPLETE BIRDS OF NORTH AMERICA, SECOND EDITION
Edited by Jonathan Alderfer with Jon L. Dunn

PUBLISHED BY THE NATIONAL GEOGRAPHIC SOCIETY

Gary E. Knell — *President and Chief Executive Officer*
John M. Fahey — *Chairman of the Board*
Declan Moore — *Executive Vice President; President, Publishing and Travel*
Melina Gerosa Bellows — *Executive Vice President; Publisher and Chief Creative Officer, Books, Kids, and Family*

PREPARED BY THE BOOK DIVISION

Hector Sierra — *Senior Vice President and General Manager*
Janet Goldstein — *Senior Vice President and Editorial Director*
Jonathan Halling — *Creative Director*
Marianne R. Koszorus — *Design Director*
Susan Tyler Hitchcock — *Senior Editor*
R. Gary Colbert — *Production Director*
Jennifer A. Thornton — *Director of Managing Editorial*
Susan S. Blair — *Director of Photography*
Meredith C. Wilcox — *Director, Administration and Rights Clearance*

STAFF FOR SECOND EDITION

Robin Terry Brown — *Project Editor*
Jennifer Seidel — *Text Editor*
Matthew Propert — *Illustrations Editor*
Linda Makarov — *Designer*
Carl Mehler — *Director of Maps*
Paul Lehman — *Chief Map Researcher and Editor*
Sven M. Dolling, Mike McNey, Gregory Ugiansky, Martin S. Walz, and Mapping Specialists, Ltd. — *Map Production*
Edward S. Brinkley, Kimball L. Garrett, and Paul Hess — *Contributing Writers*
Marshall Kiker — *Associate Managing Editor*
Judith Klein — *Production Editor*
Mike Horenstein — *Production Manager*
Galen Young — *Rights Clearance Specialist*
Katie Olsen — *Production Design Assistant*

PRODUCTION SERVICES

Phillip L. Schlosser — *Senior Vice President*
Chris Brown — *Vice President, NG Book Manufacturing*
Robert L. Barr — *Manager*
Rahsaan Jackson — *Imaging*

The National Geographic Society is one of the world's largest nonprofit scientific and educational organizations. Its mission is to inspire people to care about the planet. Founded in 1888, the Society is member supported and offers a community for members to get closer to explorers, connect with other members, and help make a difference. The Society reaches more than 450 million people worldwide each month through *National Geographic* and other magazines; National Geographic Channel; television documentaries; music; radio; films; books; DVDs; maps; exhibitions; live events; school publishing programs; interactive media; and merchandise. National Geographic has funded more than 10,000 scientific research, conservation, and exploration projects and supports an education program promoting geographic literacy. For more information, visit www.nationalgeographic.com.

For more information, please call 1-800-NGS LINE (647-5463) or write to the following address:

National Geographic Society
1145 17th Street, N.W.
Washington, D.C. 20036-4688 U.S.A.

For information about special discounts for bulk purchases, please contact National Geographic Books Special Sales: ngspecsales@ngs.org

For rights or permissions inquiries, please contact National Geographic Books Subsidiary Rights: ngbookrights@ngs.org

Copyright © 2006, 2014 National Geographic Society
All rights reserved. Reproduction of the whole or any part of the contents without permission is prohibited.

ISBN: 978-1-4262-1373-1
ISBN: 978-1-4262-1520-9 (deluxe)

430 5732
Printed in China

14/PPS/1

ACKNOWLEDGMENTS

The editors wish to thank the following individuals and institutions for their valuable assistance in the preparation of the second edition of this volume.

Bruce Anderson, David Arbour, Christian Artuso, Pierre Bannon, Louis Bevier, Dick Cannings, Russell Cannings, Steve Cardiff, Stephen Dinsmore, Cameron Eckert, Richard Erickson, Michael Force, Daniel D. Gibson, Joe Grzybowski, Peter Hamel, Margo Hearne, Matt Heindel, Steve N. G. Howell, David Irons, Oscar Johnson, Rudolf Koes, Yann Kolbeinsson, Mark Lockwood, Derek Lovitch, Carl Lundblad, Bruce Mactavish, Ron Martin, Chris McCreedy, Chet McGaugh, Robert McKernan, Martin Meyers, Charles Mills, Eric Mills, Steve Mlodinow, Kenny Nichols, Brainard Palmer-Ball, Brian Patteson, Mark Peterson, Bill Pranty, Peter Pyle, David E. Quady, Larry Rosche, Scott Seltman, David Sibley, John Sterling, Peter Taylor, Thede Tobish, Michael Todd, David Vander Pluym, Ron Weeks, Sandy Williams, Alan Wormington, and Andy Wraithmell.
Special thanks to Terry Chesser and Andy Kratter of the AOU Committee on Classification and Nomenclature for providing the accepted AOU English names for all North American bird families; and to the American Birding Association for use of their bird abundance codes.

The following individuals and institutions were extremely helpful in preparing the first edition (2006) of this volume.
George Armistead, Louis R. Bevier, Edward S. Brinkley, Jamie Cameron, Richard A. Erickson, Tom Gatz, Daniel D. Gibson, Shawneen Finnegan, Nancy Gobris, Robert A. Hamilton, Floyd Hayes, Tom and Jo Heindel, Greg Lasley, Paul Lehman, Bruce Mactavish, Curtis Marantz, Ian A. McLaren, Steven Mlodinow, Michael O'Brien, Michael A. Patten, J. Brian Patteson, Grayson Pearce, Peter Pyle, Van Remsen, Bill Schmoker, Willie Sekula, Larry Semo, Debra Shearwater, Douglas Stotz, Giles Timms, Bill Tweit, Phil Unitt, Nick Ward, Angus Wilson, and Sherrie York.
Institutions: Archbold Biological Station, Venus, FL; Burke Museum, University of Washington, Seattle; Cornell Laboratory of Ornithology, Ithaca, NY; Field Museum of Natural History, Chicago, IL; Museum of Natural Science, Louisiana State University, Baton Rouge; Museum of Vertebrate Zoology, University of California, Berkeley; Natural History Museum of Los Angeles County, Los Angeles, CA; PRBO Conservation Science, Stinson Beach, CA; San Diego Natural History Museum, San Diego, CA; and Wings, Tucson, AZ.